Developments in Environmental Modelling, 20

Numerical Ecology

SECOND ENGLISH EDITION

Developments in Environmental Modelling

1. ENERGY AND ECOLOGICAL MODELLING edited by W.J. Mitsch, R.W. Bossermann and J.M. Klopatek, 1981
2. WATER MANAGEMENT MODELS IN PRACTICE: A CASE STUDY OF THE ASWAN HIGH DAM by D. Whittington and G. Guariso, 1983
3. NUMERICAL ECOLOGY by L. Legendre and P. Legendre, 1983

4A. APPLICATION OF ECOLOGICAL MODELLING IN ENVIRONMENTAL MANAGEMENT PART A edited by S.E. Jørgensen, 1983

4B. APPLICATION OF ECOLOGICAL MODELLING IN ENVIRONMENTAL MANAGEMENT PART B edited by S.E. Jørgensen and W.J. Mitsch, 1983

5. ANALYSIS OF ECOLOGICAL SYSTEMS: STATE-OF-THE-ART IN ECOLOGICAL MODELLING edited by W.K. Lauenroth, G.V. Skogerboe and M. Flug, 1983
6. MODELLING THE FATE AND EFFECT OF TOXIC SUBSTANCES IN THE ENVIRONMENT edited by S.E. Jørgensen, 1984
7. MATHEMATICAL MODELS IN BIOLOGICAL WASTE WATER TREATMENT edited by S.E. Jørgensen and M.J. Gromiec, 1985
8. FRESHWATER ECOSYSTEMS: MODELLING AND SIMULATION by M. Straškraba and A.H. Gnauck, 1985
9. FUNDAMENTALS OF ECOLOGICAL MODELLING by S.E. Jørgensen, 1986
10. AGRICULTURAL NONPOINT SOURCE POLLUTION: MODEL SELECTION AND APPLICATION edited by A. Giorgini and F. Zingales, 1986
11. MATHEMATICAL MODELLING OF ENVIRONMENTAL AND ECOLOGICAL SYSTEMS edited by J.B. Shukla, T.G. Hallam and V. Capasso, 1987
12. WETLAND MODELLING edited by W.J. Mitsch, M. Straškraba and S.E. Jørgensen, 1988
13. ADVANCES IN ENVIRONMENTAL MODELLING edited by A. Marani, 1988
14. MATHEMATICAL SUBMODELS IN WATER QUALITY SYSTEMS edited by S.E. Jørgensen and M.J. Gromiec, 1989
15. ENVIRONMENTAL MODELS: EMISSIONS AND CONSEQUENCES edited by J. Fenhann, H. Larsen, G.A. Mackenzie and B. Rasmussen, 1990
16. MODELLING IN ECOTOXICOLOGY edited by S.E. Jørgensen, 1990
17. MODELLING IN ENVIRONMENTAL CHEMISTRY edited by S.E. Jørgensen, 1991
18. INTRODUCTION TO ENVIRONMENTAL MANAGEMENT edited by P.E. Hansen and S.E. Jørgensen, 1991
19. FUNDAMENTALS OF ECOLOGICAL MODELLING by S.E. Jørgensen, 1994

Developments in Environmental Modelling, 20

Numerical Ecology

SECOND ENGLISH EDITION

Pierre LEGENDRE, Professor

Département de sciences biologiques, Université de Montréal, Montréal (Québec) H3C 3J7, Canada

and

Louis LEGENDRE, Professor

Département de biologie, Université Laval, Québec (Québec) G1K 7P4, Canada

ELSEVIER

Amsterdam – Boston – London – New York – Oxford – Paris
San Diego – San Francisco – Singapore – Sydney – Tokyo

ELSEVIER SCIENCE B.V.
Sara Burgerhartstraat 25
P.O. Box 211, 1000 AE Amsterdam, The Netherlands

Library of Congress Cataloging in Publication Data

Legendre, Pierre, 1946-
[Écologie numérique. English]
Numerical ecology / Pierre Legendre and Louis Legendre. -- 2nd English ed.
p. cm. -- (Developments in environmental modelling ; 20)
Rev. ed. of: Ecologie numérique / Louis Legendre. 1983.
Includes bibliographical references and index.
ISBN 0-444-89249-4 (hardcover : alk. paper). -- ISBN 0-444-89250-8 (pbk. : alk. paper)
1. Ecology--Mathematics. I. Legendre, Louis. II. Legendre, Louis. Ecologie numérique. III. Legendre, Louis. Ecologie numérique. IV. Title. V. Series.
QH541.15.M34L4313 1998
577'.01'51--dc21 98-45198
CIP

ISBN: 0-444-89249-4 (Hardbound Edition)
ISBN: 0-444-89250-8 (Paperback Edition)
ISSN: 0167-8892
Second impression: 2000
Third impression: 2003

Printed and bound by CPI Antony Rowe, Eastbourne

Contents

6. Multidimensional qualitative data

7. Ecological resemblance

8. Cluster analysis

Preface

The delver into nature's aims
Seeks freedom and perfection;
Let calculation sift his claims
With faith and circumspection.

GOETHE

As a premise to this textbook on ***Numerical ecology***, the authors wish to state their opinion concerning the role of data analysis in ecology. In the above quotation, Goethe cautioned readers against the use of mathematics in the natural sciences. In his opinion, mathematics may obscure, under an often esoteric language, the natural phenomena that scientists are trying to elucidate. Unfortunately, there are many examples in the ecological literature where the use of mathematics unintentionally lent support to Goethe's thesis. This has become more frequent with the advent of computers, which facilitated access to the most complex numerical treatments. Fortunately, many other examples, including those discussed in the present book, show that ecologists who master the theoretical bases of numerical methods and know how to use them can derive a deeper understanding of natural phenomena from their calculations.

Numerical approaches can never dispense researchers from ecological reflection on observations. Data analysis must be seen as an objective and non-exclusive approach to carry out in-depth analysis of the data. Consequently, throughout this book, we put emphasis on ecological applications, which illustrate how to go from numerical results to ecological conclusions.

This book is written for the practising ecologists — graduate students and professional researchers. For this reason, it is organized both as a practical handbook and a reference textbook. Our goal is to describe and discuss the numerical methods which are successfully being used for analysing ecological data, using a clear and comprehensive approach. These methods are derived from the fields of mathematical physics, parametric and nonparametric statistics, information theory, numerical taxonomy, archaeology, psychometry, sociometry, econometry, and others. Some of these methods are presently only used by those ecologists who are especially interested in numerical data analysis; field ecologists often do not master the bases of these techniques. For this reason, analyses reported in the literature are often carried out using techniques that are not fully adapted to the data under study, leading to conclusions that are sub-optimal with respect to the field observations. When we were writing the first English edition of ***Numerical ecology*** (Legendre & Legendre, 1983a), this warning mainly concerned multivariate *versus* elementary statistics. Nowadays, most ecologists are capable of using multivariate methods; the above remark now especially applies to the analysis of autocorrelated data (see Section 1.1; Chapters 12 and 13) and the joint analysis of several data tables (Sections 10.5 and 10.6; Chapter 11).

Computer packages provide easy access to the most sophisticated numerical methods. Ecologists with inadequate background often find, however, that using high-level packages leads to dead ends. In order to efficiently use the available numerical tools, it is essential to clearly understand the principles that underlay numerical methods, and their limits. It is also important for ecologists to have guidelines for interpreting the heaps of computer-generated results. We therefore organized the present text as a comprehensive outline of methods for analysing ecological data, and also as a practical handbook indicating the most usual packages.

Our experience with graduate teaching and consulting has made us aware of the problems that ecologists may encounter when first using advanced numerical methods. Any earnest approach to such problems requires in-depth understanding of the general principles and theoretical bases of the methods to be used. The approach followed in this book uses standardized mathematical symbols, abundant illustration, and appeal to intuition in some cases. Because the text has been used for graduate teaching, we know that, with reasonable effort, readers can get to the core of numerical ecology. In order to efficiently use numerical methods, their aims and limits must be clearly understood, as well as the conditions under which they should be used. In addition, since most methods are well described in the scientific literature and are available in computer packages, we generally insist on the ecological interpretation of results; computation algorithms are described only when they may help understand methods. Methods described in the book are systematically illustrated by numerical examples and/or applications drawn from the ecological literature, mostly in English; references written in languages other than English or French are generally of historical nature.

The expression *numerical ecology* refers to the following approach. *Mathematical ecology* covers the domain of mathematical applications to ecology. It may be divided into *theoretical ecology* and *quantitative ecology*. The latter, in turn, includes a number of disciplines, among which *modelling*, *ecological statistics*, and *numerical ecology*. *Numerical ecology* is the field of quantitative ecology devoted to the numerical analysis of ecological data sets. Community ecologists, who generally use multivariate data, are the primary users of these methods. The purpose of numerical ecology is to describe and interpret the structure of data sets by combining a variety of numerical approaches. Numerical ecology differs from descriptive or inferential *biological statistics* in that it extensively uses non-statistical procedures, and systematically combines relevant multidimensional statistical methods with non-statistical numerical techniques (e.g. cluster analysis); statistical inference (i.e. tests of significance) is seldom used. Numerical ecology also differs from *ecological modelling*, even though the extrapolation of ecological structures is often used to *forecast* values in space or/and time (through multiple regression or other similar approaches, which are collectively referred to as *correlative models*). When the purpose of a study is to *predict* the critical consequences of alternative solutions, ecologists must use *predictive ecological models*. The development of models that predict the effects on some variables, caused by changes in others (see, for instance, De Neufville & Stafford, 1971), requires a deliberate causal structuring, which is based on ecological theory; it must include a validation procedure. Such models are often difficult and costly to construct. Because the ecological hypotheses that underlay causal models (see for instance Gold, 1977, Jolivet, 1982, or Jørgensen, 1983) are often developed within the context of studies using numerical ecology, the two fields are often in close contact.

Loehle (1983) reviewed the different types of models used in ecology, and discussed some relevant evaluation techniques. In his scheme, there are three types of *simulation models*: logical, theoretical, and "predictive". In a *logical model*, the representation of a system is based on logical operators. According to Loehle, such models are not frequent in ecology, and the few that exist may be questioned as to their biological meaningfulness. *Theoretical models* aim at explaining natural phenomena in a universal fashion. Evaluating a theory first requires that the model be accurately translated into mathematical form, which is often difficult to do. *Numerical models* (called by Loehle "predictive" models, *sensu lato*) are divided in two types: *application models* (called, in the present book, *predictive models*, *sensu stricto*) are based on well-established laws and theories, the laws being *applied* to resolve a particular problem; *calculation tools* (called *forecasting* or *correlative models* in the previous paragraph) do not have to be based on any law of nature and may thus be ecologically meaningless, but they may still be useful for forecasting. In forecasting models, most components are subject to adjustment whereas, in ideal predictive models, only the boundary conditions may be adjusted.

Ecologists have used quantitative approaches since the publication by Jaccard (1900) of the first association coefficient. Floristics developed from this seed, and the method was eventually applied to all fields of ecology, often achieving high levels of complexity. Following Spearman (1904) and Hotelling (1933), psychometricians and social scientists developed non-parametric statistical methods and factor analysis and, later, nonmetric multidimensional scaling (MDS). During the same period, anthropologists (e.g. Czekanowski, 1909) were interested in numerical classification. The advent of computers made it possible to analyse large data sets, using combinations of methods derived from various fields and supplemented with new mathematical developments. The first synthesis was published by Sokal & Sneath (1963), who established *numerical taxonomy* as a new discipline.

Numerical ecology combines a large number of approaches, derived from many disciplines, in a general methodology for analysing ecological data sets. Its chief characteristic is the *combined* use of treatments drawn from different areas of mathematics and statistics. Numerical ecology acknowledges the fact that many of the existing numerical methods are *complementary* to one another, each one allowing to explore a different aspect of the information underlying the data; it sets principles for interpreting the results in an integrated way.

The present book is organized in such a way as to encourage researchers who are interested in a method to also consider other techniques. The integrated approach to data analysis is favoured by numerous cross-references among chapters and the presence of sections presenting syntheses of subjects. The book synthesizes a large amount of information from the literature, within a structured and prospective framework, so as to help ecologists take maximum advantage of the existing methods.

This second English edition of ***Numerical ecology*** is a revised and largely expanded translation of the second edition of ***Écologie numérique*** (Legendre & Legendre, 1984a, 1984b). Compared to the first English edition (1983a), there are three new chapters, dealing with the analysis of semiquantitative data (Chapter 5), canonical analysis (Chapter 11), and spatial analysis (Chapter 13). In addition, new sections have been added to almost all other chapters. These include, for example, new sections (numbers given in parentheses) on: autocorrelation (1.1), statistical testing by randomization (1.2), coding (1.5), missing data (1.6), singular value decomposition (2.11), multiway contingency tables (6.3), cophenetic matrix and ultrametric property (8.3), reversals (8.6), partitioning by *K*-means (8.8), cluster validation (8.12), a review of regression methods (10.3), path analysis (10.4), a review of matrix comparison methods (10.5), the 4th-corner problem (10.6), several new methods for the analysis of data series (12.3-12.5), detection of discontinuities in multivariate series (12.6), and Box-Jenkins models (12.7). There are also sections listing available computer programs and packages at the end of several Chapters.

The present work reflects the input of many colleagues, to whom we express here our most sincere thanks. We first acknowledge the outstanding collaboration of Professors Serge Frontier (Université des Sciences et Techniques de Lille) and F. James Rohlf (State University of New York at Stony Brook) who critically reviewed our manuscripts for the first French and English editions, respectively. Many of their suggestions were incorporated into the texts which are at the origin of the present edition. We are also grateful to Prof. Ramón Margalef for his support, in the form of an influential Preface to the previous editions. Over the years, we had fruitful discussions on various aspects of numerical methods with many colleagues, whose names have sometimes been cited in the Forewords of previous editions.

During the preparation of this new edition, we benefited from intensive collaborations, as well as chance encounters and discussions, with a number of people who have thus contributed, knowingly or not, to this book. Let us mention a few. Numerous discussions with Robert R. Sokal and Neal L. Oden have sharpened our understanding of permutation methods and methods of spatial data analysis. Years of discussion with Pierre Dutilleul and Claude Bellehumeur led to the Section on spatial autocorrelation. Pieter Kroonenberg provided useful information on the relationship between singular value decomposition (SVD) and correspondence analysis (CA). Peter Minchin shed light on detrended correspondence analysis (DCA) and nonmetric multidimensional scaling (MDS). A discussion with Richard M. Cormack about the behaviour of some model II regression techniques helped us write Subsection 10.3.2. This Subsection also benefited from years of investigation of model II methods with David J. Currie. In-depth discussions with John C. Gower led us to a better understanding of the metric and Euclidean properties of (dis)similarity coefficients and of the importance of Euclidean geometry in grasping the role of negative eigenvalues in principal coordinate analysis (PCoA). Further research collaboration with Marti J. Anderson about negative eigenvalues in PCoA, and permutation tests in multiple regression and canonical analysis, made it possible to write the corresponding sections of this book; Dr. Anderson also provided comments on Sections 9.2.4, 10.5 and 11.3. Cajo J. F. ter Braak revised Chapter 11 and parts of Chapter 9, and suggested a number of improvements. Claude Bellehumeur revised Sections 13.1 and 13.2; François-Joseph Lapointe commented on successive drafts of 8.12. Marie-Josée Fortin and Daniel Borcard provided comments on Chapter 13. The ÉCOTHAU program on the Thau lagoon in southern France (led by Michel Amanieu), and the NIWA workshop on soft-bottom habitats in Manukau harbour in New Zealand (organized by Rick Pridmore and Simon Thrush of NIWA), provided great opportunities to test many of the ecological hypothesis and methods of spatial analysis presented in this book.

Graduate students at Université de Montréal and Université Laval have greatly contributed to the book by raising interesting questions and pointing out weaknesses in previous versions of the text. The assistance of Bernard Lebanc was of great value in transferring the ink-drawn figures of previous editions to computer format. Philippe Casgrain helped solve a number of problems with computers, file transfers, formats, and so on.

While writing this book, we benefited from competent and unselfish advice … which we did not always follow. We thus assume full responsibility for any gaps in the work and for all the opinions expressed therein. We shall therefore welcome with great interest all suggestions or criticisms from readers.

PIERRE LEGENDRE, Université de Montréal
LOUIS LEGENDRE, Université Laval

April 1998

Chapter

1 Complex ecological data sets

1.0 Numerical analysis of ecological data

The foundation of a general methodology for analysing ecological data may be derived from the relationships that exist between the conditions surrounding ecological observations and their outcomes. In the physical sciences for example, there often are cause-to-effect relationships between the natural or experimental conditions and the outcomes of observations or experiments. This is to say that, given a certain set of conditions, the outcome may be exactly predicted. Such totally deterministic relationships are only characteristic of extremely simple ecological situations.

Probability

Generally in ecology, a number of different outcomes may follow from a given set of conditions because of the large number of influencing variables, of which many are not readily available to the observer. The inherent genetic variability of biological material is an important source of ecological variability. If the observations are repeated many times under similar conditions, the relative frequencies of the possible outcomes tend to stabilize at given values, called the *probabilities* of the outcomes. Following Cramér (1946: 148) it is possible to state that "whenever we say that the probability of an event with respect to an experiment [or an observation] is equal to P, the concrete meaning of this assertion will thus simply be the following: in a long series of repetitions of the experiment [or observation], it is practically certain that the [relative] frequency of the event will be approximately equal to P." This corresponds to the frequency theory of probability — excluding the Bayesian or likelihood approach.

Probability distribution

Random variable

In the first paragraph, the outcomes were recurring at the individual level whereas in the second, results were repetitive in terms of their probabilities. When each of several possible outcomes occurs with a given characteristic probability, the set of these probabilities is called a *probability distribution.* Assuming that the numerical value of each outcome E_i is y_i with corresponding probability p_i, a *random variable* (or *variate*) $\mathbf{y}$ is defined as that quantity which takes on the value y_i with probability p_i at each trial (e.g. Morrison, 1990). Fig. 1.1 summarizes these basic ideas.

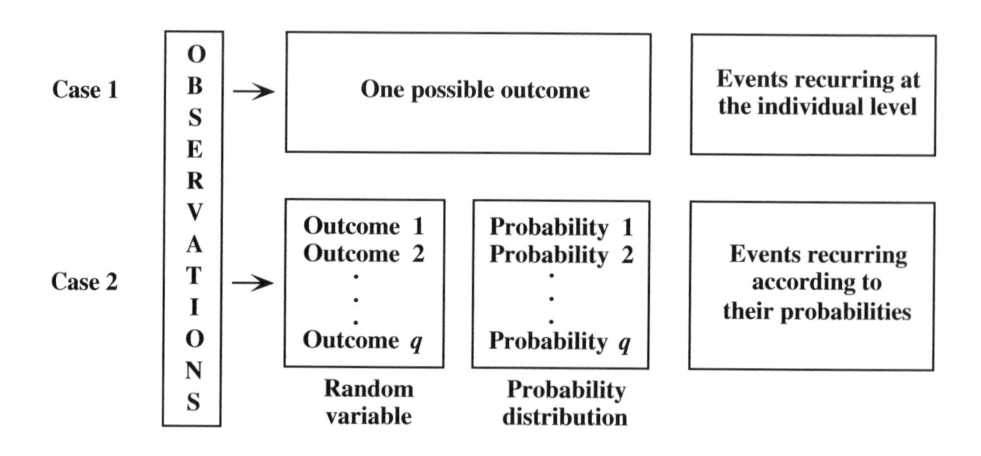

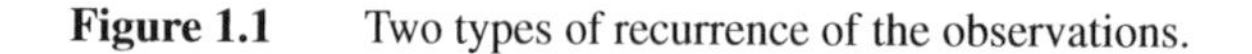

Figure 1.1 Two types of recurrence of the observations.

Of course, one can imagine other results to observations. For example, there may be *strategic* relationships between surrounding conditions and resulting events. This is the case when some action — or its expectation — triggers or modifies the reaction. Such strategic-type relationships, which are the object of *game theory*, may possibly explain ecological phenomena such as species succession or evolution (Margalef, 1968). Should this be the case, this type of relationship might become central to ecological research. Another possible outcome is that observations be *unpredictable.* Such data may be studied within the framework of *chaos theory*, which explains how natural phenomena that are apparently completely stochastic sometimes result from deterministic relationships. Chaos is increasingly used in theoretical ecology. For example, Stone (1993) discusses possible applications of chaos theory to simple ecological models dealing with population growth and the annual phytoplankton bloom. Interested readers should refer to an introductory book on chaos theory, for example Gleick (1987).

Methods of numerical analysis are determined by the four types of relationships that may be encountered between surrounding conditions and the outcome of observations (Table 1.1). The present text deals only with methods for analysing random variables, which is the type ecologists most frequently encounter.

The numerical analysis of ecological data makes use of mathematical tools developed in many different disciplines. A formal presentation must rely on a unified approach. For ecologists, the most suitable and natural language — as will become evident in Chapter 2 — is that of *matrix algebra.* This approach is best adapted to the processing of data by computers; it is also simple, and it efficiently carries information, with the additional advantage of being familiar to many ecologists.

Other disciplines provide ecologists with powerful tools that are well adapted to the complexity of ecological data. From mathematical physics comes *dimensional analysis* (Chapter 3), which provides simple and elegant solutions to some difficult ecological problems. Measuring the association among quantitative, semiquantitative or qualitative variables is based on *parametric* and *nonparametric statistical methods* and on *information theory* (Chapters 4, 5 and 6, respectively).

These approaches all contribute to the analysis of complex ecological data sets (Fig. 1.2). Because such data usually come in the form of highly interrelated variables, the capabilities of elementary statistical methods are generally exceeded. While elementary methods are the subject of a number of excellent texts, the present manual focuses on the more advanced methods, upon which ecologists must rely in order to understand these interrelationships.

In ecological spreadsheets, data are typically organized in rows corresponding to sampling sites or times, and columns representing the variables; these may describe the biological communities (species presence, abundance, or biomass, for instance) or the physical environment. Because many variables are needed to describe communities and environment, ecological data sets are said to be, for the most part, *multidimensional* (or *multivariate*). Multidimensional data, i.e. data made of several variables, structure what is known in geometry as a *hyperspace*, which is a space with many dimensions. One now classical example of ecological hyperspace is the *fundamental niche* of Hutchinson (1957, 1965). According to Hutchinson, the environmental variables that are critical for a species to exist may be thought of as orthogonal axes, one for each factor, of a multidimensional space. On each axis, there are limiting conditions within which the species can exist indefinitely; we will call upon this concept again in Chapter 7, when discussing unimodal species distributions and their consequences on the choice of resemblance coefficients. In Hutchinson's theory, the set of these limiting conditions defines a *hypervolume* called the species'

Table 1.1 Numerical analysis of ecological data.

Relationships between the natural conditions and the outcome of an observation	Methods for analysing and modelling the data
Deterministic: Only one possible result	Deterministic models
Random: Many possible results, each one with a recurrent frequency	Methods described in this book (Figure 1.2)
Strategic: Results depend on the respective strategies of the organisms and of their environment	Game theory
Uncertain: Many possible, unpredictable results	Chaos theory

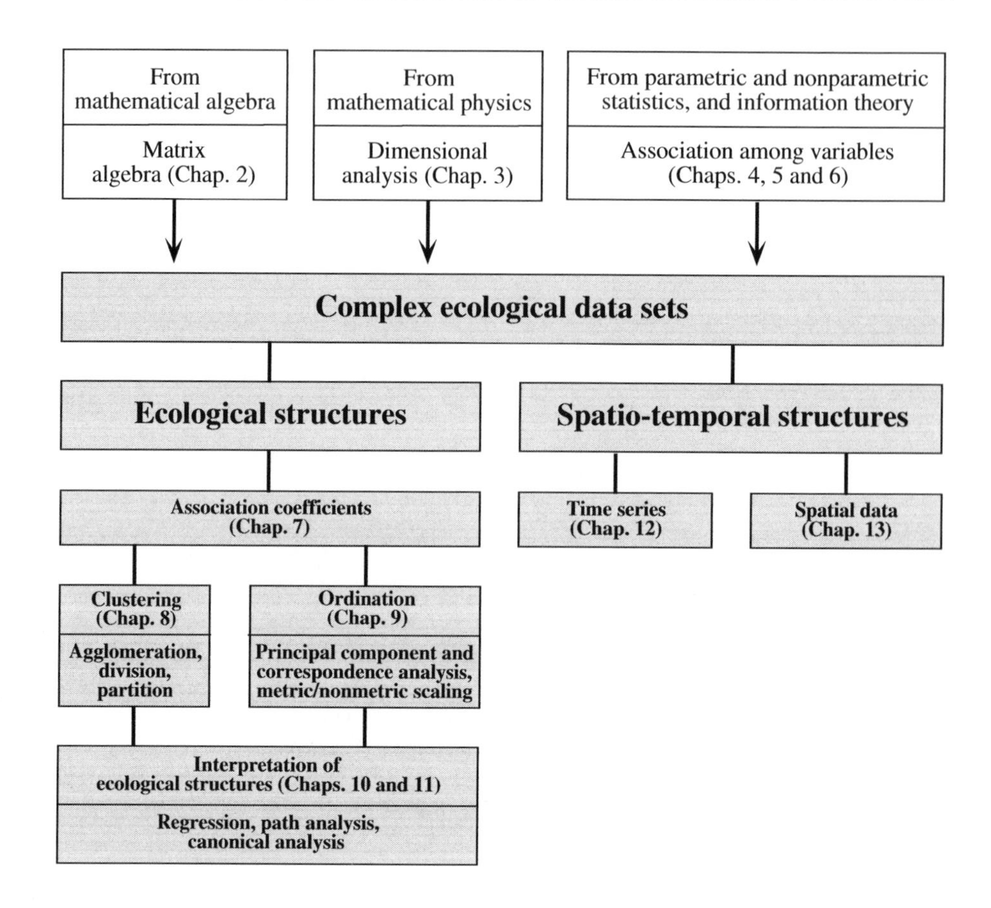

Figure 1.2 Numerical analysis of complex ecological data sets.

Fundamental niche

fundamental niche. The spatial axes, on the other hand, describe the geographical distribution of the species.

The quality of the analysis and subsequent interpretation of complex ecological data sets depends, in particular, on the compatibility between data and numerical methods. It is important to take into account the requirements of the numerical techniques when planning the sampling programme, because it is obviously useless to collect quantitative data that are inappropriate to the intended numerical analyses. Experience shows that, too often, poorly planned collection of costly ecological data, for "survey" purposes, generates large amounts of unusable data (Fig. 1.3).

The search for ecological structures in multidimensional data sets is always based on *association matrices*, of which a number of variants exist, each one leading to slightly or widely different results (Chapter 7); even in so-called association-free

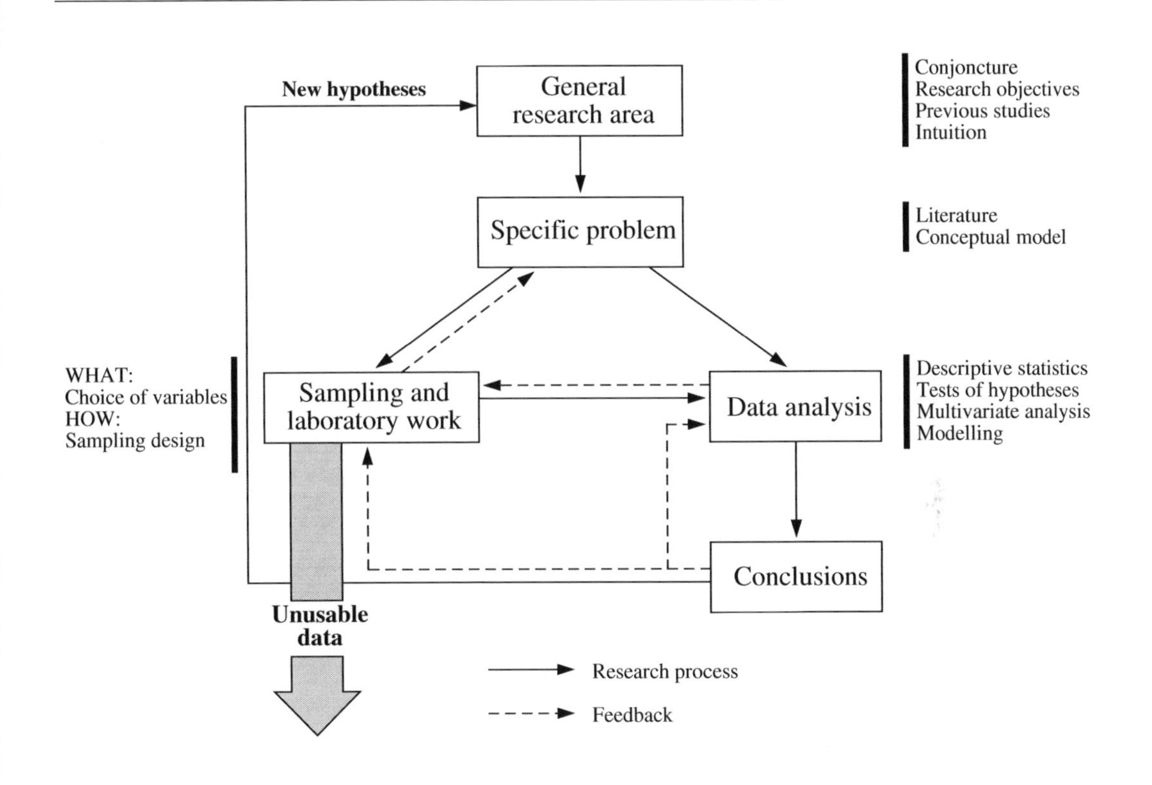

Figure 1.3 Interrelationships between the various phases of an ecological research.

methods, like principal component or correspondence analysis, or *k*-means clustering, there is always an implicit resemblance measure hidden in the method. Two main avenues are open to ecologists: (1) ecological *clustering* using agglomerative, divisive or partitioning algorithms (Chapter 8), and (2) *ordination* in a space with a reduced number of dimensions, using principal component or coordinate analysis, nonmetric multidimensional scaling, or correspondence analysis (Chapter 9). The *interpretation of ecological structures,* derived from clustering and/or ordination, may be conducted in either a direct or an indirect manner, as will be seen in Chapters 10 and 11, depending on the nature of the problem and on the additional information available.

Besides multidimensional data, ecologists may also be interested in temporal or spatial *process data,* sampled along temporal or spatial axes in order to identify time- or space-related processes (Chapters 12 and 13, respectively) driven by physics or biology. Time or space sampling requires intensive field work, which may often be automated nowadays using equipment that allows the automatic recording of ecological variables, or the quick surveying or automatic recording of the geographic positions of observations. The analysis of satellite images or information collected by airborne or shipborne equipment falls in this category. In physical or ecological

Process

applications, a *process* is a phenomenon or a set of phenomena organized along time or in space. Mathematically speaking, such ecological data represent one of the possible realizations of a random process, also called a *stochastic process*.

Design-based

Model-based

Super-population

Two major approaches may be used for inference about the population parameters of such processes (Särndal, 1978; Koch & Gillings, 1983; de Gruijter & ter Braak, 1990). In the *design-based approach*, one is interested only in the sampled population and assumes that a fixed value of the variable exists at each location in space, or point in time. A "representative" subset of the space or time units is selected and observed during sampling (for 8 different meanings of the expression "representative sampling", see Kruskal & Mosteller, 1988). Design-based (or *randomization-based*; Kempthorne, 1952) inference results from statistical analyses whose only assumption is the random selection of observations; this requires that the target population (i.e. that for which conclusions are sought) be the same as the sampled population. The probabilistic interpretation of this type of inference (e.g. confidence intervals of parameters) refers to repeated selection of observations from the same finite population, using the same sampling design. The classical (Fisherian) methods for estimating the confidence intervals of parameters, for variables observed over a given surface or time stretch, are fully applicable in the design-based approach. In the *model-based* (or *superpopulation*) *approach*, the assumption is that the target population is much larger than the sampled population. So, the value associated with each location, or point in time, is not fixed but random, since the geographic surface (or time stretch) available for sampling (i.e. the statistical population) is seen as one representation of the superpopulation of such surfaces or time stretches — all resulting from the same generating process — about which conclusions are to be drawn. Under this model, even if the whole sampled population could be observed, uncertainty would still remain about the model parameters. So, the confidence intervals of parameters estimated over a single surface or time stretch are obviously too small to account for the among-surface variability, and some kind of correction must be made when estimating these intervals. The type of variability of the superpopulation of surfaces or time stretches may be estimated by studying the spatial or temporal autocorrelation of the available data (i.e. over the statistical population). This subject is discussed at some length in Section 1.1. Ecological survey data can often be analysed under either model, depending on the emphasis of the study or the type of conclusions one wishes to derive from them.

In some instances in time series analysis, the sampling design must meet the requirements of the numerical method, because some methods are restricted to data series meeting some specific conditions, such as equal spacing of observations. Inadequate planning of the sampling may render the data series useless for numerical treatment with these particular methods. There are several methods for analysing *ecological series*. Regression, moving averages, and the variate difference method are designed for identifying and extracting general trends from time series. Correlogram, periodogram, and spectral analysis identify rhythms (characteristic periods) in series. Other methods can detect discontinuities in univariate or multivariate series. Variation in a series may be correlated with variation in other variables measured simultaneously. Finally, one may want to develop forecasting models using the Box & Jenkins approach.

Similarly, methods are available to meet various objectives when analysing spatial structures. Structure functions such as variograms and correlograms, as well as point pattern analysis, may be used to confirm the presence of a statistically significant spatial structure and to describe its general features. A variety of interpolation methods are used for mapping univariate data, whereas multivariate data can be mapped using methods derived from ordination or cluster analysis. Finally, models may be developed that include spatial structures among their explanatory variables.

For ecologists, numerical analysis of data is not a goal in itself. However, a study which is based on quantitative information must take data processing into account at all phases of the work, from conception to conclusion, including the planning and execution of sampling, the analysis of data proper, and the interpretation of results. Sampling, including laboratory analyses, is generally the most tedious and expensive part of ecological research, and it is therefore important that it be optimized in order to reduce to a minimum the collection of useless information. Assuming appropriate sampling and laboratory procedures, the conclusions to be drawn now depend on the results of the numerical analyses. It is, therefore, important to make sure in advance that sampling and numerical techniques are compatible. It follows that mathematical processing is at the heart of a research; the quality of the results cannot exceed the quality of the numerical analyses conducted on the data (Fig. 1.3).

Of course, the quality of ecological research is not a sole function of the expertise with which quantitative work is conducted. It depends to a large extent on creativity, which calls upon imagination and intuition to formulate hypotheses and theories. It is, however, advantageous for the researcher's creative abilities to be grounded into solid empirical work (i.e. work involving field data), because little progress may result from continuously building upon untested hypotheses.

Figure 1.3 shows that a correct interpretation of analyses requires that the sampling phase be planned to answer a specific question or questions. Ecological sampling programmes are designed in such a way as to capture the variation occurring along a number of axe of interest: space, time, or other ecological indicator variables. The purpose is to describe variation occurring along the given axis or axes, and to interpret or model it. Contrary to experimentation, where sampling may be designed in such a way that observations are independent of each other, ecological data are often *autocorrelated* (Section 1.1).

While experimentation is often construed as the opposite of ecological sampling, there are cases where field experiments are conducted at sampling sites, allowing one to measure rates or other processes ("manipulative experiments" *sensu* Hurlbert, 1984; Subsection 10.2.3). In aquatic ecology, for example, nutrient enrichment bioassays are a widely used approach for testing hypotheses concerning nutrient limitation of phytoplankton. In their review on the effects of enrichment, Hecky & Kilham (1988) identify four types of bioassays, according to the level of organization of the test system: cultured algae; natural algal assemblages isolated in microcosms or sometimes larger enclosures; natural water-column communities enclosed in mesocosms; whole systems. The authors discuss one major question raised by such experiments, which is whether results from lower-level systems are applicable to higher levels, and especially to natural situations. Processes estimated in experiments may be used as independent variables in empirical models accounting for survey results, while "static" survey data may be used as covariates to explain the variability observed among blocks of experimental treatments. In the future, spatial or time-series data analysis may become an important part of the analysis of the results of ecological experiments.

1.1 Autocorrelation and spatial structure

Ecologists have been trained in the belief that Nature follows the assumptions of classical statistics, one of them being the independence of observations. However, field ecologists know from experience that organisms are not randomly or uniformly distributed in the natural environment, because processes such as growth, reproduction, and mortality, which create the observed distributions of organisms, generate spatial autocorrelation in the data. The same applies to the physical variables which structure the environment. Following hierarchy theory (Simon, 1962; Allen & Starr, 1982; O'Neill *et al.*, 1991), we may look at the environment as primarily structured by broad-scale physical processes — orogenic and geomorphological processes on land, currents and winds in fluid environments — which, through energy inputs, create gradients in the physical environment, as well as patchy structures separated by discontinuities (interfaces). These broad-scale structures lead to similar responses in biological systems, spatially and temporally. Within these relatively homogeneous zones, finer-scale contagious biotic processes take place that cause the appearance of more spatial structuring through reproduction and death, predator-prey interactions, food availability, parasitism, and so on. This is not to say that biological processes are necessarily small-scaled and nested within physical processes; biological processes may be broad-scaled (e.g. bird and fish migrations) and physical processes may be fine-scaled (e.g. turbulence). The theory only purports that stable complex systems are often hierarchical. The concept of scale, as well as the expressions *broad scale* and *fine scale*, are discussed in Section 13.0.

In ecosystems, spatial heterogeneity is therefore functional, and not the result of some random, noise-generating process; so, it is important to study this type of variability for its own sake. One of the consequences is that ecosystems without spatial structuring would be unlikely to function. Let us imagine the consequences of a non-spatially-structured ecosystem: broad-scale homogeneity would cut down on diversity of habitats; feeders would not be close to their food; mates would be located at random throughout the landscape; soil conditions in the immediate surrounding of a plant would not be more suitable for its seedlings than any other location; newborn animals would be spread around instead of remaining in favourable environments; and so on. Unrealistic as this view may seem, it is a basic assumption of many of the theories and models describing the functioning of populations and communities. The view of a spatially structured ecosystem requires a new paradigm for ecologists: spatial [and temporal] structuring is a fundamental component of ecosystems. It then becomes obvious that theories and models, including statistical models, must be revised to include realistic assumptions about the spatial and temporal structuring of communities.

Autocorrelation

Spatial autocorrelation may be loosely defined as the property of random variables which take values, at pairs of sites a given distance apart, that are more similar (positive autocorrelation) or less similar (negative autocorrelation) than expected for randomly associated pairs of observations. Autocorrelation only refers to the lack of independence (Box 1.1) among the *error components* of field data, due to geographic proximity. Autocorrelation is also called *serial correlation* in time series analysis. A spatial structure may be present in data without it being caused by autocorrelation. Two models for spatial structure are presented in Subsection 1; one corresponds to autocorrelation, the other not.

Because it indicates lack of independence among the observations, autocorrelation creates problems when attempting to use tests of statistical significance that require independence of the observations. This point is developed in Subsection 1.2. Other types of dependencies (or, lack of independence) may be encountered in biological data. In the study of animal behaviour for instance, if the same animal or pair of animals is observed or tested repeatedly, these observations are not independent of one another because the same animals are likely to display the same behaviour when placed in the same situation. In the same way, paired samples (last paragraph in Box 1.1) cannot be analysed as if they were independent because members of a pair are likely to have somewhat similar responses.

Autocorrelation is a very general property of ecological variables and, indeed, of most natural variables observed along time series (temporal autocorrelation) or over geographic space (spatial autocorrelation). Spatial [or temporal] autocorrelation may be described by mathematical functions such as correlograms and variograms, called structure functions, which are studied in Chapters 12 and 13. The two possible approaches concerning statistical inference for autocorrelated data (i.e. the design- or randomization-based approach, and the model-based or superpopulation approach) were discussed in Section 1.0.

The following discussion is partly derived from the papers of Legendre & Fortin (1989) and Legendre (1993). Spatial autocorrelation is used here as the most general case, since temporal autocorrelation behaves essentially like its spatial counterpart, but along a single sampling dimension. The difference between the spatial and temporal cases is that causality is unidirectional in time series, i.e. it proceeds from $(t-1)$ to t and not the opposite. Temporal processes, which generate temporally autocorrelated data, are studied in Chapter 12, whereas spatial processes are the subject of Chapter 13.

Independence **Box 1.1**

This word has several meanings. Five of them will be used in this book. Another important meaning in statistics concerns *independent random variables*, which refer to properties of the distribution and density functions of a group of variables (for a formal definition, see Morrison, 1990, p. 7).

Independent observations — Observations drawn from the statistical population in such a way that no observed value has any influence on any other. In the time-honoured example of tossing a coin, observing a head does not influence the probability of a head (or tail) coming out at the next toss. Autocorrelated data violate this condition, their error terms being correlated across observations.

Independent descriptors — Descriptors (variables) that are not related to one another are said to be independent. *Related* is taken here in some general sense applicable to quantitative, semiquantitative as well as qualitative data (Table 1.2).

Linear independence — Two descriptors are said to be linearly independent, or *orthogonal*, if their covariance is equal to zero. A Pearson correlation coefficient may be used to test the hypothesis of linear independence. Two descriptors that are linearly independent may be related in a nonlinear way. For example, if vector **x**' is centred ($\mathbf{x}' = [x_i - \bar{x}]$), vector $[x'^2_i]$ is linearly independent of vector **x**' (their correlation is zero) although they are in perfect quadratic relationship.

Independent variable(s) of a model — In a regression model, the variable to be modelled is called the *dependent variable*. The variables used to model it, usually found on the right-hand side of the equation, are called the *independent variables* of the model. In empirical models, one may talk about *response* (or *target*) and *explanatory* variables for, respectively, the dependent and independent variables, whereas, in a causal framework, the terms *criterion* and *predictor* variables may be used. Some forms of canonical analysis (Chapter 11) allow one to model several dependent (target or criterion) variables in a single regression-like analysis.

Independent samples are opposed to *related* or *paired samples*. In related samples, each observation in a sample is paired with one in the other sample(s), hence the name *paired comparisons* for the tests of significance carried out on such data. Authors also talk of *independent* versus *matched* pairs of data. Before-after comparisons of the same elements also form related samples (matched pairs).

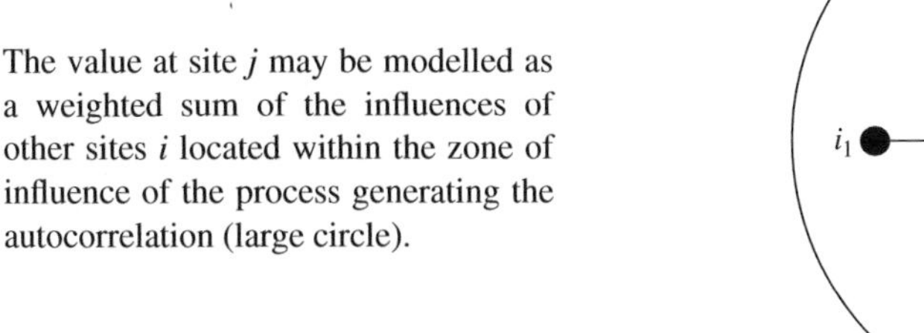

Figure 1.4 The value at site j may be modelled as a weighted sum of the influences of other sites i located within the zone of influence of the process generating the autocorrelation (large circle).

1 — Types of spatial structures

A spatial structure may appear in a variable **y** because the process that has produced the values of **y** is spatial and has generated autocorrelation in the data; or it may be caused by dependence of **y** upon one or several causal variables **x** which are spatially structured; or both. The spatially-structured causal variables **x** may be explicitly identified in the model, or not; see Table 13.3.

Autocorrelation

• Model 1: autocorrelation — The value y_j observed at site j on the geographic surface is assumed to be the overall mean of the process (μ_y) plus a weighted sum of the centred values $(y_i - \mu_y)$ at surrounding sites i, plus an independent error term ε_j:

$$y_j = \mu_y + \Sigma f(y_i - \mu_y) + \varepsilon_j \tag{1.1}$$

The y_i's are the values of **y** at other sites i located within the zone of spatial influence of the process generating the autocorrelation (Fig. 1.4). The influence of neighbouring sites may be given, for instance, by weights w_i which are function of the distance between sites i and j (eq. 13.19); other functions may be used. The total error term is $[\Sigma f(y_i - \mu_y) + \varepsilon_j]$; it contains the autocorrelated component of variation. As written here, the model assumes stationarity (Subsection 13.1.1). Its equivalent in time series analysis is the autoregressive (AR) response model (eq. 12.30).

Spatial dependence

• Model 2: spatial dependence — If one can assume that there is no autocorrelation in the variable of interest, the spatial structure may result from the influence of some explanatory variable(s) exhibiting a spatial structure. The model is the following:

$$y_j = \mu_y + f(\text{explanatory variables}) + \varepsilon_j \tag{1.2}$$

where y_j is the value of the dependent variable at site j and ε_j is an error term whose value is independent from site to site. In such a case, the spatial structure, called "trend", may be filtered out by trend surface analysis (Subsection 13.2.1), by the method of spatial variate differencing (see Cliff & Ord 1981, Section 7.4), or by some equivalent method in the case of time series (Chapter 12). The significance of the relationship of interest (e.g. correlation, presence of significant groups) is tested on the detrended data. The variables should not be detrended, however, when the spatial structure is of interest in the study. Chapter 13 describes how spatial structures may be studied and decomposed into fractions that may be attributed to different hypothesized causes (Table 13.3).

Detrending

It is difficult to determine whether a given observed variable has been generated under model 1 (eq. 1.1) or model 2 (eq. 1.2). The question is further discussed in Subsection 13.1.2 in the case of gradients ("false gradients" and "true gradients").

More complex models may be written by combining autocorrelation in variable **y** (model 1) and the effects of causal variables **x** (model 2), plus the autoregressive structures of the various **x**'s. Each parameter of these models may be tested for significance. Models may be of various degrees of complexity, e.g. *simultaneous AR model, conditional AR model* (Cliff & Ord, 1981, Sections 6.2 and 6.3; Griffith, 1988, Chapter 4).

Spatial structures may be the result of several processes acting at different spatial scales, these processes being independent of one another. Some of these — usually the intermediate or fine-scale processes — may be of interest in a given study, while other processes may be well-known and trivial, like the broad-scale effects of tides or world-wide climate gradients.

2 — Tests of statistical significance in the presence of autocorrelation

Autocorrelation in a variable brings with it a statistical problem under the model-based approach (Section 1.0): it impairs the ability to perform standard statistical tests of hypotheses (Section 1.2). Let us consider an example of spatially autocorrelated data. The observed values of an ecological variable of interest — for example, species composition — are most often influenced, at any given site, by the structure of the species assemblages at surrounding sites, because of contagious biotic processes such as growth, reproduction, mortality and migration. Make a first observation at site A and a second one at site B located near A. Since the ecological process is understood to some extent, one can assume that the data are spatially autocorrelated. Using this assumption, one can anticipate to some degree the value of the variable at site B before the observation is made. Because the value at any one site is influenced by, and may be at least partly forecasted from the values observed at neighbouring sites, these values are not stochastically independent of one another.

The influence of spatial autocorrelation on statistical tests may be illustrated using the correlation coefficient (Section 4.2). The problem lies in the fact that, when the two variables under study are positively autocorrelated, the confidence interval, estimated by the classical procedure around a Pearson correlation coefficient (whose calculation assumes independent and identically distributed error terms for all observations), is

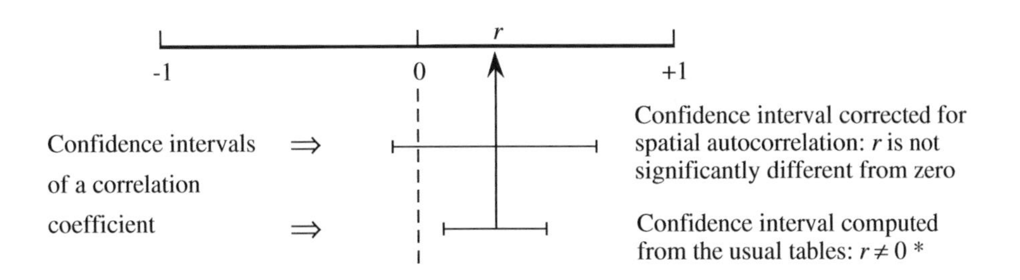

Figure 1.5 Effect of positive spatial autocorrelation on tests of correlation coefficients; * means that the coefficient is declared significantly different from zero in this example.

narrower than it is when calculated correctly, i.e. taking autocorrelation into account. The consequence is that one would declare too often that correlation coefficients are significantly different from zero (Fig. 1.5; Bivand, 1980). All the usual statistical tests, nonparametric and parametric, have the same behaviour: in the presence of positive autocorrelation, computed test statistics are too often declared significant under the null hypothesis. Negative autocorrelation may produce the opposite effect, for instance in analysis of variance (ANOVA).

The effects of autocorrelation on statistical tests may also be examined from the point of view of the *degrees of freedom*. As explained in Box 1.2, in classical statistical testing, one degree of freedom is counted for each independent observation, from which the number of estimated parameters is subtracted. The problem with autocorrelated data is their lack of independence or, in other words, the fact that new observations do not each bring with them one full degree of freedom, because the values of the variable at some sites give the observer some prior knowledge of the values the variable should take at other sites. The consequence is that new observations cannot be counted for one full degree of freedom. Since the size of the fraction they bring with them is difficult to determine, it is not easy to know what the proper reference distribution for the test should be. All that is known for certain is that positive autocorrelation at short distance distorts statistical tests (references in the next paragraph), and that this distortion is on the "liberal" side. This means that, when positive spatial autocorrelation is present in the small distance classes, the usual statistical tests too often lead to the decision that correlations, regression coefficients, or differences among groups are significant, when in fact they may not be.

This problem has been well documented in correlation analysis (Bivand, 1980; Cliff & Ord, 1981, §7.3.1; Clifford *et al.*, 1989; Haining, 1990, pp. 313-330; Dutilleul, 1993a), linear regression (Cliff & Ord, 1981, §7.3.2; Chalmond, 1986; Griffith, 1988, Chapter 4; Haining, 1990, pp. 330-347), analysis of variance (Crowder & Hand, 1990; Legendre *et al.*, 1990), and tests of normality (Dutilleul & Legendre, 1992). The problem of estimating the confidence interval for the mean when the sample data are

Degrees of freedom

Box 1.2

Statistical tests of significance often call upon the concept of degrees of freedom. A formal definition is the following: "The degrees of freedom of a model for expected values of random variables is the excess of the number of variables [observations] over the number of parameters in the model" (Kotz & Johnson, 1982).

In practical terms, the number of degrees of freedom associated with a statistic is equal to the number of its independent components, i.e. the total number of components used in the calculation minus the number of parameters one had to estimate from the data before computing the statistic. For example, the number of degrees of freedom associated with a variance is the number of observations minus one (noted $\nu = n - 1$): n components $(x_i - \bar{x})$ are used in the calculation, but one degree of freedom is lost because the mean of the statistical population is estimated from the sample data; this is a prerequisite before estimating the variance.

There is a different t distribution for each number of degrees of freedom. The same is true for the F and χ^2 families of distributions, for example. So, the number of degrees of freedom determines which statistical distribution, in these families (t, F, or χ^2), should be used as the reference for a given test of significance. Degrees of freedom are discussed again in Chapter 6 with respect to the analysis of contingency tables.

autocorrelated has been studied by Cliff & Ord (1975, 1981, §7.2) and Legendre & Dutilleul (1991).

When the presence of spatial autocorrelation has been demonstrated, one may wish to remove the spatial dependency among observations; it would then be valid to compute the usual statistical tests. This might be done, in theory, by removing observations until spatial independence is attained; this solution is not recommended because it entails a net loss of information which is often expensive. Another solution is detrending the data (Subsection 1); if autocorrelation is part of the process under study, however, this would amount to throwing out the baby with the water of the bath. It would be better to analyse the autocorrelated data as such (Chapter 13), acknowledging the fact that autocorrelation in a variable may result from various causal mechanisms (physical or biological), acting simultaneously and additively.

The alternative for testing statistical significance is to modify the statistical method in order to take spatial autocorrelation into account. When such a correction is available, this approach is to be preferred if one assumes that autocorrelation is an intrinsic part of the ecological process to be analysed or modelled.

Corrected tests rely on modified estimates of the variance of the statistic, and on corrected estimates of the effective sample size and of the number of degrees of freedom. Simulation studies are used to demonstrate the validity of the modified tests. In these studies, a large number of autocorrelated data sets are generated under the null hypothesis (e.g. for testing the difference between two means, pairs of observations are drawn at random from *the same* simulated, autocorrelated statistical distribution, which corresponds to the null hypothesis of no difference between population means) and tested using the modified procedure; this experiment is repeated a large number of times to demonstrate that the modified testing procedure leads to the nominal confidence level.

Cliff & Ord (1973) have proposed a method for correcting the standard error of parameter estimates for the simple linear regression in the presence of autocorrelation. This method was extended to linear correlation, multiple regression, and *t*-test by Cliff & Ord (1981, Chapter 7: approximate solution) and to the one-way analysis of variance by Griffith (1978, 1987). Bartlett (1978) has perfected a previously proposed method of correction for the effect of spatial autocorrelation due to an autoregressive process in randomized field experiments, adjusting plot values by covariance on neighbouring plots before the analysis of variance; see also the discussion by Wilkinson *et al.* (1983) and the papers of Cullis & Gleeson (1991) and Grondona & Cressis (1991). Cook & Pocock (1983) have suggested another method for correcting multiple regression parameter estimates by maximum likelihood, in the presence of spatial autocorrelation. Using a different approach, Legendre *et al.* (1990) have proposed a permutational method for the analysis of variance of spatially autocorrelated data, in the case where the classification criterion is a division of a territory into nonoverlapping regions and one wants to test for differences among these regions.

A step forward was proposed by Clifford *et al.* (1989), who tested the significance of the correlation coefficient between two spatial processes by estimating a modified number of degrees of freedom, using an approximation of the variance of the correlation coefficient computed on data. Empirical results showed that their method works fine for positive autocorrelation in large samples. Dutilleul (1993a) generalized the procedure and proposed an exact method to compute the variance of the sample covariance; the new method is valid for any sample size.

Major contributions to this topic are found in the literature on time series analysis, especially in the context of regression modelling. Important references are Cochrane & Orcutt (1949), Box & Jenkins (1976), Beach & MacKinnon (1978), Harvey & Phillips (1979), Chipman (1979), and Harvey (1981).

When methods specifically designed to handle spatial autocorrelation are not available, it is sometimes possible to rely on permutation tests, where the significance is determined by random reassignment of the observations (Section 1.2). Special permutational schemes have been developed that leave autocorrelation invariant; examples are found in Besag & Clifford (1989), Legendre *et al.* (1990) and ter Braak

(1990, section 8). For complex problems, such as the preservation of spatial or temporal autocorrelation, the difficulty of the permutational method is to design an appropriate permutation procedure.

The methods of clustering and ordination described in Chapters 8 and 9 to study ecological structures do not rely on tests of statistical significance. So, they are not affected by the presence of spatial autocorrelation. The impact of spatial autocorrelation on numerical methods will be stressed wherever appropriate.

3 — Classical sampling and spatial structure

Random or systematic sampling designs have been advocated as a way of preventing the possibility of dependence among observations (Cochran 1977; Green 1979; Scherrer 1982). This was then believed to be a necessary and sufficient safeguard against violations of the independence of errors, which is a basic assumption of classical statistical tests. It is adequate, of course, when one is trying to estimate the parameters of a local population. In such a case, a random or systematic sample is suitable to obtain unbiased estimates of the parameters since, *a priori*, each point has the same probability of being included in the sample. Of course, the variance and, consequently, also the standard error of the mean increase if the distribution is patchy, but their estimates remain unbiased.

Even with random or systematic allocation of observations through space, observations may retain some degree of spatial dependence if the average distance between first neighbours is shorter than the zone of spatial influence of the underlying ecological phenomenon. In the case of broad-scale spatial gradients, no point is far enough to lie outside this zone of spatial influence. Correlograms and variograms (Chapter 13), combined with maps, are used to assess the magnitude and shape of autocorrelation present in data sets.

Classical books such as Cochran (1977) adequately describe the rules that should govern sampling designs. Such books, however, emphasize only the design-based inference (Section 1.0), and do not discuss the influence of spatial autocorrelation on the sampling design. At the present time, literature on this subject seems to be only available in the field of geostatistics, where important references are: David (1977, Ch. 13), McBratney & Webster (1981), McBratney *et al.* (1981), Webster & Burgess (1984), Borgman & Quimby (1988), and François-Bongarçon (1991).

Ecologists interested in designing field experiments should read the paper of Dutilleul (1993b), who discusses how to accommodate an experiment to spatially heterogeneous conditions. The concept of spatial heterogeneity is discussed at some length in the multi-author book edited by Kolasa & Pickett (1991), in the review paper of Dutilleul & Legendre (1993), and in Section 13.0.

Heterogeneity

1.2 Statistical testing by permutation

Statistic

The role of a statistical test is to decide whether some *parameter* of the reference population may take a value assumed by hypothesis, given the fact that the corresponding statistic, whose value is estimated from a sample of objects, may have a somewhat different value. A *statistic* is any quantity that may be calculated from the data and is of interest for the analysis (examples below); in tests of significance, a statistic is called *test statistic* or *test criterion.* The assumed value of the parameter corresponding to the statistic in the reference population is given by the statistical null hypothesis (written H_0), which translates the biological null hypothesis into numerical terms; it often negates the existence of the phenomenon that the scientists hope to evidence. The reasoning behind statistical testing directly derives from the scientific method; it allows the confrontation of experimental or observational findings to intellectual constructs that are called hypotheses.

Testing is the central step of inferential statistics. It allows one to generalize the conclusions of statistical estimation to some reference population from which the observations have been drawn and that they are supposed to represent. Within that context, the problem of multiple testing is too often ignored (Box. 1.3). Another legitimate section of statistical analysis, called descriptive statistics, does not rely on testing. The methods of clustering and ordination described in Chapters 8 and 9, for instance, are descriptive multidimensional statistical methods. The interpretation methods described in Chapters 10 and 11 may be used in either descriptive or inferential mode.

1 — Classical tests of significance

Null hypothesis

Consider, for example, a correlation coefficient (which is the statistic of interest in correlation analysis) computed between two variables (Chapter 4). When inference to the statistical population is sought, the null hypothesis is often that the value of the correlation parameter (ρ, rho) in the statistical population is zero; the null hypothesis may also be that ρ has some value other than zero, given by the ecological hypothesis. To judge of the validity of the null hypothesis, the only information available is an *estimate* of the correlation coefficient, *r*, obtained from a sample of objects drawn from the statistical population. (Whether the observations adequately represent the statistical population is another question, for which the readers are referred to the literature on sampling design.) We know, of course, that a sample is quite unlikely to produce a parameter estimate which is exactly equal to the true value of the parameter in the statistical population. A statistical test tries to answer the following question: given a hypothesis stating, for example, that $\rho = 0$ in the statistical population and the fact that the estimated correlation is, say, $r = 0.2$, is it justified to conclude that the difference between 0.2 and 0.0 is due to sampling error?

The choice of the statistic to be tested depends on the problem at hand. For instance, in order to find whether two samples may have been drawn from the same

Multiple testing

Box 1.3

When several tests of significance are carried out simultaneously, the probability of a type I error becomes larger than the nominal value α. For example, when analysing a correlation matrix involving 5 variables, 10 tests of significance are carried out simultaneously. For randomly generated data, there is a probability $p = 0.40$ of rejecting the null hypothesis at least once over 10 tests, at the nominal $\alpha = 0.05$ level; this can easily be computed from the binomial distribution. So, when conducting multiple tests, one should perform a global test of significance in order to determine whether there is any significant value at all in the set.

The first approach is Fisher's method for combining the probabilities p_i obtained from k independent tests of significance. The value $-2\Sigma \ln(p_i)$ is distributed as χ^2 with $2k$ degrees of freedom if the null hypothesis is true in all k tests (Fisher, 1954; Sokal & Rohlf, 1995).

Another approach is the Bonferroni correction for k independent tests: replace the significance level, say $\alpha = 0.05$, by an adjusted level $\alpha' = \alpha/k$, and compare probabilities p_i to α'. This is equivalent to adjusting individual p-values p_i to $p'_i = kp_i$ and comparing p'_i to the unadjusted significance level α. While appropriate to test the null hypothesis for the whole set of simultaneous hypotheses (i.e. reject H_0 for the whole set of k hypotheses if the smallest unadjusted p-value in the set is less than or equal to α/k), the Bonferroni method is overly conservative and often leads to rejecting too few individual hypotheses in the set k.

Several alternatives have been proposed in the literature; see Wright (1992) for a review. For non-independent tests, Holm's procedure (1979) is nearly as simple to carry out as the Bonferroni adjustment and it is much more powerful, leading to rejecting the null hypothesis more often. It is computed as follows. (1) Order the p-values from left to right so that $p_1 \le p_2 \le \ldots \le p_i \ldots \le p_k$. (2) Compute adjusted probability values $p'_i = (k - i + 1)p_i$; adjusted probabilities may be larger than 1. (3) Proceeding from left to right, if an adjusted p-value in the ordered series is smaller than the one occurring at its left, make the smallest equal to the largest one. (4) Compare each adjusted p'_i to the unadjusted α significance level and make the statistical decision. The procedure could be formulated in terms of successive corrections to the α significance level, instead of adjustments to individual probabilities.

An even more powerful solution is that of Hochberg (1988) which has the desired overall ("experimentwise") error rate α only for independent tests (Wright, 1992). Only step (3) differs from Holm's procedure: proceeding this time from right to left, if an adjusted p-value in the ordered series is smaller than the one at its left, make the largest equal to the smallest one. Because the adjusted probabilities form a nondecreasing series, both of these procedures present the properties (1) that a hypothesis in the ordered series cannot be rejected unless all previous hypotheses in the series have also been rejected and (2) that equal p-values receive equal adjusted p-values. Hochberg's method presents the further characteristic that no adjusted p-value can be larger than the largest unadjusted p-value or exceed 1. More complex and powerful procedures are explained by Wright (1992).

For some applications, special procedures have been developed to test a whole set of statistics. An example is the test for the correlation matrix **R** (eq. 4.14, end of Section 4.2).

statistical population or from populations with equal means, one would choose a statistic measuring the difference between the two sample means $(\bar{x}_1 - \bar{x}_2)$ or, preferably, a *pivotal* form like the usual *t* statistic used in such tests; a pivotal statistic has a distribution under the null hypothesis which remains the same for any value of the measured effect (here, $\bar{x}_1 - \bar{x}_2$). In the same way, the slope of a regression line is described by the slope parameter of the linear regression equation, which is assumed, under the null hypothesis, to be either zero or some other value suggested by ecological theory. The test statistic describes the difference between the observed and hypothesized value of slope; the pivotal form of this difference is a *t* or *F* statistic.

Pivotal statistic

Alternative hypothesis

Another aspect of a statistical test is the alternative hypothesis (H_1), which is also imposed by the ecological problem at hand. H_1 is the opposite of H_0, but there may be several statements that represent some opposite of H_0. In correlation analysis for instance, if one is satisfied to determine that the correlation coefficient in the reference population (ρ) is significantly different from zero in either the positive or the negative direction, meaning that *some* linear relationship exists between two variables, then a *two-tailed* alternative hypothesis is stated about the value of the parameter in the statistical population: $\rho \neq 0$. On the contrary, if the ecological phenomenon underlying the hypothesis imposes that a relationship, if present, should have a given sign, one formulates a *one-tailed* hypothesis. For instance, studies on the effects of acid rain are motivated by the general paradigm that acid rain, which lowers the pH, has a negative effect on terrestrial and aquatic ecosystems. In a study of the correlation between pH and diversity, one would formulate the following hypothesis H_1: pH and diversity are positively correlated (i.e. low pH is associated with low diversity; H_1: $\rho > 0$). Other situations would call for a different alternative hypothesis, symbolized by H_1: $\rho < 0$.

The expressions *one-tailed* and *two-tailed* refer to the fact that, in a two-tailed test, one would look in both tails of the reference statistical distribution for values as extreme as, or more extreme than the reference value of the statistic (i.e. the one computed from the actual data). In a correlation study for instance, where the reference distribution (*t*) for the test statistic is symmetric about zero, the probability of the null hypothesis in a two-tailed test is given by the proportion of values in the *t* distribution which are, *in absolute value*, as large as, or larger than the *absolute value* of the reference statistic. In a one-tailed test, one would look only in the tail corresponding to the sign given by the alternative hypothesis; for instance, for the proportion of values in the *t* distribution which are as large as or larger than the *signed value* of the reference *t* statistic, for a test in the right-hand tail (H_1: $\rho > 0$).

In standard statistical tests, the *test statistic* computed from the data is referred to one of the usual statistical distributions printed in books or computed by some appropriate computer software; the best-known are the *z*, *t*, *F* and χ^2 distributions. This, however, can only be done if certain assumptions are met by the data, depending on the test. The most commonly encountered are the assumptions of normality of the variable(s) in the reference population, homoscedasticity (Box 1.4) and independence of the observations (Box 1.1). Refer to Siegel (1956, Chapter 2), Siegel & Castellan

(1988, Chapter 2), or Snedecor & Cochran (1967, Chapter 1), for concise yet clear classical exposés of the concepts related to statistical testing.

2 — Permutation tests

Randomization

The method of *permutation*, also called *randomization*, is a very general approach to testing statistical hypotheses. Following Manly (1997), permutation and randomization are considered synonymous in the present book, although *permutation* may also be considered to be the technique by which the principle of *randomization* is applied to data during permutation tests. Other points of view are found in the literature. For instance, Edgington (1995) considers that a randomization test is a permutation test based on randomization. A different although related meaning of *randomization* refers to the random assignment of replicates to treatments in experimental designs.

Permutation testing can be traced back to at least Fisher (1935, Chapter 3). Instead of comparing the actual value of a test statistic to a standard statistical distribution, the reference distribution is generated from the data themselves, as described below; other randomization methods are mentioned at the end of the present Section. Permutation provides an efficient approach to testing when the data do not conform to the distributional assumptions of the statistical method one wants to use (e.g. normality). Permutation testing is applicable to very small samples, like nonparametric tests. It *does not* resolve problems of independence of the observations, however. Nor does the method solve distributional problems that are linked to the hypothesis subjected to a test*. Permutation remains the method of choice to test novel or other statistics whose distributions are poorly known. Furthermore, results of permutation tests are valid even with observations that are not a random sample of some statistical population; this point is further discussed in Subsection 4. Edgington (1995) and Manly (1997) have written excellent introductory books about the method. A short account is given by Sokal & Rohlf (1995) who prefer to use the expression "randomization test". Permutation tests are used in several Chapters of the present book.

The speed of modern computers would allow users to perform any statistical test using the permutation method. The chief advantage is that one does not have to worry about distributional assumptions of classical testing procedures; the disadvantage is the amount of computer time required to actually perform a large number of permutations, each one being followed by recomputation of the test statistic. This disadvantage vanishes as faster computers come on the market. As an example, let us

---

* For instance, when studying the differences among sample means (two groups: t-test; several groups: F test of ANOVA), the classical Behrens-Fisher problem (Robinson, 1982) reminds us that two null hypotheses are tested simultaneously by these methods, i.e. equality of the means and equality of the variances. Testing the t or F statistics by permutations does not change the dual aspect of the null hypothesis; in particular, it does not allow one to unambiguously test the equality of the means without checking first the equality of the variances using another, more specific test (two groups: F ratio; several groups: Bartlett's test of equality of variances).

consider the situation where the significance of a correlation coefficient between two variables, $\mathbf{x}_1$ and $\mathbf{x}_2$, is to be tested.

Hypotheses

- H_0: The correlation between the variables in the reference population is zero ($\rho = 0$).

- For a two-tailed test, H_1: $\rho \neq 0$.

- Or for a one-tailed test, either H_1: $\rho > 0$, or H_1: $\rho < 0$, depending on the ecological hypothesis.

Test statistic

- Compute the Pearson correlation coefficient r. Calculate the pivotal statistic $t = \sqrt{n-2}\,[r/\sqrt{1-r^2}]$ (eq. 4.13; n is the number of observations) and use it as the reference value in the remainder of the test.

In this specific case, the permutation test results would be the same using either r or t as the test statistic, because t is a monotonic function of r for any constant value of n; r and t are "equivalent statistics for permutation tests", *sensu* Edgington (1995). This is not always the case. When testing a partial regression coefficient in multiple regression, for example, the test should not be based on the distribution of permuted partial regression coefficients because they are not monotonic to the corresponding partial t statistics. The partial t should be preferred because it is pivotal and, hence, it is expected to produce correct type I error.

Considering a pair of equivalent test statistics, one could choose the statistic which is the simplest to compute if calculation time would otherwise be longer in an appreciable way. This is not the case in the present example: calculating t involves a single extra line in the computer program compared to r. So the test is conducted using the usual t statistic.

Distribution of the test statistic

The argument invoked to construct a null distribution for the statistic is that, if the null hypothesis is true, all possible pairings of the two variables are equally likely to occur. The pairing found in the observed data is just one of the possible, equally likely pairings, so that the value of the test statistic for the unpermuted data should be typical, i.e. located in the central part of the permutation distribution.

- It is always the null hypothesis which is subjected to testing. Under H_0, the rows of $\mathbf{x}_1$ are seen as "exchangeable" with one another if the rows of $\mathbf{x}_2$ are fixed, or conversely. The observed pairing of $\mathbf{x}_1$ and $\mathbf{x}_2$ values is due to chance alone; accordingly, any value of $\mathbf{x}_1$ could have been paired with any value of $\mathbf{x}_2$.

• A realization of H_0 is obtained by permuting at random the values of $\mathbf{x}_1$ while holding the values of $\mathbf{x}_2$ fixed, or the opposite (which would produce, likewise, a random pairing of values). Recompute the value of the correlation coefficient and the associated t statistic for the randomly paired vectors $\mathbf{x}_1$ and $\mathbf{x}_2$, obtaining a value t^*.

• Repeat this operation a large number of times (say, 999 times). The different permutations produce a set of values t^* obtained under H_0.

• Add to these the reference value of the t statistic, computed for the unpermuted vectors. Since H_0 is being tested, this value is considered to be one that could be obtained under H_0 and, consequently, it should be added to the reference distribution (Hope, 1968; Edgington, 1995; Manly, 1997). Together, the unpermuted and permuted values form an estimate of the sampling distribution of t under H_0, to be used in the next step.

Statistical decision

• As in any other statistical test, the decision is made by comparing the reference value of the test statistic (t) to the reference distribution obtained under H_0. If the reference value of t is typical of the values obtained under the null hypothesis (which states that there is no relationship between $\mathbf{x}_1$ and $\mathbf{x}_2$), H_0 cannot be rejected; if it is unusual, being too extreme to be considered a likely result under H_0, H_0 is rejected and the alternative hypothesis is considered to be a more likely explanation of the data.

Significance level

• The significance level of a statistic is the proportion of values that are as extreme as, or more extreme than the test statistic in the reference distribution, which is either obtained by permutations or found in a table of the appropriate statistical distribution. The level of significance should be regarded as "the strength of evidence against the null hypothesis" (Manly, 1997).

3 — Numerical example

Let us consider the following case of two variables observed over 10 objects:

$\mathbf{x}_1$	–2.31	1.06	0.76	1.38	–0.26	1.29	–1.31	0.41	–0.67	–0.58
$\mathbf{x}_2$	–1.08	1.03	0.90	0.24	–0.24	0.76	–0.57	–0.05	–1.28	1.04

These values were drawn at random from a positively correlated bivariate normal distribution, as shown in Fig. 1.6a. Consequently, they would be suitable for parametric testing. So, it is interesting to compare the results of a permutation test to the usual parametric t-test of the correlation coefficient. The statistics and associated probabilities for this pair of variables, for $\nu = (n - 2) = 8$ degrees of freedom, are:

$r = 0.70156$, $t = 2.78456$, $n = 10$:
prob (one-tailed) = 0.0119, prob (two-tailed) = 0.0238.

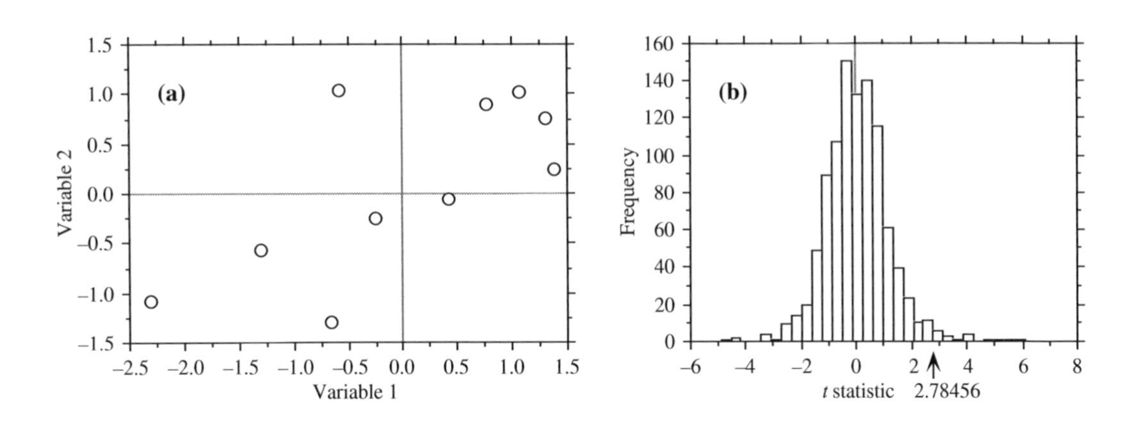

Figure 1.6 (a) Positions of the 10 points of the numerical example with respect to variables $\mathbf{x}_1$ and $\mathbf{x}_2$. (b) Frequency histogram of the (1 + 999) permutation results (*t* statistic for correlation coefficient); the reference value obtained for the points in (a), $t = 2.78456$, is also shown.

There are $10! = 3.6288 \times 10^6$ possible permutations of the 10 values of variable $\mathbf{x}_1$ (or $\mathbf{x}_2$). Here, 999 of these permutations were generated using a random permutation algorithm; they represent a random sample of the 3.6288×10^6 possible permutations. The computed values for the test statistic (t) between permuted $\mathbf{x}_1$ and fixed $\mathbf{x}_2$ have the distribution shown in Fig. 1.6b; the reference value, $t = 2.78456$, has been added to this distribution. The permutation results are summarized in the following table, where '$|t|$' is the (absolute) reference value of the t statistic ($|t| = 2.78456$) and 't^*' is a value obtained after permutation. The absolute value of the reference t is used in the table to make it a general example, because there are cases where t is negative.

	$t^* < -\lvert t\rvert$	$t^* = -\lvert t\rvert$	$-\lvert t\rvert < t^* < \lvert t\rvert$	$t^* = \lvert t\rvert$	$t^* > \lvert t\rvert$
Statistic t	8	0	974	1†	17

† This count corresponds to the reference *t* value added to the permutation results.

For a one-tailed test (in the right-hand tail in this case, since H_1: $\rho > 0$), one counts how many values in the permutational distribution of the statistic are equal to, or larger than, the reference value ($t^* \geq t$; there are 1 + 17 = 18 such values in this case). This is the only one-tailed hypothesis worth considering, because the objects are known in this case to have been drawn from a positively correlated distribution. A one-tailed test in the left-hand tail (H_1: $\rho < 0$) would be based on how many values in the permutational distribution are equal to, or smaller than, the reference value ($t^* \leq t$, which are 8 + 0 + 974 +1 = 983 in the example). For a two-tailed test, one counts all values that are as extreme as, or more extreme than the reference value *in both tails of the distribution* ($|t^*| \geq |t|$, which are 8 + 0 + 1 + 17 = 26 in the example).

Probabilities associated with these distributions are computed as follows, for a one-tailed and a two-tailed test (results using the t statistic would be the same):

One-tailed test [H_0: $\rho = 0$; H_1: $\rho > 0$]:
prob ($t^* \geq 2.78456$) = (1 + 17)/1000 = 0.018

Two-tailed test [H_0: $\rho = 0$; H_1: $\rho \neq 0$]:
prob($|t^*| \geq 2.78456$) = (8 + 0 + 1 + 17)/1000 = 0.026

Note how similar the permutation results are to the results obtained from the classical test, which referred to a table of Student t distributions. The observed difference is partly due to the small number of pairs of points ($n = 10$) sampled at random from the bivariate normal distribution, with the consequence that the data set does not quite conform to the hypothesis of normality. It is also due, to a certain extent, to the use of only 999 permutations, sampled at random among the 10! possible permutations.

4 — Remarks on permutation tests

In permutation tests, the reference distribution against which the statistic is tested is obtained by randomly permuting the data under study, without reference to any statistical population. The test is valid as long as the reference distribution has been generated by a procedure related to a null hypothesis that makes sense for the problem at hand, irrespective of whether or not the data set is representative of a larger statistical population. This is the reason why the data do not have to be a random sample from some larger statistical population. The only information the permutation test provides is whether the pattern observed in the data is likely, or not, to have arisen by chance. For this reason, one may think that permutation tests are not as "good" or "interesting" as classical tests of significance because they might not allow one to infer conclusions that apply to a statistical population.

A more pragmatic view is that the conclusions of permutation tests may be generalized to a reference population if the data set is a random sample of that population. Otherwise, they allow one to draw conclusions only about the particular data set, measuring to what extent the value of the statistic is "usual" or "unusual" with respect to the null hypothesis implemented in the permutation procedure. Edgington (1995) and Manly (1997) further argue that data sets are very often not drawn at random from statistical populations, but simply consist of observations which happen to be available for study. The generalization of results, in classical as well as permutation tests, depends on the degree to which the data were actually drawn at random, or are equivalent to a sample drawn at random, from a reference population.

Complete permutation test

For small data sets, one can compute all possible permutations in a systematic way and obtain the complete permutation distribution of the statistic; an *exact* or *complete permutation test* is obtained. For large data sets, only a sample of all possible permutations may be computed because there are too many. When designing a

Sampled permutation test

sampled permutation test, it is important to make sure that one is using a *uniform random generation algorithm*, capable of producing all possible permutations with equal probabilities (Furnas, 1984). Computer programs use procedures that produce random permutations of the data; these in turn call the 'Random' function of computer languages. Such a procedure is described in Section 5.8 of Manly's book (1997). Random permutation subprograms are also available in subroutine libraries.

The case of the correlation coefficient has shown how the null hypothesis guided the choice of an appropriate permutation procedure, capable of generating realizations of this null hypothesis. A permutation test for the difference between the means of two groups would involve random permutations of the objects between the two groups instead of random permutations of one variable with respect to the other. The way of permuting the data depends on the null hypothesis to be tested.

Some tests may be reformulated in terms of some other tests. For example, the t-test of equality of means is equivalent to a test of the correlation between the vector of observed values and a vector assigning the observations to group 1 or 2. The same value of t and probability (classical or permutational) are obtained using both methods.

Restricted permutations

Simple statistical tests such as those of correlation coefficients or differences between group means may be carried out by permuting the original data, as in the example above. Problems involving complex relationships among variables may require permuting the residuals of some *model* instead of the raw data; *model-based permutation* is discussed in Subsection 11.3.2. The effect of a nominal covariable may be controlled for by *restricted permutations*, limited to the objects within the groups defined by the covariable. This method is discussed in detail by Manly (1997). Applications are found in Brown & Maritz (1982; restrictions within replicated values in a multiple regression) and in Sokal *et al.* (1987; Mantel test), for instance.

In sampled permutation tests, adding the reference value of the statistic to the distribution has the effect that it becomes impossible for the test to produce no value "as extreme as, or more extreme than the reference value", as the standard expression goes. This way of computing the probability is biased, but it has the merit of being statistically valid (Edgington, 1995, Section 3.5). The precision of the probability estimate is the inverse of the number of permutations performed; for instance, after (999 + 1) permutations, the precision of the probability statement is 0.001.

How many permutations?

The number of permutations one should perform is always a trade-off between precision and computer time. The more permutations the better, since probability estimates are subject to error due to sampling the population of possible permutations (except in the rare cases of complete permutation tests), but it may be tiresome to wait for the permutation results when studying large data sets. In the case of the Mantel test (Section 10.5), Jackson & Somers (1989) recommend to compute 10000 to 100000 permutations in order to ensure the stability of the probability estimates. The following recommendation can be made. In exploratory analyses, 500 to 1000 permutations may be sufficient as a first contact with the problem. If the computed probability is close to the preselected significance level, run more permutations. In any case, use more permutations (e.g. 10000) for final, published results.

Interestingly, tables of critical values in nonparametric statistical tests for small sample sizes are based on permutations. The authors of these tables have computed how many cases can be found, in the complete permutation distribution, that are as extreme as, or more extreme than the computed value of the statistic. Hence, probability statements obtained from small-sample nonparametric tests are exact probabilities (Siegel, 1956).

Monte Carlo

Named after the famous casino of the principality of Monaco, Monte Carlo methods use random numbers to study either real data sets or the behaviour of statistical methods through simulations. Permutation tests are Monte Carlo methods because they use random numbers to randomly permute data. Other such methods are based on computer-intensive resampling. Among these are the jackknife (Tukey 1958; Sokal & Rohlf, 1995) and the bootstrap (Efron, 1979; Efron & Tibshirani, 1993; Manly, 1997). In these methods, the values used in each iteration to compute a statistic form a subsample of the original data. In the jackknife, each subsample leaves out one of the original observations. In the bootstrap, each subsample is obtained by resampling the original sample *with replacement*; the justification is that resampling the original sample approximates a resampling of the original population.

Jackknife
Bootstrap

As an exercise, readers are invited to figure out how to perform a permutation test for the difference between the means of two groups of objects on which a single variable has been measured, using the t statistic; this would be equivalent to a t-test. A solution is given by Edgington (1995). Other types of permutation tests are discussed in Sections 7.3, 8.9, 10.2, 10.3, 10.5, 10.6, 11.3, 12.6, 13.1 and 13.3.

1.3 Computers

Processing complex ecological data sets almost always requires the use of a computer, as much for the amount of data to be processed as for the fact that the operations to be performed are often tedious and repetitious. Work with computers is made simple by the statistical programs and packages available on microcomputers or on mainframes. For those who want to develop new methods, advanced programming languages such as S-PLUS® or MATLAB®, or the SAS® Language, may be extremely useful. One may also complement programs written in one of the standard computer languages with statistical subprograms drawn from libraries such as NAG® or IMSL®, which are available at computing centres and contain subprograms for the numerical resolution of most common numerical problems. The ease of using computers, however, has two pitfalls that ecologists must bear in mind: the fact that computations are executed does not ensure (1) that the data satisfy the conditions required by the method, or (2) that the results produced by the computer are interpreted correctly in ecological terms.

In fact, ecologists must thoroughly master the numerical methods they use. If they do not, they may end up using approaches that are incompatible with their data, or selecting techniques that do not correspond to the goal of the research, or else interpreting the results either incompletely or incorrectly. The only role of the computer is to relieve the user of the *calculations*; it makes no appraisal of the pertinence of the selected method, or of the interpretation of the results. The aim of the following chapters is to provide ecologists with a guide to the use of the many numerical methods available, as well as the bases for translating, in ecological terms, the pages of numbers produced by computers. Indeed, a computer output is not in itself a conclusion and it will never replace ecologists for interpreting results.

Ten years ago, the statistical packages most widely available were BMDP® (*Biomedical Computer Programs*), SPSS® (*Statistical Package for the Social Sciences*) and SAS® (*Statistical Analysis System*). Versions of these packages, originally developed for mainframe computers, are now available for microcomputers as well. A wide array of other programs and packages have also been developed for microcomputers, giving users a wide choice. The selection of a given package for a specific task may be guided by its availability for the user's preferred machine, the methods it contains, and its computer-friendliness. In everyday work, ecologists rely nowadays on several packages, each specialized for specific tasks such as clustering, ordination, canonical analysis, time series analysis, spatial analysis, graphics, mapping, word processing, etc. While low-end microcomputers can perform most everyday tasks of data analysis, high-end machines or mainframes retain their usefulness to analyse large data bases or for permutation-based, computer-intensive statistical testing methods.

A review of the contents of the main statistical packages available on the market is beyond the scope of the present book, and would rapidly become obsolete. Such reviews may be found in statistical Journals, for instance *The American Statistician*, *Applied Statistics*, or *Statistics and Computing*. Lists of programs for some of the more specialized fields of analysis will be provided in some chapters of this book.

Programs for all the numerical techniques described in the following chapters can be found in one or several packages. It is, therefore, much more efficient for ecologists to use these proven resources than to reprogram methods. Ecological data are, most of the time, so complex that a single analysis cannot extract all their useful information. Therefore, ecologists who judiciously use existing programs have access to a variety of numerical methods, which are needed to cover the wide range of ecological situations encountered in field studies.

1.4 Ecological descriptors

Descriptor
Variable

Any ecological study, classical or numerical, is based on *descriptors*. In the present text, the terms *descriptor* and *variable* will be used interchangeably. These refer to the attributes, or characters (also called items in the social sciences, and profiles or features in the field of pattern recognition) used to describe or compare the *objects of the study*. The *objects* that ecologists compare are the sites, quadrats, observations, sampling units, individual organisms, or subjects which are defined *a priori* by the sampling design, before making the observations (Section 2.1). Observation units are often called "samples" by ecologists; the term *sample* is only used in this book to refer to *a set of observations* resulting from a sampling action or campaign. Objects may be called individuals or OTUs (*Operational taxonomic units*) in numerical taxonomy, OGUs (*Operational geographic units*) in biogeography, cases, patterns or items in the field of pattern recognition, etc.

Object

The descriptors, used to describe or qualify the objects, are the physical, chemical, ecological, or biological characteristics of these objects that are of interest for the study. In particular, biological species are *descriptors* of sites for ecologists; in (numerical) taxonomy on the contrary, the species are the *objects* of the study, and the sites where the species are observed or collected may be used by the taxonomist as descriptors of the species. It all depends on the variable defined *a priori,* which is fixed as object for the study. In ecology, sites are compared using the species they contain, there being no possibility of choosing the species, whereas taxonomists compare populations or other taxonomic entities obtained from a number of different sites.

Descriptor

A *descriptor is a law of correspondence established by the researcher* to describe and compare, on the same basis, all the objects of the study. This definition applies to all types of descriptors discussed below (Table 1.2). The fundamental property of a descriptor, as understood in the present book, is that it distributes the objects among non-overlapping states. Each descriptor must, therefore, operate like *a law that associates with each object in the group under study one and only one element of a set of distinguishable states that belong to the descriptor.*

Descriptor state

The *states* that constitute a descriptor *must necessarily be mutually exclusive.* In other words, two different states of the same descriptor must not be applicable to the same object. Descriptors, on the contrary, do not have to be independent of one another (see Box 1.1: independent descriptors). In Chapter 6, it will be seen that the information contained in one descriptor may partially or totally overlap with the information in an other. In Chapters 8 and 9, such redundant or correlated information will be used as the basis for the clustering or ordination of ecological objects.

1 — Mathematical types of descriptor

The states which form a descriptor — that is, the qualities observed or determined on the objects — may be of a qualitative or quantitative nature, so that descriptors may be classified into several types. In ecology, a descriptor may be biological (presence, abundance, or biomass of different species), physical, chemical, geological, geographical, temporal, climatic, etc. Table 1.2 presents a classification of descriptors according to their mathematical types. This classification is, therefore, independent of the particular discipline to which the descriptors belong. The mathematical type of a descriptor determines the type of numerical processing which can be applied to it. For

Table 1.2 The different mathematical types of descriptors.

Descriptor types	Examples
Binary (two states, presence-absence)	Species present or absent
Multi-state (many states)	
Nonordered (qualitative, nominal, attributes)	Geological group
Ordered	
Semiquantitative (rank-ordered, ordinal)	Importance or abundance scores
Quantitative (metric, measurement)	
Discontinuous (meristic, discrete)	Equidistant abundance classes
Continuous	Temperature, length

example, parametric correlations (Pearson's r) may be calculated between quantitative descriptors, while nonparametric correlations (such as Kendall's τ) may be used on ordered but not necessarily quantitative descriptors, as long as their relationship is monotonic. To measure the dependence among descriptors that are not in monotonic relationship, or among qualitative descriptors, requires the use of other methods based on contingency tables (Chapter 6). Section 1.5 and Chapter 10 will show how descriptors of different mathematical types can be made compatible, in order to use them together in ecological studies.

Quantitative descriptors, which are the most usual type in ecology, are found at the bottom of Table 1.2. They include all descriptors of abundance and other quantities that can be plotted on a continuous axis of real numbers. They are called quantitative, or *metric* (Falconer, 1960), because they measure changes in a phenomenon in such a way that the difference between 1 and 2, for example, is quantitatively the same as the difference between, say, 6 and 7. Such descriptors may be further subdivided into *relative-scale* quantitative variables, where value 'zero' means the absence of the characteristic of interest, and *interval-scale* variables where the 'zero' is chosen arbitrarily. For the latter type, the fact that the 'zero' reference is chosen arbitrarily prevents comparisons of the type "this temperature (°C) is twice as high as that one". Species abundance data, or temperatures measured in Kelvin, are examples of the first type, while temperature measured in degrees Celsius, dates, or geographic directions (of wind, currents, etc.) in degrees, are examples of the second.

Relative scale

Interval scale

Continuous quantitative descriptors are usually processed as they are. If they are divided into a small number of *equidistant* classes of abundance (further discussed below), the discontinuous descriptors that are obtained may usually be processed as if they were continuous, because the distortion due to grouping is negligible for the majority of distribution types (Sneath & Sokal, 1973). Before the advent of computers, it was usual practice, in order to facilitate calculations, to divide continuous descriptors into a small number of classes. This transformation is still necessary when, due to low precision of the measurements, only a small number of classes can be distinguished, or when comparisons are sought between quantitative and semiquantitative descriptors.

Meristic variables (the result of enumeration, or counting) theoretically should be considered as discontinuous quantitative. In ecology, however, these descriptors are most often counts of the number of specimens belonging to the various species, whose range of variation is so large that they behave, for all practical purposes, as continuous variables. When they are transformed (Section 1.5), as is often the case, they become real numbers instead of integers.

In order to speed up field observations or counts in the laboratory, it is often interesting for ecologists to record observations in the form of *semiquantitative* descriptors. Usually, it is possible to estimate environmental characteristics very rapidly by ascribing them a score using a small number of ordered classes: score 1 < score 2 < score 3, etc. Ecologists may often proceed in this way without losing pertinent information, whereas precise counts would have necessitated more considerable efforts than required by the ecological phenomenon under study. For example, in studying the influence of the unevenness of the landscape on the fauna of a given area, it may be enough to describe the relief using ordered classes such as flat, undulated, rough, hilly and mountainous. In the same way, counting large numbers of organisms may be done using abundance scores instead of precise numbers of individuals. Frontier (1973), for example, established such a scoring scale to describe the variability of zooplankton. Another score scale, also developed by Frontier (1969) for counting zooplankton, was used to estimate biomass (Dévaux & Millerioux, 1976b) and diversity of phytoplankton (Dévaux & Millerioux, 1977) as well as to evaluate schools of cetaceans at sea (Frontier & Viale, 1977). Frontier & Ibanez (1974) as well as Dévaux & Millerioux (1976a) have shown that this rapid technique is as informative as classical enumeration for principal component analysis (Section 9.1). It must be noted that nonparametric statistical tests of significance, which are used on such semiquantitative descriptors, have a discriminatory power almost equal to that of their parametric equivalent. Naturally occurring semiquantitative descriptors, which give *ranks* to the objects under study, as well as quantitative descriptors divided into non-equidistant classes (which is done either to facilitate data gathering or to evidence holes in frequency distributions), are included among the semiquantitative descriptors. Method 6.4 in Subsection 1.5.4 shows how to normalize semiquantitative descriptors if they have to be used in methods of data analysis that perform better in the presence of normality. Normalized semiquantitative descriptors should only be interpreted in terms of the ordinal value that they really represent. On the other hand, methods of data analysis may often be adapted to ranked data. This is the case, for example, with principal component analysis (Lebart *et al.*, 1979; Subsection 9.1.7) and linear regression (Iman & Conover, 1979).

Qualitative descriptors often present a problem to ecologists, who are tempted to discard them, or reduce them to a series of binary variables (Section 1.5, method 9). Let us forget the cases where descriptors of this kind have been camouflaged as ordered variables by scientists who did not quite know what to do with them ...Various methods based on contingency tables (Chapter 6) may be used to compare such descriptors with one another, or to ordered descriptors divided into classes. Special resemblance coefficients (Chapter 7) allow these descriptors to be used as a basis for clustering (Chapter 8) or ordination (Chapter 9). The first paragraph of Chapter 6 gives several examples of qualitative descriptors. An important class is formed by classifications of objects, which may in turn become descriptors of these objects for subsequent analyses, since the definition of a classification (Section 8.1) corresponds to the definition of a descriptor given above.

Binary or *presence-absence* descriptors may be noted + or –, or 1 or 0. In ecology, the most frequently used type of binary descriptors is the presence or absence of species, when reliable quantitative information is not available. It is only for historical reasons that they are considered as a special class: programming the first computers was greatly facilitated by such descriptors and, as a result, several methods have been developed for processing them. Sneath & Sokal (1973) present various methods to recode variables into binary form; see also Section 1.5, transformation method 7. Binary descriptors encountered in ecology may be processed either as qualitative, semiquantitative or quantitative variables. Even though the mean and variance parameters of binary descriptors are difficult to interpret, such descriptors may be used with methods originally designed for quantitative variables — in a principal component analysis, for instance, or as independent variables in regression or canonical analysis models.

When collecting ecological data, the level of precision with which descriptors are recorded should be selected with consideration of the problem at hand. Quantitative descriptors may often be recorded either in their original form or in semiquantitative or qualitative form. The degree of precision should be chosen with respect to the following factors: (1) What is the optimal degree of precision of the descriptor for analysing this particular ecological phenomenon? (2) What type of mathematical treatment will be used? This choice may determine the mathematical types of the descriptors. (3) What additional cost in effort, time or money is required to raise the level of precision? Would it not be more informative to obtain a larger number of less precise data?

2 — Intensive, extensive, additive, and non-additive descriptors

There are other useful ways of looking at variables. Margalef (1974) classifies ecological variables as either *intensive* or *extensive*. These notions are derived from thermodynamics (Glandsdorff & Prigogine, 1971). A variable is said to be *intensive* if its value is defined independently of the size of the sampling unit in which it is measured. For example, water temperature is defined independently of the size of the bucket of water in which a thermometer would be placed: we do not say "12°C per litre" but simply "12°C". This does not mean that the *measured value* of temperature may not vary from place to place in the bucket; it may indeed, unless water is well-mixed and therefore homogeneous. Concentration of organisms (number per unit surface or volume), productivity, and other rate variables (e.g. birth, death) are also intensive because, in a homogeneous system, the same value is obtained whether the original measurements are made over 1 m^2 or over 100 m^2. In contrast, an *extensive* variable is one whose value, in a homogeneous system, changes proportionally (linear relationship) to the size of the sampling unit (transect, quadrat, or volume). It is formally defined as an integral over the sampling unit. Number of individuals and biomass in a quadrat or volume, at a given point in time, are examples of extensive variables.

Intensive

Extensive

Extensive variables have the property that the values they take in two sampling units can be added to provide a meaningful estimate of the value in the combined unit. Other variables do not have this property; either they do not vary at all (e.g. temperature in a homogeneous bucket of water, which is an intensive variable), or they vary in a nonlinear way with the size of the sampling unit. For example, species richness in a sampling unit (surface or volume) cannot be computed as the sum of the numbers of species found in two sub-units; that sum would usually be larger than the number of species actually found in the combined unit, because some species are common to the two sub-units. Species diversity (Chapter 5) also has this property. The relationship of such variables to scale is complex and depends on the distribution patterns of the species and the size of the sampling units (grain size of the measurements; He *et al.*, 1994).

Additive

Another, more statistical point of view concerns additivity. This notion is well-known in geostatistics (Olea, 1991, p. 2; Journel & Huijbregths, 1978). A variable is said to be *additive* if its values can be added while retaining the same meaning as the original variable. A good example is the number of individuals in a quadrat. Concentrations, which are intensive variables, are additive if they are referred to the same linear, surface or volume unit measure (e.g. individuals m^{-2}; kg m^{-3}) (Journel & Huijbregths, 1978, p. 199); values may be added to compute a mean for example.

Extensive variables (e.g. number of individuals) are, by definition, additive; a sum or a mean has the same meaning as the original data although, if the sampling units differ in size, the values must be weighted by the sizes of the respective sampling units for their mean to be meaningful. For intensive additive variables (e.g. temperature or concentration), only the (weighted) mean has the same meaning as the original values. Variables may be additive over either time or space (Walliser, 1977); numbers of individuals in quadrats, for example, are additive over space, but not over time if the time lag between observations is shorter than the generation time of the organisms (the same individuals would be counted several times).

Non-additive

Examples of *non-additive variables* are pH values, logarithms and ratios of random variables, indices of various kinds, and directions of vectors (wind direction, aspect of a slope, etc.). Values of non-additive variables must be transformed in some way before (and if) they could be meaningfully combined. Logarithms of counts of organisms, for instance, have to be back-transformed using antilogarithms before values can be added; for ratios, the numerator and denominator must be added separately, and the ratio recomputed from these sums. Other non-additive variables, such as species richness and diversity, simply cannot be numerically combined; values of these indices for combined sampling units must be recomputed from the combined raw data.

These notions are of prime importance when analysing spatial data (Chapter 13). To appreciate their practical usefulness, let us consider a study in which the following variables have been measured at a site in a lake or in the ocean, at different times: solar flux at water surface (W m^{-2}), temperature (°C), pH, O_2 concentration (g m^{-3}), production of phytoplankton (g C m^{-3} s^{-1}), and concentration of zooplankton (individuals m^{-3}). All variables are intensive; they all have complex physical units, except temperature (simple unit) and pH (no unit). Assuming that some form of random sampling had been conducted with constant-sized observation units, how could estimates be obtained for the whole study area? This question may be viewed from two different angles, i.e. one may be looking for a mean or for an integral value over the study area. For additive variables (i.e. all except pH), values can be computed that represent the mean over the study area. However, integrating over the study area to obtain values for total solar flux, zooplankton, etc. is not that simple, because it requires the variables to be extensive. No extensive variable can be derived from temperature or pH. In the case of variables with complex physical units, new variables may be derived with units that are appropriate for integration:

• Consider O_2 concentration. Its physical dimensions (Section 3.1) are [ML^{-3}], with units g m^{-3}. This indicates that the "mass" part (dimension [M], with unit g), which is extensive, may be integrated over a volume, for example that of the surface mixed layer over the whole study area. Also, values from different depths in the mixed layer may be vertically integrated, to provide areal concentrations (dimensions [ML^{-2}], with units g m^{-2}). The same applies to the concentration of zooplankton.

• Flux variables may be turned into variables that are additive over both space and time. Phytoplankton production (dimensions [$ML^{-3}T^{-1}$], with units g C m^{-3} s^{-1}) is a flux variable since it is expressed per unit space and time. So, the extensive "mass" part may be integrated over a volume or/and over time, e.g. the euphotic zone over the whole study area or/and for the duration of the study. Values from different depths in the euphotic zone may be vertically integrated, thus providing areal concentrations (dimensions [$ML^{-2}T^{-1}$], with units g C m^{-2} s^{-1}), which could then be integrated over time.

• Solar flux (W m^{-2}) represents a more complex case. The "power" part (W) can be integrated over space (m^2) only. However, because W = J s^{-1} (Table 3.2), it is possible to integrate the "energy" part (J) over both space and time. Since the solar flux is either W m^{-2} or J m^{-2} s^{-1}, the "power" part may be integrated over space or, alternatively, the "energy" part may be integrated over both surface (m^2) and time (s). For example, it is possible to compute solar energy over a given area during 24 h.

1.5 Coding

Coding is a technique by which original data are transformed into other values, to be used in the numerical analysis. All types of descriptors may be coded, but nonordered descriptors must necessarily be coded before they may be analysed numerically. The functions or laws of correspondence used for coding qualitative descriptors are generally discontinuous; positive integers are usually associated with the various states.

Consider the case where one needs to compute the dependence between a variable with a high degree of precision and a less precisely recorded descriptor. Two approaches are available. In the first approach, the precision of the more precise descriptor is lowered, for example by dividing continuous descriptors into classes. Computers can easily perform such transformations. Dependence is then computed using a mathematical method adapted to the descriptor with the *lowest* level of precision. In the second approach, the descriptor with the lower precision level will be given a numerical scale adjusted to the more precise one. This operation is called *quantification* (Cailliez & Pagès, 1976; Gifi, 1990); one method of quantification using canonical correspondence analysis is explained in Subsection 11.2.1. Other transformations of variables, that adjust a descriptor to another, have been developed in the regression framework; they are discussed in Section 10.3.

1 — Linear transformation

In a study where there are quantitative descriptors of different types (metres, litres, mg L^{-1}, ...), it may be useful to put them all on the same scale in order to simplify the mathematical forms of relationships. It may be difficult to find an ecological interpretation for a relationship that includes a high level of artificial mathematical complexity, where scale effects are intermingled with functional relationships. Such changes of scale may be linear (of the first order), or of some higher order.

A linear change of scale of variable y is described by the transformation $y' = b_0 + b_1 y$ where y' is the value after transformation. Two different transformations are actually included in this equation. The first one, *translation*, consists in adding or subtracting a constant (b_0 in the equation) to all data. Graphically, this consists in sliding the scale beneath the data distribution. Translation is often used to bring to zero the mean, the modal class, the weak point of a bimodal distribution, or another point of interest in the distribution. The second transformation, *expansion*, is a change of scale obtained by multiplying or dividing all observed values by a constant (b_1 in the equation). Graphically, this operation is equivalent to contracting or expanding the scale beneath the distribution of a descriptor.

Two variables that are linearly related can always be put on the same scale by a combination of a translation followed by an expansion, the values of parameters b_0 and b_1 being found by linear regression (model I or model II: Chapter 10). For example (Fig. 1.7), if a linear regression analysis shows the equation relating y_2 to y_1 to be $\hat{y}_2 = b_0 + b_1 y_1$ (where $\hat{y}_2$ represents the values estimated by the regression equation for variable y_2), then transforming y_1 into $y'_1 = b_0 + b_1 y_1$ successfully puts variable y_1 on the same scale as variable y_2, since $\hat{y}_2 = y'_1$. If one wishes to transform y_2 instead of y_1, the regression equation should be computed the other way around.

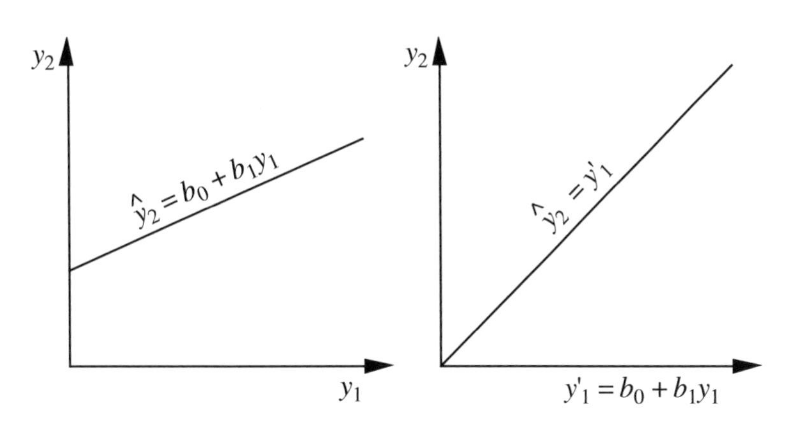

Figure 1.7 The regression parameters (b_0 and b_1) found by regressing y_2 on y_1 (left panel) may be used (right panel) to transform y_1 into y'_1 such that y'_1 is now on the same scale as y_2.

2 — *Nonlinear transformations*

The methods of multidimensional analysis described in this book are often based on covariances or linear correlations. Using them requires that the relationships among variables be made linear by an appropriate transformation. When two variables are not linearly related, their relationship may be described by a second- or higher-degree equation, or by other functional forms, depending on the situation. If the nonlinear form of the equation is derived from ecological theory, as it is often the case in population dynamics models, interpretation of the relationship poses no problem. If, however, a nonlinear transformation is chosen empirically, for reasons of mathematical elegance and without grounding in ecological theory, it may be difficult to find an ecological meaning to it.

The relationship between two variables may be determined with the help of a scatter diagram of the objects in the plane formed by the variables (Fig. 1.8). The principles of analytical geometry may then be used to recognize the type of relationship, which in turn determines the most appropriate type of transformation. A relationship frequently found in ecology is the exponential function, in which a variable y_2 increases in geometric progression with respect to y_1, according to one of the following equations:

$$y_2 = b^{(y_1)} \text{ or } y_2 = b_0 b_1^{(y_1)} \text{ or } y_2 = b_0 b_1^{(y_1 + b_2)} \text{ or else } y_2 = b_0 b_1^{(b_2 y_1)} \qquad \textbf{(1.3)}$$

Logarithmic transformation

depending on the number of constants b that shift or amplify the function. Such relationships can easily be linearized by using the logarithm of variable y_2 (called y'_2 below) instead of y_2 itself. The above relationships then become:

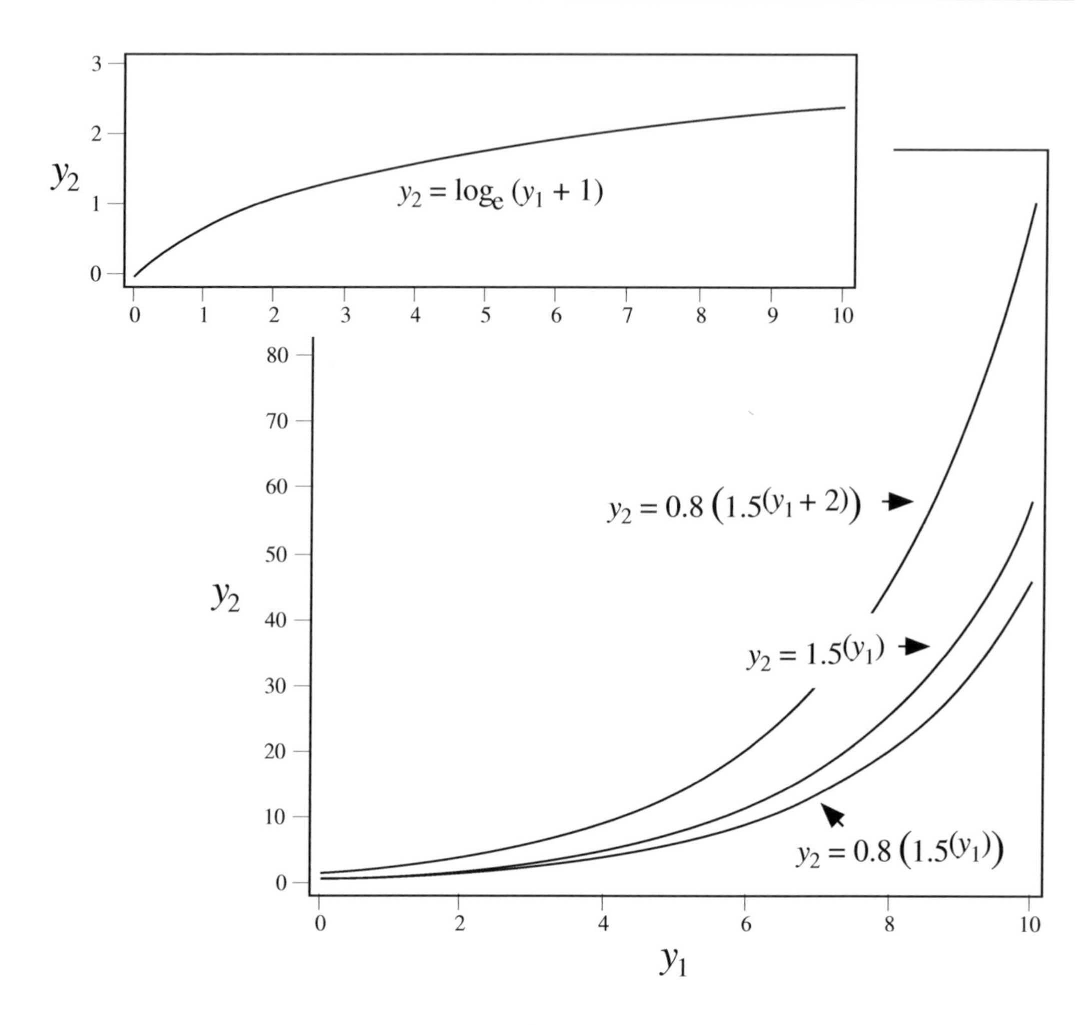

Figure 1.8 The relationship between variables may often be recognized by plotting them one against the other. In the upper panel, y_2 varies as the natural logarithm of y_1. In the lower panel, y_2 is an exponential function of y_1. These curves (and corresponding equations) may take different forms, depending on the modifying constants b (eq. 1.3).

$$y'_2 = \log(y_2) = b'y_1, \text{ or } y'_2 = b'_0 + b'_1 y_1,$$

$$\text{or } y'_2 = b'_0 + b'_1 (y_1 + b_2), \text{ or } y'_2 = b'_0 + b'_1 b_2 y_1 \quad \textbf{(1.4)}$$

where the b's are the logarithms of constants b in eq. 1.3.

If two variables display a logarithmic relationship of the form

$$y_2 = \log_b (y_1) \quad \textbf{(1.5)}$$

where b is the base of the logarithm, their relationship can be made linear by applying a $\log^{-1}$ transformation to y_2:

$$y'_2 = b^{(y_2)} = y_1 \tag{1.6}$$

When a nonlinear form may be assumed from knowledge of the ecological process involved, the corresponding equation can be used as the basis for a linearizing transformation. For instance, the nonlinear equation

$$N_t = N_0 e^{rt} \tag{1.7}$$

describes the exponential growth of a population, as observed in population explosions. In this equation, the independent variable is time (t); N_0 and N_t are the population sizes at times 0 and t, respectively; r is the Malthus parameter describing the intrinsic rate of increase of the population. This nonlinear equation indicates that N_t must be transformed into its natural logarithm. After this transformation, $\ln(N_t)$ is linearly related to t: $\ln(N_t) = \ln(N_0) + rt$.

3 — Combining descriptors

Another transformation which is often used consists in combining different descriptors by addition, subtraction, multiplication or division. In limnology, for example, the ratio (surface O_2/ bottom O_2) is often used as a descriptor. So is the Pearsall ionic ratio, all ions being in the same physical units:

$$y = \frac{\mathrm{Na} + \mathrm{K}}{\mathrm{Mg} + \mathrm{Ca}} \tag{1.8}$$

Beware, however, of the spurious correlations that may appear when comparing ratio variables to others. Jackson & Somers (1991a) describe the problem and recommend that permutation tests (Section 1.2) be employed with ratios.

It may also be required to take into account a factor of magnitude or size. For example, when observation units are of different sizes, the number of specimens of each species may be divided by the area or the volume of the unit (depending on whether the units come from an area or a volume), or by some other measure of the sampling effort. One must exert care when interpreting the results, however, since large observation units are more representative of populations than small ones.

4 — Ranging and standardization

Quantitative variables, used in ecology as environmental descriptors, are often expressed in incompatible units such as metres, mg L^{-1}, pH units, etc. In order to compare such descriptors, or before using them together in a classification or ordination procedure, they must be brought to some common scale. Among the methods available, some only eliminate size differences while others reduce both the size and variability to a common scale.

Translation, a method previously discussed, allows one to *centre* the data, eliminating size differences due to the position of the zero on the various scales. Centring is done by subtracting the mean of the observations ($\bar{y}$) from each value y_i:

$$y'_i = y_i - \bar{y} \tag{1.9}$$

For relative-scale variables (Subsection 1.4.1), dividing each y_i by the largest observed value is a way, based on expansion, to bring all values in the range [0, 1] (Cain & Harrison, 1958):

$$y'_i = y_i / y_{max} \tag{1.10}$$

For interval-scale variables, whose range may include negative values, the absolute value of the largest positive or negative value is used as divisor. The transformed values are in the interval [–1, +1].

Ranging

Other methods allow the simultaneous adjustment of the magnitude and the variability of the descriptors. The method of *ranging*, proposed by Sneath & Sokal (1973), reduces the values of a variable to the interval [0, 1] by first subtracting the minimum observed for each variable and then dividing by the range:

$$y'_i = \frac{y_i - y_{min}}{y_{max} - y_{min}} \tag{1.11}$$

Equation 1.10 is the form of ranging (eq. 1.11) to use with relative-scale variables (Subsection 1.4.1) for which y_{min} is always zero.

Standardization

The most widely used method for making descriptors compatible is to *standardize* the data (transformation into so-called "z-scores"). This method will be fully discussed in Section 4.2, dealing with correlation. Principal components (Section 9.2) are frequently computed using standardized data. Standardization is achieved by subtracting the mean (translation) and dividing by the standard deviation (s_y) of the variable (expansion):

$$z_i = \frac{y_i - \bar{y}}{s_y} \tag{1.12}$$

The position of each object on the transformed variable z_i is expressed in standard deviation units; as a consequence, it refers to the group of objects from which s_y has been estimated. The new variable z_i is called a *standardized variable*. Such a variable has three interesting properties: its mean is zero ($\bar{z} = 0$); its variance and hence its standard deviation are 1 ($s_z^2 = s_z = 1$); it is also a *dimensionless variable* (Chapter 3) since the physical dimensions (metres, mg L^{-1}, etc.) in the numerator and denominator cancel out. Transformations 1.6, 1.8 and 1.9 also produce dimensionless variables.

Beware of the "default options" of computer programs that may implicitly or explicitly suggest to standardize all variables before data analysis. Milligan & Cooper (1988) report simulation results showing that, for clustering purposes, if a transformation is needed, the ranging transformation (eqs. 1.10 and 1.11) gives results that are in general far superior to those obtained using standardization (eq. 1.12).

5 — Implicit transformation in association coefficients

When descriptors with different scales are used together to compare objects, the choice of the association coefficient (Section 7.6) may partly determine the type of transformation that must be applied to the descriptors. Some coefficients give equal weights to all variables independently of their scales while others take into account the magnitude of variation of each one. Since the amount of information (in the sense of information theory; Chapter 6) in a quantitative descriptor increases as a function of its variance, equalizing the variances before the association coefficient is computed is a way to ensure that all descriptors have the same weight. It is for ecologists to decide the kind of contribution they expect from each descriptor; again, beware of the "default options" of computer programs.

Some association coefficients require that the data be expressed as integers. Depending on the capabilities of the computer program and the degree of discrimination required, ecologists may decide to use the closest integer value, or to multiply first all values by 10 or 100, or else to apply some other simple transformation to make the data compatible with the coefficient to be computed.

6 — Normalization

Another type of transformation, called *normalizing transformation*, is performed on descriptors to make the frequency distributions of their data values look like the normal curve of errors — or, at least, as unskewed as possible. Indeed, several of the methods used in multivariate data analysis have been developed under the assumption that the variables are normally distributed. Although most of these methods do not actually require full normality (i.e. no skewness nor kurtosis), they may perform better if the distributions of values are, at least, not skewed. Skewed distributions, as in Figs. 1.8 and 1.9, are such that the variance of the distribution is controlled mostly by the few points in the extreme right tail; so, variance-partitioning methods such as principal component analysis (Chapter 9) or spectral analysis (Chapter 12) would bring out components expressing the variation of these few data points first instead of the variation of the bulk of data values. Normalizing transformations also have the property of reducing the *heteroscedasticity* of descriptors (Box 1.4).

The data analysis phase of research should always start by looking at the distributions of values for the different variables, i.e. computing basic distribution statistics (including skewness and kurtosis, eqs. 4.50-4.52), drawing histograms of frequency distributions, and testing for normality (described in Section 4.9). A normalizing transformation may have to be found for each variable separately; in other

Homoscedasticity Box 1.4

Homoscedasticity, also called *homogeneity* or *equality of the variances*, technically means that the variances of the error terms are equal for all observations. Its antonym is **heteroscedasticity** or *heterogeneity of the variances*. Homoscedasticity may actually refer to different properties of the data.

- *For a single variable*, homoscedasticity of the distribution means that, when the statistical population is sampled repeatedly, the expected value of the variance remains the same, whatever the value of the mean of the data sample. Data drawn from a normal distribution possess this property whereas data drawn from a Poisson distribution, for instance, do not since, in this type of distribution, the variance is equal to the mean.

- *In regression analysis*, homoscedasticity means that, for all values of the independent variable, the variances of the corresponding values of the dependent variable (called "error variances") are the same.

- *In t-test, analysis of variance and discriminant analysis*, homoscedasticity means that variances are equal in all groups, for each variable.

cases, one is looking for the best transformation that would normalize several variables.

• 6.1 — Ecologists often encounter distributions where a species is abundant in a few observation units (quadrats, etc.), fairly abundant in more, present in even more, and absent in many; this is in agreement with the concept of ecological niche briefly explained in Section 1.0, if the sampling programme covers a large enough area or environmental gradient. Distributions of this type are clearly not normal, being strongly skewed to the right (long tail in the higher values). Needless to say, environmental variables may also have non-normal distributions. For instance, the scales on which chemical variables are measured are conventions of chemistry which have no relation whatsoever with the processes generating these values in nature. So, any normalizing transformation is as good as the scale on which these data were originally measured.

Skewed data are often transformed by taking logarithms (below) or square roots. *Square root* is the least drastic transformation and is used to normalize data that have a Poisson distribution, where the variance is equal to the mean, whereas the *logarithmic transformation* is applicable to data that depart more widely from a normal distribution (Fig. 1.9). Several intermediate transformations have been proposed between these

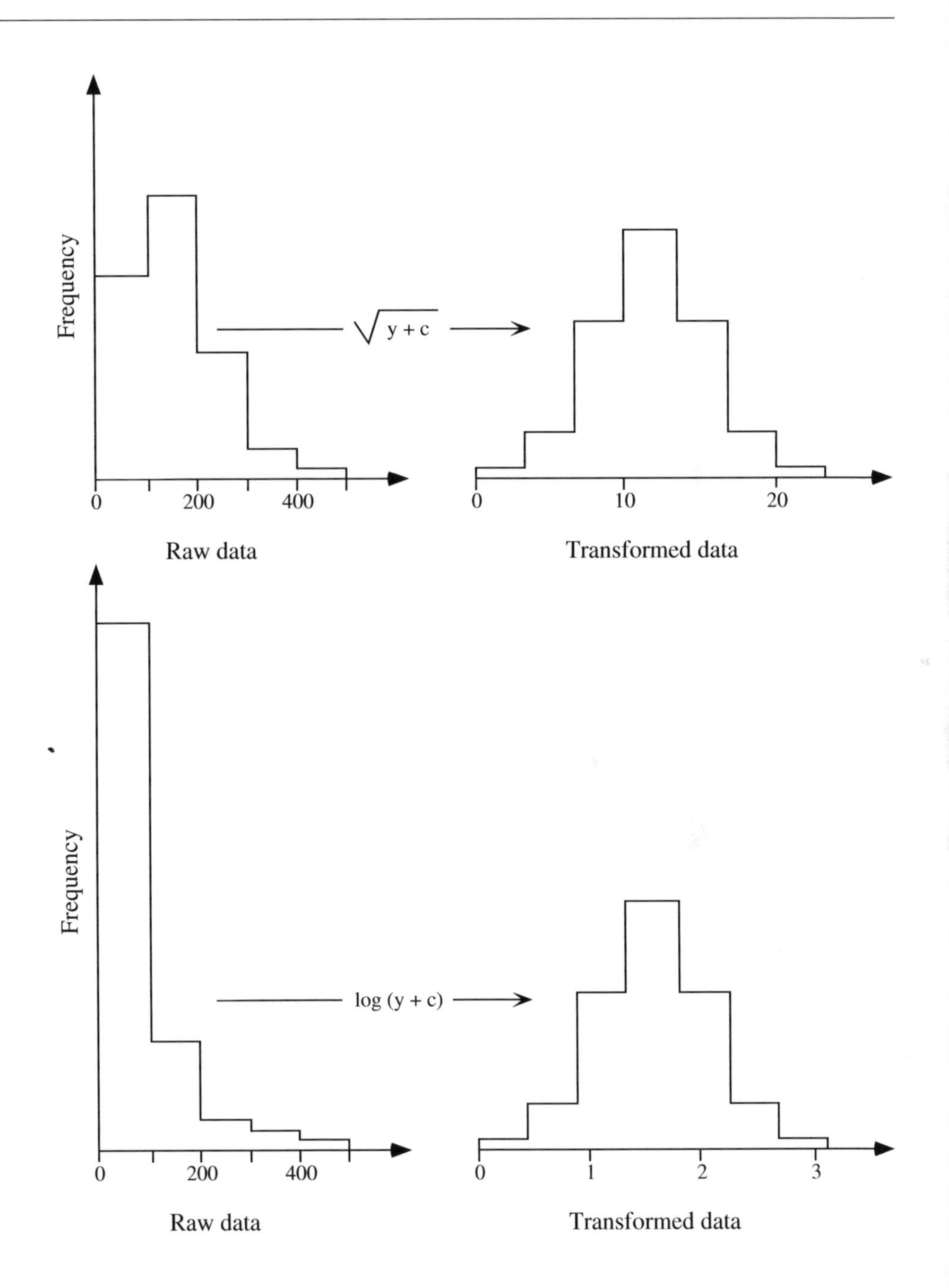

Figure 1.9 Numerical examples. Upper panel: Data that follow a Poisson distribution (left) can be normalized by the square root transformation (right). For a given species, these frequencies may represent the number of quadrats (ordinate) occupied by the number of specimens shown along the abscissa. Lower panel: Data distribution (left) that can be normalized by a logarithmic transformation (right).

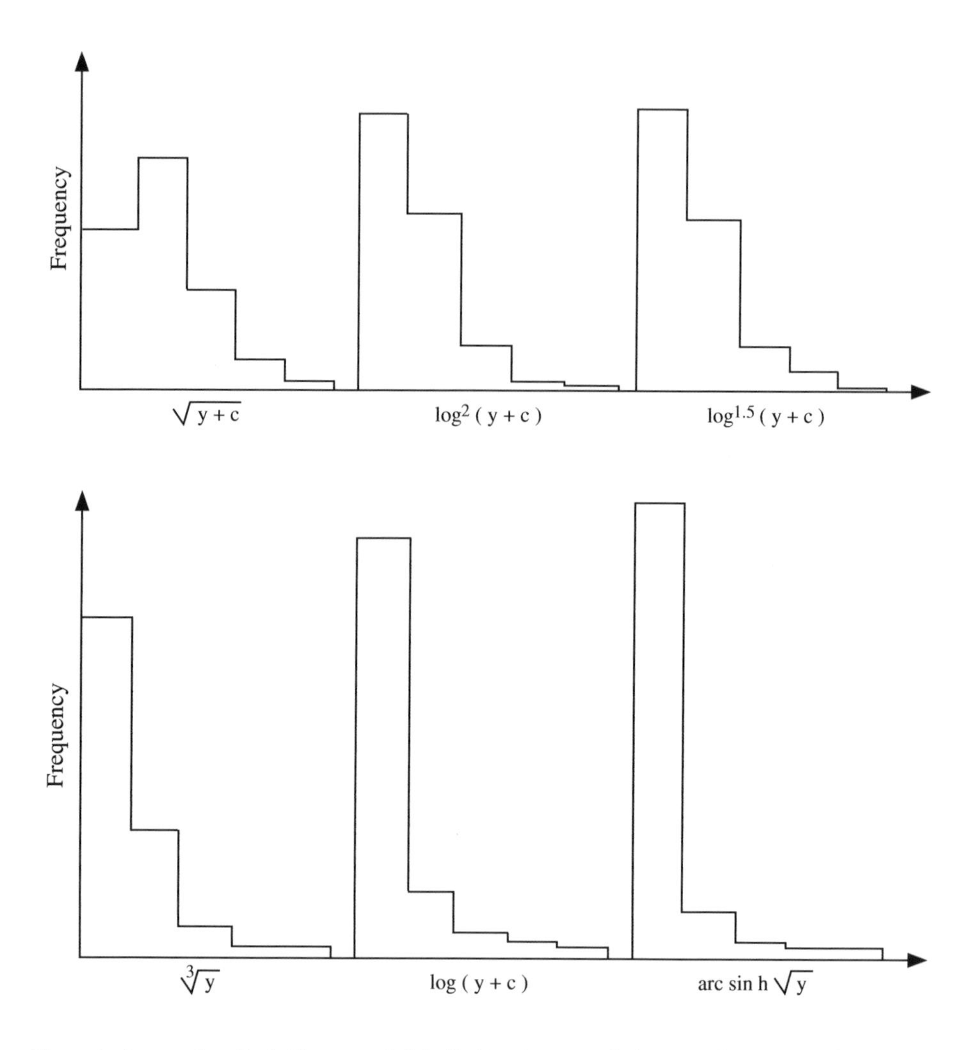

Figure 1.10 Numerical examples. Each histogram is labelled by the normalizing transformation to be used in that case.

extremes (Fig. 1.10): cubic root, $\log^2$, $\log^p$, etc. The hyperbolic transformation is useful for one particular type of data, which share the two extreme types at the same time (when the standard deviation is proportional to the mean, with many observations of a very small size which follow a Poisson distribution: Quenouille, 1950; Barnes, 1952). The *angular* or *arcsine transformation* is appropriate for percentages and proportions (Sokal & Rohlf, 1981, 1995):

$$y'_i = \arcsin\sqrt{y_i} \qquad \textbf{(1.13)}$$

In case of doubt, one may try several of these transformations and perform a test of normality (Section 4.7), or compute the skewness of the transformed data, retaining the transformation that produces the most desirable results. Alternatively, the Box-Cox method (point 6.2, below) may be used to find the best normalizing transformation.

A logarithmic transformation is computed as follows:

Logarithmic transformation

$$y'_i = \log(b_0 + b_1 y_i) \qquad \textbf{(1.14)}$$

The base of logarithm chosen has no influence on the normalising power, since transformation from one base to another is a linear change of scale (expansion, see 1 above: $\log_d y_i = \log_c y_i / \log_c d$). When the data to be transformed are all strictly positive (all $y_i > 0$), it is not necessary to carry out a translation ($b_0 = 0$). When the data contain fractional values between 0 and 1, one may multiply all values by some appropriate constant in order to avoid negative transformed values: $y'_i = \log(b_1 y_i)$. When the data to be transformed contain negative or null values, a translation must be applied first, $y'_i = \log(b_0 + y_i)$, since the logarithmic function is defined over the set of positive real numbers only. One should choose for translation a constant b_0 which is of the same order of magnitude as the significant digits of the variable to be transformed; for example, $b_0 = 0.01$ for data between 0.00 and 0.09 (the same purpose would have been achieved by selecting $b_0 = 1$ and $b_1 = 100$). For species abundance data, this rule produces the classical transformation $y'_i = \log(y_i + 1)$.

Box-Cox method

• 6.2 — When there is no *a priori* reason for selecting one or the other of the above transformations, the Box-Cox method allows one to empirically estimate what is the most appropriate exponent of the following general transformation function:

$$y'_i = (y_i^\gamma - 1)/\gamma \qquad (\text{for } \gamma \neq 0) \qquad \textbf{(1.15)}$$

and

$$y'_i = \ln(y_i) \qquad (\text{for } \gamma = 0)$$

As before, y'_i is the transformed value of observation y_i. In this transformation, the value γ is used that maximizes the following log likelihood function:

$$L = -(\nu/2)\ln(s_{y'}^2) + (\gamma - 1)(\nu/n)\sum_i \ln(y_i) \qquad \textbf{(1.16)}$$

since it is this value which yields the best transformation to normality (Box & Cox, 1964; Sokal & Rohlf, 1995). The value L that maximizes this likelihood function is found by iterative search. In this equation, $s_{y'}^2$ is the variance of the *transformed* values y'_i. When analysing several groups of observations at the same time (below), $s_{y'}^2$ is estimated instead by the within-group, or residual variance computed in a one-way ANOVA. The group size is n and ν is the number of degrees of freedom ($\nu = n - 1$ if the computation is made for a single group). All y_i values must be strictly positive numbers since logarithms are taken in the likelihood function L (eq. 1.16); all values may easily be made strictly positive by translation, as discussed in Subsection 1 above. It is interesting to note that, if $\gamma = 1$, the function is a simple linear transformation; if $\gamma = 1/2$, the function becomes the square root transformation; when $\gamma = 0$, the transformation is logarithmic; $\gamma = -1$ yields the reciprocal transformation.

Readers are invited to take a value (say 150) and transform it, using eq. 1.15, with a variety of values of γ gradually tending toward 0 (say 1, 0.1, 0.01, 0.001, etc.). Comparing the results to the logarithmic transformation will make it obvious that this transformation is indeed the limit of eq. 1.15 when γ tends towards 0.

Another log likelihood function L' is proposed by Sokal & Rohlf (1995) in order to achieve homogeneity of the variances for several groups of observations of a given variable, together with the normality of their distributions. This generalized Box-Cox transformation may also be applied to the identification of the best normalizing transformation for several species, for a given set of sampling sites.

Taylor's power law

• 6.3 — When the data distribution includes several groups, or when the same transformation is to be applied to several quantitative and dimensionally homogeneous descriptors (Chapter 3; for instance, a species abundance data table), Taylor's (1961) power law provides the basis for another general transformation which stabilizes the variances and thus makes the data *more likely* to conform to the assumptions of parametric analysis, including normality (Southwood, 1966; see also Downing, 1979 on this subject). This law relates the means and variances of the k groups through equation

$$s_{y_k}^2 = a\,(\bar{y}_k)^b \qquad \textbf{(1.17)}$$

from which constants a and b can be computed by nonlinear regression. When the latter is not available, an approximation of b may be calculated by linear regression of the logarithmic form

$$\log s_{y_k}^2 = \log a + b \log \bar{y}_k \qquad \textbf{(1.18)}$$

Having found the value of b, the variance stabilizing transformations

$$y'_i = y_i^{\left(1 - \frac{b}{2}\right)} \qquad \text{(for } b \neq 2\text{)} \qquad \textbf{(1.19)}$$

or

$$y'_i = \ln(y_i) \qquad \text{(for } b = 2\text{)}$$

are applied to the data.

Omnibus procedure

• 6.4 — The following method represents an *omnibus normalizing procedure* that should be able to normalize most kinds of data. The procedure is easy to carry out using most standard statistical packages running on microcomputers. The package

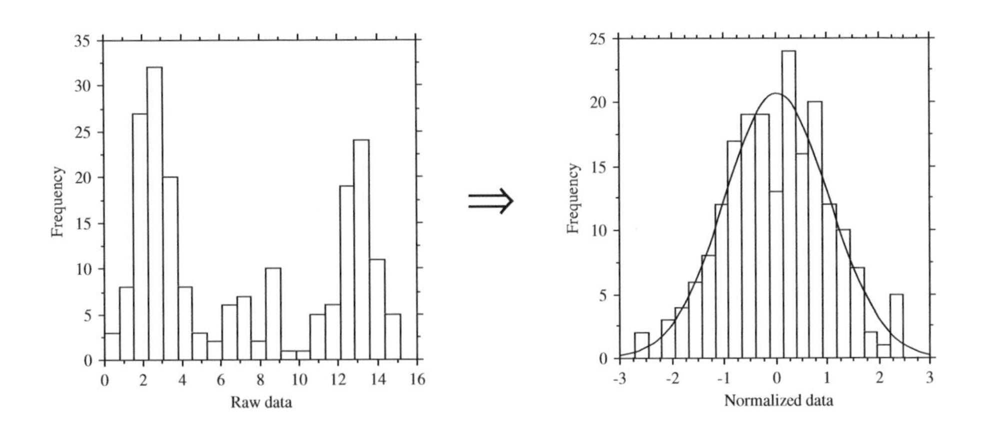

Figure 1.11 The *omnibus* procedure has been used to normalize a set of 200 data values with tri-modal distribution (left). A normal curve has been fitted to the transformed data (right).

must have a pseudo-random number generator for random normal deviates, i.e. values drawn at random from a normal distribution.

(1) Write the quantitative or semiquantitative descriptor to be normalized into a data base. Sort the data base in increasing values for that descriptor. (2) Create a new descriptor with the same number of values, using a pseudo-random normal deviate generator. Write it into another data base and sort it also in order of increasing values. (3) Copy the sorted normal deviate values into the sorted data base containing the descriptor to be normalized. Sort the data base back into the original order if necessary. (4) Use the normal deviates as a monotonic proxy for the original descriptor. Figure 1.11 shows an example of this transformation.

This procedure may be modified to handle *ex aequo* (tied) values (Section 5.3). Tied values may either receive the same normal deviate value, or they may be sorted in some random order and given neighbouring normal deviate values; one should select a solution that makes sense considering the data at hand.

Powerful as it is, this transformation only makes sense for data that have no more than ordinal value in the analysis to be conducted. The transformed data may be used in methods of data analysis that perform better in the presence of normality of the distributions. Several such methods will be studied in chapters 9 and 11. The main disadvantage is that a back-transformation is difficult. If the study requires that values of this descriptor be forecasted by a model, the data base itself will have to be used to find the original descriptor values which are the closest to the forecasted normal deviate. An interpolation may have to be made between these values.

7 — Dummy variable (binary) coding

Multistate qualitative descriptors may be binary-coded as *dummy variables*. This coding is interesting because it allows the use of qualitative descriptors in procedures such as multiple regression, discriminant analysis or canonical analysis, that have been developed for quantitative variables and in which binary variables may also be used. A multistate qualitative descriptor with s states is decomposed into $(s-1)$ dummy variables. An example is the following four-state descriptor:

States	Dummy variables			
1	1	0	0	0
2	0	1	0	0
3	0	0	1	0
4	0	0	0	1

In this example, three dummy variables are sufficient to binary-code the four states of the original nominal descriptor. Had a fourth dummy variable been included (shaded column above), its information would have been totally *linearly dependent* (Box 1.1 and Section 2.7) on the first three variables. In other words, the first three dummy variables are enough to determine the states of the multistate qualitative descriptor. Actually, any one of the four dummy variables may be eliminated to return to the condition of linear independence among the remaining ones.

Using this table, the objects are coded, in this example, by three dummy variables instead of a single 4-state descriptor. An object with state 1, for instance, would be recoded [1 0 0], an object with state 2, [0 1 0], and so on.

Other forms of binary coding have been developed for special types of variables. In phylogenetic analysis, the states of multistate characters are sometimes related by a hypothesized transformation series, going from the single hypothesized ancestral state to all the advanced states; such a series can be represented by a directed network where the states are connected by arrows representing evolutionary progression. A transformation series may be coded into binary variables using a method proposed by Kluge & Farris (1969). This same method may be applied to code the spatial relationships among localities on any geographic network. An example in freshwater ecology is a group of lakes connected by a river network (Fig. 1.12).

River network

In this example, a picture made of rivers and lakes is drawn to represent the network. A number is assigned to each river segment (which are the edges of the connected graph), while nodes represent the furcation points. In Fig. 1.12, the coding is based on the river segments; it could just as well be based on the nodes, if one felt that the nodes are the important carriers of geographic information (as in Magnan *et al.*, 1994). If the phenomenon to be modelled is, for example, fish dispersal from downstream, the arrows can be drawn going upstream, as in Fig. 1.12. In the lake-by-

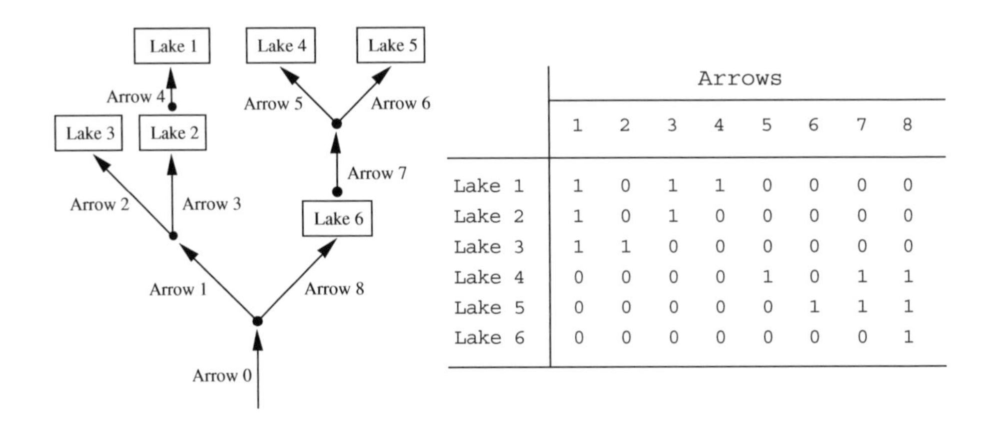

	Arrows							
	1	2	3	4	5	6	7	8
Lake 1	1	0	1	1	0	0	0	0
Lake 2	1	0	1	0	0	0	0	0
Lake 3	1	1	0	0	0	0	0	0
Lake 4	0	0	0	0	1	0	1	1
Lake 5	0	0	0	0	0	1	1	1
Lake 6	0	0	0	0	0	0	0	1

Figure 1.12 Lakes interconnected by a river network (left) can be binary-coded as shown on the right. Numbers are assigned in an arbitrary order to the edges (arrows) of the network. There is no use in representing the "root" of the network (arrow 0); all lakes would be coded '1' for that arrow.

arrow matrix, a value '1' is assigned to each arrow found downstream from the lake, meaning that the corresponding river segment is available for fish to reach that lake. All other arrows are coded '0' for the lake. The resulting matrix is a complete numerical coding of the hydrographic network information: knowing the coding procedure, one could reconstruct the picture from the matrix entries.

The coding method may be tailored to the ecological problem at hand. For a dispersion phenomenon going downstream, arrows could point the other way around; in this case, a lake would be coded '1' in the table for arrows arriving in that lake from upstream. The pattern of interconnections does not even need to be a tree-like structure; it may form a more general type of directed network, but note that no cycle is allowed. Coding the information allows the use of this type of geographical information in different types of numerical models, like multiple regression (Chapter 10) or canonical analysis (Chapter 11). In many of these methods, zeros and ones are interchangeable.

1.6 Missing data

Ecological data matrices are often plagued by missing data. They do not necessarily result from negligence on the part of the field team; most often, they are caused by the breakdown of measuring equipment during field surveys, weather events that prevent sampling sites from being visited on a given date, lost or incorrectly preserved specimens, improper sampling procedures, and so on.

Three families of solutions are available to cope with this problem for the analysis of field survey data, if one can make the assumption that the missing values occur at random in the data set. Most of the approaches mentioned below are discussed by Little & Rubin (1987), who also propose methods for estimating missing values in controlled experiments (when the missing values are only found in the outcome variable; their Chapter 2) as well as valid model-based likelihood estimation of missing values for situations where the distribution of missing values does not meet the randomness assumption stated above.

Missing values may be represented in data matrices by numbers that do not correspond to possible data values. Codes such as –1 or –9 are often used when the real data in the table are all positive numbers, as it is the case with species abundance data; otherwise, –99 or –999, or other such unambiguous codes, may be used. In spreadsheets, missing values are represented by bullets or other such symbols.

1 — Deleting rows or columns

Delete any row or column of the data matrix (Section 2.1) containing missing values. If a few rows contain most of the missing values, proceed by *rowwise* (also called *listwise*) *deletion*; conversely, if most missing values are found in a few variables only, proceed by *columnwise deletion*. This is the simplest, yet the most costly method, as it throws away the valuable information present in the remainder of these rows or columns.

2 — Accommodating algorithms to missing data

Accommodate the numerical method in such a way that the missing values are skipped during calculations. For instance, when computing resemblance coefficients among rows (Q-mode; Chapter 7) or columns (R-mode) of the data matrix, a simple method is *pairwise deletion* of missing values. This means, for example, that when computing a correlation coefficient between variables $\mathbf{y}_1$ and $\mathbf{y}_2$, if the value of the tenth object is missing for $\mathbf{y}_2$, object $\mathbf{x}_{10}$ is skipped in the computation of this correlation value. When it comes to comparing $\mathbf{y}_1$ and $\mathbf{y}_3$, if $\mathbf{x}_{10}$ has no missing data for these variables, it is then kept in the calculation for this pair of variables. However, one must be aware that covariance and correlation matrices computed in this way may be indefinite (i.e. they may have negative eigenvalues; Table 2.2). Wishart (1978, 1985) lists several such methods developed in the cluster analysis framework, that are used in the CLUSTAN clustering package.

3 — Estimating missing values

Estimate the missing values (called *imputation* by Little & Rubin, 1987). This is the best strategy when missing values are located all over the data matrix — contrary to the situation where the missing values are found in a few rows or columns only, in which case deletion of these rows or columns may be the strategy of choice. The assumption one has to make when estimating missing values is that the missing data are not grossly atypical compared to those present in the data set. Methods for estimating missing data are interesting in cases where the numerical algorithm required for analysing the data set cannot accommodate missing values. Ecologists should never imagine, however, that the estimated values are ecologically meaningful; as a consequence, they should refrain from attempting to interpret these numbers in ecological terms. Ecologists should also keep in mind that the estimation procedure has not created the missing degrees of freedom that would have accompanied observations carried out in nature or in the laboratory.

Three groups of methods are available for filling in missing values for quantitative data.

• 3.1 — The easiest way, which is often used in computer programs, is to replace missing values by the mean of the variable, estimated from the values present in the data table. When doing so, one assumes that nothing is known about the data, outside of the weak assumption mentioned above that the missing value comes from the same population as the non-missing data. Although this solution produces covariance and correlation matrices that are positive semidefinite (Section 2.10), the variances and covariances are systematically underestimated. One way around this problem is to select missing value estimates at random from some distribution with appropriate mean and variance. This is not recommended, however, when the relative positions of the objects are of interest (principal component analysis; Section 9.1). A variant of the same method is to use the median instead of the mean; it is more robust in the sense that it does not assume the distribution of values to be unskewed. It is also applicable to semiquantitative descriptors. For qualitative descriptors, use the most frequent state instead of the mean or median.

• 3.2 — Estimate the missing values by regression. Multiple linear regression (Section 10.3), with rowwise deletion of missing values, may be used when there are only a few missing values to estimate. The dependent (response) variable of the regression is the descriptor with missing value(s) while the independent (explanatory) variables are the other descriptors in the data table. After the regression equation has been computed from the objects without missing data, it can be used to estimate the missing value(s). Using this procedure, one assumes the descriptor with missing values to be linearly related to the other descriptors in the data table (unless some form of nonparametric or nonlinear multiple regression is being used) and the data to be approximately multivariate normal. This method also leads to underestimating the variances and covariances, but less so than in 3.1. An alternative approach is to use a regression program allowing for pairwise deletion of missing values in the estimation of the regression coefficients, although, in that case, a maximum likelihood estimation of the covariance matrix would be preferable (Little & Rubin, 1987, p. 152 *et seq.*).

If such a method cannot be used for estimating the covariance matrix and if the missing values are scattered throughout the data table, an approximate solution may be obtained as follows. Compute a series of simple linear regressions with pairwise deletion of missing values, and estimate the missing value from each of these simple regression equations in turn. The mean of these estimates is taken as the working estimated value. The assumptions are basically the same as in the multiple regression case (above).

To estimate missing values in qualitative (nominal) descriptors, use logistic regression (Section 10.3) instead of linear regression.

• 3.3 — Interpolate missing values in autocorrelated data. Positive autocorrelation (Section 1.1) means that near points in time or space are similar. This property allows the interpolation of missing or otherwise unknown values from the values of near points in the series. With spatial data, interpolation is the first step of any mapping procedure, and it may be done in a variety of ways (Subsection 13.2.2), including the kriging method developed by geostatisticians. The simplest such method is to assign to a missing data the value of its nearest neighbour. In time series, interpolation of missing values may be performed using the same methods; see also Shumway & Stoffer, 1982, as well as Mendelssohn & Cury, 1987, for a maximum likelihood method for estimating missing data in a time series using a state-space model.

Myers (1982, 1983, 1984) has proposed a method, called co-kriging, that combines the power of principal component analysis (Section 9.1) with that of kriging. It allows the estimation of unknown values of a data series using both the values of the same variable at neighbouring sites and the known values of other variables, correlated with the first one, observed at the same or neighbouring points in space; the spatial interrelationships of these variables are measured by a cross-variogram. This method should become very important in the future, for estimating missing data in broad-scale ecological surveys and to compute values at unobserved sites on a geographic surface.

Chapter 2 Matrix algebra: a summary

2.0 Matrix algebra

Matrix language is the algebraic form best suited to the present book. As a consequence, the following chapters will systematically use the flexible and synthetic formulation of *matrix algebra*, with which many ecologists are already acquainted.

There are many reasons why matrix algebra is especially well suited for ecology. The format of computer spreadsheets, in which ecological *data sets* are now most often recorded, is a *matrix* format. The use of *matrix notation* thus provides an elegant and compact representation of ecological information and *matrix algebra* allows operations on whole data sets to be performed. Finally, *multidimensional methods*, discussed in following chapters, are almost impossible to conceptualise and explain without resorting to matrix algebra.

Matrix algebra goes back more than one century: "After Sylvester had introduced matrices [...], it is Cayley who created their algebra [in 1851]" (translated from Bourbaki, 1960). Matrices are of great conceptual interest for theoretical formulations, but it is only with the increased use of *computers* that matrix algebra became truly popular with ecologists. The use of computers naturally enhances the use of matrix notation. Most scientific programming languages are adapted to matrix logic, some languages allowing programmers to write matrix operations directly.

Ecologists who are familiar with matrix algebra could read Sections 2.1 and 2.2 only, where the vocabulary and symbols used in the remainder of this book are defined. Other sections may be consulted whenever necessary.

The present chapter is only a *summary* of matrix algebra. Readers looking for more complete presentations of the subject should consult Bronson (1989), where numerous exercises are found; Graybill (1983), which provides applications in general statistics; or the handbook by Searle (1966), which is oriented towards biological statistics. One

Table 2.1 Ecological data matrix.

			Descriptors				
Objects	$\mathbf{y}_1$	$\mathbf{y}_2$	$\mathbf{y}_3$	…	$\mathbf{y}_j$	…	$\mathbf{y}_p$
$\mathbf{x}_1$	y_{11}	y_{12}	y_{13}	…	y_{1j}	…	y_{1p}
$\mathbf{x}_2$	y_{21}	y_{22}	y_{23}	…	y_{2j}	…	y_{2p}
$\mathbf{x}_3$	y_{31}	y_{32}	y_{33}	…	y_{3j}	…	$3p$
.	.	.	.		.		.
.	.	.	.		.		.
.	.	.	.		.		.
$\mathbf{x}_i$	y_{i1}	y_{i2}	y_{i3}	…	y_{ij}	…	y_{ip}
.	.	.	.		.		.
.	.	.	.		.		.
.	.	.	.		.		.
$\mathbf{x}_n$	y_{n1}	y_{n2}	y_{n3}	…	y_{nj}	…	y_{np}

may also consult the book of Green & Carroll (1976), which stresses the geometric interpretation of various matrix operations commonly used in statistics.

2.1 The ecological data matrix

Descriptor
Object

As explained in Section 1.4, ecological data are obtained as object-observations or sampling units which are described by a set of state values corresponding to as many descriptors, or variables. Ecological data are generally recorded in a table (spreadsheet) where each column j corresponds to a descriptor y_j (species present in the sampling unit, physical or chemical variable, etc.) and each object i (sampling site, sampling unit, locality, observation) occupies one row. In each cell (i,j) of the table is found the state taken by object i for descriptor j (Table 2.1). Objects will be denoted by a boldface, lower-case letter $\mathbf{x}$, with a subscript i varying form 1 to n, referring to object $\mathbf{x}_i$. Similarly, descriptors will be denoted by a boldface, lower case letter $\mathbf{y}$ subscripted j, with j taking values from 1 to p, referring to descriptor $\mathbf{y}_j$. When considering two set of descriptors, members of the second set will generally have subscripts k from 1 to m.

Following the same logic, the different values in a data matrix will be denoted by a doubly-subscripted *y*, the first subscript designating the object being described and the second subscript the descriptor. For example, y_{83} is the value taken by object 8 for descriptor 3.

It is not always obvious which are the objects and which are the descriptors. In ecology, for example, the different sampling sites (objects) may be studied with respect to the species found therein. In contrast, when studying the behaviour or taxonomy of organisms belonging to a given taxonomic group, the objects are the organisms themselves, whereas one of the descriptors could be the types of habitat found at different sampling sites. To unambiguously identify objects and descriptors, one must decide which is the variable defined *a priori* (i.e. the objects). When conducting field or laboratory observations, the variable defined *a priori* is totally left to the researcher, who decides how many observations will be included in the study. Thus, in the first example above, the researcher could choose the number of sampling sites needed to study their species composition. What is observed, then, are the descriptors, namely the different species present and possibly their abundances. Another approach to the same problem would be to ask which of the two sets of variables the researcher could theoretically increase to infinity; this identifies the variable defined *a priori*, or the objects. In the first example, it is the number of samples that could be increased at will — the samples are therefore the objects — whereas the number of species is limited and depends strictly on the ecological characteristics of the sampling sites. In the second example, the variable defined *a priori* corresponds to the organisms themselves, and one of their descriptors could be their different habitats (states).

The distinction between objects and descriptors is not only theoretical. One may analyse either the relationships among descriptors for the set of objects in the study (R mode analysis), or the relationships among objects given the set of descriptors (Q mode study). It will be shown that the mathematical techniques that are appropriate for studying relationships among objects are not the same as those for descriptors. For example, coefficients of correlation can only be used for studying relationships among descriptors, which are vectors with a theoretically infinite number of elements; they are in fact limited by the sampling effort. It would be incorrect to use a correlation coefficient to study the relationship between two objects for the set of descriptors, other measures of association being available for this purpose (see Section 7.3). Similarly, when using methods of multidimensional analysis, to be discussed later in this book, it is important to know which are the descriptors and which are the objects, in order to avoid methodological errors. The results of incorrectly conducted analyses — and there are unfortunately many in the literature — are not necessarily wrong because, in ecology, phenomena which are easily identified are usually sturdy enough to withstand considerable distortion. What is a pity, however, is that the more subtle phenomena, i.e. the very ones for which advanced numerical techniques are used, could very well not emerge at all from a study based on inappropriate methodology.

Linear algebra

The table of ecological data described above is an array of numbers known as a *matrix*. The branch of mathematics dealing with matrices is *linear algebra*.

Matrix **Y** is a rectangular, ordered array of numbers y_{ij}, set out in rows and columns as in Table 2.1:

$$\mathbf{Y} = [y_{ij}] = \begin{bmatrix} y_{11} & y_{12} & \dots & y_{1p} \\ y_{21} & y_{22} & \dots & y_{2p} \\ \cdot & & & \cdot \\ \cdot & & & \cdot \\ \cdot & & & \cdot \\ y_{n1} & y_{n2} & \dots & y_{np} \end{bmatrix} \quad \textbf{(2.1)}$$

Order

There are n rows and p columns. When the *order* (also known as its *dimensions* or *format*) of the matrix must be specified, a matrix of order $(n \times p)$, which contains $n \times p$ elements, is written $\mathbf{Y}_{np}$. As above, any given element of **Y** is denoted y_{ij}, where subscripts i and j identify the row and column, respectively (always in that conventional order).

In linear algebra, ordinary numbers are called *scalars*, to distinguish them from *matrices*.

The following notation will be used hereinafter: a matrix will be symbolised by a capital letter in boldface, such as **Y**. The same matrix could also be represented by its general element in italics and in brackets, such as $[y_{ij}]$, or alternatively by an enumeration of all its elements, also in italics and in brackets, as in eq. 2.1. Italics will only be used for algebraic symbols, not for actual numbers. Occasionally, other notations than brackets may be found in the literature, i.e. (y_{ij}), (y_i^j), $\{y_{ij}\}$, $\|y_i^j\|$, or $\langle iyj \rangle$.

Any subset of a matrix can be explicitly recognized. In the above matrix (eq. 2.1), for example, the following submatrices could be considered:

Square matrix

a *square matrix* $\begin{bmatrix} y_{11} & y_{12} \\ y_{21} & y_{22} \end{bmatrix}$

a *row matrix* $\begin{bmatrix} y_{11} & y_{12} & \dots & y_{1p} \end{bmatrix}$, or a *column matrix* $\begin{bmatrix} y_{12} \\ y_{22} \\ \cdot \\ \cdot \\ \cdot \\ y_{n2} \end{bmatrix}$

Matrix notation simplifies the writing of data sets. It also corresponds to the way computers work. Indeed, most programming languages are designed to input data as matrices (arrays) and manipulate them either directly or through a simple system of subscripts. This greatly simplifies programming the calculations. Accordingly, computer packages generally input data as matrices. In addition, many of the statistical models used in multidimensional analysis are based on linear algebra, as will be seen later. So, it is convenient to approach them with data already set in matrix format.

2.2 Association matrices

Two important matrices may be derived from the ecological data matrix: the association matrix among objects and the association matrix among descriptors. Any association matrix is denoted **A**, and its general element a_{ij}. Although Chapter 7 is entirely devoted to association matrices, it is important to mention them here in order to better understand the purpose of methods presented in the remainder of the present chapter.

Using data from matrix **Y** (eq. 2.1), one may examine the relationship between the first two objects $\mathbf{x}_1$ and $\mathbf{x}_2$. In order to do so, the first and second rows of matrix **Y**

$$\begin{bmatrix} y_{11} \; y_{12} \; . \; . \; . \; y_{1p} \end{bmatrix} \quad \text{and} \quad \begin{bmatrix} y_{21} \; y_{22} \; . \; . \; . \; y_{2p} \end{bmatrix}$$

are used to calculate a measure of association (similarity or distance: Chapter 7), to assess the degree of resemblance between the two objects. This measure, which quantifies the strength of the association between the two rows, is denoted a_{12}. In the same way, the association of $\mathbf{x}_1$ with $\mathbf{x}_3$, $\mathbf{x}_4$, ..., $\mathbf{x}_p$, can be calculated, as can also be calculated the association of $\mathbf{x}_2$ with all other objects, and so on for all pairs of objects. The coefficients of association for all pairs of objects are then recorded in a table, ordered in such a way that they could be retrieved for further calculations. This table is the association matrix **A** among objects:

$$\mathbf{A}_{nn} = \begin{bmatrix} a_{11} & a_{12} & . \; . \; . & a_{1n} \\ a_{21} & a_{22} & . \; . \; . & a_{2n} \\ . & & & . \\ . & & & . \\ . & & & . \\ a_{n1} & a_{n2} & . \; . \; . & a_{nn} \end{bmatrix} \qquad \textbf{(2.2)}$$

A most important characteristic of any association matrix is that it has a number of rows equal to the number of columns, this number being equal here to the number of objects n. The number of elements in the above square matrix is therefore n^2.

Similarly, one may wish to examine the relationships among descriptors. For the first two descriptors, $\mathbf{y}_1$ and $\mathbf{y}_2$, the first and second columns of matrix $\mathbf{Y}$

$$\begin{bmatrix} y_{11} \\ y_{21} \\ \cdot \\ \cdot \\ \cdot \\ y_{n1} \end{bmatrix} \quad \text{and} \quad \begin{bmatrix} y_{12} \\ y_{22} \\ \cdot \\ \cdot \\ \cdot \\ y_{n2} \end{bmatrix}$$

are used to calculate a measure of dependence (Chapter 7) which assesses the degree of association between the two descriptors. In the same way as for the objects, $p \times p$ measures of association can be calculated among all pairs of descriptors and recorded in the following association matrix:

$$\mathbf{A}_{pp} = \begin{bmatrix} a_{11} & a_{12} & \cdots & a_{1p} \\ a_{21} & a_{22} & \cdots & a_{2p} \\ \cdot & & & \cdot \\ \cdot & & & \cdot \\ \cdot & & & \cdot \\ a_{p1} & a_{p2} & \cdots & a_{pp} \end{bmatrix} \qquad \textbf{(2.3)}$$

Association matrices are most often symmetric, with elements in the upper right triangle being equal to those in the lower left triangle ($a_{ij} = a_{ji}$). Elements a_{ii} on the diagonal measure the association of a row or a column of matrix $\mathbf{Y}$ with itself. In the case of objects, the measure of association a_{ii} of an *object* with itself usually takes a value of either 1 (similarity coefficients) or 0 (distance coefficients). Concerning the association between *descriptors* (columns), the correlation a_{ii} of a descriptor with itself is 1, whereas the (co)variance provides an estimate a_{ii} of the variability among the values of descriptor i.

At this point of the discussion, it should thus be noted that the data, to which the models of multidimensional analysis are applied, are not only matrix $\mathbf{Y}_{np}$ = [objects × descriptors] (eq. 2.1), but also the two association matrices $\mathbf{A}_{nn}$ = [objects × objects] (eq. 2.2) and $\mathbf{A}_{pp}$ = [descriptors × descriptors] (eq. 2.3), as shown in Fig. 2.1.

2.3 Special matrices

Matrices with an equal number of rows and columns are called *square* matrices (Section 2.1). These, as will be seen in Sections 2.6 *et seq.*, are the only matrices for

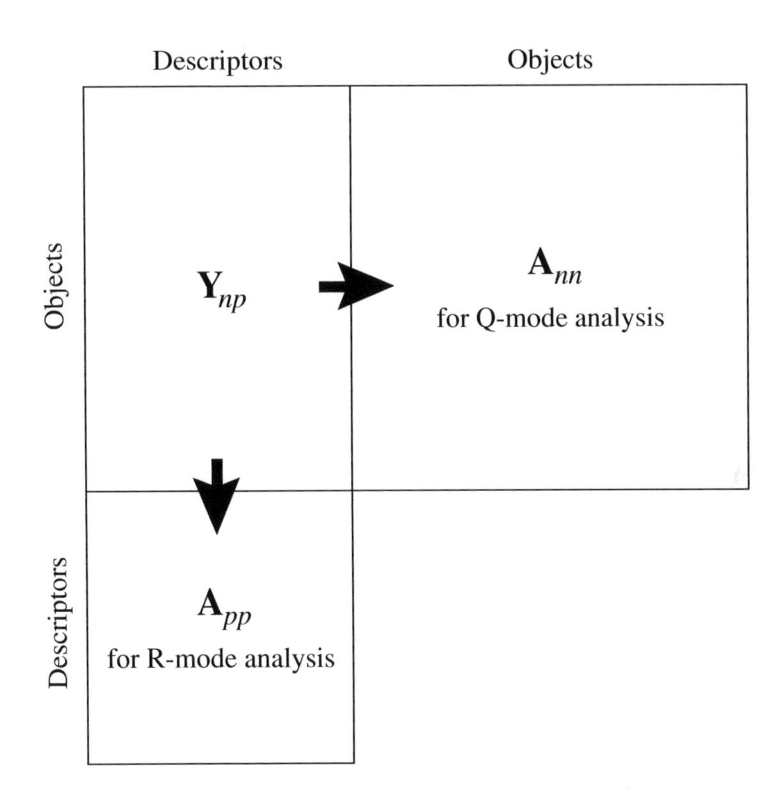

Figure 2.1 Data analysed in numerical ecology include matrix $\mathbf{Y}_{np}$ = [objects × descriptors] (eq. 2.1) as well as the two association matrices $\mathbf{A}_{nn}$ = [objects × objects] (eq. 2.2) and $\mathbf{A}_{pp}$ = [descriptors × descriptors] (eq. 2.3). The Q and R modes of analysis are defined in Section 7.1.

which it is possible to compute a determinant, an inverse, and eigenvalues and eigenvectors. As a corollary, these operations can be carried out on association matrices, which are square matrices.

Some definitions pertaining to square matrices now follow. In matrix $\mathbf{B}_{nn}$, of order $(n \times n)$ (often called "square matrix of order n" or "matrix of order n"),

$$\mathbf{B}_{nn} = [b_{ij}] = \begin{bmatrix} b_{11} & b_{12} & \dots & b_{1n} \\ b_{21} & b_{22} & \dots & b_{2n} \\ \cdot & & & \cdot \\ \cdot & & & \cdot \\ \cdot & & & \cdot \\ b_{n1} & b_{n2} & \dots & b_{nn} \end{bmatrix} \tag{2.4}$$

the *diagonal elements* are those with identical subscripts for the rows and columns (b_{ii}). They are located on the *main diagonal* (simply called *the diagonal*) which, by convention, goes from the upper left to the lower right corners. The sum of the diagonal elements is called the *trace* of the matrix.

Trace

Diagonal matrix

A *diagonal matrix* is a square matrix where all non-diagonal elements are *zero*. Thus,

$$\begin{bmatrix} 3 & 0 & 0 \\ 0 & 7 & 0 \\ 0 & 0 & 0 \end{bmatrix}$$

is a diagonal matrix. Diagonal matrices that contain on the diagonal values coming from a vector $[x_i]$ are noted $\mathbf{D}(x)$. Special examples used later in the book are the diagonal matrix of standard deviations $\mathbf{D}(\sigma)$ and the diagonal matrix of eigenvalues $\mathbf{D}(\lambda)$, also noted $\mathbf{\Lambda}$.

Identity matrix

A diagonal matrix where all diagonal elements are equal to unity is called a *unit matrix* or *identity matrix*. It is denoted $\mathbf{D}(1)$ or $\mathbf{I}$:

$$\mathbf{D}(1) = \mathbf{I} = \begin{bmatrix} 1 & 0 & \ldots & 0 \\ 0 & 1 & \ldots & 0 \\ \cdot & & & \cdot \\ \cdot & & & \cdot \\ \cdot & & & \cdot \\ 0 & 0 & \ldots & 1 \end{bmatrix} \tag{2.5}$$

This matrix plays the same role, in matrix algebra, as the number 1 in ordinary algebra, i.e. it is the neutral element in multiplication (e.g. $\mathbf{I\,B} = \mathbf{B\,I} = \mathbf{B}$).

Scalar matrix

Similarly, a *scalar matrix* is a diagonal matrix of the form

$$\begin{bmatrix} 7 & 0 & \ldots & 0 \\ 0 & 7 & \ldots & 0 \\ \cdot & & & \cdot \\ \cdot & & & \cdot \\ \cdot & & & \cdot \\ 0 & 0 & \ldots & 7 \end{bmatrix} = 7\mathbf{I}$$

All the diagonal elements are identical since a scalar matrix is the unit matrix multiplied by a scalar (here, of value 7).

Null matrix

A matrix, square or rectangular, whose elements are all zero is called a *null matrix* or *zero matrix*. It is denoted **0** or [0].*

Triangular matrix

A square matrix with all elements above (or below) the diagonal being zero is called a *lower* (or *upper*) *triangular* matrix. For example,

$$\begin{bmatrix} 1 & 2 & 3 \\ 0 & 4 & 5 \\ 0 & 0 & 6 \end{bmatrix}$$

is an upper triangular matrix. These matrices are very important in matrix algebra because their determinant (Section 2.6) is equal to the product of all terms on the main diagonal (i.e. 24 in this example). Diagonal matrices are also triangular matrices.

Transpose

The *transpose* of a matrix **B** with format $(n \times p)$ is denoted **B'** and is a new matrix of format $(p \times n)$ in which $b'_{ij} = b_{ji}$. In other words, the rows of one matrix are the columns of the other. Thus, the transpose of matrix

$$\mathbf{B} = \begin{bmatrix} 1 & 2 & 3 \\ 4 & 5 & 6 \\ 7 & 8 & 9 \\ 10 & 11 & 12 \end{bmatrix}$$

is matrix

$$\mathbf{B'} = \begin{bmatrix} 1 & 4 & 7 & 10 \\ 2 & 5 & 8 & 11 \\ 3 & 6 & 9 & 12 \end{bmatrix}$$

Transposition is an important operation in linear algebra, and also in ecology where a data matrix **Y** (eq. 2.1) is often transposed to study the relationships among descriptors after the relationships among objects have been analysed (or conversely).

* Although the concept of zero was known to Babylonian and Mayan astronomers, inclusion of the zero in a decimal system of numeration finds its origin in India, in the eighth century A.D. at least (Ifrah, 1981). The ten Western-world numerals are also derived from the symbols used by ancient Indian mathematicians. The word *zero* comes from the Arabs, however. They used the word *sifr*, meaning “empty”, to refer to a symbol designating nothingness. The term turned into *cipher*, and came to denote not only zero, but all 10 numerals. *Sifr* is at the root of the latin *zephirum*, which became *zefiro* in Italian and was then abbreviated to *zero*. It is also the root of the medieval latin *cifra*, which became *chiffre* in French where it designates any of the 10 numerals.

Symmetric matrix

A square matrix which is identical to its transpose is *symmetric*. This is the case when corresponding terms b_{ij} and b_{ji}, on either side of the diagonal, are equal. For example,

$$\begin{bmatrix} 1 & 4 & 6 \\ 4 & 2 & 5 \\ 6 & 5 & 3 \end{bmatrix}$$

is symmetric since $\mathbf{B}' = \mathbf{B}$. All symmetric matrices are square.

Non-symmetric matrix

It was mentioned in Section 2.2 that association matrices are generally symmetric. Non-symmetric (or asymmetric) matrices may be encountered, however. This happens, for example, when each coefficient in the matrix measures the ecological influence of an organism or a species on another, these influences being asymmetrical (e.g. A is a predator of B, B is a prey of A). Asymmetric matrices are also found in behaviour studies, serology, DNA pairing analysis, etc.

Skew-symmetric matrix

Matrix algebra tells us that any *non-symmetric* matrix may be expressed as the sum of two other matrices, one *symmetric* and one *skew-symmetric*, without loss of information. Consider for instance the two numbers 1 and 3, found in opposite positions (1,2) and (2,1) of the first matrix in the following numerical example:

$$\underset{\text{Non-symmetric}}{\begin{bmatrix} 1 & 1 & 2 & 2 \\ 3 & 1 & 0 & -1 \\ 1 & 2 & 1 & 0 \\ 0 & -4 & 3 & 1 \end{bmatrix}} = \underset{\text{Symmetric (average)}}{\begin{bmatrix} 1 & 2.0 & 1.5 & 1.0 \\ 2.0 & 1 & 1.0 & -2.5 \\ 1.5 & 1.0 & 1 & 1.5 \\ 1.0 & -2.5 & 1.5 & 1 \end{bmatrix}} + \underset{\text{Skew-symmetric}}{\begin{bmatrix} 0 & -1.0 & 0.5 & 1.0 \\ 1.0 & 0 & -1.0 & 1.5 \\ -0.5 & 1.0 & 0 & -1.5 \\ -1.0 & -1.5 & 1.5 & 0 \end{bmatrix}}$$

The *symmetric* part is obtained by averaging these two numbers: $(1 + 3)/2 = 2.0$. The *skew-symmetric* part is obtained by subtracting one from the other and dividing by 2: $(1 - 3)/2 = -1.0$ and $(3 - 1)/2 = +1.0$ so that, in the skew-symmetric matrix, corresponding elements on either side of the diagonal have the same absolute values but opposite signs. When the symmetric and skew-symmetric components are added, the result is the original matrix: $2 - 1 = 1$ for the upper original number, and $2 + 1 = 3$ for the lower one. Using letters instead of numbers, one can derive a simple algebraic proof of the additivity of the symmetric and skew-symmetric components. The symmetric component can be analysed using the methods applicable to symmetric matrices (for instance, metric or non-metric scaling, Sections 9.2 and 9.3), while analysis of the skew-symmetric component requires methods especially developed to assess asymmetric relationships. Basic references are Coleman (1964) in the field of sociometry and Digby & Kempton (1987, Ch. 6) in numerical ecology. An application to biological evolution is found in Casgrain *et al.* (1996). Relevant biological or ecological information may be found in the symmetric portion only and, in other instances, in the skew-symmetric component only.

2.4 Vectors and scaling

Vector

Another matrix of special interest is the *column matrix*, with format ($n \times 1$), which is also known as a *vector*. Some textbooks restrict the term 'vector' to *column matrices*, but the expression *row vector* (or simply *vector*, as in Chapter 4) may also be used for *row matrices*, with format ($1 \times p$).

A (column) vector is noted as follows:

$$\mathbf{b} = \begin{bmatrix} b_1 \\ b_2 \\ \cdot \\ \cdot \\ \cdot \\ b_n \end{bmatrix} \qquad \textbf{(2.6)}$$

A vector generally refers to a directed line segment, forming a mathematical entity on which operations can be performed. More formally, a vector is defined as an ordered n-tuple of real numbers, i.e. a set of n numbers with a specified order. The n numbers are the coordinates of a point in a n-dimensional Euclidean space, which may be seen as the end-point of a line segment starting at the origin.

For example, (column) vector [4 3]' is an ordered doublet (or 2-tuple) of two real numbers (4, 3), which may be represented in a two-dimensional Euclidean space:

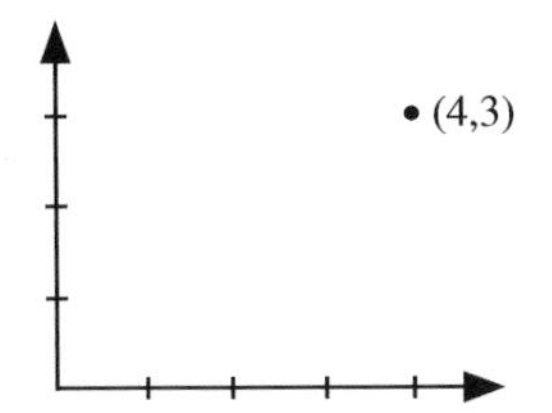

This same point (4, 3) may also be seen as the end-point of a line segment starting at the origin:

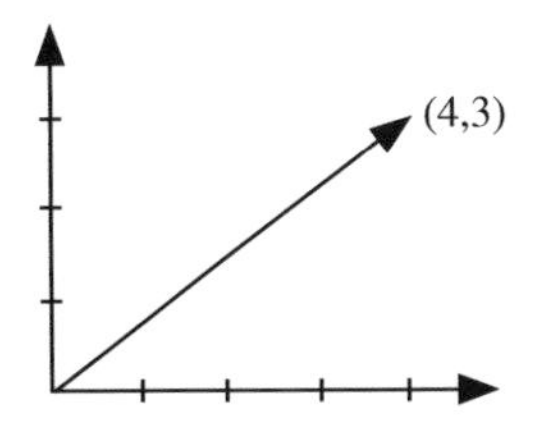

These figures illustrate the two possible representations of a vector; they also stress the *ordered* nature of vectors, since vector [3 4]' is different from vector [4 3]'.

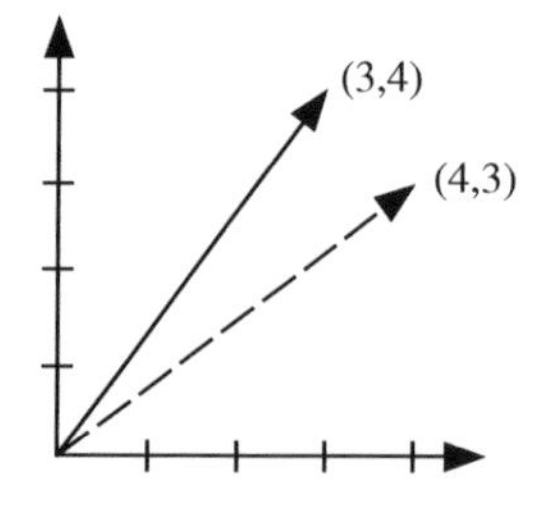

Using the Pythagorean theorem, it is easy to calculate the length of any vector. For example, the length of vector [4 3]' is that of the hypotenuse of a right triangle with base 4 and height 3:

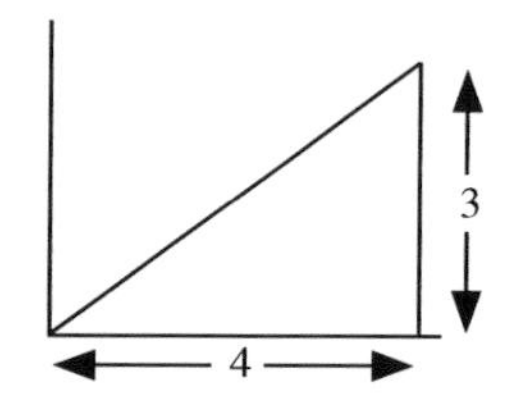

Length
Norm

The length (or *norm*) of vector [4 3]' is therefore $\sqrt{4^2 + 3^2} = 5$; it is also the length (norm) of vector [3 4]'. The norm of vector b is noted $\|\mathbf{b}\|$.

Scaling
Normalization

The comparison of different vectors, as to their directions, often requires an operation called *scaling*. In the scaled vector, all elements are divided by the same characteristic value. A special type of scaling is called *normalization*. In the normalized vector, each element is divided by the length of the vector:

$$\begin{bmatrix} 4 \\ 3 \end{bmatrix} \xrightarrow{\text{normalization}} \begin{bmatrix} 4/5 \\ 3/5 \end{bmatrix}$$

Normalized vector

The importance of normalization lies in the fact that the length of a *normalized vector* is equal to unity. Indeed, the length of vector [4/5 3/5]', calculated by means of the Pythagorean formula, is $\sqrt{(4/5)^2 + (3/5)^2} = 1$.

The example of doublet (4, 3) may be generalized to any n-tuple $(b_1, b_2, \ldots, b_n)$, which specifies a vector in n-dimensional space. The length of the vector is $\sqrt{b_1^2 + b_2^2 + \ldots + b_n^2}$, so that the corresponding normalized vector is:

$$\begin{bmatrix} b_1 / \sqrt{b_1^2 + b_2^2 + \ldots + b_n^2} \\ b_2 / \sqrt{b_1^2 + b_2^2 + \ldots + b_n^2} \\ \cdot \\ \cdot \\ \cdot \\ b_n / \sqrt{b_1^2 + b_2^2 + \ldots + b_n^2} \end{bmatrix} = \frac{1}{\sqrt{b_1^2 + b_2^2 + \ldots + b_n^2}} \begin{bmatrix} b_1 \\ b_2 \\ \cdot \\ \cdot \\ \cdot \\ b_n \end{bmatrix} \quad \textbf{(2.7)}$$

The length of any normalized vector, in the n-dimensional space, is 1.

2.5 Matrix addition and multiplication

Recording the data in table form, as is usually the case in ecology, opens the possibility of performing operations on these tables. The basic operations of matrix algebra (*algebra*, from the Arabic "al-jabr" which means *reduction*, is the theory of addition and multiplication) are very natural and familiar to ecologists.

Numerical example. Fish (3 species) were sampled at five sites in a lake, once a month during the summer (northern hemisphere). In order to get a general idea of the differences among sites, total numbers of fish caught at each site are calculated over the whole summer:

$$\begin{array}{l} \\ \text{Site 1} \\ \text{Site 2} \\ \text{Site 3} \\ \text{Site 4} \\ \text{Site 5} \\ \\ \end{array}
\begin{array}{c} \text{July} \\ \begin{bmatrix} 1 & 5 & 35 \\ 14 & 2 & 0 \\ 0 & 31 & 67 \\ 96 & 110 & 78 \\ 0 & 0 & 0 \end{bmatrix} \\ sp1 \;\; sp2 \;\; sp3 \end{array} +
\begin{array}{c} \text{August} \\ \begin{bmatrix} 15 & 23 & 10 \\ 54 & 96 & 240 \\ 0 & 3 & 9 \\ 12 & 31 & 27 \\ 8 & 14 & 6 \end{bmatrix} \\ sp1 \;\; sp2 \;\; sp3 \end{array} +
\begin{array}{c} \text{September} \\ \begin{bmatrix} 48 & 78 & 170 \\ 2 & 0 & 0 \\ 0 & 11 & 14 \\ 25 & 13 & 12 \\ 131 & 96 & 43 \end{bmatrix} \\ sp1 \;\; sp2 \;\; sp3 \end{array} =
\begin{array}{c} \text{Whole summer} \\ \begin{bmatrix} 64 & 106 & 215 \\ 70 & 98 & 240 \\ 0 & 45 & 90 \\ 133 & 154 & 117 \\ 139 & 110 & 49 \end{bmatrix} \\ sp1 \;\; sp2 \;\; sp3 \end{array}$$

This operation is known as *matrix addition*. Note that only matrices of the same order can be added together. This is why, in the first matrix, site 5 was included even if no fish had been caught there in July. Adding two matrices consists in a term-by-term addition. Matrix addition is associative and commutative; its neutral element is the null matrix **0**.

To study seasonal changes in fish productivity at each site, one possible approach would be to add together the terms in each row of each monthly matrix. However, this makes sense only if the selectivity of the fishing gear (say, a net) is comparable for the three species. Let us imagine that the efficiency of the net was 50% for species 2 and 25% for species 3 of what it was for species 1. In such a case, values in each row must be corrected before being added. Correction factors would be as follows: 1 for species 1, 2 for species 2, and 4 for species 3. To obtain estimates of total fish abundances, correction vector [1 2 4]' is first multiplied by each row of each matrix, after which the resulting values are added. Thus, for the first site in July:

Site 1 July	Correction factors	Total fish abundance Site 1, July
$\begin{bmatrix} 1 & 5 & 35 \end{bmatrix}$	$\begin{bmatrix} 1 \\ 2 \\ 4 \end{bmatrix}$	$(1 \times 1) + (5 \times 2) + (35 \times 4) = 1 + 10 + 140 = 151$

Scalar product

This operation is known in linear algebra as a *scalar product*, because it is a product of two vectors resulting in a scalar.

In physics, there is another product of two vectors, called the *external* or *vector product*, where the multiplication of two vectors results in a third one which is perpendicular to the plane formed by the first two. This product is not used in multidimensional analysis. It is however important to know that, in the literature, the expression "vector product" may be used for either this product or the scalar product of linear algebra, and that the latter is also called "inner product" or "dot product". The vector product (of physics) is sometimes called "cross product". This last expression is also used in linear algebra, for example in "sum of squares and cross products" (SSCP) which refers to the product of a matrix with its transpose.

In matrix algebra, and unless otherwise specified, multiplication follows a convention which is illustrated by the scalar product above: in this *product* of a column vector *by* a row vector, the row vector *multiplies* the column vector or, which is equivalent, the column vector *is multiplied by* the row vector. This convention, which should be kept in mind, will be followed in the remainder of the book.

The result of a scalar product is a number which is equal to the sum of the products of those elements with corresponding order numbers. The scalar product is designated by a dot, or is written <**a**,**b**>, or else there is no sign between the two terms. For example:

$$\mathbf{bc} = \mathbf{b} \bullet \mathbf{c} = \begin{bmatrix} b_1 & b_2 & \dots & b_p \end{bmatrix} \begin{bmatrix} c_1 \\ c_2 \\ \cdot \\ \cdot \\ \cdot \\ c_p \end{bmatrix} = b_1 c_1 + b_2 c_2 + \dots + b_p c_p = \text{a scalar.} \qquad \textbf{(2.8)}$$

The rules for computing scalar products are such that only vectors with the same numbers of elements can be multiplied.

In analytic geometry, it can be shown that the scalar product of two vectors obeys the relationship:

$$\mathbf{b} \bullet \mathbf{c} = (\text{length of } \mathbf{b}) \times (\text{length of } \mathbf{c}) \times \cos\theta \qquad \textbf{(2.9)}$$

Orthogonal vectors

When the angle between two vectors is $\theta = 90°$, then $\cos\theta = 0$ and the scalar product $\mathbf{b} \bullet \mathbf{c} = 0$. As a consequence, two vectors whose scalar product is zero are *orthogonal* (i.e. at right angle). This property will be used in Section 2.9 to compute eigenvectors. A matrix whose (column) vectors are all at right angle of each other is called *orthogonal*.

Numerical example. Returning to the above example, it is possible to multiply each row of each monthly matrix with the correction vector (scalar product), in order to compare total monthly fish abundances. This operation, which is the *product of a vector by a matrix*, is a simple extension of the scalar product (eq. 2.8). The product of the July matrix **B** with the correction vector **c** is written as follows:

$$\begin{bmatrix} 1 & 5 & 35 \\ 14 & 2 & 0 \\ 0 & 31 & 67 \\ 96 & 110 & 78 \\ 0 & 0 & 0 \end{bmatrix} \begin{bmatrix} 1 \\ 2 \\ 4 \end{bmatrix} = \begin{bmatrix} 1(1) + 5(2) + 35(4) \\ 14(1) + 2(2) + 0(4) \\ 0(1) + 31(2) + 67(4) \\ 96(1) + 110(2) + 78(4) \\ 0(1) + 0(2) + 0(4) \end{bmatrix} = \begin{bmatrix} 151 \\ 18 \\ 330 \\ 628 \\ 0 \end{bmatrix}$$

The product of a vector by a matrix involves calculating, for each row of matrix **B**, a scalar product with vector **c**. Such a product of a vector by a matrix is only possible if the number of *elements in the vector* is the same as the number of *columns in the matrix*. The result is no longer a scalar, but a column vector with dimension equal to the number of rows in the matrix on the left. The general formula for this product is:

$$\mathbf{B}_{pq} \bullet \mathbf{c}_q = \begin{bmatrix} b_{11} & b_{12} & \dots & b_{1q} \\ b_{21} & b_{22} & \dots & b_{2q} \\ \cdot & & & \cdot \\ \cdot & & & \cdot \\ \cdot & & & \cdot \\ b_{p1} & b_{p2} & \dots & b_{pq} \end{bmatrix} \begin{bmatrix} c_1 \\ c_2 \\ \cdot \\ \cdot \\ \cdot \\ c_q \end{bmatrix} = \begin{bmatrix} b_{11}c_1 + b_{12}c_2 + \dots + b_{1q}c_q \\ b_{21}c_1 + b_{22}c_2 + \dots + b_{2q}c_q \\ \cdot \\ \cdot \\ \cdot \\ b_{p1}c_1 + b_{p2}c_2 + \dots + b_{pq}c_q \end{bmatrix}$$

Using summation notation, this equation may be rewritten as:

$$\mathbf{B}_{pq} \bullet \mathbf{c}_q = \begin{bmatrix} \sum_{k=1}^{q} b_{1k} c_k \\ \cdot \\ \cdot \\ \cdot \\ \sum_{k=1}^{q} b_{pk} c_k \end{bmatrix} \qquad \textbf{(2.10)}$$

The product of two matrices is the logical extension of the product of a vector by a matrix. Matrix **C**, to be multiplied by **B**, is simply considered as a set of column vectors $\mathbf{c}_1$, $\mathbf{c}_2$, …; eq. 2.10 is repeated for each column. Following the same logic, the resulting column vectors are juxtaposed to form the result matrix. Matrices to be multiplied must be *conformable*, which means that the number of columns in the matrix on the left must be the same as the number of rows in the matrix on the right. For example, given

$$\mathbf{B} = \begin{bmatrix} 1 & 0 & 2 \\ 3 & 1 & 1 \\ 1 & 2 & 1 \\ -1 & 3 & 2 \end{bmatrix} \quad \text{and} \quad \mathbf{C} = \begin{bmatrix} 1 & 2 \\ 2 & 1 \\ 3 & -1 \end{bmatrix}$$

$$\mathbf{C} = [\,\mathbf{d} \quad \mathbf{e}\,]$$

the product of **B** with each of the two columns of **C** is:

$$\mathbf{Bd} = \begin{bmatrix} 1(1)+0(2)+2(3) \\ 3(1)+1(2)+1(3) \\ 1(1)+2(2)+1(3) \\ -1(1)+3(2)+2(3) \end{bmatrix} = \begin{bmatrix} 7 \\ 8 \\ 8 \\ 11 \end{bmatrix} \quad \text{and} \quad \mathbf{Be} = \begin{bmatrix} 1(2)+0(1)+2(-1) \\ 3(2)+1(1)+1(-1) \\ 1(2)+2(1)+1(-1) \\ -1(2)+3(1)+2(-1) \end{bmatrix} = \begin{bmatrix} 0 \\ 6 \\ 3 \\ -1 \end{bmatrix}$$

so that the product matrix is:

$$\mathbf{BC} = \begin{bmatrix} 7 & 0 \\ 8 & 6 \\ 8 & 3 \\ 11 & -1 \end{bmatrix}$$

Thus, the product of two conformable matrices **B** and **C** is a new matrix with the same number of rows as **B** and the same number of columns as **C**. Element d_{ij}, in row i and column j of the resulting matrix, is the scalar product of row i of **B** with column j of **C**.

The only way to master the mechanics of matrix products is to go through some numerical examples. As an exercise, readers could apply the above method to two cases which have not been discussed so far, i.e. the *product of a row vector by a column vector*, which gives a *matrix* and not a scalar, and the *product of a matrix by a row vector*, which results in a *row vector*. This exercise would help to better understand the rule of conformability.

As supplementary exercises, readers could calculate numerical examples of the eight following properties of matrix products, which will be used later in the book:

(1) $\mathbf{B}_{pq}\,\mathbf{C}_{qr}\,\mathbf{D}_{rs} = \mathbf{E}_{ps}$, of order $(p \times s)$.

(2) The existence of product **BC** does not imply that product **CB** exists, because matrices are not necessarily conformable in the reverse order; however, **C'C** and **CC'** always exist.

(3) **BC** is generally not equal to **CB**, i.e. matrix products are not commutative.

(4) $\mathbf{B}^2 = \mathbf{B} \times \mathbf{B}$ exists only if **B** is a square matrix.

(5) [**AB**]' = **B'A'** and, more generally, [**ABCD...**]' = **...D'C'B'A'**.

(6) The products **XX'** and **X'X** always give rise to symmetric matrices.

(7) In general, the product of two symmetric but different matrices **A** and **B** is not a symmetric matrix.

(8) If **B** is an orthogonal matrix (i.e. a rectangular matrix whose column vectors are orthogonal to one another), then **B'B** = **D**, where **D** is a diagonal matrix. All non-diagonal terms are zero because of the property of orthogonality, while the diagonal terms are the squares of the lengths of the column vectors. That **B'B** is diagonal does not imply that **BB'** is also diagonal. **BB'** = **B'B** only when **B** is square and symmetric.

The last type of product to be considered is that of a matrix or vector *by a scalar*. It is carried out according to the usual algebraic rules of multiplication and factoring, i.e. for matrix $\mathbf{B} = [b_{jk}]$ or vector $\mathbf{c} = [c_j]$, $d\mathbf{B} = [db_{jk}]$ and $d\mathbf{c} = [dc_j]$. For example:

$$3\begin{bmatrix} 1 & 2 \\ 3 & 4 \end{bmatrix} = \begin{bmatrix} 3 & 6 \\ 9 & 12 \end{bmatrix} \quad \text{and} \quad \begin{bmatrix} 5 \\ 6 \end{bmatrix} 2 = \begin{bmatrix} 10 \\ 12 \end{bmatrix}$$

The terms *premultiplication* and *postmultiplication* may be encountered in the literature. Product **BC** corresponds to *premultiplication* of **C** by **B**, or to *postmultiplication* of **B** by **C**. Unless otherwise specified, it is always premultiplication which is implied and **BC** simply reads: **C** is multiplied by **B**.

2.6 Determinant

It is often necessary to transform a matrix into a new one, in such a way that the information of the original matrix is preserved, while new properties which are essential to subsequent calculations are acquired. Such matrices, which are linearly derived from the original matrix, will be studied in following sections under the names *inverse matrix*, *canonical form*, etc.

The new matrix must have a minimum number of characteristics in common with the matrix from which it is linearly derived. The connection between the two matrices is a matrix function $f(\mathbf{B})$, whose properties are the following:

(1) The function must be *multilinear*, which means that it should respond linearly to any change taking place in the rows or columns of the matrix.

(2) Since the order of the rows and columns of a matrix is specified, the function should be able to detect, through *alternation of signs*, any change in the positions of rows or columns. As a corollary, if two columns (or rows) are identical, $f(\mathbf{B}) = 0$; indeed, if two identical columns (or rows) are interchanged, $f(\mathbf{B})$ must change sign but it must also remain identical, which is possible only if $f(\mathbf{B}) = 0$.

(3) Finally, there is a scalar associated with this function; it is called its *norm* or *value*. For convenience, the norm is calibrated in such a way that the value associated with the unit matrix **I** is 1, i.e. $f(\mathbf{I}) = 1$.

It can be shown that the determinant, as defined below, is the only function which has the above three properties, and that it only exists for square matrices. Therefore, it is not possible to calculate a determinant for a rectangular matrix. The determinant of matrix **B** is denoted *det* **B** or, more often, $|\mathbf{B}|$:

$$|\mathbf{B}| \equiv \begin{vmatrix} b_{11} & b_{12} & \dots & b_{1n} \\ b_{21} & b_{22} & \dots & b_{2n} \\ \cdot & & & \cdot \\ \cdot & & & \cdot \\ \cdot & & & \cdot \\ b_{n1} & b_{n2} & \dots & b_{nn} \end{vmatrix}$$

The value of function $|\mathbf{B}|$ is a scalar, i.e. a number.

What follows is the formal definition of the value of a determinant. The way to compute it in practice is explained later. The *value of a determinant* is calculated as the sum of all possible products containing one, and only one, element from each row and each column; these products receive a sign according to a well-defined rule:

$$|\mathbf{B}| = \sum \pm (b_{1j_1} b_{2j_2} \ldots b_{nj_n})$$

where indices $j_1, j_2, \ldots, j_n$, go through the $n!$ permutations of the numbers $1, 2, \ldots, n$. The sign depends on the number of inversions, in the permutation considered, relative to the regular sequence $1, 2, \ldots, n$: if the number of inversions is even, the sign is (+) and, if the number is odd, the sign is (–).

The determinant of a matrix of order 2 is calculated as follows:

$$|\mathbf{B}| = \begin{vmatrix} b_{11} & b_{12} \\ b_{21} & b_{22} \end{vmatrix} = b_{11}b_{22} - b_{12}b_{21} \qquad \textbf{(2.11)}$$

In accordance with the formal definition above, the scalar so obtained is composed of $2! = 2$ products, each product containing one, and only one, element from each row and each column.

Expansion by minors

The determinant of a matrix of order higher than 2 may be calculated using different methods, among which is the *expansion by minors*. When looking for a determinant of order 3, a determinant of order $3 - 1 = 2$ may be obtained by crossing out one row (i) and one column (j). This lower-order determinant is the *minor* associated with b_{ij}:

crossing out row 1 *and column* 2

$$\begin{vmatrix} b_{11} & b_{12} & b_{13} \\ b_{21} & b_{22} & b_{23} \\ b_{31} & b_{32} & b_{33} \end{vmatrix} \rightarrow \underbrace{\begin{vmatrix} b_{21} & b_{23} \\ b_{31} & b_{33} \end{vmatrix}}_{\textit{minor of } b_{12}} \qquad \textbf{(2.12)}$$

The minor being here a determinant of order 2, its value is calculated using eq. 2.11. When multiplied by $(-1)^{i+j}$, the minor becomes a *cofactor*. Thus, the cofactor of b_{12} is:

$$\text{cof } b_{12} = (-1)^{1+2} \begin{vmatrix} b_{21} & b_{23} \\ b_{31} & b_{33} \end{vmatrix} = - \begin{vmatrix} b_{21} & b_{23} \\ b_{31} & b_{33} \end{vmatrix} \qquad \textbf{(2.13)}$$

The expansion by minors of a determinant of order n is:

$$|\mathbf{B}| = \sum_{i=1}^{n} b_{ij} \mathrm{cof}\, b_{ij} \qquad \text{for any column } j \tag{2.14}$$

$$|\mathbf{B}| = \sum_{j=1}^{n} b_{ij} \mathrm{cof}\, b_{ij} \qquad \text{for any row } i$$

The expansion may involve the elements of any row or any column, the result being always the same. Thus, going back to the determinant of eq. 2.12, expansion by the elements of the first row gives:

$$|\mathbf{B}| = b_{11} \mathrm{cof}\, b_{11} + b_{12} \mathrm{cof}\, b_{12} + b_{13} \mathrm{cof}\, b_{13} \tag{2.15}$$

$$|\mathbf{B}| = b_{11}(-1)^{1+1} \begin{vmatrix} b_{22} & b_{23} \\ b_{32} & b_{33} \end{vmatrix} + b_{12}(-1)^{1+2} \begin{vmatrix} b_{21} & b_{23} \\ b_{31} & b_{33} \end{vmatrix} + b_{13}(-1)^{1+3} \begin{vmatrix} b_{21} & b_{22} \\ b_{31} & b_{32} \end{vmatrix}$$

Numerical example. Equation 2.15 is applied to a simple numerical example:

$$\begin{vmatrix} 1 & 2 & 3 \\ 4 & 5 & 6 \\ 7 & 8 & 10 \end{vmatrix} = 1(-1)^{1+1} \begin{vmatrix} 5 & 6 \\ 8 & 10 \end{vmatrix} + 2(-1)^{1+2} \begin{vmatrix} 4 & 6 \\ 7 & 10 \end{vmatrix} + 3(-1)^{1+3} \begin{vmatrix} 4 & 5 \\ 7 & 8 \end{vmatrix}$$

$$\begin{vmatrix} 1 & 2 & 3 \\ 4 & 5 & 6 \\ 7 & 8 & 10 \end{vmatrix} = 1(5 \times 10 - 6 \times 8) - 2(4 \times 10 - 6 \times 7) + 3(4 \times 8 - 5 \times 7) = -3$$

The amount of calculations required to expand a determinant increases very quickly with increasing order n. This is because the minor of each cofactor must be expanded, the latter producing new cofactors whose minors are in turn expanded, and so forth until cofactors of order 2 are reached. Another, faster method is normally used to calculate determinants by computer. Before describing this method, however, some properties of determinants must be examined; in all cases, *column* may be substituted for *row*.

(1) The determinant of a matrix is equal to that of its transpose, since a determinant may be computed from either the rows or columns of the matrix: $|\mathbf{A}'| = |\mathbf{A}|$.

(2) If two rows are interchanged, the sign of the determinant is reversed.

(3) If two rows are identical, the determinant is null (corollary of the second property; see beginning of the present Section).

(4) If a scalar is a factor of *one* row, it becomes a factor of the determinant (since it appears *once* in each product).

(5) If a row is a multiple of another row, the determinant is null (corollary of properties 4 and 3, i.e. factoring out the multiplier produces two identical rows).

(6) If all elements of a row are 0, the determinant is null (corollary of property 4).

(7) If a scalar c is a factor of *all* rows, it becomes a factor c^n of the determinant (corollary of property 4), i.e. $|c\mathbf{B}| = c^n |\mathbf{B}|$.

(8) If a multiple of a row is added to another row, the value of the determinant remains unchanged.

(9) The determinant of a triangular matrix (and therefore also of a diagonal matrix) is the product of its diagonal elements.

(10) The sum of the products of the elements of a row with the corresponding cofactors of a *different* row is equal to zero.

Pivotal condensation

Properties 8 and 9 can be used for rapid computer calculation of the value of a determinant; the method is called *pivotal condensation*. The matrix is first reduced to triangular form using property 8. This property allows the stepwise elimination of all terms on one side of the diagonal through combinations of multiplications by a scalar, and addition and subtraction of rows or columns. Pivotal condensation may be performed in either the upper or the lower triangular parts of a square matrix. If the lower triangular part is chosen, the upper left-hand diagonal element is used as the first pivot to modify the other rows in such a way that their left-hand terms become zero. The technique consists in calculating by how much the pivot must be multiplied to cancel out the terms in the rows below it; when this value is found, property 8 is used with this value as multiplier. When all terms under the diagonal element in the first column are zero, the procedure is repeated with the other diagonal terms as pivots, to cancel out the elements located under them in the same column. Working on the pivots from left to right insures that when values have been changed to 0, they remain so. When the whole lower triangular portion of the matrix is zero, property 9 is used to compute the determinant which is then the product of the modified diagonal elements.

Numerical example. The same numerical example as above illustrates the method:

$$\begin{vmatrix} 1 & 2 & 3 \\ 4 & 5 & 6 \\ 7 & 8 & 10 \end{vmatrix} = \underset{a}{\begin{vmatrix} 1 & 2 & 3 \\ 0 & -3 & -6 \\ 7 & 8 & 10 \end{vmatrix}} = \underset{b}{\begin{vmatrix} 1 & 2 & 3 \\ 0 & -3 & -6 \\ 0 & -6 & -11 \end{vmatrix}} = \underset{c}{\begin{vmatrix} 1 & 2 & 3 \\ 0 & -3 & -6 \\ 0 & 0 & 1 \end{vmatrix}}$$

a: (row 2 – 4 × row 1) b: (row 3 – 7 × row 1) c: (row 3 – 2 × row 2)

The determinant is the product of the diagonal elements: $1 \times (-3) \times 1 = (-3)$.

2.7 The rank of a matrix

A square matrix contains n vectors (rows or columns), which may be *linearly independent* or not (for the various meanings of "independence", see Box 1.1). Two vectors are *linearly dependent* when the elements of one are proportional to those of the other. For example:

$$\begin{bmatrix}-4\\-6\\-8\end{bmatrix} \text{ and } \begin{bmatrix}2\\3\\4\end{bmatrix} \text{ are linearly dependent, since } \begin{bmatrix}-4\\-6\\-8\end{bmatrix} = -2\begin{bmatrix}2\\3\\4\end{bmatrix}$$

Similarly, a vector is linearly dependent on two others, which are themselves linearly independent, when its elements are a linear combination of the elements of the other two. For example:

$$\begin{bmatrix}-1\\3\\4\end{bmatrix}, \begin{bmatrix}-1\\0\\1\end{bmatrix} \text{ and } \begin{bmatrix}1\\-2\\-3\end{bmatrix}$$

illustrate a case where a vector is linearly dependent on two others, which are themselves linearly independent, since

$$(-2)\begin{bmatrix}-1\\3\\4\end{bmatrix} = \begin{bmatrix}-1\\0\\1\end{bmatrix} + 3\begin{bmatrix}1\\-2\\-3\end{bmatrix}$$

Rank of a square matrix

The *rank* of a *square matrix* is defined as the number of linearly independent row vectors (or column vectors) in the matrix. For example:

$$\begin{bmatrix}-1 & -1 & 1\\3 & 0 & -2\\4 & 1 & -3\end{bmatrix}$$

(–2 × column 1) = column 2 + (3 × column 3)

or: row 1 = row 2 – row 3

rank = 2

$$\begin{bmatrix}-2 & 1 & 4\\-2 & 1 & 4\\-2 & 1 & 4\end{bmatrix}$$

(–2 × column 1) = (4 × column 2) = column 3

or: row 1 = row 2 = row 3

rank = 1

According to property 5 of determinants (Section 2.6), a matrix whose *rank* is lower than its *order* has a determinant equal to zero. Finding the rank of a matrix may therefore be based on the determinant of the lower-order submatrices it contains. The *rank* of a square matrix is the order of the largest non-zero determinant it contains; this is also the maximum number of linearly independent vectors found among the rows or the columns.

$$\begin{vmatrix} 1 & 2 & 3 \\ 4 & 5 & 6 \\ 7 & 8 & 10 \end{vmatrix} = -3 \neq 0, \text{ so that the } rank = 3$$

$$\begin{vmatrix} -1 & -1 & 1 \\ 3 & 0 & -2 \\ 4 & 1 & -3 \end{vmatrix} = 0 \qquad \begin{vmatrix} -1 & -1 \\ 3 & 0 \end{vmatrix} = 3 \qquad rank = 2$$

Rank of a rectangular matrix

It is also possible to determine the rank of a *rectangular* matrix. Several *square* submatrices may be extracted from a *rectangular* matrix, by eliminating rows or/and columns from the matrix. The *rank* of a rectangular matrix is the highest rank of all the square submatrices that can be extracted from it. A first example illustrates the case where the rank of a rectangular matrix is equal to the number of rows:

$$\begin{vmatrix} 2 & 0 & 1 & 0 & -1 & -2 & 3 \\ 1 & 2 & 2 & 0 & 0 & 1 & -1 \\ 0 & 1 & 2 & 3 & 1 & -1 & 0 \end{vmatrix} \rightarrow \begin{vmatrix} 2 & 0 & 1 \\ 1 & 2 & 2 \\ 0 & 1 & 2 \end{vmatrix} = 5 \qquad rank = 3$$

In a second example, the rank is lower than the number of rows:

$$\begin{vmatrix} 2 & 1 & 3 & 4 \\ -1 & 6 & -3 & 0 \\ 1 & 20 & -3 & 8 \end{vmatrix} \rightarrow \begin{vmatrix} 2 & 1 & 3 \\ -1 & 6 & -3 \\ 1 & 20 & -3 \end{vmatrix} = \begin{vmatrix} 2 & 1 & 4 \\ -1 & 6 & 0 \\ 1 & 20 & 8 \end{vmatrix} = \begin{vmatrix} 2 & 3 & 4 \\ -1 & -3 & 0 \\ 1 & -3 & 8 \end{vmatrix} = \begin{vmatrix} 1 & 3 & 4 \\ 6 & -3 & 0 \\ 20 & -3 & 8 \end{vmatrix} = 0$$

$$rank < 3 \rightarrow \begin{vmatrix} 2 & 1 \\ -1 & 6 \end{vmatrix} = 13 \qquad rank = 2$$

In this case, the three rows are clearly linearly dependent: (2 $\times$ row 1) + (3 $\times$ row 2) = row 3. Since it is possible to find a square matrix of order 2 that has a non-null determinant, the rank of the rectangular matrix is 2.

2.8 Matrix inversion

In algebra, division is expressed as either $c \div b$, or c/b, or $c\ (1/b)$, or $c\ b^{-1}$. In the last two expressions, division as such is replaced by multiplication with a reciprocal or inverse quantity. In matrix algebra, division of **C** by **B** does not exist. The equivalent operation is multiplication of **C** with the *inverse* or *reciprocal* of matrix **B**. The inverse of matrix **B** is denoted $\mathbf{B}^{-1}$ and the operation through which it is computed is called *inversion* of matrix **B**.

To serve its purpose, matrix $\mathbf{B}^{-1}$ must be unique and the relation $\mathbf{B}\mathbf{B}^{-1} = \mathbf{B}^{-1}\mathbf{B} = \mathbf{I}$ must be satisfied (see also the concept of generalized inverse in textbooks of advanced linear algebra; several types of generalized inverse are not unique). It can be shown that only *square matrices* have unique inverses; so, it is only for square matrices that the relation $\mathbf{B}\mathbf{B}^{-1} = \mathbf{B}^{-1}\mathbf{B}$ is satisfied. Indeed, there are *rectangular* matrices **B** for which several matrices **C** can be found, satisfying for example **CB** = **I** but not **BC** = **I**. There are also rectangular matrices for which no matrix **C** can be found such that **CB** = **I**, whereas an infinite number of matrices **C** may exist that satisfy **BC** = **I**. For example:

$$\mathbf{B} = \begin{bmatrix} 1 & 1 \\ -1 & 0 \\ 3 & -1 \end{bmatrix} \qquad \mathbf{C} = \begin{bmatrix} 1 & 3 & 1 \\ 2 & 5 & 1 \end{bmatrix} \quad \mathbf{CB} = \mathbf{I} \quad \mathbf{BC} \neq \mathbf{I}$$

$$\mathbf{C} = \begin{bmatrix} 4 & 15 & 4 \\ 7 & 25 & 6 \end{bmatrix} \quad \mathbf{CB} = \mathbf{I} \quad \mathbf{BC} \neq \mathbf{I}$$

Inverse of a square matrix

To calculate the inverse of a square matrix **B**, the *adjugate* or *adjoint matrix* of **B** is first defined. In the *matrix of cofactors* of **B**, each element b_{ij} is replaced by its cofactor (cof b_{ij}; see Section 2.6). The adjugate matrix of **B** is the *transpose* of the matrix of cofactors:

$$\underset{\text{matrix } \mathbf{B}}{\begin{bmatrix} b_{11} & b_{12} & \dots & b_{1n} \\ b_{21} & b_{22} & \dots & b_{2n} \\ \cdot & & & \cdot \\ \cdot & & & \cdot \\ \cdot & & & \cdot \\ b_{n1} & b_{n2} & \dots & b_{nn} \end{bmatrix}} \rightarrow \underset{\text{adjugate matrix of } \mathbf{B}}{\begin{bmatrix} \text{cof } b_{11} & \text{cof } b_{21} & \dots & \text{cof } b_{n1} \\ \text{cof } b_{12} & \text{cof } b_{22} & \dots & \text{cof } b_{n2} \\ \cdot & & & \cdot \\ \cdot & & & \cdot \\ \cdot & & & \cdot \\ \text{cof } b_{1n} & \text{cof } b_{2n} & \dots & \text{cof } b_{nn} \end{bmatrix}} \qquad \textbf{(2.16)}$$

In the case of second order matrices, cofactors are scalar values, e.g. cof $b_{11} = b_{22}$, cof $b_{12} = -b_{21}$, etc.

The *inverse* of matrix **B** is the adjugate matrix of **B** divided by the determinant $|\mathbf{B}|$. The product of the matrix with its inverse gives the unit matrix:

$$\underbrace{\frac{1}{|\mathbf{B}|}\begin{bmatrix} \text{cof } b_{11} & \text{cof } b_{21} & \ldots & \text{cof } b_{n1} \\ \text{cof } b_{12} & \text{cof } b_{22} & \ldots & \text{cof } b_{n2} \\ \cdot & & & \cdot \\ \cdot & & & \cdot \\ \cdot & & & \cdot \\ \text{cof } b_{1n} & \text{cof } b_{2n} & \ldots & \text{cof } b_{nn} \end{bmatrix}}_{\mathbf{B}^{-1}} \underbrace{\begin{bmatrix} b_{11} & b_{12} & \ldots & b_{1n} \\ b_{21} & b_{22} & \ldots & b_{2n} \\ \cdot & & & \cdot \\ \cdot & & & \cdot \\ \cdot & & & \cdot \\ b_{n1} & b_{n2} & \ldots & b_{nn} \end{bmatrix}}_{\mathbf{B}} = \mathbf{I} \tag{2.17}$$

All diagonal terms resulting from the multiplication $\mathbf{B}^{-1}\mathbf{B}$ (or $\mathbf{B}\mathbf{B}^{-1}$) are of the form $\sum b_{ij} \text{cof } b_{ij}$, which is the expansion by minors of a determinant (not taking into account, at this stage, the division of each element of the matrix by $|\mathbf{B}|$). Each diagonal element consequently has the value of the determinant $|\mathbf{B}|$ (eq. 2.14). All other elements of matrix $\mathbf{B}^{-1}\mathbf{B}$ are sums of the products of the elements of a row with the corresponding cofactors of a different row. According to property 10 of determinants (Section 2.6), each non-diagonal element is therefore null. It follows that:

$$\mathbf{B}^{-1}\mathbf{B} = \frac{1}{|\mathbf{B}|}\begin{bmatrix} |\mathbf{B}| & 0 & \ldots & 0 \\ 0 & |\mathbf{B}| & \ldots & 0 \\ \cdot & & & \cdot \\ \cdot & & & \cdot \\ \cdot & & & \cdot \\ 0 & 0 & \ldots & |\mathbf{B}| \end{bmatrix} = \begin{bmatrix} 1 & 0 & \ldots & 0 \\ 0 & 1 & \ldots & 0 \\ \cdot & & & \cdot \\ \cdot & & & \cdot \\ \cdot & & & \cdot \\ 0 & 0 & \ldots & 1 \end{bmatrix} = \mathbf{I} \tag{2.18}$$

Singular matrix

An important point is that $\mathbf{B}^{-1}$ exists only if $|\mathbf{B}| \neq 0$. A square matrix with a null determinant is known as a *singular* matrix and it has no inverse (but see *singular value decomposition*, Section 2.11). Matrices which can be inverted are called *nonsingular.*

Numerical example. The numerical example of Sections 2.6 and 2.7 is used again to illustrate the calculations:

$$\begin{bmatrix} 1 & 2 & 3 \\ 4 & 5 & 6 \\ 7 & 8 & 10 \end{bmatrix}$$

The determinant is already known (Section 2.6); its value is –3. The matrix of cofactors is computed, and its *transpose* (adjugate matrix) is divided by the determinant to give the inverse matrix:

$$\begin{bmatrix} 2 & 2 & -3 \\ 4 & -11 & 6 \\ -3 & 6 & -3 \end{bmatrix} \qquad \begin{bmatrix} 2 & 4 & -3 \\ 2 & -11 & 6 \\ -3 & 6 & -3 \end{bmatrix} \qquad -\frac{1}{3}\begin{bmatrix} 2 & 4 & -3 \\ 2 & -11 & 6 \\ -3 & 6 & -3 \end{bmatrix}$$

matrix of cofactors — adjugate matrix — inverse of matrix

As for the determinant (Section 2.6), various methods exist for quickly inverting matrices using computers; they are especially useful for matrices of higher rank. Description of these methods, which are available in computer packages, is beyond the scope of the present book. One such method is briefly explained here, because it is somewhat similar to the pivotal condensation presented above for determinants.

Gauss-Jordan

Inversion of matrix **B** may be conducted using the *method of Gauss-Jordan*. To do so, matrix $\mathbf{B}_{(n \times n)}$ is first augmented to the right with a same-size identity matrix **I**, thus creating a $n \times 2n$ matrix. This is illustrated for $n = 3$:

$$\begin{bmatrix} b_{11} & b_{12} & b_{13} & 1 & 0 & 0 \\ b_{21} & b_{22} & b_{23} & 0 & 1 & 0 \\ b_{31} & b_{32} & b_{33} & 0 & 0 & 1 \end{bmatrix}$$

If the augmented matrix is multiplied by matrix $\mathbf{C}_{(n \times n)}$, and if $\mathbf{C} = \mathbf{B}^{-1}$, then the resulting matrix ($n \times 2n$) has an identity matrix in its first n columns and matrix $\mathbf{C} = \mathbf{B}^{-1}$ in the last n columns.

$$\begin{bmatrix} c_{11} & c_{12} & c_{13} \\ c_{21} & c_{22} & c_{23} \\ c_{31} & c_{32} & c_{33} \end{bmatrix}\begin{bmatrix} b_{11} & b_{12} & b_{13} & 1 & 0 & 0 \\ b_{21} & b_{22} & b_{23} & 0 & 1 & 0 \\ b_{31} & b_{32} & b_{33} & 0 & 0 & 1 \end{bmatrix} = \begin{bmatrix} 1 & 0 & 0 & c_{11} & c_{12} & c_{13} \\ 0 & 1 & 0 & c_{21} & c_{22} & c_{23} \\ 0 & 0 & 1 & c_{31} & c_{32} & c_{33} \end{bmatrix}$$

This shows that, if matrix [**B**,**I**] is transformed into an equivalent matrix [**I**,**C**], then $\mathbf{C} = \mathbf{B}^{-1}$.

The Gauss-Jordan transformation proceeds in two steps.

• In the first step, the diagonal terms are used, one after the other and from left to right, as pivots to make all the off-diagonal terms equal to zero. This is done in exactly the same way as for the determinant: a factor is calculated to cancel out the target term, using the pivot, and property 8 of the determinants is applied using this factor as multiplier. The difference with determinants is that the whole row of the augmented matrix is modified, not only the part belonging to matrix **B**. If an off-diagonal zero value is encountered, then of course it is left as is, no cancellation by a multiple of the pivot being necessary or even possible. If a zero is found on the diagonal, this pivot has to be left aside for the time being (in actual programs, rows and columns are interchanged in a process called pivoting); this zero will be changed to a non-zero value during the next cycle unless the matrix is singular. Pivoting makes programming of this method a bit complex.

• Second step. When all the off-diagonal terms are zero, the diagonal terms of the former matrix **B** are brought to 1. This is accomplished by dividing each row of the augmented matrix by the value now present in the diagonal terms of the former **B** (left) portion. If the changes introduced during the first step have made one of the diagonal elements equal to zero, then of course no division can bring it back to 1 and the matrix is singular (i.e. it cannot be inverted).

A Gauss-Jordan algorithm with pivoting is available in *Numerical recipes* (Press *et al.*, 1986, pp. 28-29).

Numerical example. To illustrate the Gauss-Jordan method, the same square matrix as above is first augmented, then transformed so that its left-hand portion becomes the identity matrix:

$$(a)\quad \begin{bmatrix} 1 & 2 & 3 \\ 4 & 5 & 6 \\ 7 & 8 & 10 \end{bmatrix} \rightarrow \left[\begin{array}{ccc|ccc} 1 & 2 & 3 & 1 & 0 & 0 \\ 4 & 5 & 6 & 0 & 1 & 0 \\ 7 & 8 & 10 & 0 & 0 & 1 \end{array}\right]$$

$$(b)\quad \left[\begin{array}{ccc|ccc} 1 & 2 & 3 & 1 & 0 & 0 \\ 0 & -3 & -6 & -4 & 1 & 0 \\ 0 & -6 & -11 & -7 & 0 & 1 \end{array}\right]$$

row 2 → row 2 – 4row 1

row 3 → row 3 – 7row 1

$$(c)\quad \left[\begin{array}{ccc|ccc} 3 & 0 & -3 & -5 & 2 & 0 \\ 0 & -3 & -6 & -4 & 1 & 0 \\ 0 & 0 & 1 & 1 & -2 & 1 \end{array}\right]$$

row 1 → 3row 1 + 2row 2

row 3 → row 3 – 2row 2

$$(d)\quad \left[\begin{array}{ccc|ccc} 3 & 0 & 0 & -2 & -4 & 3 \\ 0 & -3 & 0 & 2 & -11 & 6 \\ 0 & 0 & 1 & 1 & -2 & 1 \end{array}\right]$$

row 1 → row 1 + 3row 3

row 2 → row 2 + 6row 3

$$(e)\quad \left[\begin{array}{ccc|ccc} 1 & 0 & 0 & -2/3 & -4/3 & 1 \\ 0 & 1 & 0 & -2/3 & 11/3 & -2 \\ 0 & 0 & 1 & 1 & -2 & 1 \end{array}\right]$$

row 1 → (1/3) row 1

row 2 → – (1/3) row 2

row 3 → row 3

$$(f)\quad -\frac{1}{3}\begin{bmatrix} 2 & 4 & -3 \\ 2 & -11 & 6 \\ -3 & 6 & -3 \end{bmatrix}$$

inverse of matrix **B**

The inverse of matrix **B** is the same as calculated above.

The inverse of a matrix has several interesting properties, including:

(1) $\mathbf{B}^{-1}\mathbf{B} = \mathbf{B}\mathbf{B}^{-1} = \mathbf{I}$.

(2) $|\mathbf{B}^{-1}| = 1/|\mathbf{B}|$.

(3) $[\mathbf{B}^{-1}]^{-1} = \mathbf{B}$.

(4) $[\mathbf{B}']^{-1} = [\mathbf{B}^{-1}]'$.

(5) If **B** and **C** are nonsingular square matrices, $[\mathbf{BC}]^{-1} = \mathbf{C}^{-1}\mathbf{B}^{-1}$.

(6) In the case of a symmetric matrix, since $\mathbf{B}' = \mathbf{B}$, then $[\mathbf{B}^{-1}]' = \mathbf{B}^{-1}$.

Orthonormal matrix

(7) An orthogonal matrix (Section 2.5) whose column vectors are normalized (scaled to length 1: Section 2.4) is called *orthonormal*. A square orthonormal matrix $\mathbf{B}$ has the property that $\mathbf{B}' = \mathbf{B}^{-1}$. This may be shown as follows: on the one hand, $\mathbf{B}^{-1}\mathbf{B} = \mathbf{I}$ by definition of the inverse of a square matrix. On the other hand, property 8 of matrix products (Section 2.5) shows that $\mathbf{B}'\mathbf{B} = \mathbf{D}(1)$ when the column vectors in $\mathbf{B}$ are normalized (which is the case for an orthonormal matrix); $\mathbf{D}(1)$ is a diagonal matrix of 1's, which is the identity matrix $\mathbf{I}$ (eq. 2.5). Given that $\mathbf{B}'\mathbf{B} = \mathbf{B}^{-1}\mathbf{B} = \mathbf{I}$, then $\mathbf{B}' = \mathbf{B}^{-1}$. Furthermore, combining the properties $\mathbf{B}\mathbf{B}^{-1} = \mathbf{I}$ (which is true for any square matrix) and $\mathbf{B}' = \mathbf{B}^{-1}$ shows that $\mathbf{B}\mathbf{B}' = \mathbf{I}$. For example, the matrix of normalized eigenvectors of a symmetric matrix, which is square and orthonormal (Section 2.9), has these properties.

(8) The inverse of a diagonal matrix is a diagonal matrix whose elements are the reciprocals of the original elements: $[\mathbf{D}(x_i)]^{-1} = \mathbf{D}(1/x_i)$.

Inversion is used in many types of applications, as will be seen in the remainder of this book. Classical examples of the role of inverse matrices are solving systems of equations and the calculation of regression coefficients.

System of linear equations

A system of linear equations can be represented in matrix form; for example:

$$\begin{aligned} b_1 + 2b_2 + 3b_3 &= 2 \\ 4b_1 + 5b_2 + 6b_3 &= 2 \\ 7b_1 + 8b_2 + 10b_3 &= 3 \end{aligned} \;\rightarrow\; \begin{bmatrix} 1 & 2 & 3 \\ 4 & 5 & 6 \\ 7 & 8 & 10 \end{bmatrix} \begin{bmatrix} b_1 \\ b_2 \\ b_3 \end{bmatrix} = \begin{bmatrix} 2 \\ 2 \\ 3 \end{bmatrix}$$

which may be written $\mathbf{Ab} = \mathbf{c}$. To find the values of the unknowns b_1, b_2 and b_3, vector $\mathbf{b}$ must be isolated to the left, which necessitates an inversion of the square matrix $\mathbf{A}$:

$$\begin{bmatrix} b_1 \\ b_2 \\ b_3 \end{bmatrix} = \begin{bmatrix} 1 & 2 & 3 \\ 4 & 5 & 6 \\ 7 & 8 & 10 \end{bmatrix}^{-1} \begin{bmatrix} 2 \\ 2 \\ 3 \end{bmatrix}$$

The inverse of $\mathbf{A}$ has been calculated above. Multiplication with vector $\mathbf{c}$ provides the solution for the three unknowns:

$$\begin{bmatrix} b_1 \\ b_2 \\ b_3 \end{bmatrix} = -\frac{1}{3} \begin{bmatrix} 2 & 4 & -3 \\ 2 & -11 & 6 \\ -3 & 6 & -3 \end{bmatrix} \begin{bmatrix} 2 \\ 2 \\ 3 \end{bmatrix} = -\frac{1}{3} \begin{bmatrix} 3 \\ 0 \\ -3 \end{bmatrix} = \begin{bmatrix} -1 \\ 0 \\ 1 \end{bmatrix} \qquad \begin{aligned} b_1 &= -1 \\ b_2 &= 0 \\ b_3 &= 1 \end{aligned}$$

Simple linear regression

Regression analysis is reviewed in Section 10.3. *Regression coefficients* are easily calculated for several models, using matrix inversion, so that the approach is briefly discussed here. The mathematical model for *simple linear regression* (model I: Subsection 10.3.1) is $\hat{y} = b_0 + b_1 x$.

The regression coefficients b_0 and b_1 are estimated from the observed data **x** and **y**. This is equivalent to resolving the following system of equations:

$$\begin{array}{l} y_1 = b_0 + b_1 x_1 \\ y_2 = b_0 + b_1 x_2 \\ \cdot \quad\quad \cdot \\ \cdot \quad\quad \cdot \\ \cdot \quad\quad \cdot \\ y_n = b_0 + b_1 x_n \end{array} \rightarrow \mathbf{y} = \begin{bmatrix} y_1 \\ y_2 \\ \cdot \\ \cdot \\ \cdot \\ y_n \end{bmatrix} \quad \mathbf{X} = \begin{bmatrix} 1 & x_1 \\ 1 & x_2 \\ \cdot & \cdot \\ \cdot & \cdot \\ \cdot & \cdot \\ 1 & x_n \end{bmatrix} \quad \mathbf{b} = \begin{bmatrix} b_0 \\ b_1 \end{bmatrix}$$

Least squares

Coefficients b are estimated by the *method of least squares* (Subsection 10.3.1), which minimizes the sum of squares of the differences between observed values y and values $\hat{y}$ calculated using the regression equation. In order to obtain a least-squares best fit, each member (left and right) of matrix equation $\mathbf{y} = \mathbf{Xb}$ is multiplied by the transpose of matrix **X**, i.e. $\mathbf{X'y} = \mathbf{X'Xb}$. By doing so, the rectangular matrix **X** produces a square matrix **X'X**, which can be inverted. The values of coefficients b_0 and b_1 are computed directly after inverting the square matrix [**X'X**]:

$$\mathbf{b} = [\mathbf{X'X}]^{-1}\,[\mathbf{X'y}] \quad \mathbf{(2.19)}$$

Multiple linear regression

Using the same approach, it is easy to compute coefficients $b_0, b_1, \ldots, b_p$ of a *multiple linear regression* (Subsection 10.3.3). In this type of regression, variable y is a linear function of several (p) variables x_j, so that one can write: $\hat{y} = b_0 + b_1 x_1 + \ldots + b_p x_p$. Vectors **y** and **b** and matrix **X** are defined as follows:

$$\mathbf{y} = \begin{bmatrix} y_1 \\ y_2 \\ \cdot \\ \cdot \\ \cdot \\ y_n \end{bmatrix} \quad \mathbf{X} = \begin{bmatrix} 1 & x_{11} & \ldots & x_{1p} \\ 1 & x_{21} & \ldots & x_{2p} \\ \cdot & & & \cdot \\ \cdot & & & \cdot \\ \cdot & & & \cdot \\ 1 & x_{n1} & \ldots & x_{np} \end{bmatrix} \quad \mathbf{b} = \begin{bmatrix} b_0 \\ b_1 \\ \cdot \\ \cdot \\ \cdot \\ b_p \end{bmatrix}$$

The least-squares solution is again eq. 2.19. However, readers should consult Section 10.3 for computational methods to be used in multiple linear regression when the variables x_j are strongly intercorrelated, as is often the case in ecology.

Polynomial regression

In *polynomial regression* (Subsection 10.3.4), several regression parameters b, corresponding to powers of a single variable x, are fitted to the observed data. The general regression model is $\hat{y} = b_0 + b_1 x + b_2 x^2 + \ldots + b_k x^k$. The vector of parameters, **b**, is computed in the same way. Vectors **y** and **b**, and matrix **X**, are defined as follows:

$$\mathbf{y} = \begin{bmatrix} y_1 \\ y_2 \\ \cdot \\ \cdot \\ \cdot \\ y_n \end{bmatrix} \qquad \mathbf{X} = \begin{bmatrix} 1 \; x_1 \; x_1^2 \ldots x_1^k \\ 1 \; x_2 \; x_2^2 \ldots x_2^k \\ \cdot \qquad\qquad \cdot \\ \cdot \qquad\qquad \cdot \\ \cdot \qquad\qquad \cdot \\ 1 \; x_n \; x_n^2 \ldots x_n^k \end{bmatrix} \qquad \mathbf{b} = \begin{bmatrix} b_0 \\ b_1 \\ \cdot \\ \cdot \\ \cdot \\ b_k \end{bmatrix}$$

The least-squares solution is computed using eq. 2.19. Readers should consult Section 10.3 where practical considerations concerning the calculation of polynomial regression with ecological data are discussed.

2.9 Eigenvalues and eigenvectors

There are other problems, in addition to those examined above, where the determinant and the inverse of a matrix are used to provide simple and elegant solutions. An important one in data analysis is the derivation of an orthogonal form (i.e. a matrix whose vectors are at right angles; Sections 2.5 and 2.8) for a non-orthogonal symmetric matrix. This will provide the algebraic basis for most of the methods studied in Chapters 9 and 11. In ecology, data sets generally include a large number of variables, which are associated to one another (e.g. linearly correlated; Section 4.2). The basic idea underlying several methods of data analysis is to reduce this large number of intercorrelated variables to a smaller number of composite, but linearly independent (Box 1.1) variables, each explaining a different fraction of the observed variation. One of the main goals of numerical data analysis is indeed to generate a small number of variables, each explaining a large portion of the variation, and to ascertain that these new variables explain different aspects of the phenomena under study. The present section only deals with the mathematics of the computation of *eigenvalues* and *eigenvectors*. Applications to the analysis of multidimensional ecological data are discussed in Chapters 4, 9 and 11.

Mathematically, the problem may be formulated as follows. Given a square matrix **A**, one wishes to find a diagonal matrix which is equivalent to **A**. In ecology, square matrices are most often symmetric association matrices (Section 2.2), hence the use of symbol **A**:

$$\mathbf{A} = \begin{bmatrix} a_{11} & a_{12} & \dots & a_{1n} \\ a_{21} & a_{22} & \dots & a_{2n} \\ \cdot & & & \cdot \\ \cdot & & & \cdot \\ \cdot & & & \cdot \\ a_{n1} & a_{n2} & \dots & a_{nn} \end{bmatrix}$$

In matrix **A**, the terms located above and below the diagonal characterize the degree of association of either the objects, or the ecological variables, with one another (Fig. 2.1). In the new matrix **Λ** (capital lambda) being sought, all elements outside the diagonal are null:

$$\mathbf{\Lambda} = \begin{bmatrix} \lambda_{11} & 0 & \dots & 0 \\ 0 & \lambda_{22} & \dots & 0 \\ \cdot & & & \cdot \\ \cdot & & & \cdot \\ \cdot & & & \cdot \\ 0 & 0 & \dots & \lambda_{nn} \end{bmatrix} = \begin{bmatrix} \lambda_1 & 0 & \dots & 0 \\ 0 & \lambda_2 & \dots & 0 \\ \cdot & & & \cdot \\ \cdot & & & \cdot \\ \cdot & & & \cdot \\ 0 & 0 & \dots & \lambda_n \end{bmatrix} \qquad \textbf{(2.20)}$$

This new matrix is called the *matrix of eigenvalues**. The new variables (*eigenvectors*; see below) whose association is described by this matrix **Λ** are thus linearly independent of one another. The use of the Greek letter λ (lower-case lambda) to represent eigenvalues stems from the fact that eigenvalues are actually *Lagrangian multipliers* λ, as will be shown in Section 4.4. Matrix **Λ** is known as the *canonical form* of matrix **A;** for the exact meaning of *canonical* in mathematics, see Subsection 10.2.1.

Canonical form

1 — Computation

The eigenvalues and eigenvectors of matrix **A** are found from equation

$$\mathbf{A}\mathbf{u}_i = \lambda_i \mathbf{u}_i \qquad \textbf{(2.21)}$$

which allows one to compute the different eigenvalues λ_i and their associated eigenvectors $\mathbf{u}_i$. First, the validity of eq. 2.21 must be demonstrated.

* In the literature, the following expressions are synonymous:

eigenvalue	eigenvector
characteristic root	characteristic vector
latent root	latent vector

Eigen is the German word for *characteristic*.

To do so, one uses any pair h and i of eigenvalues and eigenvectors, corresponding to matrix **A**. Equation 2.21 becomes

$$\mathbf{Au}_h = \lambda_h \mathbf{u}_h \quad \text{and} \quad \mathbf{Au}_i = \lambda_i \mathbf{u}_i \,, \quad \text{respectively.}$$

Multiplying the two equations by row vectors $\mathbf{u'}_i$ and $\mathbf{u'}_h$, respectively, gives:

$$\mathbf{u}'_i \mathbf{A} \mathbf{u}_h = \lambda_h \mathbf{u}'_i \mathbf{u}_h \quad \text{and} \quad \mathbf{u}'_h \mathbf{A} \mathbf{u}_i = \lambda_i \mathbf{u}'_h \mathbf{u}_i$$

It can be shown that, in the case of a symmetric matrix, the left-hand members of these two equations are equal: $u'_i \mathbf{A} \mathbf{u}_h = u'_h \mathbf{A} \mathbf{u}_i$; this would not be true for an asymmetric matrix, however. Using a (2×2) matrix **A**, readers can easily check that the equality holds only when $a_{12} = a_{21}$, i.e. when **A** is symmetric. So, in the case of a symmetric matrix, the right-hand members are also equal:

$$\lambda_h \mathbf{u}'_i \mathbf{u}_h = \lambda_i \mathbf{u}'_h \mathbf{u}_i$$

Since we are talking about two distinct values for λ_h and λ_i, the only possibility for the above equality to be true is that the product of vectors $\mathbf{u}_h$ and $\mathbf{u}_i$ be 0 (i.e. $\mathbf{u}'_i \mathbf{u}_h = \mathbf{u}'_h \mathbf{u}_i = 0$), which is the condition of orthogonality for two vectors (Section 2.5). It is therefore concluded that eq. 2.21

$$\mathbf{Au}_i = \lambda_i \mathbf{u}_i$$

can be used to compute vectors $\mathbf{u}_i$ which are indeed orthogonal, when matrix **A** is symmetric. In the case of a non-symmetric matrix, eigenvectors can also be calculated, but they are not orthogonal and therefore not linearly independent.

If scalars λ_i and their associated vectors $\mathbf{u}_i$ exist, then eq. 2.21 can be transformed as follows:

$$\mathbf{Au}_i - \lambda_i \mathbf{u}_i = \mathbf{0} \quad \text{(difference between two vectors)}$$

and vector $\mathbf{u}_i$ can be factorized:

$$(\mathbf{A} - \lambda_i \mathbf{I})\mathbf{u}_i = \mathbf{0} \tag{2.22}$$

Because of the nature of the elements in eq. 2.22, it is necessary to introduce a unit matrix **I** inside the parentheses, where one now finds a difference between two square matrices. According to eq. 2.22, multiplication of the square matrix $(\mathbf{A} - \lambda_i \mathbf{I})$ with column vector $\mathbf{u}_i$ must result in a null column vector (**0**).

Besides the trivial solution, where $\mathbf{u}_i$ is itself a null vector, eq. 2.22 has the following solution:

$$|\mathbf{A} - \lambda_i \mathbf{I}| = 0 \tag{2.23}$$

That is, the determinant of the difference between matrices **A** and λ_i**I** must be equal to 0 for each λ_i. Resolving eq. 2.23 provides the eigenvalues λ_i associated with matrix **A**. Equation 2.23 is known as the *characteristic* or *determinantal equation*.

Characteristic equation

Demonstration of eq. 2.23 is as follows:

1) One solution to $(\mathbf{A} - \lambda_i\mathbf{I})\mathbf{u}_i = \mathbf{0}$ is that $\mathbf{u}_i$ is the null vector: $\mathbf{u} = [0]$. This solution is trivial, since it corresponds to the centroid of the scatter of data points. A non-trivial solution must thus involve $(\mathbf{A} - \lambda_i\mathbf{I})$.

2) Solution $(\mathbf{A} - \lambda_i\mathbf{I}) = [0]$ is not acceptable either, since it implies that $\mathbf{A} = \lambda_i\mathbf{I}$ and thus that **A** be a scalar matrix, which is generally not true.

3) The solution thus requires that λ_i and $\mathbf{u}_i$ be such that the *product* $(\mathbf{A} - \lambda_i\mathbf{I})\mathbf{u}_i$ is a null vector. In other words, vector $\mathbf{u}_i$ must be orthogonal to the space corresponding to **A** after λ_i**I** has been subtracted from it; orthogonality of two vectors or matrices is obtained when their product is equal to 0 (Section 2.5). Solution $|\mathbf{A} - \lambda_i\mathbf{I}| = 0$ (eq. 2.23) means that, for each value λ_i, the rank of $(\mathbf{A} - \lambda_i\mathbf{I})$ is lower than its order, which makes the determinant equal to zero (Section 2.7). Each λ_i**I** corresponds to one dimension of matrix **A** (Section 4.4). It is easy to calculate the eigenvector $\mathbf{u}_i$, which is orthogonal to space $(\mathbf{A} - \lambda_i\mathbf{I})$ which is of lower dimension. This eigenvector is the solution to eq. 2.22, which specifies orthogonality.

For a matrix **A** of order n, the characteristic equation is a polynomial of degree n, whose solutions are the values λ_i. When these values are found, it is easy to use eq. 2.22 to calculate the eigenvector $\mathbf{u}_i$ corresponding to each eigenvalue λ_i. There are therefore as many eigenvectors as there are eigenvalues.

There are methods which enable the quick and efficient calculation of eigenvalues and eigenvectors by computer. Two of these are described in Subsection 9.1.8.

Ecologists, who are more concerned with shedding light on natural phenomena than on mathematical entities, may have found unduly technical this discussion of the computation of eigenvalues and eigenvectors. The same subject will be considered again in Section 4.4, in the context of the multidimensional normal distribution. Mastering the bases of this algebraic operation is essential to understand the methods based on *eigenanalysis* (Chapters 9 and 11), which are of prime importance to the analysis of ecological data.

2 — Numerical examples

Numerical example 1. The characteristic equation of symmetric matrix

$$\mathbf{A} = \begin{bmatrix} 2 & 2 \\ 2 & 5 \end{bmatrix}$$

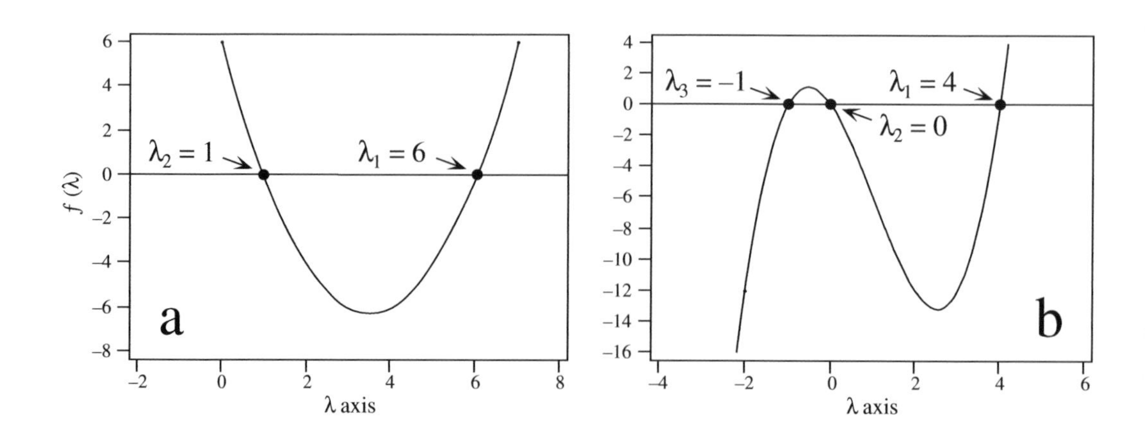

Figure 2.2 (a) The eigenvalues of numerical example 1 are the values along the λ axis where the function $\lambda^2 - 7\lambda + 6$ is zero. (b) Similarly for numerical example 2, the eigenvalues are the values along the λ axis where the function $\lambda^3 - 3\lambda^2 - 4\lambda$ is zero.

is (eq. 2.23)

$$\left| \begin{bmatrix} 2 & 2 \\ 2 & 5 \end{bmatrix} - \lambda \begin{bmatrix} 1 & 0 \\ 0 & 1 \end{bmatrix} \right| = 0$$

therefore

$$\left| \begin{bmatrix} 2 & 2 \\ 2 & 5 \end{bmatrix} - \begin{bmatrix} \lambda & 0 \\ 0 & \lambda \end{bmatrix} \right| = 0$$

and thus

$$\begin{vmatrix} 2-\lambda & 2 \\ 2 & 5-\lambda \end{vmatrix} = 0$$

The *characteristic polynomial* is found by expanding the determinant (Section 2.6):

$$(2-\lambda)(5-\lambda) - 4 = 0$$

which gives

$$\lambda^2 - 7\lambda + 6 = 0$$

from which it is easy to calculate the two values of λ which satisfy the equation (Fig. 2.2a). The two eigenvalues of **A** are therefore:

$$\lambda_1 = 6 \quad \text{and} \quad \lambda_2 = 1$$

The ordering of eigenvalues is arbitrary; it would have been equally correct to state that $\lambda_1 = 1$ and $\lambda_2 = 6$. Equation 2.22 is used to calculate eigenvectors $\mathbf{u}_1$ and $\mathbf{u}_2$ corresponding to eigenvalues λ_1 and λ_2, respectively:

for $\lambda_1 = 6$ — for $\lambda_2 = 1$

$$\left(\begin{bmatrix} 2 & 2 \\ 2 & 5 \end{bmatrix} - 6\begin{bmatrix} 1 & 0 \\ 0 & 1 \end{bmatrix}\right)\begin{bmatrix} u_{11} \\ u_{21} \end{bmatrix} = \mathbf{0} \qquad \left(\begin{bmatrix} 2 & 2 \\ 2 & 5 \end{bmatrix} - 1\begin{bmatrix} 1 & 0 \\ 0 & 1 \end{bmatrix}\right)\begin{bmatrix} u_{12} \\ u_{22} \end{bmatrix} = \mathbf{0}$$

$$\begin{bmatrix} -4 & 2 \\ 2 & -1 \end{bmatrix}\begin{bmatrix} u_{11} \\ u_{21} \end{bmatrix} = \mathbf{0} \qquad \begin{bmatrix} 1 & 2 \\ 2 & 4 \end{bmatrix}\begin{bmatrix} u_{12} \\ u_{22} \end{bmatrix} = \mathbf{0}$$

which is equivalent to the following pairs of linear equations:

$$-4u_{11} + 2u_{21} = 0 \qquad 1u_{12} + 2u_{22} = 0$$

$$2u_{11} - 1u_{21} = 0 \qquad 2u_{12} + 4u_{22} = 0$$

These sets of linear equations are always indeterminate. The solution is given by any point (vector) in the direction of the eigenvector being sought. To remove the indetermination, an arbitrary value is assigned to one of the elements u, which specifies a particular vector. For example, value $u = 1$ may be arbitrarily assigned to first term u in each set:

given that $\quad u_{11} = 1 \qquad u_{12} = 1$

it follows that $\quad -4u_{11} + 2u_{21} = 0 \qquad 1u_{12} + 2u_{22} = 0$

become $\quad -4 + 2u_{21} = 0 \qquad 1 + 2u_{22} = 0$

so that $\quad u_{21} = 2 \qquad u_{22} = -1/2$

Eigenvectors $\mathbf{u}_1$ and $\mathbf{u}_2$ are therefore:

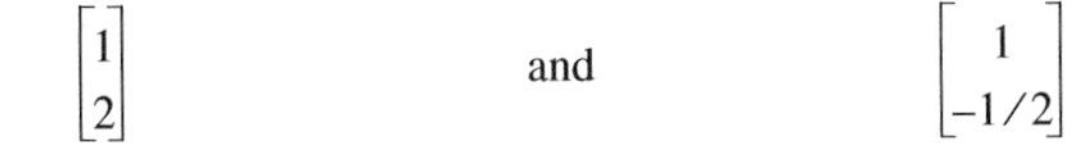

$$\begin{bmatrix} 1 \\ 2 \end{bmatrix} \quad \text{and} \quad \begin{bmatrix} 1 \\ -1/2 \end{bmatrix}$$

Values other than 1 could have been arbitrarily assigned to u_{11} and u_{12} (or, for that matter, to any other term in each vector). For example, the following vectors also satisfy the two pairs of linear equations, since these eigenvectors differ only by a multiplication by a scalar:

$$\begin{bmatrix} 2 \\ 4 \end{bmatrix} \text{ or } \begin{bmatrix} -3 \\ -6 \end{bmatrix} \qquad \begin{bmatrix} 2 \\ -1 \end{bmatrix} \text{ or } \begin{bmatrix} -4 \\ 2 \end{bmatrix}$$

This is the reason why eigenvectors are generally standardized. One method is to assign value 1 to the *largest element* of each vector, and adjust the other elements accordingly. Another standardization method, used for instance in principal component and principal coordinate analyses (Sections 9.1 and 9.2), is to make the length of each eigenvector $\mathbf{u}_i$ equal to the square root of its eigenvalue (eigenvector *scaled to* $\sqrt{\lambda_i}$).

Another, more common and more practical method, is to normalize eigenvectors, i.e. to make their lengths equal to 1. Thus, a *normalized* eigenvector is in fact *scaled to* 1, i.e. $\mathbf{u'u} = 1$. As explained in Section 2.4, normalization is achieved by dividing each element of a vector by the length of this vector, i.e. the square root of the sum of squares of all elements in the vector.

In the numerical example, the two eigenvectors

are normalized to

$$\begin{bmatrix} 1/\sqrt{5} \\ 2/\sqrt{5} \end{bmatrix} \quad \text{and} \quad \begin{bmatrix} 2/\sqrt{5} \\ -1/\sqrt{5} \end{bmatrix}$$

Since the eigenvectors are both orthogonal and normalized, they are *orthonormal* (property 7 in Section 2.8).

Had the eigenvectors been multiplied by a negative scalar, their normalized forms would now be the following:

$$\begin{bmatrix} -1/\sqrt{5} \\ -2/\sqrt{5} \end{bmatrix} \quad \text{and} \quad \begin{bmatrix} -2/\sqrt{5} \\ 1/\sqrt{5} \end{bmatrix}$$

These forms are strictly equivalent to those above.

Since matrix $\mathbf{A}$ is symmetric, its eigenvectors must be orthogonal. This is easily verified as their product is equal to zero, which is the condition for two vectors to be orthogonal (Section 2.5):

$$\mathbf{u}'_1\mathbf{u}_2 = \begin{bmatrix} 1/\sqrt{5} & 2/\sqrt{5} \end{bmatrix} \begin{bmatrix} 2/\sqrt{5} \\ -1/\sqrt{5} \end{bmatrix} = 2/5 - 2/5 = 0$$

The normalized eigenvectors may be plotted in the original system of coordinates, i.e. the Cartesian plane whose axes are the two original descriptors; the association between these descriptors is given by matrix **A.** This plot (full arrows) shows that the angle between the eigenvectors is indeed 90° (cos 90° = 0) and that their lengths are 1:

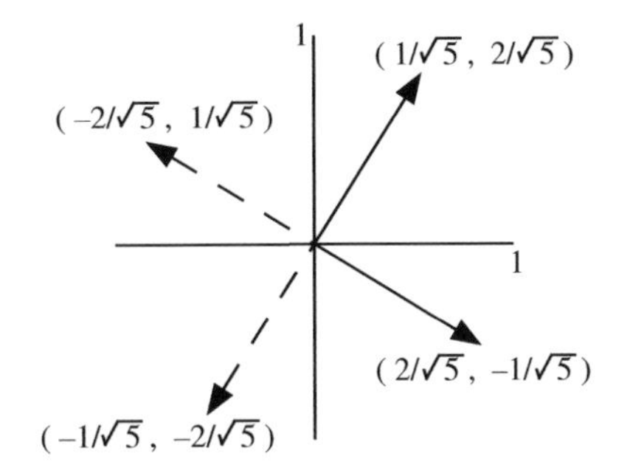

The dashed arrows illustrate the same eigenvectors with inverted signs. The eigenvectors with dashed arrows are equivalent to those with full arrows.

Resolving the system of linear equations used to compute eigenvectors is greatly facilitated by matrix inversion. Defining matrix $\mathbf{C}_{nn} = (\mathbf{A} - \lambda_n \mathbf{I})$ allows eq. 2.22 to be written in a simplified form:

$$\mathbf{C}_{nn}\mathbf{u}_n = \mathbf{0} \qquad \textbf{(2.24)}$$

Indices n designate here the dimensions of matrix **C** and vector **u**. Matrix $\mathbf{C}_{nn}$ contains all the coefficients by which a given eigenvector $\mathbf{u}_n$ is multiplied. The system of equations is indeterminate, which prevents the inversion of **C** and calculation of **u**. To remove the indetermination, it is sufficient to determine any one element of vector **u**. For example, one may arbitrarily decide that $u_1 = \alpha$ ($\alpha \neq 0$). Then,

$$\begin{bmatrix} c_{11} & c_{12} & \cdots & c_{1n} \\ c_{21} & c_{22} & \cdots & c_{2n} \\ \cdot & & & \cdot \\ \cdot & & & \cdot \\ \cdot & & & \cdot \\ c_{n1} & c_{n2} & \cdots & c_{nn} \end{bmatrix} \begin{bmatrix} \alpha \\ u_2 \\ \cdot \\ \cdot \\ \cdot \\ u_n \end{bmatrix} = \begin{bmatrix} 0 \\ 0 \\ \cdot \\ \cdot \\ \cdot \\ 0 \end{bmatrix}$$

can be written

$$\begin{bmatrix} c_{11}\alpha + c_{12}u_2 + \dots + c_{1n}u_n \\ c_{21}\alpha + c_{22}u_2 + \dots + c_{2n}u_n \\ \cdot \quad\quad\quad \cdot \\ \cdot \quad\quad\quad \cdot \\ \cdot \quad\quad\quad \cdot \\ c_{n1}\alpha + c_{n2}u_2 + \dots + c_{nn}u_n \end{bmatrix} = \begin{bmatrix} 0 \\ 0 \\ \cdot \\ \cdot \\ \cdot \\ 0 \end{bmatrix}$$

so that

$$\begin{bmatrix} c_{12}u_2 + \dots + c_{1n}u_n \\ c_{22}u_2 + \dots + c_{2n}u_n \\ \cdot \quad\quad \cdot \\ \cdot \quad\quad \cdot \\ \cdot \quad\quad \cdot \\ c_{n2}u_2 + \dots + c_{nn}u_n \end{bmatrix} = -\alpha \begin{bmatrix} c_{11} \\ c_{21} \\ \cdot \\ \cdot \\ \cdot \\ c_{n1} \end{bmatrix}$$

After setting $u_1 = \alpha$, the first column of matrix **C** is transferred to the right. The last $n - 1$ rows of **C** are then enough to define a completely determined system. The first row is removed from **C** in order to obtain a square matrix of order $n - 1$, which can be inverted. The determined system thus obtained is:

$$\begin{bmatrix} c_{22}u_2 + \dots + c_{2n}u_n \\ \cdot \quad\quad \cdot \\ \cdot \quad\quad \cdot \\ \cdot \quad\quad \cdot \\ c_{n2}u_2 + \dots + c_{nn}u_n \end{bmatrix} = -\alpha \begin{bmatrix} c_{21} \\ \cdot \\ \cdot \\ \cdot \\ c_{n1} \end{bmatrix}$$

which can be written $\quad \mathbf{C}_{(n-1)\,(n-1)}\mathbf{u}_{(n-1)} = -\alpha\mathbf{c}_{(n-1)}$ **(2.25)**

This system can be resolved by inversion of **C**, as in Section 2.8:

$$\mathbf{u}_{(n-1)} = -\alpha\mathbf{C}^{-1}_{(n-1)\,(n-1)}\mathbf{c}_{(n-1)} \qquad \textbf{(2.26)}$$

This method of computing the eigenvectors may not work, however, in the case of multiple eigenvalues (see Property 3, below). The following example provides an illustration of the computation through inversion.

Numerical example 2. The characteristic polynomial (see eq. 2.23) of the asymmetric matrix

$$\mathbf{A} = \begin{bmatrix} 1 & 3 & -1 \\ 0 & 1 & 2 \\ 1 & 4 & 1 \end{bmatrix}$$

is $\lambda^3 - 3\lambda^2 - 4\lambda = 0$, from which the three eigenvalues 4, 0 and –1 are calculated (Fig. 2.2b).

The eigenvectors are computed by inserting each eigenvalue, in turn, into eq. 2.22. For $\lambda_1 = 4$:

$$\begin{bmatrix} (1-4) & 3 & -1 \\ 0 & (1-4) & 2 \\ 1 & 4 & (1-4) \end{bmatrix} \begin{bmatrix} u_{11} \\ u_{21} \\ u_{31} \end{bmatrix} = \begin{bmatrix} 0 \\ 0 \\ 0 \end{bmatrix}$$

The above system is determined by setting $u_{11} = 1$. Using eq. 2.25 gives:

$$\begin{bmatrix} (1-4) & 2 \\ 4 & (1-4) \end{bmatrix} \begin{bmatrix} u_{21} \\ u_{31} \end{bmatrix} = -1 \begin{bmatrix} 0 \\ 1 \end{bmatrix}$$

from which it follows (eq. 2.26) that

$$\begin{bmatrix} u_{21} \\ u_{31} \end{bmatrix} = \begin{bmatrix} (1-4) & 2 \\ 4 & (1-4) \end{bmatrix}^{-1} \begin{bmatrix} 0 \\ -1 \end{bmatrix}$$

The inverse of matrix $\begin{bmatrix} -3 & 2 \\ 4 & -3 \end{bmatrix}$ is $\begin{bmatrix} -3 & -2 \\ -4 & -3 \end{bmatrix}$ so that

$$\begin{bmatrix} u_{21} \\ u_{31} \end{bmatrix} = \begin{bmatrix} -3 & -2 \\ -4 & -3 \end{bmatrix} \begin{bmatrix} 0 \\ -1 \end{bmatrix} = \begin{bmatrix} 2 \\ 3 \end{bmatrix}$$

The two other eigenvectors are computed in the same fashion, from eigenvalues $\lambda_2 = 0$ and $\lambda_3 = -1$. The resulting matrix of eigenvectors (columns) is:

$$\mathbf{U} = \begin{bmatrix} \mathbf{u}_1 & \mathbf{u}_2 & \mathbf{u}_3 \end{bmatrix} = \begin{bmatrix} 1 & 1 & 1 \\ 2 & -2/7 & -1/2 \\ 3 & 1/7 & 1/2 \end{bmatrix} \quad \text{or else} \quad \begin{bmatrix} 1 & 7 & 2 \\ 2 & -2 & -1 \\ 3 & 1 & 1 \end{bmatrix}$$

which is normalized to:

$$\mathbf{U} = \begin{bmatrix} 0.27 & 0.95 & 0.82 \\ 0.53 & -0.27 & -0.41 \\ 0.80 & 0.14 & 0.41 \end{bmatrix}$$

Readers can easily check that the above eigenvectors, which were extracted from a non-symmetric matrix, are indeed not orthogonal; none of the scalar products between pairs of columns is equal to zero.

2.10 Some properties of eigenvalues and eigenvectors

First property. — A simple rearrangement of eq. 2.21 shows that matrix **U** of eigenvectors is a transform matrix, allowing one to go from system **A** to system **Λ**. Indeed, the equation can be rewritten so as to include all eigenvalues and eigenvectors:

$$\mathbf{AU} = \mathbf{U\Lambda} \qquad \textbf{(2.27)}$$

Numerical example. Equation 2.27 can be easily verified using numerical example 2 from Section 2.9:

$$\begin{bmatrix} 1 & 3 & -1 \\ 0 & 1 & 2 \\ 1 & 4 & 1 \end{bmatrix} \begin{bmatrix} 1 & 7 & 2 \\ 2 & -2 & -1 \\ 3 & 1 & 1 \end{bmatrix} = \begin{bmatrix} 1 & 7 & 2 \\ 2 & -2 & -1 \\ 3 & 1 & 1 \end{bmatrix} \begin{bmatrix} 4 & 0 & 0 \\ 0 & 0 & 0 \\ 0 & 0 & -1 \end{bmatrix}$$

The left and right-hand sides of the equation are identical:

$$\begin{bmatrix} 4 & 0 & -2 \\ 8 & 0 & 1 \\ 12 & 0 & -1 \end{bmatrix} = \begin{bmatrix} 4 & 0 & -2 \\ 8 & 0 & 1 \\ 12 & 0 & -1 \end{bmatrix}$$

On the left-hand side of the equation, matrix **A** is postmultiplied by matrix **U** of the eigenvectors whereas, on the right-hand side, the matrix of eigenvalues **Λ** is premultiplied by **U**. It follows that **U** achieves a two-way transformation (rows, columns), from reference system **A** to system **Λ**. This transformation can go both ways, as shown by the following equations which are both derived from eq. 2.27:

$$\mathbf{A} = \mathbf{U\Lambda U}^{-1} \qquad \text{and} \qquad \mathbf{\Lambda} = \mathbf{U}^{-1}\mathbf{AU} \qquad \textbf{(2.28)}$$

A simple formula may be derived from $\mathbf{A} = \mathbf{U\Lambda U}^{-1}$, which can be used to raise matrix **A** to any power x:

$$\mathbf{A}^x = (\mathbf{U\Lambda U}^{-1})\mathbf{U\Lambda} \ldots \mathbf{U}^{-1}(\mathbf{U\Lambda U}^{-1})$$

$$\mathbf{A}^x = \mathbf{U\Lambda}(\mathbf{U}^{-1}\mathbf{U})\mathbf{\Lambda} \dots (\mathbf{U}^{-1}\mathbf{U})\mathbf{\Lambda U}^{-1}$$

$$\mathbf{A}^x = \mathbf{U\Lambda}^x\mathbf{U}^{-1}, \text{ because } \mathbf{U}^{-1}\mathbf{U} = \mathbf{I}$$

Raising a matrix to some high power is greatly facilitated by the fact that $\mathbf{\Lambda}^x$ is the matrix of eigenvalues, which is diagonal. Indeed, a diagonal matrix can be raised to any power x by raising each of its diagonal elements to power x. It follows that the last equation may be rewritten as:

$$\mathbf{A}^x = \mathbf{U}\,[\lambda_i^x]\,\mathbf{U}^{-1} \quad \textbf{(2.29)}$$

This may easily be verified using the above example.

Second property. — It was shown in Section 2.7 that, when the rank (r) of matrix $\mathbf{A}_{nn}$ is smaller than its order ($r < n$), determinant $|\mathbf{A}|$ equals 0. It was also shown that, when it is necessary to know the rank of a matrix, as for instance in dimensional analysis (Section 3.3), $|\mathbf{A}| = 0$ indicates that one must test for rank. Such a test naturally follows from the calculation of eigenvalues. Indeed, the determinant of a matrix is equal to the product of its eigenvalues:

$$|\mathbf{A}| = \prod_{i=1}^{n} \lambda_i \quad \textbf{(2.30)}$$

so that $|\mathbf{A}| = 0$ if one or several of the eigenvalues $\lambda_i = 0$. When the rank of a matrix is smaller than its order ($r < n$), this matrix has $(n - r)$ null eigenvalues. Thus, *eigenvalues can be used to determine the rank of a matrix*: the rank is equal to the *number of nonzero eigenvalues*. In the case of an association matrix among variables, the number of nonzero eigenvalues (i.e. the rank of **A**) is equal to the number of independent dimensions which are required to account for all the variance (Chapter 9).

Third property. — It was implicitly assumed, up to this point, that the eigenvalues were *all different* from one another. It may happen, however, that some (say, m) eigenvalues are equal. These are known as *multiple eigenvalues*. In such a case, the question is whether or not matrix A_{nn} has *n distinct* eigenvectors. In other words, are there *m linearly independent* eigenvectors which correspond to the *same* eigenvalue?

By definition, the determinant of $(\mathbf{A} - \lambda_i\mathbf{I})$ is null (eq. 2.23):

$$|\mathbf{A} - \lambda_i\mathbf{I}| = 0$$

which means that the rank of $(\mathbf{A} - \lambda_i\mathbf{I})$ is smaller than n. In the case of multiple eigenvalues, if there are *m distinct* eigenvectors corresponding to the m identical eigenvalues λ_i, the determinant of $(\mathbf{A} - \lambda_i\mathbf{I})$ must be null for each of these eigenvalues, but in a different way each time. When $m = 1$, the condition for $|\mathbf{A} - \lambda_i\mathbf{I}| = 0$ is for its rank to be $r = n - 1$. Similarly, in a case of *multiplicity*, the condition for $|\mathbf{A} - \lambda_i\mathbf{I}|$ to be null m times, but distinctly, is for its rank to be $r = n - m$. Consequently, for n

distinct eigenvectors to exist, the rank of $(\mathbf{A} - \lambda_i\mathbf{I})$ must be $r = n - m$, and this for any eigenvalue λ_i of multiplicity m.

Numerical example. The following matrix has eigenvalues $\lambda_1 = \lambda_2 = 1$ and $\lambda_3 = -1$:

$$\mathbf{A} = \begin{bmatrix} -1 & -2 & -2 \\ 1 & 2 & 1 \\ -1 & -1 & 0 \end{bmatrix} \quad \text{so that, for } \lambda_1 = \lambda_2 = 1, \quad (\mathbf{A} - 1\mathbf{I}) = \begin{bmatrix} -2 & -2 & -2 \\ 1 & 1 & 1 \\ -1 & -1 & -1 \end{bmatrix}$$

The rank of $(\mathbf{A} - \lambda_i\mathbf{I})$ is $r = 1$ because all three columns of this matrix are identical. Thus, for $\lambda_1 = \lambda_2 = 1$ $(m = 2)$, $n - m = 3 - 2 = 1$, so that $r = n - m$. It follows that there exist two distinct eigenvectors $\mathbf{u}_1$ and $\mathbf{u}_2$. They can indeed be found:

$$\mathbf{u}_1 = \begin{bmatrix} 1 \\ 0 \\ -1 \end{bmatrix} \quad \text{and} \quad \mathbf{u}_2 = \begin{bmatrix} 1 \\ -1 \\ 0 \end{bmatrix} \quad \text{whereas} \quad \mathbf{u}_3 = \begin{bmatrix} 2 \\ -1 \\ 1 \end{bmatrix}$$

Eigenvectors $\mathbf{u}_1$ and $\mathbf{u}_2$ both correspond to multiple eigenvalue $\lambda = 1$. Any linear combination of such eigenvectors is also an eigenvector of matrix **A** corresponding to λ. For example:

$$\mathbf{u}_1 - \mathbf{u}_2 = \begin{bmatrix} 0 \\ 1 \\ -1 \end{bmatrix} \quad \mathbf{u}_1 + 2\mathbf{u}_2 = \begin{bmatrix} 3 \\ -2 \\ -1 \end{bmatrix}$$

It can easily be verified that the above two eigenvectors, or any other linear combination of $\mathbf{u}_1$ and $\mathbf{u}_2$, are indeed eigenvectors of **A** corresponding to $\lambda = 1$. Of course, the new eigenvectors are not linearly independent of $\mathbf{u}_1$ and $\mathbf{u}_2$, so that there are still only two distinct eigenvectors corresponding to multiple $\lambda = 1$.

Numerical example. The eigenvalues of the following matrix are $\lambda_1 = 3$ and $\lambda_2 = \lambda_3 = 1$:

$$\mathbf{A} = \begin{bmatrix} 2 & -1 & 1 \\ 3 & 3 & -2 \\ 4 & 1 & 0 \end{bmatrix} \quad \text{so that, for } \lambda_2 = \lambda_3 = 1, \quad (\mathbf{A} - 1\mathbf{I}) = \begin{bmatrix} 1 & -1 & 1 \\ 3 & 2 & -2 \\ 4 & 1 & -1 \end{bmatrix}$$

The rank of $(\mathbf{A} - \lambda_i\mathbf{I})$ is $r = 2$ because any two of the three rows (or columns) of this matrix are independent of one another. Thus, for $\lambda_2 = \lambda_3 = 1$ $(m = 2)$, $n - m = 3 - 2 = 1$, so that $r \neq n - m$. The conclusion is that there do not exist two independent eigenvectors associated with the eigenvalue of multiplicity $m = 2$.

In the case of a *symmetric* matrix, it is always possible to calculate m *orthogonal* eigenvectors corresponding to multiple eigenvalues, when present. This is not necessarily true for *non-symmetric* matrices, where the number of eigenvectors may be smaller than m. Therefore, whatever their multiplicity, eigenvalues of most matrices of interest to ecologists, including association matrices (Section 2.2), have distinct

Table 2.2 Types of *symmetric* matrices and corresponding characteristics of their eigenvalues.

Symmetric matrix	Eigenvalues
All elements of matrix **A** are *real* (i.e. non-imaginary)	All eigenvalues are *real* (i.e. non-imaginary)
Matrix **A** is *positive definite*	All eigenvalues are *positive*
Matrix $\mathbf{A}_{nn}$ is *positive semidefinite* and of rank r	There are r positive and $(n - r)$ null eigenvalues
Matrix $\mathbf{A}_{nn}$ is *negative semidefinite* and of rank r	There are r negative and $(n - r)$ null eigenvalues
Matrix $\mathbf{A}_{nn}$ is *indefinite* and of rank r	There are r non-null (positive and negative) and $(n - r)$ null eigenvalues
Matrix **A** is *diagonal*	The *diagonal elements* are the eigenvalues

eigenvectors associated with them. In any case, it is unlikely that eigenvalues of matrices computed from real data be exactly equal (i.e. multiple).

Quadratic form

Fourth property. — A property of *symmetric* matrices may be used to predict the nature of their eigenvalues (Table 2.2). A symmetric matrix **A** may be combined with any vector $\mathbf{t} \neq \mathbf{0}$, in a matrix expression of the form **t'At** which is known as a *quadratic form*. This expression results in a scalar whose value leads to the following definitions:

- if **t'At** is always positive, matrix **A** is *positive definite*;
- if **t'At** can be either positive or null, matrix **A** is *positive semidefinite*;
- if **t'At** can be either negative or null, matrix **A** is *negative semidefinite*;
- if **t'At** can be either negative, null or positive, matrix **A** is *indefinite*.

2.11 Singular value decomposition

In a well-known theorem, Eckart & Young (1936) showed that any *rectangular* matrix **Y** can be decomposed as follows:

$$\mathbf{Y}(n \times p) = \mathbf{V}(n \times p)\ \mathbf{W}(\text{diagonal}, p \times p)\ \mathbf{U}'(p \times p) \qquad \textbf{(2.31)}$$

where both **U** and **V** are column-orthonormal matrices (i.e. matrices containing column vectors that are normalized and orthogonal to one another; Section 2.8) and **W** is a diagonal matrix $\mathbf{D}(w_i)$. The method is known as *singular value decomposition* (SVD). The following illustration shows more clearly the shapes of these matrices:

$$\left[\ \mathbf{Y}_{(n \times p)}\ \right] = \left[\ \mathbf{V}_{(n \times p)}\ \right] \begin{bmatrix} w_1 & 0 & 0 & \dots & 0 \\ 0 & w_2 & 0 & \dots & 0 \\ 0 & 0 & w_3 & \dots & 0 \\ \dots & \dots & \dots & \dots & \dots \\ 0 & 0 & 0 & \dots & w_p \end{bmatrix} \left[\ \mathbf{U}'_{(p \times p)}\ \right]$$

Demonstrating eq. 2.31 is beyond the scope of this book. The diagonal values w_i in **W** are non-negative; they are called the singular values of **Y**. The method is discussed in more detail by Press *et al.* (1986 and later editions), who propose computer programs for SVD*. Programs are also available in major subroutine packages.

SVD offers a way of handling matrices that are singular (Section 2.8) or numerically very close to singular. SVD may either give users a clear diagnostic of the problem, or solve it. Singularity may be encountered when solving sets of simultaneous linear equations represented by matrix equation **Ab** = **c**, where matrix **A** is square (Section 2.8), **A** and **c** are known, and **b** is unknown. **A** must be inverted in order to find **b**. **A** can always be decomposed using eq. 2.31:

$$\mathbf{A} = \mathbf{V}\ \mathbf{D}(w_i)\mathbf{U}'$$

In that case, **V**, **W** and **U** are all square matrices of the same size as **A**. Using property 5 of matrix inverses (above), the inverse of **A** is simple to compute:

$$\mathbf{A}^{-1} = [\mathbf{V}\ \mathbf{D}(w_i)\mathbf{U}']^{-1} = [\mathbf{U}']^{-1}[\mathbf{D}(w_i)]^{-1}[\mathbf{V}]^{-1}$$

---

* The *Numerical recipes* routines are available in FORTRAN and C from the following WWWeb site: <http://www.nr.com>.

Since **U** and **V** are orthonormal, their inverses are equal to their transposes (property 7), whereas the inverse of a diagonal matrix is a diagonal matrix whose elements are the reciprocals of the original elements (property 8). So, one can write:

$$\mathbf{A}^{-1} = \mathbf{U}\mathbf{D}(1/w_i)\mathbf{V}' \quad \textbf{(2.32)}$$

Singular matrix

Ill-conditioned matrix

It may happen that one or more of the w_i's are zero, so that their reciprocals are infinite; **A** is then a *singular matrix*. It may also happen that one or more of the w_i's are numerically so small that their values cannot be properly computed because of the machine's precision in floating point calculation; in that case, **A** is said to be *ill-conditioned*. When **A** is singular, the columns of **U** corresponding to the zero elements in **W** form an orthonormal basis* for the space where the system of equations has no solution, whereas the columns of **V** corresponding to the zero elements in **W** are an orthonormal basis for the space where the system has a solution. When **A** is singular or ill-conditioned, it is still possible to find one or several vectors **b** that satisfy the set of simultaneous linear equations, either exactly or approximately. How to find these solutions is explained in the book of Press *et al.* (1986), for instance.

Singular value decomposition may be applied to situations where there are more equations than unknowns (e.g. least-squares estimation of parameters, as in Section 2.8), or fewer equations than unknowns. It may also be used for eigenvalue decomposition, although it is not a general method for eigenanalysis; in particular, it does not allow one to estimate negative eigenvalues. In the present book, SVD will be used as one of the possible algorithms for principal component analysis (Subsection 9.1.9) and correspondence analysis (Subsection 9.4.1).

* A set of k linearly independent vectors form a basis for a k-dimensional vector space. Any vector in that space can be uniquely written as a linear combination of the base vectors.

Chapter 3 Dimensional analysis in ecology

3.0 Dimensional analysis

Dimensional analysis is generally not part of the curriculum of ecologists, so that relatively few are conversant with this simple but remarkably powerful tool. Yet, applications of dimensional analysis are found in the ecological literature, where results clearly demonstrate the advantage of using this mathematical approach.

"Dimensional analysis treats the *general forms of equations* that describe natural phenomena" (Langhaar, 1951). The basic principles of this discipline were established by physicists (Fourier, 1822; Maxwell, 1871) and later applied by engineers to the very important area of small-scale modelling. Readers interested in the fundamentals and engineering applications of dimensional analysis should refer, for example, to Langhaar (1951), from which are taken several of the topics developed in the present Chapter. Other useful references are Ipsen (1960), Huntley (1967), and Schneider (1994).

The use of dimensional analysis in ecology rests on the fact that a growing number of areas in ecological science use *equations*; for example, populations dynamics and ecological modelling. The study of equations is the very basis of dimensional analysis. This powerful approach can easily be used by ecologists, given the facts that it can be reduced to *a single theorem* (the Π theorem) and that many of its applications (Sections 3.1 and 3.2) only require a knowledge of elementary mathematics.

Dimensional analysis can resolve complex ecological problems in a simple and elegant manner. Readers should therefore not be surprised that ecological applications in the present Chapter are of a rather high level, since the advantage of dimensional analysis lies precisely in its ability to handle complex problems. It follows that dimensional analysis is mainly useful in those cases where it would be difficult to resolve the ecological problem by conventional approaches.

3.1 Dimensions

All fields of science, including ecology, rest on a number of abstract entities such as the mass, length, time, temperature, speed, acceleration, radioactivity, concentration, energy or volume. These entities, which can be measured, are called *quantities*. Designing a *system of units* requires to: (1) arbitrarily choose a small number of *fundamental quantities*, on which a coherent and practical system can be constructed, and (2) arbitrarily assign, to each of these quantities, *base units* chosen as reference for comparing measurements.

Various systems of units have been developed in the past, e.g. the British system and several versions of the metric system. The latter include the CGS metric system used by scientists (based on the centimetre, the gram and the second), the MKS (force) metric system used by engineers (based on the metre, the kilogram and the second, where the kilogram is the unit of *force*), and the MKS (mass) metric system (where the kilogram is the unit of *mass*). Since 1960, there is an internationally accepted version of the metric system, called the *International System of Units* (SI, from the French name *Système international d'unités*). The SI is based on seven *quantities*, to which are associated seven *base units* (Table 3.1; the mole was added to the SI in 1971 only). In addition to these seven base units, the SI recognizes two *supplementary units,* the radian (rad) and the steradian (sr), which measure planar and solid angles, respectively. All other units, called *derived units*, are combinations of the base and supplementary

International System of Units

Table 3.1 Base units of the International System of Units (SI).

Fundamental quantity	Quantity symbol*	Dimension symbol	Base unit	Unit symbol
mass	m	[M]	kilogram	kg
length	l	[L]	metre†	m
time	t	[T]	second	s
electric current	I	[I]	ampere	A
thermodynamic temperature	T ‡	[θ]	kelvin‡	K
amount of substance	n	[N]	mole	mol
luminous intensity	I_v	[J]	candela	cd

* Quantity symbols are not part of the SI, and they are not unique.

† Spelled meter in the United States of America.

‡ In ecology, temperature is generally measured on the Celsius scale, where the unit is the *degree Celsius* (°C); the quantity symbol for temperatures expressed in °C is usually t.

units. Some frequently used derived units have special names, e.g. volt, lux, joule, newton, ohm. It must be noted that: (1) *unit names* are written with small letters only, the sole exception being the degree Celsius; (2) *unit symbols* are written with small letters only, except the symbols of derived units that are surnames, whose first letter is a capital (e.g. Pa for pascal), and the litre (see Table 3.2, footnote). Unit symbols are *not* abbreviations, hence they are *never* followed by a point.*

Table 3.2 shows that derived units are not only simple products of the fundamental units, but that they are often *powers* and *combinations of powers* of these units. Maxwell (1871) used symbols such as [M], [L], [T], and [θ] to represent the quantities mass, length, time and temperature (Table 3.1). The *dimensions* of the various quantities are *products of powers* of the symbols of fundamental quantities. Thus, the dimension of an area is $[L^2]$, of a volume $[L^3]$, of a speed $[LT^{-1}]$, and of an acceleration $[LT^{-2}]$. Table 3.2 gives the exponents of the dimensional form of the most frequently encountered quantities.

Dimension

Since the various quantities are *products of powers,* going from one quantity to another is done simply by *adding* (or *subtracting) exponents* of the dimensions. For example, one calculates the dimensions of *heat conductivity* $W(mK)^{-1}$ by subtracting, from the dimension exponents of *power* W, the sum of the dimension exponents of *length* m and of *temperature* K:

$$[M^1L^2T^{-3}] / ([L^1] \times [\theta^1]) = [M^1L^{(2-1)}T^{-3}\theta^{-(1)}] = [M^1L^1T^{-3}\theta^{-1}]$$

The first three fundamental quantities (Table 3.1), mass [M], length [L], and time [T], are enough to describe any Newtonian mechanical system. Ecologists may require, in addition, temperature [θ], amount of substance [N], and luminous intensity [J]. Research in electromagnetism calls for electric current [I] and, in quantum mechanics, one uses the quantum state of the system [Ψ].

Four types of entities are recognized:

(1) *dimensional variables*, e.g. most of the quantities listed in Table 3.2;

(2) *dimensional constants*, for instance: the speed of light in vacuum $[LT^{-1}]$, $c = 2.998 \times 10^8$ m s^{-1}; the acceleration due to Earth's gravity at sea level $[LT^{-2}]$, $g = 9.807$ m s^{-2}; the number of elementary entities in a mole $N_A = 6.022 \times 10^{23}$ mol^{-1}, where N_A is the Avogadro number (note that the nature of the elementary entities in a mole must always be specified, e.g. mol C, mol photons);

* A program (for MS-DOS machines), called *The Unit Calculator*, deals with most problems involving physical measurement units. It recognizes over 600 units commonly used in science and business. This program may be used to: convert data, for instance between the American and International System of Units; carry out calculations, even with mixed units; and perform computations on the units themselves, from checking that physical equations are homogeneous to actual dimensional analysis. The program is distributed by: Applied Biomathematics, 100 North Country Road, Setauket, New York 11733, U.S.A.

Table 3.2 Dimensions, units, and names of quantities. Units follow the standards of the International System of Units (SI).

Quantity	[M]	[L]	[T]	[I]	[θ]	[N]	[J]	Units	Name*
mass	1	0	0	0	0	0	0	kg	kilogram
length	0	1	0	0	0	0	0	m	metre
time	0	0	1	0	0	0	0	s	second
electric current	0	0	0	1	0	0	0	A	ampere
temperature	0	0	0	0	1	0	0	K	kelvin
amount of substance	0	0	0	0	0	1	0	mol	mole
luminous intensity	0	0	0	0	0	0	1	cd	candela
absorbed dose	0	2	–2	0	0	0	0	$J\ kg^{-1} = Gy$	gray
acceleration (angular)	0	0	–2	0	0	0	0	$rad\ s^{-2}$	
acceleration (linear)	0	1	–2	0	0	0	0	$m\ s^{-2}$	
activity of radioactive source	0	0	–1	0	0	0	0	$s^{-1} = Bq$	becquerel
angle (planar)	0	0	0	0	0	0	0	rad	radian
angle (solid)	0	0	0	0	0	0	0	sr	steradian
angular momentum	1	2	–1	0	0	0	0	$kg\ m^2\ s^{-1}$	
angular velocity	0	0	–1	0	0	0	0	$rad\ s^{-1}$	
area	0	2	0	0	0	0	0	m^2	
compressibility	–1	1	2	0	0	0	0	Pa^{-1}	
concentration (molarity)	0	–3	0	0	0	1	0	$mol\ m^{-3}$	
current density	0	–2	0	1	0	0	0	$A\ m^{-2}$	
density (mass density)	1	–3	0	0	0	0	0	$kg\ m^{-3}$	
electric capacitance	–1	–2	4	2	0	0	0	$C\ V^{-1} = F$	farad
electric charge	0	0	1	1	0	0	0	$A\ s = C$	coulomb
electric conductance	–1	–2	3	2	0	0	0	$\Omega^{-1} = S$	siemens
electric field strength	1	1	–3	–1	0	0	0	$V\ m^{-1}$	
electric resistance	1	2	–3	–2	0	0	0	$V\ A^{-1} = \Omega$	ohm
electric potential	1	2	–3	–1	0	0	0	$W\ A^{-1} = V$	volt
energy	1	2	–2	0	0	0	0	$N\ m = J$	joule
force	1	1	–2	0	0	0	0	$kg\ m\ s^{-2} = N$	newton
frequency	0	0	–1	0	0	0	0	$s^{-1} = Hz$	hertz
heat capacity	1	2	–2	0	–1	0	0	$J\ K^{-1}$	
heat conductivity	1	1	–3	0	–1	0	0	$W(m\ K)^{-1}$	
heat flux density	1	0	–3	0	0	0	0	$W\ m^{-2}$	
illuminance	0	–2	0	0	0	0	1	$lm\ m^{-2} = lx$	lux
inductance	1	2	–2	–2	0	0	0	$Wb\ A^{-1} = H$	henry
light exposure	0	–2	1	0	0	0	1	lx s	
luminance	0	–2	0	0	0	0	1	$cd\ m^{-2}$	
luminous flux	0	0	0	0	0	0	1	cd sr = lm	lumen

* Only base units and special names of derived units are listed.

† The litre (spelled liter in the United States of America) is the *capacity* (vs. *cubic*) unit of volume. Its symbol (letter l) may be confused with digit one (1) in printed texts so that it was decided in 1979 that capital L could be used as well; $1\ m^3 = 1000\ L$.

Table 3.2 Dimensions, units, and names of quantities (continued).

Quantity	[M]	[L]	[T]	[I]	[θ]	[N]	[J]	Units	Name
magnetic field strength	0	–1	0	1	0	0	0	$A\ m^{-1}$	
magnetic flux	1	2	–2	–1	0	0	0	V s = Wb	weber
magnetic flux density	1	0	–2	–1	0	0	0	$Wb\ m^{-2} = T$	tesla
magnetic induction	1	0	–2	–1	0	0	0	$Wb\ m^{-2} = T$	tesla
magnetic permeability	1	1	–2	–2	0	0	0	$\Omega\ s\ m^{-1}$	
mass flow rate	1	0	–1	0	0	0	0	$kg\ s^{-1}$	
molality	–1	0	0	0	0	1	0	$mol\ kg^{-1}$	
molarity	0	–3	0	0	0	1	0	$mol\ m^{-3}$	
molar internal energy	1	2	–2	0	0	–1	0	$J\ mol^{-1}$	
molar mass	1	0	0	0	0	–1	0	$kg\ mol^{-1}$	
molar volume	0	3	0	0	0	–1	0	$m^3\ mol^{-1}$	
moment of force	1	2	–2	0	0	0	0	N m	
moment of inertia	1	2	0	0	0	0	0	$kg\ m^2$	
momentum	1	1	–1	0	0	0	0	$kg\ m\ s^{-1}$	
period	0	0	1	0	0	0	0	s	
permittivity	–1	–3	4	2	0	0	0	$F\ m^{-1}$	
power	1	2	–3	0	0	0	0	$J\ s^{-1} = W$	watt
pressure	1	–1	–2	0	0	0	0	$N\ m^{-2} = Pa$	pascal
quantity of light	0	0	1	0	0	0	1	lm s	
radiant intensity	1	2	–3	0	0	0	0	$W\ sr^{-1}$	
relative density	0	0	0	0	0	0	0	(no unit)	
rotational frequency	0	0	–1	0	0	0	0	s^{-1}	
second moment of area	0	4	0	0	0	0	0	m^4	
specific heat capacity	0	2	–2	0	–1	0	0	$J(kg\ K)^{-1}$	
specific latent heat	0	2	–2	0	0	0	0	$J\ kg^{-1}$	
specific volume	–1	3	0	0	0	0	0	$m^3\ kg^{-1}$	
speed	0	1	–1	0	0	0	0	$m\ s^{-1}$	
stress	1	–1	–2	0	0	0	0	$N\ m^{-2} = Pa$	pascal
surface tension	1	0	–2	0	0	0	0	$N\ m^{-1}$	
torque	1	2	–2	0	0	0	0	N m	
viscosity (dynamic)	1	–1	–1	0	0	0	0	Pa s	
viscosity (kinetic)	0	2	–1	0	0	0	0	$m^2\ s^{-1}$	
volume†	0	3	0	0	0	0	0	m^3	
volume flow rate	0	3	–1	0	0	0	0	$m^3\ s^{-1}$	
wavelength	0	1	0	0	0	0	0	m	
wave number	0	–1	0	0	0	0	0	m^{-1}	
work	1	2	–2	0	0	0	0	N m = J	joule

(3) *dimensionless variables,* such as angles, relative density (Table 3.2), or dimensionless products which will be studied in following sections;

(4) *dimensionless constants,* e.g. π, e, 2, 7; it must be noted that exponents are, by definition, dimensionless constants.

The very concept of *dimension* leads to immediate applications in physics and ecology. In physics, for example, one can easily demonstrate that the first derivative of distance with respect to time is a speed:

$$\text{dimensions of } \frac{dl}{dt} : \quad \left[\frac{\text{L}}{\text{T}}\right] = [\text{LT}^{-1}], \text{ i.e. speed.}$$

Similarly, it can be shown that the second derivative is an acceleration:

$$\text{dimensions of } \frac{d^2 l}{dt^2} = \frac{d}{dt}\left(\frac{dl}{dt}\right) : \quad \left[\frac{\text{L}}{\text{TT}}\right] = [\text{LT}^{-2}], \text{ i.e. acceleration.}$$

Note that *italics* are used for *quantity symbols* such as length (l), mass (m), time (t), area (A), and so on. This distinguishes them from *unit symbols* (roman type; Tables 3.1 and 3.2), and *dimension symbols* (roman capitals in brackets; Table 3.1).

Ecological application 3.1

Platt (1969) studied the efficiency of *primary (phytoplankton) production* in the *aquatic environment*. Primary production is generally determined at different depths in the water column, so that it is difficult to compare values observed under different conditions. The solution to this problem consists in finding a method to standardize the values, for example by transforming field estimates of *primary production* into values of *energy efficiency*. Such a transformation would eliminate the effect on production of solar irradiance at different locations and different depths. Primary production at a given depth $P(z)$ may be expressed in J m^{-3} s^{-1} [ML^{-1} T^{-3}], while irradiance at the same depth $E(z)$ is in J m^{-2} s^{-1} [MT^{-3}] (energy units).

The dimension of the ratio $P(z)/E(z)$, which defines the energy efficiency of primary production, is thus [L^{-1}]. Another property determined in the water column, which also has dimension [L^{-1}], is the *attenuation* of diffuse light as a function of depth. The *coefficient of diffuse light attenuation* (α) is defined as:

$$E(z_2) = E(z_1)\, e^{-\alpha(z_2 - z_1)}$$

where $E(z_2)$ and $E(z_1)$ are irradiances at depths z_2 and z_1, respectively. Given the fact that an exponent is, by definition, dimensionless, the dimension of α must be [L^{-1}] since that of depth z is [L].

Based on the dimensional similarity between efficiency and attenuation, and considering the physical aspects of light attenuation in the water column, Platt partitioned the attenuation coefficient (α) into physical (k_p) and biological (k_b) components, i.e. $\alpha = k_p + k_b$. The *biological attenuation coefficient* k_p may be used to estimate the attenuation of light caused by photosynthetic processes. In the same paper and in further publications by Platt & Subba Rao (1970) and Legendre (1971b), it was shown that there exists a correlation in the marine environment between k_b and the concentration of chlorophyll *a*. The above papers used the calorie as unit of energy but, according to the SI standard, this unit should no longer be used. Coherency requires here that primary production be expressed in J m^{-3} s^{-1} and irradiance in J m^{-2} s^{-1} (or W m^{-2}).

This example illustrates how a simple reflection, based on dimensions, led to an interesting development in the field of ecology.

It is therefore useful to think in terms of *dimensions* when dealing with ecological equations that contain physical *quantities*. Even if this habit is worth cultivating, it would not however, in and of itself, justify an entire chapter in the present book. So, let us move forward in the study of dimensional analysis.

3.2 Fundamental principles and the Pi theorem

It was shown in the previous section that going from one quantity to another is generally done by multiplying or dividing quantities characterized by *different dimensions*. In contrast, additions and subtractions can only be performed on quantities having the *same dimensions* — hence the fundamental principle of *dimensional homogeneity*. Any equation of the general form

Dimensional homogeneity

$$a + b + c + \ldots = g + h + \ldots$$

is dimensionally homogeneous if and only if all variables *a, b, c, ... g, h, ...* have the *same dimensions*. This property applies to all equations of a *theoretical* nature, but it does not necessarily apply to those derived *empirically*. Readers must be aware that dimensional analysis only deals with dimensionally homogeneous equations. In animal ecology, for example, the basic equation for energy budgets is:

$$dW/dt = R - T \tag{3.1}$$

where W is the mass of an animal, R its food ration, and T its metabolic expenditure rate (oxygen consumption). This equation, which describes growth dW/dt as a function of ration R and metabolic rate T, is dimensionally homogeneous. The rate of oxygen consumption T is expressed as mass per unit time, its dimensions thus being $[MT^{-1}]$, as those of food ration R. The dimensions of dW/dt are also clearly $[MT^{-1}]$. This same equation will be used in Ecological applications 3.2e and 3.3b, together with other ecological equations — all of which are dimensionally homogeneous.

In dimensional analysis, the correct identification of quantities to be included in a given equation is much more important than the exact form of the equation. Researchers using dimensional analysis must therefore have prior knowledge of the phenomenon under study, in order to identify the pertinent *dimensional variables* and *constants*. On the one hand, missing key quantities could lead to incomplete or incorrect results, or even to a deadlock. On the other hand, including unnecessary terms could overburden the solution needlessly. Hence, dimensional analysis cannot be conducted without first considering the ecological bases of the problem. A simple example, taken from hydrodynamics, will illustrate the dimensional method.

The question considered here relates to the work of many ecologists in aquatic environments, i.e. estimating the drag experienced by an object immersed in a current. Ecologists who moor current meters or other probes must consider the drag, lest the equipment might be carried away. To simplify the problem, one assumes that the immersed object is a smooth sphere and that the *velocity* of the current V is constant. The drag *force* F is then a function of: the *velocity* (V), the *diameter* of the sphere (D), the *density* of water (ρ), and its *dynamic viscosity* (η). The simplest equation relating these five quantities is:

$$F = f(V, D, \rho, \eta) \tag{3.2}$$

At first sight, nothing specifies the nature of the dependency of F on V, D, ρ, and η, except that such a dependency exists. Dimensional analysis allows one to find the form of the equation that relates F to the variables identified as governing the drag.

Dimensionless product

A number of variables are regularly encountered in hydrodynamics problems, i.e. F, V, L, ρ, η, to which one must also add g, the acceleration due to gravity. Some of these variables may be combined to form *dimensionless products*. Specialists of hydrodynamics have given names to some often-used dimensionless products:

$$\textit{Reynolds number: } Re = \frac{VL\rho}{\eta} = \frac{[\mathrm{LT^{-1}}]\,[\mathrm{L}]\,[\mathrm{ML^{-3}}]}{[\mathrm{ML^{-1}T^{-1}}]} = \frac{[\mathrm{ML^{-1}T^{-1}}]}{[\mathrm{ML^{-1}T^{-1}}]} = [1] \tag{3.3}$$

$$\textit{Newton number: } Ne = \frac{F}{\rho L^2 V^2} = \frac{[\mathrm{MLT^{-2}}]}{[\mathrm{ML^{-3}}]\,[\mathrm{L^2}]\,[\mathrm{L^2T^{-2}}]} = \frac{[\mathrm{MLT^{-2}}]}{[\mathrm{MLT^{-2}}]} = [1] \tag{3.4}$$

$$\textit{Froude number: } Fr = \frac{V^2}{Lg} = \frac{[\mathrm{L^2T^{-2}}]}{[\mathrm{L}]\,[\mathrm{T^{-2}}]} = \frac{[\mathrm{L^2T^{-2}}]}{[\mathrm{L^2T^{-2}}]} = [1] \tag{3.5}$$

Each of the above *products* is clearly *dimensionless*. It should also be noted that each product of this set is *independent* of the others, since each contains one exclusive variable, i.e. η for Re, F for Ne, and g for Fr. Finally, any other dimensionless product of these same variables would *inevitably* be a product of powers of dimensionless products from the above set. The three dimensionless products thus form a *complete set* of dimensionless products for variables F, V, L, ρ, η and g. It would obviously be possible to form other complete sets of dimensionless products using these same variables, by combining them differently.

The first important concept to remember is that of *dimensionless product.* This concept leads to the *sole* theorem of dimensional analysis, the Π theorem, which is also known as the Buckingham theorem.

Given the fundamental principle of dimensional homogeneity (see above), it follows that any equation that combines dimensionless products is dimensionally homogeneous. Thus, a *sufficient* condition for an equation to be dimensionally homogeneous is that it could be reduced to an equation combining dimensionless products. Indeed, any equation that can be reduced to an equation made of dimensionless products is dimensionally homogeneous. Buckingham (1914) did show that this condition is not only *sufficient* but also *necessary*. This leads to the Π *(Pi) theorem* (the capital Greek letter Π is the mathematical symbol for product):

Π theorem

If an equation is dimensionally homogeneous, it can be reduced to a relationship among the members of a complete set of dimensionless products.

This theorem alone summarizes the whole theory of dimensional analysis.

The power of the Π theorem is illustrated by the solution of the drag problem, introduced above. Equation 3.2 is, by definition, dimensionally homogeneous:

$$F = f(V, D, \rho, \eta)$$

It may be rewritten as:

$$f(F, V, D, \rho, \eta) = 0 \qquad \textbf{(3.6)}$$

The complete set of dimensionless products of the five variables F, V, D, ρ, η contains two products, i.e. the Reynolds (Re) and Newton (Ne) numbers (D being a length, it is a quantity of type L). Hence, eq. 3.6 may be rewritten as a relation between the members of this complete set of dimensionless products (Π theorem):

$$Ne = f(Re)$$

$$\frac{F}{\rho V^2 D^2} = f(Re) \qquad \textbf{(3.7)}$$

in which function f is, for the time being, unknown, except that it depends on the sole dimensionless variable Re.

The projected area (A) of a sphere is:

$$A = \pi\,(D/2)^2 = (1/4)\,\pi\,D^2, \text{ so that } D^2 = 4A/\pi$$

which allows one to rewrite eq. 3.7 as:

$$\frac{F}{\rho V^2 \frac{4A}{\pi}} = f(Re)$$

$$\frac{F}{\rho V^2 A} = \frac{1}{2}\left(\frac{8}{\pi}\right) f(Re)$$

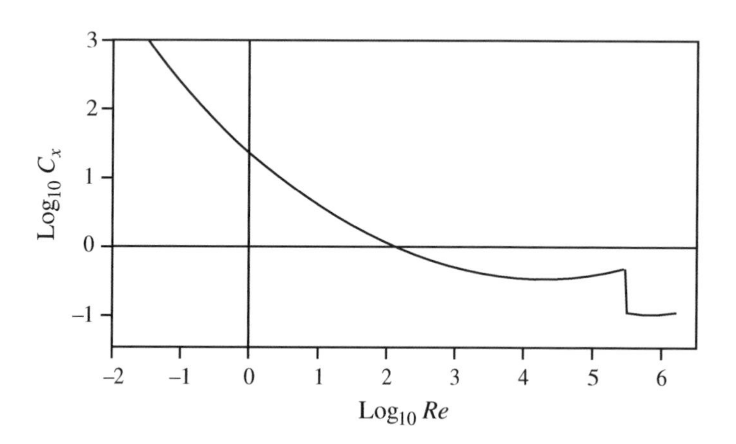

Figure 3.1 Drag coefficient on smooth spheres. Adapted from Eisner (1931).

In hydrodynamics, the term $(8/\pi) f(Re)$ is called the *drag coefficient* and is represented by C_x, so that the drag exerted on a sphere is:

$$F = (1/2)\ C_x \rho V^2 A, \text{ where } C_x = (8/\pi) f(Re) \qquad \textbf{(3.8)}$$

Since C_x is a function of the sole dimensionless coefficient *Re,* the problem is resolved by determining, in the laboratory, the experimental curve of C_x as a function of *Re*. This curve will be valid for any density (ρ) or dynamic viscosity (η) of any fluid under consideration (the same curve can thus be used for water, air, etc.) and for objects of any size, or any flow speed. The curve may thus be determined by researchers under the most suitable conditions, i.e. choosing fluids and flow speeds that are most convenient for laboratory work. As a matter of fact, this curve is already known (Fig. 3.1).

Two important properties follow from the above example.

(1) First, data for a *dimensionless graph* may be obtained under the most convenient conditions. For example, determining C_x for a sphere of diameter 3.48 m immersed in air at 14.4°C with a velocity of 15.24 m s^{-1} would be difficult and costly. In contrast, it would be much easier, in most laboratories, to determine C_x by using a sphere of diameter 0.61 m in water at 14.4°C with a speed of 5.79 m s^{-1}. In both cases, *Re* is the same so that the measured value of C_x is the same. This first property is the basis for *model testing* in engineering (Section 3.4), the sphere in air being here the *prototype* and that in water, the *model.*

Chart

(2) The dimensionless graph of Fig. 3.1 contains much more information than a set of *charts* depicting the function of the 4 variables. In a chart (Fig. 3.2), *a function of*

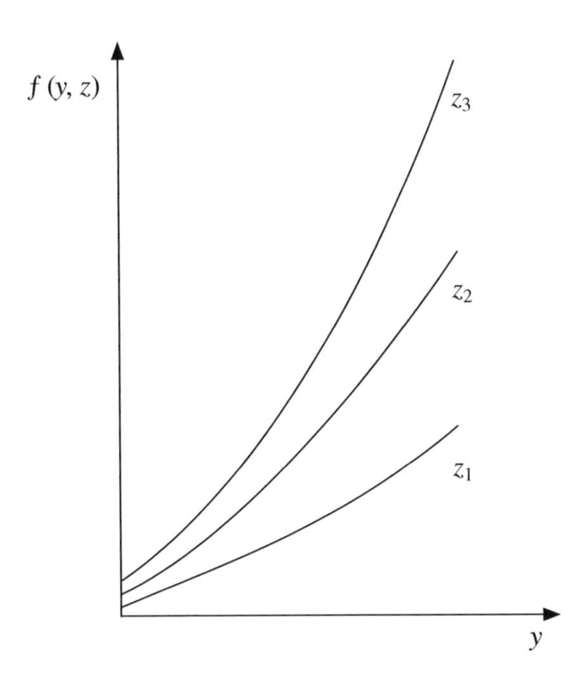

Figure 3.2 Chart representing a function of two variables. One curve is required for each value of the second variable (z_1, z_2, z_3, …)

two variables is represented by a *family of curves*, one curve being required for each value of the second variable. A *function of three variables* would be represented by a *set of sets of charts.* Hence, for four variables and assuming that there were only five values measured per variable, a total of 625 experimental points would be required, i.e. five sets of five charts each. With 25 times fewer experimental points, one can easily obtain a dimensionless graph (e.g. Fig. 3.1) which is both more exact and much more convenient.

The above physical example illustrated the great simplicity and remarkable power of dimensional analysis. Let us now examine examples from ecology.

Ecological application 3.2a

This first example belongs to the disciplines of ecology and physiology, since it concerns the dimensions of animals and their muscular dynamics. Hill (1950) compared different cetaceans, as a set of similar animals which differ in size. All these cetaceans (porpoises, dolphins, and whales), with a 5000-fold mass range, travel at high speed (ca. 7.5 m s^{-1}) which they can maintain for a long time. Table 3.3 compares the two extreme cases of the common dolphin (*Delphinus delphis*) and the blue whale (*Balaenoptera musculus*).

Table 3.3 Body characteristics of two cetaceans.

	Common dolphin	Blue whale
Maximum length (m)	2.4	30
Maximum mass (10^3 kg)	0.14	150
Mass/length^3	0.01	0.006
Area/length^2	0.45	0.40

Since these two animals can maintain a cruising speed of ca. 7.5 m s^{-1} for long periods, one may assume that they are then in a physiological steady state. The question is: how is it possible for two species with such different sizes to cruise at the same speed?

To answer this question, one must first consider the drag (F) on a streamlined body moving in a fluid. The equation is similar to eq. 3.8, except that the drag coefficient C_x is replaced here by the *friction coefficient* C_f:

$$F = 0.5\ C_f \rho V^2 A$$

where ρ is the *density* of the fluid, V the *velocity* of the body, and A its *total surface area*. For *laminar* flow, $C_f \approx 1.33\ Re^{-1/2}$ whereas, for *turbulent* flow, $C_f \approx 0.455\ (\log_{10} Re)^{-2.58}$, Re being the *Reynolds number*. Low values of Re correspond to laminar flow, where resistance to motion is relatively weak, whereas high values of Re are associated with turbulent flow, which creates stronger resistance to motion. Normally, for a streamlined body, the flow is laminar over the front portion only and is turbulent towards the back.

The *power* developed by the muscles of moving cetaceans is calculated in three steps.

• Calculation of Re, for the animal under study:

$$Re \approx 7 \times 10^5\ (\text{s m}^{-2})\ VL, \text{ in sea water at } 5°\text{C}$$

• Calculation of drag (F):

$$F = 0.5\ C_f \rho V^2 A$$

C_f being computed from Re, using the equation for either laminar or turbulent flow.

• Calculation of power (P) developed during motion:

$$P = FV$$

For the purpose of the calculation, consider (1) a dolphin with a length of 2 m, weighing 80 kg, whose surface area is 1.75 m^2 and (2) a whale 25 m long, with a mass of 100 t and surface area of 250 m^2.

(1) The value of *Re* for a dolphin moving at 7.5 m s^{-1} is of the order of 10^7, which seems to indicate highly turbulent flow. In the case of *laminar* flow,

$$C_f = 1.33 \times (10^7)^{-1/2} = 4.2 \times 10^{-4}$$

and, for *turbulent* flow,

$$C_f = 0.455\,(\log_{10} 10^7)^{-2.58} = 3 \times 10^{-3}$$

The drag (F) corresponding to these two flow regimes is:

$$F\text{ (laminar)} = 0.5\,(4.2 \times 10^{-4})\,(1028\text{ kg m}^{-3})\,(7.5\text{ m s}^{-1})^2\,(1.75\text{ m}^2) = 22\text{ N}$$

$$F\text{ (turbulent)} = 0.5\,(3 \times 10^{-3})\,(1028\text{ kg m}^{-3})\,(7.5\text{ m s}^{-1})^2\,(1.75\text{ m}^2) = 155\text{ N}$$

The *power* ($P = F \times 7.5$ m s^{-1}) that a dolphin should develop, if its motion resulted in perfectly *laminar* flow, would be 165 W and, for *turbulent* flow, 1165 W. Since the size of a dolphin is of the same order as that of a man, it is reasonable to assume that the power it can develop under normal conditions is not higher than that of an athlete, i.e. a *maximum power* of 260 W. It follows that the flow must be laminar for the 9/10 front portion of the dolphin's body, with the rear 1/10 being perhaps turbulent. This conclusion is consistent with observations made in nature on dolphins. It is assumed that the absence of turbulence along the front part of the dolphin's body comes from the fact that the animal only uses its rear section for propulsion.

(2) The blue whale also swims at 7.5 m s^{-1}, its *Re* being ca. 12.5×10^7 which corresponds to a turbulent flow regime. A *laminar* flow would lead to a value

$$C_f = 1.33 \times (12.5 \text{ x } 10^7)^{-1/2} = 1.2 \times 10^{-4}$$

and a *turbulent* flow to

$$C_f = 0.455\,(\log_{10} 12.5 \times 10^7)^{-2.58} = 2.1 \text{ x } 10^{-3}$$

The corresponding drag (F) would be:

$$F\text{ (laminar)} = 0.5\,(1.2 \times 10^{-4})\,(1028\text{ kg m}^{-3})\,(7.5\text{ m s}^{-1})^2\,(250\text{ m}^2) = 745\text{ N}$$

$$F\text{ (turbulent)} = 0.5\,(2.1 \times 10^{-3})\,(1028\text{ kg m}^{-3})\,(7.5\text{ m s}^{-1})^2\,(250\text{ m}^2) = 13\text{ kN.}$$

The *power* a whale should develop, if its motion at 7.5 m s^{-1} was accompanied by *laminar* flow, would be 5.6 kW and, in the case of *turbulent* flow, 100 kW. The maximum power developed by a 80 kg dolphin was estimated to be 260 W so that, if the maximum power of an animal was proportional to its mass, a 10^5 kg whale should be able to develop 325 kW. One should, however, take into account the fact that the available energy depends on blood flow. Since cardiac rate is proportional to (mass)$^{-0.27}$, the heart of a whale beats at a rate $(100/0.08)^{-0.27} \approx 1/7$ that of a dolphin. The *maximum power* of a whale is thus ca. 1/7 of 325 kW, i.e. 46.5 kW. This leads to the conclusion that laminar flow takes place along the 2/3 front portion of the animal and that only the 1/3 rear part can sustain turbulent flow.

Ecological application 3.2b

A second study, taken from the same paper as the previous application (Hill, 1950), deals with land animals. It has been observed that several terrestrial mammals run more or less at the same speed and jump approximately the same height, even if their sizes are very different. Table 3.4 gives some approximate maximal values. The question is to explain the small differences observed between the performances of animals with such different sizes.

Table 3.4 Performances (maximal values) of five mammals.

	Running speed (m s^{-1})	Height of jump (m)
Man	12	2
Horse	20	2
Greyhound (25 kg)	18	—
Hare	20	1.5
Deer	15	2.5

One of the explanations proposed by the author involves a relatively simple dimensional argument. The strength of tissues in the bodies of animals cannot be exceeded, during athletic performances, without great risk. For two differently sized animals, consider a pair of systems with lengths l_1 and l_2, respectively, carrying out similar movements within times t_1 and t_2, respectively. The stress at any point in these systems has dimensions $[ML^{-1}T^{-2}]$, which corresponds to the product of density $[ML^{-3}]$ with the square of speed $[L^2T^{-2}]$.

Assuming that the densities of systems are the same for the two species (i.e. $m_1 l_1^{-3} = m_2 l_2^{-3}$, which is reasonable, since the densities of bones, muscles, etc. are similar for all mammals), the stresses at corresponding points of the systems are in the ratio $(l_1^2 t_1^{-2}) : (l_2^2 t_2^{-2})$. If the two systems operate at speeds such that the stresses are the same at corresponding points, it follows that $(l_1 t_1^{-1}) = (l_2 t_2^{-1})$. In other words, the speed is the same at corresponding points of the two systems. It is therefore the strength of their tissues which would explain why athletic animals of very different sizes have the same upper limits for running speeds and jumping heights.

It is interesting to note that, over the years, the topic of maximal running speed of terrestrial mammals has been the subject of many papers, which considered at least four competing theories. These include the theory of geometric similarity, briefly explained in this example, and theories that predict an increase of maximum running speed with body mass. These are summarized in the introduction of a paper by Garland (1983), where maximum running speeds for 106 species of terrestrial mammals are analysed. The study led to several interesting conclusions, including that, even if maximal running speed is mass-independent within some mammalian orders, this is not the case when species from different orders are put together; there is then a tendency for running speed to increase with mass, up to an optimal mass of ca. 120 kg. This is quite paradoxical since, when considering mammals in general, limb bone proportions do scale consistently with geometric similarity. The author refers to Günther's (1975, p. 672) conclusion that "no single similarity criterion can provide a satisfactory quantitative explanation for every single function of an organism that can be submitted to dimensional analysis".

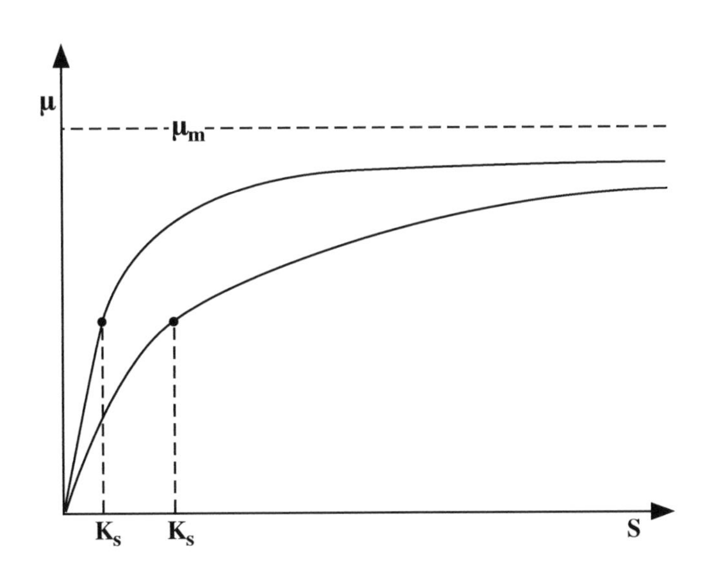

Figure 3.3 Illustration of the Michaelis-Menten equation, showing the role of parameter K_s. In the curve with higher K_s, μ approaches the asymptote μ_m more slowly than in the other curve.

Ecological application 3.2c

An example from aquatic ecology (Platt & Subba Rao, 1973) illustrates the use of dimensionless graphs. The dependence of phytoplankton growth on a given nutrient is often described by means of the Michaelis-Menten equation, borrowed from enzymology. In this equation, the *growth rate* (μ), with dimension $[T^{-1}]$, is a function of the *maximum specific* growth rate (μ_m), the *concentration* (S) of the nutrient, and the *concentration* (K_s) of nutrient at which the growth rate $\mu = 1/2\ \mu_m$:

$$\mu = \frac{1}{B}\frac{dB}{dt} = \frac{\mu_m S}{K_s + S}$$

$$[T^{-1}] = \frac{[1]}{[ML^{-3}]}\frac{[ML^{-3}]}{[T]} = \frac{[T^{-1}]\ [ML^{-3}]}{[ML^{-3}] + [ML^{-3}]}$$

where B is the concentration of phytoplankton *biomass*. This equation is that of a rectangular hyperbola, where K_s determines how fast the asymptote μ_m is approached. When K_s is high, μ approaches the asymptote μ_m slowly, which indicates a weak dependence of μ on S in the unsaturated part of the curve (Fig. 3.3).

In order to compare the effects of two different variables on phytoplankton growth, the authors defined a new entity $S_* = S/K_s$. Since this entity is dimensionless, the abscissa of the graph $\mu(S_*)$ as a function of S_* is dimensionless; $\mu(S_*)$ stands for the specific growth rate, normalized to S_*. The Michaelis-Menten equation is thus rewritten as:

$$\mu(S_*) = \frac{\mu_m S_*}{(1 + S_*)}$$

Hence, the strength of the dependence of μ on S_* is:

$$\frac{d\mu(S_*)}{dS_*} = \frac{d}{dS_*}\left(\frac{\mu_m S_*}{1 + S_*}\right) = \frac{\mu_m}{(1 + S_*)^2}$$

Using this expression, it is possible to determine the relative strength of the dependence of μ on two different variables (i and j):

$$\xi(i,j) = \frac{d\mu(S_*^i)/dS_*^i}{d\mu(S_*^j)/dS_*^j} = \frac{\mu_m^i}{\mu_m^j}\left[\frac{(1 + S_*^j)^2}{(1 + S_*^i)^2}\right]$$

Under conditions that do not limit phytoplankton growth, the maximum specific growth rate is the same for the two variables, i.e. $\mu^i_m = \mu^j_m$. In such a case, the dependence of μ on the two variables becomes:

$$\xi(i,j) = (1 + S_*^j)^2 / (1 + S_*^i)^2$$

This dimensionless approach makes it possible to compare the effects of different variables on phytoplankton growth, regardless of the dimensions of these variables. Using the above equation, one could assess, for example, the relative importance of irradiance (µmol photons $m^{-2}s^{-1}$, also denoted µEinstein $m^{-2}s^{-1}$) $[NL^{-2}T^{-1}]$ and of a nutrient $[ML^{-3}]$ for phytoplankton growth.

Characteristic value

The method described here is actually of general interest in ecology, since it shows how to approach a problem involving several variables with no common measure. In all cases, it is recommended to transform the *dimensional* variables into *dimensionless ones*. The most obvious transformation, proposed by Platt & Subba Rao (1973), consists in dividing each variable by a *characteristic value*, which has the same dimensions as the variable itself. In the case of the Michaelis-Menten equation, the characteristic value is K_s, which has the same dimensions as S. This elegant and efficient approach is also used in parametric statistics, where variables are transformed through division by their standard deviations. For this and other transformations, see Section 1.5. The approach which consists in dividing an ecologically interesting variable by another variable with the same dimensions, so as to create a dimensionless variable, is known as "scaling" (e.g. in Schneider, 1994). Scaling analysis has been used, for example, in coral reef studies (Hatcher and Firth, 1985; Hatcher *et al.*, 1987).

The following example illustrates some basic characteristics of dimensional analysis. It also stresses a major weakness of the method, of which ecologists should be aware.

Ecological application 3.2d

The study discussed here (Kierstead & Slobodkin, 1953) did not use dimensional analysis, but it provides material to which the method may usefully be applied. The authors did develop their theory for phytoplankton, but it is general enough to be used with several other types of organisms. Given a water mass containing a growing population, which loses individuals (e.g. phytoplankton cells) by diffusion and regenerates itself by multiplication, the problem is to define the minimum size of the water mass below which the growth of the population is no longer possible.

The problem is simplified by assuming that: (1) the *diffusion* (D) of organisms remains constant within the water mass, but is very large outside where the population cannot maintain itself, and (2) the water mass is *one-dimensional* (long and narrow), so that the *concentration* (c) of organisms is a function of the *position* (x) along the axis of the water mass. The equation describing the growth of the population is thus:

$$\frac{\partial c}{\partial t} = D\frac{\partial^2 c}{\partial x^2} + Kc$$

where K is the growth rate. On the right-hand side of the equation, the first term accounts for diffusion, while the second represents linear growth. A complicated algebraic solution led the authors to define a critical *length* (L_c) for the water mass, under which the population would decrease and above which it could increase:

$$L_c = \pi\sqrt{D/K}$$

It must be noted that this equation is analogous to that of the critical mass in a nuclear reactor. Associated with this critical length is a *characteristic time* (t) of the process, after which the critical length L_c becomes operative:

$$t = L_c^2/(8\pi^2 D)$$

The above results are those given in the paper of Kierstead and Slobodkin. The same problem is now approached by means of dimensional analysis, which will allow one to compare the *dimensional solution* of Platt (1981) to the algebraic solution of Kierstead and Slobodkin. In order to approach the question from a dimensional point of view, the dimensions of variables in the problem must first be specified:

x: [L] $\qquad$ K: $[T^{-1}]$

t: [T] $\qquad$ D: $[L^2T^{-1}]$

The only dimensions that are not immediately evident are those of D, but these can easily be found using the principle of dimensional homogeneity of theoretical equations.

The equation of Kierstead & Slobodkin involves three variables (c, t, x) and two constants (D, K). According to the general method developed in the previous ecological application, the variables are first *transformed* to dimensionless forms, through division by suitable *characteristic values. Dimensionless variables* C, T and X are defined using *characteristic values* c_*, t_* and x_*:

$$C = c/c_* \qquad T = t/t_* \qquad X = x/x_*$$

hence $\qquad c = C\,c_* \qquad t = T\,t_* \qquad x = X\,x_*$

Substitution of these values in the equation gives:

$$\frac{c_* \partial C}{t_* \partial T} = D \frac{c_* \partial^2 C}{x_*^2 \partial X^2} + K c_* C$$

The next step is to make all terms in the equation dimensionless, by multiplying each one by x_*^2 and dividing it by D, after eliminating from all terms the common constant c_* :

$$\left[\frac{x_*^2}{D t_*}\right] \frac{\partial C}{\partial T} = \frac{\partial^2 C}{\partial X^2} + \left[\frac{K x_*^2}{D}\right] C$$

The resulting equation thus contains three *dimensionless variables* (C, T and X) and two *dimensionless products* (in brackets).

Since the dimensions of the two products are [1], these may be transformed to isolate the characteristic values x_* and t_* :

$$\text{since} \left[\frac{x_*^2}{D t_*}\right] = [1], \text{ it follows that } [t_*] = \left[\frac{x_*^2}{D}\right]$$

$$\text{since} \left[\frac{K x_*^2}{D}\right] = [1], \text{ it follows that } [x_*^2] = \left[\frac{D}{K}\right] \text{ and thus } [x_*] = \left[\frac{D}{K}\right]^{1/2}$$

Using these relationships, the following proportionalities are obtained:

$$x_* \propto \sqrt{D/K} \text{ and } t_* \propto x_*^2/D$$

Dimensional analysis thus easily led to the same results as those obtained by Kierstead and Slobodkin (1953), reported above, except for the constant factors π and $8\pi^2$. This same example will be reconsidered in the next section (Ecological application 3.3a), where the two dimensionless products will be calculated directly.

The above example illustrates the fact that *dimensional analysis cannot generate dimensionless constants*, which is a limit of the method that must be kept in mind. Thus, in order to take advantage of the power of dimensional analysis, one must give up some precision. It is obvious that such a simple method as dimensional analysis cannot produce the same detailed results as complex algebraic developments. As mentioned above (Section 3.0), dimensional analysis deals with *general forms* of equations. Yet, starting from simple concepts, one can progress quite far into complex problems, but the final solution is only partial. As noted by Langhaar (1951): "The generality of the method is both its strength and its weakness. With little effort, a partial solution to nearly any problem is obtained. On the other hand, a complete solution is not obtained."

Ecological application 3.2e

It often happens that ecologists must synthesize published data on a given subject, either as a starting point for new research, or to resolve a problem using existing knowledge, or else as a basis for a new theoretical perspective. This is nowadays more necessary than ever, because of the explosion of ecological information. However, such syntheses are confronted to a real difficulty, which is the fact that available data are often very diversified, and must thus be unified before being used. Paloheimo & Dickie (1965) met this problem when they synthesized the mass of information available in the literature on the growth of fish as a function of food intake. As in the previous application, the authors did not themselves use dimensional analysis in their work. The dimensional solution discussed here is modified from Platt (1981).

The metabolism of fish may be described using the following relationship:

$$T = \alpha W^{\gamma}$$

where T is the rate of *oxygen consumption*, α specifies the level of *metabolic expenditure* per unit time, W is the *mass* of the fish, and γ specifies the rate of *change of metabolism* with body mass. Growth is expressed as a function of food ration (R), by means of the following equation:

$$\frac{dW}{dt} = R\,[e^{-(a+bR)}]$$

which shows that growth efficiency decreases by a constant fraction e^{-b} for each unit increase in the amount of food consumed per unit time. The value of R at maximum growth is determined, as usual, by setting the partial derivative equal to 0:

$$\frac{\partial}{\partial R}\left(\frac{dW}{dt}\right) = (1-bR)\,e^{-(a+bR)} = 0$$

Growth is thus maximum when $bR = 1$.

The basic equation for the energy budget (eq. 3.1) is:

$$\frac{dW}{dt} = R - T$$

so that

$$T = R - \frac{dW}{dt}$$

Replacing, in this last equation, dW/dt by its expression in the second equation, above, and isolating R, one obtains:

$$T = R\,[1 - e^{-(a+br)}]$$

Then, replacing T by its expression in the first equation leads to:

$$\alpha W^{\gamma} = R\,[1 - e^{-(a+br)}]$$

which is a general equation for energy budgets. This equation may be used to calculate, for any fish of mass W, the ration R required to maintain a given metabolic level. Furthermore, with an increase in ration, the term $[1 - e^{-(a+bR)}]$ tends towards 1, which indicates that the metabolism then approaches R. In other words, growth decreases at high values of R.

Values for coefficient b and food intake found in the literature are quite variable. It was shown above that the product bR determines growth. Paloheimo & Dickie therefore suggested to standardize the relationship between growth and ration in terms of bR.

Since growth is maximum when $bR = 1$, the *ration* can be brought to a common measure by expressing it in units of $1/b$. On this new *scale*, the ration (r) is defined as:

$$r = bR$$

When growth is maximum, $bR = 1$, so that $R = 1/b$. Replacing, in the general equation for the energy budget, R by $1/b$ (and bR by 1) yields:

$$\alpha W^{\gamma} = 1/b\,[1 - e^{-(a+1)}]$$

so that

$$W = \left[\frac{1 - e^{-(a+1)}}{\alpha b}\right]^{1/\gamma}$$

from which it is concluded that the *mass* should be expressed in units of $(1/\alpha b)^{1/\gamma}$, in order to bring data from the literature to a common measure. On this new *scale*, the mass (w) is defined as:

$$w = (\alpha b)^{1/\gamma}\, W$$

so that

$$\frac{w^{\gamma}}{b} = \alpha W^{\gamma} = T$$

Using the scaled ration (r) and mass (w), the general equation for energy budgets may be rewritten as:

$$\frac{w^{\gamma}}{b} = \frac{r}{b}[1 - e^{-(a+r)}]$$

and finally

$$w^{\gamma} = r\,[1 - e^{-(a+r)}]$$

In this last equation, the use of r and w brings to a common measure the highly variable values of R and W, which are available in the literature for different species or for different groups within a given fish species.

These same results could have been obtained much more easily using dimensional analysis. As with all problems of the kind, it is essential, first of all, to identify the dimensions of variables involved in the problem. The first two equations are used to identify the dimensions of all variables in the study:

$$T = \alpha W^{\gamma}$$

$$[\mathrm{MT^{-1}}] = [\mathrm{M^{(1-\gamma)}T^{-1}}]\,[\mathrm{M^{\gamma}}]$$

$$\frac{dW}{dt} = R\,[e^{-(a+bR)}]$$

$$[\mathrm{MT^{-1}}] = [\mathrm{MT^{-1}}]\,[1]^{[1] + [\mathrm{M^{-1}T}]\,[\mathrm{MT^{-1}}]}$$

The dimensions of α, which were not immediately obvious, are determined using the principle of dimensional homogeneity (i.e. same dimensions on the two sides of the equation). The dimensions of a and b are also found by applying the principle of dimensional homogeneity, taking into account the fact that an exponent is by definition dimensionless.

The problem is then to define *characteristic values* (or, more appropriately, *scale factors*) so as to obtain dimensionless ration (r), mass (w), and time (τ). Obviously, these scale factors must contain the two dimensional parameters of the above equations, α and b.

Because the product bR is dimensionless, the scale factor r for ration is:

$$r = bR$$

The cases of w and τ require the calculation of unknown exponents. These are easily found by dimensional analysis. In order to do so, unknown exponents y and z are assigned to α and b, and these unknowns are solved using the principle of dimensional homogeneity:

Calculation of w:

$$[w] = [1] = [\alpha]^y\,[b]^z\,[W]$$

$$[W]^{-1} = [\alpha]^y\,[b]^z$$

$$[\mathrm{M}^{-1}\mathrm{T}^0] = [\mathrm{M}^{(1-\gamma)}\,\mathrm{T}^{-1}]^y\,[\mathrm{M}^{-1}\mathrm{T}]^z = [\mathrm{M}^{y(1-\gamma)-z}\,\mathrm{T}^{-y+z}]$$

so that $\quad y(1-\gamma) - z = -1$

and $\quad -y + z = 0$

hence $\quad y = 1/\gamma = z$

Consequently, the scale factor w for the mass is:

$$w = (\alpha b)^{1/\gamma} W$$

Calculation of τ:

$$[\tau] = [1] = [\alpha]^y\,[b]^z\,[t]$$

$$[t]^{-1} = [\alpha]^y\,[b]^z$$

$$[\mathrm{M}^0\mathrm{T}^{-1}] = [\mathrm{M}^{y(1-\gamma)-z}\,\mathrm{T}^{-y+z}]$$

so that $\quad y(1-\gamma) - z = 0$

and $\quad -y + z = -1$

hence $\quad y = 1/\gamma$ and $z = 1/\gamma - 1$

It follows that the scale factor τ for time is:

$$\tau = \alpha^{1/\gamma}\, b^{(1/\gamma - 1)}\, t$$

$$\tau = [(\alpha b)^{1/\gamma}/b]\, t$$

These scale factors can be used to compare highly diversified data. *Ration* is then expressed in units of $(1/b)$, *mass* in units of $(\alpha b)^{-1/\gamma}$, and *time* in units of $b/(\alpha b)^{-1/\gamma}$. With this approach, it is possible to conduct generalized studies on the food intake and growth of fish as a function of time.

Other applications of dimensionless products in ecology are found, for example, in Tranter & Smith (1968), Rubenstein & Koehl (1977), and Okubo (1987). The first application analyses the performance of plankton nets, the second explores the mechanisms of filter feeding by aquatic organisms, and the third examines various aspects of biofluid mechanics, including a general relationship between the Reynolds number (*Re*) and the sizes and swimming speeds of aquatic organisms from bacteria to whales. Platt (1981) provides other examples of application of dimensional analysis in the field of biological oceanography.

Ecological applications 3.2d and 3.2e showed that dimensional analysis may be a powerful tool in ecology. They do, however, leave potential users somewhat uncertain as to how personally apply this approach to new problems. The next section outlines a general method for solving problems of dimensional analysis, which will lead to more straightforward use of the method. It will be shown that it is not even necessary to know the basic *equations* pertaining to a problem, provided that all the pertinent *variables* are identified. The above last two examples will then be reconsidered, as applications of the *systematic calculation of dimensionless products*.

3.3 The complete set of dimensionless products

As shown in the previous section, the resolution of problems using dimensional analysis involves two distinct steps: (1) the identification of variables pertinent to the phenomenon under study — these are derived from fundamental principles, for example of ecological nature — and (2) the computation of a complete set of dimensionless products. When the number of variables involved is small, complete sets of dimensionless products can be formed quite easily, as seen above. However, as the number of variables increases, this soon becomes unwieldy, so that one must proceed to a *systematic calculation of the complete set of dimensionless products*.

The physical example of the *drag on smooth spheres* (Section 3.2) will first be used to illustrate the principles of the calculation. The problem involved five variables (F, V, L, ρ, and η; see eq. 3.2), whose dimensions are written here in a *dimensional matrix*:

$$\begin{array}{c} \\ \mathrm{M} \\ \mathrm{L} \\ \mathrm{T} \end{array} \begin{array}{c} \begin{array}{ccccc} F & \eta & \rho & L & V \end{array} \\ \begin{bmatrix} 1 & 1 & 1 & 0 & 0 \\ 1 & -1 & -3 & 1 & 1 \\ -2 & -1 & 0 & 0 & -1 \end{bmatrix} \end{array} \qquad (3.9)$$

It must be kept in mind that the numbers in matrix 3.9 (i.e. dimensions) are *exponents*. *Dimensionless products* are *products* of *powers* of variables in the matrix (columns). In each product, the exponents given to the variables must be such that the result is *dimensionless*.

In other words, the systematic calculation of dimensionless products consists in finding exponents x_1, x_2, x_3, x_4 and x_5 for variables F, η, ρ, L, and V, such that a product Π, of the general form

$$\Pi = F^{x_1}\eta^{x_2}\rho^{x_3}L^{x_4}V^{x_5}$$

be dimensionless. Taking into account the respective dimensions of the five variables, the general dimensions of Π are:

$$\Pi = [MLT^{-2}]^{x_1}[ML^{-1}T^{-1}]^{x_2}[ML^{-3}]^{x_3}[L]^{x_4}[LT^{-1}]^{x_5}$$

$$\Pi = [M^{(x_1+x_2+x_3)}L^{(x_1-x_2-3x_3+x_4+x_5)}T^{(-2x_1-x_2-x_5)}]$$

The exponents of dimensions [M], [L], and [T] carry exactly the same information as the dimensional matrix (eq. 3.9). These exponents could therefore have been written directly, using matrix notation:

$$\begin{bmatrix} 1 & 1 & 1 & 0 & 0 \\ 1 & -1 & -3 & 1 & 1 \\ -2 & -1 & 0 & 0 & -1 \end{bmatrix} \begin{bmatrix} x_1 \\ x_2 \\ x_3 \\ x_4 \\ x_5 \end{bmatrix} \qquad \textbf{(3.10)}$$

where the dimensional matrix is on the left-hand side.

Since the products Π are dimensionless, the exponent of each dimension [M], [L], and [T], respectively, must be *zero*. In follows that:

$$x_1 + x_2 + x_3 = 0$$

$$x_1 - x_2 - 3x_3 + x_4 + x_5 = 0$$

$$-2x_1 - x_2 - x_5 = 0$$

or, in matrix notation:

$$\begin{bmatrix} 1 & 1 & 1 & 0 & 0 \\ 1 & -1 & -3 & 1 & 1 \\ -2 & -1 & 0 & 0 & -1 \end{bmatrix} \begin{bmatrix} x_1 \\ x_2 \\ x_3 \\ x_4 \\ x_5 \end{bmatrix} = \mathbf{0} \qquad \textbf{(3.11)}$$

Calculation of dimensionless products Π is thus achieved by simultaneously solving three equations. However, the above system of equations is *indeterminate*, since there are only three equations for five unknowns. Arbitrary values must thus be assigned to two of the unknowns, for example x_1 and x_2. The general solution is then given in terms of x_1 and x_2. The steps are as follows:

(1) Matrix equation 3.11 is rewritten so as to isolate x_1 and x_2 together with the associated first two columns of the matrix. This operation simply involves transferring all terms in x_3, x_4 and x_5 to the right-hand side of the equation:

$$\begin{bmatrix} 1 & 1 \\ 1 & -1 \\ -2 & -1 \end{bmatrix} \begin{bmatrix} x_1 \\ x_2 \end{bmatrix} = -\begin{bmatrix} 1 & 0 & 0 \\ -3 & 1 & 1 \\ 0 & 0 & -1 \end{bmatrix} \begin{bmatrix} x_3 \\ x_4 \\ x_5 \end{bmatrix} \tag{3.12}$$

It must be noted that there is now a negative sign in front of the matrix on the right-hand side. Matrix eq. 3.12 is identical to the algebraic form:

$$x_1 + x_2 = -x_3$$

$$x_1 - x_2 = 3x_3 - x_4 - x_5$$

$$-2x_1 - x_2 = x_5$$

(2) One then solves for the unknowns x_3, x_4 and x_5, using the general method of matrix inversion (Section 2.8):

$$-\begin{bmatrix} 1 & 0 & 0 \\ -3 & 1 & 1 \\ 0 & 1 & -1 \end{bmatrix}^{-1} \begin{bmatrix} 1 & 1 \\ 1 & -1 \\ -2 & -1 \end{bmatrix} \begin{bmatrix} x_1 \\ x_2 \end{bmatrix} = \begin{bmatrix} x_3 \\ x_4 \\ x_5 \end{bmatrix}$$

$$-\begin{bmatrix} 1 & 0 & 0 \\ 3 & 1 & 1 \\ 0 & 1 & -1 \end{bmatrix} \begin{bmatrix} 1 & 1 \\ 1 & -1 \\ -2 & -1 \end{bmatrix} \begin{bmatrix} x_1 \\ x_2 \end{bmatrix} = \begin{bmatrix} x_3 \\ x_4 \\ x_5 \end{bmatrix}$$

$$\begin{bmatrix} -1 & -1 \\ -2 & -1 \\ -2 & -1 \end{bmatrix} \begin{bmatrix} x_1 \\ x_2 \end{bmatrix} = \begin{bmatrix} x_3 \\ x_4 \\ x_5 \end{bmatrix} \tag{3.13}$$

(3) The simplest approach consists in successively assigning the value 1 to each unknown while setting the other equal to 0, i.e. (1) $x_1 = 1$ and $x_2 = 0$ and (2) $x_1 = 0$ and $x_2 = 1$. It follows that the first two columns of the solution matrix are a *unit matrix*:

$$\begin{array}{c} \\ \\ \Pi_1 \\ \Pi_2 \end{array} \begin{array}{ccccc} F & \eta & \rho & L & V \\ x_1 & x_2 & x_3 & x_4 & x_5 \end{array} \atop \begin{bmatrix} 1 & 0 & -1 & -2 & -2 \\ 0 & 1 & -1 & -1 & -1 \end{bmatrix} \qquad (3.14)$$

The dimensionless products of the *complete set* are therefore (as in Section 3.2):

$\Pi_1 = \dfrac{F}{\rho L^2 V^2}$, the *Newton number* (*Ne*; eq. 3.4)

$\Pi_2 = \dfrac{\eta}{\rho L V}$, the inverse of the *Reynolds number* ($1/Re$; eq. 3.3)

This example clearly shows that the systematic calculation of dimensionless products rests *solely* on recognizing the *variables* involved in the problem under consideration, without necessarily knowing the corresponding *equations*. The above solution, which was developed using a simple example, can be applied to all problems of dimensional analysis, since it has the following characteristics:

(1) Because the left-hand part of the solution matrix is an *identity matrix* (**I**), the dimensionless products Π are *independent* of one another. Indeed, given **I**, each product contains one variable which is not included in any other product, i.e. the first variable is only in Π_1, the second is only in Π_2, and so on.

(2) When partitioning the dimensional matrix, one must isolate *on the right-hand side* a matrix that can be *inverted*, i.e. a matrix whose determinant is non-zero.

(3) The *rank* (r) of the dimensional matrix is the order of the largest non-zero determinant it contains (Section 2.7). Therefore, it is always possible to isolate, *on the right-hand side*, a matrix of order r whose determinant is non-zero. The order r may however be lower than the number of rows in the dimensional matrix, as seen later.

(4) The *number* of *dimensionless products* in the *complete set* is equal to the number of variables isolated *on the left-hand side* of the dimensional matrix. It follows from item (3) that the number of dimensionless products is equal to the *total number of variables* minus the *rank of the dimensional matrix*. In the preceding example, the number of dimensionless products in the complete set was equal to the number of variables (5) minus the rank of the dimensional matrix (3), i.e. 5 – 3 = 2 dimensionless products.

(5) When the last r columns of a dimensional matrix of order r do not lead to a non-zero determinant, the columns of the matrix must be rearranged so as to obtain a non-zero determinant.

Numerical example 1. An example will help understand the consequences of the above five characteristics on the general method for the systematic calculation of the complete set of dimensionless products. The dimensional matrix is as follows:

$$\begin{array}{c} \\ \mathrm{M} \\ \mathrm{L} \\ \mathrm{T} \end{array} \begin{array}{c} \begin{array}{ccccccc} V_1 & V_2 & V_3 & V_4 & V_5 & V_6 & V_7 \end{array} \\ \begin{bmatrix} 2 & 0 & 1 & 0 & -1 & -2 & 3 \\ 1 & 2 & 2 & 0 & 0 & 1 & -1 \\ 0 & 1 & 2 & 3 & 1 & -1 & 0 \end{bmatrix} \end{array}$$

The rank (r) of this matrix is 3 (numerical example in Section 2.7), so that the number of dimensionless products of the complete set is equal to $7 - 3 = 4$. However, the determinant of the $r = 3$ last columns is zero:

$$\begin{vmatrix} -1 & -2 & 3 \\ 0 & 1 & -1 \\ 1 & -1 & 0 \end{vmatrix} = 0$$

Calculating the complete set of dimensionless products thus requires a reorganization of the dimensional matrix by rearranging, for example, the columns as follows:

$$\begin{array}{c} \\ \mathrm{M} \\ \mathrm{L} \\ \mathrm{T} \end{array} \begin{array}{c} \begin{array}{ccccccc} V_1 & V_5 & V_7 & V_4 & V_2 & V_6 & V_3 \end{array} \\ \begin{bmatrix} 2 & -1 & 3 & 0 & 0 & -2 & 1 \\ 1 & 0 & -1 & 0 & 2 & 1 & 2 \\ 0 & 1 & 0 & 3 & 1 & -1 & 2 \end{bmatrix} \end{array}$$

The solution then follows from the general method described above:

$$\begin{bmatrix} x_2 \\ x_6 \\ x_3 \end{bmatrix} = -\begin{bmatrix} 0 & -2 & 1 \\ 2 & 1 & 2 \\ 1 & -1 & 2 \end{bmatrix}^{-1} \begin{bmatrix} 2 & -1 & 3 & 0 \\ 1 & 0 & -1 & 0 \\ 0 & 1 & 0 & 3 \end{bmatrix} \begin{bmatrix} x_1 \\ x_5 \\ x_7 \\ x_4 \end{bmatrix}$$

$$\begin{bmatrix} x_2 \\ x_6 \\ x_3 \end{bmatrix} = -\begin{bmatrix} 4 & 3 & -5 \\ -2 & -1 & 2 \\ -3 & -2 & 4 \end{bmatrix} \begin{bmatrix} 2 & -1 & 3 & 0 \\ 1 & 0 & -1 & 0 \\ 0 & 1 & 0 & 3 \end{bmatrix} \begin{bmatrix} x_1 \\ x_5 \\ x_7 \\ x_4 \end{bmatrix}$$

$$\begin{bmatrix} x_2 \\ x_6 \\ x_3 \end{bmatrix} = \begin{bmatrix} -11 & 9 & -9 & 15 \\ 5 & -4 & 5 & -6 \\ 8 & -7 & 7 & -12 \end{bmatrix} \begin{bmatrix} x_1 \\ x_5 \\ x_7 \\ x_4 \end{bmatrix}$$

$$\begin{array}{c} \\ \Pi_1 \\ \Pi_2 \\ \Pi_3 \\ \Pi_4 \end{array} \begin{array}{c} \begin{matrix} V_1 & V_5 & V_7 & V_4 & V_2 & V_6 & V_3 \end{matrix} \\ \begin{bmatrix} 1 & 0 & 0 & 0 & -11 & 5 & 8 \\ 0 & 1 & 0 & 0 & 9 & -4 & -7 \\ 0 & 0 & 1 & 0 & -9 & 5 & 7 \\ 0 & 0 & 0 & 1 & 15 & -6 & -12 \end{bmatrix} \end{array}$$

Numerical example 2. This example illustrates the case of a *singular* dimensional matrix, i.e a dimensional matrix whose *rank* is less than its number of rows. This matrix has already been considered in Section 2.7:

$$\begin{array}{c} \\ \mathrm{M} \\ \mathrm{L} \\ \mathrm{T} \end{array} \begin{array}{c} \begin{matrix} V_1 & V_2 & V_3 & V_4 \end{matrix} \\ \begin{bmatrix} 2 & 1 & 3 & 4 \\ -1 & 6 & -3 & 0 \\ 1 & 20 & -3 & 8 \end{bmatrix} \end{array}$$

It was shown (Section 2.7) that the *rank* of this matrix is $r = 2$, so that it is not possible to find a combination of three columns that could be inverted. The matrix is thus *singular* (Section 2.8).

The solution consists in making the *number of rows* equal to the *rank*. This is done by eliminating any one row of the dimensional matrix, since the matrix has only two independent rows (Section 2.7). The number of dimensionless products in the complete set is thus equal to $4 - 2 = 2$.

$$\begin{array}{c} \\ \mathrm{M} \\ \mathrm{L} \end{array} \begin{array}{c} \begin{matrix} V_1 & V_2 & V_3 & V_4 \end{matrix} \\ \begin{bmatrix} 2 & 1 & 3 & 4 \\ -1 & 6 & -3 & 0 \end{bmatrix} \end{array}$$

$$\begin{array}{c} \Pi_1 \\ \Pi_2 \end{array} \begin{bmatrix} 1 & 0 & -1/3 & -1/4 \\ 0 & 1 & 2 & -7/4 \end{bmatrix}$$

It is possible to eliminate fractional exponents by multiplying each row of the solution matrix by its lowest common denominator:

$$\begin{array}{c} \Pi_1 \\ \Pi_2 \end{array} \begin{bmatrix} 12 & 0 & -4 & -3 \\ 0 & 4 & 8 & -7 \end{bmatrix}$$

Identical results would have been obtained if any other row of the dimensional matrix had been eliminated instead of row 3, since each of the three rows is a linear combination of the other two. This can easily be checked as exercise.

There now remains to discuss how to choose the ordering of variables in a dimensional matrix. This order determines the complete set of dimensionless products obtained from the calculation. The rules are as follows:

(1) The *dependent variable* is, of necessity, in the first column of the dimensional matrix, since it must be present in only one Π (the first dimensionless product is thus called the *dependent dimensionless variable*). As a consequence, this first variable can be expressed as a function of all the others, which is the goal here. For example, in eq. 3.9, the *drag F* is in the first column of the dimensional matrix since it is clearly the *dependent variable*.

(2) The other variables are then arranged in decreasing order, based on their potential for experimental variation. Indeed, a maximum amount of information will result from experimentation if those variables with a wide range of experimental variability occur in a single Π.

(3) The initial ordering of variables must obviously be changed when the last r *columns* of the dimensional matrix have a zero determinant. However, one must then still comply as well as possible with the first two rules.

Two ecological applications, already discussed in Section 3.2, will now be treated using the systematic calculation of complete sets of dimensionless products.

Ecological application 3.3a

The first example reconsiders Ecological application 3.2d, devoted to the model of Kierstead & Slobodkin (1953). This model provided equations for the *critical size* of a growing phytoplankton patch and the *characteristic time* after which this critical size becomes operative.

The dimensional matrix of variables involved in the problem includes: *length x*, *time t*, *diffusion of cells D*, and *growth rate k*. The dependent variables being x and t, they are in the first two columns of the dimensional matrix:

$$\begin{array}{c} \\ \mathrm{L} \\ \mathrm{T} \end{array} \begin{array}{c} \begin{array}{cccc} x & t & D & k \end{array} \\ \begin{bmatrix} 1 & 0 & 2 & 0 \\ 0 & 1 & -1 & -1 \end{bmatrix} \end{array}$$

The rank of the dimensional matrix being 2, the number of dimensionless products is 4 – 2 = 2. These two products are found using the general method for calculating the complete set:

$$-\begin{bmatrix} 2 & 0 \\ -1 & -1 \end{bmatrix}^{-1} \begin{bmatrix} 1 & 0 \\ 0 & 0 \end{bmatrix} = \begin{bmatrix} -1/2 & 0 \\ 1/2 & 1 \end{bmatrix}$$

$$\begin{matrix} & x & t & D & k \\ \Pi_1 & 1 & 0 & -1/2 & 1/2 \\ \Pi_2 & 0 & 1 & 0 & 1 \end{matrix} = \begin{matrix} x & t & D & k \\ 2 & 0 & -1 & 1 \\ 0 & 1 & 0 & 1 \end{matrix}$$

$$\Pi_1 = kx^2/D \text{ and } \Pi_2 = tk$$

These two dimensionless products describe, as in Ecological application 3.2d, the *critical length* x and the *characteristic time* t as:

$$x \propto \sqrt{D/k} \text{ and } t \propto 1/k \propto x^2/D$$

Ecological application 3.3b

A second example provides an easy solution to the problem which confronted Paloheimo & Dickie (1965), concerning the synthesis of data on the growth of fish with respect to food intake. The question was discussed at length in Ecological application 3.2e, which led to three scale factors, for *food ration, mass,* and *time.* These scale factors were used by the authors to compare diversified data from the ecological literature.

The solution is found directly, here, using the dimensional matrix of the six variables involved in the problem: *time t, mass W,* food *ration R,* rate of *oxygen consumption T,* rate of *metabolic expenditure* α, and coefficient b. The variables to be isolated being t, W, and R, they are in the first three columns of the dimensional matrix:

$$\begin{matrix} & t & W & R & T & \alpha & b \\ \text{M} & 0 & 1 & 1 & 1 & (1-\gamma) & -1 \\ \text{T} & 1 & 0 & -1 & -1 & -1 & 1 \end{matrix}$$

Since the *rank* of the dimensional matrix is $r = 2$, the number of dimensionless products is $6 - 2 = 4$. The four products are calculated by the method of the complete set:

$$-\begin{bmatrix} (1-\gamma) & -1 \\ -1 & 1 \end{bmatrix}^{-1} \begin{bmatrix} 0 & 1 & 1 & 1 \\ 1 & 0 & -1 & -1 \end{bmatrix} = \begin{bmatrix} 1/\gamma & 1/\gamma & 0 & 0 \\ [(1/\gamma)-1] & 1/\gamma & 1 & 1 \end{bmatrix}$$

$$\begin{matrix} & t & W & R & T & \alpha & b \\ \Pi_1 & 1 & 0 & 0 & 0 & 1/\gamma & (1/\gamma)-1 \\ \Pi_2 & 0 & 1 & 0 & 0 & 1/\gamma & 1/\gamma \\ \Pi_3 & 0 & 0 & 1 & 0 & 0 & 1 \\ \Pi_4 & 0 & 0 & 0 & 1 & 0 & 1 \end{matrix}$$

$$\Pi_1 = t\alpha^{1/\gamma}b^{(1/\gamma-1)} = [(\alpha b)^{1/\gamma}/b]t$$

$$\Pi_2 = W\alpha^{1/\gamma}b^{1/\gamma} = (\alpha b)^{1/\gamma}W$$

$$\Pi_3 = Rb = bR$$

$$\Pi_4 = Tb = bT$$

The first three dimensionless products define the three scale factors already found in Ecological application 3.2e, i.e. Π_1 for *time*, Π_2 for *mass*, and Π_3 for *ration*. Π_4 defines a scale factor for *oxygen consumption*.

Direct calculations of complete sets of dimensionless products thus led to the same results as obtained before, but operations here were more straightforward than in Section 3.2.

It should not be necessary, after these examples, to dwell on the advantage of systematically calculating the complete set of dimensionless products. In addition to providing a rapid and elegant solution to problems of dimensional analysis, the above matrix method sets researchers on the right track when tackling a problem to be investigated using the dimensional tool. The success of a dimensional study depends on: (1) adequate knowledge of the problem under study, so that *all* the pertinent variables are considered; and (2) clear ideas about which variables are functions of the others. It should be noted, as explained above, that the systematic calculation of the complete set of dimensionless products does not require prior knowledge of the fundamental equations. These, however, may be necessary to derive the dimensions of some complex variables. Dimensional analysis may be a powerful tool, provided that the ecological bases of the problem under consideration are thoroughly understood and that the objectives of the research are clearly stated.

3.4 Scale factors and models

Physical model

Given the increased awareness in society for environmental problems, major engineering projects cannot be undertaken, in most countries, before their environmental impacts have been assessed. As a consequence, an increasing number of ecologists now work within multidisciplinary teams of consultants. At the planning stage, one of the most powerful tools available to engineers, although very costly, is the small-scale *model*. *Tests* performed with such models help choose the most appropriate engineering solution. Actually, ecologists may encounter two types of model, i.e. mathematical and physical. *Mathematical models* have already been discussed in the Foreword. *Physical models* are small-scale replica of the natural environment, to which changes can be made that reproduce those planned for the real situation. Tests with physical models are generally more costly to perform than mathematical simulations, so that the latter are becoming increasingly more popular than the former. *Physical models* are often based on dimensional analysis, so that it is this type of model which is considered here. It should be noted that physical models may originate from the empirical approach of engineers, which is distinct from the dimensional approach.

In order to communicate with engineers conducting tests on small-scale models, ecologists must have some basic understanding of the principles governing model testing. In some cases, ecologists may even play a role in the study, when it is possible to integrate in the model variables of ecological significance (e.g. in a model of a harbour or estuary, such variables as salinity, sediment transport, etc.). Since small-scale models are based in part on dimensional analysis, their basic theory is thus relatively easy to understand. The actual testing, however, requires the specific knowledge and experience of model engineers. In addition to their possible involvement in applications of modelling to environmental impact studies, ecologists may at times use small-scale models to resolve problems of their own (e.g. studying the interactions between benthic organisms and sediment in a hydraulic flume). These various aspects are introduced here very briefly.

Prototype

Geometric similarity

In the vocabulary of physical modelling, the full-size system is called *prototype* and the small-size replica is called *model.* A model may be geometrically similar to the prototype, or it may be distorted. In the case of *geometric similarity,* all parts of the model have the same shapes as the corresponding parts of the prototype. In certain cases, geometric similarity would lead to errors, so that one must use a *distorted model.* In such models, one or several scales may be distorted. For example, a geometrically similar model of an estuary could result in some excessively small water depths. With such depths, the flow in the model could become subject to surface tension, which would clearly be incorrect with respect to the real flow. In the model, the depth must therefore be relatively greater than in nature, hence a distorted model.

The physical example of the *drag on smooth spheres*, already discussed in Sections 3.2 and 3.3, is now used to introduce the basic principles of scaling and small-scale modelling. Equation 3.7 describes the *drag* (F) acting on a smooth sphere of *diameter D,* immersed in a stream with *velocity V* of a fluid with density ρ and dynamic viscosity η:

$$F = \rho V^2 D^2 f(Re) \quad \textbf{(3.7)}$$

$$F = \rho V^2 D^2 f\left(\frac{VD\rho}{\eta}\right)$$

In order to experimentally determine the drag, under convenient laboratory conditions (e.g. wind tunnel or hydraulic flume), it may be appropriate to use a geometrically similar model of the sphere. Quantities pertaining to the *model* are assigned *prime indices*. If the curve of the drag coefficient for smooth spheres was not known (Fig. 3.1), the calculation of F would require that the value of the *unknown function f* be the same for both the model and the prototype. In order to do so, the test engineer should make sure that the *Reynolds numbers* for the two systems are equal:

$$Re = Re'$$

$$\frac{VD\rho}{\eta} = \frac{V'D'\rho'}{\eta'} \quad \textbf{(3.15)}$$

Scale factor

A *scale factor* is defined as the ratio of the size of the model to that of the prototype. Scale factors are therefore *dimensionless numbers*. The scale factors (K) corresponding to eq. 3.15 are:

$$K_V = V'/V \qquad K_D = D'/D \qquad K_\rho = \rho'/\rho \qquad K_\eta = \eta'/\eta$$

These scales are used to rewrite eq. 3.15 as:

$$K_V K_D K_\rho = K_\eta \tag{3.16}$$

Because $Re = Re'$, the *scale factor of the unknown function f* is equal to unity:

$$K_{f(Re)} = 1 \tag{3.17}$$

The ratio between the drag measured for the model and the real drag on the prototype is computed by combining eq. 3.7 with the above scale factors:

$$K_F = K_\rho K_V^2 K_D^2 K_{f(Re)}$$

Because of eq. 3.17, it follows that:

$$K_F = K_\rho K_V^2 K_D^2 \tag{3.18}$$

Equation 3.16 is used to find the value of K_F:

$$K_V K_D K_\rho = K_\eta \tag{3.16}$$

is squared

$$K_V^2 K_D^2 K_\rho^2 = K_\eta^2$$

from which

$$K_V^2 K_D^2 K_\rho = K_\eta^2 / K_\rho$$

and, given eq. 3.18

$$K_F = K_\eta^2 / K_\rho \tag{3.19}$$

Equation 3.19 leads to the following practical conclusions, for determining the drag on smooth spheres in the laboratory:

(1) If the model is tested using the *same fluid* as for the prototype, the *drag* measured during the test is the same as for the prototype. This follows from the fact that, if $K_\eta = 1$ and $K_\rho = 1$ (same fluid), K_F is equal to unity (eq. 3.19), hence $F' = F$.

(2) If testing is conducted using the *same fluid* as for the prototype, conservation of Re requires that the *velocity* for the model be greater than for the prototype (i.e. the model is smaller than the prototype). This follows from the fact that, when $K_\eta = 1$ and $K_\rho = 1$ (same fluid), $K_V K_D = 1$ (eq. 3.16); consequently any decrease in K_D must be compensated by a proportional increase in K_V.

(3) When it is more convenient to use *different fluids*, testing may be conducted while conserving *Re*. It has already been shown (Section 3.2) that, for example, going from a large-size prototype, in air, to a model 6 times smaller, in water, allows a reduction of the flow speed during the test by a factor of 3. The drag measured for the model would not, however, be necessarily the same as that of the prototype, since that force varies as a function of the ratio between the squares of the dynamic viscosities (K_η^2) and the densities (K_ρ) of the two fluids (eq. 3.19). Knowing this ratio (K_F), it is easy to derive the drag for the model (F) from that measured during the test (F') since:

$$F = F'/K_F$$

In more complex cases, it is sometimes necessary to simultaneously conserve two or more dimensionless products, which are incompatible. In such a situation, where a choice must be made between contradictory constraints, it rests on the test engineer to justify discrepancies in similarity and to apply theoretical corrections to compensate for them. Hence modelling, although derived from scientific concepts, becomes an art based on the experience of the researcher.

Similarity

A *general concept of similarity* follows from the previous discussion. In a Cartesian space, the *model* and the *prototype* are described by coordinates (x' y' z') and (x y z), respectively. Correspondence between the two systems is established by means of *scale factors* (K), which define *homologous* times as well as *homologous* points in the three dimensions of space:

$$t' = K_t t \qquad x' = K_x x \qquad y' = K_y y \qquad z' = K_z z$$

The *time scale factor* (K_t) would be used, for example, in the case of a flow where Δ'_t and Δ_t are the time intervals during which two homologous particles go through homologous parts of their respective trajectories. It would then be defined as

$$K_t = \Delta'_t/\Delta_t$$

Geometric similarity is defined as: $K_x = K_y = K_z = K_L$. In *distorted models*, a single length scale is usually modified, so that $K_x = K_y \neq K_z$. The ratio K_z/K_x is the *distortion factor*. It would be possible, using this same approach, to define characteristics of *kinematic similarity*, for similar motions, and of *dynamic similarity*, for systems subjected to homologous forces.

There are several types of similarity in addition to the geometric, dynamic and kinematic similarities. These include the *hydrodynamic*, *transport*, and *thermal similarities*. Readers interested in applications of dimensional analysis to the theory of biological similarity may refer to the review of Günther (1975), where the various types of physical similarity are briefly described.

Chapter

4 Multidimensional quantitative data

4.0 Multidimensional statistics

Basic statistics are now part of the curriculum of most ecologists. However, statistical techniques based on such simple distributions as the unidimensional normal distribution are not really appropriate for analysing complex ecological data sets. Nevertheless, researchers sometimes perform series of simple analyses on the various descriptors in the data set, expecting to obtain results that are pertinent to the problem under study. This type of approach is incorrect because it does not take into account the covariance among descriptors; see also Box 1.3, where the statistical problem created by multiple testing is explained. In addition, such an approach only extracts minimum information from data which have often been collected at great cost and it usually generates a mass of results from which it is difficult to draw much sense. Finally, in studies involving species assemblages, it is usually more interesting to describe the variability of the structure of the assemblage as a whole (i.e. *mensurative* variation observed through space or time, or *manipulative* variation resulting from experimental manipulation; Hurlbert, 1984) than to look at each species independently.

Fortunately, methods derived from *multidimensional statistics*, which are used throughout this book, are designed for analysing complex data sets. These methods take into account the co-varying nature of ecological data and they can evidence the structures that underlie the data set. The present chapter discusses the basic theory and characteristics of multidimensional data analysis. Mathematics are kept to a minimum, so that readers can easily reach a high level of understanding. In addition, many approaches of practical interest are discussed, including several types of linear correlation, with their statistical tests. It must be noted that this chapter is limited to linear statistics.

A number of excellent textbooks deal with detailed aspects of multidimensional statistics. For example, formal presentations of the subject are found in Muirhead (1982) and Anderson (1984). Researchers less interested in mathematical theory may

Table 4.1 Numerical example of two species observed at four sampling sites. Figure 4.1 shows that each row of the data matrix may be construed as a vector, as defined in Section 2.4.

Sampling sites (objects)	Species (descriptors) 1	2	($p = 2$)
1	5	1	
2	3	2	
3	8	3	
4	6	4	
($n = 4$)			

refer to Cooley & Lohnes (1971), Tatsuoka (1971), Press (1972), Graybill (1983), or Morrison (1990). These books describe a number of useful methods, among which the multidimensional analysis of variance. However, none of these books specifically deals with ecological data.

Multidimensional
Multivariate

Several authors use the term *multivariate* as abbreviation for *multidimensional variate* (the latter term meaning *random variable*; Section 1.0). As an adjective, *multivariate* is interchangeable with *multidimensional.*

4.1 Multidimensional variables and dispersion matrix

As stated in Section 1.0, the present textbook deals with the analysis of *random variables*. Ecological data matrices have n rows and p columns (Section 2.1). Each row is a *vector* (Section 2.4) which is, statistically speaking, one realization of a p-dimensional random variable. In other words, for example, when p species are observed at n sampling sites, the species are the p dimensions of a random variable "species" and each site is one realization of this p-dimensional random variable.

To illustrate this concept, four sampling units with two species (Table 4.1) are plotted in a two-dimensional Euclidean space (Fig. 4.1). Vector "site 1" is the doublet (5,1). It is plotted in the same two-dimensional space as the three other vectors "site i". Each row of the data matrix is a two-dimensional vector, which is one realization of the (bivariate) random variable "species". The random variable "species" is said to be two-dimensional because the sampling units (objects) contain two species (descriptors), the two dimensions being species 1 and 2, respectively.

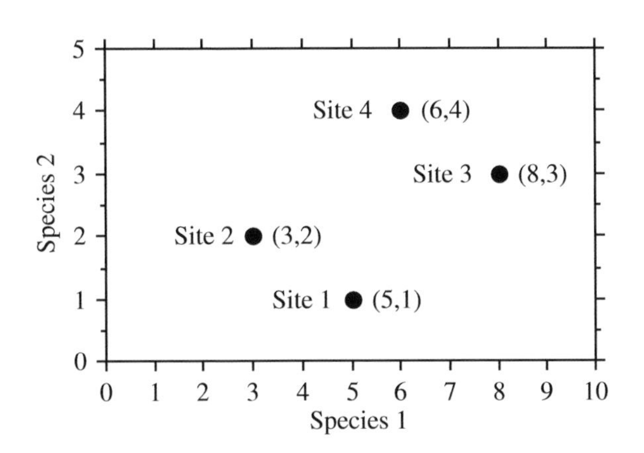

Figure 4.1 Four realizations (sampling sites from Table 4.1) of the two-dimensional random variable "species" are plotted in a two-dimensional Euclidean space.

As the number of descriptors (e.g. species) increases, the number of dimensions of the random variable "species" similarly increases, so that more axes are necessary to construct the space in which the objects are plotted. Thus, the p descriptors make up a p-dimensional random variable and the n vectors of observations are as many realizations of the p-dimensional vector "descriptors". The present chapter does not deal with *samples* of observations, which result from field or laboratory work (for a brief discussion on sampling, see Section 1.1), but it focuses instead on *populations*, which are investigated by means of the samples.

Before approaching the multidimensional normal distribution, it is necessary to define a p-dimensional random variable "descriptors":

$$\mathbf{Y} = [\mathbf{y}_1, \mathbf{y}_2, \ldots \mathbf{y}_j, \ldots \mathbf{y}_p] \qquad \textbf{(4.1)}$$

Each element $\mathbf{y}_j$ of multidimensional variable $\mathbf{Y}$ is a unidimensional random variable. Every descriptor $\mathbf{y}_j$ is observed in each of the n vectors "object", each sampling unit i providing one realization of the p-dimensional random variable (Fig. 4.2).

In ecology, the structure of *dependence* among descriptors is, in many instances, the matter being investigated. Researchers who study multidimensional data sets using univariate statistics assume that the p unidimensional $\mathbf{y}_j$ variables in $\mathbf{Y}$ are *independent* of one another (this refers to the third meaning of *independence* in Box 1.1). This is the reason why univariate statistical methods are inappropriate with most ecological data and why methods that take into account the *dependence* among descriptors must be used when analysing sets of multidimensional data. Only these methods will generate proper results when there is dependence among descriptors; it is never acceptable to replace a multidimensional analysis by a series of unidimensional treatments.

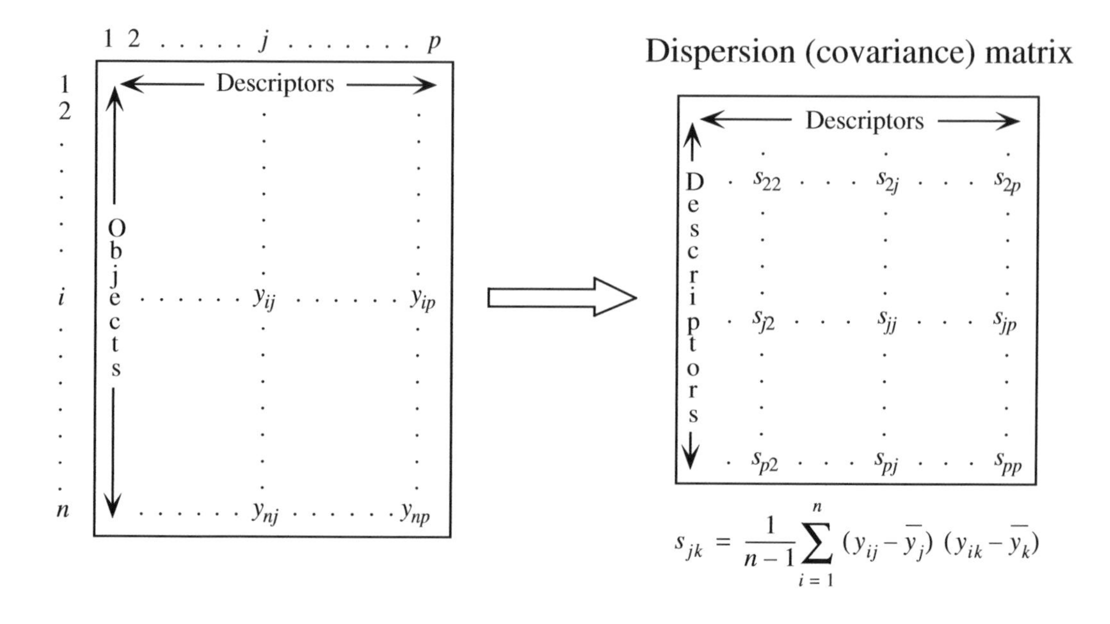

Figure 4.2 Structure of ecological data. Given their nature, ecological descriptors are *dependent* of one another. In statistics, the objects are often assumed to be *independent* observations, but this is generally not the case in ecology (Section 1.1)

Several multivariate statistical models require, however, "that successive sample observation vectors from the multidimensional population have been drawn in such a way that they can be construed as realizations of independent random vectors" (Morrison, 1990, p. 80). It has been shown in Section 1.1 that this assumption of independence among observations is most often not realistic in ecology. This major discrepancy between statistical theory and ecological data does not really matter as long as statistical models are used for descriptive purposes only, as it is generally the case in the present book. However, as explained in Section 1.1, this problem prevents utilisation of the usual tests of significance and thus statistical inference (i.e. generalization of statistics derived from the samples to parameters of populations), unless corrected tests are used.

To sum up: (1) the p descriptors in ecological data matrices are the p dimensions of a random variable "descriptors"; (2) the p descriptors are *not independent* of one another; methods of multidimensional analysis are designed to evidence the structure of dependence among descriptors; (3) each of the n sampling units is a realization of the p-dimensional vector "descriptors"; (4) most methods assume that the n sampling units are realizations of *independent* random vectors. The latter condition is generally not met in ecology, with consequences that were mentioned in the previous paragraph and discussed in Section 1.1. For the various meanings of the term *independence* in statistics, see Box 1.1.

The dependence among quantitative variables $\mathbf{y}_j$ brings up the concept of *covariance*. Covariance is the extension, to two descriptors, of the concept of *variance*. Variance is a measure of the *dispersion* of a random variable $\mathbf{y}_j$ around its mean; it is denoted σ_j^2. Covariance measures the *joint dispersion* of two random variables $\mathbf{y}_j$ and $\mathbf{y}_k$ around their means; it is denoted σ_{jk}. The *dispersion matrix* of $\mathbf{Y}$, called matrix $\mathbf{\Sigma}$ (sigma), contains the variances and covariances of the p descriptors (Fig. 4.2):

Dispersion matrix

$$\mathbf{\Sigma} = \begin{bmatrix} \sigma_{11} & \sigma_{12} & \dots & \sigma_{1p} \\ \sigma_{21} & \sigma_{22} & \dots & \sigma_{2p} \\ \cdot & \cdot & \dots & \cdot \\ \cdot & \cdot & \dots & \cdot \\ \cdot & \cdot & \dots & \cdot \\ \sigma_{p1} & \sigma_{p2} & \dots & \sigma_{pp} \end{bmatrix} \tag{4.2}$$

Matrix $\mathbf{\Sigma}$ is an *association matrix* [descriptors × descriptors] (Section 2.2). The elements σ_{jk} of matrix $\mathbf{\Sigma}$ are the covariances between all pairs of the p random variables. The matrix is symmetric because the covariance of $\mathbf{y}_j$ and $\mathbf{y}_k$ is identical to that of $\mathbf{y}_k$ and $\mathbf{y}_j$. A diagonal element of $\mathbf{\Sigma}$ is the covariance of a descriptor $\mathbf{y}_j$ with itself, which is the variance of $\mathbf{y}_j$, so that $\sigma_{jj} = \sigma_j^2$

Variance

The *estimate* of the *variance* of $\mathbf{y}_j$, denoted s_j^2, is computed on the *centred variable* $(y_{ij} - \bar{y}_j)$. Variable $\mathbf{y}_j$ is centred by subtracting the mean $\bar{y}_j$ from each of the n observations y_{ij}. As a result, the mean of the centred variable is zero. The variance s_j^2 is computed using the well-known formula:

$$s_j^2 = \frac{1}{n-1} \sum_{i=1}^{n} (y_{ij} - \bar{y}_j)^2 \tag{4.3}$$

where the sum of *squares* of the *centred data,* for descriptor j, is divided by the number of objects minus one (n –1). The summation is over the n observations of descriptor j.

Covariance

In the same way, the estimate (s_{jk}) of the *covariance* (σ_{jk}) of $\mathbf{y}_j$ and $\mathbf{y}_k$ is computed on the centred variables $(y_{ij} - \bar{y}_j)$ and $(y_{ik} - \bar{y}_k)$, using the formula of a "bivariate variance". The *covariance* s_{jk} is calculated as:

$$s_{jk} = \frac{1}{n-1} \sum_{i=1}^{n} (y_{ij} - \bar{y}_j)\,(y_{ik} - \bar{y}_k) \tag{4.4}$$

Standard deviation

When $k = j$, eq. 4.4 is identical to eq. 4.3. The positive square root of the variance is called the *standard deviation* (σ_j). Its estimate s_j is thus:

$$s_j = \sqrt{s_j^2} \tag{4.5}$$

Table 4.2 Symbols used to identify (population) parameters and (sample) statistics.

	Parameter		Statistic	
	Matrix or vector	Elements	Matrix or vector	Elements
Covariance	$\mathbf{\Sigma}$ (sigma)	σ_{jk} (sigma)	$\mathbf{S}$	s_{jk}
Correlation	$\mathbf{P}$ (rho)	ρ_{jk} (rho)	$\mathbf{R}$	r_{jk}
Mean	$\boldsymbol{\mu}$ (mu)	μ_j (mu)	$\bar{\mathbf{y}}$	$\bar{y}_j$

The symbols for matrix $\mathbf{\Sigma}$ and summation Σ should not be confused.

Contrary to the variance, which is always positive, the covariance may take any positive or negative value. In order to understand the meaning of the covariance, let us imagine that the object points are plotted in a scatter diagram, where the axes are descriptors $\mathbf{y}_j$ and $\mathbf{y}_k$. The data are then centred by drawing new axes, whose origin is at the centroid $(\bar{y}_j, \bar{y}_k)$ of the cloud of points (see Section 1.5, linear transformation: translation). A positive covariance means that most of the points are in quadrants I and III of the centred plot, where the centred values $(y_{ij} - \bar{y}_j)$ and $(y_{ik} - \bar{y}_k)$ have the same signs. This corresponds to a positive relationship between the two descriptors. The converse is true for a negative covariance, for which most of the points are in quadrants II and IV of the centred plot. In the case of a null or small covariance, the points are equally distributed among the four quadrants of the centred plot.

Parameter
Statistic

Greek and roman letters are both used here. The properties of a *population* (called *parameters*) are denoted by *greek* letters. Their *estimates* (called *statistics*), which are computed from *samples*, are symbolized by the corresponding *roman* letters. Along with these conventions are also those pertaining to matrix notation (Section 2.1). This is summarized in Table 4.2.

The dispersion matrix* $\mathbf{S}$ can be computed directly, by multiplying the *matrix of centred data* $[y - \bar{y}]$ with its transpose $[y - \bar{y}]'$:

$$\mathbf{S} = \frac{1}{n-1}[y - \bar{y}]'[y - \bar{y}] \tag{4.6}$$

* Some authors, including Lefebvre (1980), call $[y - \bar{y}]'[y - \bar{y}]$ a *dispersion matrix* and $\mathbf{S}$ a *covariance matrix*. For these authors, a covariance matrix is then a dispersion matrix divided by $(n - 1)$.

$$\mathbf{S} = \frac{1}{n-1}\begin{bmatrix} (y_{11}-\bar{y}_1) & (y_{21}-\bar{y}_1) & \dots & (y_{n1}-\bar{y}_1) \\ (y_{12}-\bar{y}_2) & (y_{22}-\bar{y}_2) & \dots & (y_{n2}-\bar{y}_2) \\ \cdot & \cdot & \dots & \cdot \\ \cdot & \cdot & \dots & \cdot \\ \cdot & \cdot & \dots & \cdot \\ (y_{1p}-\bar{y}_p) & (y_{2p}-\bar{y}_p) & \dots & (y_{np}-\bar{y}_p) \end{bmatrix} \begin{bmatrix} (y_{11}-\bar{y}_1) & (y_{12}-\bar{y}_2) & \dots & (y_{1p}-\bar{y}_p) \\ (y_{21}-\bar{y}_1) & (y_{22}-\bar{y}_2) & \dots & (y_{2p}-\bar{y}_p) \\ \cdot & \cdot & \dots & \cdot \\ \cdot & \cdot & \dots & \cdot \\ \cdot & \cdot & \dots & \cdot \\ (y_{n1}-\bar{y}_1) & (y_{n2}-\bar{y}_2) & \dots & (y_{np}-\bar{y}_p) \end{bmatrix}$$

$$\mathbf{S} = \frac{1}{n-1}\begin{bmatrix} \sum_{i=1}^{n}(y_{i1}-\bar{y}_1)^2 & \sum_{i=1}^{n}(y_{i1}-\bar{y}_1)(y_{i2}-\bar{y}_2) & \dots & \sum_{i=1}^{n}(y_{i1}-\bar{y}_1)(y_{ip}-\bar{y}_p) \\ \sum_{i=1}^{n}(y_{i2}-\bar{y}_2)(y_{i1}-\bar{y}_1) & \sum_{i=1}^{n}(y_{i2}-\bar{y}_2)^2 & \dots & \sum_{i=1}^{n}(y_{i2}-\bar{y}_2)(y_{ip}-\bar{y}_p) \\ \cdot & \cdot & \dots & \cdot \\ \cdot & \cdot & \dots & \cdot \\ \cdot & \cdot & \dots & \cdot \\ \sum_{i=1}^{n}(y_{ip}-\bar{y}_p)(y_{i1}-\bar{y}_1) & \sum_{i=1}^{n}(y_{ip}-\bar{y}_p)(y_{i2}-\bar{y}_2) & \dots & \sum_{i=1}^{n}(y_{ip}-\bar{y}_p)^2 \end{bmatrix}$$

This elegant and rapid procedure emphasizes once again the advantage of matrix algebra in numerical ecology, where the data sets are generally large.

Numerical example. Four species ($p = 4$) were observed at five stations ($n = 5$). The estimated population parameters, for the species, are the means ($\bar{y}_j$), the variances (s_j^2), and the covariances (s_{jk}). The original and centred data are shown in Table 4.3. Because $s_{jk} = s_{kj}$, the dispersion matrix is symmetric. The mean of each *centred variable* is zero.

In this numerical example, the covariance between species 2 and the other three species is zero. This does not necessarily mean that species 2 is independent of the other three, but simply that the joint *linear* dispersion of species 2 with any one of the other three is zero. This example will be considered again in Section 4.2.

The square root of the determinant of the dispersion matrix $|\mathbf{S}|^{1/2}$ is known as the *generalized variance.* It is also equal to the square root of the product of the eigenvalues of **S**.

Any dispersion matrix **S** is *positive semidefinite* (Table 2.2). Indeed, the quadratic form of **S** ($p \times p$) with any real and non-null vector **t** (of size p) is:

$$\mathbf{t'St}$$

Table 4.3 Numerical example. Calculation of centred data and covariances.

Sites	Original data	Centred data
1 2 3 4 5	$\mathbf{Y} = \begin{bmatrix} 1 & 5 & 2 & 6 \\ 2 & 2 & 1 & 8 \\ 3 & 1 & 3 & 4 \\ 4 & 2 & 5 & 0 \\ 5 & 5 & 4 & 2 \end{bmatrix}$	$[y - \bar{y}] = \begin{bmatrix} -2 & 2 & -1 & 2 \\ -1 & -1 & -2 & 4 \\ 0 & -2 & 0 & 0 \\ 1 & -1 & 2 & -4 \\ 2 & 2 & 1 & -2 \end{bmatrix}$
Means	$\bar{\mathbf{y}}' = \begin{bmatrix} 3 & 3 & 3 & 4 \end{bmatrix}$	$\overline{[y - \bar{y}]}' = \begin{bmatrix} 0 & 0 & 0 & 0 \end{bmatrix}$
$n - 1 = 4$	$\mathbf{S} = \frac{1}{n-1} [y - \bar{y}]' [y - \bar{y}] =$	$\begin{bmatrix} 2.5 & 0 & 2 & -4 \\ 0 & 3.5 & 0 & 0 \\ 2 & 0 & 2.5 & -5 \\ -4 & 0 & -5 & 10 \end{bmatrix}$

This expression may be expanded using eq. 4.6:

$$\mathbf{t'St} = \mathbf{t'} \frac{1}{n-1} [y - \bar{y}]' [y - \bar{y}] \, \mathbf{t}$$

$$\mathbf{t'St} = \frac{1}{n-1} [(y - \bar{y}) \, t]' [(y - \bar{y}) \, t] = \text{a scalar}$$

This scalar is the variance of the variable resulting from the product **Yt**. Since variance can only be positive or null, it follows that:

$$\mathbf{t'St} \geq 0$$

so that **S** is positive semidefinite. Therefore, *all the eigenvalues of* **S** *are positive or null.* This property of dispersion matrices is fundamental in numerical ecology, since it allows one to partition the variance among real *principal axes* (Sections 4.4 and 9.1).

Ideally, matrix **S** (of order p) should be estimated from a number of observations n larger than the number of descriptors p. When $n \leq p$, the rank of matrix **S** is $n - 1$ and, consequently, only $n - 1$ of its rows or columns are independent, so that $p - (n - 1)$ null eigenvalues are produced. The only practical consequence of $n \leq p$ is thus the presence of null eigenvalues in the principal component solution (Section 9.1). The first few eigenvalues of **S**, which are generally those of interest to ecologists, are not affected.

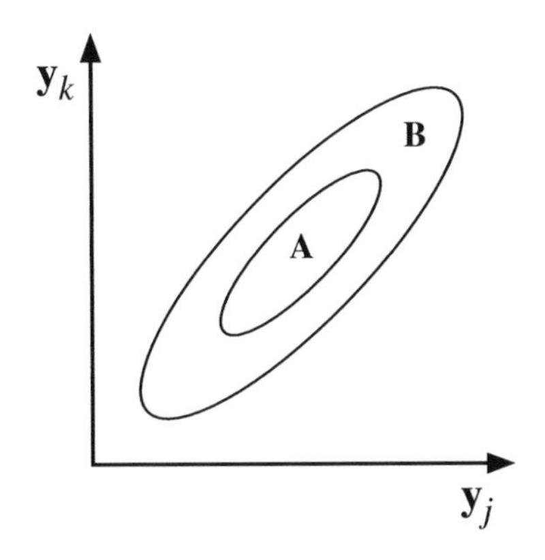

Figure 4.3 Several observations (objects), with descriptors $\mathbf{y}_j$ and $\mathbf{y}_k$, were made under two different sets of conditions (A and B). The two ellipses delineate clouds of point-objects corresponding to A and B, respectively. The covariance of $\mathbf{y}_j$ and $\mathbf{y}_k$ is twice as large for B as it is for A (larger ellipse), but the correlation between the two descriptors is the same in these two cases (i.e. the ellipses have the same shape).

4.2 Correlation matrix

The previous section has shown that the covariance provides information on the orientation of the cloud of data points in the space defined by the descriptors. That statistic, however, does not provide any information on the intensity of the relationship between variables $\mathbf{y}_j$ and $\mathbf{y}_k$. Indeed, the covariance may increase or decrease without changing the relationship between $\mathbf{y}_j$ and $\mathbf{y}_k$. For example, in Fig. 4.3, the two clouds of points correspond to different covariances (factor two in size, and thus in covariance), but the relationship between variables is identical (same shape). Since the covariance depends on the dispersion of points around the mean of each variable (i.e. their variances), determining the intensity of the relationship between variables requires to control for the variances.

The *covariance* measures the joint *dispersion* of two random variables around their means. The *correlation* is defined as a measure of the *dependence* between two random variables $\mathbf{y}_j$ and $\mathbf{y}_k$. As already explained in Section 1.5, it often happens that matrices of ecological data contain descriptors with no common scale, e.g. when some species are more abundant than others by orders of magnitude, or when the descriptors have different physical dimensions (Chapter 3). Calculating covariances on such variables obviously does not make sense, except if the descriptors are first reduced to a common scale. The procedure consists in centring all descriptors on a zero mean and reducing them to unit standard deviation (eq. 1.12). By using *standardized descriptors*, it is possible to calculate meaningful covariances, because the new variables have the same scale (i.e. unit standard deviation) and are dimensionless (see Chapter 3).

Linear correlation

The covariance of two standardized descriptors is called the linear correlation (Pearson *r*). This statistic has been proposed by the statistician Karl Pearson, so that it is named after him. Given two standardized descriptors (eq. 1.12)

$$z_{ij} = \frac{y_{ij} - \bar{y}_j}{s_j} \qquad \text{and} \qquad z_{ik} = \frac{y_{ik} - \bar{y}_k}{s_k}$$

calculating their covariance (eq. 4.4) gives

$$s(z_j,z_k) = \frac{1}{n-1}\sum_{i=1}^{n}(z_{ij} - 0)\,(z_{ik} - 0) \qquad \text{because} \qquad \bar{z}_j = \bar{z}_k = 0$$

$$s(z_j,z_k) = \frac{1}{n-1}\sum_{i=1}^{n}\left(\frac{y_{ij} - \bar{y}_j}{s_j}\right)\left(\frac{y_{ik} - \bar{y}_k}{s_k}\right)$$

$$s(z_j,z_k) = \left(\frac{1}{s_j s_k}\right)\frac{1}{n-1}\sum_{i=1}^{n}(y_{ij} - \bar{y}_j)\,(y_{ik} - \bar{y}_k)$$

$s(z_j,z_k) = \left(\frac{1}{s_j s_k}\right)s_{jk} = r_{jk}$, the *coefficient of linear correlation* between $\mathbf{y}_j$ and $\mathbf{y}_k$.

The developed formula is:

$$r_{jk} = \frac{s_{jk}}{s_j s_k} = \frac{\sum_{i=1}^{n}(y_{ij} - \bar{y}_j)\,(y_{ik} - \bar{y}_k)}{\sqrt{\sum_{i=1}^{n}(y_{ij} - \bar{y}_j)^2\sum_{i=1}^{n}(y_{ik} - \bar{y}_k)^2}} \qquad \textbf{(4.7)}$$

Correlation matrix

As in the case of dispersion (Section 4.1), it is possible to construct the *correlation matrix* of **Y**, i.e. the **P** (rho) matrix, whose elements are the coefficients of linear correlation ρ_{jk}:

$$\mathbf{P} = \begin{bmatrix} 1 & \rho_{12} & \ldots & \rho_{1p} \\ \rho_{21} & 1 & \ldots & \rho_{2p} \\ \cdot & \cdot & \ldots & \cdot \\ \cdot & \cdot & \ldots & \cdot \\ \cdot & \cdot & \ldots & \cdot \\ \rho_{p1} & \rho_{p2} & \ldots & 1 \end{bmatrix} \qquad \textbf{(4.8)}$$

The *correlation matrix is the dispersion matrix of the standardized variables.* This concept will play a fundamental role in principal component analysis (Section 9.1). It should be noted that the diagonal elements of **P** are all equal to 1. This is because the comparison of any descriptor with itself is a case of complete dependence, which leads to a correlation $\rho_j = 1$. When $\mathbf{y}_j$ and $\mathbf{y}_k$ are independent of each other, $\rho_j = 0$. However, a correlation equal to zero does not necessarily imply that $\mathbf{y}_j$ and $\mathbf{y}_k$ are independent of each other, as shown by the following numerical example. A correlation $\rho_{jk} = -1$ is indicative of a complete, but inverse dependence of the two variables.

Numerical example. Using the values in Table 4.3, matrix **R** can easily be computed. Each element r_{jk} combines, according to eq. 4.7, the covariance s_{jk} with variances s_j and s_k:

$$\mathbf{R} = \begin{bmatrix} 1 & 0 & 0.8 & -0.8 \\ 0 & 1 & 0 & 0 \\ 0.8 & 0 & 1 & -1 \\ -0.8 & 0 & -1 & 1 \end{bmatrix}$$

Matrix **R** is symmetric, as was matrix **S**. The correlation $r = -1$ between species 3 and 4 means that these species are fully, but inversely, dependent (Fig. 4.4a). Correlations $r = 0.8$ and -0.8 are interpreted as indications of strong dependence between species 1 and 3 (direct) and species 1 and 4 (inverse), respectively. The *zero* correlation between species 2 and the other three species must be interpreted with caution. Figure 4.4d clearly shows that species 1 and 2 are completely *dependent* of each other since they are related by equation $\mathbf{y}_2 = 1 + (3 - \mathbf{y}_1)^2$; the zero correlation is, in this case, a consequence of the *linear* model underlying statistic *r*. Therefore, only those correlations which are *significantly* different from zero should be considered, since a null correlation has no unique interpretation.

Since the correlation matrix is the dispersion matrix of standardized variables, it is possible, as in the case of matrix **S** (eq. 4.6), to compute **R** directly by multiplying the *matrix of standardized data* with its transpose:

$$\mathbf{R} = \frac{1}{n-1}\left[(y-\bar{y})/s_y\right]'\left[(y-\bar{y})/s_y\right] = \frac{1}{n-1}\mathbf{Z'Z} \qquad \mathbf{(4.9)}$$

Table 4.4 shows how to calculate correlations r_{jk} of the example as in Table 4.3, using this time the *standardized data*. The mean of each *standardized variable is* zero and its standard deviation is equal to unity. The *dispersion* matrix of **Z** is identical to the *correlation* matrix of **Y**, which was calculated above using the covariances and variances.

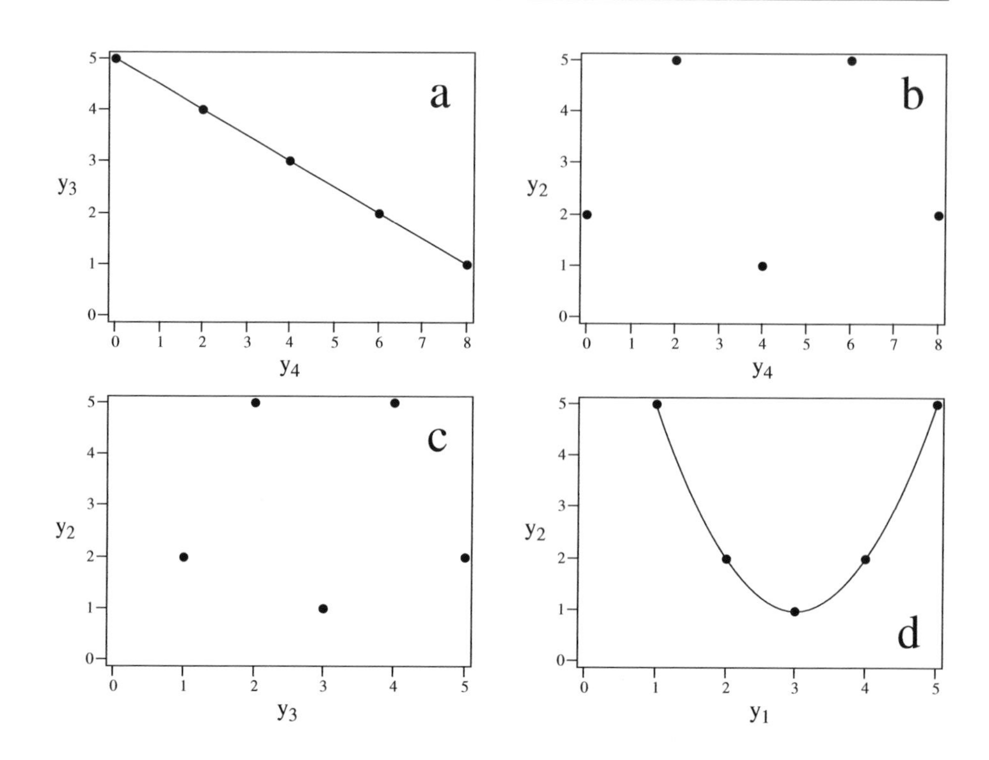

Figure 4.4 Numerical example. Relationships between species (a) 3 and 4, (b) 2 and 4, (c) 2 and 3, and (d) 2 and 1.

Matrices **Σ** and **P** are related to each other by the diagonal matrix of standard deviations of **Y**. This new matrix, which is specifically designed for relating **Σ** and **P**, is symbolized by **D**(σ) and its inverse by $\mathbf{D}(\sigma)^{-1}$:

$$\mathbf{D}(\sigma) = \begin{bmatrix} \sigma_1 & 0 & \ldots & 0 \\ 0 & \sigma_2 & \ldots & 0 \\ \cdot & \cdot & \cdots & \cdot \\ \cdot & \cdot & \cdots & \cdot \\ \cdot & \cdot & \cdots & \cdot \\ 0 & \cdot & \ldots & \sigma_p \end{bmatrix} \quad \text{and} \quad \mathbf{D}(\sigma)^{-1} = \begin{bmatrix} 1/\sigma_1 & 0 & \ldots & 0 \\ 0 & 1/\sigma_2 & \ldots & 0 \\ \cdot & \cdot & \cdots & \cdot \\ \cdot & \cdot & \cdots & \cdot \\ \cdot & \cdot & \cdots & \cdot \\ 0 & \cdot & \ldots & 1/\sigma_p \end{bmatrix}$$

Using these two matrices, one can write:

$$\mathbf{P} = \mathbf{D}(\sigma^2)^{-1/2}\ \mathbf{\Sigma}\ \mathbf{D}(\sigma^2)^{-1/2} = \mathbf{D}(\sigma)^{-1}\ \mathbf{\Sigma}\ \mathbf{D}(\sigma)^{-1} \tag{4.10}$$

Table 4.4 Numerical example. Calculation of standardized data and correlations.

Sites	Original data	Standardized data
1 2 3 4 5	$\mathbf{Y} = \begin{bmatrix} 1 & 5 & 2 & 6 \\ 2 & 2 & 1 & 8 \\ 3 & 1 & 3 & 4 \\ 4 & 2 & 5 & 0 \\ 5 & 5 & 4 & 2 \end{bmatrix}$	$\mathbf{Z} = \begin{bmatrix} -1.27 & 1.07 & -0.63 & 0.63 \\ -0.63 & -0.53 & -1.27 & 1.27 \\ 0 & -1.07 & 0 & 0 \\ 0.63 & -0.53 & 1.27 & -1.27 \\ 1.27 & 1.07 & 0.63 & -0.63 \end{bmatrix}$
Means	$\bar{\mathbf{y}}' = \begin{bmatrix} 3 & 3 & 3 & 4 \end{bmatrix}$	$\bar{\mathbf{z}}' = \begin{bmatrix} 0 & 0 & 0 & 0 \end{bmatrix}$
$n-1=4$	$\mathbf{R}(y) = \mathbf{S}(z) = \frac{1}{n-1}\mathbf{Z'Z} = \begin{bmatrix} 1 & 0 & 0.8 & -0.8 \\ 0 & 1 & 0 & 0 \\ 0.8 & 0 & 1 & -1 \\ -0.8 & 0 & -1 & 1 \end{bmatrix}$	

where $\mathbf{D}(\sigma^2)$ is the matrix of the diagonal elements of $\mathbf{\Sigma}$. It follows from eq. 4.10 that:

$$\mathbf{\Sigma} = \mathbf{D}(\sigma)\,\mathbf{P}\,\mathbf{D}(\sigma) \qquad \textbf{(4.11)}$$

Significance of r

The theory underlying tests of significance is discussed in Section 1.2. In the case of r, inference about the statistical population is in most instances through the null hypothesis H_0: $\rho = 0$. H_0 may also state that ρ has some other value than zero, which would be derived from ecological hypotheses. The general formula for testing correlation coefficients is given in Section 4.5 (eq. 4.39). The Pearson correlation coefficient r_{jk} involves two descriptors (i.e. $\mathbf{y}_j$ and $\mathbf{y}_k$, hence $m = 2$ when testing a coefficient of simple linear correlation using eq. 4.39), so that $\nu_1 = 2 - 1 = 1$ and $\nu_2 = n - 2 = \nu$. The general formula then becomes:

$$F = \nu \frac{r_{jk}^2}{1 - r_{jk}^2} \qquad \textbf{(4.12)}$$

where $\nu = n - 2$. Statistic F is tested against $F_{\alpha[1,\nu]}$.

Since the square root of a statistic $F_{[\nu_1,\nu_2]}$ is a statistic $t_{[\nu = \nu_2]}$ when $\nu_1 = 1$, r may also be tested using:

$$t = \frac{r_{jk}\sqrt{\nu}}{\sqrt{1 - r_{jk}^2}} \qquad \textbf{(4.13)}$$

The t statistic is tested against the value $t_{\alpha[\nu]}$. In other words, H_0 is tested by comparing the F (or t) statistic to the value found in a table of critical values of F (or t). Results of tests with eqs. 4.12 and 4.13 are identical. The number of degrees of freedom is $\nu = (n - 2)$ because calculating a correlation coefficient requires prior estimation of two parameters, i.e. the means of the two populations (eq. 4.7). H_0 is rejected when the probability corresponding to F (or t) is smaller than a predetermined *level of significance* (α for a two-tailed test, and $\alpha/2$ for a one-tailed test; the difference between the two types of tests is explained in Section 1.2). In principle, this test requires that the sample of observations be drawn from a population with a *bivariate normal distribution* (Section 4.3). Testing for normality and multinormality is discussed in Section 4.7, and normalizing transformations in Section 1.5. When the data do not satisfy the condition of normality, t can be tested by randomization, as shown in Section 1.2.

Test of independence of variables

It is also possible to test the *independence of all variables* in a data matrix by considering the set of all correlation coefficients found in matrix **R**. The null hypothesis here is that the $p(p - 1)/2$ coefficients are all equal to zero, H_0: $\mathbf{R} = \mathbf{I}$ (unit matrix). According to Bartlett (1954), **R** can be transformed into a X^2 (chi-square) test statistic:

$$X^2 = -[n - (2p + 11)/6] \ln |\mathbf{R}| \quad \textbf{(4.14)}$$

where $\ln |\mathbf{R}|$ is the natural logarithm of the determinant of **R**. This statistic is approximately distributed as χ^2 with $\nu = p(p - 1)/2$ degrees of freedom. When the probability associated with X^2 is significantly low, the null hypothesis of complete independence of the p descriptors is rejected. In principle, this test requires the observations to be drawn from a population with a *multivariate normal distribution* (Section 4.3). If the null hypothesis of independence of all variables is rejected, the $p(p - 1)/2$ correlation coefficients in matrix **R** may be tested individually; see Box 1.3 about multiple testing.

Other correlation coefficients are described in Sections 4.5 and 5.2. Wherever the coefficient of linear correlation must be distinguished from other coefficients, it is referred to as *Pearson's r*. In other instances, r is simply called the *coefficient of linear correlation* or *correlation coefficient*. Table 4.5 summarizes the main properties of this coefficient.

4.3 Multinormal distribution

In general, the mathematics of the normal distribution are of little concern to ecologists using unidimensional statistical methods. In the best case, data are normalized (Section 1.5) before being subjected to tests that are based on *parametric* hypotheses. It must be remembered that all *parametric tests* require the data to follow a specific *distribution,* most often the normal distribution. When the data do not obey this

Table 4.5 Main properties of the coefficient of linear correlation. Some of these properties are discussed in later sections.

Properties	Sections
1. The coefficient of linear correlation measures the *intensity of the linear relationship* between two random variables.	4.2
2. The coefficient of linear correlation between two variables can be calculated using their respective *variances* and their *covariance.*	4.2
3. The correlation matrix is the *dispersion* matrix of *standardized variables.*	4.2
4. The square of the coefficient of linear correlation is the *coefficient of determination*. It measures how much of the variance of each variable is explained by the other.	10.3
5. The coefficient of linear correlation is a *parameter* of a multinormal distribution.	4.3
6. The coefficient of linear correlation is the *geometric mean* of the *coefficients of linear regression* of each variable on the other.	10.3

condition, the results of parametric tests may be *invalid*. There also exist *nonparametric* tests, for which no reference is made to any theoretical distribution of the population, hence no use of parameters. Another advantage of nonparametric tests of significance is that they remain valid even for very small sample sizes, as are often encountered in ecological research. These tests (Chapter 5) are of great interest to ecologists, who may nevertheless attempt to normalize their data in order to have access to the powerful toolbox of parametric statistics.

Multidimensional statistics require careful examination of the main characteristics of the *multinormal* (or *multivariate normal*) *distribution*. Several of the methods described in the present chapter, and also in Chapters 9, 10 and 11, are founded on principles derived from the multinormal distribution. This is true even in cases where no test of significance is performed, which is often the case in numerical ecology (i.e. descriptive versus inferential statistics, Sections 1.1 and 4.4).

The logic of an approach centred on the multinormal distribution is based upon a theorem which is undoubtedly one of the most important of statistics. According to the *central limit theorem*, when a random variable results from several independent and additive effects, of which none has a dominant variance, then this variable tends towards a normal distribution even if the effects are not themselves normally distributed. Since ecological variables (descriptors) are often influenced by several independent random factors, the above theorem explains why the normal distribution is frequently invoked to describe ecological phenomena. This justifies a careful

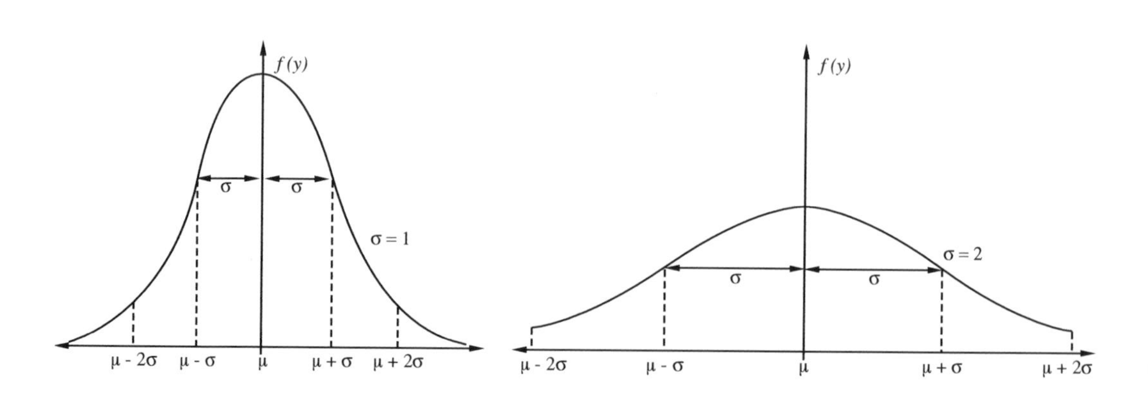

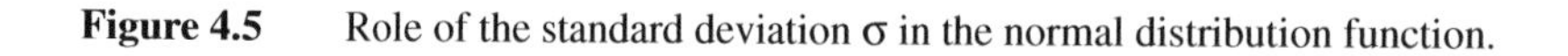
Figure 4.5 Role of the standard deviation σ in the normal distribution function.

examination of the properties of the multinormal distribution, before studying the methods for analysing multidimensional quantitative data.

Normal

The probability density of a *normal* random variable y is (Laplace-Gauss equation):

$$f(y) = \frac{1}{\sqrt{2\pi}\sigma} \exp\left[-\frac{1}{2}\left(\frac{y-\mu}{\sigma}\right)^2\right] \quad \textbf{(4.15)}$$

where exp [...] reads "e to the power [...]", e being the Napierian base (e = 2.71828...). Calculation of $f(y)$, for a given y, only requires μ and σ. The mean (μ) and standard deviation (σ) of the theoretical population completely determine the shape of the probability distribution. This is why they are called the *parameters* of the normal distribution. The curve is symmetric on both sides of μ and its exact shape depends on σ (Fig. 4.5).

The value σ determines the position of the inflexion points along the normal curve. These points are located on both sides of μ, at a distance σ, whereas μ positions the curve on the abscissa (y axis). In Fig. 4.5, the surface under each of the two curves is identical for the same number of σ units on either side of μ. The height of the curve is the probability density corresponding to the y value; for a continuous function such as that of the normal distribution, the probability of finding a value between y = a and y = b ($a < b$) is given by the surface under the curve between a and b. For example, the probability of finding a value between μ – 1.96σ and μ + 1.96σ is 0.95.

In view of examining the properties of the multinormal distribution, it is convenient to first consider the joint probability density of p *independent* unidimensional normal variables. For *each* of these p variables y_j, the probability density is given by eq. 4.15, with mean μ_j and standard deviation σ_j:

$$f(y_j) = \frac{1}{\sqrt{2\pi}\sigma_j} \exp\left[-\frac{1}{2}\left(\frac{y_j - \mu_j}{\sigma_j}\right)^2\right] \quad \textbf{(4.16)}$$

A basic law of probabilities states that the joint probability density of several independent variables is the product of their individual densities. It follows that the joint probability density for p independent variables is:

$$f(y_1, y_2, \ldots, y_p) = f(y_1) \times f(y_2) \times \ldots \times f(y_p)$$

$$f(y_1, y_2, \ldots, y_p) = \frac{1}{(2\pi)^{p/2}\sigma_1\sigma_2\ldots\sigma_p} \exp\left[-\frac{1}{2}\sum_{j=1}^{p}\left(\frac{y_j - \mu_j}{\sigma_j}\right)^2\right] \quad \textbf{(4.17)}$$

Using the conventions of Table 4.2, one defines the following matrices:

$$\mathbf{y} = \begin{bmatrix} y_1 & y_2 & \ldots & y_p \end{bmatrix} \qquad \boldsymbol{\Sigma} = \begin{bmatrix} \sigma_1^2 & 0 & \ldots & 0 \\ 0 & \sigma_2^2 & \ldots & 0 \\ \cdot & \cdot & \cdots & \cdot \\ \cdot & \cdot & \cdots & \cdot \\ \cdot & \cdot & \cdots & \cdot \\ 0 & 0 & \ldots & \sigma_p^2 \end{bmatrix} \quad \textbf{(4.18)}$$

$$\boldsymbol{\mu} = \begin{bmatrix} \mu_1 & \mu_2 & \ldots & \mu_p \end{bmatrix}$$

where $\mathbf{y}$ is the p-dimensional vector of coordinates of the point for which the probability density (i.e. the ordinate along the p-dimensional normal curve) is sought, $\boldsymbol{\mu}$ is the vector of means, and $\boldsymbol{\Sigma}$ is the dispersion matrix among the p independent variables. The determinant of a diagonal matrix being equal to the product of the diagonal elements (Section 2.6), it follows that:

$$|\boldsymbol{\Sigma}|^{1/2} = (\sigma_1\,\sigma_2\ldots\sigma_p)$$

From definitions (4.18) one may write:

$$[y - \mu]\,\boldsymbol{\Sigma}^{-1}\,[y - \mu]' = \sum_{j=1}^{p}\left(\frac{y_j - \mu_j}{\sigma_j}\right)^2$$

Using these relationships, eq. 4.17 is rewritten as:

$$f(\mathbf{y}) = \frac{1}{(2\pi)^{p/2}|\boldsymbol{\Sigma}|^{1/2}} \exp\left\{-(1/2)\,[y - \mu]\,\boldsymbol{\Sigma}^{-1}\,[y - \mu]'\right\} \quad \textbf{(4.19)}$$

Do not confuse, here, the summation symbol $\sum$ with matrix $\boldsymbol{\Sigma}$.

Multi-normal

The above equations are for the joint probability density of p *independent* unidimensional normal variables y_j. It is easy to go from there to the *multinormal distribution,* where $\mathbf{y}$ is a p-dimensional random variable whose p dimensions are *not independent*. In order to do so, one simply replaces the above matrix $\boldsymbol{\Sigma}$ by a dispersion matrix with variances and covariances, i.e. (eq. 4.2):

$$\boldsymbol{\Sigma} = \begin{bmatrix} \sigma_{11} & \sigma_{12} & \dots & \sigma_{1p} \\ \sigma_{21} & \sigma_{22} & \dots & \sigma_{2p} \\ \cdot & \cdot & \dots & \cdot \\ \cdot & \cdot & \dots & \cdot \\ \cdot & \cdot & \dots & \cdot \\ \sigma_{p1} & \sigma_{p2} & \dots & \sigma_{pp} \end{bmatrix}$$

Using this dispersion matrix $\boldsymbol{\Sigma}$, eq. 4.19 now describes the probability density $f(\mathbf{y})$ for a p-dimensional multinormal distribution.

Given eq. 4.11, eq. 4.19 may be rewritten as:

$$f(\mathbf{y}) = \frac{1}{(2\pi)^{p/2} |\mathbf{D}(\sigma)| |\mathbf{P}|^{1/2}} \exp\{-(1/2)\, [y - \mu]\, \mathbf{D}(\sigma)^{-1}\, \mathbf{P}^{-1}\, \mathbf{D}(\sigma)^{-1}\, [y - \mu]'\} \quad \textbf{(4.20)}$$

Replacing, in eq. 4.20, the p-dimensional matrix $\mathbf{Y}$ by the p-dimensional standardized matrix $\mathbf{Z}$ (eq. 1.12) gives:

$$f(\mathbf{z}) = \frac{1}{(2\pi)^{p/2} |\mathbf{P}|^{1/2}} \exp\{-(1/2)\, \mathbf{Z}\, \mathbf{P}^{-1}\, \mathbf{Z}'\} \quad \textbf{(4.21)}$$

given the fact that $[y - \mu]\, \mathbf{D}(\sigma)^{-1} = \mathbf{Z}$ and, in the case of $\mathbf{Z}$, $\mathbf{D}(\sigma) = \mathbf{I}$.

Equation 4.21 stresses a fundamental point, which was already clear in eq. 4.20: *the correlations* ρ *are parameters of the multinormal distribution,* together with the means μ and standard deviations σ. This new property of ρ is listed in Table 4.5.

Three sets of parameters are therefore necessary to specify a multidimensional normal distribution, i.e. the vector of *means* $\boldsymbol{\mu}$, the diagonal matrix of *standard deviations* $\mathbf{D}(\sigma)$, and the *correlation matrix* $\mathbf{P}$. In the unidimensional normal distribution (eq. 4.15), μ and σ were the only parameters because there is no correlation ρ for a single variable.

Bivariate normal

It is not possible to represent, in a plane, more than three dimensions. Thus, for the purpose of illustration, only the simplest case of multinormal distribution will be considered, i.e. the *bivariate normal distribution*, where:

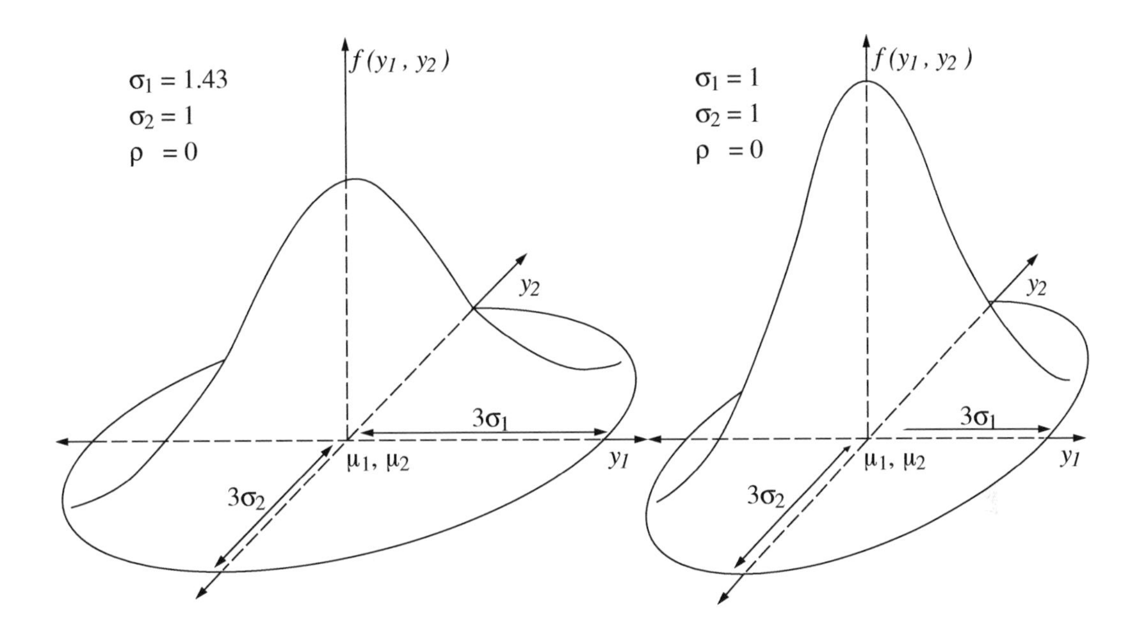

Figure 4.6 Roles of σ_1 and σ_2 in the bivariate normal distribution.

$$\boldsymbol{\mu} = \begin{bmatrix} \mu_1 & \mu_2 \end{bmatrix} \qquad \mathbf{D}(\sigma) = \begin{bmatrix} \sigma_1 & 0 \\ 0 & \sigma_2 \end{bmatrix} \qquad \mathbf{P} = \begin{bmatrix} 1 & \rho \\ \rho & 1 \end{bmatrix}$$

Since $|\mathbf{D}(\sigma)| = \sigma_1\sigma_2$ and $|\mathbf{P}| = (1-\rho^2)$ in this case, eq. 4.20 becomes:

$$f(y_1,y_2) = \frac{1}{2\pi\sigma_1\sigma_2\sqrt{1-\rho^2}} \exp\{-(1/2)\, [y-\mu]\, \mathbf{D}(1/\sigma)\, (1-\rho^2)^{-1} \begin{bmatrix} 1 & -\rho \\ -\rho & 1 \end{bmatrix} \mathbf{D}(1/\sigma)\, [y-\mu]'\}$$

$$= \frac{1}{2\pi\sigma_1\sigma_2\sqrt{1-\rho^2}} \exp\left\{-\frac{1}{2}\frac{1}{(1-\rho^2)}\left[\left(\frac{y_1-\mu_1}{\sigma_1}\right)^2 - 2\rho\left(\frac{y_1-\mu_1}{\sigma_1}\right)\left(\frac{y_2-\mu_2}{\sigma_2}\right) + \left(\frac{y_2-\mu_2}{\sigma_2}\right)^2\right]\right\}$$

Figure 4.6 shows bivariate normal distributions, with typical "bell" shapes. The two examples illustrate the roles of σ_1 and σ_2. Further examination of the multinormal mathematics is required to specify the role of ρ.

Neglecting the constant –1/2, the remainder of the exponent in eq. 4.19 is:

$$[y-\mu]\, \boldsymbol{\Sigma}^{-1}\, [y-\mu]'$$

When it is made equal to a positive constant (α), this algebraic form specifies the equation of an *ellipsoid* in the *p*-dimensional space:

$$[y - \mu]\, \mathbf{\Sigma}^{-1}\, [y - \mu]' = \alpha \qquad \textbf{(4.22)}$$

A family of such ellipsoids may be generated by varying constant α. All these ellipsoids have the multidimensional point **μ** as their common centre.

It is easy to understand the meaning of eq. 4.22 by examining the two-dimensional case. Without loss of generality, it is convenient to use the standardized variable (z_1,z_2) instead of (y_1,y_2), so that the family of *ellipses* (i.e. two-dimensional ellipsoids) be centred on the origin $\boldsymbol{\mu} = [0\ 0]$. The exponent of the *standardized bivariate normal density* is:

$$\frac{1}{1-\rho^2}\,[z_1^2 - 2\rho z_1 z_2 + z_2^2]$$

This exponent specifies, in two-dimensional space, the equation of a family of ellipses:

$$\frac{1}{1-\rho^2}\,[z_1^2 - 2\rho z_1 z_2 + z_2^2] = \alpha$$

$$z_1^2 - 2\rho z_1 z_2 + z_2^2 = \alpha\,(1-\rho^2)$$

Figure 4.7 illustrates the role played by ρ in determining the general shape of the family of ellipses. As ρ approaches zero, the shapes of the ellipses tend to become circular. In contrast, as ρ approaches +1 or –1, the ellipses tend to elongate. The sign of ρ determines the orientation of the ellipses relative to the axes.

Actually, when ρ = 0 (Fig. 4.8), the equation for the family of ellipses becomes:

$$z_1^2 - [2 \times 0 \times z_1 z_2] + z_2^2 = \alpha\,(1-0)$$

or $\quad z_1^2 + z_2^2 = \alpha$, which is the equation of a *circle*.

In contrast, when ρ = ±1, the equation becomes:

$$z_1^2 - [2 \times (\pm 1) \times z_1 z_2] + z_2^2 = \alpha\,[1 - (\pm 1)^2]$$

$$z_1^2 \mp 2 z_1 z_2 + z_2^2 = 0$$

hence $[z_1 \mp z_2]^2 = 0$, so that $z_1 \mp z_2 = 0$, and thus $z_1 = \pm z_2$,

which is the equation of a *straight line* with a positive or negative slope of 1 (±45° angle).

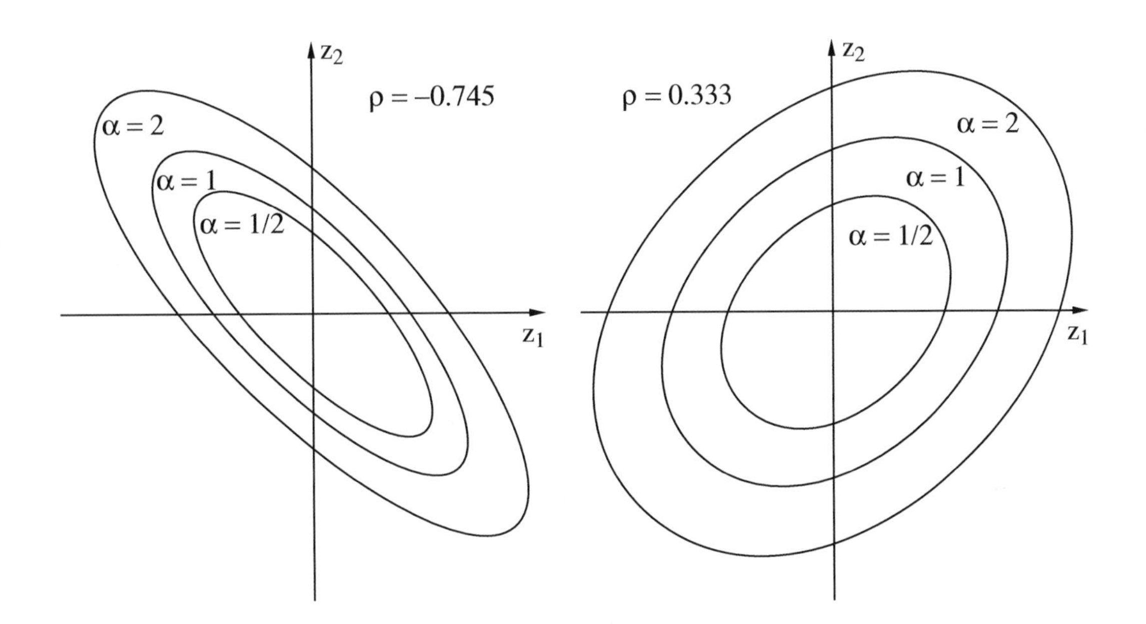

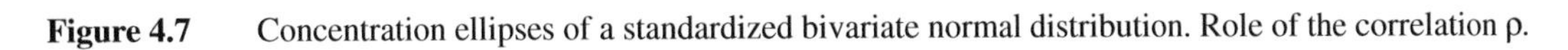
Figure 4.7 Concentration ellipses of a standardized bivariate normal distribution. Role of the correlation ρ.

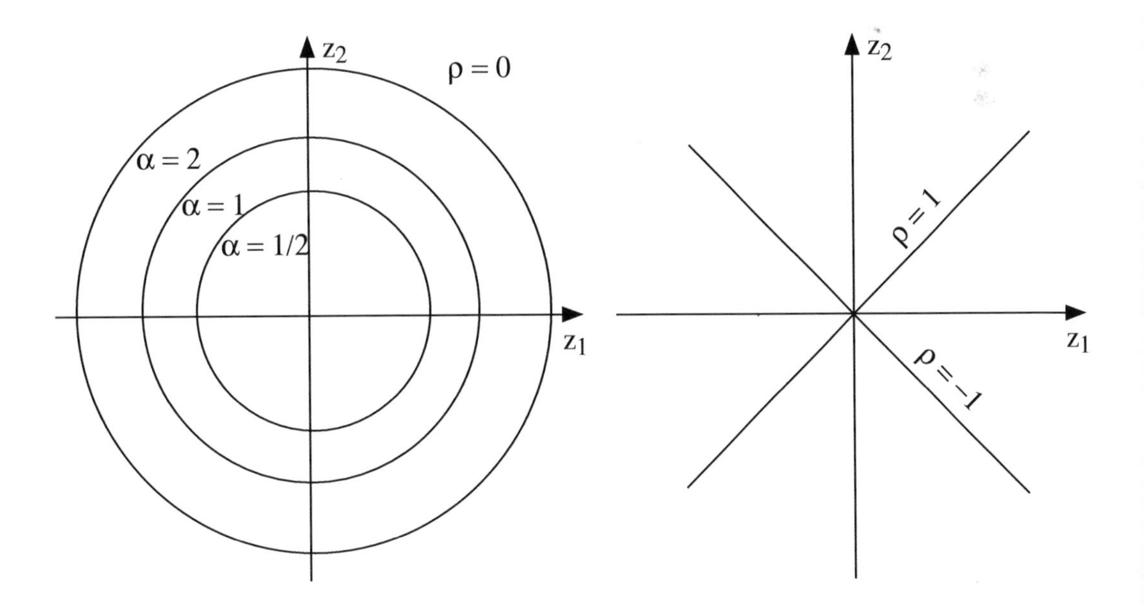

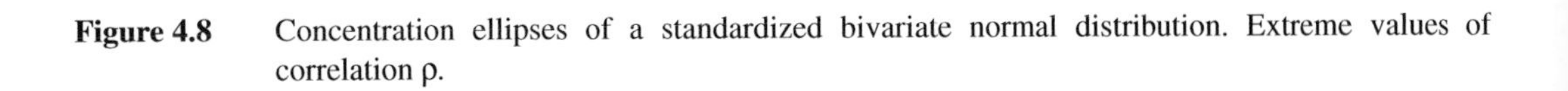
Figure 4.8 Concentration ellipses of a standardized bivariate normal distribution. Extreme values of correlation ρ.

Such a family of ellipses, called *concentration ellipses*, is comparable to a series of "contour lines" for the two-dimensional normal distribution (Fig. 4.6). Increasing the value of α corresponds to moving down along the sides of the distribution. The concentration ellipses pass through points of equal probabilities around the bivariate normal distribution. The role of ρ then becomes clear: when $\rho = 0$, the "bell" of probability densities is perfectly circular (in overhead view); as ρ increases, the "bell" of the probability densities flattens out, until it becomes unidimensional when $\rho = \pm 1$. Indeed, when there is a perfect correlation between two dimensions (i.e. $\rho = \pm 1$), a single dimension, at angle of 45° with respect to the two original variables, is sufficient to specify the distribution of probability densities.

When the number of dimensions is $p = 3$, the family of concentration ellipses becomes a family of concentration *ellipsoids* and, when $p > 3$, a family of *hyperellipsoids*. The meaning of these ellipsoids and hyperellipsoids is the same as in the two-dimensional case, although it is not possible to illustrate them.

4.4 Principal axes

Various aspects of the multinormal distribution have been examined in the previous section. One of these, namely the *concentration ellipses* (Fig. 4.7), opens a topic of great importance for ecologists. In the present section, a method will be developed for determining the *principal axes* of the concentration hyperellipsoids; for simplicity, the term ellipsoid will be used in the following discussion. The *first principal axis* is the line that passes through the greatest dimension of the ellipsoid. The next *principal axes* go through the next greatest dimensions, smaller and smaller, of the p-dimensional ellipsoid. Hence, p consecutive principal axes are determined. These principal axes will be used, in Section 9.1, as the basis for *principal component analysis*.

In the two-dimensional case, the *first principal axis* corresponds to the *major axis* of the concentration ellipse and the *second principal axis* to the *minor axis*. These two axes are perpendicular to each other. Similarly in the p-dimensional case, there are p consecutive axes, which are all perpendicular to one another in the hyperspace.

The first principal axis goes through the p-dimensional centre $\boldsymbol{\mu} = [\mu_1\ \mu_2 \ldots \mu_p]$ of the ellipsoid, and it crosses the surface of the ellipsoid at a point called here $\mathbf{y} = [y_1\ y_2 \ldots y_p]$. The values of $\boldsymbol{\mu}$ and $\mathbf{y}$ specify a vector in the p-dimensional space (Section 2.4). The length of the axis, from $\boldsymbol{\mu}$ to the surface of the ellipsoid, is calculated using Pythagoras' formula:

$$[(y_1 - \mu_1)^2 + (y_2 - \mu_2)^2 + \ldots + (y_p - \mu_p)^2]^{1/2} = ([y - \mu][y - \mu]')^{1/2}$$

Actually, this is only half the length of the axis, which extends equally on both sides of $\boldsymbol{\mu}$. The coordinates of the *first* principal axis must be such as to maximize the length of the axis. This can be achieved by maximizing the square of the half-length:

$$[y - \mu][y - \mu]'$$

Calculating coordinates corresponding to the axis with the greatest length is subjected to the constraint that the end point **y** be on the surface of the ellipsoid. This constraint is made explicit using eq. 4.22 which specifies the ellipsoid:

$$[y - \mu]\, \boldsymbol{\Sigma}^{-1}\, [y - \mu]' = \alpha$$

$$[y - \mu]\, \boldsymbol{\Sigma}^{-1}\, [y - \mu]' - \alpha = 0$$

Lagrangian multipliers (λ) are used to compute the maximum and minimum values of a function of several variables when the relationships among the variables are known. In the present case, the above two equations, for the square of the half-length of the first principal axis and for the constraint, are combined into a single function:

Principal axis

$$f(\mathbf{y}) = [y - \mu][y - \mu]' - \lambda\, \{[y - \mu]\, \boldsymbol{\Sigma}^{-1}\, [y - \mu]' - \alpha\}$$

The values that maximize this function are found by the usual method of setting the equation's partial derivative equal to 0:

$$\frac{\partial}{\partial \mathbf{y}} f(\mathbf{y}) = \mathbf{0}$$

$$\frac{\partial}{\partial \mathbf{y}} [y - \mu][y - \mu]' - \lambda \frac{\partial}{\partial \mathbf{y}} \{[y - \mu]\, \boldsymbol{\Sigma}^{-1}\, [y - \mu]' - \alpha\} = \mathbf{0}$$

Lagrangian multiplier

Scalar λ is called a *Lagrangian multiplier*. It is important to remember here that **y** is a p-dimensional *vector* $(y_1, y_2, \ldots, y_p)$, which means that the above equation is successively derived with respect to $y_1, y_2, \ldots$ and y_p. Therefore, derivation with respect to **y** represents in fact a series of p partial derivatives (∂y_j). Consequently the results of the derivation may be rewritten as a (column) vector with p elements:

$$2\, [y - \mu] - 2\, \lambda\, \boldsymbol{\Sigma}^{-1}\, [y - \mu] = \mathbf{0}$$

One may factor out $[y - \mu]$ and eliminate the constant 2:

$$(\mathbf{I} - \lambda\, \boldsymbol{\Sigma}^{-1})\, [y - \mu] = \mathbf{0}$$

Multiplying both sides of the equation by $\boldsymbol{\Sigma}$ gives:

$$(\boldsymbol{\Sigma} - \lambda \mathbf{I})\, [y - \mu] = \mathbf{0} \qquad \textbf{(4.23)}$$

The general equation defining eigenvectors (eq. 2.22) is $(\mathbf{A} - \lambda \mathbf{I})\, \mathbf{u} = \mathbf{0}$. Replacing, in this equation, **A** by $\boldsymbol{\Sigma}$ and **u** by $[y - \mu]$ gives the above equation. This leads to the conclusion that the *vector of the coordinates which specifies the first principal axis is one of the eigenvectors* $[\mathbf{y} - \mu]$ *of matrix* $\boldsymbol{\Sigma}$.

In order to find out which of the p eigenvectors of $\boldsymbol{\Sigma}$ is the vector of coordinates of the *first principal axis,* come back to the equation resulting from the partial derivation (above) and transfer the second term to the right, after eliminating the constant 2:

$$[y - \mu] = \lambda\, \boldsymbol{\Sigma}^{-1}\, [y - \mu]$$

The two sides are then premultiplied by $[y - \mu]'$:

$$[y - \mu]' [y - \mu] = \lambda [y - \mu]' \mathbf{\Sigma}^{-1} [y - \mu]$$

Since $[y - \mu]' \mathbf{\Sigma}^{-1} [y - \mu] = \alpha$ (eq. 4.22), it follows that:

$$[y - \mu]' [y - \mu] = \lambda\alpha$$

Eigenvalue

The term on the left-hand side of the equation is the square of the half-length of the first principal axis (see above). Thus, for a given value α, the length of the *first principal axis* is maximized by taking the largest possible value for λ or, in other words, the *largest eigenvalue* λ *of matrix* $\mathbf{\Sigma}$. The vector of coordinates of the *first principal axis* is therefore the eigenvector corresponding to the largest eigenvalue of $\mathbf{\Sigma}$.

Numerical example. The above equations are illustrated using the 2-dimensional data matrix from Section 9.1 (principal component analysis). The covariance matrix is:

$$\mathbf{\Sigma} = \begin{bmatrix} 8.2 & 1.6 \\ 1.6 & 5.8 \end{bmatrix}$$

There are two eigenvalues, $\lambda_1 = 9$ and $\lambda_2 = 5$, computed using eq. 2.23. To normalize the eigenvectors (written as column vectors), one arbitrarily decides that $[y - \mu]' [y - \mu] = \lambda\alpha = 1$ for each of them; in other words, $\alpha_1 = 1/9$ and $\alpha_2 = 1/5$. The normalized eigenvectors provide the coordinates of the point where each of the two principal axes crosses the surface of the ellipsoid (vectors $\mathbf{y}_1$ and $\mathbf{y}_2$):

$$\mathbf{y}_1 = \mathbf{u}_1 = \begin{bmatrix} 0.8944 \\ 0.4472 \end{bmatrix} \text{ and } \mathbf{y}_2 = \mathbf{u}_2 = \begin{bmatrix} -0.4472 \\ 0.8944 \end{bmatrix}$$

Given means $\mu_1 = \mu_2 = 0$, it can be verified, for both $\mathbf{y}_1$ and $\mathbf{y}_2$, that $[y - \mu]' [y - \mu] = 1$. This example will be further developed in Chapter 9.

In the above demonstration, $\mathbf{y}_j$ was defined as the point (vector) where principal axis j crosses the surface of the ellipsoid. Since vectors $\mathbf{y}_j$ are eigenvectors, they will be denoted $\mathbf{u}_j$ from now on, as in Sections 2.9 and 2.10, whereas $\mathbf{y}_j$ will only be used to represent descriptors.

To find the vectors of coordinates specifying the p successive principal axes,

• rank the p eigenvalues of matrix $\mathbf{\Sigma}$ in decreasing order:

$$\lambda_1 > \lambda_2 > \ldots > \lambda_p \geq 0$$

Note that the eigenvalues of a matrix $\mathbf{\Sigma}$ are all positive (end of Section 4.1);

• associate the p eigenvectors to their corresponding eigenvalues. The orientation of the p successive principal axes are given by the eigenvectors, which are associated with the p eigenvalues ranked in decreasing order. The eigenvectors of a covariance matrix $\mathbf{\Sigma}$ are orthogonal to one another because $\mathbf{\Sigma}$ is symmetric (Section 2.9). In the case of multiplicity (Section 2.10, Fourth property), the orthogonal axes may be rotated to an infinity of "principal" directions, i.e. two equal λ's result in a circle and several determine a hypersphere (multidimensional sphere) where no orientation prevails.

The next step consists in calculating a new p-dimensional variable along which the dispersion ellipses are positioned with respect to the principal axes instead of the original Cartesian system. This new variable ($\mathbf{v}$) is related to the original variables ($\mathbf{y}_j$; Section 4.1) through the following transformation:

$$\mathbf{v} = [y - \mu]\,\mathbf{U} \tag{4.24}$$

where each of the p columns in matrix $\mathbf{U}$ is the normalized eigenvector $\mathbf{u}_k$, corresponding to the k-th principal axis. Because vectors $\mathbf{u}_k$ are both orthogonal and normalized, matrix $\mathbf{U}$ is said to be *orthonormal* (Section 2.8). This transformation results in shifting the origin of the system of axes to the p-dimensional point $\boldsymbol{\mu}$, followed by a solid rotation of the translated axes into the principal axes (Fig. 4. 9), which forms matrix $\mathbf{V}$.

The dispersion matrix of $\mathbf{V}$ is:

$$\boldsymbol{\Sigma}(\mathbf{V}) = \frac{1}{(n-1)}(\mathbf{V}'\mathbf{V}) \;=\; \frac{1}{(n-1)}\mathbf{U}'\;[y-\mu]'\;[y-\mu]\;\mathbf{U} = \mathbf{U}'\boldsymbol{\Sigma}\mathbf{U}$$

where $\boldsymbol{\Sigma}$ is the dispersion matrix of the original variables $\mathbf{y}$. So, the *variance* of the k-th dimension (i.e. the k-th principal axis) is:

$$s^2(\mathbf{v}_k) = \mathbf{u}'_k\,\boldsymbol{\Sigma}\mathbf{u}_k$$

Since, by definition, $\boldsymbol{\Sigma}\mathbf{u}_k = \lambda_k\mathbf{u}_k$ (eq. 2.21) and $\mathbf{u}'_k\,\mathbf{u}_k = 1$, it follows that:

$$s^2(\mathbf{v}_k) = \mathbf{u}'_k\,\boldsymbol{\Sigma}\mathbf{u}_k = \mathbf{u}'_k\,\lambda_k\mathbf{u}_k = \lambda_k\;\mathbf{u}'_k\,\mathbf{u}_k = \lambda_k\,(1) = \lambda_k \tag{4.25}$$

with $\lambda_k \geq 0$ in all cases since $\boldsymbol{\Sigma}$ is positive definite. The *covariance* of two vectors is zero because the product of two orthogonal vectors $\mathbf{u}_k$ and $\mathbf{u}_h$ is zero (Section 2.8):

$$s(\mathbf{v}_k,\mathbf{v}_h) = \mathbf{u}'_k\,\boldsymbol{\Sigma}\mathbf{u}_h = \mathbf{u}'_k\,\lambda_h\mathbf{u}_h = \lambda_h\;\mathbf{u}'_k\,\mathbf{u}_h = \lambda_k\,(0) = 0 \tag{4.26}$$

The last two points are of utmost importance, since they are the basis for using the principal axes (and thus principal component analysis; Section 9.1) in ecology: (1) *the variance of a principal axis is equal to the eigenvalue associated with that axis* (eq. 4.25) and (2) *the p dimensions of the transformed variable are linearly independent,* since their covariances are zero (eq. 4.26).

A last point concerns the meaning of the p elements u_{jk} of each eigenvector $\mathbf{u}_k$. The values of these elements determine the rotation of the system of axes, so that they correspond to angles. Figure 4.10 illustrates, for the two-dimensional case, how the

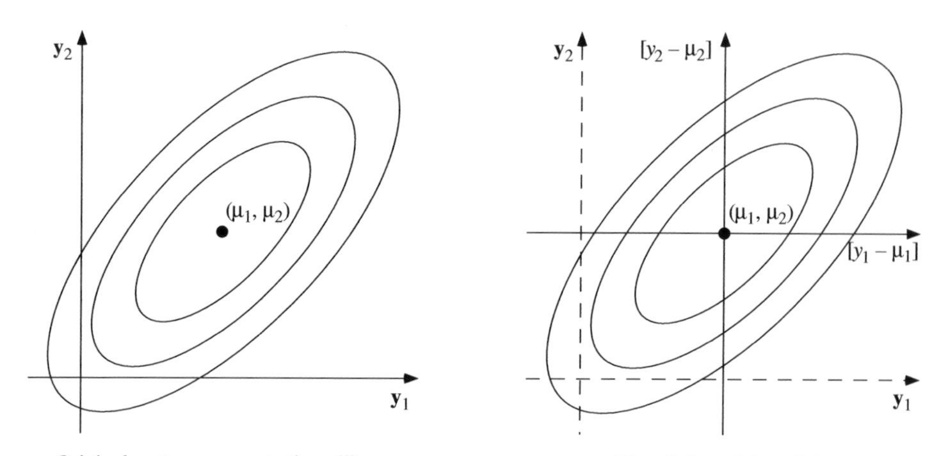

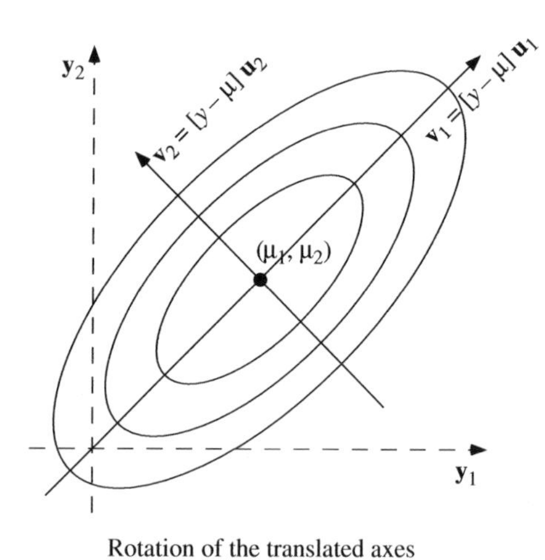

Figure 4.9 Result of the transformation $\mathbf{v} = [y - \mu]\,\mathbf{U}$ (eq. 4.24).

elements of the eigenvectors are related to the rotation angles. Using the trigonometric functions for right-angled triangles, the angular relationships in Fig. 4.10 may be rewritten as cosines:

$$\cos \alpha_{11} = \text{length } u_{11} \,/\, \text{length of vector } (u_{11}, u_{21}) = u_{11}$$

$$\cos \alpha_{21} = \text{length } u_{21} \,/\, \text{length of vector } (u_{11}, u_{21}) = u_{21}$$

$$\cos \alpha_{12} = \text{length } u_{12} \,/\, \text{length of vector } (u_{12}, u_{22}) = u_{12}$$

$$\cos \alpha_{22} = \text{length } u_{22} \,/\, \text{length of vector } (u_{12}, u_{22}) = u_{22}$$

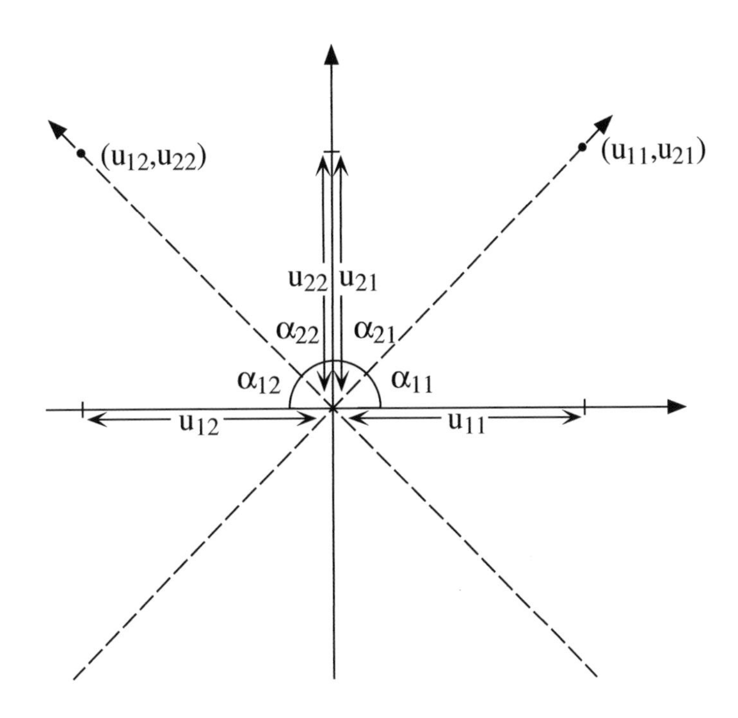

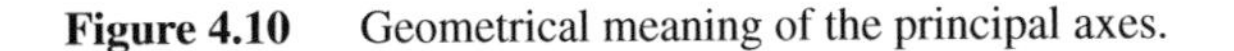

Figure 4.10 Geometrical meaning of the principal axes.

because the lengths of the *normalized* vectors (u_{11}, u_{21}) and (u_{12}, u_{22}) are 1 (Section 2.4). Eigenvector $\mathbf{u}_k$ determines the direction of the k-th main axis; it follows from the above trigonometric relationships that elements u_{jk} of normalized eigenvectors are *direction cosines*. Each direction cosine specifies the angle between an original Cartesian axis j and a principal axis k.

Direction cosine

The two-dimensional case, illustrated in Figs. 4.9 and 4.10, is the simplest to compute. The standardized dispersion matrix is of the general form:

$$\mathbf{P} = \begin{bmatrix} 1 & \rho \\ \rho & 1 \end{bmatrix}$$

When ρ is positive, the eigenvalues of $\mathbf{P}$ are $\lambda_1 = (1 + \rho)$ and $\lambda_2 = (1 - \rho)$. The *normalized* eigenvectors are:

$$\mathbf{u}_1 = \begin{bmatrix} 1/\sqrt{2} \\ 1/\sqrt{2} \end{bmatrix} \qquad \mathbf{u}_2 = \begin{bmatrix} -1/\sqrt{2} \\ 1/\sqrt{2} \end{bmatrix}$$

Therefore, the first principal axis goes through the point $(1/\sqrt{2}, 1/\sqrt{2})$, so that it cuts the first and third quadrants at a 45° angle. Its direction cosines are $\cos\alpha_{11} = 1/\sqrt{2}$ and $\cos\alpha_{12} = 1/\sqrt{2}$ which indeed specifies a 45° angle with respect to the two axes in the first quadrant. The second principal axis goes through $(-1/\sqrt{2}, 1/\sqrt{2})$, so that it cuts the second and fourth quadrants at 45°. Its direction cosines are $\cos\alpha_{21} = -1/\sqrt{2}$ and $\cos\alpha_{22} = 1/\sqrt{2}$, which determines a 45°angle with respect to the two axes in the second quadrant.

When ρ is *negative,* the eigenvalues of **P** are $\lambda_1 = (1 - \rho)$ and $\lambda_2 = (1 + \rho)$. Consequently the first principal axis goes through $(-1/\sqrt{2}, 1/\sqrt{2})$ in the second quadrant, while the second principal axis with coordinates $(1/\sqrt{2}, 1/\sqrt{2})$ cuts the first quadrant. A value $\rho = 0$ entails a case of multiplicity, since $\lambda_1 = \lambda_2 = 1$. This results in an infinite number of "principal" axes, i.e. any two perpendicular diameters would fit the concentration ellipse, which is here a circle (Fig. 4.8).

These concepts, so far quite abstract, will find direct applications to ecology in Section 9.1, dealing with principal component analysis.

4.5 Multiple and partial correlations

Section 4.2 considered, in a multidimensional context, the correlation between two variables, or two dimensions of a p-dimensional random variable. However, the multidimensional nature of ecological data offers other approaches to correlation. They are examined in the present section.

The following developments will require that the correlation matrix **R** be partitioned into four submatrices. Indices assigned to the submatrices follow the general convention on matrix indices (Section 2.1):

$$\mathbf{R} = \begin{bmatrix} \mathbf{R}_{11} & \mathbf{R}_{12} \\ \mathbf{R}_{21} & \mathbf{R}_{22} \end{bmatrix} \tag{4.27}$$

There are two possible approaches to linear correlation involving several variables or several dimensions of a multidimensional variable. The first one, which is called *multiple (linear) correlation*, measures the intensity of the relationship between a *response* variable and a linear combination of several *explanatory* variables. The second approach, called *partial (linear) correlation*, measures the intensity of the linear relationship between two variables, while taking into account their relationships with other variables.

1 — Multiple linear correlation

Multiple correlation applies to cases where there is one *response* variable and several *explanatory* variables. This situation is further studied in Section 10.3, within the context of *multiple regression*. The *coefficient of multiple determination* (R^2;

eq. 10.19) measures the fraction of the variance of $\mathbf{y}_k$ which is explained by a linear combination of $\mathbf{y}_1$, $\mathbf{y}_2$, ..., $\mathbf{y}_j$, ... and $\mathbf{y}_p$:

$$R^2_{k.12...j...p} = \frac{b_1 s_{1k} + b_2 s_{2k} + \ldots + b_j s_{jk} + \ldots + b_p s_{pk}}{s_k^2} \tag{4.28}$$

where p is here the number of explanatory variables. It is calculated in a way which is analogous to the coefficient of determination between two variables (eq. 10.8). Coefficients b are identical to those of multiple linear regression (Section 10.3). A coefficient $R^2_{k.12...j...p} = 0.73$, for example, would mean that the linear relationships of variables $\mathbf{y}_1$, $\mathbf{y}_2$, ..., $\mathbf{y}_j$, ... and $\mathbf{y}_p$ with $\mathbf{y}_k$ explain 73% of the variability of $\mathbf{y}_k$ around its mean. The *multiple correlation coefficient* (R) is the square root of the coefficient of multiple determination:

$$R_{k.12...j...p} = \sqrt{R^2_{k.12...j...p}} \tag{4.29}$$

In order to calculate R^2 using matrix algebra, a correlation matrix $\mathbf{R}$ is written for variables $\mathbf{y}_k$ and $\{\mathbf{y}_1, \mathbf{y}_2, \ldots, \mathbf{y}_j, \ldots, \mathbf{y}_p\}$, with $\mathbf{y}_k$ in the first position. Partitioning this matrix following eq. 4.27 gives, for multiple correlation:

$$\mathbf{R} = \left[\begin{array}{c|cccc} 1 & r_{k1} & r_{k2} & \cdots & r_{kp} \\ \hline r_{1k} & 1 & r_{12} & \cdots & r_{1p} \\ r_{2k} & r_{21} & 1 & \cdots & r_{2p} \\ \cdot & \cdot & \cdot & \cdots & \cdot \\ \cdot & \cdot & \cdot & \cdots & \cdot \\ \cdot & \cdot & \cdot & \cdots & \cdot \\ r_{pk} & r_{p1} & r_{p2} & \cdots & 1 \end{array}\right] = \begin{bmatrix} 1 & \mathbf{r}_{12} \\ \mathbf{r}_{21} & \mathbf{R}_{22} \end{bmatrix} \tag{4.30}$$

where $\mathbf{r}_{12} = \mathbf{r}'_{21}$ is a vector containing the correlation coefficients r_{k1}, r_{k2}, ..., r_{kp}. Using $\mathbf{r}_{12}$, $\mathbf{r}_{21}$ and $\mathbf{R}_{22}$ as defined in eq. 4.30, R^2 is calculated as:

$$R^2 = \mathbf{r}_{12}\mathbf{R}_{22}^{-1}\mathbf{r}_{21} = \mathbf{r}'_{21}\mathbf{R}_{22}^{-1}\mathbf{r}_{21} \tag{4.31}$$

Equation 4.31 is expanded using eq. 2.17:

$$R^2 = \mathbf{r}'_{21}\mathbf{R}_{22}^{-1}\mathbf{r}_{21} = \mathbf{r}'_{21}\frac{1}{|\mathbf{R}_{22}|}\begin{bmatrix} \mathrm{cof}(r_{11}) & \mathrm{cof}(r_{21}) & \cdots & \mathrm{cof}(r_{p1}) \\ \mathrm{cof}(r_{12}) & \mathrm{cof}(r_{22}) & \cdots & \mathrm{cof}(r_{p2}) \\ \cdot & \cdot & \cdots & \cdot \\ \cdot & \cdot & \cdots & \cdot \\ \cdot & \cdot & \cdots & \cdot \\ \mathrm{cof}(r_{1p}) & \mathrm{cof}(r_{2p}) & \cdots & \mathrm{cof}(r_{pp}) \end{bmatrix}\mathbf{r}_{21}$$

$$R^2 = \frac{1}{|\mathbf{R}_{22}|}(|\mathbf{R}_{22}| - |\mathbf{R}|) = 1 - \frac{|\mathbf{R}|}{|\mathbf{R}_{22}|} \qquad \textbf{(4.32)}$$

As an exercise, it is easy to check that

$$|\mathbf{R}_{22}| - |\mathbf{R}| = \mathbf{r}'_{21}\ [\text{adjugate matrix of } \mathbf{R}_{22}]\ \mathbf{r}_{21}$$

Multiple correlation

The *coefficient of multiple correlation* is calculated from eqs. 4.31 or 4.32:

$$R_{k.12\ldots j\ldots p} = \sqrt{\mathbf{r}'_{21}\mathbf{R}_{22}^{-1}\mathbf{r}_{21}} \quad \text{or} \quad R_{k.12\ldots j\ldots p} = \sqrt{1 - \frac{|\mathbf{R}|}{|\mathbf{R}_{22}|}} \qquad \textbf{(4.33)}$$

A third way of calculating R^2 is given below (eq. 4.38), at the end of Subsection 2 on partial correlation.

When two or more variables in matrix $\mathbf{R}_{22}$ are perfectly correlated (i.e. $r = 1$ or $r = -1$), the rank of $\mathbf{R}_{22}$ is smaller than its order (Section 2.7) so that $|\mathbf{R}_{22}| = 0$. Calculation of R thus requires the elimination of redundant variables from matrix $\mathbf{R}$.

Numerical example. A simple example, with three variables ($\mathbf{y}_1$, $\mathbf{y}_2$ and $\mathbf{y}_3$), illustrates the above equations. Matrix $\mathbf{R}$ is:

$$\mathbf{R} = \begin{bmatrix} 1 & 0.4 & 0.8 \\ 0.4 & 1 & 0.5 \\ 0.8 & 0.5 & 1 \end{bmatrix}$$

The *coefficient of multiple determination* $R^2_{1.23}$ is first calculated using eq. 4.31:

$$R^2_{1.23} = \begin{bmatrix} 0.4 & 0.8 \end{bmatrix} \begin{bmatrix} 1 & 0.5 \\ 0.5 & 1 \end{bmatrix}^{-1} \begin{bmatrix} 0.4 \\ 0.8 \end{bmatrix}$$

$$R^2_{1.23} = \begin{bmatrix} 0.4 & 0.8 \end{bmatrix} \begin{bmatrix} 1.33 & -0.67 \\ -0.67 & 1.33 \end{bmatrix} \begin{bmatrix} 0.4 \\ 0.8 \end{bmatrix}$$

$$R^2_{1.23} = 0.64$$

Equation 4.32 leads to an identical result:

$$R^2_{1.23} = 1 - \frac{\begin{vmatrix} 1 & 0.4 & 0.8 \\ 0.4 & 1 & 0.5 \\ 0.8 & 0.5 & 1 \end{vmatrix}}{\begin{vmatrix} 1 & 0.5 \\ 0.5 & 1 \end{vmatrix}}$$

$$R^2_{1.23} = 1 - \frac{0.27}{0.75} = 0.64$$

The linear combination of variables $\mathbf{y}_2$ and $\mathbf{y}_3$ explains 64% of the variance of $\mathbf{y}_1$. The *multiple correlation coefficient* is $R_{1.23} = 0.8$.

2 — Partial correlation

The second approach to correlation, in the multidimensional context, applies to situations where the relationship between two variables is influenced by their relationships with other variables. The *partial correlation coefficient* is related to partial multiple regression (Subsection 10.3.5). It measures what the correlation between $\mathbf{y}_j$ and $\mathbf{y}_k$ would be if other variables $\mathbf{y}_1$, $\mathbf{y}_2$, ... and $\mathbf{y}_p$, hypothesized to influence both $\mathbf{y}_j$ and $\mathbf{y}_k$, were held constant at their means. The partial correlation between variables $\mathbf{y}_j$ and $\mathbf{y}_k$, when controlling for their relationships with $\mathbf{y}_1, \mathbf{y}_2, \ldots$ and $\mathbf{y}_p$, is written $r_{jk.12 \ldots p}$.

In order to calculate the partial correlation coefficients, the set of variables is divided into two subsets. The *first* one contains the variables between which the partial correlation is to be computed while controlling for the influence of the variables in the second subset. The *second* thus contains the variables whose influence is being taken into account. Matrix $\mathbf{R}$ is partitioned as follows (eq. 4.27):

$$\mathbf{R} = \begin{bmatrix} \mathbf{R}_{11} & \mathbf{R}_{12} \\ \mathbf{R}_{21} & \mathbf{R}_{22} \end{bmatrix}$$

$\mathbf{R}_{11}$ (of order 2×2 for partial correlations) and $\mathbf{R}_{22}$ contain the correlations among variables in the first and the second subsets, respectively, whereas $\mathbf{R}_{12}$ and $\mathbf{R}_{21}$ both contain the correlations between variables from the two subsets; $\mathbf{R}_{12} = \mathbf{R}'_{21}$.

The number of variables in the second subset determines the *order* of the partial correlation coefficient. This order is the number of variables whose effects are eliminated from the correlation between $\mathbf{y}_j$ and $\mathbf{y}_k$. For example $r_{12.345}$ (third-order coefficient) means that the correlation between variables $\mathbf{y}_1$ and $\mathbf{y}_2$ is calculated while controlling for the linear effects of $\mathbf{y}_3$, $\mathbf{y}_4$, and $\mathbf{y}_5$.

The computation consists in subtracting from $\mathbf{R}_{11}$ (matrix of correlations among variables in the first subset) a second matrix containing the coefficients of multiple determination of the variables in the second subset on those in the first subset. These coefficients measure the fraction of the variance and covariance of the variables in the first subset which is explained by linear combinations of variables in the second subset. They are computed by replacing vector $\mathbf{r}_{21}$ by submatrix $\mathbf{R}_{21}$ in eq. 4.31:

$$\mathbf{R}_{12}\mathbf{R}_{22}^{-1}\mathbf{R}_{21} = \mathbf{R}'_{21}\mathbf{R}_{22}^{-1}\mathbf{R}_{21}$$

The subtraction gives the matrix of conditional correlations:

$$\mathbf{R}_{11} - \mathbf{R}_{12}\mathbf{R}_{22}^{-1}\mathbf{R}_{21} \tag{4.34}$$

It can be shown that the maximum likelihood estimate ($\mathbf{R}_{1.2}$) of the partial correlation matrix $\mathbf{P}_{1.2}$ is:

$$\mathbf{R}_{1.2} = \mathbf{D}(r_{1.2})^{-1/2}(\mathbf{R}_{11} - \mathbf{R}_{12}\mathbf{R}_{22}^{-1}\mathbf{R}_{21})\,\mathbf{D}(r_{1.2})^{-1/2} \tag{4.35}$$

where $\mathbf{D}(r_{1.2})$ is the matrix of diagonal elements of the conditional correlation matrix (eq. 4.34).

Computation for the three-dimensional case provides the algebraic formula for the partial correlation coefficients of order 1:

$$\mathbf{R} = \begin{bmatrix} 1 & r_{12} & r_{13} \\ r_{21} & 1 & r_{23} \\ r_{31} & r_{32} & 1 \end{bmatrix}$$

Coefficients pertaining to variables of the first subset ($\mathbf{y}_1$ and $\mathbf{y}_2$) are in the first two rows and columns. Using eq. 4.35 gives:

$$\mathbf{R}_{12}\mathbf{R}_{22}^{-1}\mathbf{R}_{21} = \begin{bmatrix} r_{13} \\ r_{23} \end{bmatrix} \begin{bmatrix} 1 \end{bmatrix}^{-1} \begin{bmatrix} r_{31} & r_{32} \end{bmatrix} = \begin{bmatrix} r_{13}^2 & r_{13}r_{23} \\ r_{13}r_{23} & r_{23}^2 \end{bmatrix}$$

$$\mathbf{R}_{11} - \mathbf{R}_{12}\mathbf{R}_{22}^{-1}\mathbf{R}_{21} = \begin{bmatrix} 1 & r_{12} \\ r_{21} & 1 \end{bmatrix} - \begin{bmatrix} r_{13}^2 & r_{13}r_{23} \\ r_{13}r_{23} & r_{23}^2 \end{bmatrix} = \begin{bmatrix} (1 - r_{13}^2) & (r_{12} - r_{13}r_{23}) \\ (r_{12} - r_{13}r_{23}) & (1 - r_{23}^2) \end{bmatrix}$$

$$\mathbf{R}_{1.2} = \begin{bmatrix} 1/\sqrt{1 - r_{13}^2} & 0 \\ 0 & 1/\sqrt{1 - r_{23}^2} \end{bmatrix} \begin{bmatrix} (1 - r_{13}^2) & (r_{12} - r_{13}r_{23}) \\ (r_{12} - r_{13}r_{23}) & (1 - r_{23}^2) \end{bmatrix} \begin{bmatrix} 1/\sqrt{1 - r_{13}^2} & 0 \\ 0 & 1/\sqrt{1 - r_{23}^2} \end{bmatrix}$$

$$\mathbf{R}_{1.2} = \begin{bmatrix} 1 & \dfrac{r_{12} - r_{13}r_{23}}{\sqrt{1 - r_{13}^2}\sqrt{1 - r_{23}^2}} \\ \dfrac{r_{12} - r_{13}r_{23}}{\sqrt{1 - r_{13}^2}\sqrt{1 - r_{23}^2}} & 1 \end{bmatrix} = \begin{bmatrix} 1 & r_{12.3} \\ r_{12.3} & 1 \end{bmatrix}$$

The last matrix equation provides the formula for first-order partial correlation coefficients:

$$r_{12.3} = \frac{r_{12} - r_{13} r_{23}}{\sqrt{1 - r_{13}^2} \sqrt{1 - r_{23}^2}} \quad \textbf{(4.36)}$$

The general formula, for coefficients of order p, is:

$$r_{jk.1\ldots p} = \frac{r_{jk.1\ldots(p-1)} - r_{jp.1\ldots(p-1)} r_{kp.1\ldots(p-1)}}{\sqrt{1 - r_{jp.1\ldots(p-1)}^2} \sqrt{1 - r_{kp.1\ldots(p-1)}^2}} \quad \textbf{(4.37)}$$

When there are four variables, it is possible to calculate 12 first-order and 6 second-order partial correlation coefficients. Computing a second-order coefficient necessitates the calculation of 3 first-order coefficients. For example:

$$r_{12.34} = \frac{r_{12.3} - r_{14.3} r_{24.3}}{\sqrt{1 - r_{14.3}^2} \sqrt{1 - r_{24.3}^2}} = r_{12.43} = \frac{r_{12.4} - r_{13.4} r_{23.4}}{\sqrt{1 - r_{13.4}^2} \sqrt{1 - r_{23.4}^2}}$$

It is thus possible, as the number of variables increases, to calculate higher-order coefficients. Computing a coefficient of a given order requires the calculation of three coefficients of the previous order, each of these requiring itself the calculation of coefficients of the previous order, and so on depending on the number of variables involved. Obviously, such a cascade of calculations is advantageously replaced by the direct matrix approach of eq. 4.35.

Numerical example. Partial correlations are calculated on the simple example already used for multiple correlation. Matrix **R** is:

$$\mathbf{R} = \begin{bmatrix} 1 & 0.4 & 0.8 \\ 0.4 & 1 & 0.5 \\ 0.8 & 0.5 & 1 \end{bmatrix}$$

Two subsets are formed, the first one containing descriptors $\mathbf{y}_1$ and $\mathbf{y}_2$ (between which the partial correlation is computed) and the second one $\mathbf{y}_3$ (whose influence on r_{12} is controlled for). Computations follow eqs. 4.34 and 4.35:

$$\mathbf{S}_{1.2} = \begin{bmatrix} 1 & 0.4 \\ 0.4 & 1 \end{bmatrix} - \begin{bmatrix} 0.8 \\ 0.5 \end{bmatrix} \begin{bmatrix} 1 \end{bmatrix}^{-1} \begin{bmatrix} 0.8 & 0.5 \end{bmatrix}$$

$$\mathbf{S}_{1.2} = \begin{bmatrix} 1 & 0.4 \\ 0.4 & 1 \end{bmatrix} - \begin{bmatrix} 0.64 & 0.40 \\ 0.40 & 0.25 \end{bmatrix} = \begin{bmatrix} 0.36 & 0 \\ 0 & 0.75 \end{bmatrix}$$

$$\mathbf{R}_{1.2} = \begin{bmatrix} 1.67 & 0 \\ 0 & 1.15 \end{bmatrix} \begin{bmatrix} 0.36 & 0 \\ 0 & 0.75 \end{bmatrix} \begin{bmatrix} 1.67 & 0 \\ 0 & 1.15 \end{bmatrix} = \begin{bmatrix} 1 & 0 \\ 0 & 1 \end{bmatrix}$$

Thus, the partial correlation $r_{12.3} = 0$, which was unexpected given that $r_{12} = 0.4$. It is concluded that, when their (linear) relationships with $\mathbf{y}_3$ are taken into account, descriptors $\mathbf{y}_1$ and $\mathbf{y}_2$ are (linearly) independent. Similar calculations for the other two pairs of descriptors give: $r_{13.2} = 0.76$ and $r_{23.1} = 0.33$. The meaning of these correlation coefficients will be further discussed in Subsections 4 and 5.

There is obviously a relationship between the coefficients of *multiple* and *partial* correlation. The equation linking the two types of coefficients can easily be derived; in the multiple correlation equation, p is the number of variables other than $\mathbf{y}_k$:

Nondetermination

when $p = 1$, the fraction of the variance of $\mathbf{y}_k$ which is not explained by $\mathbf{y}_1$ is the complement of the coefficient of determination $(1 - r_{k1}^2)$; this expression is sometimes called the *coefficient of nondetermination*;

when $p = 2$, the fraction of the variance of $\mathbf{y}_k$ which is not explained by $\mathbf{y}_2$, without taking into account the influence of $\mathbf{y}_1$, is $(1 - r_{k2.1}^2)$, so that the fraction of the variance of $\mathbf{y}_k$ which is not explained by $\mathbf{y}_1$ and $\mathbf{y}_2$ is $(1 - r_{k1}^2)\,(1 - r_{k2.1}^2)$.

This leads to a general expression for the fraction of the variance of $\mathbf{y}_k$ which is not explained by $\mathbf{y}_1, \mathbf{y}_2, \ldots, \mathbf{y}_j, \ldots$ and $\mathbf{y}_p$:

$$(1 - r_{k1}^2)\,(1 - r_{k2.1}^2) \ldots (1 - r_{kj.12\ldots}^2) \ldots (1 - r_{kp.12\ldots j\ldots(p-1)}^2)$$

Multiple determination

The fraction of the variance of $\mathbf{y}_k$ which is explained by $\mathbf{y}_1, \mathbf{y}_2, \ldots, \mathbf{y}_j, \ldots$ and $\mathbf{y}_p$, i.e. the *coefficient of multiple determination* (square of the *multiple correlation coefficient*), is thus:

$$R_{k.12\ldots p}^2 = 1 - [\,(1 - r_{k1}^2)\,(1 - r_{k2.1}^2) \ldots (1 - r_{kp.12\ldots p-1}^2)\,] \qquad \mathbf{(4.38)}$$

Numerical example. The same example as above is used to illustrate the calculation of the multiple correlation coefficient, using eq. 4.38:

$$R_{1.23}^2 = 1 - [\,(1 - r_{12}^2)\,(1 - r_{13.2}^2)\,]$$

$$R_{1.23}^2 = 1 - [1 - (0.4)^2]\,[1 - (0.76)^2] = 0.64$$

which is identical to the result obtained above using either eq. 4.31 or eq. 4.32.

Tables 4.6 and 4.7 summarize the main conclusions relative to the coefficients of multiple and partial correlation, respectively.

3 — Tests of statistical significance

The test of significance of the linear correlation coefficient r is discussed in Section 4.2 (eqs. 4.12-4.14). The null hypothesis H_0 is usually that the correlation coefficient is

Table 4.6 Main properties of the multiple (linear) correlation coefficient.

	Properties	Sections
1.	The multiple correlation coefficient measures the *intensity of the relationship* between a *response* variable and a *linear* combination of several *explanatory* variables.	4.5
2.	The square of the multiple correlation coefficient, called *coefficient of multiple determination*, measures the fraction of the variance of the response variable which is explained by a linear combination of the explanatory variables.	4.5
3.	The coefficient of multiple determination is the extension, to the multidimensional case, of the *coefficient of determination between two variables.*	4.5 and 10.3
4.	The multiple correlation coefficient can be computed from the matrix of correlations among *explanatory* variables and the vector of correlations between the *explanatory* and *response* variables.	4.5
5.	The multiple correlation coefficient can be computed from the determinant of the matrix of correlations among *explanatory* variables and that of the matrix of correlations among all variables involved.	4.5
6.	The multiple correlation coefficient can be computed from the product of a series of *complements of coefficients of partial determination.*	4.5

equal to zero (i.e. independence of the descriptors), but it could also be that the coefficient has some particular value other than zero. The general formula for testing correlation coefficients is:

$$F = \frac{r_{jk}^2 / \nu_1}{(1 - r_{jk}^2) / \nu_2} \qquad \textbf{(4.39)}$$

with $\nu_1 = m - 1$ and $\nu_2 = n - m$, where m is the number of variables involved in the correlation; this F statistic is tested against $F_{\alpha[\nu_1,\nu_2]}$. In the specific case of the bivariate correlation coefficient where $m = 2$, eq. 4.39 becomes eq. 4.12 or 4.13.

For the *multiple correlation coefficient* R, eq. 4.39 becomes (with $m = p$):

$$F = \frac{R_{1.2\ldots p}^2 / \nu_1}{(1 - R_{1.2\ldots p}^2) / \nu_2} \qquad \textbf{(4.40)}$$

Section 4.6 will show that the *partial correlation coefficient* is a parameter of the multinormal conditional distribution. This distribution being a special case or aspect of the multinormal distribution, partial correlation coefficients are tested in the same way

Table 4.7 Main properties of the partial (linear) correlation coefficient. Some of these properties are discussed in later sections.

Properties	Sections
1. The partial correlation coefficient measures the *intensity of the linear relationship* between two random variables while taking into account their relationships with other variables.	4.5
2. The partial correlation coefficient can be computed from the submatrix of correlations among variables *in partial relationship* (first subset), the submatrix of variables that *influence* the first subset, and the submatrix of correlations between the *two subsets* of variables.	4.5
3. The partial correlation coefficient can be computed from the *coefficients of simple correlation* between all pairs of variables involved.	4.5
4. The partial correlation coefficient is a *parameter* of the conditional distribution of multinormal variables.	4.6
5. The partial correlation coefficient can be defined as the *geometrical mean* of the *coefficients of partial regression* of each of the two variables on the other.	10.3
6. The square of the partial correlation coefficient (*coefficient of partial determination*; name seldom used) measures the fraction of the total variance of each variable which is mutually explained by the other, the influence of some other variables being taken into account.	10.3

as coefficients of simple correlation (eq. 4.12 for the F-test, or eq. 4.13 for the t-test, where $\nu = n - 2$). In the present case, one additional degree of freedom is lost for each successive *order* of the coefficient. For example, the number of degrees of freedom for $r_{jk.123}$ (third-order partial correlation coefficient) would be $\nu = (n - 2) - 3 = n - 5$. Equations 4.12 and 4.13 are, respectively:

$$F = \nu \frac{r_{jk.1\ldots p}^2}{1 - r_{jk.1\ldots p}^2} \quad \textbf{(4.12)} \qquad \text{and} \qquad t = \sqrt{\nu} \frac{r_{jk.1\ldots p}}{\sqrt{1 - r_{jk.1\ldots p}^2}} \quad \textbf{(4.13)}$$

As usual (see Sections 1.2 and 4.2), H_0 is tested by comparing the computed statistic (F or t) to a table of critical values and it is rejected when the associated probability is smaller than a predetermined *level of significance* α.

4 — Interpretation of correlation coefficients

In the ecological literature, correlation coefficients are often interpreted in terms of causal relationships among descriptors. It will now be shown that this should never be

Table 4.8 Relationships between a primary descriptor $\mathbf{y}_1$ and a dependent descriptor $\mathbf{y}_3$ in the presence of a secondary descriptor $\mathbf{y}_2$. Predictions about the relationships among linear correlation coefficients (adapted from De Neufville & Stafford, 1971) and the corresponding partial correlation coefficients.

Elementary causal models	Causal diagrams	Conditions	Predictions among simple r	partial r
Secondary descriptor $\mathbf{y}_2$ *in the middle*				
Intervening sequence: secondary descriptor $\mathbf{y}_2$ intervenes between $\mathbf{y}_1$ and $\mathbf{y}_3$	$\mathbf{y}_1 \to \mathbf{y}_2 \to \mathbf{y}_3$	$r_{12} \neq 0$ $r_{23} \neq 0$	$r_{13} = r_{12}r_{23}$	$r_{13.2} = 0$ $\lvert r_{12.3}\rvert \leq \lvert r_{12}\rvert$ $\lvert r_{23.1}\rvert \leq \lvert r_{23}\rvert$
Spurious correlation: primary descriptor $\mathbf{y}_1$ and supposedly dependent descriptor $\mathbf{y}_3$ are correlated but not causally connected	$\mathbf{y}_1 \leftarrow \mathbf{y}_2 \rightarrow \mathbf{y}_3$	$r_{12} \neq 0$ $r_{23} \neq 0$	$r_{13} = r_{12}r_{23}$	$r_{13.2} = 0$ $\lvert r_{12.3}\rvert \leq \lvert r_{12}\rvert$ $\lvert r_{23.1}\rvert \leq \lvert r_{23}\rvert$
Primary descriptor $\mathbf{y}_1$ *in the middle*				
Developmental sequence: $\mathbf{y}_1$, which is partially caused by $\mathbf{y}_2$, causes $\mathbf{y}_3$	$\mathbf{y}_2 \to \mathbf{y}_1 \to \mathbf{y}_3$	$r_{12} \neq 0$ $r_{13} \neq 0$	$r_{23} = r_{12}r_{13}$	$r_{23.1} = 0$ $\lvert r_{12.3}\rvert \leq \lvert r_{12}\rvert$ $\lvert r_{13.2}\rvert \leq \lvert r_{13}\rvert$
Double effect: primary descriptor $\mathbf{y}_1$ causes both $\mathbf{y}_2$ and $\mathbf{y}_3$	$\mathbf{y}_2 \leftarrow \mathbf{y}_1 \rightarrow \mathbf{y}_3$	$r_{12} \neq 0$ $r_{13} \neq 0$	$r_{23} = r_{12}r_{13}$	$r_{23.1} = 0$ $\lvert r_{12.3}\rvert \leq \lvert r_{12}\rvert$ $\lvert r_{13.2}\rvert \leq \lvert r_{13}\rvert$
Dependent descriptor $\mathbf{y}_3$ *in the middle*				
Double cause: both $\mathbf{y}_1$ and $\mathbf{y}_2$ independently affect $\mathbf{y}_3$	$\mathbf{y}_1 \rightarrow \mathbf{y}_3 \leftarrow \mathbf{y}_2$	$r_{13} \neq 0$ $r_{23} \neq 0$	$r_{12} = 0$	$r_{12.3} \neq 0$ $\lvert r_{13.2}\rvert \geq \lvert r_{13}\rvert$ $\lvert r_{23.1}\rvert \geq \lvert r_{23}\rvert$

done when the only information available is the correlation coefficients themselves. The matter is examined using the simple case of three linearly related descriptors $\mathbf{y}_1$, $\mathbf{y}_2$, and $\mathbf{y}_3$. Considering two causal relationships only, there are five elementary models describing the possible interactions between a *primary descriptor* $\mathbf{y}_1$ and a *response descriptor* $\mathbf{y}_3$ in the presence of a *secondary descriptor* $\mathbf{y}_2$. The five models are shown in Table 4.8. Arrows symbolize causal relationships between descriptors (e.g. $\mathbf{y}_i \to \mathbf{y}_j$: $\mathbf{y}_j$ is caused by $\mathbf{y}_i$). Using path analysis (Section 10.4), De Neufville & Stafford (1971) computed, for each causal model, the relationships that should be found among the simple linear correlation coefficients, assuming the conditions specified in the Table.

Table 4.8 also gives the *first-order partial correlation coefficients* computed from the coefficients of simple linear correlation. (1) When the correlation between the primary descriptor ($\mathbf{y}_1$) and the dependent descriptor ($\mathbf{y}_3$) is caused by the presence of a secondary variable (i.e. $\mathbf{y}_2$ in the middle: intervening sequence or spurious correlation), controlling for the effect of $\mathbf{y}_2$ through partial correlation evidences the lack of causal relationship between $\mathbf{y}_1$ and $\mathbf{y}_3$ (i.e. $r_{13.2} = 0$). The two other partial correlations ($r_{12.3}$ and $r_{23.1}$) are then smaller (i.e closer to zero) than the corresponding simple correlations (r_{12} and r_{23}). (2) When there is a direct causal relationship from the primary ($\mathbf{y}_1$) to the dependent ($\mathbf{y}_3$) descriptor (i.e. $\mathbf{y}_1$ in middle: developmental sequence or double effect), partial correlation $r_{23.1} = 0$ shows that there is no direct relationship between the secondary ($\mathbf{y}_2$) and the dependent ($\mathbf{y}_3$) descriptors. The two other partial correlations ($r_{12.3}$ and $r_{13.2}$) follow the same rule as in the previous case. (3) When the dependent descriptor ($\mathbf{y}_3$) is caused by both the primary ($\mathbf{y}_1$) and secondary ($\mathbf{y}_2$) descriptors (i.e. $\mathbf{y}_3$ in the middle: double cause), the three partial correlation coefficients are larger than the corresponding coefficients of simple correlation.

Causal model

These five elementary causal models, for the simple case where three descriptors only are involved, show how difficult it is to interpret correlation matrices, especially when several ecological descriptors are interacting in complex ways. Partial correlations may be used to help elucidate the relationships among descriptors. However, the choice of a causal model always requires hypotheses, or else the input of external ecological information. When it is possible, from *a priori* information or ecological hypotheses, to specify the causal ordering among descriptors, path analysis (Section 10.4) may be used to assess the correspondence between the data (i.e. correlations) and causal models. It must be stressed again that a causal model may never be derived from a correlation matrix, whereas a causal model is needed to interpret a correlation matrix.

Numerical example. The simple example already used for multiple and partial correlations illustrates here the problem inherent to all correlation matrices, i.e. that it is never possible to interpret correlations *per se* in terms of causal relationships. In the following matrix, the upper triangle contains the coefficients of simple correlation whereas the lower triangle contains the partial correlation coefficients:

$$\begin{bmatrix} 1 & 0.4 & 0.8 \\ 0 & 1 & 0.5 \\ 0.76 & 0.33 & 1 \end{bmatrix}$$

It may have looked as though descriptors $\mathbf{y}_1$ and $\mathbf{y}_2$ were somewhat correlated ($r_{12} = 0.4$), but the first-order partial correlation coefficient $r_{12.3} = 0$ shows that this is not the case. This numerical example corresponds to any of the first four models in Table 4.8, assuming that all the non-zero simple and partial correlation coefficients are significantly different from 0; see also the next Subsection. In the absence of external information or ecological hypotheses, there is no way of determining which pattern of causal relationships among descriptors fits the correlation matrix.

5 — *Causal modelling using correlations*

A simple form of causal modelling may be carried out on three variables. It simply involves looking at simple and partial correlation coefficients. One basic condition must be fulfilled for such a model to encompass the three variables; it is that at least two of the simple correlation coefficients be significantly different from zero. Under the assumption of linear relationships among variables, these two coefficients support two "causal arrows". In the remainder of the present Subsection, the three variables $\mathbf{y}_a$, $\mathbf{y}_b$, and $\mathbf{y}_c$ are denoted **a**, **b**, and **c** for simplicity. "Causality" refers, in statistics, to the hypothesis that changes occurring in one variable have an effect on changes in another variable; *causality resides in the hypotheses only.* Within the framework of a specific sampling design (i.e. spatial, temporal, or experimental) where variation is controlled, data are said to support the causality hypothesis if a significant portion of the variation in **b** is indeed explained by changes taking place in **a**. If the relationship is assumed to be linear, a significant linear correlation coefficient is interpreted as supporting the hypothesis of linear "causation".

Causality

Four different linear models only can be formulated, when at least two of the simple correlation coefficients among three variables are significantly different from zero. Their characteristics are specified in Fig. 4.11. Model 1 corresponds to any one of the two sequences in Table 4.8 and model 2 is the double effect. As shown above, it is not possible to distinguish between models 1 and 2 from the correlation coefficients alone, i.e. the two models are distinct only in their hypotheses. Model 3 is the double cause. Model 4 describes a triangular relationship, which may be seen as a combination of models 1 and 2. The direct and indirect effects implied in model 4 may be further analysed using path analysis (Section 10.4). Examining model 1 in some detail illustrates how the "expectations of the model" are derived.

• Significance of the simple correlations. Obviously (Table 4.8), $r_{\mathbf{ab}}$ and $r_{\mathbf{bc}}$ must be significantly different from zero for the model to hold. The model can accommodate $r_{\mathbf{ac}}$ being significant or not, although the value of $r_{\mathbf{ac}}$ should always be different from zero since $r_{\mathbf{ac}} = r_{\mathbf{ab}} r_{\mathbf{bc}}$.

• Significance of the partial correlations. The condition $r_{\mathbf{ac}} = r_{\mathbf{ab}} r_{\mathbf{bc}}$ stated in Table 4.8 implies that $r_{\mathbf{ac}} - r_{\mathbf{ab}} r_{\mathbf{bc}} = 0$ or, in other words (eq. 4.36), $r_{\mathbf{ac.b}} = 0$. In addition, for the model to hold, partial correlations $r_{\mathbf{ab.c}}$ and $r_{\mathbf{bc.a}}$ must be significantly different from 0. Indeed, $r_{\mathbf{ab.c}}$ being equal to zero would mean that $r_{\mathbf{ab}} = r_{\mathbf{ac}} r_{\mathbf{bc}}$, which would imply that **c** is in the centre of the sequence; this is not the case in the model as specified, where **b** is in the centre. The same reasoning explains the relationship $r_{\mathbf{bc.a}} \neq 0$.

• Comparison of simple correlation values. Since correlation coefficients are smaller than or equal to 1 in absolute value, the relationship $r_{\mathbf{ac}} = r_{\mathbf{ab}} r_{\mathbf{bc}}$ implies that $|r_{\mathbf{ab}}| \geq |r_{\mathbf{ac}}|$ and $|r_{\mathbf{bc}}| \geq |r_{\mathbf{ac}}|$.

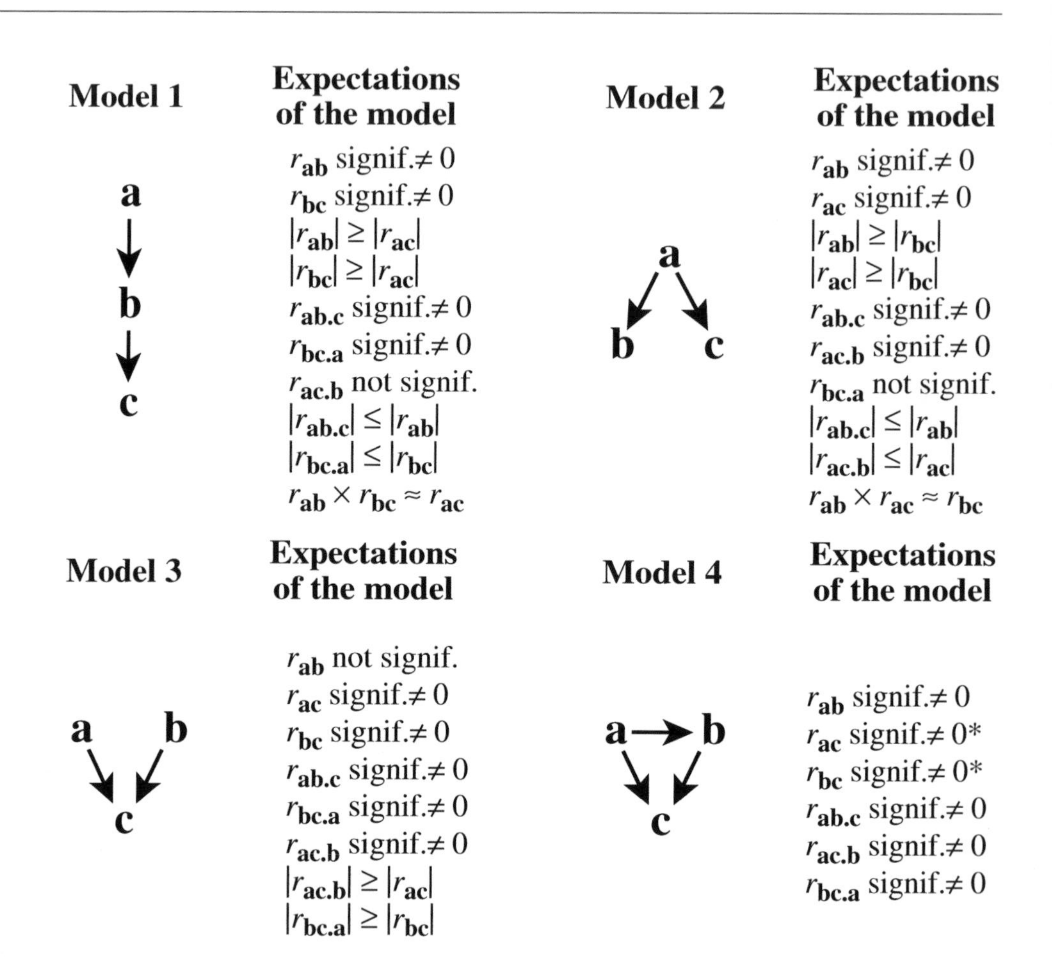

Figure 4.11 Predictions of the four possible models of causal relationships involving three variables, in terms of the expected values for the simple and partial linear correlation coefficients. 'r_{ab} signif.≠ 0' means that, under the model, the correlation must be significantly different from zero. 'r_{ab} not signif.' means that the correlation is not necessarily significantly different from zero at the pre-selected significance level. * Model 4 holds even if one, *but only one*, of these two simple correlation coefficients is not significant. Adapted from Legendre (1993).

• Comparison of partial correlation values. Consider the partial correlation formula for $r_{ab.c}$ (eq. 4.36). Is it true that $|r_{ab.c}| \leq |r_{ab}|$? The relationship $r_{ac} = r_{ab} r_{bc}$ allows one to replace r_{ac} by $r_{ab} r_{bc}$ in that equation. After a few lines of algebra, the inequality

$$|r_{ab.c}| = \frac{|r_{ab}|\,[1 - r_{bc}^2]}{\sqrt{[1 - r_{ab}^2 r_{bc}^2]\,[1 - r_{bc}^2]}} \leq |r_{ab}|$$

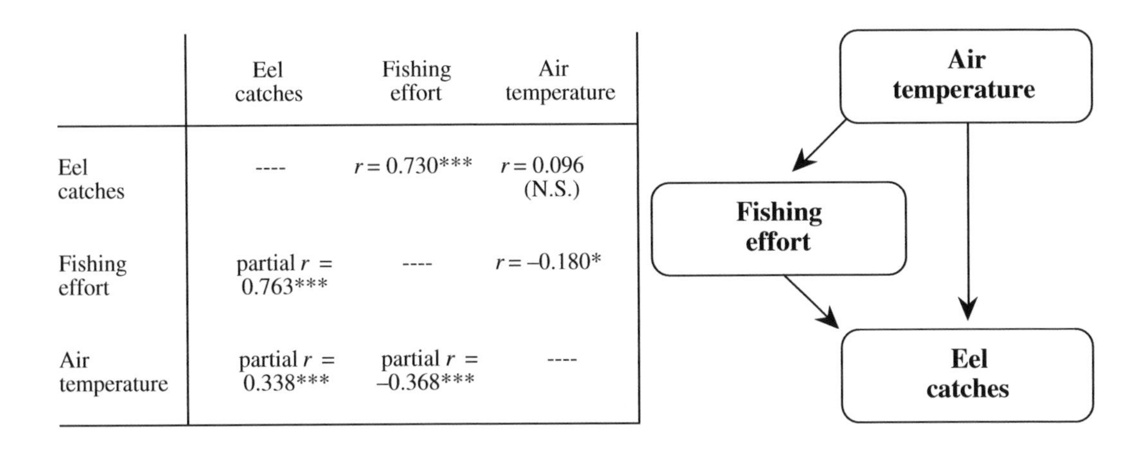

	Eel catches	Fishing effort	Air temperature
Eel catches	----	$r = 0.730$***	$r = 0.096$ (N.S.)
Fishing effort	partial $r =$ 0.763***	----	$r = –0.180$*
Air temperature	partial $r =$ 0.338***	partial $r =$ –0.368***	----

Figure 4.12 Left: simple and partial correlations among temperature, fishing effort, and eel catches using the 'capêchade' fishing gear, from Bach *et al.* (1992). Right: causal model supported by the data. *: $0.05 \geq p > 0.01$; ***: $p \leq 0.001$; N.S.: non-significant correlation ($\alpha = 0.05$).

leads to the relationship $r^2_{\mathbf{bc}}(1 - r^2_{\mathbf{ab}}) \geq 0$ which is true in all cases because $r_{\mathbf{bc}} \neq 0$ and $|r_{\mathbf{ab}}| \leq 1$. This also shows that $r_{\mathbf{ab.c}} = r_{\mathbf{ab}}$ only when $r_{\mathbf{ab}} = 1$. The same method may be used to demonstrate that $|r_{\mathbf{bc.a}}| \leq |r_{\mathbf{bc}}|$.

The (3×3) matrix of simple and partial correlations used as numerical example in the previous Subsection obeys all the conditions corresponding to models 1 and 2. Which of these two models is the correct one? This would depend on the nature of the three variables and, foremost, on the hypotheses one intends to test. The hypotheses determine the presence and direction of the arrows among variables in causal models.

Ecological application 4.5

Bach *et al.* (1992) analysed a 28-month long time series (weekly sampling) of eel catches (*Anguilla anguilla*) in the Thau marine lagoon, southern France. Fixed gears called 'capêchades', made of three funnel nets (6-mm mesh) and an enclosure, were used near the shore in less than 1.5 m of water. In the deeper parts of the lagoon, other types of gears were used: heavier assemblages of funnel nets with larger mesh sizes, called 'brandines', 'triangles' and 'gangui', as well as longlines. Various hypotheses were stated by the authors and tested using partial correlation analysis and path analysis. These concerned the influence of environmental variables on the behaviour of fish and fishermen and their effects on landings. Coefficients of linear correlation reported in the paper are used here to study the relationships among air temperature, fishing effort, and landings, for the 'capêchade' catches (Fig 4.12). The analysis in the paper was more complex; it also considered the effects of wind and moon. Linearity of the relationships was checked. The correlation coefficients support a type-4 model, stating that both effort and temperature affect the landings (temperature increases eel metabolism and thus their activity and catchability) and that the effort, represented by the number of active 'capêchade' fishermen, is affected by temperature (lower effort at high temperature, 'capêchades' being not much used from August to October). Interesting is the non-significant simple linear correlation between temperature and catches. The partial correlations indicate that this simple correlation corresponds to two effects of temperature on catches that are both significant but of opposite signs: a positive partial correlation of temperature on catches and a negative one of temperature on effort. In the paper of Bach *et al.*, partial correlation analysis was used as a first screen to eliminate variables that clearly did not influence catches. Path analysis (Section 10.4) was then used to study the direct and indirect effects of the potentially explanatory variables on catches.

Partial correlations do not provide the same information as path analysis (Section 10.4). On the one hand, partial correlations, like partial regression coefficients (Section 10.3), indicate whether a given variable has some unique (linear) relationship with some other variable, after the linear effects of all the other variables in the model have been taken into account. In path analysis, on the other hand, one is mostly interested in partitioning the relationship between predictor (explanatory, independent) and criterion (response, dependent) variables into direct and indirect components.

The above discussion was based on linear correlation coefficients. Advantages of the linear model include ease of computation and simplicity of interpretation. However, environmental processes are not necessarily linear. This is why linearity must be checked, not only assumed, before embarking in this type of computation. When the phenomena are not linear, two choices are open: either proceed with non-linear statistics (nonparametric simple and partial correlation coefficients, in particular, are available and may be used in this type of modelling), or linearize the relationships that seem promising. Monotonic relationships, identified in scatter diagrams, may usually be linearized using the transformations of Section 1.5 to one or both variables. There is no 'cheating' involved in doing so; either a monotonic relationship exists, and linearizing transformations allow one to apply linear statistics to the data; or such a relationship does not exist, and no amount of transformation will ever create one.

Simple causal modelling, as presented in this Subsection, may be used in two different types of circumstances. A first, common application is exploratory analysis, which is performed when 'weak' ecological hypotheses only can be formulated. What this means is the following: in many studies, a large number of causal hypotheses may be formulated *a priori*, some being contradictory, because processes at work in ecosystems are too numerous for ecologists to decide which ones are dominant under given circumstances. So, insofar as each of the models derived from ecological theory can be translated into hypothesized correlation coefficients, partial correlation analysis may be used to clear away those hypotheses that are not supported by the data and to keep only those that look promising for further analysis. Considering three variables, for instance, one may look at the set of simple and partial correlation coefficients and decide which of the four models of Fig. 4.11 are not supported by the data. Alternatively, when the ecosystem is better understood, one may wish to test a single set of hypotheses (i.e. a single model), to the exclusion of all others. With three variables, this would mean testing only one of the models of Fig. 4.11, to the exclusion of all others, and deciding whether that model is supported or not by the data.

Several correlation coefficients are tested in Fig. 4.11. Three simultaneous tests are performed for the simple correlation coefficients and three for the partial correlation coefficients. In order to determine whether such results could have been obtained by chance alone, some kind of global test of significance, or correction, must be performed (Box 1.3; eq. 4.14).

The simple form of modelling described here may be extended beyond the frame of linear modelling, as long as formulas exist for computing partial relationships. Examples are the partial nonparametric correlation coefficients (partial Kendall τ, eq. 5.9) and partial Mantel statistics (Subsection 10.5.2).

4.6 Multinormal conditional distribution

In Section 4.3, which deals with the multinormal distribution, an important property was demonstrated, namely that correlation coefficients are parameters of the multinormal distribution. In the same way, it will be shown here that partial correlation coefficients, described in the previous section, are parameters of a distribution, derived from the multinormal distribution, which is called the *conditional distribution of multinormal random variables* (or *multinormal conditional distribution*). The fact that ρ is a parameter of the multinormal distribution is the basis for testing the significance of simple correlation coefficients. Similarly, the fact that partial correlation coefficients are parameters of a distribution is the basis for testing their significance, using the approach explained in the previous section.

In the multidimensional context, a set of random variables is sometimes partitioned into two subsets, so as to study the distribution of the variables in the first set $(y_1,y_2,\ldots,y_p)$ while maintaining those in the second set $(y_{p+1},\ldots,y_{p+q})$ fixed. These are the conditions already described for partial correlations (eqs. 4.34 to 4.37). Such a probability distribution is called a *conditional distribution.* It can be shown that the conditional distribution of variables of the first set, given the second set of fixed variables, is:

$$g\,(y_1, y_2, \ldots, y_p | y_{p+1}, \ldots, y_{p+q}) = \frac{f\,(y_1, y_2, \ldots, y_p, y_{p+1}, \ldots, y_{p+q})}{h\,(y_{p+1}, \ldots, y_{p+q})} \quad \textbf{(4.41)}$$

where $f(y_1,y_2, \ldots, y_{p+q})$ is the joint probability density of the $(p + q)$ variables in the two subsets and $h(y_{p+1}, \ldots, y_{p+q})$ is the joint probability density of the q fixed variables (second subset). When the two subsets are *independent,* it has already been shown

(Section 4.3) that their joint probability density is the product of the densities of the two subsets, so that:

$$g(y_1, y_2, \ldots, y_p | y_{p+1}, \ldots, y_{p+q}) = \frac{f(y_1, y_2, \ldots, y_p)\, h(y_{p+1}, \ldots, y_{p+q})}{h(y_{p+1}, \ldots, y_{p+q})}$$

$$g(y_1, y_2, \ldots, y_p | y_{p+1}, \ldots, y_{p+q}) = f(y_1, y_2, \ldots, y_p) \quad \textbf{(4.42)}$$

The conditional density is then the probability density of the p random variables in the first subset; this is because variables in this subset are not influenced by those in the second subset.

In most cases of interest to ecologists, the variables under study are not independent of one another (Section 4.1). The corresponding multinormal population (i.e. with all variables intercorrelated) may be partitioned into two subsets:

$$\mathbf{Y} = [\mathbf{y}_1\ \mathbf{y}_2]$$

where $\mathbf{y}_1$ represents the p variables belonging to the first subset and $\mathbf{y}_2$ the q variables belonging to the second subset:

$$\mathbf{y}_1 = [y_1, y_2, \ldots, y_p] \quad \text{and} \quad \mathbf{y}_2 = [y_{p+1}, y_{p+2}, \ldots, y_{p+q}]$$

The multidimensional mean $\boldsymbol{\mu}$ and dispersion matrix $\boldsymbol{\Sigma}$ are partitioned in the same way:

$$\boldsymbol{\mu} = [\boldsymbol{\mu}_1\ \boldsymbol{\mu}_2] \quad \text{and} \quad \boldsymbol{\Sigma} = \begin{bmatrix} \boldsymbol{\Sigma}_{11} & \boldsymbol{\Sigma}_{12} \\ \boldsymbol{\Sigma}_{21} & \boldsymbol{\Sigma}_{22} \end{bmatrix}$$

The values of the elements in $\boldsymbol{\Sigma}_{12}$ (or $\boldsymbol{\Sigma}_{21}$), with respect to those in $\boldsymbol{\Sigma}_{11}$ and $\boldsymbol{\Sigma}_{22}$, determine the dependence between the two subsets of variables, as in eq. 11.2.

The conditional probability density of $\mathbf{y}_1$, for fixed values of $\mathbf{y}_2$, is:

$$g(\mathbf{y}_1 | \mathbf{y}_2) = \frac{f(\mathbf{y}_1, \mathbf{y}_2)}{h(\mathbf{y}_2)} \quad \textbf{(4.43)}$$

The probability densities for the *whole set* of variables $f(\mathbf{y}_1, \mathbf{y}_2)$ and for the *subset of fixed* variables $h(\mathbf{y}_2)$ are calculated as in eq. 4.19:

$$f(\mathbf{y}_1, \mathbf{y}_2) = \frac{1}{(2\pi)^{(p+q)/2} |\boldsymbol{\Sigma}|^{1/2}} \exp\{-(1/2)\,[\mathbf{y} - \boldsymbol{\mu}]\, \boldsymbol{\Sigma}^{-1}\, [\mathbf{y} - \boldsymbol{\mu}]'\} \quad \textbf{(4.44)}$$

$$h(\mathbf{y}_2) = \frac{1}{(2\pi)^{q/2} |\boldsymbol{\Sigma}_{22}|^{1/2}} \exp\{-(1/2)\,[\mathbf{y}_2 - \boldsymbol{\mu}_2]\, \boldsymbol{\Sigma}_{22}^{-1}\, [\mathbf{y}_2 - \boldsymbol{\mu}_2]'\} \quad \textbf{(4.45)}$$

Submatrices $\mathbf{\Sigma}_{11}$ and $\mathbf{\Sigma}_{22}$ being square, the determinant of the whole matrix $\mathbf{\Sigma}$ may be expressed as a function of the determinants of its four submatrices:

$$|\mathbf{\Sigma}| = |\mathbf{\Sigma}_{22}| \cdot |\mathbf{\Sigma}_{11} - \mathbf{\Sigma}_{12}\mathbf{\Sigma}_{22}^{-1}\mathbf{\Sigma}_{21}|$$

Using the above three equations, the conditional probability density of $\mathbf{y}_1$, for fixed values of $\mathbf{y}_2$, (eq. 4.43) becomes:

$$g(\mathbf{y}_1|\mathbf{y}_2) = \frac{(2\pi)^{q/2}|\mathbf{\Sigma}_{22}|^{1/2}\exp\ldots}{(2\pi)^{(p+q)/2}|\mathbf{\Sigma}_{22}|^{1/2}|\mathbf{\Sigma}_{11} - \mathbf{\Sigma}_{12}\mathbf{\Sigma}_{22}^{-1}\mathbf{\Sigma}_{21}|^{1/2}\exp\ldots}$$

$$g(\mathbf{y}_1|\mathbf{y}_2) = \frac{1}{(2\pi)^{p/2}|\mathbf{\Sigma}_{11} - \mathbf{\Sigma}_{12}\mathbf{\Sigma}_{22}^{-1}\mathbf{\Sigma}_{21}|^{1/2}}\exp\ldots \qquad \textbf{(4.46)}$$

This shows that the *dispersion matrix of the conditional probability distribution* is:

$$\mathbf{\Sigma}_{1.2} = \mathbf{\Sigma}_{11} - \mathbf{\Sigma}_{12}\mathbf{\Sigma}_{22}^{-1}\mathbf{\Sigma}_{21} \qquad \textbf{(4.47)}$$

Developing the exponent (exp...) of eq. 4.46 would show that the corresponding *mean* is:

$$\boldsymbol{\mu}_{1.2} = \boldsymbol{\mu}_1 + [y_2 - \boldsymbol{\mu}_2]\,\mathbf{\Sigma}_{22}^{-1}\mathbf{\Sigma}_{21} \qquad \textbf{(4.48)}$$

It is not easy to understand, from the above equations, what are exactly the respective roles of the two subsets of variables, $\mathbf{y}_1$ and $\mathbf{y}_2$, in the conditional distribution. Examination of the simplest case, with each of the two subsets containing *only one variable,* shows the main characteristics of this special distribution. In this simple case, $\mathbf{y}_1 = y_1$ and $\mathbf{y}_2 = y_2$, which leads to the following conditional density (eqs. 4.44 to 4.46):

$$g(y_1|y_2) = \frac{1}{(2\pi)^{1/2}(\sigma_1^2 - \sigma_{12}^2/\sigma_2^2)^{1/2}}\exp\left\{-(1/2)\left[(y_1-\mu_1) - \frac{\sigma_{12}}{\sigma_2^2}(y_2-\mu_2)\right]^2\left[\sigma_1^2 - \frac{\sigma_{12}^2}{\sigma_2^2}\right]^{-1}\right\}$$

Since $\rho = \sigma_{12}/\sigma_1\sigma_2$, it follows that:

$$\sigma_1^2 - \sigma_{12}^2/\sigma_2^2 = \sigma_1^2(1-\rho^2) \qquad \text{and} \qquad \frac{\sigma_{12}}{\sigma_2^2} = \rho\frac{\sigma_1}{\sigma_2}$$

Thus:

$$g(y_1|y_2) = \frac{1}{\sqrt{2\pi}\sqrt{\sigma_1^2(1-\rho^2)}}\exp\left\{-\frac{1}{2\sigma_1^2(1-\rho^2)}\left[(y_1-\mu_1) - \rho\frac{\sigma_1}{\sigma_2}(y_2-\mu_2)\right]^2\right\}$$

where ρ is the correlation between the two variables. This last equation may be rewritten in a simplified form, after defining the conditional mean $\mu_{1|2}$ and variance $\sigma^2_{1|2}$:

$$\mu_{1|2} = \mu_1 + \rho\frac{\sigma_1}{\sigma_2}(y_2 - \mu_2) \qquad \sigma^2_{1|2} = \sigma^2_1(1-\rho^2)$$

$$g(y_1|y_2) = \frac{1}{\sqrt{2\pi}\sigma_{1|2}}\exp\left[-\frac{1}{2}\left(\frac{y_1-\mu_{1|2}}{\sigma_{1|2}}\right)^2\right]$$

which is the conditional form of the one-dimensional normal distribution (eq. 4.15). The equation for $\mu_{1|2}$ shows that, in the conditional case, the mean $\mu_{1|2}$ is a function of the value taken by the fixed variable y_2.

Figure 4.13 illustrates the conditional distribution of a variable y_1, for four values of a fixed variable y_2. Given μ_1, μ_2, σ_1, σ_2 and ρ_{12}, the position of the conditional mean $\mu_{1|2}$ depends solely on the value taken by y_2. This property of the conditional distribution appears clearly when examining the line along which means $\mu_{1|2}$ are located (dashed line). The position of the conditional normal distribution $g(y_1|y_2)$, in the plane of axes y_1 and y_2, is determined by the position of $\mu_{1|2}$, since a normal distribution is centred on its mean.

Understanding the multidimensional case requires a little imagination. When the first subset $\mathbf{y}_1$ contains two variables (y_1 and y_2) and the second ($\mathbf{y}_2$) only one variable (y_3), each one-dimensional normal curve in Fig. 4.13 is replaced by a two-dimensional probability "bell" (Fig. 4.6). These bells represent the binormal distribution of y_1 and y_2, for values of the fixed variable y_3. This would be the case of the partial correlation $r_{12.3}$ calculated between variables y_1 and y_2 while controlling for the effect of a third one (y_3); this case has been examined in Subsection 4.5.2 (eq. 4.36).

In the more complex situation where $\mathbf{y}_1 = [y_1\ y_2]$ and $\mathbf{y}_2 = [y_3\ y_4]$, the "bells", representing the binormal distribution of y_1 and y_2, are located in a three-dimensional space instead of a plane, since the position of $\mu_{1|2}$ is then determined by the values of the two fixed variables y_3 and y_4. This would be the case of the partial correlation $r_{12.34}$, calculated between variables y_1 and y_2 while controlling for the effects of variables y_3 and y_4. It is not easy to picture higher dimensions which, however, have actual mathematical existence — and real ecological significance — since the conditional distribution of multinormal variables is that of all descriptors in partial correlation.

It was shown above (eq. 4.47) that the dispersion matrix of the conditional probability distribution is:

$$\mathbf{\Sigma}_{1.2} = \mathbf{\Sigma}_{11} - \mathbf{\Sigma}_{12}\mathbf{\Sigma}^{-1}_{22}\mathbf{\Sigma}_{21}$$

Extending eq. 4.10 to the conditional case, the partial correlation matrix is:

$$\mathbf{P}_{1.2} = \mathbf{D}(\sigma^2_{1.2})^{-1/2}\ \mathbf{\Sigma}_{1.2}\ \mathbf{D}(\sigma^2_{1.2})^{-1/2} \qquad \textbf{(4.49)}$$

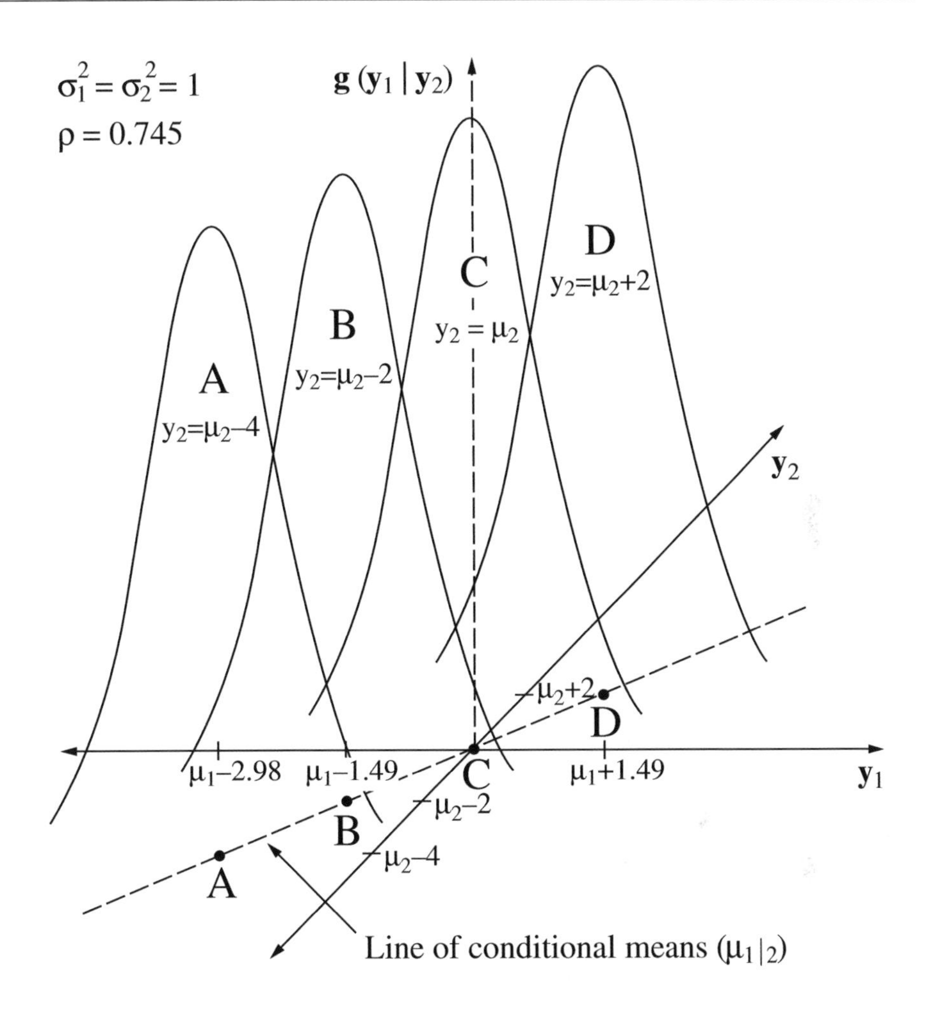

Figure 4.13 Conditional distribution of a normal variable y_1, for four values of a fixed variable y_2. Conditional means $\mu_{1|2}$ are located along the dashed line.

where $\mathbf{D}(\sigma^2_{1.2})$ is the matrix of the diagonal elements of $\mathbf{\Sigma}_{1.2}$. Equation 4.49 is the same as eq. 4.35 except for the fact that it is formulated in terms of dispersion instead of correlation, and it concerns parameters instead of statistics.

Equation 4.49 shows that *partial correlation coefficients are parameters of the multinormal conditional distribution* in the same way as simple correlation coefficients are parameters of the multinormal distribution (eqs. 4.20 and 4.21). As stated at the beginning of the present section, this property is the basis for testing partial correlation coefficients (Subsection. 4.5.3). It thus has practical significance for ecologists.

4.7 Tests of normality and multinormality

Testing the normality of empirical distributions is an important concern for ecologists who want to use linear models for analysing their data. Tests of normality are carried out in two types of circumstances. On the one hand, many tests of statistical significance, including those described in the present chapter, require the empirical data to be drawn from normally distributed populations. On the other hand, the linear methods of multivariate data analysis discussed in Chapters 9, 10, and 11 do summarize data in more informative ways if their underlying distributions are multinormal — or at least are not markedly skewed, as discussed below. Testing the normality of empirical variables is thus an important initial step in the analysis of a data set. Variables that are not normally distributed may be subjected to normalizing transformations (Section 1.5). The historical development of the tests of normality has been reviewed by D'Agostino (1982) and Dutilleul & Legendre (1992).

The problem may first be approached by plotting frequency distributions of empirical variables. Looking at these plots immediately identifies distributions that have several modes, for instance, or that are obviously too skewed, or too 'flat' or 'peaked', to have been possibly drawn from normally distributed populations.

Skewness

Next, for unimodal distributions, one may examine the parameters skewness and kurtosis. *Skewness* (α_3) is a measure of asymmetry; it is defined as the third moment of the distribution (the first moment being equal to zero, $m_1 = 0$, and the second being the variance, $m_2 = s^2$),

$$m_3 = \Sigma (x - \bar{x})^3 / (n - 1)$$

divided by the cube of the standard deviation:

$$\alpha_3 = m_3 / s_x^3 \qquad \textbf{(4.50)}$$

Kurtosis

Skewness is 0 for a normal distribution. Positive skewness corresponds to a frequency distribution with a longer 'tail' to the right than to the left, whereas a distribution with a longer 'tail' to the left than to the right shows negative skewness. *Kurtosis* (α_4 or α'_4) is a measure of flatness or peakedness; it is defined as the fourth moment of the frequency distribution,

$$m_4 = \Sigma (x - \bar{x})^4 / (n - 1)$$

divided by the standard deviation to the power 4:

$$\alpha_4 = m_4 / s_x^4 \qquad \textbf{(4.51)}$$

Since the kurtosis of a normal distribution is $\alpha_4 = 3$, authors (and computer packages) in the U. S. tradition use a modified formula for kurtosis,

$$\alpha'_4 = \alpha_4 - 3 \tag{4.52}$$

which is such that the kurtosis of a normal distribution is $\alpha_4 = 3$. Distributions flatter than the normal distribution have negative values for α'_4 whereas distributions that have more observations around the mean than the normal distribution have positive values for α'_4, indicating that they are more 'peaked'.

Although tests of significance have been developed for skewness and kurtosis, they are not used any longer because more powerful tests of goodness-of-fit are now available. For the same reason, testing the goodness-of-fit of an empirical frequency distribution to a normal distribution with same mean and variance (as in Fig 4.14a) using a chi-square test is no longer in fashion because it is not very sensitive to departures from normality (Stephens, 1974; D'Agostino, 1982), even though it is often presented in basic texts of biological statistics as a procedure of choice. The main problem is that it does not take into account the ordering of classes of the two frequency distributions that are being compared. This explains why the main statistical packages do not use it, but propose instead one or the other (or both) procedure described below.

One of the widely used tests of normality is the Kolmogorov-Smirnov test of goodness-of-fit. In Fig. 4.14b, the same data as in Fig. 4.14a are plotted as a cumulative frequency distribution. The cumulative theoretical normal distribution is also plotted on the same graph; it can easily be obtained from a published table, or by requesting in a statistical package the normal probability values corresponding to the relative cumulative frequencies. One looks for the largest deviation D between the cumulative empirical relative frequency distribution and the cumulative theoretical normal distribution. If D is larger than the critical value in the table, for a given number of observations n and significance level α, the hypothesis of normality is rejected.

K-S test

The Kolmogorov-Smirnov test of goodness-of-fit is especially interesting for small sample sizes because it does not require to lump the data into classes. When they are divided into classes, the empirical data are discontinuous and their cumulative distribution is a step-function, whereas the theoretical normal distribution to which they are compared is a continuous function. D is then formally defined as the maximum of D^- and D^+, where D^- is the maximum difference computed just before a data value and D^+ is the maximum difference computed at the data value (i.e. at the top of each step of the cumulative empirical step-function). A numerical example is given by Sokal & Rohlf (1995).

Standard Kolmogorov-Smirnov tables for the comparison of two samples, where the distribution functions are completely specified (i.e. the mean and standard deviation are stated by hypothesis), are not appropriate for testing the normality of

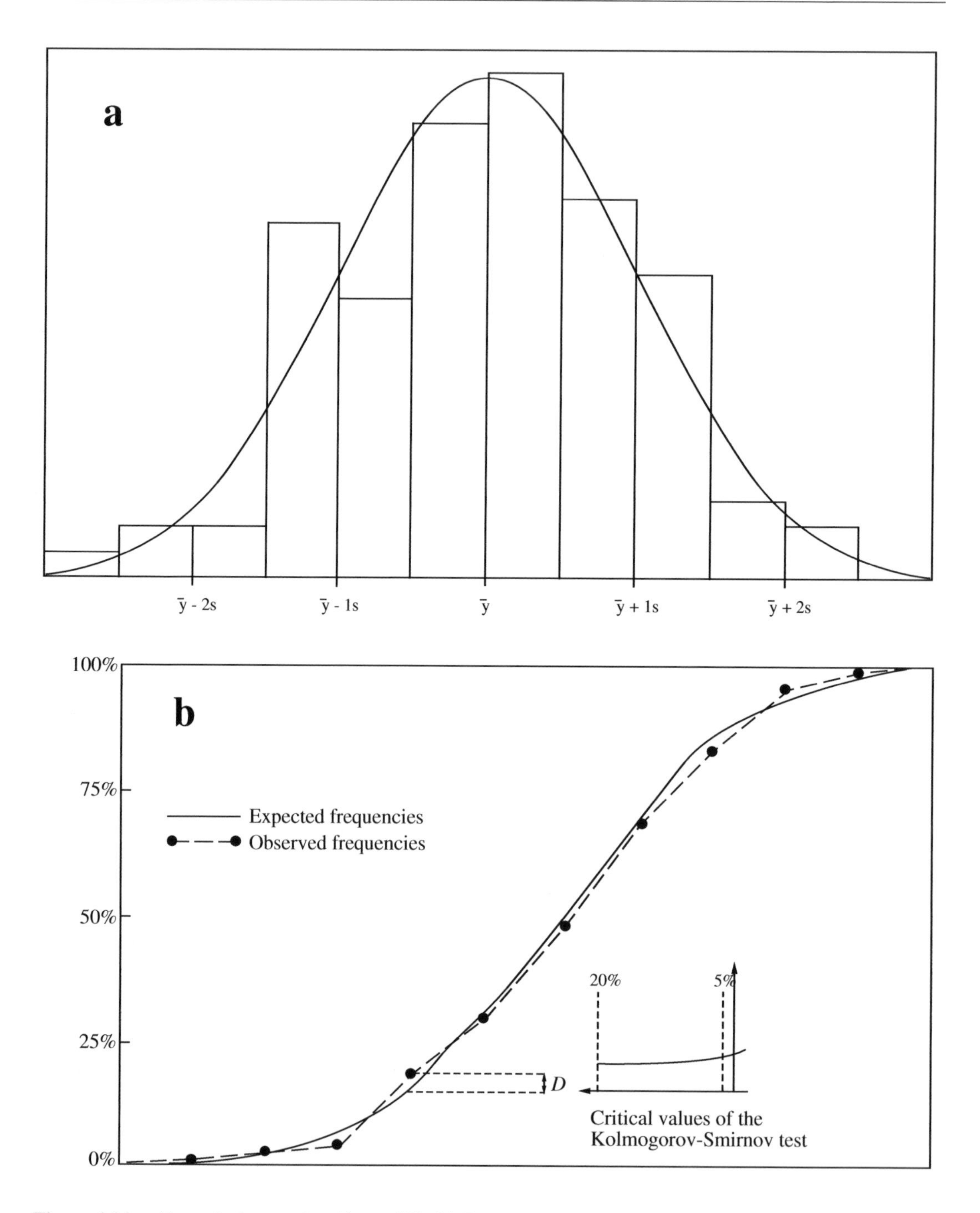

Figure 4.14 Numerical example with $n = 100$. (a) Frequency distribution and fitted theoretical normal curve, (b) relative cumulative frequencies and Kolmogorov-Smirnov test of goodness-of-fit, showing that the maximum deviation $D = 0.032$ is too small in this case to reject the hypothesis of normality.

empirical data since the mean and standard deviation of the reference normal distribution must be estimated from the observed data; critical values given in these tables are systematically too large, and thus lead to accepting too often the null hypothesis of normality. Corrected critical values for testing whether a set of observations is drawn from a normal population, that are valid for stated probabilities of type I error, have been computed by Lilliefors (1967) and, with additional corrections based on larger Monte Carlo simulations, by Stephens (1974). Critical values computed from the formulas and coefficients of Stephens are given in Table A. The same paper by Stephens evaluates other statistics to perform tests of normality, such as Cramér-von Mises W^2 and Anderson-Darling A^2 which, like D, are based on the empirical cumulative distribution function (only the statistics differ), and proposes corrections where needed for the situation where the mean and variance of the reference normal distribution are unknown and are thus estimated from the data.

Normal probability plot

The second widely used test of normality is due to Shapiro & Wilk (1965). It is based on an older graphical technique which will be described first. This technique, called *normal probability plotting*, was developed as an informal way of assessing deviations from normality. The objective is to plot the data in such a way that, if they come from a normally distributed population, they will fall along a straight line. Deviations from a straight line may be used as indication of the type of non-normality. In these plots, the values along the abscissa are either the observed or the standardized data (in which case the values are transformed to standard deviation units), while the ordinate is the percent cumulative frequency value of each point plotted on a normal probability scale. Sokal & Rohlf (1995) give computation details. Fig. 4.15 shows the same data as in Fig 4.14a, which are divided into classes, plotted on normal probability paper. The same type of plot could also be produced for the raw data, not grouped into classes. For each point, the *upper limit* of a class is used as the abscissa, while the ordinate is the percent cumulative frequency (or the cumulative percentage) of that class. Perfectly normal data would fall on a straight line passing through the point ($\bar{y}$, 50%). A straight line is fitted trough the points, using reference points based on the mean and variance of the empirical data (see the caption of Fig. 4.15); deviations from that line indicate non-normality. Alternatively, a straight line may be fitted through the points, either by eye or by regression; the mean of the distribution may be estimated as the abscissa value that has an ordinate value of 50% on that line. D'Agostino (1982) gives examples illustrating how deviations from linearity in such plots indicate the degree and type of non-normality of the data.

Shapiro-Wilk test

Shapiro & Wilk (1965) proposed to quantify the information in normal probability plots using a so-called 'analysis of variance W statistic', which they defined as the F-ratio of the estimated variance obtained from the weighted least-squares of the slope of the straight line (numerator) to the variance of the sample (denominator). The statistic is used to assess the goodness of the linear fit:

$$W = \left(\sum_{i=1}^{n} w_i x_i\right)^2 / \sum_{i=1}^{n} (x_i - \bar{x})^2 \qquad \textbf{(4.53)}$$

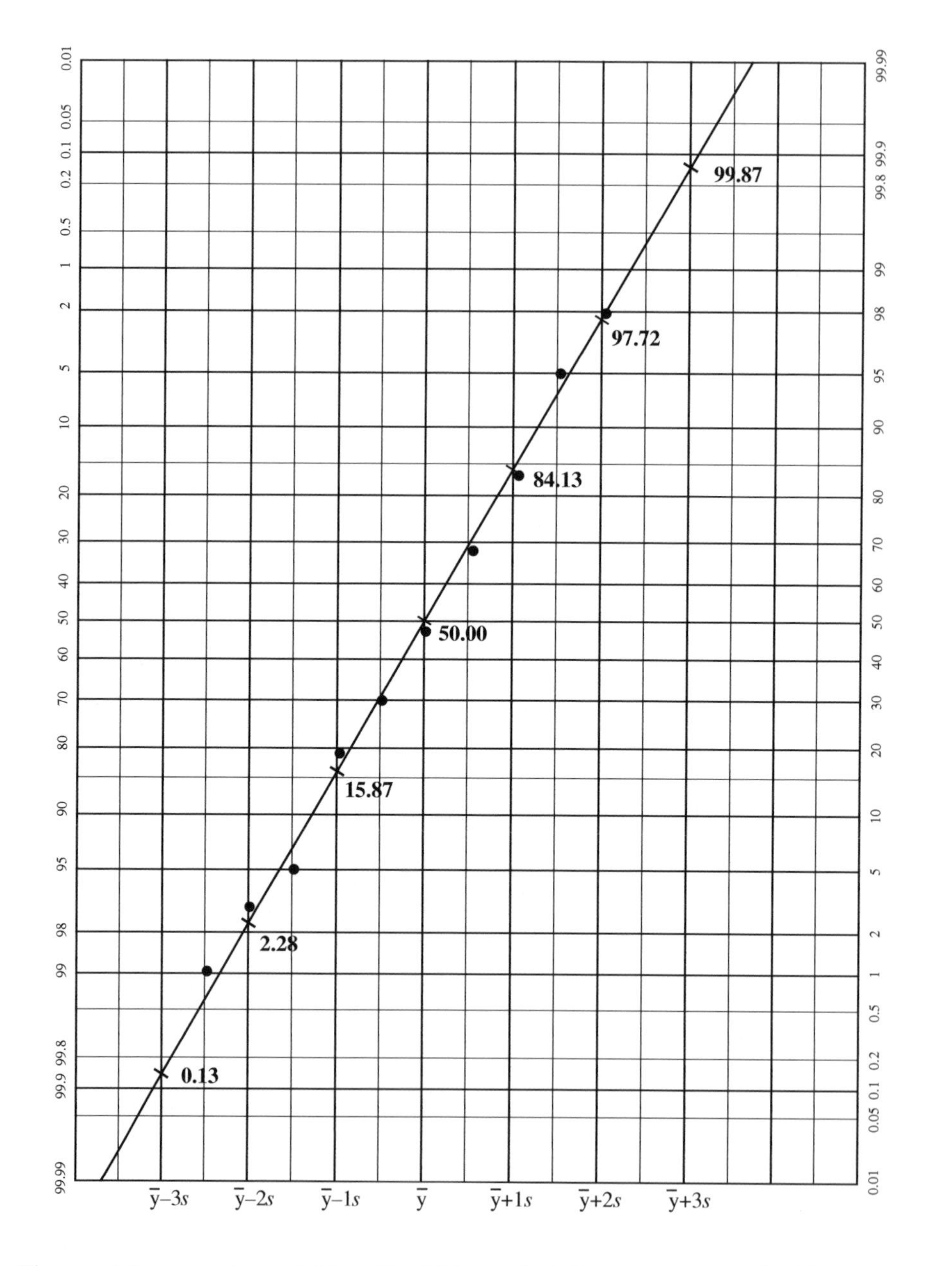

Figure 4.15 The cumulative percentages of data in Fig. 4.14a are plotted here on normal probability paper as a function of the upper limits of classes. Cumulative percentiles are indicated on the right-hand side of the graph. The last data value cannot be plotted on this graph because its cumulated percentage value is 100. The diagonal line represents the theoretical cumulative normal distribution with same mean and variance as the data. This line is positioned on the graph using reference values of the cumulative normal distribution, for example 0.13% at $\bar{y} - 3s$ and 99.87% at $\bar{y} + 3s$, and it passes through the point ($\bar{y}$, 50%). This graph contains exactly the same information as Fig. 4.14b; the difference lies in the scale of the ordinate.

where the x_i are the ordered observations ($x_1 \leq x_2 \leq \ldots \leq x_n$) and coefficients w_i are optimal weights for a population assumed to be normally distributed. Statistic W may be viewed as the square of the correlation coefficient (i.e. the coefficient of determination) between the abscissa and ordinate of the normal probability plot described above. Large values of W indicate normality (points lying along a straight line give r^2 close to 1), whereas small values indicate lack of normality. Shapiro & Wilk did provide critical values of W for sample sizes up to 50. D'Agostino (1971, 1972) and Royston (1982a, b, c) proposed modifications to the W formula (better estimates of the weights w_i), which extend its application to much larger sample sizes. Extensive simulation studies have shown that W is a sensitive *omnibus* test statistic, meaning that it has good power properties over a wide range of non-normal distribution types and sample sizes.

Which of these tests is best? Reviewing the studies on the power of tests of normality published during the past 25 years, D'Agostino (1982) concluded that the best *omnibus* tests are the Shapiro-Wilk W-test and a modification by Stephens (1974) of the Anderson-Darling A^2-test mentioned above. In a recent Monte Carlo study involving autocorrelated data (Section 1.1), however, Dutilleul & Legendre (1992) showed (1) that, for moderate sample sizes, both the D-test and the W-test are too liberal (in an asymmetric way) for high positive ($\rho > 0.4$) and very high negative ($\rho < -0.8$) values of autocorrelation along time series and for high positive values of spatial autocorrelation ($\rho > 0.2$) and (2) that, overall, the Kolmogorov-Smirnov D-test is more robust against autocorrelation than the Shapiro-Wilk W-test, whatever the sign of the first-order autocorrelation.

As stated at the beginning of the Section, ecologists must absolutely check the normality of data only when they wish to use parametric statistical tests that are based on the normal distribution. Most methods presented in this book, including clustering and ordination techniques, do not require statistical testing and hence may be applied to non-normal data. With many of these methods, however, ecological structures emerge more clearly when the data do not present strong asymmetry; this is the case, for example, with principal component analysis. Since normal data are not skewed (coefficient $\alpha_3 = 0$), testing the normality of data is also testing for asymmetry; normalizing transformations, applied to data with unimodal distributions, reduce or eliminate asymmetries. So, with multidimensional data, it is recommended to check at least the normality of variables one by one.

Some tests of significance require that the data be multinormal (Section 4.3). Section 4.6 has shown that the multidimensional normal distribution contains *conditional distributions*; it also contains *marginal distributions*, which are distributions on one or several dimensions, collapsing all the other dimensions. The normality of *unidimensional marginal distributions*, which correspond to the p individual variables in the data set, can easily be tested as described above. In a multivariate situation, however, showing that each variable does not significantly depart from normality does not prove that the multivariate data set is multinormal although, in many instances, this is the best researchers can practically do.

Test of multi-normality

Dagnelie (1975) proposed an elegant and simple way of testing the multinormality of a set of multivariate observations. The method is based on the *Mahalanobis generalized distance* (D_5; Section 7.4, eq. 7.40) which is described in Chapter 7. Generalized distances are computed, in the multidimensional space, between each object and the multidimensional mean of all objects. The distance between object $\mathbf{x}_i$ and the mean point $\bar{\mathbf{x}}$ is computed as:

$$D(\mathbf{x}_i, \bar{\mathbf{x}}) = \sqrt{[y - \bar{y}]_i \mathbf{S}^{-1} [y - \bar{y}]'_i} \qquad \textbf{(4.54)}$$

where $[y - \bar{y}]_i$ is the vector corresponding to object $\mathbf{x}_i$ in the matrix of centred data and $\mathbf{S}$ is the dispersion matrix (Section 4.1). Dagnelie's approach is that, for multinormal data, the generalized distances should be normally distributed. So, the n generalized distances (corresponding to the n objects) are put in increasing order, after which the relative cumulative frequency of each i-th distance is calculated as $(i - 0.5)/n$. The data are then plotted on a normal probability scale (Fig. 4.15), with the generalized distances on the abscissa and the relative cumulative frequencies on the ordinate. From visual examination of the plot, one can decide whether the data points are well aligned; if so, the hypothesis of multinormality of the original data may be accepted. Alternatively, the list of generalized distances may be subjected to a Shapiro-Wilk test of normality, whose conclusions are applied to the multinormality of the original multivariate data. With standardized variables $z_{ij} = (y_{ij} - \bar{y}_j)/s_j$, eq. 4.54 becomes:

$$D(\mathbf{x}_i, \bar{\mathbf{x}}) = \sqrt{\mathbf{z}_i \mathbf{R}^{-1} \mathbf{z}'_i} \qquad \textbf{(4.55)}$$

where $\mathbf{R}$ is the correlation matrix.

Chapter

5 Multidimensional semiquantitative data

5.0 Nonparametric statistics

Section 1.2 has explained that statistical testing often refers to the concepts of *parameter* and *reference population*. Section 4.3 has shown that the mean, standard deviation and correlation are *parameters* of the multinormal distribution, so that this distribution and others play a key role in testing *quantitative* data. When the data are *semiquantitative*, however, it does not make sense to compute statistics such as the mean or the standard deviation. In that case, hypothesis testing must be conducted with *nonparametric statistics*. Nonparametric tests are *distribution-free*, i.e. they do not assume that the samples were drawn from a population with a specified distribution (e.g. multinormal). Because of this, nonparametric statistics are useful not only when descriptors are semiquantitative, but also when quantitative descriptors do not conform to the multinormal distribution and researchers do not wish, or succeed, to normalize them. Many of the nonparametric tests are called *ranking tests*, because they are based on ranks of observations instead of actual quantitative values. Another advantage of nonparametric statistics is computational simplicity. Last but not least, nonparametric tests may be used with small samples, a situation that frequently occurs with ecological data. Nonparametric measures corresponding to the *mean* and *variance* (Section 4.1) are the *median* and *range*, respectively.

Nonparametric statistics cover all statistical methods developed for analysing either *semiquantitative* (rank statistics; Sections 5.2) or *qualitative* (Chapter 6) data. *Rank statistics* should always be used in the following situations:

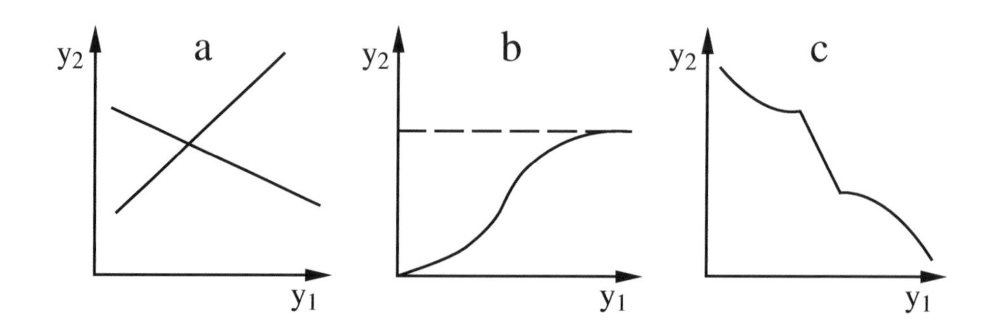

Figure 5.1 Three types of monotonic relationships between two descriptors: (a) linear (increasing and decreasing); (b) logistic (increasing monotonic); (c) atypical (decreasing monotonic).

1) One or several descriptors among those to be compared are semiquantitative.

Monotonic

2) The purpose of the study is to evidence *monotonic* relationships between quantitative descriptors. In a monotonic relationship, one of the descriptors keeps increasing or decreasing as the other increases (Fig. 5.1); the increase (or decrease) is not necessarily *linear* or smoothly *curvilinear.*

3) One or several (quantitative) descriptors are not normally distributed (see Section 4.7 for tests of normality and multinormality) and researchers do not wish to normalize them or do not succeed in doing so. Normalizing transformations are described in Subsection 1.5.6.

4) The number of observations is small.

The present Chapter first summarizes the methods available in the nonparametric approach, with reference to the corresponding parametric methods (Section 5.1). Ranking tests are then described for analysing relationships (Section 5.2) among groups of qualitative, semiquantitative, or quantitative descriptors and (Section 5.3) among groups of descriptors with mixed levels of precision (R analysis: Section 7.1). Most statistical computer packages include nonparametric procedures.

5.1 Quantitative, semiquantitative, and qualitative multivariates

As discussed in Section 1.4, ecological descriptors may be of different levels of precision (Table 1.2). Ecologists generally observe several descriptors on the same objects, so that multidimensional ecological variates may be either quantitative, semiquantitative, or qualitative, or mixed, i.e. consisting of descriptors with different precision levels. For a number of years, quantitative ecology has been based almost exclusively on quantitative descriptors and on parametric tests, even though there exist a large number of methods that can efficiently analyse semiquantitative or qualitative multivariates as well as multivariates of mixed precision levels. These methods have become increasingly popular in ecology, not only because non-quantitative descriptors often provide unique information, but also because parametric statistics cannot be tested for significance when quantitative data do not conform to a number of conditions, including multinormality. This section briefly reviews numerical methods for analysing multivariates with various levels of precision.

Table 5.1 summarizes and compares methods described elsewhere in the present book. In the same row are corresponding methods, listed under one of four column headings. The applicability of methods increases from left to right. Methods in the first (left-hand) column are restricted to *quantitative* multivariates, which must also, in most cases, be linearly related or/and multinormally distributed. Methods in the second column have been developed for *semiquantitative* descriptors exhibiting *monotonic* relationships. These methods may also be used (a) with quantitative descriptors (especially when they do not follow the conditions underlying methods in the first column) and (b) for the combined analysis of quantitative and semiquantitative descriptors. Methods in the third column were developed for the numerical analysis of *qualitative* descriptors. They may also be used for analysing quantitative or semiquantitative descriptors exhibiting nonmonotonic relationships, after partitioning these continuous descriptors into classes (see Section 6.3). Methods for qualitative descriptors thus represent a first type of techniques for multivariates of mixed precision, since they can be used for analysing together quantitative, semiquantitative, and qualitative descriptors, partitioned into classes. An alternative is to recode multiclass qualitative descriptors into dummy variables (Subsection 1.5.7) and use parametric methods (first column of the Table) on the resulting assemblage of quantitative and binary descriptors; this approach is often used in regression and canonical analyses (Chapter 10).

Other methods (right-hand column) have been developed specifically for multivariates with *mixed* levels of precision, so that these can be used for analysing together quantitative, semiquantitative and qualitative descriptors. Such methods are very general, since they may replace equivalent methods in the other three columns; the price to be paid is often greater mathematical and/or computational complexity.

There are many types of multidimensional methods (rows of Table 5.1). One interesting aspect of the Table is that there is always at least one, and often several methods for descriptors with low precision levels. Thus, ecologists should never hesitate to collect information in semiquantitative or qualitative form, since there exist numerical methods for processing descriptors with all levels of precision. However, it is always important to consider, at the very stage of the sampling design, how data will eventually be analysed, so as to avoid problems at later stages. These problems often include the local availability of specific computer programs or the lack of human resources to efficiently use sophisticated methods. Researchers could use the period devoted to sampling to acquire computer programs and improve their knowledge of

Table 5.1 Methods for analysing *multidimensional* ecological data sets, classified here according to the levels of precision of descriptors (columns). For methods concerning data series, see Table 11.1. To find pages where a given method is explained, see the Subject index (end of the book).

Quantitative descriptors	Semiquantitative descriptors	Qualitative descriptors	Descriptors of mixed precision
Difference between two samples:			
Hotelling T^2	---	Log-linear models	---
Difference among several samples:			
MANOVA	---	Log-linear models	MANOVALS
db-RDA, CCA	---	db-RDA, CCA	db-RDA
Scatter diagram	Rank diagram	Multiway contingency table	Quantitative-rank diagram
Association coefficients R:			
Covariance	---	Information, X^2	---
Pearson r	Spearman r	Contingency	---
	Kendall τ		
Partial r	Partial τ		
Multiple R	Kendall W		
Species diversity:			
Diversity measures	Diversity measures	Number of species	---
Association coeff. Q	Association coeff. Q	Association coeff. Q	Association coeff. Q
Clustering	Clustering	Clustering	Clustering
Ordination:			
Principal component a.	---	Correspondence a.	PRINCALS
Correspondence a.		HOMALS	PRINCIPALS
Principal coordinate a.			Principal coordinate a.
Nonmetric multi-dimensional scaling			Nonmetric multi-dimensional scaling
			ALSCAL, GEMSCAL
Factor analysis	---	---	FACTALS
Regression	Regression	Correspondence	Regression
simple linear (I and II)	nonparametric		logistic
multiple linear			dummy
polynomial			MORALS
partial linear			
nonlinear, logistic			
smoothing (splines, LOWESS)			
multivariate; see also canonical a.			
Path analysis	---	Log-linear models	PATHALS
		Logit models	
Canonical analysis:			
Redundancy analysis (RDA)			CORALS, OVERALS
Canonical correspondence a. (CCA)		CCA	db-RDA
Canonical correlation a. (CCorA)			
Discriminant analysis	---	Discrete discriminant a.	CRIMINALS
		Log-linear models	Logistic regression

methods. In any case, the type of data collected must take into account the local computer and human resources.

Coming back to Table 5.1, it is possible to compare groups of objects, described by quantitative multivariate data, using multidimensional analysis of variance (MANOVA). In this analysis, a test of Wilks' Λ (lambda) statistic replaces the usual F-test of one-dimensional ANOVA. When there are only two groups, another approach is Hotelling's T^2 (Section 7.4). In the case of qualitative multivariate data, the comparison may be done by adjusting log-linear models to a multiway contingency table (the relationship between contingency table analysis and analysis of variance is explained in Section 6.0; see also the caveat concerning the use of multiway contingency tables as qualitative equivalent to MANOVA, at the end of Section 6.3). Multivariate analysis of variance of descriptors with mixed levels of precision is possible using MANOVALS (a Gifi/ALSOS algorithm; Section 10.2). Multivariate analysis of variance of species presence-absence or abundance tables may be obtained using either canonical correspondence analysis (CCA, Section 11.2), or the distance-based redundancy analysis method (db-RDA) of Legendre & Anderson (1999) briefly described in Subsection 11.3.1.

The simplest approach to investigate the relationships among descriptors, considered two at a time (Fig. 5.2), is to plot the data as a scatter diagram, whose semiquantitative and qualitative equivalent are the rank-rank diagram and the contingency table, respectively. Quantitative-rank diagrams may be used to compare a quantitative to a semiquantitative descriptor (Legendre & Legendre, 1982). Two families of methods follow from these diagrams, for either *measuring* the dependence among descriptors, or *forecasting* one or several descriptors using other ones. The first family of methods is based on R-mode association coefficients (i.e. coefficients of dependence; Section 7.5), which are explained in Chapter 4 (quantitative descriptors), Chapter 5 (semiquantitative descriptors), and Chapter 6 (qualitative descriptors). It is interesting to note that measures of information and X^2 (chi-square) calculated on contingency tables (Chapter 6) are equivalent, for qualitative descriptors, to the covariance between quantitative descriptors. Methods in the second family belong to regression analysis (Section 10.4), which has a nonparametric form and whose qualitative equivalent is the analysis of correspondence in contingency tables (Section 6.4).

Various measures of species diversity are reviewed in Section 6.5. They are usually computed on quantitative species counts, but Dévaux & Millerioux (1977) have shown that this may be done just as well on semiquantitative counts. When there are no counts, the number of species present may be used to assess diversity; this is indeed the first diversity index described in the literature (Patrick, 1949; Subsection 6.5.1).

There are Q-mode association coefficients (Sections 7.3 and 7.4) adapted to descriptors of all levels of precision (see Tables 7.3 and 7.4). Some of the similarity coefficients (Chapter 7: S_{15}, S_{16}, S_{19} and S_{20}) are yet another way of combining quantitative and qualitative descriptors in multivariate data analysis. Concerning

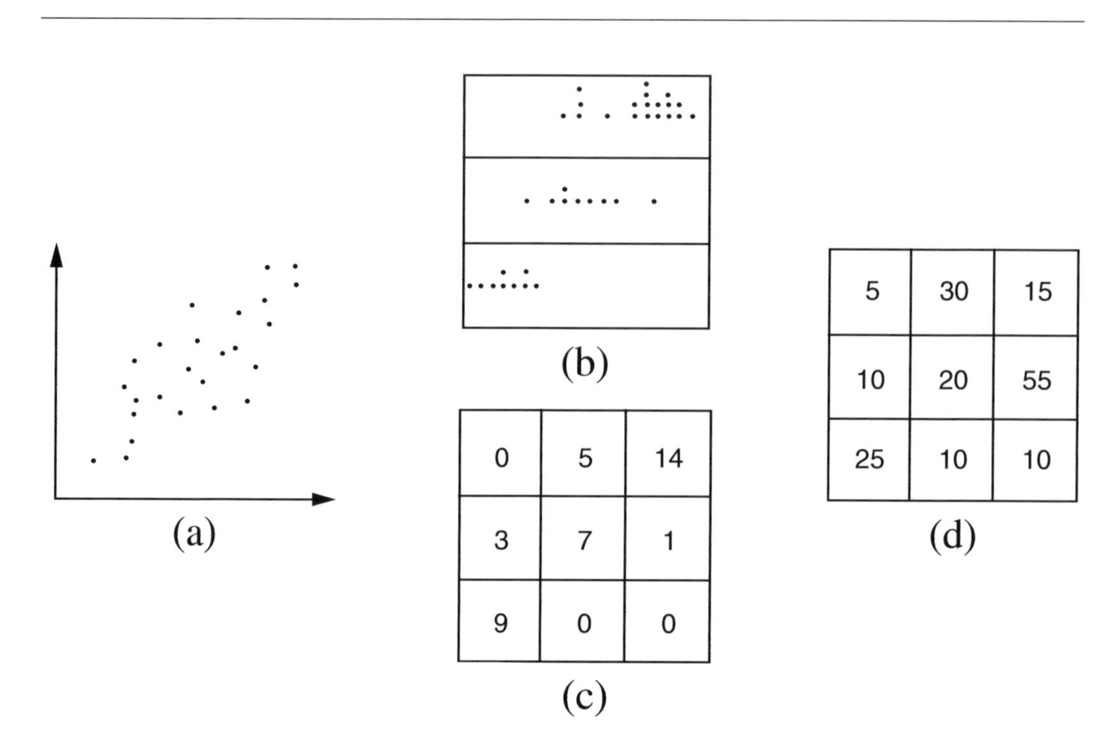

Figure 5.2 Comparison of two descriptors. (a) Scatter diagram (quantitative descriptors on both axes). (b) Quantitative-rank diagram (quantitative descriptor on the abscissa, ranked classes of a semiquantitative descriptor on the ordinate). (c) Rank-rank diagram (ranked classes of semiquantitative descriptors on both axes). (d) Two-way contingency table (nonordered classes of qualitative descriptors on both axes). From Legendre & Legendre (1982).

clustering algorithms (Chapter 8), most of them are indifferent to the precision of descriptors, since clustering is in general conducted on an association matrix, most often of type Q.

Methods of ordination in reduced space are seldom restricted to descriptors of a single level of precision, with the exception of principal component analysis which must be computed on quantitative or presence-absence data (Section 9.1). Correspondence analysis (Section 9.4) was first described for qualitative descriptors (contingency tables), but it is currently used for analysing descriptors of any precision. These must, however, be coded in such a way as to be positive and dimensionally homogenous. Principal coordinate analysis (Section 9.2) and nonmetric multidimensional scaling (Section 9.3) are indifferent to the precision of descriptors, since they are computed on an association matrix (generally Q-type). Ordination of descriptors with mixed levels of precision is also possible using the Gifi/ALSOS algorithms HOMALS, PRINCIPALS, ALSCAL, and GEMSCAL. Factor analysis

(Section 9.5) is restricted to quantitative descriptors, except when using the Gifi/ALSOS algorithm FACTALS for descriptors of mixed precision. (Algorithms PRINCIPALS, GEMSCAL and FACTALS are not discussed elsewhere in this book.)

For the interpretation of ecological structures, regression, which was briefly discussed a few paragraphs above, is the chief technique when the dependent variable is a single quantitative variable. Various forms of canonical analysis are available to interpret the structure of quantitative data sets: redundancy analysis, canonical correspondence analysis, canonical correlation analysis, and discriminant analysis (Chapter 11). Canonical correspondence analysis, in particular, allows an interpretation of the structure of species abundance or presence-absence data. For qualitative descriptors, Table 5.1 proposes methods equivalent to discriminant and path analyses. For descriptors of mixed precision, there are methods available for all types of analyses.

Table 5.1 shows that ecological data can efficiently be analysed irrespective of their levels of precision. Researchers should use ecological criteria, such as allowable effort in the field and biological meaningfulness of the decimal places to be recorded, to decide about the level of precision of their data. The strictly numerical aspects play but a secondary role.

5.2 One-dimensional nonparametric statistics

The present book is devoted to numerical methods for analysing sets of *multidimensional* ecological data. Methods for one-dimensional variables are not discussed in depth since they are the subject of many excellent textbooks. Nonparametric tests for one-dimensional descriptors are explained, among others, in the books of Siegel (1956), Hájek (1969), Siegel & Castellan (1988), and Sokal & Rohlf (1995). Because ecologists are often not fully conversant with these tests, the correspondence between approaches for quantitative, semiquantitative, and qualitative descriptors is not always clearly understood. This is why the one-dimensional methods to carry out tests of differences among groups of objects are summarized in Table 5.2.

Independent samples
Related samples

Methods in the Table are divided in two main families: those for independent samples, which are the most generally applicable, and those for related samples. Related samples are often called *matched* or *paired* samples (Box 1.1). With such samples, the effect of differences among the matched objects is eliminated from the comparison among groups. Matching may be achieved, for example, by repeating observations at the same sampling sites, or by making observations at sites representing corresponding conditions (e.g. same geological substrate, same temperature, or same depth in the water column). Related samples could be analysed using the methods for independent samples, but the information carried by the matching is then lost. Within each of the two families, methods in Table 5.2 are classified according to the number of groups (k) which are compared.

Table 5.2 Methods to carry out tests of differences among groups of objects (*one-dimensional* data) are classified here according to the levels of precision of the descriptors (columns). Most of these methods are not discussed elsewhere in the present book. Table modified from Siegel (1956) and Legendre & Legendre (1982).

Number of groups (k)	Quantitative descriptors*	Semiquantitative descriptors	Qualitative descriptors
Independent samples:			
$k = 2$	Student t (unpaired)	Mann-Whitney U test Median test Kolmogorov-Smirnov test etc.	X^2 (2 × no. states) Fisher's exact probability test Logistic regression
$k \geq 2$ (one-way)	One-way ANOVA and F-test	Kruskal-Wallis' H Extension of the median test	X^2 (k × no. states) Discriminant a.
Related samples:			
$k = 2$	Student t (paired)	Sign test Wilcoxon signed-ranks test	McNemar test (binary descriptors)
$k \geq 2$ (two-way)	Two-way ANOVA and F-tests	Friedman test	Cochran Q (binary descriptors)
$k \geq 2$ (multiway)	Multiway ANOVA and F-tests	---	---

* When quantitative data do not meet the distributional assumptions underlying parametric tests, they must be analysed using ranking tests (for semiquantitative descriptors). Another way would be to test the parametric statistics by randomization (Section 1.2).

Univariate comparison of *two independent samples* ($k = 2$), when the data are *quantitative*, is generally done by using Student's t, to test the hypothesis (H_0) of equality of the group means (i.e. that the two groups of objects were drawn from the same statistical population, or at least from populations with equal means, assuming equal standard deviations). When the data are *semiquantitative*, computing means and standard deviations would not make sense, so that the approach must be nonparametric. The Mann-Whitney U statistic first combines and ranks all objects in a single series, then allows one to test that the ranked observations come from the same statistical population (H_0). The median test, which is not as powerful as the previous one (except in cases when there are ties), is used for testing that the two groups of objects have similar medians (H_0). Other nonparametric tests consider not only the positions of the two groups along the abscissa but also the differences in dispersion and shape (e.g. skewness) of their distributions. The best-known is the Kolmogorov-Smirnov test; this is not the same test as the one described in Section 4.7 for comparing an empirical to a theoretical distribution. The method discussed here allows one to test the hypothesis (H_0) that the largest difference between the cumulative distributions of the two groups is so small that these may come from the same or identical populations. Finally, when the data are *qualitative*, the significance of differences between two groups of objects may be tested using a X^2 statistic calculated on a two-way contingency table. Section 6.2 describes contingency table analysis for the comparison of two descriptors. In the present case, the contingency table has two rows (i.e. two groups of objects) and as many columns as there are states in the quantitative descriptor. The hypothesis tested (H_0) is that the frequency distributions in the two rows are similar; this is the same as stating the more usual hypothesis of independence between rows and columns of the contingency table (see Section 6.0). When the descriptor is *binary* (e.g. presence or absence) and the number of observations in the two groups is small, it is possible to test the hypothesis (H_0) that the two groups exhibit similar proportions for the two states, using Fisher's powerful exact probability test. Logistic regression (Subsection 10.3.7) may also be used in this context; in the regression, the two groups are represented by a binary response variable while the qualitative explanatory descriptors are recoded as a series of dummy variables as in Subsection 1.5.7.

The standard parametric technique for testing that the means of *several independent samples* ($k \geq 2$) are equal, when the data are *quantitative*, is one-way analysis of variance (ANOVA). It may be considered as a generalization of the Student t–test. In one-way ANOVA, the overall variance is partitioned between two orthogonal (i.e. linearly independent; see Box 1.1) components, the first one reflecting differences among the k groups and the second one accounting for the variability among objects within the groups. The hypothesis (H_0) of equal means is rejected (F-test) when the among-groups variability is significantly larger than the within-groups component. For *semiquantitative* data, the Kruskal-Wallis' H test (also called Kruskal-Wallis' one-way ANOVA by ranks) first ranks all objects from the k groups into a single series, and then tests (H_0) that the sums of ranks calculated for the various groups are so similar that the objects are likely to have been drawn from the same or identical populations. When applied to quantitative data that are meeting all the assumptions of parametric ANOVA, Kruskal-Wallis' H is almost as efficient as the F-test. Another possibility is to extend to $k \geq 2$ groups the median test, described in the previous paragraph for $k = 2$. The latter is less efficient than Kruskal-Wallis' H because it uses less of the information in the data. As in the above case where $k = 2$, *qualitative* data can be analysed using a contingency table, but this time with $k \geq 2$ rows. Multiple logistic regression, available for example in procedure CATMOD of SAS (Subsection 10.3.7), may also be used here. Discriminant analysis could be used in the same spirit, after recoding the qualitative descriptor of interest into a series of dummy variables forming the set of discriminating variables, as in Subsection 1.5.7. See, however, the discussion on discriminant analysis *versus* logistic regression (Subsection 10.3.7 and Section 11.6).

Comparing *two related samples* ($k = 2$) is usually done, for *quantitative* data, by testing (H_0) that the mean of the differences between matched pairs of observations is null (Student t-test; the differences are assumed to be normally and independently distributed). When the data are *semiquantitative*, one can use the sign test, which first codes pairs of observations (y_i, y_k) as either (+) when $y_i > y_k$ or (–) when $y_i < y_k$, and then tests the hypothesis (H_0) that the numbers of pairs with each sign are equal; an equivalent formulation is that the proportion of pairs with either sign is equal to 0.5. This test uses information about the direction of the differences between pairs. When the relative magnitude of the differences between pairs is also known, it becomes possible to use the more powerful Wilcoxon matched-pairs signed-ranks test. Differences between pairs are first ranked according to their magnitude (absolute values), after which the sign of the difference is affixed to each rank. The null hypothesis of the test (H_0) is that the sum of the ranks having a (+) sign is equal to that of the ranks with a (–) sign. The McNemar test provides a means of comparing paired samples of *binary* data. For example, using binary observations (e.g. presence or absence) made at the same sites, before and after some event, one could test (H_0) that no overall change has occurred.

When there are *several related samples* ($k \geq 2$) and the data are *quantitative*, the parametric approach for testing (H_0) that the means of the k groups are equal is two-way analysis of variance, with or without replication. One classification criterion of the two-way ANOVA accounts for the variability among the k groups (as in one-way ANOVA above, for $k \geq 2$ independent samples) and the other for that among the related samples. Consider, for an example, 16 sites (i.e. k groups) that have been sampled at 5 depths in the water column (or at 5 different times, or using 5 different methods, etc.). The nonparametric equivalent, for *semiquantitative* data, is Friedman's *two-way analysis of variance by ranks without replication*, which is based on a two-way table similar to Table 5.7. In the two-way table, the k groups (e.g. 16 sites) are in rows and the corresponding samples (e.g. 5 depths) are in columns. Values within each column are ranked separately, and the Friedman statistic (eq. 5.10) is used to test (H_0) that the rank totals of the various rows (i.e. groups) are equal. For *binary* data, the Cochran Q test is an extension to $k \geq 2$ groups of the McNemar test, described above for $k = 2$.

Finally, when there are *several samples ($k \geq 2$), related across several classification criteria* (e.g. 16 sites all sampled at 8 different times, using each time 5 different methods), multiway ANOVA is the standard parametric method for testing the null hypothesis (H_0) that the means of the k groups are equal (F-test). In that case, there are no obvious equivalent approaches for semiquantitative or qualitative data.

5.3 Multidimensional ranking tests

Textbooks of nonparametric statistics propose a few methods only for the analysis of bi- or multivariate semiquantitative data. Section 5.1 has shown that there actually exist many numerical approaches for analysing multidimensional data, corresponding to all levels of precision (Table 5.1). These methods, which include most of those described in this book, belong to *nonparametric statistics* in a general sense, because they do not focus on the parameters of the data distributions. Within the specific realm of *ranking tests*, however, the only statistical techniques available for

Table 5.3 Numerical example. Perfect rank correlation between descriptors $\mathbf{y}_1$ and $\mathbf{y}_2$.

Objects (observation units)	Ranks of objects on the two descriptors $\mathbf{y}_1$	$\mathbf{y}_2$
$\mathbf{x}_1$	5	5
$\mathbf{x}_2$	1	1
$\mathbf{x}_3$	4	4
$\mathbf{x}_4$	2	2
$\mathbf{x}_5$	3	3

multidimensional semiquantitative data are two *rank correlation coefficients* (Spearman r and Kendall τ), which both quantify the relationship between two descriptors, and the *coefficient of concordance* (Kendall W), which assesses the relationship among several descriptors. These are described in some detail in the present section.

Spearman corr. coeff.

The Spearman r statistic, also called ρ (rho), is based on the idea that two descriptors $\mathbf{y}_1$ and $\mathbf{y}_2$ carry the same information if the largest object on $\mathbf{y}_1$ also has the highest rank on $\mathbf{y}_2$, and so on for all other objects. Two descriptors are said to be in perfect correlation when the ranks of all object are the same on both descriptors, as in the numerical example of Table 5.3. If, however, object $\mathbf{x}_1$ which has rank 5 on $\mathbf{y}_1$ had rank 2 on $\mathbf{y}_2$, it would be natural to use the difference between these ranks $d_1 = (y_{11} - y_{12}) = (5 - 2) = 3$ as a measure of the difference between the two descriptors, for this object. For the whole set of objects, differences d_i are squared before summing them, in order to prevent differences with opposite signs from cancelling each other out.

The expression for the Spearman r may be derived from the general formula of correlation coefficients (Kendall, 1948):

$$r_{jk} = \frac{\sum_{i=1}^{n} (y_{ij} - \bar{y}_j)(y_{ik} - \bar{y}_k)}{\sqrt{\sum_{i=1}^{n} (y_{ij} - \bar{y}_j)^2 \sum_{i=1}^{n} (y_{ik} - \bar{y}_k)^2}} \tag{5.1}$$

For ranked data, the average ranks $\bar{y}_j$ and $\bar{y}_k$ are equal, so that $(y_{ij} - \bar{y}_j) - (y_{ik} - \bar{y}_k) = (y_{ij} - y_{ik})$. One can write the difference between the ranks of object i on the two descriptors as $d_i = (y_{ij} - y_{ik}) = (y_{ij} - \bar{y}_j) - (y_{ik} - \bar{y}_k)$, which leads to:

$$\sum_{i=1}^{n} d_i^2 = \sum_{i=1}^{n} (y_{ij} - \bar{y}_j)^2 + \sum_{i=1}^{n} (y_{ik} - \bar{y}_k)^2 - 2\sum_{i=1}^{n} (y_{ij} - \bar{y}_j)(y_{ik} - \bar{y}_k)$$

Isolating the right-hand sum gives:

$$\sum_{i=1}^{n} (y_{ij} - \bar{y}_j)(y_{ik} - \bar{y}_k) = \frac{1}{2}\left[\sum_{i=1}^{n} (y_{ij} - \bar{y}_j)^2 + \sum_{i=1}^{n} (y_{ik} - \bar{y}_k)^2 - \sum_{i=1}^{n} d_i^2\right]$$

Using this result, eq. 5.1 is rewritten as:

$$r_{jk} = \frac{\frac{1}{2}\left[\sum_{i=1}^{n} (y_{ij} - \bar{y}_j)^2 + \sum_{i=1}^{n} (y_{ik} - \bar{y}_k)^2 - \sum_{i=1}^{n} d_i^2\right]}{\sqrt{\sum_{i=1}^{n} (y_{ij} - \bar{y}_j)^2 \sum_{i=1}^{n} (y_{ik} - \bar{y}_k)^2}} \qquad \mathbf{(5.2)}$$

The sum of ranks for each descriptor, which is the sum of the first n integers, is equal to $n(n + 1)/2$ and the sum of their squares is $\sum_{i=1}^{n} y_{ij}^2 = n(n+1)(2n+1)/6$. Since the sum of deviations from the mean rank is

$$\sum_{i=1}^{n} (y_{ij} - \bar{y}_j)^2 = \sum_{i=1}^{n} y_{ij}^2 - \frac{1}{n}\left(\sum_{i=1}^{n} y_{ij}\right)^2$$

one can write:

$$\sum_{i=1}^{n} (y_{ij} - \bar{y}_j)^2 = \frac{n(n+1)(2n+1)}{6} - \frac{1}{n}\left[\frac{n^2(n+1)^2}{4}\right] = \frac{n^3 - n}{12}$$

It follows that, when using ranks, the numerator of eq. 5.2 becomes:

$$\frac{1}{2}\left[\sum_{i=1}^{n} (y_{ij} - \bar{y}_j)^2 + \sum_{i=1}^{n} (y_{ik} - \bar{y}_k)^2 - \sum_{i=1}^{n} d_i^2\right] = \frac{1}{2}\left[\frac{n^3 - n}{12} + \frac{n^3 - n}{12} - \sum_{i=1}^{n} d_i^2\right]$$

while its denominator reduces to:

$$\sqrt{\sum_{i=1}^{n} (y_{ij} - \bar{y}_j)^2 \sum_{i=1}^{n} (y_{ik} - \bar{y}_k)^2} = \sqrt{\left(\frac{n^3 - n}{12}\right)\left(\frac{n^3 - n}{12}\right)} = \frac{n^3 - n}{12}$$

Table 5.4 Numerical example. Ranks of four objects on two descriptors, $\mathbf{y}_1$ and $\mathbf{y}_2$.

Objects (observation units)	Ranks of objects on the two descriptors $\mathbf{y}_1$	$\mathbf{y}_2$
$\mathbf{x}_1$	3	3
$\mathbf{x}_2$	4	1
$\mathbf{x}_3$	2	4
$\mathbf{x}_4$	1	2

The final formula is obtained by replacing the above two expressions in eq. 5.2. This development shows that, when using ranks, eq. 5.1 simplifies to the following formula for Spearman's r:

$$r_{jk} = \frac{\frac{1}{2}\left[\frac{n^3 - n}{12} + \frac{n^3 - n}{12} - \sum_{i=1}^{n} d_i^2\right]}{\frac{n^3 - n}{12}} = 1 - \frac{6\sum_{i=1}^{n} d_i^2}{n^3 - n} \qquad \textbf{(5.3)}$$

Alternatively, the Spearman rank correlation coefficient may be obtained in two steps: (1) replace all observations by ranks (columnwise) and (2) compute the Pearson correlation coefficient (eq. 4.7) between the ranked variables. The result is the same as obtained from eq. 5.3.

The Spearman r coefficient varies between +1 and –1, just like the Pearson r. Descriptors that are perfectly matched, in terms of ranks, exhibit values $r = +1$ (direct relationship) or $r = -1$ (inverse relationship), whereas $r = 0$ indicates the absence of a monotonic relationship between the two descriptors. (Relationships that are not monotonic, e.g. Fig. 4.4d, can be quantified using polynomial or nonlinear regression, or else contingency coefficients; see Sections 6.2 and 10.3.)

Numerical example. A small example (ranked data, Table 5.4) illustrates the equivalence between eq. 5.1 computed on ranks and eq. 5.3. Using eq. 5.1 gives:

$$r_{12} = \frac{-2}{\sqrt{5 \times 5}} = \frac{-2}{5} = -0.4$$

The same result is obtained from eq. 5.3:

$$r_{12} = 1 - \frac{6 \times 14}{4^3 - 4} = 1 - \frac{84}{60} = 1 - 1.4 = -0.4$$

Two or more objects may have the same rank on a given descriptor. This is often the case with descriptors used in ecology, which may have a small number of states or ordered classes. Such observations are said to be *tied*. Each of them is assigned the average of the ranks which would have been assigned had no ties occurred. If the proportion of tied observations is large, correction factors must be introduced into the sums of squared deviations of eq. 5.2, which become:

$$\sum_{i=1}^{n} (y_{ij} - \bar{y}_j)^2 = \frac{1}{12}\left[(n^3 - n) - \sum_{r=1}^{q} (t_{rj}^3 - t_{rj})\right]$$

and

$$\sum_{i=1}^{n} (y_{ik} - \bar{y}_k)^2 = \frac{1}{12}\left[(n^3 - n) - \sum_{r=1}^{s} (t_{rk}^3 - t_{rk})\right]$$

where t_{rj} and t_{rk} are the numbers of observations in descriptors $\mathbf{y}_j$ and $\mathbf{y}_k$ which are tied at ranks r, these values being summed over the q sets of tied observations in descriptor j and the s sets in descriptor k.

Significance of the Spearman r is usually tested against the null hypothesis H_0: $r = 0$. When $n \geq 10$, the test statistic is the same as for Pearson's r (eq. 4.13):

$$t = \sqrt{\nu}\frac{r_{ki}}{\sqrt{1 - r_{ki}^2}} \qquad \textbf{(5.4)}$$

H_0 is tested by comparing statistic t to the value found in a table of critical values of t, with $\nu = n - 2$ degrees of freedom. H_0 is rejected when the probability corresponding to t is smaller than a predetermined level of significance (α, for a two-tailed test). The rules for one-tailed and two-tailed tests are the same as for the Pearson r (Section 4.2). When $n < 10$, which is not often the case in ecology, one must refer to a special table of critical values of the Spearman rank correlation coefficient, found in textbooks of nonparametric statistics.

Kendall corr. coeff.

Kendall's τ (tau) is another rank correlation coefficient, which can be used for the same types of descriptors as Spearman's r. One major advantage of τ over Spearman's r is that the former can be generalized to a partial correlation coefficient (below), which is not the case for the latter. While Spearman's r was based on the differences between the ranks of objects on the two descriptors being compared, Kendall's τ refers to a somewhat different concept, which is best explained using an example.

Numerical example. Kendall's τ is calculated on the example of Table 5.4, already used for computing Spearman's r. In Table 5.5, the order of the objects was rearranged so as to obtain increasing ranks on one of the two descriptors (here $\mathbf{y}_1$). The Table is used to determine the degree of dependence between the two descriptors. Since the ranks are now in increasing order

Table 5.5 Numerical example. The order of the four objects from Table 5.4 has been rearranged in such a way that the ranks on $\mathbf{y}_1$ are now in increasing order

Objects (observation units)	Ranks of objects on the two descriptors $\mathbf{y}_1$	$\mathbf{y}_2$
$\mathbf{x}_4$	1	2
$\mathbf{x}_3$	2	4
$\mathbf{x}_1$	3	3
$\mathbf{x}_2$	4	1

on $\mathbf{y}_1$, it is sufficient to determine how many *pairs of ranks* are also in increasing order on $\mathbf{y}_2$ to obtain a measure of the association between the two descriptors. Considering the object in first rank (i.e. $\mathbf{x}_4$), at the top of the right-hand column, the first pair of ranks (2 and 4, belonging to objects $\mathbf{x}_4$ and $\mathbf{x}_3$) is in increasing order; a score of +1 is assigned to it. The same goes for the second pair (2 and 3, belonging to objects $\mathbf{x}_4$ and $\mathbf{x}_1$). The third pair of ranks (2 and 1, belonging to objects $\mathbf{x}_4$ and $\mathbf{x}_2$) is in decreasing order, however, so that it earns a negative score –1. The same operation is repeated for every object in successive ranks along $\mathbf{y}_1$, i.e. for the object in second rank ($\mathbf{x}_3$): first pair of ranks (4 and 3, belonging to objects $\mathbf{x}_3$ and $\mathbf{x}_1$), etc. The sum S of scores assigned to each of the $n(n-1)/2$ different pairs of ranks is then computed.

Kendall's rank correlation coefficient is defined as follows:

$$\tau_a = \frac{S}{n(n-1)/2} = \frac{2S}{n(n-1)} \qquad (5.5)$$

where S stands for "sum of scores". Kendall's τ_a is thus the sum of scores for pairs in increasing and decreasing order, divided by the total number of pairs ($n(n-1)/2$). For the example of Tables 5.4 and 5.5, τ_a is:

$$\tau_a = \frac{2(1+1-1-1-1-1)}{4 \times 3} = \frac{2(-2)}{12} = -0.33$$

Clearly, in the case of perfect agreement between two descriptors, all pairs receive a positive score, so that $S = n(n-1)/2$ and thus $\tau_a = +1$. When there is complete disagreement, $S = -n(n-1)/2$ and thus $\tau_a = -1$. When the descriptors are totally unrelated, the positive and negative scores cancel out, so that S as well as τ_a are near 0.

Equation 5.5 cannot be used for computing τ when there are tied observations. This is often the case with ecological *semiquantitative* descriptors, which may have a small number of states. The Kendall rank correlation is then computed on a contingency table (see Chapter 6) crossing two semiquantitative descriptors.

Table 5.6 Numerical example. Contingency table giving the distribution of 80 objects among the states of two semiquantitative descriptors, **a** and **b**. Numbers in the table are frequencies (*f*).

	b_1	b_2	b_3	b_4	t_j
a_1	20	10	10	0	40
a_2	0	10	0	10	20
a_3	0	0	10	0	10
a_4	0	0	0	10	10
t_k	20	20	20	20	80

Table 5.6 is a contingency table crossing two ordered descriptors. For example, descriptor **a** could represent the relative abundances of arthropods in soil enumerated on a semiquantitative scale (e.g. absent, present, abundant and very abundant), while descriptor **b** could be the concentration of organic matter in the soil, divided into 4 classes. For simplicity, descriptors are called **a** and **b** here, as in Chapter 6. The states of **a** vary from 1 to r (number of rows) while the states of **b** go from 1 to c (number of columns).

To compute τ with tied observations, S is calculated as the difference between the numbers of positive (P) and negative (Q) scores, $S = P - Q$. P is the sum of all frequencies f in the contingency table, each one multiplied by the sum of all frequencies located *lower* and on its *right*:

$$P = \sum_{j=1}^{r} \sum_{k=1}^{c} \left[f_{jk} \times \sum_{l=j+1}^{r} \sum_{m=k+1}^{c} f_{lm} \right]$$

Likewise, Q is the sum of all frequencies f in the table, each one multiplied by the sum of all frequencies *lower* and on its *left*:

$$Q = \sum_{j=1}^{r} \sum_{k=1}^{c} \left[f_{jk} \times \sum_{l=j+1}^{r} \sum_{m=1}^{k-1} f_{lm} \right]$$

Numerical example. For Table 5.6:

$P = (20 \times 40) + (10 \times 30) + (10 \times 20) + (10 \times 20) + (10 \times 10) = 1600$
$Q = (10 \times 10) + (10 \times 10) = 200$
$S = P - Q = 1600 - 200 = 1400$

Using this value S, there are two approaches for calculating τ, depending on the numbers of states in the two descriptors. When **a** and **b** have the same numbers of states ($r = c$), τ_b is computed using a formula that includes the total number of pairs $n(n-1)/2$, as in the case of τ_a (eq. 5.5). The difference with eq. 5.5 is that τ_b includes corrections for the number of pairs L_1 tied in **a** and the number of pairs L_2 tied in **b**, where

$$L_1 = \sum_{j=1}^{r} \frac{1}{2} t_j (t_j - 1) \quad \text{in which } t_j \text{ is the marginal total for row } j$$

$$L_2 = \sum_{k=1}^{c} \frac{1}{2} t_k (t_k - 1) \quad \text{in which } t_k \text{ is the marginal total for column } k.$$

The formula for τ_b is:

$$\tau_b = \frac{S}{\sqrt{\frac{1}{2} n (n-1) - L_1} \sqrt{\frac{1}{2} n (n-1) - L_2}} \qquad \textbf{(5.6)}$$

When there are no tied observations, $L_1 = L_2 = 0$ and eq. 5.6 becomes identical to eq. 5.5.

Numerical example. For Table 5.6:

$$L_1 = \frac{40 \times 39}{2} + \frac{20 \times 19}{2} + \frac{10 \times 9}{2} + \frac{10 \times 9}{2} = 1060$$

$$L_2 = \frac{20 \times 19}{2} + \frac{20 \times 19}{2} + \frac{20 \times 19}{2} + \frac{20 \times 19}{2} = 760$$

$$\tau_b = \frac{1400}{\sqrt{\frac{1}{2} (80 \times 79) - 1060} \sqrt{\frac{1}{2} (80 \times 79) - 760}} = 0.62$$

Without correction for ties, the calculated value (eq. 5.5) would have been

$$\tau_a = (2 \times 1400) / (80 \times 79) = 0.44$$

The second approach for calculating τ with tied observations should be used when **a** and **b** do not have the same number of states ($r \neq c$). The formula for τ_c uses the minimum number of states in either **a** or **b**, $min(r, c)$:

$$\tau_c = \frac{S}{\frac{1}{2} n^2 \left(\frac{min - 1}{min} \right)} \qquad \textbf{(5.7)}$$

The significance of Kendall's τ is tested by reference to the null hypothesis H_0: $r = 0$ (i.e. independence of the two descriptors). A test statistic is obtained by transforming τ into z (or t_∞) using the following formula (Kendall, 1948):

$$z = \left[|\tau| \sqrt{\frac{9n(n-1)}{2(2n+5)}} \right] - \sqrt{\frac{18}{n(n-1)(2n+5)}} \qquad (5.8)$$

When $n \geq 30$, the second term of eq. 5.8 becomes negligible (at $n = 30$, the value of this term is only 0.0178). For $n \geq 10$, the sampling distribution of τ is almost the same as the normal distribution, so that H_0 is tested using a table of z. Since z tables are one-tailed, the z statistic of eq. 5.8 may be used directly for one-tailed tests by comparing it to the value z_α read in the table. For two-tailed tests, the statistic is compared to the value $z_{\alpha/2}$ from the z table. When $n < 10$, which is seldom the case in ecology, one should refer to Table B, at the end of this book. Table B gives the critical values of τ_α for $4 \leq n \leq 50$ (one-tailed and two-tailed tests).

Spearman's r provides a better approximation of Pearson's r when the data are almost quantitative and there are but a few tied observations, whereas Kendall's τ does better when there are many ties. Computing both Spearman's r and Kendall's τ_a on the same numerical example, above, produced different numerical values (i.e. $r = -0.40$ versus $\tau_a = -0.33$). This is because the two coefficients have different underlying scales, so that their numerical values cannot be directly compared. However, given their different sampling distributions, they both reject H_0 at the same level of significance. If applied to quantitative data that are meeting all the requirements of Pearson's r, both Spearman's r and Kendall's τ have power nearly as high (about 91%; Hotelling & Pabst, 1936) as their parametric equivalent. In all other cases, they are more powerful than Pearson's r. This refers to the notion of *power* of statistical tests: a test is more powerful than another if it is more likely to detect small deviations from H_0 (i.e. smaller type II error), for constant type I error.

Power

The chief advantage of Kendall's τ over Spearman's r, as already mentioned, is that it can be generalized to a partial correlation coefficient, which cannot be done with Spearman's (Siegel, 1956: 214). The formula for a partial τ is:

$$\tau_{12.3} = \frac{\tau_{12} - \tau_{13}\tau_{23}}{\sqrt{1 - \tau_{13}^2}\sqrt{1 - \tau_{23}^2}} \qquad (5.9)$$

This formula is algebraically the same as that of first-order partial Pearson r (eq. 4.36) although, according to Kendall (1948: 103), this would be merely coincidental because the two formulae are derived using entirely different approaches. The three τ coefficients on the right-hand side of eq. 5.9 may themselves be partial τ's, thus allowing one to control for more than one descriptor (i.e. high order partial correlation coefficients). It is not possible, however, to test the significance of partial rank correlation coefficients, because quantities P_{13}, P_{23}, Q_{13} and Q_{23} used for computing

$\tau_{12.3}$ are not independent, their sum being $n(n-1)/2$ instead of n (Kendall, 1948: 122; Seigel, 1956: 229).

Rank correlation coefficients should not be used in the Q mode, i.e. for comparing objects instead of descriptors. This is also the case for the Pearson r (Section 7.5). The reasons for this are the following:

• While physical dimensions disappear when computing correlation coefficients between variables expressed in different units, the same coefficients computed between objects have complex and non-interpretable physical dimensions.

• Physical descriptors are usually expressed in somewhat arbitrary units (e.g. mm, cm, m, or km are all equally correct, in principle). Any arbitrary change in units could dramatically change the values of correlations computed between objects.

• Descriptors may be standardized first to alleviate these problems but standardization of quantitative descriptors, before rank-ordering the data within objects, changes the values along object vectors in a nonmonotonic way. The correlation between two objects is a function of the values of all the other objects in the data set.

• Consider species abundance data. At most sampling sites, several species are represented by a small number of individuals, this number being subject to stochastic variability. It follows that their ranks, in a given observation unit, may not strictly correspond to their quantitative importance in the ecosystem. A rank correlation coefficient computed between observation units would thus have high variance since it would be computed on many uncertain ranks, giving a preponderant importance to the many poorly sampled species.

• While the central limit theorem insures that means, variances, covariances, and correlations converge towards their population values when the number of objects increases, computing these same parameters in the Q mode is likely to have the opposite effect since the addition of new variables into the calculations is likely to change the values of these parameters in a non-trivial way.

Kendall coeff. of concordance

The rank correlation coefficients described above measure the correlation for pairs of descriptors, based on n objects. In contrast, Kendall's coefficient of concordance W measures the relationship among *several* rank-ordered variables for n objects. In Table 5.1, Kendall's W is listed as equivalent to the coefficient of multiple linear correlation R, but the approach is actually quite different.

The analysis is conducted on a table which contains, in each column, the ranks of the n objects on one of the p descriptors, e.g. Table 5.7. Friedman (1937) has shown that, when the number of rows and/or columns is large enough, the following statistic is approximately distributed as χ^2 with $\nu = n - 1$ degrees of freedom:

$$X^2 = \left[\frac{12}{pn(n+1)} \sum_{i=1}^{n} R_i^2\right] - 3p(n+1) \qquad \textbf{(5.10)}$$

where R_i is the sum of the ranks for row i. This is Friedman's statistic for two-way analysis of variance by ranks. Kendall's coefficient of concordance (Kendall, 1948) is a simple transform of Friedman's X^2 statistic:

$$W = \frac{X^2}{p(n-1)} \qquad \textbf{(5.11)}$$

It can be shown that the following expression is equivalent to eq. 5.11:

$$W = \frac{12 \sum_{i=1}^{n} (R_i - \bar{R})^2}{p^2 (n^3 - n)} \qquad \textbf{(5.12)}$$

Two properties are used to demonstrate the equivalence of eqs. 5.11 and 5.12. The first one is that

$$\sum_{i=1}^{n} (R_i - \bar{R})^2 = \sum_{i=1}^{n} R_i^2 - \frac{1}{n}\left(\sum_{i=1}^{n} R_i\right)^2$$

and the second is that the sum of the all R_i values in the table is $pn(n+1)/2$.

Coefficient W varies between 0 (no concordance) and 1 (maximum concordance). Its significance is tested either using eq. 5.11 directly, or after transforming W into the associated X^2 statistic:

$$X^2 = p(n-1)\mathrm{W}$$

The null hypothesis (H_0) subjected to testing is that the row sums R_i are equal or, in other words, that the p sets of ranks (or the p semiquantitative descriptors) are independent of one another. The X^2 statistic is compared to a χ^2_α value read in a table of critical values of χ^2, for $\nu = (n - 1)$. When X^2 is smaller than the critical value χ^2_α (i.e. probability larger than α), the null hypothesis that the row sums R_i are equal cannot be rejected; this leads to the conclusion that the p descriptors are independent and differ in the way they rank the n objects. On the contrary, $X^2 \geq \chi^2_\alpha$ (i.e. probability smaller than or equal to α) indicates good agreement among the descriptors in the way they rank the objects. Textbooks of nonparametric statistics provide modified formulae for X^2, for data sets with tied observations.

Table 5.7 Numerical example. Ranks of six objects on three descriptors, $\mathbf{y}_1$, $\mathbf{y}_2$, and $\mathbf{y}_3$.

Objects (observation units)	Ranks of objects on the three descriptors $\mathbf{y}_1$	$\mathbf{y}_2$	$\mathbf{y}_3$	Row sums R_i
$\mathbf{x}_1$	1	1	6	8
$\mathbf{x}_2$	6	5	3	14
$\mathbf{x}_3$	3	6	2	11
$\mathbf{x}_4$	2	4	5	11
$\mathbf{x}_5$	5	2	4	11
$\mathbf{x}_6$	4	3	1	8

Numerical example. Calculation of Kendall's coefficient of concordance is illustrated using the numerical example of Table 5.7. Data could be semiquantitative rank scores, or quantitative descriptors coded into ranks. It is important to note that the $n = 6$ objects are ranked on each descriptor (column) separately. The last column gives, for each object i, the sum R_i of its ranks on the $p = 3$ descriptors. The Friedman statistic is calculated with eq. 5.10:

$$X^2 = \left[\frac{12}{3 \times 6\,(6+1)}\,(64 + 196 + 121 + 121 + 121 + 64)\right] - [3 \times 3\,(6+1)] = 2.429$$

Using eq. 5.11, the X^2 statistic is transformed into Kendall's W:

$$W = \frac{2.429}{3\,(6-1)} = 0.162$$

Alternatively, W could have been computed using eq. 5.12:

$$W = \frac{12 \times (6.25 + 12.25 + 0.25 + 0.25 + 0.25 + 6.25)}{9\,(216-6)} = 0.162$$

A table of critical values of χ^2 indicates that $X^2 = 2.43$, for $\nu = 6 - 1 = 5$, corresponds to a probability of ca. 0.80; the probability associated with this X^2 statistic is actually 0.787. The hypothesis (H_0) that the row sums R_i are equal cannot be rejected. One concludes that the three descriptors differ in the way they rank the 6 objects.

Chapter

6 Multidimensional qualitative data

6.0 General principles

Ecologists often use variables that are neither quantitative nor ordered (Table 1.2). Variables of this type may be of physical or biological nature. Examples of qualitative physical descriptors are the colour, locality, geological substrate, or nature of surface deposits. Qualitative biological descriptors include the captured or observed species; the different states of this nonordered descriptor are the different possible species. Likewise, the presence or absence of a species cannot, in most cases, be analysed as a quantitative variable; it must be treated as a semiquantitative or qualitative descriptor. A third group of qualitative descriptors includes the results of classifications — for example, the biological associations to which the zooplankton of various lakes belong, or the chemical groups describing soil cores. Such classifications, obtained or not by clustering (Chapter 8), define qualitative descriptors and, as such, they are amenable to numerical interpretation (see Chapter 10).

The present Chapter discusses the analysis of *qualitative* descriptors; methods appropriate for bivariate or multivariate analysis are presented. Information theory is an intuitively appealing way of introducing these methods of analysis. Section 6.1 shows how the amount of information in a qualitative descriptor may be measured. This paradigm is then used in the following sections.

The comparison of qualitative descriptors is based on *contingency tables*. In order to compare pairs of qualitative descriptors, the objects are first allocated to the cells of a table with two criteria (i.e. the rows and columns). In a two-way contingency table, the number of rows is equal to the number of states of the first descriptor and the number of columns to that of the second descriptor. Any cell in the table, at the intersection of a row and a column, corresponds to one state of each descriptor; the number of objects with these two states is recorded in this cell. The analysis of *two-way contingency tables* is described in Section 6.2. When there are more than two descriptors, *multiway* (or *multidimensional*) *contingency tables* are constructed as extensions of two-way tables. Their analysis is discussed in Section 6.3. Finally, Section 6.4 deals with the *correspondence* between descriptors in a contingency table.

ANOVA hypothesis

Correlation hypothesis

Contingency table analysis is the qualitative equivalent of both *correlation analysis* and *analysis of variance*; in the particular case of a two-way contingency table, the analysis is the equivalent of a one-way ANOVA. It involves the computation of X^2 (chi-square) statistics or related measures, instead of correlation or F statistics. Two types of null hypotheses (H_0) may be tested. The first one is the independence of the two descriptors, which is the usual null hypothesis in correlation analysis (H_0: the correlation coefficient $\rho = 0$ in the statistical population). The second type of hypothesis is similar to that of the analysis of variance. In a two-way contingency table, the classification criterion of the analysis of variance corresponds to the states of one of the descriptors. The null hypothesis says that the distributions of frequencies among the states of the second descriptor (dependent variable) are the same, among the groups defined by the states of the first descriptor. In other words, the observations form a homogeneous group. For example, if the groups (classification criterion) form the columns whereas the dependent variable is in the rows, H_0 states that the frequency distributions in all columns are the same. These two types of hypotheses require the calculation of the same expected values and the same test statistics. In multiway tables, the hypotheses tested are often quite complex because they take into account interactions among the descriptors (Section 6.3).

Considering species data, the various species observed at a sampling site are the states of a qualitative multi-state descriptor. Section 6.5 will discuss *species diversity* as a measure of dispersion of this qualitative descriptor.

The mathematics used throughout this chapter are quite simple and require no prior knowledge other than the intuitive notion of probability. Readers interested in applications only may skip Section 6.1 and come back to it when necessary. To simplify the notation, the following conventions are followed throughout this chapter. When a single descriptor is considered, this descriptor is called **a** and its states have subscripts i going from 1 to q, as in Fig. 1.1. In two-way contingency tables, the descriptors are called **a** and **b**. The states of **a** are denoted a_i with subscripts i varying from 1 to r (number of rows), while the states of **b** are denoted b_j with subscripts j varying from 1 to c (number of columns).

6.1 Information and entropy

Chapters 1 and 2 have shown that the ecological information available about the objects under study is usually (or may be reformulated as) a set of biological and/or physical characteristics, which correspond to as many descriptors. Searching for groups of descriptors that behave similarly across the set of objects, or that may be used to forecast one another (R analysis, Section 7.1), requires measuring the *amount of information* that these descriptors have in common. In the simplest case of two descriptors **a** and **b** (called $\mathbf{y}_1$ and $\mathbf{y}_2$ in previous chapters), one must assess how much *information* is provided by the distribution of the objects among the states of **a**, that could be used to forecast their distribution among the states of **b**. This approach is central to the analysis of relationships among ecological phenomena.

In 1968, Ludwig von Bertalanffy wrote, in his *General System Theory* (p. 32): "Thus, there exist models, principles, and laws that apply to generalized systems or their subclasses, irrespective of their particular kind, the nature of their component elements, and the relations or 'forces' between them". This is the case with information, which can be viewed and measured in the same manner for all systems. Some authors, including Pielou (1975), think that the concepts derived from information theory are, in ecology, a model and not a homology. Notwithstanding this opinion, the following sections will discuss how to measure information for biological descriptors in terms of information to be acquired, because such a presentation provides a better understanding of the nature of information in ecological systems.

The problem thus consists in measuring the amount of information contained in each descriptor and, further, the amount of information that two (or several) descriptors have in common. If, for example, two descriptors share 100% of their information, then they obviously carry the same information. Since descriptors are constructed so as to partition the objects under study into a number of states, two descriptors have 100% of their information in common when they partition a set of objects in exactly the same way, i.e. into two equal and corresponding sets of states. When descriptors are qualitative, this correspondence does not need to follow any ordering of the states of the two descriptors. For ordered descriptors, the ordering of the correspondence between states is important and the techniques for analysing the information in common belong to correlation analysis (Chapters 4 and 5).

Entropy

The mathematical theory of information is based on the concept of *entropy*. Its mathematical formulation was developed by Shannon (Bell Laboratories) who proposed, in 1948, the well-known equation:

$$H = -\sum_{i=1}^{q} \mathrm{p}_i \log \mathrm{p}_i \qquad \textbf{(6.1)}$$

where H is a measure of the uncertainty or choice associated with a frequency distribution (vector) $\mathbf{p}$; p_i is the probability that an observation belongs to state i of the descriptor (Fig. 1.1). In practice, p_i is the proportion (or relative frequency, on a 0-1 scale) of observations in state i. Shannon recognized that his equation is similar to the equation of entropy, published in 1898 by the physicist Boltzmann as a quantitative formulation of the second law of thermodynamics, which concerns the degree of disorganization in closed physical systems. He thus concluded that H corresponds to the entropy of information systems.

Table 6.1 Contingency table (numerical example). Distribution of 120 objects on descriptors **a** and **b**.

	b_1 30	b_2 30	b_3 30	b_4 30
$a_1 = 60$	30	10	15	5
$a_2 = 30$	0	20	0	10
$a_3 = 15$	0	0	0	15
$a_4 = 15$	0	0	15	0

Negative entropy

Note that the entropy of information theory is actually the *negative entropy* of physicists. In thermodynamics, an increase in entropy corresponds to an *increase in disorder*, which is accompanied by a *decrease of information*. Strictly speaking, information is negative entropy and it is only for convenience that it is simply called entropy. *In information theory, entropy and information are taken as synonymous.*

Information

Numerical example. In order to facilitate the understanding of the presentation up to Section 6.4, a small numerical example will be used in which 120 objects are described by two descriptors (**a** and **b**) with 4 states each. The question is to determine to what extent one descriptor can be used to forecast the other. The data in the numerical example could be, for example, the benthos sampled at 120 sites of an estuary, or the trees observed in 120 vegetation quadrats. Descriptor **a** might be the dominant species at each sampling site and descriptor **b**, some environmental variable with 4 states. The following discussion is valid for any type of qualitative descriptor and also for ordered descriptors divided into classes.

Assume that the 120 observations are distributed as 60, 30, 15 and 15 among the 4 states of descriptor **a** and that there are 30 observations in each of the 4 states of descriptor **b**. The states of the observations (objects), for the two descriptors combined, are given in Table 6.1.

For each descriptor, the probability of a state is estimated by the relative frequency with which the state is found in the set of observations. Thus, the probability distributions associated with descriptors **a** and **b** are:

a_1:	60	$p(a_1) = 1/2$	b_1:	30	$p(b_1) = 1/4$
a_2:	30	$p(a_2) = 1/4$	b_2:	30	$p(b_2) = 1/4$
a_3:	15	$p(a_3) = 1/8$	b_3:	30	$p(b_3) = 1/4$
a_4:	15	$p(a_4) = 1/8$	b_4:	30	$p(b_4) = 1/4$
	120			120	

The relative frequency of a given state is the probability of observing that state when taking an object at random.

Within the framework of information theory, the entropy of a probability distribution is measured, not in kilograms, metres per second, or other such units, but in terms of decisions. The measurement of entropy must reflect how difficult it is to find, among the objects under study, one that has a given state of the descriptor. An approximate measure of entropy is the average minimum number of binary questions that are required for assigning each object to its correct state. Thus, the *amount of information* which is gained by asking binary questions, and answering them after observing the objects, is equal to the *degree of disorder* or *uncertainty* initially displayed by the frequency distribution. In this sense, the terms *entropy* and *information* may be used synonymously. A few numerical examples will help understand this measure.

(1) When all the objects exhibit the same state for a descriptor, everything is known *a priori* about the distribution of observations among the different states of the descriptor. There is a single state in this case; hence, the number of binary questions required to assign a state to an object is zero ($H = 0$), which is the minimum value of entropy.

(2) The simplest case of a descriptor with non-null entropy is when there are two states among which the objects are distributed equally:

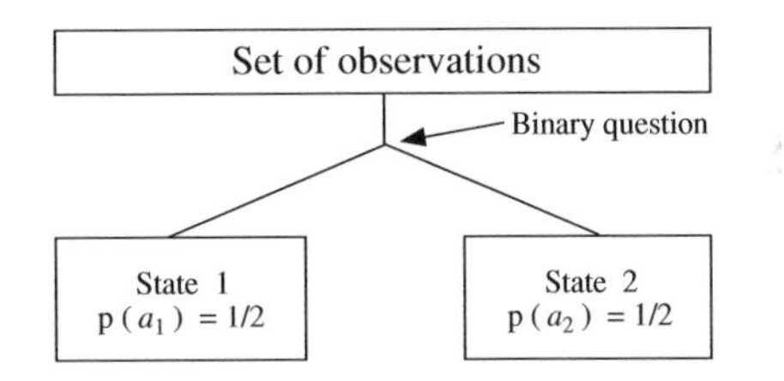

Binary question

In order to assign a state to any given object, a single binary question is necessary, of the type "Does this object belong to state 1?" If so, state 1 is assigned to the object; if not, the object belongs to state 2. The entropy associated with the descriptor is thus $H = 1$.

(3) Applying the above approach to a descriptor with four states among which the objects are distributed equally, one gets an entropy $H = 2$ since exactly two binary questions are required to assign a state to each object:

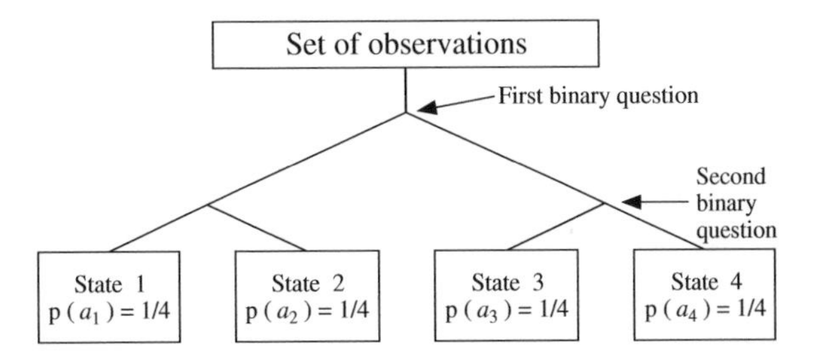

This would be the case of descriptor **b** in the numerical example of Table 6.1.

(4) For an eight-state descriptor with the objects equally distributed among the states, the binary questions are as follows:

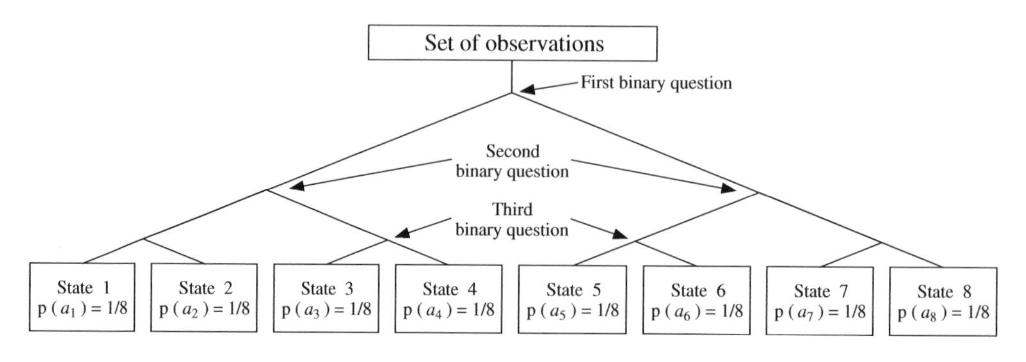

The total entropy of the descriptor is thus:

$$[3 \text{ questions} \times 8 \text{ (1/8 of the objects)}] = 3$$

and, in general, the entropy associated with a descriptor in which the objects are equally distributed among states is equal to the base 2 logarithm (if the questions are binary) of the number of states:

$\log_2 1 = 0$	$\log_2 8 = 3$
$\log_2 2 = 1$	$\log_2 16 = 4$
$\log_2 4 = 2$	etc.

Measuring the entropy from the number of binary questions is strictly equal to the logarithmic measure only when the number of states is an integer power of 2, or when the number of observations in the various states is such that binary questions divide them into equal groups (see the numerical example, below). In all other cases, the number of binary questions required is slightly larger than $\log_2$(number of states), because binary questions are then a little less efficient than in the previous case (Table 6.2). Binary questions have been used in the above discussion only to provide readers with a better understanding of entropy; the true measure is the logarithmic one. One may refer to Shannon (1948), or a textbook on information theory, for a more formal discussion of the measure of entropy.

The following example illustrates the relationship between probability and information. If an ecologist states that water in the Loch Ness is fresh, this is trivial since the probability of the event is 1 (information content null). If, however, he/she announces that she/he has captured a specimen of the famous monster, this statement contains much information because of its low probability (the dynamic aspects of Loch Ness Monster populations have been discussed by Sheldon & Kerr, 1972, Schneider & Wallis, 1973, and Rigler, 1982; see also Lehn, 1979, and Lehn & Schroeder, 1981, for a physical explanation of the Loch Ness and other aquatic monsters). Thus, information theory deals with a specific technical definition of information, which may

Table 6.2 The average minimum number of binary questions required to remove the uncertainty about the position of an object in the state-vector is equal to $\log_2$ (number of states) when the number of states is an integer power of 2 (in boldface) and the objects are equally distributed among the states. In all other cases, the number of binary questions is slightly larger than the entropy H. For example, for a three state descriptor with equal frequencies, the minimum number of binary questions is (2 questions × 2/3 of the objects) + (1 question × 1/3 of the objects) = 1.66666 binary questions.

Number of states	$\log_2$(number of states)	Average minimum number of binary questions
1	**0.00000**	**0.00000**
2	**1.00000**	**1.00000**
3	1.58496	1.66666
4	**2.00000**	**2.00000**
5	2.32193	2.40000
6	2.58496	2.66666
7	2.80735	2.85714
8	**3.00000**	**3.00000**
9	3.16993	3.22222
10	3.32193	3.40000
11	3.45943	3.54545
12	3.58496	3.66666
13	3.70044	3.76154
14	3.80735	3.85714
15	3.90689	3.93333
16	**4.00000**	**4.00000**

not correspond to the intuitive concept. A nontechnical example is that a book should contain the same amount of information before and after one has read it. From the information theory point of view, however, after one has read the book once, there is no information to be gained the next time he/she reads it (unless she/he has forgotten part of it after the first reading).

It should be clear, at this point of the discussion, that the entropy of a descriptor depends, among other characteristics, on the number of its states, among which the entropy is partitioned. In the case of the above four-state descriptor, for example, 1/4 of the entropy of the descriptor is attributed to each state, i.e. [$1/4 \log_2 4$], which is equal to [$1/4 \log_2 (1/4)^{-1}$]. The total entropy of the descriptor is thus:

$$H = \sum_{4 \text{ states}} (1/4) \log_2 (1/4)^{-1}$$

The same holds for the example of the eight-state descriptor. The entropy of each state is [$1/8 \log_2 8$] = [$1/8 \log_2 (1/8)^{-1}$], so that the total entropy of the descriptor is

$$H = \sum_{8 \text{ states}} (1/8) \log_2 (1/8)^{-1}$$

(5) Descriptor **a** in the numerical example (Table 6.1) illustrates the case of a descriptor for which the objects are not equally distributed among states. The probability distribution is [1/2, 1/4, 1/8, 1/8], which corresponds to the following scheme of binary questions:

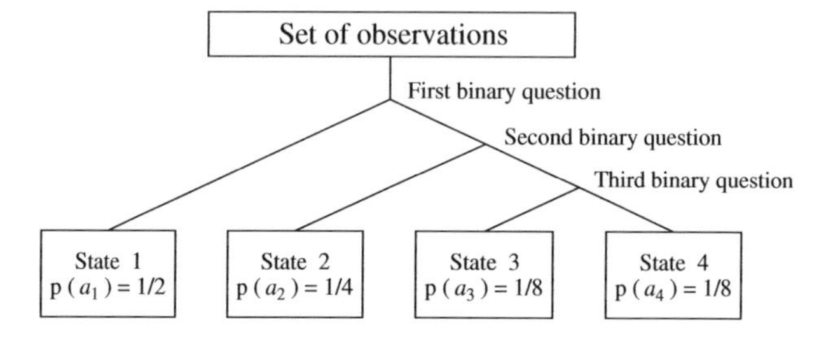

When the objects are not distributed evenly among the states, the amount of information one has *a priori* is higher than in the case of an even distribution, so that the information to be acquired by actual observation of the objects (i.e. the entropy) decreases. It follows that the entropy of the above descriptor should be $H < 2$, which is the maximum entropy for a four-state descriptor. Using binary questions, it is more economical to isolate half of the objects with the first question, then half of the remaining objects with the second question, and use a third question for the last two groups of 1/8 of the objects (see above). Since half of the objects require one question, 1/4 require 2, and the two groups of 1/8 require 3, the total entropy of this descriptor is:

$$H(\mathbf{a}) = (1/2 \times 1) + (1/4 \times 2) + (1/8 \times 3) + (1/8 \times 3) = 1.75$$

As in the previous examples, this is equal to:

$$H(\mathbf{a}) = 1/2 \log_2 2 + 1/4 \log_2 4 + 1/8 \log_2 8 + 1/8 \log_2 8$$

$$H(\mathbf{a}) = 1/2 \log_2 (1/2)^{-1} + 1/4 \log_2 (1/4)^{-1} + 1/8 \log_2 (1/8)^{-1} + 1/8 \log_2 (1/8)^{-1}$$

$$H(\mathbf{a}) = \sum_{\text{all states}} \mathrm{p}(i) \log_2 [\mathrm{p}(i)]^{-1}$$

Following the law of exponents for logarithms, exponent –1 is eliminated by writing the equation as:

$$H(\mathbf{a}) = -\sum_{\text{all states}} \mathrm{p}(i) \log_2 \mathrm{p}(i)$$

Bit
Hartley
Decit
Nat

This is exactly Shannon's formula for entropy (eq. 6.1). When the base for the logarithms is 2, the model is that of binary questions and the unit of entropy is the *bit* or *hartley* (Pinty & Gaultier, 1971). The model may be reformulated using questions with 10 answers, in which case the base of the logarithms is 10 and the unit is the *decit*. For natural logarithms, the unit is the *nat* (Pielou, 1975). These units are dimensionless, as are angles for example (Chapter 3).

Communication

Equation 6.1 may be applied to *human communications,* to calculate the information content of strings of symbols. For example, in a system of numbers with base n, there are n^N possible numbers containing N digits (in a base-10 system, there are $10^2 = 100$ numbers containing 2 digits, i.e. the numbers 00 to 99). It follows that the information content of a number with N digits is:

$$H = \log_2 n^N = N \log_2 n$$

The information per symbol (digit) is thus:

$$H/N = \log_2 n \qquad \textbf{(6.2)}$$

In the case of a binary (base 2) number, the information per symbol is $\log_2 2 = 1$ bit; for a decimal (base 10) number, it is $\log_2 10 = 3.32$ bits. A decimal digit contains 3.32 bits of information so that, consequently, a *binary* representation requires on average 3.32 times more digits than a *decimal* representation of the same number.

Alphabet

For an alphabet possessing 27 symbols (26 letters and the blank space), the information per symbol is $\log_2 27 = 4.76$ bits, assuming that all symbols have the same frequency. In languages such as English and French, each letter has a frequency of its own, so that the information per symbol is less than 4.76 bits. The information per letter is 4.03 bits in English and 3.95 bits in French. Hence, the translation from French to English should entail shorter text, which is generally the case.

English
French

Each language is characterized by a number of properties, such as the frequencies of letters, groups of letters, etc. These statistical properties, together with a defined syntax, determine a particular structure. For a given alphabet, the specific constraints of a language limit the number of messages which can actually be formulated. Thus, the number of lexical elements with 4, 5 or 6 letters is much smaller than the theoretical possible number (Table 6.3). This difference arises from the fact that every language contains a certain amount of information that is inherently embodied in its structure, which is termed *redundancy.* Without redundancy it would be impossible to detect errors slipping into communications, since any possible group of symbols would have some meaning.

In a language with n different symbols, each having a characteristic frequency ($N_1, N_2 \ldots N_n$), the total number of possible messages (P) made up of N symbols is equal to the number of *combinations*:

$$P = N! \,/\, (N_1!\, N_2! \,\ldots\, N_n!)$$

Table 6.3 Redundancy in the French language. Number of lexical elements with 4 to 6 letters (from Bourbeau *et al.*, 1984).

Number of letters	Possible number of lexical elements	Actual number of lexical elements in French
4	$26^4 \approx$ 457 000	3 558
5	$26^5 \approx$ 12 000 000	11 351
6	$26^6 \approx$ 300 000 000	24 800

The information content of a message with N symbols is:

$$H = \log_2 P = \log_2 [N! / (N_1! \, N_2! \ldots N_n!)]$$

Hence, the information per symbol is:

$$H/N = 1/N \log_2 [N! / (N_1! \, N_2! \ldots N_n!)] \tag{6.3}$$

which is the formula of Brillouin (1956). It will used later (Section 6.5) to calculate the species diversity of a sample, considered as representing a "message".

6.2 Two-way contingency tables

In order to compare two qualitative descriptors, the objects are allocated to the cells of a table with two criteria (i.e. the rows and columns). Each cell of a *two-way contingency table* (e.g. Tables 6.1 and 6.4) contains the number of observations with the corresponding pair of states of the qualitative descriptors. Numbers in the cells of a contingency table are absolute frequencies, i.e. *not* relative frequencies. The number of cells in the table is equal to the product of the number of states in the two descriptors. The first question relative to a contingency table concerns the relationship between the two descriptors: given the bivariate distribution of observations in the table, are the two descriptors related to each other, or not? This question is answered by calculating the expected frequency E for each cell of the table, according to a null hypothesis H_0, and performing a chi-square (X^2) test of the null hypothesis.

Null hypothesis

The simplest null hypothesis is that of independence of the two descriptors. E_{ij} is the number of observations that is expected in each cell (i, j) under H_0. Under this null hypothesis, E_{ij} is computed as the product of the marginal totals (i.e. the product of the

Table 6.4 Contingency table giving the observed (from Table 6.1) and expected (in parentheses) frequencies in each cell; $n = 120$. The observed frequencies that exceed the corresponding expectations are in boldface. Wilks' chi-square statistic: $X^2_W = 150.7$ ($\nu = 9$, $p < 0.001$).

	b_1 30	b_2 30	b_3 30	b_4 30
$a_1 = 60$	**30** (15)	10 (15)	15 (15)	5 (15)
$a_2 = 30$	0 (7.5)	**20** (7.5)	0 (7.5)	**10** (7.5)
$a_3 = 15$	0 (3.75)	0 (3.75)	0 (3.75)	**15** (3.75)
$a_4 = 15$	0 (3.75)	0 (3.75)	**15** (3.75)	0 (3.75)

sum of row i with the sum of column j), divided by n which is the total number of observations in the table:

Expected frequency

$$E = [(\text{row sum}) \times (\text{column sum})] / n \tag{6.4}$$

This equation generates expected frequencies whose relative distribution across the states of descriptor **a**, *within* each state of descriptor **b**, is the same as the distribution of all observed data across the states of **a**, and conversely (Table 6.4). The null hypothesis is tested using a X^2 statistic which compares the observed (O_{ij}) to the expected frequencies (E_{ij}).

In textbooks of basic statistics, the significance of relationships in two-way contingency tables is often tested using *Pearson chi-square statistic* (Pearson, 1900):

Pearson chi-square

$$X^2_P = \sum_{\text{all cells}} \frac{(O-E)^2}{E} \tag{6.5}$$

where $(O - E)$ measures the contingency of each cell. Instead of X^2_P, it is possible to compute Wilks' likelihood ratio (1935), also known as the G or *2I statistic* (Sokal & Rohlf, 1995) or G^2 (Bishop *et al.*, 1975; Dixon, 1981):

Wilks chi-square

$$X^2_W = 2 \sum_{\text{all cells}} O \ln\left(\frac{O}{E}\right) \tag{6.6}$$

where ln is the natural logarithm. For null frequencies, $\lim_{O \to 0} [O \ln(O/E)] = 0$.

Degrees of freedom

For a contingency table with r rows and c columns, the number of degrees of freedom used to determine the probability of accepting H_0, using a χ^2 table, is:

$$\nu = (r - 1)(c - 1) \tag{6.7}$$

When this probability is lower than a predetermined significance level, for example $\alpha = 0.05$, the null hypothesis (H_0) of independence of the two descriptors is rejected.

When the number of observations (n) is large (i.e. larger than ten times the number of cells, rc, in the table), the asymptotic distributions of X_P^2 and X_W^2 are χ^2. In other words, the two statistics are equivalent, when H_0 is true. There is however a problem when the number of observations is small, i.e. less than five times the number of cells. Small numbers of observations often lead to several null observed values (O_{ij}) in the contingency table, with correspondingly very low expected frequencies (E_{ij}). According to Cochran (1954) and Siegel (1956), when there is *one* value of E_{ij} smaller than 1, or when 20% or more of the expected values E_{ij} are smaller than 5, some states (rows or columns) must be grouped to increase the expected frequencies, provided that there is a logical basis to do so. It now appears that this empirical rule concerning expected frequencies is too conservative. Fienberg (1980, p. 172) cites results of simulations that lead to believe that, for $\alpha = 0.05$, the computed statistic is distributed like χ^2 (if H_0 is true) as long as all E_{ij} values are larger than 1.

Williams' correction

Concerning the choice of X_P^2 or X_W^2, there is no difference when the number of observations n is large (see the previous paragraph). When n is small, Larntz (1978) is of the opinion that X_P^2 is better than X_W^2. Sokal & Rohlf (1995) recommend using X_W^2 but suggest to correct it as proposed by Williams (1976) to obtain a better approximation of χ^2. This correction consists in dividing X_W^2 by a correction factor q_{min}. The correction factor, which is based on eq. 6.7, is computed as:

$$q_{min} = 1 + [(r^2 - 1)(c^2 - 1)/6\nu n] \tag{6.8}$$

When n is large relative to the number of cells in the contingency table, it is not necessary to apply a correction to X_W^2 since $q_{min} \approx 1$ in that case. William's correction is especially interesting when one must use X_W^2, as in the study of multiway contingency tables; the general formula for q_{min} is given in Subsection 6.3. Several computer programs allow users to compute both X_P^2 and X_W^2.

Another correction, available in some computer programs, consists in adding a small value (e.g. 0.5) to *each* observed value O_{ij} in the contingency table, when some of the O_{ij}'s are small. As indicated by Dixon (1981) and Sokal & Rohlf (1995), the effect of this correction is to lower the X^2 statistic, which makes the test more conservative. H_0 may then be rejected in a proportion of cases smaller than α, when the null hypothesis is true.

Another measure of interest to ecologists, which is related to the Wilks statistic (see below), refers to the concept of entropy (or information) discussed above. In the numerical example (Tables 6.1 and 6.4), if the correspondence between the states of descriptors **a** and **b** was perfect (i.e. descriptors completely dependent of each other), the contingency table would only have four non-zero cells — one in each row and each column. It would then be possible, using **a**, to perfectly predict the distribution of the observations among the states of **b**, and vice versa. In other words, given one state of the first descriptor, one would immediately know the state of the other descriptor. Thus, there would be no uncertainty (or entropy) concerning the distribution of the objects on **b**, after observing **a**, so that the entropy remaining in **b** after observing **a** would be null, i.e. $H(\mathbf{b}|\mathbf{a}) = 0$. On the contrary, if the descriptors were completely independent, the distribution of observations in each row of descriptor **a** would be in the same proportions as their overall distribution in **b** (found at top of Tables 6.1 and 6.4); the same would be true for the columns. $H(\mathbf{b}|\mathbf{a}) = H(\mathbf{b})$ would indicate that all the entropy contained in the distribution of **b** remains after observing **a**.

The two conditional entropies $H(\mathbf{a}|\mathbf{b})$ and $H(\mathbf{b}|\mathbf{a})$, as well as the entropy shared by the two descriptors, can be computed using the total information $H(\mathbf{a},\mathbf{b})$ and the information of each descriptor, $H(\mathbf{a})$ and $H(\mathbf{b})$, already computed in Section 6.1. $H(\mathbf{a},\mathbf{b})$ is computed on all the observed frequencies in the contingency table, using Shannon's formula (eq. 6.1):

$$H(\mathbf{a},\mathbf{b}) = -\sum_{\text{states of }\mathbf{a}} \sum_{\text{states of }\mathbf{b}} \mathrm{p}(i,j) \log \mathrm{p}(i,j) \qquad \textbf{(6.9)}$$

where $\mathrm{p}(i,j)$ is the observed frequency in each cell (i,j) of the contingency table, divided by the total number of observations n. For the example (Tables 6.1 or 6.4):

$$H(\mathbf{a},\mathbf{b}) = -\{1/4 \log_2 (1/4) + 1/6 \log_2 (1/6) + 3\,[1/8 \log_2 (1/8)] + 2\,[1/12 \log_2 (1/12)] + 1/24 \log_2 (1/24)\} = 2.84$$

$H(\mathbf{b}) = \mathrm{A} + \mathrm{B}$ and $H(\mathbf{a}) = \mathrm{B} + \mathrm{C}$, represented by circles in the Venne diagram below, have been computed in Section 6.1. $H(\mathbf{a},\mathbf{b})$ is the total information in the union of the two descriptors, A + B + C. The information (B) shared by the two descriptors is computed as:

$$\mathrm{B} = (\mathrm{A} + \mathrm{B}) + (\mathrm{B} + \mathrm{C}) - (\mathrm{A} + \mathrm{B} + \mathrm{C})$$

$$\mathrm{B} = H(\mathbf{b}) + H(\mathbf{a}) - H(\mathbf{a},\mathbf{b}) \qquad \textbf{(6.10)}$$

$$\mathrm{B} = 2.00 + 1.75 - 2.84 = 0.91$$

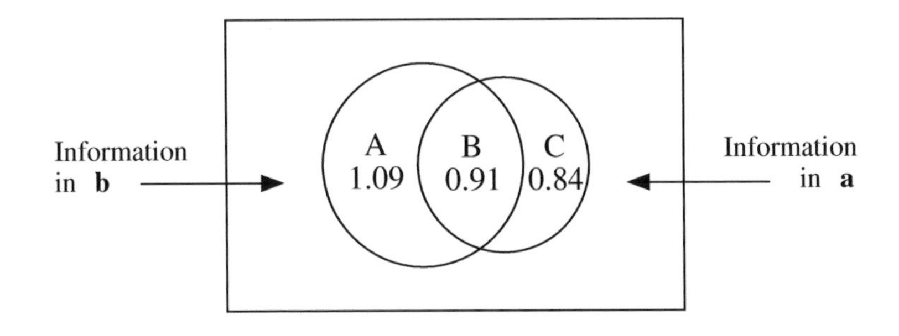

The information exclusive to each descriptor, A and C, is computed by subtraction:

$$\mathrm{A} = (\mathrm{A} + \mathrm{B} + \mathrm{C}) - (\mathrm{B} + \mathrm{C})$$

$$\mathrm{A} = H(\mathbf{b}|\mathbf{a}) = H(\mathbf{a},\mathbf{b}) - H(\mathbf{a}) \tag{6.11}$$

$$\mathrm{A} = 2.84 - 1.75 = 1.09$$

and

$$\mathrm{C} = (\mathrm{A} + \mathrm{B} + \mathrm{C}) - (\mathrm{A} + \mathrm{B})$$

$$\mathrm{C} = H(\mathbf{a}|\mathbf{b}) = H(\mathbf{a},\mathbf{b}) - H(\mathbf{b}) \tag{6.12}$$

$$\mathrm{C} = 2.84 - 2.00 = 0.84$$

There is a relationship between the reciprocal information B and Wilks' X^2_W statistic. It can be shown that $\mathrm{B} = (1/n) \sum O \ln(O/E)$ when B is computed with natural logarithms (ln), or else $\mathrm{B} \ln 2 = (1/n) \sum O \ln(O/E)$ when B is in bits. Using these relationships, it is possible to calculate the probability associated with B after transforming B into a X^2 statistic:

$$X^2_W = 2n\mathrm{B} \quad \text{(when B is in nats)} \tag{6.13}$$

$$X^2_W = 2n\mathrm{B} \ln 2 = n\mathrm{B} \ln 4 = 1.38629\, n\mathrm{B} \quad \text{(when B is in bits)} \tag{6.14}$$

Similarity

Using the measures of information A, B and C, various coefficients of reciprocal information may be computed. The *similarity* of descriptors **a** and **b** can be calculated as the amount of information that the two descriptors have in common, divided by the total information of the system:

$$S(\mathbf{a},\mathbf{b}) = \mathrm{B} \,/\, (\mathrm{A} + \mathrm{B} + \mathrm{C}) \tag{6.15}$$

$S(\mathbf{a},\mathbf{b}) = 0.91 / 2.84 = 0.32$, for the numerical example.

If the following steps of the analysis (clustering and ordination, Chapters 8 and 9) require that the measure of association between **a** and **b** be a metric, one may use the corresponding distance, defined as the sum of the information that the two descriptors possess independently, divided by the total information:

Rajski's metric

$$D(\mathbf{a},\mathbf{b}) = (\mathrm{A} + \mathrm{C}) \,/\, (\mathrm{A} + \mathrm{B} + \mathrm{C}) \tag{6.16}$$

$D(\mathbf{a},\mathbf{b}) = (1.09 + 0.84) / 2.84 = 0.68$, for the numerical example.

Obviously, $S(\mathbf{a},\mathbf{b}) + D(\mathbf{a},\mathbf{b}) = 1$.

Coherence coefficient

The measure of distance in eq. 6.16 is Rajski's metric (1961). This author has proposed another measure of similarity among descriptors, the *coherence coefficient*, which is used to assess the stochastic independence of two random variables:

$$S' = \sqrt{1 - D^2} \tag{6.17}$$

Another version of this coefficient,

$$S'' = B / (A + 2B + C) \qquad (6.18)$$

Symmetric, asymmetric uncertainty

is available in some computer programs under the name *symmetric uncertainty coefficient*. Two *asymmetric uncertainty coefficients* have also been proposed. They are used, for example, to compare the explanatory power of a descriptor with respect to several other descriptors: B / (A + B) controls for the total amount of information in **b**, whereas B / (B + C) controls for the total information in **a**.

The construction of an association matrix, containing any of the symmetric forms of coefficient described above, requires calculating $p(p-1)/2$ contingency tables; this matrix is symmetric and its diagonal is $S = 1$ or $D = 0$. *Qualitative (nonordered) descriptors* can thus be used to compute *quantitative association coefficients*, thus leading to numerical analysis of multivariate qualitative data sets. Furthermore, since quantitative or semiquantitative descriptors can be recoded into discrete states, it is possible, using uncertainty coefficients, to compute association matrices among descriptors of mixed types.

It is only through B, as discussed above, that a probability can be associated to the various uncertainty coefficients. For coefficient S above, one can state in very general terms that two descriptors are very closely related when $S(\mathbf{a},\mathbf{b}) > 0.5$; they are well associated when $0.5 > S > 0.3$; and some association exists when $S < 0.3$ without coming too close to 0 (Hawksworth *et al.*, 1968).

The probability associated with a X^2 statistic, calculated on a contingency table, assesses the hypothesis that the relationship between the two descriptors is *random*. Biological associations, for example, could be defined on the basis of relationships found to be non-random between pairs of species — the relationship being defined by reference to a pre-selected probability level (e.g. $\alpha = 0.05$ or 0.01) associated with the X^2 measuring the contingency between two species (Section 7.5). The value of X^2 may itself be used as a measure of the *strength* of the relationship between species. This is also the case for the reciprocal information measures defined above. With the same purpose in mind, it is possible to use one of the following *contingency coefficients*, which are merely transformations of a X^2 statistic on a scale from 0 to 1 (Kendall & Buckland, 1960; Morice, 1968):

Contingency coefficient

$$\text{Pearson contingency coefficient, } C = \sqrt{X^2 / (n + X^2)} \qquad (6.19)$$

$$\text{Tschuproff contingency coefficient, } T = \sqrt{X^2 / (n\sqrt{\text{degrees of freedom}})} \qquad (6.20)$$

where n is the number of observations. These contingency coefficients are not frequently used in ecology, however. They can only be used for comparing contingency tables of the same size.

Contingency tables are the main approach available to ecologists for the numerical analysis of relationships among qualitative descriptors, or else between qualitative descriptors and ordered variables divided into classes. Contingency tables are also convenient for analysing nonmonotonic relationships among ordered descriptors (a relationship is monotonic when there is a constant evolution of one descriptor with respect to the other; see Fig. 5.1). Reciprocal information and X^2 coefficients are sensitive enough that they could be used even with ordered variables, when relationships among a large number of descriptors are analysed by computer. One must simply make sure that the ordered data are divided into a sufficiently large number of classes to avoid clumping together observations that one would want to keep distinct in the results. If a first analysis indicates that redefining the boundaries of the classes could improve the interpretation of the phenomenon under study (the classes used to recode quantitative variables do not need to have the same width), ecologists should not hesitate to repeat the analysis using the recoded data. Far from being circular, this process corresponds to a progressive discovery of the structure of the information.

It is also possible to use the association coefficients described above to interpret the classifications or ordinations resulting from a first analysis of the data (Chapters 8 and 9). A classification may be compared to the descriptors from which it originates, in order to determine which descriptors are mostly responsible for it; or else, it may be compared to a new series of descriptors that could potentially explain it. Finally, one may use contingency tables to compare several classifications of the same objects, obtained through different methods. Chapter 10 deals with these higher-level analyses.

Ecological application 6.2

Legendre *et al.* (1978) analysed data from a winter aerial survey of land fauna, using contingency tables. They compared the presence or absence of tracks of different species to a series of 11 environmental descriptors. Five of these descriptors were qualitative, i.e. bioclimatic region, plant association, nature of the dominant and sub-dominant surface materials, and category of aquatic ecosystem. The others were semiquantitative, i.e. height of the trees, drainage, topography, thickness of the surface materials, abundance of streams and wetlands. The analysis identified the descriptors that determined or limited the presence of the 10 species that had been observed with sufficient frequency to permit their analysis. This allowed the authors to describe the niche of the 10 species.

6.3 Multiway contingency tables

When there are more than two descriptors, one might consider the possibility of analysing the data set using a series of two-way contingency tables, in which each pair of descriptors would be treated separately. Such an approach, however, would not take into account possible interactions among several descriptors and might thus miss part of the potential offered by the multidimensional structure of the data. This could lead to incorrect, or at least incomplete conclusions. Information on the analysis of multiway contingency tables can be found in Kullback (1959), Plackett (1974), Bishop *et al.* (1975), Upton (1978), Gokhale & Kullback (1978), Fienberg (1980) and Sokal & Rohlf (1995).

Log-linear model

The most usual approach for analysing multiway contingency tables is to adjust to the data a *log-linear model*, where the natural logarithm (ln) of the expected frequency E for each cell of the table is estimated as a sum of main effects and interactions. For example, in the case of two-way contingency tables (Section 6.2), the expected frequencies could have been computed using the following equation:

$$\ln E = [\theta] + [A] + [B] + [AB] \quad \textbf{(6.21)}$$

Symbols in brackets are the *effects*. [A] and [B] are the main effects of descriptors **a** and **b**, respectively, and [AB] is the effect resulting from the interaction between **a** and **b**. $[\theta]$ is the mean of the logarithms of the expected frequencies. In a two-way table, the hypothesis tested is that of independence between the two descriptors, i.e. H_0: [AB] = 0. The log-linear model corresponding to this hypothesis is thus:

$$\ln E = [\theta] + [A] + [B] \quad \textbf{(6.22)}$$

since [AB] = 0. The expected frequencies E computed using eq. 6.22 are exactly the same as those computed in Section 6.2 (eq. 6.4). The advantage of log-linear models becomes obvious when analysing contingency tables with more than two dimensions (or *criteria*).

For a contingency table with three descriptors (**a**, **b**, and **c**), the log-linear model containing all possible effects is:

$$\ln E = [\theta] + [A] + [B] + [C] + [AB] + [AC] + [BC] + [ABC]$$

Saturated model

Such a model is referred to as the *saturated model*. In practice, the effect resulting from the interaction among all descriptors is never included in any log-linear model, i.e. here [ABC]. This is because the expected frequencies for the saturated model are always equal to the observed frequencies ($E = O$), so that this model is useless. The general log-linear model for a three-way table is thus:

$$\ln E = [\theta] + [A] + [B] + [C] + [AB] + [AC] + [BC] \quad \textbf{(6.23)}$$

where H_0: [ABC] = 0. In other words, the logarithm of the expected frequency for each cell of the contingency table is computed here by adding, to the mean of the logarithms of the expected frequencies, one effect due to each of the three descriptors and one effect resulting from each of their two-way interactions.

Hierarchical model

Different log-linear models may be formulated by setting some of the effects equal to zero. Normally, one only considers *hierarchical models*, i.e. models in which the presence of a higher-order effect implies that all the corresponding lower effects are also included; the order of an effect is the number of symbols in the bracket. For example, in a hierarchical model, including [BC] implies that both [B] and [C] are also

Table 6.5 Possible log-linear models for a three-way contingency table. Hypotheses and corresponding models. All models include the three main effects.

Hypotheses (H_0)	Log-linear models
1.[ABC] = 0	ln E = [θ] + [A] + [B] + [C] + [AB] + [AC] + [BC]
2.[ABC] = 0, [AB] = 0	ln E = [θ] + [A] + [B] + [C] + [AC] + [BC]
3.[ABC] = 0, [AC] = 0	ln E = [θ] + [A] + [B] + [C] + [AB] + [BC]
4.[ABC] = 0, [BC] = 0	ln E = [θ] + [A] + [B] + [C] + [AB] + [AC]
5.[ABC] = 0, [AB] = 0, [AC] = 0	ln E = [θ] + [A] + [B] + [C] + [BC]
6.[ABC] = 0, [AB] = 0, [BC] = 0	ln E = [θ] + [A] + [B] + [C] + [AC]
7.[ABC] = 0, [AC] = 0, [BC] = 0	ln E = [θ] + [A] + [B] + [C] + [AB]
8.[ABC] = 0, [AB] = 0, [AC] = 0, [BC] = 0	ln E = [θ] + [A] + [B] + [C]

included. For a three-way contingency table, there are eight possible hierarchical models, corresponding to as many different hypotheses (Table 6.5). Models in the Table all include the three main effects. Each hypothesis corresponds to different types of interaction among the three variables. Methods for calculating the expected frequencies E are detailed in the general references cited above. In practice, one uses a program available in a computer package (e.g. BMDP4F, or FUNCAT in SAS), with which it is easy to estimate the expected frequencies for any hierarchical model defined by the user.

The number of degrees of freedom (ν) depends on the interactions which are included in the model. For the general hierarchical model of eq. 6.23,

$$\nu = rst - [1 + (r-1) + (s-1) + (t-1) + (r-1)(s-1) + (r-1)(t-1) + (s-1)(t-1)] \quad \textbf{(6.24)}$$

where r, s and t are the numbers of states of descriptors **a**, **b** and **c**, respectively. If there were only two descriptors, **a** and **b**, the log-linear model would not include the interaction [AB], so that eq. 6.24 would become:

$$\nu = rs - [1 + (r-1) + (s-1)] = (r-1)(s-1)$$

which is identical to eq. 6.7. In Table 6.5, model 4, for example, does not include the interaction [BC], so that:

$$\nu = rst - [1 + (r-1) + (s-1) + (t-1) + (r-1)(s-1) + (r-1)(t-1)]$$

Programs in computer packages calculate the number of degrees of freedom corresponding to each model.

It is possible to test the goodness of fit of a given model to the observed data by using one of the X^2 statistics already described for two-way tables, X^2_P or X^2_W (eqs. 6.5 and 6.6). The null hypothesis (H_0) tested is that the effects excluded from the model are null. Rejecting H_0, however, does not allow one to accept the alternative hypothesis that *all* the effects included in the model are not null. The only conclusion to be drawn from rejecting H_0 is that at least some of the effects in the model are not null. When the probability of a model is larger than the significance level α, the conclusion is that the model fits the data well.

Williams' correction

As in the case of two-way contingency tables (eq. 6.8), it is recommended to divide X^2_W by a correction factor, q_{min} (Williams, 1976), when the number of observations n is small, i.e. less than 4 or 5 times the number of cells in the table. For the general hierarchical model (eqs. 6.23 and 6.24):

$$q_{min} = 1 + (1/6\nu n)\,[r^2s^2t^2 - 1 - (r^2-1) - (s^2-1) - (t^2-1) - (r^2-1)(s^2-1) - (r^2-1)(t^2-1) - (s^2-1)(t^2-1)] \quad \textbf{(6.25)}$$

In the case of two descriptors, eq. 6.25 becomes:

$$q_{min} = 1 + (1/6\nu n)\,[r^2s^2 - 1 - (r^2-1) - (s^2-1)]$$

$$q_{min} = 1 + (1/6\nu n)\,[(r^2-1)(s^2-1)$$

which is identical to eq. 6.8. For model 4 in Table 6.5, used above as example:

$$q_{min} = 1 + (1/6\nu n)\,[r^2s^2t^2 - 1 - (r^2-1) - (s^2-1) - (t^2-1) - (r^2-1)(s^2-1) - (r^2-1)(t^2-1)]$$

This correction cannot be applied, as such, to contingency tables containing null expected frequencies (see below). The other possible correction, which consists in adding to each cell of the table a small value, e.g. 0.5, has the same effect here as in two-way contingency tables (see Section 6.2).

Ecological application 6.3a

Legendre (1987a) analysed biological oceanographic data obtained at 157 sites in Baie des Chaleurs (Gulf of St. Lawrence, eastern Canada). The data set (observations made at 5-m depth) included measurements of temperature, salinity, nutrients (phosphate and nitrate), and chlorophyll *a* (estimated from the *in vivo* fluorescence of water pumped on board the ship). As it often happens in ecology, the numerical analysis was hampered by three practical problems. (1) The measured concentrations of nutrients were often near or below the detection limit, with the result that many of them exhibited large experimental errors. (2) Relationships between variables were often nonmonotonic, i.e. they did not continuously increase or decrease but reached a maximum (or a minimum) after which they decreased (or increased). (3) Most of the

Table 6.6 Multiway contingency table analysis of oceanographic data recoded into discrete classes (Legendre, 1987a). Using a hierarchy of log-linear models, the concentrations of chlorophyll *a* (C; 4 classes) are analysed as a function of the temperature-salinity (TS) characteristics of the water masses (symbol in this Table: T; 3 classes) and the concentrations of phosphate (P; 2 classes) and nitrate (N; 2 classes). When a higher-order effect is present, all the corresponding lower-order effects are included in the model.

Effects in the model	Interpretation	ν	X^2_W
[NTP], [C]	Chl *a* is independent of the environmental variables	30	121 *
Difference	Adding [CT] to the model significantly improves the fit	9	89 *
[NTP], [CT]	Chl *a* depends on the TS characteristics	21	32
Difference	Adding [CP] to the model significantly improves the fit	3	13 *
[NTP], [CT], [CP]	Chl *a* depends on the TS characteristics and on phosphate	18	**19**
Difference	Adding [CN] does not significantly improve the fit	7	5
[NTP], [CT], [CP], [CN]	The most parsimonious model does not include a dependence of chl *a* on nitrate	11	**14**

* $p \leq 0.05$; bold X^2_W values correspond to models with $p > 0.05$ of fitting the data

variables were intercorrelated, so that no straightforward interpretation of phytoplankton (i.e. chlorophyll *a*) concentrations was possible in terms of the environmental variables. Since multiway contingency table analysis can handle these three types of problems, it was decided to partition the (ordered) variables into discrete classes and analyse the transformed data using hierarchical log-linear models.

The initial model in Table 6.6 (line 1) only includes the interaction among the three environmental variables, with no effect of these on chl *a*. This initial model does not fit the data well. Adding the interaction between chl *a* and the temperature-salinity (TS) characteristics significantly improves the fit (i.e. there is a significant difference between models; line 2). The resulting model could be accepted (line 3), but adding the interaction between chl *a* and phosphate further improves the fit (significant difference, line 4) and the resulting model fits the data well (line 5). Final addition of the interaction between chl *a* and nitrate does not improves the fit (difference not significant, line 6). The most parsimonious model (line 5) thus shows a dependence of chl *a* concentrations on the TS characteristics and phosphate. The choice of the initial model, for this example, is explained in Ecological application 6.3b.

There are 8 hierarchical models associated with a three-way contingency table, 113 with a four-way table, and so forth, so that the choice of a single model, among all those possible, rapidly becomes a major problem. In fact, it often happens that several models could fit the data well. Also, in many instances, the fit to the data could be improved by adding supplementary terms (i.e. effects) to the model. However, this improved fit would result in a more complex ecological interpretation because of the added interaction(s) among descriptors. It follows that the choice of a model generally involves a compromise between goodness of fit and simplicity of interpretation. Finally, even if it was possible to test the fit of all possible models to the data, this would not be an acceptable approach since these tests would not be independent. One must therefore use some other strategy to choose the "best" model.

There are several methods to select a model which are both statistically acceptable and ecologically parsimonious. These methods are described in the general references mentioned at the beginning of this Section. In practice, since none of the methods is totally satisfactory, one could simply use, with care, those included in the available computer package.

Partitioning the X^2_W

1) A first method consists in *partitioning* the X^2_W statistics associated with a hierarchy of log-linear models. The hierarchy contains a series of models, which are made progressively simpler (or more complex) by removing (or adding) one effect at a time. It can be shown that the difference between the X^2_W statistics of two successive models in the hierarchy is itself a X^2_W statistic, which can therefore be tested. The corresponding number of degrees of freedom is the difference between those of the two models. The approach is illustrated using Ecological application 6.3a (Table 6.6). The initial model (line 1) does not fit the data well. The difference (line 2) between it and the next model is significant, but the second model in the hierarchy (line 3) still does not fit the data very well. The difference (line 4) between the second and third models is significant and the resulting model (line 5) fits the data well. The difference (line 6) between the third model and the next one being non-significant, the most parsimonious model in the hierarchy is that on line 5. The main problem with this method is that one may find different "most parsimonious" models depending on the hierarchy chosen *a priori*. Partitioning X^2 statistics is possible only with X^2_W, not X^2_P.

Stepwise selection

2) A second family of approaches lies in the *stepwise forward selection* or *backward elimination* of terms in the model. As always with stepwise methods (see Section 10.3), (a) it may happen that forward selection lead to models quite different from those resulting from backward elimination, and (b) the tests of significance must be interpreted with caution because the computed statistics are not independent. Stepwise methods thus only provide guidance, which may be used for limiting the number of models to be considered. It often happens that models other than those identified by the stepwise approach are found to be more parsimonious and interesting, and to fit the data just as well (Fienberg 1980: 80).

Effect screening

3) Other methods simultaneously consider all possible effects. An example of *effect screening* (Brown 1976) is given in Dixon (1981). The approach is useful for reducing the number of models to be subsequently treated, for example, by the method of hierarchical partitioning of X^2_W statistics (see method 1 above).

When analysing multiway contingency tables, ecologists must be aware of a number of possible practical problems, which may sometimes have significant impact on the results. These potential problems concern the cells with zero expected frequencies, the limits imposed by the sampling design, the simultaneous analysis of descriptors with mixed levels of precision (i.e. qualitative, semiquantitative, and quantitative), and the use of contingency tables for the purpose of explanation or forecasting.

Cells with $E = 0$

1) Multiway contingency tables, in ecology, often include cells with expected frequencies $E = 0$. There are two types of zero expected frequencies, i.e. those resulting from sampling and those that are of structural nature.

Sampling zeros are caused by random variation, combined with small sample size relative to the number of cells in the multiway contingency table. Such zeros would normally disappear if the size of the sample was increased. The presence of cells with null observations ($O = 0$) may result, when calculating specific models, in some expected frequencies $E = 0$. This is accompanied by a reduction in the number of degrees of freedom. For example, according to eq. 6.24, the number of degrees of freedom for the initial model in Table 6.6 (line 1) should be $\nu = 33$, since this model includes four main effects [C], [N], [P], and [T] and interactions [NP], [NT], [PT], and [NPT]; however, the presence of cells with null observations ($O = 0$) leads to cells with $E = 0$, which reduces the number of degrees of freedom to $\nu = 30$. Rules to calculate the reduction in the number of degrees of freedom are given in Bishop *et al.* (1975: 116 *et seq.*) and Dixon (1981: 666). In practice, computer programs generally take into account the presence of zero expected frequencies when computing the number of degrees of freedom for multiway tables. The problem does not occur with two-way contingency tables because cells with $E = 0$ are only possible, in the two-way configuration, if all the observations in the corresponding row or column are null, in which case the corresponding state is automatically removed from the table.

Structural zeros correspond to combinations of states that cannot occur *a priori* or by design. For example, in a study where two of the descriptors are sex (female, male) and sexual maturity (immature, mature, gravid), the expected frequency of the cell "gravid male" would *a priori* be $E = 0$. Another example would be combinations of states which have not been sampled, either by design or involuntarily (e.g. lack of time, or inadequate planning). Several computer programs allow users to specify the cells which contain structural zeros, before computing the expected frequencies.

2) In principle, the methods described here for multiway contingency tables can only be applied to data resulting from *simple random sampling* or *stratified sampling* designs. Fienberg (1980: 32) gives some references in which methods are described for analysing qualitative descriptors within the context of *nested sampling* or a *combination of stratified and nested sampling* designs. Sampling designs are described in Cochran (1977), Green (1979), and Thompson (1992), for example.

Mixed precision

3) Analysing together *descriptors with mixed levels of precision* (e.g. a mixture of qualitative, semiquantitative, and quantitative descriptors) may be done using multiway contingency tables. In order to do so, continuous descriptors must first be partitioned into a small number of classes. Unfortunately, there exists no general approach to do so. When there is no specific reason for setting the class limits, it has been suggested, for example, to partition continuous descriptors into classes of equal width, or containing an equal number of observations. Alternatively, Cox (1957) describes a method which may be used for partitioning a normally distributed descriptor into a predetermined number of classes (2 to 6). For the specific case discussed in the next paragraph, where there is one response variable and several explanatory variables, Legendre & Legendre (1983b) describe a method for partitioning the ordered explanatory variables into classes in such a way as to maximize the relationships to the response variable. It is important to be aware that, when analysing the contingency table, different ways of partitioning continuous descriptors may sometimes lead to different conclusions. In practice, the number of classes of each descriptor should be as small as possible, in order to minimize the problems discussed above concerning the calculation of X_W^2 (see eqs. 6.8 ad 6.25 for correction factor q_{min}) and the presence of sampling zeros. Another point is that contingency table analysis considers the different states of any descriptor to be nonordered. When some of the descriptors are in fact ordered (i.e. originally semiquantitative or quantitative), the information pertaining to the ordering of states may be used when adjusting log-linear models (see for example Fienberg 1980: 61 *et seq.*).

4) There is an analogy between *log-linear models* and *analysis of variance* since the two approaches use the concepts of effects and interactions. This analogy is superficial, however, since analysis of variance aims at assessing the effects of explanatory factors on a single response variable, whereas log-linear models have been developed to describe structural relationships among several descriptors corresponding to the dimensions of the table.

5) It is possible to use contingency table analysis for interpreting a *response variable* in terms of several interacting *explanatory variables*. In such a case, the following basic rules must be followed. (1) Any log-linear model fitted to the data must include by design the term for the highest-order interaction among all *explanatory variables*. In this way, all possible interactions among the explanatory variables are included in the model, because of its hierarchical nature. (2) When interpreting the model, one should not discuss the interactions among the explanatory variables. They are incorporated in the model for the reason given above, but no test of significance is performed on them. In any case, one is only interested in the interactions between the explanatory and response variables. An example follows.

Ecological application 6.3b

The example already discussed in application 6.3a (Legendre, 1987a) aimed at interpreting the horizontal distribution of phytoplankton in Baie des Chaleurs (Gulf of St. Lawrence, eastern

Canada), in terms of selected environmental variables. In such a case, where a single response variable is interpreted as a function of several potentially explanatory variables, all models considered must include by design the highest-order interaction among the explanatory variables. Thus, all models in Table 6.6 include the interaction [NPT]. The simplest model in the hierarchy (line 1 in Table 6.6) is that with effects [NPT] and [C]. In this simplest model, there is no interaction between chlorophyll and any of the three environmental variables, i.e. the model does not include [CN], [CP] or [CT]. When interpreting the model selected as best fitting the data, one should not discuss the interaction among the explanatory variables, because the presence of [NPT] prevents a proper analysis of this interaction. Table 6.6 then leads to the interpretation that the horizontal distribution of phytoplankton depends on the TS characteristics of water masses and on phosphate concentration.

Logistic regression

When the *qualitative response variable* is *binary*, one may use the *logistic linear* (or *logit*) *model* instead of the log-linear model. Fitting such a model to data is also called *logistic regression* (Subsection 10.3.7). In logistic regression, the explanatory descriptors do not have to be divided into classes; they may be discrete or continuous. This type of regression is available in various computer packages and some programs allow the *response variable* to be *multi-state*. Efficient use of logistic regression requires that *all* the explanatory descriptors be potentially related to the response variable. This method may also replace discriminant analysis in cases discussed in Subsection 10.3.7 and Section 11.6.

There are many cases where multiway contingency tables have been successfully used in ecology. Examples are found in Fienberg (1970) and Schoener (1970) for the habitat of lizards, Jenkins (1975) for the selection of trees by beavers, Legendre & Legendre (1983b) for marine benthos, and Fréchet (1990) for cod fishery.

6.4 Contingency tables: correspondence

Once it has been established that two or more qualitative descriptors in a contingency table are not independent (Sections 6.2 and 6.3), it is often of interest to identify the cells of the table that account for the existing relationship between descriptors. These cells, which show how the descriptors are related, define the *correspondence* between the rows and columns of the contingency table. By comparison with parametric and nonparametric statistics (Chapters 4 and 5), the measures of contingency described in Sections 6.2 and 6.3 are, for qualitative descriptors, analogous to the *correlation* between ordered descriptors, whereas correspondence would be analogous to *regression* (Section 10.3) because it makes it possible to forecast the state of one descriptor using another descriptor. *Correspondence analysis* (Section 8.4) is another method that allows, among other objectives, the identification of the relationships between the rows and columns of a contingency table. This can be achieved directly through the approach described in the present section.

In a contingency table where the descriptors are not independent (i.e. the null hypothesis of independence has been rejected), the cells of interest to ecologists are those in which the observed frequencies (O_{ij}) are very different from the corresponding expected frequencies (E_{ij}). Each of these cells corresponds to a given state for each descriptor in the contingency table. The fact that $O_{ij} \neq E_{ij}$ is indicative of a stronger interaction, between the states in question, than expected under the null hypothesis used for computing E. For example, hypothesis H_0 in Table 6.4 is that of independence between descriptors **a** and **b**. This hypothesis being rejected ($p < 0.001$), one may identify in the contingency table the observed frequencies O_{ij} that are higher than the corresponding expected frequencies E_{ij}. Values $O_{ij} > E_{ij}$ (bold-face type in Table 6.4) give an indication of how **a** and **b** are related. These values may be located anywhere in the table, since contingency table analysis does not take into account the ordering of states.

Conditional probability distribution

A mathematically identical result is reached using the concept of entropy of Sections 6.1 and 6.2. The entropy of a single descriptor (Section 6.1) is called the *unconditional entropy* of its probability distribution. When analysing contingency tables, one may also calculate the *conditional probability distribution* of, say, descriptor **b** for each state of descriptor **a**. A particular example of such a distribution, the conditional multinormal distribution, has been discussed in Section 4.7.

Section 6.1 has shown that the entropy of a descriptor is equal to the amount of information that can be gained by observing the distribution of objects among the states of this descriptor. It follows that the entropy of descriptor **b** is the maximum amount of information which may be obtained concerning **b**. Let us assume that, instead of **b**, it is **a** which is observed in order to learn something about **b**. Information can be obtained only insofar as **a** and **b** have information in common. If the two descriptors have no information in common, the sum of the conditional entropies of **b** is equal to its unconditional entropy, as calculated in Section 6.1. If **a** and **b** have information in common, the entropy of **b** is lowered by an amount equal to the information shared by the two descriptors (i.e. the reciprocal information B, Section 6.2), as shown by this Venne diagram:

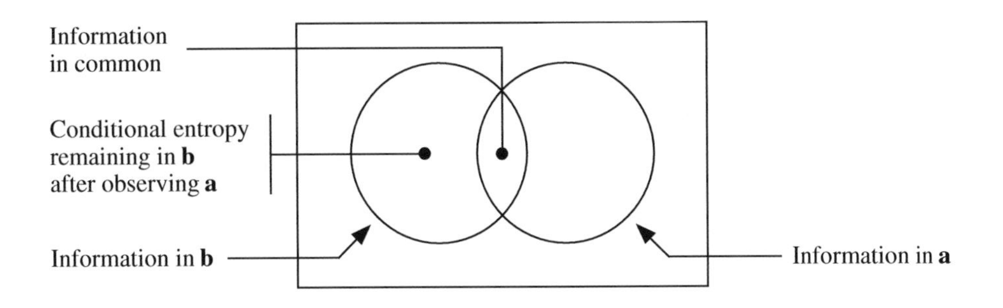

The first step in analysing the correspondence in a two-way contingency table, using conditional entropy, is to construct a table in which the frequencies of the objects, for each state of descriptor **a**, are distributed among the states of **b**. The result is a table of the *conditional probabilities of* **b** *for each state of* **a** (Table 6.7). Each row

Table 6.7 Conditional probability distributions of **b**, for each state of descriptor **a**. The conditional probabilities that exceed the corresponding unconditional probabilities (heading row) are in bold-face type. These are the same as the cells with bold-face values in Table 6.4

	p(b_1)	p(b_2)	p(b_3)	p(b_4)	
	0.25	0.25	0.25	0.25	Σ
a_1	**0.50**	0.17	0.25	0.08	1.00
a_2	0.00	**0.67**	0.00	**0.33**	1.00
a_3	0.00	0.00	0.00	**1.00**	1.00
a_4	0.00	0.00	**1.00**	0.00	1.00

of the table contains the probabilities of selecting an object having the various states of descriptor **b**, assuming that this object has the state of descriptor **a** corresponding to the row. The first row, for example, contains all the objects with dominant species a_1, and it gives their probability distribution on descriptor **b**. The sum of each row is 1, since each frequency in the contingency table (Table 6.1, for the example) has been divided by the (marginal) total of the corresponding row.

To analyse a table of conditional probabilities, one identifies the conditional probabilities that are larger than the corresponding unconditional probabilities, in the column headings. Under the null hypothesis of independence of the two descriptors, each conditional probability distribution (i.e. each row of the table) should be approximately equal to the unconditional distribution (heading row). Thus, in Table 6.7, b_1 has a probability of 0.50 among the objects with dominant species a_1, while this same state b_1 has a probability of 0.25 among all objects. A similar table of conditional probabilities could have been computed with descriptor **a** conditional on **b**; the cells of interest, and thus the ecological conclusions, would have been exactly the same.

Test of $O_{ij} = E_{ij}$

It is also possible to *test* the significance of the difference between O_{ij} and E_{ij} in each cell of the contingency table. Ecologists may be interested in any difference, whatever its sign, or only in cases where O_{ij} is significantly higher than E_{ij} (preference) or significantly lower (avoidance, exclusion).

Bishop *et al.* (1975: 136 *et seq.*) describe three statistics for measuring the difference between O and E. They may be used for two-way or multiway contingency tables. The three statistics are the components of X_P^2, components of X_W^2, and Freeman-Tukey deviates:

$$\text{component of } X_P^2 : (O - E) / (\sqrt{E}) \quad \textbf{(6.26)}$$

$$\text{component of } X_W^2 : 2\,O \ln(O/E) \quad \textbf{(6.27)}$$

$$\text{Freeman-Tukey deviate: } \sqrt{O} + \sqrt{O+1} - \sqrt{4E+1} \quad \textbf{(6.28)}$$

These statistics are available in various computer packages. A critical value has been proposed by Bishop *et al.* (1975) for testing the significance of statistics 6.26 and 6.28:

$$\sqrt{X^2_{[\nu,\alpha]} / (\text{no.cells})}$$

E_{ij} is said to be significantly different from O_{ij} when the absolute value of the statistic, for cell (i, j), is larger than the critical value. According to Sokal & Rohlf (1995), however, the above critical value often results in a type I error much greater than the nominal α level. These authors use instead the following approximate criterion to test Freeman-Tukey deviates:

$$\sqrt{\nu\, X^2_{[1,\alpha]} / (\text{no.cells})} \quad \textbf{(6.29)}$$

In cells where the (absolute) value of the Freeman-Tukey deviate is larger than the criterion, it is concluded that $E_{ij} \neq O_{ij}$ at significance level α. Neu *et al.* (1974) recommend to test only the cells where $5 \leq E_{ij} \leq (n - 5)$. It is also recommended to apply a Bonferroni or Holm correction (Box 1.3) to significance level α in order to account for multiple testing. An example is provided in Table 6.8.

Alternatively, the following statistic (adapted from Neu *et al.*, 1974) may be computed:

$$Z = \frac{|O - E|}{\sqrt{O\,(1 - O/n)}} \quad \textbf{(6.30)}$$

where n is the total number of observations in the contingency table. When statistic Z is larger than the critical value $z_{[1-(\alpha/2\text{ no. cells})]}$ read from a table of standard normal deviates, it is concluded that O_{ij} is significantly different from E_{ij} at probability level $\alpha/2$ (one-tailed test); the further division by the number of cells is the Bonferroni correction (Box 1.3). Statistics higher than the critical value z, in Table 6.9, are in boldface type. As is often the case, the conclusions drawn from Tables 6.8 and 6.9 are not the same.

Comparing Tables 6.4 and 6.7 to Tables 6.8 and 6.9 shows that considering only the cells where $O_{ij} > E_{ij}$ may lead to conclusions which, without necessarily being

Table 6.8 Statistics (Freeman-Tukey deviates, eq. 6.28) for testing the significance of individual cells in a contingency table. The observed and expected values are in Table 6.4. Absolute values larger than the criterion (eq. 6.29) $[9\ \chi^2_{[1,0.05]}\ /\ 16]^{1/2} = [9 \times 3.84\ /\ 16]^{1/2} = 1.47$ are in boldface type. A Bonferroni-corrected criterion $[9\ \chi^2_{[1,0.05/16]}\ /\ 16]^{1/2} = [9 \times 9.5\ /\ 16]^{1/2} = 2.31$ would have led to the same conclusions with the present example. Values in boldface print identify the cells of the table in which the number of observations (O_{ij}) significantly ($p < 0.05$) differs (higher or lower) from the corresponding expected frequencies (E_{ij}). The overall null hypothesis (H_0: complete independence of descriptors **a** and **b**) was rejected first (Table 6.4) before testing the significance of the observed values in individual cells of the contingency table.

	b_1	b_2	b_3	b_4
a_1	**3.23**	–1.33	0.06	**–3.12**
a_2	**–4.57**	**3.49**	**–4.57**	0.91
a_3	–3.00 *	–3.00 *	–3.00 *	3.87 *
a_4	–3.00 *	–3.00 *	3.87 *	–3.00 *

* No test because $E_{ij} < 5$ (Table 6.4).

incorrect, are subject to some risk of error. For example, dominant species a_2 may well not be indicative of environmental condition b_4 as suggested in Table 6.7. Tables 6.8 and 6.9 also show that dominant species a_1 is significantly under-represented in environmental condition b_4, suggesting that this condition is strongly adverse to the species.

Ecological application 6.4

Legendre *et al.* (1982) explored the relationship between the abundance of phytoplankton and vertical stability of the water column in a coastal embayment of Hudson Bay (Canadian Arctic). Surface waters are influenced by the plume of the nearby Great Whale River. There were intermittent phytoplankton blooms from mid-July through mid-September. In order to investigate the general relationship between phytoplankton concentrations (chlorophyll *a*) and the physical conditions, chl *a* and salinity data from 0 and 5 m depths were allocated to a contingency table (Table 6.10). The null hypothesis of independence being rejected, the correspondence between the two descriptors rests in three cells. (1) At high salinities (> 22 mg L^{-1}), there is a significantly small number of high chl *a* (> 1.5 mg m^{-3}) observations. At intermediate salinities (18-22 mg L^{-1}), (2) high chl *a* observations are significantly numerous, whereas (3) low chl *a* observations are significantly infrequent. At low salinities (< 18 mg L^{-1}), the numbers observed are not significantly different from the frequencies expected under the null hypothesis of independence.

Table 6.9 Statistics (eq. 6.30) for testing the significance of individual cells in a contingency table. The observed and expected values are in Table 6.4. Values larger than $z_{[1\,-\,0.05\,/\,(2\times4\times4)]} = z_{0.9984} =$ 2.95 are in boldface. They identify cells in which the number of observations (O_{ij}) significantly ($p < 0.05$) differs (higher or lower) from the corresponding expected frequency (E_{ij}).

	b_1	b_2	b_3	b_4
a_1	**3.16**	1.65	0.00	**4.57**
a_2	---	**3.06**	---	0.83
a_3	--- *	--- *	--- *	3.11 *
a_4	--- *	--- *	3.11 *	--- *

--- Statistic not computed because the denominator is 0. * No test because $E_{ij} < 5$ (Table 6.4).

Table 6.10 shows that, on the one hand, high chl *a* concentrations were positively associated with intermediate salinities, whereas they were much reduced in waters of high salinity. On the other hand, low chl *a* concentrations were characteristically infrequent in waters of intermediate salinities. The overall interpretation of these results, which also took into account estimates of the vertical stability of the water column (Richardson number), was as follows: (1) strong vertical mixing led to high salinities at the surface; this mixing favoured nutrient replenishment, but dispersed phytoplankton biomass over the water column; (2) low salinity conditions were not especially favourable nor adverse to phytoplankton, i.e. stratification was favourable, but dilution by water from the nearby river was adverse; (3) intermediate salinities were associated with intermittent conditions of stability; under such conditions, both the high nutrient concentrations and the stability of the water column were favourable to phytoplankton growth and accumulation. Intermittent summer blooms thus occurred upon stabilization of the water column, as a combined result of wind relaxation and fortnightly tides.

6.5 Species diversity

Sections 4.1 and 4.3 have shown that the distribution of a quantitative variable is characterized by its *dispersion* around the mean. The parametric and nonparametric measures of dispersion are the *variance* (eq. 4.3) and the *range*, respectively. These two measures do not apply to qualitative variables, for which the *number of states* (q) may be used as a simple measure of dispersion. However, this measure does not take advantage of the fact that, in several instances, the frequency distribution of the observations in the various states is known (Section 6.1). When the relative

Table 6.10 Contingency table: chlorophyll a concentrations as a function of salinity in the surface waters of Manitounuk Sound (Hudson Bay, Canadian Arctic). In each cell: observed (O_{ij}) and expected (E_{ij}, in parentheses) frequencies, and statistic (eq. 6.30) to test the hypothesis that $O_{ij} = E_{ij}$ ($\alpha = 0.05$). Statistics in bold-face print are larger than $z_{[1-0.05/12]} = 2.64$, indicating that $O_{ij} \neq E_{ij}$. Total no. observations $n = 207$. $X_W^2 = 33.78$ ($\nu = 2$, $p < 0.001$); hence the hypothesis of independence between chl a and salinity is rejected.

Chlorophyll a (mg m^{-3})	Salinity (mg L^{-1}) 6-18	18-22	22-26
1.5-6.1 (high values)	2 (3.29) *	22 (8.09) **3.14**	7 (19.62) **4.85**
0-1.5 (low values)	20 (18.71) 0.30	32 (45.91) **2.67**	124 (111.38) 1.79

* Statistic not computed nor tested because $E_{ij} < 5$.

frequencies of the states are available, eq. 6.1 may be used to measure the dispersion of a qualitative variable:

$$H = -\sum_{i=1}^{q} p_i \log p_i$$

where p_i is the relative frequency or proportion (on a 0-1 scale) of observations in state i. This formula can be rewritten as:

$$H = \frac{1}{n}\sum_{i=1}^{q} -(\log n_i - \log n)\, n_i$$

where n is the total number of organisms and n_i is the number of organisms belonging to species i. The latter equation is similar to the formula for the variance of n objects divided into q classes:

$$s_y^2 = \frac{1}{n-1}\sum_{i=1}^{q} (y_i - \bar{y})\, f_i$$

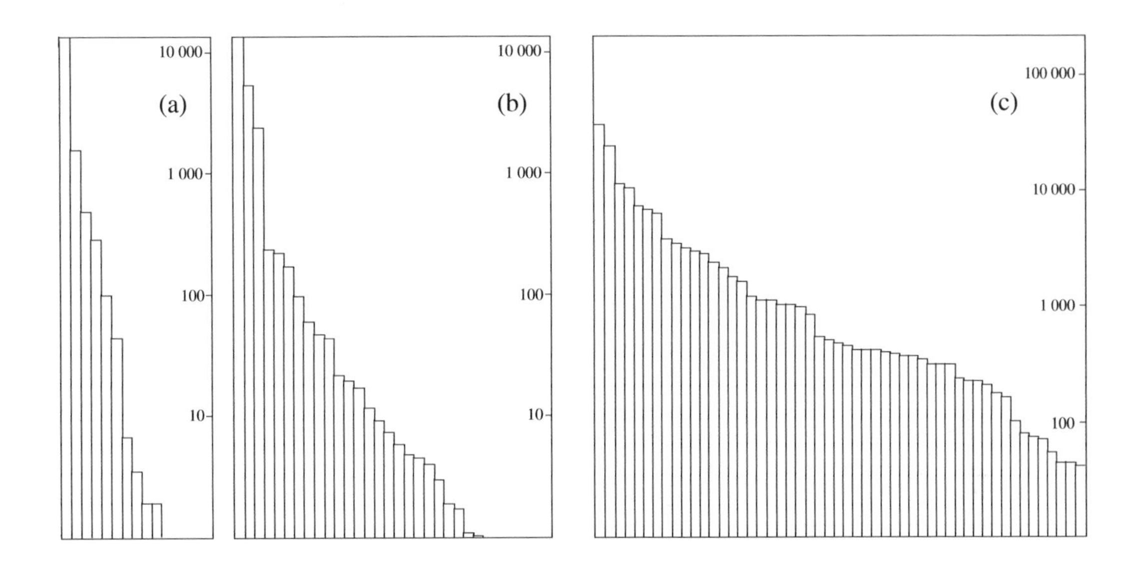

Figure 6.1 Fish catches (abundances) in (a) the Barents Sea, (b) the Indian Ocean, and (c) the Red Sea. Along the abscissa, species are arranged in order of decreasing frequencies. The ordinates of histograms are logarithmic. Adapted from Margalef (1974).

where f_i is the frequency of the ith class. In ecology, H is widely used to measure the *diversity* of a species assemblage; it is generally computed for each sampling site separately. In species diversity studies, the qualitative descriptor is the list of the q species present and each state corresponds to a species name. Both the number of species q and the entropy H belong to the same family of generalized entropies (eq. 6.31, below).

In assemblages of biological species, there are generally several species represented by a single or a few individuals, and a few species that are very abundant. The few abundant species often account for many more individuals than all the rare species together. Figure 6.1 shows, in order of decreasing frequencies, the abundances of fish species caught in the Barents Sea, the Indian Ocean, and the Red Sea. Diversity indices must be applicable to any type of species assemblage regardless of the shape of the abundance distribution. One parameter of the distribution is clearly the *number of species*; another is the *shape of the distribution*. An alternative approach is to combine these two parameters in a single index, using the entropy measure H for example. Species diversity may thus be defined as a *measure of species composition, in terms of both the number of species and their relative abundances*. It is a synthetic biotic index which captures multidimensional information relative to the species composition of an assemblage or a community.

Species diversity

Diversity indices characterize species composition at a given site and a given time. These indices are used by ecologists for various purposes, which may be of theoretical or practical nature. In theoretical ecology, measurements of diversity may be used to compare different communities, or else the same community at different times. For example, temporal changes in species diversity is a major characteristic of ecological succession (Margalef, 1968, 1974; Gutierrez & Fey, 1980). Diversity may also be compared to other characteristics that may change within communities, such as productivity, maturity, stability, and spatial heterogeneity. In studies that encompass several communities, species diversity may be compared to chemical, geomorphological, or climatological environmental variables; see for example Strömgren *et al.* (1973). In a more practical context, spatial or temporal changes in diversity are often used to assess the effects of pollution on biological communities; see for instance Wilhm & Dorris (1968). More recently, the rapid disappearance of species in several regions and ecosystems have caused international action to assess and preserve biodiversity.

In principle, diversity should not be computed on taxonomic levels other than species. Similarly, species are the basic units in the niche, competition, and succession theories. This is because the resources of an ecosystem are apportioned among the local populations (demes) of the species present in the system, each species representing a separate genetic pool. Attempts at measuring diversity at supraspecific levels generally produce confusing or trivial results.

It is generally not useful to measure species diversity of a whole community (e.g. primary, secondary, and tertiary producers and decomposers), because of the different roles played by various species in an ecosystem. It is better (Hurlbert, 1971; Pielou, 1975) to restrict the study of species diversity (and of the underlying theoretical phenomena, e.g. competition, succession) to a single *taxocene*. A taxocene is a set of species belonging to a given supraspecific taxon that make up a natural ecological community or, in other words, that represent a taxonomic segment of a community or association (Chodorowski, 1959; Hurlbert, 1971). The supraspecific taxon must be such that its member species are about the same size, have similar life histories, and compete over both ecological and evolutionary time for a finite amount of similar resources (Deevey, 1969). A taxocene occupies a limited segment in space and in the environment. For these reasons, the following information about the reference population should accompany any measure of diversity: (1) the spatial boundaries of the region or volume within which the population is found and a description of the sampling method; (2) the temporal limits within which the observations have been made; (3) the taxocene under study (Hurlbert, 1971; Pielou, 1975).

Taxocene

Sampling sites may harbour species that differ much in size or role in the environment. This may occur, for example, when all plants in quadrats (ligneous and herbaceous) are counted, or when species at different developmental stages are collected (e.g. counting saplings as equivalent to adult trees). Comparisons of diversity indices with production or environmental variables may be easier in such cases if species diversity is computed, not on the numbers of individuals, but instead on measures of biomass (Wilhm, 1968) or dry mass, productivity (Dickman, 1968), fecundity, or any other appropriate measure of energy transfer.

Species diversity indices may be used to compare successive observations from the same community (time series: O mode, Fig. 7.1) or sampling sites from different areas (Q mode). Coefficients in Chapter 7 compare objects by combining paired information available for each species. In contrast, diversity indices pool the multispecies information into a single value for each observation, before comparing them.

Over the years, several formulae have been proposed in the ecological literature for measuring species diversity. The present section describes only the few indices that are usually found in the modern literature. Species diversity has been the subject of detailed discussions, for example by Pielou (1969, 1975), Margalef (1974), Peet (1974), and Legendre & Legendre (1983a).

1 — Diversity

Hill (1973a) and Pielou (1975) noted that the three diversity indices mostly used by ecologists are specific cases of the *generalized entropy* formula of Rényi (1961)

$$H_a = \frac{1}{1-a} \log \sum_{i=1}^{q} \mathrm{p}_i^a \qquad \textbf{(6.31)}$$

where: $a = 0, 1, \ldots$; q is the number of species; p_i is the relative frequency or proportion of species i. Hill (1973a) prefers the corresponding diversity numbers:

Diversity number

$$N_a = \exp H_a \qquad \textbf{(6.32)}$$

It can be shown that the first three entropies (order 0 to 2) and corresponding diversity numbers are:

$$\text{(a) } H_0 = \log q \qquad \text{(b) } N_0 = q \qquad \textbf{(6.33)}$$

$$\text{(a) } H_1 = -\sum \mathrm{p}_i \log \mathrm{p}_i = H \qquad \text{(b) } N_1 = \exp H \qquad \textbf{(6.34)}$$

$$\text{(a) } H_2 = -\log \sum \mathrm{p}_i^2 = -\log(\text{concentration}) \qquad \text{(b) } N_2 = \text{concentration}^{-1} \qquad \textbf{(6.35)}$$

Hill (1973a) noted that increasing the order a diminishes the relative weights of rare species in the resulting index. In a review of the topic, Peet (1974) proposed other ways of creating families of diversity indices. Let us examine the first three orders of eq. 6.31 in more detail.

Number of species

1) *Entropy of order a = 0* — The *number of species q* (eq. 6.33b) is the index of diversity most often used in ecology. It goes back to Patrick (1949):

$$Diversity = q \qquad \textbf{(6.36)}$$

It is more sensitive to the presence of rare species than higher-order indices. The number of species can also be seen as a component of other diversity indices (e.g. *H*, Subsection 2).

As the size of the sampling units increases, additional rare species appear. This is a problem with all diversity indices and it is at its worst in eq. 6.36. It is incorrect to compare the diversities of sampling units having different sizes because diversity measures are not additive (Subsection 1.4.2). This point has been empirically shown by He *et al.* (1996). This problem can be resolved by calculating the numbers of species that the sampling units would contain if they all had the same size, for example 1000 organisms. This may be done using Sanders' (1968) *rarefaction method*, whose formula was corrected by Hurlbert (1971). The formula computes the expected number of species q' in a standardized sampling unit of n' organisms, from a nonstandard sampling unit containing q species, a total of n organisms, and n_i organisms belonging to each species i:

$$E(q') = \sum_{i=1}^{q} \left[1 - \frac{\binom{n-n_i}{n'}}{\binom{n}{n'}} \right] \qquad \textbf{(6.37)}$$

where $n' \le (n - n_1)$, n_1 being the number of individuals in the most abundant species (y_1), and the terms in parentheses are combinations. For example:

$$\binom{n}{n'} = \frac{n!}{n'!\,(n-n')!}$$

Shannon's entropy

2) *Entropy of order a = 1* — Margalef (1958) proposed to use Shannon's entropy *H* (eqs. 6.1 and 6.34a) as an index of species diversity.

$$H = -\sum_{i=1}^{q} \mathrm{p}_i \log \mathrm{p}_i$$

The properties of *H* as a measure of diversity are the following:

• $H = 0$ (minimum value), when the sampling unit contains a single species; *H* increases with the number of species.

• For a given number of species, *H* is maximum when the organisms are equally distributed among the *q* species: $H = \log q$. For a given number of species, *H* is lower when there is stronger dominance in the sampling unit by one or a few species (e.g. Figs. 6.1a and b). The actual value of H depends on the base of logarithms (2, e, 10, or other). This base must always be reported since it sets the scale of measurement.

• Like the variance, diversity can be partitioned into different components. It follows that the calculation of diversity can take into account not only the proportions of the different species but also those of genera, families, etc. Partitioning diversity into a component for *genera* and a component for *species within genera* allows one to examine two adaptive levels among the environmental descriptors. Such partitioning can be done using eqs. 6.10-6.12. Total diversity, $H = \mathrm{A} + \mathrm{B} + \mathrm{C}$, which is calculated using the proportions of species without taking into account those of genera, is equal to the diversity due to genera, $H(G) = \mathrm{A} + \mathrm{B}$, plus that due to species within genera, $H(S \mid G) = \mathrm{C}$, which is calculated as the sum of the species diversities in each genus, weighted by the proportions of genera. The formula is:

$$H = H(G) + H(S \mid G) \qquad \textbf{(6.38)}$$

This same calculation may be extended to other systematic categories. Considering, for example, the categories family (F), genus (G), and species (S), diversity can be partitioned into the following hierarchical components:

$$H = H(F) + H(G \mid F) + H(E \mid G,F) \qquad \textbf{(6.39)}$$

Using this approach, Lloyd *et al.* (1968) measured hierarchical components of diversity for communities of reptiles and amphibians in Borneo.

Most diversity indices share the above first two properties, but only the indices derived from eq. 6.31 have the third one (Daget, 1980). The probabilistic interpretation of H refers to the *uncertainty* about the identity of an organism chosen at random in a sampling unit. The uncertainty is small when the sampling unit is dominated by a few species or when the number of species is small. These two situations correspond to low H.

In principle, H should only be used when a sample is drawn from a theoretically infinite population, or at least a population large enough that sampling does not modify it in any noticeable way. In cases of samples drawn from small populations, or samples whose representativeness is unknown, it is theoretically better, according to Pielou (1966), to use Brillouin's formula (1956), proposed by Margalef (1958) for computing diversity H. This formula was introduced in Section 6.1 to calculate the information per symbol in a message (eq. 6.3):

$$H = (1/n) \log[n! \,/\, (n_1!\, n_2! \,\ldots\, n_i! \,\ldots\, n_q!)]$$

where the n_i is the number of individuals in species i and n is the total number of individuals in the collection. Brillouin's H corresponds to sampling *without* replacement (and is thus more exact) whereas Shannon's H applies to sampling *with* replacement. In practice, H computed with either formula is the same to several decimal places, unless samples are so small that they should not be used to estimate species diversity in any case. Species diversity cannot, however, be computed on measures of biomass or energy transfer using Brillouin's formula.

3) *Entropy of order a = 2* — Simpson (1949) proposed an index of species diversity based on the probability that two interacting individuals of a population belong to the same species. This index is frequently used in ecology. When randomly drawing, without replacement, two organisms from a sampling unit containing q species and n individuals, the probability that the first organism belong to species i is n_i/n and that the second also belong to species i is $(n_i - 1)/(n - 1)$. The combined probability of the two events is the product of their separate probabilities. Simpson's *concentration* index is the probability that two randomly chosen organisms belong to the same species, i.e. the sum of combined probabilities for the different species:

Concen-tration

$$Concentration = \sum_{i=1}^{q} \frac{n_i (n_i - 1)}{n\ (n-1)} = \frac{\sum_{i=1}^{q} n_i (n_i - 1)}{n (n-1)}$$

When n is large, n_i is almost equal to $(n_i - 1)$, so that the above equation becomes:

$$Concentration = \sum_{i=1}^{q} \left(\frac{n_i}{n}\right)^2 = \sum_{i=1}^{q} p_i^2 \qquad \textbf{(6.40)}$$

which corresponds to eq. 6.35a. This index may be computed from numbers of individuals, or from measures of biomass or energy transfer. The higher is the probability that two organisms be conspecific, the smaller is the diversity of the sampling unit. For this reason, Greenberg (1956) proposed to measure species diversity as:

$$Diversity = 1 - concentration \qquad \textbf{(6.41)}$$

which is also the *probability of interspecific encounter* (Hurlbert, 1971). Pielou (1969) has shown that this index is an unbiased estimator of the diversity of the population from which the sample has been drawn. However, eq. 6.41 is more sensitive than H to changes in abundance of the few very abundant species, so that Hill (1973a) recommended to use instead:

$$Diversity = concentration^{-1} \qquad \textbf{(6.42)}$$

which is identical to eq. 6.35b. Hill (1973a) also showed that this index is linearly related to exp H (eq. 6.34b). Examples to the same effect are also given by Daget (1980).

2 — Evenness, equitability

Several authors, for example Margalef (1974), prefer to directly interpret species diversity, as a function of physical, geographical, biological, or temporal variables, whereas others consider that species diversity is made of two components which should be interpreted separately. These two components are the *number of species* and the *evenness* of their frequency distribution. Although the concept of evenness had been introduced by Margalef (1958), it was formally proposed by Lloyd & Ghelardi (1964) for characterizing the *shape* of distributions such as in Fig. 6.1, where the component "number of species" corresponds to the length of the abscissa. In the literature "evenness" and "equitability" are synonyms terms (Lloyd & Ghelardi, 1964; see also the review of Peet, 1974). Several indices of evenness have been proposed.

1) The simplest approach to evenness consists in comparing the measured diversity to the corresponding maximum value. When using H (eqs. 6.1 and 6.34a), diversity takes its maximum value when all species are equally represented. In such a case:

$$H_{max} = -\sum_{i=1}^{q} \frac{1}{q}\log\frac{1}{q} = \log q \qquad \textbf{(6.43)}$$

Pielou's evenness

where q *is* the number of species. Evenness (J) is computed as (Pielou, 1966):

$$J = H/H_{max} = \left(-\sum_{i=1}^{q} p_i \log p_i\right) / \log q \qquad \textbf{(6.44)}$$

which is a ratio, whose value is independent of the base of logarithms used for the calculation. Using the terms defined by Hill (1973a; see eqs. 6.31-6.35), Daget (1980) rewrote eq. 6.44 as the ratio of entropies of orders 1 (eq. 6.34a) and 0 (eq. 6.33a):

$$J = H_1 / H_0 \qquad \textbf{(6.45)}$$

Equations 6.44 and 6.45 show that diversity H combines the *number of species* (q) and the *evenness* of their distribution (J):

$$H = JH_{max} = J\log q \qquad \textbf{(6.46)}$$

Hurlbert's evenness

2) Hurlbert (1971) proposed an evenness index based on the minimum and maximum values of diversity. Diversity is minimum when one species is represented by $(n - q + 1)$ organisms and the $(q - 1)$ others by a single organism. According to Hurlbert, the following indices are independent of q:

$$J = (D - D_{min}) / (D_{max} - D_{min}) \qquad \textbf{(6.47)}$$

$$1 - J = (D_{max} - D) / (D_{max} - D_{min}) \qquad \textbf{(6.48)}$$

Patten's redundancy

Equation 6.48 was proposed by Patten (1962) as a measure of *redundancy* (see Section 6.1). The two indices can be computed for any diversity index D.

Broken stick model

3) Instead of dividing the observed diversity by its maximum value, Lloyd & Ghelardi (1964) proposed to use a model based on the *broken stick distribution* (Barton & David, 1956; MacArthur, 1957). To generate this distribution, a set of individuals is taken as equivalent to a stick of unit length which is broken randomly into a number of pieces (i.e. in the present case, the number of species q). The divisor in the evenness formula is the diversity computed from the lengths of the pieces of the randomly broken stick. The expected lengths (E) of the pieces of the broken stick (species) y_i are given, in decreasing order, by the successive terms of the following series (Pielou, 1975), corresponding to the successive values $i = 1, 2, \ldots, q$, for a given number of species q:

$$E(y_i) = q^{-1} \sum_{x=i}^{q} x^{-1} \quad \textbf{(6.49)}$$

For example, for $q = 3$ species, eq. 6.49 gives the following lengths for species $i = 1$ to 3: 0.6111, 0.2778, and 0.1111, respectively. Diversity of this series is computed using the formula for H (eq. 6.1 or 6.34a):

$$M = -\sum_{i=1}^{q} E(y_i) \log E(y_i) \quad \textbf{(6.50)}$$

The evenness index of Lloyd & Ghelardi (1964),which they called *equitability*, is similar to eq. 6.44, with M being used instead of $H_{\max}$:

$$J = H / M \quad \textbf{(6.51)}$$

In the same paper, Lloyd & Ghelardi proposed another evenness index:

$$J = q' / q \quad \textbf{(6.52)}$$

where q is the observed number of species and q' is the number of species for which the broken stick model predicts the observed diversity H, i.e. $H(q) = M(q')$. Table C, at the end of this book, which is taken from Lloyd & Ghelardi (1964), makes it easy to get M or q'. Values computed with eq. 6.51 or 6.52 are usually, but not always, smaller than one. Indeed, it happens that biological populations are more diversified than predicted by the broken stick model.

Functional evenness

4) Troussellier & Legendre (1981) described an *index of functional evenness*, for studying bacterial assemblages. In such assemblages, the species level is often poorly defined. The index bypasses the step of species identification, using instead as data the set of binary biochemical (and other) descriptors that characterize the microbial

isolates. The authors have shown that their index has the usual properties of an evenness measure. Functional evenness E of a bacterial sampling unit is defined as:

$$E = \frac{I}{I_{max}} = \frac{1}{c \log 0.5} \sum_{i=1}^{c} [p_i \log p_i + (1 - p_i) \log (1 - p_i)] \qquad \textbf{(6.53)}$$

where I and I_{max} are measures of information, c is the number of binary descriptors used, and p_i is the proportion of positive responses to test i.

Evenness indices 6.44, 6.47, 6.51, and 6.52 all suffer from the problem that they depend on field estimation of the number of species in the population; in other words, q is not a fixed and known value but a random variable. Because the true value of q is not known and cannot be estimated from the data, there is no formula for computing a standard error (and, thus, a confidence interval) for these estimates of J. This point has been stressed by Pielou (1975) for eq. 6.44. This is not the case with eq. 6.53, where the denominator of E is a constant ($I_{max} = c \log 0.5$ where c is the number of binary descriptors used in the calculation). Several methods may be used for computing the confidence interval of E (e.g. the jackknife, briefly described at the end of Subsection 1.2.4). Legendre *et al.* (1984b) provided examples where the computation of confidence intervals for E, measured during biodegradation experiments, showed that significant changes had taken place, at some point in time, in the structure of the bacterial assemblages involved in the biodegradation processes.

In changing environments, the ecological interpretation of the two components of diversity could be carried out, for example, along the lines proposed by Legendre (1973). (1) The *number of species* may be a function of the stability of the environment. Indeed, a more stable environment entails a higher degree of organization and complexity of the food web (Margalef, 1958), so that such an environment contains more niches and, thus, more species. The number of species is proportional to the number of niches since, by definition, the realized niche of a species is the set of environmental conditions that this species does not share with any other sympatric species (Hutchinson, 1957, 1965). This approach has the advantage of linking species diversity to environmental diversity. (2) The *evenness of species distribution* may be inversely related to the overall biological activity in the studied environment; the lower the evenness, the higher the biological activity (e.g. production, life cycles, energy flows among trophic levels). On a seasonal basis, another factor may contribute to lower the evenness. In an environment where interspecific competition is low (high evenness), seasonal reduction of resources or deterioration of weather conditions could induce stronger competition and thus favour some species over others, which would decrease the evenness. The same is often observed in cases of pollution.

Chapter

7 Ecological resemblance

7.0 The basis for clustering and ordination

For almost a century, ecologists have collected quantitative observations to determine the resemblance between either the objects under study (sites) or the variables describing them (species or other descriptors). Measuring the association (Section 2.2) between objects (Q mode) or descriptors (R mode) is the first, and sometimes the only step in the numerical analysis of ecological data. The various modes of analysis are discussed in Section 7.1. It may indeed happen that examining the association matrix suffices to elucidate the structure and thus answer the question at the origin of the investigation.

The present Chapter provides a review of the main measures of association available to ecologists. Section 7.2 introduces the three types of association coefficients and the measures pertaining to each type — similarity, distance, and dependence — are described in Sections 7.3 to 7.5, respectively. In order to help ecologists choose from among this plurality of coefficients, Section 7.6 summarizes criteria for choosing a coefficient; the criteria are presented in the form of identification keys. Ecologists who do not wish to study the theory that underlies the measures of association may directly go to Section 7.6, after making themselves familiar with the terminology (Sections 7.1 and 7.2). When necessary, they may then refer to the paragraphs of Sections 7.3 to 7.5 describing the measures of interest.

Clustering

Ordination

In the following chapters, measures of resemblance between objects or descriptors will be used to cluster the objects or descriptors (Chapter 8) or to produce scatter diagrams in spaces of reduced dimensionality (Chapter 9). The clustering of objects (or descriptors) is an operation by which the set of objects (or descriptors) is partitioned in two or more subsets (clusters), using pre-established rules of agglomeration or division. Ordination in reduced space is an operation by which the objects (or descriptors) are positioned in a space that contains fewer dimensions that in the original data set; the positions of the objects or descriptors with respect to one another may also be used to cluster them. Both operations are often carried out on association matrices, as described in the following sections.

7.1 Q and R analyses

As noted by Cattell (1952), the ecological data matrix may be studied from two main viewpoints. One may wish to look at relationships among either the objects or the descriptors. The important point here is that these modes of analysis are based on different measures of association. The different types of coefficients are described in Section 7.2. Measuring the dependence between descriptors is done using coefficients like Pearson's r correlation coefficient (eq. 4.7, Section 4.2), so that studying the data matrix based on such coefficients is called *R analysis*. By opposition, studying the data matrix to uncover relationships among objects is called *Q analysis* (Fig. 2.2).

R mode
Q mode

Cattell (1966) had also observed that the *data box* (objects × descriptors × time instances; Fig. 7.1) may be looked at from other viewpoints than simply Q and R. He defined six modes of analysis:

O: among time instances, based on all observed descriptors (a single object);
P: among descriptors, based on all observed times (a single object);
Q: among objects, based on all observed descriptors (a single instance);
R: among descriptors, based on all observed objects (a single instance);
S: among objects, based on all observed times (a single descriptor);
T: among time instances, based on all observed objects (a single descriptor).

In the present chapter, the discussion of association coefficients will deal with the two basic modes only, i.e. Q measures (computed among objects) and R measures (computed among descriptors); objects and descriptors are defined in Section 1.4.

O-mode studies are conducted using Q measures; see, for example, Section 12.6. Similarly, P-mode studies are generally carried out with the usual R-type coefficients. When the data set forms a time series, however, P studies are based on special R-type coefficients which are discussed in Chapter 12: cross-covariance, cross-correlation, co-spectrum, coherence.

S- and T-mode studies mostly belong to autecology, i.e. studies involving a single species. S-mode comparisons among objects use the same coefficients as in P-mode analysis. Studying the relationship between "descriptor y observed at site x_1" and "the same descriptor y observed at site x_2" is analogous to the comparison of two descriptors along a time axis.

In T-mode studies, a variable is observed across several objects (sites, etc.) and comparing it at different instances through time. Statistical tests of hypothesis for related samples are often applicable to these problems; see Table 5.2. In other cases,

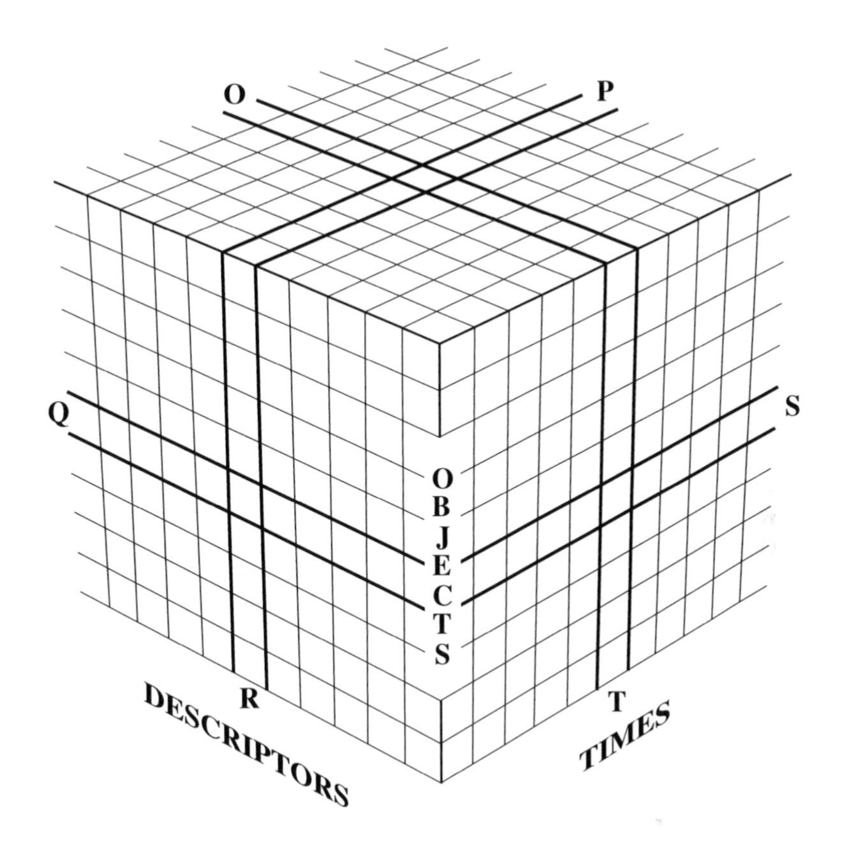

Figure 7.1 The three-dimensional data box (objects × descriptors × times). Adapted from Cattell (1966).

the two time instances to be compared are considered to define two descriptors, as in the S mode, so that normal R-type measures may be used. Environmental impact studies form an important category of T-mode problems; ecologists should look at the literature on BACI designs when planning such studies (Before/After - Control/Impact: Green, 1979; Bernstein & Zalinski, 1983; Stewart-Oaten *et al.*, 1986; Underwood, 1991, 1992, 1994).

Q or R?

It is not always obvious whether an analysis belongs to the Q or R mode. As a further complication, in the literature, authors define the mode based either on the association matrix or on the purpose of the analysis. Principal component analysis (Section 9.1), for instance, is based on a dispersion matrix among descriptors (R mode?), but it may be used for ordination of either the objects (Q mode?) or the descriptors (R mode?). In order to prevent confusion, in the present book, any study starting with the computation of an *association matrix among objects* is called a *Q analysis* whereas studies starting with the computation of an *association matrix among descriptors* are referred to as *R analyses*. In Chapter 9 for example, it is

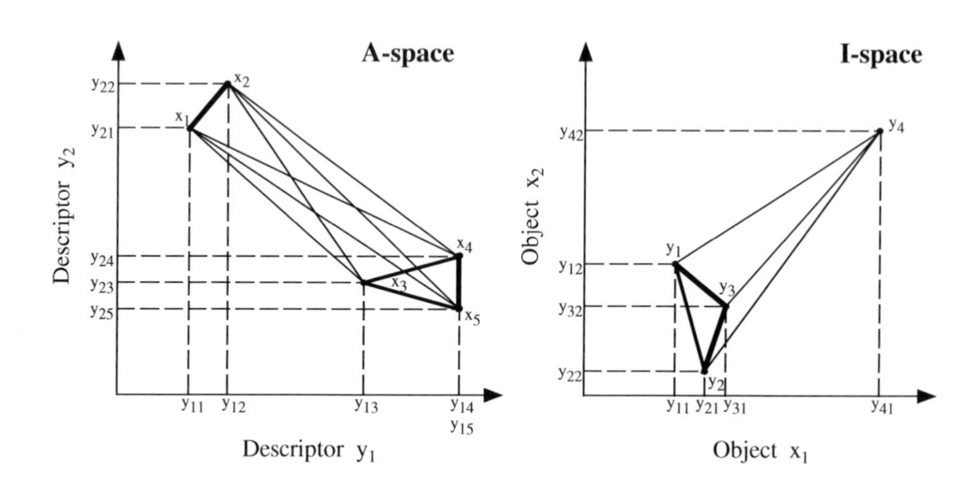

Figure 7.2 On the left, representation of five objects in an A-space with two descriptors. The thickness of the lines that join the objects is proportional to their degree of resemblance with regard to the two descriptors, i.e. their proximity in the A-space. On the right, a similar representation of four descriptors in an I-space consisting of two objects.

possible to obtain an ordination of objects in low-dimension space using either the R method of principal component analysis or the Q method of principal coordinate analysis. Interestingly, these two analyses lead to the same ordination of the objects when the principal coordinates are computed using coefficient D_1 (Section 7.4), although the results of Q and R analyses are not always reducible to each other.

A-space

Following the terminology of Williams & Dale (1965), the space of descriptors (attributes) is called "A-space". In this space, the objects may be represented along axes which correspond to the descriptors. For example, sites may be positioned in an A-space whose axes are the abundances of various species. A low-dimension space obtained using the ordination methods described in Chapter 9, in which the objects are represented, is also an A-space. Symmetrically, the space of reference in which the descriptors are positioned relative to axes corresponding to objects (or individuals) is called "I-space". For example, species could be positioned in an I-space whose axes would be the different sites. An I-space representation is more artificial than an A-space plot because, in I-space, values on the x_i axes are y_{ij} values (Table 2.1) and are therefore states of the observed descriptors. Figure 7.2 illustrates the A- and I-spaces.

I-space

The number of dimensions that can be represented on paper is limited to two or eventually three. Hence, one generally imagines distances between objects or descriptors as embedded in a 2- or 3-dimensional space. Section 7.4 will show that such models can be extended to a large number of dimensions. Distances and similarities computed in the present chapter will, in most instances, be based on measurements made in high-dimensional spaces.

In addition to the methods described in the present and following chapters, there exist approaches allowing the analysis of the whole data box instead of subsets of it, as was the case in the six modes described above. Examples are found in Williams & Stephenson (1973), Williams *et al.* (1982), Cailliez & Pagès (1976), Marcotorchino & Michaud (1979), Kroonenberg (1983: three-way principal component analysis*), and Carlier & Kroonenberg (1996: three-way correspondence analysis).

Euclidean space

The A- and I-spaces are called *metric* or *Euclidean* because the reference axes are quantitative and metric. Even though it might be interesting to plot objects in A-space, or descriptors in I-space, in order to detect their clustering structure, the clustering methods described in Chapter 8 are free from the restrictions inherent to metric spaces. As a matter of fact, clustering may be achieved without reference to any particular space, metric or not. This makes the two dimensions of figures, drawn on paper, available for illustrating other aspects of the clustering, e.g. the tree of the successive partitions of the set of objects.

7.2 Association coefficients

The most usual approach to assess the resemblance among objects or descriptors is to first condense all (or the relevant part of) the information available in the ecological data matrix (Section 2.1) into a square matrix of association among the objects or descriptors (Section 2.2). In most instances, the association matrix is *symmetric*. Non-symmetric matrices can be decomposed into symmetric and skew-symmetric components, as described in Section 2.3; the components may then be analysed separately. In Chapters 8 and 9, objects or descriptors will be clustered or represented in reduced space after analysing an association matrix. It follows that *the structure resulting from the numerical analysis is that of the association matrix; the results of the analysis do not necessarily reflect all the information originally contained in the ecological data matrix.* This stresses the importance of choosing an appropriate measure of association. This choice determines the issue of the analysis. Hence, it must take into account the following considerations:

• The nature of the study (i.e. the initial question and the hypothesis) determines the kind of ecological structure to be evidenced through an association matrix, and consequently the type of measure of resemblance to be used.

• The various measures available are subject to different mathematical constraints. The methods of analysis to which the association matrix will be subjected (clustering,

* Program 3WAYPACK (Kroonenberg, 1996) for three-way principal component analysis is available from Pieter M. Kroonenberg, Department of Educational Sciences, Leiden University, Wassenaarseweg 52, NL-2333 AK Leiden, The Netherlands. Other three-mode software is described on the WWWeb page: <http://www.fsw.leidenuniv.nl/~kroonenb/>.

ordination) often require measures of resemblance with specific mathematical properties.

• One must also consider the computational aspect, and thus preferably choose a measure which is available in a computer package (Section 7.7) or can easily be programmed.

Ecologists are, in principle, free to define and use any measure of association suitable to the ecological phenomenon under study; mathematics impose few constraints to this choice. This is why so many association coefficients are found in the literature. Some of them are of wide applicability whereas others have been created for specific needs. Several coefficients have been rediscovered by successive authors and may be known under various names. Reviews of some coefficients may be found in Cole (1949, 1957), Goodman & Kruskal (1954, 1959, 1963), Dagnelie (1960), Sokal & Sneath (1963), Williams & Dale (1965), Cheetham & Hazel (1969), Sneath & Sokal (1973), Clifford & Stephenson (1975), Orlóci (1978), Daget (1976), Blanc *et al.* (1976), Gower (1985), and Gower & Legendre (1986).

Dependence
Similarity
Distance

In the following sections, *association* will be used as a general term to describe any measure or coefficient used to quantify the resemblance or difference between objects or descriptors, as proposed by Orlóci (1975). With *dependence* coefficients, used in the R mode, zero corresponds to no association. In Q-mode studies, *similarity* coefficients between objects will be distinguished from *distance* (or *dissimilarity*) coefficients. Similarities are *maximum* when the two objects are identical and *minimum* when the two objects are completely different; distances follow the opposite rule. Figure 7.2 (left) clearly shows the difference between the two types of measures: the length of the line between two objects is a measure of their distance, whereas its thickness, which decreases as the two objects get further apart, is proportional to their similarity. If needed, a similarity can be transformed into a distance, for example by computing its one-complement. For a similarity measure varying between 0 and 1, as is generally the case, the corresponding distance may be computed as:

$$D = 1 - S, \quad D = \sqrt{1 - S}, \quad \text{or} \quad D = \sqrt{1 - S^2}$$

Distances, which in some cases are not bound by a pre-determined upper value, may be normalized, using eqs. 1.10 or 1.11:

$$D_{norm} = \frac{D}{D_{max}} \quad \text{or} \quad D_{norm} = \frac{D - D_{min}}{D_{max} - D_{min}}$$

where D_{norm} is the distance normalized between [0, 1] whereas D_{max} and D_{min} are the maximum and minimum values taken by the distance coefficient, respectively. Normalized distances can be used to compute similarities, by reversing the transformations given above:

$$S = 1 - D_{norm}, \quad S = 1 - D_{norm}^2, \quad \text{or} \quad S = \sqrt{1 - D_{norm}^2}$$

The following three sections describe the coefficients that are most useful with ecological data. Criteria to be used as guidelines for choosing a coefficient are discussed in Section 7.6. Computer programs are briefly reviewed in Section 7.7.

7.3 Q mode: similarity coefficients

The largest group of coefficients in the literature is the similarities. These coefficients are used to measure the association between *objects*. In contrast to most distance coefficients, similarity measures are never metric (definition at the beginning of Section 7.4) since it is always possible to find two objects, A and B, that are more similar than the sum of their similarities with another, more distant, object C. It follows that similarities cannot be used directly to position objects in a metric space (ordination; Chapter 9); they must be converted into distances. Clustering (Chapter 8), on the other hand, can be conducted on either a distance or a similarity matrix.

Similarity coefficients were first developed for binary data, representing presence-absence data, or answers to yes-no questions. They were later generalized to multi-state descriptors, when computers made it possible. Another major dichotomy among similarity coefficients concerns how they deal with double-zeros or negative matches. This dichotomy is so fundamental with ecological data that it is discussed first.

Double-zero problem

The double-zero problem shows up in ecology because of the special nature of species descriptors. Species are known to have unimodal distributions along environmental gradients (Whittaker, 1967). Hutchinson's (1957) niche theory states that species have ecological preferences, meaning that they are more likely to be found at sites where they encounter appropriate living conditions; the distribution of a species has its mode at this optimum value. A species becomes rare and eventually absent as one departs from optimal conditions. If a species is present at two sites, this is an indication of the similarity of these sites; but if a species is absent from two sites, it may be because the two sites are both above the optimal niche value for that species, or both are below, or else one site is above and the other is below that value. One cannot tell which of these circumstances is the correct one.

Asymmetrical coefficient

It is thus preferable to abstain from drawing any ecological conclusion from the absence of a species at two sites. In numerical terms, this means to skip double zeros altogether when computing similarity or distance coefficients using species presence-absence or abundance data. Coefficients of this type are called *asymmetrical* because they treat zeros in a different way than other values. On the other hand, the presence of a species at one of two sites and its absence at the other are considered as a difference between these sites.

Symmetrical coefficient

In *symmetrical* coefficients, the state *zero* for two objects is treated in exactly the same way as any other pair of values, when computing a similarity. These coefficients can be used in cases where the state *zero* is a valid basis for comparing two objects and represents the same kind of information as any other value. This obviously excludes the special case where *zero* means “lack of information”. For example, finding that two lakes in winter have 0 mgL^{-1} of dissolved oxygen in the hypolimnion provides valuable information concerning their physical and chemical similarity and their capacity to support species.

Several reasons may preside to the absence of a species from a site, besides unimodal distributions and niche optimality. Species are known to occur in suboptimal environments, especially when they are subjected to competition. So, the niche of a species may be present in one (or both) of two sites but be occupied by substitute species. Absence may also be the result of the dispersion pattern of a species, historical events, or, more simply, stochastic sampling variation.

It often occurs that a large number of species are present in a data matrix. There may be several sites where only a small number of species are found. If sampling has been carried out along an environmental gradient, the species that are present may not be the same from site to site. Including double-zeros in the comparison between sites would result in high values of similarity for the many pairs of sites holding only a few species; this would not reflect the situation adequately. Thus, when analysing species presence-absence or abundance data, ecologists normally use asymmetrical coefficients, in which double absences are not counted as indications of resemblance.

The remainder of this Section distinguishes between binary and quantitative similarity coefficients and, for each type, those that use double-zeros or exclude them. It ends with a description of probabilistic coefficients. Tables 7.3 and 7.4 summarize the use of the various similarity coefficients in ecology.

1 — Symmetrical binary coefficients

In the simplest cases, the similarity between two sites is based on presence-absence data. Binary descriptors may describe the presence or absence of environmental conditions (here) or species (next Subsection). Observations may be summarized in a 2×2 frequency table:

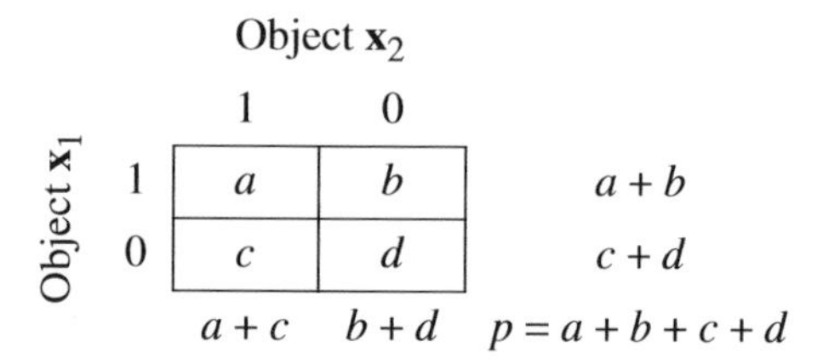

where a is the number of descriptors for which the two objects are coded 1, d is the number of descriptors coding the two objects 0, whereas b and c are the numbers of descriptors for which the two objects are coded differently; and p is the total number of descriptors. An obvious way of computing the similarity between two objects is to count the number of descriptors that code the objects in the same way and divide this by the total number of descriptors:

$$S_1(\mathbf{x}_1, \mathbf{x}_2) = \frac{a+d}{p} \quad \textbf{(7.1)}$$

Coefficient S_1* is called the *simple matching coefficient* (Sokal & Michener, 1958). When using this coefficient, one assumes that there is no difference between double-0 and double-1. This is the case, for instance, when any one of the two states of each descriptor could be coded 0 or 1 indifferently. A variant of this measure is the *coefficient of Rogers & Tanimoto* (1960) in which differences are given more weight than resemblances:

$$S_2(\mathbf{x}_1, \mathbf{x}_2) = \frac{a+d}{a+2b+2c+d} \quad \textbf{(7.2)}$$

Sokal & Sneath (1963) proposed four other measures that include double-zeros. They have their counterparts in coefficients that exclude double-zeros, in the next Subsection:

$$S_3(\mathbf{x}_1, \mathbf{x}_2) = \frac{2a+2d}{2a+b+c+2d} \quad \textbf{(7.3)}$$

counts resemblances as being twice as important as differences;

$$S_4(\mathbf{x}_1, \mathbf{x}_2) = \frac{a+d}{b+c} \quad \textbf{(7.4)}$$

compares the resemblances to the differences, in a measure that goes from 0 to infinity;

$$S_5(\mathbf{x}_1, \mathbf{x}_2) = \frac{1}{4}\left[\frac{a}{a+b} + \frac{a}{a+c} + \frac{d}{b+d} + \frac{d}{c+d}\right] \quad \textbf{(7.5)}$$

compares the resemblances to the marginal totals;

$$S_6(\mathbf{x}_1, \mathbf{x}_2) = \frac{a}{\sqrt{(a+b)\,(a+c)}}\;\frac{d}{\sqrt{(b+d)\,(c+d)}} \quad \textbf{(7.6)}$$

is the product of the geometric means of the terms relative to a and d, respectively, in coefficient S_5.

* Coefficient numbers are used in the SIMIL program of *The R Package* to identify resemblance coefficients; see Section 7.7 and Table 13.4. For this reason, coefficient numbers have not been changed since the first edition of this book (*Écologie numérique*, Masson, Paris, 1979). Coefficients added since the 1979 edition have been numbered sequentially.

Among the above coefficients, S_1 to S_3 are of more general interest for ecologists, but the others may occasionally prove useful to adequately handle special descriptors. Three additional measures are available in the NT-SYS computer package (Section 7.7): the *Hamann coefficient*:

$$S = \frac{a + d - b - c}{p} \tag{7.7}$$

the *Yule coefficient*:

$$S = \frac{ad - bc}{ad + bc} \tag{7.8}$$

and *Pearson's* ϕ (phi):

$$\phi = \frac{ad - bc}{\sqrt{(a + b)\ (c + d)\ (a + c)\ (b + d)}} \tag{7.9}$$

where the numerator is the determinant of the 2×2 frequency table. ϕ is actually the square root of the X^2 (chi-square) statistic for 2×2 tables (eq. 7.60). In ecology, coefficients of this type are mostly used in R-mode analyses. These last indices are described in detail in Sokal & Sneath (1963).

2 — Asymmetrical binary coefficients

Coefficients that parallel those above are available for comparing sites using species presence-absence data, where the comparison must exclude double-zeros. The best-known measure is Jaccard's (1900, 1901, 1908) *coefficient of community*. It is often referred to simply as *Jaccard's coefficient*:

$$S_7(\mathbf{x}_1, \mathbf{x}_2) = \frac{a}{a + b + c} \tag{7.10}$$

in which all terms have equal weight. As a variant, *Sørensen's coefficient* (1948) gives double weight to double presences:

$$S_8(\mathbf{x}_1, \mathbf{x}_2) = \frac{2a}{2a + b + c} \tag{7.11}$$

because (see above) one may consider that the presence of a species is more informative than its absence. Absence may be due to various factors, as discussed above; it does not necessarily reflect differences in the environment. Double-presence, on the contrary, is a strong indication of resemblance. Note, however, that S_8 is monotonic to S_7. This property means that if the similarity for a pair of objects computed with S_7 is higher than that of another pair of objects, the same will be true when using S_8. In other words, S_7 and S_8 only differ by their scales. Before Sørensen,

Dice (1945) had used S_8 under the name *coincidence index* in an R-mode study of species associations; this question is further discussed in Section 7.5.

The distance version of this coefficient, $D_{13} = 1 - S_8$, is a semimetric, as shown in the example that follows eq. 7.56. A consequence is that principal coordinate analysis of a S_8 or D_{13} resemblance matrix is likely to produce negative values. Solutions to this problem are discussed in Subsection 9.2.4. The easiest way is to base the principal coordinate analysis on square-root-transformed distances $D = \sqrt{1 - S_8}$ instead of $D = 1 - S_8$ (Table 7.2).

Another variant of S_7 gives triple weight to double presences:

$$S_9(\mathbf{x}_1, \mathbf{x}_2) = \frac{3a}{3a + b + c} \qquad \textbf{(7.12)}$$

The counterpart to the coefficient of Rogers & Tanimoto (S_2), in the above Subsection, was proposed by Sokal & Sneath (1963). This coefficient gives double weight to differences in the denominator:

$$S_{10}(\mathbf{x}_1, \mathbf{x}_2) = \frac{a}{a + 2b + 2c} \qquad \textbf{(7.13)}$$

Russell & Rao (1940) suggested a measure which allows the comparison of the number of double presences, in the numerator, to the total number of species found at all sites, including species that are absent (d) from the pair of sites considered:

$$S_{11}(\mathbf{x}_1, \mathbf{x}_2) = \frac{a}{p} \qquad \textbf{(7.14)}$$

Kulczynski (1928) proposed a coefficient opposing double-presences to differences:

$$S_{12}(\mathbf{x}_1, \mathbf{x}_2) = \frac{a}{b + c} \qquad \textbf{(7.15)}$$

Among their coefficients for presence-absence data, Sokal & Sneath (1963) mention the binary version of Kulczynski's coefficient S_{18} for quantitative data:

$$S_{13}(\mathbf{x}_1, \mathbf{x}_2) = \frac{1}{2}\left[\frac{a}{a + b} + \frac{a}{a + c}\right] \qquad \textbf{(7.16)}$$

where double-presences are compared to the marginal totals $(a + b)$ and $(a + c)$.

Ochiai (1957) used, as measure of similarity, the geometric mean of the ratios of a to the number of species in each site, i.e. the marginal totals $(a + b)$ and $(a + c)$:

$$S_{14}(\mathbf{x}_1, \mathbf{x}_2) = \sqrt{\frac{a}{(a+b)}\ \frac{a}{(a+c)}} = \frac{a}{\sqrt{(a+b)\ (a+c)}} \quad \textbf{(7.17)}$$

This measure is the same as S_6 except for the part that concerns double-zeros (d).

Faith (1983) suggested the following coefficient, in which disagreements (presence in one site and absence in the other) are given a weight opposite to that of double presences. The value of S_{26} decreases with an increasing number of double-zeros:

$$S_{26}(\mathbf{x}_1, \mathbf{x}_2) = \frac{a + d/2}{p} \quad \textbf{(7.18)}$$

3 — Symmetrical quantitative coefficients

Ecological descriptors often have more than two states. Binary coefficients in Subsection 1, above, may sometimes be extended to accommodate multi-state descriptors. For example, the simple matching coefficient may be used as follows with multi-state descriptors:

$$S_1(\mathbf{x}_1, \mathbf{x}_2) = \frac{\text{agreements}}{p} \quad \textbf{(7.19)}$$

where the numerator contains the number of descriptors for which the two objects are in the same state. For example, if a pair of objects was described by the following 10 multi-state descriptors:

	Descriptors										
Object $\mathbf{x}_1$	9	3	7	3	4	9	5	4	0	6	
Object $\mathbf{x}_2$	2	3	2	1	2	9	3	2	0	6	
Agreements	0 +	1 +	0 +	0 +	0 +	1 +	0 +	0 +	1 +	1	= 4

the value of S_1 computed for the 10 multi-state descriptors would be:

$$S_1(\mathbf{x}_1, \mathbf{x}_2) = 4 \text{ agreements}/10 \text{ descriptors} = 0.4$$

It is possible to extend in the same way the use of all binary coefficients to multi-state descriptors. Coefficients of this type often result in a loss of valuable information, however, especially in the case of ordered descriptors for which two objects can be compared on the basis of the *amount of difference* between states.

Gower (1971a) proposed a general coefficient of similarity which can combine different types of descriptors and process each one according to its own mathematical type. Although the description of this coefficient may seem a bit complex, it can be easily translated into a short computer program. The coefficient initially takes the following form (see also the final form, eq. 7.20):

$$S_{15}(\mathbf{x}_1, \mathbf{x}_2) = \frac{1}{p}\sum_{j=1}^{p} s_{12j}$$

Partial similarity

The similarity between two objects is the average, over the p descriptors, of the similarities calculated for all descriptors. For each descriptor j, the *partial similarity value* s_{12j} between objects $\mathbf{x}_1$ and $\mathbf{x}_2$ is computed as follows.

- For *binary* descriptors, $s_j = 1$ (agreement) or 0 (disagreement). Gower proposed two forms for this coefficient. The form used here is symmetrical, giving $s_j = 1$ to double-zeros. The other form, used in Gower's asymmetrical coefficient S_{19} (Subsection 4), gives $s_j = 0$ to double-zeros.

- *Qualitative* and *semiquantitative* descriptors are treated following the simple matching rule stated above: $s_j = 1$ when there is agreement and $s_j = 0$ when there is disagreement. Double-zeros are treated as in the previous paragraph.

- *Quantitative* descriptors (real numbers) are treated in an interesting way. For each descriptor, one first computes the difference between the states of the two objects $|y_{1j} - y_{2j}|$, as in the case of distance coefficients belonging to the Minkowski metric group (Section 7.4). This value is then divided by the largest difference (R_j) found for this descriptor across all sites in the study — or, if one prefers, in a reference population*. Since this ratio is actually a normalized distance, it is subtracted from 1 to transform it into a similarity:

$$s_{12j} = 1 - [|y_{1j} - y_{2j}| / R_j]$$

Missing values

Kronecker delta

Gower's coefficient may be programmed to include an additional element of flexibility: no comparison is computed for descriptors where information is *missing* for one or the other object. This is obtained by a value w_j, called *Kronecker's delta*, describing the presence or absence of information: $w_j = 0$ when the information about y_j is missing for one or the other object, or both; $w_j = 1$ when information is present for both objects. The final form of *Gower's coefficient* is the following:

$$S_{15}(\mathbf{x}_1, \mathbf{x}_2) = \frac{\sum_{j=1}^{p} w_{12j} s_{12j}}{\sum_{j=1}^{p} w_{12j}} \qquad \textbf{(7.20)}$$

* In most applications, the largest difference R_j is calculated for the data table under study. In epidemiological studies, for example, one may proceed to the analysis of a subset of a much larger data base. To insure consistency of the results in all the partial studies, it is recommended to calculate the largest differences (the "range" statistic of data bases) observed throughout the whole data base for each descriptor j and use these as values R_j when computing S_{15} or S_{19}.

Coefficient S_{15} produces similarity values between 0 and 1 (maximum similarity).

One last touch of complexity, which was not suggested in Gower's paper but is added here, provides weighting to the various descriptors. Instead of *0 or 1*, one can assign to w_j a value *between 0 and 1* corresponding to the weight one wishes each descriptor to have in the analysis. Descriptors with weights close to 0 contribute little to the final similarity value whereas descriptors with higher weights (closer to 1) contribute more. Giving a weight of 0 to a descriptor is equivalent to removing it from the analysis. A missing value automatically changes the weight w_j to 0.

The following numerical example illustrates the computation of coefficient S_{15}. In the example, two sites are described by eight quantitative environmental descriptors. Values R_j (the range of values among all objects, for each descriptor y_j) given in the table have been calculated for the whole data base prior to computing coefficient S_{15}. Weights w_{12j} are only used in this example to eliminate descriptors with missing values (Kronecker delta function):

	Descriptors j								Sum
Object $\mathbf{x}_1$	2	2	–	2	2	4	2	6	
Object $\mathbf{x}_2$	1	3	3	1	2	2	2	5	
w_{12j}	1	1	0	1	1	1	1	1	= 7
R_j	1	4	2	4	1	3	2	5	
$\lvert y_{1j} - y_{2j} \rvert$	1	1	–	1	0	2	0	1	
$\lvert y_{1j} - y_{2j} \rvert / R_j$	1	0.25	–	0.25	0	0.67	0	0.20	
$w_{12j} s_{12j}$	0	0.75	0	0.75	1	0.33	1	0.80	= 4.63

thus $S_{15}(\mathbf{x}_1, \mathbf{x}_2) = 4.63/7 = 0.66$.

When computing S_{15}, one may decide to handle semiquantitative descriptors as if they were quantitative, in order to use differences between states in the final similarity assessment. It is important in such cases to make sure that distances between adjacent states are comparable in magnitude. For example, with ordered (semiquantitative) descriptors coded from 1 to 3, $\lvert y_{1j} - y_{2j} \rvert$ can be used only if the difference between states 1 and 2 can be thought of as almost equal to that between states 2 and 3. If there is too much difference, values $\lvert y_{1j} - y_{2j} \rvert$ are not comparable and semiquantitative descriptors should not be used in that way in coefficient S_{15}.

Another general coefficient of similarity was proposed by Estabrook & Rogers (1966). The similarity between two objects is, as in S_{15}, the sum of the partial similarities by descriptors, divided by the number of descriptors for which there is information for the two objects. In the original publication, the authors used state 0 to mean "no information available", but any other convention would be acceptable. The general form of this coefficient is therefore the same as Gower's coefficient (eq. 7.20):

$$S_{16}(\mathbf{x}_1, \mathbf{x}_2) = \frac{\sum_{j=1}^{p} w_{12j} s_{12j}}{\sum_{j=1}^{p} w_{12j}} \qquad \textbf{(7.21)}$$

As in S_{15}, the w_j parameters may be used as weights (between 0 and 1) instead of only playing the roles of Kronecker deltas. The coefficient of Estabrook & Rogers differs from S_{15} in the computation of the partial similarities s_j.

Partial similarity

In the paper of Estabrook & Rogers (1966), the state values were positive integers and the descriptors were either ordered or unordered. The partial similarity between two objects for a given descriptor j is computed using a monotonically decreasing function of partial similarity. On an empirical basis, and among all functions of this type, the authors proposed to use the following function of two numbers d and k:

$$s_{12j} = f(d_{12j}, k_j) = \frac{2(k+1-d)}{2k+2+dk} \qquad \text{when } d \leq k$$
$$s_{12j} = f(d_{12j}, k_j) = 0 \qquad \text{when } d > k \qquad \textbf{(7.22)}$$

where d is the distance between the states of the two objects $\mathbf{x}_1$ and $\mathbf{x}_2$ for descriptor j, i.e. the same value $|y_{1j} - y_{2j}|$ as in Gower's coefficient, and k is a parameter determined *a priori* by the users for each descriptor, describing how far non-null partial similarities are permitted to go. Parameter k is equal to the largest difference d for which the partial similarity s_{12j} (for descriptor j) is allowed to be different from 0. Values k for the various descriptors may be quite different from one another. For example, for a descriptor coded from 1 to 4, one might decide to use $k = 1$ for this descriptor; for another descriptor with code values from 1 to 50, $k = 10$ could be used. In order to fully understand the partial similarity function s_{12j} (eq. 7.22), readers are invited to compute by hand s_{12j} for some descriptors in the following numerical example. Values k, which are usually small numbers, are given for each descriptor in the table:

	Descriptors j						$S_{16}(\mathbf{x}_1, \mathbf{x}_2)$
Object $\mathbf{x}_1$	2	1	3	4	2	1	
Object $\mathbf{x}_2$	2	2	4	3	2	3	
k_j	1	0	1	2	1	1	
	↓	↓	↓	↓	↓	↓	
$s_{12j} = f(d_{12j}, k_j)$	1.0 +	0 +	0.4 +	0.5 +	1.0 +	0	$= 2.9 / 6 = 0.483$

Values taken by the partial similarity function for the first values of k are given in Table 7.1. Values in the table show that, if $k = 0$ for all descriptors, S_{16} is identical to the simple-matching coefficient for multistate descriptors (eq. 7.19).

Table 7.1 Values of the partial similarity function $f(d, k)$ in coefficients S_{16} and S_{20}, for the most usual values of k (adapted from Legendre & Rogers, 1972: 594).

k	*d*							
	0	1	2	3	4	5	6	7
0	1	0.00	0.00	0.00	0.00	0.00	0.00	0.00
1	1	0.40	0.00	0.00	0.00	0.00	0.00	0.00
2	1	0.50	0.20	0.00	0.00	0.00	0.00	0.00
3	1	0.55	0.28	0.12	0.00	0.00	0.00	0.00
4	1	0.57	0.33	0 18	0.08	000	0.00	0.00
5	1	0.59	0.36	0.22	0.13	0.05	0.00	0.00

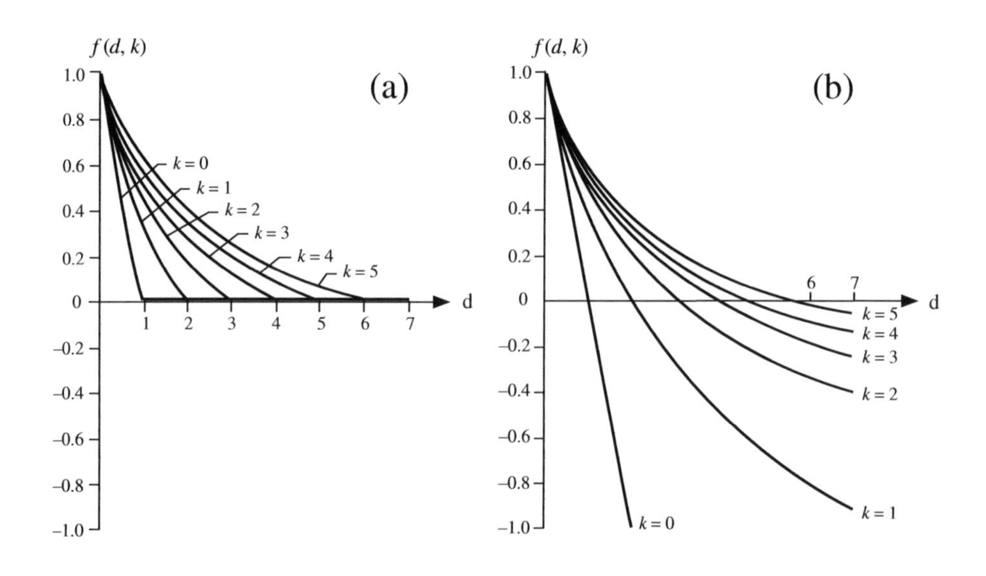

Figure 7.3 Coefficients S_{16} and S_{20}: change in $f(d, k)$ as a function of d, for six values of k, (a) under the condition $f(d, k) = 0$ when $d > k$; (b) without this condition. Adapted from Legendre & Chodorowski (1977).

These same values of function $f(d, k)$ are shown in Fig. 7.3a, which illustrates how the function decreases with increasing d. It is easy to see that function $f(d, k)$, which was originally defined by Estabrook & Rogers for discontinuous descriptors (coded only with integers: 0, 1, 2, 3, …), can actually be used with real-number descriptors since the function only requires that d and k be differences, i.e. natural numbers. Figure 7.3a also raises the question: could $f(d, k)$ take negative values? To accomplish that, Legendre & Chodorowski (1977) proposed to simply leave out the second line of eq. 7.22 stating that $f(d, k) = 0$ when $d > k$. This is shown in Fig. 7.3b, where the function decreases over the interval $[0, \infty)$, taking negative values when $d > (k + 1)$. Such a measure follows the same general idea as in the coefficients presented at the end of Subsection 1 (eqs. 7.7 to 7.9) where differences are subtracted from resemblances. This contributes to further separate dissimilar objects when the similarity matrix is subjected to clustering (Chapter 8).

Estabrook & Rogers (1966) proposed another partial similarity measure s_{12j}, to be used with coefficient S_{16} in situations where function $f(d, k)$ (eq. 7.22) does not adequately describe the relationships between objects for some descriptor j. The approach consists in providing the computer program with a small "do-it yourself" matrix that describes the partial similarities between all states of descriptor j. This partial similarity matrix replaces $s_{12j} = f(d_{12j}, k)$ in eq. 7.21 for descriptor j; the partial similarity of the other descriptors is computed with eq. 7.22. Partial similarity matrices may be constructed for any or all descriptors if needed.

Ecological application 7.3a

Partial similarity matrix

In a study of terrestrial fauna, Legendre *et al.* (1978) used the following partial similarity matrix (matrix unpublished) for the descriptor "category of aquatic ecosystem", which was one of the descriptors of the ecosystems where tracks of animal had been recorded in the snow blanket:

	1	2	3	4	5	6	7	8	9
1. <5% water	1.0								
2. 5-15% water	0.4	1.0							
3. >15% water, lakes < 250 ha	0.0	0.4	1.0						
4. >15% water, lakes 250-500 ha	0.0	0.3	0.8	1.0					
5. >15% water, lakes 500-1000 ha	0.0	0.2	0.7	0.8	1.0				
6. >15% water, lakes 1000-2500 ha	0.0	0.2	0.6	0.7	0.8	1.0			
7. >15% water, lakes > 2500 ha	0.0	0.2	0.5	0.6	0.7	0.8	1.0		
8. >5% rivers	0.0	0.0	0.0	0.0	0.0	0.0	0.0	1.0	
9. Bordering large rivers	0.0	0.0	0.0	0.0	0.0	0.0	0.0	0.4	1.0

The upper triangle of the matrix is not given; it is symmetric to the lower one. The diagonal may also be left out because the partial similarity of a state with itself must be 1. The matrix means that a site from an area with less than 5% of its surface covered by water is given a partial similarity $s_j = 0.4$ with another site from an area with 5 to 15% of its surface covered by water. State 1 has partial similarity with state 2 only; lake systems only have partial similarities with other lake systems, the similarity decreasing as the difference in lake areas increases; and rivers only have partial similarities when compared to other rivers. Partial similarity matrices are especially useful with descriptors that are nonordered, or only partly ordered as is the case here.

Partial similarity matrix

Partial similarity matrices represent a powerful way of using unordered or partly ordered descriptors in multivariate data analyses. They may prove useful in the following cases, in particular:

- When, from the user's point of view, function $f(d, k)$ (eq. 7.22) does not adequately describe the partial similarity relationships.

- When the descriptor states are not fully ordered. For example, in a study on ponds, the various states of descriptor "water temperature" may be followed by state "dry pond", which is quite different from a lack of information.

- If some states are on a scale different from that of the other states. For example, 0-10, 10-20, 20-30, 30-40, and then 50-100, 100-1000, and >1000.

- With nonordered or only partly ordered descriptors (including "circular variables" such as directions of the compass card or hours of the day), if one considers that pairs of sites coded into different states are partly similar, as in Ecological application 7.3a.

Partial similarity matrices may be used even in analyses where one wishes to use some other similarity function than S_{16}. Proceed as follows:

- Consider a descriptor whose inter-state relationships should be described by a "do-it-yourself" partial similarity matrix. Compute a similarity matrix S_{16} among the objects of the study *using this descriptor alone*; similarities among states are described by the partial similarity matrix.

- Carry out a principal coordinate analysis of the similarity matrix (Section 9.2). Use a correction for negative eigenvalues if any are present (Subsection 9.2.4). A good approximate solution may also usually be obtained using nonmetric multidimensional scaling (Section 9.3).

- The principal coordinates or multidimensional scaling axes form a new set of Euclidean descriptors that fully account for the variation among objects described by the partial similarity matrix. They can be combined with other quantitative descriptors in subsequent data analyses.

4 — Asymmetrical quantitative coefficients

Subsection 3 started with an extension of coefficient S_1 to multi-state descriptors. In the same way, binary coefficients from Subsection 2 can be extended to accommodate multi-state descriptors. For example, Jaccard's coefficient becomes

$$S_7(\mathbf{x}_1, \mathbf{x}_2) = \frac{\text{agreements}}{p - \text{double-zeros}} \qquad \textbf{(7.23)}$$

where the numerator is the number of species with *the same* abundance state at the two sites. This form can be used when species abundances are coded in a small number of classes and one wishes to strongly contrast differences in abundance. In other cases, using such a coefficient would obviously cause the loss of part of the information carried by species abundances.

Other measures are more efficient than eq. 7.23 in using species abundance information. They are divided in two categories: the coefficients for raw data and the measures for normalized data.

• As discussed in Subsections 1.5.6, the distribution of abundances of a species across an ecological gradient is often strongly skewed, so that normalization of species abundances often calls for square root, double square root, or logarithmic transformations. Another way to obtain approximately normal data is to use a scale of relative abundances with boundaries forming a geometric progression, for example a scale from 0 (absent) to 7 (very abundant).

• Abundances thus normalized reflect the role of each species in the ecosystem better than the raw abundance data, since a species represented by 100 individuals at a site does not have a role 10 times as important in the ecological equilibrium as another species represented by 10 individuals. The former is perhaps twice as important as the latter; this is the ratio obtained after applying a base-10 logarithmic transformation (assuming that numbers 100 and 10 at the site are representative of true relative abundances in the population).

Some coefficients lessen the effect of the largest differences and may therefore be used with raw species abundances, whereas others compare the different abundance values in a more linear way and are thus better adapted to normalized data.

In the group of coefficients to be used with raw species abundances, the best-known is a coefficient attributed to the Polish mathematician H. Steinhaus by Motyka (1947). This measure has been rediscovered a number of times; its one-complement is known as the Odum or Bray-Curtis coefficient (eq. 7.57). It is sometimes incorrectly attributed to anthropologist Czekanowski (1909 and 1913; Czekanowski's distance coefficient is described in the next Section, eq. 7.46). This coefficient compares two sites ($\mathbf{x}_1$, $\mathbf{x}_2$) in terms of the minimum abundance of each species:

$$S_{17}(\mathbf{x}_1, \mathbf{x}_2) = \frac{W}{(A+B)/2} = \frac{2W}{(A+B)} \qquad \mathbf{(7.24)}$$

where W is the sum of the minimum abundances of the various species, this minimum being defined as the abundance at the site where the species is the rarest. A and B are the sums of the abundances of all species at each of the two sites or, in other words, the total number of specimens observed or captured at each site, respectively. Consider the following numerical example:

	Species abundances						A	B	W
Site $\mathbf{x}_1$	7	3	0	5	0	1	16		
Site $\mathbf{x}_2$	2	4	7	6	0	3		22	
Minimum	2	3	0	5	0	1			11

$$S_{17}(\mathbf{x}_1, \mathbf{x}_2) = \frac{2 \times 11}{16 + 22} = 0.579$$

This measure is closely related to Sørensen's coefficient (S_8): if presence-absence data are used instead of species counts, S_{17} becomes S_8 (eq. 7.11).

The distance version of this coefficient, $D_{14} = 1 - S_{17}$, is a semimetric, as shown in the example that follows eq. 7.57. A consequence is that principal coordinate analysis of a S_{17} or D_{14} resemblance matrix is likely to produce negative values. Solutions to this problem are discussed in Subsection 9.2.4. One of the possible solutions is to base the principal coordinate analysis on square-root-transformed distances $D = \sqrt{1 - S_{17}}$ instead of $D = 1 - S_{17}$ (Table 7.2).

Kulczynski's coefficient (1928) also belongs to the group of measures that are suited to raw abundance data. The sum of minima is first compared to the grand total at each site; then the two values are averaged:

$$S_{18}(\mathbf{x}_1, \mathbf{x}_2) = \frac{1}{2}\left(\frac{W}{A} + \frac{W}{B}\right) \qquad \textbf{(7.25)}$$

For presence-absence data, S_{18} becomes S_{13} (eq. 7.16). For the numerical example above, coefficient S_{18} is computed as follows:

$$S_{18}(\mathbf{x}_1, \mathbf{x}_2) = \frac{1}{2}\left(\frac{11}{16} + \frac{11}{22}\right) = 0.594$$

Coefficients S_{17} and S_{18} always produce values between 0 and 1, although Kulczynski (1928) multiplied the final value by 100 to obtain a percentage. Kulczynski's approach, which consists in computing the average of two comparisons, seems more arbitrary than Steinhaus' method, in which the sum of minima is compared to the mean of the two site sums. In practice, values of these two coefficients are almost monotonic.

The following coefficients belong to the group adapted to "normalized" abundance data; this means here unskewed frequency distributions. These coefficients parallel S_{15} and S_{16} of the previous Subsection. Concerning coefficient S_{19}, Gower (1971a) had initially proposed that his general coefficient S_{15} should exclude double-zeros from the comparison (Subsection 3); this makes it well-suited for quantitative species abundance data. Since the differences between states are computed as $|y_{1j} - y_{2j}|$ and are thus linearly related to the measurement scale, this coefficient should be used with previously normalized data. The general form is:

$$S_{19}(\mathbf{x}_1, \mathbf{x}_2) = \frac{\sum_{j=1}^{p} w_{12j} s_{12j}}{\sum_{j=1}^{p} w_{12j}}, \text{ where} \qquad \textbf{(7.26)}$$

Partial similarity

- $s_{12j} = 1 - [|y_{1j} - y_{2j}|/R_j]$, as in S_{15},
- and $w_{12j} = 0$ when y_{1j} or y_{2j} = absence of information, or
  when y_{1j} and y_{2j} = absence of the species ($y_{1j} + y_{2j} = 0$);
- while $w_{12j} = 1$ in all other cases.

With species abundance data, values of w_j could be made to vary between 0 and 1, as in coefficient S_{15}, either to reflect the biomasses or the biovolumes of the different species, or to compensate for selective effects of the sampling gear.

Legendre & Chodorowski (1977) proposed a general coefficient of similarity that parallels S_{16}. This measure uses a slightly modified version of the partial similarity function $f(d, k)$ (eq. 7.22), or else a matrix of partial similarities as in Ecological application 7.3a. Since S_{20} processes all differences d in the same way, irrespective of whether they correspond to high or low values in the scale of abundances, it is better to use this measure with unskewed abundance data. The only difference between S_{16} and S_{20} is in the way in which double-zeros are handled. The general form of the coefficient is the sum of partial similarity values for all the species, divided by the total number of species in the combined two sites:

$$S_{20}(\mathbf{x}_1, \mathbf{x}_2) = \frac{\sum_{j=1}^{p} w_{12j} s_{12j}}{\sum_{j=1}^{p} w_{12j}} \text{, where} \tag{7.27}$$

Partial similarity

- $$s_{12j} = f(d_{j12}, k_j) \begin{cases} = \dfrac{2(k+1-d)}{2k+2+dk} & \text{when } d \leq k \quad \text{(I)} \\ = 0 & \text{when } d > k \quad \text{(II)} \\ = 0 & \text{when } y_{j1} \text{ or } y_{j2} = 0 \text{ (i.e. } y_{j1} \times y_{j2} = 0) \quad \text{(III)} \end{cases}$$
- or else $s_{12j} = f(y_{1j}, y_{2j})$ as given by a partial similarity matrix, as in Ecological application 7.3a, in which $s_{12j} = 0$ when y_{1j} or $y_{2j} = 0$,
- and $w_{12j} = 0$ when y_{1j} or y_{2j} = absence of information, or
  when y_{1j} and y_{2j} = absence of the species ($y_{1j} + y_{2j} = 0$),
- while $w_{12j} = 1$ in all other cases. Else, w_{12j} may receive a value between 0 and 1, as explained above for S_{19}.

In summary, the properties of coefficient S_{20} are the following:

• when d_j is larger than k_j, the partial similarity between sites is $s_{12j} = 0$ for species j;

• when $d_j = 0$, then $s_{12j} = 1$(see $f(d, k)$, part I), except when $y_{1j} = 0$ or $y_{2j} = 0$ (see $f(d, k)$, part III);

• $f(d, k)$ decreases with increasing d, for a given k;

• $f(d, k)$ increases with increasing k, for a given d;

• when $y_{1j} = 0$ or $y_{2j} = 0$, the partial similarity between sites is $s_{12j} = 0$ for species j, even if d_{12j} is not larger than k_j (see $f(d, k)$, part III);

• when $k_j = 0$ for all species j, S_{20} is the same as Jaccard's coefficient (S_7) for multi-state descriptors (eq. 7.23).

The above properties correspond to the opinion that ecologists may have on the problem of partial similarities between unskewed abundances of species. Depending on the scale chosen (0 to 5 or 0 to 50, for example), function $f(d, k)$ can be used to contrast to various degrees the differences between species abundances, by increasing or decreasing k_j, for each species j if necessary. An example of clustering using this measure of similarity is presented in Ecological application 8.2.

The last quantitative coefficient that excludes double-zeros is called the χ^2 *similarity*. It is the complement of the χ^2 metric (D_{15}; Section 7.4):

$$S_{21}(\mathbf{x}_1, \mathbf{x}_2) = 1 - D_{15}(\mathbf{x}_1, \mathbf{x}_2) \qquad \textbf{(7.28)}$$

The discussion of how species that are absent from two sites are excluded from the calculation of this coefficient is deferred to the presentation of D_{15}.

5 — *Probabilistic coefficients*

Probabilistic measures form a special category. These coefficients are based on statistical estimation of the significance of the relationship between objects.

When the data are frequency counts, a first probabilistic measure would be obtained by computing a Pearson chi square statistic (X_P^2, eq. 6.5) between pairs of sites, instead of the χ^2 similarity (S_{21}) described above. The complement of the probability associated with X_P^2 which would provide a probabilistic measure of the similarity between sites. The number of degrees of freedom (ν) should exclude the number of double-zeros:

$$\nu = \left[\left(\sum_{j=1}^{p} w_{12j} \right) - 1 \right] \qquad \textbf{(7.29)}$$

where w_{12j} is Kronecker's delta for each species j, as in Gower's coefficient S_{19}. The *probabilistic* X_P^2 *similarity* could be defined as:

$$S_{22} = 1 - \text{prob}\,[X_P^2(\mathbf{x}_1, \mathbf{x}_2)] \qquad \textbf{(7.30)}$$

This coefficient does not seem to have been used in ecology. Other measures are based on the probabilities of the various states of each descriptor. Some are derived from information theory (Chapter 4). In ecology, the best-known measure of this type is the *information statistic* of Lance & Williams (1966b), which is discussed in Subsection 8.5.11.

Following a different approach, *Goodall's probabilistic coefficient* (1964, 1966a) takes into account the frequency distribution of the various states of each descriptor in the whole set of objects. Indeed, it is less likely for two sites to both contain the same rare species than a more frequent species. In this sense, agreement for a rare species should be given more importance than for a frequent species, when estimating the similarity between sites. Goodall's probabilistic index, which had been originally developed for taxonomy, is especially useful in ecological classifications, because abundances of species in different sites are stochastic functions (Sneath & Sokal, 1973: 141). Orlóci (1978) suggests to use it for clustering sites (Q mode). The index has also been used in the R mode, for clustering species and identifying associations (Subsection 7.5.2).

The probabilistic coefficient of Goodall is based on the probabilities of the various states of each descriptor. The resulting measure of similarity is itself a probability, namely the complement of the probability that the resemblance between two sites is due to chance. In the next chapter, this index will be used for probabilistic clustering, in which sites in the same cluster are linked to one another at a specified level of probability (Subsection 8.9.2).

The probabilistic index, as formulated by Goodall (1966a), is a general taxonomic measure in which binary and quantitative descriptors can be used together. The coefficient as presented here follows the modifications of Orlóci (1978) and is limited to the clustering of sites based on species abundances. It also takes into account the remarks made at the beginning of the present section concerning double-zeros. The resulting measure is therefore a simplification of Goodall's original coefficient, oriented towards the clustering of sites. The computational steps are as follows:

(a) A partial similarity measure s_j is first calculated for all pairs of sites and for each species j. Because there are n sites, the number of partial similarities s_j to compute, for each species, is $n(n-1)/2$. If the species abundances have been normalized, one may choose either the partial similarity measure $s_{12j} = 1 - [|y_{1j} - y_{2j}|/R_j]$ from Gower's S_{19} coefficient or function s_{12j} from coefficient S_{20}, which were both described above. In all cases, double-zeros must be excluded. This is done by multiplying the partial similarities s_j by Kronecker delta w_{12j}, whose value is 0 upon occurrence of a double-zero. For raw species abundance data, Steinhaus' similarity

S_{17}, computed for a single species at a time, may be used as the partial similarity measure. The outcome of this first step is a partial similarity matrix, containing as many *rows* as there are species in the ecological data matrix (p) and $n(n-1)/2$ *columns*, i.e. one column for each pair of sites; see the numerical example below.

(b) In a second table of the same size, for each species j and each of the $n(n-1)/2$ pairs of sites, one computes the proportion of partial similarity values belonging to species j that are larger than or equal to the partial similarity of the pair of sites being considered; the s_j value under consideration is itself included in the calculation of the proportion. The larger the proportion, the less similar are the two sites with regard to the given species.

(c) The above proportions or probabilities are combined into a site × site similarity matrix, using Fisher's method, i.e. by computing the product Π of the probabilities relative to the various species. Since none of the probabilities is equal to 0, there is no problem in combining these values, but one must assume that the probabilities of the different species are independent vectors. If there are correlations among species, one may use, instead of the original descriptors of species abundance (Orlóci, 1978: 62), a matrix of component scores from a principal coordinate or correspondence analysis of the original species abundance data (Sections 9.2 and 9.4).

(d) There are two ways to define Goodall's similarity index. In the first approach, the products Π are put in increasing order. Following this, the similarity between two sites is calculated as the proportion of products that are larger than or equal to the product for the pair of sites considered:

$$S_{23}(\mathbf{x}_1, \mathbf{x}_2) = \frac{\sum_{pairs} \mathrm{d}}{n(n-1)/2} \quad \text{where} \quad \begin{cases} \mathrm{d} = 1 \text{ if } \Pi \geq \Pi_{12} \\ \mathrm{d} = 0 \text{ if } \Pi < \Pi_{12} \end{cases} \qquad \textbf{(7.31)}$$

(e) In the second approach, the χ^2 value corresponding to each product is computed under the hypothesis that the probabilities of the different species are independent vectors:

$$\chi^2_{12} = -2 \ln \Pi_{12}$$

which has $2p$ degrees of freedom (p is the number of species). The similarity index is the complement of the probability associated with this χ^2, i.e. the complement of the probability that a χ^2 value taken at random exceeds the observed χ^2 value:

$$S_{23}(\mathbf{x}_1, \mathbf{x}_2) = 1 - \text{prob}\ (\chi^2_{12}) \qquad \textbf{(7.32)}$$

It should be clear to the reader that the value of Goodall's index for a given pair of sites may vary depending on the sites included in the computation, since it is based on the rank of the partial similarity for that pair of sites among all pairs. This makes Goodall's measure different from the other coefficients discussed so far.

The following numerical example illustrates the computation of Goodall's index. In this example, five ponds are characterized by the abundances of eight zooplankton species. Data are on a scale of relative abundances, from 0 to 5 (data from Legendre & Chodorowski, 1977).

Species	Ponds					Range R_j
	212	214	233	431	432	
1	3	3	0	0	0	3
2	0	0	2	2	0	2
3	0	2	3	0	2	3
4	0	0	4	3	3	4
5	4	4	0	0	0	4
6	0	2	0	3	3	3
7	0	0	0	1	2	2
8	3	3	0	0	0	3

(a) Gower's matrix of partial similarities has 6 rows and $n(n-1)/2 = 10$ columns which correspond to the 10 pairs of ponds:

Species	Pairs of ponds									
	212	212	212	212	214	214	214	233	233	431
	214	233	431	432	233	431	432	431	432	432
1	1	0	0	0	0	0	0	0	0	0
2	0	0	0	0	0	0	0	1	0	0
3	0.33	0	0	0.33	0.67	0.33	1	0	0.67	0.33
4	0	0	0.25	0.25	0	0.25	0.25	0.75	0.75	1
5	1	0	0	0	0	0	0	0	0	0
6	0.33	0	0	0	0.33	0.67	0.67	0	0	1
7	0	0	0.50	0	0	0.50	0	0.50	0	0.50
8	1	0	0	0	0	0	0	0	0	0

(b) In the next table, one computes, for each pair of sites and each row (species), the proportion of partial similarity values in the row that are larger than or equal to the partial similarity of the pair of sites being considered. The value under consideration is itself included in the proportion. For example, for the pair of ponds (214, 233), the third species has a similarity of 0.67. In the third row, there are 3 values out of 10 that are larger than or equal to 0.67. Thus the ratio associated with the pair (214, 233) in the table is 0.3.

Species	Pairs of ponds									
	212	212	212	212	214	214	214	233	233	431
	214	233	431	432	233	431	432	431	432	432
1	0.1	1	1	1	1	1	1	1	1	1
2	1	1	1	1	1	1	1	0.1	1	1
3	0.7	1	1	0.7	0.3	0.7	0.1	1	0.3	0.7
4	1	1	0.7	0.7	1	0.7	0.7	0.3	0.3	0.1
5	0.1	1	1	1	1	1	1	1	1	1
6	0.5	1	1	1	0.5	0.3	0.3	1	1	0.1
7	1	1	0.4	1	1	0.4	1	0.4	1	0.4
8	0.1	1	1	1	1	1	1	1	1	1

(c) The next table is a site × site symmetric matrix, in which are recorded the products of the terms in each column of the previous table

Ponds	Ponds				
	212	214	233	431	432
212	–				
214	0.00035	–			
233	1.00000	0.15000	–		
431	0.28000	0.05880	0.01200	–	
432	0.49000	0.02100	0.09000	0.00280	–

(d) The first method for computing the similarity consists in entering, in a site × site matrix, the proportions of the above products that are larger than or equal to the product corresponding to each pair of sites. For example, the product corresponding to pair (212, 431) is 0.28. In the table, there are 3 values out of 10 that are larger than or equal to 0.28, hence the similarity S_{23} (212, 431) = 0.3 (eq. 7.31).

Ponds	Ponds				
	212	214	233	431	432
212	–				
214	1.0	–			
233	0.1	0.4	–		
431	0.3	0.6	0.8	–	
432	0.2	0.7	0.5	0.9	–

(e) If the chosen similarity measure is the complement of the probability associated with χ^2 (eq. 7.32), the following table is obtained. For example, to determine the similarity for pair (212, 431), the first step is to compute χ^2 (212, 431) = –2 ln(0.28) = 2.5459, where 0.28 was the product associated with this pair in the table at step (d). The value of χ^2 (212, 431) is 2.5459 and the number of degrees of freedom is $2p$ = 16, so that the corresponding probability is 0.9994. The similarity is the complement of this probability: S_{23} (212, 431) = 1 – 0.99994 = 0.00006.

Ponds	Ponds				
	212	214	233	431	432
212	–				
214	0.54110	–			
233	0.00000	0.00079	–		
431	0.00006	0.00869	0.08037	–	
432	0.00000	0.04340	0.00340	0.23942	–

Even though the values in the last two tables are very different, the differences are only in term of scale; measures S_{23} computed with eqs. 7.31 and 7.32 are monotonic to each other. Section 8.9 shows how to use similarities computed with eq. 7.32 for the probabilistic clustering of objects or species.

Another, simpler probabilistic similarity coefficient among sites has been proposed by palaeontologists Raup & Crick (1979) for species presence-absence data; this is the level of measurement usually favoured in palaeoecology. Consider the number of species in common to sites h and i; this is statistic a_{hi} of the binary coefficients of Section 7.3. The null hypothesis here is H_0: there is no association between sites h and i because species are independent of one another and each one is distributed at random among the sites (hypothesis of random sprinkling of species). The association between sites, measured by a_{hi}, is tested as follows using permutations:

Permutation test

1. Compute the reference value of the number of species in common, a_{hi}, for each pair of sites h and i.

2. Permute at random each vector of species occurrences, independently of the other species, as in permutation model 1 of Section 10.6. Compute the number of species in common, a^*_{hi}, for each pair of sites under permutation.

3. Repeat step 2 a large number of times to obtain the distribution of a^*_{hi}. Add the reference value a_{hi} to the distribution.

4. For each pair of sites, compare a_{hi} to the reference distribution and calculate the probability $p(a_{hi})$ that $a^*_{hi} \geq a_{hi}$, using the procedure of Subsection 1.2.2.

This description is based upon the papers of Raup & Crick (1979) and McCoy *et al.* (1986). However, calculation of the probability in the upper tail of the reference distribution is modified, in the above description, to agree with Subsection 1.2.2. The probability $p(a_{hi})$ is expected to be near 0 for sites h and i showing high association, i.e. with more species in common than expected under the null hypothesis. A value near 0.5 indicates that the data support the null hypothesis of "random sprinkling of species".

One could also test in the lower tail of the distribution, looking for pairs of sites that are significantly dissimilar. The probability to calculate would be: $p(a^*_{hi} \leq a_{hi})$. Significantly dissimilar sites would suggest that some process may have influenced the selection of species, so that fewer species are common to the two sites than expected under the null hypothesis.

Using the probability calculated in the upper tail, a probabilistic similarity measure of association between sites $\mathbf{x}_1$ and $\mathbf{x}_2$ is defined here as follows:

$$S_{27}(\mathbf{x}_1, \mathbf{x}_2) = 1 - p(a_{12}) \quad \textbf{(7.33)}$$

This measure of similarity is different from that used by Raup & Crick (1979). Now that computer power is widely available, it is recommended to use 999 or 9999 permutations instead of the 50 permutations recommended by Raup & Crick in 1979.

7.4 Q mode: distance coefficients

Distance coefficients are functions which take their maximum values (often 1) for two objects that are entirely different, and 0 for two objects that are identical over all descriptors. Distances, like similarities, (Section 7.3), are used to measure the association between *objects*. Distance coefficients may be subdivided in three groups. The first group consists of *metrics* which share the following four properties:

Metric properties

1) minimum 0: if $a = b$, then $D(a, b) = 0$;

2) positiveness: if $a \neq b$, then $D(a, b) > 0$;

3) symmetry: $D(a, b) = D(b, a)$;

4) triangle inequality: $D(a, b) + D(b, c) \geq D(a, c)$. In the same way, the sum of two sides of a triangle drawn in Euclidean space is necessarily equal to or larger than the third side.

The second group of distances are the *semimetrics* (or *pseudometrics*). These coefficients do not follow the triangle inequality axiom. These measures cannot directly be used to order points in a *metric* or *Euclidean space* because, for three points (a, b and c), the sum of the distances from a to b and from b to c may be smaller than the distance between a and c. A numerical example is given is Subsection 2. Some authors prefer to talk about *dissimilarity* coefficients as the general expression and use the term *distance* only to refer to coefficients that satisfy the four metric properties.

Table 7.2 Some properties of distance coefficients calculated from the similarity coefficients presented in Section 7.3. These properties (from Gower & Legendre, 1986), which will be used in Section 9.2, only apply when there are no missing data.

Similarity	$D = 1 - S$ metric, etc.	$D = 1 - S$ Euclidean	$D = \sqrt{1-S}$ metric	$D = \sqrt{1-S}$ Euclidean
$S_1 = \dfrac{a+d}{a+b+c+d}$ (simple matching; eq. 7.1)	metric	No	Yes	Yes
$S_2 = \dfrac{a+d}{a+2b+2c+d}$ (Rogers & Tanimoto; eq. 7.2)	metric	No	Yes	Yes
$S_3 = \dfrac{2a+2d}{2a+b+c+2d}$ (eq. 7.3)	semimetric	No	Yes	No
$S_4 = \dfrac{a+d}{b+c}$ (eq. 7.4)	nonmetric	No	No	No
$S_5 = \dfrac{1}{4}\left[\dfrac{a}{a+b}+\dfrac{a}{a+c}+\dfrac{d}{b+d}+\dfrac{d}{c+d}\right]$ (eq. 7.5)	semimetric	No	No	No
$S_6 = \dfrac{a}{\sqrt{(a+b)\,(a+c)}}\dfrac{d}{\sqrt{(b+d)\,(c+d)}}$ (eq. 7.6)	semimetric	No	Yes	Yes
$S_7 = \dfrac{a}{a+b+c}$ (Jaccard; eq. 7.10)	metric	No	Yes	Yes
$S_8 = \dfrac{2a}{2a+b+c}$ (Sørensen; eq. 7.11)	semimetric	No	Yes	Yes
$S_9 = \dfrac{3a}{3a+b+c}$ (eq. 7.12)	semimetric	No	No	No
$S_{10} = \dfrac{a}{a+2b+2c}$ (eq. 7.13)	metric	No	Yes	Yes
$S_{11} = \dfrac{a}{a+b+c+d}$ (Russell & Rao; eq. 7.14)	metric	No	Yes	Yes
$S_{12} = \dfrac{a}{b+c}$ (Kulczynski; eq. 7.15)	nonmetric	No	No	No

The third group of distances consists of *nonmetrics*. These coefficients may take negative values, thus violating the property of positiveness of metrics.

All similarity coefficient from Section 7.3 can be transformed into distances, as mentioned in Section 7.2. Some properties of distance coefficients resulting from the transformations $D = (1 - S)$ and $D = \sqrt{1-S}$ are discussed in Table 7.2. Stating that a distance coefficient is *not* metric or Euclidean actually means that an example can be found; it does not mean that the coefficient is never metric or Euclidean. A coefficient is *likely* to be metric or Euclidean when the binary form of the coefficient (name given in the Table) is known to be metric or Euclidean, and test runs have never turned up cased to the contrary. A coefficient is said to be Euclidean if the distances are fully embeddable in an Euclidean space; principal coordinate analysis (Section 9.2) of such a distance matrix does not produce negative eigenvalues.

Euclidean coefficient

Table 7.2 Continued.

Similarity	$D = 1 - S$ metric, etc.	$D = 1 - S$ Euclidean	$D = \sqrt{1-S}$ metric	$D = \sqrt{1-S}$ Euclidean
$S_{13} = \frac{1}{2}\left[\frac{a}{a+b} + \frac{a}{a+c}\right]$ (eq. 7.16)	semimetric	No	No	No
$S_{14} = \frac{a}{\sqrt{(a+b)(a+c)}}$ (Ochiai; eq. 7.17)	semimetric	No	Yes	Yes
$S_{15} = \sum w_j s_j / \sum w_j$ (Gower; eq. 7.20)	metric	No	Yes	Likely* (S_1)
$S_{16} = \sum w_j s_j / \sum w_j$ (Estabrook & Rogers; eq. 7.21)	metric	No	Yes	Likely* (S_1)
$S_{17} = \frac{2W}{A+B}$ (Steinhaus; eq. 7.24)	semimetric	No	Likely* (S_8)	Likely* (S_8)
$S_{18} = \frac{1}{2}\left[\frac{W}{A} + \frac{W}{B}\right]$ (Kulczynski; eq. 7.25)	semimetric	No	No* (S_{13})	No* (S_{13})
$S_{19} = \sum w_j s_j / \sum w_j$ (Gower; eq. 7.26)	metric	No	Yes	Likely
$S_{20} = \sum w_j s_j / \sum w_j$ (Legendre & Chodorowski; 7.27)	metric	No	Yes	Likely* (S_7)
$S_{21} = 1 - \chi^2\ metric$ (eq. 7.28)	metric	Yes	Yes	Yes
$S_{22} = 1 - p(\chi^2)$ (eq. 7.30)	semimetric	No	–	–
$S_{23} = 2(\sum d)/n(n-1)$ (Goodall; eq. 7.31) or	semimetric	No	–	–
$S_{23} = 1 - p(\chi^2)$ (Goodall; eq. 7.32)	semimetric	No	–	–
$S_{26} = (a + d/2)/p$ (Faith, 1983; eq. 7.18)	metric	–	Yes	–

* These results follow from the properties of the corresponding binary coefficients (coefficient numbers given), when continuous variables are replaced by binary variables.
– Property unknown for this coefficient.

In the case of ordered descriptors, some additional distance measures are described here, in addition to those in Table 7.2. How to use the various distance coefficients is summarized in Tables 7.3 and 7.4.

1 — Metric distances

Metric distances have been developed for quantitative descriptors, but they have occasionally been used with semiquantitative descriptors. Some of these measures (D_1, D_2, D_5 to D_8, D_{12}) process the double-zeros in the same way as any other value of the descriptors; refer to the discussion of the double-zero problem at the beginning of

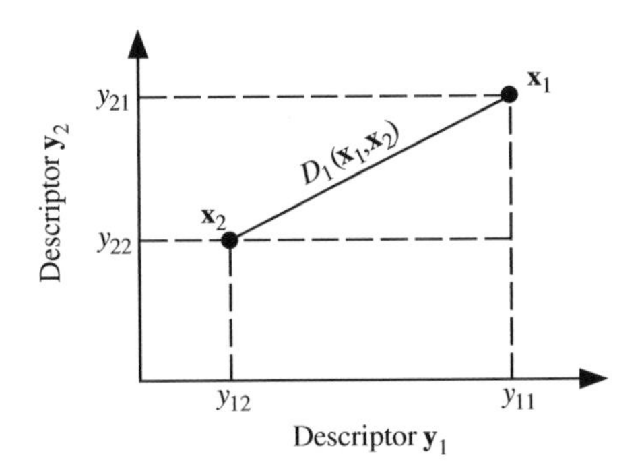

Figure 7.4 Computation of the Euclidean distance (D_1) between objects $\mathbf{x}_1$ and $\mathbf{x}_2$ in 2-dimensional space.

Section 7.3. These coefficients should not be used, in general, with species abundances, as will be seen in the paradox described below. Coefficients D_3, D_4, D_9, to D_{11}, and D_{15} to D_{17}, on the contrary, are well adapted to species abundance data.

The most common metric measure is the *Euclidean distance*. It is computed using Pythagoras' formula, from site-points positioned in a p-dimensional space called a *metric* or *Euclidean space*:

$$D_1(\mathbf{x}_1, \mathbf{x}_2) = \sqrt{\sum_{j=1}^{p} (y_{1j} - y_{2j})^2} \tag{7.34}$$

When there are only two descriptors, this expression becomes the measure of the hypotenuse of a right-angled triangle (Fig. 7.4; Section 2.4):

$$D_1(\mathbf{x}_1, \mathbf{x}_2) = \sqrt{(y_{11} - y_{21})^2 + (y_{12} - y_{22})^2}$$

The square of D_1 may also be used for clustering purpose. One should notice, however, that D_1^2 is a semimetric, which makes it less appropriate than D_1 for ordination:

$$D_1^2(\mathbf{x}_1, \mathbf{x}_2) = \sum_{j=1}^{p} (y_{1j} - y_{2j})^2 \tag{7.35}$$

The Euclidean distance does not have an upper limit, its value increasing indefinitely with the number of descriptors. The value also depends on the scale of each descriptor, to such an extent that changing the scale of some descriptors may result in measures that are not monotonic to each other. The problem may be avoided by using standardized variables (eq. 1.12) instead of the original data, or by restricting the use of D_1 and other distances of the same type (D_2, D_6, D_7 and D_8) to dimensionally homogeneous data matrices (Chapter 3).

Williams & Stephenson (1973) and Stephenson *et al.* (1974) proposed a model for analysing three-dimensional ecological data matrices (sites × species × times) based on the Euclidean distance among vectors of data centred on their means (eq. 1.9).

Species abundance paradox

The Euclidean distance, used as a measure of resemblance among sites on the basis of species abundances, may lead to the following paradox: two sites without any species in common may be at a smaller distance than another pair of sites sharing species. This paradox is illustrated by a numerical example from Orlóci (1978: 46):

Sites	Species		
	$\mathbf{y}_1$	$\mathbf{y}_2$	$\mathbf{y}_3$
$\mathbf{x}_1$	0	1	1
$\mathbf{x}_2$	1	0	0
$\mathbf{x}_3$	0	4	4

From these data, the following distances are calculated between sites:

Sites	Sites		
	$\mathbf{x}_1$	$\mathbf{x}_2$	$\mathbf{x}_3$
$\mathbf{x}_1$	0	1.732	4.243
$\mathbf{x}_2$	1.732	0	5.745
$\mathbf{x}_3$	4.243	5.745	0

Thus the Euclidean distance between sites $\mathbf{x}_1$ and $\mathbf{x}_2$, which have no species in common, is smaller than the distance between $\mathbf{x}_1$ and $\mathbf{x}_3$ which share species $\mathbf{y}_2$ and $\mathbf{y}_3$. In general, double-zeros lead to reduction of distances. This must be avoided with species abundance data. For environmental descriptors, on the contrary, double-zeros may well be a valid basis for comparing sites. The Euclidean distance should therefore not be used for comparing sites on the basis of species abundances. The main difficulty in ecology concerning the Euclidean distance arises from the fact that a frequently used method, i.e. principal component analysis, orders objects in the multidimensional space of descriptors using D_1. The ensuing problems are discussed in Section 9.1.

Various modifications have been proposed to deal with the drawbacks of the Euclidean distance applied to species abundances. First, the effect of the number of descriptors may be tempered by computing an *average distance*:

$$D_2^2(\mathbf{x}_1, \mathbf{x}_2) = \frac{1}{p}\sum_{j=1}^{p} (y_{1j} - y_{2j})^2 \text{ or } D_2(\mathbf{x}_1, \mathbf{x}_2) = \sqrt{D_2^2} \quad \textbf{(7.36)}$$

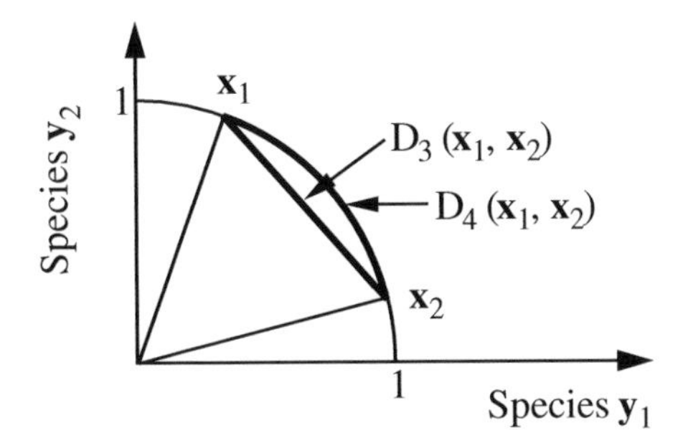

Figure 7.5 Computation of the chord distance D_3 and geodesic metric D_4 between sites $\mathbf{x}_1$ and $\mathbf{x}_2$.

Orlóci (1967b) proposed to use the *chord distance*, which has a maximum value of $\sqrt{2}$ for sites with no species in common and a minimum of 0 when two sites share the same species in the same *proportions*, without it being necessary for these species to be represented by the same *numbers of individuals* at the two sites. This measure is the Euclidean distance computed after scaling the site vectors to length 1 (normalization of a vector, Section 2.4). After normalization, the Euclidean distance computed between two objects (sites) is equivalent to the length of a chord joining two points within a segment of a sphere or hypersphere of radius 1. If there are only two species involved, the normalization places the sites on the circumference of a 90° sector of a circle with radius 1 (Fig. 7.5). The chord distance may also be computed directly from non-normalized data through the following formula:

$$D_3(\mathbf{x}_1, \mathbf{x}_2) = \sqrt{2\left(1 - \frac{\sum_{j=1}^{p} y_{1j} y_{2j}}{\sqrt{\sum_{j=1}^{p} y_{1j}^2 \sum_{j=1}^{p} y_{2j}^2}}\right)} \qquad \textbf{(7.37)}$$

The inner part of this equation is actually the cosine of the angle (θ) between the two site vectors, normalized or not (eq. 2.9). So the chord distance formula may be written:

$$D_3 = \sqrt{2(1 - \cos\theta)} \qquad \textbf{(7.38)}$$

The chord distance is maximum when the species at two sites are completely different. In such a case, the normalized site vectors are at 90° from each other on the circumference of a 90° sector of a circle (when there are only two species), or on the surface of a segment of a hypersphere (for p species), and the distance between the two sites is $\sqrt{2}$. This measure solves the problem caused by sites having different total abundances of species as well as the paradox explained above for D_1. Indeed, with D_3, the distances between pairs of sites for the numerical example are:

Sites	Sites		
	$\mathbf{x}_1$	$\mathbf{x}_2$	$\mathbf{x}_3$
$\mathbf{x}_1$	0	1.414	0.000
$\mathbf{x}_2$	1.414	0	1.414
$\mathbf{x}_3$	0.000	1.414	0

The chord distance is a metric. Since double-zeros do not influence the chord distance, it can be used to compare sites described by species abundances.

A transformation of the previous measure, known as the *geodesic metric*, measures the length of the arc at the surface of the hypersphere of unit radius (Fig. 7.5):

$$D_4(\mathbf{x}_1, \mathbf{x}_2) = \text{arc cos}\left[1 - \frac{D_3^2(\mathbf{x}_1, \mathbf{x}_2)}{2}\right] \quad \textbf{(7.39)}$$

In the numerical example, pairs of sites ($\mathbf{x}_1$, $\mathbf{x}_2$) and ($\mathbf{x}_2$, $\mathbf{x}_3$), with no species in common, are at an angle of 90°, whereas pair of sites ($\mathbf{x}_1$, $\mathbf{x}_2$), which share two of the three species, are at smaller angle (88°).

Mahalanobis (1936) has developed a generalized distance which takes into account the correlations among descriptors and is independent of the scales of the various descriptors. This measure computes the distance between two points in a space whose axes are not necessarily orthogonal, in order to take into account the correlations among descriptors. Orlóci (1978: 48) gives the formula for computing distances between individual sites but, in practice, the *Mahalanobis generalized distance* is only used for comparing *groups of sites*. For two groups of sites, $\mathbf{w}_1$ and $\mathbf{w}_2$, containing n_1 and n_2 sites, respectively, and described by the same p variables, the square of the generalized distance is given by the following matrix formula:

$$D_5^2(\mathbf{w}_1, \mathbf{w}_2) = \overline{\mathbf{d}_{12}}\mathbf{V}^{-1}\overline{\mathbf{d}'_{12}} \quad \textbf{(7.40)}$$

In this equation, $\overline{\mathbf{d}_{12}}$ is the vector (length $= p$) of the differences between the means of the p variables in the two groups of sites. $\mathbf{V}$ is the pooled within-group dispersion matrix of the two groups of sites, estimated from the matrices of sums of squares and cross products between group-centred descriptors for each of the two groups, added up term by term and divided by $(n_1 + n_2 - 2)$, as in discriminant analysis (Table 11.8) and multivariate analysis of variance:

$$\mathbf{V} = \frac{1}{n_1 + n_2 - 2}\left[(n_1 - 1)\mathbf{S}_1 + (n_2 - 1)\mathbf{S}_2\right] \quad \textbf{(7.41)}$$

where $\mathbf{S}_1$ and $\mathbf{S}_2$ are the dispersion matrices for each of the two groups*. Whereas vector $\overline{\mathbf{d}}_{12}$ measures the difference between the p-dimensional means of two groups (p descriptors), $\mathbf{V}$ takes into account the covariances among descriptors. This formula may be adapted to calculate the distance between a single object and a group.

If one wishes to test D_5 for significance, the within-group dispersion matrices must be homogeneous (homoscedasticity, Box 1.4). Homoscedasticity of matrices $\mathbf{S}_1$ and $\mathbf{S}_2$ may be tested using Kullback's test (eq. 11.34). The test of significance also assumes multinormality of the within-group distributions (Sections 4.3 and 4.7) although the generalized distance tolerates some degree of deviation from this condition. The Mahalanobis generalized distance is the distance preserved among group means in a canonical space of discriminant functions (Section 11.5).

To perform the test of significance, the generalized distance is transformed into Hotelling's T^2 (1931) statistic, using the following equation:

$$T^2 = \frac{n_1 n_2}{(n_1 + n_2)} D_5^2 \qquad \textbf{(7.42)}$$

The F statistic is computed as follows:

$$F = \frac{n_1 + n_2 - (p+1)}{(n_1 + n_2 - 2)\, p} T^2 \qquad \textbf{(7.43)}$$

with p and $[n_1 + n_2 - (p + 1)]$ degrees of freedom. Statistic T^2 is a generalization of Student's t statistic to the multidimensional case. It allows one to test the hypothesis that two groups originate from populations with similar centroids. The final generalization to several groups, called Wilks Λ (lambda), is discussed in Section 11.5 (eq. 11.35).

Coefficients D_2 and D_5 are related to the Euclidean distance D_1, which is the second degree ($r = 2$) of Minkowski's metric:

$$D_6(\mathbf{x}_1, \mathbf{x}_2) = \left[\sum_{j=1}^{p} \left| y_{1j} - y_{2j} \right|^r \right]^{1/r} \qquad \textbf{(7.44)}$$

Forms of this metric with $r > 2$ are seldom used in ecology because powers higher than 2 give too much importance to the largest differences $\left| y_{1j} - y_{2j} \right|$. For the exact opposite reason, exponent $r = 1$ is used in many instances. The basic form,

---

* Procedure NEIGHBOUR of the SAS package uses the total dispersion matrix $\mathbf{T}$ (Table 11.8), instead of $\mathbf{V}$, to compute Mahalanobis distances.

$$D_7(\mathbf{x}_1, \mathbf{x}_2) = \sum_{j=1}^{p} |y_{1j} - y_{2j}| \quad \textbf{(7.45)}$$

is known as the *Manhattan metric*, *taxicab metric*, or *city-block metric*. This refers to the fact that, for two descriptors, the distance between two sites is the distance on the abscissa (descriptor y_1) plus the distance on the ordinate (descriptor y_2), like the distance travelled by a taxicab around blocks in a city with an orthogonal plan like Manhattan. This metric presents the same problem for double-zeros as in the Euclidean distance and thus leads to the same paradox.

The *mean character difference* ("durchschnittliche Differenz", in German), proposed by anthropologist Czekanowski (1909),

$$D_8(\mathbf{x}_1, \mathbf{x}_2) = \frac{1}{p} \sum_{j=1}^{p} |y_{1j} - y_{2j}| \quad \textbf{(7.46)}$$

has the advantage over D_7 of not increasing with the number of descriptors (p). It may be used with species abundances if one modifies eq. 7.46 to exclude double-zeros from the computation of $|y_{1j} - y_{2j}|$ and p is replaced by (p – no. double-zeros).

Whittaker's *index of association* (1952) is well adapted to species abundance data, because each species is first transformed into a fraction of the total number of individuals at the site, before the subtraction. The complement of this index is the following distance:

$$D_9(\mathbf{x}_1, \mathbf{x}_2) = \frac{1}{2} \sum_{j=1}^{p} \left| \frac{y_{1j}}{\sum_{j=1}^{p} y_{1j}} - \frac{y_{2j}}{\sum_{j=1}^{p} y_{2j}} \right| \quad \textbf{(7.47)}$$

The difference is zero for a species when its *proportions* are identical in the two sites. An identical result is obtained by computing, over all species, the sum of the smallest fractions calculated for the two sites:

$$D_9(\mathbf{x}_1, \mathbf{x}_2) = \left[1 - \sum_{j=1}^{p} \min \left(\frac{y_j}{\sum y} \right) \right] \quad \textbf{(7.48)}$$

The Australians Lance & Williams (1967a) give several variants of the Manhattan metric, including their *Canberra metric* (Lance & Williams, 1966c):

$$D_{10}(\mathbf{x}_1, \mathbf{x}_2) = \sum_{j=1}^{p} \left[\frac{|y_{1j} - y_{2j}|}{(y_{1j} + y_{2j})} \right] \quad \textbf{(7.49)}$$

which must exclude double-zeros in order to avoid indetermination. This measure has no upper limit. It can be shown that, in D_{10}, a given difference between abundant species contributes less to the distance than the same difference found between rarer species (Section 7.6). As an ecological *similarity* measure, Stephenson *et al.* (1972) and Moreau & Legendre (1979) used the one-complement of the Canberra metric, after scaling it between 0 and 1:

$$S(\mathbf{x}_1, \mathbf{x}_2) = 1 - \frac{1}{p} D_{10} \tag{7.50}$$

Another version of coefficient D_{10}, scaled and based on the Euclidean distance, has been used for taxonomic purposes by Clark (1952) under the name *coefficient of divergence*:

$$D_{11}(\mathbf{x}_1, \mathbf{x}_2) = \sqrt{\frac{1}{p} \sum_{j=1}^{p} \left(\frac{y_{1j} - y_{2j}}{y_{1j} + y_{2j}} \right)^2} \tag{7.51}$$

D_{11} is to D_{10} what D_2 is to D_7. Because, in D_{11}, the difference for each descriptor is first expressed as a fraction, before squaring the values and summing them, this coefficient may be used with species abundance data. As in coefficient D_8,however, double-zeros must be excluded from the computation and their number subtracted from p. Unless one intends to use this coefficient as a basis for ordination (Chapter 9), it is better, for species abundance data, to use the semimetric D_{14} described below.

Another coefficient, which is related to D_{11}, was developed by Pearson (1926) for anthropological studies under the name *coefficient of racial likeness*. Using this coefficient, it is possible to measure a distance between groups of sites, like with the Mahalanobis generalized distance D_5, but without eliminating the effect of correlations among descriptors:

$$D_{12}(\mathbf{w}_1, \mathbf{w}_2) = \sqrt{\frac{1}{p} \sum_{j=1}^{p} \left[\frac{(\bar{y}_{1j} - \bar{y}_{2j})^2}{(s_{1j}^2 / n_1) + (s_{2j}^2 / n_2)} \right] - \frac{2}{p}} \tag{7.52}$$

for two groups of sites $\mathbf{w}_1$ and $\mathbf{w}_2$ containing respectively n_1 and n_2 sites; $\bar{y}_{ij}$ is the mean of descriptor j in group i and s_{ij}^2 is the corresponding variance.

Other measures, related to χ^2, are available to calculate the distance among sites using species abundances or other frequency data; no negative value is allowed in the data. The first of these coefficients is called the χ^2 *metric*. The sum of squares of differences is calculated between profiles of conditional probabilities in two rows (or columns) of a frequency table, weighting each term of the sum of squares by the inverse of the frequency of the column (or row) in the overall table. This measure has been used by Roux & Reyssac (1975) to calculate distances among sites described by species abundances.

In order to compute the χ^2 metric, the data matrix must be transformed into a matrix of conditional probabilities; see Table 6.7. If the metric is computed between the rows of the matrix, the conditional probabilities are computed by rows. The elements of the matrix become new terms y_{ij}/y_{i+} where y_{i+} is the sum of frequencies in row i. In the numerical example below, rows of the data matrix, on the left-hand side, are sites and columns are species. The matrix of conditional probabilities (by rows) on the right is used below to compute the association between rows (sites):

$$\mathbf{Y} = \begin{bmatrix} 45 & 10 & 15 & 0 & 10 \\ 25 & 8 & 10 & 0 & 3 \\ 7 & 15 & 20 & 14 & 12 \end{bmatrix} \begin{bmatrix} y_{i+} \end{bmatrix} = \begin{bmatrix} 80 \\ 46 \\ 68 \end{bmatrix} \rightarrow \begin{bmatrix} y_{ij}/y_{i+} \end{bmatrix} = \begin{bmatrix} 0.563 & 0.125 & 0.188 & 0.000 & 0.125 \\ 0.543 & 0.174 & 0.217 & 0.000 & 0.065 \\ 0.103 & 0.221 & 0.294 & 0.206 & 0.176 \end{bmatrix}$$

$$\begin{bmatrix} y_{+j} \end{bmatrix} = \begin{bmatrix} 77 & 33 & 45 & 14 & 25 \end{bmatrix} \quad 194$$

The distance between the first two rows of the right-hand matrix could be computed using the formula of the Euclidean distance D_1 (eq. 7.34). The equation of the distance between the two sites would then be

$$D(\mathbf{x}_1, \mathbf{x}_2) = \sqrt{\sum_{j=1}^{p} \left(\frac{y_{1j}}{y_{1+}} - \frac{y_{2j}}{y_{2+}} \right)^2}$$

With this equation, however, the most abundant species would contribute predominantly to the sum of squares. Instead, the χ^2 metric is computed using a weighted expression:

$$D_{15}(\mathbf{x}_1, \mathbf{x}_2) = \sqrt{\sum_{j=1}^{p} \frac{1}{y_{+j}} \left(\frac{y_{1j}}{y_{1+}} - \frac{y_{2j}}{y_{2+}} \right)^2} \qquad \textbf{(7.53)}$$

where y_{+j} is the sum of frequencies in column j. Although this measure has no upper limit, it produces distances smaller than 1 in most cases. For the numerical example, computation of D_{15} between the first two sites (rows) gives:

$$D_{15}(\mathbf{x}_1, \mathbf{x}_2) =$$

$$\left[\frac{(0.563 - 0.543)^2}{77} + \frac{(0.125 - 0.174)^2}{33} + \frac{(0.188 - 0.217)^2}{45} + \frac{(0 - 0)^2}{14} + \frac{(0.125 - 0.065)^2}{25} \right]^{1/2}$$

$$= 0.015$$

The fourth species, which is absent from the first two sites, cancels itself out. This is how the χ^2 metric excludes double-zeros from the calculation.

The two sites were compared above using (weighted) profiles of species conditional probabilities. D_{15} may also be used to measure the distance between species from the (weighted) distribution profiles among the various sites. The conditional probability matrix is then computed by columns $[y_{ij}/y_{.j}]$ before applying the above formula, interchanging columns for rows.

A related measure is called the χ^2 *distance* (Lebart & Fénelon, 1971). It differs from the χ^2 metric in that the terms of the sum of squares are divided by the *probability* (relative frequency) of each row in the overall table instead of its absolute frequency. In other words, it is identical to the χ^2 metric multiplied by $\sqrt{y_{++}}$ where y_{++} is the sum of all frequencies in the data table:

$$D_{16}(\mathbf{x}_1, \mathbf{x}_2) = \sqrt{\sum_{j=1}^{p} \frac{1}{y_{+j}/y_{++}} \left(\frac{y_{1j}}{y_{1+}} - \frac{y_{2j}}{y_{2+}} \right)^2} = \sqrt{y_{++}} \sqrt{\sum_{j=1}^{p} \frac{1}{y_{+j}} \left(\frac{y_{1j}}{y_{1+}} - \frac{y_{2j}}{y_{2+}} \right)^2} \quad \textbf{(7.54)}$$

The χ^2 distance is the distance preserved in correspondence analysis (Section 9.4), when computing the association between species or between sites. More generally, it is used for computing the association between the rows or columns of a contingency table. This measure has no upper limit.

The small numerical example already used to illustrate the paradox associated with the Euclidean distance (D_1) computed for species abundances is used again here to contrast D_{16} with D_1.

$$\mathbf{Y} = \begin{bmatrix} 0 & 1 & 1 \\ 1 & 0 & 0 \\ 0 & 4 & 4 \end{bmatrix} \overset{[y_{i+}]}{\begin{bmatrix} 2 \\ 1 \\ 8 \end{bmatrix}} \quad \rightarrow [y_{ij}/y_{i+}] = \begin{bmatrix} 0 & 0.5 & 0.5 \\ 1 & 0 & 0 \\ 0 & 0.5 & 0.5 \end{bmatrix}$$

$$[y_{+j}] = \begin{bmatrix} 1 & 5 & 5 \end{bmatrix} \quad 11$$

Computing D_{16} between the first two rows (sites) gives:

$$D_{16}(\mathbf{x}_1, \mathbf{x}_2) = \left[\frac{(0-1)^2}{1/11} + \frac{(0.5-0)^2}{5/11} + \frac{(0.5-0)^2}{5/11} \right]^{1/2} = 3.477$$

The distances between all pairs of sites are:

Sites	Sites		
	$\mathbf{x}_1$	$\mathbf{x}_2$	$\mathbf{x}_3$
$\mathbf{x}_1$	0	3.477	0
$\mathbf{x}_2$	3.477	0	3.477
$\mathbf{x}_3$	0	3.477	0

Comparison with results obtained for D_1, above, shows that the problem then caused by the presence of double-zeros does not exist here. Distance D_{16} may therefore be used directly with sites described by species abundances, contrary to D_1.

A coefficient related to D_{15} and D_{16} is the Hellinger distance, described by Rao (1995). The formula for the Hellinger distance is:

$$D_{17}(\mathbf{x}_1, \mathbf{x}_2) = \sqrt{\sum_{j=1}^{p} \left[\sqrt{\frac{y_{1j}}{y_{1+}}} - \sqrt{\frac{y_{2j}}{y_{2+}}}\right]^2} \quad (7.55)$$

Its properties are briefly discussed near the end of Section 7.6. Like D_{16}, this distance has no upper limit. Rao (1995) recommends this measure as a basis for a new ordination method. One can obtain a similar ordination by computing this distance function among the objects and going to principal coordinate analysis.

2 — *Semimetrics*

Some distance measures do not follow the fourth property of metrics, i.e. the triangle inequality axiom described at the beginning of this Section. As a consequence, they do not allow a proper ordination of sites in a full Euclidean space. They may, however, be used for ordination by principal coordinate analysis after correction for negative eigenvalues (Subsection 9.2.4) or by nonmetric multidimensional scaling (Section 9.3). These measures are called *semimetrics* or *pseudometrics*. Some semimetrics derived from similarities are identified in Table 7.2. Other such measures are presented here.

The distance corresponding to Sørensen's coefficient S_8 was described by Watson *et al.* (1966) under the name *nonmetric coefficient*:

$$D_{13}(\mathbf{x}_1, \mathbf{x}_2) = 1 - \frac{2a}{2a+b+c} = \frac{b+c}{2a+b+c} \quad (7.56)$$

a, b and c were defined at the beginning of Subsection 7.3.1. The following numerical example shows that D_{13} does not obey the triangle inequality axiom:

Sites	Species				
	$\mathbf{y}_1$	$\mathbf{y}_2$	$\mathbf{y}_3$	$\mathbf{y}_4$	$\mathbf{y}_5$
$\mathbf{x}_1$	1	1	1	0	0
$\mathbf{x}_2$	0	0	0	1	1
$\mathbf{x}_3$	1	1	1	1	1

Distances between the three pairs of sites are:

$$D_{13}(\mathbf{x}_1, \mathbf{x}_2) = \frac{3+2}{0+3+2} = 1.00$$

$$D_{13}(\mathbf{x}_1, \mathbf{x}_3) = \frac{0+2}{(2\times 3)+0+2} = 0.25$$

$$D_{13}(\mathbf{x}_2, \mathbf{x}_3) = \frac{0+3}{(2\times 2)+0+3} = 0.43$$

hence, 1.00 > 0.25 + 0.43, contrary to the triangle inequality axiom.

Among the measures for species abundance data, the coefficients of Steinhaus S_{17} and Kulczynski S_{18} are semimetrics when transformed into distances (Table 7.2). In particular, $D_{14} = 1 - S_{17}$ was first described by Odum (1950), who called it the *percentage difference*, and then by Bray & Curtis (1957):

$$D_{14}(\mathbf{x}_1, \mathbf{x}_2) = \frac{\sum_{j=1}^{p} |y_{1j} - y_{2j}|}{\sum_{j=1}^{p} (y_{1j} + y_{2j})} = 1 - \frac{2W}{A+B} \qquad \textbf{(7.57)}$$

Contrary to the Canberra metric D_{10}, differences between abundant species contribute the same to D_{14} as differences between rare species. This may be seen as a desirable property, for instance when using normalized species abundance data. Bloom (1981) compared the Canberra metric, the percentage difference and other indices to a theoretical standard. He showed that only D_{14} (or S_{17}) accurately reflects the true resemblance along its entire 0 to 1 scale, whereas D_{10}, for example, underestimated the resemblance over much of its 0 to 1 range.

The following numerical example, from Orlóci (1978: 59), shows that D_{14} does not obey the triangle inequality axiom:

Quadrats	Species				
	y_1	y_2	y_3	y_4	y_5
$\mathbf{x}_1$	2	5	2	5	3
$\mathbf{x}_2$	3	5	2	4	3
$\mathbf{x}_3$	9	1	1	1	1

Distances between the three pairs of sites are:

$$D_{14}(\mathbf{x}_1, \mathbf{x}_2) = \frac{1+0+0+1+0}{17+17} = 0.059$$

$$D_{14}(\mathbf{x}_1, \mathbf{x}_3) = \frac{7+4+1+4+2}{17+13} = 0.600$$

$$D_{14}(\mathbf{x}_2, \mathbf{x}_3) = \frac{6+4+1+3+2}{17+13} = 0.533$$

hence, $0.600 > 0.059 + 0.533$, contrary to the triangle inequality axiom. If the numbers of specimens are the same in all sites (sums of rows), however, then D_{14} is a metric. It follows that, when numbers of specimens are quite different from site to site, ordination by principal coordinate analysis based upon D_{14} (or S_{17} and S_{18}) matrices are likely to produce negative eigenvalues. How a meaningful ordination of sites may be obtained in such conditions is described at the beginning of the present Subsection.

7.5 R mode: coefficients of dependence

The main purpose of R-mode analysis is to investigate the relationships among *descriptors*; R matrices may also be used, in some cases, for the ordination of *objects* e.g. in principal component or discriminant analysis (Sections 9.1 and 10.5). Following the classification of descriptors in Table 1.2, dependence coefficients will be described for quantitative, semiquantitative, and qualitative descriptors. This will be followed by special measures to assess the dependence between species, to be used for the identification of biological associations (Section 8.9).

Most dependence coefficients are amenable to *statistical testing*. For such coefficients, it is thus possible to associate a matrix of probabilities with the R matrix, if required by subsequent analyses. While it is not always legitimate to apply statistical tests of significance, it is never incorrect to compute a dependence coefficient. For example, there is no objection to computing a Pearson correlation coefficient for any pair of metric variables, but these same variables must be normally distributed (Sections 4.2 and 4.3) and the sites must be independent realizations (Sections 1.1 and

1.2) to legitimately test the significance of the coefficient using the standard test. Furthermore, a test of significance only allows one to reject or not a specific hypothesis concerning the value of the statistic (here, the coefficient of resemblance) whereas the coefficient itself measures the intensity of the relationship between descriptors. Table 7.5 summarizes the use of R-mode coefficients with ecological variables.

1 — Descriptors other than species abundances

Why the resemblance between species abundance descriptors must be measured using special coefficients is explained at the beginning of the next Subsection. Measures of resemblance in the present Subsection are used for comparing descriptors for which double-zeros provide unequivocal information (for a discussion of double-zeros in ecology, see the beginning of Section 7.3).

The resemblance between *quantitative descriptors* can be computed using parametric measures of dependence, i.e. measures based on parameters of the frequency distributions of the descriptors. These measures are the covariance and the Pearson correlation coefficient; they have been described in Chapter 5. They are only adapted to descriptors whose relationships are *linear.*

The *covariance* s_{jk} between descriptors j and k is computed from centred variables $(y_{ij} - \bar{y}_j)$ and $(y_{ik} - \bar{y}_k)$ (eq. 4.4). The range of values of the covariance has no *a priori* upper or lower limits. The variances and covariances among a group of descriptors form their dispersion matrix **S** (eq. 4.6).

Pearson's *correlation coefficient* r_{jk} is their covariance of descriptors j and k computed from standardized variables (eqs. 1.12 and 4.7). The coefficients of correlations among a group of descriptors form the correlation matrix **R** (eq. 4.8). Correlation coefficients range in values between –1 and +1. The significance of individual coefficients (the null hypothesis being generally H_0: $r = 0$) is tested using eq. 4.13, whereas eq. 4.14 is used to test the complete independence among all descriptors.

Q-mode correlation

Some authors have used Pearson's r for Q-mode analyses, after interchanging the positions of objects and descriptors in the data matrix. Lefebvre (1980) calls this Q measure the *resemblance coefficient*. There are at least five objections to this:

- In the R mode, Pearson's r is a dimensionless coefficient (Chapter 3). When the descriptors are not dimensionally homogeneous, the Q-mode correlation coefficient, which combines all descriptors, has complex dimensions that cannot be interpreted.

- In most cases, one may arbitrarily rescale quantitative descriptors (e.g. multiplying one by 100 and dividing another by 10). In the R mode, the value of r remains unchanged after rescaling, whereas doing so in the Q mode may change the value of resemblance between objects in unpredictable and nonmonotonic fashion.

• In order to avoid the two previous problems, it has been suggested to standardize the descriptors (eq. 1.12) before computing correlations in the Q mode. Consider two objects $\mathbf{x}_1$ and $\mathbf{x}_2$: their similarity should be independent of the other objects in the study; removing objects from the data set should not change it. Any change in object composition of the data set changes the standardized variables, however, and so it affects the value of the correlation computed between $\mathbf{x}_1$ and $\mathbf{x}_2$. Hence, standardization does not solve the problems.

• Even with dimensionally homogeneous data (e.g. counts of different species), the second objection still holds. In addition, in the R mode, the central limit theorem (Section 4.3) predicts that, as the number of objects increases, the means, variances, and covariances (or correlations) converge towards their values in the statistical population. In the Q mode, on the contrary, adding new descriptors (their positions have been interchanged with that of objects in the data matrix) causes major variations in the resemblance coefficient if these additional descriptors are not perfectly correlated to those already present.

• If correlation coefficients could be used as a general measure of resemblance in the Q mode, they should be applicable in particular to the simple case of the description of the proximities among sites, computed from their geographic coordinates X and Y on a map; the correlations obtained from this calculation should reflect in some way the distances among the sites. This is not the case: correlation coefficients computed among sites from their geographic coordinates are all +1 or –1. As an exercise, readers are encouraged to compute an example of their own.

It follows that the measures designed for R-mode analysis should not be used in the Q mode. Sections 7.3 and 7.4 describe several Q-mode coefficients, whose properties and dimensions are already known or easy to determine.

The resemblance between *semiquantitative descriptors* and, more generally between any pair of *ordered* descriptors whose relationship is *monotonic* may be determined using nonparametric measures of dependence (Chapter 5). Since *quantitative* descriptors are ordered, nonparametric coefficients may be used to measure their dependence, as long as they are monotonically related.

Two *nonparametric correlation coefficients* have been described in Section 5.3: Spearman's r and Kendall's τ (tau). In Spearman's r (eq. 5.3), quantitative values are replaced by ranks before computing Pearson's r formula. Kendall's τ (eqs. 5.5 to 5.7) measures the resemblance in a way that is quite different from Pearson's r. Values of Spearman's r and Kendall's τ range between –1 and +1. The significance of individual coefficients (the null hypothesis being generally H_0: $r = 0$) is tested using eq. 5.4 (Spearman's r) or 5.8 (Kendall's τ).

As with Pearson's r above, rank correlation coefficients should not be used in the Q mode. Indeed, even if quantitative descriptors are standardized, the same problem arises as with Pearson's r, i.e. the Q measure for a pair of objects is a function of all objects in the data set. In addition, in most biological sampling units, several species are represented by small numbers of individuals. Because these small numbers are subject to large stochastic variation, the ranks of the corresponding species are uncertain in the reference ecosystem. As a consequence, rank correlations between sites would be subject to important random variation because their values would be based on large numbers of uncertain ranks. This is equivalent to giving preponderant weight to the many poorly sampled species.

The importance of *qualitative descriptors* in ecological research is discussed in Section 6.0. The measurement of resemblance between pairs of such descriptors is based on two-way contingency tables (Sections 6.2 and 6.3), whose analysis is generally conducted using X^2 (chi-square) statistics. Contingency table analysis is also the major approach available for measuring the dependence between *quantitative* or *semiquantitative* ordered descriptors that are not monotonically related. The minimum value of X^2 is zero, but it has no *a priori* upper limit. Its formulae (eqs. 6.5 and 6.6) and test of significance are explained in Section 6.2. X^2 may be transformed into contingency coefficients (eqs. 6.19 and 6.20), whose values range between 0 and +1.

Two-way contingency tables may also be analysed using measurements derived from information theory. In this case, the amounts of information (B) shared by two descriptors j and k and exclusive to each one (A and C) are first computed. These quantities may be combined into similarity measures, such as $S(j, k) = B/(A + B + C)$ (eq. 6.15; see also eqs. 6.17 and 6.18), or into distance coefficients such as $D(j, k) = (A + C)/(A + B + C)$ (eq. 6.16). The analysis of multiway contingency tables (Section 6.3) is based on the Wilks X^2 statistic (eq. 6.6).

A *qualitative* descriptor (including a *classification*; Chapter 8) can be compared to a *quantitative* descriptor using *one-way analysis of variance* (one-way ANOVA; Table 5.2 and accompanying text). The classification criterion for this ANOVA is the qualitative descriptor. As long as the assumptions underlying analysis of variance are met (i.e. normality of within-group distributions and homoscedasticity, Box 1.4), the significance of the relationship between the descriptors may be tested. If the quantitative descriptor does not obey these assumptions or the comparison is between a *quantitative* and a *semiquantitative* descriptor, *nonparametric one-way analysis of variance* (Kruskal-Wallis H test; Table 5.2) is used instead of parametric ANOVA.

2 — *Species abundances: biological associations*

Analysing species abundance descriptors causes the same problem in the R as in the Q mode, i.e. what to do with double-zeros? The problem often occurs with ecological data because biological assemblages generally consist of a small number of dominant species and a large number of rare species (Section 6.5). Since biological association studies are generally based on all species in a given taxocene (term defined in Section 6.5), the data matrix most often contains a large number of zeros. As a consequence, measurements of dependence among species, upon which rests the identification of biological associations, is based on many pairs of zeros for which there is no clear biological interpretation. The dubious interpretation of double zeros has been discussed at the beginning of Section 7.3.

The present Subsection explains how to measure the degree of association, or dependence, between *species descriptors*. How these measures can be used for the identification of biological associations is discussed in Section 8.9, where an operational concept of *biological association* is defined. Some of these measures have been discussed by Southwood (1966), in reference to the study of insect populations.

Since species abundances are generally quantitative, it is appropriate to first examine the parametric measures of dependence in view of identifying species associations. If the abundance data are first normalized, using for example transformation $y' = \log(y + 1)$ (eq. 1.12), it may seem appropriate to measure the association between species using the covariance (eq. 4.4) or Pearson's linear correlation coefficient (eq. 4.7). The ecological literature on species associations contains many examples of this approach and of analyses based on dispersion or correlation matrices (e.g. principal component analysis). This is incorrect because covariances or Pearson's correlations use zeros as any another quantitative value. Normalizing the data does not minimize the basic problem, which is that the presence of many double-zeros, without any clear biological meaning, significantly distorts the dispersion ellipse of the sites with respect to the "species" axes. In order to minimize this effect, several options are available to ecologists who only have access to the coefficients available in regular computer packages:

(1) Eliminate from the study the less frequent species, so as to reduce the number of double-zeros. The absence of these species from the following analyses is of little consequence to the study of ecologically meaningful species associations (see Subsection 1.6.1).

(2) Eliminate all zeros from the comparisons, by declaring that zeros are missing values (see Subsection 1.6.2).

(3) Eliminate double-zeros only from the computations of the covariance or correlation matrix; this must generally be programmed separately. The resulting dispersion (or correlation) matrix may then be fed to a computer package for the remaining operations, such as the computation of principal components (Section 9.1).

Note that correspondence analysis (Section 9.4) is a form of principal component analysis which preserves the χ^2 distance (D_{16}; eq. 7.54) instead of the Euclidean distance (D_1; eq. 7.34). Because D_{16} excludes double-zeros whereas D_1 includes them, correspondence analysis is better suited than principal component analysis to the study of species associations.

Fager (1957) pointed out that associations must group species which are almost always part of one another's biological environment. Because the covariance or correlation coefficients measure the (linear) relationship between fluctuations in the abundances of two species, it follows that, if two species do not covary in a linear way (even though they are always present together), Pearson's correlation coefficient will not show any relationship (i.e. r not significantly different from zero). This problem with covariance or correlation coefficients calls for an operational definition of associations (see Section 8.9) and stresses again the fact that low correlation values do not have an unequivocal meaning (see Section 4.2). A low correlation does not indicate that two species are unrelated; this is a major drawback for the identification of associations (Section 8.9). The same applies to nonparametric correlation coefficients which are used with classes of relative species abundances.

Moving away from correlation coefficients, another approach consists in applying Goodall's probabilistic coefficient (S_{23}; eqs. 7.31 or 7.32) in the R mode to species abundances, as was done by Legendre (1973). This probabilistic coefficient allows one to set an "objective" limit to species associations; indeed, one may then use a probabilistic definition of an association, such as: "all species that are related at a probability level $p \geq 0.95$ are members of the association". With this coefficient, it is also possible to use a probabilistic clustering model (Section 8.9). In the R mode, Goodall's coefficient has the following meaning: given p species and n sites, the similarity of a pair of species is defined as the complement $(1 - p)$ of the probability that any pair of species chosen at random would be as similar as, or more similar than the two species under consideration. Goodall's similarity coefficient is computed as in Subsection 7.3.5, with species interchanged with sites. At step (a), if all species vectors have been normalized (for example using the transformation $y' = \log(y + 1)$; eq. 1.12), the partial similarity of Gower's coefficient S_{19} (eq.7.26)

Probabilistic association

$$s_{i12} = 1 - [|y_{i1} - y_{i2}| / R_i]$$

may be used to describe the similarity between species y_1 and y_2 at site i. R_i is the range of variation of the normalized species abundances at site i. It is useless to standardize the data in this case since differences between sites are scaled with respect to R_i for each species.

Biological associations may also be defined on the basis of the *co-occurrence of species* instead of the relationships between fluctuations in abundances. Indeed, the definition of association may refer to the sole concept of co-occurrence, as was the case with Fager's approach above. This is because quantitative data may not accurately reflect the proportions of the various species in the environment, because of problems with sampling, preservation, identification or counting, or simply because the concept of individuality is not clear (e.g. plants multiplying through rhizomes; colonial algae or animals) or the comparison of individuals does not make ecological sense (e.g. the baobab and the surrounding herbaceous plants). Finally, spatio-temporal aggregration of organisms may obscure the true quantitative relationships among species, as in the case of plankton patches or reindeer herds. It follows that associations are often defined on the sole basis of the presence or absence of species.

There are many approaches in the literature for measuring the association between species based on binary data. These coefficients are based on the following 2×2 frequency table:

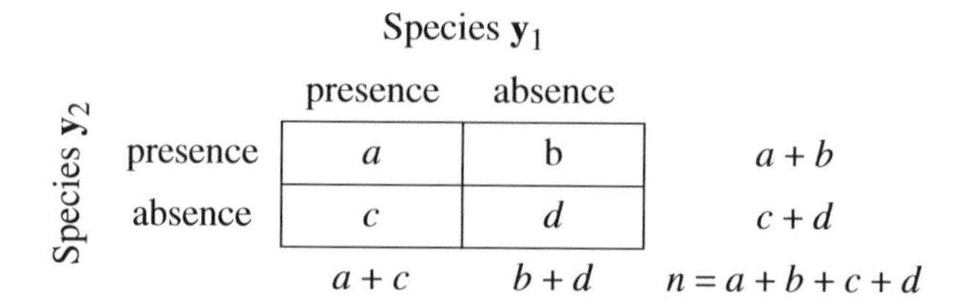

		Species $\mathbf{y}_1$ presence	Species $\mathbf{y}_1$ absence	
Species $\mathbf{y}_2$	presence	a	b	$a+b$
	absence	c	d	$c+d$
		$a+c$	$b+d$	$n=a+b+c+d$

where a and d are numbers of sites in which the two species are present and absent, respectively, whereas b and c are the numbers of sites in which only one of the two species is present; n is the total number of sites. The measures of association between species always exclude the number of double absences, d.

Among the many binary coefficients, described in Section 7.3, that exclude double-zeros, at least two have been used for assessing association between species. Jaccard's *coefficient of community* (eq. 7.10) has been used by Reyssac & Roux (1972):

$$S_7(\mathbf{y}_1, \mathbf{y}_2) = \frac{a}{a+b+c}$$

The corresponding distance has been used by Thorrington-Smith (1971) for the same purpose:

$$D = 1 - S_7(\mathbf{y}_1, \mathbf{y}_2) = \frac{b+c}{a+b+c} \qquad \textbf{(7.58)}$$

The coincidence index (eq. 7.11)

$$S_8(\mathbf{y}_1, \mathbf{y}_2) = \frac{2a}{2a+b+c}$$

was originally defined for studying species associations (Dice, 1945).

A more elaborate coefficient was proposed by Fager & McGowan (1963), to replace Fager's probabilistic coefficient (1957) proposed earlier:

$$S_{24}(\mathbf{y}_1, \mathbf{y}_2) = \frac{a}{\sqrt{(a+b)\,(a+c)}} - \frac{1}{2\sqrt{a+c}} \qquad (c \geq b) \qquad \textbf{(7.59)}$$

The first part of the coefficient is the same as S_{14}, i.e. the geometric mean of the proportions of co-occurrence for each of the two species; the second part is a correction for small sample size.

As a probabilistic coefficient for presence-absence data, Krylov (1968) proposed to use the probability associated with the chi-square statistic of the above 2×2 frequency table to test the null hypothesis that two species are distributed independently of each other among the various sites. Rejecting H_0 leads gives support to the alternative hypothesis of association between the two species. In the case of a 2×2 contingency table, and using Yate's correction factor for small samples, the X^2 formula is:

$$X^2 = \frac{n\left[|ad - bc| - (n/2)\right]^2}{(a+b)\,(c+d)\,(a+c)\,(b+d)} \tag{7.60}$$

The number of degrees of freedom for the test of significance is ν = (no. rows – 1) × (no. columns – 1) = 1. The X^2 statistic could also be tested by permutation (Section 1.2). Given that associations should be based on positive relationships between pairs of species (negative relationships reflecting competition), Krylov proposed to set $S(\mathbf{y}_1, \mathbf{y}_2) = 0$ when the expected value of co-occurrence, $E = (a + b)\,(a + c) / n$, is larger than or equal to the observed frequency ($E \geq a$). Following the test, two species are considered associated if the probability (p) associated to their X^2 value is smaller than a pre-established significance threshold, for example $\alpha = 0.05$. The similarity measure between species is the complement of this probability:

$$S_{25}(\mathbf{y}_1, \mathbf{y}_2) = 1 - \mathrm{p}(X^2), \text{ with } \nu = 1, \qquad \text{when } (a + b)(a + c) / n < a$$

$$S_{25}(\mathbf{y}_1, \mathbf{y}_2) = 0 \qquad \text{when } (a + b)(a + c) / n \geq a \tag{7.61}$$

When the number of sites n is smaller than 20 or a, b, c or d are smaller than 5, Fisher's exact probability formula should be used instead of X^2. This formula can be found in most textbooks of statistics

The same formula can be derived from Pearson's ϕ (phi) (eq. 7.9), given that $X^2 = n\phi^2$. Pearson's ϕ is also called the *point correlation coefficient* because it is the general correlation coefficient (eq. 5.1) computed from presence-absence values.

7.6 Choice of a coefficient

Criteria for choosing a coefficient are summarized in Tables 7.3 to 7.5. In these Tables, the coefficients are identified by the names and numbers used in Sections 7.3 to 7.5. The three Tables distinguish between coefficients appropriate for "species" descriptors and those for other types of descriptors.

Levels 5, 7 and 8 of Table 7.3 require some explanation. Coefficients found in these levels are classified with respect to two criteria, i.e. (a) standardization (or not) of each object-vector prior to the comparison and (b) relative importance given by the coefficient to the abundant or rare species. This defines various types of coefficients.

Type 1 coefficients. Consider two objects, each represented by a vector of species abundances, to be compared using a Q-mode measure. With type 1 coefficients, if there is a given difference between sites for some abundant species and the same difference for a rare species, the two species contribute equally to the similarity or distance between sites. A small numerical example illustrates this property for the percentage difference (D_{14}), which is the complement of Steinhaus' similarity (S_{17}):

Species:	$\mathbf{y}_1$	$\mathbf{y}_2$	$\mathbf{y}_3$
Site $\mathbf{x}_1$	100	40	20
Site $\mathbf{x}_2$	90	30	10
$\lvert y_{1j} - y_{2j} \rvert$	10	10	10
$(y_{1j} + y_{2j})$	190	70	30

Using eq. 7.57 shows that each of the three species contributes 10/290 to the total distance between the two sites. With some coefficients (D_3, D_4, D_9), the standardization of the site-vectors, which is automatically done prior to the computation of the coefficient, may make the result unclear as to the importance given to each species. With these coefficients, the property of "equal contribution" is found only when the two site-vectors are equally important, the importance being measured in different ways depending on the coefficient (see the note at the foot of Table 7.3).

Type 2a coefficients. With coefficients of this type, a difference between values for an abundant species contributes less to the distance (and, thus, more to the similarity) than the same difference for a rare species. The *Canberra metric* (D_{10}) belongs to this type. For the above numerical example, calculation of D_{10} (eq. 7.49) shows that species $\mathbf{y}_1$, which is the most abundant, contributes 10/190 to the distance, $\mathbf{y}_2$ contributes 10/70, whereas the contribution of $\mathbf{y}_1$, which is the rarest species, is the largest of the three (10/30). The total distance is $D_{10} = 0.529$. The *coefficient of divergence* (D_{11}; eq. 7.51) also belongs to this type.

Type 2b coefficients. Coefficients of this type behave similarly to the previous ones, except that the importance of each species is calculated with respect to the whole data set, instead of the two site-vectors being compared. The χ^2 *metric* (D_{15}) is representative of this. In eq. 7.53 and accompanying example, the squared difference between conditional probabilities, for a given species, is divided by y_{+j} which is the total number of individuals belonging to this species in all sites. If this number is large, it reduces the contribution of the species to the total distance between two rows (sites) more than would happen in the case of a rarer species. *Gower's coefficient* (S_{19}; eq 7.26) has the same behaviour (unless special weights w_{12j} are used for some species), since the importance of each species is determined from its range of variation through all sites. The coefficient of Legendre & Chodorowski (S_{20}; eq 7.27) also belongs to this type when parameter k in the partial similarity function s_{12j} for each species is made proportional to its range of variation through all sites.

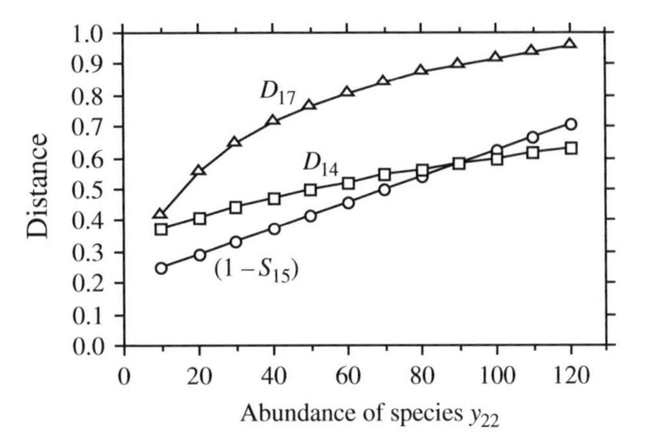

Figure 7.6 Results from an *ordered comparison case series* (OCCAS) where the abundance of species y_{22} varies from 10 to 120 by steps of 10. The values taken by coefficients $(1 - S_{15})$, D_{14}, and D_{17} are shown.

Legendre *et al.* (1985) suggested that it is more informative to compare dominant or well-represented species than rare taxa, because the latter are generally not well sampled. This provides an approach for choosing a coefficient. In immature communities, most of the species are represented by small numbers of individuals, so that only a few species are well sampled, whereas, in mature communities, several species exhibit intermediate or high frequencies of occurrence. When calculating similarities between species from immature communities, a reasonable approach may thus be to give more weight to the few well-sampled species (type 2 coefficients) whereas, for sites from mature communities, type 1 coefficients may be more appropriate.

OCCAS

Another way of choosing a resemblance coefficient is to construct artificial data representing contrasting situations that the similarity or distance measure should be able to differentiate. Computing several candidate coefficients for the test data will indicate which coefficient is the most appropriate for data of the same type. In this spirit, Hajdu (1981) constructed series of test cases, that he called *ordered comparison case series* (OCCAS), corresponding to linear changes in the abundances of two species along different types of simulated environmental gradients. The results are distances between sites, computed using different coefficients, for linearly changing species composition.

To illustrate the method, consider one of Hajdu's OCCAS with two species. For these species, site 1 had frequencies $y_{11} = 100$ and $y_{12} = 0$; site 2 had frequency $y_{21} = 50$ whereas y_{22} varied from 10 to 120. Figure 7.6 shows the results for three coefficients: $(1 - S_{15})$ has a completely linear behaviour across the values of y_{22}, D_{14} is not quite linear, and D_{17} is strongly curvilinear.

Non-linearity

Resolution

An ideal coefficient should change linearly when plotted against a series of test cases corresponding to a linear change in species composition, as simulated in OCCAS runs. Hajdu (1981) proposed a measure of *non-linearity*, defined as the *standard deviation of the changes* in values of distance between adjacent test cases along the series. A good distance coefficient should also change substantially along the series and reach its maximum when the species composition becomes maximum. *Resolution* was defined as the *mean change* occurring in distances between adjacent test cases along the series. High linearity is desirable in ordination methods whereas high resolution is desirable in cluster analysis. The ratio of non-linearity over resolution defines a coefficient of variation which should be small for a "good" overall resemblance coefficient.

Resolutions are only comparable among coefficients that are [0, 1]-bounded; as a consequence, this measure should not be used to compare coefficients, such as D_1, D_2, D_{10}, D_{15}, D_{16}, and D_{17}, which do not have an upper bound. Non-linearity near 0 is always a good property, but, again, higher values are only comparable for coefficients that are [0, 1]-bounded. Coefficients of variation are comparable because the scale of variation of each specific coefficient is taken into account in the calculation.

Gower & Legendre (1986) used Hajdu's OCCAS to study the behaviour of several similarity and distance coefficients and to make recommendations about their use. They studied 15 coefficients for binary data (all of which are described in the present Chapter) and 10 coefficients for quantitative data (5 of them are described here). Among the binary coefficients, S_{12} (eq. 7.15) and the coefficient of Yule (eq. 7.8) were strongly non-linear and should be avoided; all the other coefficients (S_1, S_2, S_3, S_5, S_6, S_7, S_8, S_{10}, S_{13}, S_{14}, as well as eqs. 7.7 and 7.9) behaved well. The coefficients for quantitative data included in that study were S_{15}, $D_{14} = 1 - S_{17}$, D_2, D_{10} and D_{11}. Coefficients D_2 and S_{15}, which are adapted to physical descriptors (Table 7.4), behaved well. D_2 is a standardized form of the Euclidean distance D_1; they both have the same behaviour. All coefficients adapted to species abundance data (Table 7.3) that were included in the study (D_{10}, D_{11}, D_{14}) behaved well and are recommended. Coefficients S_{15} and D_{10} had perfect linearity in all specific OCCAS runs; they are thus the best of their kinds *for linear ordination*.

A later analysis of coefficient $D = \sqrt{D_{14}} = \sqrt{1 - S_{17}}$ showed that its non-linearity was very similar to $D_{14} = 1 - S_{17}$; the resolution of $\sqrt{D_{14}}$ was slightly lower than that of D_{14}. Both forms are thus equally suitable for ordination whereas D_{14} may be slightly preferable for clustering purposes.The square root transformation of D_{14}, used in the latter part of Numerical example 2 in Section 9.2, is one of the ways to avoid negative eigenvalues in principal coordinate ordination. Another comparative analysis involving the chi-square metric and related forms (D_{15}, D_{16}, and D_{17}) showed that the best of this group *for linear ordination* is the Hellinger distance (D_{17}) which has the lowest coefficient of variation (best compromise between linearity and resolution), despite the fact that it is strongly non-linear. Other properties of resemblance coefficients have been investigated by Bloom (1981), Wolda (1981) and Hubálek (1982).

Table 7.3 Choice of an association measure among objects (Q mode), to be used with species descriptors (asymmetrical coefficients). For explanation of levels 5, 7 and 8, see the accompanying text.

1) Descriptors: presence-absence or ordered classes on a scale of relative abundances (no partial similarities computed between classes) **see 2**

- 2) Metric coefficients: *coefficient of community* (S_7) and variants (S_{10}, S_{11})
- 2) Semimetric coefficients: variants of the coef. community (S_8, S_9, S_{13}, S_{14})
- 2) Nonmetric coefficient: Kulczynski (S_{12}) (non-linear: not recommended)
- 2) Probabilistic coefficient: S_{27}

1) Descriptors: quantitative or semiquantitative (states defined in such a way that partial similarities can be computed between them) **see 3**

- 3) Data: raw abundances **see 4**
  - 4) Coefficients without associated probability levels **see 5**
    - 5) No standardization by object; the same difference for either abundant or rare species, contributes equally to the similarity between sites: *coefficients of Steinhaus* (S_{17}) and *Kulczynski* (S_{18})
    - 5) Standardization by object-vector; differences for abundant species (in the whole data set) contribute more than differences between rare species to the similarity (less to the distance) between sites: χ^2 *similarity* (S_{21}), χ^2 *metric* (D_{15}), χ^2 *dist.* (D_{16}), *Hellinger dist.* (D_{17})
  - 4) Probabilistic coefficient: probabilistic χ^2 similarity (S_{22})
- 3) Data: normalized abundances (or, at least, distributions not skewed) or classes on a scale of relative abundances (e.g. 0 to 5, 0 to 7). [Normalization, Subsection 1.5.6, is useful when abundances cover several orders of magnitude.] **see 6**
  - 6) Coefficients without associated probability levels **see 7**
    - 7) No standardization by object **see 8**
      - 8) The same difference for either abundant or rare species, contributes equally to the similarity between sites: *coefficients of Steinhaus* (S_{17}) and *Kulczynski* (S_{18}), *mean character difference* (D_8), *percentage difference* (D_{14})
      - 8) Differences for abundant species (for the two sites under consideration) contribute more than differences between rare species to the similarity (less to the distance) between sites: *Canberra metric* (D_{10}), *coefficient of divergence* (D_{11}). Both have low resolution: not recommended for clustering
      - 8) Differences for abundant species (in the whole data set) contribute more than differences between rare species to the similarity (less to the distance) between sites: *asymmetrical Gower coefficient* (S_{19}), *coefficient of Legendre & Chodorowski* (S_{20})
    - 7) Standardization by object-vector; if objects are of equal importance*, same contributions for abundant or rare species to the similarity between sites: *chord distance* (D_3), *geodesic metric* (D_4), *complement of index of association* (D_9)
  - 6) Probabilistic coefficient: *Goodall coefficient* (S_{23})

* D_3 and D_4: importance is quantified as the length of vector $\sqrt{\sum_i y_{ij}^2}$

D_9: importance is quantified as the total number of individuals in vector $\sum_i y_{ij}$

Table 7.4 Choice of an association measure among objects (Q mode), to be used with chemical, geological physical, etc. descriptors (symmetrical coefficients, using double-zeros).

1) Association measured between individual objects	**see 2**
2) Descriptors: presence-absence or multistate (no partial similarities computed between states)	**see 3**
3) Metric coefficients: *simple matching* (S_1) and derived coefficients (S_2, S_6)	
3) Semimetric coefficients: S_3, S_5	
3) Nonmetric coefficient: S_4	
2) Descriptors: multistate (states defined in such a way that partial similarities can be computed between them)	**see 4**
4) Descriptors: quantitative and dimensionally homogeneous	**see 5**
5) Differences enhanced by squaring: *Euclidean distance* (D_1) and *average distance* (D_2)	
5) Differences mitigated: *Manhattan metric* (D_7), *mean character difference* (D_8)	
4) Descriptors: not dimensionally homogeneous; weights (equal or not, according to values w_j used) given to each descriptor in the computation of association measures	**see 6**
6) Descriptors are qualitative (no partial similarities computed between states) and quantitative (partial similarities based on the range of variation of each descriptor): *symmetrical Gower coefficient* (S_{15})	
6) Descriptors are qualitative (possibility of using matrices of partial similarities between states) and semiquantitative or quantitative (partial similarity function for each descriptor): *coefficient of Estabrook & Rogers* (S_{16})	
1) Association measured between groups of objects	
7) Removing the effect of correlations among descriptors: *Mahalanobis generalized distance* (D_5)	
7) Not removing the effect of correlations among descriptors: *coefficient of racial likeness* (D_{12})	

Table 7.5 Choice of a dependence measure among descriptors (R mode).

1) Descriptors: species abundances — see 2
   2) Descriptors: presence-absence — see 3
      3) Coefficients without associated probability levels:
      3) Probabilistic coefficient:
   2) Descriptors: multistate
      4) Data are raw abundances: χ^2 *similarity* (S_{21}), χ^2 *metric* (D_{15}), χ^2 *distance* (D_{16}), *Hellinger distance* (D_{17}), *Spearman r, Kendall* τ — see 4
      4) Data are normalized abundances — see 5
         5) Coefficients without associated probability levels: *covariance* or *Pearson r,* after elimination of as much double-zeros as possible, *Spearman r, Kendall* τ
         5) Probabilistic coefficients: *probabilities associated to Pearson r, Spearman r* or *Kendall* τ, *Goodall coefficient* (S_{23})
1) Descriptors: chemical, geological, physical, etc. — see 6
   6) Coefficients without associated probability levels — see 7
      7) Descriptors are quantitative and linearly related: *covariance, Pearson r*
      7) Descriptors are ordered and monotonically related: *Spearman r, Kendall* τ
      7) Descriptors are qualitative or ordered but not monotonically related: χ^2, *reciprocal information coefficient, symmetric uncertainty coefficient*
   6) Probabilistic coefficients — see 8
      8) Descriptors are quantitative and linearly related: *probabilities associated to Pearson r*
      8) Descriptors are ordered and monotonically related: *probabilities associated to Spearman r* and *Kendall* τ
      8) Descriptors are ordered and monotonically related: *probabilities associated to* χ^2

7.7 Computer programs and packages

Only the largest general-purpose commercial statistical packages, such as SAS, SPSS, SYSTAT, JMP, or STATISTICA, offer clustering among their methods for data analysis, as well as capacities to compute some resemblance coefficients. The smaller commercial packages offer no such facility. Among the Q-mode coefficients, one always finds the Euclidean distance; the squared Euclidean, Manhattan, Chebychev and Minkowski distances may also be found, as well as the simple matching coefficient for multistate nominal data. For R-mode analyses, one finds Pearson's *r* in most programs, or related measures such as the cosine of the angle between variables, dot product, or covariance; nonparametric correlation coefficients may also be found.

At least four programs strongly emphasize resemblance coefficients and clustering methods. They are: NTSYS-PC[*], developed by F. J. Rohlf, originally for numerical taxonomy studies; CLUSTAN[†], developed by D. Wishart; PATN[‡], developed by Lee Belbin; and *The R Package*[**] (Legendre & Vaudor, 1991; see Table 13.4) which offers all the coefficients described in the present Chapter.

[*] How to obtain NTSYS-PC is described in Table 13.4.

[†] The CLUSTAN package was written by David Wishart. It may be ordered from CLUSTAN Limited, 16 Kingsburgh Road, Edinburgh EH12 6DZ, Scotland. See also the WWWeb site <http://www.clustan.com/>.

[‡] PATN, developed by Lee Belbin, is available from CSIRO Division of Wildlife and Ecology, P.O. Box 84, Lyneham ACT 2614, Australia; technical information is obtained from Fiona Vogt at the same address.
See also the WWWeb site <http://www.dwe.csiro.au/local/research/patn/patn0.htm>.

[**] How to obtain *The R Package* is described in Table 13.4.

Chapter 8 Cluster analysis

8.0 A search for discontinuities

Humans have always tried to classify the animate and inanimate objects that surround them. Classifying objects into collective categories is a prerequisite to naming them. It requires the recognition of discontinuous subsets in an environment which is sometimes discrete, but most often continuous.

To cluster is to recognize that objects are sufficiently similar to be put in the same group and to also identify distinctions or separations between groups. Measures of similarity between objects (Q mode) or descriptors (R mode) have been discussed in Chapter 7. The present Chapter considers the different criteria that may be used to decide whether objects are similar enough to be allocated to a group; it also shows that different clustering strategies correspond to different definitions of a what a cluster is.

Few ecological theories predict the existence of discontinuities in nature. Evolutionary theory tells taxonomists that discontinuities exist between species, which are the basic units of evolution, as a result of reproductive barriers; taxonomists use classification methods to reveal these discontinuities. For the opposite reason, taxonomists are not surprised to find continuous differentiation at the sub-species level. In contrast, the world that ecologists try to understand is most often a continuum. In numerical ecology, methods used to identify clusters must therefore be more contrasting than in numerical taxonomy. Ecologists who have applied taxonomic clustering methods directly to data, without first considering the theoretical applicability of such methods, have often obtained disappointing results. This has led many ecologists to abandon clustering methods altogether, hence neglecting the rich potential of similarity measures, described in Chapter 7, and to rely instead on factor analysis and other ordination methods. These are not always adapted to ecological data and, in any case, they don't aim at bringing out partitions, but gradients.

Given a sufficiently large group of objects, ecological clustering methods should be able to recognize clusters of similar objects while ignoring the few intermediates which often persist between clusters. Indeed, one cannot expect to find discontinuities when clustering sampling sites unless the physical environment is itself discontinuous, or unless sampling occurred at opposite ends of a gradient, instead of within the gradient (Whittaker, 1962: 88). Similarly, when looking for associations of species, small groups of densely associated species are usually found, with the other species gravitating around one or more of the association nuclei.

Typology

The result of clustering ecological objects sampled from a continuum is often called a *typology* (i.e. a system of types). In such a case, the purpose of clustering is to identify various *object types* which may be used to describe the structure of the continuum; it is thus immaterial to wonder whether these clusters are "natural" or unique.

For readers with no practical experience in clustering, Section 8.2 provides a detailed account of single linkage clustering, which is simple to understand and is used to introduce the principles of clustering. The review of other methods includes a survey of the main dichotomies among existing methods (Section 8.4), followed by a discussion of the most widely available methods of interest to ecologists (8.5, 8.7 and 8.8). Theoretical aspects are discussed in Sections 8.3 and 8.6. Section 8.9 deals with clustering algorithms useful in identifying biological associations, whereas Section 8.10 gives an overview of seriation, a method useful to cluster non-symmetric resemblance matrices. A review of clustering statistics, methods of cluster validation, and graphical representations, completes the chapter (Sections 8.11 to 8.13). The relationships between clustering and other steps of data analysis are depicted in Fig. 10.3.

Several, but not all statistical packages offer clustering capabilities. All packages with clustering procedures offer at least a Lance & Williams algorithm capable of carrying out the clustering methods listed in Table 8.8. Many also have a K-means partitioning algorithm. Few offer proportional-link linkage or additional forms of clustering. Some methods are available in specialized packages only: clustering with constraints of temporal (Section 12.6) or spatial contiguity (Section 13.3); fuzzy clustering (e.g. Bezdek, 1987); or clustering by neural network algorithms (e.g. Fausett, 1994). The main difference among packages lies in the list of resemblance coefficients available (Section 7.7). Ecologists should consider this point when selecting a clustering package.

While most packages nowadays illustrate clustering results in the form of dendrograms, some programs use "skyline plots", which are also called "trees" or "icicle plots". These plots contain the same information as dendrograms but are rather odd to read and interpret. The way to transform a skyline plot into a dendrogram is explained in Section 8.13.

Despite the versatility of clustering methods, one should remember that not all problems are clustering problems. Before engaging in clustering, one should be able to justify why one believes that discontinuities exist in the data; or else, explain that one has a practical need to divide a continuous swarm of objects into groups.

8.1 Definitions

Clustering
Partition

Clustering is an operation of multidimensional analysis which consists in partitioning the collection of objects (or descriptors) in the study. A *partition* (Table 8.1) is a division of a set (collection) into subsets, such that each object or descriptor belongs to one and only one subset for that partition (Legendre & Rogers, 1972). The classification of objects (or descriptors) that results from clustering may include a single partition, or several hierarchically nested partitions of the objects (or descriptors), depending on the clustering model that has been selected.

From this definition, it follows that the subsets of any level of partition form a series of *mutually exclusive* cells, among which the objects (or descriptors) are distributed. This definition *a priori* excludes all classification models in which classes have elements in common (overlapping clusters) or in which objects have fractional degrees of membership in different clusters (fuzzy partitions: Bezdek, 1987); these models have not been used in ecology yet. This limitation is such that a "hard" or "crisp" (*versus* fuzzy) partition has the same definition as a descriptor (Section 1.4). Each object is characterized by a state (its cluster) of the classification and it belongs to only one of the clusters. This property will be useful for the interpretation of classifications (Chapter 10), since any partition may be considered as a qualitative descriptor and compared as such to any other descriptor. A clustering of objects defined in this way imposes a discontinuous structure onto the data set, even if the objects have originally been sampled from a continuum. This structure results from the grouping into subsets of objects that are sufficiently similar, given the variables considered, and from the observation that different subsets possess unique recognizable characteristics.

Table 8.1 Example of hierarchically nested partitions of a group of objects (e.g. sampling sites). The first partition separates the objects by the environment to which they belong. The second partition, hierarchically nested into the first, recognizes clusters of sites in each of the two environments.

Partition 1	Partition 2	Sampling sites
	Cluster 1	7, 12
Observations in environment A	Cluster 2	3, 5, 11
	Cluster 3	1, 2, 6
	Cluster 4	4, 9
Observations in environment B	Cluster 5	8, 10, 13, 14

Clustering has been part of ecological tradition for a long time. It goes back to the Polish ecologist Kulczynski (1928), who needed to cluster ecological observations; he developed a method quite remote from the clustering algorithms discussed in the paragraphs to come. His technique, called seriation, consists in permuting the rows and columns of an association matrix in such a way as to maximize the values on the diagonal. The method is still used in phytosociology, anthropology, social sciences, and other fields; it is described in Section 8.10 where an analytical solution to the problem is presented.

Most clustering (this Chapter) and ordination (Chapter 9) methods proceed from association matrices (Chapter 7). Distinguishing between clustering and ordination is somewhat recent. While ordination in reduced space goes back to Spearman (factor analysis: 1904), most modern clustering methods have only been developed since the era of second-generation computers. The first programmed method, developed for biological purposes, goes back to 1958 (Sokal & Michener)*. Before that, one simply plotted the objects in a scatter diagram with respect to a few variables or principal axes; clusters were then delineated manually (Fig. 8.1) following a method which today would be called centroid (Section 8.4), based upon the Euclidean distances among points. This empirical clustering method still remains the best approach when the number of variables is small and the structure to be delineated is not obscured by intermediate objects between clusters.

Clustering is a family of techniques which is undergoing rapid development. In their report on the literature they reviewed, Blashfield & Aldenderfer (1978) mentioned that they found 25 papers in 1964 that contained references to the basic texts on clustering; they found 136 papers in 1970, 294 in 1973, and 501 in 1976. The number has been growing ever since. Nowadays, hundreds of mathematicians and researchers from various application fields are collaborating within 10 national or multinational *Classification Societies* throughout the world, under the umbrella of the *International Federation of Classification Societies* founded in 1985.

The commonly-used clustering methods are based on easy-to-understand mathematical constructs: arithmetic, geometric, graph-theoretic, or simple statistical models (minimizing within-group variance), leading to rather simple calculations on the similarity or dissimilarity values. It must be understood that most clustering methods are heuristic; they create groups by reference to some concept of what a group embedded in some space should be like, but without reference, in most case, to the processes occurring in the application field — ecology in the present book. They have been developed first by the schools of numerical taxonomists and numerical ecologists, later joined by other researchers in the physical sciences and humanities.

* Historical note provided by Prof. F. James Rohlf: "Actually, Sokal & Michener (1958) did not use a computer for their very large study. They used an electromechanical accounting machine to compute the raw sums and sums of products. The coefficients of correlation and the cluster analysis itself were computed by hand with the use of mechanical desk calculators. Sneath did use a computer in his first study."

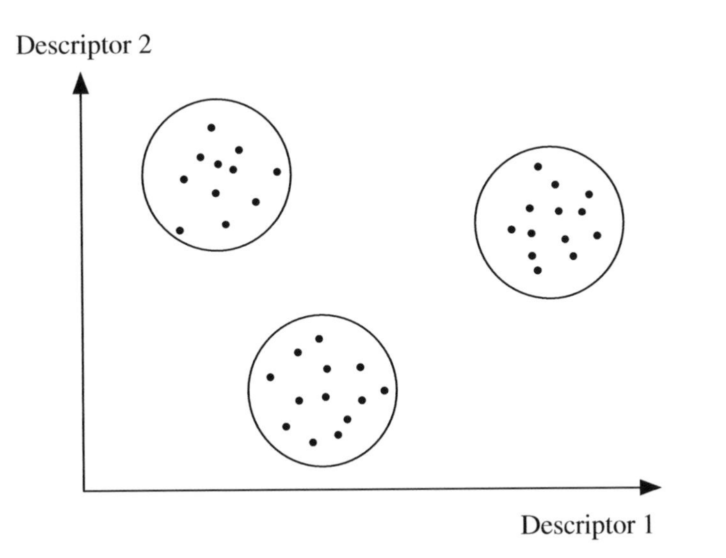

Figure 8.1 Empirically delineating clusters of objects in a scatter diagram is easy when there are no intermediate objects between the groups.

Clusters are delineated on the basis of statements such as: "$\mathbf{x}_1$ is closer to $\mathbf{x}_2$ than it is to $\mathbf{x}_3$", whereas other methods rest on probabilistic models of the type: "Chances are higher that $\mathbf{x}_1$ and $\mathbf{x}_2$ pertain to the same group than $\mathbf{x}_1$ and $\mathbf{x}_3$". In all cases, clustering models make it possible to link the points without requiring prior positioning in a graph (i.e. a metric space), which would be impractical in more than three dimensions. These models thus allow a graphical representation of other interesting relationships among the objects in the data set, for example the dendrogram of their interrelationships. Chapter 10 will show how it is possible to combine clustering and ordination, computed with different methods, to obtain a more complete picture of the data structure.

Descriptive, synoptic clustering

The choice of a clustering method is as critical as the choice of an association measure. It is important to fully understand the properties of clustering methods in order to correctly interpret the ecological structure they bring out. Most of all, the methods to be used depend upon the type of clustering sought. Williams *et al.* (1971) recognized two major categories of methods. In a *descriptive clustering*, misclassifying objects is to be avoided, even at the expense of creating single object clusters. In a *synoptic clustering*, all objects are forced into one of the main clusters; the objective is to construct a general conceptual model which encompasses a reality wider than the data under study. Both approaches are useful.

When two or more clustering models seem appropriate to a problem, ecologists should apply them all to the data and compare the results. Clusters that repeatedly come out of all runs (based on appropriate methods) are the robust solutions to the clustering problem. Differences among results must be interpreted in the light of the known properties of clustering models, which are explained in the following sections.

8.2 The basic model: single linkage clustering

For natural scientists, a simple-to-understand clustering method (or *model*) is *single linkage* (or *nearest neighbour*) clustering (Sneath, 1957). Its logic seems natural, so that it is used to introduce readers to the principles of clustering. Its name, *single linkage*, distinguishes it from other clustering models, called complete or intermediate linkage, detailed in Section 8.4. The algorithm for single linkage clustering is sequential, agglomerative, and hierarchical, following the nomenclature of Section 8.3. Its starting point is any association matrix (similarity or distance) among the objects or descriptors to be clustered. One assumes that the association measure has been carefully chosen, following the recommendations of Section 7.6. To simplify the discussion, the following will deal only with objects, although the method is equally applicable to an association matrix among descriptors.

The method proceeds in two steps:

• First, the association matrix is rewritten in order of decreasing similarities (or increasing distances), heading the list with the two most similar objects of the association matrix, followed by the second most similar pair, and proceeding until all the measures comprised in the association matrix have been listed.

• Second, the clusters are formed hierarchically, starting with the two most similar objects, and then letting the objects clump into groups, and the groups aggregate to one another, as the similarity criterion is relaxed. The following example illustrates this method.

Ecological application 8.2

Five ponds characterized by 38 zooplankton species have been studied by Legendre & Chodorowski (1977). The data were counts, recorded on a relative abundance scale from 0 = absent to 5 = very abundant. These ponds have already been used as example for the computation of Goodall's coefficient (S_{23}, Chapter 7; only eight zooplankton species were used in that example). These five ponds, with others, have been subjected to single linkage clustering after computing similarity coefficient S_{20} with parameter $k = 2$. The symmetric similarity matrix is represented by its lower triangle. The diagonal is trivial because it contains 1's by construct.

Ponds	Ponds				
	212	214	233	431	432
212	—				
214	0.600	—			
233	0.000	0.071	—		
431	0.000	0.063	0.300	—	
432	0.000	0.214	0.200	0.500	—

The first clustering step consists in rewriting these similarities in decreasing order:

S_{20}	Pairs formed
0.600	212-214
0.500	431-432
0.300	233-431
0.214	214-432
0.200	233-432
0.071	214-233
0.063	214-431
0.000	212-233
0.000	212-431
0.000	212-432

Link

As the similarity levels drop, pairs of objects are formed. These pairs are called "links"; they serve to link the objects or groups into a chain, as discussed below.

Connected subgraphs are one of the many possible graphical representations of cluster formation (Fig. 8.2a). As the similarity decreases, clusters are formed, following the list of links in the table of ordered similarities above. Only the similarity levels at which clusters are modified by addition of objects are represented here. The first link is formed between ponds 212 and 214 at $S = 0.6$, then between 431 and 432 at $S = 0.5$. Pond 233 joins this second cluster nucleus at $S = 0.3$. Finally these two clusters merge at $S = 0.214$ by a link which is formed between ponds 214 and 432. The clustering may stop at this point since all ponds now belong to the same cluster. If the similarity criterion was relaxed down to $S = 0$, links would form between members of the cluster up to a point where all ponds would be linked to one another. This part of the clustering is of no interest in single linkage clustering, but these links will be of interest in the other forms of linkage clustering below.

Dendrogram

Edge

Node

A dendrogram (Fig. 8.2b) is another, more commonly-used representation of hierarchical clustering results. Dendrograms only display the clustering topology and object labels, not the links between objects. Dendrograms are made of branches ("edges") that meet at "nodes" which are drawn at the similarity value where fusion of branches takes place. For graphical convenience, vertical lines are used in Fig. 8.2b to connect branches at the similarity levels of the nodes; the lengths of these lines are of no consequence. Branches could be directly connected to nodes. The branches furcating from a node may be switched ("swivelled") without affecting the information contained in a dendrogram.

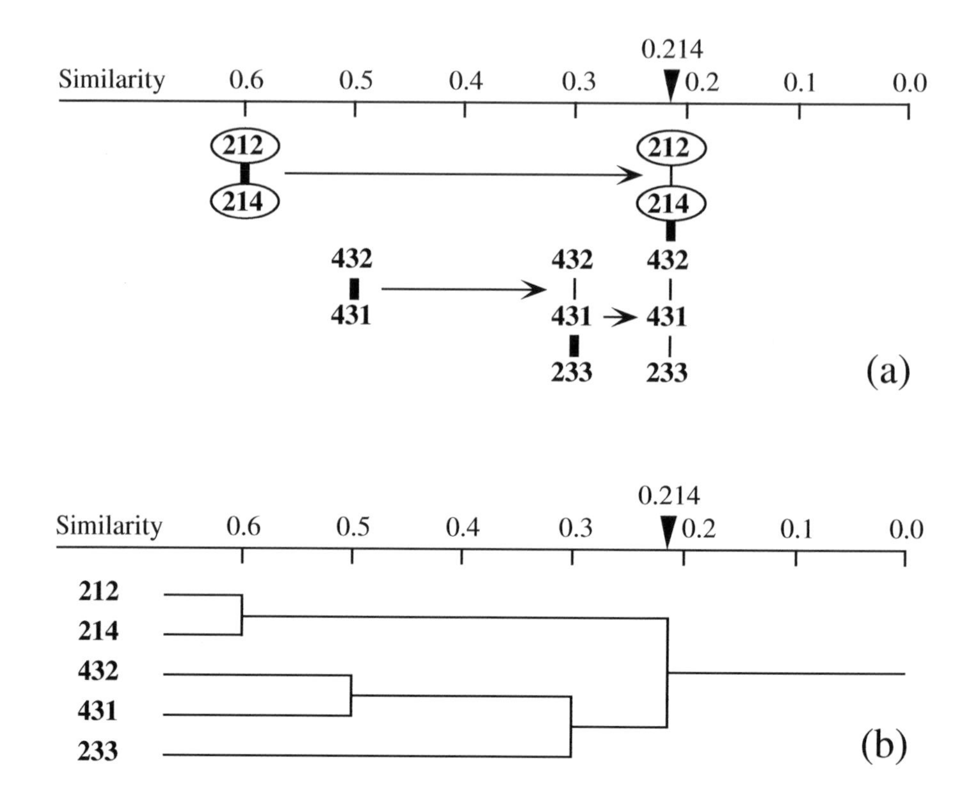

Figure 8.2 Illustrations of single linkage agglomerative clustering for the ponds of the example. (a) Connected subgraphs: groups of objects are formed as the similarity level is relaxed from left to right. Only the similarity levels where clusters are modified by addition of objects are represented. New links between ponds are represented by heavy lines; thin lines are used for links formed at previous (higher) similarity levels. Circled ponds are non-permanent; the others are permanent. (b) Dendrogram for the same cluster analysis.

The clustering results have been interpreted by Legendre & Chodorowski (1977) with respect to the conditions prevailing in the ponds. In their larger study, all non-permanent ponds (including 212 and 214) formed a cluster, while the permanent ponds (including 233, 431 and 432) formed a distinct group, on the basis of zooplankton assemblages.

Single linkage rule

From this example, it should be clear that the rule for assigning an object to a cluster, in single linkage clustering, requires an object to display a similarity at least equal to the considered level of partition with *at least one object already member of the cluster.* In complete linkage hierarchical clustering, the assignment rule differs and requires the object to display the given level of similarity with *all* the objects already members of the cluster. The chaining rule used in single linkage clustering may be stated as follows: at each level of partition, two objects must be allocated to the same subset if their degree of similarity is equal to or higher than that of the partitioning level considered. The same rule can be formulated in terms of dissimilarities (distances) instead: two objects must be allocated to the same subset if their dissimilarity is less than or equal to that of the partitioning level considered.

Estabrook (1966) discussed single linkage clustering using the language of graph theory. The exercise has didactic value. A cluster is defined through the following steps:

Link

a) For any pair of objects $\mathbf{x}_1$ and $\mathbf{x}_2$, a *link* is defined between them by a relation G_c:

$$\mathbf{x}_1 \; G_c \; \mathbf{x}_2 \text{ if and only if } S(\mathbf{x}_1, \mathbf{x}_2) \geq c$$

$$\text{or equally, if } D(\mathbf{x}_1, \mathbf{x}_2) \leq (1 - c)$$

Index c in clustering relation G_c is the similarity level considered. At a similarity level of 0.55, for instance, ponds 212 and 214 of the example are in relation $G_{0.55}$ since $S(212, 214) \geq 0.55$. This definition of a link has the properties of symmetry ($\mathbf{x}_1 \; G_c \; \mathbf{x}_2$ if and only if $\mathbf{x}_2 \; G_c \; \mathbf{x}_1$) and reflexivity ($\mathbf{x}_i \; G_c \; \mathbf{x}_i$ is always true since $S(\mathbf{x}_i, \mathbf{x}_i) = 1.0$). A group of links for a set of objects, such as defined by relation G_c, is called an *undirected graph.*

Undirected graph

Chain
Chaining

b) The chaining which characterizes single linkage clustering may be described by a G_c-chain. A G_c-chain is said to extend from $\mathbf{x}_1$ to $\mathbf{x}_2$ if there exist other points $\mathbf{x}_3, \mathbf{x}_4, \ldots, \mathbf{x}_i$ in the collection of objects under study, such that $\mathbf{x}_1 \; G_c \; \mathbf{x}_3$ and $\mathbf{x}_3 \; G_c \; \mathbf{x}_4$ and ... and $\mathbf{x}_i \; G_c \; \mathbf{x}_2$. For instance, at similarity level $c = 0.214$ of the example, there exists a $G_{0.214}$-chain from pond 212 to pond 233, since there are intermediate ponds such that $212 \; G_{0.214} \; 214$ and $214 \; G_{0.214} \; 432$ and $432 \; G_{0.214} \; 431$ and $431 \; G_{0.214} \; 233$. The number of links in a G_c-chain defines the *connectedness* of a cluster (Subsection 8.11.1).

c) There only remains to delineate the clusters resulting from single linkage chaining. For that purpose, an equivalence relation R_c ("member of the same cluster") is defined as follows:

$$\mathbf{x}_1 \; R_c \; \mathbf{x}_2 \text{ if and only if there exists a } G_c\text{-chain from } \mathbf{x}_1 \text{ to } \mathbf{x}_2 \text{ at similarity level } c.$$

In other words, $\mathbf{x}_1$ and $\mathbf{x}_2$ are assigned to the same cluster at similarity level c if there exists a chain of links joining $\mathbf{x}_1$ to $\mathbf{x}_2$. Thus, at level $S = 0.214$ in the example, ponds 212 and 233 are assigned to the same cluster ($212 \; R_{0.214} \; 233$) because there exists a $G_{0.214}$-chain from 212 to 233. The relationship "member of the same cluster" has the following properties: (1) it is reflexive ($\mathbf{x}_i \; R_c \; \mathbf{x}_i$) because G_c is reflexive; (2) the G_c-chains may be reversed because G_c is symmetric; as a consequence, $\mathbf{x}_1 \; R_c \; \mathbf{x}_2$ implies that $\mathbf{x}_2 \; R_c \; \mathbf{x}_1$; and (3) it is transitive because, by G_c-chaining, $\mathbf{x}_1 \; R_c \; \mathbf{x}_2$ and $\mathbf{x}_2 \; R_c \; \mathbf{x}_3$ implies that $\mathbf{x}_1 \; R_c \; \mathbf{x}_3$. Each cluster thus defined is a connected subgraph, which means that the objects of a cluster are all connected in their subgraph; in the graph of all the objects, distinct clusters (subgraphs) have no links attaching them to one another.

Connected subgraph

Single linkage clustering provides an accurate picture of the relationships between pairs of objects but, because of its propensity to chaining, it may be undesirable for ecological analysis. This means that the presence of an object midway between two compact clusters, or a few intermediates connecting the two clusters, is enough to turn them into a single cluster. Of course, clusters do not chain unless intermediates are present; so, the occurrence of chaining provides information about the data. To describe this phenomenon, Lance & Williams (1967c) wrote that this method contracts the reference space. Picture the objects as laying in A-space (Fig. 7.2). The presence of a cluster increases the probability of inclusion, by chaining, of neighbouring objects into the cluster. This is as if the distances between objects were smaller in that region of the space; see also Fig. 8.23a.

Section 10.1 will show how to take advantage of the interesting properties of single linkage clustering by combining it with ordination results, while avoiding the undue influence of chaining on the clustering structure.

Chain of primary connections

Minimum spanning tree

Ecologists who have access to several computer programs for single linkage clustering should rely on those that permit recognition of the first connection making an object a member of a cluster, or allowing two clusters to fuse. These similarities form a *chain of primary* (Legendre, 1976) or *external connections* (Legendre & Rogers, 1972), also called *dendrites* by Lukaszewicz (1951). They are very useful when analysing clusters drawn in an ordination space (Section 10.1). Dendrites are also called a *network* (Prim, 1957), a *Prim network* (Cavalli-Sforza & Edwards, 1967), a *minimum spanning tree* (Gower & Ross, 1969), a *shortest spanning tree*, or a *minimum-length tree* (Sneath & Sokal, 1973). If these dendrites are drawn on a scatter diagram of the objects, one can obtain a non-hierarchical clustering of the objects by removing the last (weakest) similarity links. Such graphs are illustrated in Figs. 10.1 and 10.2, drawn on top of an ordination in a space of principal coordinates; they may also be drawn without any reference space.

8.3 Cophenetic matrix and ultrametric property

Any classification or partition can be fully described by a cophenetic matrix. This matrix is used for comparing different classifications of the same objects.

1 — Cophenetic matrix

Cophenetic similarity

The *cophenetic similarity* (or *distance*) of two objects $\mathbf{x}_1$ and $\mathbf{x}_2$ is defined as the similarity (or distance) level at which objects $\mathbf{x}_1$ and $\mathbf{x}_2$ become members of the same cluster during the course of clustering (Jain & Dubes, 1988), as depicted by connected subgraphs or a dendrogram (Fig. 2a, b). Any dendrogram can be uniquely represented by a matrix in which the similarity (or distance) for a pair of objects is their cophenetic similarity (or distance). Consider the single linkage clustering dendrogram of Fig. 8.2. The clustering levels, read directly on the dendrogram, lead to the following matrices of similarity (**S**) and distance (**D**, where $D = 1 - S$):

S	212	214	233	431	432
212	—		(upper triangle symmetric to lower)		
214	0.600	—			
233	0.214	0.214	—		
431	0.214	0.214	0.300	—	
432	0.214	0.214	0.300	0.500	—

D	212	214	233	431	432
212	—		(upper triangle symmetric to lower)		
214	0.400	—			
233	0.786	0.786	—		
431	0.786	0.786	0.700	—	
432	0.786	0.786	0.700	0.500	—

Cophenetic matrix

Such a matrix is often called a *cophenetic matrix* (Sokal & Rohlf, 1962; Jain & Dubes, 1988). The ordering of objects in the cophenetic matrix is irrelevant; any order that suits the researcher is acceptable. The same applies to dendrograms; the order of the objects may be changed at will, provided that the dendrogram is redrawn to accommodate the new ordering.

For a *partition* of the data set (as in the K-means method, below), the resulting groups of objects are not related through a dendrogram. A cophenetic matrix may nevertheless be obtained. Consider the groups (212, 214) and (233, 431, 432) obtained by cutting the dendrogram of Fig. 8.2 at similarity level $S = 0.25$, ignoring the hierarchical structure of the two clusters. The cophenetic matrices would be:

S	212	214	233	431	432
212	—		(upper triangle symmetric to lower)		
214	1	—			
233	0	0	—		
431	0	0	1	—	
432	0	0	1	1	—

D	212	214	233	431	432
212	—		(upper triangle symmetric to lower)		
214	0	—			
233	1	1	—		
431	1	1	0	—	
432	1	1	0	0	—

2 — *Ultrametric property*

If there are no *reversals* in the clustering (Fig. 8.16), a classification has the following *ultrametric property* and the cophenetic matrix is called ultrametric:

$$D(\mathbf{x}_1, \mathbf{x}_2) \leq \max [D(\mathbf{x}_1, \mathbf{x}_3), D(\mathbf{x}_2, \mathbf{x}_3)] \qquad \textbf{(8.1)}$$

for every triplet of objects $(\mathbf{x}_1, \mathbf{x}_2, \mathbf{x}_3)$ in the study. Cophenetic distances also possess the four *metric properties* of Section 7.4. The ultrametric property may be expressed in terms of similarities:

$$S(\mathbf{x}_1, \mathbf{x}_2) \geq \min [S(\mathbf{x}_1, \mathbf{x}_3), S(\mathbf{x}_2, \mathbf{x}_3)] \qquad \textbf{(8.2)}$$

As an exercise, readers can verify that the five properties apply to all triplets of similarities and distances in the above matrices.

8.4 The panoply of methods

Clustering algorithms have been developed using a wide range of conceptual models and for studying a variety of problems. Sneath & Sokal (1973) propose a classification of clustering procedures. Its main dichotomies are now briefly described.

1 — Sequential versus simultaneous algorithms

Most clustering algorithms are sequential in the sense that they proceed by applying a recurrent sequence of operations to the objects. The agglomerative single linkage clustering of Section 8.2 is an example of a sequential method: the search for the equivalence relation R_c is repeated at all levels of similarity in the association matrix, up to the point where all objects are in the same cluster. In *simultaneous* algorithms, which are less frequent, the solution is obtained in a single step. Ordination techniques (Chapter 9), which may be used for delineating clusters, are of the latter type. This is also the case of the direct complete linkage clustering method presented in Section 8.9. The *K*-means (Section 8.8) and other non-hierarchical partitioning methods may be computed using sequential algorithms, although these methods are neither agglomerative nor divisive (next paragraph).

2 — Agglomeration versus division

Among the sequential algorithms, *agglomerative* procedures begin with the discontinuous partition of all objects, i.e. the objects are considered as being separate from one another. They are successively grouped into larger and larger clusters until a single, all-encompassing cluster is obtained. If the continuous partition of all objects is used instead as the starting point of the procedure (i.e. a single group containing all objects), *divisive* algorithms subdivide the group into sub-clusters, and so on until the discontinuous partition is reached. In either case, it is left to users to decide which of the intermediate partitions is to be retained, given the problem under study. Agglomerative algorithms are the most developed for two reasons. First, they are easier to program. Second, in clustering by division, the erroneous allocation of an object to a cluster at the beginning of the procedure cannot be corrected afterwards (Gower, 1967) unless a special procedure is embedded in the algorithm to do so.

3 — Monothetic versus polythetic methods

Divisive clustering methods may be monothetic or polythetic. *Monothetic* models use a single descriptor as basis for partitioning, whereas *polythetic* models use several descriptors which, in most cases, are combined into an association matrix (Chapter 7) prior to clustering. Divisive monothetic methods proceed by choosing, for each partitioning level, the descriptor considered to be the best for that level; objects are then partitioned following the state to which they belong with respect to that descriptor. For example, the most appropriate descriptor at each partitioning level could be the one that best represents the information contained in all other descriptors, after measuring the reciprocal information between descriptors (Subsection 8.6.1). When a single partition of the objects is sought, monothetic methods produce the clustering in a single step.

4 — Hierarchical versus non-hierarchical methods

In *hierarchical* methods, the members of inferior-ranking clusters become members of larger, higher-ranking clusters. Most of the time, hierarchical methods produce non-overlapping clusters, but this is not a necessity according to the definition of "hierarchy" in the dictionary or the usage recognized by Sneath & Sokal (1973). Single linkage clustering of Section 8.2 and the methods of Sections 8.5 and 8.6 are hierarchical. *Non-hierarchical methods* are very useful in ecology. They produce a single partition which optimizes within-group homogeneity, instead of a hierarchical series of partitions optimizing the hierarchical attribution of objects to clusters. Lance & Williams (1967d) restrict the term "clustering" to the non-hierarchical methods and call the hierarchical methods "classification". Non-hierarchical methods include *K*-means partitioning, the ordination techniques (Chapter 9) used as clustering methods, the primary connection diagrams (dendrites) between objects with or without a reference space, the methods of similarity matrix seriation of Section 8.10, and one of the algorithms of Section 8.9 for the clustering of species into biological associations. These methods should be used in cases where the aim is to obtain a direct representation of the relationships among objects instead of a summary of their hierarchy. Hierarchical methods are easier to compute and more often available in statistical data analysis packages than non-hierarchical procedures.

Most hierarchical methods use a resemblance matrix as their starting point. This prevents their use with very large data sets because the resemblance matrix, with its $n(n-1)/2$ values, becomes extremely large and may exceed the handling capacity of computers. Jambu & Lebeaux (1983) have described a fast algorithm for the hierarchical agglomeration of very large numbers of objects (e.g. $n = 5000$). This algorithm computes a fraction only of the $n(n-1)/2$ distance values. Rohlf (1978, 1982a) has also developed a rather complex algorithm allowing one to obtain single linkage clustering after computing only a small fraction of the distances.

5 — Probabilistic versus non-probabilistic methods

Probabilistic methods include the clustering model of Clifford & Goodall (1967) and the parametric and nonparametric methods for estimating density functions in multivariate space.

In the method of Clifford & Goodall (1967), clusters are formed in such a way that the within-group association matrices have a given probability of being homogeneous. This clustering method is discussed at length in Subsection 8.9.2, where it is recommended, in conjunction with Goodall's similarity coefficient (S_{23}, Chapter 7), for the clustering of species into biological associations.

Sneath & Sokal (1973) describe other dichotomies for clustering methods, which are of lesser interest to ecologists. These are: global or local criteria, direct or iterative solutions, equal or unequal weights, and adaptive or non-adaptive clustering.

8.5 Hierarchical agglomerative clustering

Most methods of hierarchical agglomeration can be computed as special cases of a general model which is discussed in Subsection 8.5.9.

1 — Single linkage agglomerative clustering

In single linkage agglomeration (Section 8.2), two clusters fuse when the two objects closest to each other (one in each cluster) reach the similarity of the considered partition. (See also the method of simultaneous single linkage clustering described in Subsection 8.9.1). As a consequence of chaining, results of single linkage clustering are sensitive to noise in the data (Milligan, 1996), because noise changes the similarity values and may thus easily modify the order in which objects cluster. The origin of single linkage clustering is found in a collective work by Florek, Lukaszewicz, Perkal, Steinhaus, and Zubrzycki, published by Lukaszewicz in 1951.

2 — Complete linkage agglomerative clustering

Complete linkage rule

Opposite to the single linkage approach is *complete linkage agglomeration*, also called *furthest neighbour sorting*. In this method, first proposed by Sørensen (1948), the fusion of two clusters depends on the most distant pair of objects instead of the closest. Thus, an object joins a cluster only when it is linked (relationship G_c, Section 8.2) to all the objects already members of that cluster. Two clusters can fuse only when all objects of the first are linked to all objects of the second, and vice versa.

Coming back to the ponds of Ecological application 8.2, complete linkage clustering (Fig. 8.3) is performed on the table of ordered similarities of Section 8.2. The pair (212, 214) is formed at $S = 0.6$ and the pair (431, 432) at $S = 0.5$. The next clustering step must wait until $S = 0.2$, since it is only at $S = 0.2$ that pond 233 is finally linked (relationship G_c) to both ponds 431 and 432. The two clusters hence formed cannot fuse, because it is only at similarity zero that ponds 212 and 214 become linked to all the ponds of cluster (233, 431, 432). $S = 0$ indicating, by definition, distinct entities, the two groups are not represented as joining at that level.

In the compete linkage strategy, as a cluster grows, it becomes more and more difficult for new objects to join to it because the new objects should bear links with all the objects already in the cluster before being incorporated. For this reason, the growth of a cluster seems to move it away from the other objects or clusters in the analysis. According to Lance & Williams (1967c), this is equivalent to dilating the reference

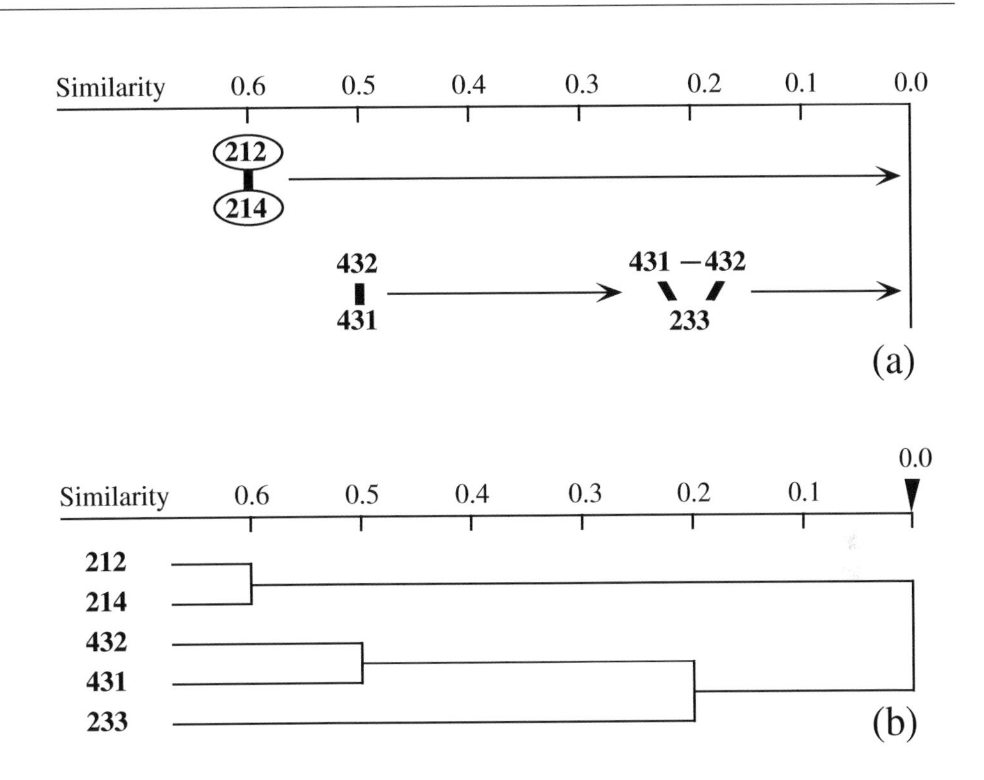

Figure 8.3 Complete linkage clustering of the ponds of Ecological application 8.2. Symbols as in Fig. 8.2.

space in the neighbourhood of that cluster; see also Fig. 8.23c and related text. This effect is opposite to what was found in single linkage clustering, which contracted the reference space. In reference space A (Fig. 7.2), complete linkage produces maximally linked and rather spherical clusters, whereas single linkage may produce elongated clusters with loose chaining. Complete linkage clustering is often desirable in ecology, when one wishes to delineate clusters with clear discontinuities.

The intermediate (next Subsection) and complete linkage clustering models have one drawback when compared to single linkage. In all cases where two incompatible candidates present themselves at the same time to be included in a cluster, algorithms use a preestablished and often arbitrary rule, called a "right-hand rule", to choose one and exclude the other. This problem does not exist in single linkage. An example is when two objects or two clusters could be included in a third cluster, while these two objects or clusters have not completed the linkage with each other. For this problem, Sørensen (1948) recommends the following: (1) choose the fusion leading to the largest cluster; (2) if equality persists, choose the fusion that most reduces the number of clusters; (3) as a last criterion, choose the fusion that maximizes the average similarity within the cluster.

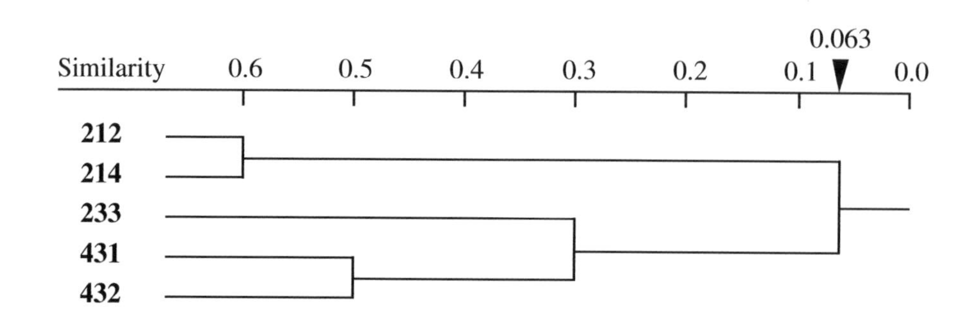

Figure 8.4 Intermediate linkage clustering, using the proportional link linkage criterion (Co = 50%), for the ponds of Ecological application 8.2 (dendrogram only).

3 — Intermediate linkage clustering

Between the chaining of single linkage and the extreme space dilation of complete linkage, the most interesting solution in ecology may be a type of linkage clustering that approximately conserves the metric properties of reference space A; see also Fig. 8.23b. If the interest only lies in the clusters shown in the dendrogram, and not in the actual similarity links between clusters shown by the subgraphs, the average clustering methods of Subsections 4 to 7 below could be useful since they also conserve the metric properties of the reference space.

Connectedness

Proportional link linkage

In intermediate linkage clustering, the fusion criterion of an object or a cluster with another cluster is considered satisfied when a given proportion of the total possible number of similarity links is reached. For example, if the criterion of connectedness (Co) is 0.5, two clusters are only required to share 50% of the possible links to fuse; in other words, the fusion is authorized when $£/n_1n_2 \geq Co$ where £ is the actual number of *between-group* links at sorting level L, while n_1 and n_2 are the numbers of objects in the two clusters, respectively. This criterion has been called *proportional link linkage* by Sneath (1966). Fig. 8.4 gives the results of proportional link linkage clustering with Co = 50% for the pond example.

Sneath (1966) has described three other ways of defining intermediate linkage clustering criteria: (1) by *integer link linkage*, which specifies the number of links required for the fusion of two groups (fusion when £ is larger than or equal to a fixed integer, *or else* when $£ = n_1n_2$; (2) by their *absolute resemblance*, based on the sum of similarities between the members of two clusters (the sum of between-group similarities, ΣS_{12}, must reach a given threshold before the fusion occurs); or (3) by their *relative resemblance*, where the sum of similarities between the two clusters, ΣS_{12}, is divided by the number of between-group similarities, n_1n_2 (fusion occurs at level L when the ratio $\Sigma S_{12}/n_1n_2$ is greater than cL, where c is an arbitrary constant.)

Table 8.2 Average clustering methods discussed in Subsections 8.5.4 to 8.5.7.

	Arithmetic average	Centroid clustering
Equal weights	4. Unweighted arithmetic average clustering (UPGMA)	6. Unweighted centroid clustering (UPGMC)
Unequal weights	5. Weighted arithmetic average clustering (WPGMA)	7. Weighted centroid clustering (WPGMC)

When c equals 1, the method is called *average linkage clustering*. These strategies are not combinatorial in the sense of Subsection 8.5.9.

4 — Unweighted arithmetic average clustering (UPGMA)

Average clustering

There are four methods of *average clustering* that conserve the metric properties of reference space A. These four methods were called "average linkage clustering" by Sneath & Sokal (1973), although they do not tally the links between clusters. As a consequence they are not object-linkage methods in the sense of the previous three subsections. They rely instead on average similarities among objects or on centroids of clusters. The four methods have nothing to do with Sneath's (1966) "average linkage clustering" described in the previous paragraph, so that we prefer calling them "average clustering". These methods (Table 8.2) result from the combinations of two dichotomies: (1) arithmetic average versus centroid clustering and (2) weighting versus non-weighting.

The first method in Table 8.2 is the *unweighted arithmetic average clustering* (Rohlf, 1963), also called "UPGMA" ("Unweighted Pair-Group Method using Arithmetic averages") by Sneath & Sokal (1973) or "group-average sorting" by Lance & Williams (1966a and 1967c). It is also called "average linkage" by SAS, SYSTAT and some other statistical packages, thus adding to the confusion pointed out in the previous paragraph. The highest similarity (or smallest distance) identifies the next cluster to be formed. Following this event, the method computes the arithmetic average of the similarities or distances between a candidate object and each of the cluster members or, in the case of a previously formed cluster, between all members of the two clusters. All objects receive equal weights in the computation. The similarity or distance matrix is updated and reduced in size at each clustering step. Clustering proceeds by agglomeration as the similarity criterion is relaxed, just as it does in single linkage clustering.

For the ponds of Section 8.2, UPGMA clustering proceeds as shown in Table 8.3 and Fig. 8.5. At step 1, the highest similarity value in the matrix is

Table 8.3 Unweighted arithmetic average clustering (UPGMA) of the pond data. At each step, the highest similarity value is identified (italicized boldface value) and the two corresponding objects or groups are fused by averaging their similarities as described in the text (boxes).

Objects	212	214	233	431	432
212	—				**Step 1**
214	***0.600***	—			
233	0.000	0.071	—		
431	0.000	0.063	0.300	—	
432	0.000	0.214	0.200	0.500	—
212-214		—			**Step 2**
233		0.0355	—		
431		0.0315	0.300	—	
432		0.1070	0.200	***0.500***	—
212-214		—			**Step 3**
233		0.0355	—		
431-432		0.06925	***0.250***	—	
212-214		—			**Step 4**
233-431-432		***0.058***	—		

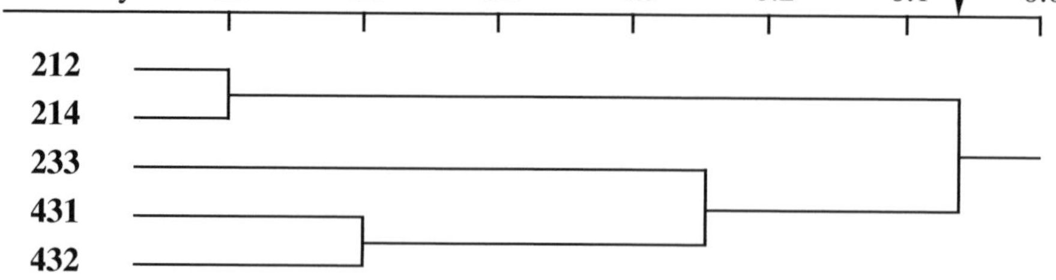

Figure 8.5 Unweighted arithmetic average clustering (UPGMA) of the ponds from Ecological application 8.2. This type of clustering only produces a dendrogram. It cannot be represented by connected subgraphs since it is not a linkage clustering as in Figs. 8.2 and 8.3.

$S(212, 214) = 0.600$; hence the two objects fuse at level 0.600. As a consequence of this fusion, the similarity values of these two objects with each of the remaining objects in the study must be averaged (values in the inner boxes in the Table, step 1); this results in a reduction of the size of the similarity matrix. Considering the reduced matrix (step 2), the highest similarity value is $S = 0.500$; it indicates that objects 431 and 432 fuse at level 0.500. Again, this similarity value is obtained by averaging the boxed values; this produces a new reduced similarity matrix for the next step. In step 3, the largest similarity is 0.250; it leads to the fusion of the already-formed group (431, 432) with object 233 at level 0.250. In the example, this last fusion is the difficult point to understand. Before averaging the values, each one is multiplied by the number of objects in the corresponding group. There is one object in group (233) and two in group (431, 432), so that the fused similarity value is calculated as $[(0.0355 \times 1) + (0.06925 \times 2)]/3 = 0.058$. This is equivalent to averaging the six boxed similarities in the top matrix (larger box) with equal weights; the result would also be 0.058. So, this method is "unweighted" in the sense that it gives equal weights to the original similarities. To achieve this at step 3, one has to use weights that are equal to the number of objects in the groups. At step 4, there is a single remaining similarity value; it is used to perform the last fusion at level 0.058. In the dendrogram, fusions are drawn at the identified levels.

Because it gives equal weights to the original similarities, the UPGMA method assumes that the objects in each group form a representative sample of the corresponding larger groups of objects in the reference population under study. For that reason, UPGMA clustering should only be used in connection with simple random or systematic sampling designs if the results are to be extrapolated to some larger reference population.

Unlike linkage clustering methods, information about the relationships between pairs of objects is lost in methods based on progressive reduction of the similarity matrix, since only the relationships among groups are considered. This information may be extracted from the original similarity matrix, by making a list of the strongest similarity link found, at each fusion level, between the objects of the two groups. For the pond example, the *chain of primary connections* corresponding to the dendrogram would be made of the following links: (212, 214) for the first fusion level, (431, 432) for the second level, (233, 431) for the third level, and (214, 432) for the last level (Table 8.3, step 1). The topology obtained from UPGMA clustering may differ from that of single linkage; if this had been the case here, the chain of primary connections would have been different from that of single linkage clustering.

5 — Weighted arithmetic average clustering (WPGMA)

It often occurs in ecology that groups of objects, representing different regions of a territory, are of unequal sizes. Eliminating objects to equalize the clusters would mean discarding valuable information. However, the presence of a large group of objects, which are more similar *a priori* because of their common origin, may distort the UPGMA results when a fusion occurs with a smaller group of objects. Sokal &

Michener (1958) proposed a solution to this problem, called *weighted arithmetic average clustering* ("WPGMA" in Sneath & Sokal, 1973: "Weighted Pair-Group Method using Arithmetic averages"). This solution consists in giving equal weights, when computing fusion similarities, to the two *branches* of the dendrogram that are about to fuse. This is equivalent, when computing a fusion similarity, to giving different weights to the original similarities, i.e. down-weighting the largest group. Hence the name of the method.

Table 8.4 and Fig. 8.6 describe the WPGMA clustering sequence for the pond data. In this example, the only difference with UPGMA is the last fusion value. It is computed here by averaging the two similarities from the previous step: (0.0355 + 0.06925)/2 = 0.052375. Weighted arithmetic average clustering increases the separation of the two main clusters, compared to UPGMA. This gives sharper contrast to the classification.

6 — Unweighted centroid clustering (UPGMC)

Centroid

The *centroid* of a cluster of objects may be imagined as the type-object of the cluster, whether that object actually exists or is only a mathematical construct. In A-space (Fig. 7.2), the coordinates of the centroid of a cluster are computed by averaging the coordinates of the objects in the group.

Unweighted centroid clustering (Lance & Williams, 1967c; "UPGMC" in Sneath & Sokal, 1973: "Unweighted Pair-Group Centroid Method") is based on a simple geometric approach. Along a decreasing scale of similarities, UPGMC proceeds to the fusion of objects or clusters presenting the highest similarity, as in the previous methods. At each step, the members of a cluster are replaced by their common centroid (i.e. "mean point"). The centroid is considered to represent a new object for the remainder of the clustering procedure; in the next step, one looks again for the pair of objects with the highest similarity, on which the procedure of fusion is repeated.

Gower (1967) proposed the following formula for centroid clustering, where the similarity of the centroid (**hi**) of the objects or clusters **h** and **i** with a third object or cluster **g** is computed from the similarities $S(\mathbf{h}, \mathbf{g})$, $S(\mathbf{i}, \mathbf{g})$, and $S(\mathbf{h}, \mathbf{i})$:

$$S(\mathbf{hi}, \mathbf{g}) = \frac{w_h}{w_h + w_i} S(\mathbf{h}, \mathbf{g}) + \frac{w_i}{w_h + w_i} S(\mathbf{i}, \mathbf{g}) + \frac{w_h w_i}{(w_h + w_i)^2} [1 - S(\mathbf{h}, \mathbf{i})] \qquad \mathbf{(8.3)}$$

were the w's are weights given to the clusters. To simplify the symbols, letters **g**, **h**, and **i** are used here to represent three objects considered in the course of clustering; **g**, **h**, and **i** may also represent centroids of clusters obtained during previous clustering steps. Gower's formula insures that the centroid **hi** of objects (or clusters) **h** and **i** is geometrically located on the line between **h** and **i**. In classical centroid clustering, the numbers of objects $n_\mathbf{h}$ and $n_\mathbf{i}$ in clusters **h** and **i** are taken as values for the weights w_h and w_i; these weights are 1 at the start of the clustering because there is then a single

Table 8.4 Weighted arithmetic average clustering (WPGMA) of the pond data. At each step, the highest similarity value is identified (italicized boldface value) and the two corresponding objects or groups are fused by averaging their similarities (boxes).

Objects	212	214	233	431	432
212	—				**Step 1**
214	***0.600***	—			
233	0.000	0.071	—		
431	0.000	0.063	0.300	—	
432	0.000	0.214	0.200	0.500	—
212-214		—			**Step 2**
233		0.0355	—		
431		0.0315	0.300	—	
432		0.1070	0.200	***0.500***	—
212-214		—			**Step 3**
233		0.0355	—		
431-432		0.06925	***0.250***	—	
212-214		—			**Step 4**
233-431-432		***0.05238***	—		

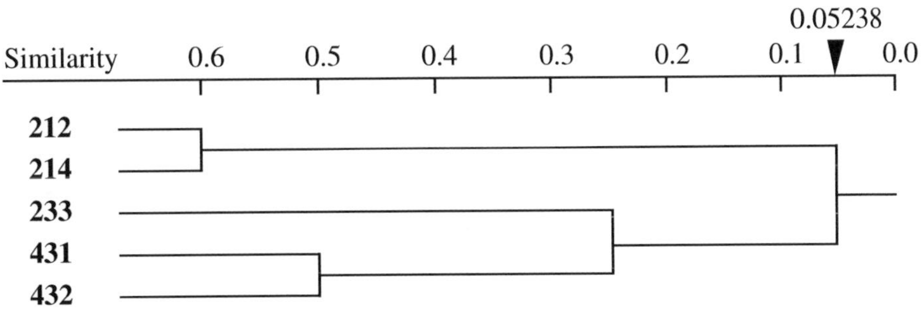

Figure 8.6 Weighted arithmetic average clustering (WPGMA) of the ponds from Ecological application 8.2. This type of clustering only produces a dendrogram. It cannot be represented by connected subgraphs since it is not a linkage clustering as in Figs. 8.2 and 8.3.

object per cluster. If initial weights are attached to individual objects, they may be used instead of 1's in eq. 8.3.

Centroid clustering may lead to reversals (Section 8.6). Some authors feel uncomfortable about reversals since they violate the ultrametric property; such violations make dendrograms more difficult to draw. A reversal is found with the pond example (Table 8.5, Fig. 8.7): the fusion similarity found at step 4 is higher than that of step 3. The last fusion similarity (step 4), for instance, is calculated as follows:

$$S[(233, 431\text{-}432), (212\text{-}214)] = \frac{1}{3} \times 0.1355 + \frac{2}{3} \times 0.29425 + \frac{2}{3^2}(1 - 0.375) = 0.38022$$

As indicated above, the geometric interpretation of UPGMC clustering is the fusion of objects into cluster centroids. Figure 8.8 presents the four clustering steps depicted by the dendrogram, drawn in an A-space (Fig. 7.2) reduced to two dimensions through principal coordinate analysis (Section 9.2) to facilitate representation. At the end of each step, a new cluster is formed and its centroid is represented at the *centre of mass* of the cluster members (examine especially steps 3 and 4).

Unweighted centroid clustering may be used with any measure of similarity, but Gower's formula above only retains its geometric properties for similarities corresponding to metric distances (Table 7.2). Note also that in this clustering procedure, the links between clusters do not depend upon identifiable pairs of objects; this was also the case with clustering methods 4 and 5 above. Thus, if the chain of primary connections is needed, it must be identified by some other method.

The assumptions of this model with respect to representativeness of the observations are the same as in UPGMA, since equal weights are given to all objects during clustering. So, UPGMC should only be used in connection with simple random or systematic sampling designs if the results are to be extrapolated to some larger reference population. When the branching pattern of the dendrogram displays asymmetry (many more objects in one branch than in the other), this can be attributed to the structure of the reference population if the sampling design was random.

In order to obtain clusters that were well separated even though the objects came from an ecological continuum, Flos (1976) provided his computer program with a strip (non-clustering zone) between centroids. The width of the zone was set by the user at the beginning of the calculations. Points found within that zone were not included in any cluster. At the end of the formation of clusters, the unclassified objects were allocated to the closest cluster centroid.

7 — Weighted centroid clustering (WPGMC)

Weighted centroid clustering was proposed by Gower (1967). It plays the same role with respect to UPGMC as WPGMA (method 5) plays with respect to UPGMA (method 4). When many observations of a given type have been included in the set to

Table 8.5 Unweighted centroid clustering (UPGMC) of the pond data. At each step, the highest similarity value is identified (italicized boldface value) and the two corresponding objects or groups are fused using eq. 8.3.

Objects	212	214	233	431	432
212	—				**Step 1**
214	***0.600***	—			
233	0.000	0.071	—		
431	0.000	0.063	0.300	—	
432	0.000	0.214	0.200	0.500	—
212-214		—			**Step 2**
233		0.1355	—		
431		0.1315	0.300	—	
432		0.2070	0.200	***0.500***	—
212-214		—			**Step 3**
233		0.1355	—		
431-432		0.29425	***0.375***	—	
212-214		—			**Step 4**
233-431-432		***0.3802***	—		

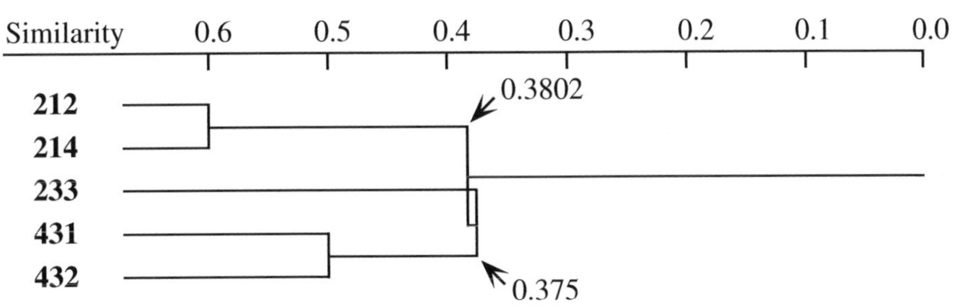

Figure 8.7 Unweighted centroid clustering (UPGMC) of the ponds from Ecological application 8.2. This type of clustering only produces a dendrogram. The reversal in the structure of the dendrogram is explained in Section 8.6.

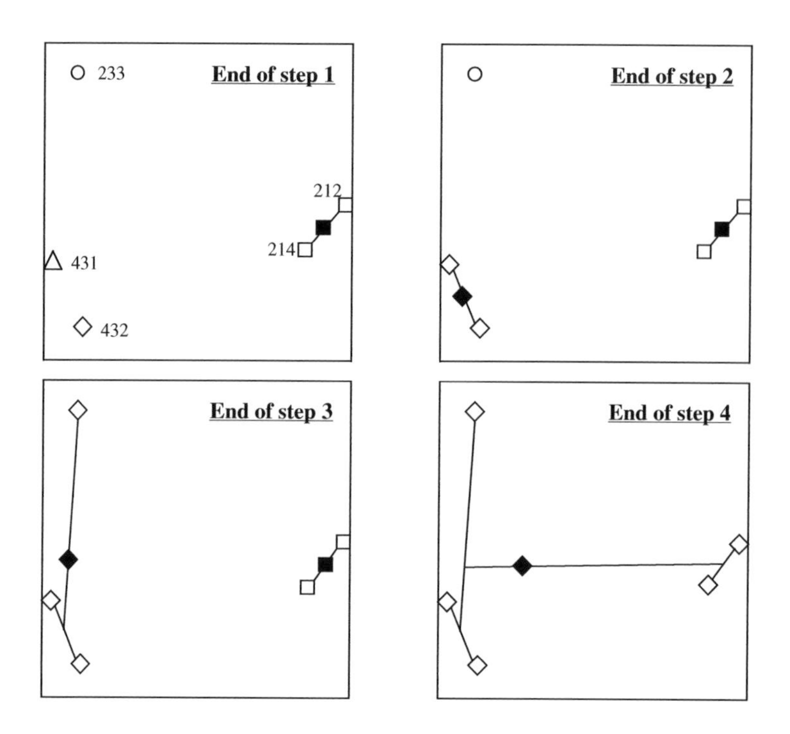

Figure 8.8 The four UPGMC clustering steps of Fig. 8.7 are drawn in A-space. Objects are represented by open symbols and centroids by dark symbols; object identifiers are shown in the first panel only. Distinct clusters are represented by different symbols. The first two principal coordinates, represented here, account for 87.4% of the variation of the full A-space.

be clustered, next to other types which were not as well-sampled (sampling design other than simple random or systematic), the positions of the centroids may be biased towards the over-represented types, which in turn could distort the clustering. In *weighted centroid clustering*, which Sneath & Sokal (1973) call "WPGMC" ("Weighted Pair-Group Centroid Method"), this problem is corrected by giving equal weights to two clusters on the verge of fusing, independently of the number of objects in each cluster. To achieve this, eq. 8.3 is replaced by the following formula (Gower, 1967):

$$S(\mathbf{hi}, \mathbf{g}) = \frac{1}{2}[S(\mathbf{h}, \mathbf{g}) + S(\mathbf{i}, \mathbf{g})] + \frac{1}{4}[1 - S(\mathbf{h}, \mathbf{i})] \quad \mathbf{(8.4)}$$

The five ponds of Ecological application 7.2 are clustered as described in Table 8.6 and Fig. 8.9. The last fusion similarity (step 4), for example, is calculated as follows:

$$S[(233, 431\text{-}432), (212\text{-}214)] = \frac{1}{2}[0.1355 + 0.29425] + \frac{1}{4}(1 - 0.375) = 0.371125$$

Table 8.6 Weighted centroid clustering (WPGMC) of the pond data. At each step, the highest similarity value is identified (italicized boldface value) and the two corresponding objects or groups are fused using eq. 8.4.

Objects	212	214	233	431	432
212	—				**Step 1**
214	***0.600***	—			
233	0.000	0.071	—		
431	0.000	0.063	0.300	—	
432	0.000	0.214	0.200	0.500	—
212-214		—			**Step 2**
233		0.1355	—		
431		0.1315	0.300	—	
432		0.2070	0.200	***0.500***	—
212-214		—			**Step 3**
233		0.1355	—		
431-432		0.29425	***0.375***	—	
212-214		—			**Step 4**
233-431-432		***0.37113***	—		

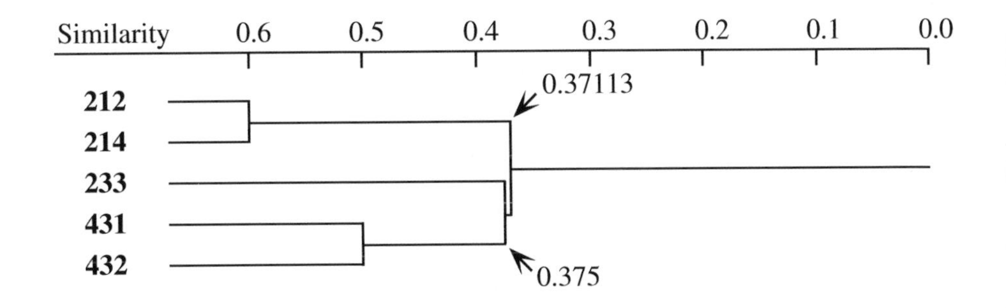

Figure 8.9 Weighted centroid clustering (WPGMC) of the ponds from Ecological application 8.2. This type of clustering only produces a dendrogram.

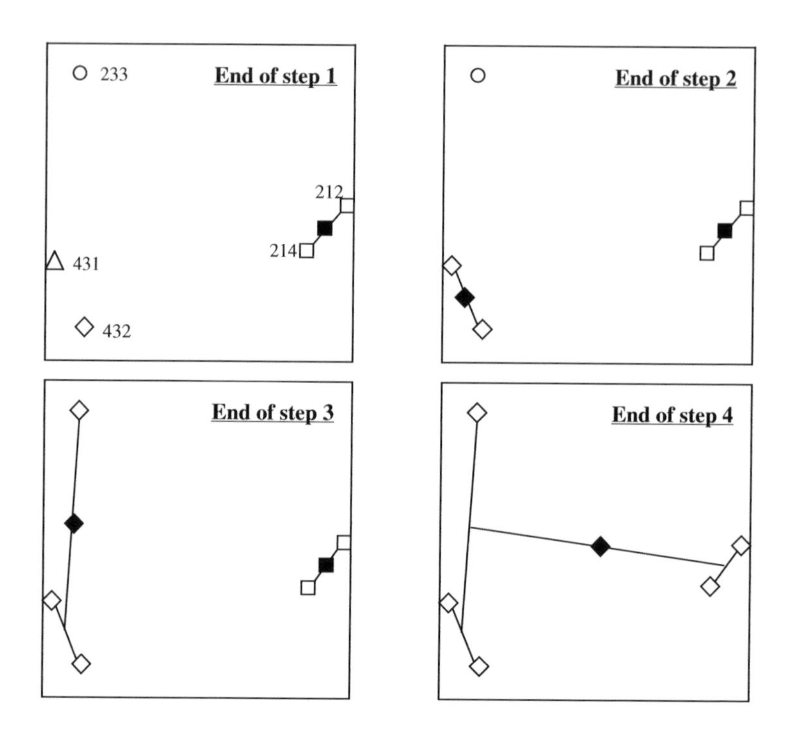

Figure 8.10 The four WPGMC clustering steps of Fig. 8.9 are drawn in A-space. Objects are represented by open symbols and centroids by dark symbols; object identifiers are shown in the first panel only. Distinct clusters are represented by different symbols. The first two principal coordinates, represented here, account for 87.4% of the variation of the full A-space.

This value is the level at which the next fusion takes place. Note that no reversal appears in this result, although WPGMC may occasionally produce reversals, like UPGMC clustering.

As indicated above, the geometric interpretation of WPGMC clustering is the fusion of objects into cluster centroids. Fig. 8.10 presents the four clustering steps depicted by the dendrogram, in A-space (Fig. 7.2) reduced to two dimensions through principal coordinate analysis (Section 9.2) to facilitate representation. At the end of each step, a new cluster is formed and its centroid is represented at the *geometric centre* of the last line drawn (examine especially steps 3 and 4 and compare to Fig. 8.8).

In the R mode, weighted centroid clustering does not make sense if the measure of association is Pearson's r. Correlations are cosine transformations of the angles between descriptors; these cannot be combined using eq. 8.4.

8 — *Ward's minimum variance method*

Ward's (1963) minimum variance method is related to the centroid methods (Subsections 6 and 7 above) in that it also leads to a geometric representation in which cluster centroids play an important role. To form clusters, the method minimizes an *objective function* which is, in this case, the same "squared error" criterion as that used in multivariate analysis of variance.

Objective function

At the beginning of the procedure, each objects is in a cluster of its own, so that the distance of an object to its cluster's centroid is 0; hence, the sum of all these distances is also 0. As clusters form, the centroids move away from actual object coordinates and the sums of the squared distances from the objects to the centroids increase. At each clustering step, Ward's method finds the pair of objects or clusters whose fusion increases as little as possible the sum, over all objects, of the *squared distances* between objects and cluster centroids. The distance of object $\mathbf{x}_i$ to the centroid $\mathbf{m}$ of its cluster is computed using the Euclidean distance formula (eq. 7.33) over the various descriptors $\mathbf{y}_j$ ($j = 1 \dots p$):

$$\sum_{j=1}^{p} [y_{ij} - m_j]^2$$

The centroid $\mathbf{m}$ of a cluster was defined at the beginning of Subsection 8.5.6. The sum of squared distances of all objects in cluster k to their common centroid is called "error" in ANOVA, hence the symbol e_k^2:

Squared error

$$e_k^2 = \sum_{i=1}^{n_k} \sum_{j=1}^{p} [y_{ij}^{(k)} - m_j^{(k)}]^2 \qquad \textbf{(8.5)}$$

where $y_{ij}^{(k)}$ is the value of descriptor $\mathbf{y}_j$ for an object i member of group (k) and $m_j^{(k)}$ is the mean value of descriptor j over all members of group k. Alternatively, the within-cluster sums of squared errors e_k^2 can be computed as the mean of the squared distances among cluster members:

$$e_k^2 = \left[\sum_{h, i=1}^{n_k} D_{hi}^2 \right] / n_k \qquad \textbf{(8.6)}$$

where the D_{hi}^2 are the squared distances among objects in cluster k (Table 8.7) and n_k is the number of objects in that cluster. Equations 8.5 and 8.6 both allow the calculation of the squared error criterion. The equivalence of these two equations is stated in a theorem whose demonstration, for the univariate case, is found in Kendall & Stuart (1963, parag. 2.22). Numerical examples illustrating the calculation of eqs. 8.5 and 8.6 are given at the end of Section 8.8 (K-means partitioning).

Table 8.7 Ward's minimum variance clustering of the pond data. Step 1 of the table contains *squared distances* computed as $D^2 = (1 - S)^2$ from the similarity values in Tables 8.3 to 8.6. At each step, the lowest squared distance is identified (italicized boldface value) and the two corresponding objects or groups are fused using eq. 8.10.

Objects	212	214	233	431	432
212	—				**Step 1**
214	***0.16000***	—			
233	1.00000	0.86304	—		
431	1.00000	0.87797	0.49000	—	
432	1.00000	0.61780	0.64000	0.25000	—
212-214		—			**Step 2**
233		1.18869	—		
431		1.19865	0.49000	—	
432		1.02520	0.64000	***0.25000***	—
212-214		—			**Step 3**
233		0.18869	—		
431-432		0.54288	***0.67000***	—	
212-214		—			**Step 4**
233-431-432		***1.67952***	—		

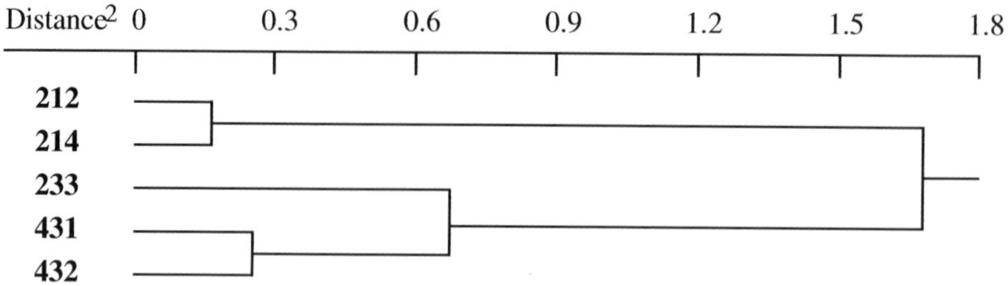

Distance^2 0 0.3 0.6 0.9 1.2 1.5 1.8

212
214
233
431
432

Figure 8.11 Ward's minimum variance clustering of the ponds from Ecological application 8.2. The scale of this dendrogram is the squared distances computed in Table 8.7.

The sum of squared errors E_K^2, for all K clusters corresponding to given partition, is the criterion to be minimized at each clustering step:

Sum of squared errors

$$E_K^2 = \sum_{k=1}^{K} e_k^2 \qquad \textbf{(8.7)}$$

At each clustering step, two objects or clusters **h** and **i** are merged into a new cluster **hi**, as in previous sections. Since changes occurred only in the groups **h**, **i**, and **hi**, the change in the overall sum of squared errors, $\Delta E_{\mathbf{hi}}^2$, may be computed from the changes that occurred in these groups only:

$$\Delta E_{\mathbf{hi}}^2 = e_{\mathbf{hi}}^2 - e_{\mathbf{h}}^2 - e_{\mathbf{i}}^2 \qquad \textbf{(8.8)}$$

It can be shown that this change depends only on the distance between the centroids of clusters **h** and **i** and on their numbers of objects $n_{\mathbf{h}}$ and $n_{\mathbf{i}}$ (Jain & Dubes, 1988):

$$\Delta E_{\mathbf{hi}}^2 = \frac{n_{\mathbf{h}} n_{\mathbf{i}}}{n_{\mathbf{h}} + n_{\mathbf{i}}} \sum_{j=1}^{p} [m_j^{(\mathbf{h})} - m_j^{(\mathbf{i})}]^2 \qquad \textbf{(8.9)}$$

So, one way of identifying the next fusion is to compute the $\Delta E_{\mathbf{hi}}^2$ statistic for all possible pairs and select the pair which generates the smallest value of this statistic for the next merge. Another way is to use the following updating formula to compute the fusion distances between the new cluster **hi** and all other objects or clusters **g**, in the agglomeration table (Table 8.7):

$$D^2(\mathbf{hi}, \mathbf{g}) = \frac{n_{\mathbf{h}} + n_{\mathbf{g}}}{n_{\mathbf{h}} + n_{\mathbf{i}} + n_{\mathbf{g}}} D^2(\mathbf{h}, \mathbf{g}) + \frac{n_{\mathbf{i}} + n_{\mathbf{g}}}{n_{\mathbf{h}} + n_{\mathbf{i}} + n_{\mathbf{g}}} D^2(\mathbf{i}, \mathbf{g}) - \frac{n_{\mathbf{g}}}{n_{\mathbf{h}} + n_{\mathbf{i}} + n_{\mathbf{g}}} D^2(\mathbf{h}, \mathbf{i}) \qquad \textbf{(8.10)}$$

Squared distances are used instead of similarities in eq. 8.10 and in Table 8.7.

Dendrograms for Ward's clustering may be represented along a variety of scales. They all lead to the same clustering topology.

- Fig. 8.11 uses the same scale of squared distances as Table 8.7. This is the solution advocated by Jain & Dubes (1988) and other authors.

- One can easily compute the *square roots* of the fusion distances of Table 8.7 and draw the dendrogram accordingly. This solution, illustrated in Fig. 8.12a, removes the distortions created by squaring the distances. It is especially suitable when one wants to compare the fusion distances of Ward's clustering to the original distances, either graphically (Shepard-like diagrams, Fig. 8.23) or numerically (cophenetic correlations, Subsection 8.11.2).

TESS

• The sum of squared errors E_K^2 (eq. 8.7) is used in some computer programs as the clustering scale. This statistic is also called the *total error sum of squares* (TESS) by Everitt (1980) and other authors. This solution is illustrated in Fig. 8.12b.

• The SAS package (1985) recommends two scales for Ward's clustering. The first one is the proportion of variance (R^2) accounted for by the clusters at any given partition level. It is computed as the total sum of squares (i.e. the sum of squared distances from the centroid of all objects) minus the within-cluster squared errors E_K^2 of eq. 8.7 for the given partition, divided by the total sum of squares. R^2 decreases as clusters grow. When all the objects are lumped in a single cluster, the resulting one-cluster partition does not explain any of the objects' variation so that $R^2 = 0$. The second scale recommended by SAS is called the *semipartial* R^2. It is computed as the between-cluster sum of squares divided by the (corrected) total sum of squares. This statistic increases as the clusters grow.

Like the K-means partitioning method (Section 8.8), Ward's agglomerative clustering can be computed from either a raw data table using eq. 8.8, or a matrix of squared distances through eq. 8.10. The latter is the most usual approach in computer programs. It is important to note that distances are computed as (squared) Euclidean

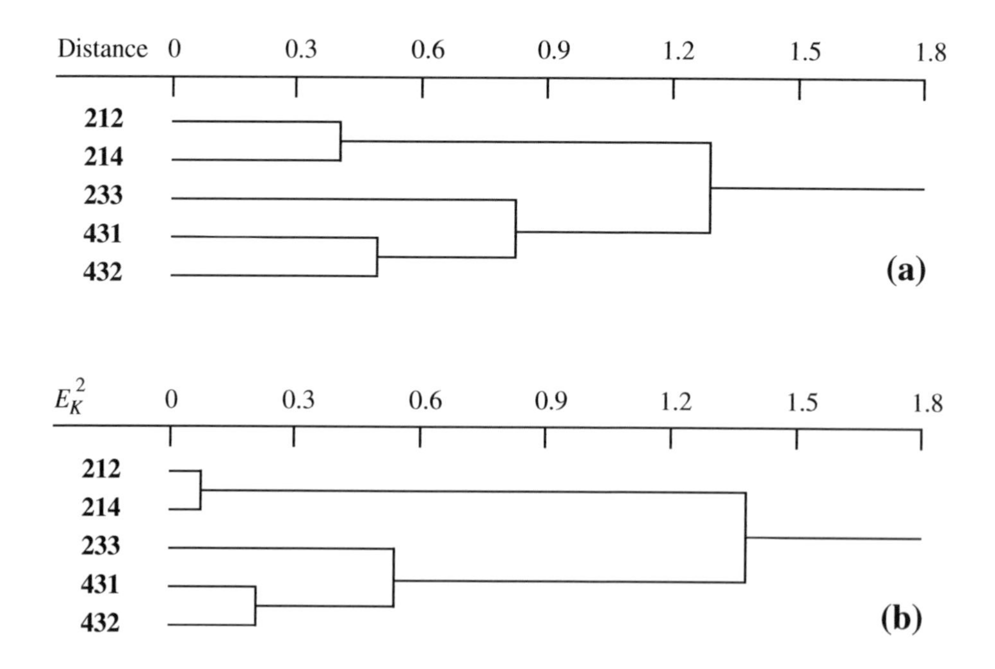

Figure 8.12 Ward's minimum variance clustering of the ponds from Ecological application 8.2. The scale of dendrogram (a) is the square root of the squared distances computed in Table 8.7; in dendrogram (b), it is the E_K^2 (or TESS) statistic.

distances in Ward's method. So, unless the descriptors are such that Euclidean distances (D_1 in Chapter 7) are an appropriate model for the relationships among objects, one should not use a Ward's algorithm based or raw data. It is preferable in such cases to compute a distance matrix using an appropriate coefficient (Tables 7.3 to 7.5), followed by clustering of the resemblance matrix in A-space by a distance-based Ward algorithm. Section 9.2 will show that resemblance matrices can also be used for plotting the positions of objects in A-space, "as if" the distances were Euclidean.

Because of the squared error criterion used as the objective function to minimize, clusters produced by the Ward minimum variance method tend to be hyperspherical, i.e. spherical in multidimensional A-space, and to contain roughly the same number of objects if the observations are evenly distributed through A-space. The same applies to the centroid methods of the previous subsections. This may be seen as either an advantage or a problem, depending on the researcher's conceptual model of a cluster.

9 — General agglomerative clustering model

Lance & Williams (1966a, 1967c) have proposed a general model that encompasses all the agglomerative clustering methods presented up to now, except intermediate linkage (Subsection 3). The general model offers the advantage of being translatable into a single, simple computer program, so that it is used in most statistical packages that offer agglomerative clustering. The general model allows one to select an agglomerative clustering model by choosing the values of four parameters, called α_h, α_i, β, and γ, which determine the clustering strategy. This model only outputs the branching pattern of the clustering tree (the dendrogram), as it was the case for the methods described in Subsections 8.5.4 to 8.5.8. For the linkage clustering strategies (Subsections 8.5.1 to 8.5.3), the list of links responsible for cluster formation may be obtained afterwards by comparing the dendrogram to the similarity matrix.

Combinatorial method

The model of Lance & Williams is limited to *combinatorial* clustering methods, i.e. those for which the similarity $S(\mathbf{hi}, \mathbf{g})$ between an external cluster $\mathbf{g}$ and a cluster $\mathbf{hi}$, resulting from the prior fusion of clusters $\mathbf{h}$ and $\mathbf{i}$, is a function of the three similarities $S(\mathbf{h}, \mathbf{g})$, $S(\mathbf{i}, \mathbf{g})$, and $S(\mathbf{h}, \mathbf{i})$ and also, eventually, the numbers $n_\mathbf{h}$, $n_\mathbf{i}$, and $n_\mathbf{g}$ of objects in clusters $\mathbf{h}$, $\mathbf{i}$, and $\mathbf{g}$, respectively (Fig. 8.13). Individual objects are considered to be single-member clusters. Since the similarity of cluster $\mathbf{hi}$ with an external cluster $\mathbf{g}$ can be computed from the above six values, $\mathbf{h}$ and $\mathbf{i}$ can be condensed into a single row and a single column in the updated similarity matrix; following that, the clustering proceeds as in the Tables of the previous Subsections. Since the new similarities at each step can be computed by *combining* those from the previous step, it is not necessary for a computer program to retain the original similarity matrix or data set. Non-combinatorial methods do not have this property. For similarities, the general model for combinatorial methods is the following:

$$S(\mathbf{hi}, \mathbf{g}) = (1 - \alpha_h - \alpha_i - \beta) + \alpha_h S(\mathbf{h}, \mathbf{g}) + \alpha_i S(\mathbf{i}, \mathbf{g}) + \beta S(\mathbf{h}, \mathbf{i}) - \gamma |S(\mathbf{h}, \mathbf{g}) - S(\mathbf{i}, \mathbf{g})| \quad \textbf{(8.11)}$$

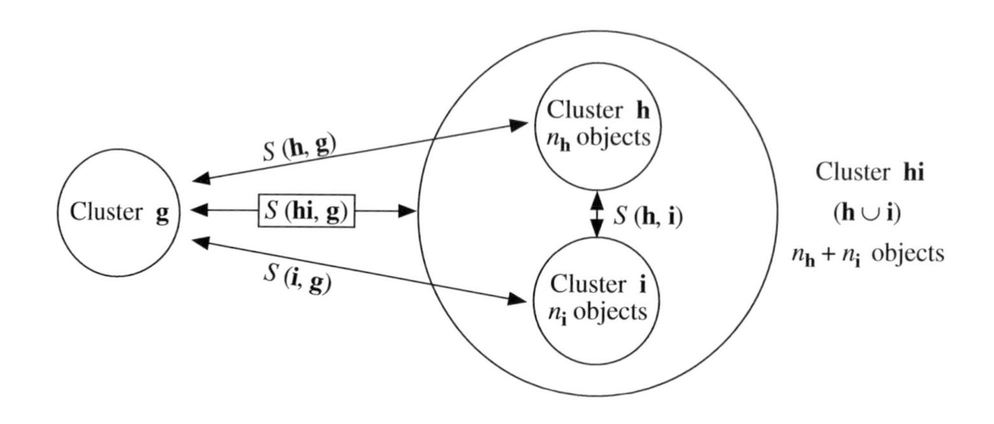

Figure 8.13 In combinatorial clustering methods, the similarity between a cluster **hi**, resulting from the fusion of two previously formed clusters **h** and **i**, and an external cluster **g** is a function of the three similarities between (**h** and **i**), (**h** and **g**), and (**i** and **g**), and of the number of objects in **h**, **i**, and **g**.

When using distances, the combinatorial equation becomes:

$$D(\mathbf{hi}, \mathbf{g}) = \alpha_h D(\mathbf{h}, \mathbf{g}) + \alpha_i D(\mathbf{i}, \mathbf{g}) + \beta D(\mathbf{h}, \mathbf{i}) + \gamma |D(\mathbf{h}, \mathbf{g}) - D(\mathbf{i}, \mathbf{g})| \qquad \textbf{(8.12)}$$

Clustering proceeds in the same way for all combinatorial agglomerative methods. As the similarity decreases, a new cluster is obtained by the fusion of the two most similar objects or groups, after which the algorithm proceeds to the fusion of the two corresponding rows and columns in the similarity (or distance) matrix using eq. 8.11 or 8.12. The matrix is thus reduced by one row and one column at each step. Table 8.8 gives the values of the four parameters for the most commonly used combinatorial agglomerative clustering strategies. Values of the parameters for some other clustering strategies are given by Gordon (1996a).

In the case of equality between two mutually exclusive pairs, the decision may be made on an arbitrary basis (the so-called "right-hand rule" used in most computer programs) or based upon ecological criteria (as, for example, Sørensen's criteria reported at the end of Subsection 8.5.2, or those explained in Subsection 8.9.1).

In several strategies, $\alpha_h + \alpha_i + \beta = 1$, so that the term $(1 - \alpha_h - \alpha_i - \beta)$ becomes zero and disappears from eq. 8.11. One can show how the values chosen for the four parameters make the general equation correspond to each specific clustering method. For single linkage clustering, for instance, the general equation becomes:

$$S(\mathbf{hi}, \mathbf{g}) = \frac{1}{2}[S(\mathbf{h}, \mathbf{g}) + S(\mathbf{i}, \mathbf{g}) + |S(\mathbf{h}, \mathbf{g}) - S(\mathbf{i}, \mathbf{g})|]$$

Table 8.8 Values of parameters α_h, α_i, β, and γ in Lance and Williams' general model for combinatorial agglomerative clustering. Modified from Sneath & Sokal (1973) and Jain & Dubes (1988).

Clustering method	α_h	α_i	β	γ	Effect on space A
Single linkage	1/2	1/2	0	–1/2	Contracting*
Complete linkage	1/2	1/2	0	1/2	Dilating*
UPGMA	$\frac{n_h}{n_h + n_i}$	$\frac{n_i}{n_h + n_i}$	0	0	Conserving*
WPGMA	1/2	1/2	0	0	Conserving
UPGMC	$\frac{n_h}{n_h + n_i}$	$\frac{n_i}{n_h + n_i}$	$\frac{-n_h n_i}{(n_h + n_i)^2}$	0	Conserving
WPGMC	1/2	1/2	–1/4	0	Conserving
Ward's	$\frac{n_h + n_g}{n_h + n_i + n_g}$	$\frac{n_i + n_g}{n_h + n_i + n_g}$	$\frac{-n_g}{n_h + n_i + n_g}$	0	Conserving
Flexible	$\frac{1 - \beta}{2}$	$\frac{1 - \beta}{2}$	$-1 \le \beta < 1$	0	Contracting if $\beta \approx 1$ Conserving if $\beta \approx -.25$ Dilating if $\beta \approx -1$

* Terms used by Sneath & Sokal (1973).

The last term (absolute value) completes the smallest of the two similarities $S(\mathbf{h}, \mathbf{g})$ and $S(\mathbf{i}, \mathbf{g})$, making it equal to the largest one. Hence, $S(\mathbf{hi}, \mathbf{g}) = \max[S(\mathbf{h}, \mathbf{g}), S(\mathbf{i}, \mathbf{g})]$. In other words, the similarity between a newly-formed cluster **hi** and some other cluster **g** becomes equal to the largest of the similarity values previously computed between the two original clusters (**h** and **i**) and **g**.

Intermediate linkage clustering is not a combinatorial strategy. All along the clustering procedure, it is necessary to refer to the original association matrix in order to calculate the connectedness of pairs of clusters. This is why it cannot be obtained using the Lance & Williams general agglomerative clustering model.

10 — Flexible clustering

Lance & Williams (l966a. 1967c) proposed to vary the parameter β (eq. 8.11 or 8.12) between –1 and +1 to obtain a series of intermediates solutions between single linkage chaining and the space dilation of complete linkage. The method is also called *beta-flexible clustering* by some authors. Lance & Williams (*ibid.*) have shown that, if the other parameters are constrained a follows:

$$\alpha_h = \alpha_i = (1 - \beta)/2 \qquad \text{and} \qquad \gamma = 0$$

the resulting clustering is ultrametric (no reversals; Section 8.6).

When β is close to 1, strong chaining is obtained. As β decreases and becomes negative, space dilation increases. The space properties are conserved for small negative values of β, near –0.25. Figure 8.14 shows the effect of varying β in the clustering of 20 objects. Like weighted centroid clustering, flexible clustering is compatible with all association measures except Pearson's r.

Ecological application 8.5a

Pinel-Alloul *et al.* (1990) studied phytoplankton in 54 lakes of Québec to determine the effects of acidification, physical and chemical characteristics, and lake morphology on species assemblages. Phytoplankton was enumerated into five main taxonomic categories (microflagellates, chlorophytes, cyanophytes, chrysophytes, and pyrrophytes). The data were normalized using the generalized form of the Box-Cox method that finds the best normalizing transformation for all species (Subsection 1.5.6). A Gower (S_{19}) similarity matrix, computed among lakes, was subjected to flexible clustering with parameter $\beta = -0.25$. Six clusters were found, which were roughly distributed along a NE-SW geographic axis and corresponded to increasing concentrations of total phytoplankton, chlorophytes, cyanophytes, and microflagellates. Explanation of the phytoplankton-based lake typology was sought by comparing it to the environmental variables (Section 10.2.1).

11 — Information analysis

The Q-mode clustering method called *information analysis* was developed for ecological purposes by Williams *et al.* (1966) and Lance & Williams (1966b). It does not go through the usual steps of similarity calculation followed by clustering. It is a direct method of clustering, based on information measures.

Entropy

Shannon's formula (eq. 6.1) can be used to measure the diversity or information in a frequency or probability distribution:

$$H = -\sum_{j=1}^{p} \mathrm{p}_j \log \mathrm{p}_j$$

Information analysis is a type of unweighted centroid clustering, adapted to species data. At each step, the two objects or clusters causing the smallest gain in within-group diversity (or information) are fused. As a consequence, the clusters are as homogeneous as possible in terms of species composition.

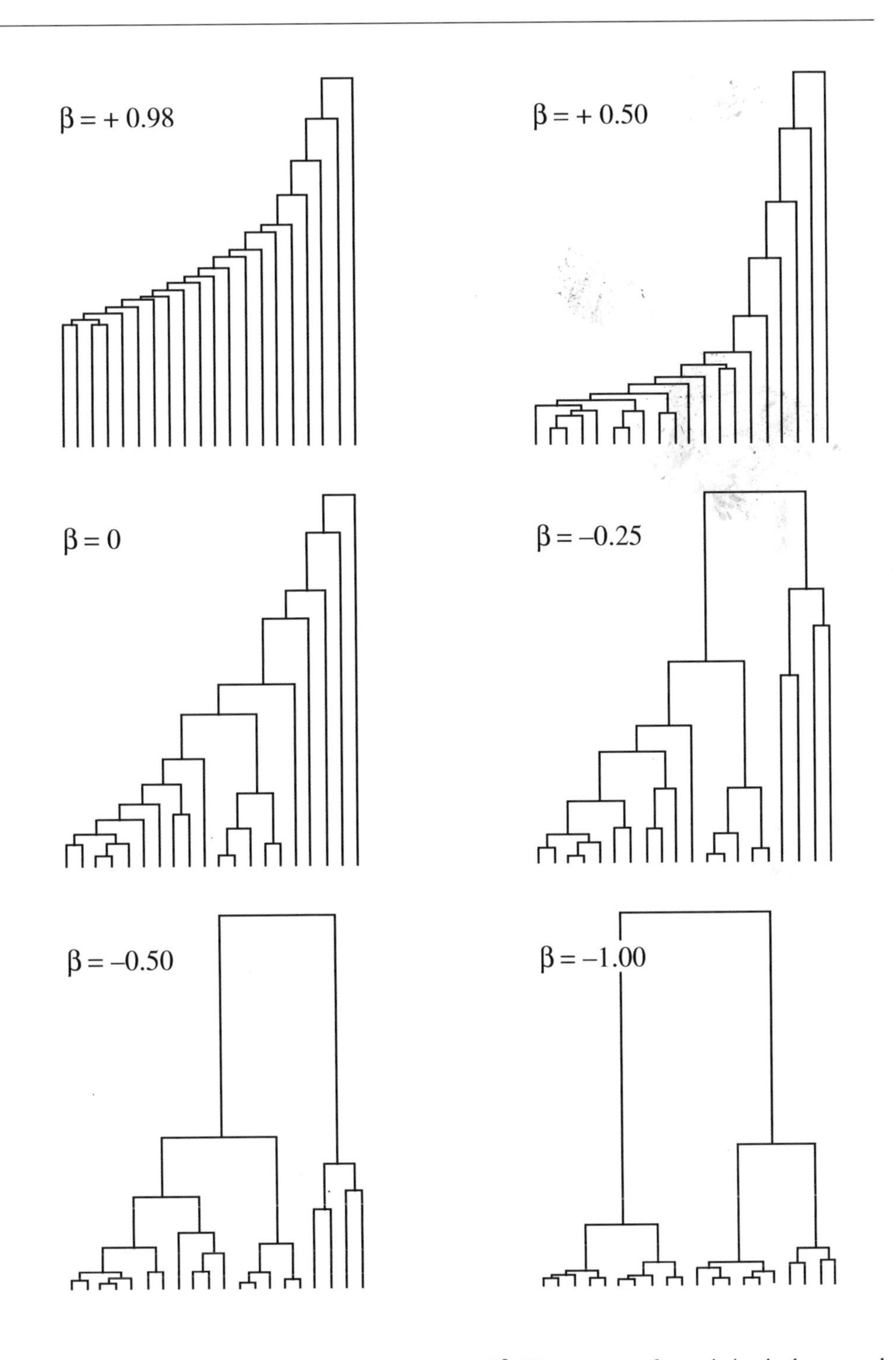

Figure 8.14 Flexible clustering of 20 objects for six values of β. The measure of association is the squared Euclidean distance D_1^2. Adapted from Lance & Williams (1967c: 376).

The method could be applied to species abundance data, divided into a small number of classes but, in practice, it is mostly used with presence-absence data. The information measure described below is not applicable to raw abundance data because the number of different states would then vary from one species to another, which would give them different weights in the overall measure.

To illustrate the method, the pond zooplankton counts used in Chapter 7 (coefficient S_{23}) are transformed here into presence-absence data (see also Ecological application 8.2):

Species j	Ponds					p_j	$(1 - p_j)$
	212	214	233	431	432		
1	1	1	0	0	0	0.4	0.6
2	0	0	1	1	0	0.4	0.6
3	0	1	1	0	1	0.6	0.4
4	0	0	1	1	1	0.6	0.4
5	1	1	0	0	0	0.4	0.6
6	0	1	0	1	1	0.6	0.4
7	0	0	0	1	1	0.4	0.6
8	1	1	0	0	0	0.4	0.6

Total information in this group of ponds is computed using an information measure derived from the following reasoning (Lance & Williams, 1966b). The entropy of each species presence-absence descriptor j is calculated on the basis of the probabilities of presence p_j and absence $(1 - p_j)$ of species j, which are written in the right-hand part of the table. The probability of presence is estimated as the number of ponds where species j is present, divided by the total number of ponds *in the cluster under consideration* (here, the group of five ponds). The probability of absence is estimated likewise, using the number of ponds where species j is absent. The entropy of species j is therefore:

$$H(j) = -[p_j \log p_j + (1 - p_j) \log(1 - p_j)] \quad \text{for } 0 < p_j < 1 \qquad \textbf{(8.13)}$$

The base of the logarithms is indifferent, as long as the same base is used throughout the calculations. Natural logarithms are used throughout the present example. For the first species, $H(1)$ would be:

$$H(1) = -[0.4 \ln(0.4) + 0.6 \ln(0.6)] = 0.673$$

The information of the conditional probability table can be calculated by summing the entropies per species, considering that all species have the same weight. Since the measure of *total information* in the group must also take into account the number of objects in the cluster, it is defined as follows:

$$I = -n \sum_{j=1}^{p} [p_i \log p_j + (1 - p_j) \log (1 - p_j)] \quad \text{for } 0 < p_j < 1 \qquad \textbf{(8.14)}$$

where p is the number of species represented in the group of n objects (ponds). For the group of 5 ponds above,

$$I = -5\,[8\,(-0.673)] = 26.920$$

If I is to be expressed as a function of the number a_j of ponds with species j present, instead of a function of probabilities $p_j = a_j/n$, it can be shown that the following formula is equivalent to eq. 8.14:

$$I = np \log n - \sum_{j=1}^{p} [a_i \log a_j + (n - a_j) \log (n - a_j)] \qquad \textbf{(8.15)}$$

I is zero when all ponds in a group contain the same species. Like entropy H, I has no upper limit; its maximum value depends on the number of species present in the study.

At each clustering step, three series of values are considered: (a) the total information I in each group, which is 0 at the beginning of the process since each object (pond) then forms a distinct cluster; (b) the value of I for all possible combinations of groups taken two at a time; (c) the increase of information ΔI resulting from each possible fusion. As recommended by Sneath & Sokal (1973), all these values can be placed in a matrix, initially of dimension $n \times n$ which decreases as clustering proceeds. For the example data, values (a) of information in each group are placed on the diagonal, values (b) in the lower triangle, and values (c) of ΔI in the upper triangle, in italics.

Ponds	Ponds				
	212	214	233	431	432
212	0	*2.773*	*8.318*	*9.704*	*9.704*
214	2.773	0	*8.318*	*9.704*	*6.931*
233	8.318	8.318	0	*4.159*	*4.159*
431	9.704	9.704	4.159	0	*2.773*
432	9.704	6.931	4.159	2.773	0

The ΔI for two groups is found by subtracting the corresponding values I, on the diagonal, from the value I of their combination in the lower triangle. Values on the diagonal are 0 in this first calculation matrix, so that values in the upper triangle are the same as in the lower triangle, but this will not be the case in subsequent matrices.

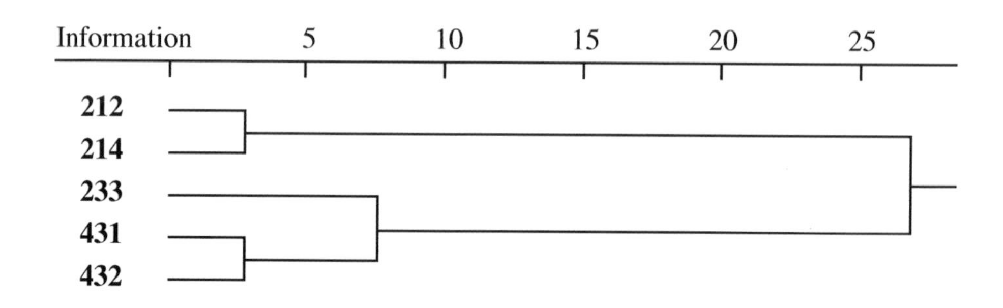

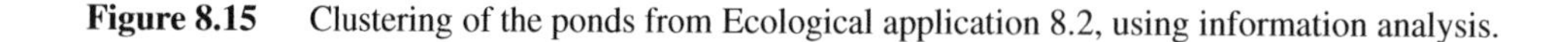

Figure 8.15 Clustering of the ponds from Ecological application 8.2, using information analysis.

The first fusion is identified by the lowest ΔI value found in the upper triangle. This value is 2.773 for pairs (212, 214) and (431, 432), which therefore fuse. A new matrix of I values is computed:

Groups	Groups		
	212 214	233	431 432
212-214	2.773	*10.594*	*15.588*
233	13.367	0	*4.865*
431-432	21.134	7.638	2.773

This time, the ΔI values in the upper triangle differ from the I's in the lower triangle since there are I values on the diagonal. The ΔI corresponding to group (212, 214, 431, 432), for example, is computed as: 21.134 – 2.773 – 2.773 = 15.588. The lowest value of ΔI is for the group (233, 431, 432), which therefore fuses at this step.

For the last clustering step, the only I value to calculate in the lower triangle is for the cluster containing the five ponds. This value is 26.920, as computed above from eq. 8.14. ΔI is then 26.920 – 2.773 – 7.638 = 16.509.

Groups	Groups	
	212 214	233 431-432
212-214	2.773	*16.509*
233-431-432	26.920	7.638

The last fusion occurs at I = 26.920; computing ΔI would not have been necessary in this case. The values of I can be used as the scale for a dendrogram summarizing the clustering steps (Fig. 8.15).

According to Williams *et al.* (1966), information analysis minimizes chaining and quickly delineates the main clusters, at least with ecological data. Field (1969) pointed out, however, that information analysis bases the similarity between objects on double absences as well as double presences. This method may therefore not be appropriate when a gradient has been sampled and the data matrix contains many zeros; see Section 7.3 and Subsection 9.4.5 for a discussion of this problem.

Efficiency coefficient

The inverse of ΔI is known as the *efficiency coefficient* (Lance & Williams, 1966b). An analogue to the efficiency coefficient may be computed for dendrograms obtained using other agglomerative clustering procedures. In that case, the efficiency coefficient is still computed as $1/\Delta$, where Δ represents the amount by which the information in the classification is reduced due to the fusion of groups. The reduction is computed as the entropy in the classification before a fusion level minus the entropy after that fusion. In Fig. 8.2 for instance, the partition at $S = 0.40$ contains three groups of 2, 2, and 1 objects respectively; using natural logarithms, Shannon's formula (eq. 6.1) gives $H = 1.05492$. The next partition, at $S = 0.25$, contains two groups of 2 and 3 objects respectively; Shannon's formula gives $H = 0.67301$. The difference is $\Delta = 0.38191$, hence the efficiency coefficient $1/\Delta = 2.61843$ for the dendrogram fusion level $S = 0.3$.

When $1/\Delta I$ is high, the procedure clusters objects that are mostly alike. The efficiency coefficient does not monotonically decrease as the clustering proceeds. With real data, it may decrease, reach a minimum, and increase again. If $1/\Delta I$ is plotted as a function of the successive fusion levels, the minima in the graph indicate the most important partitions. If one wants to select a single cutting level through a dendrogram, this graph may help in deciding which partition should be selected. In Fig. 8.2 for example, one would choose the value $1/\Delta I = 1.48586$, which corresponds to the last fusion level ($S = 0.214$), as the most informative partition. The efficiency coefficient is not a rigorous decision criterion, however, since no test of statistical significance is performed.

8.6 Reversals

Reversals may occasionally occur in the clustering structure when using UPGMC or WPGMC (Subsections 8.5.6 and 8.5.7), or with some unusual combinations of parameters in the general agglomerative model of Lance & Williams (Subsection 8.5.9). As an example, a reversal was produced in Fig. 8.7. Two types of situations lead to reversals:

- When $\mathbf{x}_1$ and $\mathbf{x}_2$ cluster first, because they represent the closest pair, although the distance from $\mathbf{x}_3$ to the centroid $\mathbf{c}_{12}$ is smaller than the distance from $\mathbf{x}_1$ to $\mathbf{x}_2$ (Fig. 8.16a).

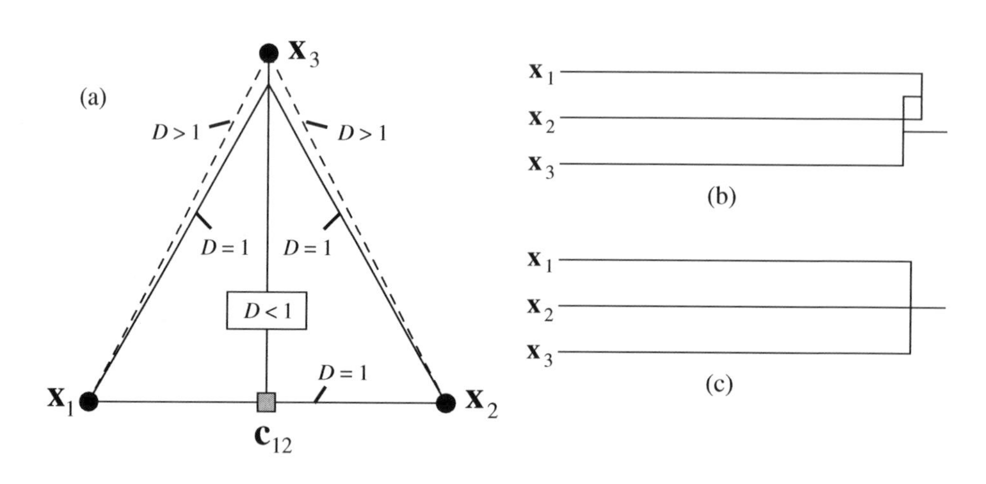

Figure 8.16 A reversal may occur in situations such as (a), where $\mathbf{x}_1$ and $\mathbf{x}_2$ cluster first because they represent the closest pair, although the distance from $\mathbf{x}_3$ to the centroid $\mathbf{c}_{12}$ is smaller than the distance from $\mathbf{x}_1$ to $\mathbf{x}_2$. (b) The result is usually depicted by a non-ultrametric dendrogram. (c) The reversal may also be interpreted as a trichotomy.

- When $D(\mathbf{x}_1, \mathbf{x}_2) = D(\mathbf{x}_1, \mathbf{x}_3) = D(\mathbf{x}_2, \mathbf{x}_3)$. In such a situation, most computer programs use an arbitrary rule ("right-hand rule") and first cluster two of the three objects. A reversal appears when the third object is added to the cluster.

When this happens, the cophenetic matrix (Subsection 8.3.1) violates the ultrametric property (Subsection 8.3.2) and the dendrogram is more difficult to draw than in the no-reversal cases (Fig. 8.16b). However, departures from ultrametricity are never large in practice. For this reason, a reversal may be interpreted as nearly equivalent to a trichotomy in the hierarchical structure (Fig. 8.16c). They may also indicate true trichotomies, as discussed above; this can be checked by examination of the similarity or distance matrix.

A clustering method is said to be *monotonic* (i.e. without reversals) if:

$$S(\mathbf{x}_1 \cup \mathbf{x}_2, \mathbf{x}_3) \leq S(\mathbf{x}_1, \mathbf{x}_2)$$

or

$$D(\mathbf{x}_1 \cup \mathbf{x}_2, \mathbf{x}_3) \geq D(\mathbf{x}_1, \mathbf{x}_2)$$

Assuming that $\alpha_h > 0$ and $\alpha_i > 0$ (Table 8.8), necessary and sufficient conditions for a clustering method to be monotonic in all situations are the following:

$$\alpha_h + \alpha_i + \beta \geq 1$$

and

$$\gamma \geq -\min(\alpha_h, \alpha_i)$$

(Milligan, 1979; Jain & Dubes, 1988). Some authors use the term *classification* only for hierarchies *without reversals* or for single non-overlapping partitions of the objects (Section 8.8).

8.7 Hierarchical divisive clustering

Contrary to the agglomerative methods of Section 8.5, hierarchical divisive techniques use the whole set of objects as the starting point. They divide it into two or several subgroups, after which they consider each subgroup and divide it again, until the criterion chosen to end the divisive procedure is met (Lance & Williams, 1967b).

In practice, hierarchical divisive clustering can only be achieved in the monothetic case, or when working in an ordination space. In monothetic divisive methods, the objects are divided, at each step of the procedure, according to the states of a single descriptor. This descriptor is chosen because it best represents the whole set of descriptors (next subsection). Polythetic algorithms have been developed, but it will be seen that they are not satisfactory.

An alternative is to use a partitioning method (Section 8.8) for all possible numbers of groups from $K = 2$ to $K = (n - 1)$ and assemble the results into a graph. There is no guarantee, however, that the groups will be nested and form a hierarchy, unless the biological or ecological processes that have generated the data are themselves hierarchical.

1 — Monothetic methods

Association analysis

The clustering methods that use only one descriptor at a time are less than ideal, even when the descriptor is chosen after considering all the others. The best-known monothetic method in ecology is Williams & Lambert's (1959) *association analysis*, originally described for species presence-absence data. Association analysis may actually be applied to any binary data table, not only species. The problem is to identify, at each step of the procedure, which descriptor is the most strongly associated with all the others. First, X^2 (chi-square) values are computed for 2×2 contingency tables comparing all pairs of descriptors in turn. X^2 is computed using the usual formula:

$$X^2 = n\,(ad - bc)^2/[(a + b)\,(c + d)\,(a + c)\,(b + d)]$$

The formula may include Yates' correction for small sample sizes, as in similarity coefficient S_{25}. The X^2 values relative to each descriptor k are summed up:

$$\sum_{j=1}^{p} X_{jk}^2 \quad \text{for } j \neq k \qquad \textbf{(8.16)}$$

The largest sum identifies the descriptor that is the most closely related to all the others. The first partition is made along the states of this descriptor; a first cluster is made of the objects coded 0 for the descriptor and a second cluster for the objects coded 1. The descriptor is eliminated from the study and the procedure is repeated, separately for each cluster. Division stops when the desired number of clusters is reached or when the sum of X^2 values no longer reaches a previously set threshold.

This method has been adapted by Lance & Williams (1968) to the information statistic I of Subsection 8.5.11. Lance & Williams (1965) have also suggested using the point correlation coefficient $\varphi = \sqrt{X^2/n}$ (Subsection 7.3.1) instead of X^2. This may prevent aberrant or unique objects in the study from determining the first partitions. This is analogous to the problem encountered with the higher powers of Minkowski's metric (D_6), which could give too much weight to the largest differences; this problem was less severe when using power 1, which is the Manhattan metric (D_7). One then looks for the descriptor that maximizes the sum $\Sigma\varphi_{jk}$ ($j \neq k$; see eq. 8.16). Gower (1967) suggested maximizing the multiple correlation of species k with all the other species, instead of the sum of simple correlations. This is equivalent to looking for the best multiple regression function of a descriptor with respect to all the others. Gower also provided the mathematical bases for an algorithm designed to generate all the results from a single matrix inversion.

The principles of association analysis may be applied to descriptors with multiple states (semiquantitative or qualitative), by computing X^2 values between descriptors with the usual X^2 formula. Raw species abundance data should not be analysed in this way, however, because the large number of different abundance values then makes the contingency table meaningless. Using correlation coefficients instead of X^2 statistics would not help, because of the problem of the double zeros discussed at length in section 7.5.

Legendre & Rogers (1972) proposed a monothetic divisive method similar to association analysis, in which the choice of the descriptor best representing all the others is made with the help of an information statistic computed on contingency tables. For each descriptor k, two quantities developed by Christanson (*in* Brill *et al.*, 1972) are computed: SUMRAT (k) and SAMRAT (k) ("sum of ratios"). SUMRAT (k) is the sum of the fractions representing the amount of information that k has in common with each descriptor j ($j \neq k$), divided by the amount of information in j. In SAMRAT (k), the divisor is the amount of information in k instead of j. Using the symbolism of Section 6.2:

$$\text{SUMRAT}\,(k) = \sum_{j=1}^{p} \frac{H(k) - H(k|j)}{H(j)} \quad \text{for } j \neq k \tag{8.17}$$

$$\text{SAMRAT}(k) = \sum_{j=1}^{p} \frac{H(k) - H(k|j)}{H(k)} \quad \text{for } j \neq k \tag{8.18}$$

which can be recognized as sums of asymmetric uncertainty coefficients, Σ B/(B + C) and Σ B/(A + C), respectively (Section 6.2). SUMRAT (k) and SAMRAT (k) both have the property of being high when k has much information in common with the other descriptors in the study. The descriptor that best represents the divisive power of all descriptors should have the highest SUMRAT and SAMRAT values. However, SUMRAT (k) and SAMRAT (k) are also influenced by the number of states in k, which may unduly inflate $H(k)$, thus causing SUMRAT (k) to increase and SAMRAT (k) to decrease. This factor must be taken into account if there is conflict between the indications provided by SUMRAT and SAMRAT as to the descriptor that best represents the whole set. This peculiarity of the method requires the user's intervention at each division step, in the present state of development of the equations.

Since the information measures on which SUMRAT and SAMRAT are based are at the same exponent level as X^2 (Section 6.2), one could compute instead:

$$\text{SUMRAT}(k) = \sum_{j=1}^{p} \sqrt{\frac{H(k) - H(k|j)}{H(j)}} \quad \text{for } j \neq k \tag{8.19}$$

$$\text{SAMRAT}(k) = \sum_{j=1}^{p} \sqrt{\frac{H(k) - H(k|j)}{H(k)}} \quad \text{for } j \neq k \tag{8.20}$$

thus minimizing the effect of single objects on the first partitions, as indicated above.

Williams & Lambert (1961) have suggested using association analysis in the R mode for identifying species associations. This approach does not seem, however, to be based on an acceptable operational concept of association (see Section 8.9).

2 — *Polythetic methods*

There is no satisfactory algorithm for the hierarchical division of objects based on the entire set of descriptors.

The method of Edwards & Cavalli-Sforza (1965) tries all possible divisions of the set of objects into two clusters, looking for the division that maximizes the distance between the centroids. Like the K-means method of Section 8.8, it can only be applied to quantitative descriptors. Using sums of squared distances to centroids, one first computes SS, which is the sum of squares of the Euclidean distances of all objects to the centroid of the whole set of objects; this is the total sum of squares of single classification analysis of variance. Then, for each possible partition of the objects into two groups **h** and **i**, the sums of squares of the distances to the centroids are computed within each cluster, using eq. 8.5, to obtain $SS(\mathbf{h})$ and $SS(\mathbf{i})$, respectively. The distance between the two clusters is therefore $SS - SS(\mathbf{h}) - SS(\mathbf{i})$. This is the quantity to be maximized for the first partition. Then each cluster is considered in turn and the operation is repeated to obtain subsequent divisions.

This method may seem attractive but, apart from the theoretical objections that he raised about it, Gower (1967) calculated that, before obtaining the first partition of a cluster of 41 objects, 54 000 years of computing time would be required using a computer with an access time of 5 microseconds, to try all $(2^{40} - 1)$ possible partitions of 41 objects into two groups. The problem remains with modern computers, even though they have an access time of 10 to 100 nanoseconds.

Dissimilarity analysis

The *dissimilarity analysis* of Macnaughton-Smith *et al.* (1964) first looks for the object which is the most different from all the others and removes it from the initial cluster. One by one, the most different objects are removed. Two groups are defined: the objects removed and the remaining ones, between which a distance is calculated. Objects are removed up to the point where the distance between clusters can no longer be increased. Each of the two clusters thus formed is subdivided again, using the same procedure. The first partition of a cluster of n objects requires at most $3n^2/4$ operations instead of the $(2^{n-1} - 1)$ operations required by the previous method. Other authors have developed special measures of distance to be used in dissimilarity analysis, such as Hall's (1965) *singularity index* and Goodall's (1966b) *deviant index.* Although attractive, dissimilarity analysis may produce strange results when many small clusters are present in the data, in addition to major clusters of objects. Actually, there is always the danger that a division into major clusters may also separate the members of some minor cluster, which cannot be fused again unless special procedures are included in the algorithm for this purpose (Williams & Dale, 1965).

3 — Division in ordination space

Efficient polythetic hierarchical divisive clustering can be obtained by partitioning the objects according to the axes of an ordination space. Using principal component analysis (PCA, Section 9.1), the set of objects may be partitioned in two groups: those that have positive values along the first PCA axis and those that have negative values. The PCA analysis is repeated for each of the groups so obtained and a new partition of each group is performed. The process is repeated until the desired level of resolution is obtained (Williams, 1976b).

Following a similar suggestion by Piazza & Cavalli-Sforza (1975), Lefkovitch (1976) developed a hierarchical classification method for very large numbers of objects, based on principal coordinate analysis (PCoA, Section 9.2). The dendrogram is constructed from the successive principal coordinate axes, the signs of the objects on the coordinate axes indicating their membership in one of the two groups formed at each branching step. The objects are partitioned in two groups according to their signs along the first PCoA axis; each group is then divided according to the positions of the objects along the second axis; and so on. This differs from the method used with PCA above, where the analysis is repeated for each group before a new division takes place.

To calculate the principal coordinates of a large number of objects, Lefkovitch proposed to first measure the similarity among objects by an equation which, like the covariance or correlation, is equivalent to the product of a matrix with its transpose. He described such a measure, applicable if necessary to combinations of binary, semiquantitative, and quantitative descriptors. The association matrix among objects is obtained by the matrix product **YY'** (order $n \times n$). In many problems where there are many more objects than descriptors, computation of the eigenvalues and eigenvectors of the association matrix among *descriptors,* **Y'Y**, represents an important saving of computer time because **Y'Y** (order $p \times p$) is much smaller than **YY'** (order $n \times n$). After Rao (1964) and Gower (1966), Lefkovitch showed that the principal coordinates **V** of the association matrix among *objects* can then be found, using the relation **V** = **YU** where **U** is the matrix of the principal coordinates among *descriptors.* The principal coordinates thus calculated allow one to position the objects, numerous as they may be, in the reduced space. Principal coordinates can be used for the binary hierarchical divisive classification procedure that was Lefkovitch's goal.

Another way of obtaining Lefkovitch's classification is to compute the principal coordinates from the ($n \times n$) similarity matrix among objects, using the TWWS algorithm described in Subsection 9.2.6. This algorithm allows one to quickly obtain a small number of principal coordinates for a fairly large number of objects. The first few principal coordinates, calculated in this way, could be used as the basis for Lefkovitch's clustering method.

A divisive algorithm of the same type is used in TWINSPAN (below). It is based upon an ordination obtained by correspondence analysis instead of PCA or PCoA.

4 — TWINSPAN

TWINSPAN stands for *Two Way INdicator SPecies ANalysis*, although its purpose is *not* primarily to identify indicator species. The method is widely used in vegetation science. The TWINSPAN procedure*, proposed by Hill (1979), classifies the objects by hierarchical division and constructs an ordered two-way table from the original species-by-sites data table; Hill calls it a *dichotomized ordination analysis*. TWINSPAN produces a tabular matrix arrangement that approximates the results of a Braun-Blanquet phytosociological analysis table.

Pseudo-species

To model the concept of *differential species* (i.e. species with clear ecological preferences), which is qualitative, TWINSPAN creates *pseudospecies.* Each species is recoded into a set of dummy variables (pseudospecies) corresponding to relative abundance levels; these classes are cumulative. If the pseudospecies cutting levels are 1%, 11%, 26%, 51%, and 76%, for instance, a relative abundance of 18% at a site will fill the first and second dummy pseudospecies vectors with "1" (= presence). Cutting levels are arbitrarily decided by users. A (sites × pseudospecies) data table is created.

* Available as part of the package PC-ORD (distribution: see footnote in Section 9.3). TWINSPAN also available from Micro-computer Power: <http://www.microcomputerpower.com>.

The TWINSPAN procedure is rather complex. A detailed description is given by Kent & Coker (1992). It may be summarized as follows.

1. Objects are divided in two groups according to their signs along the first correspondence analysis axis (CA, Section 9.4) obtained from analysing the original (sites × species) data table. This is called the primary ordination.

Indicator value

2. Pseudospecies make it possible to turn species relative abundances into measures of their *indicator values* for that binary partition, on a scale from –1 (indicative of the group of objects found on the arbitrarily selected left-hand side of the CA axis) to +1 (indicative of the group on the right-hand side). The indicator value actually describes the preference of a pseudospecies for one or the other side of the partition.

3. A refined ordination is calculated, based on the species indicator values.

4. The pseudospecies indicator values are integrated to construct an indicator index for each species: the indicator value of a species is the code number (1 to 5 in the example above) of the pseudospecies with the highest indicator value, after removing signs.

Fidelity

5. Prior to classifying the species, a new table of *fidelity* values is calculated. The fidelity of a species to a particular group of sites is the degree to which the species in confined to that group. Classification of the species is achieved in a similar way as for sites, based on the table of fidelity values.

6. After taking care of misclassifications, borderline cases, and other problems, a final division of the sites is obtained.

7. Preferential pseudospecies are now tabulated. A pseudospecies is considered as preferential to one side or the other of the partition if it is more than twice as likely to occur on one side than on the other.

After a first binary division of the sites has been obtained, each subset is divided in smaller subsets by repeating steps 1 to 7. This goes on until groups become very small. Typically, groups of 4 objects or less are not partitioned any further.

The main output of TWINSPAN is a two-way table of species × sites reorganized to provide information about the dichotomies as well as the indicator values of the various species. When preparing that table, groups in the hierarchy are swivelled, where necessary, to make adjacent groups as similar as possible. In this way, the final table presents both a classification and an ordination of the objects; its interpretation is straightforward. A dendrogram can easily be drawn, if required, from the TWINSPAN output table. Additional tables provide information about each partition of the sites.

TWINSPAN has been criticized by Belbin and McDonald (1993) on two grounds: (1) The method assumes the existence of a strong gradient dominating the data structure, so that it may fail to identify secondary gradients or other types of structure in data sets. (2) The cutting points along the dominant axis are rather arbitrary; instead of selecting large gaps in the data, sites that may be very close in species composition may become separated.

There are other problems with TWINSPAN when it comes to identifying clusters of species or computing indicator values.

• Firstly, when identifying clusters of species or computing indicator values, one cannot introduce some other classification of the sites in the program; only the classifications produced by TWINSPAN, which are based on correspondence analysis (CA) or detrended correspondence analysis (DCA, Subsection 9.4.5), may be used to delineate species groups.

• Secondly, the pseudospecies concept is based on species relative abundances. The relative abundance of a species depends on the absolute abundances of the other species present at the site. Such relative frequencies may be highly biased when sampling or harvesting mobile organisms; all species are not sampled with the same efficiency because of differences in behaviour. There is always a distortion between the estimated (i.e. observed) and real relative frequencies of species at any given site. A species abundance value observed at a site should only be compared to abundance values for the same species at other sites.

• Finally, whereas simple CA is well suited for studying species abundances observed at several sites (ter Braak 1985), DCA has recently been severely criticized (Subsection 9.4.5). Jackson and Somers (1991b) have shown that results obtained with DCA vary depending on the number of segments used to remove the arch effect. Therefore, several runs with different numbers of segments must be done to find stable factorial axes and interpretable results.

8.8 Partitioning by *K*-means

Partitioning consists in finding a single partition of a set of objects (Table 8.1). Jain & Dubes (1988) state the problem in the following terms: given n objects in a p–dimensional space, determine a partition of the objects into K groups, or clusters, such that the objects within each cluster are more similar to one another than to objects in the other clusters. The number of groups, K, is determined by the user. This problem has first been stated in statistical terms by Fisher (1958), who proposed solutions (with or without constraint of contiguity; see Sections 12.6 and 13.2) for a single variable.

The difficulty is to define what “more similar” means. Several criteria have been suggested; they can be divided into global and local criteria. A *global criterion* would be, for instance, to represent each cluster by a type-object (on *a priori* grounds, or using the centroids of Subsections 8.5.6 and 8.5.7) and assign each object to the nearest type-object. A *local criterion* uses the local structure of the data to delineate clusters; groups are formed by identifying high-density regions in the data. The

K–means method, described in the next paragraphs, is the most commonly used of the latter type.

Objective function

In *K*-means, the objective function that the partition to discover should minimize is the same as in Ward's agglomerative clustering method (Subsection 8.5.8): the total error sum of squares (E_K^2, or TESS). The major problem encountered by the algorithms is that the solution on which the computation eventually converges depends to some extent on the initial positions of the centroids. This problem does not exist in Ward's method, which proceeds iteratively by hierarchical agglomeration. However, even though Ward's algorithm guarantees that the *increase* in sum of squared errors ($\Delta E_{\mathbf{hi}}^2$, eq. 8.8) is minimized at each step of the agglomeration (so that any order of entry of the objects should lead to the same solution, except in cases of equal distances where a "right-hand" programming rule may prevail), there is no guarantee that any given Ward's partition is optimal in terms of the E_K^2 criterion — surprising at this may seem. This same problem occurs with all stepwise statistical methods.

Local minimum

The problem of the final solution depending on the initial positions of the centroids is known as the "local minimum" problem in algorithms. The concept is illustrated in Fig. 8.17, by reference to a *solution space*. It may be explained as follows. Solutions to the *K*-means problem are the different ways to partition *n* objects into, say, $K = 4$ groups. If a single object is moved from one group to another, the corresponding two solutions will have slightly different values for the criterion to be minimized (E_K^2). Imagine that all possible solutions form a "space of solutions". The different solutions can be plotted as a graph with the E_K^2 criterion as the ordinate. It is not necessary to accurately describe the abscissa to understand the concept; it would actually be a multidimensional space. A *K*-means algorithm starts at some position in this space, the initial position being assigned by the user (see below). It then tries to navigate the space to find the solution that minimizes the objective criterion (E_K^2). The space of solutions is not smooth, however. It may contain *local minima* from which the algorithm may be unable to escape. When this happens, the algorithm has not found the overall minimum and the partition is not optimal in terms of the objective criterion.

Overall minimum

Several solutions may be used to help a *K*-means algorithm converge towards the overall minimum of the objective criterion E_K^2. They involve either selecting specific objects as "group seeds" at the beginning of the run, or attributing the objects to the *K* groups in some special way. Here are some commonly-used approaches:

• Provide an initial configuration corresponding to an (ecological) hypothesis. The idea is to start the algorithm in a position in the solution space which is, hopefully, close to the final solution sought. This ideal situation is seldom encountered in real studies, however.

• Provide an initial configuration corresponding to the result of a hierarchical clustering, obtained from a space-conserving method (Table 8.8). One simply chooses the partition into *K* groups found on the dendrogram and lists the objects pertaining to

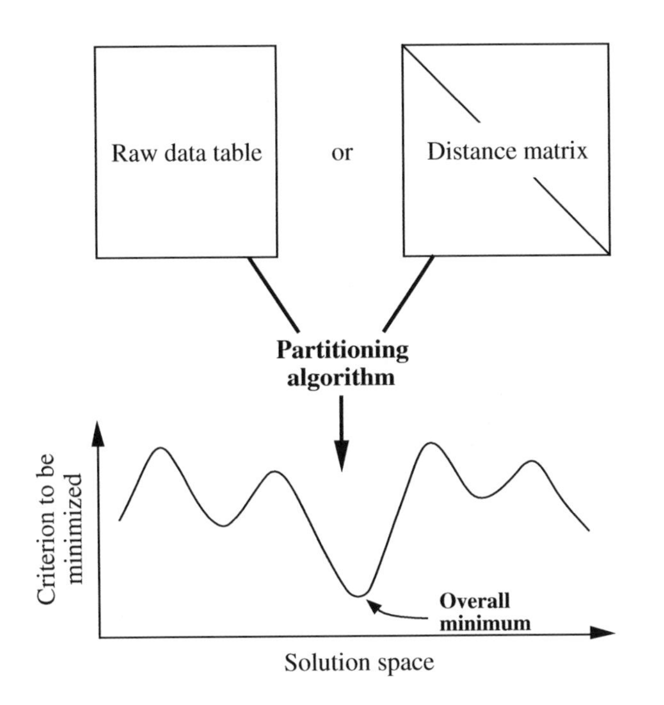

Figure 8.17 *K*-means algorithms search the space of solutions, trying to find the overall minimum (arrow) of the objective criterion to be minimized, while avoiding local minima (troughs).

each group. The *K*-means algorithm will then be asked to rearrange the group membership and look for a better overall solution (lower E_K^2 statistic).

• If the program allows it, select as "group seed", for each of the *K* groups to be delineated, some object located near the centroid of that group. For very large problems, Lance & Williams (1967d) suggested to use as starting point the result of a hierarchical clustering of a *random subset of the objects*, using as "group seeds" either the centroids of *K* clusters, or objects located near these centroids.

• Attribute the objects at random to the various groups. All *K*-means computer programs offer this option. Find a solution and note the E_K^2 value. It is possible, of course, that the solution found corresponds to a local minimum of E_K^2. So, repeat the whole procedure a number of times (for example, 100 times), starting every time from a different random configuration. Retain the solution that minimizes the E_K^2 statistic. One is more confident that this solution corresponds to the overall minimum when the corresponding value of E_K^2 is found several times across the runs.

Several algorithms have been proposed to solve the *K*–means problem, which is but one of a family of problems known in computer sciences as the *NP–complete* or

NP-hard

NP–hard problems[*]. In all these problems, the only way to be sure that the optimal solution has been found is to try all possible solutions in turn. This is impossible, of course, for any real-size problems, even with modern-day computers, as explained in Subsection 8.7.2. Classical references to K-means algorithms are Anderberg (1973), Hartigan (1975), Späth (1975, 1980), Everitt (1980), and Jain & Dubes (1988). Milligan & Cooper (1987) have reviewed the most commonly used algorithms and compared them for structure recovery, using artificial data sets. One of the best algorithms available is the following; it frequently converges to the solution representing the overall minimum for the E_K^2 statistic. It is a very simple alternating least-squares algorithm, which iterates between two steps:

- Compute cluster centroids and use them as new cluster seeds.
- Assign each object to the nearest seed.

At the start of the program, K observations are selected as "group seeds". Each iteration reduces the sum of squared errors E_K^2, if possible. Since only a finite number of partitions are possible, the algorithm eventually reaches a partition from which no improvement is possible; iterations stop when E_K^2 can no longer be improved. The FASTCLUS procedure of the SAS package, mentioned here because it can handle very large numbers of objects, uses this algorithm. Options of the program can help deal with outliers if this is a concern. The SAS (1985) manual provides more information on the algorithm and available options.

This algorithm was originally proposed in a pioneering paper by MacQueen (1967) who gave the method its name: K-means. Lance & Williams made it popular by recommending it in their review paper (1967d). In the MacQueen paper, group centroids are recomputed after each addition of an object; this is also an option in SAS. MacQueen's algorithm contains procedures for the fusion of clusters, if centroids become very close, and for creating new clusters if an object is very distant from existing centroids.

K-means partitioning may be computed from either a table of raw data or a distance matrix, because the total error sum of squares E_K^2 is equal to the sum of squares of the distances from the points to their respective centroids (eq. 8.5; Fig. 8.18a) and to the sum (over groups) of the mean squared within-group distances (eq. 8.6; Fig. 8.18b). It is especially advantageous to compute it on raw data when the number of objects is large because, in such a situation, the distance matrix may become very cumbersome or even impossible to store and search. In contrast, when using a table of original data, one only needs to compute the distance of each object to each group centroid, rather than to all other objects.

[*] *NP* stands for *Non-deterministic Polynomial.* In theory, these problems can be solved in polynomial time (i.e. some polynomial function of the number of objects) on a (theoretical) non-deterministic computer. NP-hard problems are probably not solvable by efficient algorithms.

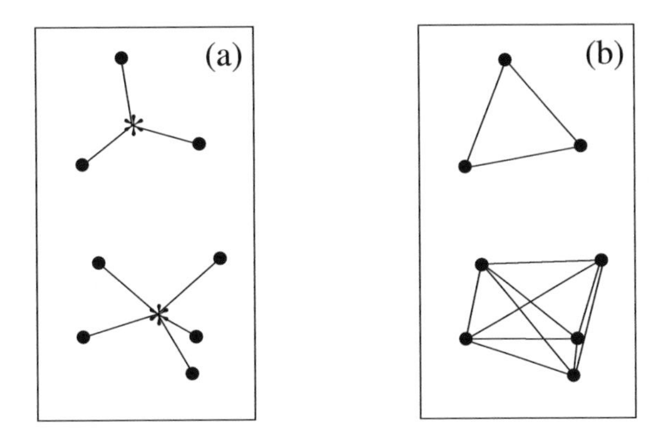

Figure 8.18 The total error sum of squares (E_K^2, TESS) is equal (a) to the sum of squares of the distances from the points to their respective centroids. (b) It is also equal to the sum (over groups) of the mean squared within-group distances.

The disadvantage of using a table of raw data is that the only distance function among points, available during *K*-means partitioning, is the Euclidean distance (D_1, Chapter 7) in A-space. This is not suitable with species counts and other ecological problems (Chapter 7). When the Euclidean distance is unsuitable, one may first compute a suitable similarity or distance measure among objects (Table 7.3 and 7.4); run the resemblance matrix through a metric or nonmetric scaling procedure (principal coordinate analysis, Section 9.2; nonmetric multidimensional scaling, Section 9.3); obtain a new table of coordinates for the objects in A-space; and run *K*-means partitioning using this table.

Following are two numerical examples that illustrate the behaviour of the E_K^2 criterion (eq. 8.5 and 8.6).

Numerical example 1. For simplicity, consider a single variable. The best partition of the following five objects (dark squares) in two clusters (boldface horizontal lines) is obviously to put objects with values 1, 2 and 3 in one group, and objects with values 6 and 8 in the other:

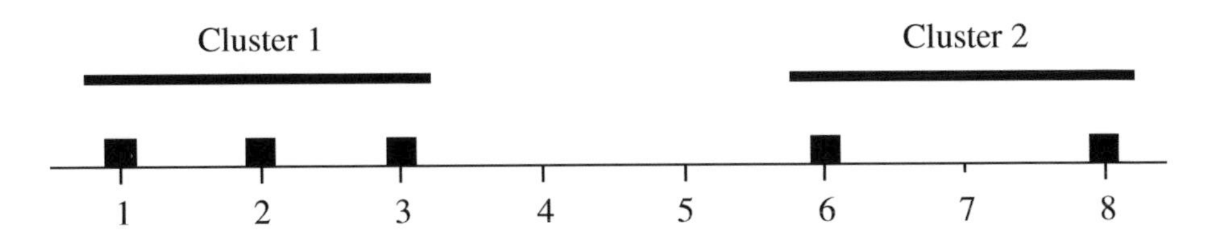

This example is meant to illustrate that the E_K^2 criterion can be computed from either raw data (eq. 8.5) or distances among objects (eq. 8.6). Using raw data (left-hand column, below), the group centroids are at positions 2 and 7 respectively; deviations from the centroids are calculated for each object, squared, and added within each cluster. Distances among objects (right-hand column, below) are easy to calculate from the object positions along the axis; the numbers of objects (n_k), used in the denominators, are 3 for cluster 1 and 2 for cluster 2.

$e_1^2 = (1^2 + 0^2 + (-1)^2) = 2$	$e_1^2 = (2^2 + 1^2 + 1^2)/3 = 2$
$e_2^2 = (1^2 + (-1)^2) = 2$	$e_2^2 = 2^2/2 = 2$
$E_K^2 = 4$	$E_K^2 = 4$

Numerical example 2. Considering a single variable again, this example examines the effect on the E_K^2 statistic of changing the cluster membership. There are six objects and they are to be partitioned into $K = 2$ clusters. The optimal solution is that represented by the boldface horizontal lines:

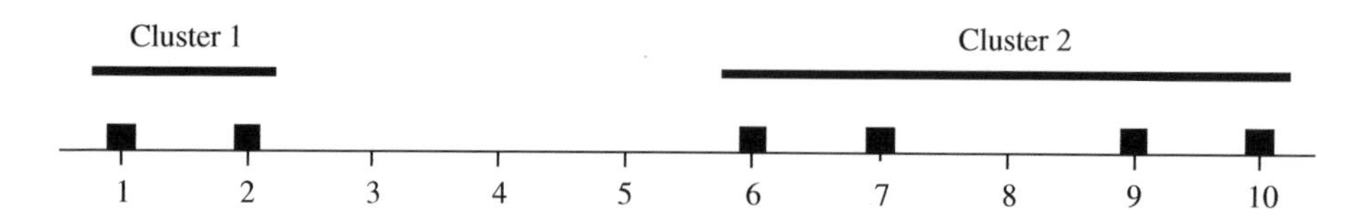

Calculations are as above. Using raw data (left-hand column, below), the group centroids are at positions 0.5 and 8 respectively; deviations from the centroids are calculated for each object, squared, and added within each cluster. Distances among objects (right-hand column, below) are easy to calculate from the object positions along the axis; the numbers of objects (n_k), used in the denominators, are 2 for cluster 1 and 4 for cluster 2.

$e_1^2 = (0.5^2 + (-0.5)^2) = 0.5$	$e_1^2 = 1^2/2 = 0.5$
$e_2^2 = (2^2 + 1^2 + (-1)^2 + (-2)^2) = 10.0$	$e_2^2 = (1^2 + 3^2 + 4^2 + 2^2 + 3^2 + 1^2)/4 = 10.0$
$E_K^2 = 10.5$	$E_K^2 = 10.5$

Consider now a sub-optimal solution where the clusters would contain the objects located at positions (1, 2, 6, 7) and (9, 10), respectively. The centroids are now at positions 4 and 9.5 respectively. Results are the following:

$e_1^2 = (3^2 + 2^2 + (-2)^2 + (-3)^2) = 26.0$	$e_1^2 = (1^2 + 5^2 + 6^2 + 4^2 + 5^2 + 1^1)/4 = 26.0$
$e_2^2 = (0.5^2 + (-0.5)^2) = 0.5$	$e_2^2 = 1^2/2 = 0.5$
$E_K^2 = 26.5$	$E_K^2 = 26.5$

This example shows that the E_K^2 criterion quickly increases when the cluster membership departs from the optimum.

In some studies, the number of clusters K to be delineated is determined by the ecological problem, but this it is not often the case. The problem of determining the most appropriate number of clusters has been extensively discussed in the literature. Over 30 different methods, called "stopping rules", have been proposed to do so. The efficiency coefficient, described in the last paragraph of Section 8.5, is one of them. Milligan & Cooper (1985; see also Milligan, 1996) compared them through an extensive series of simulations using artificial data sets with known numbers of clusters. Some of these rules recover the correct number of clusters in most instances, but others are appallingly inefficient. SAS, for instance, has implemented two among the best of these rules: a pseudo-F statistic and the cubic clustering criterion.

8.9 Species clustering: biological associations

Most of the methods discussed in the previous sections may be applied to clustering descriptors as well as objects. When searching for species associations, however, it is important to use clustering methods that model as precisely as possible a clearly formulated concept of association. The present section attempts (1) to define an operational concept of association and (2) to show how species associations can be identified in that framework.

Several concepts of species association have been developed since the nineteenth century; Whittaker (1962) wrote a remarkable review about them. These concepts are not always operational, however. In other words, they cannot always be translated into a series of well-defined analytical steps which would lead to the same result if they were applied by two independent researchers, using the same data. In general, the concept of association refers to a group of species that are "significantly" found together, without this implying necessarily any positive interaction among these species. In other words, an association is simply a group of species (or of taxa pertaining to some systematic category) recognized as a cluster following the application of a clearly stated set of rules.

Several procedures have been proposed for the identification of species associations. Quantitative algorithms have progressively replaced the empirical methods, as they have in other areas of science. All these methods, whether simple or elaborate, have two goals: first, identify the species that occur together and, second, minimize the likelihood that the co-occurrences so identified be fortuitous. The search for valid associations obviously implies that the sampling be random and planned in accordance with the pattern of variability under study (e.g. geographical, temporal, vertical, experimental). This pattern defines the framework within which the groups of species, found repeatedly along the sampling axes, are called associations; one then speaks of association of species over geographic space, or in time, etc. The criterion is the recurrence of a group of species along the sampling axes under study.

Ecologists are interested in associations of species as a conceptual framework to synthesize environmental characteristics. When associations have been found, one can concentrate on finding the ecological requirements common to most or all species of an association instead of having to describe the biology and habitat of each species individually. In an inverse approach, species associations may be used to predict environmental characteristics. Associations may be better predictors of environmental conditions because they are less subject to sampling error than individual species. In other studies, trophic groups or size classes may be used for the same purpose.

The ecological interpretation of species associations is a subject open to discussion. Research examining the quantitative relationships between species associations and their environment, in a multidimensional framework such as Hutchinson's (1957) fundamental niche, should enrich this discussion with appropriate data and provide some idea of the kind of variability to expect in species responses. These relationships could be discovered using the techniques for the interpretation of ecological structures summarized in Chapter 10. Among the association concepts that are best adapted to multidimensional analysis, the following one may be taken as starting point for an ecological discussion: associations are composed of species that have *similar reactions* to properties of the environment (Fager & McGowan, 1963). In this way, associations are characterized by their internal stability along the sampling axes. Associated species are thus responding in a *related fashion* to environmental changes (Legendre, 1973), which implies that variations in the abundance of associated species are under similar environmental control. Consequently, an association would also be a simplification of the responses of individual species to environmental controls. This viewpoint is quite remote from the organismic concepts of association, but it lends itself more easily to quantitative analysis. According to this model, the recurrence of associations is an indication that the abstraction, called association, corresponds to some fundamental properties of the interaction between species and their environment. This analysis of the species-environment relationships would be incomplete if it did not include the recognition that the environment of a species is not solely composed of physical variables, but also of the other species with which it interacts in a positive or negative fashion. This is in agreement with the notion that species frequently found together may have evolved mechanisms of biological accommodation to one another, in addition to adaptations to their common environment. This phenomenon has been determinant when selecting measures of similarity for studying species associations (Subsection 7.5.2).

Species association

According to Fager (1963), recurrent organized systems of biological organisms are characterized by structural similarity in terms of species presence and abundance. This language belongs to the quantitative approach. Following the above discussion, a simple and operational definition is proposed: *a species association is a recurrent group of co-occurring species* (Legendre & Legendre, 1978). Associations of taxa belonging to categories other than species may also be defined, if the fundamental units that interact in the environment belong to some other taxonomic level.

Using this definition, one can choose some clustering methods that are appropriate to delineate species associations. Appropriate measures of resemblance have already been described in the previous chapter (Table 7.5).

A great variety of clustering methods have been used for the identification of associations, although the choice of a given method often appears to have been based on the availability of a program on the local computer instead of a good understanding of the properties and limitations of the various techniques. An alternative to standard clustering techniques was proposed by Lambshead & Paterson (1986) who used numerical cladistic methods to delineate species associations. Among the ordination methods, principal component and correspondence analyses do not generally produce good clusters of species, even though these analyses may be very useful to investigate other multivariate ecological problems (Chapter 9). One of the problems of principal component analysis is that of the double zeros; difficulties with this method in the search for species associations have been discussed in detail in Subsection 7.5.2.

After selecting the most appropriate coefficient of dependence for the data at hand (Table 7.5), one must next make a choice among the usual hierarchical clustering methods discussed in the previous sections of this chapter, including TWINSPAN (Subsection 8.7.4). Partitioning by *K*-means (Section 8.8) should also be considered. In addition, there are two specialized partitioning methods for probabilistic coefficients, described below. When the analysis aims at identifying hierarchically-related associations using non-probabilistic similarities, the hierarchical clustering methods are appropriate. When one simply looks for species associations without implying that they should form a hierarchy, partitioning methods are in order. Hierarchical clustering may also be used in this case but one must decide, using a dendrogram or another graphical representation, which level of partition in the hierarchy best corresponds to the ecological situation to be described (see definition above). One must take into account the level of detail required and the limits of significance or interpretability of the species clusters thus found. In any case, space-conserving or space-dilating methods should be preferred to single linkage, especially when one is trying to delimit functional groups of species from data sampled along an ecological continuum. After the main association structure has been identified with the help of more robust clustering methods, however, single linkage can play a role in bringing out the fine relationships, such as between the members of associations and their satellite species.

An alternative to clustering is to test the significance of the pattern of species co-occurrence across sites, using some appropriate testing procedure. Jackson *et al.* (1992) discussed several null models that may be used for this purpose.

Ecological application 8.9a

Thorrington-Smith (1971) identified 237 species of phytoplankton in water samples from the West Indian Ocean. 136 of the species were clustered into associations by single linkage hierarchical clustering of a Jaccard (S_7) association matrix among species. The largest of the 11 associations contained 50 species; its distribution mostly corresponded to the equatorial subsurface water. This association was dominant at all sites and may be considered typical of the endemic flora of the West Indian Ocean. Other phytoplankton associations represented seasonal or regional differences, or characterized currents or nutrient-rich regions. Since phytoplankton associations did not lose their identities even when they were mixed, the study of associations in zones of water mixing seemed a good way of tracing back the origins of water masses.

1 — Non-hierarchical complete linkage clustering

The first specialized partitioning method for discovering species associations is Fager's (1957) *non-hierarchical complete linkage clustering.* It is well-adapted to probabilistic measures of dependence among species and to other measures of dependence for which a critical or significance level can be set. This method differs from hierarchical complete linkage clustering in that one looks for clusters formed at a stated level of similarity without taking into account the hierarchical cluster structure that may have been be found at higher similarity levels. The working level is usually $S \geq 0.95$ or $S \geq 0.99$ for probabilistic coefficients. With the non-probabilistic measure S_{24} (Chapter 7), Fager & McGowan (1963) used $S \geq 0.5$ as the clustering threshold.

Fager and Krylov rules

The authors developed a computer program which makes the method operational, but it is possible to implement it without a special program. Select a threshold similarity level and draw a graph (as in Fig. 8.2a) of the objects with link edges corresponding to all values of $S \geq$ (threshold). Then, delineate the species associations on the graph as the groups meeting the complete-linkage criterion, i.e. the groups in which all objects are linked to all others at the stated similarity level (Subsection 8.5.2). In case of conflicts, use the following decision rules.

1. Complete-linkage clusters of species, obtained by this method, must be independent of one another, i.e. they must have no species in common. Between two possible species partitions, *form first the clusters containing as many species as possible.* For instance, if a cluster of 8 species has two species in common with another cluster of 5 species, create clusters of 8 and 3 species instead of clusters of 6 and 5 species. Krylov (1968) adds that no association should be recognized that contains less than three species.

If non-independent clusters remain (i.e. clusters with objects in common), consider rules 2 and 3, in that order.

2. Between several non-independent clusters *containing the same number of species,* choose the partition that maximizes the size of the resulting independent clusters. For example, if there are three clusters of 5 species each where clusters 1 and 2 have one species in common and clusters 2 and 3 also have one species in common, select clusters 1 and 3 with five species each, leaving 3 species into cluster 2. One thus creates three clusters with membership 5, 3, and 5, instead of three clusters with membership 4, 5, and 4.

3a. If the above two criteria do not solve the problem, between two or more non-independent clusters having about the same number of species, select the one *found at the largest number of sites* (Fager, 1957). One has to go back to the original data matrix in order to use this criterion.

3b. Krylov (1968) suggested replacing this last criterion with the following one: among alternative species, the species to include in a cluster is the one *that has the least affinity* with all the other species that are not members of that cluster, i.e. the species that belongs to the cluster more exclusively. This criterion may be decided from the graph of link edges among species (above).

This form of non-hierarchical complete linkage clustering led Fager (1957), Fager & McGowan (1963), and Krylov (1968) to identify meaningful and reproducible plankton associations. Venrick (1971) explains an interesting additional step of Fager's computer program; this step answers an important problem of species association studies. After having recognized independent clusters of completely linked species, the program associates the remaining species, by single linkage clustering, to one or several of the main clusters. These *satellite species* do not have to be associated with all members of a given association. They may also be satellites of several associations. This reflects adequately the organizational complexity of biological communities.

This last point shows that *overlapping* clustering methods could be applied to the problem of delineating species associations. The mathematical bases of these methods have been established by Jardine & Sibson (1968, 1971) and Day (1977).

Ecological application 8.9b

Fager's non-hierarchical complete linkage clustering was used by Legendre & Beauvais (1978) to identify fish associations in 378 catches from 299 lakes of northwestern Québec. Their computer program provided the list of all possible complete linkage clusters formed at a user-selected similarity level. Species associations were determined using the criteria listed above. The similarity between species was established by means of the probabilistic measure S_{25} (Subsection 7.5.2), based on presence-absence data.

At similarity level $S_{25} \geq 0.989$, the program identified 25 non-independent species clusters, involving 26 of the 29 species in the study. Each subgroup of at least three species could eventually become an association since the clustering method was complete linkage. Many of these clusters overlapped. The application of Fager's decision rules (with rule 3b of Krylov) led to the identification of five fish associations, each one completely formed at the similarity level indicated to the right. Stars indicate the internal strength of the associations (*** all links ≥ 0.999, ** all links ≥ 0.99, * all links ≥ 0.95).

1)	Lake whitefish	*Coregonus clupeaformis*	$S_{25} \geq 0.999$ ***
	Longnose sucker	*Catostomus catostomus*	
	Lake trout	*Salvelinus namaycush*	
	Round whitefish	*Prosopium cylindraceum*	
	Lake chub	*Couesius plumbeus*	
2)	Northern pike	*Esox lucius*	$S_{25} \geq 0.995$ **
	White sucker	*Catostomus commersoni*	
	Walleye	*Stizostedion vitreum*	
	Shallowwater cisco	*Coregonus artedii*	
	Yellow perch	*Perca fluviatilis*	

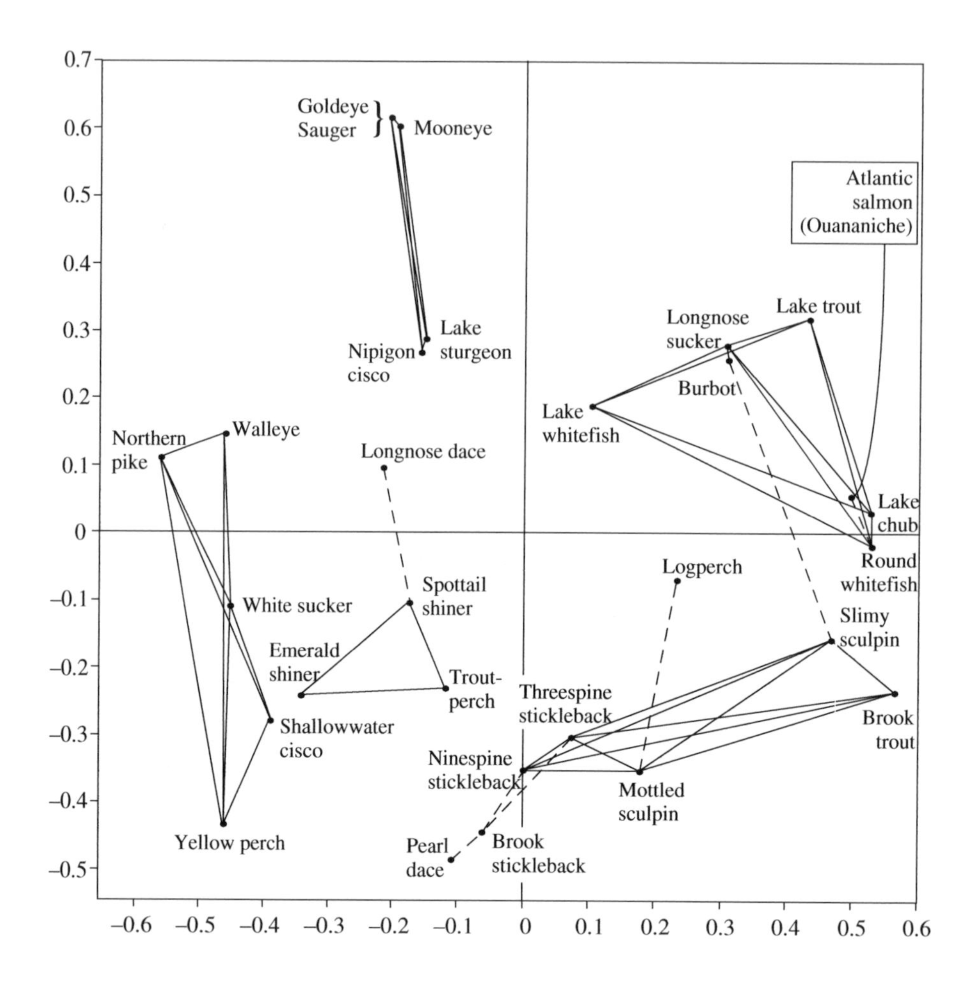

Figure 8.19 Fish associations are drawn on a two-dimensional principal coordinate ordination of the species. Axes I (abscissa) and II (ordinate) explain together 55% of the variability among species. Full lines link species that are members of associations, which were identified by non-hierarchical complete linkage clustering at $S \geq 0.989$. Dashed lines attach satellite species to the most closely related species member of an association. Redrawn from Legendre & Beauvais (1978).

3)	Brook trout	*Salvelinus fontinalis*	$S_{25} \geq 0.991$ **
	Ninespine stickleback	*Pungitius pungitius*	
	Mottled sculpin	*Cottus bairdi*	
	Threespine stickleback	*Gasterosteus aculeatus*	
	Slimy sculpin	*Cottus cognatus*	

4) Nipigon cisco	*Coregonus nipigon*	$S_{25} \geq 0.991$ **
Lake sturgeon	*Acipenser fulvescens*	
Goldeye	*Hiodon alosoides*	
Mooneye	*Hiodon tergisus*	
Sauger	*Stizostedion canadense*	
5) Trout-perch	*Percopsis omiscomaycus*	$S_{25} \geq 0.989$ *
Spottail shiner	*Notropis hudsonius*	
Emerald shiner	*Notropis atherinoides*	

The six remaining species were attached as satellites, by single linkage chaining, to the association containing the closest species. Figure 8.19 shows the species associations drawn on a two-dimensional principal coordinate ordination of the species. Three of these associations can be interpreted ecologically. Association 1 was characteristic of the cold, clear, low-conductivity lakes of the Laurentide Shield. Association 2 characterized lakes with warmer and more turbid waters, found in the lowlands. Association 4 contained species that were all at the northern limit of their distributions; they were found in the southern part of the study area.

2 — *Probabilistic clustering*

Clifford & Goodall (1967) developed a probabilistic clustering procedure to be used in conjunction with Goodall's probabilistic index (S_{23}). Lance & Williams (1967d) state that this is one of the few completely internally-consistent clustering procedures. Clifford & Goodall proposed this method for Q mode studies, but it is especially interesting for clustering species into associations. Because of its computational requirements (the similarity matrix has to be recomputed at each iteration), there seems to be little advantage in using this model when another algorithm is applicable.

The probabilistic clustering approach allows one to find, for a given data set, all the clusters whose association matrices have a given *probability* of being homogeneous. Among all clusters thus formed, the largest one is chosen. If a tie occurs (two or more clusters are equally the largest), the most homogeneous one is kept. Notice the resemblance between this association concept and that of Fager described above. The first cluster is removed from the main data matrix, a similarity matrix is recomputed for the remainder of the data matrix, and the clustering operation is carried out again. In the same way, a new similarity matrix is computed for the species belonging to the selected cluster in order to determine whether it contains sub-clusters.

The detailed mechanism of this clustering procedure is summarized in Fig. 8.20 and explained below with the help of an example taken from Legendre (1971a), which consists of 10 species of marine phytoplankton identified in 13 water samples collected at a single site. The clustering of all 69 species identified in the 13 water samples is discussed in Ecological application 8.9c below.

Goodall's (S_{23}) similarity matrix for the 10 species forms Table 8.9. Note again that the matrix was computed using all 69 species present; 10 of them are extracted here from that matrix to illustrate the method. First, one must check whether the matrix is homogeneous; if this is the case, it cannot be divided into associations in the sense of the present method. To carry out the test, one examines the probability of finding by chance, in the matrix, the largest similarity value

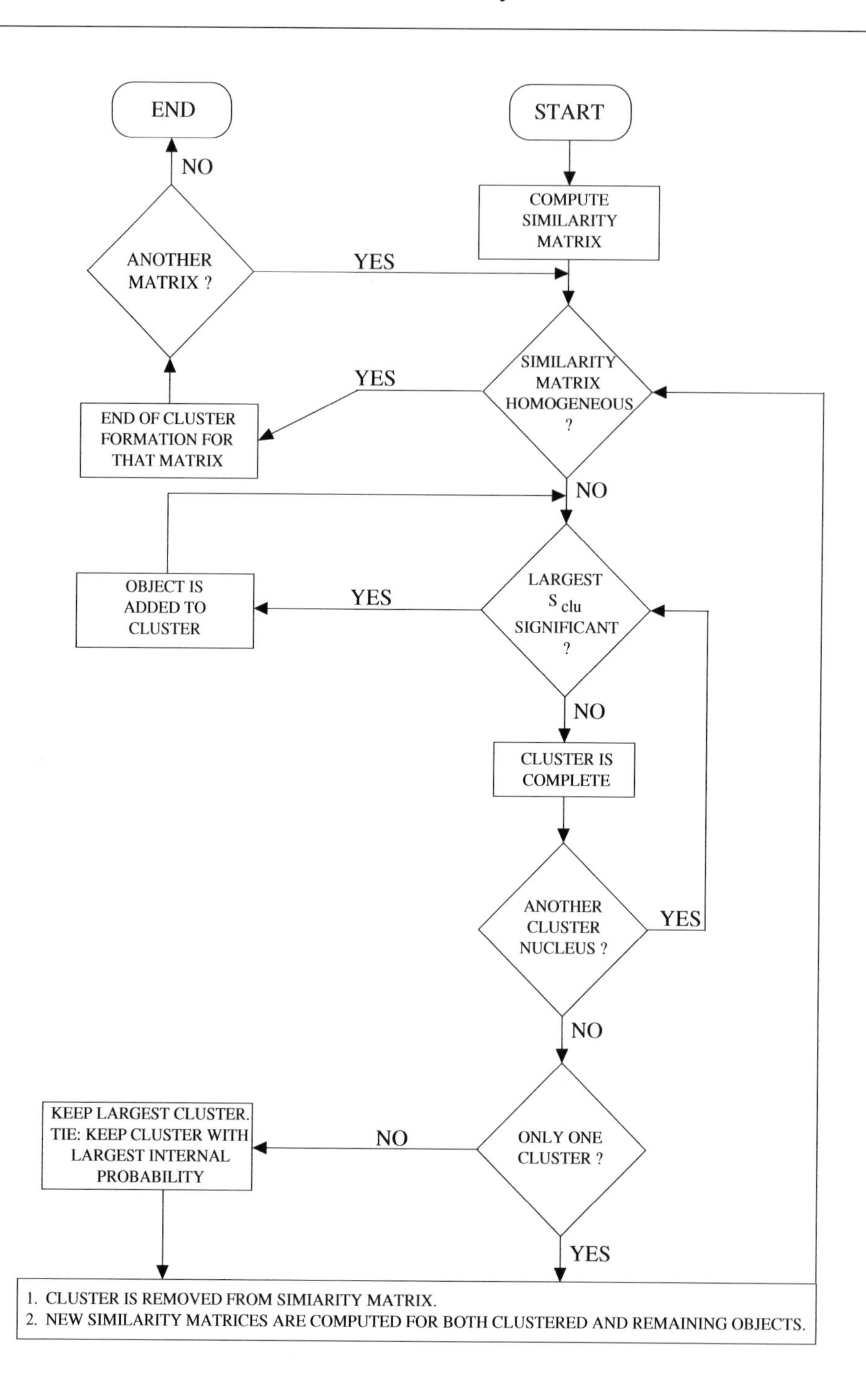

Figure 8.20 Flow chart of the procedure for probabilistic clustering of species into associations. From Legendre (1973).

Table 8.9 Similarity matrix among 10 phytoplankton species (from Legendre, 1971a) calculated using Goodall's index S_{23}. Similarities higher than the minimum significant value are in italics.

Species	1	2	3	4	5	6	7	8	9
2	*0.999996*								
3	0.999873	0.999873							
4	0.999222	0.999222	0.992320						
5	0.013925	0.013925	0.003509	0.261027					
6	0.059115	0.059115	0.025796	0.959650	*0.999994*				
7	0.000020	0.000020	0.000000	0.229895	0.999914	0.976088			
8	0.489400	0.489400	0.550799	0.420580	0.968154	0.921217	0.991849		
9	0.958317	0.958317	0.750147	0.948276	0.000293	0.032500	0.804223	0.562181	
10	0.679196	0.679196	0.726537	0.699797	0.565639	0.986997	0.719274	0.998504	0.029591

The minimum significant value ($\alpha = 0.05$) is 0.99998, considering the 69 species in the whole data set.

S_{max} which is actually found in it. If the probability is lower than a pre-established significance level α, for example 0.05, the hypothesis of homogeneity is rejected for this matrix and the associative procedure can begin. Assuming that the similarities are independent of one another and are thus distributed uniformly (distribution where the frequencies of all similarity values are equal), the probability sought, p, is that of finding a value at least as high as S_{23max} in at least one of the $p(p-1)/2$ distinct cells of the similarity matrix:

$$\text{p} = 1 - S_{max}^{\left[\frac{p(p-1)}{2}\right]}$$

The rationale of this equation* involves the complement of the probability of observing similarity values smaller than S_{max} in each of the $p(p-1)/2$ cells of the similarity matrix. This is also the probability that the hypothesis of matrix homogeneity is supported by the data.

Is p smaller than or equal to the predetermined significance level α? Reversing the equation, the hypothesis of matrix homogeneity is to be rejected when

$$S_{max} \geq (1-\alpha)^{\left[\frac{2}{p(p-1)}\right]}$$

* In Clifford & Goodall (1967: 505) and in Legendre (1971a), this equation is incorrect.

Table 8.10 Species associations are formed by agglomeration around two nuclei, (1, 2) and (5, 6), following Clifford & Goodall (1967). At each clustering step (rows), the weakest similarity between a non-clustered species and all members of the cluster is found in Table 8.9. The largest of these similarities (in italics) is then compared to the minimum significant value (S_{min} in boldface). If this similarity is larger than S_{min}, the corresponding species is included in the association.

Steps	Species 1	2	3	4	5	6	7	8	9	10	S_{min}
1	cluster	cluster	*0.999873*	0.999222	0.013925	0.059115	0.000020	0.489400	0.958317	0.679196	**0.9723**
2	cluster	cluster	cluster	*0.992320*	0.003509	0.025796	0.000000	0.489400	0.750147	0.679196	**0.9081**
3	cluster	cluster	cluster	cluster	0.003509	0.025796	0.000000	0.420580	*0.750147*	0.679196	**0.8324**
	Internal probability of the cluster = 2.999×10^{-22}										
1	0.013925	0.013925	0.003509	0.261027	cluster	cluster	*0.976088*	0.921217	0.000293	0.565639	**0.9723**
2	0.000020	0.000020	0.000000	0.229895	cluster	cluster	cluster	*0.921217*	0.000293	0.565639	**0.9081**
3	0.000020	0.000020	0.000000	0.229895	cluster	cluster	cluster	cluster	0.000293	*0.565639*	**0.8324**
	Internal probability of the cluster = 2.523×10^{-16}										

where α is the significance level and p is the number of species in the matrix. For $\alpha = 0.05$, the $p = 69$ species from which the values in Table 8.9 are calculated result in a maximum value of 0.99998. By this criterion, the pairs of species (1, 2) and (5, 6) are significantly linked at the 5% probability level. The similarity matrix is not homogeneous and these two pairs of species (which could overlap) are used as nuclei for the formation of clusters by agglomeration.

The next step in the clustering procedure (Table 8.10) consists in determining which species from outside the cluster is the most strongly related to the species that are already in the cluster. This species is the next potential candidate for clustering. If the weakest link with all species in a cluster is larger than the minimum significant value, the candidate species is included in the cluster. By testing the weakest link, one makes sure that the new members of a cluster are significantly related to all the other species in that cluster. Therefore Clifford and Goodall's probabilistic method follows the complete linkage clustering philosophy, as was the case with Fager's method above. The probability of the weakest similarity value S_{min} actually observed between a candidate species and the members of a cluster is (Clifford & Goodall, 1967):

$$\text{p} = 1 - [1 - (1 - S_{min})^{g}]^{p-g}$$

Table 8.11 Similarity matrix for the 6 non-clustered species of the example (upper panel; part of a 65 × 65 matrix), and clustering (lower panel).

Species	Species 1	2	3	4	9
2	*0.999998*				
3	0.999921	0.999921			
4	0.999432	0.999432	0.991196		
9	0.964934	0.964934	0.723616	0.957422	
10	0.696301	0.696301	0.752215	0.713782	0.003661

The minimum significant value ($\alpha = 0.05$) is 0.99998, considering the 65 unclustered species.

Steps	Species 1	2	3	4	9	10	S_{min}
1	cluster	cluster	*0.999921*	0.999432	0.964934	0.696301	**0.9715**
2	cluster	cluster	cluster	*0.991196*	0.723616	0.696301	**0.9061**
3	cluster	cluster	cluster	cluster	*0.723616*	0.696301	**0.8297**

where g is the number of species already in the cluster. Therefore, at significance level α, the lowest similarity S_{min} between a species and all the members of a cluster must be

$$S_{min} \geq 1 - \left[1 - (1-\alpha)^{\frac{1}{p-g}}\right]^{1/g}$$

for that species to be included in the cluster. This formula is such that S_{min} decreases as the size of the cluster increases, thus allowing the largest clusters to be more heterogeneous.

The clustering steps are shown in Table 8.10 for the cluster nuclei (1, 2) and (5, 6) found during the first clustering step. In each case, the clustering procedure stops when no further significant similarities are found. The two clusters could have had species in common, although it is not the case here. At the end of this first clustering cycle, only one of the clusters is retained, i.e. the one containing more species; in cases of equality (as is the case here), the cluster that has the highest internal probability is selected. The internal probability of a cluster is the product of the complements of the similarities $(1 - S)$ between all members of the cluster. It is shown in Table 8.10 for the two clusters formed. The four species (5, 6, 7, 8) forming the cluster with the

Table 8.12 Similarity matrix for the two clusters of 4 species. In both cases, the highest similarity value S_{max} (in italics) does not reach the minimum significant value, so that the similarity matrices are considered homogeneous and the clustering procedure stops.

Species	Species 1	2	3
2	*0.914072*		
3	0.698359	0.698359	
4	0.280874	0.280874	0.073669

The minimum significant value ($\alpha = 0.05$) is 0.99149 for these 4 species.

Species	Species 5	6	7
6	*0.987365*		
7	0.430935	0.330135	
8	0.142089	0.357724	0.659724

The minimum significant value ($\alpha = 0.05$) is 0.99149 for these 4 species.

highest internal probability are now removed from the main cluster of 69 species and *new similarity matrices are computed* for the cluster of 4 species (Table 8.12) and the remaining 65 species (Table 8.11 for the remaining 6 species of the present example).

The similarity matrix for the remaining species is analysed in the same way as the original matrix. This results in another complete-linkage cluster of 4 species (Table 8.11). In this example, the new cluster is identical to the cluster rejected at the end of the first clustering cycle, but this need not always be the case.

The two clusters of 4 species are homogeneous (Table 8.12); consequently they contain no sub-clusters. Since one may not wish to recognize clusters containing less than 3 species, as in Fager's method, the clustering is now complete. It has been pursued with the 61 non-clustered species in the ecological application below.

The species associations delineated by this clustering method are thus clusters of species agglomerated by complete linkage, using a criterion of statistical significance. These associations may or may not be grouped into larger significant associations. At the end of the procedure, a variable number of species remain unclustered. They can be attached by single linkage to the closest cluster, as was done in Fager's method (Subsection 8.9.1).

Ecological application 8.9c

Legendre (1973) collected 104 water samples for enumeration of phytoplankton at two sites in Baie des Chaleurs (Gulf of Saint Lawrence, Québec) during two consecutive summers (1968 and 1969). Sampling site 110 was located near the centre a cyclonic gyre; site 112N was located outside the gyre. For each site and each sampling season, phytoplankton associations were determined, for depths 0 and 10 m, using Clifford & Goodall's method. These associations were not based on spatial recurrence as in the previous application; they represented, instead,

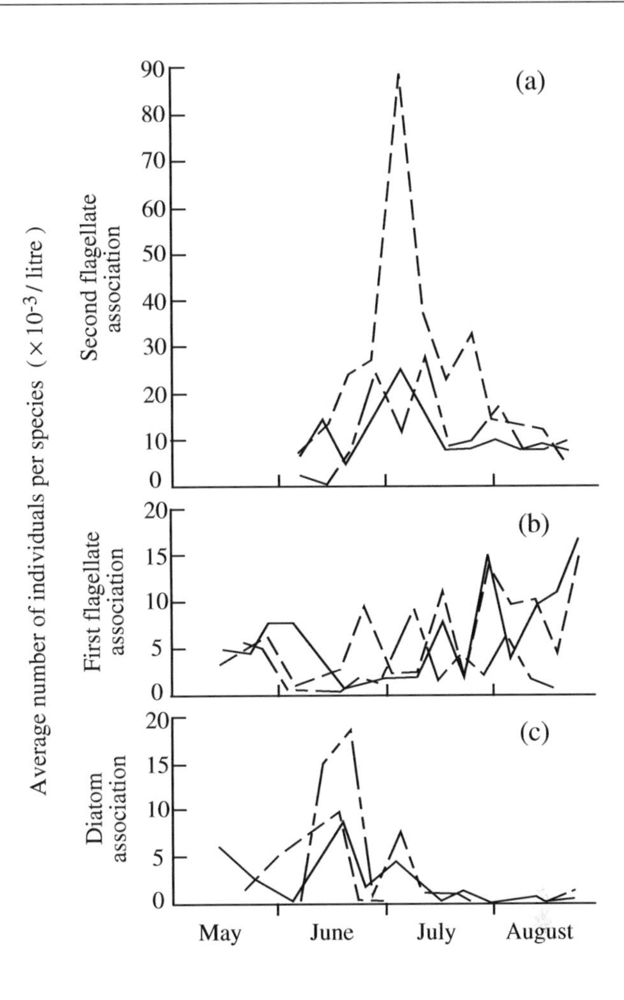

Figure 8.21 Temporal variation in abundance of the three main phytoplankton associations at sites 110 and 112N of Baie des Chaleurs, detected over the whole summer using significance level $\alpha = 0.05$. (a) 1968. Sites: —, 110 (0 m); ---, 110 (10 m); –•–, 112N (0 m). (b) 1969. Sites: —, 112N (0 m); ---, 112N (10 m); –•–, 110 (0 m). (c) —, 1969, site 112N (10 m); ---, 1969, site 110 (0 m); –•–, 1968, site 112N (0 m). From Legendre (1973).

temporal recurrence in the co-occurrence of species. Three major associations were found. One exclusively contained diatoms; the other two were mostly composed of flagellates. To facilitate ecological interpretation, each association was identified within each water sample and plotted as the average number of phytoplankters per species of the association and per litre, for each water sample in which it was found. Results are shown in Fig. 8.21.

The first flagellate association (Fig. 8.21b), which included the cluster of species 5 to 8 of the above example (Table 8.12), was observed mainly in 1969 where it seemed to increase slowly from the beginning to the end of the summer. This association was also present during short periods during the summer of 1968. In all instances where this association was present, the abundance of nitrate was higher than when only the second association was detected. Surface waters in 1968 contained little nitrate compared to 1969, where large quantities were observed at the beginning of the summer. The second flagellate association (Fig. 8.21a) was using nitrate perhaps more efficiently, which might have resulted in it becoming dominant in 1968. It was not detected in 1969.

The diatom association (Fig. 8.21c), which included species 1 to 4 clustered in Table 8.12, was related to silicate, which showed a minimum around mid-June. This condition was well correlated with the observed diatom peak. The boundary condition seemed to be temperature-dependent since the diatom association disappeared at ca. 12°C. This boundary condition accounted for the difference between the two sampling sites (the association being present at site 110 during the first third of the summer only), since the critical temperature of 12°C was reached at site 110 at least 3 weeks sooner than at 112N. The diatom association showed a similar behaviour during both summers.

The link between the two groups of associations could be the strong positive seasonal correlation between nitrate and silicate (Legendre, 1971b). This relationship was overridden in the case of diatoms by a temperature limit.

The search for associations, which were identifiable during part of the summer only, showed that associations that were distinct on a seasonal basis could be combined during shorter periods. This suggests that phytoplankton associations identified along time are an abstraction, resulting from the action of environmental processes. Indeed, two autocorrelated environmental variables are more likely to fluctuate in the same way when a shorter time period is considered. The fact that statistically significant clusters of species could be detected indicated, however, that the environment itself presented clusters of niches, at least during some periods.

The above application shows that the results of numerical methods (i.e. the identification of species associations by clustering, in this case) may contribute to a better understanding of ecological processes when the numerical analysis is followed by a careful ecological interpretation. Numerical methods do not represent a "new ecology" by themselves; they may only favour its emergence.

3 — Indicator species

The identification of characteristic or indicator species is traditional in ecology and biogeography. Field studies describing sites or habitats usually mention one or several species that characterize each habitat. The most widespread method for identifying indicator species is TWINSPAN (Hill, 1979; Subsection 8.7.4). There is clearly a need for the identification of characteristic or indicator species in the fields of monitoring, conservation, and management, as discussed below. Because indicator species add ecological meaning to groups of sites discovered by clustering, they provide criteria to compare typologies derived from data analysis, to identify where to stop dividing clusters into subsets, and to point out the main levels in a hierarchical classification of sites. *Indicator species* differ from *species associations* in that they are indicative of particular groups of sites. Good indicator species should be found mostly in a single group of a typology and be present at most of the sites belonging to that group. This duality is of ecological interest; yet it is seldom exploited in indicator species studies.

Dufrêne & Legendre (1997) present an alternative to TWINSPAN in the search for indicator species and species assemblages characterizing groups of sites. Like TWINSPAN, the new method is *asymmetric*, meaning that species are analysed on the basis of a *prior* partition of the sites. The first novelty of the method is that it derives indicator species from any hierarchical or non-hierarchical classification of the objects (sampling sites), contrary to TWINSPAN where indicator species can only be derived for classifications obtained by splitting sites along correspondence analysis (CA) or detrended correspondence analysis (DCA) axes (Subsection 8.7.4). The second novelty lies in the way the indicator value of a species is measured for a group of sites. The *indicator value index* (INDVAL) is based only on within-species abundance and occurrence comparisons; its value is not affected by the abundances of other species. The significance of the indicator value of each species is assessed by a randomization procedure (Section 1.2).*

Specificity
Fidelity

The *indicator value* (INDVAL) index is defined as follows. For each species j in each cluster of sites k, one computes the product of two values, A_{kj} and B_{kj}. A_{kj} is a measure of *specificity* whereas B_{kj} is a measure of *fidelity*:

$$A_{kj} = Nindividuals_{kj} \,/\, Nindividuals_{+k}$$

$$B_{kj} = Nsites_{kj} \,/\, Nsites_{k+}$$

$$INDVAL_{kj} = 100\, A_{kj}\, B_{kj} \qquad \textbf{(8.21)}$$

In the formula for A_{kj}, $Nindividuals_{kj}$ is the mean abundance of species j across the sites pertaining to cluster k and $Nindividuals_{+k}$ is the sum of the mean abundances of species j within the various clusters. The *mean* number of individuals in each cluster is used, instead of summing the individuals across all sites of a cluster, because this removes any effect of variations in the number of sites belonging to the various clusters. Differences in abundance among sites of a cluster are not taken into account. A_{kj} is maximum when species j is present in cluster k only. In the formula for B_{kj}, $Nsites_{kj}$ is the number of sites in cluster k where species j is present and $Nsites_{k+}$ is the total number of sites in that cluster. B_{kj} is maximum when species j is present at all sites of cluster k. Quantities A and B must be combined by multiplication because they represent independent information (i.e. specificity and fidelity) about the distribution of species j. Final multiplication by 100 produces a percentage.

The indicator value of species j for a partition of sites is the largest value of $INDVAL_{kj}$ observed over all clusters k of that partition:

Indicator value

$$INDVAL_j = \max\,[INDVAL_{kj}] \qquad \textbf{(8.22)}$$

* A FORTRAN program (INDVAL) is available from the following WWWeb site to compute the indicator value index and perform the randomization testing procedure: <http://www.biol.ucl.ac.be/ecol/html/Outils/Tools.IndVal.html>. INDVAL is also available in the package PC-ORD. Distribution: see footnote in Section 9.3.

The index is maximum (100%) when the individuals of species j are observed at all sites belonging to a single cluster. A random reallocation procedure of sites among the site groups is used to test the significance of $INDVAL_j$ (Section 1.2). The index can be computed for any given partition of sites, or for all levels of a hierarchical classification of sites. Dufrêne & Legendre (1997) provide a numerical example of the calculation procedure.

Ecological application 8.9d

In order to illustrate the indicator value method, Dufrêne & Legendre (1997) used a large data set of Carabid beetle distributions in open habitats of Belgium (189 species collected in pitfall traps, for a total of 39 984 specimens). The data represented 123 year-catch cycles at 69 locations; a year-catch cycle cumulates catches at a site during a full year; 54 sites were studied during two years and 15 sites were sampled during a single year. The Q-mode typology of sites was computed by K-means partitioning of a data matrix representing the ordination axes of a principal coordinate analysis (Section 9.2), computed from a Steinhaus similarity matrix among sites (eq. 7.24) calculated from log-transformed species abundance data. Similarities among sites, which had been computed using an asymmetrical similarity coefficient, were thus turned into Euclidean coordinates prior to K-means partitioning. The clusters produced by K-means were not forced to be hierarchically nested. Despite of that, the results of K-means partitioning showed a strong hierarchical structure for $K = 2$ to 10 groups. This allowed the authors to represent the relationships among partitions as a dendrogram. The $K = 10$ level corresponded to the main types of habitat, recognized *a priori*, where sampling had been conducted.

Indicator values were computed for each species and partitioning level. Some species were found to be stenotopic (narrow niches) while others were eurytopic (species with wide niches, present in a variety of habitats). Others characterized intermediate levels of the hierarchy. The best indicator species ($INDVAL > 25\%$) were assembled into a two-way indicator table; this tabular representation displayed the hierarchical relationships among species.

Results of the indicator value method were compared to TWINSPAN. Note that the partitions of sites used in the two methods were not the same; the TWINSPAN typology was obtained by partitioning correspondence analysis ordination axes (Subsection 8.7.4). TWINSPAN identified, as indicators, pseudospecies pertaining to very low cut-off levels. These species were not particularly useful for prediction because they were simply known to be present at all sites of a group. Several species identified by TWINSPAN as indicators also received a high indicator value from the *INDVAL* procedure, for the same or a closely related habitat class. The *INDVAL* method identified several other indicator species, with rather high indicator values, that also contributed to the specificity of the groups of sites but had been missed by TWINSPAN. So, the *INDVAL* method appeared to be more sensitive than TWINSPAN to the fidelity and specificity of species.

McGeoch & Chown (1998) found the indicator value method important to conservation biology because it is conceptually straightforward and allows researchers to identify bioindicators for any combination of habitat types or areas of interest, e.g. existing conservation areas, or groups of sites based on the outcome of a classification procedure. In addition, it may be used to identify bioindicators for groups of sites classified using the target taxa, as in Ecological application 8.9d, or using non-target taxa (e.g. insect bioindicators of plant community classifications).

Because each *IndVal* index is calculated independently of other species in the assemblage, comparisons of indicator values can be made between taxonomically unrelated taxa, taxa in different functional groups, or those in different communities. Comparisons across taxa are robust to differences in abundance that may or may not be due to differences in catchability or sampling methods. The method is also robust to differences in the numbers of sites between site groups, to differences in abundance among sites within a particular group, and to differences in the absolute abundances of very different taxa which may show similar trends.

When a group of sites corresponds to a delimited geographic area, distribution maps for the indicator species of that group should help identify the core conservation areas for these species, even when little other biological information is available. McGeoch & Chown (1998) also consider the *indicator measure of a species absence* to be of value. The species absence *IndVal* provides a method for improving the objectivity with which species transient to an assemblage can be identified. Species with high values for this absence index may also be of ecological interest as indicators of peculiar ecological conditions where the species is seldom or never present.

Taxa proposed as bioindicators are often merely the favourite taxa of their proponents; ornithologists prefer birds, lepidopterists butterflies, and coleopterists beetles. According to McGeoch & Chown (1998), *IndVal* provides an objective method for addressing this problem by enabling assessment of the relative merits of different taxa for a given study area. The species that do emerge from this procedure as the most useful indicators of a group of sites should prove useful in practical conservation for monitoring site changes.

Borcard (1996) and Borcard & Vaucher-von Ballmoos (1997) present applications of the indicator value method to the identification of the Oribatid mite species that characterize well-defined zones in a peat bog of the Swiss Jura. The indicator values of beetle species characterizing different types of forests have also been studied by Barbalat & Borcard (1997).

8.10 Seriation

Before clustering methods were developed, the structure of a similarity matrix was often studied by *matrix rearrangement* (Orlóci, 1978). In this approach, the order of the objects is modified in such a way as to concentrate the highest similarities near the main diagonal of the similarity matrix. This is very similar to a method called *seriation* in archaeology, where the rows and columns of a *rectangular* matrix of (artefacts × descriptors) are rearranged in such a way as to bring the highest values near the main diagonal, in order to discover the temporal seriation of the artefacts; see Kendall (1988) for a review. This technique was developed by anthropologists Petrie (1899) and Czekanowski (1909) and was applied to ecology by Kulczynski (1928). Traditionally, the rearranged matrices are represented as *trellis diagrams* (Fig. 8.22 is

Trellis diagram

an example), in which half of the matrix is represented by shades of gray corresponding to similarity values. Seriation is available in the clustering package CLUSTAN, referenced in Section 7.7.

At the end of the seriation procedure (Fig. 8.22), the high similarities, which are now found close to the diagonal, indicate clusters of objects. Permutation of the rows and columns was traditionally carried out by hand, although this rapidly becomes tedious with matrices exceeding about ten objects. The various methods of ordination (Chapter 9) would be other ways to find the order of the objects, to be used in a trellis diagram, from an (objects × descriptors) data matrix.

An analytical solution to seriation was proposed in 1950 by Beum & Brundage. This solution can be applied to a similarity matrix among objects presented in any order; the ordering is, of course, the same for the rows and columns. In each column, compute the product of each similarity by the *rank of the row*, add up these values, and divide them by the *sum* of values in the column. These values are used as weights to reorder the objects (rows and columns). The procedure is repeated until convergence is reached, i.e. until the weights do not change the ordering any more. The algorithm may, at times, end up alternating between two equally optimum final solutions.

Non-symmetric matrix

For the analysis of a usual similarity matrix, the clustering methods discussed in the previous sections are more specific and efficient than seriation, since they directly produce clusters of objects. A dendrogram provides a partial order for the objects, which may be used to seriate the diagonal of the similarity matrix, if needed. Seriation is an interesting approach because it allows the analysis of non-symmetric as well as symmetric matrices. Non-symmetric matrices, in which $S(\mathbf{x}_1, \mathbf{x}_2) \neq S(\mathbf{x}_2, \mathbf{x}_1)$, are rather rare in ecology. They may, however, be encountered in cases where the similarity is a direct measure of the influence of an organism on another, or in behavioural studies where the attraction of an organism for another can be used as a similarity measure. They are becoming more common in numerical taxonomy (serological data, DNA pairing data, etc.). These matrices could be decomposed into symmetric and skew-symmetric components, as described in Subsection 2.3, before analysis by clustering and/or ordination methods. Rohlf's (1970) adaptive hierarchical clustering may also be used to analyse non-symmetric resemblance matrices.

Ecological application 8.10a

Kulczynski (1928) studied the phytosociology of a region in the Carpathian Mountains, southeastern Poland. He recognized 37 plant associations, listed the species found in each, and computed a similarity matrix among them. Part of that similarity matrix is reproduced in Fig. 8.22, after seriation. When the largest similarity values are brought near the diagonal, the associations clearly form a series, from association 22 (*Varietum pinetosum czorsztynense*) to association 13 (*Seslerietum variae normale*). The blocs of higher (darker) values near the diagonal allow one to recognize two main groups of associations: (22, 21) and (15, 14, 17, 18); association 11 also has a high similarity with association 15.

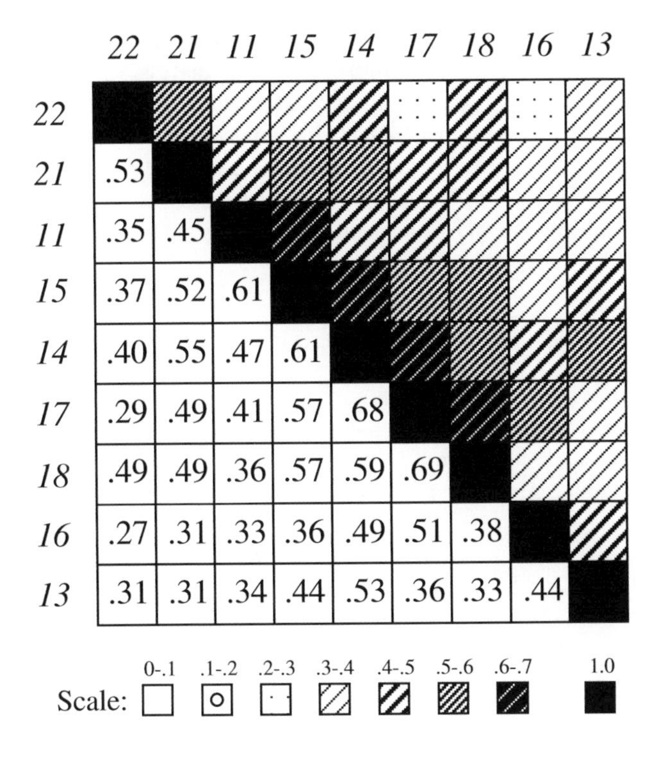

Figure 8.22 Similarity matrix (lower half) and trellis diagram (upper half) for part of Kulczynski's (1928) plant associations of the Carpathian Mountains. Numbers in italics, in the margins, identify the associations; the latin names can be found in Kulczynski's paper. In the trellis diagram, the similarities are represented by shadings, as indicated underneath the matrix.

Ecological application 8.10b

Wieser (1960) studied the meiofauna (small benthic metazoans) at three sites (6 or 7 cores per site) in Buzzards Bay, Massachusetts, USA. After representing the resemblance among cores as a similarity matrix (using Whittaker's index of association, $1 - D_9$) and a trellis diagram, he found that although the three sites differed in species composition, the two sandy sites were more similar to each other than they resembled the third site where the sediment contained high concentrations of fine deposits.

The classical study reported in Ecological application 8.10b has triggered other applications of trellis diagrams in benthic ecology. Among these is Sanders' (1960) representation of an ecological time series, also from Buzzards Bay, using a trellis diagram. Inspired by these applications to benthic ecology, Guille (1970) and Soyer (1970) used the method of trellis diagrams to delineate benthic communities (macrofauna and harpacticoid copepods, respectively) along the French Catalonian coast of the Mediterranean Sea, near Banyuls-sur-Mer.

Wieser's (1960) study offers an opportunity to come back to the warning of Section 8.0, that not all problems of data analysis belong to the clustering approach. Nowadays, one would not have to seriate or cluster the sites before comparing the species to the sediment data. One could compare two similarity matrices, one for species and one for sediment data, using the Mantel test or the ANOSIM procedure (Section 10.5), or else directly compare the species abundance table to the sediment data table using canonical analysis (Chapter 11).

8.11 Clustering statistics

This section is devoted to clustering statistics, the correlation between a cophenetic matrix and the original similarity or distance matrix, and the problem of cluster validation.

1 — Connectedness and isolation

Clustering statistics describe the connectedness within clusters or their degree of isolation. Some of these measures are described here.

The basic statistic of a cluster k is its *number of objects*, n_k. In linkage clustering, when the number of objects is compared to the number of links among them, a measure of density of the cluster in A-space is obtained. Density increases with the *degree of connectedness* of a cluster. Connectedness can be measured as follows (Estabrook, 1966):

Connectedness

$$Co = \frac{\text{number of links in a cluster}}{\text{maximum possible number of links}} \tag{8.23}$$

where the maximum possible number of links is $n_k(n_k - 1)/2$, with n_k being the number of objects in cluster k. This measure of connectedness varies between 0 and 1. Day (1977) proposed other related measures. One of them is the *cohesion index* which considers only the links that exceed the minimum number of links necessary for the cluster to be connected. If this minimum number is called m, the cohesion index can be written as follows:

Cohesion index

$$\frac{\text{No. links} - m}{[n_k(n_k - 1)/2] - m} \tag{8.24}$$

For single linkage clustering, the minimum number of links necessary for n_k objects to be connected is $n_k - 1$, so that the cohesion index becomes:

$$\frac{2(\text{No. links} - n + 1)}{(n - 1)(n - 2)} \tag{8.25}$$

which is Estabrook's (1966) normalized connectedness index. Another measure of cluster density could be the minimum similarity within a cluster, or else the average similarity (Estabrook, 1966).

Isolation

The degree of isolation of clusters can be measured, in metric space A, as the distance between the two closest objects in different clusters. It may also be measured as the average distance between all the objects of one cluster and all the objects of another, or else as the ratio of the distance between the two closest objects to the distance between the centroids of the two clusters. All these measures are ways of quantifying distances between clusters; a clustering or ordination *of clusters* can be computed using these distances. In the context of linkage clustering without reference to a metric space A, Wirth *et al.* (1966) used, as a measure of isolation, the difference between the similarity at which a cluster is formed and the similarity at which it fuses with another cluster.

2 — Cophenetic correlation and related measures

Pearson's correlation coefficient, computed between the values in a cophenetic matrix (Subsection 8.3.1) and those in the original resemblance matrix (excluding the values on the diagonal), is called *cophenetic correlation* (Sokal & Rohlf, 1962), *matrix correlation* (Sneath & Sokal, 1973) or *standardized Mantel* (1967) *statistic* (Subsection 10.5.1). It measures the extent to which the clustering result corresponds to the original resemblance matrix. When the clustering perfectly corresponds to the coefficients in the original matrix, the cophenetic correlation is 1.

Besides the cophenetic correlation, which compares the original similarities to those in a cophenetic matrix, matrix correlations are useful in four other situations:

• To compare any pair of resemblance matrices, such as the original similarity matrix of Section 8.2 and a matrix of distances among the objects in a space of reduced dimension (Chapter 9).

• To compare two similarity or distance matrices obtained by computing different resemblance measures on the same data.

• To compare the results of two clustering methods applied to a resemblance matrix.

• To compare various clustering levels in a dendrogram. The ultrametric matrix of a given clustering level contains only zeros and ones in that case, as shown in Subsection 8.3.1.

Correlations take values between –1 and +1. The cophenetic correlation is expected to be positive if the original similarities are compared to cophenetic similarities (or distances to distances) and negative if similarities are compared to distances. The higher the absolute value of the cophenetic correlation, the better the correspondence between the two matrices being compared. Ecologists might prefer to use a nonparametric correlation coefficient (Kendall's τ or Spearman's r) instead of Pearson's r, if the interest lies more in the geometric structure of the dendrogram than the actual lengths of its branches.

A cophenetic correlation cannot be tested for significance because the cophenetic matrix is not independent of the original similarity matrix; one comes from the other through the clustering algorithm. In order to test the significance of a cophenetic correlation, one would have to pretend that, under H_0, the two matrices may be independent of each other, i.e. that the clustering algorithm is likely to have a null efficiency. On the contrary, the similarity between two hierarchical classifications of *different data sets* about the same objects, measured by matrix correlation or some other measure of consensus (Rohlf, 1974, 1982b), can be tested for significance (Section 10.2, Fig. 10.4).

Gower distance

Other coefficients have been proposed to measure the goodness-of-fit between matrices. For instance, Gower's (1983) distance is the sum of the squared differences between values in the cophenetic similarity matrix and the original similarity matrix:

$$D_{\text{Gower}} = \sum_{i,j} (original\ s_{ij} - cophenetic\ s_{ij})^2 \qquad \textbf{(8.26)}$$

This measure, also called *stress 1* (Kendall, 1938), takes values in the interval $[0, \infty)$; it is used as a measure of goodness-of-fit in nonmetric multidimensional scaling (eq. 9.28). Small values indicate high fit. Like the cophenetic correlation, this measure only has relative value when comparing clustering results obtained from the same original similarity matrix. Several other such functions are listed in Rohlf (1974).

Modified Rand index

Other measures have been proposed for comparing different *partitions* of the same objects. Consider in turn all pairs of objects and determine, for each one, whether the two objects are placed in the same group, or not, by the partition. One can construct a 2×2 contingency table, similar to the one shown at the beginning of Subsection 7.3.1, comparing the pair assignments made by two partitions. The simple matching coefficient (eq. 7.1), computed on this contingency table, is often called the Rand index (1971). Hubert & Arabie (1985) have suggested a modified form that corrects the Rand index as follows: if the relationship between two partitions is comparable to that of partitions picked at random, the corrected Rand index returns a value near 0. The *modified Rand index* is widely used for comparing partitions.

Shepard-like diagram

A Shepard diagram is a scatter plot comparing distances in a space of reduced dimension, obtained by ordination methods, to distances in the original association matrix (Fig. 9.1). This type of diagram has been proposed by Shepard (1962) in the paper where he first described nonmetric multidimensional scaling (Section 9.3). Shepard-like diagrams can be constructed to compare the similarities (or distances) of the cophenetic matrix (Section 8.3) to the similarities (or distances) of the original resemblance matrix (Fig. 8.23). Such a plot may help choose between parametric and nonparametric cophenetic correlation coefficients. If the relationship between original

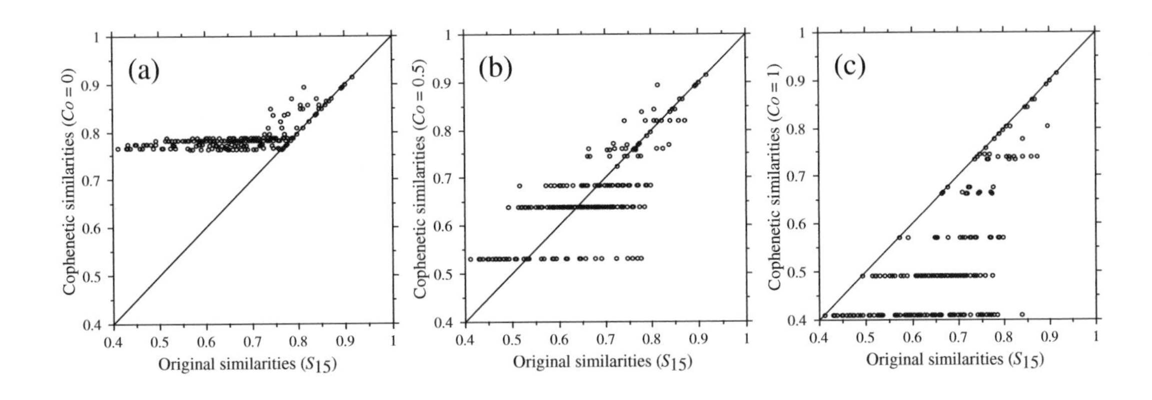

Figure 8.23 Shepard-like diagrams comparing cophenetic similarities to original similarities for 21 lakes clustered using (a) single linkage ($Co = 0$, cophenetic $r = 0.64$, $\tau = 0.45$), (b) proportional link linkage ($Co = 0.5$, cophenetic $r = 0.75$, $\tau = 0.58$), and (c) complete linkage clustering ($Co = 1$, cophenetic $r = 0.68$, $\tau = 0.51$). Co is the connectedness of the linkage clustering method (Subsection 8.5.3). There are 210 points (i.e. 210 similarity pairs) in each graph. The diagonal lines are visual references.

and cophenetic similarities is curvilinear in the Shepard-like diagram, as it is the case in Figs. 23a and c, a nonparametric correlation coefficient should be used.

Fig. 8.23 also helps understand the space-contraction effect of single linkage clustering, where the cophenetic similarities are always larger than or equal to the original similarities; the space-conservation effect of intermediate linkage clustering with connectedness values around $Co = 0.5$; and the space-dilation effect of complete linkage clustering, in which cophenetic similarities can never exceed the original similarities. There are $(n - 1)$ clustering levels in a dendrogram. This limits to $(n - 1)$ the number of different values that can be found in a cophenetic matrix and, hence, along the ordinate of a Shepard-like diagram. This is why points form horizontal bands in Fig. 8.23.

Following are three measures of goodness-of-fit between the single linkage clustering results and the original similarity matrix, for the pond example:

Pearson r cophenetic correlation = 0.941
Kendall τ_b cophenetic correlation = 0.774
Gower distance = 0.191

8.12 Cluster validation

Users of clustering methods may wonder whether the result of a clustering program run is valid or not, i.e. whether the clusters are "real", or simply artefacts of the clustering algorithm. Indeed, clustering algorithms may produce misleading results. On the one hand, most hierarchical clustering (or partitioning) algorithms will give rise to a hierarchy (or a partition), whether the objects are, or not, hierarchically interrelated (or pertaining to distinct clusters). On the other hand, different clustering algorithms may produce markedly different results because clustering methods impose different models onto the data, as shown in the present Chapter. Finally, different clustering methods are variously sensitive to noise (error) in the data. A simulation study comparing several clustering and partitioning methods under different levels of noise can be found in Milligan (1980); see also the review paper of Milligan (1996).

It is important to validate the results of cluster analyses. One has to show that a clustering structure is unusual in some sense, i.e. that it departs from what may be expected from unstructured data. Unfortunately, most of the validation methods summarized below are not presently available in standard clustering packages. Readers are referred to Chapter 4 of Jain & Dubes (1988) for details, and to the review papers of Perruchet (1983a, b), Bock (1989, 1996), Gordon (1994, 1996a, 1996b) and Milligan (1996). Lapointe (1998) provides a review of the validation methods used in phylogenetic studies.

Validation may be carried out in nonstatistical or statistical ways. Statistical ways involve tests of hypotheses, whereas nonstatistical assessment accepts weaker evidence for the presence of clusters. Commonly-used nonstatistical methods are:

• Plot the clusters onto an ordination diagram and look for separation of the clusters (Section 10.1). This method is often used to assess the degree of refinement of hierarchical clustering results that one should consider for interpretation.

• Compare the results of several clustering algorithms, either informally (using visual examination, identify the partition levels that are found in most or all trees being compared) or formally (calculate consensus indices or construct a compromise "consensus" tree: below).

Different issues can be considered in cluster validation:

• The most general hypothesis is that of complete absence of classification structure in the data. In principle, such tests should be carried out before cluster analysis is attempted. Several methods have been proposed to assess the positions of the objects distributed in multidimensional space (random position hypothesis) and test for either uniform or unimodal distributions (i.e. greater density of objects near the centre of the distribution). There are also tests that are carried out on graphs linking the objects, and others that involve only the object labels.

• Other methods are available to test (1) for the presence of a hierarchical structure in the data, (2) for partitions (are there distinct clusters in the data? how many clusters?), or (3) for the validity of individual clusters.

For any one of these hypotheses, validation may be carried out at different conceptual levels.

1. *Internal validation using* **Y** — *Internal validation* methods allow the assessment of the *consistency* of a clustering topology. Internal validation consists in using the original data (i.e. matrix **Y** containing the data originally used for clustering) to assess the clustering results. One approach is to resample the original data set. One repeatedly draws subsets of objects at random, using the jackknife or bootstrap methods (Subsection 1.2.4), to verify that the original clusters of objects are found by the clustering method for the different subsets. Nemec & Brinkhurst (1988) present an ecological application of this method to species abundance data. Another approach is to randomize the original data set, or generate simulated data with similar distribution parameters, and compute the classification a large number of times to obtain a null distribution for some clustering statistic of interest; one may use one of the statistics discussed in Subsection 8.11.2, or the *U* statistic of Gordon (1994) described at the end of Subsection 10.5.3. The test of cluster fusion in chronological clustering (Subsection 12.6.4) is an example of an internal validation criterion. Using simulations, Milligan (1981) compared 30 internal validation criteria that may be used in this type of study. One *must not*, however, use a standard hypothesis testing procedure such as ANOVA or MANOVA on the variables used to determine the clusters. This approach would be incorrect because the *alternative hypothesis* of the test would be constructed to fit the group structure computed from the very data that would now be used for testing the null hypothesis. As a consequence, such a test would almost necessarily (subject to type II error) result in significant differences among the groups.

2. *External validation comparing* **Y** *to* **X** — *External validation* methods involve the comparison of two different data tables. The clustering results derived from data matrix **Y** are compared to a matrix of explanatory variables, which is called **X** in the context of regression (Chapter 10) and canonical analysis (Chapter 11). Comparisons can be made at different levels. One may compare a partition of the objects to matrix **X** using discriminant analysis (Table 10.1; Section 11.5). Else, the whole hierarchical tree structure may be coded using binary variables (Baum, 1992; Ragan, 1992), in the same way as nested factors in ANOVA; this matrix is then compared to the explanatory matrix **X** using redundancy analysis (Subsections 11.1). A third way is to compare the cophenetic matrix (Subsection 2 above) that represents the hierarchical tree structure to a similarity or distance matrix computed from matrix **X**, using a Mantel test (Subsection 10.5.1; Hubert & Baker, 1977). Contrary to the cophenetic correlations considered in Subsection 8.11.2, testing is legitimate here because matrix **X** is independent of the data matrix **Y** used to construct the classification.

3. *External validation comparing two or several matrices* **Y**, *same variables* — Confirmation of the presence of a clustering structure in the data can be obtained by

repeating the cluster analysis using different sets of objects (data matrices $\mathbf{Y}_1$, $\mathbf{Y}_2$, etc.) and comparing the results. The first case is that where replicate data are available. For instance, if lakes can be selected at random from different geographic regions, one could conduct independent cluster analyses of the regions using one lake per region (different lakes being used in the separate runs), followed by a comparison of the resulting partitions or dendrograms. Methods are available for comparing independently-obtained dendrograms representing the same objects (Fig. 10.4 and references in Section 10.2). Alternatively, the classification of regions obtained from the first set of lakes (matrix $\mathbf{Y}_1$) could be taken as a model to be validated, using discriminant analysis, by comparing it to a second, independent set of lakes (matrix $\mathbf{Y}_2$) representing the same regions.

In *replication analysis*, external validation is carried out for data that are not replicate observations of the same objects. One finds a classification using matrix $\mathbf{Y}_1$, determines group centroids, and assigns the data points in $\mathbf{Y}_2$ to the nearest centroid (McIntyre & Blashfield, 1980). Then, the data in $\mathbf{Y}_2$ are clustered without considering the result from $\mathbf{Y}_1$. The independently obtained classification of $\mathbf{Y}_2$ is compared to the first one using some appropriate measure of consensus (point 4, below).

In studies where data are costly to obtain, this approach is, in most cases, not appealing to researchers who are more interested in using all the available information in a single cluster analysis, instead of dividing the data set into two or several analyses. This approach is only feasible when the objects are numerous.

4. *External validation comparing two or several matrices* **Y**, *same objects* — Several groups of variables may be available about the same objects; one may wish to conduct separate cluster analyses on them. An example would be sites where data are available about several groups of arthropods (e.g. matrices $\mathbf{Y}_1$ = mites, $\mathbf{Y}_2$ = insects, and $\mathbf{Y}_3$ = spiders), besides physical or other variables of the environment which would form a matrix **X** of explanatory variables. Classifications may be obtained independently for each matrix **Y**. Measures of resemblance between trees, called *consensus indices* (Rohlf, 1982b), may be calculated. The cophenetic correlation coefficient of the previous Subsection can be used as a consensus index; other indices are available, that only take the classification topologies into account. Alternatively, one may compute a compromise tree, called a *consensus tree*, which represents the areas of agreement among trees. Several criteria have been proposed for constructing consensus trees: majority rule, strict consensus, average consensus, etc. (Leclerc & Cucumel, 1987). Tests of significance are available for comparing independently-obtained dendrograms that describe relationships among the same objects (Fig. 10.4 and references in Section 10.2).

8.13 Cluster representation and choice of a method

This section summarizes the most usual graphical representations of clustering results. More complete reviews of the subject are found in Sneath & Sokal (1973) and Chambers & Kleiner (1982).

Hierarchical clustering results are represented, in most cases, by dendrograms or by plots of connected subgraphs. The construction of these graphs has been explained in Sections 8.2 and 8.5. The branches of dendrograms may point upwards or downwards, but they are more often represented horizontally because this is an easier way of fitting a dendrogram of a large number of objects into a page. The abscissa is graduated in similarities or distances; the branching pattern indicates the similarity or distance of bifurcating branches. Usually, the names of the objects (or descriptors), or their code numbers, are written at the tips of the branches. The ordinate (on horizontal dendrograms) has no specified ordering, except in TWINSPAN. Bifurcating branches are not fixed; they may be swivelled as required by the presentation of results, without altering the nature of the ultrametric information in the dendrogram.

Dendrogram

Dendrograms clearly illustrate the clusters formed at each partition level, but they do not allow the identification of the exact similarity links among objects. With some clustering methods, this information is not directly available and must be found *a posteriori* when needed. In any case, for a synoptic clustering which only aims at recognizing major clusters of objects, connecting links are not required.

Connected subgraphs

Series of connected subgraphs, as in Fig. 8.3 and 8.4, may be used to represent all the information of the similarity or distance matrix. Complex information may be represented by different types of lines; colours may also be used. When they become numerous, objects can be placed at the rim of a circle; similarity links are drawn as lines between them. In each subgraph, the relative positions of the objects are of little importance. They are merely arranged in such a way as to simplify the paths of the links connecting them. The objects may have been positioned beforehand in a two-dimensional ordination space, which may be obtained by principal coordinate analysis or nonmetric scaling of the association matrix (Sections 9.2 and 9.3). Figures of connected subgraphs, informative as they may be, are quite time consuming to draw and difficult to publish.

Skyline plot
Tree
Icicle plot

Some programs still use "skyline plots" (Ward, 1963, Wirth *et al.*, 1966), which may also be called "trees" or "icicle plots". These plots may be imagined as negatives of dendrograms. They contain the same information as dendrograms, but they are rather odd to read and interpret. In Fig. 8.24a for instance (UPGMA clustering of the pond data, Fig. 8.5), the object names are sitting on the lines *between* columns of X's; the ordinate of the plot is a scale of similarities or distances. Since the value $S = 1$ ($D = 0$) is at the bottom of the graph, this is where the hierarchical agglomeration begins. The first clustering step is materialized by the first horizontal row of X's, at distance $S = 0.6$ ($D = 0.4$), which joins objects 212 and 214. It is drawn like the lintel

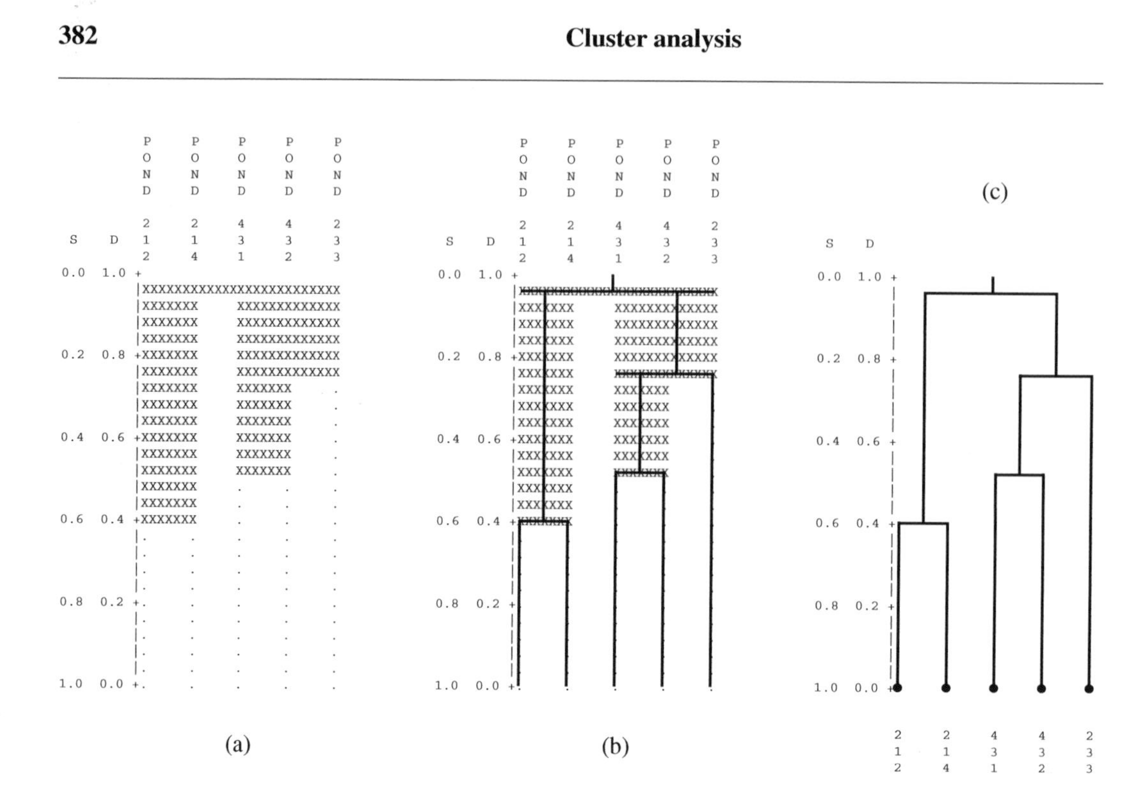

Figure 8.24 A skyline plot (a) can be transformed into a dendrogram (c) by going through the drawing steps described in (b).

of a door. The surface above the lintel of X's is filled with X's; these are without meaning. The next clustering step is at distance $S = D = 0.5$; it consists in a row of X's joining ponds 431 and 432. The third clustering step is more interesting. Note how a new lintel of X's, at $S = 0.25$ ($D = 0.75$), goes from pond 233, and right across the column of X's already joining ponds 431 and 432. The final clustering step is at $S = 0.058$ ($D = 0.942$). This new lintel crosses the last remaining gap, uniting the two columns of X's corresponding to the two already-formed clusters.

A skyline plot can directly be transformed into a dendrogram (Fig. 8.24b, c). Working from the bottom to the top, proceed as follows:

- Identify lintels and draw lines across the column of X's. The lines should not extend beyond the row of X's.

- When all the horizontal lines have been drawn, draw vertical lines from the *middle* of a lower lintel up to the one above. Erase the overhanging part of the upper lintel. Repeat the operation for the next lintel up.

- The result is a standard dendrogram (Fig. 8.23c).

Seriated matrix

A hierarchical clustering (or a partition) can also be represented as a seriated similarity matrix, as in Fig. 8.22. The order in which the objects appear in the dendrogram can be used as a first approximation of the seriation order. Of course, branches of the dendrogram may be swivelled to obtain a better ordering of the similarity values.

Chapter 10 shows how to superimpose clustering results onto an ordination of the same objects. This is often the only way to clarify the structure when ecological objects form an almost perfect continuum. When it comes to representing the results of a partition, the objects are represented in an ordination space and envelopes are drawn around points corresponding to the various clusters.

Table 8.13 summarizes, in a comparative way, the various clustering methods discussed in the present Chapter. Some advantages and disadvantages of each method are pointed out.

Table 8.13 Synoptic summary of the clustering methods presented in Chapter 8.

Method	Pros & cons	Use in ecology
Hierarchical agglomeration: linkage clustering	Pairwise relationships among the objects are known.	
Single linkage	Computation simple; contraction of space (chaining); combinatorial method.	Good complement to ordination.
Complete linkage (see also: species associations)	Dense nuclei of objects; space expansion; many objects cluster at low similarity; arbitrary rules to resolve conflicts; combinatorial method.	To increase the contrast among clusters.
Intermediate linkage	Preservation of reference space A; non-combinatorial: not included in Lance & Williams' general model.	Preferable to the above in most cases where only one clustering method is to be used.
Hierarchical agglomeration: average clustering	Preservation of reference space A; pairwise relationships between objects are lost; combinatorial method.	
Unweighted arithmetic average (UPGMA)	Fusion of clusters when the similarity reaches the mean inter-cluster similarity value.	For a collection of objects obtained by simple random or systematic sampling.
Weighted arithmetic average (WPGMA)	Same, with adjustment for group sizes.	Preferable to the previous method in all other sampling situations.
Unweighted centroid (UPGMC)	Fusion of clusters with closest centroids; may produce reversals.	For simple random or systematic samples of objects.
Weighted centroid (WPGMC)	Same, with adjustment for group sizes; may produce reversals.	Preferable to the previous method in all other sampling situations.
Ward's method	Minimizes the within-group sum of squares.	When looking for hyperspherical clusters in space A.
Hierarchical agglomeration: flexible clustering	The algorithm allows contraction, conservation, or dilation of space A; pairwise relationships between objects are lost; combinatorial method.	This method, as well as all the other combinatorial methods, are implemented using a simple algorithm.
Hierarchical agglomeration: information analysis	Minimal chaining; only for Q-mode clustering based upon presence-absence of species.	Use is unclear: similarities reflect double absences as well as double presences.

Table 8.13 Continued.

Method	Pros & cons	Use in ecology
Hierarchical division	Danger of incorrect separation of members of minor clusters near the beginning of clustering.	
Monothetic	Division of the objects following the states of the "best" descriptor.	Useful only to split objects into large clusters, inside which clustering may depend on different phenomena.
Polythetic	For small number of objects only.	Impossible to compute for sizable data sets.
Division in ordination space	Binary division along each axis of ordination space; no search is done for high concentrations of objects in space A.	Efficient algorithms for large data sets, when a coarse division of the objects is sought.
TWINSPAN	Dichotomized ordination analysis; ecological justification of several steps unclear.	Produces an ordered two-way table classifying sites and species.
***K*-means partitioning**	Minimizes within-group sum of squares; different rules may suggest different optimal numbers of clusters.	Produces a partition of the objects into *K* groups, *K* being determined by the user.
Species associations	Non-hierarchical methods; clustering at a pre-selected level of similarity or probability.	Concept of association based on co-occurrence of species (for other concepts, use the hierarchical methods).
Non-hierarchical complete linkage	For all measures of dependence among species; species associated by complete linkage (no overlap); satellite species joined by single linkage (possible overlap).	Straightforward concept; easy application to any problem of species association.
Probabilistic clustering	Theoretically very consistent algorithm; test of significance on species associations; limited to similarities computed using Goodall's probabilistic coefficient.	Not often used because of heavy computation.
Seriation	One-dimensional ordination along the main diagonal of the similarity matrix.	Especially useful for non-symmetric association matrices.
Indicator species		
TWINSPAN	Only for classifications of sites obtained by splitting CA axes; ecological justification of several steps unclear.	Gives indicator values for the pseudospecies.
Indicator value index	For any hierarchical or non-hierarchical classification of sites; *IndVal* for a species is not affected by the other species in the study.	Gives indicator values for the species under study; the *IndVal* index is tested by permutation.

Chapter 9 Ordination in reduced space

9.0 Projecting data sets in a few dimensions

Ordination (from the Latin *ordinatio* and German *Ordnung*) is the arrangement of units in some order (Goodall, 1954). This operation is well-known to ecologists. It consists in plotting object-points along an axis representing an ordered relationship, or forming a scatter diagram with two or more axes. The ordered relationships are usually quantitative, but it would suffice for them to be of the type "larger than", "equal to", or "smaller than" (semiquantitative descriptors) to permit meaningful ordinations. Gower (1984) points out that the term *ordination*, used in multivariate statistics, actually comes from ecology where it refers to the representation of objects (sites, stations, relevés, etc.) as points along one or several axes of reference.

Ordination

In ecology, several descriptors are usually observed for each object under study. In most instances, ecologists are interested in characterizing the main trends of variation of the objects with respect to all descriptors, not only a few of them. Looking at scatter plots of the objects with respect to all possible pairs of descriptors is a tedious approach, which generally does not shed much light on the problem at hand. In contrast, the multivariate approach consists in representing the scatter of objects in a multidimensional diagram, with as many axes as there are descriptors in the study. It is not possible to draw such a diagram on paper with more than two or eventually three dimensions, however, even though it is a perfectly valid mathematical construct. For the purpose of analysis, ecologists therefore project the multidimensional scatter diagram onto bivariate graphs whose axes are known to be of particular interest. The axes of these graphs are chosen to represent a large fraction of the variability of the multidimensional data matrix, in a space with reduced (i.e. lower) dimensionality relative to the original data set. Methods for *ordination in reduced space* also allow one to derive quantitative information on the quality of the projections and study the relationships among descriptors as well as objects.

Table 9.1 Domains of application of the ordination methods presented in this chapter.

Method	Distance preserved	Variables
Principal component analysis (PCA)	Euclidean distance	Quantitative data, linear relationships (beware of double-zeros)
Principal coordinate analysis (PCoA), metric (multidimensional) scaling, classical scaling	Any distance measure	Quantitative, semiquantitative, qualitative, or mixed
Nonmetric multidimensional scaling (NMDS, MDS)	Any distance measure	Quantitative, semiquantitative, qualitative, or mixed
Correspondence analysis (CA)	χ^2 distance	Non-negative, dimensionally homogeneous quantitative or binary data; species abundance or presence/absence data
Factor analysis *sensu stricto*	Euclidean distance	Quantitative data, linear relationships (beware of double-zeros)

Factor analysis

Ordination in reduced space is often referred to as *factor* (or *inertia*) *analysis* since it is based on the extraction of the eigenvectors or *factors* of the association matrix. In the present book, the expression *factor analysis* will be restricted to the methods discussed in Section 9.5. Factor analysis *sensu stricto* is mainly used in the social sciences; it aims at representing the covariance structure of the descriptors in terms of a hypothetical causal model.

The domains of application of the techniques discussed in the present chapter are summarized in Table 9.1. Section 9.1 is devoted to principal component analysis, a powerful technique for ordination in reduced space which is, however, limited to sets of quantitative descriptors. Results are also sensitive to the presence of double-zeros. Sections 9.2 and 9.3 are concerned with principal coordinate analysis (metric scaling) and nonmetric multidimensional scaling, respectively. Both methods project, in reduced space, the distances among objects computed using some appropriate association measure (S or D; Chapter 7); the descriptors may be of any mathematical kind. Section 9.4 discusses correspondence analysis, a most useful ordination technique for species presence/absence or abundance data. Finally, and as mentioned above, Section 9.5 summarizes the principles of factor analysis *sensu stricto*. The presentation of the various forms of canonical analysis, which are also eigenvector-based techniques (like PCA, PCoA, and CA), is deferred to Chapter 11.

It often happens that the structure of the objects under study is not continuous. In such a case, an ordination in reduced space, or a scatter diagram produced using two important variables, may be sufficient to make the grouping structure of the objects obvious. Ordination methods may thus be used, sometimes, to delineate clusters of objects (Section 8.1); see however the remarks of Section 8.9 about the use of ordination methods in the study of species associations. More generally, ordinations may always be used as complements to cluster analyses. The reason is that clustering investigates pairwise distances among objects, looking for fine relationships, whereas ordination in reduced space considers the variability of the whole association matrix and thus brings out general gradients. Different methods for superimposing the results of clustering onto ordinations of the same objects are described in Section 10.1.

Reduced space

Ecologists generally use ordination methods to study the relative positions of objects in reduced space. An important aspect to consider is the representativeness of the representation in reduced space, which usually has $d = 2$ or 3 dimensions. To what extent does the reduced space preserve the distance relationships among objects? To answer this, one can compute the distances between all pairs of objects, both in the multidimensional space of the original p descriptors and in the reduced d-dimensional space. The resulting values are plotted in a scatter diagram such as Fig. 9.1. When the projection in reduced space accounts for a high fraction of the variance, the distances between projections of the objects in reduced space are quite similar to the original distances in multidimensional space (case a). When the projection is less efficient, the distances between objects in reduced space are much smaller than in the original space. Two situations may then occur. When the objects are at *proportionally* similar distances in the two spaces (case b), the projection is still useful even if it accounts for a small fraction of the variance. When, however, the relative positions of objects are not the same in the two spaces (case c), the projection is useless. Ecologists often disregard the interpretation of ordinations when the reduced space does not account for a high fraction of the variance. This is not entirely justified, since a projection in reduced space may be informative even if that space only accounts for a small fraction of the variance (case b).

Shepard diagram

The scatter diagram of Fig. 9.1, which is often referred to as a Shepard diagram (Shepard, 1962; diagrams in Shepard's paper had their axes transposed relative to Fig. 9.1), may be used to estimate the representativeness of ordinations obtained using any reduced-space ordination method. In principal component analysis (Section 9.1), the distances among objects, in both the multidimensional space of original descriptors and the reduced space, are calculated using Euclidean distances (D_1, eq. 7.34). The **F** matrix of principal components (eq. 9.4 below) gives the coordinates of the objects in the reduced space. In principal coordinate analysis (Section 9.2) and nonmetric multidimensional scaling (Section 9.3), Euclidean distances among the objects in reduced space are compared to distances D_{hi} found in matrix **D** used as the basis for computing the ordination. In correspondence analysis (Section 9.4), it is the χ^2 distance (D_{16}, eq. 54) among objects which is used on the abscissa (Table 9.1). Shepard-like diagrams can also be constructed for cluster analysis (Fig. 8.23).

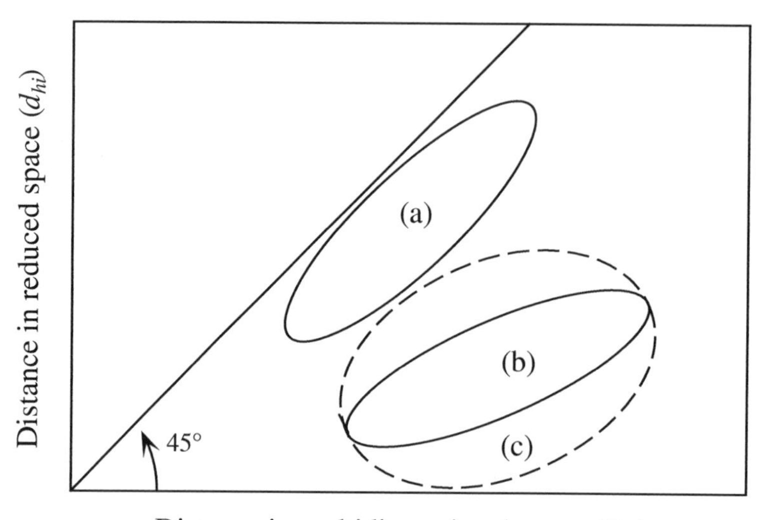

Figure 9.1 Shepard diagram. Three situations encountered when comparing distances among objects, in the p-dimensional space of the p original descriptors (abscissa) *versus* the d-dimensional reduced space (ordinate). The figure only shows the contours of the scatters of points. (a) The projection in reduced space accounts for a high fraction of the variance; the relative positions of objects in the d-dimensional reduced space are similar to those in the p-dimensional space. (b) The projection accounts for a small fraction of the variance, but the relative positions of the objects are similar in the two spaces. (c) Same as (b), but the relative positions of the objects differ in the two spaces. Adapted from Rohlf (1972). Compare to Fig. 8.23.

The following sections discuss the ordination methods most useful to ecologists. They are written to be easily understood by ecologists, so that they may not entirely fulfil the expectations of statisticians. Many programs are available to carry out ordination analysis; several of them are described by Michael Palmer*. For detailed discussions on the theory or computing methods, one may refer to ter Braak (1987c) and Morrison (1990), among other works. Important references about correspondence analysis are Benzécri and coll. (1973), Hill (1974), Greenacre (1983), and ter Braak (1987c). Gower (1984 and 1987) reviewed the ordination methods described in this chapter, plus a number of other techniques developed by psychometricians. Several of these are progressively finding their way into numerical ecology. They include methods of metric scaling other than principal coordinate analysis, multidimensional unfolding, orthogonal Procrustes analysis (the Procrustes statistic m^2 is described in Subsection 10.5.4) and its generalized form, scaling methods for several distance matrices, and a method for ordination of non-symmetric matrices.

* WWWeb site: <http://www.okstate.edu/artsci/botany/ordinate/software.htm>.

Ordination vocabulary*

Box 9.1

Major axis. Axis in the direction of maximum variance of a scatter of points.

First principal axis (of the concentration ellipsoid in a multinormal distribution; Section 4.4, Fig. 4.9). Line passing through the greatest dimension of the ellipsoid; major axis of the ellipsoid.

Principal components. New variates (*variates* = random variables) specified by the axes of a rigid rotation of the original system of coordinates, and corresponding to the successive directions of maximum variance of the scatter of points. The principal components give the positions of the objects in the new system of coordinates.

Principal-component axes (also called *principal axes* or *component axes*). System of axes resulting from the rotation described above.

*Adapted from Morrison (1990, pp. 87 and 323-325).

9.1 Principal component analysis (PCA)

Section 4.4 has shown that, in a multinormal distribution, the first principal axis is the line that goes through the greatest dimension of the concentration ellipsoid describing the distribution. In the same way, the following principal axes (orthogonal to one another, i.e. at right angles to one another, and successively shorter) go through the following greatest dimensions of the p-dimensional ellipsoid. A maximum of p principal axes may be derived from a data table containing p variables (Fig. 4.9). The principal axes of a dispersion matrix **S** are found by solving (eq. 4.23):

$$(\mathbf{S} - \lambda_k \mathbf{I})\, \mathbf{u}_k = \mathbf{0} \qquad \textbf{(9.1)}$$

whose characteristic equation

$$|\mathbf{S} - \lambda_k \mathbf{I}| = 0 \qquad \textbf{(9.2)}$$

Eigenvalue
Eigenvector

is used to compute the *eigenvalues* λ_k. The eigenvectors $\mathbf{u}_k$ associated with the λ_k are found by putting the different λ_k values in turn into eq. 9.1. These eigenvectors are the *principal axes* of dispersion matrix **S** (Section 4.4). The eigenvectors are normalized (i.e. scaled to unit length, Section 2.4) before computing the *principal components,* which give the coordinates of the objects on the successive principal axes. Principal component analysis (PCA) is due to Hotelling (1933). The method and several of its

Principal components

implications for data analysis are clearly presented in the seminal paper of Rao (1964). PCA possesses the following properties, which make it a powerful instrument for the analysis of ecological data:

1) Since any dispersion matrix **S** is symmetric, its principal axes $\mathbf{u}_k$ are *orthogonal* to one another. In other words, they correspond to *linearly independent directions* in the concentration ellipsoid of the distribution of objects (Section 2.9).

2) The eigenvalues λ_k of a dispersion matrix **S** give the amount of *variance* corresponding to the successive principal axes (Section 4.4).

3) Because of the first two properties, principal component analysis *can often summarize, in a few dimensions,* most of the variability of a dispersion matrix of a large number of descriptors. It also provides a measure of the amount of variance explained by these few independent principal axes.

The present Section shows how to compute the relationships among objects and among descriptors, as well as the relationships between the principal axes and the original descriptors. A simple numerical example is developed, involving five objects and two quantitative descriptors:

$$\mathbf{Y} = \begin{bmatrix} 2 & 1 \\ 3 & 4 \\ 5 & 0 \\ 7 & 6 \\ 9 & 2 \end{bmatrix} \quad \text{After centring on the column means,} \quad [y - \bar{y}] = \begin{bmatrix} -3.2 & -1.6 \\ -2.2 & 1.4 \\ -0.2 & -2.6 \\ 1.8 & 3.4 \\ 3.8 & -0.6 \end{bmatrix}$$

In practice, principal component analysis is never used for two descriptors only; in such a case, the objects can simply be represented in a two-dimensional scatter diagram (Fig. 9.2a). A two-dimensional example is used here for simplicity, in order to show that the main result of principal component analysis is to rotate the axes, using the centroid of the objects as pivot.

1 — Computing the eigenvectors

The dispersion matrix of the above descriptors (eq. 4.6) is:

$$\mathbf{S} = \frac{1}{n-1}[y - \bar{y}]'[y - \bar{y}] = \begin{bmatrix} 8.2 & 1.6 \\ 1.6 & 5.8 \end{bmatrix}$$

The corresponding characteristic equation (eq. 2.23) is:

$$\left|\mathbf{S} - \lambda_k \mathbf{I}\right| = \left| \begin{bmatrix} 8.2 & 1.6 \\ 1.6 & 5.8 \end{bmatrix} - \begin{bmatrix} \lambda_k & 0 \\ 0 & \lambda_k \end{bmatrix} \right| = 0$$

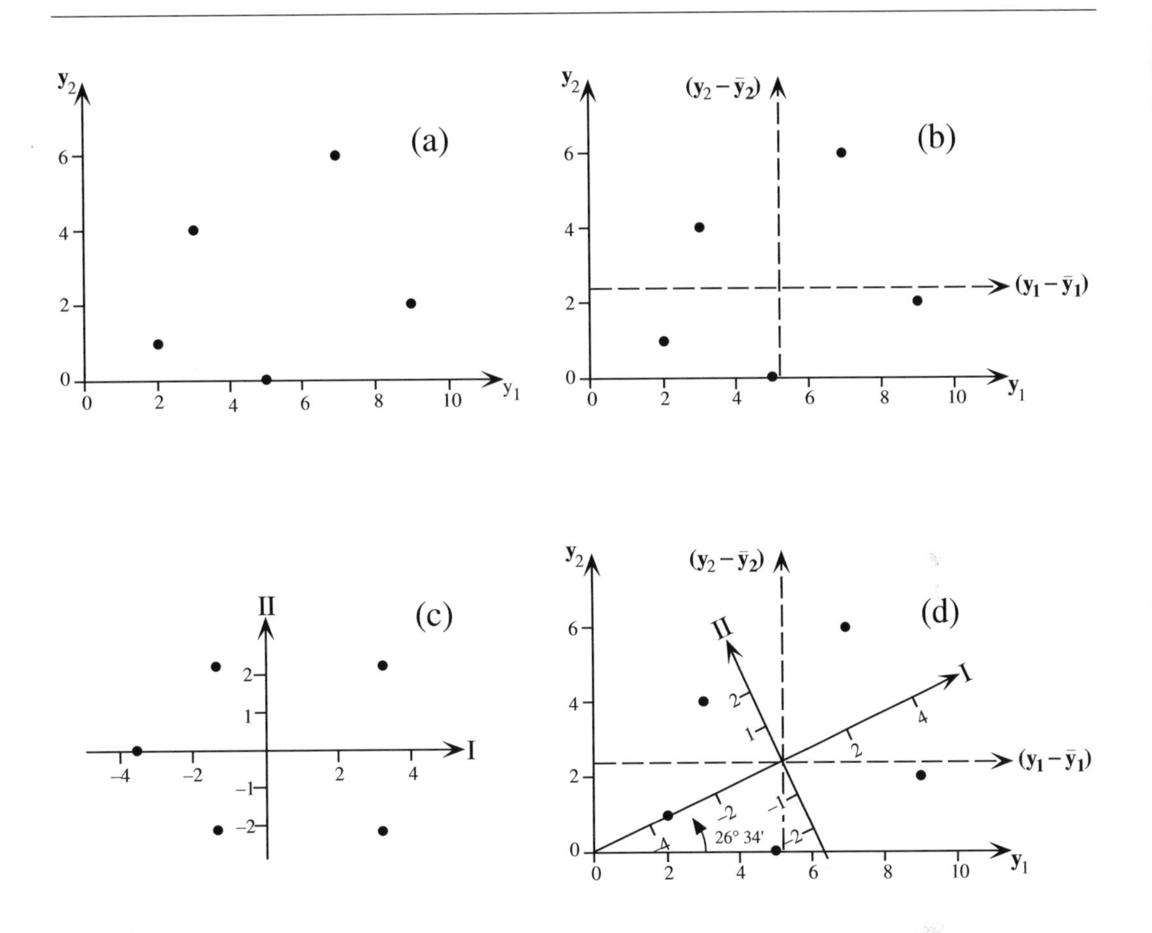

Figure 9.2 Numerical example of principal component analysis. (a) Five objects are plotted with respect to descriptors $\mathbf{y}_1$ and $\mathbf{y}_2$. (b) After centring the data, the objects are now plotted with respect to $(\mathbf{y}_1 - \bar{\mathbf{y}}_1)$ and $(\mathbf{y}_2 - \bar{\mathbf{y}}_2)$, represented by dashed axes. (c) The objects are plotted with reference to principal axes I and II, which are centred with respect to the scatter of points. (d) The two systems of axes (b and c) can be superimposed after a rotation of 26°34'.

It has two eigenvalues, $\lambda_1 = 9$ and $\lambda_2 = 5$. The total variance remains the same, but it is partitioned in a different way: the sum of the variances on the main diagonal of matrix **S** is $(8.2 + 5.8 = 14)$, while the sum of the eigenvalues is $(9 + 5 = 14)$. $\lambda_1 = 9$ accounts for 64.3% of the variance and λ_2 makes up for the difference (35.7%). There are as many eigenvalues as there are descriptors. The successive eigenvalues account for progressively smaller fractions of the variance. Introducing, in turn, the λ_k's in matrix equation 9.1:

$$(\mathbf{S} - \lambda_k \mathbf{I})\, \mathbf{u}_k = \mathbf{0}$$

provides the eigenvectors associated with the eigenvalues. Once these vectors have been normalized (i.e. set to unit length, $\mathbf{u}'\,\mathbf{u} = 1$) they become the *columns* of matrix **U**:

$$\mathbf{U} = \begin{bmatrix} 0.8944 & -0.4472 \\ 0.4472 & 0.8944 \end{bmatrix}$$

If a different sign had been arbitrarily assigned to one of the terms of matrix **U**, when calculating the eigenvectors, this would have resulted in mirror images for Figs. 9.2c and d. These images would be as good at representing the data as Fig. 9.2c and d.

Ortho-gonality

It is easy to check the orthogonality of the two eigenvectors: their cross-product $\mathbf{u}'_1\mathbf{u}_2 = (0.8944 \times (-0.4472)) + (0.4472 \times 0.8944) = 0$. Moreover, Section 4.4 has shown that the elements of **U** are direction cosines of the angles between the original descriptors and the principal axes. Using this property, one finds that the system of principal axes specifies a rotation of $(\text{arc cos } 0.8944) = 26°34'$ of the system of reference defined by the original descriptors.

2 — Computing and representing the principal components

Loading

Principal component

The elements of the eigenvectors are also weights, or *loadings* of the original descriptors, in the linear combination of descriptors from which the principal components are computed. The *principal components* give the positions of the objects with respect to the new system of principal axes. Thus the position of object $\mathbf{x}_i$ on the first principal axis is given by the following function, or linear combination:

$$f_{i1} = (y_{i1} - \bar{y}_1)\,u_{11} + \ldots + (y_{ip} - \bar{y}_p)\,u_{p1} = [y - \bar{y}]_i\mathbf{u}_1 \qquad \textbf{(9.3)}$$

The values $(y_{ij} - \bar{y}_j)$ are the coordinates of object i on the various centred descriptors j while the values u_{j1} are the loadings of the descriptors on the first eigenvector. The positions of all objects with respect to the system of principal axes is given by matrix **F** of the transformed variables. It is also called the *matrix of component scores*:

Matrix of principal components

$$\mathbf{F} = [y - \bar{y}]\,\mathbf{U} \qquad \textbf{(9.4)}$$

where **U** is the matrix of eigenvectors and $[y - \bar{y}]$ is the matrix of centred observations. The system of principal axes is centred with respect to the scatter of point-objects. This would not be the case if **U** had been multiplied by **Y** instead of the centred matrix, as in some special forms of principal component analysis (*non-centred PCA*). For the numerical example, the principal components are computed as follows:

$$\mathbf{F} = \begin{bmatrix} -3.2 & -1.6 \\ -2.2 & 1.4 \\ -0.2 & -2.6 \\ 1.8 & 3.4 \\ 3.8 & -0.6 \end{bmatrix} \begin{bmatrix} 0.8944 & -0.4472 \\ 0.4472 & 0.8944 \end{bmatrix} = \begin{bmatrix} -3.578 & 0 \\ -1.342 & 2.236 \\ -1.342 & -2.236 \\ 3.130 & 2.236 \\ 3.130 & -2.236 \end{bmatrix}$$

Since the two columns of the matrix of component scores are the coordinates of the five objects with respect to the principal axes, they can be used to plot the objects with respect to principal axes I and II (Fig. 9. 2c). It is easy to verify (Fig. 9.2d) that, in this two-descriptor example, the objects are positioned by the principal components in the same way as in the original system of descriptor-axes. Principal component analysis has simply rotated the axes by 26° 34' in such a way that the new axes correspond to the two main components of variability. When there are more than two descriptors, as it is usually the case in ecology, principal component analysis still only performs a rotation of the system of descriptor-axes, but now in multidimensional space. In that case, principal components I and II define the plane allowing the representation of the largest amount of variance. The objects are projected on this plane in such a way as to preserve, as much as possible, the relative Euclidean distances they have in the multidimensional space of the original descriptors.

Euclidean distance

The relative positions of the objects in the rotated p-dimensional space of principal components are the same as in the p-dimensional space of the original descriptors (Fig. 9.2d). This means that *the Euclidean distances among objects* (D_1, eq. 7.34) *have been preserved through the rotation of axes.* This important property of principal component analysis is noted in Table 9.1.

R^2-like ratio

The quality of the representation in a reduced Euclidean space with m dimensions only ($m \le p$) may be assessed by the ratio:

$$\left(\sum_{k=1}^{m} \lambda_k \right) \Big/ \left(\sum_{k=1}^{p} \lambda_k \right) \qquad \textbf{(9.5)}$$

This ratio is the equivalent of a coefficient of determination (R^2, eq. 10.8) in regression analysis. The denominator of eq. 9.5 is actually equal to the trace of matrix **S** (sum of the diagonal elements). Thus, with the current numerical example, a representation of the objects, along the first principal component only, would account for a proportion 9/(9+5) = 0.643 of the total variance in the data matrix.

When the observations have been made along a temporal or spatial axis, or on a geographic plane (i.e. a map giving the coordinates of the sampling sites), one may plot the principal components along the sampling axis, or on the geographic map. Figure 9.20 is an example of such a map, for the first ordination axis of a detrended correspondence analysis. The same approach may be used with the results of a principal component analysis, or any other ordination method.

3 — Contributions of descriptors

Principal component analysis provides the information needed to understand the role of the original descriptors in the formation of the principal components. It may also be used to show the relationships among original descriptors in the reduced space. The role of descriptors in principal component analysis is examined in this Subsection under various aspects: matrix of eigenvectors, projection in reduced space (matrix $\mathbf{U\Lambda}^{1/2}$), and projection in reduced space (matrix **U**).

1. *The matrix of eigenvectors* — In Subsection 1 (above), the relationships among the *normalized eigenvectors* (which are the columns of the square matrix **U**) were studied using an expression of the form **U'U**. For the numerical example:

$$\mathbf{U'U} = \begin{bmatrix} 0.8944 & 0.4472 \\ -0.4472 & 0.8944 \end{bmatrix} \begin{bmatrix} 0.8944 & -0.4472 \\ 0.4472 & 0.8944 \end{bmatrix} = \begin{bmatrix} 1 & 0 \\ 0 & 1 \end{bmatrix} = \mathbf{I}$$

The diagonal terms of **U'U** result from the multiplication of the eigenvectors with themselves. These values are equal to their $(\text{length})^2$, here equal to unity because the vectors are scaled to 1. The nondiagonal terms, resulting from the multiplication of two different eigenvectors, are equal to zero because the eigenvectors are orthogonal. This result would be the same for any matrix **U** of normalized eigenvectors computed from a symmetric matrix. Matrix **U** is a *square orthonormal matrix* (Section 4.4); several properties of such matrices are described in Section 2.8.

Orthonormal matrix

In the same way, the relationships among *descriptors*, which correspond to the rows of matrix **U**, can be studied through the product **UU'**. The diagonal and non-diagonal terms of **UU'** have the same meaning as in **U'U**, except that they now concern the relationships among descriptors. The relationships among the rows of a square orthonormal matrix are the same as among the columns (Section 2.8, property 7), so that:

$$\mathbf{UU'} = \mathbf{I} \qquad \textbf{(9.6)}$$

The descriptors are therefore of unit lengths in the multidimensional space and they lie at 90° of one another (orthogonality).

Principal component analysis is simply a rotation, in the multidimensional space, of the original system of axes (Figs. 9.2 and 9.3a, for a two-dimensional space). It therefore follows that, after the analysis (rotation), the original descriptor-axes are still at 90° of one another. Furthermore, normalizing the eigenvectors simultaneously normalizes the descriptor-axes (the lengths of row and column vectors are given outside the matrix):

$$\mathbf{U} = \begin{bmatrix} u_{11} & \dots & u_{1p} \\ \cdot & & \cdot \\ \cdot & & \cdot \\ \cdot & & \cdot \\ u_{p1} & \dots & u_{pp} \end{bmatrix} \begin{matrix} \sqrt{\Sigma u_{1k}^2} = 1 \\ \cdot \\ \cdot \\ \cdot \\ \sqrt{\Sigma u_{pk}^2} = 1 \end{matrix} \qquad \textbf{(9.7)}$$

$$\sqrt{\Sigma u_{j1}^2} = 1 \quad \dots \quad \sqrt{\Sigma u_{jp}^2} = 1$$

There is a second approach to the study of the relationships among descriptors. It consists in scaling the eigenvectors in such a way that the cosines of the angles between descriptor-axes be proportional to their *covariances.* In this approach, the angles between descriptor-axes are between 0° (maximum positive covariance) and 180° (maximum negative covariance); an angle of 90° indicates a null covariance (orthogonality). This result is achieved by scaling each eigenvector k to a length equal to its standard deviation $\sqrt{\lambda_k}$ *. Using this scaling for the eigenvectors, the Euclidean distances among objects are not preserved.

Using the diagonal matrix $\mathbf{\Lambda}$ of eigenvalues (eq. 2.20), the new matrix of eigenvectors can be directly computed by means of expression $\mathbf{U\Lambda}^{1/2}$. For the numerical example:

$$\mathbf{U\Lambda}^{1/2} = \begin{bmatrix} 0.8944 & -0.4472 \\ 0.4472 & 0.8944 \end{bmatrix} \begin{bmatrix} \sqrt{9} & 0 \\ 0 & \sqrt{5} \end{bmatrix} = \begin{bmatrix} 2.6833 & -1.0000 \\ 1.3416 & 2.0000 \end{bmatrix}$$

In this scaling, the relationships among descriptors are the same as in the dispersion matrix $\mathbf{S}$ (on which the analysis is based), since

$$(\mathbf{U\Lambda}^{1/2})\,(\mathbf{U\Lambda}^{1/2})' = \mathbf{U\Lambda U}' = \mathbf{U\Lambda U}^{-1} = \mathbf{S} \qquad \textbf{(9.8)}$$

Equation $\mathbf{U\Lambda U}^{-1} = \mathbf{S}$ is derived directly from the general equation of eigenvectors $\mathbf{SU} = \mathbf{U\Lambda}$ (eq. 2.27). In other words, the new matrix $\mathbf{U\Lambda}^{1/2}$ is of the following form (the lengths of the row and column vectors are given outside the matrix):

$$\mathbf{U\Lambda}^{1/2} = \begin{bmatrix} u_{11}\sqrt{\lambda_1} & \dots & u_{1p}\sqrt{\lambda_p} \\ \cdot & & \cdot \\ \cdot & & \cdot \\ \cdot & & \cdot \\ u_{p1}\sqrt{\lambda_1} & \dots & u_{pp}\sqrt{\lambda_p} \end{bmatrix} \begin{matrix} \sqrt{\Sigma\,(u_{1k}\sqrt{\lambda_k})^2} = s_1 \\ \cdot \\ \cdot \\ \cdot \\ \sqrt{\Sigma\,(u_{pk}\sqrt{\lambda_k})^2} = s_p \end{matrix} \qquad \textbf{(9.9)}$$

$$\sqrt{\Sigma\,(u_{j1}\sqrt{\lambda_1})^2} = \sqrt{\lambda_1} \quad \dots \quad \sqrt{\Sigma\,(u_{jp}\sqrt{\lambda_p})^2} = \sqrt{\lambda_p}$$

This equation shows that, when the eigenvectors are scaled to the lengths of their respective standard deviations $\sqrt{\lambda_k}$, the lengths of the descriptor-axes are $\sqrt{s_j^2} = s_j$ (i.e. their standard deviations) in multidimensional space. The product of two descriptor-axes, which corresponds to their angle in the multidimensional space, is therefore equal to their *covariance* s_{jl}.

* In some computer packages, the principal component procedure only scales the eigenvectors to length $\sqrt{\lambda}$ and only provides a plot of the descriptor-axes (no plot of the objects is available).

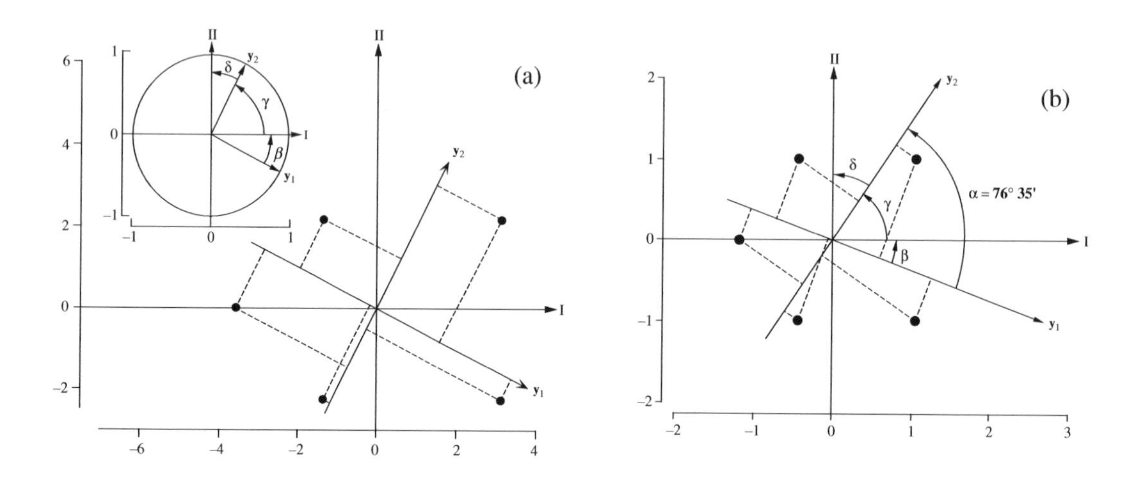

Figure 9.3 Numerical example from Fig. 9.2. Distance and correlation biplots are discussed in Subsection 9.1.4. (a) Distance biplot. The eigenvectors are scaled to lengths 1. Inset: descriptors (matrix **U**). Main graph: descriptors (matrix **U**; arrows) and objects (matrix **F**; dots). The interpretation of the object-descriptor relationships is not based on their proximity, but on orthogonal projections (dashed lines) of the objects on the descriptor-axes or their extensions. (b) Correlation biplot. Descriptors (matrix $\mathbf{U\Lambda}^{1/2}$; arrows) with a covariance angle of 76°35'. Objects (matrix **G**; dots). Projecting the objects orthogonally on a descriptor (dashed lines) reconstructs the values of the objects along that descriptors, to within a multiplicative constant.

2. *Projection of descriptors in reduced space (matrix* $\mathbf{U\Lambda}^{1/2}$*)* — Ecologists using principal component analysis are not interested in the whole multidimensional space but only in a simplified *projection* of the objects in a *reduced space* (generally a two-dimensional plane). Elements $u_{jk}\sqrt{\lambda_k}$ of the eigenvectors, scaled to $\sqrt{\lambda_k}$, are the coordinates of the projections of the descriptors j on the different principal axes k. They are scaled in such a way that the projections of descriptor-axes can be drawn in the reduced space formed by the principal axes (Fig. 9.3b). The descriptors are represented by arrows since they are *axes*. In a reduced-dimension plane, projections of descriptor-axes are shorter than or equal to their lengths in the multidimensional space. In the particular case of Fig. 9.3b, the lengths are the same in the projection plane as in the original space because the latter only has two dimensions.

In the reduced-space plane, the angles between descriptors are *projections* of their true covariance angles. It is thus important to consider only the descriptors that are well represented in the projection plane. To do so, one must recognize, in the multidimensional space, the descriptors that form small angles with the reduced plane; they are the descriptors whose projections approach their real lengths s in the multidimensional space. Since the length of the *projection* of a descriptor-axis j is equal to or shorter than s_j, one must choose a criterion to assess the value of representations in the projection plane.

If a descriptor j was equally associated with each of the p principal axes, all elements of row j (which is of length s_j) of matrix $\mathbf{U\Lambda}^{1/2}$ would be equal, their values being $s_j\sqrt{1/p}$. The length of the descriptor-axis would be $[p(s_j\sqrt{1/p})^2]^{1/2} = s_j$ in multidimensional space. The length of the projection of this descriptor-axis in a reduced space with d dimensions would therefore be $s_j\sqrt{d/p}$. The latter expression defines, in the d-dimensional space, a measure of the *equilibrium contribution of a descriptor* to the various axes of the whole multidimensional space.

Equilibrium contribution

This measure may be compared to the actual length of a descriptor in reduced space, to help judge whether the contribution of this descriptor to the reduced space is larger or smaller than it would be under the hypothesis of an equal contribution to all principal axes. For the above numerical example, the lengths of the rows of matrix $\mathbf{U\Lambda}^{1/2}$, in two-dimensional space, are:

$$\text{length of the first descriptor} = \sqrt{2.6833^2 + (-1.0000)^2} = 2.8636$$

$$\text{length of the second descriptor} = \sqrt{1.3416^2 + 2.0000^2} = 2.4083$$

Because this simple numerical example has two dimensions only, these lengths are equal to their equilibrium contributions in the two-dimensional space. This is easily verified, using the variances of the descriptors, which are already known (Subsection 9.1.1):

$$\text{equilibrium projection of first descriptor} = s_1\sqrt{2/2} = \sqrt{8.2}\sqrt{2/2} = 2.8636$$

$$\text{equilibrium projection of second descriptor} = s_2\sqrt{2/2} = \sqrt{5.8}\sqrt{2/2} = 2.4083$$

In general, because ecological data sets are multidimensional, the lengths of descriptors in the reduced space are not equal to their equilibrium contributions.

The angular interpretation of the product of two descriptor-axes follows directly when the descriptors are scaled to unit lengths; this is done by dividing the elements of each row j by its length s_j. In practice, this is achieved by using the diagonal matrix $\mathbf{D}(s)$ of the standard deviations (Section 4.2), from which matrix $\mathbf{D}(s)^{-1}\mathbf{U\Lambda}^{1/2}$ is calculated. The relationship among descriptors is:

$$[\mathbf{D}(s)^{-1}\mathbf{U\Lambda}^{1/2}]\,[\mathbf{D}(s)^{-1}\mathbf{U\Lambda}^{1/2}]' = \mathbf{D}(s)^{-1}\,\underbrace{\mathbf{U\Lambda U'}}_{\mathbf{S}}\,\mathbf{D}(s)^{-1} = \mathbf{D}(s)^{-1}\,\mathbf{S}\,\mathbf{D}(s)^{-1} = \mathbf{R} \qquad \mathbf{(9.10)}$$

The correlation matrix $\mathbf{R}$ is connected to the dispersion matrix $\mathbf{S}$ by the diagonal matrix of standard deviations $\mathbf{D}(s)$ (eq. 4.10).

The cosine of the angle α_{jl} between two descriptors $\mathbf{y}_j$ and $\mathbf{y}_l$, in multidimensional space, is therefore related to their *correlation* (r_{jl}); it can actually be shown that $\cos(\alpha_{jl}) = r_{jl}$. This angle is the same as that of the *covariance*, because standardization of the rows to unit lengths has only changed the lengths of the descriptor-axes and not their positions in multidimensional space. For the numerical example, the correlation between the two descriptors is equal to $1.6/\sqrt{8.2 \times 5.8} = 0.232$. The angle corresponding to this correlation is (arc cos 0.232) = 76°35', which is indeed the same as the angle of the covariance in Fig. 9.3b.

In the same way, the angle between a descriptor j and a principal axis k, in multidimensional space, is the arc cosine of the correlation between descriptor j and principal component k. This correlation is the element jk of the new matrix of eigenvectors:

$$u_{jk}\sqrt{\lambda_k}/s_j \tag{9.11}$$

In other words, the correlation is calculated by weighting the element of the eigenvector by the ratio of the standard deviation of the principal component to that of the descriptor. For the numerical example, these correlations and corresponding angles are computed using matrix $\mathbf{U\Lambda}^{1/2}$ (calculated above) and the standard deviations of the two descriptors ($s_1 = 2.8636$, $s_2 = 2.4083$):

$$[u_{jk}\sqrt{\lambda_k}/s_j] = \begin{bmatrix} 0.9370 & -0.3492 \\ 0.5571 & 0.8305 \end{bmatrix} \xrightarrow{\text{arc cos}} \begin{bmatrix} 20°26' & 110°26' \\ 56°09' & 33°51' \end{bmatrix}$$

The values of angles in Fig. 9.3b are thus: $\beta = 20°26'$, $\gamma = 56°09'$, $\delta = 33°51'$. These correlations may be used to study the contributions of the descriptors to the various components, the scale factors of the descriptors being removed. The highest correlations (absolute values), in the correlation matrix between descriptors and components, identify the descriptors that contribute most to each eigenvector. The significance of the correlations between descriptors and components cannot be tested using a standard statistical test for Pearson correlation coefficients, however, because the principal components are linear combinations of the descriptors themselves.

When the descriptor-axes of $\mathbf{U\Lambda}^{1/2}$ are standardized to unit lengths, as in eq. 9.10, drawing their projections in the principal space is not recommended. This is because the rescaled eigenvectors are not necessarily orthogonal and may be of any lengths:

$$[\mathbf{D}(s)^{-1}\,\mathbf{U\Lambda}^{1/2}]'\,[\mathbf{D}(s)^{-1}\,\mathbf{U\Lambda}^{1/2}] \neq \mathbf{I} \tag{9.12}$$

The representations of these principal axes are therefore not necessarily at right angles.

The projections of the descriptor-axes of matrix $\mathbf{U\Lambda}^{1/2}$ may be examined, in particular, with respect to the following points:

• The projection coordinates of a descriptor-axis specify the position of the apex of this descriptor-axis in the reduced space. It is recommended to use arrows to represent projections of descriptor-axes. Some authors call them point-descriptors or point-variables and represent them by *points* in the reduced space. This representation is ambiguous and misleading. It is acceptable only if the nature of the point-descriptors is respected; they actually are *apices of descriptor-axes,* so that the relationships among them are defined in terms of angles, not proximities (correlations; Figs. 9.3 and 9.6).

• The projection $u_{jk}\sqrt{\lambda_k}$ of a descriptor-axis j on a principal axis k shows its covariance with the principal axis and, consequently, its positive or negative contribution to the position of the objects along the axis. It follows that a principal axis may often be qualified by the names of the descriptors that are mostly contributing, and in a preferential way, to its formation. Thus, in Fig. 9.6, principal axis I is formed mainly by descriptors 6 to 10 and axis II by descriptors 1 to 4.

• Note the descriptors whose projected lengths reach or exceed the values of their respective equilibrium contributions. Descriptor-axes that are clearly shorter than this value contribute little to the formation of the reduced space under study and, therefore, contribute little to the structure that may be found in the projection of *objects* in that reduced space.

• The correlation among descriptors is given by *the angle between descriptor-axes, not by the proximity between the apices of axes.* In the reduced space, groups of descriptor-axes that form small angles with one another, or angles close to 180° ($\cos 180° = -1$, i.e. perfect negative correlation), can be identified. One must remember, however, that projections of correlation angles in a reduced space do not render the complete correlations among variables. Thus, it may be better to cluster descriptors, in the reduced-space plot, with respect to the multidimensional space, by carrying out a cluster analysis on the original dispersion matrix using the clustering methods of Chapter 8.

• Objects can be projected at right angles onto the descriptor-axes to approximate their values along the descriptors, as shown in Subsection 4 (correlation biplot) and Fig. 9.3b. The distances among objects in such plots *are not* approximations of their Euclidean distances, however.

3. *Projection of descriptors in reduced space (matrix* **U***)* — The projections of the descriptor-axes of matrix **U** differ from those of matrix $\mathbf{U\Lambda}^{1/2}$. Indeed, when the eigenvectors have not been scaled to lengths equal to their standard deviations $\sqrt{\lambda}$, the descriptor-axes are of unit lengths and at right angles in multidimensional space (Fig. 9.3a). The angles between descriptor-axes and principal axes are projections of the *rotation angles*. For the numerical example, the angles between descriptors and principal axes are computed as above, using matrix **U**:

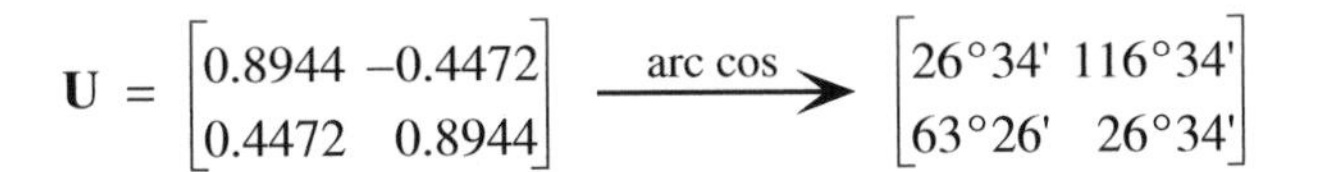

$$\mathbf{U} = \begin{bmatrix} 0.8944 & -0.4472 \\ 0.4472 & 0.8944 \end{bmatrix} \xrightarrow{\text{arc cos}} \begin{bmatrix} 26°34' & 116°34' \\ 63°26' & 26°34' \end{bmatrix}$$

The values of angles in the inset of Fig. 9.3a are thus: $\beta = 26°34'$, $\gamma = 63°26'$, $\delta = 26°34'$. Contrary to the previous case, it is not possible to represent here the

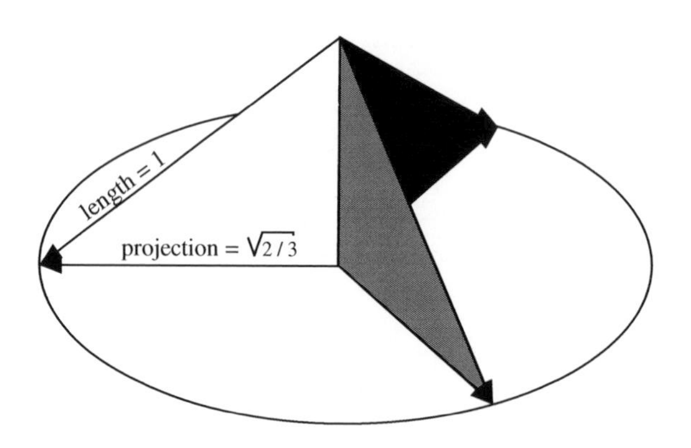

Figure 9.4 Equilibrium projection, in a plane, of three orthogonal vectors with unit lengths.

correlations among descriptors, since descriptors are always orthogonal (i.e. at right angles) in this representation, where the eigenvectors are scaled to 1.

Projection u_{jk} of a descriptor-axis j on a principal axis k is *proportional* to *the covariance* of that descriptor with the principal axis. The proportionality factor is different for each principal axis, so that it is not possible to compare the projection of a descriptor on one axis to its projection on another axis. It is correct, however, to compare the projections of different descriptor-axes on the same principal axis. It can be shown that an isogonal projection (with respectively equal angles) of p orthogonal axes of unit lengths gives a length $\sqrt{d/p}$ to each axis in d-dimensional space. In Fig. 9.4, the equilibrium projection of each of the three orthogonal unit axes, in two-dimensional space, has a length of $\sqrt{2/3}$. This formula is almost identical to the equilibrium contribution of descriptors for eigenvectors scaled to length $\sqrt{\lambda}$ (i.e. $s_j\sqrt{d/p}$, see above), with the difference that s_j is replaced here by 1 because the length of each descriptor-axis is 1. This is explained by the fact that an isogonal projection results in an equal association of all descriptor-axes with the principal axes.

Equilibrium circle

An *equilibrium circle of descriptors,* with radius $\sqrt{d/p}$, may be drawn as reference to assess the contribution of each descriptor to the formation of the reduced space. Such a circle is drawn in the inset of Fig. 9.3a; its radius is $\sqrt{2/2} = 1$ because, in the numerical example, both the reduced space and the total space are two-dimensional. If one was only interested in the equilibrium contribution of descriptors to the first principal axis, the one-dimensional "circle" would then have a "radius" of $\sqrt{1/2} = 0.7071$. For the example, the projection of the first descriptor on the first principal axis is equal to 0.8944 (examine matrix **U** and Fig. 9.3a), so that this descriptor significantly contributes to the formation of axis I. This is not the case for the second descriptor, whose projection on the first axis is only 0.4472.

Table 9.2 Principal component analysis. Main properties for centred descriptors.

Centred descriptor j	Scaling of eigenvectors	
	$\sqrt{\lambda_k}$	1
Total length	s_j	1
Angles in reduced space	projections of covariances (correlations)	90°, i.e. rigid rotation of the system of axes
Length of equilibrium contribution	$s_j\sqrt{d/p}$	circle with radius $\sqrt{d/p}$
Projection on principal axis k	$u_{jk}\sqrt{\lambda_k}$ i.e. covariance with component h	u_{jk} i.e. proportional to the covariance with h
Correlation with principal component k	$u_{jk}\sqrt{\lambda_k}/s_j$	$u_{jk}\sqrt{\lambda_k}/s_j$

The main properties of a principal component analysis of centred descriptors are summarized in Table 9.2.

4 — Biplots

The previous two Subsections have shown that, in principal component analysis, both the descriptor-axes and object-vectors can be plotted in the reduced space. This led Jolicoeur & Mosimann (1960), and subsequently Gabriel (1971, 1982), to plot these projections together in the same diagram, called biplot.

Scalings in PCA

Two types of biplots may be used to represent PCA results (Gabriel, 1982; ter Braak, 1994): *distance biplots* are made of the juxtaposition of matrices **U** (eigenvectors scaled to lengths 1) and **F** (eq. 9.4, where each principal component k is scaled to variance = λ_k), whereas *correlation biplots* use matrix $\mathbf{U\Lambda}^{1/2}$ for descriptors (each eigenvector k is scaled to length $\sqrt{\lambda_k}$) and a matrix $\mathbf{G} = \mathbf{F\Lambda}^{-1/2}$ for objects whose columns have unit variances. Matrices **F** and **U**, or **G** and $\mathbf{U\Lambda}^{1/2}$, can be used together in biplots because the products of the eigenvectors with the object score matrices reconstruct the original (centred) matrix **Y** perfectly: $\mathbf{FU'} = \mathbf{Y}$ and $\mathbf{G}(\mathbf{U\Lambda}^{1/2})' = \mathbf{Y}$. Actually, the eigenvectors and object score vectors may be multiplied by any constant without changing the interpretation of a biplot.

Distance biplot

• Distance biplot (Fig. 9.3a) — The main features of a distance biplot are the following: (1) Distances among objects in the biplot are approximations of their

Euclidean distances in multidimensional space. (2) Projecting an object at right angle on a descriptor approximates the position of the object along that descriptor. (3) Since descriptors have lengths 1 in the full-dimensional space (eq. 9.7), the length of the projection of a descriptor in reduced space indicates how much it contributes to the formation of that space. (4) The angles among descriptor vectors are meaningless.

Correlation biplot

• Correlation biplot (Fig. 9.3b) — The main features of a correlation biplot are the following: (1) Distances among objects in the biplot *are not* approximations of their Euclidean distances in multidimensional space. (2) Projecting an object at right angle on a descriptor approximates the position of the object along that descriptor. (3) Since descriptors have lengths s_j in full-dimensional space (eq. 9.9), the length of the projection of a descriptor in reduced space is an approximation of its standard deviation. (4) The angles between descriptors in the biplot reflect their correlations. When the relationships among objects are important for interpretation, this type of biplot is inadequate; use a distance biplot in this case.

For the numerical example, the positions of the objects in the correlation biplot are computed as follows:

$$\mathbf{G} = \mathbf{F\Lambda}^{-1/2} = \begin{bmatrix} -3.578 & 0 \\ -1.342 & 2.236 \\ -1.342 & -2.236 \\ 3.130 & 2.236 \\ 3.130 & -2.236 \end{bmatrix} \begin{bmatrix} 0.3333 & 0 \\ 0 & 0.4472 \end{bmatrix} = \begin{bmatrix} -1.193 & 0.000 \\ -0.447 & 1.000 \\ -0.447 & -1.000 \\ 1.044 & 1.000 \\ 1.044 & -1.000 \end{bmatrix}$$

In this particular example, the relationships between objects and descriptors are fully represented in two-dimensional space. Figure 9.3b shows that projecting the objects at right angles on a descriptor reconstructs the values of the objects along that descriptors, to within a multiplicative constant. Let us stress again that the distances among objects in a correlation biplot *are not* approximations of their Euclidean distances.

The object or descriptor coordinates must often be multiplied by a constant to produce a clear visual display. In Fig. 9.5 for instance, the scatter of objects would be too large if it was plotted in the same system of coordinates as the descriptors, or the lengths of the descriptor arrows would be too short for visual appraisal. Ecologists may be tempted to interpret the relationships between objects and descriptors in terms of their proximity in the reduced space, whereas the correct interpretation implies the projection of objects on the descriptor-axes (centred with respect to the scatter of points) or on their extensions (Fig. 9.3a); in Fig. 9.2a for example, it would not come to mind to interpret the relationship between the objects and descriptor $\mathbf{y}_1$ in terms of the distance between the object-points and the apex (head of the arrow) of axis $\mathbf{y}_1$. Projections of objects onto an axis specify the coordinates of the objects with respect to the descriptor-axis.

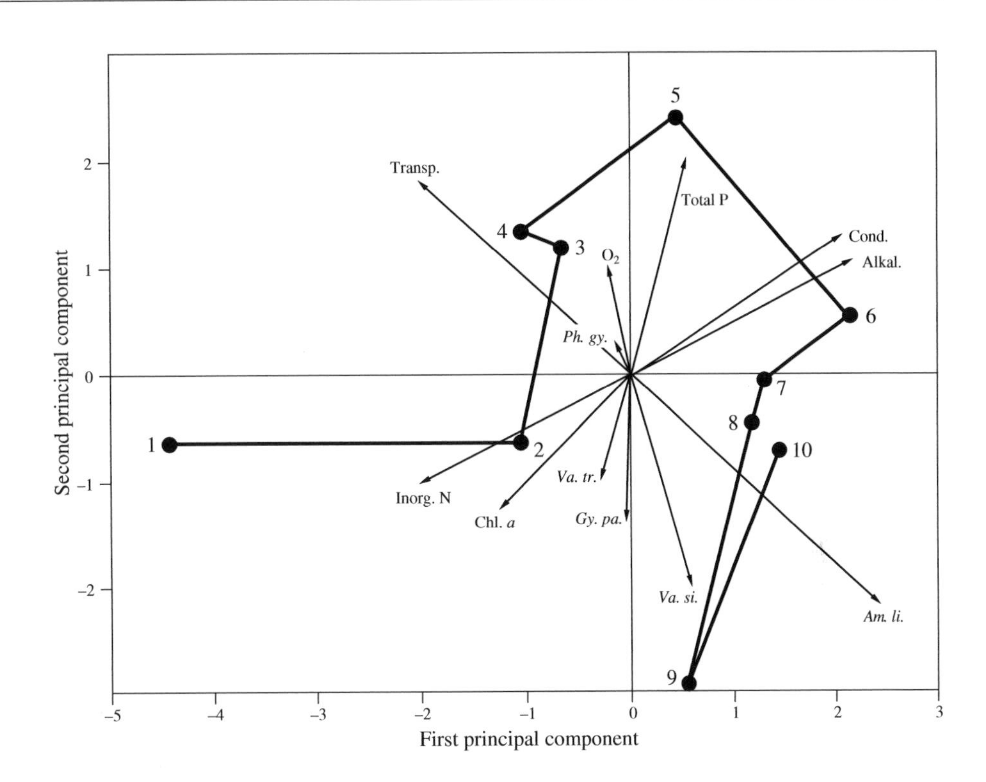

Figure 9.5 Principal component analysis of a time series of 10 observations, from Rivière du Sud (nos. 1 through 10: 30 May to 10 October 1979). The distance biplot shows the relationships between the biological and environmental descriptors, on the one hand, and the time series of responses of the site, on the other. The environmental descriptors are: conductivity (Cond.), transparency (Transp.), alkalinity (Alcal.), dissolved oxygen (O_2), total phosphorus (Total P), inorganic nitrogen (Inorg. N) and chlorophyll *a* (Chl. a). Species of benthic gastropods are: *Amnicola limosa*, *Gyraulus parvus*, *Physa gyrina*, *Valvata tricarinata*, and *Valvata sincera*. The species and environmental descriptor scores were multiplied by 5 before plotting. Modified from Legendre *et al.* (1984a).

Ecological application 9.1a

Legendre *et al.* (1984a) studied gastropod assemblages at two neighbouring sites, one in mesotrophic Richelieu River and the other in its hypereutrophic tributary (Rivière du Sud), near Montréal (Québec). The purpose of the study was to test the hypothesis that physical and chemical conditions influenced the benthic gastropod assemblages. Principal component analysis was used to describe the succession of species within each community, from May through October 1979. This example is not a typical principal component analysis since the observations form a time series; the analysis of multivariate time series is discussed in more detail in Section 12.6. Another, more straightforward approach to the analysis of such data would be canonical correspondence analysis (Section 11.2), which only became available in 1986. The present application illustrates the use of biplots for interpreting ecological data.

Principal component analysis of a data set containing both environmental descriptors and species abundances is a delicate enterprise (Dagnelie, 1965). The problem in this study was not the presence of double-zeros in the data (the comparison of species abundances involved only 4% of double zeros), but the fact that the analysis should obey the following conditions: (1) environmental descriptors must be scaled by standardization since they are not dimensionally homogenous; (2) species abundances should not be scaled because scaling them would give to rare species the same weights as those of dominant species; (3) overall, the analysis must be balanced, i.e. the two types of descriptors must have equal weights. These conditions were met as follows: (1) after normalization, the values of the seven environmental descriptors were standardized (eq. 1.12); the sum of their standard deviations is thus 7. (2) After normalization [$y'_i = \log(y_i + 1)$], each species abundance value was divided by the sum of the standard deviations of all species and multiplied by 7, in order to make the sum of the standard deviations of species equal to 7. As a result, the total weights of the two groups of descriptors in the analysis were equal.

Results for the hypereutrophic environment (Rivière du Sud) are shown in Fig. 9.5. The data set included five descriptors of species abundances and seven environmental variables. The first two principal components accounted for more than 60% of the variance. The chronological sequence of observations is represented by a line. The biplot allowed the authors to characterize each sampling time, by reference to the succession of species and to the changes in environmental descriptors. Conclusions were drawn about the effects of nutrients on the succession of gastropods. The study also compared the two sampled environments using canonical correlation analysis (Section 11.4).

5 — Principal components of a correlation matrix

Even though principal component analysis is defined for a dispersion matrix **S** (Section 4.5), it can also be carried out on a correlation matrix **R** since correlations are covariances of standardized descriptors (Section 4.2). In an **R** matrix, all diagonal elements are equal to one. It follows that the sum of eigenvalues, which corresponds to the total variance of the dispersion matrix, is equal to the order of **R**, which is given by the number of descriptors *p*. Before computing the principal components, it is a sound practice to check that $\mathbf{R} \neq \mathbf{I}$ (eq. 4.14).

Principal components extracted from correlation matrices are not the same as those computed from dispersion matrices. [Beware: some computer packages only allow the computation of principal components from correlation matrices; this may be inappropriate in many studies.] Consider the basic equation for the eigenvalues and eigenvectors, $(\mathbf{S} - \lambda_k \mathbf{I})\, u_k = 0$. The sum of the eigenvalues of **S** is equal to the sum of variances s^2, whereas the sum of eigenvalues of **R** is equal to *p*, so that the eigenvalues of the two matrices, and therefore also their eigenvectors, are necessarily different. This is due to the fact that distances between objects are not the same in the two cases.

In the case of correlations, the descriptors are standardized. It follows that the distances among objects are independent of measurement units, whereas those in the space of the original descriptors vary according to measurement scales. When the descriptors are all of the same kind and order of magnitude, and have the same units, it is clear that the **S** matrix must be used. In that case, the eigenvectors, on the one hand,

and the correlation coefficients between descriptors and components, on the other hand, provide complementary information. The former give the loadings of descriptors and the latter quantify their relative importance. When the descriptors are of different natures, it may be necessary to use matrix **R** instead of **S**. In Section 11.5, differences of the same type will be found between discriminant functions and identification functions.

S or **R** matrix?

Ecologists who wish to study the relationships among objects in a reduced space of principal components may base their decision of conducting the analysis on **S** or **R** on the answer to the following question:

• If one wanted to cluster the objects in the reduced space, should the clustering be done with respect to the original descriptors (or any transformation of these descriptors; Section 1.3), thus preserving their differences in magnitude? Or, should all descriptors contribute equally to the clustering of objects, independently of the variance exhibited by each one? In the second instance, one should proceed from the correlation matrix. An alternative in this case is to transform the descriptors by "ranging", using eq. 1.10 for relative-scale descriptors and eq. 1.11 for interval-scale descriptors, and carry out the analysis on matrix **S**.

Another way to look at the same problem was suggested by Gower (1966):

• Consider that the Euclidean distance (eq. 7.34) is the distance preserved among objects through principal component analysis. Is it with the raw data (covariances) or with the standardized data (correlations) that the spatial configuration of the objects, in terms of Euclidean distances, is the most interesting for interpretation? In the first case, choose matrix **S** as the basis for PCA; in the second case, use matrix **R**.

The principal components of a correlation matrix are computed from matrix **U** of the eigenvectors of **R** and the matrix of standardized observations:

$$\mathbf{F} = \left[\frac{y - \bar{y}}{s_y}\right]\mathbf{U} \qquad \textbf{(9.13)}$$

Principal component analysis is still only a rotation of the system of axes (Subsection 9.1.2). However, since the descriptors are now *standardized*, the objects are not positioned in the same way as if the descriptors had simply been *centred* (i.e. principal components computed from matrix **S**, above).

As far as the representation of descriptors in the reduced space computed from matrix **R** is concerned, the conclusions of Subsection 9.1.3, which concern matrix **S**, can be used here, after replacing *covariance* by *correlation*, s_{jl} by r_{jl}, and *dispersion matrix* **S** by *correlation matrix* **R**.

The variances, and therefore also the standard deviations, of the *standardized* descriptors are equal to one, which leads to some special properties for the $\mathbf{U\Lambda}^{1/2}$ matrix. First, $\mathbf{D}(s) = \mathbf{I}$, so that $\mathbf{U\Lambda}^{1/2} = \mathbf{D}(s)^{-1}\mathbf{U\Lambda}^{1/2}$, i.e. the coefficients $u_{jk}\sqrt{\lambda_k}$ are

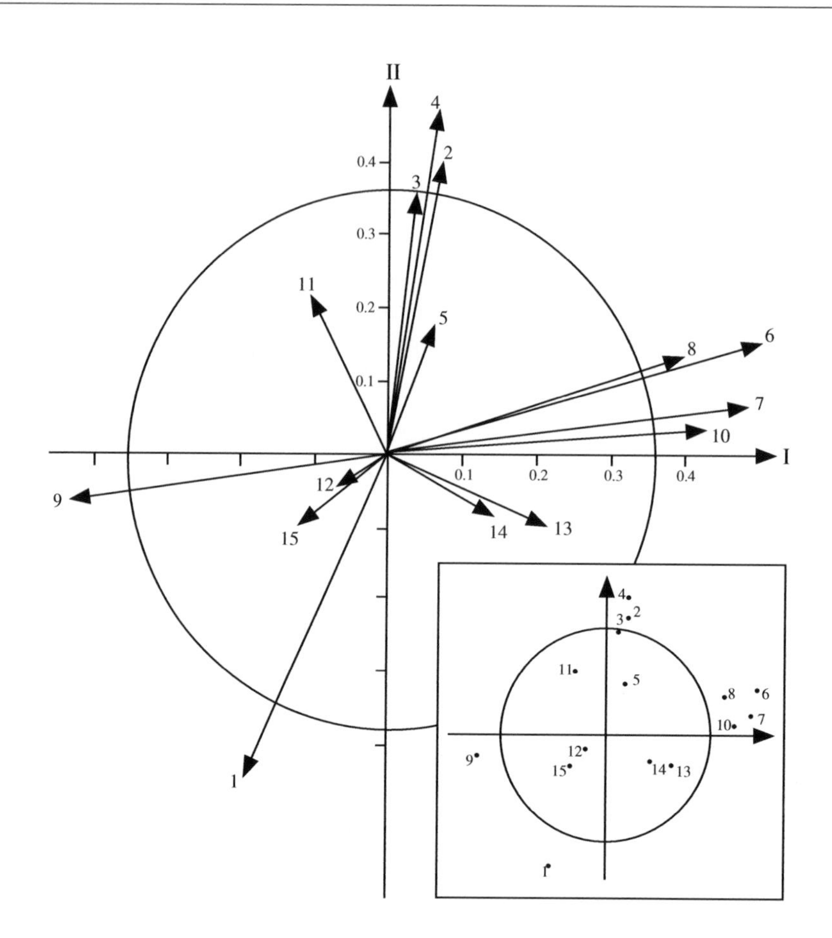

Figure 9.6 Fifteen descriptors plotted in the plane determined by the first two principal axes. The coordinates of each descriptor are the first two elements of the corresponding row of matrix $\mathbf{U\Lambda}^{1/2}$ (i.e. the eigenvectors of $\mathbf{R}$ scaled to $\sqrt{\lambda}$). The circle of equilibrium descriptor contribution is drawn at $\sqrt{2/15} = 0.365$. The inset figure shows the same descriptor-axes using only the apices of the vectors. This representation is often encountered in the ecological literature; it must be avoided because of possible confusion with point-objects.

the correlation coefficients between descriptors j and components k. In addition, the equilibrium contribution corresponding to each descriptor, in the reduced space of $\mathbf{U\Lambda}^{1/2}$, is $s_j\sqrt{d/p} = \sqrt{d/p}$ (since $s_j = 1$). It is therefore possible to judge whether the contribution of each descriptor to the reduced space is greater or smaller than expected under the hypothesis of an equal contribution to all principal axes, by comparing the lengths of their projections to an equilibrium circle with radius $\sqrt{d/p}$ (Fig. 9.6).

The main properties for standardized descriptors are summarized in Table 9.3, which parallels Table 9.2 for centred descriptors.

Table 9.3 Principal component analysis. Main properties for standardized descriptors.

Standardized descriptor j	**Scaling of eigenvectors**	
	$\sqrt{\lambda_k}$	1
Total length	1	1
Angles in reduced space	projections of correlations	90°, i.e. rigid rotation of the system of axes
Radius of equilibrium contribution circle	$\sqrt{d/p}$	$\sqrt{d/p}$
Projection on principal axis k	$u_{jk}\sqrt{\lambda_k}$ i.e. correlation with component h	u_{jk} i.e. proportional to the correlation with h
Correlation with principal component k	$u_{jk}\sqrt{\lambda_k}$	$u_{jk}\sqrt{\lambda_k}$

6 — *The meaningful components*

The successive principal components correspond to progressively smaller fractions of the total variance. One problem is therefore to determine how many components are meaningful in ecological terms or, in other words, what should be the number of dimension of the reduced space. The best approach may be to study the representativeness of the projections in reduced space for two, three, or more dimensions, using Shepard diagrams (Fig. 9.1). However, principal component analysis being a form of variance partitioning, researchers may wish to test the significance of the variance associated with the successive principal axes.

There are a number of classical statistical approaches to this question, such as Bartlett's (1950) test of sphericity. These approaches have been reviewed by Burt (1952) and Jackson (1993). The problem is that these formal tests require normality of all descriptors, a condition which is rarely met by ecological data.

Kaiser-Guttman criterion

There is an empirical rule suggesting that one should only interpret a principal components if the corresponding eigenvalue λ is larger than the mean of the λ's. In the particular case of standardized data, where **S** is a correlation matrix, the mean of the λ's is 1 so that, according to the rule, only the components whose λ's are larger than 1 should be interpreted. This is the so-called Kaiser-Guttman criterion. Ibanez (1973) has provided a theoretical framework for this empirical rule. He showed that, if a variable made of randomly selected numbers is introduced among the descriptors, it is impossible to interpret the eigenvectors that follow the one on which this random-number variable has the highest loading. One can show that this random-number variable, which has covariances near zero with all the other descriptors, introduces in the analysis an eigenvalue equal to 1 if the descriptors have been standardized. For non-standardized descriptors, this eigenvalue is equal to the mean of the λ's if the variance of the random-number variable is made equal to the mean variance of the other descriptors.

Broken stick

Frontier (1976) proposed to compare the list of decreasing eigenvalues to the decreasing values of the broken stick model (Subsection 6.5.2). This comparison is based on the following idea. Consider the variance shared among the principal axes to be a resource embedded in a stick of unit length. If principal component analysis had divided the variance at random among the principal axes, the fractions of total variation explained by the various axes would be about the same as the relative lengths of the pieces obtained by breaking the unit stick at random into as many pieces as there are axes. If a unit stick is broken at random into $p = 2, 3, \ldots$ pieces, the expected values (E) of the relative lengths of the successively smaller pieces (j) are given by eq. 6.49:

$$E\,(\text{piece}_j) = \frac{1}{p}\sum_{x=j}^{p}\frac{1}{x}$$

The expected values are equal to the mean lengths that would be obtained by breaking the stick at random a large number of times and calculating the mean length of the longest pieces, the second longest pieces, etc. A stick of unit length may be broken at random into p pieces by placing on the stick $(p - 1)$ random break points selected using a uniform [0, 1] random number generator. Frontier (1976) has computed the percentage of variance associated with successive eigenvalues, under the broken stick null model, for 2 to 20 eigenvalues (Table D, end of this book).

Coming back to the eigenvalues, it would be meaningless to interpret the principal axes that explain a fraction of the variance as small as or smaller than that predicted by the broken stick null model. The test may be carried out in two ways. One may compare individual eigenvalues to individual predictions of the broken stick model (Table D) and select for interpretation only the eigenvalues that are larger than the values predicted by the model. Or, to decide whether eigenvalue λ_k should be interpreted, one may compare the *sum of eigenvalues*, from 1 to k, to the sum of the values from 1 to k predicted by the model. This test usually recognizes the first two or three principal components as meaningful; this corresponds to the experience of ecologists.

After an empirical study using a variety of matrix types, using simulated and real ecological data, Jackson (1993) concluded that two methods consistently pointed to the correct number of ecologically meaningful components in data sets: the broken-stick model and a bootstrapped eigenvalue-eigenvector method proposed in his paper.

Chapter 10 will discuss how to use explanatory variables to ecologically interpret the first few principal components that are considered to be meaningful according to one of the criteria mentioned in the present Subsection.

7 — Misuses of principal components

Given the power of principal component analysis, ecologists have sometimes used it in ways that exceed the limits of the model. Some of these limits may be transgressed without much consequences, while others are more critical. The most common errors are: the use of descriptors for which a measure of covariance makes no sense and the interpretation of relationships between descriptors, in reduced space, based on the relative positions of the apices of axes instead of the angles between them.

Principal component analysis was originally defined for data with *multinormal distributions* (Section 4.4), so that its optimal use (Cassie & Michael, 1968) calls for normalization of the data (Subsection 1.5.6). Deviations from normality do not necessarily bias the analysis, however (Ibanez, 1971). It is only important to make sure that the descriptors' distributions are reasonably unskewed. Typically, in analyses conducted with strongly skewed distributions, the first few principal components only separate a few objects with extreme values from the remaining objects, instead of displaying the main axes of variation of all objects in the study.

Principal components are computed from the eigenvectors of a dispersion matrix. This means that the method is to be used on a matrix of covariances (or possibly correlations) with the following properties: (a) matrix **S** (or **R**) has been computed among descriptors (b) that are quantitative and (c) for which valid estimates of the covariances (or correlations) may be obtained. These conditions are violated in the following cases:

1) Technically, a dispersion matrix cannot be estimated using a number of observations n smaller than or equal to the number of descriptors p. When $n \leq p$, since only $(n - 1)$ independent comparisons can be made (because there are $n - 1$ degrees of freedom), the resulting matrix of order p has $(n - 1)$ independent rows or columns only. In such a case, the matrix has $[p - (n - 1)]$ null eigenvalues; indeed, in order to position n objects while respecting their distances, one only requires $n - 1$ dimensions. To get a statistically valid estimate of the dispersion matrix, the number of objects n must be larger than p. However, the first few eigenvectors, which are the only ones used in most cases, are little affected when the matrix is not of full rank, so that this problem should not lead to incorrect interpretations of ordinations in reduced space.

2) Some authors have transposed the original data matrix and computed correlations among the *objects* (i.e. Q mode) instead of among the descriptors (R mode). Their aim was to position the descriptors in the reduced space of the objects. There are several reasons why this operation is incorrect, the least being its uselessness since principal component analysis provides information about the relationships among both objects and descriptors. In addition, the covariances or correlations thus

estimated do not make sense, for the reasons given in Subsection 7.5.1. Among these arguments, the following one is of utmost importance in PCA: calculation of correlations implies a standardization of the vectors; the two steps of standardization (eq. 1.12) only makes sense for dimensionally homogeneous data. Standardizing a vector of objects across non-homogeneous descriptors violates this condition; a simple and legitimate change of scale in one of the descriptors (for example moving the position of the decimal point) could completely change the results of the analysis.

A different problem arises from the fact that, in the literature, the expression "components in Q mode" may sometimes designate a rightful analysis conducted on an R matrix. This expression comes from the fact that one can use principal component analysis primarily as a method for positioning objects in reduced space. The meanings of "Q mode" and "R mode" are variable in the literature; their meanings in numerical ecology are defined in Section 7.1.

Rao (1964), Gower (1966), and Orlóci (1967a) have shown that, *as a computational technique*, principal components can be found by computing the eigenvalues and eigenvectors of a Q matrix. The steps are the following:

• Compute matrix $\mathbf{C}_{np} = [y - \bar{y}] / \sqrt{n-1}$ of centred variables scaled to $\sqrt{n-1}$; this matrix is such that $\mathbf{C'C} = \mathbf{S}_{pp}$, which is the usual variance-covariance matrix.

• Compute the scalar cross-product matrix $\mathbf{Q}_{nn} = \mathbf{CC'}$ instead of $\mathbf{S}_{pp} = \mathbf{C'C}$.

• Determine the non-zero eigenvalues of $\mathbf{Q}$ and their associated eigenvectors.

• Scale each eigenvector k to the corresponding value $\sqrt{\lambda_k}$.

• The eigenvalues of matrix $\mathbf{Q}$ are the same as those of matrix $\mathbf{S}$ and the scaled eigenvectors are the principal components of $\mathbf{S}$ (i.e. matrix $\mathbf{F}$). This computational technique is very different from the approach criticised in the previous paragraph.

3) Covariances and correlations are defined for quantitative descriptors only (Section 7.5). This implies, in particular, that one must not use multistate qualitative descriptors in analyses based upon covariance matrices; means and variances computed from non-ordered states are meaningless.

Precision of data

Spearman correlation

Principal component analysis is very robust, however, to variations in the *precision of data*. Variables may be recoded into a few classes without noticeable change to the results (Frontier & Ibanez, 1974; Dévaux & Millerioux, 1976a). The correlation coefficients calculated using semiquantitative data are equivalent to Spearman's rank correlation coefficients (eq. 5.3). In a discussion of principal component analysis computed using semiquantitative data, Lebart *et al.* (1979) give, for various numbers of objects and descriptors, values above which the λ's of the first two principal components may be taken as significant. Gower (1966) has also shown that, with binary descriptors, principal component analysis positions the objects, in the multidimensional space, at distances that are the square roots of complements of simple matching coefficients S_1 (eq. 7.1).

4) When calculated over data sets with many double-zeros, coefficients such as the covariance or correlation lead to ordinations that produce inadequate estimates of the distances among sampling sites. The problem arises from the fact that the principal-component rotation preserves the Euclidean distance among objects (Fig. 9.2d; Table 9.1). The double-zero problem has been discussed in the context of the measures of association that are appropriate for species abundances (Sections 7.3 to 7.5). With this type of data, principal component analysis should only be used when the sampling sites cover very short gradients. In such a case, the species present at the various sites should be mostly the same, the sites differing by the *abundances* of the various species. Correspondence analysis (Section 9.4), or else metric or nonmetric scaling (Sections 9.2 and 9.3), should be used when studying longer gradients.

This last remark explains to a large extent why, in the ecological literature, principal component analysis has often not provided interesting results, for example in studies of species associations (e.g. Margalef & Gonzalez Bernaldez, 1969; Ibanez, 1972; Reyssac & Roux, 1972). This problem had also been noted by Whittaker & Gauch (1973).

Attempts to interpret the *proximities between the apices* of species-axes in the reduced space, instead of considering the angles separating these descriptor-axes (e.g. Fig. 9.6), may also have clouded the relationships among species in principal component analysis results.

• On the one hand, examination of the angles allows one to identify cases of negative association among species; see for instance the opposition between *Physa gyrina* and *Valvata sincera* in Fig. 9.5. The apices of negatively associated species vectors are not close to one another in principal space.

• On the other hand, in studies involving many species, the first two principal axes usually do not account for a large fraction of the variability. The species that require a higher-dimensional space are poorly represented in two dimensions. These species usually form an unstructured swarm of points around the origin of the system of axes, from which little can be concluded. This problem is common to all studies of species associations using ordination methods.

Fortunately, principal component analysis is useful for studying other questions of ecological interest, as will be seen in the following applications.

Table 9.4 summarizes, with reference to the text, the various questions that may be addressed in the course of a principal component analysis.

Table 9.4 Questions that can be addressed in the course of a principal component analysis and the answers found in Section 9.1.

Before starting a principal component analysis	*Pages*
1) Are the descriptors appropriate?	
⇒ Quantitative descriptors; multinormality; not too many zeros. In principle, there should be more objects than descriptors	411-413
2) Are the descriptors dimensionally homogeneous?	
⇒ If YES: conduct the analysis on the dispersion matrix	403, 407
⇒ If NO: conduct the analysis on the correlation matrix	406-409
3) Purpose of the ordination in reduced space:	
⇒ To preserve and display the relative positions of the objects: scale the eigenvectors to unit lengths	394, 396
⇒ Distance biplots: scale eigenvectors to unit lengths; compute $\mathbf{F} = \mathbf{YU}$	403
⇒ To display the correlations among descriptors: scale the eigenvectors to $\sqrt{\lambda}$	397-400
⇒ Correlation biplots: scale eigenvectors to $\sqrt{\lambda}$; compute $\mathbf{G} = \mathbf{F}\mathbf{\Lambda}^{1/2}$ (beware: Euclidean distances among objects are not preserved)	403-404
While examining the results of a principal component analysis	
1) How informative is a representation of the objects in an m-dimensional reduced space?	
⇒ Compute eq. 9.5	395
2) Are the distances among objects well preserved in the reduced space?	
⇒ Compare Euclidean distances using a Shepard diagram	389-390
3) Which are the significant eigenvalues?	
⇒ Test: is λ_k larger than the mean of the λ's?	409-410
⇒ Test: is the percentage of the variance corresponding to λ_k larger than the corresponding value in the broken stick model?	410
4) What are the descriptors that contribute the most to the formation of the reduced space?	
⇒ Represent descriptors by arrows	398, 400-401, 404
⇒ Compute the equilibrium contribution of descriptors and, when appropriate, draw the circle	399, 401-402, 408
⇒ Compute correlations between descriptors and principal axes	400, 403, 407-409
5) How to represent the objects in the reduced space?	
⇒ $\mathbf{F} = [y - \bar{y}]\,\mathbf{U}$	394, 403

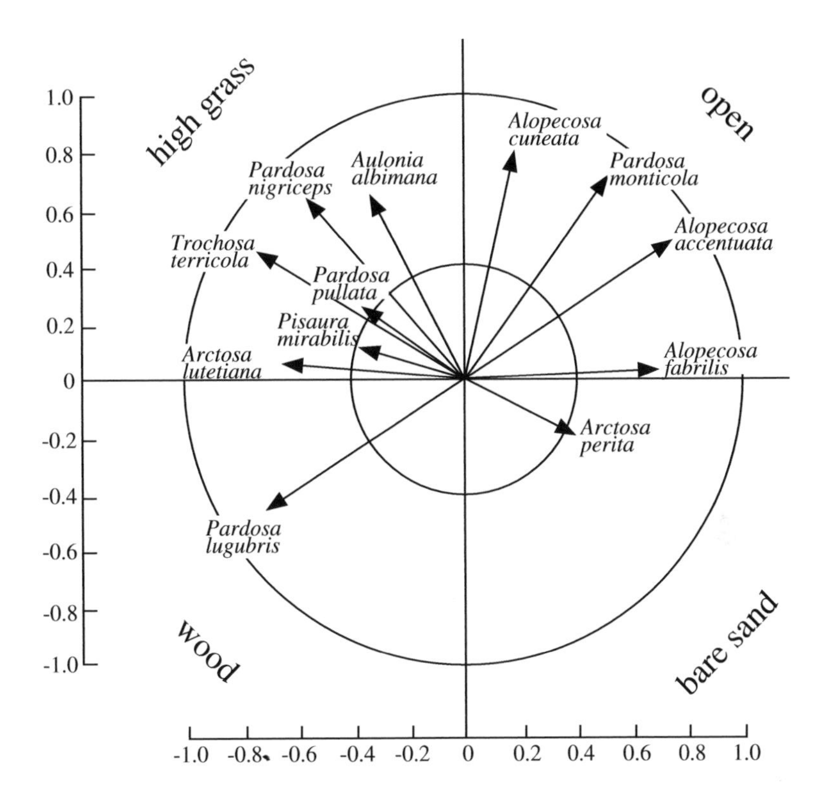

Figure 9.7 Positions of species-axes in the plane of the first two principal axes. Large circle: unit radius (maximum length for a descriptor, in the case of a principal component analysis conducted on an **R** matrix). Small circle: equilibrium circle of descriptors, with radius $\sqrt{2/p} = 0.41$. Modified from Aart (1973).

8 — Ecological applications

Ecological application 9.1b

From 1953 to 1960, 100 pitfall traps, located in a sand dune area north of the Hague (The Netherlands), were sampled weekly during 365 weeks. In these 36 500 relevés, approximately 425 animal species were identified, about 90% of them being arthropods. Aart (1973) studied the wolf spiders (*Lycosidea* and *Pisauridae*: 45 030 specimens) to assess how lycosids shared the multidimensional space of resources (see Chapter 1 for the concept of niche). In the analysis reported here, values from the different weeks were added up, which resulted in a data matrix of 100 pitfall traps × 12 species. Principal component analysis was applied to the data matrix, which contained about 30% zero values; two of the 14 species were eliminated because they had been found only twice and once, respectively.

Figure 9.7 shows the contributions of the different descriptors (species) to the formation of principal axes I and II as well as the projections, in the plane of these two axes, of the correlation

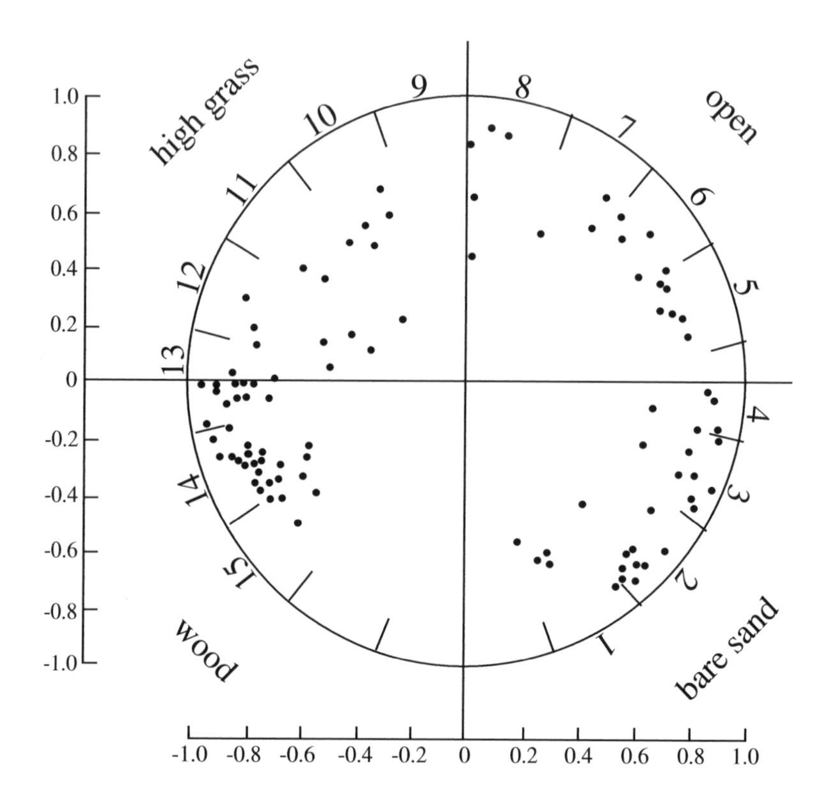

Figure 9.8 Positions of the 100 trap-objects plotted in the reduced space of the first two principal components. Modified from Aart (1973).

angles between descriptors. The equilibrium circle was added to the original figure of Aart. Its radius is $\sqrt{2/p} = 0.41$, since the analysis was conducted on the correlation matrix (each descriptor thus has unit variance); the eigenvectors were scaled to $\sqrt{\lambda}$ (matrix $\mathbf{U\Lambda}^{1/2}$). A circular gradient, from bare sand to woods, ordered the species as a function of their habitats (see also Ecological application 9.5). Only the first two principal components were considered to be meaningful.

The 100 traps were also positioned in the space of the first two principal components. The result (Fig. 9.8) clearly displays a circular gradient of habitats that follows vegetation conditions (numbered 1 to 15 along the circle margin). This gradient goes from bare sand with lichens and *Corynephorus canescens,* via the same vegetation plus *Ammophilia arenaria*, to open areas with scattered *Salix repens*, *Rubus* and short grass vegetation (*Carex arenaria, Galium* and *Onosis*). Next are dune meadows dominated by *Calamagrostis epigejos*, leading to a park-like landscape with bushes of *Betula* and *Crataegus*, and ending with woods of *Betula* and *Populus tremula*.

The 15 divisions of the circular gradient in Fig. 9.8 may be taken to represent as many biotopes. Examining the average numbers of catches for the different species in each of the biotopes confirmed the preference of each lycosid species for a specific type of vegetation.

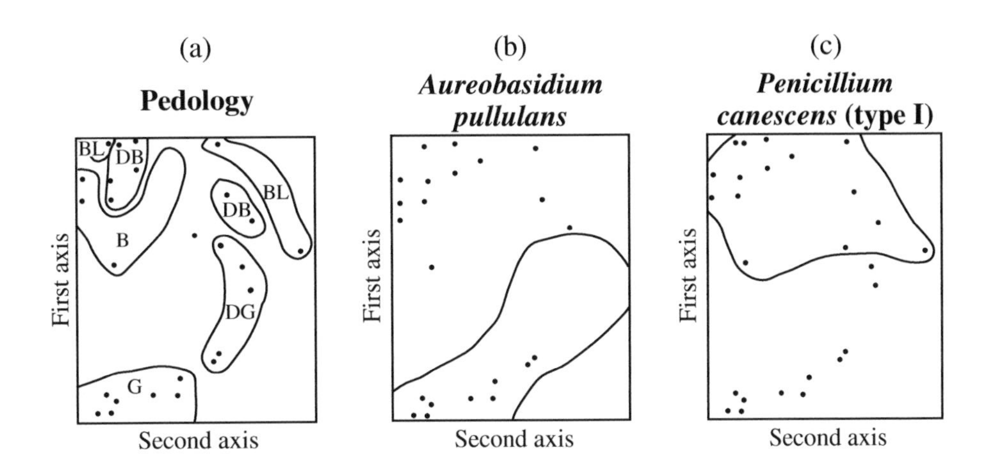

Figure 9.9 Principal component analysis computed from presence-absence of 51 soil microfungi. (a) Pedological information drawn on the ordination of the 26 sampling sites, plotted in the reduced space of the first two principal components. From north to south, soil types are: G = grey, DG = dark grey, BL = black, DB = dark brown, B = brown. (b) and (c) Distributions of the sites where two species of the microflora were present. Modified from Morrall (1974).

Ecological application 9.1c

A study of soil microfungi living in association with the aspen *Populus tremuloides* Michx. provides another type of utilization of principal component analysis. This study by Morrall (1974) covered 26 stations with 6 sites each, scattered throughout Saskatchewan (Canada). It evidenced relationships between the distributions of some species and soil types.

Among the 205 species or taxonomic entities that were identified, 51 were included in the ordination study. The others were not, considering criteria aimed at eliminating rare taxa which could have been either ephemeral constituents of the soil microflora or even contaminants of the laboratory cultures. Observations were transformed into presence-absence data.

Following principal component analysis, the 26 sampling sites were plotted in the reduced space of the first two principal components, onto which information about the nature of the soils was superimposed (Fig. 9.9a). Soils of Saskatchewan may be classified into 5 types, i.e. (G) the grey wooded soils of the northern boreal forest, followed southward by the dark grey (DG) transition soils and the black soils (BL). Further south are dark brown soils (DB), which give way to the brown soils (B) of the grassland. Since the principal components were calculated on presence-absence data, the distribution of sites in the reduced space is expected to correspond to that of the species. The author tested this for the most abundant species in the study, by plotting, in the reduced space, distributions of the sites where they were present; two examples are given in Figs. 9.9b and c. In this way, the author could compare these distributions to that of soil types.

9 — *Algorithms*

A number of standard subprograms are available for computing the eigenvalues and eigenvectors of a real, symmetric matrix, such as covariance matrix **S**. The most widely used method is called Householder reduction; it is very efficient for problems in which *all* eigenvalues and eigenvectors must be computed. Subprograms implementing it can be found in standard subroutine packages available at computing centres, or in books such as Press *et al.* (1986 and later editions; these programs are also available on a diskette).

Householder

Clint & Jennings (1970) published a pioneering piece of work describing how to compute a *subset* only of the eigenvalues and corresponding eigenvectors of a real symmetric matrix, using an iterative method. Hill (1973b) used this idea to develop a "reciprocal averaging" algorithm for correspondence analysis; Hill's work will be further discussed in Section 9.4. Building on these bases, ter Braak (1987c) proposed a "two-way weighted summation algorithm" for principal component analysis. This algorithm is described in detail here for three reasons: (1) it is closely linked to the basic equations of the PCA method, so that it may help readers understand them; (2) it is very easy to program; (3) using it, one can compute the first few components only, when these are the ones of interest. The algorithm is summarized in Table 9.5. The next sections will also refer to it.

TWWS algorithm

The example worked out in Table 9.6 should help understand how the algorithm manages to compute the principal components, the eigenvectors, and the eigenvalues, using the numerical example of Section 9.1. The procedure starts with the matrix of centred data $[y - \bar{y}]$, found in the upper left-hand corner of Table 9.6 (boxed values).

To estimate principal component I, arbitrary scores are first assigned to the rows of the centred data matrix (Table 9.6, column R0); values $[f_{i1}] = [1\ 2\ 3\ 4\ 5]'$ are used here. Any other initial choice would lead to the same estimate for the first principal component $[f_{i1}]$, although the number of iterations necessary to reach it may differ. The only choice to avoid is to make all initial f_{i1} values equal. From these, column scores are found by multiplying the transpose of the data matrix by the row scores (Table 9.6, row C1):

$$[\text{column scores}_{1j}] = [y - \bar{y}]'\,[f_{i1}] \tag{9.14}$$

At the end of the iteration process, the column scores will provide estimates of the first column of matrix **U**. The rationale for this operation comes from the basic equation of eigenanalysis (eq. 2.27) as applied to matrix **S**:

$$\mathbf{S\,U} = \mathbf{U\,\Lambda}$$

Replacing **S** by its value in the definition of the covariance matrix (eq. 4.6), $\mathbf{S} = (n-1)^{-1}\,[y - \bar{y}]'\,[y - \bar{y}]$, one obtains:

$$[y - \bar{y}]'\,[y - \bar{y}]\,\mathbf{U} = (n-1)\,\mathbf{U\,\Lambda}$$

Table 9.5 Two-way weighted summation (TWWS) algorithm for PCA. Modified from ter Braak (1987c).

a) Iterative estimation procedure

Step 1: Consider a table of n objects (rows) $\times p$ variables (columns).
Centre each variable (column) on its mean.

Decide how many eigenvectors are needed and, for each one. **DO** the following:

Step 2: Take the row order as the arbitrary initial object scores (1, 2, …).
Set the initial eigenvalue estimate to 0.

Iterative procedure begins

Step 3: Compute new variable loadings: $colscore(j) = \Sigma\, y(i,j) \times rowscore(i)$

Step 4: Compute new object scores: $rowscore(i) = \Sigma\, y(i,j) \times colscore(j)$

Step 5: For the second and higher-order axes, make the object scores uncorrelated with all previous axes (Gram-Schmidt orthogonalization procedure: see *b* below).

Step 6: Scale the vector of object scores to length 1 (normalization procedure *c*, below).

Step 7: Upon convergence, the eigenvalue is $S/(n-1)$ where n is the number of objects. So, at the end of each iteration, $S/(n-1)$ provides an estimate of the eigenvalue. If this estimate does not differ from that of the previous iteration by more than a small quantity ("tolerance", set by the user), go to step 8. If the difference is larger than the tolerance value, go to step 3.

End of iterative procedure

Step 8: Normalize the eigenvector (variable loadings), i.e. scale it to length 1 (procedure *c*, below).
Rescale the principal component (object scores) to variance = eigenvalue.

Step 9: If more eigenvectors are to be computed, go to step 2. If not, continue with step 10.

Step 10: Print out the eigenvalue, % variance, cumulative % variance, eigenvector (variable loadings), and principal component (object scores).

b) Gram-Schmidt orthogonalization procedure

DO the following, in turn, for all previously computed principal components k:

Step 5.1: Compute the scalar product $SP = \Sigma\, [rowscore(i) \times v(i,k)]$ of the current object score vector estimate with previous component k, where vector $v(i,k)$ contains the object scores of component k, scaled to length 1. This product varies between 0 (if the vectors are orthogonal) to 1.

Step 5.2: Compute new values of $rowscore(i)$ such that vector *rowscore* becomes orthogonal to vector $v(i,k)$: $rowscore(i) = rowscore(i) - (SP \times v(i,k))$.

c) Normalization procedure

Step 6.1: Compute the sum of squares of the object scores: $S^2 = \Sigma\, rowscore(i)^2$; $S = \sqrt{S^2}$.

Step 6.2: Compute the normalized object scores: $rowscore(i) = rowscore(i)/S$.

Since $\mathbf{F} = [y - \bar{y}]\ \mathbf{U}$ (eq. 9.4), it follows that:

$$[y - \bar{y}]'\,\mathbf{F} = (n - 1)\ \mathbf{U}\ \mathbf{\Lambda}$$

Hence, the column scores obtained from eq. 9.14 are the values of the first eigenvector (first column of matrix **U**) multiplied by eigenvalue λ_1 (which is a diagonal element of matrix **Λ**) and by $(n - 1)$.

From the first estimate of column scores, a new estimate of row scores is computed using eq. 9.4, $\mathbf{F} = [y - \bar{y}]\ \mathbf{U}$:

$$[\text{row scores}_{i1}] = [y - \bar{y}]\ [u_{i1}] \qquad \textbf{(9.15)}$$

The algorithm alternates between estimating row scores and column scores until convergence. At each step, the row scores (columns called R in Table 9.6) are scaled to length 1 in order to prevent the scores from becoming too large for the computer to handle, which they may easily do. Before this normalization, the length of the row score vector, divided by $(n - 1)$, provides the current estimate of the eigenvalue. This length actually measures the amount of "stretching" the row score vector has incurred during an iteration.

This description suggests one of several possible stopping criteria (Table 9.5, step 7): if the estimate of the eigenvalue has not changed, during the previous iteration, by more than a preselected tolerance value, the iteration process is stopped. Tolerance values between 10^{-10} and 10^{-12} produce satisfactory estimates when computing all the eigenvectors of large matrices, whereas values between 10^{-6} and 10^{-8} are sufficient to compute only the first two or three eigenvectors.

At the end of the iterative estimation process (Table 9.5, step 8),

- the eigenvector (Table 9.6, line C13) is normalized (i.e. scaled to unit length), and

- the principal component is scaled to length $\sqrt{(n-1)\,\lambda_1}$. This makes its variance equal to its eigenvalue.

Note that the eigenvalues, eigenvectors, and principal components obtained using this iterative procedure are the same as in Sections 9.1 and 9.2, except for the signs of the second eigenvector and principal component, which are all changed. One may arbitrarily change all the signs of an eigenvector and the corresponding principal component, since signs result from an arbitrary decision made when computing the eigenvectors (Section 2.9). This is equivalent to turning the ordination diagram by 180° if signs are changed on both the first and second principal components, or looking at it from the back of the page, or in a mirror, if signs are changed for one axis only.

To estimate the second principal component, eigenvalue, and eigenvector, row scores are again assigned arbitrarily at the beginning of the iterative process. In Table 9.6 (bottom part), the same values were actually chosen as for axis I, as stated in

Table 9.6 Estimation of axes I (top) and II (bottom) for the centred data of the numerical example (values in boxes), using the "two-way weighted summation" algorithm (TWWS, Table 9.5). Iterations 1 to 13: estimates of the row scores (R1 to R13) and column scores (C1 to C13).

Objects ⇓	Var. 1	Var. 2	R0 (arbitrary)	R1	R1 length=1	R2	R2 length=1	R3	R3 length=1 ...	R9	R9 length=1 ...	R13	R13 length=1	R13 scaled (var=λ)
x_1	**−3.2**	**−1.6**	1	−64.000	−0.586	−21.103	−0.593	−21.352	−0.595	−21.466	−0.596	−21.466	−0.596	−3.578
x_2	**−2.2**	**1.4**	2	−34.000	−0.311	−9.745	−0.274	−9.037	−0.252	−8.080	−0.224	−8.053	−0.224	−1.342
x_3	**−0.2**	**−2.6**	3	−14.000	−0.128	−6.082	−0.171	−6.977	−0.195	−8.019	−0.223	−8.047	−0.224	−1.341
x_4	**1.8**	**3.4**	4	46.000	0.421	16.633	0.467	17.653	0.492	18.752	0.521	18.780	0.522	3.130
x_5	**3.8**	**−0.6**	5	66.000	0.605	20.297	0.570	19.713	0.550	18.813	0.523	18.786	0.522	3.131
Estimates of $\lambda_1 \Rightarrow$					27.295		8.895		8.967		9.000		9.000	
C1	18.000	4.000												
C2	5.642	1.905												
C3	5.544	2.257												
C4	5.473	2.449												
C5	5.428	2.554												
C6	5.401	2.612												
C7	5.386	2.644												
C8	5.377	2.661												
C9	5.373	2.671												
C10	5.370	2.677												
C11	5.368	2.680												
C12	5.368	2.681												
C13	5.367	2.682												
C13 length=1	0.895	0.447												

Objects ⇓	Var. 1	Var. 2	R0 (arbitrary)	R1	R1 ortho*	R1 length=1	R2	R2 ortho*	R2 length=1	R2 scaled (var=λ)
x_1	**−3.2**	**−1.6**	1	−64.000	0.001	0.000	0.002	0.001	0.000	0.000
x_2	**−2.2**	**1.4**	2	−34.000	−9.995	−0.500	−9.999	−10.000	−0.500	−2.236
x_3	**−0.2**	**−2.6**	3	−14.000	9.996	0.500	10.001	10.000	0.500	2.236
x_4	**1.8**	**3.4**	4	46.000	−9.996	−0.500	−10.002	−10.001	−0.500	−2.236
x_5	**3.8**	**−0.6**	5	66.000	9.994	0.500	9.998	9.999	0.500	2.236
Estimates of $\lambda_2 \Rightarrow$						4.998			5.000	
C1	18.000	4.000								
C2	2.000	−4.000								
C2 length=1	0.447	−0.894								

* ortho: scores are made orthogonal to R13 found in the upper portion of the Table.

step 2 of the algorithm (Table 9.5). Iterations proceed in the same way as above, with the exception that, during each iteration, the row scores are made orthogonal to the final estimate obtained for the first principal component (column R13 in the upper portion of Table 9.6). This follows from the basic rule that principal components must be linearly independent (i.e. orthogonal) of one another. For the third principal axis and above, the vector estimating row scores is made orthogonal, in turn, to *all* previously computed principal components. As a result, the vector is orthogonal to all previously computed principal components.

The algorithm converges fairly rapidly, even with small tolerance values. For the example of Table 9.6, it took 13 iterations to reach convergence for axis I, and 2 iterations only for axis II, using a tolerance value of 10^{-6}; double-precision real numbers (REAL*8) were used where needed. With a tolerance value of 10^{-10}, it took 21 and 2 iterations, respectively. The initial, arbitrary values assigned to the row scores also have an (unpredictable) effect on the number of iterations; e.g. with a different set of initial values [2 5 4 3 1], it took 14 iterations instead of 13 to reach convergence for the first axis (tolerance = 10^{-6}).

Supplementary object and variable

Supplementary objects or variables may easily be incorporated in the calculations using this algorithm. These are objects or variables that have not been used to compute the eigenvalues and eigenvectors of the ordination space, but whose positions are sought with respect to the original set of objects and variables that were used to compute the eigenvalues and eigenvectors. In Ecological application 9.1a, for example, the principal component analysis could have been computed using the species abundance data, and the environmental descriptors could have been added to the analysis as supplementary variables; in addition, the observations from Richelieu River could have been added as supplementary objects. Preliminary transformations are required: (1) supplementary variables must be centred on their respective means; (2) on the other hand, for each variable, supplementary objects must be centred using the mean value of that variable calculated for the original set of objects. When the algorithm has reached convergence for an axis, using the original set of objects, it is a simple matter to compute the column scores of the supplementary variables using eq. 9.14 and the row scores of the supplementary objects using eq. 9.15. The final step consists in applying to the supplementary variable scores the scaling that was applied to the terms of the eigenvector corresponding to the original set of variables and, to the supplementary object scores, the scaling that was applied to the original set of objects.

SVD

There is another way of computing principal components. It involves *singular value decomposition* (SVD, Section 2.11). One of the properties of SVD leads to principal component analysis, although this is not the most commonly used method for PCA. SVD is also an excellent approach to correspondence analysis (Section 9.4).

The relationship with principal component analysis is the following. Let us assume that the column vectors of **Y** are centred on their respective means and compute the product **Y'Y** using the singular value decomposition of **Y** (eq. 2.31):

$$\mathbf{Y'Y} = \mathbf{UW'}\ (\mathbf{V'V})\ \mathbf{W}\ \mathbf{U'} \qquad \textbf{(9.16)}$$

Since **V** is orthonormal (Section 2.11), **V'V** = **VV'** = **I** and one gets:

$$\mathbf{Y'Y} = \mathbf{U}\ \mathbf{W'W}\ \mathbf{U'} \qquad \textbf{(9.17)}$$

The eigenvalues (forming the diagonal matrix **Λ**) and eigenvectors (written out in matrix **U**) of a square matrix **A** obey the relationship (eq. 2.28):

$$\mathbf{A} = \mathbf{U}\boldsymbol{\Lambda}\mathbf{U}^{-1}$$

If the vectors in **U** are normalized, **U** is an orthonormal matrix with the property $\mathbf{U}^{-1} = \mathbf{U'}$ (property 7 of inverses, Section 2.8). Equation 2.28 may be rewritten as:

$$\mathbf{A} = \mathbf{U}\boldsymbol{\Lambda}\mathbf{U'} \qquad \textbf{(9.18)}$$

Matrix **U** of eqs. 9.16 and 9.17 is the same as matrix **U** of eq. 9.18, i.e. it is the matrix of eigenvectors of **A** = [**Y'Y**]. The diagonal matrix [**W'W**], containing squared singular values, is the diagonal matrix **Λ** of eigenvalues of **A** = [**Y'Y**].

In this book*, principal component analysis has been defined as the eigenanalysis of the covariance matrix $\mathbf{S} = (n-1)^{-1}\ \mathbf{Y'Y}$ (for matrix **Y** centred), so that the eigenvalues of [**Y'Y**], found in the previous paragraph, are larger than the eigenvalues of standard principal component analysis by a factor $(n-1)$. The singular values of **Y** (eq. 2.31) are the square roots of the eigenvalues of [**Y'Y**] because **Λ** = [**W'W**].

Matrix **F** of the principal components of **Y** (centred) can be calculated from eq. 9.4 (**F** = **YU**). Combining it to eq. 2.31 (**Y** = **VWU'**) shows that **F** may also be computed from the left-hand column-orthonormal matrix **V** of the singular value decomposition of **Y**, as follows:

$$\mathbf{F} = \mathbf{VW} \quad \text{or} \quad \mathbf{F} = \mathbf{V}\boldsymbol{\Lambda}^{-1/2} \qquad \textbf{(9.19)}$$

Equation 9.19 amounts to multiplying the elements in the columns of **V** by the corresponding singular values, found on the diagonal of **W**. The next paragraph shows that **V** is also the matrix of eigenvectors of [**YY'**].

The principal component analysis of **Y** may be computed from matrix [**YY'**] instead of [**Y'Y**], as shown in Subsection 9.1.7. This can be done using SVD. Compute the SVD of the square matrix [**YY'**]:

$$\mathbf{YY'} = \mathbf{VW}\ (\mathbf{U'U})\ \mathbf{W'}\ \mathbf{V'} = \mathbf{V}\ \mathbf{WW'}\ \mathbf{V'}$$

* Some authors define PCA as the eigenanalysis of the dispersion matrix **Y'Y** (for **Y** centred).

The eigenvalues of [$\mathbf{YY'}$] are the diagonal values of matrix $\mathbf{WW'}$; they are the same as the diagonal elements of $\mathbf{W'W}$ which are the eigenvalues of $\mathbf{Y'Y}$. Matrix $\mathbf{V}$ now contains the eigenvectors of [$\mathbf{YY'}$]. Section 9.2 (principal coordinate analysis) will show that the eigenvectors of matrix [$\mathbf{YY'}$] (which is called $\mathbf{\Delta}_1$ in that Section), scaled to a length equal to the square root of their respective eigenvalues, give the coordinates of the objects in the ordination space. This scaling is obtained by multiplying the elements in the column vectors of $\mathbf{V}$ by the corresponding singular values. The coordinates of the objects in matrix [$\mathbf{VW}$] are thus the same as in matrix $\mathbf{F}$ of principal component analysis (eq. 9.4).

When there are as many, or more variables than there are objects (i.e. $p \geq n$), eigenvalues and eigenvectors can still be computed using any of the methods described above: Householder reduction, the TWWS algorithm, or singular value decomposition. The covariance matrix is positive semidefinite in such cases, so that null eigenvalues are expected (Table 2.2). When p is much larger than n and all eigenvalues and eigenvectors must be computed, important savings in computer time may be made by applying Householder reduction or singular value decomposition to matrix [$\mathbf{YY'}$], which is of size ($n \times n$), instead of [$\mathbf{Y'Y}$] which is ($p \times p$) and is thus much larger. The eigenvalues of [$\mathbf{YY'}$] are the same as the non-zero eigenvalues of [$\mathbf{Y'Y}$]. Matrix $\mathbf{U}$ of the eigenvectors of [$\mathbf{Y'Y}$] can be found from matrix $\mathbf{V}$ of the eigenvectors of [$\mathbf{YY'}$] using the transformation $\mathbf{U} = \mathbf{Y'VW}^{-1}$ or $\mathbf{U} = \mathbf{Y'V\Lambda}^{-1/2}$, where $\mathbf{Y}$ is centred by columns. Matrix $\mathbf{F}$ of the principal components is found using eq. 9.19. The rationale justifying to carry out the calculations on [$\mathbf{YY'}$] is presented in the previous paragraph.

Negative eigenvalues may occur in principal component analysis due to missing values. Pairwise deletion of missing data (Subsection 1.6.2), in particular, creates covariances computed with different numbers of degrees of freedom; this situation may make the covariance matrix indefinite (Table 2.2). SVD is not appropriate in this case because the square roots of the negative eigenvalues would be complex numbers; singular values cannot be negative nor complex. A Householder algorithm should be used in such a case.

9.2 Principal coordinate analysis (PCoA)

Chapter 7 discussed various measures of resemblance that are especially useful in ecological analysis with non-quantitative descriptors or with data sets containing many double-zeros (e.g. species presence/absence or abundance data). In such cases, the Euclidean distance (among objects) and the corresponding measures of covariance or correlation (among descriptors) do not provide acceptable models, so that the method of principal component analysis is not adequate for ordination.

Euclidean representation

Gower (1966) has described a method to obtain a Euclidean representation (i.e. a representation in a Cartesian coordinate system) of a set of objects whose relationships are measured by any similarity or distance coefficient chosen by users. This method, known as *principal coordinate analysis* (abbreviated PCoA), *metric multidimensional scaling* (in contrast to the nonmetric method described in Section 9.3), or *classical scaling* by reference to the pioneering work of Torgerson (1958), allows one to position objects in a space of reduced dimensionality while preserving their distance relationships as well as possible; see also Rao (1964).

Mixed precision

The interest of the PCoA method lies in the fact that it may be used with all types of variables — even data sets with variables of mixed levels of precision, provided that a coefficient appropriate to the data has been used to compute the resemblance matrix (e.g. S_{15} or S_{16}, Chapter 7). It will be shown that, if the distance matrix is metric (no violation of the triangle inequality), the relationships among objects can, in most cases, be fully represented in Euclidean space. In the case of violations of the triangle inequality, or when problems of "non-Euclideanarity" occur with metric distances (Gower, 1982; Fig. 9.10), negative eigenvalues are produced. In most cases, this does not impair the quality of the Euclidean representation obtained for the first few principal coordinates. It is also possible to transform the resemblance matrix, or use an alternative resemblance measure, to eliminate the problem. These matters are discussed in Subsection 3.

Euclidean model

One may look at principal coordinates as the equivalent of principal components. Principal components, on the one hand, are linear combinations of the original (or standardized) descriptors; *linear* is the key concept. Principal coordinates, on the other hand, are also functions of the original variables, but mediated through the similarity or distance function that has been used. In any case, PCoA can only embed (i.e. fully represent), in Euclidean space, the Euclidean part of a distance matrix. This is not a property of the data, but a result of the Euclidean model which is forced upon the data because the objective is to draw scatter diagrams on flat sheets of paper. By doing so, one must accept that whatever is non-Euclidean cannot be drawn on paper. This may be viewed as the problem of fitting Euclidean to possibly non-Euclidean distances; there is a remaining fraction in some cases (see Numerical example 3 in Subsection 5).

The method of nonmetric multidimensional scaling (MDS, Section 9.3) also obtains ordinations of objects from any resemblance matrix. It is better than principal coordinate analysis at compressing the distance relationships among objects into, say, two or three dimensions. By construction, MDS always obtains a Euclidean representation, even from non-Euclidean-embeddable distances. MDS is, however, a computer-intensive technique requiring far more computing time than PCoA. For large distance matrices, principal coordinate analysis is faster in most cases.

1 — Computation

Gower (1966) explained how to compute the principal coordinates of a resemblance matrix:

• The initial matrix must be a distance matrix $\mathbf{D} = [D_{hi}]$. It is also possible to carry out the calculations on a similarity matrix $\mathbf{S} = [S_{hi}]$; the results are detailed in Subsection 3. It is better, however, to first transform the **S** matrix into a **D** matrix.

• Matrix **D** is transformed into a new matrix **A** by defining:

$$a_{hi} = -\frac{1}{2}D_{hi}^2 \quad \textbf{(9.20)}$$

The purpose of this transformation is explained in Subsection 3.

• Matrix **A** is centred to give matrix $\mathbf{\Delta}_1 = [\delta_{hi}]$, using the following equation:

$$\delta_{hi} = a_{hi} - \bar{a}_h - \bar{a}_i + \bar{a} \quad \textbf{(9.21)}$$

where $\bar{a}_h$ and $\bar{a}_i$ are the means of the row and column corresponding to element a_{hi} of matrix **A**, respectively, whereas $\bar{a}$ is the mean of all a_{hi}'s in the matrix. This centring has the effect of positioning the origin of the new system of axes at the centroid of the scatter of objects, without altering the distances among objects. Since the sums of the rows and columns of $\mathbf{\Delta}_1$ is null, $\mathbf{\Delta}_1$ has at least one null eigenvalue.

Euclidean distance

In the particular case of distances computed using the Euclidean distance coefficient (D_1, eq. 7.34), it is possible to obtain the Gower-centred matrix $\mathbf{\Delta}_1$ directly, i.e. without calculating a matrix **D** of Euclidean distances and going through eqs. 9.20 and 9.1, because $\mathbf{\Delta}_1 = \mathbf{YY'}$ (for **Y** centred). This may be verified using numerical examples. In this particular case, $\mathbf{\Delta}_1$ is a positive semidefinite matrix (Table 2.2).

• The eigenvalues and eigenvectors are computed and the latter are scaled to lengths equal to the square roots of the respective eigenvalues:

$$\sqrt{\mathbf{u'}_k\mathbf{u}_k} = \sqrt{\lambda_k}$$

Due to the centring, matrix $\mathbf{\Delta}_1$ always has at least one zero eigenvalue. The reason is that at most $(n - 1)$ real axes are necessary for representing n points in Euclidean space. There may be more than one zero eigenvalue in cases where the distance matrix is degenerate, i.e. if the objects can be represented in fewer than $(n - 1)$ dimensions. In practice, there are c positive eigenvalues and c real axes forming the Euclidean representation of the data, the general rule being that $c \leq n - 1$.

Degenerate **D** matrix

With the Euclidean distance (D_1), when there are more objects than descriptors ($n > p$), the maximum value of c is p; when $n \leq p$, then $c \leq n - 1$. Take as example a set of three objects or more, and two descriptors ($n > p$). The objects, as many as they are, may be represented in a two-dimensional space — for example, the scatter diagram of the two descriptors. Consider now the situation where there are two objects and two descriptors ($n \leq p$); the two objects only require one dimension for representation.

• After scaling, if the eigenvectors are written as columns (e.g. Table 9.7), the *rows* of the resulting table are the *coordinates of the objects* in the space of principal coordinates, without any further transformation. Plotting the points on, say, the first two principal coordinates (or more) produces a reduced-space ordination diagram of the objects in two (or more) dimensions.

Table 9.7 Principal coordinates of the objects (rows) are obtained by scaling the eigenvectors to $\sqrt{\lambda}$.

	Eigenvalues			
	λ_1	λ_2	...	λ_c
Objects	**Eigenvectors**			
$\mathbf{x}_1$	u_{11}	u_{12}	...	u_{1c}
$\mathbf{x}_2$	u_{21}	u_{22}	...	u_{2c}
.	.			.
$\mathbf{x}_h$	u_{h1}	u_{h2}	...	u_{hc}
.	.			.
$\mathbf{x}_i$	u_{i1}	u_{i2}	...	u_{ic}
.	.			.
$\mathbf{x}_n$	u_{n1}	u_{n2}	...	u_{nc}
Length: $\sqrt{\sum_i u_{ik}^2}$ =	$\sqrt{\lambda_1}$	$\sqrt{\lambda_2}$	...	$\sqrt{\lambda_c}$
Centroid: $\bar{u}_k$ =	0	0	...	0

2 — Numerical example

Readers may get a better feeling of what principal coordinate analysis does by comparing it to principal component analysis. Consider a data matrix $\mathbf{Y}$ on which a principal component analysis (PCA) has been computed, with resulting eigenvalues, eigenvectors, and principal components. If one also computed a Euclidean distance matrix $\mathbf{D} = [D_{hi}]$ among the same n objects, the eigenvectors obtained by principal coordinate analysis would be exactly the same as the principal components. The eigenvalues of the PCoA are equal to the eigenvalues one would obtain from a PCA conducted on the cross-product matrix $[y - \bar{y}]' [y - \bar{y}]$; these are larger than the eigenvalues of a PCA conducted on the covariance matrix $\mathbf{S}$ by factor $(n - 1)$, because $\mathbf{S} = (1/(n - 1)) [y - \bar{y}]' [y - \bar{y}]$. Since PCA has been defined, in this book, as the eigenanalysis of the covariance matrix **S**, the same PCA eigenvalues can be obtained by carrying out a principal coordinate analysis on the Euclidean distance matrix among objects and dividing the resulting PCoA eigenvalues by $(n - 1)$. If one is only interested in the *relative* magnitude of the eigenvalues, this scaling may be ignored.

The previous paragraph does not mean that principal coordinate analysis is limited to Euclidean distance matrices. It can actually be computed for *any* distance matrix. If the distances cannot readily be embedded in Euclidean space, negative eigenvalues may be obtained, with consequences described below (Subsection 4).

Numerical example 1. The numerical example for principal component analysis (Section 9.1) is used here to illustrate the main steps in the computation of principal coordinates. The example also shows that computing principal coordinates from a matrix of Euclidean distances $\mathbf{D} = [D_{hi}]$ gives the exact same results as principal component analysis of the raw data, with the exception of the variable loadings. Indeed, information about the original variables is not passed on to the PCoA algorithm. Since PCoA is computed from a distance matrix, it cannot give back the loadings of the variables. A method for computing them *a posteriori* is described in Subsection 5.

1) The matrix of Euclidean distances among the 5 objects of data matrix **Y** used to illustrate Section 9.1 is:

$$\mathbf{D} = \begin{bmatrix} 0.00000 & 3.16228 & 3.16228 & 7.07107 & 7.07107 \\ 3.16228 & 0.00000 & 4.47214 & 4.47214 & 6.32456 \\ 3.16228 & 4.47214 & 0.00000 & 6.32456 & 4.47214 \\ 7.07107 & 4.47214 & 6.32456 & 0.00000 & 4.47214 \\ 7.07107 & 6.32456 & 4.47214 & 4.47214 & 0.00000 \end{bmatrix}$$

2) Matrix $\mathbf{\Delta}_1$ obtained by Gower's centring (eqs. 9.20 and 9.21) is:

$$\mathbf{\Delta}_1 = \begin{bmatrix} 12.8 & 4.8 & 4.8 & -11.2 & -11.2 \\ 4.8 & 6.8 & -3.2 & 0.8 & -9.2 \\ 4.8 & -3.2 & 6.8 & -9.2 & 0.8 \\ -11.2 & 0.8 & -9.2 & 14.8 & 4.8 \\ -11.2 & -9.2 & 0.8 & 4.8 & 14.8 \end{bmatrix}$$

The trace (sum of the diagonal elements) of this matrix is 56. This is $(n - 1) = 4$ times the trace of the covariance matrix computed in PCA, which was 14. Note that matrix $\mathbf{\Delta}_1$ could have been obtained directly from data matrix **Y** centred by columns, as mentioned in Subsection 9.2.1: $\mathbf{\Delta}_1 = \mathbf{YY}'$. Readers can verify this property numerically for the example.

3) The eigenvalues and eigenvectors of matrix $\mathbf{\Delta}_1$, scaled to $\sqrt{\lambda}$, are given in Table 9.8. There are only two eigenvalues different from zero; this was to be expected since the distances have been computed on two variables only, using the Euclidean distance formula $(c = p = 2)$. The values of the principal coordinates, which are the standardized eigenvectors of the PCoA, are exactly the same as the principal components (Table 9.6). Measures of resemblance other than the Euclidean distance may produce a different number of eigenvalues and principal coordinates and they would, of course, position the objects differently.

Table 9.8 Principal coordinates computed for the numerical example developed in Section 9.1. Compare with PCA results in Table 9.6.

	Eigenvalues	
	λ_1	λ_2
Objects	**Eigenvectors**	
$\mathbf{x}_1$	–3.578	0.000
$\mathbf{x}_2$	–1.342	–2.236
$\mathbf{x}_3$	–1.342	2.236
$\mathbf{x}_4$	3.130	–2.236
$\mathbf{x}_5$	3.130	2.236
Eigenvalues of PCoA	36.000	20.000
PCoA eigenvalues/$(n-1)$ = eigenvalues of corresponding PCA	9.000	5.000
Length: $\sqrt{\sum_i u_{ik}^2}$ =	$6.000 = \sqrt{36}$	$4.472 = \sqrt{20}$

PCA and PCoA

While the numerical example illustrates the fact that a PCoA computed on a Euclidean distance matrix gives the same results as a PCA conducted on the original data, the converse is also true: taking the coordinates of the objects in the full space (all eigenvectors) obtained from a PCoA and using them as input for a principal component analysis, the eigenvalues of the PCA will be the same (to a factor $n-1$) as those of the original PCoA and the principal components will be identical to the principal coordinates. All the signs of any one component could have been inverted, though, as explained in Subsection 9.1.8; signs depend on an arbitrary decision made during execution of a computer program. Because of this, users of ordination methods are free to change all the signs of any principal component before presenting the results, if this suits them better.

3 — Rationale of the method

Gower (1966) has shown that the distance relationships among objects are preserved in the full-dimensional principal coordinate space. His proof is summarized as follows.

• In the total space of the principal coordinates (i.e. all eigenvectors), the distance between objects h and i can be found by computing the Euclidean distance between rows h and i of Table 9.7:

$$D'_{hi} = \left[\sum_{k=1}^{c} (u_{hk} - u_{ik})^2\right]^{1/2} = \left[\sum_{k=1}^{c} u_{hk}^2 + \sum_{k=1}^{c} u_{ik}^2 - 2\sum_{k=1}^{c} u_{hk}u_{ik}\right]^{1/2} \quad \textbf{(9.22)}$$

• Since the eigenvectors are scaled in such a way that their lengths are $\sqrt{\lambda_k}$ (in other words, $\mathbf{U}$ is scaled here to $\mathbf{\Lambda}^{1/2}$), the eigenvectors have the property that $\mathbf{\Delta}_1 = \mathbf{UU'}$ (eq. 9.8 for matrix $\mathbf{S}$). One can thus write:

$$\mathbf{\Delta}_1 = [\delta_{hi}] = \mathbf{u}_1\mathbf{u}'_1 + \mathbf{u}_2\mathbf{u}'_2 + \ldots + \mathbf{u}_c\mathbf{u}'_c$$

from which it can be shown, following eq. 9.22, that:

$$D'_{hi} = [\delta_{hh} + \delta_{ii} - 2\delta_{hi}]^{1/2}$$

Readers can verify this property on the above numerical example.

• Since $\delta_{hi} = a_{hi} - \bar{a}_h - \bar{a}_i + \bar{a}$ (eq. 9.21), replacing the values of δ in the right-hand member of the previous equation gives:

$$\delta_{hh} + \delta_{ii} - 2\delta_{hi} = a_{hh} + a_{ii} - 2a_{hi}$$

hence

$$D'_{hi} = [a_{hh} + a_{ii} - 2a_{hi}]^{1/2}$$

The transformation of $\mathbf{A}$ into $\mathbf{\Delta}_1$ is not essential. It is simply meant to eliminate one of the eigenvalues, which could be the largest and would only account for the distance between the centroid and the origin.

• The transformation of the matrix of original distances D_{hi} into $\mathbf{A}$ is such that distances are preserved in the course of the calculations. Actually, a_{hi} may be replaced by its value $-0.5\ D_{hi}^2$, from which one can show that

$$D'_{hi} = \left[-\frac{1}{2}D_{hh}^2 - \frac{1}{2}D_{ii}^2 + D_{hi}^2\right]^{1/2}$$

and, since $D_{hh} = D_{ii} = 0$ (property of distances),

$$D'_{hi} = [D_{hi}^2]^{1/2}$$

Principal coordinate analysis thus preserves the original distances, regardless of the formula used to compute them. If the distances have been calculated from similarities, $D_{hi} = 1 - S_{hi}$ will be preserved in the full-dimensional principal coordinate space. If the transformation of similarities into distances was done by $D_{hi} = \sqrt{1 - S_{hi}}$ or $D_{hi} = \sqrt{1 - S_{hi}^2}$, then it is these distances that are preserved by the PCoA. As a corollary, these various representations in principal coordinate space should be as different from one another as are the distances themselves.

Gower (1966) has also shown that principal coordinates can be directly computed from a similarity matrix **S** instead of a distance matrix **D**. If principal coordinates are calculated from matrix **S** after centring (eq. 9.21, *without applying* eq. 9.20 first; make sure that the diagonal of matrix **S** contains 1's and not 0's before centring), distances D'_{hi} among the reconstructed point-objects in the full-dimensional principal coordinate space are not the same as distances $D_{hi} = (1 - S_{hi})$; they are distortions of the distances D_{hi} such that $D'_{hi} = \sqrt{2}\sqrt{D_{hi}}$. As a consequence, if **S** was derived from a **D** matrix through the transformation $S_{hi} = 1 - D_{hi}^2$, then the distances D'_{hi} among the reconstructed point-objects in the full-dimensional principal coordinate space are too large by a factor $\sqrt{2}$, compared to the original distances D_{hi}, but without distortion: $D'_{hi} = \sqrt{2}D_{hi}$. The above holds only if **S** is positive semidefinite, that is, if there are no negative eigenvalues.

To summarize, principal coordinate analysis produces a representation of objects in Euclidean space which preserves the distance relationships computed using any measure selected by users. This is a major difference with PCA, where the distance among objects is always, by definition, the Euclidean distance (Table 9.1). In PCoA, the representation of objects in the reduced space of the first few principal coordinates forms the best possible Euclidean approximation of the original distances, in the sense that the sum of squares of the projection distances of the objects onto the selected subspace is minimum (Gower, 1982). The quality of a Euclidean representation in a space of principal coordinates can be assessed using a Shepard diagram (Fig. 9.1).

Contrary to principal component analysis, the relationships between the principal coordinates and the original descriptors are not provided by a principal coordinate analysis. Indeed the descriptors, from which distances were computed among objects, do not play any role during the calculation of a PCoA. However, computing covariances or linear correlations between the principal coordinates (matrix $\mathbf{U\Lambda}^{1/2}$, Table 9.7) and the original descriptors is a straightforward operation:

$$\mathbf{S}_{pc} = \frac{1}{(n-1)} [y - \bar{y}]'\mathbf{U}\,\mathbf{\Lambda}^{1/2} \tag{9.23}$$

$$\mathbf{R}_{pc} = \frac{1}{(n-1)} \left[\frac{y - \bar{y}}{s}\right]'\mathbf{U} \tag{9.24}$$

In these equations, matrix **U** is that of normalized eigenvectors (lengths 1). The equations assume, of course, that the descriptors in matrix $\mathbf{Y} = [y]$ are quantitative; as in multiple regression analysis (Subsection 10.3.3), multistate qualitative descriptors may be recoded into dummy variables (Subsection 1.5.7) before estimating the covariances or correlations. The rows of matrices $\mathbf{S}_{pc}$ and $\mathbf{R}_{pc}$ correspond to the p descriptors and their columns correspond to the c principal coordinates.

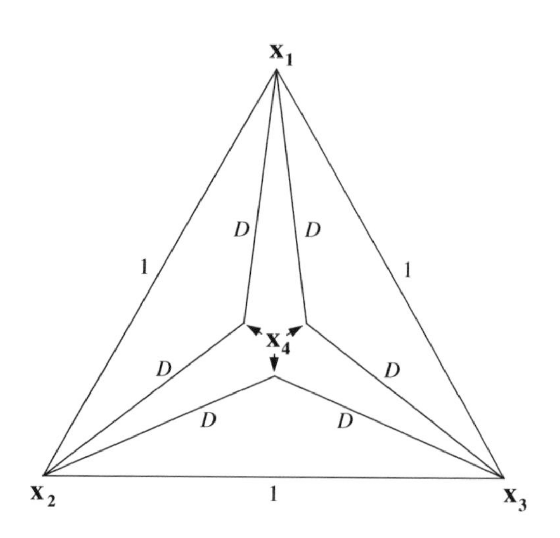

Figure 9.10 Here is a case where the triangle inequality is not violated, yet no Euclidean representation of the 4 points ($\mathbf{x}_1$ to $\mathbf{x}_4$) is possible because distances D, that are all equal, are too small for the inner points ($\mathbf{x}_4$) to join. Assuming that the outer edges are all of length 1, the triangle inequality would be violated if D was smaller than 0.5. $D = 1/\sqrt{3}$ would allow the three representations of $\mathbf{x}_4$ to meet at the centroid. If $D > 1/\sqrt{3}$, the Euclidean representation of the four points, $\mathbf{x}_1$ to $\mathbf{x}_4$, forms a three-dimensional pyramid (Gower, 1982).

4 — Negative eigenvalues

There are resemblance matrices that do not allow a full representation of the distance relationships among objects in Euclidean space (i.e. a set of real Cartesian coordinates).

• Problems of Euclidean representation may result from the use of a distance measure that violates the triangle inequality. Such distances are called *semimetric* and *nonmetric* in Section 7.4 and in Table 7.2.

• These problems may also result from an imbalance in the distance matrix, due to the handling of missing values. See for instance how missing values are handled in coefficients S_{15}, S_{16}, S_{19}, and S_{20} of Chapter 7, using Kronecker delta functions.

Non-Euclide-anarity

• Some *metric* distance matrices present problems of "non-Euclideanarity", as described by Gower (1982, 1985). Figure 9.10 illustrates such a case; the closing of all individual triangles (triangle inequality condition, Section 7.4) is a necessary, but not a sufficient condition to guarantee a full Euclidean representation of a set of objects. This "non–Euclideanarity", when present, translates itself into negative eigenvalues. For instance, most of the metric distances resulting from the transformation of a similarity coefficient, using the formula $D = 1 - S$, are non-Euclidean (Table 7.2). This does not mean that all distance matrices computed using these coefficients are non-Euclidean, but that cases can be found where PCoA produces negative eigenvalues. Among the metric coefficients described in Subsection 7.4.1, D_1, D_2, and D_{11} have been proved Euclidean whereas D_7 and D_{10} are known to be non-Euclidean (Gower & Legendre, 1986). The χ^2 distance D_{16} is also Euclidean; this is the distance preserved by correspondence analysis (Section 9.4) which is known not to produce negative eigenvalues. The forms derived from D_{16}, i.e. D_{15} and S_{21}, are also Euclidean. The "Euclideanarity" of the other metric distances described in Chapter 7 is not known.

Table 7.2 shows that, for many coefficients, the distances $D_{hi} = \sqrt{1 - S_{hi}}$ are Euclidean even though the distances $D_{hi} = 1 - S_{hi}$ are not necessarily Euclidean. The transformation $D_{hi} = \sqrt{1 - S_{hi}}$ should thus be preferred in view of computing PCoA. This transformation may resolve the negative eigenvalue problem even for coefficients that are known to be semimetric. This is the case, for instance, with coefficients S_8, S_{17}, and its one-complement $D_{14} = 1 - S_{17}$, which are all widely used by ecologists to analyse tables of species abundance data. A square root transformation of $D_{14} = 1 - S_{17}$ eliminates negative eigenvalues in principal coordinate analysis; see numerical example 2 in Subsection 5. While it is difficult to demonstrate mathematically that a coefficient is always Euclidean, simulations have not turned up non-Euclidean cases for $\sqrt{1 - S_{17}}$ (Anderson, 1996). In support of this statement, Gower & Legendre (1986) have shown that coefficient S_8, which is the binary form of S_{17}, is Euclidean when transformed into $D = \sqrt{1 - S_8}$.

When one does not wish to apply a square root transformation to the distances, or when negative eigenvalue problems persist in spite of a square root transformation, Gower & Legendre (1986) have shown that the problem of "non–Euclideanarity", and of the negative eigenvalues that come with it, can be solved by adding a (large enough) constant to all values of a distance matrix that would not lend itself to full Euclidean representation. No correction is made along the diagonal, though, because the distance between an object and itself is always zero. Actually, adding some large constant would make the negative eigenvalues disappear and produce a fully Euclidean representation, but it would also create extra dimensions (and eigenvalues) to express the additional variance so created. In Fig. 9.11, for instance, adding a large value, like 0.4, to all six distances among the four points in the graph would create a pyramid, requiring three dimensions for a full Euclidean representation, instead of two.

The problem is to add just the right amount to all distances in the matrix to eliminate all negative eigenvalues and produce a Euclidean representation of the distance relationships among objects, without creating unnecessary extra dimensions. Following Gower & Legendre (1986, Theorem 7*), this result can be obtained by adding a constant c to either the squared distances D_{hi}^2 or the original distances D_{hi}. This gives rise to two methods for adjusting the original distances and correcting for their non-Euclidean behaviour.

* The present Subsection corrects two misprints in theorem 7 of Gower & Legendre (1986).

• *Correction method 1* (derived from the work of Lingoes, 1971) — Add a constant to all squared distances D_{hi}^2, except those on the diagonal, creating a new matrix $\hat{\mathbf{D}}$ of distances $\hat{D}_{hi}$ through the following transformation:

$$\hat{D}_{hi} = \sqrt{D_{hi}^2 + 2c_1} \qquad \text{for } h \neq i \qquad \textbf{(9.25)}$$

then proceed to the transformation of $\hat{\mathbf{D}}$ into matrix $\hat{\mathbf{A}}$ using eq. 9.20. The two operations may be combined into a single transformation producing the new matrix $\hat{\mathbf{A}} = [\hat{a}_{hi}]$ directly from the original distances D_{hi}:

$$\hat{a}_{hi} = -\frac{1}{2}\hat{D}_{hi}^2 = -\frac{1}{2}(D_{hi}^2 + 2c_1) = -\frac{1}{2}D_{hi}^2 - c_1 \qquad \text{for } h \neq i$$

Then, proceed with eq. 9.21 and recompute the PCoA. The constant to be added, c_1, is the absolute value of the *largest negative eigenvalue* obtained by analysing the original matrix $\mathbf{\Delta}_1$. Constant c_1 is also used, below, in the corrected formula for assessing the quality of a representation in reduced space (eq. 9.27). After correction, all non-zero eigenvalues are augmented by a value equal to c_1, so that the largest negative eigenvalue is now shifted to value 0. As a consequence, the corrected solution has two null eigenvalues, or more if the matrix is degenerate. The constant c_1 is the smallest value that will produce the desired effect. Any value larger than c_1 will also eliminate all negative eigenvalues and make the system fully Euclidean, but it may also create a solution requiring more dimensions.

• *Correction method 2* (proposed by Cailliez, 1983) — Add a constant c_2 to all elements D_{hi} of matrix $\mathbf{D}$, except those on the diagonal, creating a new matrix $\hat{\mathbf{D}}$ of distances $\hat{D}_{hi}$ through the transformation:

$$\hat{D}_{hi} = D_{hi} + c_2 \qquad \text{for } h \neq i \qquad \textbf{(9.26)}$$

then proceed to the transformation of $\hat{\mathbf{D}}$ into matrix $\hat{\mathbf{A}}$ using eq. 9.20. The two operations may be combined into a single transformation producing the new matrix $\hat{\mathbf{A}} = [\hat{a}_{hi}]$ directly from the original distances D_{hi}:

$$\hat{a}_{hi} = -\frac{1}{2}(D_{hi} + c_2)^2 \qquad \text{for } h \neq i$$

Then, proceed with eq. 9.21 and recompute the PCoA. The constant to be added, c_2, is equal to the *largest positive eigenvalue* obtained by analysing the following special matrix, which is of order $2n$:

$$\begin{bmatrix} \mathbf{0} & 2\Delta_1 \\ -\mathbf{I} & -4\Delta_2 \end{bmatrix}$$

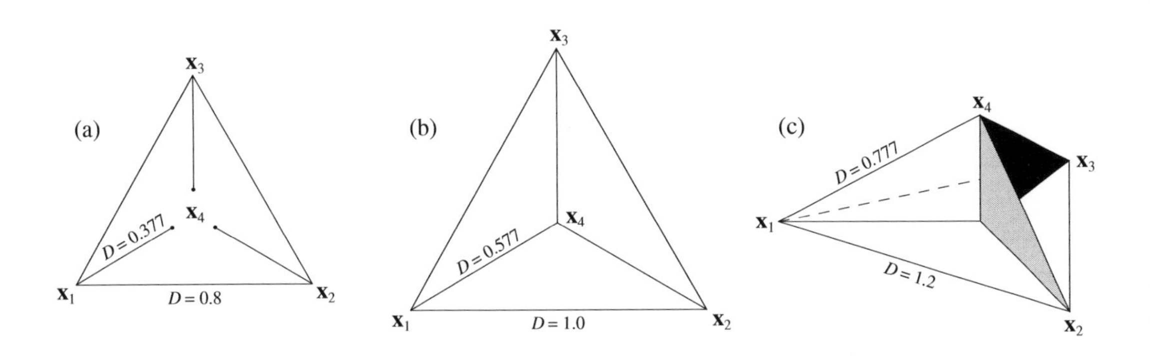

Figure 9.11 (a) Distances among four points constructed in such a way that the system cannot be represented in Euclidean space because the three lines going towards point $\mathbf{x}_4$ do not meet. (b) By adding a constant to all distances ($c_2 = 0.2$ in the present case), correction method 2 makes the system Euclidean; in this example, the distances can be associated with a representation of the points in 2-dimensional space. (c) When increasing the distances further (adding again 0.2 to each distance in the present case), the system remains Euclidean but requires more dimensions for representation (3 dimensions in this example).

where $\mathbf{0}$ is a null matrix, $\mathbf{I}$ is an identity matrix, $\mathbf{\Delta}_1$ is the centred matrix defined by eqs. 9.20 and 9.21, and $\mathbf{\Delta}_2$ is a matrix containing values ($-0.5D_{hi}$) centred using eq. 9.21. The order of each of these matrices is n. Beware: the special matrix is asymmetric. Press *et al.* (1986, Sections 11.5-11.6) give an algorithm to compute the eigenvalues of such a matrix. The corrected solution has two null eigenvalues, or more if the matrix is degenerate. The constant c_2 is the smallest value that will produce the desired effect; any value larger than c_2 will also eliminate all negative eigenvalues and make the system fully Euclidean, but the solution may require more dimensions. Fig. 9.11a-b shows the effect of adding constant c_2 to a non-Euclidean group of four points, whereas Fig. 9.11c shows the effect of adding some value larger than c_2.

The two correction methods do not produce the exact same Euclidean representation. This may be understood by examining the consequences of adding c_2 to the distances in $\mathbf{D}$. When $\hat{\mathbf{D}}$ is transformed into $\hat{\mathbf{A}}$ (eq. 9.20), $(D_{hi} + c_2)$ becomes:

$$\hat{a}_{hi} = -0.5\,(D_{hi} + c_2)^2 = -0.5\,(D_{hi}^2 + 2c_2 D_{hi} + c_2^2) \qquad \text{for } h \neq i$$

The effect on $\hat{a}_{hi}$ does not only depend on the value of c_2; it varies with each value D_{hi}. This is clearly not the same as subtracting a constant from all a_{hi} values (i.e. correction method 1). The eigenvectors resulting from one or the other correction also differ from those resulting from a PCoA without correction for negative eigenvalues. The two correction methods, and PCoA without correction, thus correspond to different partitions of the variation because the total variance, given by the trace of centred matrix $\mathbf{\Delta}_1$, differs among methods.

How large are constants c_1 and c_2 for coefficient $D_{14} = 1 - S_{17}$ which is important for the analysis of species abundance data? To answer this question, Legendre & Anderson (1999) simulated species abundance data tables. After computing distance D_{14}, the correction constants (c_1 for method 1, c_2 for method 2) increased nearly linearly with the ratio (*number of sites/number of species*). In extremely species-poor ecosystems, corrections were the largest; for instance, with a ratio 20:1 (e.g. 200 sites, 10 species), c_1 was near 0.4 and c_2 was near 0.8. When the ratio was near 1 (i.e. number of sites ≈ number of species), c_1 was about 0.06 and c_2 was about 0.2. In species-rich ecosystems, corrections were small, becoming smaller as the species richness increased for a constant number of sites; with a ratio 1:2 for example (e.g. 100 sites, 200 species), c_1 was near 0.02 and c_2 was about 0.1. Results also depended to some extent on the data generation parameters.

To summarize, the methods for eliminating negative eigenvalues, when they are present, all involve making the small distances larger, compared to the large distances, in order to allow all triangles to close (Figs. 9.10, 9.11a and b), even those that did not meet the triangle inequality axiom (Section 7.4). The first approach consists in taking the square root of all distances; this reduces the largest distances more than the small ones. The other two approaches (described as correction methods 1 and 2 above) involve adding a constant to all non-diagonal distances; small distances are proportionally more augmented than large distances. In correction method 1, a constant ($2c_1$) is added to the squared distances D^2_{hi} whereas in method 2 a constant (c_2) is added to the distances D_{hi} themselves.*

Numerical example 2. Consider the numerical example used in Chapter 7 to demonstrate the semimetric nature of the Odum/Bray & Curtis coefficient. The data matrix contains 3 objects and 5 species. Matrix **D** (computed using D_{14}), matrix $\mathbf{A} = [-0.5D^2_{hi}]$, and matrix $\mathbf{\Delta}_1$ are:

$$\mathbf{D} = \begin{bmatrix} 0.00000 & 0.05882 & 0.60000 \\ 0.05882 & 0.00000 & 0.53333 \\ 0.60000 & 0.53333 & 0.00000 \end{bmatrix} \quad \mathbf{A} = \begin{bmatrix} 0.00000 & -0.00173 & -0.18000 \\ -0.00173 & 0.00000 & -0.14222 \\ -0.18000 & -0.14222 & 0.00000 \end{bmatrix} \quad \mathbf{\Delta}_1 = \begin{bmatrix} 0.04916 & 0.03484 & -0.08401 \\ 0.03484 & 0.02398 & -0.05882 \\ -0.08401 & -0.05882 & 0.14283 \end{bmatrix}$$

The trace of $\mathbf{\Delta}_1$ is 0.21597. The eigenvalues are: $\lambda_1 = 0.21645$, $\lambda_2 = 0.00000$, and $\lambda_3 = -0.00049$. The sum of the eigenvalues is equal to the trace.

For correction method 1, value $c_1 = 0.00049$ is subtracted from all non-diagonal values of **A** to give $\hat{\mathbf{A}}$, which is centred (eq. 9.21) to give the corrected matrix $\mathbf{\Delta}_1$:

$$\hat{\mathbf{A}} = \begin{bmatrix} 0.00000 & -0.00222 & -0.18049 \\ -0.00222 & 0.00000 & -0.14271 \\ -0.18049 & -0.14271 & 0.00000 \end{bmatrix} \quad \mathbf{\Delta}_1 = \begin{bmatrix} 0.04949 & 0.03468 & -0.08417 \\ 0.03468 & 0.02430 & -0.05898 \\ -0.08417 & -0.05898 & 0.14315 \end{bmatrix}$$

* A program (DISTPCOA: FORTRAN source code and compiled versions for Macintosh and Windows95 or WindowsNT) is available from the following WWWeb site to carry out principal coordinate analysis, including the corrections for negative eigenvalues described here: <http://www.fas.umontreal.ca/BIOL/legendre/>.

The trace of the corrected matrix $\mathbf{\Delta}_1$ is 0.21694. The corrected eigenvalues are: $\lambda_1 = 0.21694$, $\lambda_2 = 0.00000$, and $\lambda_3 = 0.00000$. This Euclidean solution is one-dimensional.

For correction method 2, value $c_2 = 0.00784$, which is the largest eigenvalue of the special matrix, is added to all non-diagonal elements of matrix $\mathbf{D}$ to obtain $\hat{\mathbf{D}}$, which is then transformed into $\hat{\mathbf{A}}$ (eq. 9.20) and centred (eq. 9.21) to give the corrected matrix $\mathbf{\Delta}_1$:

$$\hat{\mathbf{D}} = \begin{bmatrix} 0.00000 & 0.06667 & 0.60784 \\ 0.06667 & 0.00000 & 0.54118 \\ 0.60784 & 0.54118 & 0.00000 \end{bmatrix} \quad \hat{\mathbf{A}} = \begin{bmatrix} 0.00000 & -0.00222 & -0.18474 \\ -0.00222 & 0.00000 & -0.14644 \\ -0.18474 & -0.14644 & 0.00000 \end{bmatrix} \quad \mathbf{\Delta}_1 = \begin{bmatrix} 0.05055 & 0.03556 & -0.08611 \\ 0.03556 & 0.02502 & -0.06058 \\ -0.08611 & -0.06058 & 0.14669 \end{bmatrix}$$

The trace of the corrected matrix $\mathbf{\Delta}_1$ is 0.22226. The corrected eigenvalues are: $\lambda_1 = 0.22226$, $\lambda_2 = 0.00000$, and $\lambda_3 = 0.00000$. This Euclidean solution is also one-dimensional.

Using the square root of coefficient D_{14}, matrices $\mathbf{D}$, $\mathbf{A}$ and $\mathbf{\Delta}_1$ are:

$$\mathbf{D} = \begin{bmatrix} 0.00000 & 0.24254 & 0.77460 \\ 0.24254 & 0.00000 & 0.73030 \\ 0.77460 & 0.73030 & 0.00000 \end{bmatrix} \quad \mathbf{A} = \begin{bmatrix} 0.00000 & -0.02941 & -0.30000 \\ -0.02941 & 0.00000 & -0.26667 \\ -0.30000 & -0.26667 & 0.00000 \end{bmatrix} \quad \mathbf{\Delta}_1 = \begin{bmatrix} 0.08715 & 0.04662 & -0.13377 \\ 0.04662 & 0.06492 & -0.11155 \\ -0.13377 & -0.11155 & 0.24532 \end{bmatrix}$$

The trace of $\mathbf{\Delta}_1$ is 0.39739. The eigenvalues are: $\lambda_1 = 0.36906$, $\lambda_2 = 0.02832$, and $\lambda_3 = 0.00000$. No negative eigenvalue is produced using this coefficient. This Euclidean solution is two-dimensional.

If negative eigenvalues are present in a full-dimensional PCoA solution and no correction is made to the distances to eliminate negative eigenvalues, problems of interpretation arise. Since the eigenvectors $\mathbf{u}_k$ are scaled to length $\sqrt{\lambda_k}$, it follows that the axes corresponding to negative eigenvalues are not real, but complex. Indeed, in order for the sum of squares of the u_{ik}'s in an eigenvector $\mathbf{u}_k$ to be negative, the coordinates u_{ik} must be imaginary numbers. When some of the axes of the reference space are complex, the distances cannot be fully represented in Euclidean space, as in the example of Figs. 9.10 and 9.11a.

It is, however, legitimate to investigate whether the Euclidean approximation corresponding to the positive eigenvalues (i.e. the non-imaginary principal coordinates) provides a good representation, when no correction for negative eigenvalues is applied. Cailliez & Pagès (1976) show that such a representation is meaningful as long as the largest negative eigenvalue is smaller, in absolute value, than any of the m positive eigenvalues of interest for representation in reduced space (usually, the first two or three). When there are no negative eigenvalues, the quality of the representation in a reduced Euclidean space with m dimensions can be assessed, as in principal component analysis (eq. 9.5), by the R^2-like ratio:

R^2-like ratio

$$\left(\sum_{k=1}^{m} \lambda_k\right) \Big/ \left(\sum_{k=1}^{c} \lambda_k\right)$$

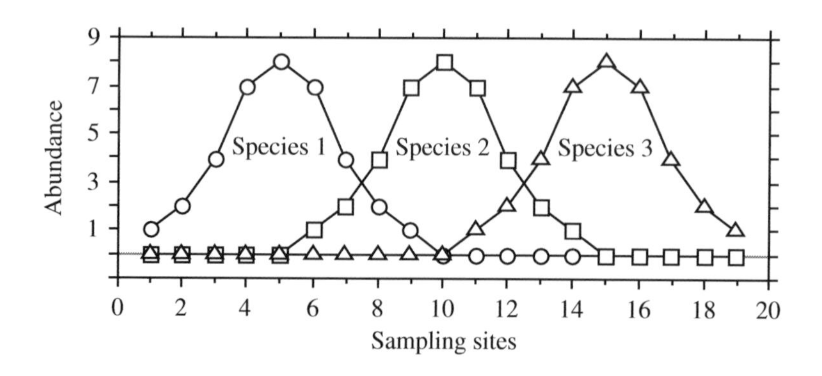

Figure 9.12 Distributions of three species at 19 sampling sites along a hypothetical environmental gradient These artificial data are given in the upper panel of Table 9.9.

where c is the number of positive eigenvalues. This comes from the fact that the eigenvalues of a PCoA are the same (to a factor $n - 1$) as those of a PCA performed on the coordinates of the same points in the full-dimensional space of the principal coordinates (Subsection 9.2.2). Cailliez & Pagès (1976) further suggest that, when negative eigenvalues are present, a correct estimate of the quality of a reduced-space representation can be obtained by the corrected R^2-like ratio:

Corrected R^2-like ratio

$$\frac{\left(\sum_{k=1}^{m} \lambda_k\right) + m|\lambda_n|}{\left(\sum_{k=1}^{n} \lambda_k\right) + (n-1)|\lambda_n|} \quad \textbf{(9.27)}$$

where m is the dimensionality of the reduced space, n is the order of the distance matrix (total number of objects), and $|\lambda_n|$ is the absolute value of the largest negative eigenvalue. Equation 9.27 gives the same value as if correction method 1, advocated above, was applied to the distance matrix, the PCoA was recomputed, and the quality of the representation was calculated using eq. 9.5. All non-zero eigenvalues would be augmented by a value equal to $c_1 = |\lambda_n|$, producing the same changes to the numerator and denominator as in eq. 9.27.

5 — *Ecological applications*

Principal coordinate analysis is an ordination method of great interest to ecologists because the nature of ecological descriptors often makes it necessary to use other measures of resemblance than the Euclidean distance preserved by principal component analysis (Table 9.1). Ordination methods such as principal coordinate analysis and nonmetric multidimensional scaling (Section 9.3) provide Euclidean representations of point-objects for any distance measure selected by the users.

Table 9.9 Upper panel: artificial data illustrated in Fig. 9.12. Lower panel: first row of each of four distance matrices, comparing site 1 to itself and the 18 other sites. The distance matrices are not fully printed to save space; readers can compute them from the data in the upper panel. Values in the lower panel are rounded to a single decimal place.

Sampling sites	1	2	3	4	5	6	7	8	9	10	11	12	13	14	15	16	17	18	19
Species 1	1	2	4	7	8	7	4	2	1	0	0	0	0	0	0	0	0	0	0
Species 2	0	0	0	0	0	1	2	4	7	8	7	4	2	1	0	0	0	0	0
Species 3	0	0	0	0	0	0	0	0	0	0	1	2	4	7	8	7	4	2	1
D_1 (Euclidean)	0.0	1.0	3.0	6.0	7.0	6.1	3.6	4.1	7.0	8.1	7.1	4.6	4.6	7.1	8.1	7.1	4.1	2.2	1.4
$D_{14} = (1 - S_{17})$	0.0	0.3	0.6	0.8	0.8	0.8	0.7	0.7	0.8	1.0	1.0	1.0	1.0	1.0	1.0	1.0	1.0	1.0	1.0
$\sqrt{D_{14}}$	0.0	0.6	0.8	0.9	0.9	0.9	0.8	0.8	0.9	1.0	1.0	1.0	1.0	1.0	1.0	1.0	1.0	1.0	1.0
D_{16} (χ^2 distance)	0.0	0.0	0.0	0.0	0.0	0.3	0.8	1.6	2.1	2.4	2.3	2.2	2.2	2.3	2.4	2.4	2.4	2.4	2.4

Gradient

Numerical example 3. A data set was created (Fig. 9.12; Table 9.9) to represent the abundances of three hypothetical species at 19 sites across an environmental gradient along which the species were assumed to have unimodal distributions (Whittaker, 1967). Four Q-mode distance measures were computed among sites to illustrate some properties of principal coordinate analysis. The same data will be used again in Subsection 9.4.5 to create examples of horseshoes and arches.

• The Euclidean distance D_1 is a symmetrical coefficient. It is not ideal for species abundance data; it is only used here for comparison. A principal coordinate analysis of this matrix led to 19 eigenvalues: three positive (50, 41, and 9% of the variation) and 16 null. This was expected since the original data matrix contained three variables.

• Distance D_{14} is widely used for species abundance data. Like its one-complement S_{17}, it excludes double-zeros. Principal coordinate analysis of this distance matrix led to 19 eigenvalues: 11 positive, one null, and 7 negative. The distance matrix was corrected using method 1 of Subsection 4, which makes use of the largest negative eigenvalue. Re-analysis led to 17 positive and two null eigenvalues, the largest one accounting for 31% of the variation. The distance matrix was also corrected using method 2 of Subsection 4, which makes use of the largest eigenvalue of the special matrix. Re-analysis also led to 17 positive and two null eigenvalues, the largest one accounting for 34% of the variation.

• Principal coordinate analysis was also conducted using the square root of coefficient D_{14}. The analysis led to 18 positive, one null, and no negative eigenvalues, the largest one accounting for 35% of the variation.

• A fourth distance matrix was computed using the χ^2 distance D_{16}, which is also a coefficient excluding double-zeros. Principal coordinate analysis produced 19 eigenvalues: two positive (64 and 36% of the variation) and 17 null. The χ^2 distance (D_{16}) is the coefficient preserved in correspondence analysis (Section 9.4). This analysis also produces one dimension less than the original number of species, or even fewer in the case of degenerate matrices.

This example shows that different distance (or similarity) measures may lead to very different numbers of dimensions of the Euclidean representations. In the analyses reported here, the numbers of dimensions obtained were 3 for distance D_1, 11 for D_{14} (not counting the complex axes corresponding to negative eigenvalues), 17 (after correction of D_{14} by the largest negative eigenvalue), 18 for the square root of D_{14}, and 2 for D_{16}.

Only the Euclidean distance and derived coefficients lead to a number of principal axes equal to the original number of variables. Other coefficients may produce fewer, or more axes. The dimensionality of a principal coordinate space is a function of the number of original variables, mediated through the similarity or distance measure that was selected for the analysis.

There are many applications of principal coordinate analysis in the ecological literature. This method may be used in conjunction with clustering techniques; an example is presented in Ecological application 10.1. Direct applications of the method are summarized in Ecological applications 9.2a and 9.2b. The application of principal coordinate analysis to the clustering of large numbers of objects is discussed in Subsection 8.7.3.

Ecological application 9.2a

Field & Robb (1970) studied the molluscs and barnacles from a rocky shore (21 quadrats) in False Bay, South Africa, in order to determine the influence of factors *emergence* and *wave* on these communities. Quadrats 1 to 10, on a transect parallel to the shoreline, differed in their exposure to wave action; quadrats 11 to 21, located on a transect at right angle to the shoreline, covered the spectrum between the mean high and mean low waters of spring tides. 79 species were counted, one reaching 10 864 individuals in a single quadrat. When going up the shore, quadrats had progressively larger numbers of individuals and smaller numbers of species. This illustrates the principle according to which an increasing environmental stress (here, the *emergence* factor) results in decreasing diversity. It also shows that the few species that can withstand a high degree of stress do not encounter much interspecific competition and may therefore become very abundant.

The same principal coordinate ordination could have been obtained by estimating species abundances with a lesser degree of precision, e.g. using classes of abundance. Table 7.3 gives the association measures that would have been appropriate for such data.

Species abundances (y'_{ij}) were first normalized by logarithmic transformation $y''_{ij} = \ln(y'_{ij} + 1)$, then centred ($y_{ij} = y''_{ij} - \bar{y}_i$), to form matrix $\mathbf{Y} = [y_{ij}]$ containing the data to be analysed. Scalar products among quadrat vectors were used as the measure of similarity:

$$\mathbf{S}_{n \times n} = \mathbf{Y}_{n \times p} \mathbf{Y}'_{p \times n}$$

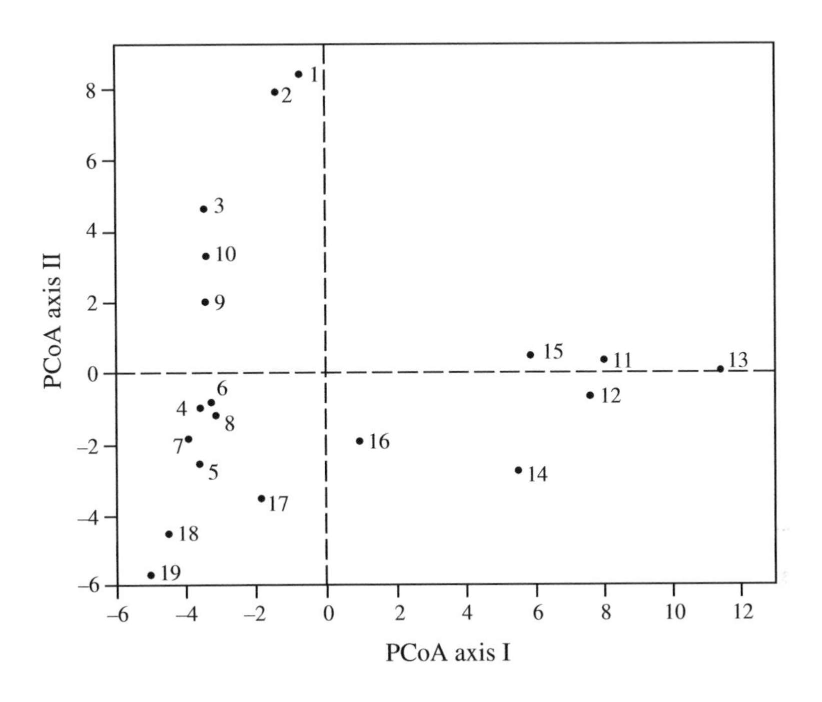

Figure 9.13 Ordination of quadrats 1 to 19 in the space of the first two principal coordinates (PCoA axes I and II). Modified from Field & Robb (1970).

Principal coordinates were computed using a variant procedure proposed by Orlóci (1966). Fig. 9.13 displays the ordination of quadrats 1 to 19 in the space of the first two principal coordinates. The ordination was also calculated including quadrats 20 and 21 but, since these came from the highest part of the shore, they introduced so much variation in the analysis that the factor *emergence* dominated the first two principal coordinates. For the present Ecological application, only the ordination of quadrats 1 to 19 is shown. The authors looked for a relationship between this ordination and the two environmental factors, by calculating Spearman's rank correlation coefficients (eq. 5.3) between the ranks of the quadrats on each principal axis and their ranks on the two environmental factors. This showed that the first principal axis had a significant correlation with elevation with respect to the shoreline (*emergence*), whereas the second axis was significantly related to *wave action*. The authors concluded that PCoA is well adapted to the study of ecological gradients, provided that the data set is fairly homogeneous. (Correspondence analysis, described in Section 9.4, would have been another appropriate way of obtaining an ordination of these quadrats.)

Ecological application 9.2b

Ardisson *et al.* (1990) investigated the spatio-temporal organization of epibenthic communities in the Estuary and Gulf of St. Lawrence, an area ca. 1150 × 300 km. Quantitative data were

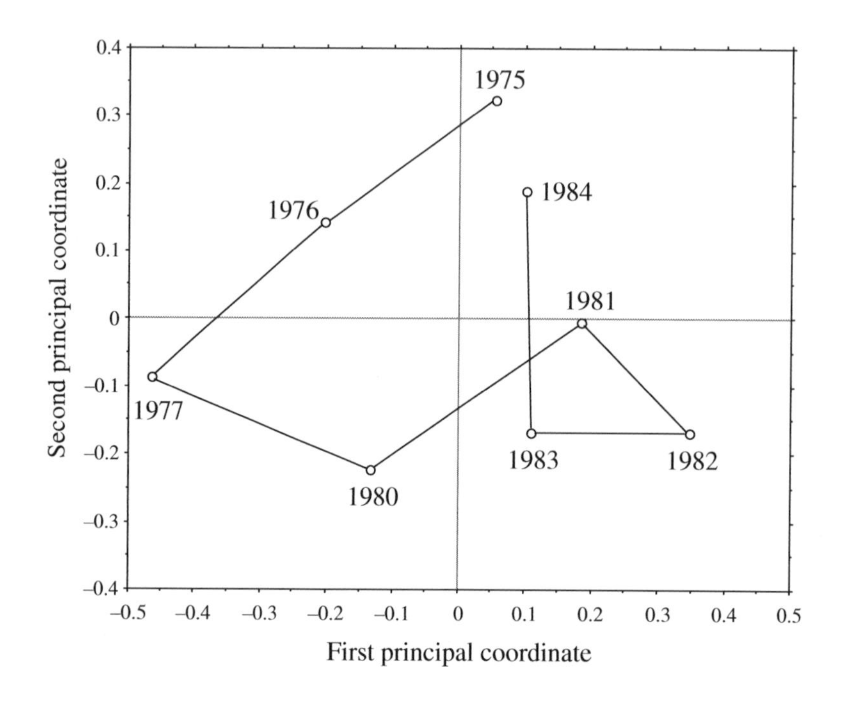

Figure 9.14 Among-year variability illustrated in the space of the first two principal coordinates, obtained from analysing a matrix of Mantel statistics comparing yearly similarity matrices. Recomputed from the Mantel statistic values provided in Fig. 8 of Ardisson *et al.* (1990).

obtained over 8 years, between 1975 and 1984, from 161 collectors (navigation buoys) moored yearly from May through November.

Mantel statistic

Each year was represented by a data table of 161 sites (buoys) × 5 dominant species (dry biomass). A similarity matrix among sites was computed for each year separately, using the asymmetrical form of the Gower similarity coefficient (S_{19}). The eight yearly matrices were compared to one another using the Mantel statistic (Subsection 10.5.1). A principal coordinate analysis (Fig. 9.14) was conducted on the resulting matrix of Mantel statistics, to determine whether year-to-year differences were random or organized. The among-year pattern of dispersion suggested the existence of a cycle of variation whose length was about equal to the duration of the study. This is supported by the fact that all Mantel tests conducted between consecutive years were highly significant ($p \leq 0.001$). This cycle might represent the response of the Estuary-Gulf system, as an integrated unit, to external inputs of auxiliary energy, although the specific causal process, physical or biotic, remains unknown.

This last Ecological application showed that the usefulness of principal coordinate analysis is not limited to projecting, in reduced space, classical resemblance matrices among objects. In the example, the relationships among data tables, as expressed by Mantel statistics, were represented in a Euclidean space using PCoA. The method may

Table 9.10 Computing principal coordinates.

a) Centre the distance matrix

Centre the distance matrix following Gower's method (eqs. 9.21 and 9.22).

b) Compute the eigenvalues and eigenvectors

Use any eigenanalysis subroutine.

If the "two-way weighted summation" algorithm (TWWS) is chosen, use steps 2 to 6 of Table 9.5, including orthogonalization and normalization, as well as the stopping rule of step 7. Modify the algorithm (Table 9.5) as follows:

• Step 3: the column scores are made equal to the row scores.

• Step 7: the eigenvalue is estimated by S obtained from the last normalization (Step 6.1), at the end of the iteration loop. No division by $(n - 1)$ is necessary.

c) Final scaling

Scale each eigenvector k to length $\sqrt{\lambda_k}$ to obtain the principal coordinates.

Eigenvalues obtained from PCoA are larger than those from PCA by a factor $(n - 1)$.

actually be applied to any type of symmetric resemblance matrix. This includes cases where the measures of resemblance are obtained directly from observation (e.g. interaction matrices in behavioural studies) or from laboratory work (DNA hybridisation results, serological data, etc.). If the resulting matrix is non-symmetric, it may be decomposed into a symmetric and a skew-symmetric component (Section 2.3) which can be analysed separately by PCoA. Either the upper or the lower triangular portion of the skew-symmetric matrix is analysed in such a case; the two analyses produce identical results.

6 — Algorithms

Principal coordinates are easy to compute for any distance matrix, using the standard eigenanalysis subprograms mentioned in Subsection 9.1.8 (e.g. Householder reduction). Follow the steps summarized in Table 9.10.

One may prefer to use, here again, an algorithm that allows the computation of the first few eigenvectors only. Indeed, it may be painfully long to wait for a program to compute all eigenvectors of a very large matrix, when only the first few are needed to plot a reduced-space ordination diagram. The "two-way weighted summation" algorithm (TWWS) of ter Braak, detailed in Table 9.5, is a good way of obtaining them. Two changes must be made, however (Table 9.10):

• Since the analysis involves a square symmetric distance matrix, the row and column scores provide the same ordination. So, in Step 3 of the algorithm (Table 9.5), the column scores are simply made equal to the row scores.

• The eigenvalue associated with each principal coordinate is estimated by the S value obtained during the last normalization (Step 6.1), at the end of the iteration loop (Steps 3 to 7); no division by $(n-1)$ is necessary. The reason is that the eigenvector itself is sought, this time, and not the principal component.

If the TWWS algorithm is used to compute a large number of principal coordinates, or all $(n-1)$ of them, a decision rule must be included to stop the program when an eigenvalue estimate becomes "dangerously" close to zero; if a null eigenvalue occurred, there would be division by 0 during normalization (Step 6.2, Table 9.5) and the computation would be interrupted. Cases where the number of positive eigenvalues c is smaller than $(n-1)$ have been described in Subsection 9.2.1. Negative eigenvalues, if present, are produced by the algorithm mixed among the positive values, in decreasing order of their absolute values; they are identifiable by the fact that the corresponding eigenvector elements change signs at each iteration.

9.3 Nonmetric multidimensional scaling (MDS)

All reduced-space ordination methods start from an ordination (scaling) of the objects in full-dimensional space and attempt to represent them in a few dimensions while preserving, as well as possible, the distance relationships among the objects. There are cases where the exact preservation of distances is not of primary importance, the priority being instead to represent the objects in a small and specified number of dimensions — usually two or three. In such cases, the objective is to plot dissimilar objects far apart in the ordination space and similar objects close to one another. This is called the *preservation of ordering relationships* among objects. The method to do so is called *nonmetric multidimensional scaling* (NMDS, or simply MDS). It was devised by psychometricians Shepard (1962, 1966) and Kruskal (1964a, b). Programs for MDS were originally distributed by Bell Laboratories in New Jersey, where the method originated; see Carroll (1987) for a review. The method is now available in several major (SPSS, SAS, SYSTAT, etc.) and specialized computer packages*. A useful reference is the book of Kruskal & Wish (1978). Relationships between MDS and other forms of ordination have been explained by Gower (1987). Extensions of MDS to several matrices, weighted models, the analysis of preference data, etc. are discussed by Young (1985) and Carroll (1987). A form of *hybrid scaling*, combining metric and nonmetric scaling criteria, was proposed by Faith *et al.* (1987); it is further explained in Belbin (1991) and is available in packages DECODA and PATN.

Like principal coordinate analysis (PCoA), MDS is not limited to Euclidean distance matrices; it can produce ordinations of objects from any distance matrix. The method can also proceed with missing distance estimates — actually, the more missing distances there are, the easier the computations — as long as there are enough measures left to position each object with respect to a few of the others. This feature makes it a method of choice for the analysis of matrices obtained by direct observation (e.g. behaviour studies) or laboratory assays, where missing pairwise distances often occur. Some programs can handle non-symmetric distance matrices, for which they provide a compromise solution between distances in the upper and lower triangular parts of the matrix. Contrary to PCA, PCoA, or CA, which are eigenvector methods, MDS calculations do not maximize the variability associated with individual axes of the ordination; MDS axes are arbitrary, so that plots may arbitrarily be rotated, centred, or inverted. Reasons for this will become clear from the presentation of the method.

Consider a distance matrix $\mathbf{D}_{n \times n} = [D_{hi}]$ computed using a measure appropriate to the data at hand (Chapter 7). Matrix $\mathbf{D}$ may also result from direct observations, e.g. affinities among individuals or species found in serological, DNA pairing, or behavioural studies; these matrices may be non-symmetric. Nonmetric multidimensional scaling of matrix $\mathbf{D}$ may be summarized in the following steps.

1) Specify the number m of dimensions chosen *a priori* for scaling the objects. The output will provide coordinates of the n objects on m axes. If several configurations for different numbers of dimensions are sought — say, 2, 3, 4, and 5 dimensions, they must be computed separately. Several programs actually allow solutions to cascade from high to low numbers of dimensions — for instance from 4 to 3 to 2 to 1.

2) Construct an initial configuration of the objects in m dimensions, to be used as a starting point for the iterative adjustment process of steps 3 to 7. The way this initial configuration is chosen is critical because the solution on which the algorithm eventually converges depends to some extent on the initial positions of the objects. The

* MDS programs for microcomputers are found in the following commercially available packages (list not exhaustive):

• DECODA is a package for vegetation studies written by Peter R. Minchin, School of Botany, The University of Melbourne. It is distributed by Anutech Pty Ltd., Canberra, A.C.T. 0200, Australia.

• How to obtain NTSYS is described in Table 13.4.

• How to obtain PATN, is described in the footnote of p. 302.

• PRIMER was developed by M. R. Carr and K. R. Clarke at Plymouth Marine Laboratory, Prospect Place, West Hoe, Plymouth PL1 3DH, Great Britain.

• PC-ORD is available from MjM Software, P.O. Box 129, Gleneden Beach, Oregon 97388, USA. Besides MDS, PC-ORD contains programs for PCA, CA, and DCA (this Chapter), diversity indices (Section 6.5), agglomerative cluster analysis (Section 8.5), TWINSPAN (Subsection 8.7.4), indicator value analysis (Subsection 8.9.3), Mantel test (Section 10.5), CCA (Section 11.3), species-area curves, and other procedures of interest to ecologists. WWWeb site: <http://ourworld.compuserve.com/homepages/MJMSoftware/pcordwin.htm>.

Local minimum

same problem was encountered with *K*-means partitioning (Section 8.8); the "space of solutions" may contain several local minima besides the overall minimum (Fig. 8.17). The most commonly used solutions to this problem are the following:

- Run the program several times, starting from different random initial placements of the objects. The solution minimizing the objective function (step 5) is retained.

- Initiate the run from an ordination obtained using another method, e.g. PCoA.

- If the data are thought to be spatially structured and the geographic positions of the objects are known, these geographic positions may be used as the starting configuration for MDS of a matrix **D** computed from the data.

- Work step by step from higher to lower dimensionality. Compute, for instance, a first MDS solution in 5 dimensions from a random initial placement of the objects. Note the stress value (eqs. 9.28 to 9.30), which should be low because the high number of dimensions imposes little constraint to the distances. Use 4 of the 5 dimensions so obtained as the initial configuration for a run in 4 dimensions, and so forth until the desired number (m) of ordination dimensions is reached.

3) Calculate a matrix of fitted distances d_{hi} in the ordination space, using one of Minkowski's metrics (D_6, eq. 7.44). Most often, one chooses the second degree of Minkowski's metric, which is the Euclidean distance. (a) In the first iteration, distances d_{hi} are computed from the initial (often random) configuration. (b) In subsequent iterations, the configuration is that obtained in step 6.

Shepard diagram

4) Consider the Shepard diagram (Figs. 9.1 and 9.15) comparing the fitted distances d_{hi} to the empirical (i.e. original) distances D_{hi}. Regress d_{hi} on D_{hi}. Values forecasted by the regression line are called $\hat{d}_{hi}$. The choice of the type of regression is left to the users, given the choices implemented in the computer program. Usual choices are the linear, polynomial, or monotone regressions (also called "nonparametric", although there are other types of nonparametric regression methods).

Monotone regression is a step-function which is constrained to always increase from left to right (Fig. 9.15b); this is a common choice in MDS. A monotone regression is equivalent to a linear regression performed after monotonic transformation of the original distances D_{hi}, so as to maximize the linear relationship between D_{hi} and d_{hi}. The regression is fitted by least squares.

Tied values

If there are tied values among the empirical distances, Kruskal (1964a, b) has defined two approaches that may be followed in monotone regression. Ties are likely to occur when the empirical distances D_{hi} are computed from a table of raw data using one of the coefficients of Chapter 7; they are less likely to occur when distances result from direct observations. In Fig. 9.15b, for instance, there are ties for several of the values on the abscissa; the largest number of ties is found at $D = D_{max} = 1$.

• In Kruskal's *primary approach*, one accepts the fact that, if an empirical distance D_{hi} corresponds to different fitted values d_{hi}, it also corresponds to different forecasted values $\hat{d}_{hi}$. Hence the monotone regression line is allowed to go straight up in a column of tied values, subject to the constraint that the regression line is not allowed to decrease compared to the previous values D. The monotone regression line is not a mathematical function in that case, however. In order to insure monotonicity, the only constraint on the $\hat{d}_{hi}$ values is:

$$\text{when } D_{gi} < D_{hi}, \text{ then } \hat{d}_{gi} \le \hat{d}_{hi}$$

• In the *secondary approach*, the forecasted value $\hat{d}_{hi}$ is the same for all fitted distances d_{hi} that are tied to a given empirical distance value D_{hi}. To insure monotonicity, the constraints on the $\hat{d}_{hi}$ values are:

$$\text{when } D_{gi} < D_{hi}, \text{ then } \hat{d}_{gi} \le \hat{d}_{hi}$$

$$\text{when } D_{gi} = D_{hi}, \text{ then } \hat{d}_{gi} = \hat{d}_{hi}$$

In this approach, the least-squares solution for $\hat{d}_{hi}$ is the mean of the tied d_{hi}'s when considering a single value D_{hi}. The vertical difference in the diagram between d_{hi} and $\hat{d}_{hi}$ is used as the contribution of that point to the stress formula, below. In Fig. 9.15b, the secondary approach is applied to all tied values found for $D_{hi} < (D_{max} = 1)$, and the primary approach when $D_{hi} = D_{max} = 1$.

Computer programs may differ in the way they handle ties. This may cause major discrepancies between reported stress values corresponding to the final solutions, although the final configurations of points are usually very similar from program to program, except when different programs identify distinct final solutions having very similar stress values.

A reduced-space scaling would be perfect if all points in the Shepard diagram fell exactly on the regression line (straight line, smooth curve, or step-function); the rank order of the fitted distances d_{hi} would be exactly the same as that of the original distances D_{hi} and the value of the objective function (step 5) would be zero.

Objective function

5) Measure the goodness-of-fit of the regression using an objective function. All objective functions used in MDS are based on the sum of the squared differences between fitted values d_{hi} and the corresponding values $\hat{d}_{hi}$ forecasted by the regression function; this is the usual sum of squared residuals of regression analysis (least-squares criterion, Subsection 10.3.1). Several variants have been proposed and are used in MDS programs:

$$\textit{Stress} \text{ (formula 1)} = \sqrt{\sum_{h,i} (d_{hi} - \hat{d}_{hi})^2 / \sum_{h,i} d_{hi}^2} \qquad \textbf{(9.28)}$$

$$Stress\text{ (formula 2)} = \sqrt{\sum_{h,i} (d_{hi} - \hat{d}_{hi})^2 / \sum_{h,i} (d_{hi} - \bar{d})^2} \quad \textbf{(9.29)}$$

$$Sstress = \sqrt{\sum_{h,i} (d_{hi}^2 - \hat{d}_{hi}^2)^2} \quad \textbf{(9.30)}$$

The denominators in the two *Stress* formulas (eq. 9.28 and 9.29) are scaling terms that make the objective functions dimensionless and produce *Stress* values between 0 and 1. These objective functions may use the square root, or not, without changing the issue; a configuration that minimizes these objective functions would also minimize the non-square-rooted forms. Other objective criteria, such as *Strain*, have been proposed. All objective functions measure how far the reduced-space configuration is from being monotonic to the original distances D_{hi}. Their values are only relative, measuring the decrease in lack-of-fit between iterations of the calculation procedure.

Steepest descent

6) Improve the configuration by moving it slightly in a direction of decreasing stress. This is done by a numerical optimization algorithm called *method of steepest descent*; the method is explained, for instance, in *Numerical Recipes* (Press *et al.*, 1986) and in Kruskal (1964b). The direction of steepest descent is the direction in the space of solutions along which stress is decreasing most rapidly. This direction is found by analysing the partial derivatives of the stress function (Carroll, 1987). The idea is to move points in the ordination plot to new positions that are likely to decrease the stress most rapidly.

7) Repeat steps 3 to 6 until the objective function reaches a small, predetermined value (tolerated lack-of-fit), or until convergence is achieved, i.e. until it reaches a minimum and no further progress can be made. The coordinates calculated at the last passage through step 6 become the coordinates of the n objects in the m dimensions of the multidimensional scaling ordination.

8) Most MDS programs rotate the final solution using principal component analysis, for easier interpretation.

In most situations, users of MDS decide that they want a representation of the objects in two or three dimensions, for illustration or other purpose. In some cases, however, one wonders what the “best” number of dimensions would be for a data set, i.e. what would be the best compromise between a summary of the data and an accurate representation of the distances. As pointed out by Kruskal & Wish (1978), determining the dimensionality of an MDS ordination is as much a substantive as a statistical question. The substantive aspects concern the interpretability of the axes, ease of use, and stability of the solution. The statistical aspect is easier to approach since stress may be used as a guide to dimensionality. Plot the *stress* values as a function of *dimensionality* of the solutions, using one of the stress formulas above (eqs. 9.28 - 9.30). Since stress decreases as dimensionality increases, choose for the final solution the dimensionality where the change in stress becomes small.

For species count data, Faith *et al.* (1987) have shown, through simulations, that the following strategy yields informative ordination results: (1) standardize the data by dividing each value by the maximum abundance for that species in the data set; (2) use the Steinhaus (S_{17}) or the Kulczynski (S_{18}) similarity measure; (3) compute the ordination by MDS.

Besides the advantages mentioned above for the treatment of nonmetric distances or non-symmetric matrices (see also Sections 2.3 and 8.10 on this topic), Gower (1966) pointed out that MDS can summarize distances in fewer dimensions than principal coordinate analysis (i.e. lower stress in, say, two dimensions). Results of the two methods may be compared by examining Shepard diagrams of the results obtained by PCoA and MDS, respectively. If the scatter of points in the PCoA Shepard diagram is narrow, as in Fig. 9.1a or b, the reduced-space ordination is useful in that it correctly reflects the relative positions of the objects. If it is wide or nearly circular (Fig. 9.1c), the ordination diagram is of little use and one may try MDS to find a more satisfactory solution in a few dimensions. A PCoA solution remains easier to compute in most cases, however, because it does not require multiple runs, and it is obtained using a direct eigenanalysis algorithm instead of an iterative procedure.

Numerical example 2 continued. The Bray & Curtis distance matrix (D_{14}) computed in Table 9.9 was subjected to MDS analysis using the package DECODA (see above). This MDS program uses Stress formula 1 (eq. 9.28). Repeated runs, using $m = 2$ dimensions but different random starting configurations, produced very similar results; the best one had a stress value of 0.0181 (Fig. 9.15a).

Kruskal's secondary approach, explained with computation step 4 above, was used in Fig. 9.15b for all tied values found when $D_{hi} < D_{max}$, while the primary approach was used when $D_{hi} = D_{max} = 1$. The rationale for this follows from the fact that the empirical distances D_{hi} are blocked by an artificial ceiling D_{max} of the distance function, over which they cannot increase. So, pairs of sites tied at distance $D_{max} = 1$, for which d_{hi} is larger than the previous value $\hat{d}$, are not expected to be the same distance apart in the ordination. Hence these values should not contribute to the stress despite their ties.

Using $\sqrt{D_{14}}$ as the distance measure, instead of D_{14}, produced an identical ordination, since MDS is invariant to monotonic transformations of the distances. The stress value did not change either, because the square root transformation of D_{14} affects only the abscissa of Fig. 9.15b, whereas the stress is computed along the ordinate. The arch effect found in Fig. 9.17 (Subsection 9.4.5) does not appear in Fig. 9.15a. The horizontal axis of the MDS ordination reproduces the original gradient almost perfectly in this example.

Points in an MDS plot may be rotated, translated, inverted, or scaled *a posteriori* in any way considered appropriate to achieve maximum interpretability or to illustrate the results. This may be done either by hand or, for example, through canonical analysis of the MDS axes with respect to a set of explanatory variables (Chapter 11).

With the present data, a one-dimensional ordination (stress = 0.1089) perfectly reconstructed the gradient of sites 1 to 19; the same ordination was always obtained when repeating the run from different random starting configurations and cascading from 3 to 2 to 1 dimensions. This configuration, and the low stress value, were hardly ever obtained when performing the MDS ordination directly in one dimension, without the cascading procedure.

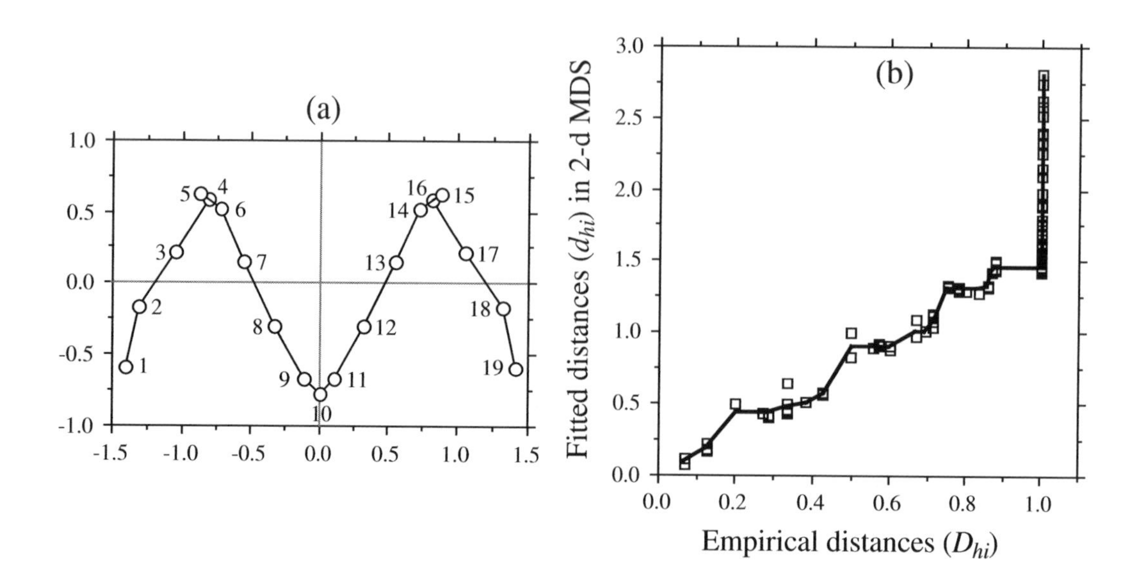

Figure 9.15 (a) MDS ordination (2-dimensional) of the D_{14} distance matrix in Table 9.9. Sampling sites are numbered as in Fig. 9.12 and Table 9.9. (b) Shepard diagram of the final solution showing the monotone regression line fitted by nonparametric regression. The scatter about the line is measured by a stress function (eq. 9.28 to 9.30).

Ecological application 9.3

Sprules (1980) used nonmetric multidimensional scaling to analyse seasonal changes in zooplankton assemblages at a site located in Lake Blelham, in the Lake District of northern England, and in two experimental enclosures built in that lake. The three sites were surveyed on a weekly basis from June to December 1976. MDS was preferred to PCA because the responses of species to environmental gradients could not be assumed to be linear.

For each site, points in the MDS ordination diagram were connected in chronological order to reflect the seasonal changes in faunal composition. The plot (not reproduced here) is therefore of the same type as Fig. 12.23. In one of the enclosures, the assemblage oscillated about a mean value without any clear cycle; small-size species dominated the assemblage. In the other enclosure and the lake, changes were more directional; at these sites, predators were more abundant. Based on available evidence, Sprules concluded that the differences observed between the two patterns of seasonal change were related to differences in predation intensity, quality of food available to herbivores, and nutrient dynamics.

Several ecological applications of nonmetric multidimensional scaling are found in the ecological literature. Two papers are especially interesting: Whittington & Hughes (1972; Ordovician biogeography from the analysis of trilobite fauna), and Fasham (1977; comparison of nonmetric multidimensional scaling, principal coordinate analysis and correspondence analysis for the ordination of simulated coenoclines and coenoplanes). Ecological application 12.6a features a MDS plot.

9.4 Correspondence analysis (CA)

Correspondence analysis (CA) was developed independently by several authors. It was first proposed for the analysis of contingency tables by Hirschfeld (1935), Fisher (1940), Benzécri (1969), and others. In a historical review of the subject, Nishisato (1980) traces its origin back to 1933. It was applied in ecology to the analysis of sites × species tables by Roux & Roux (1967), Hatheway (1971), Ibanez & Séguin (1972), Hill (1973b, 1974), Orlóci (1975), and others. Its use was generalized to other types of data tables by Benzécri and his collaborators (Escofier-Cordier, 1969; Benzécri and coll., 1973). Other important books on correspondence analysis are those of Nishisato (1980), Greenacre (1983), ter Braak (1988), and van Rijckevorsel & de Leeuw (1988). In the course of its history, the method was successively designated under the English names *contingency table analysis* (Fisher, 1940), RQ-*technique* (Hatheway, 1971), *reciprocal averaging* (Hill, 1973b), *correspondence analysis* (Hill, 1974), *reciprocal ordering* (Orlóci, 1975), *dual scaling* (Nishisato, 1980), and *homogeneity analysis* (Meulman, 1982), while it is known in French as *analyse factorielle des correspondances* (Cordier, 1965; Escofier-Cordier, 1969).

Contingency table

Correspondence analysis was first proposed for analysing two-way contingency tables (Section 6.4). In such tables, the states of a first descriptor (rows) are compared to the states of a second descriptor (columns). Data in each cell of the table are frequencies, i.e. numbers of objects coded with a combination of states of the two descriptors. These frequencies are positive integers or zeros. The most common application of CA in ecology is the analysis of species data (presence-absence or abundance values) at different sampling sites (Subsection 4). The rows and columns of the data table then correspond to sites and species, respectively. Such a table is analogous to a contingency table because the data are frequencies.

Frequencies

In general, correspondence analysis may be applied to any data table that is dimensionally homogeneous (i.e. the physical dimensions of all variables are the same; Chapter 3) and only contains positive integers or zeros. The χ^2 distance (D_{16}), which is a coefficient excluding double-zeros (Chapter 7), is used to quantify the relationships among rows and columns (Table 9.1).

Correspondence analysis can also be conducted on contingency tables that compare two *groups* of descriptors. For example, the rows of the table could be different species, each divided into a few classes of abundance, and the columns, different descriptors of the physical environment with, for each, a number of columns equal to the number of its states. Each site (object) then contributes to several frequencies of the table, but this does not invalidate the results because of the transformations described in the next subsection. A better way of comparing species to environmental data is canonical correspondence analysis (CCA, Section 11.2). CCA does not require that the species and environmental data be recoded into a few classes.

Correspondence analysis is primarily a method of ordination. As such, it is similar to principal component analysis; it preserves, in the space of the principal axes (i.e. after rotation), the Euclidean distance between *profiles of weighted conditional probabilities*. This is equivalent to preserving the χ^2 distance (D_{16}, eq. 7.52) between the rows or columns of the contingency table. Relationships between correspondence analysis and principal component analysis will be further described in the next subsections.

1 — Computation

This description of correspondence analysis will proceed in three steps. (1) First, the contingency table will be transformed into a table of contributions to the Pearson chi-square statistic after fitting a null model to the contingency table. (2) Singular value decomposition will be applied to that table and the eigenvalues and eigenvectors will be computed, as in PCA. (3) Further matrix operations will lead to the various tables needed for plotting useful diagrams. Besides its role as an ordination method, CA may be used for studying the proximities between the rows (or the columns) of the contingency table, as well as the correspondence between rows and columns.

Consider a contingency table with r rows and c columns, as in Section 6.2. Assume that the table is written in such a way that $r \geq c$; the table may be transposed to meet this condition, since the rows and columns of a contingency table play identical roles. Symbolism is as follows:

- Absolute frequencies are represented by f_{ij} and relative frequencies ("probabilities" or "proportions") by p_{ij}.

- p_{ij} is the frequency f_{ij} in cell ij divided by the sum f_{++} of the f_{ij}'s over the whole table. The table containing the relative frequencies p_{ij} is called $\mathbf{Q}$; its size is $(r \times c)$.

- Row weight p_{i+} is equal to f_{i+}/f_{++}, where f_{i+} is the sum of values in row i. Vector $[p_{i+}]$ is of size (r).

- Likewise, column weight p_{+j} is equal to f_{+j}/f_{++}, where f_{+j} is the sum of values in column j. Vector $[p_{+j}]$ is of size (c).

The steps are as follows:

1) The Pearson chi-square statistic, χ^2_P (eq. 6.5), is a sum of squared χ_{ij} values, computed for every cell ij of the contingency table. Each χ_{ij} value is the standardized residual of a frequency f_{ij} after fitting a null model to the contingency table. The null model states that there is no relationship between the rows and columns of the table (eq. 6.4). Simple algebra shows that the χ_{ij} value for each cell is:

Contribution to chi-square

$$\chi_{ij} = \frac{O_{ij} - E_{ij}}{\sqrt{E_{ij}}} = \sqrt{f_{++}} \left[\frac{p_{ij} - p_{i+}p_{+j}}{\sqrt{p_{i+}p_{+j}}} \right] \tag{9.31}$$

Correspondence analysis is based upon a matrix called $\bar{\mathbf{Q}}$ $(r \times c)$ in this book:

$$\bar{\mathbf{Q}} = [\bar{q}_{ij}] = \left[\frac{p_{ij} - p_{i+}p_{+j}}{\sqrt{p_{i+}p_{+j}}}\right] \qquad \textbf{(9.32)}$$

Values $\bar{q}_{ij}$, which are at the basis of correspondence analysis, only differ from the χ_{ij} values by a constant, so that $\bar{q}_{ij} = \chi_{ij} / \sqrt{f_{++}}$. This difference causes all the eigenvalues to be smaller than or equal to 1, as will be shown below. Values $\bar{q}_{ij}$ could also be calculated directly from the f_{ij}'s:

$$\bar{q}_{ij} = \frac{f_{ij}f_{++} - f_{i+}f_{+j}}{f_{++}\sqrt{f_{i+}f_{+j}}} \qquad \textbf{(9.33)}$$

The sum of squares of all values in matrix $\bar{\mathbf{Q}}$, $\Sigma \bar{q}_{ij}^2$, measures the total inertia in $\bar{\mathbf{Q}}$. It is also equal to the sum of all eigenvalues to be extracted by eigenanalysis of $\bar{\mathbf{Q}}$.

SVD

2) Singular value decomposition (SVD, eq. 2.31) is applied to matrix $\bar{\mathbf{Q}}$, with the following result (symbolism is slightly modified compared to Section 2.11):

$$\bar{\mathbf{Q}}\,(r \times c) = \hat{\mathbf{U}}\,(r \times c)\;\mathbf{W}(\text{diagonal}, c \times c)\;\mathbf{U}'(c \times c) \qquad \textbf{(9.34)}$$

where both $\mathbf{U}$ and $\hat{\mathbf{U}}$ are column-orthonormal matrices (i.e. matrices containing column vectors that are normalized and orthogonal to one another; Section 4.4) and $\mathbf{W}$ is a diagonal matrix $\mathbf{D}(w_i)$. The diagonal values w_i in $\mathbf{W}$, which are non-negative, are the singular values of $\bar{\mathbf{Q}}$.

Because $\bar{\mathbf{Q}} = \hat{\mathbf{U}}\mathbf{W}\mathbf{U}'$ (eq. 9.34), the multiplication $\bar{\mathbf{Q}}'\bar{\mathbf{Q}}$ gives the following result:

$$\bar{\mathbf{Q}}'\bar{\mathbf{Q}}\,(c \times c) = \mathbf{U}\mathbf{W}'\,(\hat{\mathbf{U}}'\hat{\mathbf{U}})\,\mathbf{W}\,\mathbf{U}' \qquad \textbf{(9.35)}$$

Since $\hat{\mathbf{U}}$ is orthonormal, $\hat{\mathbf{U}}'\hat{\mathbf{U}} = \hat{\mathbf{U}}\hat{\mathbf{U}}' = \mathbf{I}$, so that:

$$\bar{\mathbf{Q}}'\bar{\mathbf{Q}} = \mathbf{U}\,\mathbf{W}'\mathbf{W}\,\mathbf{U}' \qquad \textbf{(9.36)}$$

Equation 2.28 shows that the eigenvalues (forming diagonal matrix $\mathbf{\Lambda}$) and eigenvectors (matrix $\mathbf{U}$) of a square matrix $\mathbf{A}$ obey the relationship:

$$\mathbf{A} = \mathbf{U}\mathbf{\Lambda}\mathbf{U}^{-1}$$

If the vectors in $\mathbf{U}$ are normalized, $\mathbf{U}$ is an orthonormal matrix with the property $\mathbf{U}^{-1} = \mathbf{U}'$. As a consequence, eq. 2.28 may be rewritten as

$$\mathbf{A} = \mathbf{U}\mathbf{\Lambda}\mathbf{U}' \qquad \textbf{(9.37)}$$

It follows that the diagonal matrix $[\mathbf{W}'\mathbf{W}]$, which contains squared singular values on its diagonal, is the diagonal matrix $\mathbf{\Lambda}(c \times c)$ of the eigenvalues of $\bar{\mathbf{Q}}'\bar{\mathbf{Q}}$. Similarly, the orthonormal matrix **U** of eqs. 9.35 and 9.36 is the same as matrix **U** of eq. 9.37; it is the matrix of eigenvectors of $\bar{\mathbf{Q}}'\bar{\mathbf{Q}}$ $(c \times c)$, containing the loadings of the *columns* of the contingency table. A similar reasoning applied to matrix $\bar{\mathbf{Q}}\bar{\mathbf{Q}}'$ $(r \times r)$ shows that the orthonormal matrix $\hat{\mathbf{U}}$ produced by singular value decomposition is the matrix of eigenvectors of $\bar{\mathbf{Q}}\bar{\mathbf{Q}}'$, containing the loadings of the *rows* of the contingency table.

The relationship between eq. 9.34 and eigenvalue decomposition (eq. 2.22) is the same as in principal component analysis (Subsection 9.1.9). Prior to eigenvalue decomposition, a square matrix of sums of squares and cross products $\bar{\mathbf{Q}}'\bar{\mathbf{Q}}$ must be computed; this matrix may be called a "covariance matrix" in some general sense. This is similar to using matrix **Y'Y** for eigenvalue decomposition in PCA; **Y'Y** is a covariance matrix **S** except for the division by $(n-1)$. In PCA, however, matrix **Y** was centred on the column means prior to computing **Y'Y** whereas, in CA, matrix $\bar{\mathbf{Q}}$ is centred by the operation $(O_{ij} - E_{ij})$ (eqs. 9.31 and 9.32). This operation does not make the sums of the rows and columns equal to zero.

Eigen-analysis

Results identical to those of SVD would be obtained by applying eigenvalue analysis (eq. 2.22 and 9.1), either to the covariance matrix $\bar{\mathbf{Q}}'\bar{\mathbf{Q}}$, which would produce the matrices of eigenvalues **Λ** and eigenvectors **U**, or to matrix $\bar{\mathbf{Q}}\bar{\mathbf{Q}}'$, which would provide the matrices of eigenvalues **Λ** and eigenvectors $\hat{\mathbf{U}}$. Actually, it is not necessary to repeat the eigenanalysis to obtain **U** and $\hat{\mathbf{U}}$, because:

$$\hat{\mathbf{U}}\,(r \times c) = \bar{\mathbf{Q}}\mathbf{U}\,\boldsymbol{\Lambda}^{-1/2} \qquad \textbf{(9.38)}$$

and

$$\mathbf{U}(c \times c) = \bar{\mathbf{Q}}'\hat{\mathbf{U}}\,\boldsymbol{\Lambda}^{-1/2} \qquad \textbf{(9.39)}$$

In the sequel, all matrices derived from **U** will be without a hat and all matrices derived from $\hat{\mathbf{U}}$ will bear a hat.

Singular value decomposition, or eigenvalue analysis of matrix $\bar{\mathbf{Q}}'\bar{\mathbf{Q}}$, always yields one null eigenvalue. This is due to the centring in eq. 9.32, where $(p_{i+}\, p_{+j})$ is subtracted from each value p_{ij}. Thus, there are $(c-1)$ positive eigenvalues when $r \geq c$, so that the part of matrix **U** which is considered for interpretation is of size $c{\times}(c{-}1)$. Likewise, the part of $\hat{\mathbf{U}}$ which is considered is of size $r{\times}(c{-}1)$.

The analysis, by either SVD or eigenvalue decomposition, is usually performed on matrix $\bar{\mathbf{Q}}$ with $r \geq c$, for convenience. The reason is that not all SVD programs can handle matrices with $r < c$. In addition, when using eigenanalysis programs, computations are shorter when performed on the smallest of the two possible covariance matrices, both solutions leading to identical results. If one proceeds from a matrix such that $r < c$, the first $r{-}1$ eigenvalues are the same as in the analysis of the transposed matrix, the remaining eigenvalues being zero.

Consider now the uncentred matrix $\tilde{\mathbf{Q}}\,(r \times c)$, where $(p_{i+}\, p_{+j})$ is not subtracted from each term p_{ij} in the numerator:

$$\tilde{\mathbf{Q}} = [\tilde{q}_{ij}] = \left[\frac{p_{ij}}{\sqrt{p_{i+}p_{+j}}}\right] \qquad \textbf{(9.40)}$$

What would happen if the analysis was based on matrix $\tilde{\mathbf{Q}}$ instead of $\bar{\mathbf{Q}}$ (eq. 9.32)? The only difference is that using $\tilde{\mathbf{Q}}$ would produce one extra eigenvalue; all the other results would be identical. This extra eigenvalue is easy to recognize because its value is 1 in correspondence analysis. This eigenvalue is meaningless because it only reflects the distance between the centre of mass of the data points in the ordination space and the origin of the system of axes. In other words, it corresponds to the lack of centring of the scatter of points on the origin (Hill, 1974); it explains none of the dispersion (Lebart & Fénelon, 1971). There are computer programs that do not make the centring; the first eigenvalue ($\lambda_1 = 1$) and eigenvector thus obtained must be discarded. All other programs, that carry out the calculations on matrix $\bar{\mathbf{Q}}$, produce one eigenvalue less than min$[r, c]$; if the data table $\mathbf{Q}$ is such that $r \geq c$, correspondence analysis yields $(c-1)$ non-null and positive eigenvalues.

Alternatively, what would happen if the analysis was based on the matrix of χ_{ij} values (eq. 9.31) instead of matrix $\bar{\mathbf{Q}}$? Since values $\chi_{ij} = \sqrt{f_{++}}\,\bar{q}_{ij}$ it follows that the total variance in matrix $[\chi_{ij}]$ is larger than that of matrix $\bar{\mathbf{Q}}$ by a factor $(\sqrt{f_{++}})^2 = f_{++}$, so that all eigenvalues of matrix $[\chi_{ij}]$ would be larger than those of $\bar{\mathbf{Q}}$ by a factor f_{++}. The normalized eigenvectors in matrices $\mathbf{U}$ and $\hat{\mathbf{U}}$ would remain unaffected. When the analysis is carried out on matrix $\bar{\mathbf{Q}}$, all eigenvalues are smaller than or equal to 1, which is more convenient.

Joint plot

3) Matrices $\mathbf{U}$ and $\hat{\mathbf{U}}$ may be used to plot the positions of the row and column vectors in two separate scatter diagrams. For *joint plots*, various scalings of the row and column scores have been proposed. First, matrices $\mathbf{U}$ and $\hat{\mathbf{U}}$ can be weighted by the inverse of the square roots of the column and row scores, written out in diagonal matrices $\mathbf{D}(p_{+j})^{-1/2}$(size $c \times c$) and $\mathbf{D}(p_{i+})^{-1/2}$(size $r \times r$), respectively:

$$\mathbf{V}(c \times c) = \mathbf{D}(p_{+j})^{-1/2}\,\mathbf{U} \qquad \textbf{(9.41)}$$

$$\hat{\mathbf{V}}(r \times c) = \mathbf{D}(p_{i+})^{-1/2}\,\hat{\mathbf{U}} \qquad \textbf{(9.42)}$$

Discarding the null eigenvalue, the part of matrix $\mathbf{V}$ to consider for interpretation is of size $c \times (c-1)$ and the part of matrix $\hat{\mathbf{V}}$ to consider is of size $r \times (c-1)$.

Matrix $\mathbf{F}$, which gives the positions of the *rows* of the contingency table in the correspondence analysis space, is obtained from the transformed matrix of eigenvectors $\mathbf{V}$, which gives the positions of the *columns* in that space. This is done by applying the usual equation for component scores (eq. 9.4) to data matrix $\mathbf{Q}$, with division by the row weights:

$$\mathbf{F}(r \times c) = \hat{\mathbf{V}}\,\mathbf{\Lambda}^{1/2} \qquad \textbf{(9.43a)}$$

or

$$\mathbf{F}(r \times c) = \mathbf{D}(p_{i+})^{-1}\mathbf{Q}\mathbf{V} \qquad \textbf{(9.43b)}$$

In the same way, matrix $\hat{\mathbf{F}}$, which gives the positions of the *columns* of the contingency table in the correspondence analysis space, is obtained from the transformed matrix of eigenvectors $\hat{\mathbf{V}}$, which gives the positions of the *rows* in that space. The equation is the same as above, except that division here is by the column weights:

$$\hat{\mathbf{F}}\,(c{\times}c) = \mathbf{V}\boldsymbol{\Lambda}^{1/2} \qquad \textbf{(9.44a)}$$

or

$$\hat{\mathbf{F}}\,(c{\times}c) = \mathbf{D}(\mathrm{p}_{+j})^{-1}\,\mathbf{Q}'\hat{\mathbf{V}} \qquad \textbf{(9.44b)}$$

Discarding the null eigenvalue, the part of matrix **F** to consider for interpretation is of size $r{\times}(c{-}1)$ and the part of matrix $\hat{\mathbf{F}}$ to consider is of size $c{\times}(c{-}1)$. With this scaling, matrices **F** and **V** form a pair such that the rows (given by matrix **F**) are at the centroid (centre of mass or "barycentre", from the Greek βαρυς, pronounced "barus", heavy) of the columns in matrix **V**. In the same way, matrices $\hat{\mathbf{F}}$ and $\hat{\mathbf{V}}$ form a pair such that the columns (given by matrix $\hat{\mathbf{F}}$) are at the centroids of the rows in matrix $\hat{\mathbf{V}}$. This property is illustrated in the numerical example below.

Scatter diagrams may be drawn using different combinations of the matrix scalings described above. Scaling types 1 and 2 are the most commonly used by ecologists when analysing species presence-absence or abundance data (ter Braak, 1990).

Scalings in CA

• Scaling type 1 — Draw a joint plot with the rows (matrix **F**) at the centroids of the columns (matrix **V**). For sites × species data tables where sites are rows and species are columns, this scaling is the most appropriate if one is primarily interested in the ordination of sites. In matrix **F**, distances among sites preserve their χ^2 distances (D_{16}) (ter Braak, 1987c; Numerical example, Subsection 2).

• Scaling type 2 — Draw a joint plot with the columns (matrix $\hat{\mathbf{F}}$) at the centroids of the rows (matrix $\hat{\mathbf{V}}$). This scaling puts the species at the centroids of sites in the graph. For sites × species data tables where sites are rows and species are columns, this is the most appropriate scaling if one is primarily interested in the relationships among species. This is because, in matrix $\hat{\mathbf{F}}$, distances among species preserve their χ^2 distances (see Numerical example, Subsection 2).

• Scaling type 3 — Assuming that the rows of the data matrix are objects (sites) and the columns are descriptors (species), *use separate diagrams* to plot matrix **U** for the column scores and matrix **F** for the row scores. The eigenvectors in matrix **U** are normalized, as in PCA. The scaling of **F** is such that the Euclidean distances among the rows of **F** are equal to the χ^2 distances (D_{16}) among objects of the original data table; this property is illustrated in the example below. This is often the only choice available in correspondence analysis programs that are not ecologically oriented. This scaling is not appropriate for joint plots. [When the data matrix subjected to the analysis has the variables (species) as rows and the objects (sites) as columns, one should plot, *in separate diagrams*, matrix $\hat{\mathbf{U}}$ for the row scores and matrix $\hat{\mathbf{F}}$ for the column scores in order to obtain the same representation as above. With this type of input, matrices **U** and **F** are meaningless because **U** contains normalized site scores which do not have any interesting distance-preservation properties.]

Table 9.11 Numerical example. Contingency table between two descriptors (roman type). Alternatively, the Table could be a site-by-species data table (italics).

Descr. 2 – The species is:		Rare or absent (0) *Species 1*	Abundant (+) *Species 2*	Very abundant (++) *Species 3*	**Row sums**
Descr. 1 Temperature	*Sites*				
Cold (1)	*1*	10	10	20	40
Medium (2)	*2*	10	15	10	35
Warm (3)	*3*	15	5	5	25
Column sums		35	30	35	100

Other possible, but less often used scaling methods are discussed by ter Braak (1987c, 1990), including CANOCO's* scaling method 3 which is offered as an intermediate between the first two scaling options described above.

2 — *Numerical example*

The following example illustrates the calculations involved in correspondence analysis. In this numerical example, a species (abundance score, 3 classes) has been observed at 100 sites. The temperature (or any other environmental factor) at each site is coded from 1 (cold) to 3 (warm). The contingency table (Table 9.11) contains the number of sites at which each combination of the two descriptors was encountered. Subsection 4 will show that the same calculations may be conducted on a site $\times$ species data table, so that Table 9.11 can alternatively be seen as a numerical example for the latter, as indicated in italics in the Table. The data table is of small size (3×3), so as to allow readers to repeat the calculations.

Matrix **Q** contains the proportions p_{ij} and the marginal distributions p_{i+} and p_{+j} of the rows and columns, respectively. Identifiers of the rows and columns are given in parentheses, following Table 9.11:

$$\begin{array}{lcc} & (0)\ (+)\ (++) & [p_{i+}] \\ \mathbf{Q} = [p_{ij}] = \begin{array}{c}(1)\\(2)\\(3)\end{array} & \begin{bmatrix} 0.10 & 0.10 & 0.20 \\ 0.10 & 0.15 & 0.10 \\ 0.15 & 0.05 & 0.05 \end{bmatrix} & \begin{bmatrix} 0.40 \\ 0.35 \\ 0.25 \end{bmatrix} \\ [p_{+j}] = & \begin{bmatrix} 0.35 & 0.30 & 0.35 \end{bmatrix} & \end{array}$$

* CANOCO is widely used for PCA, CA, and canonical analysis. See note at end of Section 11.0.

Matrix $\bar{\mathbf{Q}}$ is computed following eq. 9.32:

$$\bar{\mathbf{Q}} = [\bar{q}_{ij}] = \left[\frac{p_{ij} - p_{i+}p_{+j}}{\sqrt{p_{i+}p_{+j}}}\right] = \begin{array}{c} \\ (1) \\ (2) \\ (3) \end{array} \begin{array}{c} \begin{array}{ccc} (0) & (+) & (++) \end{array} \\ \begin{bmatrix} -0.10690 & -0.05774 & 0.16036 \\ -0.06429 & 0.13887 & -0.06429 \\ 0.21129 & -0.09129 & -0.12677 \end{bmatrix} \end{array}$$

and matrix $\tilde{\mathbf{Q}}$ following eq. 9.40:

$$\tilde{\mathbf{Q}} = [\tilde{q}_{ij}] = \left[\frac{p_{ij}}{\sqrt{p_{i+}p_{+j}}}\right] = \begin{array}{c} \\ (1) \\ (2) \\ (3) \end{array} \begin{array}{c} \begin{array}{ccc} (0) & (+) & (++) \end{array} \\ \begin{bmatrix} 0.26726 & 0.28868 & 0.53452 \\ 0.28571 & 0.46291 & 0.28571 \\ 0.50709 & 0.18257 & 0.16903 \end{bmatrix} \end{array}$$

The eigenvalues of $\bar{\mathbf{Q}}'\bar{\mathbf{Q}}$ are $\lambda_1 = 0.09613$ (70.1%), $\lambda_2 = 0.04094$ (29.9%), and $\lambda_3 = 0$ (because of the centring). The first two eigenvalues are also eigenvalues of $\tilde{\mathbf{Q}}'\tilde{\mathbf{Q}}$, its third eigenvalue being 1 (because $\tilde{\mathbf{Q}}$ is not centred; eq. 9.40). The normalized eigenvectors of $\bar{\mathbf{Q}}'\bar{\mathbf{Q}}$, corresponding to λ_1 and λ_2, are (in columns):

$$\mathbf{U} = \begin{array}{c} \\ (0) \\ (+) \\ (++) \end{array} \begin{array}{c} \begin{array}{cc} (\lambda_1) & (\lambda_2) \end{array} \\ \begin{bmatrix} 0.78016 & -0.20336 \\ -0.20383 & 0.81145 \\ -0.59144 & -0.54790 \end{bmatrix} \end{array}$$

The normalized eigenvectors of $\bar{\mathbf{Q}}\bar{\mathbf{Q}}'$ are (in columns):

$$\hat{\mathbf{U}} = \begin{array}{c} \\ (0) \\ (+) \\ (++) \end{array} \begin{array}{c} \begin{array}{cc} (\lambda_1) & (\lambda_2) \end{array} \\ \begin{bmatrix} -0.53693 & -0.55831 \\ -0.13043 & 0.79561 \\ 0.83349 & -0.23516 \end{bmatrix} \end{array}$$

The third eigenvector is of no use and is therefore not given. Most programs do not compute it.

In scaling type 1 (Fig. 9.16a), the states in the rows of the data matrix (1, 2, 3, called "rows" hereinafter), whose coordinates will be stored in matrix **F**, are to be plotted at the centroids of the column states (0, +, ++, called "columns" hereinafter). The scaling for the columns is obtained using eq. 9.41:

$$\mathbf{V} = \mathbf{D}(p_{+j})^{-1/2}\,\mathbf{U} = \begin{array}{c} \\ (0) \\ (+) \\ (++) \end{array} \begin{array}{c} \begin{array}{cc} (\lambda_1) & (\lambda_2) \end{array} \\ \begin{bmatrix} 1.31871 & -0.34374 \\ -0.37215 & 1.48150 \\ -0.99972 & -0.92612 \end{bmatrix} \end{array}$$

To put the rows (matrix **F**) at the centroids of the columns (matrix **V**), the position of each row along an ordination axis is computed as the mean of the column positions, weighted by the

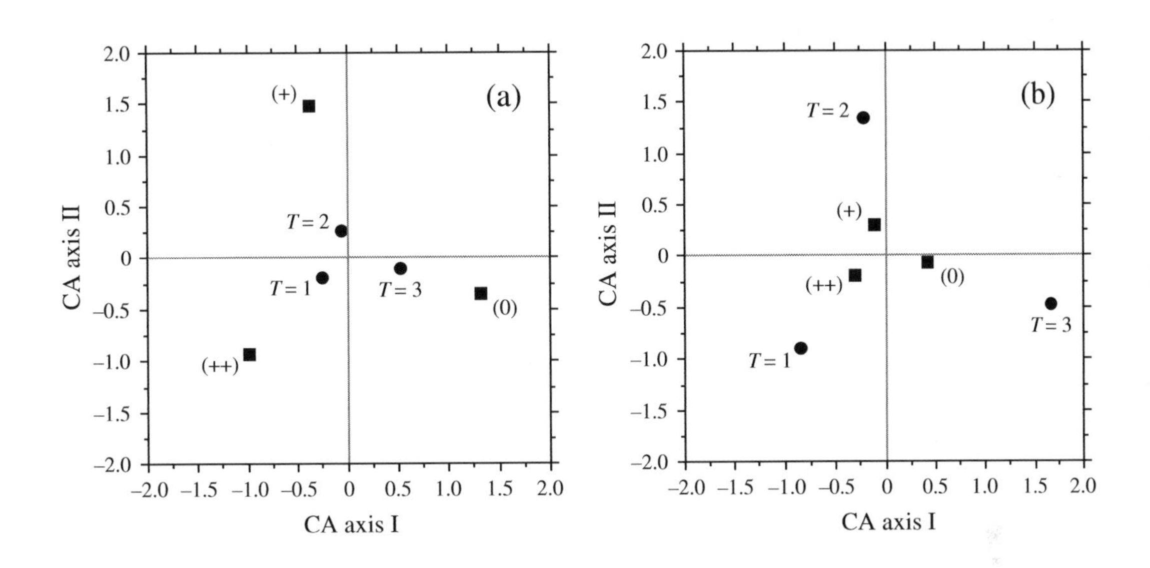

Figure 9.16 Correspondence analysis joint plots. (a) Scaling type 1: the rows of the contingency table (circles, matrix **F**) are at the centroids (barycentres) of the columns (squares, matrix **V**). (b) Scaling type 2: the columns (squares, matrix $\hat{\mathbf{F}}$) are at the centroids (barycentres) of the rows (circles, matrix $\hat{\mathbf{V}}$).

relative frequencies of the observations in the various columns of that row. Consider the first row of the data table (Table 9.11), for example. The relative frequencies (or *conditional probabilities*, Section 6.4) of the three columns in that row are 0.25, 0.25, and 0.50. Multiplying matrix **V** by that vector provides the coordinates of the first row in the ordination diagram:

$$[0.25\ \ 0.25\ \ 0.50]\begin{bmatrix} 1.31871 & -0.34374 \\ -0.37215 & 1.48150 \\ -0.99972 & -0.92612 \end{bmatrix} = [-0.26322\ \ -0.17862]$$

These coordinates put the first row at the centroid of the columns in Fig. 9.16a; they are stored in the first row of matrix **F**. The row-conditional probabilities for the whole data table are found using the matrix operation $\mathbf{D}(p_{i+})^{-1}\mathbf{Q}$, so that matrix **F** is computed using eq. 9.43b:

$$\mathbf{F} = \mathbf{D}(p_{i+})^{-1}\mathbf{QV} = \begin{matrix} \\ (1) \\ (2) \\ (3) \end{matrix} \begin{matrix} \begin{matrix} (\lambda_1) & \quad (\lambda_2) \end{matrix} \\ \begin{bmatrix} -0.26322 & -0.17862 \\ -0.06835 & 0.27211 \\ 0.51685 & -0.09517 \end{bmatrix} \end{matrix}$$

Using the formulae for the Euclidean (D_1, eq. 7.34) and χ^2 (D_{16}, eq. 7.54) distances, one can verify that the Euclidean distances among the rows of matrix **F** are equal to the χ^2 distances among the rows of the original data table (Table 9.11):

$$\mathbf{D} = \begin{matrix} (1) \\ (2) \\ (3) \end{matrix} \overset{\begin{matrix} (1) & (2) & (3) \end{matrix}}{\begin{bmatrix} 0 & & \\ 0.49105 & 0 & \\ 0.78452 & 0.69091 & 0 \end{bmatrix}}$$

Matrix **F** thus provides a proper ordination of the rows of the original data matrix (temperatures in the numerical example).

In scaling type 2 (Fig. 9.16b), the columns, whose coordinates will be stored in matrix $\hat{\mathbf{F}}$, are to be plotted at the centroids of the rows (matrix $\hat{\mathbf{V}}$). The scaling for matrix $\hat{\mathbf{V}}$ is obtained using eq. 9.42:

$$\hat{\mathbf{V}} = \mathbf{D}(\mathrm{p}_{i+})^{-1/2}\,\hat{\mathbf{U}} = \begin{matrix} (1) \\ (2) \\ (3) \end{matrix} \overset{\begin{matrix} (\lambda_1) & (\lambda_2) \end{matrix}}{\begin{bmatrix} -0.84896 & -0.88276 \\ -0.22046 & 1.34482 \\ 1.66697 & -0.47032 \end{bmatrix}}$$

To put the columns (matrix $\hat{\mathbf{F}}$) at the centroids of the rows (matrix $\hat{\mathbf{V}}$), the position of each column along an ordination axis is computed as the mean of the row positions, weighted by the relative frequencies of the observations in the various rows of that column. Consider the first column of the data table (Table 9.11), for example. The relative frequencies of the three rows in that column are (10/35 = 0.28571), (10/35 = 0.28571) and (15/35 = 0.42857). Multiplying matrix $\hat{\mathbf{V}}$ by that vector provides the coordinates of the first column in the ordination diagram:

$$[0.28571\;\;0.28571\;\;0.42857] \begin{bmatrix} -0.84896 & -0.88276 \\ -0.22046 & 1.34482 \\ 1.66697 & -0.47032 \end{bmatrix} = [0.40887\;\;-0.06955]$$

These coordinates put the first column at the centroid of the rows in Fig. 9.16a; they are stored in the first row of matrix $\hat{\mathbf{F}}$. The column-conditional probabilities for the whole data table are found using the matrix operation $\mathbf{D}(\mathrm{p}_{+j})^{-1}\mathbf{Q}'$, so that matrix $\hat{\mathbf{F}}$ is computed using eq. 9.44a or 9.44b:

$$\hat{\mathbf{F}} = \mathbf{V}\mathbf{\Lambda}^{1/2} = \mathbf{D}(\mathrm{p}_{+j})^{-1}\mathbf{Q}'\hat{\mathbf{V}} = \begin{matrix} (0) \\ (+) \\ (++) \end{matrix} \overset{\begin{matrix} (\lambda_1) & (\lambda_2) \end{matrix}}{\begin{bmatrix} 0.40887 & -0.06955 \\ -0.11539 & 0.29977 \\ -0.30997 & -0.18739 \end{bmatrix}}$$

Using the formulae for the Euclidean (D_1, eq. 7.34) and χ^2 (D_{16}, eq. 7.54) distances, one can verify that the Euclidean distances among the rows of matrix $\hat{\mathbf{F}}$ are equal to the χ^2 distances among the columns of the original data table (Table 9.11):

$$\mathbf{D} = \begin{matrix} & (0) & (+) & (++) \\ (0) & 0 & & \\ (+) & 0.64128 & 0 & \\ (++) & 0.72843 & 0.52458 & 0 \end{matrix}$$

Matrix $\hat{\mathbf{F}}$ thus provides a proper ordination of the columns of the original data matrix (species abundance classes in the numerical example).

3 — Interpretation

The relationship between matrices $\mathbf{V}$ and $\hat{\mathbf{V}}$, which provide the ordinations of the columns and rows of the contingency (or species data) table, respectively, is found by combining eqs. 9.38, 9.41, and 9.42 in the following expression:

$$\hat{\mathbf{V}}\,\mathbf{\Lambda}^{1/2} = \mathbf{D}(\mathrm{p}_{i+})^{-1/2}\bar{\mathbf{Q}}\,\mathbf{D}(\mathrm{p}_{+j})^{1/2}\mathbf{V} \qquad \textbf{(9.45)}$$

This equation means that the ordination of the rows (matrix $\hat{\mathbf{V}}$) is related to the ordination of the columns (matrix $\mathbf{V}$), along principal axis h, by the value $\sqrt{\lambda_h}$ which is a measure of the "correlation" between these two ordinations. Value $(1 - \lambda_h)$ actually measures the difficulty of ordering, along principal axis h, the rows of the contingency table from an ordination of the columns, or the converse (Orlóci, 1978). The highest eigenvalue (0.096 in the above numerical example), or its square root ($\sqrt{\lambda_1} = 0.31$), is consequently a measure of the dependence between two unordered descriptors, to be added to the measures described in Chapter 6. Williams (1952) discusses different methods for testing the significance of $R^2 = \lambda$.

Examination of joint plots (e.g. Fig. 9.16) allows one to draw conclusions about the ecological relationships displayed by the data. With scaling type 1, (a) the distances among rows (or sites in the case of a species × sites data table) in reduced space approximate their χ^2 distances and (b) the rows (sites) are at the centroids of the columns (species). Positions of the centroids are calculated using weights equal to the relative frequencies of the columns (species); columns (species) that are absent from a row (site) have null weights and do not contribute to the position of that row (site). Thus, the ordination of rows (sites) is meaningful. In addition, any row (site) found near the point representing a column (species) is likely to have a high contribution of that column (species); for binary (or species presence-absence) data, the row (site) is more likely to possess the state of that column (or contain that species).

With scaling type 2, it is the distances among columns (species) in reduced space that approximate their χ^2 distances, whereas columns (species) are at the centroids of the rows (sites). Consequently (a), the ordination of columns (species) is meaningful, and (b) any column (species) that lies close to the point representing a row (site) is more likely to be found in the state of that row (site), or with higher frequency (abundance) than in rows (sites) that are further away in the joint plot.

For species presence-absence or abundance data, insofar as a species has a unimodal (i.e. bell-shaped) response curve along the axes of ecological variation corresponding to the ordination axes, the optimum for that species should be close to the point representing it in the ordination diagram and its frequency of occurrence or abundance should decrease with distance from that point. Species that are absent at most sites often appear at the edge of the scatter plot, near the point representing a site where they happen to be present — by chance, or because they are favoured by some rare condition occurring at that site. Such species have little influence on the analysis because their numerical contributions are small (column sums in Table 9.11). Finally, species that lie near the centre of the ordination diagram may have their optimum in this area of the plot, or have two or several optima (bi- or multi-modal species), or else be unrelated to the pair of ordination axes under consideration. Species of this last group may express themselves along some other axis or axes. Close examination of the raw data table may be required in this case. It is the species found away from the centre of the diagram, but not near the edges, that are the most likely to display clear relationships with the ordination axes (ter Braak, 1987c).

In Fig. 9.16 (a and b), the first CA axis (70.1% of the variance) orders the abundances in a direction opposite to that of temperatures. Both graphs associate abundance (0) to the highest temperature (3), abundance (+) to the intermediate temperature (2), and abundance (++) to the lowest temperature (1). An analysis of the correspondence between rows and columns of the contingency table following the methods described in Section 6.4 would have shown the same relationships.

4 — Site × species data tables

Correspondence analysis has been applied to data tables other than contingency tables. Justification is provided by Benzécri and coll. (1973). Notice, however, that the elements of a table to be analysed by correspondence analysis must be *dimensionally homogeneous* (i.e. same physical units, so that they can be added) and *non-negative* (≥ 0, so that they can be transformed into probabilities or proportions). Several data sets already have these characteristics, such as (bio)mass values, concentrations, financial data (in \$, £, etc.), or species abundances.

Other types of data may be recoded to make the descriptors dimensionally homogeneous and positive; the most widely used data transformations are discussed in Section 1.5. For descriptors with different physical units, the data may, for example, be standardized (which makes them dimensionless; eq. 1.12) and made positive by translation, i.e. by subtracting the highest negative value; or divided by the maximum or by the range of values (eqs. 1.10 and 1.11). Data may also be recoded into ordered classes. Regardless of the method, recoding is then a critical step of correspondence analysis. Consult Benzécri and coll. (1973) on this matter.

Several authors, mentioned at the beginning of this Section, have applied correspondence analysis to the analysis of site × species matrices containing species presence/absence or abundance data. This generalization of the method is based on the following sampling model. If sampling had been designed in such a way as to collect individual organisms (which is usually not the case, the sampled elements being, most often, sampling sites), each organism could be described by two descriptors: the site where it was collected and the taxon to which it belongs. These two descriptors may be written out to an *inflated data table* which has as many rows as there are individual organisms. The more familiar site × species data table would then be the contingency table resulting from crossing the two descriptors, sites and taxa. It could be analysed using any of the methods applicable to contingency tables. Most methods involving tests of statistical significance cannot be used, however, because the hypothesis of independence of the individual organisms, following the model described above, is not met by species presence-absence or abundance data collected at sampling sites.

Inflated data table

Niche theory tells us that species have ecological preferences, meaning that they are found at sites where they encounter favourable conditions. This statement is rooted in the idea that species have unimodal distributions along environmental variables (Fig. 9.12), more individuals being found near some environmental value which is "optimal" for the given species. This has been formalised by Hutchinson (1957) in his *fundamental niche* model. Furthermore, Gause's (1935) competitive exclusion principle suggests that, in their micro-evolution, species should have developed non-overlapping niches. These two principles indicate together that species should be roughly equally spaced in the n-dimensional space of resources. This model has been used by ter Braak (1985) to justify the use of correspondence analysis on presence-absence or abundance data tables; he showed that the χ^2 distance preserved through correspondence analysis (Table 9.1) is an appropriate model for species with unimodal distributions along environmental gradients.

Niche

Let us follow the path travelled by Hill (1973b), who rediscovered correspondence analysis while exploring the analysis of vegetation variation along environmental gradients; he called his method "reciprocal averaging" before realizing that this was correspondence analysis (Hill, 1974). Hill started from the simpler method of *gradient analysis*, proposed by Whittaker (1960, 1967) to analyse site × species data tables. Gradient analysis uses a matrix **Y** (site × species) and an initial vector **v** of values v_j which are ascribed to the various species j as indicators of the physical *gradient* to be evidenced. For example, a score (scale from 1 to 10) could be given to the each species for its preference with respect to soil moisture. These coefficients are used to calculate the positions of the sites along the gradient. The score $\hat{v}_i$ of a site i is calculated as the average score of the species ($j = 1 \dots p$) present at that site, using the formula:

Reciprocal averaging

$$\hat{v}_i = \frac{\sum_{j=1}^{p} y_{ij} v_j}{y_{i+}} \qquad \textbf{(9.46)}$$

where y_{ij} is the abundance of species j at site i and y_{i+} is the sum of the organisms at this site (i.e. the sum of values in row i of matrix **Y**).

Gradient analysis produces a vector $\hat{\mathbf{v}}$ of the positions of the sites along the gradient under study. Hill (1973b, 1974) suggested to continue the analysis, using now vector $\hat{\mathbf{v}}$ of the ordination of sites to compute a new ordination ($\mathbf{v}$) of the species:

$$v_j = \frac{\sum_{i=1}^{n} y_{ij}\hat{v}_i}{y_{+j}} \tag{9.47}$$

in which y_{+j} is the sum of values in column j of matrix **Y**. Alternating between $\mathbf{v}$ and $\hat{\mathbf{v}}$ (scaling the vectors at each step as shown in step 6 of Table 9.12) defines an iterative procedure that Hill (1973b) called "reciprocal averaging". This procedure converges towards a unique unidimensional ordination of the species and sites, which is independent of the values initially given to the v_j's; different initial guesses as to the values v_j may however change the number of steps required to reach convergence. Being aware of the work of Clint & Jennings (1970), Hill realized that he had discovered an eigenvalue method for gradient analysis, hence the title of his 1973b paper. It so happens that Hill's method produces the barycentred vectors $\mathbf{v}$ and $\hat{\mathbf{v}}$ for species and sites, that correspond to the first eigenvalue of a correspondence analysis. Hill (1973b) showed how to calculate the eigenvalue (λ) corresponding to these ordinations and how to find the other eigenvalues and eigenvectors. He thus created a simple algorithm for correspondence analysis (described in Subsection 7).

When interpreting the results of correspondence analysis, one should keep in mind that the simultaneous ordination of species and sites aims at determining how useful the ordination of species is, as a whole, for predicting the ordination of the sites. In other words, it seeks the predictive value of one ordination with respect to the other. Subsection 3 has shown that, for any given dimension h, $(1 - \lambda_h)$ measures the difficulty of ordering, along principal axis h, the row states of the contingency table from an ordination of the column states, or the converse. The interpretation of the relationship between the two ordinations must be done with reference to this statistic.

When it is used as an ordination method, correspondence analysis provides an ordination of the sites which is somewhat similar to that resulting from a principal component analysis of the correlation matrix among species (standardized data). This is to be expected since the first step in the calculation actually consists in weighting each datum by the sums (or the relative frequencies) of the corresponding row and column (eq. 9.32 and 9.33), which eliminates the effects due to the large variances that certain rows or columns may have. In the case of steep gradients (i.e. many zeros in the data matrix), correspondence analysis should produce a better ordination than PCA (Hill, 1973b). This was also shown by Gauch *et al.* (1977) using simulated and experimental floristic data. This result logically follows from the fact that the χ^2 distance (D_{16}) is a coefficient that excludes double-zeros from the estimation of resemblance. This is not the case with the Euclidean distance (eq. 7.33), which is the distance preserved in principal component analysis. For this reason, correspondence analysis is recommended for reduced-space ordination of species abundances when the data contain a large number of null values; this situation is encountered when sampling environmental gradients that are long enough for species to replace one another.

For clustering species into associations, correspondence analysis does not seem to escape the problems encountered with principal component analysis (Reyssac & Roux, 1972; Ibanez & Séguin, 1972; Binet *et al.*, 1972). Causes for this were discussed in Section 9.1. Undoubtedly, the most serious problem arises from the fact that the species are multidimensional descriptor-axes, which are projected in a low-dimensional space by both PCA and CA. This explains the tendency for the species to form a more or less uniformly dense scatter centred on the origin. It may nevertheless be interesting to superimpose a clustering of species, determined using the methods of Section 8.9, on a reduced-space ordination obtained by correspondence analysis.

When sites (objects) and species (descriptors) are plotted together, the joint plot must be interpreted with due consideration of the remarks in Subsection 9.1.4; species are the variables in the joint plot and a correspondence analysis of a site × species data table is but a variant of a principal component analysis. The practice which consists in only explaining the sites by the neighbouring species on the plot often gives good results, although it overlooks possible indications of avoidance of sites by certain species. An interesting complement to correspondence analysis is the direct analysis of the site × species table using the method of Section 6.4. This method is better at evidencing all the correspondences between sites and species (attraction and avoidance). Applying contingency table analysis to sites × species tables is justified by the same logic that allows correspondence analysis to be applied to such data matrices.

5 — Arch effect

Environmental gradients often support a succession of species. Since the species that are controlled by environmental factors (*versus* population dynamics, historical events, etc.) generally have unimodal distributions along gradients, the effect of gradients on the distance relationships among sites, calculated on species presence-absence or abundance data, is necessarily nonlinear. The three species in Fig. 9.12 have unimodal distributions; each one shows a well-defined mode along the gradient represented by sites 1 to 19. Ordination methods aim at rendering this non-linear phenomenon in an Euclidean space, in particular as two-dimensional plots. In such plots, non-linearities end up being represented by curves, called *arches* or *horseshoes*. The shapes of these curves depend on the distance function used by different ordination methods to model the relationships among sites. While most ecologists are content with interpreting the ordination plots for the information they display about distances among sites, some feel that they should try to reconstruct the original gradient underlying the observed data. Hence their concern with *detrending*, which is an operation carried out on the ordination axes of correspondence analysis whereby the arch is unbent to let the gradient appear as a linear arrangement of the sites.

Detrending

Imagine an alternative Fig. 9.12, representing environmental factors varying along the environmental gradient of sites 1 to 19, with some variables increasing linearly from left to right (e.g. altitude) and others decreasing (e.g. temperature, humidity). A principal component analysis of such data would clearly render the original gradient along PCA axis I, because the Euclidean distances (D_1) from site 1 to sites 2, 3, etc., which are implicit in PCA, would increase from one end of the transect to the other. The second axis would only display the residual variation in the data (the "error" component). Readers are invited to make up such a data set, adding random error to the data, and see for themselves what the result would be. Contrary to that, Euclidean distances calculated on the species data of Fig. 9.12, between site 1 and sites 2, 3, etc., do not increase monotonically from one end of the gradient to the other. These distances, which form the first row of the Euclidean distance matrix among sites, are reported in the bottom panel of Table 9.9. Distances from site 1 increase up to site 5, after which they decrease; they increase again up to site 10, then decrease; they increase up to site 15 and decrease again. The other rows of the Euclidean distance matrix display equally complex patterns; they are not shown in Table 9.9 to save space. The PCA algorithm is facing the task of representing these complex patterns in at most three dimensions because PCA ordinations cannot have more axes than the number of original variables (i.e. three species in Fig. 9.12). The result is illustrated in Fig. 9.17a and b. The most dramatic effect is found at the ends of the transect, which are folded inwards along axis I. This is because the Euclidean distance formula considers the extreme sites to be very near each other (small distance). This shape is called a *horseshoe*. Fig. 9.17b shows that the end sites also go "down" along the third axis. In most instances in correspondence analysis, extremities of the gradient are not folded inwards in the plot (but see Wartenberg *et al.*, 1987, Fig. 3, for a case where this occurs); a bent ordination plot with extremities not folded inwards is called an *arch*.

Horseshoe

Arch

The presence in ordination plots of a *bow* (Swan, 1970), *horseshoe* (Kendall, 1971), or *arch* (Gauch, 1982) had already been noted by plant ecologist Goodall (1954). Benzécri and coll. (1973) discuss the arch under the name *Guttman effect.* Several authors have explained the nature of this mathematical construct, which occurs when the taxonomic composition of the sites progressively changes along an environmental gradient. For correspondence analysis, ter Braak (1987c) discusses how this effect results from the fact that all ordination axes try to maximally separate the species while remaining uncorrelated with one another. When a single axis (the first one) is enough to order the sites and species correctly, a second axis, which is independent of the first, can be obtained by folding the first axis in the middle and bringing the ends together; hence the arch effect. Subsequent independent ordination axes can be obtained by folding the first axis in three parts, four, etc., until $\min[(r-1), (c-1)]$ axes have been produced. *Detrended correspondence analysis* (DCA; Hill & Gauch, 1980; Gauch, 1982) aims at eliminating the arch effect.

Fig. 9.17c helps in understanding the meaning of CA joint plots. This joint plot has been produced using scaling type 1 to preserve the χ^2 distances among sites and make this plot comparable to the other ordinations shown in Fig. 9.17. The ordination is two-dimensional since the data set contains three species. The species (black squares) occupy the edges of a triangle; since they are the descriptors of the analysis, heavy lines are drawn joining them to the centre of the plot. Sites 1-5, 10, and 15-19, which have only one species present, occupy the same position as the point representing that species, because sites are at the barycentres (centroids) of the species; CA does not spread apart sites that possess a single and same species, even in different amounts. Sites 6-9 and 11-14, which possess two species in various combinations, lie on a line between the two species; their positions along that line depend on the relative abundances of the two species at each site. No site has three species in this example, so that no point lies inside the triangular shape of the scatter of sites. Considering site 1 (lower left in Fig. 9.17c), examine its distances to all the other sites at the bottom of Table 9.9: they increase from site 6 to 10, after which they remain constant. This corresponds to the relative positions of the sites in the Figure. Had the example contained more species along the gradient, the site points would have exhibited a rounded shape.

The PCA ordination (Fig. 9.17a, b) is identical to the ordination that would have been obtained from PCoA of a matrix of Euclidean distances among sites. In the same way, the ordination of sites in the CA plot (Fig. 9.17c), that used scaling type 1, is similar to a PCoA ordination obtained from a matrix of χ^2 distances (D_{16}) among sites. The ordinations obtained from distance coefficients D_{14} (Bray & Curtis or Odum distance) and $\sqrt{D_{14}}$ are also of interest because of the favour these coefficients have among ecologists. They are displayed in Fig. 9.17e-h. The ordinations produced by these coefficients are quite similar to each other and present a horseshoe, but not as pronounced as in PCA because these coefficients exclude double-zeros from the calculations. In Fig. 9.17e (coefficient D_{14}), sites 6 to 14 form an arch depicting the three-species gradient, with arms extending in a perpendicular direction (Fig. 9.17f) to account for the dispersion of sites 1 to 5 and 15 to 19, each group containing one species only. The ordination produced by coefficient $\sqrt{D_{14}}$ is very similar to the above (Fig. 9.17g-h). There are two advantages to $\sqrt{D_{14}}$ over D_{14}, though: $\sqrt{D_{14}}$ never produces negative eigenvalues and, in the present case at least, the ordination explains more variation than D_{14} in two or three dimensions.

DCA

Two main approaches have been proposed to remove arches in correspondence analysis: detrending by segments and by polynomials. These methods lead to *detrended correspondence analysis* (DCA).

• When detrending by segments (Hill & Gauch, 1980), axis I is divided into a number of "segments" and, within each one, the mean of the scores along axis II is made equal to zero; in other words, data points in each segment are moved along axis II to make their mean coincide with the abscissa. Fig. 9.18b shows the result of detrending the ordination of Fig. 9.17c, using the three segments defined in Fig. 9.18a. The bottom line is that scores along detrended axis II are meaningless. Proximities among points should in no case be interpreted ecologically, because segmenting generates large differences in scores for points that are near each other in the original ordination but happen to be on either side of segment divisions (Fig. 9.18). The number of segments is arbitrary; different segmentations lead to different ordinations along axis II.

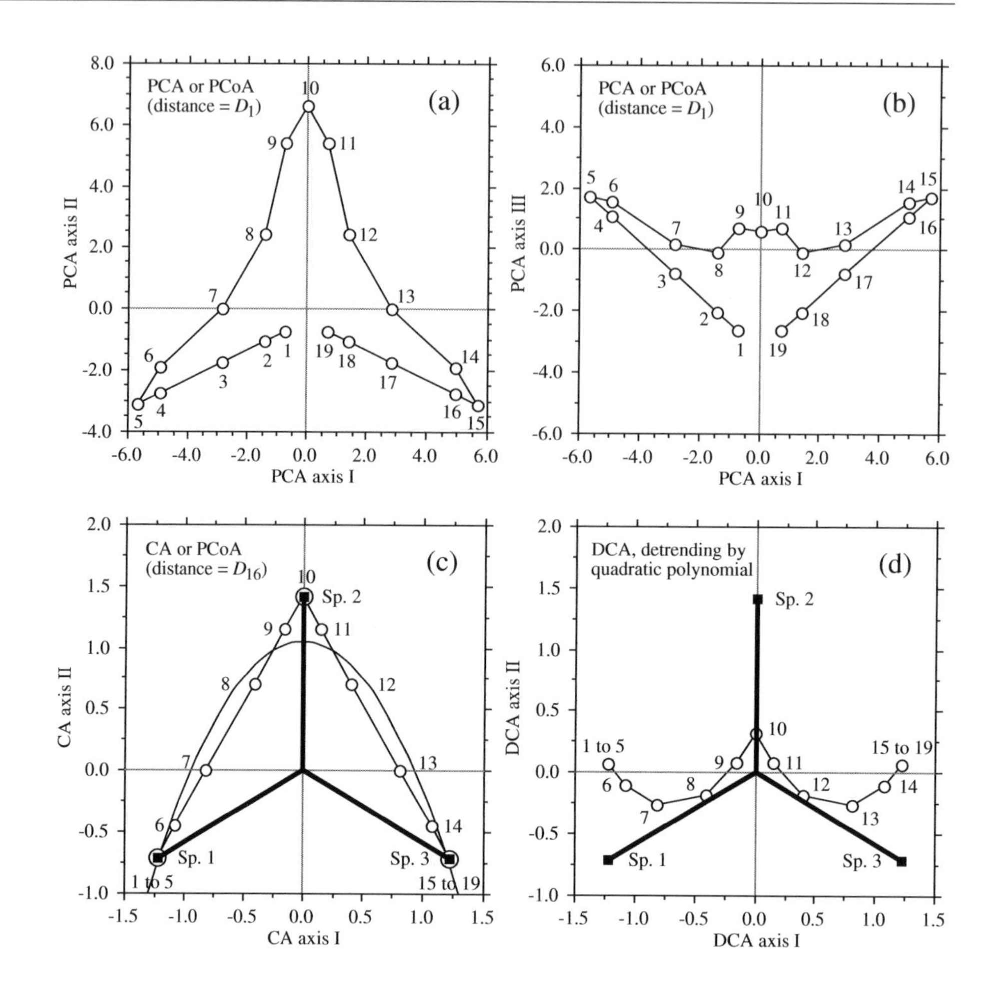

Figure 9.17 Ordinations of the data from Fig. 9.12 and Table 9.9. Circles are sites and squares are species. Principal component analysis: (a) PCA axes I and II ($\lambda_1 = 50.1\%$, $\lambda_2 = 40.6\%$), (b) axes I and III ($\lambda_1 = 50.1\%$, $\lambda_3 = 9.3\%$). (c) Correspondence analysis (scaling type 1), CA axes I and II ($\lambda_1 = 58.1\%$, $\lambda_2 = 41.9\%$). A quadratic polynomial function of axis I is also shown (convex curve): (axis II) = 1.056 – 1.204 (axis I)^2. (d) Detrended correspondence analysis (scaling type 1, detrending by quadratic polynomial), DCA axes I and II ($\lambda_1 = 58.1\%$, $\lambda_2 = 1.6\%$). (c) and (d) Heavy lines, representing the species axes, are drawn from the centres of the plots.

The method is only used with a fairly large number of segments. Programs DECORANA (Hill, 1979) and CANOCO use a minimum of 10 and a maximum of 46 segments, 26 being the 'recommended' number (i.e. the default value). This obviously requires a number of data points larger than in the numerical example of Fig. 9.17. With Hill's iterative algorithm for CA (Table 9.12), detrending by segments is done at the end of each iteration, but the final site scores are derived from the species scores

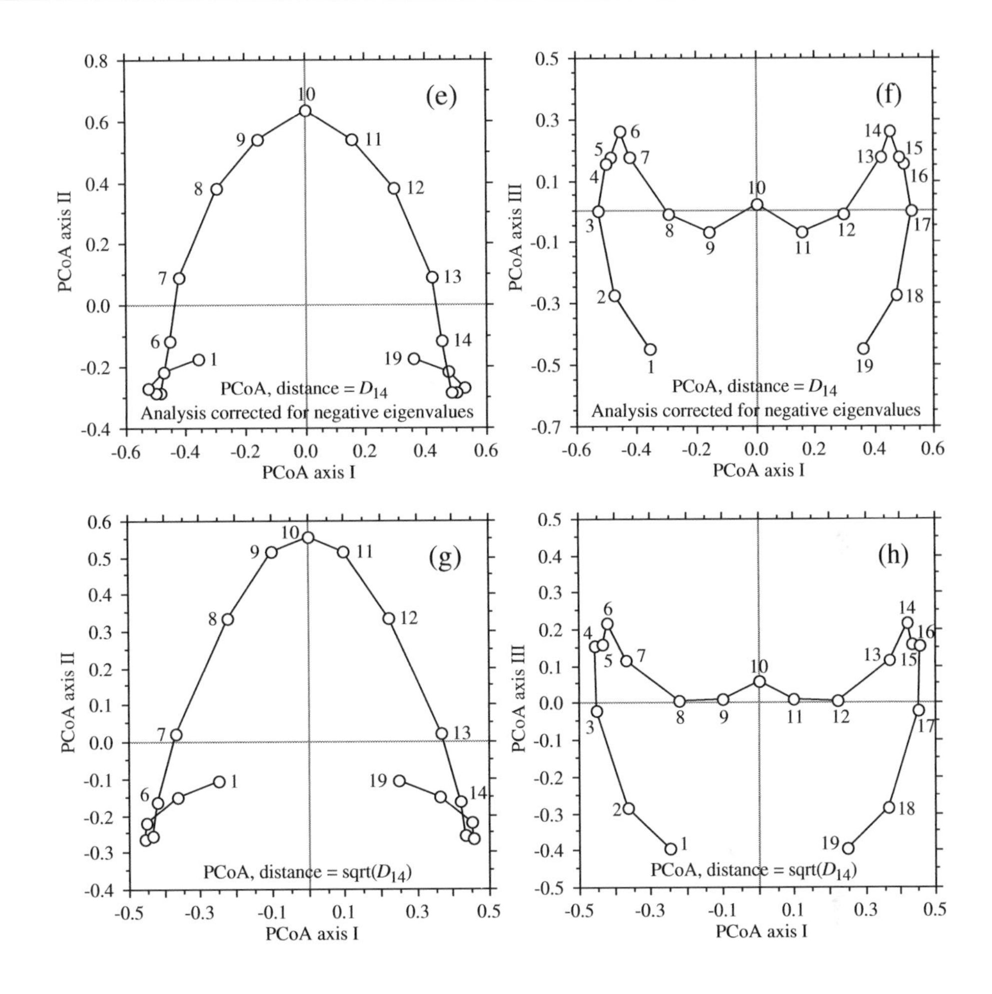

Figure 9.17 **(continued)** Principal coordinate ordinations of the data from Fig. 9.12 and Table 9.9. Distance D_{14}, analysis corrected for negative eigenvalues: (e) PCoA axes I and II ($\lambda_1 = 30.8\%$, $\lambda_2 = 18.6\%$), (f) axes I and III ($\lambda_1 = 30.8\%$, $\lambda_3 = 8.3\%$). Distance $\sqrt{D_{14}}$: (g) PCoA axes I and II ($\lambda_1 = 34.5\%$, $\lambda_2 = 22.9\%$), (h) axes I and III ($\lambda_1 = 34.5\%$, $\lambda_3 = 10.5\%$).

without detrending in order to insure that the site scores are simply weighted averages of the species scores. For the third DCA axis, detrending is carried out with respect to the first and second axes, and so on for subsequent axes.

In order to deal with the contraction of the ends of the gradient when the sites are projected onto the first axis, nonlinear rescaling of the axes is often performed following detrending. An extreme case is represented by Fig. 9.17c where sites 1 to 5 and 15 to 19 each occupy a single point along axis I. To equalize the breadths of the species response curves, the axis is divided into small segments and segments with

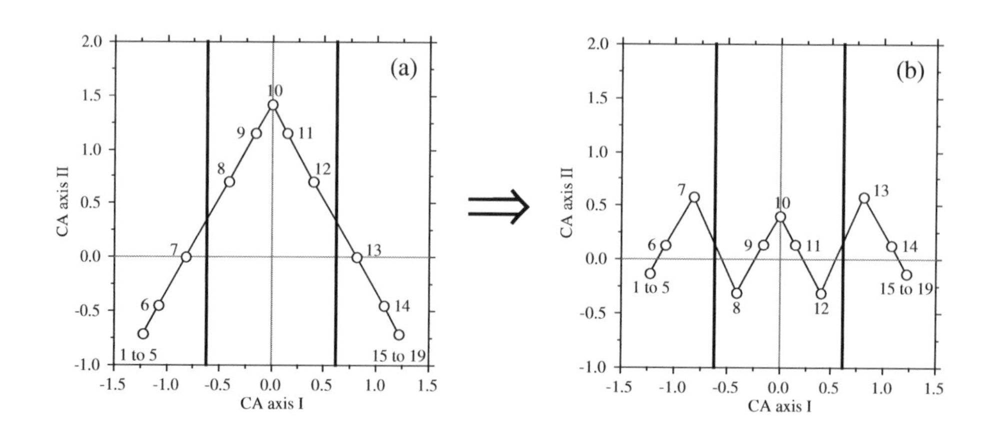

Figure 9.18 Detrending by segments. (a) Three arbitrarily defined segments are delimited by vertical lines in the CA ordination (from Fig. 9.17c). (b) After detrending, the mean of the points in each segment is zero.

small within-group variances are expanded, whereas segments with large within-group variances are contracted (Hill, 1979). Fig. 5.5 of ter Braak (1987c) provides a good illustration of the process; ter Braak (1987c) advises *against* the routine use of nonlinear rescaling.

After detrending by segments and nonlinear rescaling of the axes, the DCA ordination has the interesting property that the axes are scaled in units of the average standard deviation (SD) of species turnover (Gauch, 1982). Along a regular gradient, a species appears, rises to its modal value, and disappears over a distance of about 4 SD; similarly, a complete turnover in species composition occurs, over the sites, in about 4 SD units. A half-change in species composition occurs within about 1 to 1.4 SD units. Thus the length of the first DCA axis is an approximate measure of the length of the ecological gradient, in species turnover units. In this respect, DCA with nonlinear rescaling of the axes is a useful method to estimate the lengths of ecological gradients. The length of a gradient revealed by a pilot study may help determine the *extent* (Section 13.0) to be given to a subsequent full-scale study.

Length of gradient

• Detrending by polynomials (Hill & Gauch, 1980; ter Braak, 1987c) directly follows from the fact that an arch is produced when a gradient of sufficient length is present in data. When a sufficient number of species are present and replace each other along the gradient, the second CA axis approaches a quadratic function of the first one (i.e. a second-degree polynomial), and so on for the subsequent axes. This is clearly not the case with the data of Table 9.9, which consist of three species only. Fig. 9.17c shows that the 'arch' is reduced to a triangular shape in that case.

The arch effect is removed by imposing, in the CA algorithm (orthogonalization procedure in Table 9.12), the constraint that axis II be uncorrelated not only to axis I, but also to its square, its cube, and so on; the degree of the polynomial function is chosen by the user. In the same way, axis III is made uncorrelated to the 1st, 2nd, 3rd … k-th degree polynomial of axes I and II. And so forth. When detrending is sought, detrending by polynomial is an attractive method. The result is a continuous function of the previous axes, without the discontinuities generated by detrending-by-segments. However, detrending by polynomials imposes a specific model onto the data, so that the success of the operation depends on how closely the polynomial model corresponds to the data. Detrending by polynomial does not solve the problem of compression of the sites at the ends of the ordination axes.

Detrending by quadratic polynomial was applied to the test data. Fig. 9.17c shows the quadratic polynomial (convex curve; among the terms of the quadratic polynomial, only the (axis I)^2 term was significant) that was fitted to the CA ordination, which has a triangular shape in the present example. Detrending involves computing and plotting the vertical (residual) distances between the data points and the fitted polynomial. The detrended ordination is shown in Fig. 9.17d. The regression residuals display an elegant but meaningless shape along axis II.

When the data are controlled by a single environmental gradient, detrending is useless. Only the first ordination axis is meaningful, subsequent axes being meaningless linear combinations of the first.

The controversy about detrending has been raging in the literature over the past 10 years. Key papers are those of Wartenberg *et al.* (1987), Peet *et al.* (1988), and Jackson & Somers (1991b). Wartenberg *et al.* (1987) argued that the arch is an important and inherent attribute of the distances among sites, not a mathematical artifact. The only effect of DCA is to flatten the distribution of points onto axis I without affecting the ordination of sites along that axis. They also pointed out that detrending-by-segments is an arbitrary method for which no theoretical justification has been offered. Similarly, the nonlinear rescaling procedure assumes that, on average, each species appears and disappears at the same rate along the transect and that the parametric variance is an adequate measure of that rate; these assumptions have not been substantiated. Despite these criticisms, Peet *et al.* (1988) still supported DCA on the ground that detrending and rescaling may facilitate ecological interpretation. They called for improved algorithms for detrending and rescaling. Jackson & Somers (1991b) showed that the DCA ordination of sites greatly varies with the number of segments one arbitrarily decides to use, so that the ecological interpretation of the results may vary widely, as do the correlations one can calculate with environmental variables. One should always try different numbers of segments when using DCA.

Simulation studies involving DCA have been conducted on artificial data representing unimodal species responses to environmental gradients in one (*coenoclines*) or two (*coenoplanes*) dimensions, following the method pioneered by Swan (1970). Kenkel & Orlóci (1986) report that DCA did not perform particularly well in recovering complex gradients. Using Procrustes statistics (Subsection 10.5.4) as measures of structure recovery, Minchin (1987) showed that DCA did not perform well with complex response models and non-regular sampling schemes. Furthermore, DCA may remove real structures that appear as curved in ordination diagrams and should legitimately contribute to the second axis. This last criticism would apply to detrending by polynomials as well as detrending by segments. Both studies concurred that MDS is the best technique for recovering complex gradients, not DCA.

Present evidence indicates that detrending should be avoided, except for the specific purpose of estimating the lengths of gradients; such estimates remain subject to the assumptions of the model being true. In particular, DCA should be avoided when analysing data that represent complex ecological gradients. Most ordination techniques are able to recover simple, one-dimensional environmental gradients. When there is a single gradient in the data, detrending is useless since the gradient is best represented by CA axis I.

Satisfactory mathematical solutions to the problem of detrending remain to be found. In the meantime, ordination results should be interpreted with caution and in the light of the type of distance preserved by each method.

6 — Ecological applications

Ecological application 9.4a

Cadoret *et al.* (1995) investigated the species composition (presence/absence and abundance) of chaetodontid fish assemblages off Moorea Island, French Polynesia, in order to describe the spatial distribution of the butterflyfishes and to determine their relationships with groups of benthic organisms. Sampling was conducted in four areas around the island: (a) Opunohu Bay, (b) Cook Bay, (c) the Tiahura transect across the reef in the northwestern part of the island, and (d) the Afareaitu transect across the reef in the eastern part of the island.

Correspondence analysis (Fig. 9.19) showed that the fish assemblages responded to the main environmental gradients that characterized the sampling sites. For areas c and d (transects across the reef), axis I corresponded to a gradient from the coastline to the ocean; from left to right, in the plot, are the sites of the fringing reef, the shallow (found only in sector c), the barrier reef, and the outer slope. Sites from the bays (areas a and b) are also found in the left-hand part of the graph. Axis II separates the sites located in the upper reaches of Opunohu Bay (a11, a12 and a13, in the upper-left of the plot) from all the others. This application will be further developed in Section 11.2, to identify species assemblages and evidence the relationships between species and environmental variables, using canonical correspondence analysis.

Ecological application 9.4b

In a study on the vegetation dynamics of southern Wisconsin, Sharpe *et al.* (1987) undertook a systematic field survey of all forest tracts in two townships. Detrended correspondence analysis was used to display the relationships among stands with respect to species composition. The scores of the first ordination axes were used to construct three-dimensional maps. Mapping of the first axis (Fig. 9.20) shows that the scores were generally low in the southern and central portions of the area and increased towards the west and north. Since the first axis showed a trend from forest tracts dominated by *Acer saccharum* to oak-dominated forests (not shown), Fig. 9.20 indicates that stands dominated by *A. saccharum* were located in the south-central portion of the

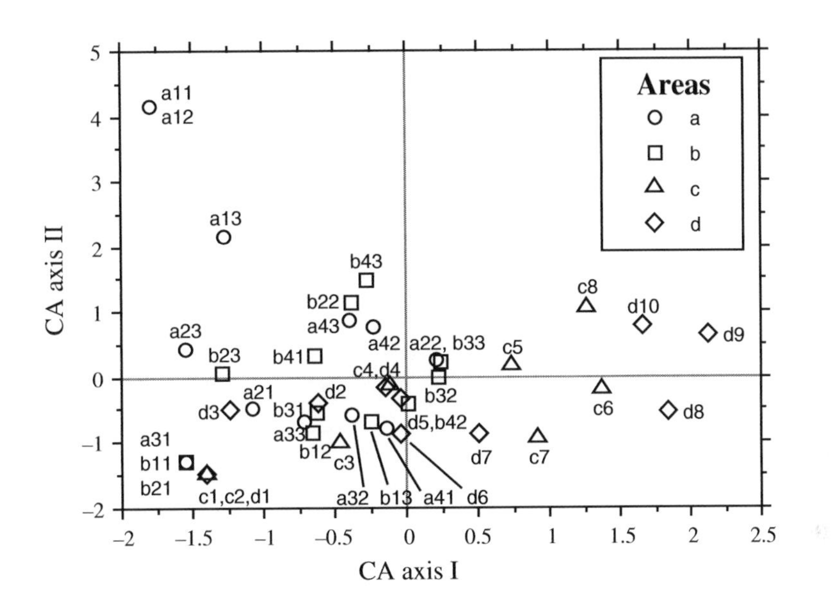

Figure 9.19 Correspondence analysis (CA): ordination of sampling sites with respect to axes I and II from presence/absence observations of butterflyfishes (21 species) around Moorea Island. Axes I and II explain together 29% of the variation among sites. Species vectors are not drawn; they would have overloaded the plot. Modified from Cadoret *et al.* (1995).

area, whereas oak-dominated stands were to the west, north and, to a lesser extent, east. Such a mapping, using a 3- or 2-dimensional representation, is often a useful way of displaying synthetic information provided by the scores of objects along the first ordination axes.

Maps, like that of Fig. 9.20, may be produced for the ordination scores computed by any of the methods described in the present Chapter; see Section 13.2.

7 — Algorithms

There are several computer programs available for correspondence analysis. They do not all provide the same ordination plots, though, because the site and species score vectors may be scaled differently. A simple, empirical way for discovering which of the matrices described in Subsection 1 are actually computed by a program is to run the small numerical example of Subsection 2. The main variants are:

1) General-purpose data analysis programs, which compute the eigenvalues and eigenvectors using traditional eigenanalysis algorithms. They usually output matrices **U** of the eigenvectors (ordination of the columns, i.e. the species, when the analysis is conducted on a site × species table) and **F** (ordination of rows, i.e. sites). Some programs also output matrix $\hat{\mathbf{F}}$ (ordination of the species at the centroids of sites).

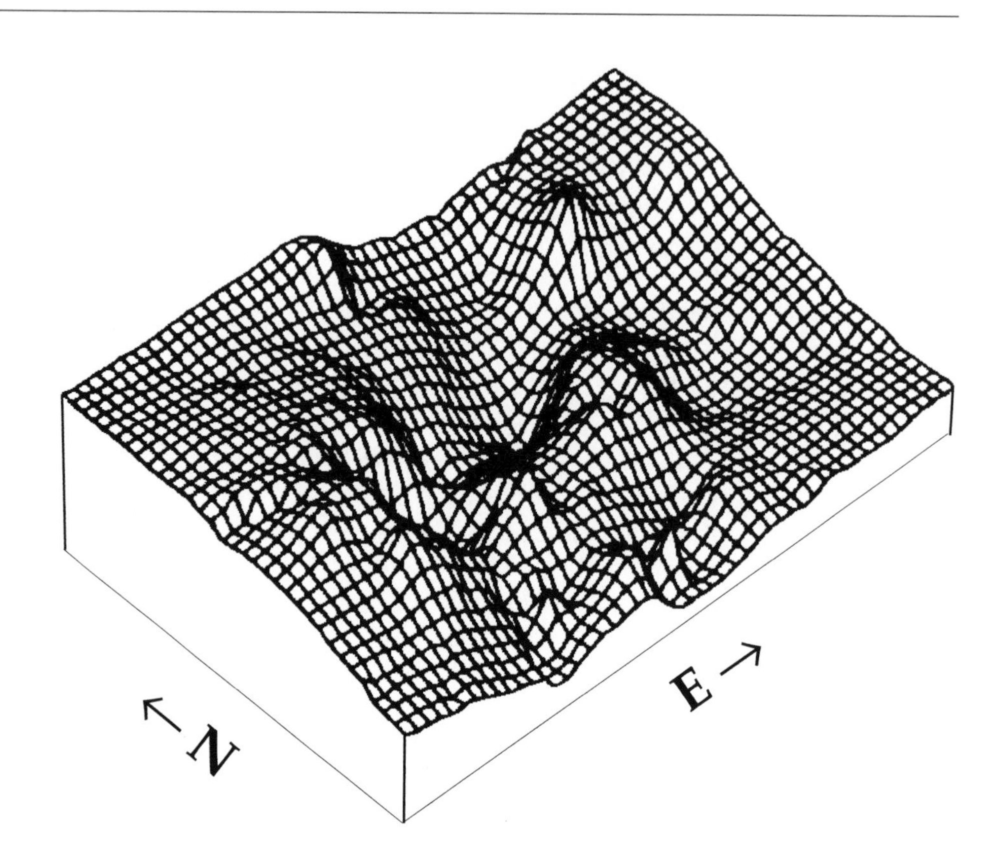

Figure 9.20 Three-dimensional map of the scores of the first ordination axis (detrended correspondence analysis), based on trees observed in 92 forest tracts of southern Wisconsin, U.S.A. (survey area: 11 × 17 km). Modified from Sharpe *et al.* (1987).

2) Ecologically-oriented programs, which often use the *two-way weighted averaging* iterative algorithm (Hill' *reciprocal averaging* method), although this is by no means a requirement. They allow for several types of scalings, including types 1 and 2 discussed in Subsection 9.4.1 (e.g. CANOCO; ter Braak, 1988b, 1988c, 1990; ter Braak & Smilauer, 1998).

TWWA algorithm

Table 9.12 presents Hill's *two-way weighted averaging* (TWWA) algorithm, as summarized by ter Braak (1987c). There are three main differences with the TWWS algorithm for PCA presented in Table 9.5: (1) variables are centred in PCA, not in CA. (2) In CA, the centroid of the site scores is not zero and must thus be estimated (step 6.1) (3) In CA, summations are standardized by the row sum, column sum, or grand total, as appropriate. This produces shrinking of the ordination scores at the end of each iteration in CA (step 6.4), instead of stretching in PCA.

Table 9.12 Two-way weighted averaging (TWWA) algorithm for correspondence analysis. From Hill (1973b) and ter Braak (1987c).

a) Iterative estimation procedure

Step 1: Consider a table **Y** with n rows (sites) × p columns (species).
Do NOT centre the columns (species) on their means.

Determine how many eigenvectors are needed. For each one, **DO** the following:

Step 2: Take the row order as the arbitrary initial site scores. (1, 2, …)
Set the initial eigenvalue estimate to 0. In what follows, y_{i+} = row sum for site i, y_{+j} = column sum for species j, and y_{++} = grand total for the data table **Y**.

Iterative procedure begins

Step 3: Compute new species loadings: $colscore(j) = \Sigma\, y(i,j) \times rowscore(i)/y_{+j}$

Step 4: Compute new site scores: $rowscore(i) = \Sigma\, y(i,j) \times colscore(j)/y_{i+}$

Step 5: For the second and higher-order axes, make the site scores uncorrelated with all previous axes (Gram-Schmidt orthogonalization procedure: see *b* below).

Step 6: Normalize the vector of site scores (procedure *c*, below) and obtain an estimate of the eigenvalue. If this estimate does not differ from the previous one by more than the tolerance set by the user, go to step 7. If the difference is larger than the tolerance, go to step 3.

End of iterative procedure

Step 7: If more eigenvectors are to be computed, go to step 2. If not, continue with step 8.

Step 8: The row (site) scores correspond to matrix $\hat{\mathbf{V}}$. The column scores (species loadings) correspond to matrix $\hat{\mathbf{F}}$. Matrices $\hat{\mathbf{F}}$ and $\hat{\mathbf{V}}$ provide scaling type 2 (Subsection 9.4.1). Scalings 1 or 3 may be calculated if required. Print out the eigenvalues, % variance, species loadings, and site scores.

b) Gram-Schmidt orthogonalization procedure

DO the following, in turn, for all previously computed components k:

Step 5.1: Compute the scalar product $SP = \Sigma\, (y_{i+} \times rowscore(i) \times v(i,k)/y_{++})$ of the current site score vector estimate with the previous component k. Vector $v(i,k)$ contains the site scores of component k scaled to length 1. This product is between 0 (if the vectors are orthogonal) and 1.

Step 5.2: Compute new values of $rowscore(i)$ such that vector *rowscore* becomes orthogonal to vector $v(i,k)$: $rowscore(i) = rowscore(i) - (SP \times v(i,k))$.

c) Normalization procedure[†]

Step 6.1: Compute the centroid of the site scores: $z = \Sigma\, (y_{i+} \times rowscore(i)/y_{++})$.

Step 6.2: Compute the sum of squares of the site scores: $S^2 = \Sigma\, (y_{i+} \times (rowscore(i) - z)^2/y_{++})$; $S = \sqrt{S^2}$.

Step 6.3: Compute the normalized site scores: $rowscore(i) = (rowscore(i) - z)/S$.

Step 6.4: At the end of each iteration, S, which measures the amount of shrinking during the iteration, provides an estimate of the eigenvalue. Upon convergence, the eigenvalue is S.

[†] Normalization in CA is such that the *weighted* sum of squares of the elements of the vector is equal to 1.

Householder
SVD

Alternative algorithms for CA are Householder reduction and singular value decomposition. SVD was used to describe the CA method in Subsection 9.4.1; it directly provides the eigenvalues (they are actually the squares of the singular values) as well as matrices **U** and $\hat{\mathbf{U}}$. The various scalings for the row and column scores may be obtained by simple programming. Efficient algorithms for singular value decomposition are available in Press *et al.* (1986 and later editions).

9.5 Factor analysis

In the social sciences, analysis of the relationships among the descriptors of a multidimensional data matrix is often carried out by *factor analysis*. The name *factor analysis* was first used by Spearman (1904), who proposed that the correlations among a set of intelligence-test scores could be explained by a common factor combined with several other factors reflecting the qualities of individual tests. There are fundamental differences between principal component analysis, described above (Section 9.1), and factor analysis. The primary aim of principal component analysis is to account for a maximum amount of the variance in the data, whereas the goal of *factor analysis* is to account for the *covariance* among descriptors. To do this, factor analysis assumes that the observed descriptors are linear combinations of hypothetical underlying (or latent) variables (i.e. the *factors*). As a result, factor analysis explains the covariance structure of the descriptors in terms of a hypothetical causal model (i.e. the observed descriptors are caused by the underlying factors), whereas principal component analysis summarizes the data set by means of linear combinations of the descriptors. Readers may refer to Kim & Mueller (1978a 1978b) for a simple introduction to factor analysis and to Mulaik (1972) for a more detailed discussion.

There are two types of factor analysis. *Exploratory factor analysis*, on the one hand, computes the minimum number of hypothetical factors needed to account for the observed covariation among descriptors. *Confirmatory factor analysis*, on the other hand, may be used to test hypotheses concerning the causal structure between the underlying factors and the observed descriptors. Exploratory factor analysis is sometimes used by ecologists, simply because it is widely available in computer packages. With the exception of the *varimax* rotation, described below, factor analysis is not currently used in ecology. The purpose of the present section is not to recommend the use of exploratory or confirmatory factor analysis in ecology, but mainly to provide basic information to ecologists who might encounter computer programs or results obtained using these methods. Confirmatory factor analysis, which is briefly described at the end of the present section, is not currently used in ecology. It could be used as an easy way for testing simple, causal ecological models.

In factor analysis, one assumes that there are two types of underlying factors, i.e. the *common factors*, which are involved in the causation of more than one observed descriptor, and the *unique* (or *specific*) *factors*, which are each causally

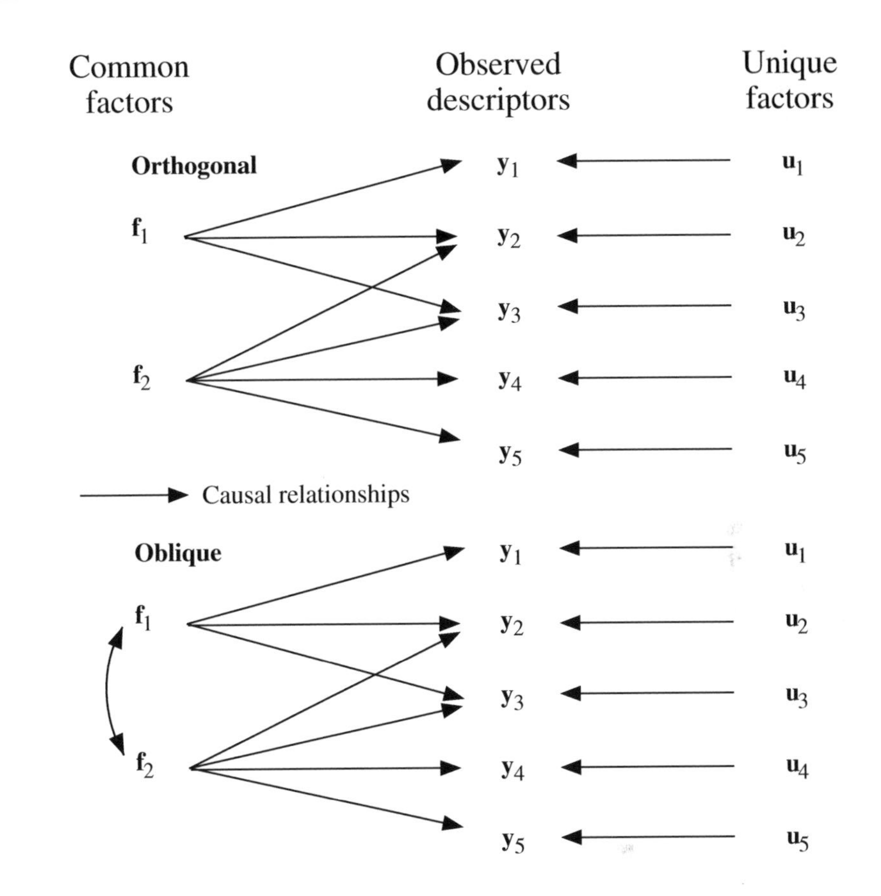

Figure 9.21 Path models (Section 10.4) for a hypothetical situation with five descriptors and two common factors, showing the difference between orthogonal rotation (i.e. common factors uncorrelated) and oblique rotation (i.e. common factors correlated). Adapted from Kim & Mueller (1978a).

related to only one observed descriptor (Fig. 9.21). Principal component analysis uses the overall variance of the data set, whereas factor analysis considers only the variance pertaining to the common factors. This is because, in factor analysis, *unique factors* do not contribute to the covariation among the observed descriptors; only the *common factors* do so. A fundamental assumption of factor analysis is that the observed covariation among descriptors results from their sharing common factors, not from being direct causes of one another. To be of practical use, the number of common factors must be much smaller than the number of descriptors.

Exploratory factor analysis

The first step of *exploratory factor analysis* is the extraction of initial factors from an appropriate covariance matrix. These initial factors are somewhat analogous to the eigenvectors computed in principal component analysis, but they concern here only that part of the variance of each descriptor (called *communality*) which is accounted for by the common factors. The fraction of the variance pertaining to unique factors (called *uniqueness*) is therefore excluded. Several methods may be used for extracting the initial factors. They are: principal axis factoring (or principal factoring), least squares (i.e. principal axis factoring with iterative estimation of communalities; minimum residual method or Minres), maximum likelihood (or canonical factoring), alpha factoring, and image analysis. These methods are based on different assumptions, but they all result in a small number of (orthogonal) common factors which account for the covariation among a much larger number of observed descriptors.

When looking at the extracted initial factors, one usually finds that several factors have significant loadings on any given descriptor; this is also the case in principal component analysis. The number of such factors is the *factorial complexity.* As a result, the causal structure between factors and descriptors at this stage of the analysis is quite complex, since each descriptor is more or less explained by all factors. The next step of the analysis is to seek a *simple structure*, i.e. a solution where causal relationships between the underlying common factors and the observed descriptors is simpler. This is achieved through *rotation.*

The purpose of rotation is not to improve the fit of factors to the observed data; the rotated factors explain the exact same amount of covariance among the descriptors as the initial factors. The sole purpose of rotation is to achieve a simple structure, that is a solution where certain factor loadings are maximized while others are minimized. There exist two general types of rotations, i.e. orthogonal and oblique. In *orthogonal rotations*, the causal underlying factors cannot be covariates whereas, in *oblique rotations*, the factors can be correlated (Fig. 9.21). In a sense, orthogonal rotations are special cases of oblique rotations (i.e. cases where the correlations among the common factors are null). However, the existence of correlations among the common factors makes the interpretation of oblique solutions much more complicated than that of orthogonal ones.

Orthogonal rotation

Several criteria have been proposed to carry out orthogonal rotations. The *quartimax* method (Burt, 1952) minimizes the sum of squares of the products of the loadings of the descriptors onto factors. This is done in such a way that each descriptor has a high loading for only one factor and low loadings for all the others. While quartimax simplifies the rows of the loading matrix, the *varimax* rotation (Kaiser, 1958) simplifies the columns by maximizing the variance of the squared loadings in each column. The *equimax* rotation simplifies both the rows and the columns. A variant of equimax is called *biquartimax.* Another method of orthogonal rotation, which minimizes the entropy, has been proposed by McCammon (1970). In ecology, the most often used rotation is varimax. Examples are found in Ibanez & Séguin (1972), Parker (1975), and Ecological application 9.5.

Oblique rotation

Various oblique rotations have been proposed; several are available in computer packages. They include: *oblimax* (Saunders, 1961), *quartimin, oblimin, covarimin* and *biquartimin* (Carroll, 1957), *binormamin* (Kaiser & Dickman, 1959), *radial rotation,*

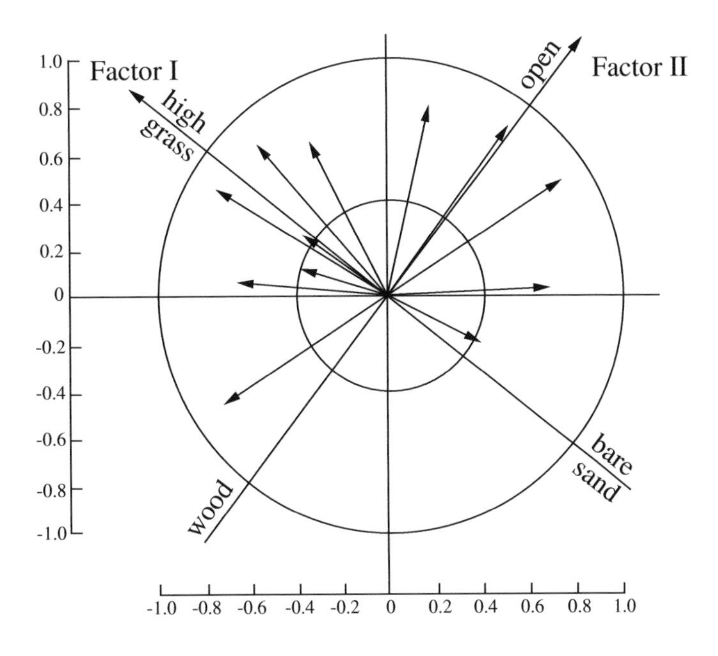

Figure 9.22 Varimax rotation of the axes of Fig. 9.7. Modified from Aart (1973).

prolonged vectors (Thurstone, 1947), *orthoblique* (Harris & Kaiser, 1964), *maxplane* (Cattell & Muerle, 1960; Eber, 1966), and *promax* (Hendrickson & White, 1964). A comparative review summarizing most of the oblique rotations is found in Ibanez (1976).

It should now be clear that factor analysis aims primarily at analysing the covariation among descriptors. It follows that, in most models, the relative positions of the objects cannot be calculated directly, whereas this is often of primary interest to ecologists. In factor analysis, coordinates of the objects in factor space are called *factor scales*. Kim & Mueller (1978b, p. 60 *et. seq.*) review several methods for estimating the factor scales, i.e. regression estimates, estimates based on the least squares criterion, Bartlett's method of minimizing the error variance, estimates with orthogonality constraints, and factor-based scales.

Ecological application 9.5

In order to facilitate ecological interpretation, the axes of Fig. 9.7 (resulting from a principal component analysis, so that the present application is not a typical factor analysis) were subjected to a varimax orthogonal rotation. In Fig. 9.22, factor I may be interpreted as a gradient from bare sand to high grass, via sparse vegetation, and factor II as a gradient from woods to open areas with no trees or shrubs, via sparsely distributed shrubs (Aart, 1973).

Confirmatory factor analysis

In contrast to the above, *confirmatory factor analysis* may be used to test specific expectations concerning the number of factors and their loadings on the various descriptors. Hypotheses tested may specify the number of common factors involved, the nature of their relationships (i.e. orthogonal or oblique), the general structure of the factor loadings, equality constraints on factor loadings, fixed values for individual factor loadings for each descriptor, as well as the correlation structure among residuals. These hypotheses are normally based on empirical or conceptual preliminary information about the causal mechanisms involved and on expectations about the magnitude of the different parameters to be estimated. Higher-order confirmatory factor analysis models allow a set of exogenous hypothetical variables, called the higher-order factors, to directly influence lower-order common factors.

An elementary introduction to confirmatory factor analysis is found in Schumacker & Lomax (1996). A more advanced treatment is provided by Bollen (1989). The most widely used program for confirmatory factor analysis is LISREL[*]. Another interesting program is EQS[†].

[*] LISREL, written by K. G. Jöreskog and D. Sörbom, is distributed by Scientific Software International (SIS), 7383 N. Lincoln Ave., Suite 100, Chicago, Illinois 60646-1704, USA. WWWeb site: <http://ssicentral.com/>.

[†] Program EQS, written by P. M. Bentler and E. J. C. Wu, is distributed by Multivariate Software, Inc., 4924 Balboa Blvd., Suite 368, Encino, California 91316, USA. WWWeb site: <http://www.mvsoft.com/>.

Chapter

10 *Interpretation of ecological structures*

10.0 Ecological structures

The previous chapters explained how to use the techniques of clustering and ordination to investigate relationships among objects or descriptors. What do these analyses contribute to the understanding of ecological phenomena? Ecological applications in Chapters 8 and 9 have shown how clustering and ordination can synthesize the variability of the data and present it in a format which is easily amenable to interpretation. It often happens, however, that researchers who are using these relatively sophisticated methods do not go beyond the description of the structure of multidimensional data matrices, in terms of clusters or gradients. The descriptive phase must be followed by interpretation, which is conducted using either the descriptors that were analysed in order to bring out the structure, or other ecological descriptors which have not yet been involved in the analysis.

Structure

From the previous chapters, it should be clear that the *structure* of a data matrix is the organization of the objects, or descriptors, along gradients in a continuum, or in the form of subsets (clusters). This organization characterizes the data matrix, and it is derived from it. The first phase of multidimensional analysis (i.e. clustering or/and ordination) thus consists in characterizing the data matrix in terms of a simplified structure. In a second phase, ecologists may use this structure to interpret the phenomenon which underlies the data matrix. To do so, analyses are conducted to quantify the relationships between the structure of the data matrix and potentially explanatory descriptors. The methods which are most often used for interpreting ecological structures are described in the present Chapter and in Chapter 11.

During interpretation, one must assume that the analysis of the structure has been conducted with care, using measures of association which are appropriate to the objects and/or descriptors of the data matrix (Chapter 7) as well as analytical methods that correspond to the objectives of the study. Ordination (Chapter 9) is used when gradients are sought, and clustering (Chapter 8) when one is looking for a partition of the objects or descriptors into subsets. When the gradient is a function of a single or a pair of ordered descriptors, the ordination may be plotted in the original space of the descriptors. When the gradient results from the combined action of several descriptors, however, the ordination must be carried out in a reduced space, using the methods discussed in Chapter 9. It may also happen that an ordination is used as a basis for visual clustering. Section 10.1 discusses the combined use of clustering and ordination to optimize the partition of objects or descriptors.

Explanation

Forecasting

Prediction

The interpretation of structures, in ecology, has three main objectives: (1) *explanation* (often called *discrimination*) of the structure of one or several descriptors, using those descriptors at the origin of the structure or, alternatively, a set of other descriptors that may potentially explain the structure; (2) *forecasting* of one or several descriptors (which are the response, or dependent variables: Box 1.1), using a number of other descriptors (called the explanatory, or independent variables); (3) *prediction* of one or several descriptors, using descriptors that can be experimentally manipulated, or that naturally exhibit environmental variation. It must be noted that the terms *forecasting* and *prediction*, which are by no means equivalent (Subsection 10.2.2), are often confused in the ecological and statistical literatures. Each of the above objectives covers a large number of numerical methods, which correspond to various levels of precision of the descriptors involved in the analysis.

Section 10.2 reviews the methods available for interpretation. The next sections are devoted to some of the methods introduced in Section 10.2. Regression and other scatterplot smoothing methods are discussed in Section 10.3. Section 10.4 deals with path analysis, used to assess causal relationships among quantitative descriptors. Section 10.5 discusses methods developed for the comparison of resemblance matrices. Various forms of canonical analysis are presented in Chapter 11.

10.1 Clustering and ordination

Section 8.2 showed that single linkage clustering accurately accounts for the relationships between closely similar objects. However, due to its tendency to chaining, single linkage agglomeration is not very suitable for investigating ecological questions. Because ecological data generally form a continuum in A-space (Fig. 7.2), single linkage clustering is best used in conjunction with an ordination of the objects. In the full multidimensional ordination space, distances among the main clusters of objects are the same as in the original A-space (Section 9.1), or they are some appropriate Euclidean representation of them (Sections 9.2 to 9.4). However, when only the first two or three dimensions are considered, ordinations in reduced space may misrepresent the structure by projecting together clusters of objects which are distinct in higher dimensions. Clustering methods allow one to separate clusters whose projections in reduced space may sometimes obscure the relationships.

Several authors (e.g. Gower & Ross, 1969; Rohlf, 1970; Schnell, 1970; Jackson & Crovello, 1971; Legendre, 1976) have independently proposed to take advantage of the characteristics of clustering and ordination by combining the results of the two types of analyses on the same diagram. The same similarity or distance matrix (Tables 7.3 to 7.5) is often used for the ordination and cluster analyses. Any clustering method may be used, as long as it is appropriate to the data. If linkage clustering is chosen, it is easy to draw the links between objects onto the ordination diagram, up to a given level of similarity. One may also identify the various similarity levels by using different colours or streaks (for example: solid line for $1.0 \geq S > 0.8$, dashed for $0.8 \geq S > 0.6$, dotted for $0.6 \geq S > 0.4$, etc., or any other convenient combination of codes or levels). If a divisive method or centroid clustering was used, a polygon or envelope may be drawn, on the ordination diagram, around the members of a given cluster. This is consistent with the opinion of Sneath & Sokal (1973), who suggest to always simultaneously carry out clustering and ordination on a set of objects. Field *et al.* (1982) express the same opinion about marine ecological data sets. It is therefore recommended, as routine procedure in ecology, to represent clustering results onto ordination diagrams.

The same approach may be applied to cluster analyses of descriptors. Clustering may be conducted on a dependence matrix among descriptors — especially species — in the same way as on an association matrix among objects. An ordination of species may be obtained using correspondence analysis (Section 9.4), which preserves the χ^2 distance among species, or through principal coordinates (Section 9.2) or nonmetric multidimensional scaling (Section 9.3) if some other measure of dependence among species is preferred (Table 9.1). With physical or chemical descriptors of the environment, the method of choice is principal component analysis of the correlation matrix (Section 9.1); descriptors are represented by arrows in the ordination diagram. Before clustering, negative correlations among descriptors should be made positive because they are indicative of resemblance on an inverted scale.

When superimposed onto an ordination, single linkage clustering becomes a most interesting procedure in ecology. Single linkage clustering is the best complement to an ordination due to its contraction of the clustering space (Section 8.2). Drawing single linkage results onto an ordination diagram provides both the correct positions for the main clusters of objects (from the ordination) and the fine relationships between closely similar objects (from the clustering). It is advisable to only draw the chain of primary connections (Section 8.2) on the ordination diagram because it reflects the changes in the composition of clusters. Otherwise, the groups of highly similar objects may become lost in the multitude of links drawn on the ordination diagram. Ecological application 10.1 provides an example of this procedure.

Jackson & Crovello (1971) suggested to indicate the directions of the links on the ordination diagram (Fig. 10.1). This information may be useful when delineating clusters, although the strengths of the links may be considered more important than their directions. In such diagrams, each link of the primary chain is drawn with an arrow. On a link from $\mathbf{x}_1$ to $\mathbf{x}_2$, an arrow pointing towards $\mathbf{x}_2$ indicates that object $\mathbf{x}_1$ has

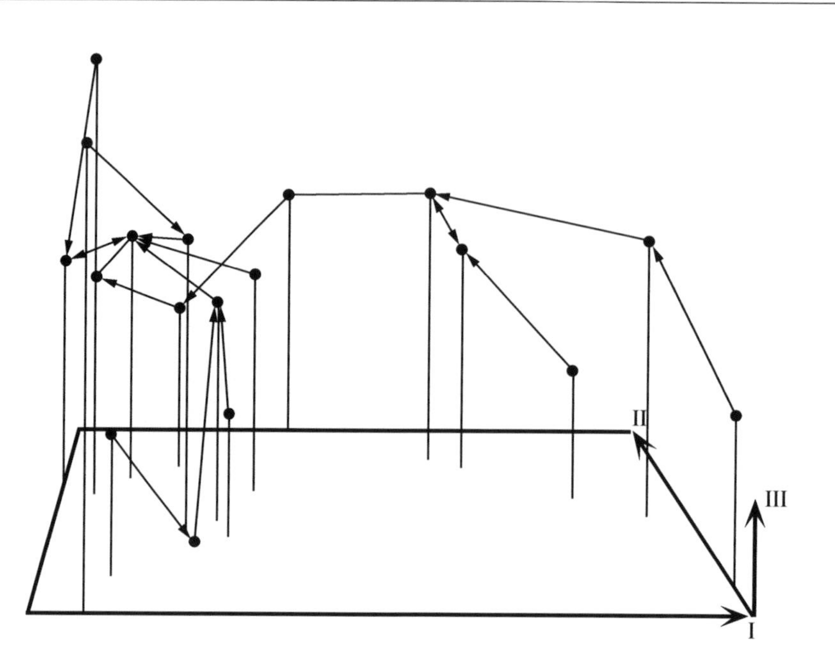

Figure 10.1 Three-dimensional ordination of objects (dots), structured by the primary connections of a single linkage clustering. The arrows (excluding those of the principal axes I to III) specify the directions of the relationships between nearest neighbours; see text. Modified from Jackson & Crovello (1971).

$\mathbf{x}_2$ as its closest neighbour in multidimensional A-space (i.e. in the association matrix among objects). When $\mathbf{x}_2$ also has $\mathbf{x}_1$ as its closest neighbour, the arrow goes both ways. When $\mathbf{x}_2$ has $\mathbf{x}_3$ as its closest neighbour, the arrow from $\mathbf{x}_2$ points towards $\mathbf{x}_3$. New links formed between objects that are already members of the same cluster do not have arrows. These links may be removed to separate the clusters.

Ecological application 10.1

Single linkage clustering was illustrated by Ecological application 8.2 taken from a study of a group of ponds, based upon zooplankton. The same example (Legendre & Chodorowski, 1977) is used again here. Twenty ponds were sampled on islands of the St. Lawrence River, east and south of Montréal (Québec). Similarity coefficient S_{20} (eq. 7.27) was computed with $k = 2$. The matrix of similarities among ponds was used to compute single linkage clustering and an ordination in reduced space by principal coordinate analysis. In Fig. 10.2, the chain of primary connections is superimposed onto the ordination, in order to evidence the clustering structure. The ponds are first divided into a cluster of periodic ponds, which are dry during part of the year (encircled), and a cluster of permanent ponds. Ponds with identification numbers beginning with the same digit (which indicates the region) tend to be close to one another and to cluster first with one another. The second digit refers to the island on which a pond was located.

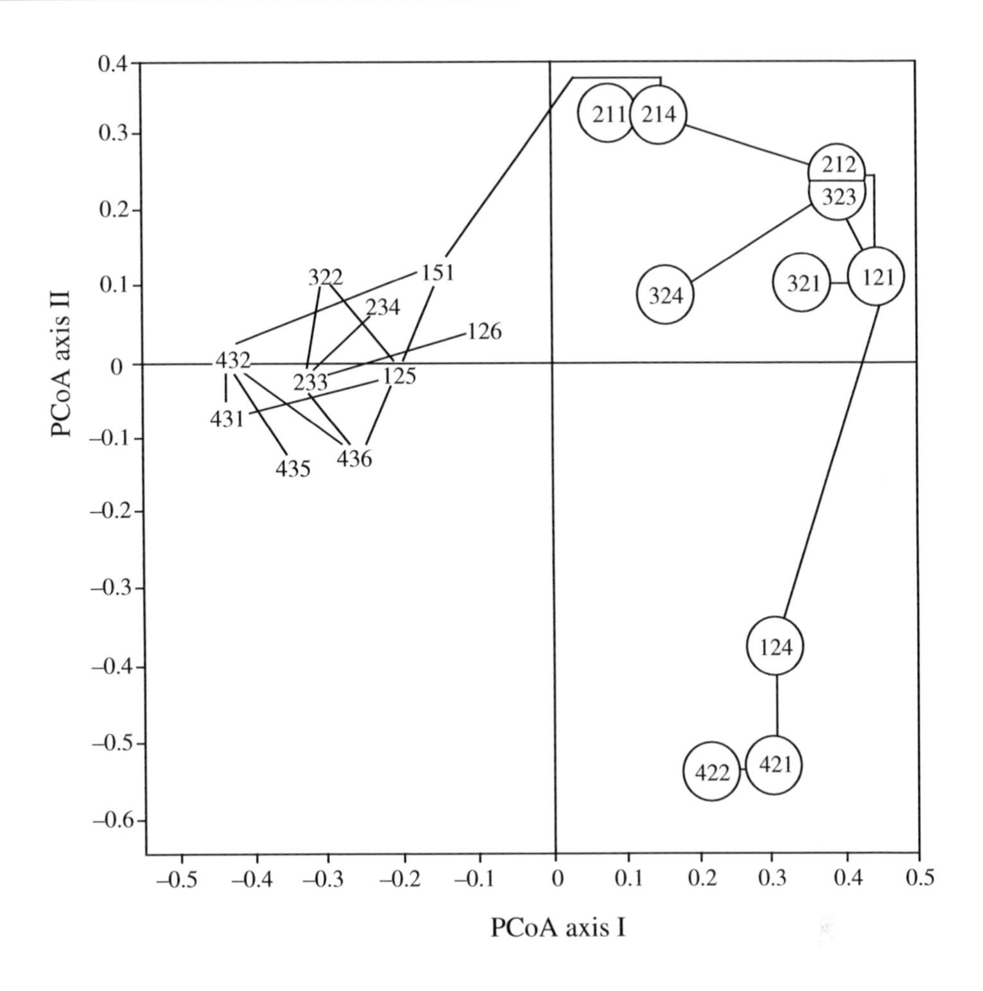

Figure 10.2 Comparison of 20 ponds on the basis of their zooplankton fauna. Ordination in a space of principal coordinates (principal axes I and II), and superimposition of the chain of primary connections obtained by single linkage clustering. The encircled ponds are periodic; the others are permanent. Adapted from Legendre & Chodorowski (1977).

When no clear clustering structure is present in the data but groups are still needed, for management purpose for instance, arbitrary groups may be delineated by drawing a regular grid on the reduced-space ordination diagram. This grid may be orthogonal (i.e. square or rectangular), or polar as in Fig. 9.8. Another method is to divide the objects according to the quadrants of the ordination in reduced space (in 2^d groups for a d-dimensional space); the result is the hierarchic classification scheme of Lefkovitch (1976) described in Subsection 9.2.5.

Figure 10.3 summarizes the steps involved in producing a cluster analysis and an ordination from a resemblance matrix. Description of the data structure is clearer when the clustering results are drawn onto the ordination. In order to assess to what extent the clustering and the ordination correspond to the resemblance matrix from which they originate, these representations may be compared to the original resemblance matrix using matrix correlation or related methods (Subsection 8.11.2).

10.2 The mathematics of ecological interpretation

The present Section summarizes the numerical methods available for the interpretation of ecological structures. They include some of the methods discussed in Chapters 4 to 9 and also other, more specialized techniques. The most widely used of these techniques (regression, path analysis, matrix comparison, the 4th-corner method, and canonical analysis) are discussed in Sections 10.3 to 10.6 and in Chapter 11. A few other methods are briefly described in the present Section.

The numerical methods presented in this Section are grouped into three subsections, which correspond to the three main objectives of ecological interpretation, set in Section 10.0: explanation, forecasting, and prediction. For each of these objectives, there is a summary table (Tables 10.1 to 10.3) which allows researchers to choose the method(s) best suited to their ecological objectives and the nature of their data. There are computer programs available for all methods mentioned in this chapter; several of these are found in the most currently used packages.

Ecological interpretation, and especially the *explanation* and *forecasting* of the structure of several descriptors (i.e. multivariate data), may be conducted following two approaches, which are the indirect and direct comparison schemes (Fig. 10.4).

Indirect comparison

Indirect comparison proceeds in two steps. The structure (ordination axes, or clusters) is first identified from a set of descriptors of prime interest in the study. In a second step, the structure is interpreted using either (a) the descriptors that were analysed in the first step to identify the structure, or (b) another set of descriptors assumed to help explain the structure. In his chapter on ordination analysis, ter Braak (1987c) referred to this form of analysis as *indirect gradient analysis* because is was mostly concerned with the study of environmental gradients.

Direct comparison

In *direct comparison*, one simultaneously analyses the response and explanatory data tables in order to identify what they have in common. For two tables of quantitative or binary data, canonical analysis, which combines and is an extension of regression and ordination methods, offers an interesting approach. It allows one to bring out the ordination structure common to two data sets; ter Braak (1987c) refers to this approach as *direct gradient analysis*.

Other forms of direct comparison analysis are available. One may compare similarity or distance matrices, derived from the original data matrices, using the

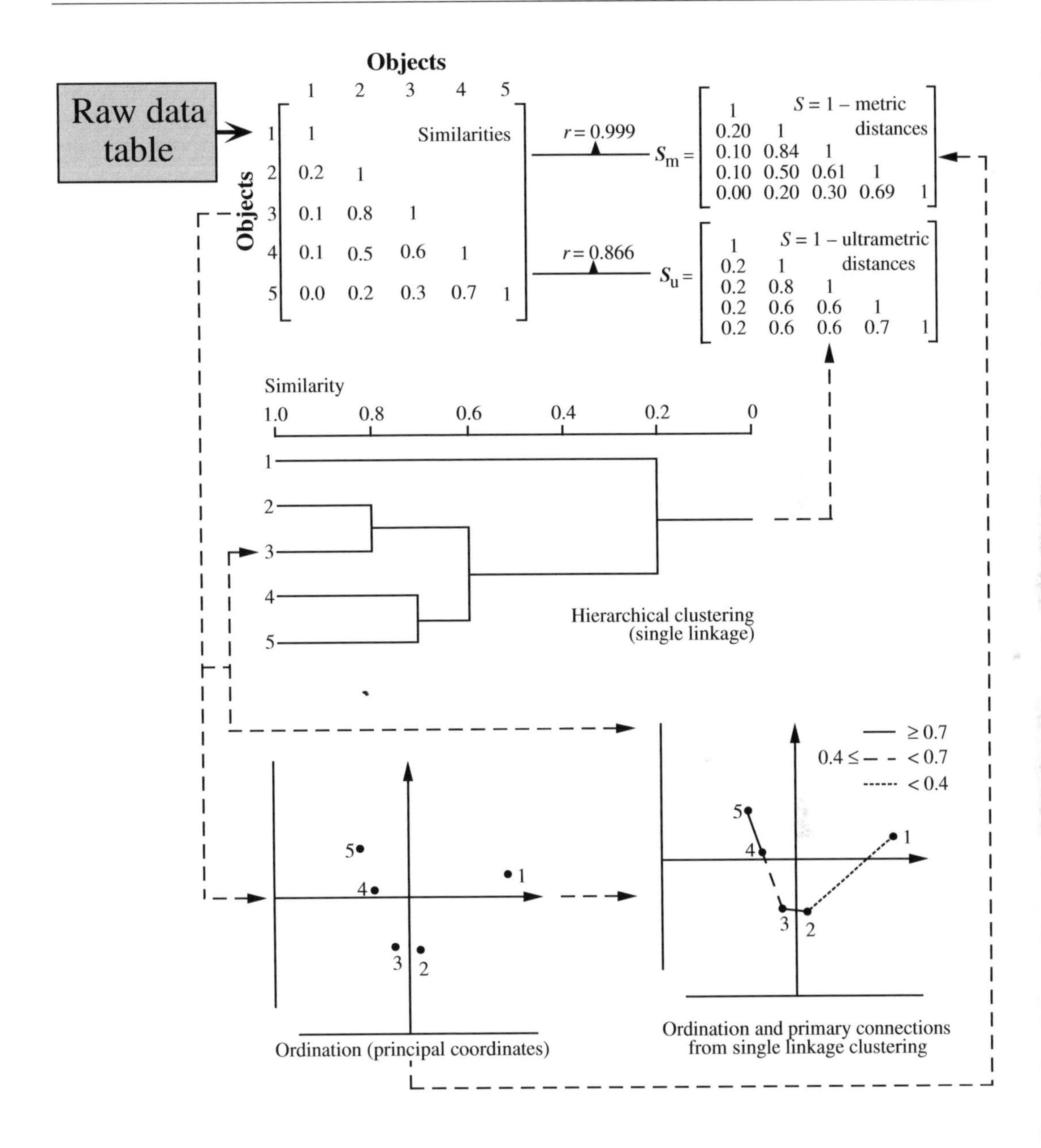

Figure 10.3 Identification of the structure of five objects, using clustering and ordination. Bottom right: the chain of primary connections is superimposed on a 2-dimensional ordination, as in Figs. 10.1 and 10.2. Top: the reduced-space ordination and the clustering results are compared to the resemblance matrix from which they originate. Upper right (top): a matrix of metric distances (or its complement $\mathbf{S_m} = [1 - D_m]$) is computed from the reduced-space ordination, and compared to the original similarities using matrix correlation ($r = 0.999$ is a rather high score). Upper right (below): a cophenetic matrix (Section 8.3) is computed from the dendrogram, and compared to the original similarities using matrix correlation ($r = 0.866$).

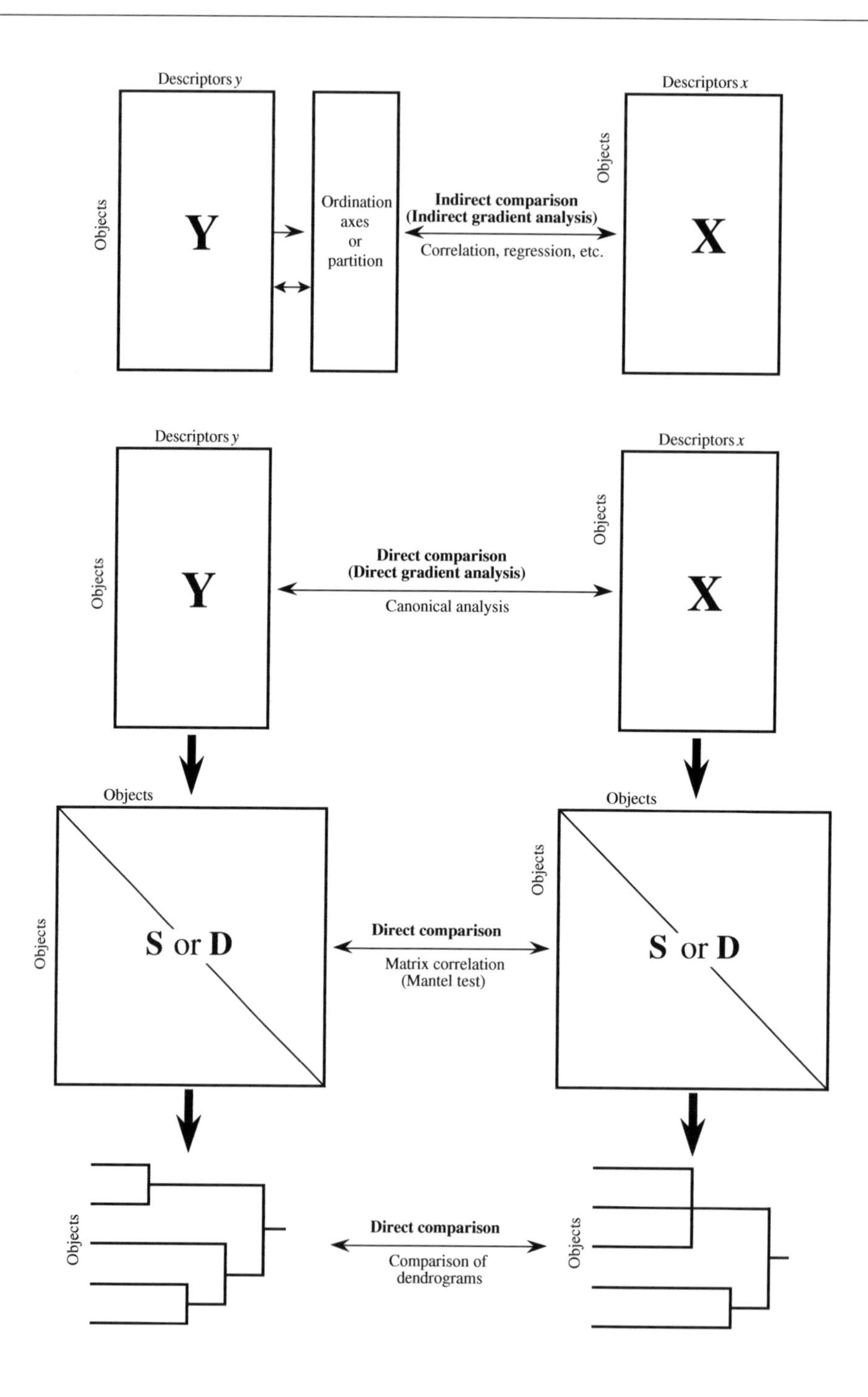

Figure 10.4 Indirect and direct comparison approaches for analysing and interpreting the structure of ecological data. Single thin arrow: inference of structure. Double arrow: interpretation strategy.

techniques of matrix comparison (Section 10.5). One may also directly compare dendrograms derived from resemblance matrices, using *consensus indices*. Two main approaches have been developed to test the significance of consensus statistics: (1) a probability distribution derived for a given consensus statistic may be used, or (2) a specific test may be carried out to assess the significance of the consensus statistic, in which the reference distribution is found by permuting the two dendrograms under study in some appropriate way (Lapointe & Legendre, 1995). Readers are referred to the papers of Day (1983, 1986), Shao & Rohlf (1983), Shao & Sokal (1986), Lapointe & Legendre (1990, 1991, 1992a, 1992b, 1995), and Steel & Penny (1993), where these methods are described. Lapointe & Legendre (1994) used the three forms of direct comparison analysis (i.e. comparison of raw data, distance matrices, and dendrograms; Fig. 10.4) on five data sets describing the same objects. They showed that all methods essentially led to similar conclusions, with minor differences.

Consensus index

Permutation test

The interpretation of a structure, using the descriptors from which it originates, makes it possible to identify which descriptors mainly account for the structuring of the objects. In some methods of ordination (e.g. principal component analysis, correspondence analysis), the eigenvectors readily identify the important descriptors. Other types of ordination, or the clustering techniques, do not directly provide this information, which must therefore be found *a posteriori* using methods of indirect comparison. This type of interpretation does not allow one to perform formal tests of significance. The reason is that the structure under study is derived from the very same descriptors that are now used to interpret it; it is thus not independent of them.

Interpretation of a structure using external information (data table **X** in Fig. 10.4) is central to numerical ecology. This approach is used, for example, to diagnose abiotic conditions (response data table **Y**) from the available biological descriptors (explanatory data table **X**) or, alternatively, to forecast the responses of species assemblages (table **Y**) using available environmental descriptors (table **X**). In the same way, it is possible to compare two groups of biological descriptors, or two tables of environmental data. Until the mid-1980's, the indirect comparison scheme was favoured because of methodological problems with the classical technique of canonical correlations, which was then the only one available in computer packages to analyse two sets of descriptors. With the availability of new computer programs and methods, the direct comparison scheme is becoming increasingly popular in the ecological literature.

In the indirect comparison approach, the first set of descriptors is reduced to a single or a few one-dimensional variables (i.e. a partition resulting from clustering, or one or several ordination axes, the latter being generally interpreted one at the time). It follows that the methods of interpretation for univariate descriptors may also be used for indirect comparisons. This is the approach used in Tables 10.1 and 10.2.

1 — Explaining ecological structures

Table 10.1 summarizes the methods available for *explaining* the structure of one or several ecological descriptors. The purpose here is data exploration, not hypothesis testing. The first dichotomy of the Table separates methods for univariate descriptors (used also in the indirect comparison approach) from those for multivariate data.

Methods used for explaining the structure of *univariate descriptors* belong to three major groups: (1) measures of dependence, (2) discriminant functions, (3) and methods for qualitative descriptors. Methods used for explaining the structure of *multivariate descriptors* belong to two major types: (4) canonical analysis methods and (5) matrix comparison methods. The following paragraphs briefly review these five groups of methods, paying special attention to those that are not discussed elsewhere in this book.

1. Various coefficients have been described in Chapters 4 and 5 to measure the *dependence* between two descriptors exhibiting monotonic relationships (i.e. the parametric and nonparametric *correlation coefficients*). When there are more than two descriptors, one may use the *coefficients of partial correlation* or the *coefficient of concordance* (Section 5.3). The *coefficient of multiple correlation* (R^2), which is derived from multiple regression (*multiple linear regression* and *dummy variable regression*), may be used when the response descriptor is quantitative. *Dummy variable regression* is the same as multiple regression, but conducted on explanatory variables that are qualitative or of mixed levels of precision; the qualitative variables are coded as *dummy variables*, as explained in Subsection 1.5.7. Finally, in *logistic regression*, it is possible to compute partial correlation coefficients between the response and each explanatory variables. These different types of regression are briefly discussed in Subsection 2, in relation with Table 10.2, and in more detail in Section 10.3.

2. Explaining the structure of a qualitative descriptor is often called *discrimination*, when the aim of the analysis is to identify explanatory descriptors that would allow one to discriminate among the various states of the qualitative descriptor. *Discriminant analysis* may be used when (1) the explanatory (or *discriminant*) descriptors are quantitative, (2) their distributions are not too far from normal, and (3) the within-group dispersion matrices are reasonably homogeneous. Discriminant analysis is described in Section 11.5. Its use with species data is discussed in Section 11.6, where alternative strategies are proposed.

3. When both the descriptor to be explained and the explanatory descriptors are qualitative, one may use *multidimensional contingency table analysis*. It is then imperative to follow the rules, given in Section 6.3, concerning the models to use when a distinction is made between the explained and explanatory descriptors. When the response variable is binary, *logistic regression* may be a better choice than multidimensional contingency table analysis. An additional advantage is that logistic regression allows one to use explanatory variables presenting a mixture of precision

Table 10.1 Numerical methods for *explaining* the structure of descriptors, using either the descriptors from which the structure originates, or other, potentially explanatory descriptors. In parentheses, identification of the Section where a method is discussed. Tests of significance cannot be performed when the structure of a descriptor is explained by the descriptors at the origin of that structure.

1) Explanation of the structure of a *single* descriptor, or *indirect comparison* .... **see 2**
- 2) Structure of a quantitative or a semiquantitative descriptor .... **see 3**
  - 3) Explanatory descriptors are quantitative or semiquantitative .... **see 4**
    - 4) To *measure* the dependence between descriptors .... **see 5**
      - 5) Pairs of descriptors: *Pearson r*, for quantitative descriptors exhibiting linear relationships (4.2); *Kendall* τ or *Spearman r*, for quantitative or semiquantitative descriptors exhibiting monotonic relationships (5.2)
      - 5) A single quantitative descriptor as a function of several others: *coefficient of multiple determination* R^2 (4.5)
      - 5) Several descriptors exhibiting monotonic relationships: *coefficient of concordance W* (5.2)
    - 4) To *interpret* the structure of a single descriptor: *partial Pearson r*, for quantitative descriptors exhibiting linear relationships (4.5); *partial Kendall* τ, for descriptors exhibiting monotonic relationships (5.2)
  - 3) Explanatory descriptors of mixed precision: R^2 of *dummy variable regression* (10.3)
  - 3) Estimation of the dependence between descriptors of the sites and descriptors of the species (any precision level): *the 4th-corner method* (10.6)
- 2) Structure of a qualitative descriptor (*or* of a classification) .... **see 6**
  - 6) Explanatory descriptors are quantitative: *discriminant analysis* (11.5)
  - 6) Explanatory descriptors are qualitative: *multidimensional contingency table analysis* (6.3); *discrete discriminant analysis* (10.2)
  - 6) Explanatory descriptors are of mixed precision: *logistic regression* (in most computer programs, the explained descriptor is binary; 10.3)

1) Explanation of the structure of a *multivariate* data table .... **see 7**
- 7) *Direct comparison* .... **see 8**
  - 8) Structure of quantitative descriptors explained by quantitative descriptors: *redundancy analysis* (variables in linear relationships; 11.1); *canonical correspondence analysis* (species data, unimodal distributions; 11.2)
  - 8) The response and the explanatory data tables are transformed into resemblance matrices, using *S* or *D* functions appropriate to their mathematical types: *matrix comparison* (*Mantel test, Procrustes analysis*: 10.5)
  - 8) Classifications are computed for the two data tables .... **see 9**
    - 9) Partitions are compared: *contingency table analysis* (6.2), or *modified Rand index* (8.11)
    - 9) Dendrograms are compared (10.2, Fig. 10.4)
- 7) *Indirect comparison* .... **see 10**
  - 10) Ordination in reduced space: each axis is treated in the same way as a single quantitative descriptor .... **see 2**
  - 10) Clustering: each partition is treated as a qualitative descriptor .... **see 2**

levels. For qualitative variables, the equivalent of discriminant analysis is called *discrete discriminant analysis*. Goldstein & Dillon (1978) describe models used for this analysis and provide FORTRAN programs.

4. The standard approach for comparing two sets of descriptors is *canonical analysis* (Chapter 11). The classical method in parametric statistics is *canonical correlation analysis* (Section 11.4), which may be thought of as two principal component analyses — one on each of the two sets — followed by rotation of the principal axes so as to make them correspond, i.e. maximizing their correlations. Canonical correlations are restricted to quantitative descriptors where the relationships between the two data sets are linear; they may also include binary descriptors, just as in multiple regression. There are two problems with this method in the context of the explanation of ecological structures. (1) The solution of canonical correlations, even when mathematically valid, may not necessarily lead to interesting results because the highest correlations may well be found between axes which are of minor importance for the two data sets. It may be simpler to conduct a principal component analysis that includes both sets of descriptors, whose results would be easier to interpret than those of a canonical correlation analysis. Ecological application 9.1a is an example of this approach. (2) In most instances in ecology, one is not interested so much in correlating two data sets as to explain one using the other. In other words, the questions to be answered focus on one of the two sets, which is thought of as the response, or dependent data set, while the other is the explanatory, or independent data table. The solution to these two problems is found in an indirect comparison approach, where one asks how much of the structure of the response data set is explained by the explanatory data table. Two variants of canonical analysis are now available to do so: *redundancy analysis* and *canonical correspondence analysis* (Sections 11.1 and 11.2). The main difference between the two methods is the same as between principal component and correspondence analyses (Table 9.1).

5. Raw data tables may be turned into similarity or distance matrices (Fig. 10.4) when one wishes to express the relationships among objects through a specific measure of resemblance, or because the descriptors are of mixed types; similarity coefficients are available to handle mixed-type data (S_{15}, S_{16}, S_{19}, S_{20}, Chapter 7). Two resemblance matrices concerning the same objects may be compared using *matrix correlation* (Subsection 8.11.2), that is, by computing a parametric or nonparametric correlation coefficient between corresponding values in these two matrices (excluding the main diagonals). Furthermore, when the two resemblance matrices are independent of each other, i.e. they originate from different data sets, the matrix correlation may be tested for significance using the *Mantel test* (Section 10.5). In the same way, classifications of objects may be computed from resemblance matrices (Fig. 10.4); two classifications may be compared using appropriate techniques. (1) If one is concerned with specific partitions resulting from hierarchical classifications, or if a non-hierarchical method of classification has been used, one may compare two partitions using *contingency table analysis*, since partitions are equivalent to qualitative descriptors, or the *modified Rand index* (Subsection 8.11.2). (2) If one is interested in the relationships depicted by whole dendrograms, cophenetic matrices corresponding

to the two dendrograms may be compared and tested for significance using the methods mentioned in the paragraphs where Fig. 10.4 is described. An interesting application of these methods is the comparison of a dendrogram computed from data to a dendrogram taken from the literature.

6. Consider a (site × species) table containing presence-absence data, for which supplementary variables are known for the sites (e.g. habitat characteristics) and for the species (e.g. biological or behavioural traits). The *4th-corner method*, described in Section 10.6, offers a way to estimate the dependence between the supplementary variables of the rows and those of the columns, and to test the resulting correlation-like statistics for significance.

2 — Forecasting ecological structures

Forecasting model
Predictive model

It is useful to recall here the distinction between *forecasting* and *prediction* in ecology. Forecasting models extend, into the future or to different situations, structural relationships among descriptors that have been quantified for a given data set. A set of relationships among variables, which simply describe the changes in one or several descriptors in response to changes in others as computed from a "training set", make up a *forecasting* model. In contrast, when the relationships are assumed to be causal and to describe a process, the model is *predictive*. A condition to successful forecasting is that the values of all important variables that have not been observed (or controlled, in the case of an experiment) be about the same in the new situation as they were during the survey or experiment. In addition, forecasting does not allow extrapolation beyond the observed range of the explanatory variables. *Forecasting models* (also called *correlative models*) are frequently used in ecology, where they are sometimes misleadingly called "predictive models". Forecasting models are useful, provided that the above conditions are fulfilled. In contrast, predictive models describe known or assumed causal relationships. They allow one to estimate the effects, on some variables, of changes in other variables; they will be briefly discussed at the beginning of the next Subsection.

Methods in Table 10.2 are used to *forecast* descriptors. As in Table 10.1, the first dichotomy in the Table distinguishes the methods that allow one to forecast a single descriptor (*response* or *dependent* variable) from those that may be used to simultaneously forecast several descriptors. Forecasting methods belong to five major groups: (1) regression models, (2) identification functions, (3) canonical analysis methods, and (4) matrix comparison methods.

1. Methods belonging to *regression* models are numerous. Several regression methods include measures of dependence that have already been mentioned in the discussion of Table 10.1: *multiple linear regression* (the explanatory variables must be quantitative), *dummy variable regression* (i.e. multiple regression conducted on explanatory variables that are qualitative or of mixed levels of precision; the qualitative variables are then coded as *dummy variables*, as explained in Subsection 1.5.7), and *logistic regression* (the explanatory variables may be of mixed

Table 10.2 Numerical methods to *forecast* one or several descriptors (response or dependent variables) using other descriptors (explanatory or independent variables). In parentheses, identification of the Section where a method is discussed.

1) Forecasting the structure of a *single* descriptor, or *indirect comparison* .... **see 2**
  2) The response variable is quantitative .... **see 3**
    3) The explanatory variables are quantitative............................. **see 4**
      4) Null or low correlations among explanatory variables: *multiple linear regression* (10.3); *nonlinear regression* (10.3)
      4) High correlations among explanatory variables (collinearity): *ridge regression* (10.3); *regression on principal components* (10.3)
    3) The explanatory variables are of mixed precision: *dummy variable regression* (10.3)
  2) The response variable is qualitative (*or* a classification) .... **see 5**
    5) Response: two or more groups; explanatory variables are quantitative (but qualitative variables may be recoded into dummy variables): *identification functions in discriminant analysis* (11.5)
    5) Response: binary (presence-absence); explanatory variables are quantitative (but qualitative variables may be recoded into dummy var.): *logistic regression* (10.3)
  2) The response and explanatory variables are quantitative, but they display a nonlinear relationship: *nonlinear regression* (10.3)
1) Forecasting the structure of a *multivariate* data table............................ **see 6**
  6) *Direct comparison*.. **see 7**
    7) The response as well as the explanatory variables are quantitative: *redundancy analysis* (variables linearly related; 11.1); *canonical correspondence analysis* (species presence-absence or abundance data; unimodal distributions; 11.2)
    7) Forecasting a resemblance matrix, or a cophenetic matrix representing a dendrogram, using several other explanatory resemblance matrices: *multiple regression on resemblance matrices* (10.5)
  6) *Indirect comparison* .... **see 8**
    8) Ordination in reduced space: each axis is treated in the same way as a single quantitative descriptor .... **see 2**
    8) Clustering: each partition is treated as a qualitative descriptor .... **see 2**

levels of precision; the response variable is qualitative; most computer programs are limited to the binary case*). Section 10.3 provides a detailed description of several regression methods.

2. *Identification functions* are part of multiple discriminant analysis (Section 11.5), whose discriminant functions were briefly introduced in the previous Subsection.

* In the SAS computer package, the standard procedure for logistic regression is LOGIST. One may also use CATMOD, which makes it possible to forecast a multi-state qualitative descriptor.

These functions allow the assignment of any object to one of the states of a qualitative descriptor, using the values taken by several quantitative variables (i.e. the explanatory or discriminant descriptors). As already mentioned in the previous Subsection, the distributions of the discriminant descriptors must not be too far from normality, and their within-group dispersion matrices must be reasonably homogeneous (i.e. about the same among groups).

3. Canonical analysis, and especially *redundancy analysis* and *canonical correspondence analysis*, which were briefly discussed in the previous Subsection (and in more detail in Sections 11.1 and 11.2), allow one to model a data table from the descriptors of a second data table; these two data tables form the "training set". Using the resulting model, it is possible to forecast the position of any new observation among those of the "training set", e.g. along environmental gradients. The new observation may represent some condition which may occur in the future, or at a different but comparable location.

4. Finally, resemblance (**S** or **D**) and cophenetic matrices representing dendrograms may be interpreted in the regression framework, against an array of other resemblance matrices, using *multiple regression on resemblance matrices* (Subsection 10.5.2). The permutational tests of significance for the regression parameters (R^2 and partial regression coefficients) are performed in the manner of either the Mantel test or the double-permutation test, depending on the nature of the dependant matrix (an ordinary similarity or distance matrix, or a cophenetic matrix).

3 — Ecological prediction

As explained in the Foreword, numerical modelling does not belong to numerical ecology *sensu stricto*. However, some methods of numerical ecology may be used to analyse causal relationships among a small number of descriptors, thus linking numerical ecology to predictive modelling. Contrary to the *forecasting* or *correlative models* (previous Subsection), *predictive models* allow one to foresee how some variables of interest would be affected by changes in other variables. Prediction is possible when the model is based on causal relationships among descriptors (i.e. not only correlative evidence). Causal relationships are stated as hypotheses (theory) for modelling; they may also be validated through experiments in the laboratory or in the field. In *manipulative experiments*, one observes the responses of some descriptors to user-determined changes in others, by reference to a *control*. Besides manipulative experiments, which involve two or more treatments, Hurlbert (1984) recognizes *mensurative experiments* which involve measurements made at one or more points in space or time and allow one to test hypotheses about patterns in space (Chapter 13) and/or time (Chapter 12). The numerical methods in Table 10.3 allow one to explore a network of causal hypotheses, using the observed relationships among descriptors. The design of experiments and analysis of experimental results are discussed by Mead (1988) who offers a statistically-oriented presentation, and by Underwood (1997) in a book emphasizing ecological experiments.

Predictive model

Experiment

Table 10.3 Numerical methods for analysing causal relationships among ecological descriptors, with the purpose of *predicting* one or several descriptors using other descriptors. In parentheses, identification of the Section where a method is discussed. In addition, forecasting methods (Table 10.2) may be used for prediction when there are reasons to believe that the relationships between explanatory and response variables are of causal nature.

1) The causal relationships among descriptors are given by hypothesis .... **see 2**
   2) Quantitative descriptors; linear causal relationships: *causal modelling using correlations* (4.5); *path analysis* (10.4)
   2) Qualitative descriptors: *logit* and *log-linear models* (6.3)
   2) Modelling from resemblance matrices: *causal modelling on resemblance matrices* (10.5)

1) Hidden variables (latent variables, factors) are assumed to cause the observed structure of the descriptors: *confirmatory factor analysis* (9.5)

One may hypothesize that there exist causal relationships among the observed descriptors or, alternatively, that the observed descriptors are caused by underlying hidden variables. Depending on the hypothesis, the methods for analysing causal relationships are not the same (Table 10.3). Methods appropriate to the first case belong to the family of *path analysis*; the second case leads to *confirmatory factor analysis*. The present Chapter only discusses the former since the latter was explained in Section 9.5. In addition to these methods, techniques of forecasting (Table 10.2) may be used for predictive purposes when there are reasons to believe that the relationships between explanatory and response variables are of causal nature.

Fundamentals of *path analysis* are presented in Section 10.4. Path analysis is an extension of multiple linear regression and is thus limited to quantitative or binary descriptors (including qualitative descriptors recoded as dummy variables: Subsection 1.5.7). In summary, path analysis is used to decompose and interpret the relationships among a small number of descriptors, assuming (a) a *(weak) causal order* among descriptors, and (b) that the relationships among descriptors are *causally closed*. *Causal order* means, for example, that y_2 possibly (but not necessarily) affects y_3 but that, under no circumstance, y_3 would affect y_2 through the same process. Double causal "arrows" are allowed in a model only if different mechanisms may be hypothesized for the reciprocal relationships. Using this assumption, it is possible to set a causal order between y_2 and y_3. The assumption of *causal closure* implies independence of the residual causalities, which are the unknown factors responsible for the residual variance (i.e. the variance not accounted for by the observed descriptors). Path analysis is restricted to a small number of descriptors. This is not due to computational problems, but to the fact that the interpretation becomes complex when the number of descriptors in a model becomes large.

When the analysis involves three descriptors only, the simple method of *causal modelling using correlations* may be used (Subsection 4.5.5). For three resemblance matrices, causal modelling may be carried out using the results of Mantel and partial Mantel tests, as described in Subsection 10.5.2 and Section 13.6.

For qualitative descriptors, Fienberg (1980; his Chapter 7) explains how to use *logit* or *log-linear models* (Section 6.3) to determine the signs of causal relationships among such descriptors, by reference to diagrams similar to the path diagrams of Section 10.4.

10.3 Regression

The purpose of regression analysis is to describe the relationship between a *dependent* (or *response*) *random* variable* (y) and a set of *independent* (or *explanatory*) *variables*, in order to forecast or predict the values of y for given values of the independent variables $x_1, x_2, \ldots, x_p$. Box 1.1 gives the terminology used to refer to the dependent and independent variables of a regression model in an empirical or causal framework. The explanatory variables may be either random* or controlled (and, consequently, known *a priori*). On the contrary, the response variable must of necessity be a random variable. That the explanatory variables be random or controlled will be important when choosing the appropriate computation method (model I or II).

Model

A *mathematical model* is simply a mathematical formulation (algebraic, in the case of regression models) of a relationship, or set of relationships among variables, whose parameters have to be estimated, or that are to be tested; in other words, it is a simplified mathematical description of a real-life system. Regression, with its many variants, is the first type of modelling method presented in this Chapter for analysing ecological structures. It is also used as a platform to help introduce the principles of structure analysis. The same principles will apply to more advanced forms, collectively referred to as canonical analysis, that are discussed in Chapter 11.

Regression modelling may be used for description, inference, or forecasting/prediction:

1. Description aims at finding the best functional relationship among variables in the model, and estimating its parameters, based on available data. In mathematics, a function $y = f(x)$ is a rule of correspondence, often written as an equation, that associates with each value of x one and only one value of y. A well-known functional

* A random variable is a variable whose values are assumed to result from some random process (p. 1); these values are not known before observations are made. A random variable is *not* a variable consisting of numbers drawn at random; such variables, usually generated with the help of a pseudo-random number generator, are used by statisticians to assess the properties of statistical methods under some hypothesis.

relationship in physics is Einstein's equation $E = mc^2$, which describes the amount of energy E associated with given amounts of mass m; the scalar value c^2 is the parameter of the model, where c is the speed of light in vacuum.

2. Inference means generalizing the results of a set of observations to the whole target population, as represented by a sample drawn from that population. Inference may consist in estimating the confidence intervals within which the true values of the statistical population parameters are likely to be found, or testing *a priori* hypotheses about the values of model parameters in the statistical population. (1) The ecological hypotheses may simply concern the *existence* of a relationship (i.e. the slope is different from 0), and/or it may state that the intercept is different from zero. The test consists in finding the *two-tailed* probability of observing the slope (b_1) and/or intercept (b_0) values which have been estimated from the sample data, given the null hypothesis (H_0) stating that the slope (β_1) and/or intercept (β_0) parameters are zero in the statistical population. These tests are described in manuals of elementary statistics. (2) In other instances, the ecological hypothesis concerns the sign that the relationship should have. One then tests the *one-tailed* null statistical hypotheses (H_0) that the intercept and/or slope parameters in the statistical population are zero, against alternative hypotheses (H_1) that they have the signs (positive or negative) stated in the ecological hypotheses. For example, one might want to test Bergmann's law (1847), that the body mass of homeotherms, within species or groups of closely related species, *increases* with latitude. (3) There are also cases where the ecological hypothesis states specific values for the parameters. Consider for instance the isometric relationship specifying that mass should increase as the cube of the length in animals, or in log form: $\log(mass) = b_0 + 3\log(length)$. Length-to-mass relationships found in nature are most often allometric, especially when considering a multi-species group of organisms. Reviewing the literature, Peters (1983) reported allometric slope values from 1.9 (algae) to 3.64 (salamanders).

3. Forecasting (or prediction) consists in calculating values of the response variable using a regression equation. Forecasting (or prediction) is sometimes described as *the* purpose of ecology. In any case, ecologists agree that empirical or hypothesis-based regression equations are helpful tools for management. This objective is achieved by using the equation that minimizes the residual mean square error, or maximizes the coefficient of determination (r^2 in simple regression; R^2 in multiple regression).

A study may focus on one or two of the above objectives, but not necessarily all three. Satisfying two or all three objectives may call upon different methods for computing the regressions. In any case, these objectives differ from that of correlation analysis, which is to support the existence of a relationship between two random variables, without reference to any functional or causal link between them (Box 10.1).

This Section does not attempt to present regression analysis in a comprehensive way. Interested readers are referred to general texts of (bio)statistics such as Sokal & Rohlf (1995), specialized texts on regression analysis (e.g. Draper & Smith, 1981), or textbooks such as those of Ratkowski (1983) or Ross (1990) for nonlinear estimation.

Correlation or regression analysis? Box 10.1

Regression analysis is a type of modelling. Its purpose is either to find the best functional model relating a response variable to one or several explanatory variables, in order to test hypotheses about the model parameters, or to forecast or predict values of the response variable.

The purpose of correlation analysis is quite different. It aims at establishing whether there is *interdependence*, in the sense of the coefficients of dependence of Chapter 7, between two random variables, without assuming any functional or explanatory-response or causal link between them.

In model I simple linear regression, where the explanatory variable of the model is controlled, the distinction is easy to make; in that case, a correlation hypothesis (i.e. interdependence) is meaningless. Confusion comes from the fact that the coefficient of determination, r^2, which is essential to estimate the forecasting value of a regression equation and is automatically reported by most regression programs, happens to be the square of the coefficient of linear correlation.

When the two variables are random (i.e. not controlled), the distinction is more tenuous and depends on the intent of the investigator. If the purpose is modelling (as broadly defined in the first paragraph of this Box), model II regression is the appropriate type of analysis; otherwise, correlation should be used to measure the interdependence between such variables. In Sections 4.5 and 10.4, the same confusion is rampant, since correlation coefficients are used as an *algebraic tool* for choosing among causal models or for estimating path coefficients.

The purpose here is to survey the main principles of regression analysis and, in the light of these principles, explain the differences among the regression models most commonly used by ecologists: simple linear (model I and model II), multiple linear, polynomial, partial, nonlinear, and logistic. Some smoothing methods will also be described. Several other types of regression will be mentioned, such as dummy variable regression, ridge regression, multivariate linear regression, and monotone or nonparametric regression.

Regression

Incidentally, the term *regression* has a curious origin. It was coined by the anthropologist Francis Galton (1889, pp. 95-99), a cousin of Charles Darwin, who was studying the relationship between the heights of parents and offspring. Galton observed "that the Stature of the adult offspring … [is] … more *mediocre* than the stature of their Parents", or in other words, closer to the population mean; so, Galton

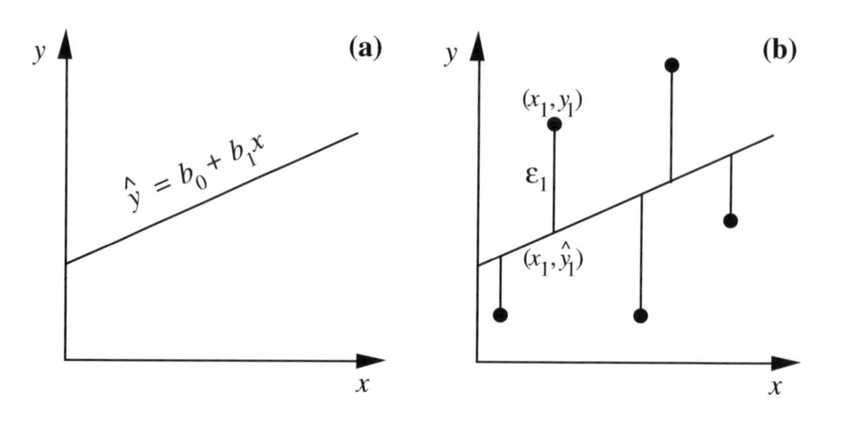

Figure 10.5 (a) Linear regression line, of equation $\hat{y} = b_0 + b_1x$, fitted to the scatter of points shown in b. (b) Graphical representation of regression residuals ε_i (vertical lines); ε_1 is the residual for point 1 with coordinates (x_1, y_1).

said, they *regressed* (meaning *going back*) towards the population mean. He called the slope of this relationship "the ratio of 'Filial Regression' ". For this historical reason, the slope parameter is now known as the regression coefficient.

1 — Simple linear regression: model I

Linear regression is used to compute the parameters of a first-degree equation relating variables y and x. The expression *simple linear regression* applies to cases where there is a single explanatory variable x. The equation (or model) for simple linear regression has the form:

$$\hat{y} = b_0 + b_1x \tag{10.1}$$

This corresponds to the equation of a straight line (hence the name *linear*) that crosses the scatter of points in some optimal way and allows the computation of an estimated value $\hat{y}$ (ordinate of the scatter diagram) for any value of x (abscissa; Fig. 10.5a). Parameter b_0 is the estimate of the intercept of the regression line with the *ordinate*; it is also called the y-intercept. Parameter b_1 is the slope of the regression line; it is also called the *regression coefficient*. In the Subsection on polynomial regression, a distinction will be made between linearity in parameters and linearity in response to the explanatory variables.

Intercept
Slope

When using this type of regression, one must be aware of the fact that a *linear model* is imposed. In other words, one assumes that the relationship between variables may be adequately described by a straight line and that the vertical dispersion of observed values above and below the line is the result of a random process. The difference between the observed and estimated values along y, noted $\varepsilon_i = (y_i - \hat{y}_i)$

for every observation i, may be either positive or negative since the observed data points may lie above or below the regression line. ε_i is called the *residual* value of observation y_i after fitting the regression line (Fig. 10.5b). Including ε_i in the equation allows one to describe exactly the ordinate value y_i of each point (x_i, y_i) of the data set; y_i is equal to the value $\hat{y}_i$ predicted by the regression equation plus the residual ε_i:

$$y_i = \hat{y}_i + \varepsilon_i = b_0 + b_1 x_i + \varepsilon_i \qquad \textbf{(10.2)}$$

This equation is the *linear model* of the relationship. $\hat{y}_i$ is the predicted, or *fitted* value corresponding to each observation i. The model assumes that the only deviations from the linear functional relationship $y = b_0 + b_1 x$ are vertical differences ("errors") ε_i on values y_i of the response variable, and that there is no "error" associated with the estimation of x. "Error" is the traditional term used by statisticians for deviations of all kind due to random processes, and not only measurement error. In practice, when it is known by hypothesis — or found by studying a scatter diagram — that the relationship between two variables is not linear, one may either try to linearise it (Section 1.5), or else use polynomial or nonlinear regression methods to model the relationship (Subsections 4 and 6, below).

Model I

Besides the supposition that the variables under study are linearly related, *model I regression* makes the following additional assumptions about the data:

1. The explanatory variable x is controlled, or it is measured without error. (The concepts of random and controlled variables have been briefly explained above.)

2. For any given value of x, the y's are independently and normally distributed. This does not mean that the response variable y must be normally distributed, but instead that the "errors" ε_i are normally distributed about a mean of zero. One also assumes that the ε_i's have the same variance for all values of x in the range of the observed data (homoscedasticity: Box 1.3).

So, model I regression is appropriate to analyse results of controlled experiments, and also the many cases of field data where a response random variable y is to be related to sampling variables under the control of the researcher (e.g. location in time and space, volume of water filtered). The next Subsection will show how to use model II regression to analyse situations where these assumptions are not met.

Least squares

In simple linear regression, one is looking for the straight line with equation $\hat{y} = b_0 + b_1 x$ that minimizes the sum of squares of the vertical residuals, ε_i, between the observed values and the regression line. This is the *principle of least squares*, first proposed by the mathematician Adrien Marie Le Gendre from France, in 1805, and later by Karl Friedrich Gauss from Germany, in 1809; these two mathematicians were interested in problems of astronomy. This sum of squared residuals, $\Sigma(y_i - \hat{y}_i)^2$, offers the advantage of providing a unique solution, which would not be the case if one chose to minimize another function — for example $\Sigma|y_i - \hat{y}_i|$. It can also be shown

OLS

that the straight line that meets the *ordinary least-squares* (OLS) criterion passes through the centroid, or centre of mass $(\bar{x}, \bar{y})$ of the scatter of points, whose

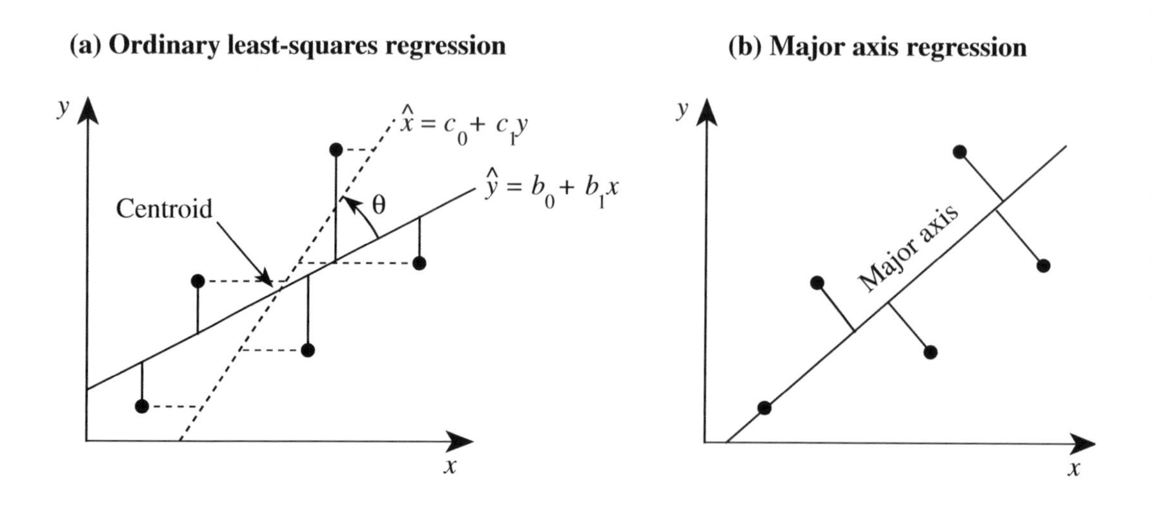

Figure 10.6 (a) Two least-squares regression equations are possible in the case of two random variables (called x and y here, for simplicity). When regressing y on x, the sum of *vertical* squared deviations is minimized (full lines); when regressing x on y, the sum of *horizontal* squared deviations is minimized (dashed lines). Angle θ between the two regression lines is computed using eq. 10.5. (b) In major axis regression, the sum of the squared Euclidean distances to the regression line is minimized.

coordinates are the means $\bar{x}$ and $\bar{y}$. The formulae for parameters b_0 and b_1 of the line meeting the least-squares criterion are found using partial derivatives. The solution is:

$$b_1 = s_{xy}/s_x^2 \quad \text{and} \quad b_0 = \bar{y} - b_1\bar{x} \qquad \textbf{(10.3)}$$

where s_{xy} and s_x^2 are estimates of covariance and variance, respectively (Section 4.1). These formulae, written in full, are found in textbooks of introductory statistics. Least-squares estimates of b_0 and b_1 may also be computed directly from the **x** and **y** data vectors, using matrix eq. 2.19. Least-squares estimation provides the line of best fit for parameter estimation and forecasting when the explanatory variable is controlled.

Regressing y on x does not lead to the same least-squares equation as regressing x on y. Figure 10.6a illustrates this for two random variables; they would thus represent a case for model II regression, discussed in the next Subsection. Even when x is a random variable, the variables will continue to be called x and y (instead of y_1 and y_2) to keep the notation simple. Although the covariance s_{xy} is the same for the regression coefficient of y on x ($b_{1(y \cdot x)}$) and that of x on y ($c_{1(x \cdot y)}$), the denominator of the slope equation (eq. 10.3) is s_x^2 when regressing y on x, whereas it is s_y^2 when regressing x on y. Furthermore, the means $\bar{x}$ and $\bar{y}$ play inverted roles when estimating the two intercepts, $b_{0(y \cdot x)}$ and $c_{0(x \cdot y)}$. This emphasizes the importance of clearly defining the explanatory and response variables when performing regression.

The two least-squares regression lines come together only when all observation points fall on the same line (correlation = 1). According to eq. 4.7, $r_{xy} = s_{xy}/s_x s_y$. So, when $r = 1$, $s_{xy} = s_x s_y$ and, since $b_{1(y \cdot x)} = s_{xy}/s_x^2$ (eq. 10.3), then $b_{1(y \cdot x)} = s_x s_y/s_x^2 = s_y/s_x$. Similarly, the slope $c_{1(x \cdot y)}$, which describes the same line in the transposed graph, is $s_x/s_y = 1/b_{1(y \cdot x)}$. In the more general case where r is not equal to 1, $c_{1(x \cdot y)} = r_{xy}^2/b_{1(y \cdot x)}$. When the two regression lines are drawn on the same graph, assuming that the variables have been standardized prior to the computations, there is a direct relationship between the Pearson correlation coefficient r_{xy} and angle θ between the two regression lines:

$$\theta = 90° - 2\tan^{-1} r, \quad \text{or} \quad r = \tan\left(\frac{90° - \theta}{2}\right) \tag{10.4}$$

If $r = 0$, the scatter of points is circular and angle $\theta = 90°$, so that the two regression lines are at a right angle; if $r = 1$, the angle is 0°. Computing angle θ for non-standardized variables, as in Fig. 10.6a, is a bit more complicated:

$$\theta = 90° - [\tan^{-1}(r\, s_x/s_y) + \tan^{-1}(r\, s_y/s_x)] \tag{10.5}$$

Coefficient of deter-mination

The *coefficient of determination* r^2 measures how much of the variance of each variable is explained by the other. This coefficient has the same value for the two regression lines. The amount of explained variance for y is the variance of the fitted values $\hat{y}_i$. It is calculated as:

$$s_{\hat{y}}^2 = \Sigma(\hat{y}_i - \bar{y})^2/(n-1) \tag{10.6}$$

whereas the total amount of variation in variable y is

$$s_y^2 = \Sigma(y_i - \bar{y})^2/(n-1)$$

It can be shown that the coefficient of determination, which is the ratio of these two values (the two denominators $(n - 1)$ cancel out), is equal to the square of the Pearson correlation coefficient r. With two random variables, the regression of y on x makes as much sense as the regression of x on y. In this case, the coefficient of determination may be computed as the product of the two regression coefficients:

$$r^2 = b_{1(y \cdot x)}\, c_{1(x \cdot y)} \tag{10.7}$$

In other words, the coefficient of correlation is the geometric mean of the coefficients of linear regression of each variable on the other: $r = (b_{1(y \cdot x)}\, c_{1(x \cdot y)})^{1/2}$. It may also be computed as the square of r in eq. 4.7:

$$r^2 = \frac{(s_{xy})^2}{s_x^2\, s_y^2} \tag{10.8}$$

Coefficient of non-determination

A value $r^2 = 0.81$, for instance, means that 81% of the variation in y is explained by x, and vice versa. In Section 10.4, the quantity $(1 - r^2)$ will be called the *coefficient of nondetermination*; it measures the proportion of the variance of a response variable that is not explained by the explanatory variable(s) of the model.

When x is a controlled variable, one must be careful not to interpret the coefficient of determination in terms of interdependence, as one would for a coefficient of correlation, in spite of their algebraic closeness and the fact that the one may, indeed, be directly calculated from the other (Box 10.1).

2 — Simple linear regression: model II

Model II

When both the response and explanatory variables of the model are random (i.e. *not* controlled by the researcher), there is error associated with the measurements of x and y. Such situations are often referred to as *model II regression*; they are not the regression equivalent of model II ANOVA, though. Examples are:

• In microbial ecology, the concentrations of two substances produced by bacterial metabolism have been measured. One is of economical interest, but difficult to measure with accuracy, whereas the other is easy to measure. Determining their relationship by regression allows ecologists to use the second substance as a proxy for the first.

• In aquatic ecology, *in vivo* fluorescence is routinely used to estimate phytoplankton chlorophyll *a*. The two variables must be determined to establish their relationship; they are both random and measured with error.

• In comparative growth studies, one may use length as an indicator for the mass of individuals pertaining to a given taxonomic group.

In all these cases, one may be interested in estimating the parameters of the functional relationship, or to use one variable to forecast the other. In model II regression, different computational procedures may be required for description and inference, as opposed to forecasting. Other applications follow.

• In freshwater sediment, one may be interested to compare the rate of microbial anaerobic methane production to total particulate carbon, in two environments (e.g. two lakes) in which several sites have been studied, but that differ in some other way. Since total particulate carbon and methane production have been measured with error in the field, rates are given by the slopes of model II regression equations; the confidence intervals of these slopes may serve to compare the two environments.

• Deterministic models are often used to describe ecological processes. In order to test how good a model is at describing reality, one uses field data about the control variables incorporated in the model, and compares the predictions of the model to the observed field values of the response variable. Since both sets of variables (control, response) are random, the predictions of the model are just as random as the field response variable values, so that they should be compared using model II regression. The hypothesis is one-tailed in this case; indeed, a model may be said to accurately reflect the field process only if its predictions are *positively* correlated with the field observations. Theory and examples are provided by Mesplé *et al.* (1996).

In the examples above, one may be interested to estimate the parameters of the equation that describes the functional relationship between pairs of random variables, in order to quantify the underlying physiological or ecological processes. Model II regression should be used for parameter estimation, in such cases, since the slope found by ordinary least squares (OLS) is biased by the presence of measurement error in the explanatory variable. OLS should be used only when x is fixed by experiment or is a random variable measured without error, or with little error compared to y (see recommendation 1 at the end of the Subsection), or else when the objective of the study is forecasting (recommendation 6, also discussed below).

To give substance to the above assertion, let us consider again the relationship between length and mass of adult animals of a given species. Let us further assume that the relationship is isometric ($mass = c \cdot length^3$) for the species under study; this equation would correspond to the case where all individuals, short or long, have the same shape (fatness). The same functional equation, in log form, is $\log(mass) = b_0 + 3\log(length)$, where b_0 is the log of parameter c. Since individual measurements are each subject to a large number of small genetic and environmental influences, presumably additive in their effects and uncorrelated among individuals, it is expected that both length and mass include random deviations from the functional equation; measurement errors must be added to this inherent variability. In such a system, the slope of the OLS regression line of $\log(mass)$ on $\log(length)$ would be smaller than 3 (Fig. 10.7; Ecological application 10.2), which would lead one to conclude that the species displays allometric growth, with longer individuals thinner than short ones. On the contrary, the slope of the regression line of $\log(length)$ on $\log(mass)$, computed in the transposed space, would produce a slope smaller than 1/3; its inverse, drawn in Fig. 10.7, is larger than 3; this slope would lead to the opposite conclusion, i.e. that shorter individuals are thinner than long ones. This apparent paradox is simply due to the fact that OLS regression is inappropriate to describe the functional relationship between these variables.

When the purpose of regression analysis is merely to compute fitted values for forecasting or prediction purposes, the OLS method should be used. The reason is simple: by definition (see Subsection 1), ordinary least squares is the method that produces fitted values with the smallest error, defined as $\Sigma(y_i - \hat{y}_i)^2$. So, OLS is also one of the methods to be used in model II situations, when the purpose is as stated above.

Several methods have been proposed to estimate model II regression parameters, and the controversy about which is the best still goes on in the literature. The four following methods are the most popular — although, surprisingly, the major statistical

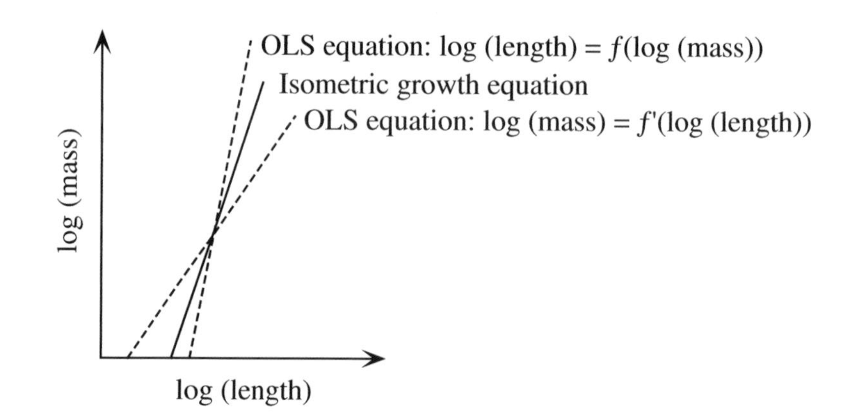

Figure 10.7 Isometric growth is depicted by the functional relationship $\log(mass) = b_0 + 3\log(length)$. The ordinary least-squares (OLS) regression line of $\log(mass)$ on $\log(length)$ would suggest allometric growth of one type, while the OLS regression line of $\log(length)$ on $\log(mass)$ would suggest allometric growth of the opposite type.

packages are still ignoring them (except, of course, method 5, i.e. OLS). For methods 1 to 4 described below, slope estimates can easily be calculated with a pocket calculator, from values of the means, variances, and covariance, computed with standard statistical software.

Methods 1, 2, and 5 are special cases of the *structural relationship*, in which it is assumed that there is error ε_i on y and δ_i on x, ε_i and δ_i being independent of each other. As stated above, "error" means deviation of any kind due to a random process, not only measurement error. The maximum likelihood (ML) estimate of the slope for such data is (Madansky, 1959; Kendall & Stuart, 1966):

ML slope formula

$$b_{ML} = \frac{s_y^2 - \lambda s_x^2 + \sqrt{(s_y^2 - \lambda s_x^2)^2 + 4\lambda s_{xy}^2}}{2s_{xy}} \qquad \textbf{(10.9)}$$

where s_y^2 and s_x^2 are the estimated variances of y and x, respectively, s_{xy} is their covariance, and λ is the ratio $\sigma_\varepsilon^2/\sigma_\delta^2$ of the variances of the two error terms.

When λ is large, another form may prove easier to compute than eq. 10.9; it is derived from the property that the slope of the regression line of y on x is the inverse of the slope of the regression of x on y in the case of symmetric regression lines. After the proper substitutions, equation 10.9 becomes:

$$b_{ML} = \frac{2s_{xy}}{s_x^2 - (s_y^2/\lambda) + \sqrt{[s_x^2 - (s_y^2/\lambda)]^2 + (4s_{xy}^2/\lambda)}} \qquad \textbf{(10.10)}$$

Major axis

• 1 — In *major axis regression* (MA), the estimated regression line is the first principal component of the scatter of points (see principal component analysis, Section 9.1). The quantity which is minimized is the sum, over all points, of the squared *Euclidean distances* between the points and the regression line (Fig. 10.6b), instead of *vertical distances* as in OLS (Fig. 10.6a). In this method, one assumes that the two error variances (σ_ε^2 on y and σ_δ^2 on x) are equal, so that their ratio $\lambda = 1$. This assumption is strictly met, for example, when both variables have been measured using the same instrument and all of the error is measurement error (McArdle, 1988). The slope of the major axis is estimated by the following formula (Pearson, 1901; Jolicoeur, 1973; Sokal & Rohlf, 1995) which is a special case of eq. 10.9, with $\lambda = 1$:

$$b_{MA} = \frac{s_y^2 - s_x^2 + \sqrt{(s_y^2 - s_x^2)^2 + 4(s_{xy})^2}}{2s_{xy}} \tag{10.11}$$

The positive square root is used in the numerator. A second equation is obtained by using the negative square root; it estimates the slope of the minor axis, which is the second principal component, of the bivariate scatter of points. When the covariance is near 0, b_{MA} is estimated using eq. 10.10 (with $\lambda = 1$) instead of eq. 10.11, in order to avoid numerical indetermination.

The slope of the major axis may also be calculated using estimates of the slope of the OLS regression line, b_{OLS}, and of the correlation coefficient, r_{xy}:

$$b_{MA} = \frac{d \pm \sqrt{d^2 + 4}}{2} \quad \text{where} \quad d = \frac{(b_{OLS})^2 - r_{xy}^2}{r_{xy}^2 \times b_{OLS}}$$

The positive root of the radical is used when the correlation coefficient is positive, and conversely (Currie *et al.*[*]).

Just as with principal component analysis, this method is mostly useful in situations where both variables are expressed in the same physical units, or are dimensionless (naturally, or after standardization or some other transformation). Many natural ecological variables are not in the same physical units. Major axis regression has been criticized because, in that case, the slope estimated by major axis regression is not invariant under an arbitrary change of scale such as expansion (Section 1.5) and, after a change of scale, b_{MA} cannot be directly calculated using the change-of-scale factor. In these conditions, the actual *value* of the slope may be meaningless (Teissier, 1948; Kermack and Haldane, 1950; Ricker, 1973; McArdle, 1988) or difficult to interpret. By comparison, the slopes of the OLS, SMA, and RMA regression lines are

[*] Currie, D. J., P. Legendre & A. Vaudor. Regression in the correlation situation. Paper under review.

not invariant either to change-of-scale transformations, but the slopes of the transformed data can easily be calculated using the change-of-scale factor. For example, after regressing a variable in g onto a variable in cm, if the OLS slope is b_1 (in g/cm), then after rescaling the explanatory variable from cm to m, the OLS slope becomes $b'_1 = b_1 \times 100$.

Major axis regression may be used with variables in different physical units (1) when the purpose of the analysis is to compare slopes (e.g. at different sampling sites) computed from variables measured in an identical way. It may also be useful (2) when the purpose of the study is to compare the slope of empirical data to a value given by hypothesis.

Permutation test

Significance of b_{MA} estimates can be tested by permutation (Section 1.2); the values of one or the other variable are permuted a large number of times and slope estimates are computed using eq. 10.11. The test should be carried out on the lesser of the two slopes: b_1 of *y* on *x*, or $b'_1 = 1/b_1$ of *x* on *y*. If the objective is simply to assess the relationship between the two variables under study, the correlation coefficient should be tested for significance instead of the slope of a model II regression line.

When the variances s_y^2 and s_x^2 are equal, the slope estimated by eq. 10.11 is ±1, the sign being that of the covariance, whatever the value of s_{xy}. As in the case of SMA (below), permutations produce slopes estimates of +1 or –1 in equal numbers, with a resulting probability near 0.5 whatever the value of the correlation. This result is meaningless. The practical consequence is that, if the slope estimate b_{MA} is to be tested by permutations, variables should not be standardized (eq. 1.12).

C.I. of MA slope

Alternatively, one may compute the confidence interval of the slope at a predetermined confidence level and check whether the value 0 (or, for that matter, any other value of interest) lies inside or outside the confidence interval. Computation of the confidence interval involves several steps; the formulae are given in Jolicoeur & Mosimann (1968), Jolicoeur (1990), and Sokal & Rohlf (1995, pp. 589-591), among others. When both *n* and the ratio of the eigenvalues of the bivariate distribution (see principal component analysis, Section 9.1) are small, limits of the confidence interval cannot be computed because it covers all 360° of the plane. Such a confidence interval always includes slope 0, as well as any other value. For example, when $n = 10$, the ratio of the eigenvalues must be larger than 2.21 for the 95% confidence interval to be real; for $n = 20$, the ratio must be larger than 1.63; and so on.

It frequently happens in ecology that a scatter plot displays a bivariate lognormal distribution; the univariate frequency distributions of such variables are positively skewed, with longer tails in the direction of the higher values. Such distributions may be normalized by applying a log transformation (Subsection 1.5.6; Fig. 1.9). This transformation also solves the problem of dimensionally heterogeneous variables and makes the estimate of the major axis slope invariant over expansion (multiplication or division by a constant: Section 1.5) — but not over translation. One should verify, of course, that the log-transformed data conform to a bivariate normal distribution before proceeding with major axis regression.

This property can easily be demonstrated as follows. Consider a model II functional equation describing the linear relationship of two log-transformed variables x and y:

$$\log(y) = b_0 + b_1 \log(x)$$

If x and y are divided by constants c_1 and c_2 respectively (expansion), one obtains new variables $x' = x/c_1$ and $y' = y/c_2$, so that $x = c_1 x'$ and $y = c_2 y'$. The functional equation becomes:

$$\log(c_2 y') = b_0 + b_1 \log(c_1 x')$$

$$\log(y') + \log(c_2) = b_0 + b_1 \log(c_1) + b_1 \log(x')$$

$$\log(y') = [b_0 + b_1 \log(c_1) - \log(c_2)] + b_1 \log(x')$$

which may be rewritten as

$$\log(y') = b_0' + b_1 \log(x')$$

where $b_0' = [b_0 + b_1 \log(c_1) - \log(c_2)]$ is the new intercept, while the slope of $\log(x')$ is still b_1. So, under log transformation, the slope b_1 is invariant for any values of expansion coefficients c_1 and c_2; it differs, of course, from the major axis regression coefficient (slope) of the untransformed variables.

Dividing x and y by their respective standard deviations, s_x and s_y, is an expansion which makes the two variables dimensionless. It thus follows that the major axis slope of the original log-transformed data is the same as that of the log of the standardized (dimensionless) data. This also applies to other standardization methods such as division by the maximum value or the range (eqs. 1.10 and 1.11).

Readers who prefer numerical examples can easily check the above derivation by computing a principal component analysis on a small data set containing two log-transformed variables only, with or without expansion (multiplication or division by a constant prior to the log transformation). The angles between the original variables and the first principal component are easily computed as the $\cos^{-1}$ of the values in the first normalized eigenvector (Subsection 9.1.3); the slopes of the major axis regression coefficients of $y = f(x)$ and $x = f(y)$, which are the tangents (tan) of these angles, remain the same over such a transformation.

• 2 — Regression using variables that are not dimensionally homogeneous produces results that vary with the scales of the variables. If the physical dimensions are arbitrary (e.g. measurements that may indifferently be recorded in mm, cm, m, or km), the slope estimate is also arbitrary. In ordinary least squares regression (OLS), the slope and confidence interval values change proportionally to the measurement units. For example, multiplying all y values by 10 produces a slope estimate ten times larger, whereas multiplying all x values by 10 produces a slope estimate 10 times smaller. This is not the case with MA; the major axis slope does not scale proportionally to the units of measurement. For this reason, it may be desirable to make the variables dimensionally homogeneous prior to model II regression.

Standard major axis

Standard major axis regression (SMA) is a way to make the variables dimensionally homogeneous prior to the regression. It is computed as follows:

- Standardize variables x and y using eq. 1.12.
- Compute MA regression on the standardized variables. The slope estimate is always +1 or –1; the sign is that of the covariance s_{xy} or correlation coefficient r_{xy}.
- Back-transform the slope estimate to the original units by multiplying it by (s_y/s_x).

To summarize, the slope of the *standard major axis* (SMA), or *reduced major axis*, is computed as the ratio (Teissier, 1948):

$$b_{\text{SMA}} = \sqrt{s_y^2/s_x^2} = \pm(s_y/s_x) \qquad \textbf{(10.12)}$$

This formula is obtained from eq. 10.9 by assuming that the error variances σ_ε^2 and σ_δ^2 of y and x, respectively, are identically proportional to their respective variances σ_y^2 and σ_x^2; in other words, $\sigma_\varepsilon^2/\sigma_y^2 = \sigma_\delta^2/\sigma_x^2$. This assumption is unlikely to be strictly true with real data, except in cases where both variables are counts (e.g. animal numbers) or log-transformed data (McArdle, 1988). Replacing variances σ_y^2 and σ_x^2 by their unbiased estimates s_y^2 and s_x^2 gives the following value to λ in eq. 10.9:

$$\lambda = \sigma_\varepsilon^2/\sigma_\delta^2 = \sigma_y^2/\sigma_x^2 = s_y^2/s_x^2$$

Equation 10.9 then simplifies to eq. 10.12. Since the square root $\sqrt{s_y^2/s_x^2}$ is either positive or negative, the slope estimate is given the sign of the Pearson correlation coefficient, which is the same as that of the covariance s_{xy} in the denominator of eq. 10.9 or that of the OLS slope estimate. The b_{SMA} estimate is also the geometric mean of the OLS regression coefficient of y on x and the *reciprocal* of the regression coefficient of x on y; this is why the method is also called *geometric mean regression*, besides a variety of other names.

From equations 4.7 (Pearson coefficient of linear correlation r), 10.3 (b_{OLS}) and 10.12 (b_{SMA}), one can show that

$$b_{\text{SMA}} = b_{\text{OLS}}/r_{xy} \quad \text{when} \quad r_{xy} \neq 0 \qquad \textbf{(10.13)}$$

So, one can easily compute b_{SMA} from eq. 10.13, using values of b_{OLS} and r_{xy} provided by an OLS regression program. This equation also shows that, when the variables are highly correlated ($r \to 1$), $b_{\text{SMA}} \to b_{\text{OLS}}$. When they are not, b_{SMA} is always larger than b_{OLS} for positive values of r, and smaller for negative values of r; in other words, b_{OLS} is always closer to 0 than b_{SMA}.

When $r_{xy} = 0$, the b_{SMA} estimate obtained from eq. 10.12, which is the ratio of the standard deviations, is meaningless. It does not fall to zero when the correlation is zero, except in the trivial case where s_y is zero (Jolicoeur, 1975, 1990). Since the b_{SMA} estimate is independent of the presence of a significant covariance between x and y (eq. 10.12), ecologists should always compute a Pearson correlation coefficient and test it for significance prior to computing the slope of a standard major axis regression line. If r is not significantly different from zero, b_{SMA} should not be computed nor used for forecasting.

The slope of the standard major axis cannot be tested for significance. There are two reasons for this.

Permutation test

• Consider permutation testing. The b_{SMA} slope estimate is $\pm s_y/s_x$ but, for all permuted data, s_y/s_x is a constant. Giving the signs of the permuted covariances to the permuted slope estimates inevitably produces a probability near 0.5 of obtaining, by permutation, a value as extreme as or more extreme than the estimate b_{SMA}.

• The confidence interval of the slope b_{SMA}, described below, is inappropriate to test the null hypothesis $\beta = 0$ because the ratio s_y/s_x cannot be zero unless s_y is equal to zero. This is a trivial case, unsuitable for regression analysis (Sokal & Rohlf, 1995).

McArdle (1988) suggests that the solution to this problem is to test the correlation coefficient r_{xy} for significance instead of testing b_{SMA}.

C.I. of SMA slope

When needed, an approximate confidence interval $[b_1, b_2]$ can be computed for b_{SMA} as follows (Jolicoeur & Mosimann, 1968):

$$b_1 = b_{\mathrm{SMA}} [\sqrt{(B+1)} - \sqrt{B}]$$

$$b_2 = b_{\mathrm{SMA}} [\sqrt{(B+1)} + \sqrt{B}]$$

where

$$B = t^2 (1 - r^2) / (n - 2)$$

and t is a two-tailed Student's $t_{\alpha/2}$ value for significance level α and $(n - 2)$ degrees of freedom.

• 3 — An alternative transformation to make the variables dimensionally homogeneous is *ranging* (eq. 1.11). This transformation does not make the variances equal and thus does not lead to the problems encountered with SMA regression. It leads to *ranged major axis regression* (RMA, described here and in Currie *et al.*; see footnote p. 507) which proceeds as follows:

Ranged major axis

• Transform the y and x variables into y' and x', respectively, using eq. 1.11. For relative-scale variables (Subsection 1.4.1), which have zero as their natural minimum, the ranging transformation is carried out using eq. 1.10.

• Compute MA regression between the ranged variables y' and x'.

• Back-transform the estimated slope and confidence interval limits to the original units by multiplying them by the ratio of the ranges, $(y_{max} - y_{min})/(x_{max} - x_{min})$.

The RMA slope estimator has several desirable properties when variables x and y are not expressed in the same units. The slope estimator scales proportionally to the units of x and y. The estimator is not insensitive to the covariance, as is the case for SMA. Finally, it is possible to test the hypothesis that an RMA slope estimate is equal to a stated value, in particular 0 or 1. As in MA, this may be done either by permutations, or by comparing the confidence interval of the slope to the hypothetical value of interest. Thus, whenever MA regression cannot be used because of incommensurable units, RMA regression can be used. There is no reason, however, to use RMA with variables that are expressed in the same units.

Prior to RMA, one should check for the presence of outliers, using a scatter diagram of the objects. Outliers cause important changes to the estimates of the ranges of the variables. Outliers that are not aligned with the bulk of the objects may thus have an undesirable influence on the slope estimate. RMA should not be used in the presence of such outliers.

Bartlett three groups

• 4 — *Bartlett's three-group method* consists in arbitrarily dividing the data into three groups along the abscissa; the first and third groups must be of the same size. Discarding the middle group, the centroids of the first and third groups, $(\bar{x}_1, \bar{y}_1)$ and $(\bar{x}_3, \bar{y}_3)$, are computed. The slope of the straight line that goes through the centroids of the two extreme groups is calculated from these values: $b_1 = (\bar{y}_3 - \bar{y}_1) / (\bar{x}_3 - \bar{x}_1)$. The formula for computing the confidence interval of this regression coefficient is given by Sokal & Rohlf (1995), among others. The main handicap of this method is that the regression lines are not the same depending on whether the grouping is carried out based on x or y; deviation from the true major axis may be large.

OLS method

• 5 — The *ordinary least squares* (OLS) method is derived from eq. 10.10 by assuming that there is no error on x, so that the error variance on x, σ_δ^2, is zero and thus $\lambda = \sigma_\varepsilon^2/\sigma_\delta^2 = \infty$. After simplification, the OLS slope is equal to (eq. 10.3)

$$b_{\text{OLS}} = s_{xy}/s_x^2$$

With all methods of model II regression, an estimate of the intercept, b_0, can be computed from b_1 and the centroid of the scatter of points $(\bar{x}, \bar{y})$, using eq. 10.3. The same equation may be used to calculate approximate estimates of the confidence limits of the intercept. Call $b_{1\text{–INF}}$ and $b_{1\text{–SUP}}$, respectively, the lower and upper limits of the confidence interval of the slope computed either for the major axis, the standard major axis, or Bartlett's method. Put $b_{1\text{–INF}}$ into eq. 10.3 to obtain one of the confidence limits of the intercept, $b_{0\text{–Lim1}}$, at the same confidence level, e.g. 95%: $b_{0\text{-Lim1}} = \bar{y} - b_{1\text{-INF}}\bar{x}$. $b_{0\text{–Lim1}}$ is the upper limit of the confidence interval of the intercept if the centroid is on the right of the abscissa, and the lower limit if it is on its

C.I. of intercept

left. The other confidence limit of the intercept, $b_{0\text{-Lim2}}$, is calculated in the same way: $b_{0\text{-Lim2}} = \bar{y} - b_{1\text{-SUP}}\bar{x}$.

The first three methods (MA, SMA, RMA) have the property that the slope of the regression $y = f(x)$ is the reciprocal of the slope of $x = f(y)$. This property of symmetry is desirable here since there is no functional distinction between x and y in a model II situation. OLS regression does not have this property (Fig. 10.6a). In Bartlett's three-group method, symmetry does not hold unless the data points are ranked exactly in the same way on x and on y, which is a rare occurrence with real data. If the three groups formed are different, the two slopes are not the reciprocal of each other.

Since users of model II regression techniques are never certain that the assumptions of the various methods are met in data sets (i.e. MA: $\sigma_\varepsilon^2 = \sigma_\delta^2$ so that $\lambda = \sigma_\varepsilon^2/\sigma_\delta^2 = 1$; SMA: $\lambda = \sigma_y^2/\sigma_x^2$; OLS: $\sigma_\delta^2 = 0$ so that $\lambda = \sigma_\varepsilon^2/\sigma_\delta^2 = \infty$), McArdle (1988) produced an extensive series of simulations to investigate the influence of the error variances, σ_ε^2 for y and σ_δ^2 for x, on the efficiency (i.e. precision of the estimation) of the MA, SMA and OLS methods, measuring how variable the estimated slopes are under various conditions. He found the following:

• When $\sigma_y^2 = \sigma_x^2$ in the reference population from which the data sets are drawn, the true slope that must be estimated by MA or SMA regression is 1. In this case, the standard major axis was more efficient than the major axis, producing less variable slope estimates with simulated data.

• With true slopes departing from 1, the results were mixed. When the true slope was larger than 1, MA was more efficient than SMA at estimating the true slope when the error rate on x (i.e. $\sigma_\delta^2/\sigma_x^2$) was the largest, whereas SMA did better when the error rate on y (i.e. $\sigma_\varepsilon^2/\sigma_y^2$) was the largest. When the true slope was smaller than 1, the situation was reversed.

• When the error rate on y was more than three times that on x, OLS was the most efficient estimator of the slope (smallest variance among the estimates).

• The error curves for MA and SMA had minima. The minimum occurred at $\lambda = 1$ for MA and $\lambda = \sigma_y^2/\sigma_x^2$ for SMA, as predicted by theory.

In another series of simulations, Jolicoeur (1990) investigated the effects of small sample sizes and low correlations on slope estimations by MA and SMA.

• He found that the 95% confidence interval of MA remained largely accurate under all simulation conditions. In other words, with simulated data, the true slope was included in the 95% confidence interval nearly precisely 95% of the time.

• He also showed that when n was very small or the correlation was weak, the 95% confidence interval of SMA was narrower than its nominal value, this occurring even under conditions where the true slope was 1 (ideal, in principle, for SMA). In other words the true slope was not included in the nominal 95% confidence interval in more than 5% of the cases. This phenomenon became important when, for instance, $n = 10$ and $|r| < 0.6$, or $n = 20$ and $|r| < 0.4$.

This may be explained as follows. In the SMA method, slope estimates are always attracted towards the value +1 when they are positive and towards –1 when they are negative; this bias of the SMA method is further described in the next paragraph. When simulating data with low r and small n, data sets are more likely to be produced that have, by chance occurrence, a slope of opposite sign to that of the distribution from which they are drawn. In that case, the slope estimated by SMA departs further than it should from the mean of the population of simulated slopes and is found near the middle of the opposite quadrant. As a consequence, the distribution of slopes of the simulated data becomes very wide and the 95% confidence interval includes the true slope in fewer than 95% of the simulated cases.

Currie *et al.* (see footnote p. 507) investigated the relationship between slope estimate formulas, comparing MA to OLS and MA to SMA in the *correlation situation* defined as that where researchers are interested in describing the slope of the bivariate relationship displayed by two correlated random variables, i.e. variables that are not controlled or error-free. In their view, the regression line in the correlation situation is generally interpreted as a mathematical descriptor of the central tendency of a bivariate distribution. To be consistent with this interpretation of the slope, they set forth four criteria that the regression line should possess:

Criteria

1. Symmetry: the estimator of the slope of the regression $y = f(x)$ should be the reciprocal of that of $x = f(y)$. By this criterion, OLS and Bartlett's methods are inappropriate in the correlation situation. MA, SMA and RMA meet this criterion.

2. A slope estimator should vary proportionally to changes in the units of measurement. Geometrically, this means that the regression line should always be located in the same way with respect to the scatter of points, irrespective of the vertical or horizontal stretching of the data. This criterion is always met by SMA, RMA, and OLS. Cases where this criterion is not critical for MA have been discussed above.

3. The regression line must describe the central tendency of the joint distribution of two variables. This is essentially an operational definition of the major axis. For dimensionless data, b_{OLS} and b_{SMA} may fall quite far from the major axis.

Using an equation relating b_{OLS} to b_{MA} and r, Currie *et al.* plotted the relationship between b_{OLS} and b_{MA} at different degrees of correlation. They found that OLS was unbiased when the true slope was 0 ($r = 0$). Bias increased as the slope departed from zero in either the positive or negative directions, the degree of under-estimation depending upon the strength of the correlation; when $r = 1$, $b_{OLS} = b_{MA}$.

Using an equation relating b_{SMA} to b_{MA} and r, they plotted the relationship between b_{SMA} and b_{MA} at different degrees of correlation. They found that SMA was unbiased when the true slope was ±1, that is, when the scatter of points formed a 45° angle with the abscissa in either the positive or negative directions. Bias increased as the slope increased towards $\pm\infty$ or decreased towards 0. This was true for all values of $0 < r < 1$ (b_{SMA} is meaningless when $r = 0$; when $r = 1$, $b_{SMA} = b_{MA} = b_{OLS}$).

Whenever b_{OLS} or b_{SMA} depart from b_{MA}, they are inappropriate by this criterion. RMA may be used if it is the user's opinion that the relationship between dimensionally heterogeneous variables is adequately represented by the scatter diagram in the dimension-free space of ranged variables. The RMA regression line has the properties of an MA regression line in that space.

4. It must be possible to test the hypothesis that the slope is equal to a stated value, in particular 0 or 1. This criterion is met by OLS, MA, and RMA. SMA is unsuitable by this criterion.

Recommendations

Considering the results of these various studies, the following recommendations are offered to ecologists who have to estimate the slope of linear relationships when both variables are random and measured with error.

1. If the magnitude of the error on y is known to be much larger than that on x (i.e. $\sigma_\varepsilon^2/\sigma_y^2$ more than three times $\sigma_\delta^2/\sigma_x^2$), use OLS (McArdle, 1988). Otherwise, proceed as follows.

2. Check whether the data are approximately bivariate normal, either by looking at a scatter diagram or by performing a formal test of significance (Section 4.7). If not, attempt transformations to make them bivariate normal. For data that are reasonably bivariate normal, consider recommendations 3 to 5. If the data cannot be made to be reasonably bivariate normal, see recommendation 6.

3. Use major axis (MA) regression if both variables are expressed in the same physical units (untransformed variables that were originally measured in the same units) or are dimensionless (e.g. log-transformed variables forming a bivariate normal distribution), if it can reasonably be assumed that $\lambda = \sigma_\varepsilon^2/\sigma_\delta^2 \approx 1$; see example before eq. 10.11.

When no information is available on the $\sigma_\varepsilon^2/\sigma_\delta^2$ ratio and there is no reason to believe that λ would differ from 1, MA may be used provided that the results are interpreted with caution. MA produces unbiased slope estimates and accurate confidence intervals (Jolicoeur, 1990).

MA may also be used with dimensionally heterogeneous variables when the purpose of the analysis is (1) to compare two slopes computed for a pair of variables measured in an identical way (e.g. at two sampling sites), or (2) to test the hypothesis that the major axis does not significantly differ from a value given by hypothesis.

4. For variables that are not in the same units, if MA cannot be used (previous paragraph), use ranged major axis regression (RMA), i.e. transform the variables by ranging, compute major axis regression, and back-transform the slope estimator by multiplication by the ratio of the ranges. Prior to RMA, one should check for the presence of outliers, using a scatter diagram of the objects.

5. Standard major axis (SMA) regression may be used if it can reasonably be assumed that $(\sigma_\varepsilon^2/\sigma_y^2) \approx (\sigma_\delta^2/\sigma_x^2)$; examples are given after eq. 10.12. One should first test the significance of the correlation coefficient (r) to determine if the hypothesis of a relationship is supported. No regression equation should be computed when this condition is not met. This remains a less-than-ideal solution since SMA slope estimates cannot be tested for significance. Confidence intervals should also be used with caution: as the slope departs from ±1, the SMA slope estimate is increasingly biased and the confidence interval includes the true value less and less often. Even when the slope is near ±1, the confidence interval is too narrow if n is very small or if the correlation is weak.

6. If the distribution is not bivariate normal and the data cannot be transformed to satisfy that condition (e.g. bi- or multi-modal distributions), one should wonder whether the slope of a regression line is really an adequate model to describe the functional relationship between the two variables. If it is, the next step is to determine how to compute it. Since the distribution is not bivariate normal, there seems little reason to apply models such as MA or SMA, which primarily describe the first principal component of a bivariate normal distribution. So, OLS is recommended to estimate the parameters of the regression line. The significance of these parameters should be tested by permutation, however, because the distributional assumptions of parametric testing are not satisfied (Edgington, 1987). If a straight line is not the necessary model, polynomial or nonlinear regression should be considered (Subsections 4 and 6 below).

7. When the purpose of the study is not to estimate functional parameters, but simply to forecast or predict values of y for given x's, use OLS in all cases.

8. Observations may be compared to the predictions of a statistical or deterministic model (e.g. simulation model) in order to assess the quality of the model. If the model contains random variables measured with error, use MA for the comparison. If the model fits the data well, the slope is expected to be 1 and the intercept 0. A slope that significantly differs from 1 indicates a difference between observed and simulated values which is proportional to the observed values. An intercept which significantly differs from 0 indicates a systematic and constant difference between observations and simulations (Mesplé *et al.*, 1996).

9. With all methods, the confidence intervals are large when n is small; they become smaller as n goes up to about 60, after which they change much more slowly. Model II regression should ideally be applied to data sets containing 60 observations or more.

Ecological application 10.2

Laws & Archie (1981) re-analysed data published in two previous papers which had quantified the relationships between the log of respiration rates and the log of biomass for zooplankton, under various temperature conditions. The authors of the original papers had computed the OLS slopes and confidence intervals (model I regression) of the biomass-respiration relationships for each temperature condition. They had come to the conclusions (1) that the *surface law*, which states that the slope of the log-log relationship should fall between 0.66 and 1.00, was not verified by the data, and (2) that the slope significantly varied as a function of temperature. Based on the same data, Laws & Archie recomputed the slopes using the standard major axis method. They found that all slopes were larger than estimated by OLS (same phenomenon as in Fig. 10.7) and that none of them was significantly outside the 0.66 to 1.00 interval predicted by the surface law. Furthermore, comparing the slopes of the different temperature data sets at $\alpha = 2\%$, they found that they did not differ significantly from one another.

3 — Multiple linear regression

When there are several explanatory variables $x_1, x_2, \ldots, x_p$, it is possible to compute a regression equation where the response variable y is a linear function of all explanatory variables x_j. The multiple linear regression model is a direct extension of simple linear regression:

$$y_i = b_0 + b_1 x_{i1} + b_2 x_{i2} + \ldots + b_p x_{ip} + \varepsilon_i \quad \textbf{(10.14)}$$

for object i. Equation 10.14 leads to the well-known formula for the fitted values:

$$\hat{y}_i = b_0 + b_1 x_{i1} + b_2 x_{i2} + \ldots + b_p x_{ip} \quad \textbf{(10.15)}$$

Using ordinary least squares (OLS), the vector of regression parameters $\mathbf{b} = [b_j]$ is easily computed from matrix eq. 2.19: $\mathbf{b} = [\mathbf{X'X}]^{-1}\,[\mathbf{X'y}]$. If an intercept ($b_0$) must be estimated, a column of 1's is added to matrix $\mathbf{X}$ of the explanatory variables.

Standard minor axis

Equation 10.15 provides a model I estimation, which is valid when the x_j variables have been measured without error. This is the only method presently available in commercial statistical packages and, for this reason, it is the multiple regression model most widely used by ecologists. McArdle (1988) proposed a multiple regression method, the *standard minor axis*, to be used when the explanatory variables of the model are random (i.e. with measurement error or natural variability). McArdle's standard minor axis is the multivariate equivalent of the standard major axis (SMA) method described in the previous Subsection.

Orthogonal distance regression

Another approach is *orthogonal distance regression* (ODR), computed through generalized least squares. The method minimizes the sum of the squares of the orthogonal distances between each data point and the curve described by the model equation; this is the multivariate equivalent of the major axis regression (MA) method described in the previous Subsection. ODR is used extensively in econometrics. Boggs & Rogers (1990) give entry points to the numerous papers that have been published on the subject in the computer science and econometric literature and they propose an extension of the method to nonlinear regression modelling. They also give references to ODRPACK*, a public-domain collection of FORTRAN subprograms for "weighted

* ODRPACK is available from the following WWWeb site: <http://www.netlib.org/odrpack/>.

orthogonal distance regression" which allows estimation of the parameters that minimize the sum of squared weighted orthogonal distances from a set of observations to the curve or surface determined by the parameters.

When the same multiple regression model is to be computed for several response variables in turn, regression coefficients can be estimated by ordinary least squares for all the response variables simultaneously, using a single matrix expression:

$$\hat{\mathbf{B}} = [\mathbf{X'X}]^{-1}[\mathbf{X'Y}] \qquad \mathbf{(10.16)}$$

Multivariate linear regression

(Finn, 1974). In this expression, which is the multivariate equivalent of eq. 2.19, **X** is the matrix of explanatory variables, **Y** is the matrix of response variables, and $\hat{\mathbf{B}}$ is the matrix of regression coefficients. The coefficients found using this equation are the same as those obtained from multiple regressions computed in separate runs for each response variable. The procedure is called *multivariate linear regression.*

Partial regression coefficient

Two types of regression coefficients may be obtained in regression analysis. Ordinary regression coefficients are computed on the original variables, whereas standard regression coefficients are computed on standardized variables **X** and **y**. The first ones, represented by symbols b, are useful when the regression equation is to be used to compute estimated values of y for objects that have not been used to estimate the parameters of the regression equation, and for which observed x values are available. This is the case, for instance, when a regression model is validated using a new set of observations: estimates $\hat{y}$ are computed from the regression equation to be validated, using the observed values of the explanatory variables x_j, and they are compared to the observed y's, to assess how efficient the regression model is at calculating y for new data. In contrast, the standard regression coefficients, often represented by symbols b', are useful as a means of assessing the relative importance of the explanatory variables x_j included in the regression equation. The variables with the highest standard regression coefficients are also those that contribute the most to the estimated $\hat{y}$ values. The relationship between coefficients b and b' obtained by ordinary least squares estimation is: $b'_{yx_j} = b_{yx_j} s_{x_j} / s_y$, where b_{yx_j} is the partial regression coefficient for the explanatory variable x_j. Both the ordinary and standard regression coefficients in multiple regression are *partial regression coefficients*. The term 'partial' means that each regression coefficient is a measure, standardized or not, of the rate of change variable y would have per unit of variable x_j, if all the other explanatory variables in the study were held constant. The concept of partial regression is further developed in Subsection 7 below. Partial regression coefficient may be tested by randomization using methods similar to those described in Subsection 11.3.2 for canonical analysis.

Collinearity

When the explanatory variables x_j of the model are uncorrelated, multiple regression is a straightforward extension of simple linear regression. In experimental work, controlled variables may satisfy this condition if the experiment has been planned with care. With observational data, however, the explanatory variables used in multiple regression models are most often collinear (i.e. correlated to one another), and it will be seen that strong collinearity may affect the ability to correctly estimate model parameters. How to deal with this problem will depend again on the purpose of the analysis. If one is primarily interested in forecasting, the objective is to maximize the coefficient of multiple determination (called R^2 in multiple regression); collinearity of the explanatory variables is not a concern. For description or inference, however, the primary interest is to correctly estimate the parameters of the model; the effect of multicollinearity on the estimates of the model parameters must then be minimized.

The effect of collinearity on the estimates of regression parameters may be described as follows. Let us assume that one is regressing y on two explanatory variables x_1 and x_2. If x_1 is orthogonal to x_2, the variables form a well-defined Cartesian plane. If y is represented as an axis orthogonal to that plane, a multiple linear regression equation corresponds to a plane in the three-dimensional space; this plane represents the variation of y as a linear function of x_1 and x_2. If x_1 is strongly correlated (i.e. collinear) to x_2, the axes of the base plane form an acute angle instead of being at a right angle. In the limit situation where $r\ (x_1, x_2) = 1$, they become a single axis. With such correlated explanatory variables, the angles determined by the slope coefficients (b_1 and b_2), which set the position of the regression plane in the x_1–x_2–y space, are more likely to be unstable; their values may change depending on the random component ε_i in y_i. In other words, two samples drawn from the same statistical population may be modelled by regression equations with very different parameters — even to the point that the signs of the regression coefficients may change.

Simulation is the easiest way to illustrate the effect of collinearity on the estimation of regression parameters. One hundred pairs of data points were generated in the [0, 5] range using a uniform random number generator. Since the values were generated at random, vectors $\mathbf{x_1}$ and $\mathbf{x_2}$ produced in this way, each with 100 values, should be orthogonal. Actually, the correlation between them was –0.019. The control data set was completed by computing variable $\mathbf{y_1}$ as the sum of $\mathbf{x_1}$ and $\mathbf{x_2}$, to which a random component was added in the form of an error term drawn at random from a normal distribution with mean 0 and standard deviation 2:

$$y_i = x_{i1} + x_{i2} + \varepsilon_i$$

For the test data set, two *correlated* explanatory variables $\mathbf{w_1}$ and $\mathbf{w_2}$ were created by multiplying matrix $\mathbf{X}$, containing vectors $\mathbf{x_1}$ and $\mathbf{x_2}$, by the square root of a correlation matrix stating that the correlation between $\mathbf{x_1}$ and $\mathbf{x_2}$ should be 0.8:

$$\mathbf{W} = [\mathbf{w_1}, \mathbf{w_2}] = \mathbf{X}\mathbf{R}^{1/2} \quad \text{where} \quad \mathbf{X} = [\mathbf{x_1}, \mathbf{x_2}] \text{ and } \mathbf{R}^{1/2} = \begin{bmatrix} 1.0 & 0.8 \\ 0.8 & 1.0 \end{bmatrix}^{1/2} = \begin{bmatrix} \sqrt{0.8} & \sqrt{0.2} \\ \sqrt{0.2} & \sqrt{0.8} \end{bmatrix}$$

$\mathbf{R}^{1/2}$ is computed using the first property of Section 2.10; Cholesky factorization of $\mathbf{R}$ may be used instead of square root decomposition. The new variables are such that, if $\mathbf{x_1}$ and $\mathbf{x_2}$ are orthogonal, the correlation matrix of the variables in matrix $\mathbf{W}$ is $\mathbf{R}$. The correlation between $\mathbf{w_1}$ and $\mathbf{w_2}$ turned out to be 0.794, which is very close to 0.8. The test data set was completed by computing a variable $\mathbf{y_2}$ from $\mathbf{w_1}$, $\mathbf{w_2}$, with the same error term as above:

$$y_2 = w_{i1} + w_{i2} + \varepsilon_i$$

Table 10.4 Parameters of the multiple regression equations for two data sets, each divided into five groups of 20 'objects'. Top: control data set where variables x_1 and x_2 are uncorrelated. Bottom: test data set with $r(\mathbf{x_1}, \mathbf{x_2}) \approx 0.8$. Two statistics are provided to describe the variation in the estimates of the slope parameters. The intercept estimates (b_0) are the same in the two panels.

$\hat{y}_1 = b_0 + b_1x_1 + b_2x_2 \quad \Rightarrow$	b_0	b_1	b_2
Group 1	–1.317	1.061	1.174
Group 2	–0.763	1.377	1.063
Group 3	–1.960	0.632	1.810
Group 4	0.645	0.730	1.133
Group 5	1.159	0.839	0.782
Range of slope estimates = Max – Min		0.745	1.028
Standard deviation of slope estimates		0.298	0.378

$\hat{y}_2 = b_0 + b_1w_1 + b_2w_2 \quad \Rightarrow$	b_0	b_1	b_2
Group 1	–1.317	0.962	1.213
Group 2	–0.763	1.515	0.813
Group 3	–1.960	–0.152	2.482
Group 4	0.645	0.498	1.400
Group 5	1.159	0.922	0.795
Range of slope estimates = Max – Min		1.667	1.687
Standard deviation of slope estimates		0.620	0.689

Each data matrix was divided into five independent groups of 20 'objects' each, and multiple regression equations were computed; the groups were independent of one another since the generated data were not autocorrelated. Results are given in Table 10.4. Note the high variability of the slope estimates obtained for the test data sets (lower panel, with collinearity in the explanatory variables). In one case (group 3), the sign of the estimate of b_1 was changed.

Parsimony
Ockham's razor

When trying to find the 'best' possible model describing an ecological process, another important aspect is the principle of parsimony, also called 'Ockham's razor'. This principle, formulated by the English logician and philosopher William Ockham (1290-1349), professor at Oxford University, states that

Pluralites non est ponenda sine necessitate

which literally translates: "Multiplicity ought not to be posited without necessity". In other words, unnecessary assumptions should be avoided when formulating hypotheses. Following this principle, parameters should be used with parsimony in modelling, so that any parameter that does not significantly contribute to the model (e.g. by increasing the R^2 coefficient in an important way) should be eliminated.

Indeed, any model containing as many parameters as the number of data points can be adjusted to perfectly fit the data. The corresponding 'cost' is that there is no degree of freedom left to test its significance, hence the 'model' cannot be extended to any other situation.

When the explanatory variables of the model are orthogonal to one another (no collinearity, for example among the controlled factors of well-planned factorial experiments), applying Ockham's razor is easy. One then removes from the model any variable whose contribution (slope parameter) is not statistically significant. Tests of significance for the partial regression coefficients (i.e. the individual b's) are described in standard textbooks of statistics. The task is not that simple, however, with observational data, which often display various degrees of collinearity. The problem is that significance may get 'diluted' among collinear variables contributing in the same way to the explanation of a response variable y. Consider for instance a data set where an explanatory variable x_1 makes a significant contribution to a regression model; introducing a copy of x_1 in the calculation is usually enough to make the contribution of each copy non-significant, simply as the result of the collinearity that exists between copies. Hocking (1976) compared a number of methods proposed for selecting variables in linear regression exhibiting collinearity.

Some statistical programs offer procedures that allow one to test all possible models with k explanatory variables, and select the one where R^2 is the highest *and* where all partial regression coefficients are significant. When such a procedure is not available and one does not want to manually test all possible models, heuristic methods that have been developed for selecting the 'best' subset of explanatory variables may be used, although with caution. The explanatory variables with the strongest contributions may be chosen by backward elimination, forward selection, or stepwise procedure. The three strategies do not necessarily lead to the same selection of explanatory variables.

Backward elimination

• The *backward elimination procedure* is easy to understand. All variables are initially included and, at each step, the variable that contributes the least to explaining the response variable (usually that with the smallest partial correlation) is removed, until all explanatory variables remaining in the model have a significant partial regression coefficient. Some programs express the selection criterion in terms of a F-to-remove (F statistic for testing the significance of the partial regression coefficient), or a p-to-remove criterion (same, but expressed in terms of probability), instead of the value of the partial correlation.

Forward selection

• The *forward selection procedure* starts with no explanatory variable in the model. The variable entered is the one that produces the largest increase in R^2, provided this increase is significantly different from zero, using a pre-determined α level. The procedure is iteratively repeated until no more explanatory variable can be found that produces a significant increase in R^2. Calculations may be simplified by computing partial correlations for all variables not yet in the model, and only testing the significance of the largest partial correlation. Again, some programs base the final decision for including an explanatory variable on a F-to-enter value, which is equivalent to using the actual probability values. The major problem with forward selection is that all variables included at previous steps are kept in the model, even though some of them may finally contribute little to the R^2 after incorporation of some other variables.

Stepwise procedure

• The latter problem may be alleviated by the *stepwise procedure,* which alternates between forward selection and backward elimination. After each step of forward inclusion, the significance of all the variables in the model is tested, and those that are not significant are excluded before the next forward selection step.

In any case, a problem common to all stepwise inclusion procedures remains: when a model with, say, k explanatory variables has been selected, the procedure offers no guarantee that there does not exist another subset of k explanatory variables, with significant partial correlations, that would explain together more of the variation of y (larger R^2) than the subset selected by stepwise procedure. Furthermore, Sokal & Rohlf (1995) warn users that, after repeated testing, the probability of type I error is far greater than the nominal value α. The stepwise approach to regression can only be recommended in empirical studies, where one must reduce the number of explanatory variables in order to simplify data collection during the next phase of field study.

There are other ways to counter the effects of multicollinearity in multiple regression. Table 10.4 shows that collinearity has the effect of inflating the variance of regression coefficients, with the exception of the intercept b_0. Regression on principal components and ridge regression are helpful when the objective is forecasting or prediction. Both have the same effects: on the one hand, they reduce the variance of the regression coefficients, which leads to better predictions of the response variable; on the other hand, the regression coefficients they produce are biased, but they are still better estimates of the 'true' regression coefficients than those obtained by OLS. In other words, the price to pay for reducing the inflation of variance is some bias in the estimates of the regression coefficients. This may provide better forecasting or prediction than the OLS solution since, as a consequence of the larger variance in the regression coefficients, multicollinearity tends to increase the variance of the forecasted or predicted values (Freund & Minton, 1979).

Regression on principal components

• *Regression on principal components* consists of the following steps: (1) perform a principal component analysis on the matrix of the explanatory variables **X**, (2) compute the multiple regression of y on the principal components instead of the original explanatory variables, and (3) find back the contributions of the explanatory variables by multiplying matrix **U** of the eigenvectors with the vector of regression coefficients **b** (using either the unstandardized or the standardized coefficients) of y on the principal components. One obtains a new vector **b**' of contributions of the original variables to the regression equation:

$$\mathbf{U}_{(p \times k)} \mathbf{b}_{(k \times 1)} = \mathbf{b}'_{(p \times 1)} \tag{10.17}$$

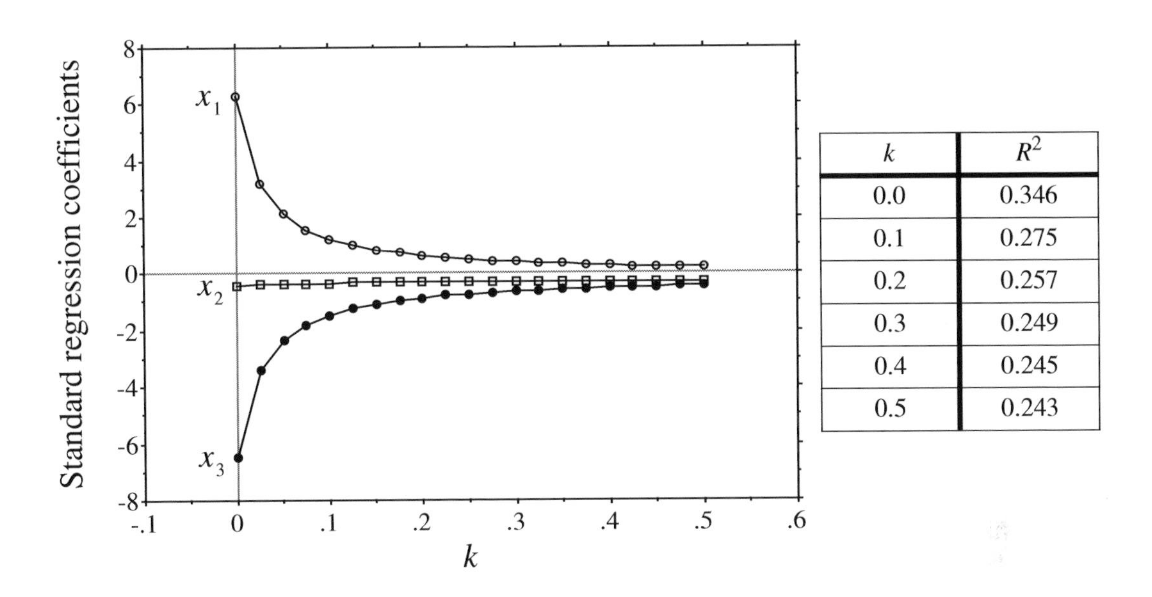

k	R^2
0.0	0.346
0.1	0.275
0.2	0.257
0.3	0.249
0.4	0.245
0.5	0.243

Figure 10.8 'Ridge trace' diagram showing the estimates of the standardized regression coefficients for explanatory variables x_1 to x_3 as a function of k. Table: decrease of R^2 as a function of k.

where p is the number of explanatory variables in the analysis and k is the number of principal components retained for step (3). This procedure* does not necessarily solve the problem of multicollinearity, although it is true that the regression is performed on principal components which are not collinear by definition. Consider the following case: if all p eigenvectors are kept in matrix $\mathbf{U}$ for step (3), one obtains exactly the same regression coefficients as in ordinary multiple regression. There is gain in stability of the regression coefficients only by eliminating some of the principal components from eq. 10.17. One may either eliminate the eigenvectors with the smallest eigenvalues or, better, use only in eq. 10.17 the principal components that significantly contribute to explain y. By doing so, the regression coefficient estimates become biased, of course. In problems involving a small number of explanatory variables, regression on principal components may be difficult to use because the number of principal components is small, so that eliminating one of them from the analysis may result in a large drop in R^2. Ecological application 12.7 provides an example of regression on principal components.

Ridge regression

• *Ridge regression*, developed by Hoerl (1962) and Hoerl & Kennard (1970a, b), approaches the problem in a different way; another important paper on the subject is

* Regression on principal components is available in some computer packages, e.g. BMDP.

Marquardt & Snee (1975). Instead of the usual matrix eq. 2.19 $\mathbf{b} = [\mathbf{X'X}]^{-1}[\mathbf{X'Y}]$, the regression coefficients are estimated using a modified equation,

$$\mathbf{b} = [\mathbf{X'X} + k\mathbf{I}]^{-1}[\mathbf{X'Y}] \text{ , where } k > 0. \qquad \textbf{(10.18)}$$

Hence, the method consists in increasing the diagonal terms (variances) of the covariance matrix $[\mathbf{X'X}]$ by a constant positive quantity k. This reduces the variance of the regression coefficients while creating a bias in the resulting estimates. So, users are left with the practical problem of choosing a value for k which is optimal in some sense. This is accomplished by computing regression coefficient estimates for a series of values of k, and plotting them (ordinate) as a function of k (abscissa); this plot is called the 'ridge trace', for historical reasons (Hoerl, 1962). After studying the plot, one chooses a value of k which is as small as possible, but large enough that the regression coefficient estimates change little after it. Since ridge regression is usually computed on standardized variables, no intercept is estimated. A number of criteria have been proposed by Obenchain (1977) to help choose the value of k. These criteria must be used with caution, however, since they often do not select the same value of k as the optimal one. An example of a 'ridge trace' diagram is presented in Fig. 10.8. The data set consists of a response variable y and three collinear explanatory variables x_1 to x_3; their correlation matrix is as follows:

	y	x_1	x_2	x_3
y	1			
x_1	–0.40	1		
x_2	–0.44	0.57	1	
x_3	–0.41	**0.99**	0.56	1

The leftmost regression coefficient estimates in Fig. 10.8 ($k = 0$) are the standardized OLS multiple regression coefficients. Going from left to right in the Figure, the regression coefficients stabilize after a sharp decrease or increase. One may decide that setting the cut-off point at $k = 0.20$ would be an appropriate compromise between small k and stable regression coefficients. Boudoux & Ung (1979) and Bare & Hann (1981) provide applications of ridge regression to forestry; in both papers, some regression coefficients change signs with increasing k. An application of ridge regression to modelling heterotrophic bacteria in a sewage lagoon ecosystem is presented by Troussellier *et al.* (1986, followed-up by Troussellier & Legendre, 1989).

The coefficient of multiple determination R^2, also called the *unadjusted* coefficient of multiple determination, is the square of the multiple correlation coefficient R (Section 4.5); it varies between 0 and 1. It is the most commonly used statistic to measure the forecasting potential of a multiple regression equation. As in simple linear regression, it is the ratio of the sum of squares of distances of the estimated values $\hat{y}$ to the mean $\bar{y}$, to the sum of squares of distances of the original response variable values y to the mean $\bar{y}$:

$$R^2 = \frac{\Sigma (\hat{y}_i - \bar{y})^2}{\Sigma (y_i - \bar{y})^2} \qquad \textbf{(10.19)}$$

It measures the proportion of the variation of y about its mean which is explained by the regression equation. Another measure, the *adjusted coefficient of multiple determination* R_a^2, takes into account the respective numbers of degrees of freedom of the numerator and denominator of R^2:

$$R_a^2 = 1 - (1 - R^2)\left(\frac{n-1}{n-p}\right) \qquad \textbf{(10.20)}$$

where n is the number of observations and p is the number of parameters b in the model, including intercept b_0 if any. This statistic is now available in a number of statistical packages. The purpose of the correction is to produce a statistic suitable for comparing regression equations fitted to different data sets, with different numbers of objects and explanatory variables. According to Draper & Smith (1981), however, R_a^2 is not very good at doing that.

Multiple regression may make use of a variety of explanatory variables:

• Binary descriptors may be used as explanatory variables in multiple regression, as well as quantitative variables. This means that multistate qualitative variables can also be used, insofar as they are recoded into a set of binary dummy variables, as described in Subsection 1.5.7. This method is often referred to as *dummy variable regression.*

• Geographic information may be used in multiple regression models in different ways. On the one hand, X (longitude) and Y (latitude) information form perfectly valid quantitative descriptors if they are recorded as axes of a Cartesian plane. Geographic data in the form of degrees-minutes-seconds should be recoded, however, because they are not decimal. The X and Y coordinates may be used either alone, or in the form of a polynomial (X, Y, X^2, XY, Y^2, etc.). Regression using such explanatory variables is referred to as *trend surface analysis* in Chapter 13. On the other hand, if replicate observations are available for each site, the grouping of observations, which is also a kind of geographic information, may be introduced in a multiple regression equation as a qualitative multistate descriptor, recoded into a set of dummy variables.

• Finally, one should be aware of the fact that any analysis of variance may be reformulated as a regression analysis. Consider one-way ANOVA for instance: the classification criterion can be written down as a multistate qualitative variable and, as such, recoded as a set of dummy variables (Subsection 1.5.7) on which multiple regression may be performed. The analysis of variance table obtained by multiple regression is identical to that produced by ANOVA. This equivalence is discussed in more detail by ter Braak & Looman (1987), in an ecological framework. Draper & Smith (1981, Chapter 9) and Searle (1987) discuss in some detail how to apply multiple regression to various analysis of variance configurations.

4 — Polynomial regression

Several solutions have been proposed to the problem of fitting, to a response variable y, a nonlinear function of a single explanatory variable x. An elegant and easy solution is to use a polynomial of x, whose terms are treated as so many explanatory variables in a multiple-regression-like procedure. In this method, y is expressed as a polynomial function of x:

Polynomial model

$$\hat{y} = b_0 + b_1 x + b_2 x^2 + \ldots + b_k x^k \qquad \textbf{(10.21)}$$

Such an equation is linear in its parameters (if one considers the terms x^2, …, x^k as so many explanatory variables), although the response of y to the explanatory variable x is nonlinear. The degree of the equation, which is its highest exponent, determines the shape of the curve, each degree above 1 (straight line) allowing an additional inflexion point. This subject is discussed in detail in manuals of analytic geometry. Increasing the degree of the equation always increases its adjustment to the data (R^2). If one uses as many parameters b as there are data points, one can fit the data perfectly ($R^2 = 1$); the cost is that there are no degrees of freedom left to test the relationship, hence the "model" cannot be extended to any other situation. In any case, a high-degree polynomial is of little interest in view of the principle of parsimony (Ockham's razor) discussed at the beginning of this section, i.e. the best model is the simplest one that adequately describes the relationship.

So, the problem left to ecologists is to find the most parsimonious polynomial equation which has the highest possible coefficient of determination and where all regression coefficients are significant. The methods for selecting variables, described above for multiple regression, may be used profitably here in a two-stage process.

• First, one must determine the degree of the polynomial. To do that, start with an equation of a degree high enough (5 or 6, for instance) that the last terms are not significant. Then, remove terms *one by one*, from the highest degree down, until a significant term is found. Do not remove several terms at a time; the term in x^4, for example, may make the term in x^3 non-significant, because of collinearity, in an equation containing both, but x^3 may become significant after x^4 has been removed.

Monomial

• When the highest monomials (*monomial*: each term of a polynomial expression) have been eliminated, use one of the standard selection procedures described above to find the most effective subset of monomials (highest R^2) in which all terms are statistically significant. The best procedure, of course, is one that tests all possible models with fewer terms. Short of that, backward, forward, or stepwise procedures may be used, with caution. The final equation does not necessarily possess all successive monomials from degree 1 up.

Polynomial regression procedures are directly available in some statistical packages (beware: in most packages, polynomial regression procedures do not allow one to remove the monomials of degree lower than k if x^k is retained in the equation).

When no such packaged procedure is available, one can easily construct a series of variables corresponding to the successive monomials. Starting with a variable x, it is easy to square it to construct x^2, and so on, creating as many new variables as deemed necessary. These are then introduced in a multiple regression procedure. See also the last paragraph of Section 2.8.

One must be aware of the fact that the successive terms of an ordinary polynomial expression are not linearly independent of one another. Starting for instance with a variable x made of the successive integers 1 to 10, variables x^2, x^3, and x^4 computed from it display the following correlations:

	x	x^2	x^3	x^4
x	1			
x^2	0.975	1		
x^3	0.928	0.987	1	
x^4	0.882	0.961	0.993	1

The problem of multicollinearity is severe with such data. Centring variable x on its mean before computing the polynomial is good practice. It reduces the linear dependency of x^2 on x (it actually eliminates it when the x values are at perfectly regular intervals, as in the present example), and somewhat alleviates the problem for the higher terms of the polynomial. This may be enough when the objective is descriptive. If, however, it is important to estimate the exact contribution (standard regression coefficient) of each term of the polynomial in the final equation, the various monomials (x, x^2, etc.) should be made orthogonal to one another before computing the regression equation. Orthogonal monomials may be obtained, for example, through the Gram-Schmidt procedure described in Table 9.5 and in textbooks of linear algebra (for instance Lipschutz, 1968).

Orthogonal monomials

Numerical example. Data from the ECOTHAU program (Ecology of the Thau lagoon, southern France; Amanieu *et al.*, 1989) are used to illustrate polynomial regression. Salinity (response variable y) was measured at 20 sites in the brackish Thau lagoon (Mediterranean Sea) on 25 October 1988. The lagoon is elongated in a SW-NE direction. The explanatory variable x is the projection of the positions of the sampling sites on the long axis of the lagoon, as determined by principal component analysis of the site coordinates. Being a principal component, variable x is automatically centred. The other terms of a 6th-degree polynomial were computed from it. Using the selection procedure described above, a polynomial equation was computed to describe the spatial distribution of salinity in the lagoon (Fig. 10.9). The higher-degree terms (x^6, then x^5) were eliminated one by one until it was found that term x^4 was significant. Then, all possible models involving x, x^2, x^3, and x^4 were computed. Among the models in which all terms were significant, the one with x and x^4 had the highest coefficient of determination.

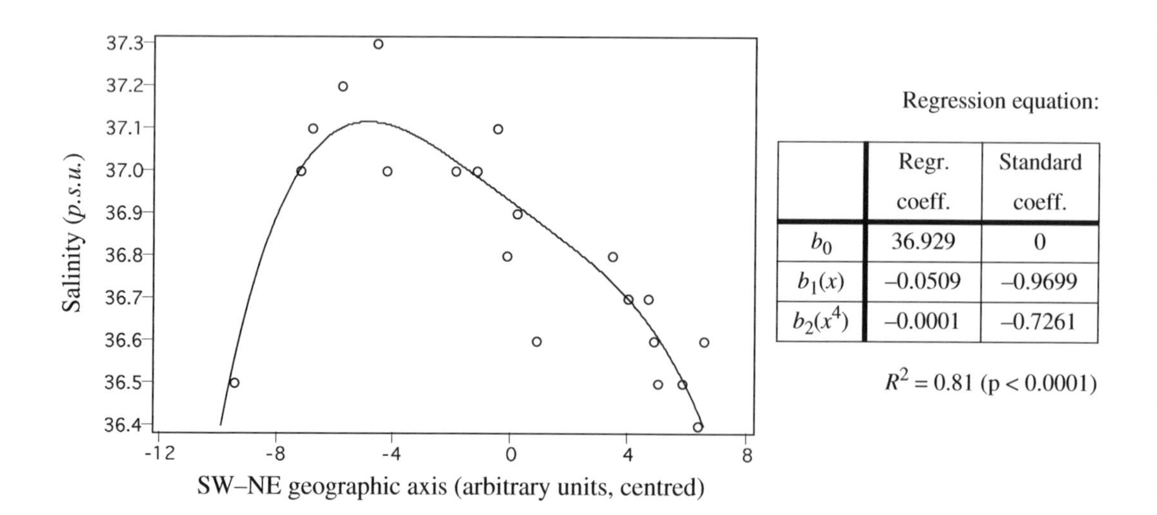

	Regr. coeff.	Standard coeff.
b_0	36.929	0
$b_1(x)$	–0.0509	–0.9699
$b_2(x^4)$	–0.0001	–0.7261

$R^2 = 0.81$ ($p < 0.0001$)

Figure 10.9 Polynomial regression equation describing the structure of salinity (*practical salinity units*) in the Thau lagoon (Mediterranean Sea), along its main geographic axis, on 25 October 1988.

5 — Partial linear regression

There are situations where two or more complementary sets of hypotheses may be invoked to explain the variation of an ecological variable. For example, the abundance of a species could vary as a function of biotic and abiotic factors. Regression modelling may be used to study one set of factors, or the other, or the two together. Partial regression is a way of estimating how much of the variation of the response variable can be attributed exclusively to one set of factors, once the effect of the other set has been taken into account. The purpose may be to measure the amount of variation (R^2) that can be attributed exclusively to one or the other set of explanatory variables, or else to estimate the vector of fitted values corresponding to the exclusive effect of one set of variables. When the purpose of the study is simply to assess the unique contribution of each explanatory variable, there is no need for partial regression analysis since the multiple linear regression coefficients, which are *partial regression coefficients*, already provide that information.

Consider three data sets. Vector **y** is the response variable and matrices **X** and **W** contain the explanatory variables. Assume that one wishes to model the relationship between **y** and **X**, while controlling for the effects of the variables in matrix **W**, called the matrix of *covariables*. There are two ways of doing this, i.e. a long and a short way. The long way provides the justification for the short. One proceeds as follows:

• First, compute multiple regressions of **y** and of each variable in **X** against matrix **W**. Calculate the *residuals* of all these regressions. For vector **y** for instance, the residuals,

Table 10.5 Data collected at 20 sites in the Thau lagoon on 25 October 1988. There are two response variables (Bna and Ma), three environmental variables (NH_4, phaeopigments, and bacterial production), and three spatial variables (the X and Y geographic coordinates measured with respect to arbitrary axes and centred on their respective means, plus the geographic variable X^2). The variables are further described in the text. The code names of these variables in the present Section are y, x_1 to x_3, and w_1 to w_3, respectively.

Site No.	Bna	Ma y	NH_4 x_1	Phaeo. *a* x_2	Prod. x_3	X w_1	Y w_2	X^2 w_3
1	4.615	10.003	0.307	0.184	0.274	–8.75	3.7	76.5625
2	5.226	9.999	0.207	0.212	0.213	–6.75	2.7	45.5625
3	5.081	9.636	0.140	0.229	0.134	–5.75	1.7	33.0625
4	5.278	8.331	1.371	0.287	0.177	–5.75	3.7	33.0625
5	5.756	8.929	1.447	0.242	0.091	–3.75	2.7	14.0625
6	5.328	8.839	0.668	0.531	0.272	–2.75	3.7	7.5625
7	4.263	7.784	0.300	0.948	0.460	–1.75	0.7	3.0625
8	5.442	8.023	0.329	1.389	0.253	–0.75	–0.3	0.5625
9	5.328	8.294	0.207	0.765	0.235	0.25	–1.3	0.0625
10	4.663	7.883	0.223	0.737	0.362	0.25	0.7	0.0625
11	6.775	9.741	0.788	0.454	0.824	0.25	2.7	0.0625
12	5.442	8.657	1.112	0.395	0.419	1.25	1.7	1.5625
13	5.421	8.117	1.273	0.247	0.398	3.25	–4.3	10.5625
14	5.602	8.117	0.956	0.449	0.172	3.25	–2.3	10.5625
15	5.442	8.487	0.708	0.457	0.141	3.25	–1.3	10.5625
16	5.303	7.955	0.637	0.386	0.360	4.25	–5.3	18.0625
17	5.602	10.545	0.519	0.481	0.261	4.25	–4.3	18.0625
18	5.505	9.687	0.247	0.468	0.450	4.25	–2.3	18.0625
19	6.019	8.700	1.664	0.321	0.287	5.25	–0.3	27.5625
20	5.464	10.240	0.182	0.380	0.510	6.25	–2.3	39.0625

called ε_i above, are computed as $(y_i - \hat{y}_i)$ and make up vector $\mathbf{y}_r$. The residuals of the regressions of the variables in $\mathbf{X}$ are computed in the same way; they form matrix $\mathbf{X}_r$.

• Then, compute a multiple regression, with $\mathbf{y}_r$ as the response variable and $\mathbf{X}_r$ as the matrix of explanatory variables, to obtain the partial regression equation.

Numerical example 1. The example data set (Table 10.5) is from the ECOTHAU research program mentioned in the numerical example of Subsection 4 (Amanieu *et al.*, 1989). It contains two bacterial variables (Bna, the concentration of colony-forming units of aerobic heterotrophs growing on bioMérieux nutrient agar, with low NaCl concentration; and Ma, the concentration of aerobic heterotrophs growing on marine agar at 34 gL^{-1} salinity); three environmental variables (NH_4 in the water column, in $\mu molL^{-1}$; phaeopigments from degraded chlorophyll *a*, in μgL^{-1}; and bacterial production, determined by incorporation of tritiated thymidine in bacterial DNA, in $nmolL^{-1}d^{-1}$); and three spatial variables of the sampling sites on the nodes of an arbitrarily located grid (the X and Y geographic coordinates, in km, each centred on its mean, and X^2, which was found to be important for explaining the response variables). All bacterial and environmental variables were log-transformed using $\ln(x + 1)$. One of the bacterial variables, Ma, is used here as the response variable **y**; the three environmental variables form the matrix of explanatory variables **X**; the three spatial variables make up matrix **W** of the covariables. Table 10.5 will be used again in Chapter 13. A multiple regression of **y** against **X** and **W** together was computed first as a reference; the regression equation is the following:

$$\hat{y} = 9.64 - 0.90x_1 - 1.34x_2 + 0.54x_3 + 0.10w_1 + 0.14w_2 + 0.02w_3 \quad (R^2 = 0.5835)$$

The vector of fitted values was also computed; this vector will be plotted as fraction [a + b + c] in Fig. 10.12. Since the total sum of squares in **y** is 14.9276 [$SS = s_y^2 \times (n-1)$], the coefficient of multiple determination R^2 allows the computation of the sum of squares corresponding to the vector of fitted values $\hat{\mathbf{y}}$: $14.9276 \times 0.5835 = 8.7108$. This value may also be derived by directly computing the sum of squares of the values in the fitted vector.

Residuals of the regressions (not shown here) of variables y, x_1, x_2, and x_3 on matrix **W** were computed as described above and assembled in vector $\mathbf{y}_r$ and matrix $\mathbf{X}_r$. The multiple regression model of $\mathbf{y}_r$ on $\mathbf{X}_r$ has the following equation:

$$\hat{y}_r = 0.00 - 0.90x_{r1} - 1.34x_{r2} + 0.54x_{r3} \quad (R^2 = 0.3197)$$

The vector of fitted values was also computed; this vector will be plotted as fraction [a] in Fig. 10.12. Note that the three slope coefficients in this partial regression equation are exactly the same as in the previous multiple regression equation. This gives substance to the statement of Subsection 3, that regression coefficients obtained from multiple linear regression are *partial regression coefficients* in the sense of the present Subsection. Since the sum of squares of deviations from the mean (SS) of vector $\mathbf{y}_r$ is 9.1380, the R^2 coefficient allows the computation of the sum of squares corresponding to the vector of fitted values: $9.1380 \times 0.3197 = 2.9213$. This R^2 value is the square of the partial correlation coefficient (Section 4.5) between **y** and matrix **X** when controlling for the effect of **W**.

The exact same regression coefficients would be obtained by regressing **y** (instead of $\mathbf{y}_r$) on matrix $\mathbf{X}_r$. The only difference is that this new equation would have an intercept whereas the above equation has none. The multiple regression equation for variable Ma of the example would be:

$$\hat{y} = 8.90 - 0.90x_1 - 1.34x_2 + 0.54x_3 \quad (R^2 = 0.1957)$$

As a consequence, the fitted values would differ by the value of the intercept, i.e. 8.90. The R^2 coefficients differ because variable **y** does not have the same variance as $\mathbf{y}_r$. The advantage of using $\mathbf{y}_r$ instead of **y** is that $R = 0.5654$, which is the square root of the coefficient of determination $R^2 = 0.3197$, is the true partial correlation coefficient (in the sense of Subsection 4.5.2) between **y** and **X**, when controlling for the effect of **W**. The value $R^2 = 0.1957$, obtained from the regression of **y**, directly provides the fraction of the variation of **y** accounted for by the partial linear model of the three explanatory variables **x**. This is the value that will be needed in the decomposition of the variation, below. Otherwise the two analyses are equivalent.

If the study requires it, the partial regression equation of **y** on **W**, while controlling for the effects of **X**, may be computed in the same way. The result is the following:

$$\hat{y}_r = 0.00 + 0.10w_{r1} + 0.14w_{r2} + 0.02w_{r3} \quad (R^2 = 0.2002)$$

The vector of fitted values was also computed; this vector will be plotted as fraction [c] in Fig. 10.12. Since the sum of squares of vector $\mathbf{y}_r$ is 7.7729, the R^2 coefficient allows the computation of the sum of squares corresponding to the vector of fitted values: $7.7729 \times 0.2002 = 1.5562$.

The short way of achieving the same result as above is simply to compute the multiple regression of **y** against **X** and **W** together. Using that equation (copied from above)

$$\hat{y} = 9.64 - 0.90x_1 - 1.34x_2 + 0.54x_3 + 0.10w_1 + 0.14w_2 + 0.02w_3 \quad (R^2 = 0.5835)$$

one directly obtains the two partial regression equations by deleting the terms corresponding to the variables controlled for. With the present example, the partial regression model of **y** on **X**, when controlling for the effects of **W**, is obtained by deleting all terms w and the intercept:

$$\hat{y}_r = -0.90x_1 - 1.34x_2 + 0.54x_3 \quad (R^2 = 0.3197)$$

The R^2 coefficient is obtained by computing the correlation between vectors $\hat{\mathbf{y}}_r$ and **y** and squaring it. In the same way, the partial regression model of **y** on **W**, when controlling for the effects of **X**, is obtained by deleting all terms x, and the intercept:

$$\hat{y}_r = 0.10w_1 + 0.14w_2 + 0.02w_3 \quad (R^2 = 0.2002)$$

The R^2 coefficient is obtained again by computing the correlation between vectors $\hat{\mathbf{y}}_r$ and **y** and squaring it. Sums of squares are obtained by multiplying the values of R^2 of the partial models by the total sum of squares in **y**.

Variation partitioning

Coefficients of determination computed using the above three equations provide all the information required to partition the variation of variable **y** among the explanatory data sets **X** and **W**. Partial regression assumes that the effects are additive. Figure 10.10 sets a nomenclature for the four fractions of variation, called [a] to [d]. Following this convention, the above results are assembled in the following table (rounded values):

Fractions of variation	Sums of squares	Proportions of variation of y (R^2)
[a]	2.9213	0.1957
[b]	4.2333	0.2836
[c]	1.5562	0.1042
[d]	6.2167	0.4165
	14.9276	1.0000

If one is simply interested in partitioning the variation of vector **y** among the explanatory data sets **X** and **W**, without estimating the vectors of fitted values that

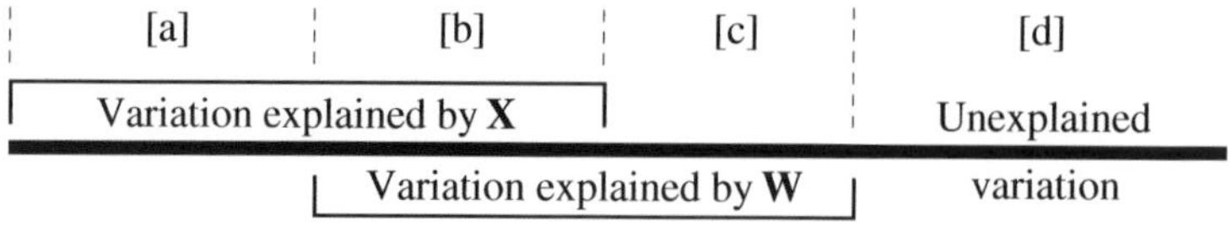

[a] [b] [c] [d]

Variation explained by **X** | Unexplained variation

Variation explained by **W**

Figure 10.10 Partition of the variation of a response variable **y** among two sets of explanatory variables **X** and **W**. The length of the horizontal line corresponds to 100% of the variation in **y**. Fraction [b] is the intersection of the linear effects of **X** and **W** on **y**. Adapted from Legendre (1993).

correspond exclusively to one or the other data set (fractions [a] and [c]), there is a simpler way to obtain the information, considering the ease with which multiple regressions is computed using a statistical package:

- Compute the multiple regression of **y** against **X** and **W** together. The corresponding R^2 measures the fraction of information [a + b + c], which is the sum of the fractions of variation [a], [b], and [c] defined in Fig. 10.10. For the example data set, $R^2 = 0.5835$, as already shown.

- Compute the multiple regression of **y** against **X**. The corresponding R^2 measures [a + b], which is the sum of the fractions of variation [a] and [b]. For the example, this $R^2 = 0.4793$. The vector of fitted values corresponding to fraction [a + b], which is required to plot Fig. 10.12 (below), is also computed.

- Compute the multiple regression of **y** against **W**. The corresponding R^2 measures [b + c], which is the sum of the fractions of variation [b] and [c]. For the example, this $R^2 = 0.3878$. The vector of fitted values corresponding to fraction [b + c], which is required to plot Fig. 10.12 (below), is also computed.

- If needed, fraction [d] may be computed by subtraction. It is equal to 1 – [a + b + c], or 1 – 0.5835 = 0.4165 for the example data set.

- Fraction [b] of the variation may be obtained by subtraction, in the same way as the quantity B used for comparing two qualitative descriptors in Section 6.2:

$$[b] = [a + b] + [b + c] - [a + b + c]$$

For the example data set,

$$[b] = 0.4793 + 0.3878 - 0.5835 = 0.2836$$

Note that no fitted vector can be estimated for fraction [b], which is obtained by subtraction and not by estimation of an explicit parameter in the regression model. For this reason, fraction [b] may be negative and, as such, it is not a rightful measure of variance; this is why it is referred to by the looser term *variation*.

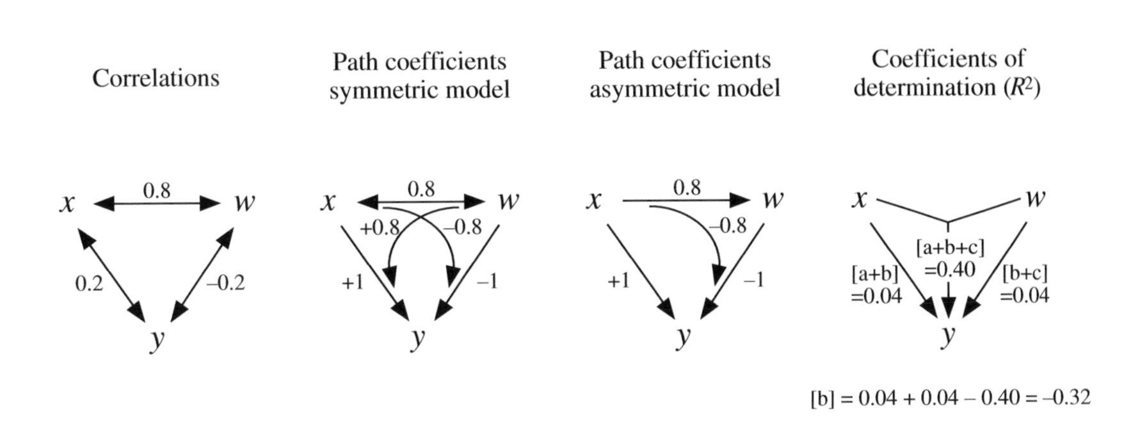

Figure 10.11 Correlations, path coefficients, and coefficients of determination for Numerical example 2.

Negative [b]

A negative fraction [b] simply indicates that two variables (or groups of variables **X** and **W**), together, explain the response variable better than the sum of the individual effects of these variables. This is due to the fact that, through the model, one or the other of the (groups of) explanatory variables **X** and **W** — or both — have effects of opposite signs on the response variable **y**, which lowers the total covariance of **X**, **W**, or both, with **y**. Further conditions for a negative fraction [b] to appear are the presence of strong direct effects of **X** and **W** on **y** and strong correlations between **X** and **W** (non-orthogonality).

Numerical example 2. From three variables measured over 50 objects, the following correlations are obtained: $r(x, w) = 0.8$, $r(x, y) = 0.2$ and $r(w, y) = -0.2$; w, x, and y have the same meaning as in the previous numerical example; $r(x, y)$ and $r(w, y)$ are not statistically significant at the $\alpha = 0.05$ level. In anticipation of Section 10.4, one may use path analysis to compute the direct and indirect causal covariation relating the explanatory variables w and x to the response variable y. One can also compute the coefficient of determination of the model $y = f(x, w)$; its value is $R^2 = 0.40$. From these values, the partition of the variation of y can be achieved: [a + b + c] = R^2 of the model = 0.4; [a + b] = $r^2(w, y) = 0.04$; [b + c] = $r^2(x, y) = 0.04$. Hence, [b] = [a + b] + [b + c] – [a + b + c] = –0.32, [a] = [a + b] – [b] = 0.36, and [c] = [b + c] – [b] = 0.36. How is this possible?

Carrying on the path analysis (Fig. 10.11), and assuming a symmetric model of relationships (i.e. w affects x and x affects w), the direct effect of x on y, $p_{xy} = 1.0$, is positive and highly significant, but it is counterbalanced by a strong negative indirect covariation of –0.8 going through w. In the same way, $p_{wy} = -1.0$ (which is highly significant), but this direct effect is counterbalanced by a strong positive indirect covariation of +0.8 going through x. As a result, and although they both have the maximum possible value of 1.0 for direct effects on the response variable y, both w and x turn out to have non-significant total correlations with y. In the present variation partitioning model, this translates into small amounts of explained variation [a + b] = 0.04 and [b + c] = 0.04, and a negative value for fraction [b]. If an asymmetric model of relationship had been assumed (e.g. w affects x but x does not affect w), essentially the same conclusion would have been reached from path analysis.

The above decomposition of the variation of a response vector **y** between two sets of explanatory variables **X** and **W** was described by Whittaker (1984), for the simple case where there is a single regressor in each set **X** and **W**. Whittaker showed that the various fractions of variation may be represented as vectors in space, and that the value of fraction [b] [noted G(12:) by Whittaker, 1984] is related to angle θ between the two regressors through the following formula:

$$1 - 2\cos^2(\theta/2) \leq [\mathrm{b}] \leq 2\cos^2(\theta/2) - 1 \qquad \textbf{(10.22)}$$

θ is related to the coefficient of linear correlation (eq. 10.4). This formula has three interesting properties. (1) If the two regressors are orthogonal ($r = 0$), then $2\cos^2(\theta/2) = 1$, so that $0 \leq [\mathrm{b}] \leq 0$ and consequently $[\mathrm{b}] = 0$. Turning the argument around, the presence of a non-zero fraction [b] indicates that the two explanatory variables are not orthogonal; there are also, of course, instances where [b] is zero with two non-orthogonal regressors. (2) If the two regressors are identical, or at least pointing in the same direction ($\theta = 0°$), then $-1 \leq [\mathrm{b}] \leq 1$. It follows that the proportion of variation of **y** that is accounted for by either regressor (fraction [b]) may be, in some cases, as large as 1, i.e. 100%. (3) The formula allows for negative values of [b], as shown in Numerical example 2.

In conclusion, fraction [b] represents the fraction of variation of **y** that may indifferently be attributed to **X** or **W**. The interpretation of a negative [b] is that the two processes, represented in the analysis by the data sets **X** and **W**, are competitive; in other words, they have opposite effects, one process hindering the contribution of the other. One could use eq. 6.15, $S = [\mathrm{b}]/[\mathrm{a + b + c}]$, to quantify how similar **X** and **W** are in explaining **y**. Whittaker (1984) also suggested that if **X** and **W** represent two factors of an experimental design, [b] may be construed as a measure of the effective balance (i.e. orthogonality) of the design; [b] is 0 in a balanced design.

Whittaker's representation may be used even when regressors **X** and **W** are multivariate data sets. Figure 10.12 illustrates the angular relationships among the fitted vectors corresponding to the fractions of variation of the above example. One plane is needed for vectors {[a], [b + c], and [a + b + c]} in which [a] is orthogonal and additive to [b + c]; another plane is needed for vectors {[c], [a + b], and [a + b + c]} where [c] is orthogonal and additive to [a + b]. However, the sets {[a], [b + c]} and {[c], [a + b]} belong to different planes, which intersect along vector [a + b + c]; so, the whole set of fitted vectors is embedded in a three-dimensional space when there are two explanatory data sets; this is independent of the number of variables in each set. The vector of residuals corresponding to fraction [d] is orthogonal to all the fitted vectors and lies in a fourth dimension. Whittaker (1984) gives examples involving more than two explanatory data sets. The graphical representation of the partitioned fitted vectors in such cases requires spaces with correspondingly more dimensions.

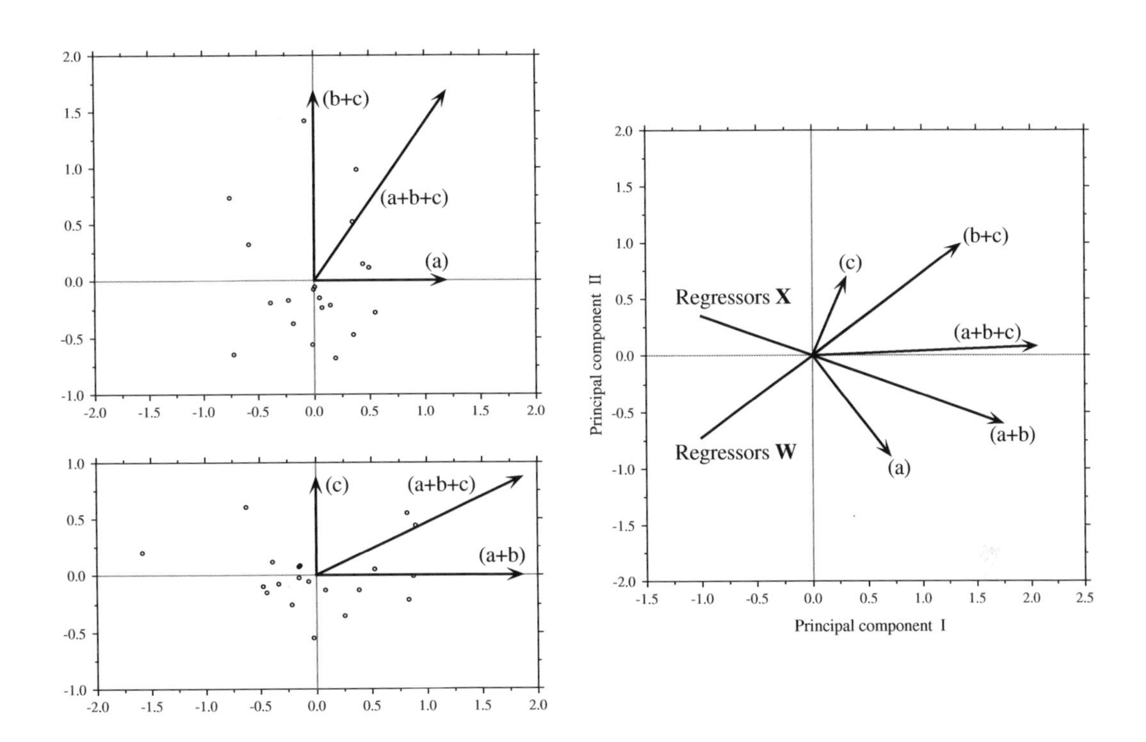

Figure 10.12 Numerical example of partial regression analysis: representation of the fitted vectors in the regression space, as suggested by Whittaker (1984). Vectors are represented with lengths proportional to their standard deviations. Upper left: scatter diagram of objects along orthogonal vectors [a] and [b + c]. Vector [a + b + c], also shown, is obtained by adding vectors [a] and [b + c]. Lower left: same for orthogonal vectors [c] and [a + b]. Right: all five fitted vectors are represented in a compromise plane obtained by principal component analysis (PCA axes I and II, which explain 96.7% of the variation). [a] is still orthogonal to [b + c], and [c] to [a + b], but these orthogonal relationships are slightly deformed by the projection in two dimensions.

Ecological application 10.3

Birks (1996) used partial regressions to analyse the mountain plant species richness in 75 grid squares covering Norway (109 species in total), in order to test whether the nunatak hypothesis was necessary to explain the present distribution of these plants. The nunatak, or refugial hypothesis, holds that apparent anomalies in present-day species distributions are explained by survival through glaciations on ice-free mountain peaks or rocky outcrops (called 'nunataks') projecting above continental glaciers. Implicit in this hypothesis is that the presumed refugial species have poor dispersal power. According to the nunatak hypothesis, one would expect a concentration of rare plants in either the glacial refuges or their vicinity. Hence, a variable describing unglaciated areas (3 abundance classes for occurrence of presumed unglaciated areas) was introduced in the analysis to represent "history". The alternative hypothesis, called the "tabula rasa", holds that present-day distributions are well-explained by the environmental control model (Whittaker, 1956; Bray & Curtis, 1957). To materialise this hypothesis in the analysis, Birks used 10 explanatory variables that described bedrock geology, geography, topography, and climate. "Geography" was introduced in the analysis in the form of a third-degree polynomial of the geographic coordinates, which allowed a representation of the geographic variation of species richness by a cubic trend-surface of latitude and longitude, as explained in Subsection 13.2.1; the terms of the polynomial representing latitude and longitude^2 were retained by a forward selection procedure.

(1) Birks (1996) first used a form of stepwise multiple regression, adding variables in a specified order, to determine the importance of unglaciated areas in explaining mountain plant species richness. In the "ecology first" analysis, history (i.e. variable "unglaciated areas") was introduced last in the analysis; it added about 0.1% to the explained variation, whereas the environmental variables explained together 84.9% of the variation. In the "history first" analysis, history was entered first; it only explained 7.6% of the variation, which was not a significant contribution. (2) The contribution of "history" did not improve in partial regression analyses, when controlling for either land area per grid square alone, or land area, latitude and longitude. Modern ecological variables such as bedrock geology, climate, topography, and geography were considerably more effective explanatory variables of species richness than "history". (3) In order to find out whether "history" made a unique statistically significant contribution to the variance of the species richness, when the effects of the other variables were controlled for, Birks carried out the variance decomposition described above using partial regression analyses. Fraction [a], corresponding to the influence of all environmental variables independent of "history", explained 77.4% of the variation of species richness; fraction [b], in which "environment" covaried with "history", explained 7.5%; fraction [c], "history" independent of "environment", explained 0.1%; the unexplained variation, fraction [d], was 15.0%. Fraction [b] is likely to result from the spatial coincidence of unglaciated areas with high elevation, western coastal areas, and certain types of bedrock, all these being included among the environmental variables.

(4) In another paper, Birks (1993) used partial canonical correspondence analysis, instead of partial regression analysis, to carry out the same type of analysis (including variance decomposition) on a table of grid cells × species presence/absence. Again, the results suggested that there was no statistically significant contribution from unglaciated areas in explaining present-day distribution patterns when the effects of modern topography, climate, and geology were considered first.

These two papers (Birks 1993, 1996) show that the hypothesis of survival in glacial nunataks is unnecessary to explain the present-day patterns of species distribution and richness of Norwegian mountain plants. Following Ockham's razor principle (Subsection 10.3.3), this unnecessary assumption should be avoided when formulating hypotheses intended to explain present-day species distributions.

6 — Nonlinear regression

In some applications, ecologists know from existing theory the algebraic form of the nonlinear relationship between a response variable and one or several explanatory variables. An example is the logistic equation, which describes population growth in population dynamics:

Logistic equation

$$N_t = \frac{K}{1 + e^{(a - rt)}} \qquad \textbf{(10.23)}$$

This equation gives the population size (N_t) of a species at time t as a function of time (t) and three parameters a, r, and K, which are adjusted to the data; r is the Malthus parameter describing the natural rate of increase of the population, and K is the support capacity of the ecosystem. Nonlinear regression allows one to estimate the parameters (a, r, and K in this example) of the curve that best fits the data, for a user-selected function. This type of modelling does not assume linear relationships among variables; the equation to be fitted is provided by users.

Nonlinear regression is available in several statistical packages and subroutine libraries. The most usual objective functions to minimize are (1) the usual least-squares criterion $\Sigma(y_i - \hat{y}_i)^2$, or (2) the sum of squared Euclidean distances of the points to the regression function. These criteria are illustrated in Fig. 10.6b. The parameters of the best-fitting equation are found by iterative adjustment; users usually have the choice among a variety of rules for stopping the iterative search process. Common choices are: when the improvement in R^2 becomes smaller than some preselected value, or when some preselected maximum number of iterations is reached, or when the change in all parameters becomes smaller than a given value. Good references on this topic are Hollander & Wolfe (1973), Ratkowsky (1983), Ross (1990), and Huet *et al.* (1992).

Consider the Taylor equation relating the means $\bar{y}$ and variances s_y^2 of several groups of data:

$$s_{y_k}^2 = a\bar{y}_k^b \quad \textbf{(1.17)}$$

One must decide whether the equation should be fitted to the data by nonlinear regression, or to the corresponding logarithmic form (eq. 1.18) by linear regression. Look at the data in the original mean-variance space and in the transformed log(mean)-log(variance) space, and choose the form for which the data are homoscedastic.

Other often-encountered functions are the exponential, hyperbolic, Gaussian, and trigonometric (for periodic phenomena; see Section 12.4), and other growth models for individuals or populations.

Monotone regression

As an alternative to linear or nonlinear regression, Conover (1980, his Section 5.6) proposes *monotone regression* which may be used when (1) the relationship is monotonic (increasing or decreasing), (2) the purpose is forecasting or prediction rather than parameter estimation, and (3) one does not wish to carefully model the functional relationship; see also Iman & Conover (1983, their Section 12.6). Monotone regression consists in assigning ranks to the x and y observations and computing a linear regression on these ranks. Simple, natural rules are proposed to reassign real-number values to the forecasted/predicted values obtained from the rank-based equation for given values of x. Monotone regression is sometimes called *nonparametric regression*. A specialized form of monotone regression is used in MDS algorithms (Section 9.3).

7 — *Logistic regression*

Binary variables form an important category of response variables that ecologists may wish to model. In process studies, one may wonder whether a given effect will be present under a variety of circumstances. Population ecologists are also often interested in determining the factors responsible for the presence or absence of a species. When the explanatory variables of the model are qualitative, modelling may call upon log-linear models computed on multivariate contingency tables (Section 6.3). When the explanatory variables are quantitative, or represent a mixture of quantitative and qualitative data, logistic regression is the approach of choice.

In logistic regression, the response variable is binary (presence-absence, or 1-0; see example below). A linear model of quantitative explanatory variables would necessarily produce some forecasted/predicted values larger than 1 and some values smaller than 0. Consider Fig. 10.13, which illustrates the example developed below. A linear regression line fitting the data points would have a positive slope and would span outside the vertical [0, 1] interval, so that the equation would forecast ordinate values smaller than 0 (for small x) and larger than 1 (for large x); these would not make sense since the response variable can only be 0 or 1.

If one tries to predict the *probability* of occurrence of an event (for example the presence of a species), instead of the event itself (0 or 1 response), the model should be able to produce real-number values in the range [0, 1]. The logistic equation (eq. 10.23) described in Subsection 6 provides a sigmoid model for such a response between limit values (Fig. 10.13). It is known to adequately model several ecological, physiological and chemical phenomena. Since the extreme values of the probabilistic response to be modelled are 0 and 1, then $K = 1$, so that eq. 10.23 becomes:

$$\mathrm{p} = \frac{1}{1 + \mathrm{e}^{-z}} \qquad \textbf{(10.24)}$$

where p is the probability of occurrence of the event. z is a linear function of the explanatory variable(s):

$$z = b_0 + b_1 x \qquad \text{for a singe predictor } x \qquad \textbf{(10.25a)}$$

or

$$z = b_0 + b_1 x_1 + b_2 x_2 + \ldots + b_p x_p \qquad \text{for several predictors} \qquad \textbf{(10.25b)}$$

Note that there are other, equivalent algebraic forms for the logistic equation. A form equivalent to eq. 10.24 is: $\mathrm{p} = \mathrm{e}^{z}/(1 + \mathrm{e}^{z})$.

For the error part of the model, the ε_i values cannot be assumed to be normally distributed and homoscedastic, as it is the case in linear regression, since the response variable can only take two values (presence or absence). The binomial distribution is the proper model in such a case, or the multinomial distribution for multistate

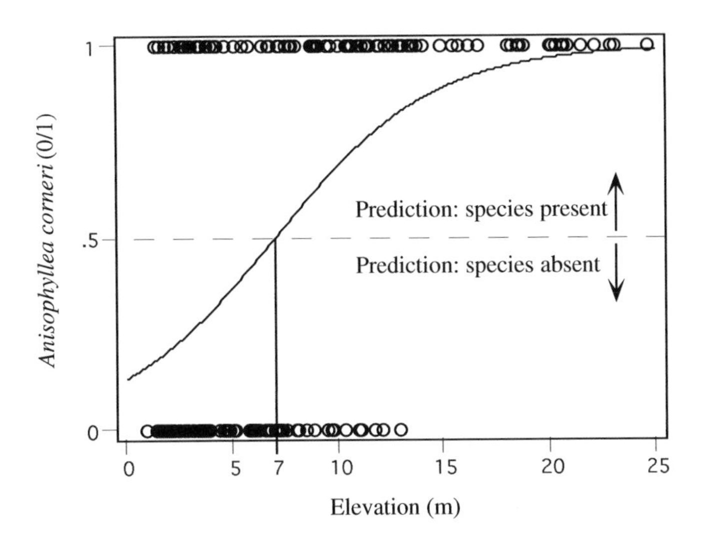

Figure 10.13 Logistic regression equation fitted to presence/absence of *Anisophyllea corneri*, as a function of elevation, in 200 forest quadrats.

qualitative response variables, as allowed in some computer software (e.g. CATMOD in SAS). The parameters of the model cannot be estimated by ordinary least-squares since the error term is not normally distributed. This is done instead by maximum likelihood. Logistic regression is a special case of the generalized linear model (GLM: McCullagh & Nelder, 1983); least-squares regression is another special case of GLM.

Maximum likelihood

According to the *maximum likelihood principle*, the best values for the parameters of a model are those for which the likelihood is maximum. The likelihood L of a set of parameter estimates is defined as the probability of observing the values that have actually been observed, given the model and the parameter estimates. This probability is expressed as a function of the parameters:

$$L = \text{p(observed data} \mid \text{model, parameters)}$$

So, one iteratively searches for parameter estimates that maximize the likelihood function.

Numerical example. Data describing the structure of a tree community, sampled over a 50-ha plot in the Pasoh forest, Malaysia, were studied by He *et al.* (1994, 1996, 1997). The plot was established to monitor long-term changes in a primary tropical forest. The precise locations of the 334077 individual trees and shrubs at least 1 cm in diameter at breast height (dbh) were determined (825 species in total) and a few environmental variables were recorded at the centres of 20 × 20 m quadrats. The present example uses the presence or absence of one species, *Anisophyllea corneri*, in each quadrat. One hundred quadrats were selected at random in the plot among those where *A. corneri* was present, and also 100 among the quadrats where it was absent, for a total of 200 quadrats. Results of the logistic regression study presented below were reported by He *et al.* (1997).

Stepwise logistic regression is used here to model the presence or absence of the species with respect to *slope* and *elevation* (i.e. altitude, in metres, measured by reference to the lowest part of the forest plot floor), using the SPSS software package. Following the calculations, *elevation* was included in the model for its significant contribution, whereas *slope* was left out. The linear part of the fitted model (from eqs. 10.24 and 10.25) is:

$$z = -1.8532 + 0.2646 \times elevation$$

Significance of the regression coefficients was tested using the Wald statistic, which is the square of the ratio of a regression coefficient to its standard error; this statistic is distributed like χ^2. Both the intercept and slope coefficients of the model were significant ($p < 0.001$).

As explained above, the probability of the observed values of the response variable, for given values of the parameters, is called the likelihood. Since a probability is in the range [0, 1], its natural logarithm is a negative number. It is customary to multiply it by –2 to obtain $-2\ln(L)$, noted –2LL, which is a measure of how badly a model fits the data; –2LL = 0 in the case of a perfect fit. This value presents the advantage of being distributed like χ^2, so that it may be tested for significance. The significance of the model is tested using the following table:

	χ^2	ν	$p(\chi^2)$
Intercept only	277.259	199	0.0002
Difference	59.709	1	< 0.0001
Intercept + *elevation*	217.549	198	0.1623
Difference	1.616	1	0.2057
Intercept + *elevation* + *slope*	215.933	197	0.1690
Goodness of fit	183.790	198	0.7575

Parameters are added to the model, one by one, as long as they improve the fit. The procedure is the same as in log-linear models (e.g. Table 6.6).

• For a model with an intercept only, –2LL = 277.259. The hypothesis to be tested is that –2LL = 277.259 is not significantly different from 0, which would be the value of –2LL for a model fitting the data perfectly. Degrees of freedom are computed as the number of observations (200) minus the number of fitted parameters (a single one, up this point). The significant χ^2 statistic ($p < 0.05$) indicates that the model does not fit the data well.

• Inclusion of *elevation* adds a second parameter to the model; this parameter was fitted iteratively and the resulting value of –2LL was 217.549 at convergence, i.e. when –2LL did not change by more than a small preselected value. Since the probability associated with the χ^2 statistic is large, one cannot reject the null hypothesis that the model fits the data. The difference in χ^2 between the two models (277.259 – 217.549 = 59.709) is tested with 1 degree of freedom. The significant probability ($p < 0.05$) shows that *elevation* brings a significant contribution to the likelihood of the model.

• Inclusion of *slope* adds a third parameter to the model. The resulting model also fits the data well ($p > 0.05$), but the difference in χ^2 between the two models (217.549 – 215.933 = 1.616) is not significant ($p = 0.2057$), indicating that *slope* does not significantly contribute to increase the likelihood of the model. Hence, *slope* is left out of the final model.

The last row of the table tests a goodness-of-fit statistic which compares the observed values (0 or 1 in logistic regression) to the probabilities forecasted by the model, which includes the intercept and *elevation* in the present example (Norusis, 1990, p. 52). This statistic (183.790) is distributed like χ^2 and has the same number of degrees of freedom as the χ^2 statistic for the complete model. In the present example, this statistic is not significant ($p > 0.05$), which leads to conclude that there is no significant discrepancy between the forecasted values and the data.

Putting back the observed values of the explanatory variable(s) into the model (eq. 10.25) provides estimates of z. For instance, one of the quadrats in the example data had *elevation* = 9.5 m, so that

$$z = -1.8532 + 0.2646 \times 9.5 = 0.6605$$

Incorporating this value into eq. 10.24 provides the following probability that *A. corneri* be present in the quadrat:

$$p = \frac{1}{1 + e^{-0.6605}} = 0.659$$

Since $p > 0.5$, the forecast is that the species should be found in this quadrat. In general, if $p < 0.5$, the event is unlikely to occur whereas it is likely to occur if $p > 0.5$. (Flip a coin if a forecasted value is required in a case where $p = 0.5$ exactly.) With the present equation, the breaking point between forecasted values of 0 and 1 (i.e. the point where $p = 0.5$) corresponds to an *elevation* of 7 m. The logistic curve fitted to the *A. corneri* data is shown in Fig. 10.13.

Classification table

Forecasted values may be used to produce a classification (or "confusion") table, as in discriminant analysis (Section 11.5), in which the forecasted values are compared to observations. For the example data, the classification table is:

Observed	*Forecasted*		*Percent*
	0	1	*correct*
0	78	22	78%
1	35	65	65%
		Total	71.5%

Since most values are in the diagonal cells of the table, one concludes that the logistic regression equation based solely on elevation is successful at forecasting the presence of *A. corneri* in the quadrats.

Gaussian logistic model

A Gaussian logistic equation may be used to model the unimodal response of a species to an environmental gradient. Fit the logistic equation with a quadratic response function $z = b_0 + b_1 x + b_2 x^2$, instead of eq. 10.25a, to obtain a Gaussian logistic model; the response function for several predictors (eq. 10.25b) may be modified in the same way. See ter Braak & Looman (1987) for details.

Discriminant analysis (Section 10.5) has often been used by ecologists to study niches of plants or animals, before logistic regression became widely available in computer packages. Williams (1983) gives examples of such works. The problem with discriminant analysis is that it constructs a linear model of explanatory variables, such that forecasted values are not limited to the [0, 1] range. Negative values and values higher than 1 may be produced, which are ecologically unrealistic for presence-absence data. This problem does not exist with logistic regression, which is available in several major statistical packages such as BMDP, SPSS, SAS, GLIM, GENSTAT, STATISTICA, SYSTAT, and others. This question is further discussed in Section 11.6.

In procedure CATMOD of SAS, the concept of logistic regression is extended to multi-state qualitative response variables. Trexler & Travis (1993) provide an application of logistic regression to an actual ecological problem, including selection of the most parsimonious model; they also discuss the relative merits of various alternatives to the logistic model.

8 — Splines and LOWESS smoothing

There are instances where one is only interested in estimating an empirical relationship between two variables, without formally modelling the relationship and estimating parameters. In such instances, smoothing methods may be the most appropriate, since they provide an empirical representation of the relationship, efficiently and at little cost in terms of time spent specifying a model. Since they fit the data locally (i.e. within small windows), smoothing methods are useful when the relationship greatly varies in shape along the abscissa. This is the opposite of the parametric regression methods, where a single set of parameters is used to adjust the same function to all data points (global fit). Smoothing methods are far less sensitive to exceptional values and outliers than regression, including polynomial regression. Several numerical methods are available for smoothing.

Moving average

A simple way to visualize an empirical relationship is the method of moving averages, described in more detail in Section 12.2. Define a 'window' of a given width, position it at one of the margins of the scatter diagram, and compute the mean ordinate value (y) of all the observations in the window. Move the window by small steps along the abscissa, recomputing the mean every time, until the window reaches the opposite margin of the scatter diagram. Plot the window means as a function of the positions of the window centres along the abscissa. Link the mean estimates by line segments. This empirical line may be used to estimate y as a function of x.

Piecewise polynomial fitting by "splines" is a more advanced form of local smoothing. In its basic form, spline estimation consists in dividing the range of the explanatory variable x (which is also the width of the scatter diagram) into a number of intervals, which are generally of equal widths and separated by *knots*, and adjusting a polynomial of order k to the data points within each segments, using polynomial regression (Subsection 4). To make sure that the transitions between spline segments are smooth at the junction points (knots), one imposes two constraints: (1) that the values of the function be equal on the left and right of the knots, and (2) that the (k–1) first derivatives of the curves be also equal on the left and right of the knots. Users of the method have to make arbitrary decisions about (1) the level k of the polynomials to be used for regression (a usual choice is cubic splines) and (2) the number of segments along the abscissa. If a large enough number of intervals is used, the spline function can be made to fit every data point. A smoother curve is obtained by using fewer knots. It is recommended to choose the interval width in such a way as to have at least 5 or 6 data points per segment (Wold, 1974). Knots should be positioned at or near inflexion points, where the behaviour of the curve changes (see example below). A large body of literature exists about splines. Good introductory texts are Chambers (1977), de Boor (1978), Eubanks (1988), and Wegman & Wright (1983). The simplest text is Montgomery & Peck (1882, Section 5.2.2); it inspired the explanation of the method that follows.

When the positions of the knots are known (i.e. decided by users), a cubic spline model *with no continuity restriction* is written as:

$$\hat{y} = \sum_{j=0}^{3} b_{0j} x^j + \sum_{k=1}^{h} \sum_{j=0}^{3} b_{kj} (x - t_k)_+^j \qquad \textbf{(10.26)}$$

In this equation, the parameters b_{0j} in the first sum correspond to a cubic polynomial equation in x. The parameters b_{kj} in the second sum allow the curve segments to be disconnected at the positions of the knots. There are h knots, and their positions along the abscissa are represented by t_k; the knots are ordered in such a way that $t_1 < t_2 < \ldots < t_h$. This equation, written out in full, is the following for a single knot (i.e. $h = 1$) located at position t:

$$\hat{y} = b_{00} + b_{01} x + b_{02} x^2 + b_{03} x^3 + b_{10} (x - t)_+^0 + b_{11} (x - t)_+^1 + b_{12} (x - t)_+^2 + b_{13} (x - t)_+^3$$

The expression $(x - t_k)_+$ takes the value $(x - t_k)$ when $x - t_k > 0$ (i.e. if the given value x is to the right of the knot), and 0 when $x - t_k \leq 0$ (for values of x on the knot or to the left of the knot). The constraint of continuity is implemented by giving the value zero to all terms b_{kj}, except the last one. In eq. 10.26, it is these parameters that allow the relationship to be described by discontinuous curves; by removing them, eq. 10.26 becomes a cubic splines equation with continuity constraint:

Cubic splines

$$\hat{y} = \sum_{j=0}^{3} b_{0j} x^j + \sum_{k=1}^{h} b_k (x - t_k)_+^3 \qquad \textbf{(10.27)}$$

which has a single parameter b_k for each knot. Written in full, eq. 10.27 is the following for two knots (i.e. $h = 2$) located at positions $t_1 = -5$ and $t_2 = +4$, as in the numerical example below:

$$\hat{y} = b_{00} + b_{01}x + b_{02}x^2 + b_{03}x^3 + b_1 (x+5)^3_+ + b_2 (x-4)^3_+$$

This approach is not the one used in advanced spline smoothing packages because it has some numerical drawbacks, especially when the number of knots is large. It is, however, the most didactic, because it shows spline smoothing to be an extension of OLS polynomial regression. Montgomery & Peck (1982) give detailed computational examples and show how to test the significance of the difference in R^2 between models with decreasing numbers of knots, or between a spline model and a simple polynomial regression model. They finally show that *piecewise linear regression* — that is, fitting a continuous series of straight lines through a scatter of points — is a natural extension of the spline eq. 10.27 in which the exponent is limited to 1.

LOWESS

LOWESS refers to *Locally Weighted Scatterplot Smoothing* (Cleveland, 1979). This method is an extension of moving averages in the sense that, for each value x_i along the abscissa, a value $\hat{y}_i$ is estimated from the data present in a window around x_i. The number of data points included in the moving window is a proportion f, determined by users, of the total number of observations; a commonly-used first approximation for f is 0.5. The higher this proportion, the smoother the line of fitted values will be. For the end values, all observed points in the window come from the same side of x_i; this prevents the line s from becoming flat near the ends. Estimation proceeds in two steps:

• First, a weighted simple linear regression is computed for the points within the window and an estimate $\hat{y}_i$ is obtained. Weights, given to the observation points by a 'tricube' formula, decrease from the focal point x_i outwards. Points outside the window receive a zero weight. This regression procedure is repeated for all values x_i for which estimates are sought.

• The second step is to make these first estimates more robust, by reducing the influence of exceptional values and outliers. Residuals are computed from the fitted values and, from these, new weights are calculated that give more importance to the points with low residuals. Weighted linear regression is repeated, using as weights the products of the new weights with the original neighbourhood weights. This second step may be repeated until the recomputed weights display no more changes.

Trexler & Travis (1993) give a detailed account of the LOWESS method, together with a full example, and details on two techniques for choosing the most appropriate value for f. The simplest approach is to start with a (low) initial value, and increase it until a non-random pattern along x appears in the residuals; at that point, f is too large. Other important references are Chambers *et al.* (1983) and Cleveland (1985).

Numerical example. Consider again the dependence of salinity on the position along a transect, as modelled in Fig. 10.9. This same relationship may be studied using cubic splines and LOWESS (Fig. 10.14). For splines smoothing, the arbitrary rule stated above (5 or 6 points at

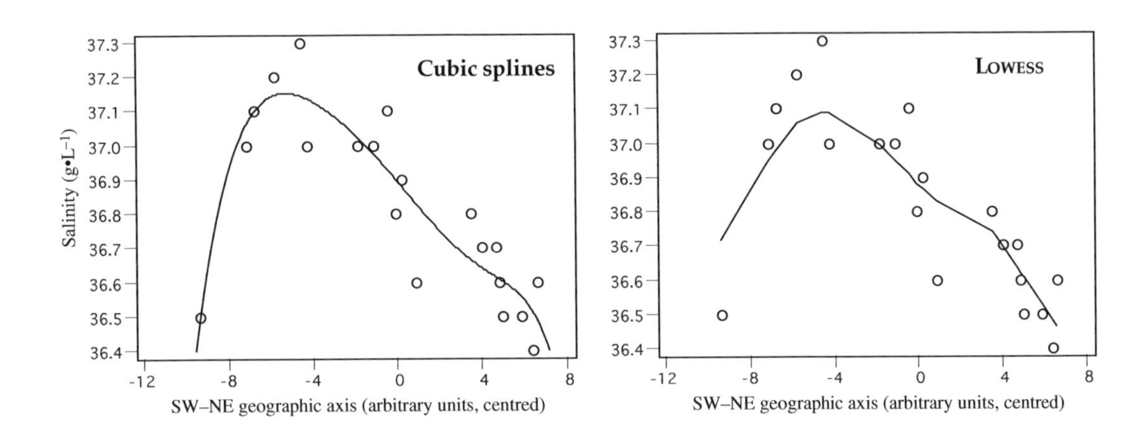

Figure 10.14 Cubic splines and LOWESS scatter diagrams describing the relationship of salinity to the position of the sites along the main geographic axis of the Thau lagoon, on 25 October 1988. Cubic splines were computed with knots at abscissa –5 and +4. For LOWESS (computed using SYSTAT), the proportion of the points included in each smoothing window was $f = 0.5$.

least per interval) leads to 3 or 4 intervals. Fig. 10.9 indicates, on the other hand, that there are at least three regions in the scatter of points, which can be delimited by knots located at approximately –5 and +4 along the abscissa. The computed spline regression equation which follows has $R^2 = 0.841$:

$$\hat{y} = 37.500 + 0.291x + 0.072x^2 + 0.006x^3 - 0.005\,(x+5)_+^3 - 0.007\,(x-4)_+^3$$

The difference in explained variation between this spline model and a cubic polynomial model ($R^2 = 0.81$, Fig. 10.9) is not significant.

The LOWESS curve also clearly suggests the presence of three distinct physical processes which determine the values of salinity along the long axis of the lagoon, i.e. from abscissa –10 to about –5, the central portion, and the right-hand portion from abscissa 4 and on.

Other smoothing methods are available in computer software, such as negative exponentially weighted smoothing (the influence of neighbouring points decreases exponentially with distance); inverse squared distance smoothing, described in Subsection 13.2.2 (eq. 13.20 with $k = 2$); distance-weighted least-squares smoothing (the surface is allowed to bend locally to fit the data); and step smoothing (a step function is fitted to the data).

10.4 Path analysis

Section 4.5 showed that causal relationships among descriptors cannot be unambiguously derived from the sole examination of correlation coefficients, whether simple, multiple, or partial. Several causal models may account for the same correlation matrix. In the case of *prediction* (*versus forecasting*, see Subsection 10.2.3), however, causal (and not only correlative) relationships among descriptors must be established with reasonable certainty. *Path analysis* is an extension of *multiple linear regression* (Subsection 10.3.3) which allows the decomposition and interpretation of *linear* relationships among a (small) number of descriptors. It is thus possible to formally state *a priori hypotheses* concerning the causal relationships among descriptors and, using path analysis, to examine their consequences given the coefficients of regression and correlation computed among these descriptors.

Path analysis was developed by Wright (1921, 1960). The number of ecological applications is growing. There are also many interesting examples in population genetics and the social sciences. Introductory presentations are found in Nie *et al.* (1975, p. 383 *et seq.*)* and Sokal & Rohlf (1995, p. 634 *et seq.*). The present Section provides a summary of path analysis, and concludes on some ecological applications.

Causal order

Path diagram

Causal closure

As mentioned in Section 10.2, path analysis is based on two fundamental assumptions. (1) Researchers must determine a *causal order* among the variables. This causal order may be derived from ecological theory, or established experimentally (for a brief discussion of experiments, see Subsection 10.2.3). The assumption is that of *weak causal* ordering, i.e. y_1 *may* (or may not) affect y_2 but y_2 *cannot* affect y_1. In *path diagrams* (see Figs. 10.15 to 10.18 and Table 4.8), the causal ordering is represented by arrows, e.g. $y_1 \rightarrow y_2$. (2) No single model can account for all the observed variance. Path models thus include *residual variables* u_i which represent the unknown factors responsible for the residual variance (i.e. the variance not accounted for by the observed descriptors). The assumption of *causal closure* implies the independence of the residual causal variables; in other words, one assumes the existence of residual variables such that $u_1 \rightarrow y_1$ and $u_2 \rightarrow y_2$, whereas $u_1 \rightarrow y_2$ or $u_2 \rightarrow y_1$ is not allowed.

Numeral example. A simple example, with three variables exhibiting causal relationships, is used to illustrate the main features of path analysis. It is adapted from Nie *et al.* (1975, p. 386 *et seq.*). The example considers hypothesized relationships among water temperature, picophytoplankton (algae < 2 μm), and microzooplankton (e.g. ciliates) grazing on the picophytoplankton. In the model, it is assumed that water temperature (y_3) directly affects the growth of microzooplankton (y_1) and picophytoplankton (y_2), whose abundance, in turn, affects that of microzooplankton. Following the terminology of Sokal & Rohlf (1995, Section 16.3), y_2 and y_3 are *predictor* (or explanatory) variables while y_1 is the *criterion* (or response) variable. Figure 10.15 illustrates this hypothetical network of causal relationships in schematic form. Since the three variables probably do not explain all the observed variance, the model also

* Recent SPSS manuals unfortunately do not provide an introduction to path analysis.

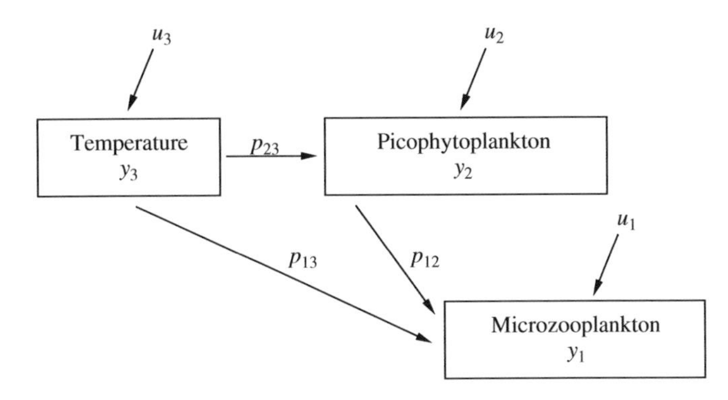

Figure 10.15 Path diagram for three linearly related descriptors. Adapted from Nie *et al.* (1975).

includes residual variables u_1 to u_3. The *causal ordering* of Fig. 10.15 is summarized in the following system of linear equations:

$$y_3 = u_3$$

$$y_2 = p_{23}y_3 + u_2$$

$$y_1 = p_{13}y_3 + p_{12}y_2 + u_1$$

Path coefficient

where parameters p_{ij} are the *path coefficients*. All variables are centred on their respective means. The hypothesis of causal closure implies that:

$$s(u_1,u_2) = s(u_1,u_3) = s(u_2,u_3) = 0$$

because the residual causes are independent; s represents covariances.

The path coefficients are estimated using multiple linear regression (Subsection 10.3.3):

$$\hat{y}_2 = p_{23}y_3$$

$$\hat{y}_1 = p_{13}y_3 + p_{12}y_2$$

There are no intercepts (coefficients p_0) in the regression equations because the data are centred. For a model with n descriptors, one can estimate all path coefficients using at most $(n - 1)$ regression equations. Each descriptor is predicted from the descriptors with immediately higher causal order. Two regression equations are needed to calculate the three path coefficients in Fig. 10.15. Let us use the following values for the path coefficients (Fig. 10.16) and coefficients of determination (R^2) of the numerical example:

$$\hat{y}_2 = 0.5y_3 \qquad R^2 = 0.25$$

$$\hat{y}_1 = 0.4y_3 + 0.2y_2 \qquad R^2 = 0.28$$

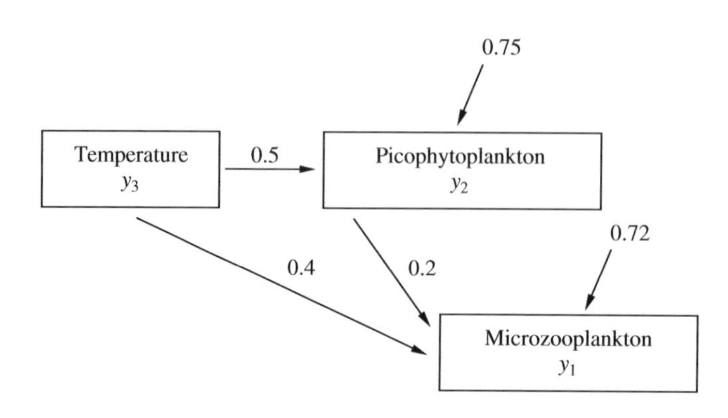

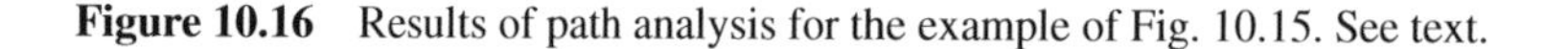

Figure 10.16 Results of path analysis for the example of Fig. 10.15. See text.

and the following correlation coefficients among the descriptors:

$$r_{12} = 0.4 \qquad r_{13} = 0.5 \qquad r_{23} = 0.5$$

The correlation r_{13} depends on both the direct relationship between y_1 and y_3 and the indirect relationship *via* y_2 (Figs. 10.15 and 10.16). Path analysis makes it possible to interpret the correlation r_{13} within the framework of the above model of causal relationships. Because the regressions which provide the estimates of the path coefficients are computed using *standardized* variables (eq. 1.12), it follows (Sokal & Rohlf, 1995, eq. 16.6) that

$$\begin{aligned} r_{13} &= p_{13} + r_{23}p_{12} \\ &= 0.4 + 0.5 \times 0.2 \\ &= 0.4 + 0.1 = 0.5 \end{aligned}$$

The correlation between y_3 (predictor variable) and y_1 (criterion variable) includes the direct contribution of y_3 to y_1 (path coefficient p_{13}), and also the common causes behind the correlations between y_3 and y_1. More generally, the correlation between a predictor variable y_i and a criterion variable y_1 includes the direct contribution of y_i to y_1, plus the common causes behind the correlations between y_i and any other variable that has a direct effect on y_1. These various contributions may either increase (as in the present example) or decrease the correlation between the predictor and criterion variables. The correlation coefficient r_{13} thus includes both a direct (0.4) and an indirect component (0.1).

Coefficient of nondetermination

*Coefficients of nondetermination** are used to estimate the fraction of the variance that is not explained by the model (Fig. 10.16):

$$r^2(u_2, y_2) = 1 - R^2_{2.3} = 1 - 0.25 = 0.75$$

$$r^2(u_1, y_1) = 1 - R^2_{1.23} = 1 - 0.28 = 0.72$$

* The *coefficient of nondetermination* is $(1 - R^2)$; $\sqrt{1 - R^2}$ is called the *coefficient of alienation*.

Table 10.6 Decomposition of bivariate covariation among the (standardized) variables of Fig. 10.16. Adapted from Nie *et al.* (1975).

Bivariate relationships	Total covariation	Causal covariation Direct	Indirect	Total	Noncausal covariation
	(A)	(B)	(C)	(D = B+C)	(A–D)
y_2y_3	$r_{23} = 0.5$	0.5	0.0	0.5	0.0
y_1y_3	$r_{13} = 0.5$	0.4	0.1	0.5	0.0
y_1y_2	$r_{12} = 0.4$	0.2	0.0	0.2	0.2

One concludes that 75% of the variance of the picophytoplankton (y_2) and 72% of the variance of the microzooplankton (y_1) are not explained by the causal relationships stated in the model. The same results are obtained using the following general formula (Sokal & Rohlf, 1995):

$$r^2(u_1, y_1) = 1 - \left[\sum_i p_{1i}^2 + 2\sum_{ij} p_{1i}p_{1j}r_{ij}\right]$$

$$= 1 - [(p_{12}^2 + p_{13}^2) + 2(p_{12}p_{13}r_{23})$$

$$= 1 - [(0.04 + 0.16) + 2(0.2 \times 0.4 \times 0.5)]$$

$$= 1 - [0.20 + 2(0.04)]$$

$$= 1 - 0.28 = 0.72$$

The above results may be summarized in a single table. In the numerical example (Table 10.6), 0.1/0.5 = 20% of the covariation between microzooplankton (y_1) and temperature (y_3) is through picophytoplankton (y_2). In addition, 0.2/0.4 = 50% of the observed relationship between microzooplankton (y_1) and picophytoplankton (y_2) is not causal, and thus spurious according to the path model of Figs. 10.15 and 10.16. Such spurious correlations (Table 4.8) occur when two descriptors are caused by a third one whose values have not been observed in the study.

Path analysis may be applied to more than three variables. As the number of variables increases, interpretation of the results becomes more complex and the number of possible models increases rapidly. In practice, path analysis is restricted to exploring the causal structure of relatively simple systems. This analysis is very useful in many ecological situations, if only because it forces researchers to explicitly state their hypotheses about the causal relationships among descriptors. The method helps assess the consequences of hypotheses, given the observed covariation among descriptors. Other methods, mentioned in Table 10.3, must be used when the descriptors do not exhibit linear relationships, or when they are not quantitative.

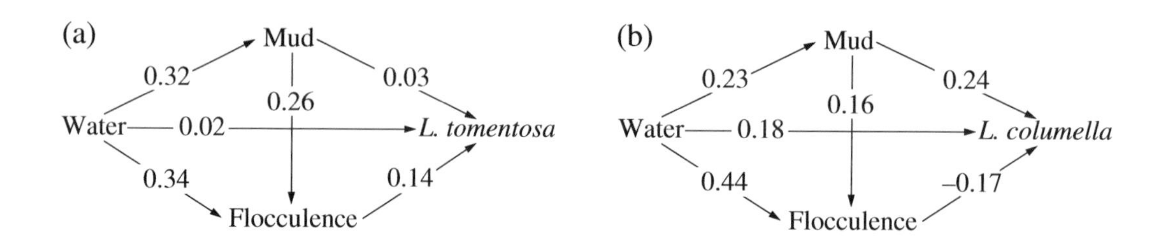

Figure 10.17 Path diagrams of the hypothesized effects of water, mud and flocculence on population densities of two pulmonate snails in marsh microhabitats. After Harris & Charleston (1977).

The following Ecological application (10.4a) concerns freshwater ecology. Other applications of path analysis may be found, for example, in the fields of bacterial ecology (Troussellier *et al.*, 1986), biological oceanography (Gosselin *et al.*, 1986; Legendre *et al.*, 1991), and plant ecology (Hermy, 1987; Kuusipalo, 1987).

Ecological application 10.4a

Harris & Charleston (1977) used path analysis to compare the microhabitats of two pulmonate snails, *Lymnaea tomentosa* and *L. columella*. The two species live in freshwater marshes; there are no obvious differences in the physical or chemical features of their respective habitats. Path analysis was used to examine, for each of the two species, the hypothetical model of causal relationships represented in schematic form in Fig. 10.17. In this model, water was assumed to affect snail numbers directly, and also *via* mud and flocculence, since both factors are partly determined by the amount of water present. The amount of mud was also expected to influence snails directly; however, larger mud areas are less likely to contain vegetation and are thus more likely to be flocculent, hence the indirect path from mud to snails *via* flocculence.

Results of path analysis (Fig. 10.17) suggest major differences between the microhabitats of the two species. Overall, increasing water cover has a direct (positive) effect on *L. columella*; in addition, flocculent mud appears to favour *L. tomentosa* whereas *L. columella* seem to prefer firm mud. The effects of water and mud on *L. columella* are thus direct, whereas they are indirect on *L. tomentosa* (i.e. *via* flocculence). The tentative hypothesis generated by the path diagrams must be further tested by observations and experiments. However, designing experiments to test the role played by the consistency of mud, while controlling for other (confounding) variables, would require considerable ingenuity.

Ecological application 10.4b

Nantel & Neumann (1992) studied ectomycorrhizal-basidiomycete fungi at 11 forest sites in southern Québec, with the purpose of examining the relationships between fungi, woody vegetation (referred to as "trees"), and environmental conditions. Matrices representing the three data sets (11 sites) were compared using Mantel tests (Subsection 10.5.1). Similarity matrices were computed for each data set, using appropriate similarity coefficients (S_{17} for fungi and trees, S_{16} for abiotic conditions). Since the standardized Mantel statistic (r_M) is computationally

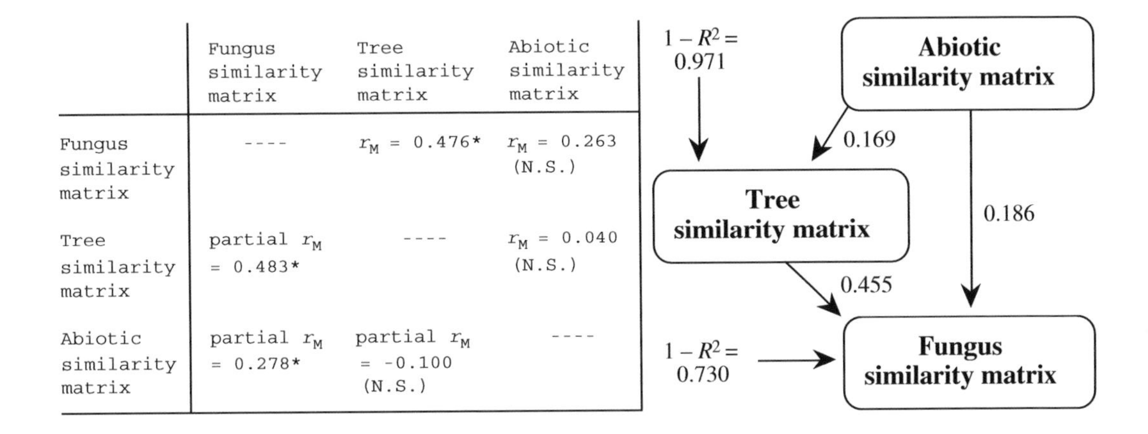

	Fungus similarity matrix	Tree similarity matrix	Abiotic similarity matrix
Fungus similarity matrix	----	r_M = 0.476*	r_M = 0.263 (N.S.)
Tree similarity matrix	partial r_M = 0.483*	----	r_M = 0.040 (N.S.)
Abiotic similarity matrix	partial r_M = 0.278*	partial r_M = -0.100 (N.S.)	----

Figure 10.18 Relationships among fungi, trees, and environmental conditions. Left: Mantel statistics r_M (above the diagonal) and partial Mantel statistics (below). Tests of significance: *, significant result; N.S., result not significant at the Bonferroni-corrected 0.05 level. Right: path diagram computed from the Mantel r_M coefficients. The figures on the arrows are the path coefficients.

equivalent to a Pearson r correlation coefficient, the Mantel statistic values were used to carry out a path analysis in order to test a model of interdependence among the three matrices (Fig. 10.18). Partial t statistics (eq. 4.13 of Subsection 4.5.3) could not be used for testing in this case because the Mantel statistics had been computed from similarities, which were not independent of one another. Partial Mantel tests were used instead to determine the significance of the path coefficients. The study showed that, while both the trees and abiotic conditions were significantly related to the ectomycorrhizal basidiomycete community structure, the influence of the host trees was far greater. The influence of abiotic conditions on the fungus and tree communities was further studied using canonical correspondence analysis (Section 11.2).

Another application of Mantel statistics to path analysis, also coupled with partial Mantel tests (Subsection 10.5.1), is presented by Leduc *et al.* (1992), who tried to untangle the spatial component from the relationships between environmental conditions and the distributions of trees and saplings in a hardwood forest.

10.5 Matrix comparisons

Regression and path analysis are restricted to the interpretation of univariate response variables. Other methods are required to perform direct comparison analyses when the response descriptors form a multivariate data table. As indicated in Fig. 10.4, the analysis may focus on either the original raw data tables, using canonical analysis (Chapter 11), or similarity or distance matrices derived from the raw data tables, using the techniques of matrix comparison described in the present Section. Three main approaches are discussed here: the Mantel test and derived forms (partial Mantel test, multiple regression on distance matrices); the analysis of similarities (ANOSIM); and Procrustes analysis.

1 — Mantel test

One method for comparing two similarity or distance matrices, computed about the same objects, is the Mantel test (1967). The data tables used to compute the two resemblance matrices must have been obtained independently of each other (i.e. different variables). One of the matrices may actually reflect a hypothesis instead of real data, as will be shown below.

Let us consider a set of n sampling sites. To fix ideas, matrix **Y** may contain similarities about species composition among sites; matrix **X** may be computed from characteristics of the environment (e.g. physics or chemistry of the water or soil, or geomorphology) for the same sites, *listed in the same order.* The choice of an appropriate similarity or distance measure is discussed in Chapter 7 (see Tables 7.3 and 7.4). In the applications discussed in Chapter 13, one of the matrices contains geographic distances among sites.

z_M statistic

The basic form of the Mantel statistic, called z_M, is the sum of cross-products of the (*unstandardized*) values in the two similarity or distance matrices, excluding the main diagonal which only contains trivial values (1's for similarities, 0's for distances) for which no estimation has been computed (Fig. 10.19). A second approach is to *standardize* the values in each of the two vectors of similarities or distances before computing the Mantel statistic. The cross-product statistic, divided by the number of distances in each half-matrix minus 1 [i.e. $(n(n-1)/2)-1$], is bounded between -1 and $+1$; it behaves like a correlation coefficient and is called r_M. A third approach is to transform the actual distances into ranks (Dietz, 1983) before computing the standardized Mantel statistic; this is equivalent to computing a Spearman correlation coefficient (Section 5.3) between the corresponding values of matrices **Y** and **X**.

r_M statistic

Permutation test

Mantel statistics are tested by permutation (Section 1.2). The permutations actually concern the n objects forming the similarity or distance matrices and not the $[n(n-1)/2]$ values in each half-matrix. The reason is that there exist liaisons among the values in a similarity or distance matrix. The best-known is the triangle inequality for metric coefficients (property 4 of metrics, Section 7.4). As a consequence of that property, when two distances among three points are known, the third distance cannot take any possible value but is bounded to be larger than or equal to the difference between the first two distances, and smaller than or equal to their sum. Semimetrics may violate this property, but not by much. The testing procedure for Mantel statistics is summarized in Box 10.2.

The permutation test leads to exactly the same probability with statistics z_M or r_M because all cross-product results, permuted or not, are affected in the same way by linear transformations (such as standardization, eq. 1.12) of one or the other series of

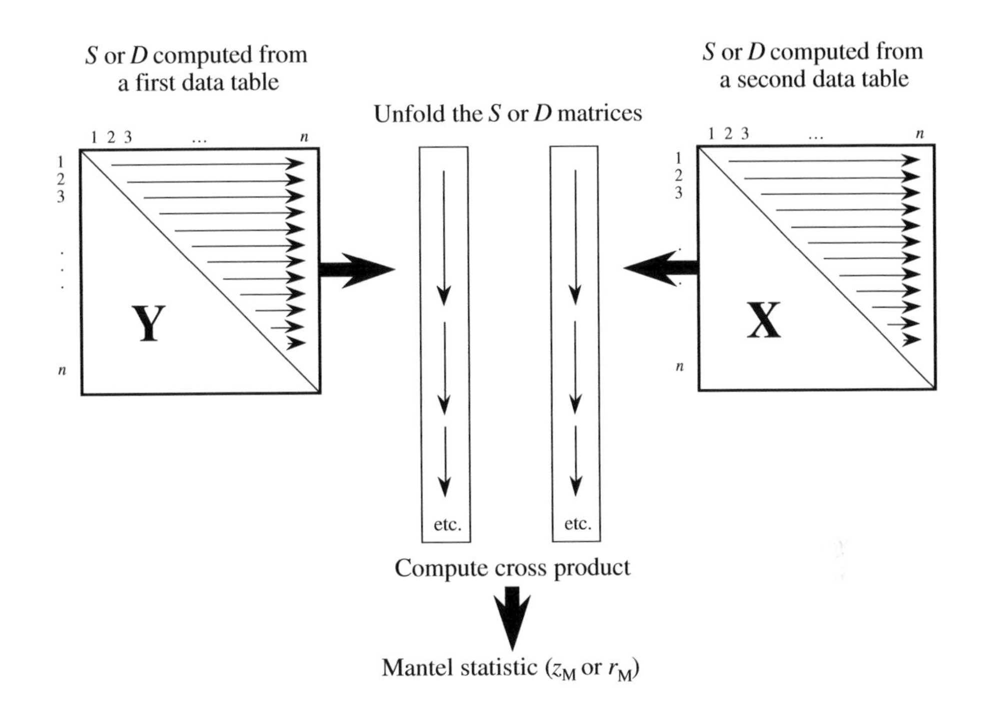

Figure 10.19 The Mantel statistic is the sum of cross products of the corresponding values in two similarity or distance matrices. Values in the vectors representing the unfolded matrices (i.e. written out as vectors) may be standardized before computing the statistic (r_M), or not (z_M), or transformed into ranks.

similarities or distances. This is a most important property of the Mantel test. Thanks to it, the arbitrary values used in *model matrices* (below) are not an issue because any pair of chosen contrasting values leads to the same probability.

Mantel tests are usually one-tailed since, in most cases, ecologists have a strong hypothesis that the two matrices being compared are correlated with a given sign. The hypothesis is often that the two similarity or distance matrices are positively related, which leads to a test of significance in the upper tail of the reference distribution. This is certainly the case when comparing similarities based on species composition among sites to similarities based on characteristics of the environment. When comparing a similarity to a distance matrix, one generally expects a negative relationship to be found, if any, so that the test is then in the lower tail of the reference distribution.

Examples of Mantel tests are found in Ecological application 10.4b, Upton & Fingleton (1985), Legendre & Fortin (1989), Sokal & Rohlf (1995), and in the papers cited near the end of the present Subsection.

Theory of the Mantel test

Box 10.2

Hypotheses

H_0: The distances among objects in matrix **Y** are not (linearly) correlated with the corresponding distances in **X**. When **X** contains geographic distances (Chapter 13), H_0 reads as follows: the variable (or multivariate data) in **Y** is not structured as a gradient.

H_1: The distances among points in matrix **Y** are (linearly) correlated to the distances in **X**.

Test statistics

• Mantel (1967) statistic: $z_M = \sum_{i=1}^{n-1} \sum_{j=i+1}^{n} x_{ij} y_{ij}$ where i and j are row and column indices.

• Standardized Mantel statistic: $r_M = \frac{1}{d-1} \sum_{i=1}^{n-1} \sum_{j=i+1}^{n} \left(\frac{x_{ij} - \bar{x}}{s_x} \right) \left(\frac{y_{ij} - \bar{y}}{s_y} \right)$

$d = [n(n-1)/2]$ is the number of distances in the upper triangular part of each matrix.

Distribution of the test statistic

• According to H_0, the vector of values observed for any one object could have been observed for any other object; in other words, the objects are the permutable units. A realization of H_0 is obtained by permuting the objects (rows) in one of the original data matrices, bringing with them their vectors of values for the observed variables, and recomputing the **S** or **D** matrix.

• An equivalent result is obtained by permuting at random the rows of matrix **Y** and the corresponding columns. Either **Y** or **X** can be permuted at random, with the same net effect.

• Repeating the above operation, the different permutations produce a set of values of the Mantel statistic, z_M or r_M, obtained under H_0. These values estimate the sampling distribution of the Mantel statistic under H_0.

Statistical decision

As in any other statistical test, the decision to reject H_0 or not is made by comparing the actual value of the auxiliary variable (z_M or r_M) to the reference distribution obtained under H_0. If the actual value of the Mantel statistic is one likely to have been obtained under the null hypothesis (no relationship between **Y** and **X**), then H_0 is accepted; if it is too extreme to be considered a likely result under H_0, then H_0 is rejected. See Section 1.2 for details.

Remarks

• The z_M or the r_M statistics may be transformed into another statistic, called t by Mantel (1967), which is asymptotically normal. It is tested by referring to a table of the standard normal distribution. It provides a good approximation of the probability **when *n* is large**.

• Like Pearson's correlation coefficient, the Mantel statistic formula is a linear model that brings out the linear component of the relationship between the values in two distance matrices. Strong nonlinearity may prevent relationships from being identified in the Mantel test. This led Dietz (1983) to suggest the use of the Spearman or Kendall nonparametric correlation coefficients as statistics in the Mantel test.

Clarke & Ainsworth (1993) used Mantel tests based on Spearman correlations (see Box 10.2, Remarks) to measure the agreement between distance matrices computed from species abundances and environmental (abiotic) variables, respectively. They proposed to repeat the analysis for all possible subsets of variables in the abiotic data set in order to circumscribe the subset of environmental variables that best matches the biotic data. This subset is the one leading to the highest nonparametric Mantel statistic. The method is available as procedure BIO-ENV in package PRIMER (Clarke & Warwick, 1994).

The Mantel test is only valid if matrix **X** is independent of the resemblance measures in **Y**, i.e. **X** should not come from an analysis of **Y**. The Mantel test has two chief domains of application in community ecology:

1. It may be used to compare two resemblance matrices computed from empirical data tables (e.g. field observations), as described above. For the test to be valid, **X** must be computed from a different data set than that used to compute **Y**.

Model matrix

2. The Mantel test may also be used to assess the goodness-of-fit of data to an *a priori* model. The test compares the empirical resemblance matrix (**S** or **D**) to a *model matrix* (also called *pattern* or *design matrix*, e.g. Fig. 10.20). This matrix is constructed to represent the model to be tested; in other words, it depicts the alternative hypothesis of the test. If the model is a classification of the objects into groups, the Mantel test is equivalent to a nonparametric multivariate analysis of variance. The way to code the model matrix in this case is explained in Ecological application 10.5a and in Sokal & Rohlf (1995, Section 18.3). Model matrices may take complicated forms for hypotheses stated in quantitative rather than qualitative terms. The model matrix may, for instance, represent the hypothesis that a gradient is present in the data.

Here are examples of the two domains of application described above:

• Species abundance data are recorded at various sites and used to cluster the sites in two groups. The two groups are transcribed into a model matrix **X** and tested against a similarity matrix **Y** computed from the environmental descriptors collected at the same sites as for matrix **X**, to test the hypothesis of environmental control of community differentiation. Clustering the sites is equivalent to asking the species where they draw the line between communities. The procedure is valid as long as the test involves independently obtained matrices **Y** and **X**.

• In order to find out whether there are differences in habitat preferences between males and females of a species, individuals are collected and sexed, and descriptors of their habitats are recorded. **Y** is a similarity matrix among individuals, computed from the observed habitat descriptors. **X** is a model matrix composed of contrasting dummy variables which describe the hypothesis of ecological differentiation between sexes, obtained independently of the habitat data. See Fig. 10.20 for an example.

The Mantel test cannot be used to check the conformity to a matrix **Y** of a model derived from the same data, e.g. testing the conformity of **Y** to a group structure obtained by clustering matrix **Y**. In such a case, the model matrix **X**, which depicts the *alternative hypothesis* of the test, would describe a structure made to fit the very data that would now be used for testing the null hypothesis. The hypothesis (**X**) would not then be independent of the data (**Y**) used to test it. Such a test would be incorrect; it would almost always reject the null hypothesis and support the conformity of **Y** to **X**.

Goodness-of-fit Mantel tests have been used in vegetation studies to investigate hypotheses related to questions like the concept of climax (McCune & Allen, 1985) and the environmental control model (Burgman, 1987, 1988). Hypotheses of niche segregation have been tested for trees by Legendre & Fortin (1989), and for animals by Hudon & Lamarche (1989; see Ecological application 10.5a). Somers & Green (1993) used Mantel tests based on Spearman correlation coefficients (see Box 10.2, Remarks) to assess the relationship between crayfish catches in six Ontario lakes and five model matrices corresponding to different ecological hypotheses.

Ecological application 10.5a

Hudon & Lamarche (1989) studied the stomach contents of American lobsters (*Homarus americanus*) and rock crabs (*Cancer irroratus*) captured at the same sites at Îles-de-la-Madeleine in the Gulf of St. Lawrence (Québec). The two species feed upon the same prey and their stomach contents generally reflect the relative abundances of prey species in the habitat. Gause's competitive exclusion principle suggests that, if two species are found together, there should be some difference in their feeding habits, for the two species to coexist on a permanent basis. To test the hypothesis of niche segregation, the authors compared the stomach contents of 42 lobsters and 103 crabs captured at Cap Irving, and of 136 lobsters and 59 crabs caught at Gros Cap, using a goodness-of-fit Mantel tests. For each location, a Gower (S_{19}) similarity matrix was computed among the stomach content vectors, based upon the abundances of 61 prey taxa grouped into 18 taxa. This matrix was compared to a model corresponding to the alternative hypothesis of the study, i.e. the existence of differences between the two species (Fig. 10.20). In the model matrix, stomach contents are similar within species ($S = 1$ in the two "within-group" parts of the matrix) and different between species ($S = 0$ in the "between-group" portion). (Any other pair of values, one large and one small, for instance –1 and +1, would have served the same purpose. This is because all possible pairs of contrasting model values give the same result after standardization of the matrices.) The Mantel statistics (values not reported in the paper) were tested using the normal approximation (Box 10.2) and found to be significant ($p < 0.001$), showing the good fit of the niche segregation model to the data. The authors concluded that lobsters and crabs were able to cohabit probably because they used the substrate and food resources in different ways.

This test is equivalent to a multivariate analysis of variance (MANOVA) or a discriminant analysis. The parametric forms of these analyses could not be used in this case because the authors did not wish the absence of a prey from two stomach contents to be interpreted as an indication of resemblance. The similarity coefficient used by the authors, which treats the zeros in an asymmetrical way (Chapter 7), was considered a better model of resemblance among stomach contents than the Euclidean distance which is implicit in standard parametric analyses. The diversity of resemblance measures in Chapter 7 allows ecologists to choose resemblance coefficients appropriate to the data and the problem at hand.

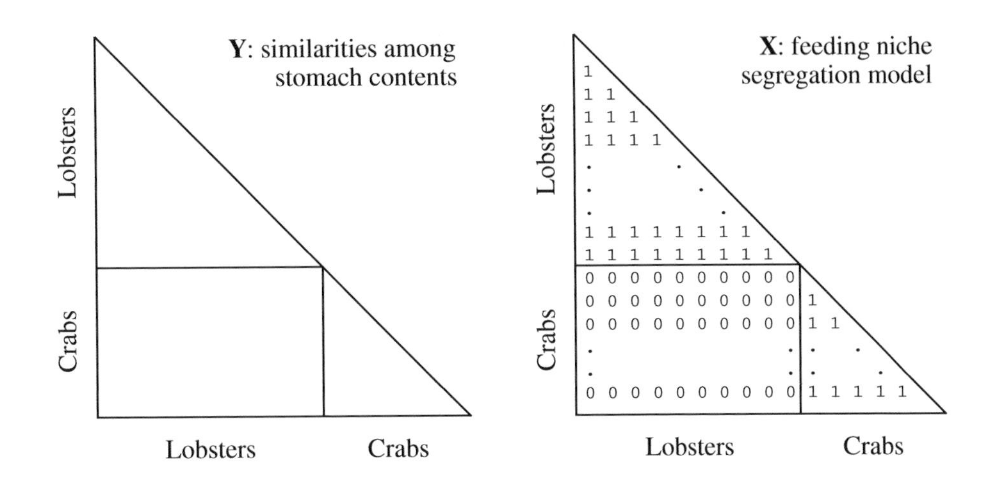

Figure 10.20 Goodness-of-fit Mantel test comparing the actual similarity matrix among stomach contents (**Y**) to a model matrix (**X**), to test the hypothesis of niche segregation between lobsters and crabs.

The problem in this Ecological application could also be studied using canonical analysis (Chapter 11). Indeed, the classification criterion could be coded into the explanatory matrix **X** of canonical analysis as easily as in matrix **X** of the Mantel test. If the Euclidean distance is adequate to represent distances among objects in the response data matrix, redundancy analysis (RDA, Section 11.1) may be used for such a test. If the χ^2 distance is adequate, canonical correspondence analysis may be used (CCA, Section 11.2). If the distances among objects in **Y** are better modelled by some other resemblance measure, the Mantel test and the distance-based RDA method of Legendre & Anderson (1999; Subsection 11.3.1) offer alternatives to RDA and CCA.

The ANOSIM method discussed in Subsection 10.5.3 is another way of testing the same type of hypothesis. Other techniques derived from the Mantel test are presented in the next Subsection and in Chapter 13 (multivariate Mantel correlogram).

2 — More than two matrices

Researchers often collect data about different types of variables that may all be useful to model the variations of the community structure. These variables may include, for example, lake morphometry; pedology; soil, sediment or water chemistry; geology; predator species assemblage; prey species assemblage; and "geography" (i.e. geographic coordinates of the sampling sites). How is it possible to use several matrices of explanatory variables in the same analysis? The simultaneous analysis of three or more similarity or distance matrices is examined here and in Subsection 13.6.1. The canonical analysis of three (sites × descriptors) data tables is described in Sections 11.3 and 13.5. The two techniques discussed in the present Subsection are related to partial correlations and multiple regression, respectively.

Smouse *et al.* (1986) proposed to compute partial correlations involving similarity or distance matrices. Consider distance matrices **A**, **B**, and **C** computed from three multivariate data tables, using a distance measure appropriate to each case. The partial Mantel statistic, $r_M(\mathbf{AB.C})$, estimating the correlation between matrices **A** and **B** while controlling for the effect of **C**, is computed in the same way as a partial correlation coefficient (eq. 4.36), except that the calculation is based here on standardized Mantel statistics r_M (Box 10.2) instead of Pearson correlations *r*. For symmetric distance matrices, only the upper (or lower) triangular portions are used in the calculations.

Partial Mantel test

The difficult aspect is testing the significance of the partial Mantel statistic. The first method proposed by Smouse *et al.* (1986) is to permute matrix **A** (or matrix **B**). This procedure holds **B** and **C** constant (i.e. unpermuted) with respect to each other, thus preserving their relationship. The testing procedure is as follows:

Permutation test

1. Compute the standardized Mantel statistics $r_M(\mathbf{AB})$, $r_M(\mathbf{AC})$ and $r_M(\mathbf{BC})$. Combine these values using the equation for first-order partial correlation coefficients (eq. 4.36) to obtain the reference value of the test statistic, $r_M(\mathbf{AB.C})$.

2. Permute **A** at random using matrix permutation (Box 10.2), obtaining $\mathbf{A}^*$.

3. Compute $r_M(\mathbf{A^*B})$ and $r_M(\mathbf{A^*C})$. Using the value $r_M(\mathbf{BC})$ calculated in step 1, compute $r_M(\mathbf{A^*B.C})$ (eq. 4.36) to obtain a value r_M^* of the partial correlation statistic under permutation.

4. Repeat steps 2 and 3 a large number of times to obtain the distribution of r_M^* under permutation. Add the reference value $r_M(\mathbf{AB.C})$ to the distribution.

5. Calculate the probability as in Section 1.2.

Note that only two of the three correlation coefficients must be computed after each permutation of **A**, i.e. $r_M(\mathbf{A^*B})$ and $r_M(\mathbf{A^*C})$. There is no need to recompute $r_M(\mathbf{BC})$ since neither **B** nor **C** are permuted. Mantel's normal approximation (Box 10.2, Remarks) cannot be used to test the significance of partial Mantel statistics computed in this way. Tests must be carried out using permutations.

An alternative procedure was proposed by Smouse *et al.* (1986) to calculate and test partial Mantel statistics. One computes matrix $\mathbf{Res}_{\mathbf{A|C}}$ containing the residuals of the linear regression of the values of **A** over the values of **C**, and also matrix $\mathbf{Res}_{\mathbf{B|C}}$ of the residuals of the linear regression of the values of **B** on those of **C**. The reference value of the test statistic, $r_M(\mathbf{AB.C})$, is obtained by computing the standardized Mantel statistic between $\mathbf{Res}_{\mathbf{A|C}}$ and $\mathbf{Res}_{\mathbf{B|C}}$; this method of calculation produces the exact same value $r_M(\mathbf{AB.C})$ as above. Using matrix permutation, $\mathbf{Res}_{\mathbf{A|C}}$ is permuted at random (Box 10.2) to obtain a permuted residual matrix $\mathbf{Res}^*_{\mathbf{A|C}}$. The standardized Mantel statistic is computed between $\mathbf{Res}^*_{\mathbf{A|C}}$ and $\mathbf{Res}_{\mathbf{B|C}}$, producing a value $r_M^*(\mathbf{AB.C})$ of the test statistic under permutation. The permutation of $\mathbf{Res}_{\mathbf{A|C}}$ and computation of $r_M^*(\mathbf{AB.C})$ are repeated a large number of times to obtain the distribution of $r_M^*(\mathbf{AB.C})$ under permutation. The reference value $r_M(\mathbf{AB.C})$ is added to the distribution and the probability is calculated as in Section 1.2. The advantage of this method is that it is shorter to compute than the full partial correlation formula.

The alternative procedure of Smouse *et al.* (1986) is similar to a procedure proposed by Kennedy (1995) to test partial regression coefficients in multiple regression. Simulations carried out by Anderson & Legendre (1999), described in Subsection 11.3.2, showed that the method of Kennedy suffers from inflated type I error in tests of partial regression coefficients, especially with small sample sizes. A similar simulation study carried out by P. Legendre (to be reported elsewhere) showed that the alternative procedure of Smouse *et al.* (1986) for partial Mantel tests also suffers from inflated type I error, especially with small sample sizes. Hence this procedure cannot be recommended.

Causal modelling

Partial Mantel tests are not always easy to interpret. Legendre & Troussellier (1988) have shown the consequences of all possible three-matrix causal models on the significance of Mantel and partial Mantel statistics. The models (and their predictions) are the same as those illustrated in Fig. 4.11 for three simple variables. This approach leads to a form of *causal modelling on resemblance matrices* (Legendre, 1993).

This type of analysis has been used mostly to study the distribution of organisms (matrix **A**) with respect to environmental variables (matrix **B**) while considering the spatial positions of the sampling sites (matrix **C**). In such applications, spatial autocorrelation in the data causes the tests of significance to be too liberal (Subsection 1.1.2). Oden & Sokal (1992) have shown, however, that the partial Mantel analysis method developed by Smouse *et al.* (1986) is quite resilient in the presence of spatial autocorrelation, being unlikely to reject the null hypothesis falsely when a conservative critical value is used. Applications of this method to the analysis of spatial ecological data are reviewed in Subsection 13.6.1.

Multiple regression on **S** or **D** matrices

One may also want to model a response multivariate data table (e.g. community structure) as dependent upon a variety of explanatory data sets, such as those mentioned at the beginning of the present Subsection. *Multiple regression on resemblance matrices* has been suggested by a number of authors (Hubert & Golledge, 1981; Smouse *et al.*, 1986; Manly, 1986; Krackhardt, 1988). Legendre *et al.* (1994) described an appropriate testing procedure. The parameters of the multiple regression model are obtained using a procedure similar to that of the Mantel test (Fig. 10.21). The response matrix **Y** is unfolded into a vector **y**; likewise, each explanatory matrix **X** is unfolded into a vector **x**. A multiple regression is computed in which **y** is a function of the vectors $\mathbf{x}_j$. The parameters of this regression (the coefficient of multiple determination R^2 and the partial regression coefficients) are tested by permutations, as follows. When the response matrix **Y** is an ordinary distance or similarity matrix, the permutations of the corresponding vector **y** are carried out in the way of the Mantel permutational test (Subsection 10.5.1). When it is an ultrametric matrix representing a dendrogram (Subsection 8.3.1), the double-permutation method of Lapointe and Legendre (1990, 1991) is used. When it is a path-length matrix representing an additive tree (i.e. a cladogram in phylogenetic studies), the triple-permutation method (Lapointe and Legendre, 1992a) is used. Vectors $\mathbf{x}_j$ representing the explanatory matrices are kept fixed with respect to one another during the permutations. Selection of explanatory matrices may be done by forward selection, backward elimination, or a stepwise procedure, described in Legendre *et al.* (1994).

Permutation test

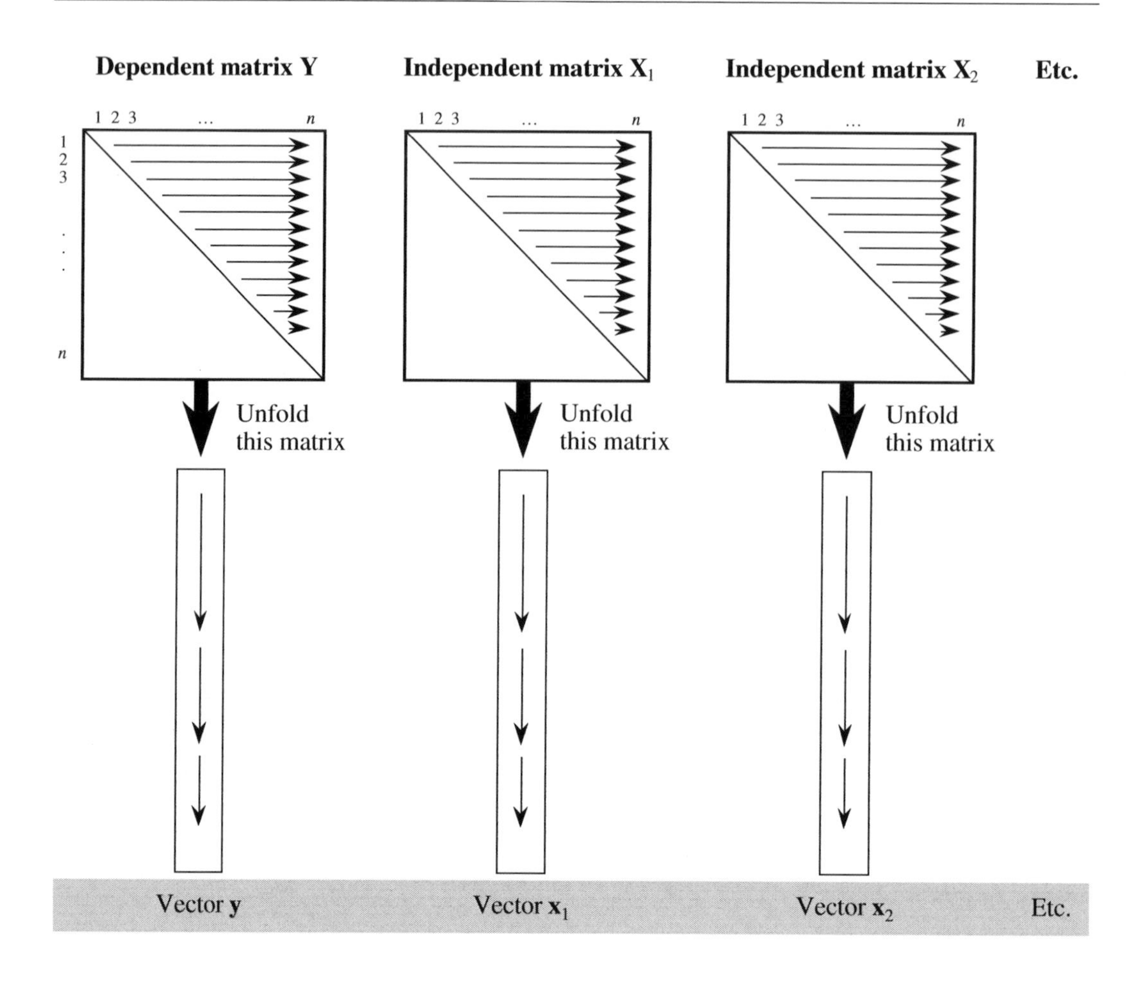

Figure 10.21 Multiple regression is computed on the vectors resulting from unfolding matrices **Y** (response) and $\mathbf{X}_1$, $\mathbf{X}_2$, etc. (explanatory).

3 — ANOSIM test

Focusing on problems of analysis of variance that involve community composition, Clarke (1988, 1993) developed a parallel approach to the goodness-of-fit Mantel tests. Clarke's method, called ANOSIM (*ANalysis Of SIMilarities*), is implemented in the PRIMER package, referred to at the beginning of Section 9.3. Program ANOSIM includes one-way and two-way analyses (crossed or nested) for replicated data, whereas program ANOSIM2 covers two-way analyses without replication (Clarke & Warwick, 1994). After a brief presentation of Clarke's statistic, below, the similarities and differences between the ANOSIM and Mantel approaches will be stressed.

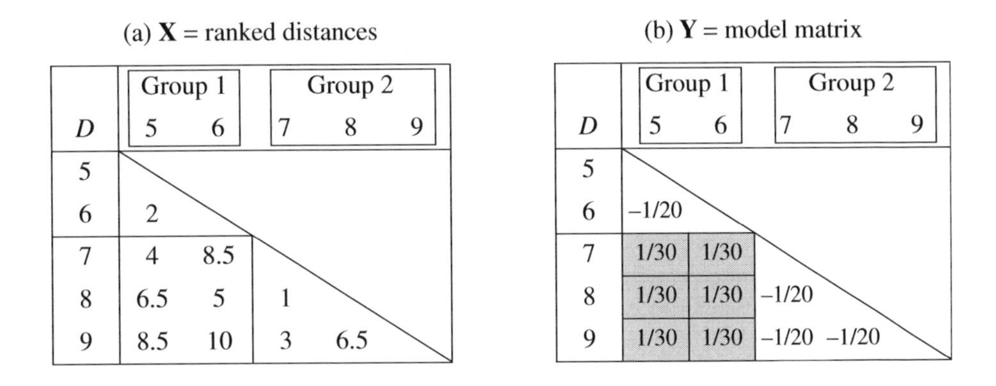

(a) **X** = ranked distances

D	5	6	7	8	9
	Group 1		Group 2		
5					
6	2				
7	4	8.5			
8	6.5	5	1		
9	8.5	10	3	6.5	

(b) **Y** = model matrix

D	5	6	7	8	9
	Group 1		Group 2		
5					
6	–1/20				
7	1/30	1/30			
8	1/30	1/30	–1/20		
9	1/30	1/30	–1/20	–1/20	

Figure 10.22 (a) Distances from the numerical example in Fig. 12.21a are transformed into ranks, the most similar pair receiving rank 1. (b) Weighting required to compute Clarke's statistic as a Mantel statistic.

Consider the situation illustrated in Fig. 10.22a. The distances, taken from Fig. 12.21, have been transformed into ranks, the least dissimilar pair (i.e. the most similar) receiving rank 1. Tied values in Fig. 12.21a are given mean rank values, as usual in nonparametric statistics. Object are arbitrarily numbered 5, 6, 7, 8, 9. The objects are assumed to form two groups, defined here on *a priori* bases; the two groups are not supposed to result from clustering as in Fig. 12.21. The two *a priori* groups are (5, 6) and (7, 8, 9). The null hypothesis is of the ANOVA type:

H_0: There are no differences between the two (or more) groups.

In Fig. 10.22a, does one find the kind of variation among distance values that one may expect if the data correspond to the null hypothesis? Clarke (1988, 1993) proposed the following statistic to assess the differences among groups:

$$R = \frac{\bar{r}_B - \bar{r}_W}{n(n-1)/4} \qquad \textbf{(10.28)}$$

where $\bar{r}_B$ is the mean of the ranks in the *between*-group submatrix (i.e. the rectangle, in Fig. 10.22a, crossing groups 1 and 2), $\bar{r}_W$ is the mean of the ranks in all *within*-group submatrices (i.e. the two triangles in the Figure), and n is the total number of objects. In the present example, $\bar{r}_B = 7.083$ and $\bar{r}_W = 3.125$, so that $R = 0.79167$ (eq. 10.28).

Using ranks instead of the original distances is not a fundamental requirement of the method. It comes from a (reasonable) recommendation, by Clarke and co-authors, that the test statistic should reflect the patterns formed among sites represented by multidimensional scaling plots (MDS, Section 9.3), which preserve rank-transformations of distances. The R statistic is tested by permutations of the objects, as

Permutation test

explained in Box 10.2. The denominator of eq. 10.28 is chosen in such a way that $R = 1$ if all the lowest ranks are in the "within-group" submatrices, and $R = 0$ if the high and low ranks are perfectly mixed across the "within" and "between" submatrices. R is unlikely to be substantially smaller than 0; this would indicate that the similarities within groups are systematically lower than among groups.

Clarke (1988, 1993) actually applied the method to the analysis of several groups. This is also the case in the nonparametric ANOVA-like example of the Mantel test proposed by Sokal & Rohlf (1995). The statistic (eq. 10.28) can readily handle the more-than-two-group case: $\bar{r}_B$ is then the mean of the ranks in *all* between-group submatrices, whereas $\bar{r}_W$ is the mean of the ranks in *all* within-group submatrices.

Both ANOSIM and the goodness-of-fit Mantel test (Fig. 10.20) assume that, if H_0 is false, distances within all distinct groups are comparable in that they are smaller than the among-group distances. This should be checked before proceeding with ANOSIM. Strong heteroscedasticity among groups (i.e. the presence of dense and more dispersed groups) may actually violate this condition and increase the type I error of the test. In other words, the Mantel or ANOSIM tests may find significant differences among groups of objects exhibiting different dispersions, even if they come from statistical populations having identical centroids; see the footnote in Subsection 1.2.2 about the Behrens-Fisher problem.

Equation 10.28 may be reformulated as a Mantel cross-product statistic z_M (Box 10.2). To achieve this, define a model matrix containing positive constants in the "between-group" portion and negative constants in the "within-group" parts:

- the "between" values (shaded area in Fig. 10.22b) are chosen to be the inverse of the number of between-group distances (1/6 in this example), divided by the denominator of eq. 10.28, i.e. $[n(n-1)/4]$ (which is 5 in the present example);

- similarly, the "within" values in Fig. 10.22b are chosen to be the inverse, with negative signs, of the number of distances in all within-group submatrices (–1/4 in the example), also divided by $[n(n-1)/4]$ (= 5 in the present example).

The coding is such that the sum of values in the half-matrix is zero. The unstandardized Mantel statistic (Box 10.2), computed between matrices **X** and **Y** of Fig. 10.22, is $z_M = 0.79167$. This result is identical to Clarke's ANOSIM statistic.

Since the permutation method is the same in the Mantel and ANOSIM procedures, the tests should produce similar probabilities. They may differ slightly in practice because different programs, and even different runs of the same program, may produce different sequences of permutations of the objects. Actually, Subsection 10.5.1 has shown that *any binary coding* of the "within" and "between" submatrices of the model matrix should lead to the same probabilities. Of course, interchanging the small and large values produces a change of sign of the statistic and turns an upper-tail test into a lower-tail test. The only substantial difference between the Mantel goodness-of-fit and

ANOSIM tests is one of tradition: Clarke (1988, 1993) and the PRIMER package (Clarke & Warwick, 1994) transform the distances into ranks before computing eq. 10.28. Since Clarke's R is equivalent to a Mantel statistic computed on ranked distances, it is thus analogous to a Spearman correlation coefficient (eqs. 5.1 and 5.3).

Mann-Whitney's U statistic could also be used for analysis-of-variance-like tests of significance performed on distance matrices. This has been suggested by Gordon (1994) in a different context, i.e. as a way of measuring the differentiation of clusters produced by clustering procedures (internal validation criterion), as reported in Section 8.12. In Gordon's method, distances are divided in two subsets, i.e. the within-group (W) and between-group (B) distances — just like in Clarke's method. A U statistic is computed between the two subsets. U is closely related to the Spearman rank correlation coefficient (eqs. 5.1 and 5.3); a U test of a variable against a dummy variable representing a classification in two groups is equivalent to a Spearman correlation test (same probability). Since Clarke's statistic is also equivalent to a Spearman correlation coefficient, the Mann-Whitney U statistic should lead to the exact same probability as the Clarke or Mantel statistics, if U was used as the statistic in a Mantel-like permutation test. [Using the U statistic as an internal validation criterion, as proposed by Gordon (1994), is different. On the one hand, the grouping of data into clusters is obtained from the distance matrix which is also used for testing; this is not authorized in an analysis-of-variance approach. On the other hand, Gordon's Monte Carlo testing procedure differs from the Mantel permutation test.]

4 — Procrustes analysis

Procrustes was an infamous inn-keeper in the Greek mythology. His pastime was to seize travellers, tie them to an iron bed, and either cut off their legs if they were taller than the bed, or stretch the victims if they were too short, till they fitted in.

Procrustes analysis, proposed by Gower (1971b, 1975, 1987), is primarily an ordination technique. The purpose is to find a compromise ordination for two data matrices concerning the same objects (raw data or distances), using a rotational-fit algorithm that minimizes the sum of squared distances between corresponding points of the two matrices. In the resulting ordination, each object has two representations, one from each matrix, so that the scatter diagram allows one to visualize the differences between the two original matrices. In *orthogonal Procrustes*, two matrices are considered and fitted using rigid-body motions (translation, rotation, and mirror reflection). The extension of the method to more than two matrices is called *generalized Procrustes analysis*. Details are given in the references given above. Other ordination methods involving several data matrices are described by Gower (1987) and Carroll (1987).

The present Subsection focuses on the residual sum-of-squares statistic of orthogonal Procrustes analysis, which is a goodness-of-fit statistic called m^2 by Gower. It is computed as follows:

Procrustes statistic m^2

$$m_{12}^2 = \text{Trace}(\mathbf{Y}_1\mathbf{Y}_1') - \frac{(\text{Trace}\mathbf{W})^2}{\text{Trace}(\mathbf{Y}_2\mathbf{Y}_2')} \quad \mathbf{(10.29)}$$

where $\mathbf{Y}_1$ and $\mathbf{Y}_2$ are the two rectangular matrices of raw data to be analysed, with column vectors centred on their respective means, and $\mathbf{W}$ is a diagonal matrix which is the solution to the singular value decomposition $(\mathbf{Y}_1'\mathbf{Y}_2) = \mathbf{VWU}'$ (eq. 2.31). When $\mathbf{Y}_1$ and $\mathbf{Y}_2$ do not contain the same number of descriptors (columns), the narrower matrix is completed with columns of zeros.

Equation 10.29 is not symmetric. The m_{12}^2 value resulting from fitting $\mathbf{Y}_2$ to $\mathbf{Y}_1$ differs from m_{21}^2. If the matrices $\mathbf{Y}_1$ and $\mathbf{Y}_2$ (with column vectors centred on their respective means) are transformed, before the analysis, to have unit sums-of-squares, the traces of the two cross-product matrices become 1. To accomplish this, compute the trace of $(\mathbf{Y}_1\mathbf{Y}_1')$ and divide each value of $\mathbf{Y}_1$ by the square root of the trace. The trace of $(\mathbf{Y}_1\mathbf{Y}_1')$, which is the same as the trace of $(\mathbf{Y}_1'\mathbf{Y}_1)$, is easily computed as the sum of squares of all values in $\mathbf{Y}_1$. Do the same for $\mathbf{Y}_2$. For matrices $\mathbf{Y}_1$ and $\mathbf{Y}_2$ transformed in this way, the two Procrustes statistics are now identical:

$$m_{12}^2 = m_{21}^2 = 1 - (\text{Trace}\mathbf{W})^2 \quad \mathbf{(10.30)}$$

Permutation test PROTEST

Jackson (1995) suggested to use the symmetric orthogonal Procrustes statistic m_{12}^2 (eq. 10.30) as a measure of concordance, or similarity, between two data matrices representing, in particular, species abundances and environmental variables. The statistic is tested by permutation, as described in Box 10.2 for the Mantel test. Jackson (1995) called this procedure *Procrustean randomization test* (PROTEST). He provided examples of applications to ecological data: benthic invertebrates, lake morphometry, lake water chemistry, and geographic coordinates, for 19 lakes in Ontario, Canada. What Jackson actually compared were, for each data set, the first two ordination axes from correspondence analysis (CA, for benthic invertebrates) or principal component analysis (PCA, for lake morphometry and chemistry); geographic coordinates were left untransformed.

Although Procrustes analysis was originally proposed by Gower as a method for comparing raw data matrices, it may also be used to compare similarity or distance matrices, provided that these matrices are converted back to rectangular data tables by principal coordinate analysis (PCoA, Section 9.2) or nonmetric multidimensional scaling (MDS, Section 9.3). Procrustes comparison can then take full advantage of the diversity of similarity and distance functions available to researchers (Chapter 7). PROTEST is an alternative to the Mantel test, using a different statistic.

Simulation studies comparing type I error and power of the various methods of matrix comparison presented in this Section are needed.

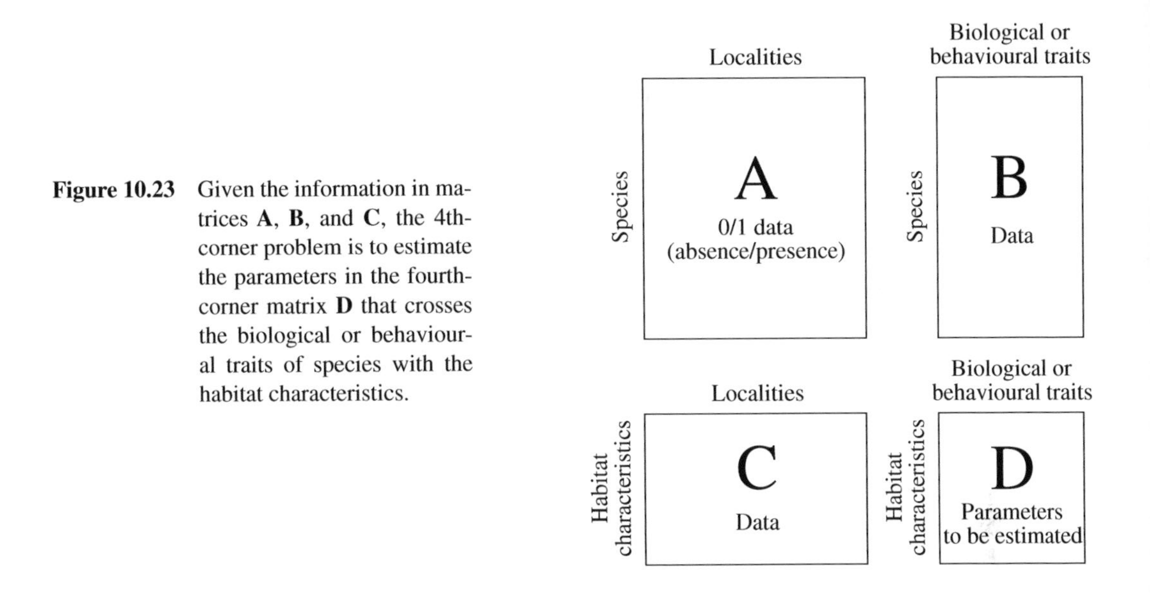

Figure 10.23 Given the information in matrices **A**, **B**, and **C**, the 4th-corner problem is to estimate the parameters in the fourth-corner matrix **D** that crosses the biological or behavioural traits of species with the habitat characteristics.

10.6 The 4th-corner problem

How do the biological and behavioural characteristics of species determine their relative locations in an ecosystem?

This question, which stems from niche theory, has been neglected so far by ecologists because they lacked an appropriate method of analysis. Observation of species in nature leads ecologists to formulate hypotheses in that respect. Testing such hypotheses requires (1) a way of detecting relationships between species traits and habitat characteristics, and (2) of testing the significance of these relationships. Legendre *et al.* (1997) proposed a method to do this.

Consider a table **A** ($p \times n$) containing data on the presence or absence of p species at n sites (Fig. 10.23); the standard ecological data matrix (Table 2.1) is transposed for convenience here. A second table **B** ($p \times q$) describes q biological or behavioural traits of the same p species. A third table **C** ($m \times n$) contains information about m habitat characteristics (environmental variables) at the n sites. How does one go about associating the q biological and behavioural traits to the m habitat characteristics? To help find a solution, let us translate the problem into matrix form:

$$\begin{bmatrix} \mathbf{A}_{(p \times n)} & \mathbf{B}_{(p \times q)} \\ \mathbf{C}_{(m \times n)} & \mathbf{D}_{(m \times q)} \end{bmatrix} \qquad \textbf{(10.31)}$$

Using this representation, the problem may now be stated as follows:

• How does one go about estimating the parameters in matrix **D** ($m \times q$) where the q biological and behavioural traits are related to the m habitat characteristics?

• Are these parameters significant in some sense, i.e. are they different from 0 (no relationship) or from the value they could take in a randomly organized environment?

Because of the above matrix representation, the underlying statistical problem is that of estimating the parameters in the fourth-corner matrix **D**; it is thus referred to as the fourth-corner problem. While the data in matrix **A** are necessarily of the presence/absence type, data in matrices **B** and **C** may be either quantitative or qualitative (nominal). The paper of Legendre *et al.* (1997) describes solutions to accommodate the different types of variables.

1 — Comparing two qualitative variables

The first situation considered here concerns two qualitative variables, one from matrix **B** (behaviour), the other from matrix **C** (habitat). Any qualitative variable can be expanded into a series of binary variables, one for each state (Subsection 1.5.7).

Numerical example. In test cases 1 and 2 (Table 10.7), **A** is a matrix of presence-absence of species at two sites. **B** and **C** hold supplementary variables (qualitative, two states) for the rows and columns of **A**, respectively. To fix ideas, assume that the variable in **B** describes two feeding habits (herbivorous, carnivorous) and **C** is the nature of the substrate at two sampling sites on a coral reef (live coral, turf). This example is used to describe the approach for qualitative variables (Subsection 1) and introduce the method for significance testing (Subsection 2).

Matrices **A**, **B**, and **C** are all needed to estimate the parameters in **D**. The three matrices can be combined by multiplication around the set of four matrices while preserving matrix compatibility:

clockwise: $$\mathbf{D} = \mathbf{C}\,\mathbf{A}'\,\mathbf{B} \qquad \textbf{(10.32)}$$

or counter-clockwise: $$\mathbf{D}' = \mathbf{B}'\,\mathbf{A}\,\mathbf{C}' \qquad \textbf{(10.33)}$$

Inflated data table

For the two test cases of the numerical example, matrix **D** is shown in Table 10.7. Equations 10.32 and 10.33 have an equivalent in traditional statistics. If the data in **A**, **B**, and **C** are frequencies, they can be combined to form an "inflated data table". Matrix **D**, which results from crossing the two columns of the inflated table, is a contingency table as shown in Table 10.8; values d in matrix **D** are frequencies or pseudo-frequencies (see Ecological application 10.6). So, a solution that naturally comes to mind for significance testing is to compute a χ^2 statistic, using either Pearson's (eq. 6.5) or Wilks' formula (eq. 6.6, also called the G statistic). The G statistic is used here; it is the first type of 4th-corner statistic.

For large contingency tables **D**, relationships among descriptor states could be visualized using correspondence analysis (Section 9.4).

Table 10.7 Test cases for qualitative variables. Matrix **A** is (10 species × 2 sites), **B** is (10 species × 2 feeding habits), and **C** is (2 habitat types × 2 sites). So, **D** is (2 habitat types × 2 feeding habits). Probabilities (p) are one-tailed, assuming that the sign of the relationship is stated in the hypothesis. The hypothesis is indicated by a sign in each cell of matrix **D**, + meaning that the actual value is expected to be in the upper tail and – that it is expected to be in the lower tail. Probabilities were calculated after 9999 random permutations. E = exact probabilities; see text.

Test case 1

A:	Site 1	Site 2	**B**: Herbiv.	Carniv.
Sp. 1	1	0	0	1
Sp. 2	0	1	0	1
Sp. 3	1	0	0	1
Sp. 4	1	0	0	1
Sp. 5	1	0	0	1
Sp. 6	0	1	1	0
Sp. 7	0	1	1	0
Sp. 8	0	1	1	0
Sp. 9	0	1	1	0
Sp. 10	0	1	1	0

C:	Site 1	Site 2	**D**: Herbiv.	Carniv.
Live coral	1	0	0 – p = 0.029 E = 0.031	4 + p = 0.189 E = 0.188
Turf	0	1	5 + p = 0.029 E = 0.031	1 – p = 0.189 E = 0.188

Contingency statistic:

G = 8.4562, p (9999 permutations) = 0.021

Test case 2

A:	Site 1	Site 2	**B**: Herbiv.	Carniv.
Sp. 1	1	1	0	1
Sp. 2	1	1	0	1
Sp. 3	1	1	0	1
Sp. 4	1	1	0	1
Sp. 5	1	1	0	1
Sp. 6	1	1	1	0
Sp. 7	1	1	1	0
Sp. 8	1	1	1	0
Sp. 9	1	1	1	0
Sp. 10	1	1	1	0

C:	Site 1	Site 2	**D**: Herbiv.	Carniv.
Live coral	1	0	5 p = 1.000 E = 1.000	5 p = 1.000 E = 1.000
Turf	0	1	5 p = 1.000 E = 1.000	5 p = 1.000 E = 1.000

Contingency statistic:

G = 0.0000, p (9999 permutations) = 1.000

2 — *Test of statistical significance*

In 4th-corner problems, one cannot test the G statistics in the usual manner because, in the general case (although not in test case 1 of Table 10.7), several species are observed at any one sampling so that the rows of the inflated table are not independent of one another; several rows of that matrix result from observations at a single site. To solve the problem, G is tested by permutations (Section 1.2). The procedure is as follows.

Permutation test

Table 10.8 Inflated data table (left); there is one row in this table for each species "presence" ("1") in matrix **A** of test case 1 (Table 10.7). The contingency table (matrix **D**, right) is constructed from the inflated table.

Inflated data table

Occurrences in test case 1	Feeding habits from **B**	Habitat types from **C**
Sp. 1 @ Site 1	Carnivorous	Live coral
Sp. 2 @ Site 2	Carnivorous	Turf
Sp. 3 @ Site 1	Carnivorous	Live coral
Sp. 4 @ Site 1	Carnivorous	Live coral
Sp. 5 @ Site 1	Carnivorous	Live coral
Sp. 6 @ Site 2	Herbivorous	Turf
Sp. 7 @ Site 2	Herbivorous	Turf
Sp. 8 @ Site 2	Herbivorous	Turf
Sp. 9 @ Site 2	Herbivorous	Turf
Sp. 10 @ Site 2	Herbivorous	Turf

Contingency table

D:	Herbivorous	Carnivorous
Live coral	0	4
Turf	5	1

Hypotheses

• H_0: the species (reef fish in the numerical examples) are distributed at random among the sampling sites. Various null models are detailed in the next Subsection.

• H_1: the species are not distributed at random among the sampling sites.

Test statistic

Compute a χ^2 statistic (G here) on the contingency table (matrix **D**) and use it as reference value for the remainder of the test.

Distribution of the test statistic

Under H_0, the species found at any one site could have been observed at any other one. Where the species have actually been observed is due to chance alone. So, a realization of H_0 is obtained by permuting at random the values in matrix **A**, using one of the methods described in the next Subsection. After each permutation of matrix **A**, recompute the χ^2 statistic on **D**.

• Repeat the permutation a large number of times (say, 999 or 9999 times). The different permutations produce a set of values of the χ^2 statistic, obtained under H_0.

• Add to this set the reference value of the statistic, computed for the unpermuted data matrix. Together, the unpermuted and permuted values (for a total of 1000 values, 10000 values, etc.) form an estimate of the sampling distribution of χ^2 under H_0.

Statistical decision

As in any other statistical test, the decision is made by comparing the reference value of the χ^2 statistic to the distribution obtained under H_0. If the reference value of χ^2 is one likely to have been obtained under the null hypothesis, H_0 is not rejected. If it is too extreme (i.e. located out in a tail) to be considered a likely result under H_0, the H_0 is rejected.

Individual values d in matrix **D** can also be tested for significance, as shown below in the Numerical example and the Ecological application.

3 — Permutational models

Permutations may be conducted in different ways, depending on the ecological hypotheses to be tested against observations. Technically, the fourth-corner statistical method can accommodate any of the permutation models described below, as well as constrained permutations for spatial or temporal autocorrelation (e.g. Legendre *et al.*, 1990; ter Braak, 1990). The random component is clearly the field information about the species found at the sampling sites, i.e. matrix **A**. It is thus matrix **A** that should be permuted (randomized) for the purpose of hypothesis testing. This may be done in various ways (Fig. 10.24).

Model 1: Environmental control over individual species — The environmental control model (Whittaker 1956, Bray and Curtis 1957, Hutchinson 1957) states that species are found at sites where they encounter favourable living conditions. Species do that independently of one another, contrary to the randomly-located species assemblage model (next). Realizations of this null hypothesis are generated by permuting at random the values (0's and 1's) within each row vector of matrix **A**; this is done independently from row to row. In this model, species associations are not functional; they simply result from the co-occurrence of species at particular sites, driven by environmental control (Section 8.9). The number of sites occupied by any given species in a row of matrix **A** is fixed because it is considered to reflect such characteristics of the species as abundance, intraspecific competition, and territoriality, as well as ecological plasticity. If all parts of the environment were equally suitable for all species, as stated by H_0, they could eventually all be present at any given site. The present permutation model allows for this, whereas the species assemblage model (next) does not. The alternative hypothesis is that *individual species* find optimal living conditions at the sites where they are actually found. This permutational model is also used in probabilistic similarity coefficient S_{27} (Subsection 7.3.5).

Model 2: Environmental control over species assemblages — Permutation of whole columns is appropriate to test the null ecological hypothesis that the species

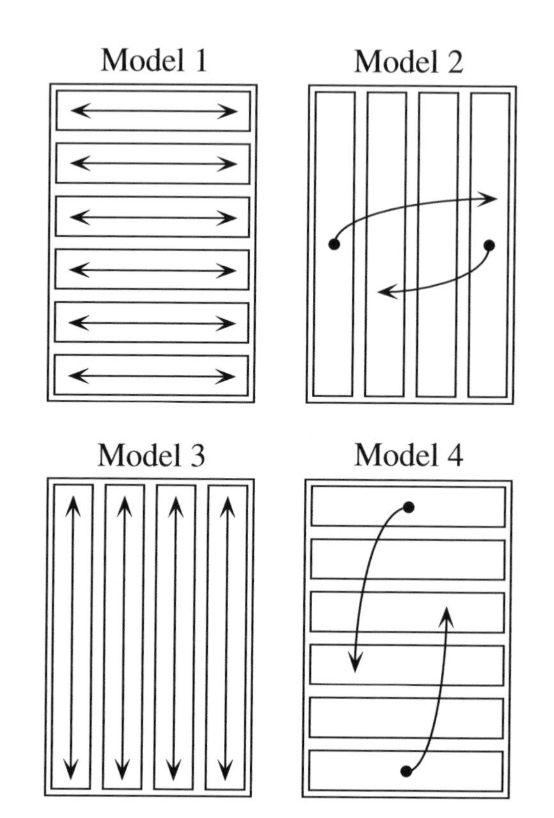

Figure 10.24 Permutations of matrix **A** may be performed in different ways which correspond to different null ecological models.

(1) The occurrence of a species on the reef is constant, but positions are random; permute at random within rows.

(2) Positions of species assemblages are random; permute whole columns (assemblages).

(3) Lottery hypothesis: the species that arrives first occupies a site; permute at random within columns.

(4) Species have random attributes; permute whole rows.

composition found at any one site could have been observed at any other site. The alternative hypothesis is that species assemblages are dependent upon the habitat characteristics of the sites where they are actually found. In the context of the fourth-corner problem, this alternative hypothesis is an extension of the environmental control model (previous paragraph) to species assemblages, implying strong biotic ties among the species that are found together. Permutation of whole vectors of species composition (which are the columns of table **A** here) is equivalent to the method of designed-based permutation of raw data used in tests of significance in canonical analysis (Subsection 11.1.4); the other permutation frameworks used in canonical analysis, i.e. model-based permutation of residuals under a null or a full model (Section 11.3), are modifications of model 2.

Model 3: Lottery — A third method consists in permuting values within the columns of matrix **A**, this being done independently from column to column. The null hypothesis states that there is a fixed number of niches at any one site and that species invade them through some form of lottery, the identity (species) of an individual settling at a site being a chance event. The lottery model has been advocated by Sale (1978) for coral reef fishes; he argued that the main determinant of species composition at various sites on coral reefs is chance, coupled with an over-abundance of juveniles available for settlement. Instead of assuming the number of occurrences of any one species to be fixed, it is the number of niches available for settlement which is assumed to be fixed. The alternative hypothesis is here that some species have competitive advantage over chance settlers in given habitats.

Model 4: Random species attributes — A fourth method would be to permute whole rows at random with respect to one another. The corresponding null hypothesis is that species have random biological and behavioural attributes. This model is not appropriate to study the relationship between behaviour and habitat because the linkage between species and their behavioural and biological characteristics is fixed. It may be appropriate to other types of problems involving three data matrices.

Numerical example. Let us examine how the method behaves when applied to the data sets introduced in the numerical example of Subsection 1. The first test case (Table 10.7, left) was constructed to suggest that herbivores are found on turf while carnivores are more ubiquitously distributed. Globally, the G statistic indicates a significant relationship ($\alpha = 0.05$) between behavioural states and types of habitat ($p = 0.0207$ after 9999 random permutations under model 1 above). The expected values in the various cells of matrix **D** determine the tail in which each frequency d of the contingency table is to be tested for significance; this value is taken to be the mean frequency expected from all possible permutations of matrix **A**, given the permutation model that has been selected. Looking at individual values d, herbivores are clearly positively associated with turf and negatively with coral ($p = 0.0287$, computed from the random permutation results), while carnivores are not significantly associated with either live coral or turf ($p = 0.1890$). These probabilities are very close to the exact probabilities calculated for the same data, which are the values obtained from a complete permutation procedure (E in the Table). Values of exact probabilities E are computed as follows: consider all possible permutations that result from independently permuting the rows of matrix **A** (permutation model 1); count how many of these would produce values equal to, or more extreme than the observed value in each given cell of matrix **D**. This value may differ slightly from the random permutational probability. Globally, the testing procedure for the relationship between behaviour and habitat behaved as expected in this example, and the random permutation procedure produced values quite close to the exact probabilities.

The second test case (Table 10.7, right) illustrates a situation where the null hypothesis is true in all cases, matrix **A** indicating all 10 species to be present everywhere. Indeed, the testing procedure finds all permutation statistics to be equal to the unpermuted ones, so that the probability of the data under the null hypothesis is 1 everywhere. The procedure once more behaved correctly.

4 — Other types of comparison among variables

Variables in matrices **B** and **C** are not always qualitative. Through lines of reasoning similar to that of Subsection 10.6.1, involving inflated data matrices (as in Table 10.8), 4th-corner statistics can be formulated to accommodate other types of comparisons among variables:

• To compare a quantitative variable in **B** to a quantitative variable in **C**, a Pearson correlation coefficient may be computed between the columns of the inflated matrix.

A correlation coefficient is directly obtained from the 4th-corner equation **D** = **CA'B** if the columns of the inflated data table are first standardized and the scalar product is divided by the number of rows of the inflated table minus 1.

• When comparing a quantitative variable in **B** to a qualitative variable coded into dummy variables (Subsection 1.4.7) in **C**, or the converse, the 4th-corner matrix product (eq. 10.32) is equivalent to computing an overall F statistic for the pair of variables, as explained in Legendre *et al.* (1997); the cells of matrix **D** contain measures of within-group homogeneity. Correlations may also be computed between the quantitative variable on the one hand, and each of the dummy variables coding for the qualitative variable.

Each of these statistics may be tested for significance using the permutational procedure described above.

The 4th-corner method* offers a way of analysing the relationships between *supplementary variables* associated with the rows and columns of a binary (presence-absence) data table. Other types of problems could be studied using this method. Here are two examples.

• In biogeography, consider a matrix **A** of presence/absence of species; a matrix **B** describing the extensiveness of the species' distributions, their migratory behaviour, etc.; and a matrix **C** of habitat characteristics (environmental variables), as above. The question is again to relate habitat to species characteristics.

• In the study of feeding behaviour, consider a matrix **A** with rows that are *individuals* while columns correspond to sites. The prey ingested by each individual are found in matrix **B** (in columns). Matrix **C** may contain either microhabitat environmental variables, or prey availability variables. The question is to determine feeding preferences: choice of prey *versus* availability, or choice of prey *versus* microhabitat conditions. Problems of the same type are found in such fields as sociology, marketing, political science, and the like.

Ecological application 10.6

Development of the 4th-corner method was motivated by the study of a fish assemblage (280 species) surveyed along a one-km transect across the coral reef of Moorea Island, French Polynesia (Legendre *et al.*, 1997). Biological and behavioural characteristics of the species were used as descriptors (supplementary variables) for the rows, and characteristics of the environment for the columns of the fish presence-absence data table **A**. Parameters of the relationship between habitat characteristics (distance from the beach, water depth, and substrate variables) and biological and behavioural traits of the species (feeding habits, ecological niche categories, size classes, egg types, activity rhythms) were estimated and tested for significance

---

* 4THCORNER is a FORTRAN program available from the following WWWeb site to carry out the calculations and perform tests of significance: <http://www.fas.umontreal.ca/BIOL/legendre/>.

Table 10.9 Contingency table comparing feeding habits (7 states) to materials covering reef bottom (8 proportions). First row in each cell: pseudo-frequency resulting from the matrix operation **D** = **CA'B**; lower row, probability adjusted using Holm's procedure; *: $p \leq 0.05$. Probabilities before correction resulted from 9999 random permutations. Sign indicates whether a statistic is above (+) or below (–) the expected value, estimated as the mean of the permutation results.

	Herbiv-orous	Omniv-orous	Sessile invertebrates	Carniv. 1 diurnal	Carniv. 2 nocturnal	Fish only	Copepod eater
Stone slab	6.20–	5.84+	3.72–	8.42–	5.18+	0.96+	2.40–
p	0.429	0.232	1.535	2.650	2.650	2.650	2.650
Sand	81.22–	54.26–	43.34–	94.38–	35.90–	8.94–	26.26–
p	0.039*	0.799	0.006*	0.006*	0.006*	0.799	0.039*
Coral debris	34.96+	20.22–	24.32+	46.74+	25.60+	4.48+	12.08–
p	1.976	1.976	0.006*	0.009*	0.645	2.650	2.650
Turf, dead cor.	45.46+	27.88+	28.28+	57.58+	33.58+	6.20+	15.76+
p	0.207	2.650	0.081	0.013*	0.029*	1.976	2.650
Live coral	49.86+	28.50+	29.20+	58.28+	40.82+	6.22+	21.06+
p	0.006*	1.976	0.006*	0.006*	0.006*	1.976	0.006*
Large algae	44.66–	37.50+	28.12–	59.68–	32.26–	6.34–	19.20–
p	0.006*	2.650	0.105	0.048*	0.140	2.650	2.650
Calcar. algae	29.12+	16.32+	16.08+	31.00+	26.02+	4.50+	11.32+
p	0.006*	1.030	0.079	0.122	0.006*	0.207	0.036*
Other substrate	2.52+	1.48+	1.94+	2.92+	1.64+	0.36+	0.92+
p	0.105	2.650	0.006*	0.795	1.734	1.976	1.976

using permutations. Results were compared to predictions made independently by reef fish ecologists, in order to assess the method as well as the pertinence of the variables subjected to the analysis.

Table 10.9 summarizes the comparison of reef bottom materials to feeding habits. This is an interesting case: the eight "reef bottom materials" variables are relative frequencies; each one represents the proportion of the habitat covered by a category of substrate material, so that non-integer pseudo-frequencies are obtained in the contingency table where the variables are crossed (Table 10.9). The permutation testing procedure allows data in matrices **B** and **C** to be relative or absolute frequencies. Probabilities remain the same under any linear transformation of the frequency values, even though the value of the G statistic is changed. This would not be allowed by a standard test whose outcome would be read from a χ^2 table.

The relationship is globally significant (G = 15.426, p(G) = 0.0001 after 9999 random permutations following model 1 of Subsection 3 above); 20 of the 56 4th-corner statistics d were significant (*) after applying Holm's correction for multiple testing (Box 1.3). Compared to the null hypothesis, fish are under-represented on sand and large algae, and are unrelated to stone slab. In addition, herbivores are over-represented on live coral and calcareous algae. Grazers of sessile invertebrates and as carnivores of types 1 and 2 are over-represented on coral debris, turf and dead coral, live coral, calcareous algae, and "other substrate" (large echinoderms, sponges, anemones, alcyonarians); this includes all areas where herbivores are found. Copepod eaters are over-represented on live coral and calcareous algae. Omnivores and specialist piscivores (fish-only diet) do not exhibit significant relationships to substrate.

Distance from the beach and size of fish species (adult individuals) are quantitative variables. The 4th-corner statistic that crosses these two variables is thus correlation-like; its value is $r = 0.0504$, with a probability of 0.001 after 999 random permutations. There is thus a weak but significant correlation, indicating that larger fish are found farther away from the beach than smaller ones. Other comparisons between biological-behavioural and habitat variables are presented in the published paper.

Chapter

11 Canonical analysis

11.0 Principles of canonical analysis

Canonical analysis is the simultaneous analysis of two, or eventually several data tables. It allows ecologists to perform a *direct comparison* of two data matrices ("direct gradient analysis"; Fig. 10.4, Table 10.1). Typically, one may be interested in the relationship between a first table describing species composition and a second table of environmental descriptors, observed *at the same locations*; or, say, a table about the chemistry of lakes and another about drainage basin geomorphology.

In *indirect comparison* (indirect gradient analysis; Section 10.2, Fig. 10.4), the matrix of explanatory variables **X** does not intervene in the calculation producing the ordination of **Y**. Correlation or regression of the ordination vectors on **X** are computed *a posteriori*. In *direct comparison analysis*, on the contrary, matrix **X** intervenes in the calculation, forcing the ordination vectors to be maximally related to combinations of the variables in **X**. This description applies to all forms of canonical analysis and in particular to the asymmetric forms described in Sections 11.1 to 11.3. There is a parallel in cluster analysis, when clustering results are constrained to be consistent with temporal (Subsection 12.6.4) or spatial relationships (Subsection 13.3.2) among observations, which are inherent to the sampling design. When using a constraint (clustering, ordination), the results should differ from those of unconstrained analysis and be, hopefully, more readily interpretable. Thus, direct comparison analysis allows one to directly test *a priori* ecological hypotheses by (1) bringing out *all* the variance of **Y** that is related to **X** and (2) allowing formal tests of these hypotheses to be performed, as detailed below. Further examination of the unexplained variability may help generate new hypotheses, to be tested using new field observations (Section 13.5).

Canonical form

In mathematics, a *canonical form* (from the Greek κανων, pronounced "kanôn", rule) is the simplest and most comprehensive form to which certain functions, relations, or expressions can be reduced without loss of generality. For example, the canonical form of a covariance matrix is its matrix of eigenvalues. In general, methods of canonical analysis use eigenanalysis (i.e. calculation of eigenvalues and eigenvectors), although some extensions of canonical analysis have been described that use multidimensional scaling (MDS) algorithms (Section 9.3).

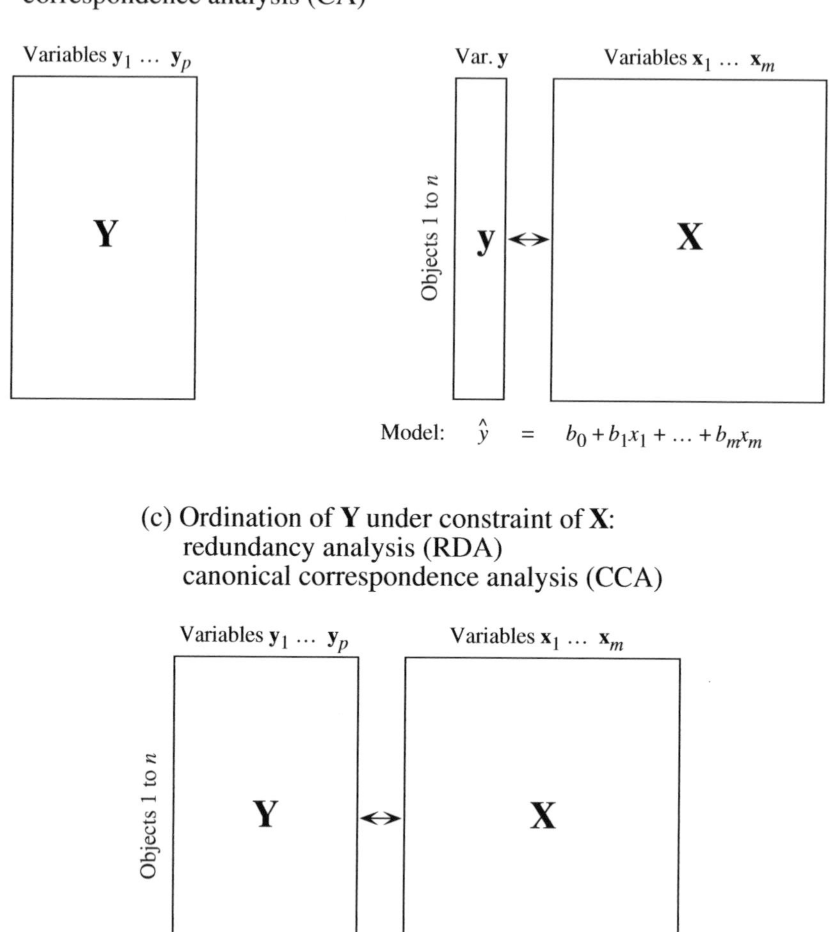

Figure 11.1 Relationships between (a) ordination, (b) regression, and (c) the asymmetric forms of canonical analysis (RDA and CCA). In (c), each canonical axis of **Y** is constrained to be a linear combination of the explanatory variables **X**.

Canonical analysis combines the concepts of ordination and regression. It involves a response matrix **Y** and an explanatory matrix **X** (names used throughout this chapter). Like the other ordination methods (Chapter 9; Fig. 11.1a), canonical analysis produces (usually) orthogonal axes from which scatter diagrams may be plotted.

Canonical analysis — in particular redundancy analysis (RDA, Section 11.1) and canonical correspondence analysis (CCA, Section 11.2) — is related to multiple regression analysis. In Subsection 10.3.3, multiple regression was described as a method for modelling a response variable **y** using a set of explanatory variables assembled into a data table **X**. Another aspect of regression analysis must be stressed: while the original response variable **y** provides, by itself, an ordination of the objects in one dimension, the vector of fitted values (eq. 10.15)

$$\hat{y}_i = b_0 + b_1 x_{i1} + b_2 x_{i2} + \ldots + b_p x_{ip}$$

creates a new one-dimensional ordination of the same objects (Fig. 11.1b). The ordinations corresponding to **y** and $\hat{\mathbf{y}}$ differ; the square of their correlation is the coefficient of determination of the multiple regression model (eq. 10.19):

$$R^2 = [r(\mathbf{y}, \hat{\mathbf{y}})]^2 \qquad \textbf{(11.1)}$$

So, multiple regression creates a correspondence between ordinations **y** and $\hat{\mathbf{y}}$, because ordination $\hat{\mathbf{y}}$ is constrained to be optimally and linearly related to the variables in **X**. This property is shared with canonical analysis. The constraint is optimal in the least-square sense, meaning that the linear multiple regression maximizes R^2.

Canonical analysis combines properties of these two families of methods (i.e. ordination and regression; Fig. 11.1c). It produces ordinations of **Y** that are constrained to be related in some way to a second set of variables **X**. The way in which the relationship between **X** and **Y** is established differs among methods of canonical analysis.

Problems of canonical analysis may be represented by the following partitioned covariance matrix, resulting from the fusion of the **Y** and **X** data sets; the joint dispersion matrix $\mathbf{S}_{\mathbf{Y+X}}$ contains blocks that are identified as follows for convenience:

$$\mathbf{S}_{\mathbf{Y+X}} = \left[\begin{array}{ccc|ccc} s_{y_1,y_1} & \cdots & s_{y_1,y_p} & s_{y_1,x_1} & \cdots & s_{y_1,x_m} \\ \cdot & & \cdot & \cdot & & \cdot \\ \cdot & & \cdot & \cdot & & \cdot \\ \cdot & & \cdot & \cdot & & \cdot \\ s_{y_p,y_1} & \cdots & s_{y_p,y_p} & s_{y_p,x_1} & \cdots & s_{y_p,x_m} \\ \hline s_{x_1,y_1} & \cdots & s_{x_1,y_p} & s_{x_1,x_1} & \cdots & s_{x_1,x_m} \\ \cdot & & \cdot & \cdot & & \cdot \\ \cdot & & \cdot & \cdot & & \cdot \\ \cdot & & \cdot & \cdot & & \cdot \\ s_{x_m,y_1} & \cdots & s_{x_m,y_p} & s_{x_m,x_1} & \cdots & s_{x_m,x_m} \end{array}\right] = \begin{bmatrix} \mathbf{S}_{\mathbf{YY}} & \mathbf{S}_{\mathbf{YX}} \\ \mathbf{S}_{\mathbf{XY}} & \mathbf{S}_{\mathbf{XX}} \end{bmatrix} = \begin{bmatrix} \mathbf{S}_{\mathbf{YY}} & \mathbf{S}_{\mathbf{YX}} \\ \mathbf{S'}_{\mathbf{YX}} & \mathbf{S}_{\mathbf{XX}} \end{bmatrix} \qquad \textbf{(11.2)}$$

Submatrices $\mathbf{S_{YY}}$ (order $p \times p$) and $\mathbf{S_{XX}}$ ($m \times m$) concern each of the two sets of descriptors, respectively, whereas $\mathbf{S_{YX}}$ ($p \times m$) and its transpose $\mathbf{S'_{YX}} = \mathbf{S_{XY}}$ ($m \times p$) account for the covariances among the descriptors of the two groups, as in eq. 4.27.

• In redundancy analysis (RDA, Section 11.1), each canonical ordination axis corresponds to a direction, in the multivariate scatter of objects (matrix **Y**), which is maximally related to a linear combination of the explanatory variables **X**. A canonical axis is thus similar to a principal component (Box 9.1). Two ordinations of the objects are obtained along the canonical axes: (1) linear combinations of the **Y** variables (matrix **F**, eq. 11.12), as in PCA, and (2) linear combinations of the fitted $\hat{\mathbf{Y}}$ variables (matrix **Z**, eq. 11.13), which are thus also linear combinations of the **X** variables. RDA preserves the Euclidean distance among objects in matrix $\hat{\mathbf{Y}}$ containing values of **Y** fitted by regression to the explanatory variables **X** (Fig. 11.2); variables in $\hat{\mathbf{Y}}$ are therefore linear combinations of the **X** variables.

• Canonical correspondence analysis (CCA, Section 11.2) is similar to RDA. The difference is that it preserves the χ^2 distance (as in correspondence analysis), instead of the Euclidean distance among objects. Calculations are a bit more complex since matrix $\hat{\mathbf{Y}}$ contains fitted values obtained by weighted linear regression of matrix $\bar{\mathbf{Q}}$ of correspondence analysis (eq. 9.32) on the explanatory variables **X**. As in RDA, two ordinations of the objects are obtained.

• In canonical correlation analysis (CCorA, Section 11.4), the canonical axes maximize the correlation between linear combinations of the two sets of variables **Y** and **X**. This is obtained by maximizing the among-variable-group covariance (or correlation) in eq. 11.2 with respect to the within-variable-group covariance (or correlation; eq. 11.22). Two ordinations of the objects are obtained again.

• In canonical discriminant analysis (Section 11.5), the objects are divided into k groups, described by a qualitative descriptor. The method maximizes the dispersion of the centroids of the k groups. This is obtained by maximizing the ratio of the among-object-group dispersion over the pooled within-object-group dispersion (eq. 11.31).

The application of the various methods of canonical analysis to ecological data sets has already been discussed in Section 10.2. In summary, canonical correlation analysis (CCorA) is used to find axes of maximum linear correlation between two data tables. When one of the data sets (**Y**) is to be explained by another (**X**), the asymmetric forms of canonical analysis should be used; the methods of choice are redundancy analysis (RDA) and canonical correspondence analysis (CCA). RDA is used when the **X** variables display linear relationships with the **Y** variables whereas CCA should be used in all cases where correspondence analysis (CA, Section 9.4) would be appropriate for an ordination analysis of matrix **Y** alone. Discriminant analysis may be used when the target data set contains a single qualitative variable **y** representing a classification of the objects; in ecology, it should be used mostly to discriminate among groups of sites that are characterized by descriptors of the physical environment (Section 11.6).

Canonical analysis has become an instrument of choice for ecological analysis. A 1994 bibliography of ecological papers on the subject already contained 379 titles (Birks *et al.*, 1994). CCorA and discriminant analysis are readily available in most major statistical packages. For RDA and CCA, one must rely on specialized ordination packages. The most widely used program is CANOCO* (ter Braak, 1988b). A closely related procedure, called ACPVI (principal component analysis with instrumental variables), is available in the ADE-4 package† (Thioulouse *et al.*, 1996).

11.1 Redundancy analysis (RDA)

Redundancy

Redundancy analysis (RDA) is the direct extension of multiple regression to the modelling of multivariate response data. *Redundancy* is synonymous with *explained variance* (Gittins, 1985). The analysis is asymmetric: **Y** is the table of response variables and **X** is the table of explanatory variables. Looking at the matter from a descriptive perspective, one would say that the ordination of **Y** is constrained in such a way that the resulting ordination vectors are linear combinations of the variables in **X**. The difference between RDA and canonical correlation analysis (CCorA, Section 11.4) is the same as that between simple linear regression and linear correlation analysis. RDA may also be seen as an extension of principal component analysis (Section 9.1), because the canonical ordination vectors are linear combinations of the response variables **Y**. This means that each ordination vector is a one-dimensional projection of the distribution of the objects in a space that preserves the Euclidean distances (D_1, Chapter 7) among them. These ordination vectors differ, of course, from the principal components that could be computed on the **Y** data table, because they are also constrained to be linear combinations of the variables in **X**.

Redundancy analysis was first proposed by Rao (1964); in his 1973 book (p. 594-595), he proposed the problem as an exercise at the end of his Chapter 8 on multivariate analysis. The method was later rediscovered by Wollenberg (1977).

The eigenanalysis equation for redundancy analysis

$$(\mathbf{S_{YX}}\, \mathbf{S}^{-1}_{\mathbf{XX}}\, \mathbf{S'_{YX}} - \lambda_k \mathbf{I})\, \mathbf{u}_k = \mathbf{0} \qquad \textbf{(11.3)}$$

* CANOCO, which contains procedures for both RDA and CCA, was written by C. J. F. ter Braak who also developed CCA. Distribution: see Table 13.4, p. 784.

The package PC-ORD contains a procedure for CCA. Distribution: see footnote in Section 9.3.

RDACCA is a FORTRAN program for RDA and CCA written by P. Legendre. It is distributed free of charge from the WWWeb site: <http://www.fas.umontreal.ca/BIOL/legendre/>. It uses the direct eigenanalysis methods described in Subsections 11.1.1(for RDA) and 11.2.1 (for CCA).

† The ADE-4 package (for Macintosh and Windows) was written by D. Chessel and J. Thioulouse at Université de Lyon, France. It is distributed free of charge from the following WWWeb site: <http://biomserv.univ–lyon1.fr/ADE–4.html>.

may be derived through multiple linear regression, followed by principal component decomposition (Fig. 11.2). This way of looking at the calculations makes it intuitively easy to understand what RDA actually does to the data. It also shows that the computations can be carried out using any standard general statistical package for micro-computer or mainframe, provided that multiple regression and principal component analysis are available; the procedure is also easy to program using advanced languages such as MATLAB or S-PLUS. RDA is appropriate when the response data table **Y** could be analysed, alone, by principal component analysis (PCA); in other words, when the **y** variables are linearly related to one another and the Euclidean distance is deemed appropriate to describe the relationships among objects in factorial space. The data matrices must be prepared as follows, prior to RDA.

1. The table of response variables **Y** is of size $(n \times p)$, where n is the number of objects and p is the number of variables. Centre the response variables on their means, or standardize them by column if the variables are not dimensionally homogeneous (e.g. a mixture of temperatures, concentrations, pH values, etc.), as one would do prior to PCA. Centring at this early step simplifies several of the equations from 11.4 to 11.12 in which, otherwise, the centring of the columns of matrix **Y** should be specified.

2. Table **X** of the explanatory variables is of size $(n \times m)$ with $m \leq n$. The variables are centred on their respective means for convenience; centring the variables in **X** and **Y** has the effect of eliminating the regression intercepts, thus simplifying the interpretation without loss of pertinent information. The **X** variables may also be standardized (eq. 1.12). This is not a necessary condition for a valid redundancy analysis, but removing the scale effects of the physical dimensions of the explanatory variables (Subsection 1.5.4) turns the regression coefficients into standard regression coefficients which are comparable to one another. The amount of explained variation, as well as the fitted values of the regression, remain unchanged by centring or standardization of the variables in **X**. In the program CANOCO, for instance, standardization is automatically performed for the explanatory variables (matrix **X**) when computing RDA or CCA.

The distributions of the variables should be examined at this stage, as well as bivariate plots within and between the sets **Y** and **X**. Transformations (Section 1.5) should be applied as needed to linearize the relationships and make the distributions more symmetric, reducing the effect of outliers.

If **X** and **Y** are made to contain the same data (i.e. $\mathbf{X} = \mathbf{Y}$), eq. 11.3 becomes $(\mathbf{S_{YY}} - \lambda_k \mathbf{I})\, \mathbf{u}_k = \mathbf{0}$, which is the equation for principal component analysis (eq. 9.1). The result of RDA is then a principal component analysis of that data table.

1 — The algebra of redundancy analysis

The following algebraic development describes how to arrive at eq. 11.3 through multiple regression and principal component analysis. The steps are (Fig. 11.2): (1) regress each variable in **Y** on all variables in **X** and compute the fitted values;

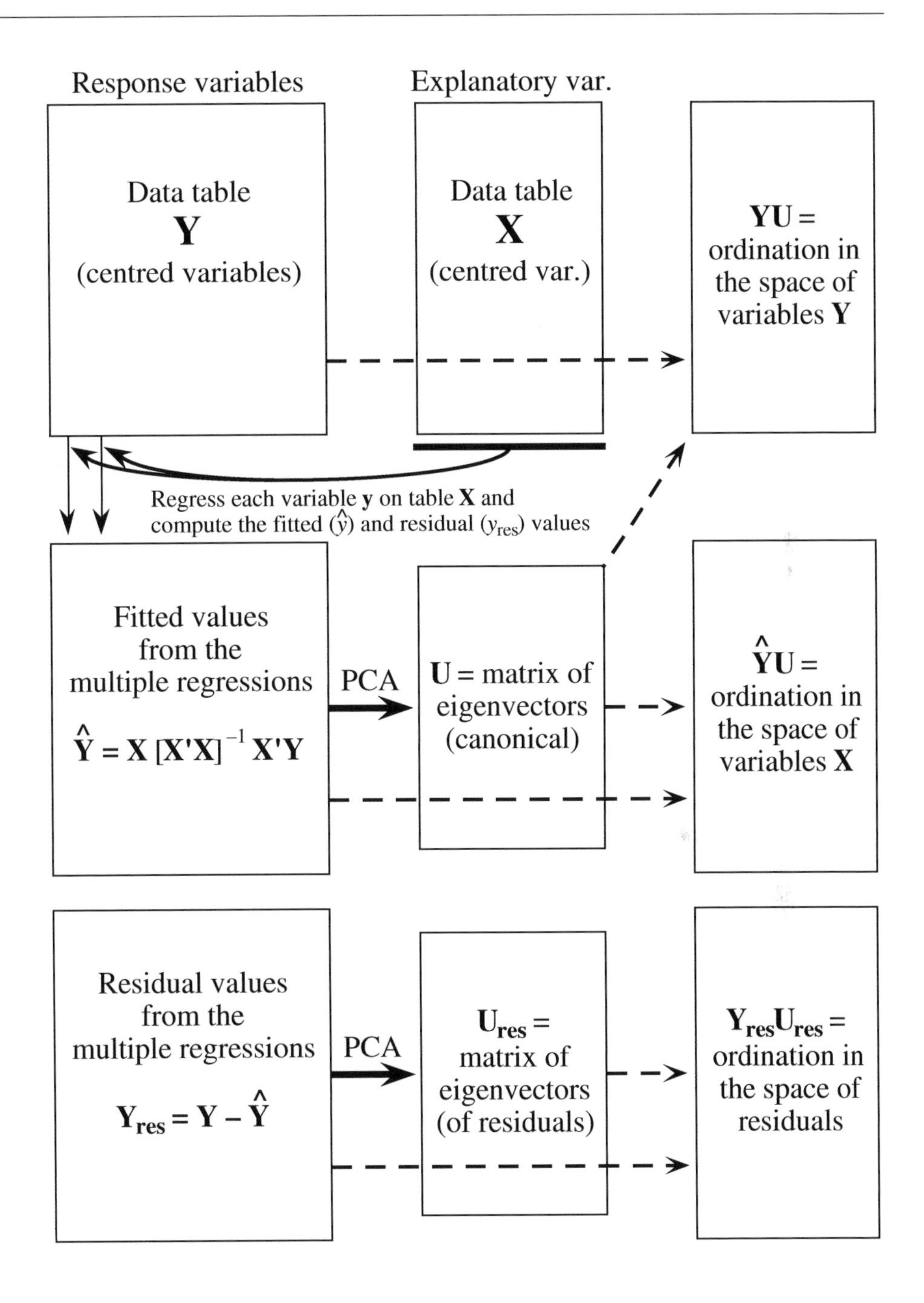

Figure 11.2 Redundancy analysis may be understood as a two-step process: (1) regress each variable in **Y** on all variables in **X** and compute the fitted values; (2) carry out a principal component analysis of the matrix of fitted values to obtain the eigenvalues and eigenvectors. Two ordinations are obtained, one (**YU**) in the space of the response variables **Y**, the other ($\hat{\mathbf{Y}}\mathbf{U}$) in the space of the explanatory variables **X**. Another PCA ordination can be obtained using the matrix of residuals.

(2) carry out a principal component analysis on the matrix of fitted values to obtain the eigenvalues and eigenvectors.

1) For *each* response variable in table **Y**, compute a multiple linear regression on all variables in table **X**. This may be done using any general-purpose statistical package. The matrix equation corresponding to each regression is (eq. 2.19):

$$\mathbf{b} = [\mathbf{X'X}]^{-1}\,\mathbf{X'y}$$

so that the matrix equation corresponding to the whole set of regressions (i.e. for all response variables) is

$$\mathbf{B} = [\mathbf{X'X}]^{-1}\,\mathbf{X'Y} \qquad \textbf{(11.4)}$$

where **B** is the matrix of regression coefficients of all response variables **Y** on the regressors **X**. Computing all linear regressions simultaneously has been called *multivariate linear regression* by Finn (1974) and is available, for instance, in the SAS procedure GLM.

In multiple regression, the fitted values $\hat{y}$ are computed as:

$$\hat{\mathbf{Y}} = \mathbf{X}\,\mathbf{B} \qquad \textbf{(11.5)}$$

This is the multivariate extension of eq. 10.1. The whole table of fitted values, $\hat{\mathbf{Y}}$, may be computed in a single matrix operation in this way. Using **B** estimated by eq. 11.4, eq. 11.5 becomes:

$$\hat{\mathbf{Y}} = \mathbf{X}\,[\mathbf{X'}\,\mathbf{X}]^{-1}\,\mathbf{X'}\,\mathbf{Y} \qquad \textbf{(11.6)}$$

Because variables **X** and **Y** are centred on their respective means, there is no intercept parameter in the **B** vectors. The $\hat{\mathbf{Y}}$ vectors are centred, as is always the case in ordinary linear regression. If $m = n$, **X** is square; in that case, the multiple regressions always explain the variables in matrix **Y** entirely, so that $\hat{\mathbf{Y}} = \mathbf{Y}$. Using property 5 of matrix inverses (Section 2.8), one can indeed verify that eq. 11.6 gives $\hat{\mathbf{Y}} = \mathbf{Y}$ when **X** is square.

2) The covariance matrix corresponding to the table of fitted values $\hat{\mathbf{Y}}$ is computed from eq. 4.6:

$$\mathbf{S}_{\hat{\mathbf{Y}}'\hat{\mathbf{Y}}} = [1/(n-1)]\;\hat{\mathbf{Y}}'\hat{\mathbf{Y}} \qquad \textbf{(11.7)}$$

Replacing $\hat{\mathbf{Y}}$ by the expression from eq. 11.6, eq. 11.7 becomes:

$$\mathbf{S}_{\hat{\mathbf{Y}}'\hat{\mathbf{Y}}} = [1/(n-1)]\,\mathbf{Y'}\,\mathbf{X}\,[\mathbf{X'}\,\mathbf{X}]^{-1}\,\mathbf{X'}\,\mathbf{X}\,[\mathbf{X'}\,\mathbf{X}]^{-1}\,\mathbf{X'}\,\mathbf{Y} \qquad \textbf{(11.8)}$$

This equation reduces to:

$$\mathbf{S}_{\hat{\mathbf{Y}}'\hat{\mathbf{Y}}} = \mathbf{S}_{\mathbf{YX}}\,\mathbf{S}^{-1}{}_{\mathbf{XX}}\,\mathbf{S'}_{\mathbf{YX}} \qquad \textbf{(11.9)}$$

where $\mathbf{S}_{\mathbf{YY}}$ is the $(p \times p)$ covariance matrix among the response variables, $\mathbf{S}_{\mathbf{XX}}$ the $(m \times m)$ covariance matrix among the regressors (it is actually a matrix $\mathbf{R}_{\mathbf{XX}}$ if all the **X** variables have been standardized, as suggested above), and $\mathbf{S}_{\mathbf{YX}}$ is the $(p \times m)$ covariance matrix among the variables of the two sets; its transpose $\mathbf{S'}_{\mathbf{YX}} = \mathbf{S}_{\mathbf{XY}}$ is of size $(m \times p)$. If the **Y** variables had also been standardized, this equation would read $\mathbf{R}_{\mathbf{YX}}\,\mathbf{R}^{-1}{}_{\mathbf{XX}}\,\mathbf{R'}_{\mathbf{YX}}$, which is the equation for the coefficient of multiple determination (eq. 4.31).

3) The table of fitted values $\hat{\mathbf{Y}}$ is subjected to principal component analysis to reduce the dimensionality of the solution. This corresponds to solving the eigenvalue problem:

$$(\mathbf{S}_{\hat{\mathbf{Y}}'\hat{\mathbf{Y}}} - \lambda_k \mathbf{I})\,\mathbf{u}_k = \mathbf{0} \qquad \textbf{(11.10)}$$

which, using eq. 11.9, translates into:

$$(\mathbf{S}_{\mathbf{YX}}\,\mathbf{S}^{-1}{}_{\mathbf{XX}}\,\mathbf{S'}_{\mathbf{YX}} - \lambda_k \mathbf{I})\,\mathbf{u}_k = \mathbf{0} \qquad \textbf{(11.11)}$$

This is the equation for redundancy analysis (eq. 11.3); it may also be obtained from the equation for canonical correlation analysis (eq. 11.22), by defining $\mathbf{S}_{11} = \mathbf{S}_{\mathbf{YY}} = \mathbf{I}$ (Rao, 1973; ter Braak, 1987c). Different programs may express the eigenvalues in different ways: raw eigenvalues, fraction of total variance in matrix **Y**, or percentage; see Tables 11.2 and 11.4 for examples.

The matrix containing the normalized canonical eigenvectors $\mathbf{u}_k$ is called **U**. The eigenvectors give the contributions of the descriptors of $\hat{\mathbf{Y}}$ to the various canonical axes. Matrix **U**, of size $(p \times p)$, contains only $\min[p, m, n-1]$ eigenvectors with non-zero eigenvalues, since the number of canonical eigenvectors cannot exceed the minimum of p, m and $(n-1)$:

• It cannot exceed p which is the size of the reference space of matrix **Y**. This is obvious in multiple regression, where matrix **Y** contains a single variable; the ordination given by the fitted values $\hat{y}$ is, consequently, one-dimensional.

• It cannot exceed m which is the number of variables in **X**. Consider an extreme example: if **X** contains a single explanatory variable ($m = 1$), regressing all p variables in **Y** on this single regressor produces p fitted vectors $\hat{y}$ which all point in the same direction of the space; a principal component analysis of matrix $\hat{\mathbf{Y}}$ of these fitted vectors can only produce one common (canonical) variable.

• It cannot exceed $(n-1)$ which is the maximum number of dimensions required to represent n points in Euclidean space.

The canonical coefficients in the normalized matrix $\mathbf{U}$ give the contributions of the variables of $\hat{\mathbf{Y}}$ to the canonical axes. They should be interpreted as in PCA. For biplots (discussed below), matrix $\mathbf{U}$ can be rescaled in such a way that the length of each eigenvector is $\sqrt{\lambda_k}$, using eq. 9.9.

4) The ordination of objects in the space of the response variables $\mathbf{Y}$ can be obtained directly from the centred matrix $\mathbf{Y}$, using the standard equation for principal components (eq. 9.4) and matrix $\mathbf{U}$ of the eigenvectors $\mathbf{u}_k$ found in eq. 11.11:

$$\mathbf{F} = \mathbf{Y}\mathbf{U} \tag{11.12}$$

Site scores

The ordination vectors (columns of $\mathbf{F}$) defined in eq. 11.12 are called the vectors of "site scores". They have variances that are close, but not equal to the corresponding eigenvalues. How to represent matrix $\mathbf{F}$ in biplot diagrams is discussed in point 8 (below).

5) Likewise, the ordination of objects in space $\mathbf{X}$ is obtained as follows:

$$\mathbf{Z} = \hat{\mathbf{Y}}\mathbf{U} = \mathbf{X}\mathbf{B}\mathbf{U} \tag{11.13}$$

Fitted site scores

As stated above, the vectors in matrix $\hat{\mathbf{Y}}$ are centred on their respective means. The right-hand part of eq. 11.13, obtained by replacing $\hat{\mathbf{Y}}$ by it value in eq. 11.5, shows that this ordination is a linear combinations of the $\mathbf{X}$ variables. For that reason, these ordination vectors (columns of matrix $\mathbf{Z}$) are also called "fitted site scores", or "sample scores which are linear combinations of environmental variables" in program CANOCO. The ordination vectors, as defined in eq. 11.13, have variances equal to the corresponding eigenvalues. The representation of matrix $\mathbf{Z}$ in biplot diagrams is discussed in point 8 (below).

The "site scores" of eq. 11.12 are obtained by projecting the original data (matrix $\mathbf{Y}$) onto axis k; they approximate the observed data, which contain residuals ($\mathbf{Y} = \hat{\mathbf{Y}} + \mathbf{Y}_{res}$, Fig. 11.2). On the other hand, the "fitted site scores" of eq. 11.13 are obtained by projecting the fitted values of the multiple regressions (matrix $\hat{\mathbf{Y}}$) onto axis k; they approximate the fitted data. Either set may be used in biplots. The practical difference between "site scores" and "fitted site scores" is further discussed in the second example below and in the numerical example of Section 13.4.

6) The correlation r_k between the ordination vectors in spaces $\mathbf{Y}$ (from eq. 11.12) and $\mathbf{X}$ (from eq. 11.13) for dimension k is called the "species-environment correlation" in program CANOCO. It measures how strong the relationship is between the two data sets, as expressed by each canonical axis k. It should be interpreted with caution because a canonical axis with high species-environment correlation may explain but a small fraction of the variation in $\mathbf{Y}$, which is given by the amount (or proportion) of variance of matrix $\mathbf{Y}$ explained by each canonical axis; see example in Table 11.2.

7) The last important information needed for interpretation is the contribution of the explanatory variables **X** to the canonical ordination axes. Either the regression or the correlation coefficients may be considered:

• Matrix **C** of the canonical coefficients,

$$\mathbf{C} = \mathbf{B}\,\mathbf{U} \qquad \textbf{(11.14)}$$

gives directly the weights of the explanatory variables **X** in the formation of the matrix of fitted site scores. The ordination of objects in the space of the explanatory variables can be found directly by computing **XC**; these vectors of site scores are the same as in eq. 11.13. The coefficients in the columns of matrix **C** are identical to the regression coefficients of the ordination scores from eq. 11.13 on the matrix of standardized explanatory variables **X**; they may thus be interpreted in the same way.

• Correlations may also be computed between the variables in **X**, on the one hand, and the ordination vectors, in either space **Y** (from eq. 11.12) or space **X** (from eq. 11.13), on the other. The correlations between **X** and the ordination vectors in space **X** are used to represent the explanatory variables in biplots.

Biplot

8) In RDA, *biplot diagrams* may be drawn that contain two sets of points, as in PCA (Subsection 9.1.4), or three sets: site scores (matrices **F** or **Z**, from eqs. 11.12 and 11.13), response variables from **Y,** and explanatory variables from **X**. Each pair of sets of points forms a biplot. Biplots help interpret the ordination of objects in terms of **Y** and **X**. When there are too many objects, or too many variables in **Y** or **X**, separate ordination diagrams may be drawn and presented side by side. The construction of RDA biplot diagrams is explained in detail in ter Braak (1994); his conclusions are summarized here. As in PCA, two main types of scalings may be used (Table 9.2):

Scalings in RDA

• RDA scaling type 1 — The eigenvectors in matrix **U**, representing the scores of the response variables along the canonical axes, are scaled to lengths 1*. The site scores in space **X** are obtained from equation $\mathbf{Z} = \hat{\mathbf{Y}}\mathbf{U}$ (eq. 11.13); these vectors have variances equal to λ_k. The site scores in space **Y** are obtained from equation $\mathbf{F} = \mathbf{YU}$; the variances of these vectors are usually slightly larger than λ_k because **Y** contains both the fitted and residual components and has thus more total variance than $\hat{\mathbf{Y}}$. Matrices **Z** and **U**, or **F** and **U**, can be used together in biplots because the products of the eigenvectors with the site score matrices reconstruct the original matrices perfectly:

* In CANOCO 3.1, RDA scaling –1 produces a matrix **U** with vectors ("species scores") scaled to lengths $\sqrt{n}$ (or $\sqrt{p}$ in CANOCO 4.0), instead of 1, if all species and site weights are equal. In both versions of CANOCO, the site scores in space **X** (matrix **Z**) are scaled to lengths $\sqrt{n\lambda_k}$ (or, in other words, to sums of squares of $n\lambda_k$); the site scores in space **Y** (matrix **F**) have lengths slightly larger than $\sqrt{n\lambda_k}$. For RDA, CANOCO expresses the eigenvalues as fractions of the total variance in **Y**. As a result, site scores in matrices **F** and **Z**, as described in the present Section (z_{here}), are related to site scores given by CANOCO (z_{CANOCO}) through the formula: $z_{\text{CANOCO}} = z_{\text{here}}\ \sqrt{n/((n-1)\ \text{Total variance in}\,\mathbf{Y})}$. Changing the scaling of species and site score vectors by any multiplicative constant does not change the interpretation of a biplot.

$\mathbf{ZU}' = \hat{\mathbf{Y}}$ and $\mathbf{FU}' = \mathbf{Y}$, as in PCA (Subsection 9.1.4). A quantitative explanatory variable **x** may be represented in the biplot using the correlations of **x** with the fitted site scores. Each correlation is multiplied by $\sqrt{\lambda_k / \text{Total variance in}\mathbf{Y}}$ where λ_k is the eigenvalue of the corresponding axis *k*; this correction accounts for the fact that, in this scaling, the variances of the site scores differ among axes. The correlations were obtained in calculation step 7 above.

The consequences of this scaling, for PCA, are summarized in the right-hand column of Table 9.2. This scaling, called *distance biplot*, allows the interpretation to focus on the ordination of objects because the distances among objects approximate their Euclidean distances in the space of response variables (matrix **Y**).

Distance biplot

The main features of a distance biplot are the following: (1) Distances among objects in a biplot are approximations of their Euclidean distances. (2) Projecting an object at right angle on a response variable **y** approximates the value of the object along that variable, as in Fig. 9.3a. (3) The angles among variables **y** are meaningless. (4) The angles between variables **x** and **y** in the biplot reflect their correlations. (5) Binary explanatory **x** variables may be represented as the centroids of the objects possessing state "1" for that variable. Examples are given in Subsection 2. Since a centroid represents a "mean object", its relationship to a variable **y** is found by projecting it at right angle on the variable, as for an object. Distances among centroids, and between centroids and individual objects, approximate Euclidean distances.

• RDA scaling type 2 — Alternatively, one obtains response variable scores by rescaling the eigenvectors in matrix **U** to lengths $\sqrt{\lambda_k}$, using the transformation $\mathbf{U\Lambda}^{1/2}$*. The site scores in space **X** obtained for scaling 1 (eq. 11.13) are rescaled to unit variances using the transformation $\mathbf{Z\Lambda}^{-1/2}$. The site scores in space **Y** obtained for scaling 1 are rescaled using the transformation $\mathbf{F\Lambda}^{-1/2}$; the variances of these vectors are usually slightly larger than 1 for the reason explained in the case of scaling 1. Matrices **Z** and **U**, or **F** and **U**, as rescaled here, can be used together in biplots because the products of the eigenvectors with the site score matrices reconstruct the original matrices perfectly: $\mathbf{ZU}' = \hat{\mathbf{Y}}$ and $\mathbf{FU}' = \mathbf{Y}$, as in PCA (Subsection 9.1.4). A quantitative explanatory variable **x** may be represented in the biplot using the correlations of **x** with the fitted site scores, obtained in calculation step 7 above.

---

* In CANOCO 3.1, RDA scaling –2 produces a matrix **U** with vectors ("species scores") scaled to lengths $\sqrt{n\lambda_k}$ (or $\sqrt{p\lambda_k}$ in CANOCO 4.0), instead of $\sqrt{\lambda_k}$, if all species and site weights are equal. For RDA, CANOCO expresses the eigenvalues as fractions of the total variance in **Y**. As a result, the values in matrix **U** as described here (u_{here}) are related to the "species scores" of CANOCO 3.1 ($u_{\text{CANOCO 3.1}}$) through the formula: $u_{\text{CANOCO 3.1}} = u_{\text{here}} \sqrt{n / \text{Total variance in}\mathbf{Y}}$, or $u_{\text{CANOCO 4.0}} = u_{\text{here}} \sqrt{p / \text{Total variance in}\mathbf{Y}}$ in CANOCO 4.0. In both versions of CANOCO, the site scores in space **X** (matrix **Z**) are scaled to lengths $\sqrt{n}$ instead of $\sqrt{n-1}$; the site scores in space **Y** (matrix **F**) have lengths slightly larger than $\sqrt{n}$. Site scores in matrices **F** and **Z**, as described in the present Section (z_{here}), are related to site scores given by CANOCO (z_{CANOCO}) through the formula: $z_{\text{CANOCO}} = z_{\text{here}} \sqrt{n/(n-1)}$. Changing the scaling of species and site score vectors by any multiplicative constant does not change the interpretation of a biplot.

The consequences of this scaling, for PCA, are summarized in the central column of Table 9.2. This scaling, called *correlation biplot*, is appropriate to focus on the relationships among response variables (matrix **Y**).

Correlation biplot

The main features of a correlation biplot are the following: (1) Distances among objects in the biplot *are not* approximations of their Euclidean distances. (2) Projecting an object at right angle on a response variable **y** approximates the value of the object along that variable. (3) The angles between variables (from sets **X** and **Y**) in the biplot reflect their correlations. (4) Projecting an object at right angle on a variable **x** approximates the value of the object along that variable. (5) Binary explanatory variables may be represented as described above. Their interpretation is done in the same way as in scaling type 1, except for the fact that the distances in the biplot among centroids, and between centroids and individual objects, does not approximate Euclidean distances.

The type of scaling depends on the emphasis one wants to give to the biplot, i.e. display of distances among objects or of correlations among variables. When most explanatory variables are binary, scaling type 1 is probably the most interesting; when most of the variables in set **X** are quantitative, one may prefer scaling type 2. When the first two eigenvalues are nearly equal, both scalings lead to nearly the same biplot.

9) Redundancy analysis usually does not completely explain the variation in the response variables (matrix **Y**). During the regression step (Fig. 11.2), regression residuals may be computed for each variable **y**; the residuals are obtained as the difference between observed values y_{ij} and the corresponding fitted values $\hat{y}_{ij}$ in matrix $\hat{\mathbf{Y}}$. The matrix of residuals ($\mathbf{Y}_{\text{res}}$ in Fig. 11.2) is also a matrix of size $(n \times p)$. Residuals may be analysed by principal component analysis, leading to $\min[p, n-1]$ non-canonical eigenvalues and eigenvectors (Fig. 11.2, bottom). So, the full analysis of matrix **Y** (i.e. the analysis of fitted values and residuals) may lead to more eigenvectors than a principal component analysis of matrix **Y**: there is a maximum of $\min[p, m, n-1]$ non-zero canonical eigenvalues and corresponding eigenvectors, plus a maximum of $\min[p, n-1]$ non-canonical eigenvalues and eigenvectors, the latter being computed from the matrix of residuals (Table 11.1). When the variables in **X** are good predictors of the variables in **Y**, the canonical eigenvalues may be larger than the first non-canonical eigenvalues, but this need not always be the case. If the variables in **X** are not good predictors of **Y**, the first non-canonical eigenvalues, computed on the residuals, may be larger than their canonical counterparts.

In the trivial case where **Y** contains a single response variable, redundancy analysis is nothing but multiple linear regression analysis.

2 — *Numerical examples*

As a first example, consider again the data set presented in Table 10.5. The first five variables are assembled into matrix **Y** and the three spatial variables make up matrix **X**. Calculations performed as described above, or using the iterative algorithm

Table 11.1 Maximum number of non-zero eigenvalues and corresponding eigenvectors that may be obtained from canonical analysis of a matrix of response variables $\mathbf{Y}(n \times p)$ and a matrix of explanatory variables $\mathbf{X}(n \times m)$ using redundancy analysis (RDA) or canonical correspondence analysis (CCA).

	Canonical eigenvalues and eigenvectors	Non-canonical eigenvalues and eigenvectors
RDA	$\min[p, m, n-1]$	$\min[p, n-1]$
CCA	$\min[(p-1), m, n-1]$	$\min[(p-1), n-1]$

described in the next subsection, lead to the same results (Table 11.2). There are $\min[5, 3, 19] = 3$ canonical eigenvectors in this example and 5 non-canonical PCA axes computed from the residuals. This is a case where the first non-canonical eigenvalue is larger than any of the canonical eigenvalues. The ordination of objects along the canonical axes (calculation steps 4 and 5 in the previous Subsection) as well as the contribution of the explanatory variables to the canonical ordination axes (calculation step 6) are not reported, but the correlations between these two sets of ordinations are given in the Table; they are rather weak. The sum of the three canonical eigenvalues accounts for only 32% of the variation in response matrix **Y**.

A second example has been constructed to illustrate the calculation and interpretation of redundancy analysis. Imagine that fish have been observed at 10 sites along a transect running from the beach of a tropical island, with water depths going from 1 to 10 m (Table 11.3). The first three sites are on sand and the others alternate between coral and "other substrate". The first six species avoid the sandy area, possibly because little food is available there, whereas the last three are ubiquitous. The sums of abundances for the 9 species are in the last row of the Table. Species 1 to 6 come in three successive pairs, with distributions forming opposite gradients of abundance between sites 4 and 10. Species 1 and 2 are not associated to a single type of substrate. Species 3 and 4 are found in the coral areas only while species 5 and 6 are found on other substrates only (coral debris, turf, calcareous algae, etc.). The distributions of abundance of the ubiquitous species (7 to 9) have been produced using a random number generator, fitting the frequencies to a predetermined sum; these species will only be used to illustrate CCA in Section 11.2.

RDA was computed using the first six species as matrix **Y**, despite the fact that CCA (Subsection 11.2) is probably more appropriate for these data. Comparison of Tables 11.4 and 11.5, and of Figs. 11.3 and 11.5, allows, to a certain extent, a comparison of the two methods. The analysis was conducted on centred **y** variables because species abundances do not require standardization. When they are not

Table 11.2 Results of redundancy analysis (selected output). Matrix **Y** contained the first five variables of Table 10.5 and matrix **X**, the last three.

	Canonical axes			Non-canonical axes				
	I	II	III	IV	V	VI	VII	VIII
Eigenvalues (with respect to total variance in **Y** = 1.40378)								
	0.3374	0.1126	0.0027	0.5577	0.2897	0.0605	0.0261	0.0171
Fraction of total variance in **Y**								
	0.2404	0.0802	0.0019	0.3973	0.2064	0.0431	0.0186	0.0122
Correlations between the ordination vectors in spaces **Y** and **X**								
	0.6597	0.5588	0.1404					
Normalized eigenvectors (the rows correspond to the five variables in matrix **Y**)								
1	0.0578	0.8320	0.4855	0.2760	0.6362	0.6882	0.1495	–0.1519
2	–0.9494	0.0719	–0.0062	0.9215	0.0109	–0.3170	–0.2106	0.0770
3	0.0148	0.4997	–0.8615	–0.2642	0.7212	–0.4100	–0.3932	0.2957
4	0.3074	0.0244	–0.0786	–0.0300	–0.2692	0.4897	–0.7855	0.2643
5	0.0231	0.2289	0.1265	0.0632	–0.0513	0.1339	0.4021	0.9021

dimensionally homogeneous, the **y** variables should be standardized before RDA; this may be done by requesting that the calculations be carried out on the correlation matrix, or standardization may be done prior to canonical analysis. Three of the environmental variables form matrix **X**: depth (quantitative variable) and two of the three binary variables coding for the three-state qualitative variable "substrate type". Including all three binary "substrate" variables would make one of them linearly dependent on the other two (Subsection 1.5.7); the covariance matrix [**X'X**] would then be singular (Section 2.8), which would prevent calculation of the regression coefficients. It is not necessary to eliminate one of the dummy variables when using programs for canonical analysis such as CANOCO (version 3 and above); the last dummy variable is automatically eliminated from the calculations leading to eigenanalysis, but its position in the ordination diagram is estimated in the final calculations. Scaling type 1 was selected for the biplot in order to focus the interpretation on the ordination of the objects; this is especially interesting in this case because most of the explanatory variables (matrix **X**) are binary.

Table 11.3 Artificial data set representing observations (e.g. fish abundances) at 10 sites along a tropical reef transect. The variables are further described in the text.

Site No.	Sp. 1	Sp. 2	Sp. 3	Sp. 4	Sp. 5	Sp. 6	Sp. 7	Sp. 8	Sp. 9	Depth (m)	Substrate type Coral	Sand	Other
1	1	0	0	0	0	0	2	4	4	1	0	1	0
2	0	0	0	0	0	0	5	6	1	2	0	1	0
3	0	1	0	0	0	0	0	2	3	3	0	1	0
4	11	4	0	0	8	1	6	2	0	4	0	0	1
5	11	5	17	7	0	0	6	6	2	5	1	0	0
6	9	6	0	0	6	2	10	1	4	6	0	0	1
7	9	7	13	10	0	0	4	5	4	7	1	0	0
8	7	8	0	0	4	3	6	6	4	8	0	0	1
9	7	9	10	13	0	0	6	2	0	9	1	0	0
10	5	10	0	0	2	4	0	1	3	10	0	0	1
Sum	60	50	40	30	20	10	45	35	25				

Results of the analysis are presented in Table 11.4; programs such as CANOCO provide more output tables than presented here. The data could have produced 3 canonical axes and up to 6 non-canonical eigenvectors. In this example, only 4 of the 6 non-canonical axes had variance larger than 0. An overall test of significance (Subsection 11.3.2) showed that the canonical relationship between matrices **X** and **Y** is very highly significant ($p = 0.001$ after 999 permutations; permutation of residuals using CANOCO). The canonical axes explain 66%, 22% and 8% of the response table's variance, respectively; they are all significant ($p < 0.05$) and display strong species-environment correlations ($r = 0.999$, 0.997, and 0.980, respectively).

In Table 11.4, the eigenvalues are first given with respect to the total variance in matrix **Y**, as is customary in principal component analysis. They are also presented as proportions of the total variance in **Y** as is the practice in program CANOCO in the case of PCA and RDA. The species and sites are scaled for a distance biplot (RDA scaling type 1, Subsection 11.1.1). The eigenvectors (called "species scores" in CANOCO) are normalized to length 1. The site scores (matrix **F**) are obtained from eq. 11.12. They provide the ordination of the objects in the space of the original matrix **Y**. These ordination axes are not orthogonal to one another because matrix **Y** contains the "residual" components of the multiple regressions (Fig. 11.2). The "site scores that are linear combinations of the environmental variables", or "fitted site sores" (matrix **Z**, not printed in Table 11.4), are obtained from eq. 11.13. They provide the ordination of the objects in the space of matrix $\hat{\mathbf{Y}}$ which contains the fitted values of the multiple regressions (Fig. 11.2). These ordination axes are orthogonal to one another because

Table 11.4 Results of redundancy analysis of data in Table 11.3 (selected output). Matrix **Y**: species 1 to 6. Matrix **X**: depth and substrate classes.

	Canonical axes			Non-canonical axes			
	I	II	III	IV	V	VI	VII
Eigenvalues (with respect to total variance in **Y** = 112.88889)							
	74.52267	24.94196	8.87611	4.18878	0.31386	0.03704	0.00846
Fraction of total variance in **Y** (these are the eigenvalues of program CANOCO for RDA)							
	0.66014	0.22094	0.07863	0.03711	0.00278	0.00033	0.00007
Cumulative fraction of total variance in **Y** accounted for by axes 1 to *k*							
	0.66014	0.88108	0.95971	0.99682	0.99960	0.99993	1.00000
Normalized eigenvectors ("species scores"): mat. **U** for the canonical and $\mathbf{U}_{res}$ for the non-canonical portions							
Species 1	0.30127	–0.64624	–0.39939	–0.00656	–0.40482	0.70711	–0.16691
Species 2	0.20038	–0.47265	0.74458	0.00656	0.40482	0.70711	0.16690
Species 3	0.74098	0.16813	–0.25690	–0.68903	–0.26668	0.00000	0.67389
Species 4	0.55013	0.16841	0.26114	0.58798	0.21510	0.00000	0.68631
Species 5	–0.11588	–0.50594	–0.29319	0.37888	–0.66624	0.00000	0.12373
Species 6	–0.06292	–0.21535	0.25679	–0.18944	0.33312	0.00000	–0.06187
Site scores ("sample scores"): matrices **F** for the canonical and non-canonical portions, eqs. 11.12 and 9.4							
Site 1	–6.82791	5.64392	–1.15219	0.24712	1.14353	0.23570	0.01271
Site 2	–7.12919	6.29016	–0.75280	0.00000	0.00000	–0.47140	0.00000
Site 3	–6.92880	5.81751	–0.00823	–0.24712	–1.14353	0.23570	–0.01271
Site 4	–4.00359	–6.97190	–4.25652	2.14250	–0.28230	0.00000	0.00141
Site 5	13.63430	0.85534	–3.96242	–3.80923	–0.14571	0.00000	0.10360
Site 6	–4.03654	–5.82821	–1.12541	0.71417	–0.09410	0.00000	0.00047
Site 7	12.11899	1.03525	0.13651	0.22968	0.08889	0.00000	–0.22463
Site 8	–4.06949	–4.68452	2.00570	–0.71417	0.09410	0.00000	–0.00047
Site 9	11.34467	1.38328	3.97855	3.57956	0.05682	0.00000	0.12103
Site 10	–4.10243	–3.54082	5.13681	–2.14250	0.28230	0.00000	–0.00141
Correlations of environmental variables with site scores from eq. 11.12							
Depth	0.42204	–0.55721	0.69874				
Coral	0.98708	0.15027	0.01155				
Sand	–0.55572	0.81477	–0.14471				
Other subs.	–0.40350	–0.90271	0.12456				
Biplot scores of environmental variables							
Depth	0.34340	–0.26282	0.20000				
Coral	0.80314	0.07088	0.00330				
Sand	–0.45216	0.38431	–0.04142				
Other subs.	–0.32831	–0.42579	0.03565				
Centroids of sites with code "1" for BINARY environmental variables, in ordination diagram							
Coral	12.36599	1.09129	0.05088				
Sand	–6.96197	5.91719	–0.63774				
Other subs.	–4.05301	–5.25636	0.44014				

the eigenanalysis (PCA in Fig. 11.2) has been conducted on matrix $\hat{\mathbf{Y}}$. Both the "site scores" (matrix **F**) and "fitted site scores" (matrix **Z**) may be used in RDA biplots.*

Correlations of the environmental variables with the ordination vectors can be obtained in two forms: either with respect to the "site scores" (eq. 11.12) or with respect to the "fitted site scores" (eq. 11.13). The latter set of correlations is used to draw biplots containing the sites as well as the variables from **Y** and **X** (Fig. 11.3). There were three binary variables in Table 11.3. Each such variable may be represented by the centroid of the sites possessing state "1" for that variable (or else, the centroid of the sites possessing state "0"). These three variables are represented by both arrows (correlations) and symbols (centroids) in Fig. 11.3 to show the difference between these representations; in real-case studies, one chooses one of the representations.

The following question may arise when the effect of some environmental variables on the dependent variables **Y** is already well known (e.g. the effect of altitude on vegetation along a mountain slope, or the effect of depth on plankton assemblages): what would the *residual* ordination of sites (or the *residual* correlations among variables) be like if one could control for the linear effect of such well-known environmental variables? An approximate answer may be obtained by looking at the structure of the *residuals* obtained by regressing the original variables on the variables representing the well-known factors. With the present data set, for instance, one could examine the residual structure, after controlling for depth and substrate, by plotting ordination biplots of the *non-canonical axes* in Table 11.4. These axes correspond to a PCA of the table of residual values of the multiple regressions (Fig. 11.2).

3 — *Algorithms*

There are different ways of computing RDA. One may go through the multiple regression and principal component analysis steps described in Fig. 11.2, or calculate the matrix corresponding to $\mathbf{S_{YX}}\ \mathbf{S^{-1}_{XX}}\ \mathbf{S'_{YX}}$ in eq. 11.3 and decompose it using a standard eigenvalue-eigenvector algorithm.

Alternatively, ter Braak (1987c) suggested to modify his iterative algorithm for principal component analysis (Table 9.5), by incorporating a regression analysis at the end of each iteration, as illustrated in Fig. 11.4. Because it would be cumbersome to repeat a full multiple regression calculation at the end of each iteration and for each

---

* To obtain a distance biplot based upon the covariance matrix using program CANOCO (version 3 or later), one should centre the response variables (no standardization) and emphasize the ordination of sites by choosing scaling –1 in the "long dialogue" option. In the Windows version of CANOCO 4.0, focus on inter-site distances and do not post-transform the species scores. CANOCO prints the eigenvalues as proportions of the total variation in matrix **Y**. The scalings of eigenvalues and eigenvectors produced by CANOCO are described in the footnotes of Subsection 11.1.1. Changing the scaling of species and site score vectors by any multiplicative constant does not change the interpretation of a biplot.

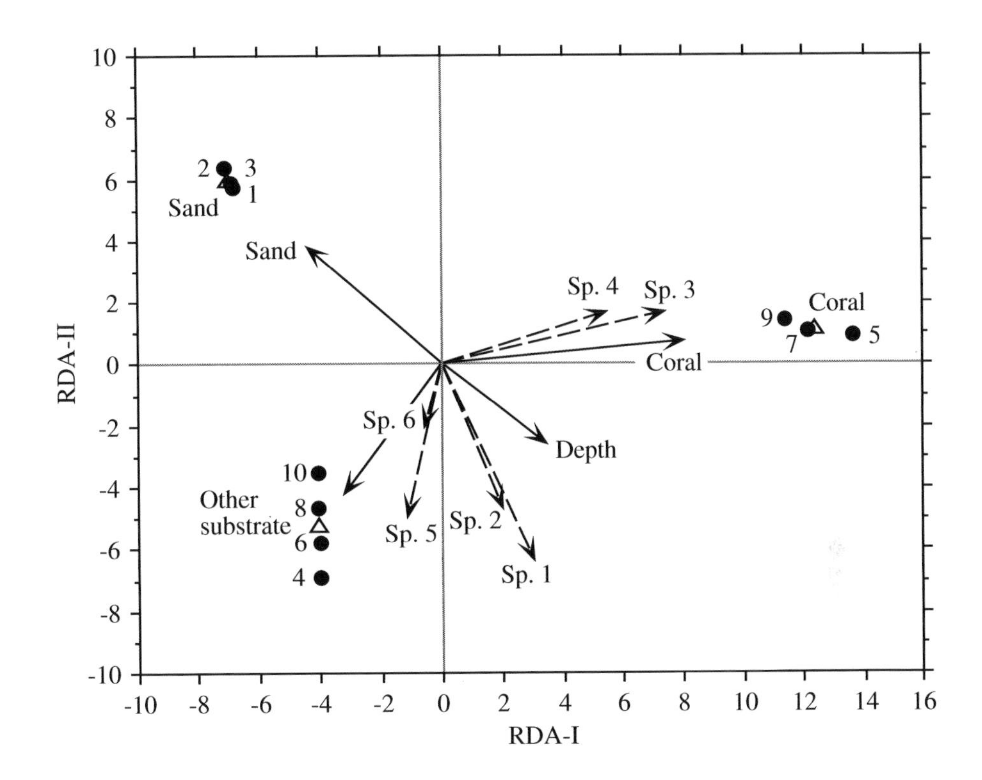

Figure 11.3 RDA ordination biplot of the artificial data presented in Table 11.3; the numerical results of the analysis are in Table 11.4. Dots are the sampling sites; numbers represent both the site number identifiers and depths (in m). Dashed arrows are the species. Full-line arrows represent the "biplot scores of environmental variables". The lengths of all arrows have been multiplied by 10 for clarity of the diagram. The "centroids of sites with code 1 for [the three] binary environmental variables" are represented by triangles. Binary environmental variables are usually represented by *either* arrows *or* symbols, not both as in this diagram.

canonical eigenvector, a short cut can be used. Vector **b** of regression coefficients is obtained from eq. 2.19:

$$\mathbf{b} = [\mathbf{X'X}]^{-1}\,[\mathbf{X'y}]$$

Only the $[\mathbf{X'y}]$ portion must be recomputed during each iteration of the estimation of the canonical eigenvectors; the $[\mathbf{X'X}]^{-1}$ part, which is the most cumbersome to calculate, is constant during the whole redundancy analysis run so that it needs to be computed only once.

The iterative procedure presents two advantages: (1) with large problems, one is satisfied, in most instances, with computing a few axes only, instead of having to estimate all eigenvalues and eigenvectors. The iterative procedure was developed to do

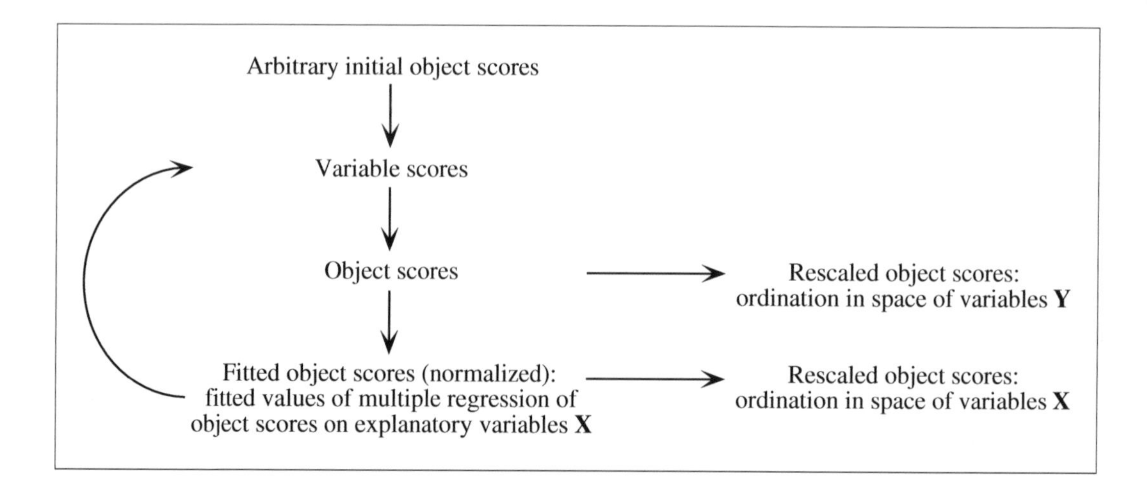

Figure 11.4 Two-way weighted summation algorithm (from Table 9.5), modified to compute redundancy analysis. Two types of ordinations are produced, one in the space of the **Y** variables and the other in the space of the **X** variables. Translated from Borcard & Buttler (1997).

that. (2) With smaller problems, the canonical *and* non-canonical axes can all be computed at once; one does not have to carry out a separate calculation on the matrix of residuals to obtain the non-canonical axes. The main disadvantage of the procedure is the possibility of numerical instability *when a large number of successive axes are computed.* For large problems in which all the canonical axes are needed, this procedure also requires more calculation time than regular eigenanalysis.

11.2 Canonical correspondence analysis (CCA)

Canonical correspondence analysis is a canonical ordination method developed by ter Braak (1986, 1987a, 1987c) and implemented in the program CANOCO (ter Braak, 1988b, 1988c, 1990; ter Braak & Smilauer, 1998). It is the canonical form of correspondence analysis. Any data table that could be subjected to correspondence analysis forms a suitable *response matrix* **Y** for CCA; this is the case, in particular, for species presence-absence or abundance tables (Section 9.4).

1 — The algebra of canonical correspondence analysis

The mathematics of CCA is essentially the same as that of redundancy analysis. Differences involve the diagonal matrices of row totals $\mathbf{D}(f_{i+})$ and row relative frequencies $\mathbf{D}(p_{i+})$, as defined in Section 9.4 for correspondence analysis; f_{i+} is the sum of values in row i of matrix **Y** whereas p_{i+} is f_{i+} divided by the grand total f_{++} of all values in **Y**.

Inflated data matrix

The calculations are modified in such a way as to simulate an analysis carried out on *inflated data matrices* $\mathbf{Y}_{infl}$ and $\mathbf{X}_{infl}$ constructed in a way similar to the inflated data table of Subsection 9.4.4. Assume that **Y** contains species presence-absence or abundance data and **X** contains environmental data. A *single species presence*, from the original table of species abundances, is placed in each row of $\mathbf{Y}_{infl}$. An inflated matrix $\mathbf{Y}_{infl}$ usually has many more rows than matrix **Y**. In the corresponding inflated matrix $\mathbf{X}_{infl}$, the row vectors of environmental data are repeated as required to make every species presence (in $\mathbf{Y}_{infl}$) face a copy of the appropriate vector of environmental data (in $\mathbf{X}_{infl}$). Modifications to the RDA algorithm are the following:

- The dependent data table is not matrix **Y** centred by variables (columns) as in RDA. CCA uses matrix $\bar{\mathbf{Q}}$ of the contributions to chi-square, also used in correspondence analysis. $\bar{\mathbf{Q}}$ is derived from matrix **Y** using eq. 9.32. Matrix **Q** of the relative frequencies is also computed ($\mathbf{Q} = (1/f_{++})\mathbf{Y}$); it is used in the scaling operations.

- Matrix **X** is standardized using weights $\mathbf{D}(f_{i+})$. To achieve this, compute the mean and standard deviation for each column of the inflated matrix $\mathbf{X}_{infl}$, which contains f_{++} rows, and use them to standardize the environmental data. Use the maximum likelihood estimator for the variance instead of the usual estimator (eq. 4.3); in other words, divide the sum of squared deviations from the mean by the number of rows of matrix $\mathbf{X}_{infl}$ (which is equal to f_{++}), instead of the number of rows minus 1.

- Weighted multiple regression is used instead of a conventional multiple regression. The weights, given by diagonal matrix $\mathbf{D}(p_{i+})^{1/2}$, are applied to matrix **X** everywhere it occurs in the multiple regression equations, which become:

$$\mathbf{B} = [\mathbf{X'}\,\mathbf{D}(p_{i+})\,\mathbf{X}]^{-1}\,\mathbf{X'}\,\mathbf{D}(p_{i+})^{1/2}\,\bar{\mathbf{Q}}$$

and

$$\hat{\mathbf{Y}} = \mathbf{D}(p_{i+})^{1/2}\,\mathbf{X}\,\mathbf{B}$$

The equation for computing $\hat{\mathbf{Y}}$ is then:

$$\hat{\mathbf{Y}} = \mathbf{D}(p_{i+})^{1/2}\,\mathbf{X}\,[\mathbf{X'}\,\mathbf{D}(p_{i+})\,\mathbf{X}]^{-1}\,\mathbf{X'}\,\mathbf{D}(p_{i+})^{1/2}\,\bar{\mathbf{Q}} \tag{11.15}$$

The matrix of residuals is computed as $\mathbf{Y_{res}} = \bar{\mathbf{Q}} - \hat{\mathbf{Y}}$. This is the equivalent, for CCA, of the equation $\mathbf{Y_{res}} = \mathbf{Y} - \hat{\mathbf{Y}}$ used in Fig. 11.2 for RDA.

- Eigenvalue decomposition (eqs. 11.10 and 11.11) is carried out on matrix $\mathbf{S}_{\hat{Y}'\hat{Y}}$ which, in this case, is simply the matrix of sums of squares and cross products, without division by the number of degrees of freedom — as in correspondence analysis:

$$\mathbf{S}_{\hat{Y}'\hat{Y}} = \hat{\mathbf{Y}}'\hat{\mathbf{Y}} \tag{11.16}$$

One can show that $\mathbf{S}_{\hat{\mathbf{Y}}'\hat{\mathbf{Y}}}$, computed as described, is equal to $\mathbf{S}_{\bar{\mathbf{Q}}\mathbf{X}}\mathbf{S}_{\mathbf{XX}}^{-1}\mathbf{S}'_{\bar{\mathbf{Q}}\mathbf{X}}$ (eq. 11.3 and 11.11) if the covariance matrices $\mathbf{S}_{\bar{\mathbf{Q}}\mathbf{X}}$ and $\mathbf{S}_{\mathbf{XX}}$ are computed with weights on $\mathbf{X}$, given by matrix $\mathbf{D}(p_{i+})^{1/2}$, and without division by the number of degrees of freedom.

With these modifications, CCA is computed using eq. 11.3, obtaining matrices $\mathbf{\Lambda}$ of eigenvalues and $\mathbf{U}$ of eigenvectors. Canonical correspondence analysis is thus a weighted form of redundancy analysis, applied to dependent matrix $\bar{\mathbf{Q}}$. It approximates chi-square distances among the rows (objects) of the dependent data matrix, subject to the constraint that the canonical ordination vectors be maximally related to weighted linear combinations of the explanatory variables. The equations are also described in Section 5.9.5 of ter Braak (1987c). The method is perfectly suited to analyse the relationships between species presence/absence or abundance data matrices and tables of environmental variables. The number of canonical and non-canonical axes expected from the analysis are given in Table 11.1. Tests of significance are available, in CCA and RDA, for the total canonical variation and for individual eigenvalues (Subsection 11.3.2).

• The normalized matrix $\hat{\mathbf{U}}$ is obtained using eq. 9.38:

$$\hat{\mathbf{U}} = \bar{\mathbf{Q}}\mathbf{U}\mathbf{\Lambda}^{-1/2}$$

In CCA, matrix $\hat{\mathbf{U}}$ as defined here does not contain the loadings of the rows of $\hat{\mathbf{Y}}$ on the canonical axes. It contains instead the loadings of the rows of $\bar{\mathbf{Q}}$ on the ordination axes, as in CA. It will be used to find the site scores (matrices $\mathbf{F}$ and $\hat{\mathbf{V}}$) in the space of the original variables $\mathbf{Y}$. The site scores in the space of the fitted values $\hat{\mathbf{Y}}$ will be found using $\mathbf{U}$ instead of $\hat{\mathbf{U}}$.

Scalings in CCA

• Matrix $\mathbf{V}$ of species scores (for scaling type 1) and matrix $\hat{\mathbf{V}}$ of site scores (for scaling type 2) are obtained from $\mathbf{U}$ and $\hat{\mathbf{U}}$ using the transformations described for correspondence analysis (Subsection 9.4.1):

eq. 9.41 (species scores, scaling 1): $\mathbf{V} = \mathbf{D}(p_{+j})^{-1/2}\mathbf{U}$

and eq. 9.42 (site scores, scaling 2): $\hat{\mathbf{V}} = \mathbf{D}(p_{i+})^{-1/2}\hat{\mathbf{U}}$

or combining eqs. 9.38 and 9.42: $\hat{\mathbf{V}} = \mathbf{D}(p_{i+})^{-1/2}\bar{\mathbf{Q}}\mathbf{U}\mathbf{\Lambda}^{-1/2}$

Scalings 1 and 2 are the same as in correspondence analysis (Subsection 9.4.1). Matrices $\mathbf{F}$ (site scores for scaling type 1) and $\hat{\mathbf{F}}$ (species scores for scaling type 2) are found using eqs. 9.43a and 9.44a:

$$\mathbf{F} = \hat{\mathbf{V}}\mathbf{\Lambda}^{1/2} \quad \text{and} \quad \hat{\mathbf{F}} = \mathbf{V}\mathbf{\Lambda}^{1/2}$$

Equations 9.43b and 9.44b cannot be used here to find $\mathbf{F}$ and $\hat{\mathbf{F}}$ because the eigenanalysis has been conducted on a covariance matrix (eq. 11.16) computed from the matrix of fitted values $\hat{\mathbf{Y}}$ (eq. 11.15) and not from $\mathbf{Q}$. The site scores which are linear combinations of the environmental variables, corresponding to eq. 11.13 of RDA, are found from $\hat{\mathbf{Y}}$ using the following equations:

For scaling type 1: $$\mathbf{Z}_{\text{scaling 1}} = \mathbf{D}(\text{p}_{i+})^{-1/2}\,\hat{\mathbf{Y}}\,\mathbf{U} \quad \textbf{(11.17)}$$

For scaling type 2: $$\mathbf{Z}_{\text{scaling 2}} = \mathbf{D}(\text{p}_{i+})^{-1/2}\,\hat{\mathbf{Y}}\,\mathbf{U}\mathbf{\Lambda}^{-1/2} \quad \textbf{(11.18)}$$

With scaling type 1, biplots can be drawn using either $\mathbf{F}$ and $\mathbf{V}$, or $\mathbf{Z}_{\text{scaling 1}}$ and $\mathbf{V}$. With scaling type 2, one can use either $\hat{\mathbf{V}}$ and $\hat{\mathbf{F}}$, or $\mathbf{Z}_{\text{scaling 2}}$ and $\hat{\mathbf{F}}$. The construction and interpretation of CCA biplots is discussed by ter Braak & Verdonschot (1995).

• Residuals can be analysed by applying eigenvalue decomposition (eq. 11.10) to matrix $\mathbf{Y}_{\text{res}}$, producing matrices of eigenvalues $\mathbf{\Lambda}$ and normalized eigenvectors $\mathbf{U}$. Matrix $\hat{\mathbf{U}}$ is obtained using eq. 9.38: $\hat{\mathbf{U}} = \bar{\mathbf{Q}}\,\mathbf{U}\mathbf{\Lambda}^{-1/2}$. Species and site scores are obtained for scaling types 1 and 2 (eqs. 9.41, 9.42, 9.43a and 9.44a) using the matrices of row and column sums $\mathbf{D}(\text{p}_{i+})^{-1/2}$ and $\mathbf{D}(\text{p}_{+j})^{-1/2}$ of the original matrix $\mathbf{Y}$.

A little-known application of CCA is worth mentioning here. Consider a qualitative environmental variable and a table of species presence-absence or abundance data. How can one “quantify” the qualitative states, i.e. give them values along a quantitative scale which would be related in some optimal way to the species data? CCA provides an easy answer to this problem. The species data form matrix $\mathbf{Y}$; the qualitative variable, recoded as a set of dummy variables, is placed in matrix $\mathbf{X}$. Compute CCA and take the fitted site scores (“site scores which are linear combinations of environmental variables”): they provide a quantitative rescaling of the qualitative variable, maximizing the weighted linear correlation between the dummy variables and matrix $\bar{\mathbf{Q}}$. In the same way, RDA may be used to rescale a qualitative variable with respect to a table of quantitative variables of the objects if linear relationships can be assumed.

McCune (1997) warns users of CCA against inclusion of noisy or irrelevant explanatory variables in the analysis. They may lead to misleading interpretations.

2 — Numerical example

Table 11.3 will now be used to illustrate the computation and interpretation of CCA. The 9 species are used as matrix $\mathbf{Y}$. Matrix $\mathbf{X}$ is the same as in Subsection 11.1.2. Results are presented in Table 11.5 and Fig. 11.5; programs such as CANOCO provide more output tables than presented here. There was a possibility of 3 canonical and 8 non-canonical axes. Actually, the last 2 non-canonical axes have zero variance. An overall test of significance (Subsection 11.3.2) showed that the canonical relationship between matrices $\mathbf{X}$ and $\mathbf{Y}$ is very highly significant ($p = 0.001$ after 999 permutations, by permutation of residuals under a full model; Subsection 11.3.2). The canonical axes explain 47%, 24% and 10% of the response table’s variance, respectively. They are all significant ($p < 0.05$) and display strong row-weighted species-environment correlations ($r = 0.998$, 0.940, and 0.883, respectively).

Table 11.5 Results of canonical correspondence analysis of the data in Table 11.3 (selected output). Matrix **Y**: species 1 to 9; **X**: depth and 3 substrate classes. Non-canonical axes VIII and IX not shown.

	Canonical axes			Non-canonical axes			
	I	II	III	IV	V	VI	VII
Eigenvalues (their sum is equal to the total inertia in matrix $\bar{\mathbf{Q}}$ of species data = 0.78417)							
	0.36614	0.18689	0.07885	0.08229	0.03513	0.02333	0.00990
Fraction of the total variance in $\bar{\mathbf{Q}}$							
	0.46691	0.23833	0.10055	0.10494	0.04481	0.02975	0.01263
Cumulative fraction of total inertia in $\bar{\mathbf{Q}}$ accounted for by axes 1 to k							
	0.46691	0.70524	0.80579	0.91072	0.95553	0.98527	0.99791
Eigenvectors ("species scores"): matrices $\hat{\mathbf{F}}$ for the canonical and the non-canonical portions (eq. 9.44a)							
Species 1	–0.11035	–0.28240	–0.20303	0.00192	0.08223	0.08573	–0.01220
Species 2	–0.14136	–0.30350	0.39544	0.14127	0.02689	0.14325	0.04303
Species 3	1.01552	–0.09583	–0.19826	0.10480	–0.13003	0.02441	0.04647
Species 4	1.03621	–0.10962	0.22098	–0.22364	0.24375	–0.02591	–0.05341
Species 5	–1.05372	–0.53718	–0.43808	–0.22348	0.32395	0.12464	–0.11928
Species 6	–0.99856	–0.57396	0.67992	0.38996	–0.29908	0.32845	0.21216
Species 7	–0.25525	0.17817	–0.20413	–0.43340	–0.07071	–0.18817	0.12691
Species 8	–0.14656	0.85736	–0.01525	–0.05276	–0.35448	–0.04168	–0.19901
Species 9	–0.41371	0.70795	0.21570	0.69031	0.14843	–0.33425	–0.00629
Site scores ("sample scores"): matrices $\hat{\mathbf{V}}$ for the canonical and the non-canonical portions (eq. 9.42)							
Site 1	–0.71059	3.08167	0.21965	1.24529	1.07293	–0.50625	0.24413
Site 2	–0.58477	3.00669	–0.94745	–2.69965	–2.13682	0.81353	0.47153
Site 3	–0.76274	3.15258	2.13925	3.11628	2.30660	–0.69894	–1.39063
Site 4	–1.11231	–1.07151	–1.87528	–0.66637	1.10154	1.43517	–1.10620
Site 5	0.97912	0.06032	–0.69628	0.61265	–0.98301	0.31567	0.57411
Site 6	–1.04323	–0.45943	–0.63980	–0.28716	0.57393	–1.44981	1.70167
Site 7	0.95449	0.08470	0.13251	0.42143	0.11155	–0.39424	–0.67396
Site 8	–0.94727	0.10837	0.52611	0.00565	–1.26273	–1.06565	–1.46326
Site 9	1.14808	–0.49045	0.47835	–1.17016	1.00599	0.07350	0.08605
Site 10	–1.03291	–1.03505	2.74692	1.28084	–0.36299	1.98648	1.05356
Correlations of environmental variables with site scores							
Depth	0.18608	–0.60189	0.65814				
Coral	0.99233	–0.09189	–0.04614				
Sand	–0.21281	0.91759	0.03765				
Other subs.	–0.87958	–0.44413	0.02466				
Correlations of environmental variables with fitted site scores (for biplots)							
Depth	0.18636	–0.64026	0.74521				
Coral	0.99384	–0.09775	–0.05225				
Sand	–0.21313	0.97609	0.04263				
Other subs.	–0.88092	–0.47245	0.02792				
Centroids of sites with code "1" for BINARY environmental variables, in ordination diagram							
Coral	1.02265	–0.10059	–0.05376				
Sand	–0.66932	3.06532	0.13387				
Other subs.	–1.03049	–0.55267	0.03266				

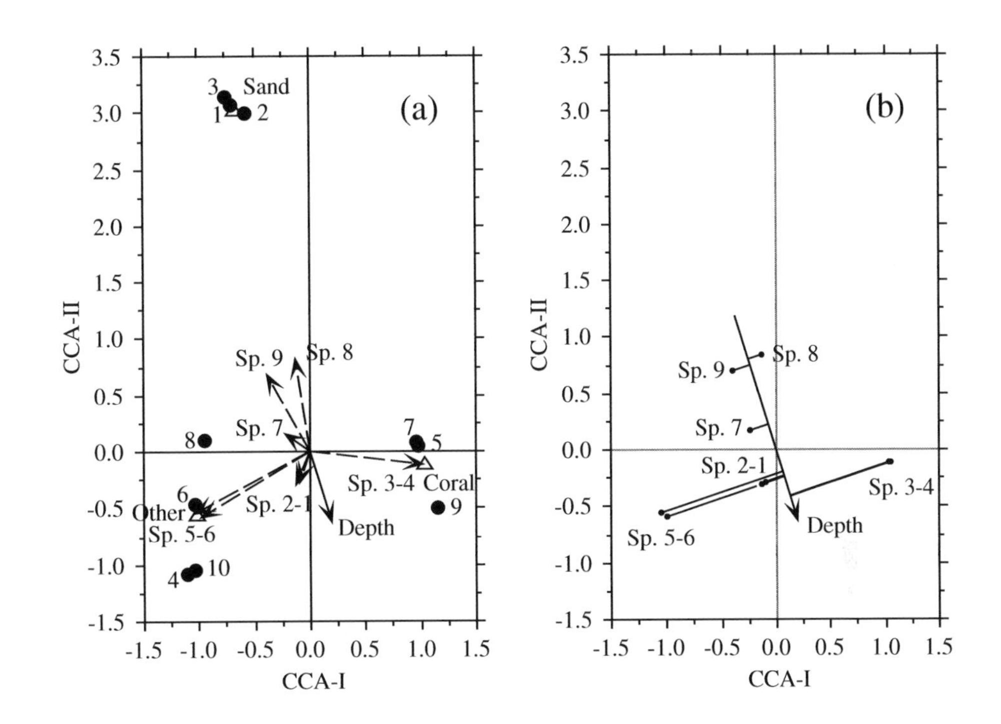

Figure 11.5 CCA ordination biplot of the artificial data in Table 11.3; the numerical results of the analysis are in Table 11.5. (a) Biplot representing the species (dashed arrows), sites (dots, with site identifiers which also correspond to depths in m) and environmental variables (full arrow for depth, triangles for the three binary substrate variables). (b) Ranking of the species along a quantitative environmental variable (depth in the present case) is inferred by projecting the species onto the arrow representing that variable.

Scaling type 2 (from Subsection 9.4.1) was used, in this example, to emphasize the relationships among species. As a result, the species (matrix $\hat{\mathbf{F}}$) are at the centroids of the sites (matrix $\hat{\mathbf{V}}$) in Fig. 11.5a and distances among species approximate their chi-square distances. Species 3 and 4 characterize the sites with coral substrate, whereas species 5 and 6 indicate the sites with "other substrate". Species 1 and 2, which occupy an intermediate position between the sites with coral and other substrate, are not well represented in the biplot of canonical axes I and II; axis III is needed to adequately represent the variance of these species. Among the ubiquitous species 7 to 9, two are well represented in the subspace of canonical axes I and II; their arrows fall near the middle of the area encompassing the three types of substrate. The sites are not perfectly ordered along the depth vector; the ordering of sites along this variable mainly reflects the difference in species composition between the shallow sandy sites (1, 2 and 3) and the other sites.

Figure 11.5b shows how to infer the ranking of species along a quantitative environmental variable. Depth is used in this example. The graphical method simply consists in projecting (at right angle) the species onto the arrow representing that variable. This gives an approximation of the weighted averages of the species with respect to environmental variables. Ecologists like to interpret this ranking as representing the niche optima for the species under study. It is important to realize that three rather strong assumptions have to be made when attempting such an interpretation:

- that the various species have unimodal distributions along the environmental variable of interest (subsection 9.4.4);

- that the species distributions are under environmental control (Whittaker, 1956; Bray & Curtis, 1957), so that the mode of each species is at its optimum along the various environmental variables; and

- that the gradient under study is long enough to allow each species to go from some less-than-optimum low frequency to its high-frequency optimum, and back to some past-optimum low frequency.

In the data of the present example (Table 11.2), only species 1, 3 and 5 were constructed to approximately correspond to these criteria. Species 7, which may also look like it has a unimodal distribution, has actually been constructed using a pseudo-random number generator.

To investigate the similarities among sites or the relationships among species after controlling for the linear effects of depth and type of substrate, one could draw ordination biplots of the *non-canonical axes* in Table 11.5. These axes correspond to a correspondence analysis of the table of regression residuals, as in Fig. 11.2.

3 — Algorithms

CCA may be computed following the same three routes as RDA (Subsection 11.1.3). One may go through the weighted multiple regression steps followed by eigenanalysis of matrix $\mathbf{S}_{\hat{\mathbf{Y}}'\hat{\mathbf{Y}}}$. Alternatively, one may choose to estimate the matrix corresponding to $\mathbf{S}_{\bar{\mathbf{Q}}\mathbf{X}}\mathbf{S}_{\mathbf{X}\mathbf{X}}^{-1}\mathbf{S}'_{\bar{\mathbf{Q}}\mathbf{X}}$ (taking into account the modifications described in Subsection 1) and proceed to eigenanalysis. Finally, one may use the modified iterative algorithm proposed by ter Braak (1986, 1987a) (Table 11.6) and implemented in program CANOCO. The advantages and disadvantages of this procedure have been discussed in Subsection 11.1.3. (By removing the weights y_{i+} and y_{+j} from Table 11.6, one obtains the iterative algorithm for RDA. In this case, however, take the normalization from the iterative algorithm for PCA, Table 9.5, instead of Table 9.12 where the iterative algorithm for CA is outlined.)

Table 11.6 Two-way weighted averaging (TWWA) algorithm for canonical correspondence analysis (CCA). From ter Braak (1986, 1987c). The regression steps, by which CCA differs from CA, are identified by arrows.

Step 1: Consider a table **Y** of n sites (rows) × p species (columns) and a standardized matrix **X** of explanatory variables, as in Subsection 11.2.1. Do NOT centre the species on their means.

Determine how many eigenvectors are needed. For each one, **DO** the following:

Step 2: Take the row order as the arbitrary initial fitted site scores (vector *fitted-rowscore*). Set the initial eigenvalue estimate to 0. In what follows, y_{i+} = row sum for site i, y_{+j} = column sum for species j, and y_{++} = grand total for the data table **Y**.

Iterative procedure begins

Step 3: Compute new species loadings: $colscore(j) = \Sigma\, y(i,j) \times fitted\text{-}rowscore(i)/y_{+j}$

Step 4: Compute new site scores: $rowscore(i) = \Sigma\, y(i,j) \times colscore(j)/y_{i+}$

⇒ Step 5: For the non-canonical axes, skip this and go directly to step 7. For the canonical axes, regress the site scores (vector *rowscore*) on **X** using weighted regression; the weights are given by the diagonal matrix of row sums $\mathbf{D}(y_{i+})^{1/2}$. The equation used here for the regression coefficients is:

$\mathbf{c} = [\mathbf{X}'\,\mathbf{D}(y_{i+})\,\mathbf{X}]^{-1}\,[\mathbf{X}'\,\mathbf{D}(y_{i+})\,\mathbf{x}^*]$ where $\mathbf{x}^*$ is the *rowscore* vector[1].

The $[\mathbf{X}'\,\mathbf{D}(y_{i+})\,\mathbf{X}]^{-1}$ part of the regression procedure may have been calculated beforehand, once and for all. What remains to calculate here is $[\mathbf{X}'\,\mathbf{D}(y_{i+})\,\mathbf{x}^*]$.

⇒ Step 6: Calculate the fitted values ($\hat{\mathbf{y}} = \mathbf{Xc}$) and use them as new site scores (vector *fitted-rowscore*)[1].

Step 7: For the second and higher axes, make the fitted site scores uncorrelated with all previous axes (Gram-Schmidt orthogonalization procedure: Table 9.12).

Step 8: Normalize the fitted site scores to obtain an estimate of the eigenvalue (normalization procedure: Table 9.12). If this estimate does not differ from the previous one by more than a small quantity ("tolerance"), go to step 9.

End of iterative procedure

Step 9: If more eigenvectors are to be computed, go to step 2. If not, continue with step 10.

Step 10: **Scaling type 2** (Subsection 9.4.1) — The row (site) scores (vector *rowscore*) give the ordination of the sites in the space of the original species data (matrix $\hat{\mathbf{V}}$). The column scores (species loadings) are in vector *colscore*; they correspond to matrix $\hat{\mathbf{F}}$. The fitted site scores (vector *fitted-rowscore*) give the ordination of the sites in the space of the explanatory variables **X** (eq. 11.18).

Scaling type 1 (Subsection 9.4.1) — Matrices **F** and **V** are obtained from eqs. 9.43a and 9.44a. The fitted site scores are vectors *fitted-rowscore*, computed above, multiplied by the square root of the corresponding eigenvalue, as shown by combining eqs. 11.17 and 11.18.

Step 11: Print out the eigenvalues, % variance, species loadings, site scores, and fitted site scores.

[1] These equations differ slightly from the corresponding portions of eq. 11.15 because the analysis is based here upon matrix **Y** instead of $\bar{\mathbf{Q}}$.

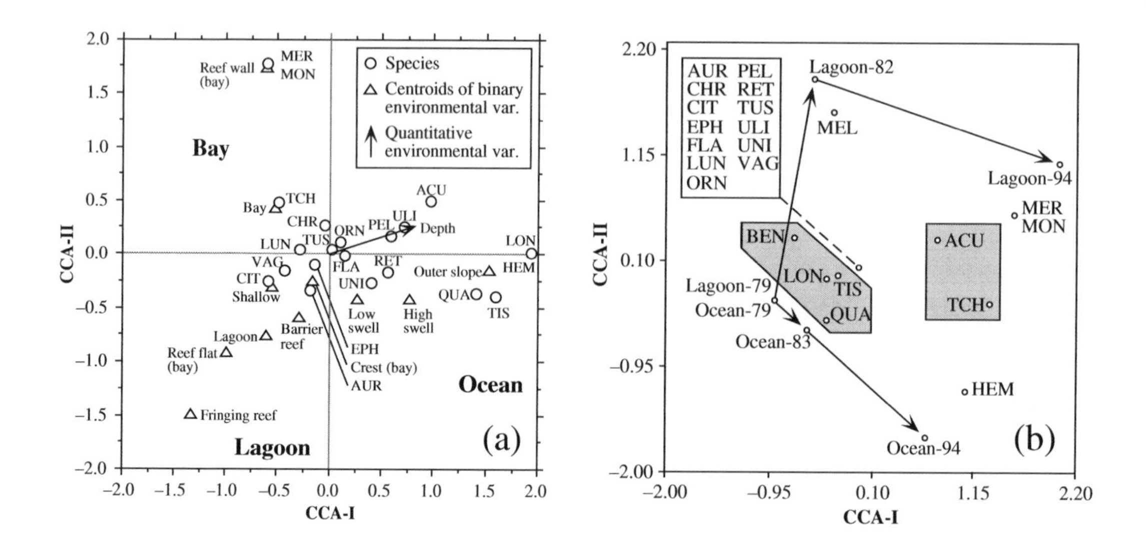

Figure 11.6 (a) CCA ordination diagram: presence/absence of 21 Chaetodontid fish species at 42 sampling sites around Moorea Island, French Polynesia, against environmental variables. The species (names abbreviated to 3 letters) are represented by circles instead of arrows for readability of the diagram. Axis I: 14.6% of the variation (p = 0.001 after 999 permutations); axis II: 7.4% (p = 0.010). Redrawn from the original data of Cadoret *et al.* (1995). (b) CCA ordination diagram for relevés from the lagoon and the outer slope, at three time periods over 15 years. Arrows show the sampling sequence in each environment. Thirteen species found in all relevés are located in the centre of the ordination diagram. Modified from Cadoret *et al.* (1995).

Ecological application 11.2a

Ecological application 9.4a described the spatial distribution of chaetodontid fish assemblages (butterflyfishes) around a tropical island, using correspondence analysis. This application is continued here. Cadoret *et al.* (1995) next described the relationships between the fish species (quantitative relevés) and some environmental variables, using canonical correspondence analysis. The environmental variables are: the type of environment (qualitative descriptor: bay, lagoon, or outer slope of the reef on the ocean side), geomorphology (qualitative: reef flat, crest, and reef wall of the fringing reefs of bays; fringing reef, shallow, barrier reef, and outer slope for transect sites), depth (quantitative: from 0.5 to 35 m), and exposure to swell (qualitative: low, high, or sites located in bays).

The ordination of sampling sites by CCA was virtually identical to that in Fig. 9.18; this indicates that the first CA axes are closely related to the environmental variables. The canonical axes accounted together for 35% of the variation in the species data (p = 0.001 after 999 permutations). The description of the ordination of sites presented in Ecological application 9.4a may be compared to Fig. 11.6a. This Figure shows which types of environment are similar in their chaetodontid species composition and which species are associated with the various types of environment. It indicates that the reef flats of the fringing reefs of the bays are similar in species composition to the fringing reefs of the lagoon; likewise, the crests of the fringing reefs of the bays are similar to the barrier reefs in the lagoon. Species composition along the reef walls of bays and that on the outer slopes differ, however, from all the other types of environment. The ecology of the most important species is discussed in the paper.

CCA has also been used in that paper to analyse the temporal changes of the fish assemblages along one of the transects, called Tiahura, which had been repeatedly studied over the previous 15 years. One of the objectives of the study was to determine whether the chaetodontid assemblages were stable or changing with time. In this analysis, matrix **Y** contained presence/absence data for 23 butterflyfish species in each of 6 relevés made either in the lagoon (in 1979, 1982 and 1994) or on the outer slope (in 1979, 1983 and 1994) along the transect. Matrix **X** contained 3 binary variables dividing the relevés by location (lagoon, outer slope: 1 dummy variable) and time (3 time periods: 2 dummy variables). Matrix **X** explains 67% of the variation in the species presence-absence data table **Y**. The hypothesis of "no change with time" cannot be properly tested for significance since there is no replicate sampling available. Figure 11.6b shows, however, that the species composition has been mostly stable in the lagoon and on the outer slope of Moorea during the past 15 years, although a few coral-eating species were encountered less frequently in 1982-83 and 1994 (BEN, LON, TIS and QUA, in the shaded polygon), following changes in the coral community, while two species have been observed almost exclusively in the last sampling year (ACU and TCH, in the shaded rectangle). It is interesting to note that, while the species composition was the same in the lagoon and the outer slope in 1979, compositions diverged slightly in one way in the lagoon and in a different way on the seaward side, by gain and loss of species.

Ecological application 11.2b

Canonical correspondence analysis is widely used in palaeoecology, together with regression and calibration, to infer past ecological conditions (climatic, limnological, etc.) from palaeo-assemblages of species. This vast and growing literature has been summarized by Birks (1995); references to these methods can also be found in the bibliography assembled by Birks *et al.* (1994), under the headings *limnology*, *palaeoecology*, *palaeolimnology*, etc.

One of the classical papers on the subject is that of Birks *et al.* (1990a). Palaeolimnological reconstruction involves two main steps: modelling from a *training data set*, followed by the construction of forecasting models that are then applied to the palaeo-data. In this paper, diatoms were used to reconstruct past water chemistry. The training data set consisted of diatom assemblages comprising 287 species, from present-day surface samples from 138 lakes in England, Norway, Scotland, Sweden, and Wales. Data were also available on pH, conductivity, Ca, Mg, K, SO_4, Cl, alkalinity, total Al, and DOC. Data from more lakes were available for subsets of these variables. CCA was used to relate species composition to water chemistry. The first two canonical eigenvalues were significant and displayed strong species-environment correlations ($r = 0.95$ and 0.84, respectively). The first axis expressed a significant diatom gradient which was strongly positively correlated with alkalinity and its close correlates, Ca and pH, and negatively but less strongly-correlated with total Al; the second axis corresponded to a significant gradient strongly correlated with DOC. This result indicated that pH (or alkalinity), Al, and DOC are potentially reconstructible from fossil diatom assemblages.

The fossil data set contained 101 slices of a sediment core from a small lake, the Round Loch of Glenhead, in Galloway, southwestern Scotland. The data series covered the past 10000 years. The fossil data (292 diatom taxa) were included in the CCA as passive objects (called *supplementary objects* in Subsection 9.1.8) and positioned in the ordination provided by canonical axes I and II. All fossil objects were well-fitted in that space (they had low squared residual distances), indicating that the pattern of variation in diatom composition can be linked to the modern chemical variables.

Reconstruction of past surface-water chemistry involved two steps. First, the training set was used to model, by regression, the responses of modern diatoms to the chemical variables of interest (one variable at a time). Secondly, the modelled responses were used to infer past chemistry from the composition of fossil diatom assemblages; this phase is called *calibration* (ter Braak, 1987b; ter Braak & Prentice, 1988). Extensive simulations led Birks *et al.* (1990b) to prefer weighted averaging (WA) over maximum likelihood (ML) regression and calibration. Consider pH in lakes, for example. *WA regression* simply consists in applying eq. 9.47 to estimate the pH optimum of each taxon of the training set as the weighted average of all the pH values for lakes in which this taxon occurs, weighted by the taxon's relative abundance. *WA calibration* consists in applying eq. 9.46 to estimate the pH of each lake as the weighted average of the pH optima of all the taxa present. Taxa with a narrow pH tolerance or amplitude may, if required, be given greater weight in WA regression and calibration than taxa with a wide pH tolerance (Birks *et al.*, 1990b).

Application of eqs. 9.47 and 9.46 to data resulted in shrinkage of the range of pH scores. Shrinkage occurred for the same reason as in the TWWA algorithm for correspondence analysis; in step 6.4 of that algorithm (Table 9.12), the eigenvalue was actually estimated from the amount of shrinkage incurred by the site scores after each iteration through eqs. 9.47 and 9.46 (steps 3 and 4). Deshrinking may be done in at least two ways; the relative merits of the two methods are discussed by Birks *et al.* (1990b).

• Deshrinking by classical regression proceeds in two steps. (1) The pH values inferred by WA regression and calibration ($\hat{x}_i$) are regressed on the observed values x_i for the training set, using the linear regression model $\hat{x}_i = b_0 + b_i x_i + \varepsilon_i$. (2) The parameters of that model are then used to deshrink the $\hat{x}_i$ values, using the equation: final $\hat{x}_i = (\hat{x}_i - b_o)/b_1$. This method was used to deshrink the inferred pH values.

• Another way of deshrinking, advocated by ter Braak & van Dam (1989) for palaeolimnological data, is to use "inverse regression" of x_i on $\hat{x}_i$ (ter Braak, 1987b). Inverse regression was used to deshrink the inferred Al and DOC values.

Training sets containing different numbers of lakes were used to infer pH, total Al, and DOC. Past values of these variables were then reconstructed from the palaeo-assemblages of diatoms, using the pH optima estimated above (eq. 9.47) for the various diatom species, followed by deshrinking. Reconstructed values were plotted against depth and time, together with error estimates obtained by bootstrapping. The past history of the Round Loch of Glenhead over the past 10000 years was discussed in the paper.

This approach involving CCA, WA regression and WA calibration is now widely used in palaeolimnology to reconstruct, for example, surface-water temperatures from fossil chironomid assemblages, lake salinity from diatom assemblages, lake-water phosphorous concentrations from diatom assemblages, and surface-water chlorophyll *a* concentrations from fossil diatom assemblages. The WA regression and WA calibration method was further improved by ter Braak & Juggins (1993). Birks (1995) provides a review of quantitative palaeoenvironmental reconstructions, both theory and applications, whereas ter Braak (1995) gives a theoretical comparison of recent reconstruction methods. Programs are available to carry out the calculations (ter Braak & Juggins, 1993; Line *et al.*, 1994).

11.3 Partial RDA and CCA

Partial canonical analysis is the extension of partial linear regression to the case where the response data table **Y** is multivariate. The logic is the same as explained in Subsection 10.3.5 and will not be detailed further. Partial canonical analysis includes partial RDA (Davies & Tso, 1982) and partial CCA (ter Braak, 1988a). As in partial linear regression, the explanatory variables are seen as belonging to two groups: **X** for the explanatory variables to be included in the model and **W** for the covariables (also called covariates) whose effect on **Y** is to be controlled.*

1 — Applications

Partial canonical analysis may be used to investigate a variety of problems. Here are some examples.

• Consider the case where matrix **W** contains variables whose effects on **Y** are well known. One wants to control for these well-known effects when analysing the effect on **Y** of a set of variables of interest, **X**. For instance, one may want to control for a well-known effect of a gradient of salinity while analysing the effect of nutrients on phytoplankton assemblages.

• After conducting a standard canonical analysis as in Subsections 11.1.2 and 11.2.2, one may want to isolate the effect of a single explanatory variable. Using all the other explanatory variables as covariates leads to a single canonical axis expressing the partial effect of the variable of interest on **Y**. The corresponding canonical eigenvalue divided by the total inertia in **Y** measures the fraction of the variation of **Y** accounted for by that factor. In partial CCA, the coordinates of the species along this axis provide a ranking along the given environmental variable, conditional on the ranking explained by the covariables; this ranking must be interpreted with caution, as discussed in Subsection 11.2.2. Measures of species "tolerance" (i.e. standard deviation for each species along an axis) provided by program CANOCO may help in the interpretation.

• Partial canonical analysis may be used, instead of MANOVA, to analyse multivariate response data (matrix **Y**) in balanced experimental designs, including tests of significance for main effects and interaction terms. For a single experimental factor, the analysis can be conducted using simple RDA or CCA. For two or more factors and their interactions, partial RDA or CCA must be used. When the response data consist of a matrix of species presence-absence or abundance data, one can use either partial CCA to preserve the chi-square distance among sites, or the distance-based RDA method (db-RDA) of Legendre & Anderson (1999) to preserve some other distance.

* In program CANOCO, which is widely used to carry out partial canonical analysis, **X** is called the "matrix of environmental data" and **W** is called the "matrix of covariables".

In distance-based redundancy analysis (db-RDA), a resemblance matrix, **S** or **D**, is computed among the sites using a similarity measure appropriate to species data. Principal coordinate analysis (PCoA, Section 9.2) is applied to this matrix to obtain new Euclidean axes (matrix **Y**) fully representing the relationships among the sites; a correction for negative values, using method 1 of Subsection 9.2.4, may be required. The experimental factors and their interactions are coded as orthogonal dummy variables. RDA is applied to the new matrix **Y** to test the significance of the factor (or interaction) coded into matrix **X**, with all the other factors (and interactions) coded into a matrix of covariables **W**.

• For sampling conducted at different occasions, the effect of time can be estimated by canonical analysis, as in Ecological application 11.2. One may also wish to remove the effect of the times of sampling. This can be done by using the variable(s) describing the sampling occasions as covariables in the analysis. Time may be represented by dummy variables, or by a quantitative variable whose effect on **Y** is assumed to be linear, or by a sine transformation of Julian days, etc. The effect of time will effectively be removed if the sampling times (days, weeks, years, ...) only affect the means of the response variables and nothing else. If there is an interaction between sampling time and the other environmental or spatial variables of interest in the analysis, the effect of time cannot be removed through this simple approach. In the presence of an interaction, the interaction terms must remain in the analysis for the model to be valid.

• In the same way, the effect of the sampling locations can be controlled for. Sampling locations may be represented by dummy variables, or by a trend-surface polynomial of the geographic coordinates of the sites (Chapter 13). The caveat of the previous paragraph concerning interactions applies here as well.

• In Chapter 13, partial canonical analysis is used to partition the variation in a table **Y** between environmental (**X**) and spatial components (**W**). A numerical example and an ecological application of partial canonical analysis are presented in that context.

2 — Tests of significance

Contrary to ordinary ordination methods (Chapter 9), eigenvalues can be tested for significance in canonical analysis, using the method of permutations (Section 1.2) because a null hypothesis can be formulated about the relationship between matrices **Y** and **X**. Tests may be used whenever one has explicit hypotheses about the relationship that may exist between the response (**Y**) and explanatory (**X**) data tables. Hypotheses may imply specific quantitative environmental variables, or mixtures of quantitative and binary variables, cast into matrix **X**, as in the numerical example of Table 11.3. Hypotheses may also be of the analysis-of-variance type, involving either a single classification criterion (factor) or several factors and their interactions, each recoded as dummy variables. Examples of the use of RDA to test multivariate hypotheses are found in Sabatier *et al.* (1989), ter Braak & Wiertz (1994), Verdonschot & ter Braak (1994), and Legendre & Anderson (1999).

Tests of significance are usually performed at two levels in RDA and CCA:

• The most general test concerns the null hypothesis of independence between **Y** and **X**. The statistic is the sum of all canonical eigenvalues; it is tested using an F ratio (eq. 11.19). The alternative hypothesis states that the sum of all canonical eigenvalues is larger (one-tailed test) than could be obtained from matrices with permuted rows, using either permutation of the raw data or permutation of residuals (below). This is an overall test of the relationship between **Y** and **X**. The sum of all canonical eigenvalues, divided by the total variation in matrix **Y**, gives the proportion of variation of **Y** explained by **X**, like a coefficient of determination (R^2) in multiple regression.

• Individual canonical eigenvalues can be tested for significance. This test differs from the overall test in that it tries to identify axes that *individually* explain more canonical variation than could be obtained from permuted matrices. Following this test, one may decide to use only the significant axes in ordination diagrams. The null hypothesis is the same as above, i.e. independence between **Y** and **X**. The alternative hypothesis is that λ_1 explains more of the variation of **Y** than matrices with permuted rows would.*

In program CANOCO, only the first canonical eigenvalue can be tested for significance (eq. 11.20). To test the second canonical eigenvalue, one must control for the effect of the first one. This is done by turning the first canonical axis into a covariable, whose effect is removed from the analysis before repeating the calculations and permutation testing. The operation may be repeated to test further canonical eigenvalues, one at a time. The program allows users to perform this operation easily.

Without covariables

In RDA and CCA, F statistics, described below, may be tested under two different frameworks (ter Braak, 1990). These frameworks, which also apply to regression analysis, generally lead to the same probabilities, although these may differ in some cases. They are compared in the study of Anderson & Legendre (1999) reported below. They are described, first, for canonical analysis without covariables:

Permutation of raw data

• Permutation of raw data — The null hypothesis is that of *exchangeability of the rows* of **Y** with respect to the observations in **X**. This is implemented by permuting the rows of matrix **Y** (or, alternatively, the rows of matrix **X**) at random and recomputing the redundancy analysis. Repeat the permutation and computation a large number of times and test the statistic for significance as in Section 1.2. When permuting the raw data, the error associated with each observation "travels with it" (Edgington, 1995).

* Release 2.1 of program CANOCO was directly testing the first eigenvalue against its permutational distribution (ter Braak, 1988c). In the 3.10 release of CANOCO (ter Braak, 1990), this was modified to an F-type statistic (eqs. 11.19 and 11.20) whereby the eigenvalue is divided by the residual error, calculated after fitting the explanatory variables (and also the covariables in partial redundancy analysis). The actual value of the F statistic is tested against the distribution of values obtained from the permutations.

Permutation of residuals

• Permutation of residuals — Here, the residuals of a linear (or other) model are the permutable units (Kempthorne, 1952). In multiple regression and canonical analysis, the null hypothesis is that of *exchangeability of the residuals* of the response variables after fitting some explanatory variables using a linear model. Tests of significance using permutation of residuals have only asymptotically exact significance levels (i.e. as n becomes large).

When there are no covariables in the study, permutation of residuals may be implemented as follows. Compute the canonical analysis of **Y** on **X** to obtain the reference value of F (eq. 11.19) for the original data. Consider matrix $\mathbf{Res}_{\mathbf{Y|X}}$ of the residuals of the regressions of variables **Y** on **X** (Fig. 11.2, bottom). Permute the rows of $\mathbf{Res}_{\mathbf{Y|X}}$ at random, obtaining $\mathbf{Res}^*_{\mathbf{Y|X}}$, and recompute the redundancy analysis of $\mathbf{Res}^*_{\mathbf{Y|X}}$ on **X**. Repeat the permutation and analysis. Test the F statistic for significance as in Section 1.2. The null hypothesis in this reduced form of the test is that an RDA of **Y** on **X** explains no more of the variation of **Y** than an RDA of the permuted residuals on **X**. This is the *permutation of residuals under a full model* without covariables, found in the central column of Table 11.7.

In the presence of covariables, there are two ways of permuting residuals: *under a reduced model* or *under a full model*. The two methods are described below.

With covariables

Tests of significance in partial canonical analysis, involving a matrix of covariables **W**, pose special problems. A partial F is used as the reference statistic when testing the significance of either the sum of all canonical eigenvalues or the first canonical eigenvalue alone (ter Braak, 1990). In the F statistic, the numerator (eigenvalue or sum of eigenvalues) is divided by the residual error. It thus becomes an *asymptotically pivotal statistic*, i.e. a statistic whose distribution under the null hypothesis remains the same for any value of the numerator, as n tends towards infinity. The importance of using a pivotal test statistic, such as F, as opposed to derivatives such as sums of squares or eigenvalues, is discussed in Section 1.2 and in Manly (1997).

Pivotal statistic

• For the sum of all canonical eigenvalues,

$$F = \frac{\text{sum of all canonical eigenvalues}/m}{\text{RSS}/(n-m-q-1)} \qquad \textbf{(11.19)}$$

where the "sum of all canonical eigenvalues" in the numerator measures the relationship between **Y** and **X** after controlling for **W**. RSS is the residual sum of squares, computed as [total inertia in **Y** after fitting the covariables, minus sum of all canonical eigenvalues after fitting the covariables]. n is the number of objects, m the number of environmental variables in **X**, and q is the number of covariables in **W**. The "total inertia in **Y** after fitting the covariables" is the sum of all eigenvalues of a non-canonical ordination (PCA or CA) of **Y** after controlling for **W**. The same statistic F is used when there are no covariables in the study; in that case $q = 0$.

To understand the construction of the F statistic, consider the simpler problem of partial regression illustrated in Fig. 10.10. The "sum of all canonical eigenvalues" in the numerator is equivalent to fraction [a] in the Figure. RSS is equivalent to [d]; it is computed as "total inertia in **Y** after fitting the covariables", which is [a + d], minus the "sum of all canonical eigenvalues after fitting the covariables", which is [a]. So, the test statistic is essentially $F = [a]/[d]$, leaving aside the degrees of freedom of the numerator and denominator.

- For the first eigenvalue,

$$F = \frac{\lambda_1}{\text{RSS}_1 / (n - m - q - 1)} \tag{11.20}$$

where RSS_1 is the residual sum of squares, computed as [total inertia in **Y** after fitting the covariables, minus λ_1].

Simulation results, reported below, indicate that, in most situations, permutation of either the raw data or residuals are adequate to carry out tests of significance in multiple regression and canonical analysis. Restricted permutation, which is briefly described in Subsection 1.2.4, is also appropriate if **W** contains a qualitative variable; the permutations are then limited to the objects within the groups defined by that variable (Edgington, 1995; Manly, 1997; Subsection 1.2.4).

In permutation of raw data, the variance associated with the permuted data corresponds to all the variance in matrix **Y**; this is fraction [a + b + c + d] of Fig. 10.10 and Table 11.7. Compare this to the variance associated with the permuted portions in the methods of permutation of residuals (below).

Two methods of permutation of residuals may be used to test the significance of the sum of all canonical eigenvalues (eq. 11.19). These procedures can readily be adapted to test the first eigenvalue only, using eq. 11.20. They may also be used to test the significance of partial regression coefficients in multiple regression, as shown by the simulations reported below, after the description of the methods.

1) Permutation of residuals under a reduced model — This method is ter Braak's adaptation to canonical analysis of the permutation method proposed by Freedman & Lane (1983) for partial regression coefficients. The method is called permutation "under the null model" by ter Braak (1990), or "under the reduced model" by Cade and Richards (1996). The "reduced regression model" contains only the covariables of **W**.

1. Compute the canonical analysis of **Y** on **X** and **W** together to obtain the reference value of the F statistic (eq. 11.19) for the unpermuted data.

2. Compute matrix $\mathbf{Fit}_{\mathbf{Y|W}}$ of the fitted values and matrix $\mathbf{Res}_{\mathbf{Y|W}}$ of the residuals of the regressions of variables **Y** on **W**.

3. Permute the rows of $\mathbf{Res}_{\mathbf{Y|W}}$ to obtain $\mathbf{Res}^*_{\mathbf{Y|W}}$. Compute matrix $\mathbf{Ynew} = \mathbf{Fit}_{\mathbf{Y|W}} + \mathbf{Res}^*_{\mathbf{Y|W}}$. The values of $\mathbf{Fit}_{\mathbf{Y|W}}$ remain fixed (i.e. unpermuted).

4. Compute the canonical analysis of **Ynew** against **X** and **W** together to obtain a value for the F statistic under permutation (called F^*) using eq. 11.19.

5. Repeat steps 3 and 4 a large number of times to obtain the distribution of F^*. Add the reference value of F to the distribution.

6. Calculate the probability as in Section 1.2.

In this method, the variance associated with the permuted portion of the data, $\mathbf{Res}_{\mathbf{Y}|\mathbf{W}}$, is fraction [a + d] of Fig. 10.10 and Table 11.7.

2) Permutation of residuals under a full model — This method was developed by ter Braak (1990, 1992) to conduct tests of significance in the CANOCO program of canonical analysis, version 3 and later. He was inspired by results obtained by Hall & Titterington (1989) in the context of bootstrapping; bootstrapping is briefly described at the end of Section 1.2. The "full regression model" contains all the explanatory variables and covariables of **X** and **W**. Proceed as follows:

1. Compute the canonical analysis of **Y** on **X** and **W** together to obtain the reference value of the F statistic (eq. 11.19) for the unpermuted data, as well as matrix $\mathbf{Fit}_{\mathbf{Y}|\mathbf{XW}}$ of the fitted values and matrix $\mathbf{Res}_{\mathbf{Y}|\mathbf{XW}}$ of the residuals.

2. Permute the rows of $\mathbf{Res}_{\mathbf{Y}|\mathbf{XW}}$ to obtain $\mathbf{Res}^*_{\mathbf{Y}|\mathbf{XW}}$. Compute matrix **Ynew** = $\mathbf{Fit}_{\mathbf{Y}|\mathbf{XW}} + \mathbf{Res}^*_{\mathbf{Y}|\mathbf{XW}}$. The values of $\mathbf{Fit}_{\mathbf{Y}|\mathbf{XW}}$ remain fixed (i.e. unpermuted).

3. Compute the canonical analysis of **Ynew** against **X** and **W** together (unpermuted) to obtain the residual sum of squares $\text{RSS}_{\mathbf{XW}}$ computed as [total inertia in **Ynew** – sum of all canonical eigenvalues].

4. Compute the canonical analysis of **Ynew** against **W** alone (unpermuted) to obtain the residual sum of squares $\text{RSS}_{\mathbf{W}}$ computed as [total inertia in **Ynew** – sum of all canonical eigenvalues].

5. Calculate a value for the F statistic under permutation (called F^*) as follows (ter Braak, 1992):

$$F^* = \frac{(\text{RSS}_{\mathbf{W}} - \text{RSS}_{\mathbf{XW}})/m}{\text{RSS}_{\mathbf{XW}}/(n-m-q-1)}$$

To understand the construction of the F^* statistic, consider Fig. 10.10 again. $\text{RSS}_{\mathbf{W}}$ corresponds to [a + d] whereas $\text{RSS}_{\mathbf{XW}}$ is fraction [d]. So $(\text{RSS}_{\mathbf{W}} - \text{RSS}_{\mathbf{XW}})$ in the numerator is [a] and $\text{RSS}_{\mathbf{XW}}$ in the denominator is [d].

6. Repeat steps 2 to 5 a large number of times to obtain the distribution of F^*. Add the reference value of F to the distribution.

7. Calculate the probability as in Section 1.2.

In this method, the variance associated with the permuted portion of the data, $\mathbf{Res}_{\mathbf{Y}|\mathbf{XW}}$, is $\text{RSS}_{\mathbf{XW}}$ which corresponds to fraction [d] of Fig. 10.10 and Table 11.7.

Which methods are adequate for permutation tests? Using Monte Carlo simulations, Anderson & Legendre (1999) compared empirical type I error and power of different permutation techniques for a test of significance of a single partial regression coefficient. Their results are relevant to canonical analysis because partial RDA with a single variable in matrices **Y**, **X** and **W** is identical to a multiple regression; permutation tests of significance carried out on such data using multiple regression or partial RDA are strictly equivalent. As a consequence, methods that are inappropriate to test single partial regression coefficients in multiple regression would also be inadequate for partial canonical analysis.

Anderson & Legendre limited their study to methods that maintain the covariance structure between **X** and **W** constant throughout the permutations; this property is called *ancillarity*, which means *relatedness*. They compared permutation of the raw data to three methods of permutation of residuals. Two of these are the methods of permutation of residuals under a reduced and a full model, described above. The normal-theory *t*-test was also included in the comparison. Their study showed that

• when the error in the data strongly departed from normality, permutation tests had more power than parametric *t*-tests;

• type I error and power were asymptotically equivalent for permutation of raw data or permutation of residuals under the reduced or the full model;

• when the covariable contained an extreme outlier, permutation of raw data resulted in unstable (often inflated) type I error. The presence of outliers in the covariable did not adversely affect the tests based on permutation of residuals. Thus, permutation of raw data cannot be recommended when the covariable contains outliers; permutation of residuals should be used.

The method of permutation of residuals under a reduced model (Freedman & Lane, 1983) described above is closely related to another method proposed by Kennedy (1995) to test partial regression coefficients. Although the two methods, under permutation, give the same value for the estimate of a partial regression coefficient, they do not give the same value for the *t* statistic. Tests of regression coefficients by permutation require, however, the use of a pivotal statistic when covariables are involved (Kennedy 1995, Manly 1997). In their simulation study, Anderson & Legendre (1999) found that the Kennedy method had inflated type I error, especially with small sample sizes. The reason for the discrepancy between the two methods is described in more detail in Anderson & Legendre (1999). The method of Kennedy is similar to the alternative procedure proposed by Smouse *et al.* (1986) for partial Mantel tests, described in Subsection 10.5.2; the latter is not recommended there, just as the Kennedy method is not recommended here.

Methods of permutation of raw data or residuals are compared in Table 11.7 in terms of the permuted portions of variation, in the presence or absence of a matrix of covariables **W**. *Without covariables*, permutation of raw data involves fraction [a + d] whereas permutation of residuals involves [d]. No residual can be computed under a reduced model in the absence of covariables; the method becomes a permutation of raw data. *With covariables*, permutation of residuals may involve the residuals of a reduced model of the covariables only (fraction [a + d]), or the residuals of a full model of the explanatory variables and covariables, in which case the permutation involves fraction [d].

Table 11.7 Tests of statistical significance in canonical analysis. Comparison of the methods of permutation of raw data or residuals in terms of the permuted fractions of variation, in the presence or absence of a matrix of covariables **W**. Fractions of variation are noted as in Fig. 10.10: [a] is the variation of matrix **Y** explained by **X** alone, [c] the variation explained by **W** alone, [b] the variation explained jointly by **X** and **W**, and [d] the residual variation.

	Without covariables	With matrix **W** of covariables
	[a] Explained by **X** ; [d] Unexplained variation	[a] [b] Explained by **X** ; [b] [c] Explained by **W** ; [d] Unexplained variation
Permute raw data	Permute [a + d]	Permute [a + b + c + d][1]
Permute residuals:		
• reduced model	Equivalent to permuting raw data	Permute [a + d]
• full model	Permute [d]	Permute [d]

[1] Permutation of raw data may result in unstable (often inflated) type I error when the covariable contains outliers. This does not occur, however, when using restricted permutations of raw data within groups of a qualitative covariable, which gives an exact test. See text.

11.4 Canonical correlation analysis (CCorA)

Canonical correlation analysis (CCorA; Hotelling, 1936), differs from redundancy analysis (RDA) in the same way as linear correlation differs from simple linear regression. In CCorA, the two matrices under consideration are treated in a symmetric way whereas, in RDA, the **Y** matrix is considered to be dependent on an explanatory matrix **X**. The algebraic consequence is that, in CCorA, the matrix whose eigenvalues and eigenvectors are sought (eq. 11.22) is constructed from all four parts of eq. 11.2 whereas, in the asymmetric RDA, eq. 11.3 does not contain the $\mathbf{S}_{YY}$ portion.

The brief discussion below is only meant to show the general principles of the CCorA method. Ecologists interested in delving deeper into the method will find detailed accounts of the theory in Kendall & Stuart (1966), and computation procedures in Anderson (1958). Gittins (1985) presents a comprehensive review of the theory and applications of CCorA in ecology. Now that RDA (Section 11.1) and CCA (Section 11.2) are available, CCorA has limited applications; RDA and CCA correspond better than CCorA to the way most two-matrix problems are formulated.

In CCorA, the objects (sites) under study are described by two sets of quantitative descriptors; for example, a first set $\mathbf{Y}_1$ of p chemical and a second set $\mathbf{Y}_2$ of m geomorphological descriptors of the sampling sites; or, a first set $\mathbf{Y}_1$ of p species and a second set $\mathbf{Y}_2$ of m descriptors of the physical environment. The dispersion matrix $\mathbf{S}$ of these $p + m$ descriptors is therefore made of four blocks, as in eq. 11.2:

$$\mathbf{S} = \begin{bmatrix} \mathbf{S}_{11} & \mathbf{S}_{12} \\ \mathbf{S'}_{12} & \mathbf{S}_{22} \end{bmatrix} \qquad \textbf{(11.21)}$$

The algebra that follows applies equally well to $\mathbf{S}$ matrices defined as variance-covariance matrices (e.g. $\mathbf{S_{YY}} = (1/(n-1))\,\mathbf{Y'Y}$) or matrices of sums of squares and cross products (e.g. $\mathbf{S_{YY}} = \mathbf{Y'Y}$). Submatrices $\mathbf{S}_{11}$ (order $p \times p$) and $\mathbf{S}_{22}$ (order $m \times m$), refer, respectively, to one of the two sets of descriptors, whereas $\mathbf{S}_{12}$ (order $p \times m$) and its transpose $\mathbf{S'}_{12} = \mathbf{S}_{21}$ (order $m \times p$) account for the interactions between the two sets of descriptors. Numbers (1, 2) are used here to designate matrices, instead of letters (**X**, **Y**) as in eq. 11.2, to emphasize the fact that the two data matrices ($\mathbf{Y}_1$, $\mathbf{Y}_2$) play equivalent roles in CCorA.

The problem consists in maximizing the between-set dispersion with respect to the within-set dispersion. The expression to be optimized is $\mathbf{S}_{12}\mathbf{S}_{22}^{-1}\mathbf{S'}_{12}\mathbf{S}_{11}^{-1}$ since $\mathbf{S}_{12}\mathbf{S'}_{12}/\mathbf{S}_{11}\mathbf{S}_{22}$ does not exist in matrix algebra. Finding solutions to this optimization problem calls for eigenvalues and eigenvectors. Canonical correlations are obtained by solving the characteristic equation:

$$\left|\mathbf{S}_{12}\mathbf{S}_{22}^{-1}\mathbf{S'}_{12}\mathbf{S}_{11}^{-1} - \lambda_k\mathbf{I}\right| = 0 \qquad \textbf{(11.22)}$$

which corresponds to one of the following equations, resulting from the multiplication of both members of eq. 11.22 by either $\mathbf{S}_{11}$ or $\mathbf{S}_{22}$:

$$\left|\mathbf{S}_{12}\mathbf{S}_{22}^{-1}\mathbf{S'}_{12} - \lambda_k\mathbf{S}_{11}\right| = 0 \qquad \textbf{(11.23)}$$

or

$$\left|\mathbf{S'}_{12}\mathbf{S}_{11}^{-1}\mathbf{S}_{12} - \lambda_k\mathbf{S}_{22}\right| = 0 \qquad \textbf{(11.24)}$$

Canonical correlations r_k are the square roots of the eigenvalues λ_k ($\lambda_k = r_k^2$). The same λ_k values are found using either equation. The next step is to calculate the eigenvectors of the two equation systems, corresponding to each eigenvalue. The two eigenvectors give the linear combinations of the two sets of original descriptors ($\mathbf{Y}_1$, $\mathbf{Y}_2$) corresponding to each eigenvalue. For eigenvalue λ_k, the eigenvectors $\mathbf{u}_k$ and $\mathbf{v}_k$ are computed using the following matrix equations:

$$(\mathbf{S}_{12}\mathbf{S}_{22}^{-1}\mathbf{S'}_{12} - \lambda_k\mathbf{S}_{11})\,\mathbf{u}_k = 0 \qquad \textbf{(11.25)}$$

and

$$(\mathbf{S'}_{12}\mathbf{S}_{11}^{-1}\mathbf{S}_{12} - \lambda_k\mathbf{S}_{22})\,\mathbf{v}_k = 0 \qquad \textbf{(11.26)}$$

For convenience, the eigenvectors are normalized, as in discriminant analysis (Section 11.5), by dividing each $\mathbf{u}_k$ by the scalar resulting from $(\mathbf{u'}_k\mathbf{S}_{11}\mathbf{u}_k)^{1/2}$ and each $\mathbf{v}_k$ by $(\mathbf{v'}_k\mathbf{S}_{22}\mathbf{v}_k)^{1/2}$, which makes the variances of the canonical variates equal to 1. The two matrices of canonical ordination scores,

$$\mathbf{T}_u = \mathbf{Y}_1\mathbf{U} = [y_{i1}u_{1k} + \ldots + y_{ip}u_{pk}] \text{ for matrix } \mathbf{Y}_1 \text{ with } p \text{ descriptors} \qquad \textbf{(11.27)}$$

and

$$\mathbf{T}_v = \mathbf{Y}_2\mathbf{V} = [y_{i1}v_{1k} + \ldots + y_{im}v_{mk}] \text{ for matrix } \mathbf{Y}_2 \text{ with } m \text{ descriptors,} \qquad \textbf{(11.28)}$$

contain the coordinates (ordination scores) of the objects in the two systems of principal axes. Column vectors (ordination scores) in $\mathbf{T}_u$ are uncorrelated with one another; the same is true for the column vectors in $\mathbf{T}_v$.

CCorA produces a total of $\max[p, m]$ eigenvalues and canonical correlations. When $p + m < n - 1$, if p is greater than m, there are p eigenvalues but $(p - m)$ of them are null. If the m positive eigenvalues are distinct, there is only one possible pair of eigenvectors $\mathbf{u}_k$ and $\mathbf{v}_k$ for each of them. When $p + m \geq n - 1$, $p + m - (n - 1)$ of the eigenvalues (and canonical correlations) are equal to 1; some computer programs may refuse to carry out the calculations in such a case. CCorA cannot handle matrices with more variables (columns) in any one of the sets than there are sites (rows) minus 1 because covariance matrices $\mathbf{S}_{11}$ and $\mathbf{S}_{22}$ must be inverted (eq. 11.22). This is a commonly encountered problem with species abundance tables; rare species must often be dropped from the analysis to satisfy the requirements of the method.

When one of the matrices only contains one descriptor ($p = 1$ and $m > 1$, for example), there is only one positive eigenvalue. The canonical correlation problem reduces to the problem of finding the linear combination of variables in $\mathbf{Y}_2$ that is maximally correlated with the single variable $\mathbf{y}_1$; this is simply a problem of multiple regression (Subsection 10.3.3). The general equation for eigenvalues (eq. 11.22) then simplifies to:

$$\lambda = r^2 = \mathbf{s}_{12}\mathbf{S}_{22}^{-1}\mathbf{s'}_{12} / s_1^2 \qquad \textbf{(11.29)}$$

where $\mathbf{s}_{12}$ is a vector of covariances. This equation corresponds to that of multiple correlation (eq. 4.31), previously given in terms of $\mathbf{r}$ instead of $\mathbf{s}$. Finally, when the two sets contain only one descriptor each ($p = m = 1$), eq. 11.29 becomes:

$$\lambda = r^2 = s_{12}s_{22}^{-1}s_{12}s_{11}^{-1} = \frac{(s_{12})^2}{s_{11}s_{22}} = \frac{(s_{12})^2}{s_1^2 s_2^2} \qquad \textbf{(11.30)}$$

which is the formula for the square of Pearson's simple linear correlation (eq. 4.7).

If canonical correlations are computed from a correlation matrix instead of a dispersion matrix, the interpretation must take into account the fact that the canonical ordination scores now concern standardized descriptors instead of the original ones.

The interpretation of canonical correlation analyses is more difficult than that of other multidimensional analyses. The main use of this technique is to explore the structure of multidimensional data and reveal the correlations that can be found between linear functions of two groups of descriptors (Kendall & Stuart, 1966). The detailed study of pairs of eigenvectors is usually restricted to the first few canonical correlations, although Blackith & Reyment (1971) give an example taken from Blackith & Albrecht (1959) where the lowest canonical correlations were of interest; the corresponding canonical eigenvectors made it possible to isolate a "phase" vector in locusts which was independent of the "size" vector. When using CCorA, one should remember that high canonical correlations do not necessarily mean that the corresponding vectors of ordination scores $\mathbf{T}_u$ and $\mathbf{T}_v$ explain a large fraction of the variation in $\mathbf{Y}_1$ or $\mathbf{Y}_2$. *Redundancy coefficients* are used in CCorA to measure the proportion of the variance of $\mathbf{Y}_1$ (or $\mathbf{Y}_2$) which is explained by a linear combination of the variables in $\mathbf{Y}_2$ (or $\mathbf{Y}_1$); they should always be computed together with canonical correlations to help interpret them.

Ecological application 11.4

Some authors produced interesting results using canonical correlations. For example, Aart & Smeenk-Enserink (1975) obtained high canonical correlations between the eigenvectors of species and those that summarized the environmental descriptors, in a study of the spatial distribution of lycosid spiders in the Netherlands. The survey analysed in this study was conducted 10 years after that of Ecological application 9.1b. This classical study is an early application of canonical analysis to ecological problems. Nowadays, the analysis would likely be carried out using RDA (Section 11.1) or CCA (Section 11.2), to account for the fact that the species are the response variables in this problem; this question is discussed in Section 11.6. The Aart & Smeenk-Enserink spider data set has been reanalysed by ter Braak (1986) using CCA.

At the 28 sites included in the analysis, the descriptors were the abundances of 12 spider species, normalized by logarithmic transformation $\log(y + 1)$, and 15 environmental variables characterizing the light, vegetation, and soil. Among all the environmental descriptors that had been observed, only those that were linearly correlated with the species descriptors were chosen, so as to ensure the linearity of the relationship between the two sets of descriptors. Calculations were conducted on the correlation matrix instead of the covariance matrix; this was equivalent to standardizing the descriptors before the analysis.

Results show strong correlations between the canonical ordination vectors produced by the two sets of descriptors: the first four canonical correlations are larger than 0.98. The authors found a high resemblance between the contributions of the 12 species to the first two canonical axes (Fig. 11.7a) and the first two principal components (not illustrated here: practically the same graph). The results led them to conclude that the principal axes of species variation could be interpreted from the contributions of the environmental descriptors (Fig. 11.7b) to the canonical axes.

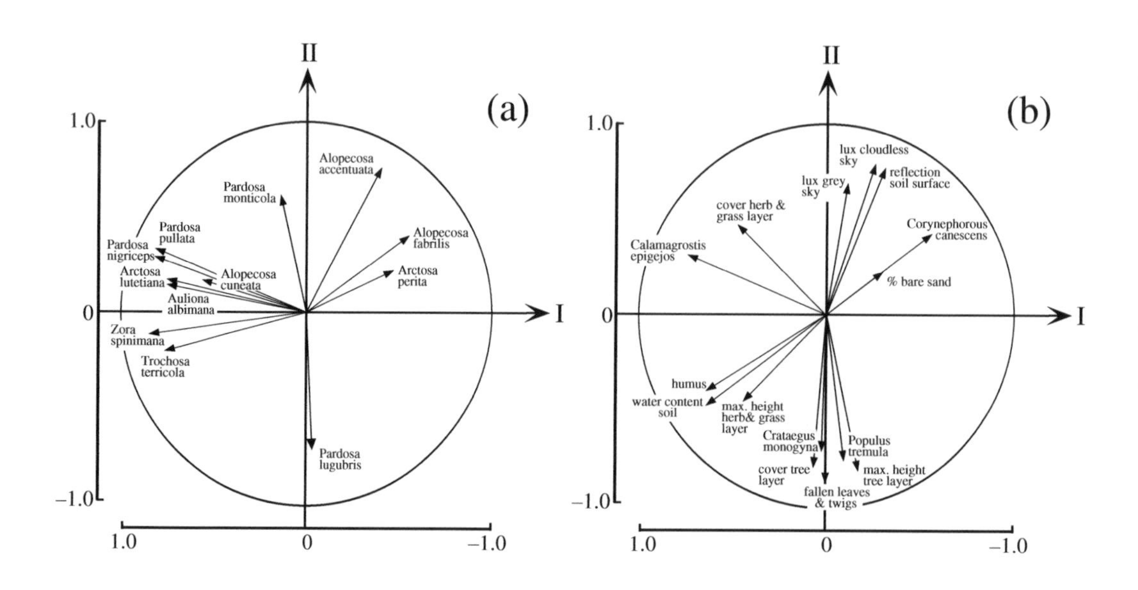

Figure 11.7 Contribution of the descriptors to the first two canonical axes. (a) The 12 species of spiders. (b) The 15 environmental descriptors. The figure was redrawn using data published by the authors. Adapted from Aart & Smeenk-Enserink (1975).

Other interesting applications of CCorA to ecology are presented by Gittins (1985), van der Meer (1991), and Varis (1991).

Dolédec & Chessel (1994) proposed a method called *co-inertia analysis*. This is a form of CCorA which maximizes the covariance between the two sets of projected object scores, instead of the correlation. They applied the method to study the relationships between macroinvertebrate communities in ponds, on the one hand, and environmental variables on the other.

11.5 Discriminant analysis

A usual step in ecological analysis is to start with an already known grouping of the objects (considered to be a qualitative response variable **y** in this form of analysis) and try to determine to what extent a set of quantitative descriptors (seen as the explanatory variables **X**) can actually explain this grouping. In this type of analysis, the grouping is known at the start of the analysis. It may be the result of a cluster analysis computed from a *different* data set, or reflect an ecological hypothesis to be tested. The problem thus no longer consists in delineating groups, as in cluster analysis, but in interpreting them.

Discriminant analysis is a method of linear modelling, like analysis of variance, multiple linear regression, and canonical correlation analysis. It proceeds in two steps. (1) First, one tests for differences in the explanatory variables (**X**), among the predefined groups. This part of the analysis is identical to the overall test performed in MANOVA. (2) If the test supports the alternative hypothesis of significant differences among groups in the **X** variables, the analysis proceeds to find the linear combinations (called *discriminant functions* or *identification functions*) of the **X** variables that best discriminate among the groups.

Like one-way analysis of variance, discriminant analysis considers a single classification criterion (i.e. division of the objects into groups) and allows one to test whether the explanatory variables can discriminate among the groups. Testing for differences among group means in discriminant analysis is identical to ANOVA for a single explanatory variable and to MANOVA for multiple variables (**X**).

When it comes to modelling, i.e. finding the linear combinations of the variables (**X**) that best discriminate among the groups, discriminant analysis is a form of "inverse analysis" (ter Braak, 1987b), where the classification criterion is considered to be the response variable (**y**) whereas the quantitative variables are explanatory (matrix **X**). In ANOVA, on the contrary, the objective is to account for the variation in a response quantitative descriptor **y** using one or several classification criteria (explanatory variables, **X**).

Like multiple regression, discriminant analysis estimates the parameters of a linear model of the explanatory variables which may be used to forecast the response variable (states of the classification criterion). While inverse multiple regression would be limited to two groups (expressed by a single binary variable **y**), discriminant analysis can handle several groups. Discriminant analysis is a canonical method of analysis; its link to canonical correlation analysis (CCorA) will be explained in Subsection 1, after some necessary concepts have been introduced.

After the overall test of significance, the search for discriminant functions may be conducted with two different purposes in mind. One may be interested in obtaining a linear equation to allocate new objects to one of the states of the classification criterion (identification), or simply in determining the relative contributions of various explanatory descriptors to the distinction among these states (discrimination).

Discriminant analysis is also called *canonical variate analysis* (CVA). The method was originally proposed by Fisher (1936) for the two-group case ($g = 2$). Fisher's results were extended to $g \geq 2$ by Rao (1948, 1952). Fisher (1936) illustrated the method using a famous data set describing the morphology (lengths and widths of sepals and petals) of 150 specimens of irises (Iridaceae) belonging to three species. The data had originally been collected in the Gaspé Peninsula, eastern Québec

(Canada), by the botanist Edgar Anderson of the Missouri Botanical Garden who allowed Fisher to publish and use the raw data. Fisher showed how to use these morphological measurements to discriminate among the species. The data set is sometimes — erroneously — referred to as "Fisher's irises".

The analysis is based upon an explanatory data matrix **X** of size $(n \times m)$, where n objects are described by m descriptors. **X** is meant to discriminate among the groups defined by a separate classification criterion vector (**y**). As in regression analysis, the explanatory descriptors must in principle be quantitative, although qualitative descriptors coded as dummy variables may also be used (Subsection 1.5.7). Other methods are available for discrimination using non-quantitative descriptors (Table 10.1). The objects, whose membership in the various groups of **y** is known before the analysis is undertaken, may be sites, specimens, quadrats, etc.

One possible approach would be to examine the descriptors one by one, either by hand or using analyses of variance, and to note those which have states that characterize one or several groups. This information could be transformed into an identification key, for example. It often occurs, however, that no single descriptor succeeds in separating the groups completely. The next best approach is then to search for a linear combination of descriptors that provides the most efficient discrimination among groups. Figure 11.8 shows an idealized example of two groups (A and B) described by two descriptors only. The groups cannot be separated on either of the two axes taken alone. The solution is a new discriminant descriptor **z**, drawn on the figure, which is a linear combination of the two original descriptors. Along **z**, the two groups of objects are perfectly separated. Note that discriminant axis **z** is parallel to the direction of greatest variability *between groups*. This suggests that the weights u_j used in the discriminant function could be the elements of the eigenvectors of a between-group dispersion matrix. The method can be generalized to several groups and many descriptors.

Discriminant function

• *Discriminant functions* (also called standardized discriminant functions) are computed from *standardized descriptors*. The coefficients of these functions are used to assess the relative contributions of the descriptors to the final discrimination.

Identification function

• *Identification functions* (also called unstandardized discriminant functions) are computed from the *original descriptors* (not standardized). They may be used to compute the group to which a new object is most likely to belong. Discriminant analysis is seldom used for this purpose in ecology; it is widely used in that way in taxonomy.

When there are only two groups of objects, the method is called *Fisher's*, or *simple discriminant analysis* (a single function is needed to discriminate between two clusters), whereas the case with several groups is called *multiple discriminant analysis* or *canonical variate analysis*. The simple discriminant analysis model (two groups) is a particular case of multiple discriminant analysis, so that it will not be developed here. The solution can be entirely derived from the output of a multiple regression using a

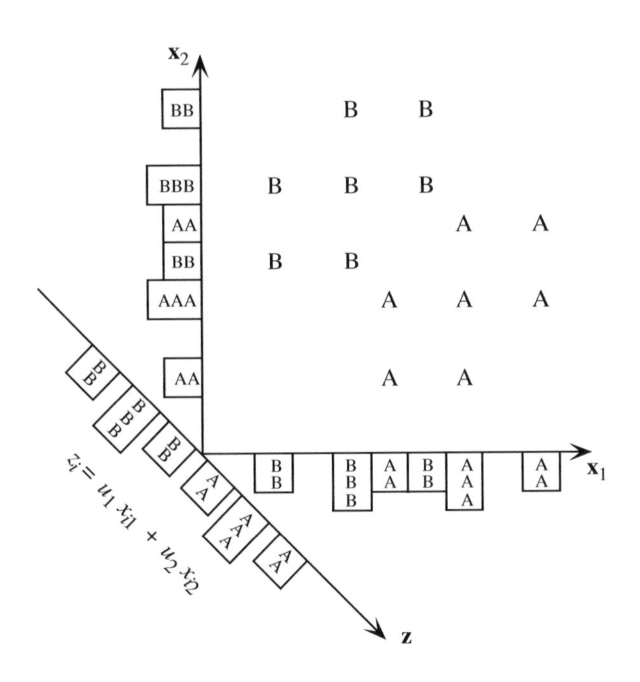

Figure 11.8 Two groups, A and B, with 6 objects each, cannot be separated on either descriptor-axis $\mathbf{x}_1$ or $\mathbf{x}_2$ (histograms on the axes). They are perfectly separated, however, by a discriminant axis **z**. The position of each object i is calculated along **z** using the equation $z_i = (\cos 45^\circ)\, x_{i1} - (\cos 45^\circ)\, x_{i2}$. Adapted from Jolicoeur (1959).

dummy variable defining the two groups (used as the dependent variable **y**) against the table of explanatory variables **X**.

Analysis of variance is often used for screening variables prior to discriminant analysis: each variable in matrix **X** is tested for its capacity to discriminate among the groups of the classification criterion **y**. Figure 11.8 shows however that there is a danger in this approach; any single variable may not discriminate groups well although it may have high discriminating power in combination with other variables. One should be careful when using univariate analysis to eliminate variables. If the analysis requires that poorly discriminating variables be eliminated, one should use stepwise discriminant analysis instead, which allows users to identify a subset of good discriminators. Bear in mind, though, that stepwise selection of explanatory variables does not guarantee that the "best" set of explanatory variables is necessarily going to be found. This is equally true in discriminant analysis and regression analysis (Subsection 10.3.3).

Table 11.8 Discriminant analysis is computed on either dispersion matrices (right-hand column) or matrices of sums of squares and cross-products (centre). Matrices in the right-hand column are simply those in the central column divided by their respective numbers of degrees of freedom. The size of all matrices is $(m \times m)$.

	Matrices of sums of squares and cross-products	Dispersion matrices
Total dispersion	$\mathbf{T}$	$\mathbf{S} = \mathbf{T}/n-1$
Pooled within-group dispersion	$\mathbf{W} = \mathbf{W}_1 + \ldots + \mathbf{W}_g$	$\mathbf{V} = \mathbf{W}/n-g$
Among-group dispersion	$\mathbf{B} = \mathbf{T} - \mathbf{W}$	$\mathbf{A} = \mathbf{B}/g-1$

1 — The algebra of discriminant analysis

The problem consists in finding linear combinations of the discriminant descriptors in matrix **X** that maximize the differences among groups while minimizing the variation within the groups. As in regression analysis, the descriptors must be quantitative or binary since they are combined into a linear function. If necessary, they must have already been transformed to meet the condition of multinormality. The discriminant analysis model is robust to departures from this condition, but the statistical tests assume *within-group normality* of each descriptor.

Computations are carried out on either dispersion matrices or matrices of sums of squares and cross-products of centred descriptors (Table 11.8). These matrices are constructed much in the same way as in the analysis of variance. **T** is the matrix of scalar products of the centred descriptors, $[x - \bar{x}]$, for the all objects irrespective of the groups: $\mathbf{T} = [x - \bar{x}]'[x - \bar{x}]$ (total sums of squares and cross-products). When divided by the total number of degrees of freedom $n - 1$, it becomes the total dispersion matrix **S** used in principal component analysis.

Matrix **W**, which pools the sums of squares within all groups, is computed by adding up matrices $\mathbf{W}_1$ to $\mathbf{W}_g$ of the sums of squares and cross-products for each of the g groups. Each matrix $\mathbf{W}_j$ is computed from descriptors that have been *centred for the objects of that group only*, just as in ANOVA. In other words, matrix $\mathbf{W}_j$ is the product $[x - \bar{x}]'[x - \bar{x}]$ for the objects that belong to group j only. Dividing the pooled within-group matrix **W** by the within-group number of degree of freedom, $n - g$, produces the pooled within-group dispersion matrix **V**.

Matrix **B** of the sums of squares among groups is computed by subtracting the pooled within-group matrix **W** from the total matrix of sums of squares **T**. Since

$\mathbf{B} = \mathbf{T} - \mathbf{W}$, the number of degrees of freedom by which $\mathbf{B}$ must be divided to obtain the among-group dispersion matrix $\mathbf{A}$ is: $(n - 1) - (n - g) = g - 1$.

The solution to the problem of maximizing the variation among groups while minimizing that within groups calls for the eigenvalues and eigenvectors of a matrix corresponding to the ratio of the among-group dispersion $\mathbf{A}$ to the pooled within-group dispersion $\mathbf{V}$. Since $\mathbf{A}/\mathbf{V}$ does not exist in matrix algebra, $\mathbf{V}^{-1}\mathbf{A}$ is computed instead. The maximization problem is stated by the following matrix equation:

$$(\mathbf{V}^{-1}\mathbf{A} - \lambda_k\mathbf{I})\,\mathbf{u}_k = \mathbf{0} \qquad \textbf{(11.31)}$$

which has the same form as the basic equation for principal component analysis (eq. 9.1). Equation 11.24 indicates that the eigenvectors $\mathbf{u}_k$ will not be orthogonal in the reference system of the original descriptors. Indeed, matrix $\mathbf{V}^{-1}\mathbf{A}$ from which the eigenvectors are calculated is not symmetric; this condition leads to non-orthogonal eigenvectors (Section 2.9). If both members of eq. 11.31 are premultiplied by $\mathbf{V}$, it becomes:

$$(\mathbf{A} - \lambda_k\mathbf{V})\,\mathbf{u}_k = \mathbf{0} \qquad \textbf{(11.32)}$$

The number of discriminant axes needed for the ordination of g groups is $(g - 1)$.

When the non-orthogonal eigenvectors are plotted at right angles, they straighten the reference space and, with it, the ellipsoids of the within-group scatters of objects. As a result, if the eigenvectors are normalized in an appropriate manner, the within-group scatters of objects can be made circular (Fig. 11.9), insofar as the within-group cross-product matrices $\mathbf{W}_j$ are homogeneous (same dispersion in all groups). This result is obtained by dividing each eigenvector $\mathbf{u}_k$ by the square root of $\mathbf{u}'_k\mathbf{V}\mathbf{u}_k$ which is a scalar. In matrix form, this operation is:

$$\mathbf{C} = \mathbf{U}\,(\mathbf{U}'\mathbf{V}\mathbf{U})^{-1/2} \qquad \textbf{(11.33)}$$

Matrix $\mathbf{C}$ contains the normalized eigenvectors defining the *canonical space* of the discriminant analysis. After this transformation, the variance among group centroids is maximized even if the group dispersion matrices are not homogeneous. This leads to the conclusion that the principal axes describe the dispersion *among groups*. The first principal axis indicates the direction of largest variation among group centroids, and so on for the successive canonical axes, after the reference space has been straightened up to make each group spherical. The SAS and STATISTICA packages, among others, offer the normalization of eq. 11.33.

Other methods for normalizing the eigenvectors are found in the literature: to length 1 or $\sqrt{\lambda}$. Some statistical packages unfortunately compute the positions of the objects along the canonical axes (matrix $\mathbf{F}$, eq. 11.39) directly from matrix $\mathbf{U}$ of the eigenvectors normalized to length 1. In that case, the group dispersions remain nonspherical; it is then difficult to compare the eigenvectors because they describe a

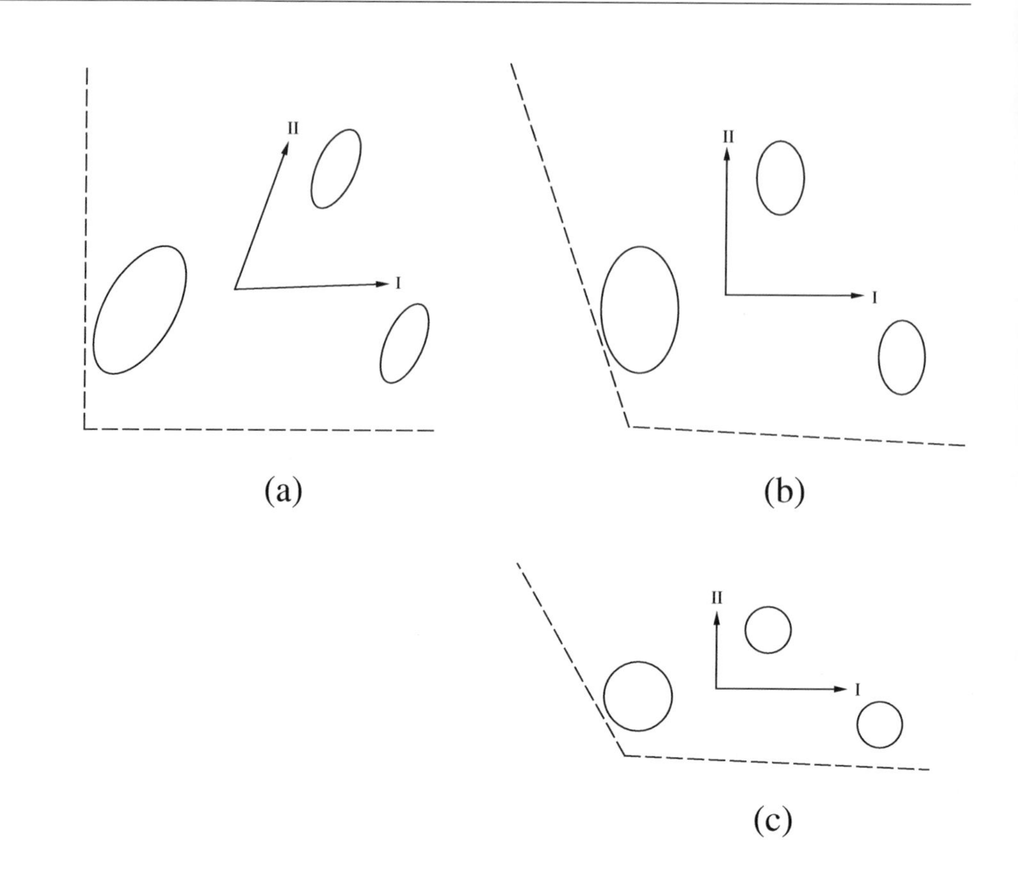

Figure 11.9 Basic principles of multiple discriminant analysis. Dashed: two original descriptors. Full lines: canonical axes. The within-group dispersion matrices are homogeneous in this example. (a) The canonical axes are not orthogonal in the reference space of the original descriptors. (b) When they are used as the orthogonal reference space, the ellipsoids of the within-group scatters of objects are straightened up. (c) Normalizing the eigenvectors by dividing each $\mathbf{u}_k$ by $(\mathbf{u}'_k \mathbf{V} \mathbf{u}_k)^{1/2}$ makes the within-group dispersions circular if they are homogeneous.

combination of within-group and among-group dispersion. It is not always easy to understand, from the documentation, what a specific statistical program does. A simple way to decide what kind of normalization is used, is to run the small example presented in Subsection 11.5.3.

Spherical within-group dispersions are obtained only if the condition of homogeneity of the within-group dispersion matrices is fulfilled. Even if discriminant analysis is moderately robust to departures from this condition, it remains advisable to examine whether this condition is met. Several statistics have been developed to test the hypothesis of homogeneity of the within-group dispersion matrices. One of them is Kullback's statistic (1959) which is approximately distributed as χ^2:

$$\chi^2 = \sum_{j=1}^{g} \frac{(n_j - 1)}{2} \ln \frac{|\mathbf{V}|}{|\mathbf{V}_j|} \tag{11.34}$$

with $(g - 1)\, m\, (m + 1)/2$ degrees of freedom, where n_j is the number of objects in group j, $|\mathbf{V}|$ is the determinant of the pooled within-group dispersion matrix **V**, and $|\mathbf{V}_j|$ is the determinant of the within-group dispersion matrix of group j. When the test value is larger than the critical χ^2 value, the hypothesis of homogeneity is rejected.

Another useful test statistic is Wilks' Λ (lambda) (1932), which measures to what extent the groups differ in the positions of their centroids. Λ is computed as the ratio of the determinants of the matrices of sums of squares and cross-products **W** and **T**:

$$\Lambda = \frac{|\mathbf{W}|}{|\mathbf{T}|} \tag{11.35}$$

This ratio may assume values in the range from 0 (maximum dispersion of the centroids) to 1 (no dispersion among groups). It can be transformed to a χ^2 statistic with $m(g - 1)$ degrees of freedom (Bartlett, 1938):

$$\chi^2 = -\left[(n-1) - \frac{1}{2}(m+g) \right] \ln \Lambda \tag{11.36}$$

Wilks' Λ can also be transformed into an F statistic following Rao (1951). It is a generalization of Student's t-test to several groups and several explanatory variables. Another multidimensional generalization of t, for two groups, is Hotelling's T^2, which has been discussed with reference to the Mahalanobis generalized distance (eqs. 7.40 and 7.41).

As explained above, discrimination among g groups requires a maximum of $(g - 1)$ discriminant functions. To test the significance of the $(g - k - 1)$ eigenvalues that remain after examining the first k, Wilks' ratio is computed as the product:

$$L = \prod_{j=k+1}^{g-1} \frac{1}{1 + \lambda_j} \tag{11.37}$$

Transformation of this statistic to χ^2, as above (eq. 11.36), allows one to estimate the significance of the discriminant power of the axes remaining after accepting the first k eigenvalues as significant (Bartlett, 1948):

$$\chi^2 = \left[(n-1) - \frac{1}{2}(m+g) \right] \ln \left[\prod_{j=k+1}^{g-1} (1 + \lambda_j) \right] \tag{11.38}$$

with $(m-k)\ (g-k-1)$ degrees of freedom. (The log of L in eq. 11.37 is equal to minus the log of the product of the $(1+\lambda_j)$ terms in the denominator.) When the last $(g-k-1)$ canonical eigenvalues, taken together, do not reach the chosen critical χ^2 value, the null hypothesis that the centroids of the groups do not differ on the remaining $(g-k-1)$ discriminant functions cannot be rejected. This indicates that the discriminant power is limited to the first k functions.

The last step of the computation is to find the positions of the objects in the space of the canonical axes. The matrix of discriminant scores **F** is obtained by multiplying the matrix of centred data with the matrix of normalized eigenvectors **C**:

$$\mathbf{F} = [x-\bar{x}]\,\mathbf{C} \tag{11.39}$$

Since the matrix of centred data $[x-\bar{x}]$ is used in eq. 11.39, the origin of the discriminant axes is located at the centroid of all objects, as in Fig. 11.9. It is common practice to also compute the positions of the centroids of the g groups of objects in canonical space, by multiplying the matrix of the original group centroids (computed from data centred over all objects in the analysis) with matrix **C**. The centroid of a group is a point whose coordinates are made of the mean values of the objects of that group for all descriptors. The matrix of group centroids therefore has g rows and m columns.

As in principal component analysis, equation $\mathbf{F} = [x-\bar{x}]\,\mathbf{C}$ contains a set of functions that provide the position (or *score*) of each object, i, on each canonical axis k:

$$f_{ik} = (x_{i1}-\bar{x}_1)\,c_{1k} + \ldots + (x_{ip}-\bar{x}_p)\,c_{pk} \tag{11.40}$$

The columns of matrix **F** are called *canonical variates* in discriminant analysis.

The columns of matrix **C** are called *discriminant functions*, or "standardized discriminant function coefficients", when the descriptors are *standardized* at the beginning of the study. Discriminant functions are used to assess the relative importance of the original descriptors in the discrimination among groups after eliminating, by standardization, the differences in variance among descriptors. If the analysis is carried out on the *non-standardized* descriptors, the columns of matrix **C** are called *identification functions* or "unstandardized discriminant function coefficients".

Identification functions serve to place new objects in the canonical space. To do so, values of the various descriptors of a new object are centred using the same descriptor means as in eq. 11.40 and the centred values are multiplied by the weights c_{jk}. This provides the position of this object on the canonical axes. By plotting the point representing this object in the canonical ordination space together with the original set of objects, it is possible to identify the group to which the new object is most likely to belong.

There are other ways of assigning objects to groups. *Classification functions*[*] are linear equations that can be used for that purpose. A separate classification function is computed as follows for each group j:

$$\text{Classification function for group } j = -0.5\ \bar{\mathbf{x}}'_j \mathbf{V}^{-1} \bar{\mathbf{x}}_j + \mathbf{V}^{-1} \bar{\mathbf{x}}_j \qquad \textbf{(11.41)}$$

where $\mathbf{V}$ is the pooled within-group dispersion matrix (Table 11.8) and $\bar{\mathbf{x}}_j$ is the vector describing the centroid of group j for all m variables of matrix $\mathbf{X}$. Each classification function looks like a multiple regression equation; eq. 11.41 provides the weights ($\mathbf{V}^{-1}\bar{\mathbf{x}}_j$) to apply to the various descriptors of matrix $\mathbf{X}$ combined in the linear equation, as well as a constant ($-0.5\ \bar{\mathbf{x}}'_j \mathbf{V}^{-1} \bar{\mathbf{x}}_j$). The classification score of each object is calculated for each of the g classification functions; an object is assigned to the group for which it receives the highest classification score. Another way is to compute Mahalanobis distances (eq. 7.38) of the objects from each of the group centroids. An object is assigned to the group to which it is the closest.

Classification table

A *classification table* (also called *classification matrix* or *confusion table*) can be constructed; this is a contingency table comparing the original assignment of objects to groups (usually in rows) to the assignment made by the classification functions (in columns). From this table, users can determine the number and percentage of cases correctly classified by the discriminant functions.[†]

To obtain matrices $\mathbf{V}$ and $\mathbf{A}$, matrices $\mathbf{W}$ and $\mathbf{B}$ were divided by their respective numbers of degrees of freedom (Table 11.8). These divisions may be avoided by carrying out the canonical calculations directly on matrices $\mathbf{W}$ and $\mathbf{B}$. Solving the matrix equation

$$(\mathbf{B} - l_k \mathbf{W})\, \mathbf{u}_k = 0 \qquad \textbf{(11.42)}$$

provides eigenvectors $\mathbf{u}_k$ which are the same as in eqs. 11.31 and 11.32. The eigenvalues l_k are smaller than the λ_k by a constant ratio:

$$l_k = \frac{g-1}{n-g}\, \lambda_k \qquad \textbf{(11.43)}$$

which leaves unchanged the percentage of the variance of $\mathbf{W}^{-1}\mathbf{B}$ explained by each eigenvector. The eigenvectors obtained from matrix $\mathbf{W}$ are rescaled using the following formula:

---

[*] This terminology is unfortunate. In biology, classification consists in establishing groups, using clustering methods for instance (Chapter 8), whereas identification is to assign objects to preestablished groups.

[†] In the SAS package, procedure DISCRIM only computes the reclassification of the objects using either Mahalanobis distances or classification functions (called "linear discriminant functions" in SAS). Procedure CANDISC allows users to obtain raw and standardized canonical coefficients (eq. 11.44).

$$\mathbf{C} = \mathbf{U}\left(\mathbf{U}'\frac{\mathbf{W}}{n-g}\mathbf{U}\right)^{-1/2} = \mathbf{U}\,(\mathbf{U}'\mathbf{V}\mathbf{U})^{-1/2} \qquad \textbf{(11.44)}$$

The relationship between discriminant analysis and canonical correlation analysis (CCorA) can now be described (Gittins, 1985). Consider that the classification criterion of discriminant analysis is expressed in matrix $\mathbf{Y}_2$ containing dummy variables coded as in Subsection 1.5.7 while the quantitative variables are in matrix $\mathbf{Y}_1 = \mathbf{X}$. Apply CCorA to $\mathbf{Y}_1$ and $\mathbf{Y}_2$ using the form where the various matrices $\mathbf{S}$ are matrices of sums of squares and cross products (i.e. dispersion matrices without the final division by the degrees of freedom, e.g. $\mathbf{S}_{11} = \mathbf{Y}'_1\mathbf{Y}_1$). One can show that $\mathbf{S}_{12}\mathbf{S}_{22}^{-1}\mathbf{S}'_{12}$ of eqs. 11.23 and 11.25, calculated in this way, is the among-group dispersion matrix $\mathbf{B}$ of Table 11.8, while $\mathbf{S}_{11}$ is the total dispersion matrix $\mathbf{T}$. So eq. 11.25 may be rewritten as:

$$(\mathbf{B} - r_k^2\mathbf{T})\,\mathbf{w}_k = \mathbf{0} \qquad \textbf{(11.45)}$$

The eigenvalues of this equation are noted r_k^2 and the eigenvectors are noted $\mathbf{w}_k$ to differentiate them from the eigenvalues l_k and eigenvectors $\mathbf{u}_k$ of eq. 11.42. One can show that the eigenvalues of eqs. 11.42 and 11.45 are related by the formulae:

$$r_k^2 = \frac{l_k}{1 + l_k} \quad \text{and} \quad l_k = \frac{r_k^2}{1 - r_k^2}$$

and that the eigenvectors are related as follows:

$$\mathbf{w}_k = (1 + l_k)^{1/2}\,\mathbf{u}_k \quad \text{and} \quad \mathbf{u}_k = (1 - r_k^2)^{1/2}\,\mathbf{w}_k$$

2 — *Numerical example*

Discriminant analysis is illustrated by means of a numerical example in which seven objects, allocated to three groups, are described by two descriptors. The calculation of *identification functions* is shown first (raw data), followed by *discriminant functions* (standardized data). Normally, these data should not be submitted to discriminant analysis since the variances of the group matrices are not homogeneous; they are used here to illustrate the steps involved in the computation. The data set is the following:

Groups =	1			2		3		Means
$\mathbf{X}'$ =	1	2	2	8	8	8	9	5.42857
	2	2	1	7	6	3	3	3.42857

The centred data for the objects and the group centroids are the following:

$$\text{Groups} = \quad \underbrace{\qquad 1 \qquad}\quad \underbrace{\qquad 2 \qquad}\quad \underbrace{\qquad 3 \qquad}$$

$$[\mathbf{X}\ \text{centred}]' = [x-\bar{x}]' = \begin{bmatrix} -4.429 & -3.429 & -3.429 & 2.571 & 2.571 & 2.571 & 3.571 \\ -1.429 & -1.429 & -2.429 & 3.571 & 2.571 & -0.429 & -0.429 \end{bmatrix}$$

$$[\text{Centroids}]' = \begin{bmatrix} -3.762 & 2.571 & 3.071 \\ -1.762 & 3.071 & -0.429 \end{bmatrix}$$

The matrix of sums of squares and cross-products is:

$$\mathbf{T} = [x-\bar{x}]'[x-\bar{x}] = \begin{bmatrix} 75.71429 & 32.71429 \\ 32.71429 & 29.71429 \end{bmatrix}$$

The pooled within-groups matrix $\mathbf{W}$ is computed by adding up the three group matrices of sums of squares and cross-products $\mathbf{W}_1$, $\mathbf{W}_2$ and $\mathbf{W}_3$:

$$\mathbf{W} = \begin{bmatrix} 0.66667 & -0.33333 \\ -0.33333 & 0.66667 \end{bmatrix} + \begin{bmatrix} 0 & 0 \\ 0 & 0.5 \end{bmatrix} + \begin{bmatrix} 0.5 & 0 \\ 0 & 0 \end{bmatrix} = \begin{bmatrix} 1.16667 & -0.33333 \\ -0.33333 & 1.16667 \end{bmatrix}$$

The determinants of matrices $\mathbf{W}$ and $\mathbf{T}$ are 1.25000 and 1179.57, respectively. The ratio of these two values is Wilks' Λ (eq. 11.35: $\Lambda = 0.00106$). The matrix of sums of squares between groups is computed as:

$$\mathbf{B} = \mathbf{T} - \mathbf{W} = \begin{bmatrix} 74.54762 & 33.04762 \\ 33.04762 & 28.54762 \end{bmatrix}$$

The characteristic equation $|\mathbf{B} - l\mathbf{W}| = 0$ is used to calculate the two eigenvalues:

$$l_1 = 106.03086 \Rightarrow \lambda_1 = \frac{(7-3)}{(3-1)} \times 106.03086 = 212.06171\ (93.13\%)$$

$$l_2 = 7.81668 \Rightarrow \lambda_2 = 2 \times 7.81668 = 15.63336\ (6.87\%)$$

In this example, canonical axes 1 and 2 explain 93.13 and 6.87% of the among-group variation, respectively. The two eigenvalues are used to compute the eigenvectors, by means of matrix equation $(\mathbf{B} - l_k\mathbf{W})\ \mathbf{u}_k = \mathbf{0}$. These eigenvectors, normalized to length 1, are the columns of matrix $\mathbf{U}$:

$$\mathbf{U} = \begin{bmatrix} 0.81202 & -0.47849 \\ 0.58363 & 0.87809 \end{bmatrix}$$

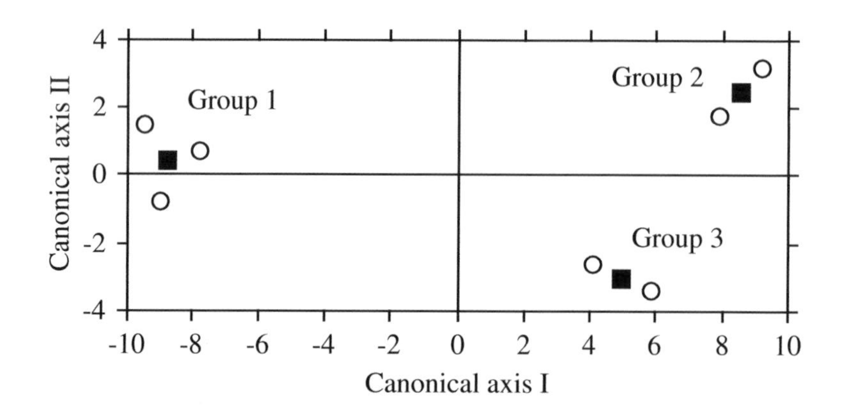

Figure 11.10 Ordination diagram of the seven objects (circles) and group centroids (squares) of the example in the canonical discriminant space.

The vectors are not orthogonal since $\mathbf{u}_1' \mathbf{u}_2 = 0.12394$. In order to bring the eigenvectors to their final lengths, the following scaling matrix is computed:

$$\left(\mathbf{U}'\frac{\mathbf{W}}{n-g}\mathbf{U}\right)^{1/2} = \begin{bmatrix} 0.46117 & 0 \\ 0 & 0.60141 \end{bmatrix}$$

The component terms of each eigenvector $\mathbf{u}_j$ are *divided* by the corresponding diagonal term from this matrix, to obtain the final vectors:

$$\mathbf{C} = \begin{bmatrix} 1.76077 & -0.79562 \\ 1.26553 & 1.46006 \end{bmatrix}$$

Multiplication of the centred matrices of the raw data and centroids by $\mathbf{C}$ gives the positions of the objects and centroids in canonical space (Fig. 11.10):

$$\text{Groups} = \quad \underline{\qquad 1 \qquad} \quad \underline{\qquad 2 \qquad} \quad \underline{\qquad 3 \qquad}$$

$$[\mathbf{X}\text{ centred}]\,\mathbf{C} = [x - \bar{x}]\,\mathbf{C} = \begin{bmatrix} -9.606 & -7.845 & -9.111 & 9.047 & 7.783 & 3.985 & 5.747 \\ 1.438 & 0.642 & -0.818 & 3.169 & 1.708 & -2.672 & -3.466 \end{bmatrix}'$$

$$[\text{Centroids}]\,\mathbf{C} = \begin{bmatrix} -8.854 & 8.415 & 4.866 \\ 0.420 & 2.438 & -3.069 \end{bmatrix}'$$

One can verify that, in canonical space, the among-group dispersion matrix $\mathbf{A}$ is equal to the matrix of eigenvalues and that the pooled within-groups dispersion matrix $\mathbf{V}$ is the identity matrix **I**. Beware: some computer programs calculate the discriminant scores as **XU** instead of [**X** centred] **U** or [**X** centred] **C**.

The *classification functions*, computed from eq. 11.41, are the following for descriptors x_1 and x_2 of the example:

Group 1: $\text{Score}_i = -13.33333 + 8.00000\, x_{i1} + 8.00000\, x_{i2}$

Group 2: $\text{Score}_i = -253.80000 + 36.80000\, x_{i1} + 32.80000\, x_{i2}$

Group 3: $\text{Score}_i = -178.86667 + 34.93333\, x_{i1} + 20.26667\, x_{i2}$

The scores of the 7 objects i, computed from these functions, are the following:

Object number	Observed group	Function 1	Function 2	Function 3	Assigned to group
1	1	10.66667	–151.40000	–103.40000	1
2	1	18.66667	–114.60000	–68.46667	1
3	1	10.66667	–147.40000	–88.73334	1
4	2	106.66667	270.20000	242.46666	2
5	2	98.66667	237.40000	222.19999	2
6	3	74.66667	139.00000	161.39998	3
7	3	82.66667	175.80000	196.33331	3

Each object is assigned (right-hand column) to the group corresponding to the function giving it the highest score. The *classification table* can now be constructed; this is a contingency table comparing the original group assignment of the objects (from the second column in table above) to the assignment made from the classification functions (last column in table above):

Observed group	Assigned to group			Total and % correct
	1	2	3	
1	3	0	0	3 (100%)
2	0	2	0	2 (100%)
3	0	0	2	2 (100%)
Total	3	2	2	7 (100%)

In order to compute *discriminant functions*, the descriptors are standardized at the start of the analysis:

$$\text{Groups} = \quad \underline{\quad 1 \quad} \quad \underline{\quad 2 \quad} \quad \underline{\quad 3 \quad}$$

$$[\mathbf{X}\ \text{standardized}]' = \left[\frac{x-\bar{x}}{s_x}\right]' = \begin{bmatrix} -1.247 & -0.965 & -0.965 & 0.724 & 0.724 & 0.724 & 1.005 \\ -0.642 & -0.642 & -1.091 & 1.605 & 1.155 & -0.193 & -0.193 \end{bmatrix}$$

$$[\text{Centroids}]' = \begin{bmatrix} -1.059 & 0.724 & 0.865 \\ -0.792 & 1.380 & -0.193 \end{bmatrix}$$

The remaining calculations are the same as for the identification functions (above):

$$\mathbf{T} = \left[\frac{x-\bar{x}}{s_x}\right]'\left[\frac{x-\bar{x}}{s_x}\right] = \begin{bmatrix} 6.00000 & 4.13825 \\ 4.13825 & 6.00000 \end{bmatrix}$$

$$\mathbf{W} = \begin{bmatrix} 0.05283 & -0.04217 \\ -0.04217 & 0.13462 \end{bmatrix} + \begin{bmatrix} 0 & 0 \\ 0 & 0.10096 \end{bmatrix} + \begin{bmatrix} 0.03962 & 0 \\ 0 & 0 \end{bmatrix} = \begin{bmatrix} 0.09246 & -0.04217 \\ -0.04217 & 0.23558 \end{bmatrix}$$

$$\mathbf{B} = \mathbf{T} - \mathbf{W} = \begin{bmatrix} 5.90755 & 4.18042 \\ 4.18042 & 5.76441 \end{bmatrix}$$

$$l_1 = 106.03086 \quad \Rightarrow \quad \lambda_1 = \frac{(7-3)}{(3-1)} \times 106.03086 = 212.06171\ (93.13\%)$$

$$l_2 = 7.81668 \quad \Rightarrow \quad \lambda_2 = 2 \times 7.81668 = 15.63336\ (6.87\%)$$

The amounts of among-group variation explained by the canonical axes (93.13 and 6.87%) are the same as those obtained above with the unstandardized data.

$$\mathbf{U} = \begin{bmatrix} 0.91183 & -0.65630 \\ 0.41057 & 0.75450 \end{bmatrix}$$

$$\left(\mathbf{U}'\frac{\mathbf{W}}{n-g}\mathbf{U}\right)^{1/2} = \begin{bmatrix} 0.14578 & 0 \\ 0 & 0.23221 \end{bmatrix} \quad \Rightarrow \quad \mathbf{C} = \begin{bmatrix} 6.25473 & -2.82627 \\ 2.81631 & 3.24918 \end{bmatrix}$$

$$\text{Groups} = \quad \underbrace{\quad 1 \quad}\ \underbrace{\quad 2 \quad}\ \underbrace{\quad 3 \quad}$$

$$[\mathbf{X}\ \text{standardized}]\,\mathbf{C} = \begin{bmatrix} -9.606 & -7.845 & -9.111 & 9.047 & 7.783 & 3.985 & 5.747 \\ 1.438 & 0.642 & -0.818 & 3.169 & 1.708 & -2.672 & -3.466 \end{bmatrix}'$$

$$[\text{Centroids}]\,\mathbf{C} = \begin{bmatrix} -8.854 & 8.415 & 4.866 \\ 0.420 & 2.438 & -3.069 \end{bmatrix}'$$

The raw and standardized data produce exactly the same ordination of the objects and group centroids.

The classification functions computed for standardized descriptors differ from those reported above for raw data, but the classification table is the same in both cases.

Computer packages usually have an option for variable selection using forward entry, backward elimination, or stepwise selection, as in multiple regression (Subsection 10.3.3). These procedures are useful for selecting only the descriptors that significantly contribute to discrimination, leaving the others out of the analysis. This option must be used with caution. As it is the case with any stepwise computation method, the step-by-step selection of s successively most discriminant descriptors does not guarantee that they form the most discriminant set of s descriptors.

The following Ecological application is an example of multiple discriminant analysis among groups of observations, using physical and chemical descriptors as discriminant variables. Steiner *et al.* (1969) have applied discriminant analysis to the agronomic interpretation of aerial photographs, based upon a densimetric analysis of different colours. Other interesting ecological applications are found in Gittins (1985).

Ecological application 11.5

Sea ice is an environment with a rich and diversified biota. This is because ice contains a network of brine cells and channels in which unicellular algae, heterotrophic bacteria, protozoa, and small metazoa can develop and often reach very high concentrations. Legendre *et al.* (1991) investigated the environmental factors controlling the growth of microscopic algae in the sea ice of southeastern Hudson Bay, Canadian Arctic.

Ice cores were taken at eight sites along a transect that extended from the mouth of the Great Whale River to saline waters 25 km offshore. Ice thickness ranged from 98 to 125 cm. The cores were used to determine the crystallographic structure of the ice, at 2 cm intervals from the top to the bottom of each core, together with several chemical and biological variables (nutrients, algal pigments, and taxonomic composition of algal assemblages) along the cores. The chemical and biological variables were determined on melted 10-cm thick sections of the cores; using crystallographic information, the chemical and biological data were transformed into values per unit of brine volume. The rate of ice growth for each 10-cm interval of each core was calculated

Table 11.9 Standardized canonical coefficients for the first two canonical variates.

Discriminant variable	Canonical variate 1	Canonical variate 2
Nitrate	–0.63	0.69
Phosphate	0.55	–0.08
Silicate	0.29	0.44
Rate of ice growth	0.89	0.54

by combining the mean daily air temperatures since the start of ice formation with the ice thickness at the date of sampling. Data on taxonomic composition of the algal assemblages in the brine cells were analysed as follows: (1) Similarities (χ^2 similarity; S_{21} eq. 7.28) were computed among all pairs of core sections, on the basis of their taxonomic composition. (2) The similarity matrix was subjected to flexible clustering (Subsection 8.5.10) to identify groups of core sections that were taxonomically similar. (3) Discriminant analysis was used to determine which environmental variables best accounted for differences among the groups of core sections. Chlorophyll *a* is not a descriptor of the environment but of the ice algae, so that it was not used as discriminant variable; it is, however, the response variable in the path analysis mentioned below. Another approach to this question would have been to look directly at the relationships between the physical and chemical data and the species, using CCA.

Cluster analysis evidenced five groups among the 10-cm ice sections. The groups were distributed at various depths in the cores, sometimes forming clusters of up to 5 adjacent ice sections from within the same core. Discriminant analysis was conducted on standardized descriptors. The first canonical variate accounted for 62% of the variation among groups, and the second one 29%.

The standardized canonical coefficients for the first two canonical variates (Table 11.8) indicate that the environmental descriptors that best accounted for the among-group variation were the rate of ice growth (first variate) and nitrate (second variate). Figure 11.11 shows the position of the centroids of the 5 groups of core sections, plotted in the space of the first two canonical axes, with an indication of the role played by the environmental variables in discriminating among the groups of core sections. According to the Figure, the groups of core sections are distributed along two gradients, one dominated by ice growth rate (with groups 1, 3 and 5 in faster-growing ice) and the other by nitrate (with group 1 in low-nitrate and group 5 in high-nitrate environments). These results are consistent with those of a path analysis (Section 10.4) conducted on the same data, showing that algal biomass (chl *a*) was inversely related to the rate of ice growth. The paper concluded that slower ice growth favoured the colonization of brine cells by microalgae (path analysis) and that the rate of ice growth had a selective effect on taxa, with nutrient limitation playing a secondary role in some brine cells (discriminant analysis).

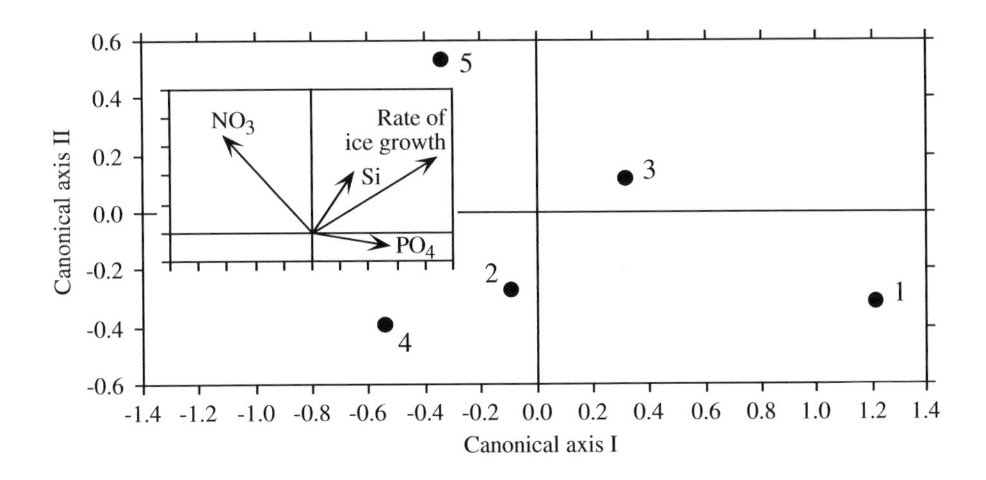

Figure 11.11 Centroids of the five groups of taxonomically similar core sections plotted along the first two canonical axes. Insert: contributions (from Table 11.9) of the four environmental variables (arrows) to the formation of the canonical axes. The groups of core sections are distributed along two gradients, one dominated by ice growth (groups 4, 2 and 3), the other by nitrate (groups 1, 3 and 5). Modified from Legendre *et al.* (1991).

11.6 Canonical analysis of species data

In early numerical ecology papers, canonical correlation analysis and discriminant analysis were used to analyse tables of species presence/absence or abundance. In many applications, the assumptions of linearity and the algebraic constraints imposed by the models make these methods unsuitable for such data. RDA and CCA provide alternatives that are often more appropriate. Let us consider different types of problems that may involve species data.

(1) The first case involves a matrix **Y** of species presence-absence or abundance data and a matrix **X** of habitat characteristics. One may wish to find support for the ecological hypothesis of environmental control of the species distributions (Whittaker, 1956; Bray & Curtis, 1957) and/or describe in what way the species are related to the environmental variables. The analysis is not symmetric; the species clearly form the response variables, to be explained by the environmental variables. Hence a symmetric form of analysis such as CCorA is not appropriate; one should rely instead on asymmetric forms of analysis such as RDA or CCA. When the ecological gradient represented in the data is short, RDA is appropriate; when it is long, generating many zeros in the species data table, CCA is the method of choice.

It often happens that matrix **Y** contains many species, more than there are sites (i.e. objects). CCorA is unable to analyse such data because it cannot handle more variables in any one of the data sets than there are sites minus 1. Rare species would have to be dropped from the analysis to satisfy the requirements of the method. In RDA and CCA, on the contrary, there is no upper limit to the number of species that can be analysed.

When **X** contains dummy variables describing types of habitat (classification criterion) coded as in Subsection 1.5.7, RDA or CCA may be used to test the hypothesis that groups of sites, identified *a priori*, do not differ from one another in species composition. The question is of the same type as in multivariate analysis of variance. When **X** codes for factors of an experiment, CCA or the db-RDA method, briefly described in Subsection 11.2.1, may be used.

(2) Ecologists may wish to use the environmental variables in **X** to forecast a classification criterion (**y**) representing the presence or absence of a single species, a group of species, or one of several dominant species at various locations. Although this is a discrimination problem, discriminant analysis is often inappropriate for such data.

For two groups (e.g. presence or absence of a species), discriminant analysis, like linear regression, creates a linear function generating forecasted values that may be smaller than 0 or larger than 1. Forecasted responses smaller than 0 make no sense in the case of a species classification criterion. For the presence or absence of a single species, logistic regression (Subsection 10.3.7) is an appropriate model, because forecasted responses are in the range [0, 1]. One should worry, however, of unimodal distributions of species along environmental gradients (Subsection 9.4.4): a species may be absent *under both low and high values* of an environmental variable. One should plot scatter diagrams of the presence/absence of the target species against each of the environmental variables in matrix **X**. When a unimodal response is detected, a quadratic polynomial function of that response variable should be used in the logistic model (see Gaussian logistic response, Subsection 10.3.7).

(3) Species may represent the explanatory variables (matrix **X**). What are the species assemblages that characterize different types of habitat? In such cases, the types of habitat form the classification criterion **y**.

This is typically a discriminant analysis problem. However, because of algebraic constraints, not all species data sets are suitable for discriminant analysis. The size of the pooled within-group dispersion matrix **V** is $(m \times m)$, where m is the number of variables in **X**. Because **V** must be inverted (eq. 11.31), it must be constructed from a matrix **X** containing more objects (n) than descriptors (m). Actually, Williams & Titus (1988) recommend that the total number of observations *per group* be at least three times the number of variables in matrix **X**; ter Braak (1987c) recommends that n be much larger than the number of species (m) plus the number of groups (g). This problem, which was also encountered with CCorA, often prevents the use of discriminant analysis with species presence/absence or abundance data. Else, rare species must be dropped from the analysis in order to satisfy the algebraic requirements of the method.

RDA may be used as a form of inverse analysis related to discriminant analysis. The classification criterion (e.g. types of habitat) is made of a set of dummy variables, written into response matrix **Y**; the species data are the explanatory variables **X**. The condition of more objects (n) than species (m) must also be satisfied in this analysis.

Matrix **X**, in which each species is represented by a vector, may be transformed prior to RDA or discriminant analysis, by replacing the m species vectors by $(m - 1)$ ordination axes obtained by correspondence analysis (CA) of the raw species data. An alternative is to compute a similarity or distance matrix among sites using the species data and obtain new axes by principal coordinate analysis (PCoA). CA or PCoA axes might relate to the environmental descriptors better than the original species data.

(4) Ecologists may wish to use the species data in **X** to predict or reconstruct one or more environmental variables in **Y**. This case, which is related to Ecological application 11.2b, is like CCA but with **X** and **Y** interchanged. A neat solution, which circumvents the too-many-species-problem, is Weighted Averaging Partial Least Squares (WA-PLS), which extends PLS regression in the correspondence analysis framework (ter Braak, 1995).

Chapter 12 *Ecological data series*

12.0 Ecological series

The use and analysis of *data series* is increasingly popular in ecology, especially as equipment becomes available to automatically measure and record environmental variables. Ecological data series may concern either continuous or discrete (discontinuous) variables, which may be sampled over time or along transects in space.

Stochastic process

A data series is a sequence of observations that are *ordered* along a temporal or spatial axis. As mentioned in Section 1.0, a series is one of the possible realizations of a *stochastic process*. A *process* is a phenomenon (response variable), or a set of phenomena, which is organized along some independent axis. In most cases, the independent axis is time, but it may also be space, or a trajectory through both time and space (e.g. sampling during a cruise). *Stochastic processes* generally exhibit three types of component, i.e. deterministic, systematic, and random. Methods for the numerical analysis of data series are designed to characterize the deterministic and systematic components present in series, given the probabilistic environment resulting from the presence of random components.

The most natural axis along which processes may be studied is *time* because temporal phenomena develop in an irreversible way, and independently of any decision made by the observer. The temporal evolution of populations or communities, for example, provides information which can unambiguously be interpreted by ecologists. Ecological variability is not a characteristic limited to the time domain, however; it may also be studied across space. In that case, the decisions to be made concerning the observation axis and its direction depends on the working hypothesis. In ecology, the distinction between space and time is not always straightforward. At a fixed sampling location, for example, time series analysis may be used to study the spatial organization of a moving system (e.g. migrating populations, plankton in a current), whereas a spatial series is required to assess temporal changes in that same

Eulerian
Lagrangian

system. The first approach (i.e. at a fixed point in space) is called *Eulerian*, whereas the second (i.e. at a fixed point within a moving system) is known as *Lagrangian*.

Periodic phenomena

Ecologists are often interested in *periodic* changes. This follows in part from the fact that many ecological phenomena are largely determined by geophysical rhythms; there are also rhythms that are endogenous to organisms or ecosystems. The geophysical cycles of glaciations, for example, or, at shorter time scales, the solar (i.e. seasons, days) or lunar (tides) periods, play major roles in ecology. Concerning endogenous rhythms, considerations of non-linear statistical mechanics (Yates *et al.*, 1972) arising from thermodynamic considerations about nonlinear irreversible phenomena (Glansdorff & Prigogine, 1971) suggest that regular fluctuations (called free oscillations, for they are independent of any geophysical forcing) are a fundamental characteristic of biological systems. This question is briefly discussed by Platt & Denman (1975) in a review paper on spectral analysis in ecology. Endogenous rhythms in fish are discussed in some detail in Ali (1992).

The analysis of data series often provides unique information concerning ecological phenomena. The quality of the results depends to a large extent, however, on the *sampling design*. As a consequence, data series must be sampled following well-defined rules, in order (1) to preserve the spatio-temporal variability, which is often minimized on purpose in other types of ecological sampling design, and (2) to take into account the various conditions prescribed by the methods of numerical analysis. These conditions will be detailed later in the present Chapter. An even more demanding framework prevails for *multidimensional series*, which result from sampling several variables simultaneously. Most numerical methods require that the series be made up of *large numbers of observations* ($n > 100$, or even $n > 1000$) for the analysis to have enough statistical power to provide conclusive results, especially when large random fluctuations are present. Long series require extensive sampling. This is often carried out, nowadays, using equipment that automatically measures and records the variables of ecological interest. There also exist a few methods especially designed for the analysis of short time series; they are discussed below.

Observational window

Lag
Period
Frequency
Wavelength
Wavenumber

The most fundamental constraint in periodic analysis is the *observational window*. The width of this window is determined by the number of observations in the data series (n) and the interval (time or distance) between successive observations. This interval is called the *lag*, Δ; for the time being, it is assumed to be uniform over the whole data series. These two characteristics set the time or space domain that can be "observed" when analysing data series (Table 12.1). For temporal data, one refers to either the *period* (T) or the *frequency* ($f = 1/T$) whereas, for spatial data, the corresponding concepts are the *wavelength* (λ) and the *wavenumber* ($1/\lambda$). The length of the series (Δn) sets, for temporal data, the *fundamental period* ($T_0 = \Delta n$) or *fundamental frequency* ($f_0 = 1/T_0 = 1/\Delta n$) and, for spatial data, the *fundamental wavelength* ($\lambda_0 = \Delta n$) or *fundamental wavenumber* ($1/\lambda_0 = 1/\Delta n$). *Harmonic periods* and *wavelengths* are *integral fractions* of the fundamental period and wavelength, respectively ($T_i = T_0/i$ and $\lambda_i = \lambda_0/i$, where $i = 1, 2, \dots n$), whereas *harmonic frequencies* and *wavenumbers* are *integral multiples* of the fundamental frequency and

Table 12.1 Characteristics of the observational window in periodic analysis. Strictly speaking, the length of a data series is $(n - 1)\Delta$ but, for simplicity, one assumes that the series is long, hence $(n - 1) \approx n$.

Harmonic i	Period (T_i) Wavelength (λ_i)	Frequency (f_i) Wavenumber (i)	
1	$n\Delta$	$1/n\Delta$	Fundamental value, i.e. the whole series
2	$n\Delta/2$	$2/n\Delta$	Limit of observational window
.	.	.	
.	.	.	
i	$n\Delta/i$	$i/n\Delta$	ith harmonic
.	.	.	
.	.	.	
$n/2$	2Δ	$1/2\Delta$	Limit of window: Nyquist frequency

wave number, respectively ($f_i = if_0$ and $1/\lambda_i = i/\lambda_0$). Concerning the actual limits of the observational window, the *longest* period or wavelength that can be statistically investigated is, at best, equal to *half the length* of the series ($\Delta n/2$). For example, in a study on circadian (24-h) rhythms, the series must have a *minimum* length of two days (better 4 days or more). Similarly, in an area where spatial structures are of the order of 2 km, a transect must cover *at least* 4 km (better 8 km or more). Similarly, the *shortest* period or wavelength that can be resolved is equal to *twice the interval* between observations (2Δ). In terms of frequencies, the highest possible frequency that can be resolved, $1/2\Delta$, is called the *Nyquist frequency*. For example, if one is interested in hourly variations, observations must be made *at least* every 30 min. In space, in order to resolve changes at the metre scale, observations must be collected along a transect *at least* every 50 cm, or closer.

Nyquist frequency

To summarize the above notions concerning the observational window, let us consider a variable observed every month during one full year. The data series would allow one to study periods ranging between (2×1 month = 2 months) and (12 months/2 = 6 months). Periods shorter than 2 months and longer than 6 months are outside the observational window. In other words, statistical analysis cannot resolve frequencies higher than 1/(2 months) = 0.5 cycle month^{-1} = 6 cycles year^{-1} (Nyquist frequency), or lower than 1/(6 months) = 0.167 cycle month^{-1} = 2 cycles year^{-1}. The longest period (or lowest frequency) of the observational window is easy to understand, by reference to the usual notion of degrees of freedom (Box 1.2).

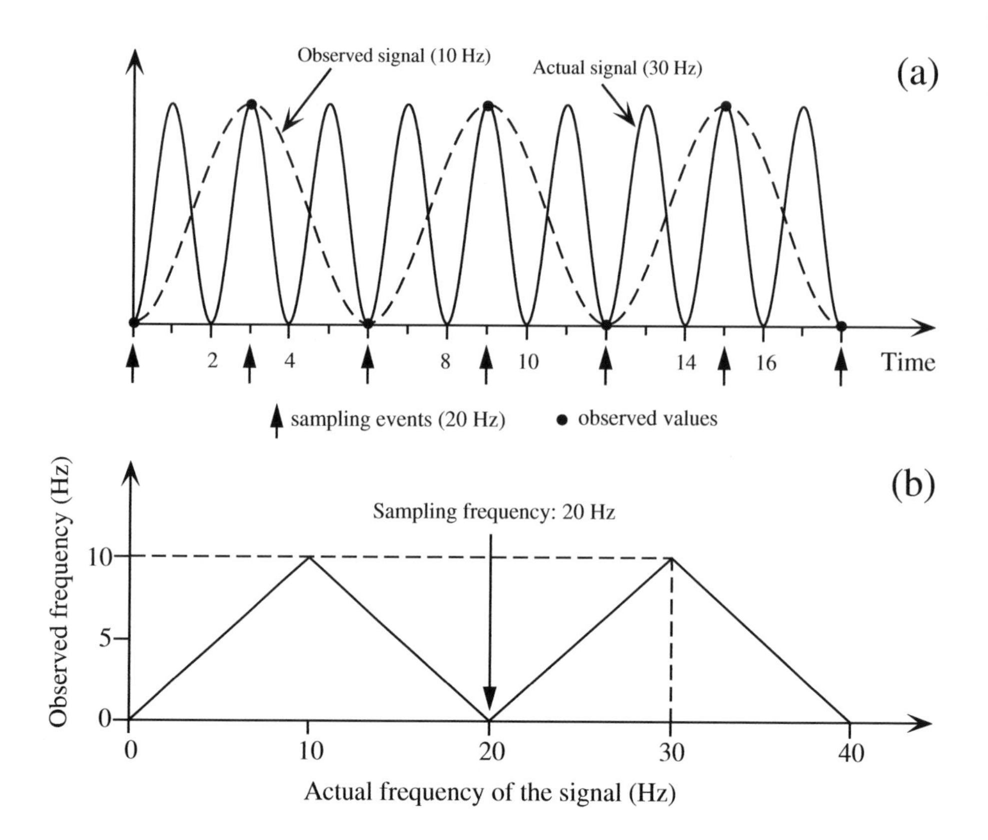

Figure 12.1 Aliasing. (a) The artificial signal detected in the data series (dashed line) is caused by observations (dots) made at a frequency lower than twice that present in the series under study (solid line). On the abscissa, 1 time unit = 1/60 s. (b) With a sampling frequency of 20 Hz, the observed frequency (ordinate) varies between 0 and 10 Hz, as the actual frequency of the signal increases (abscissa). The observed frequency would be identical to the frequency of the signal (no aliasing) only if the latter was ≤ 10 Hz, which is half the 20 Hz sampling frequency. In the example, the frequency of the signal is 30 Hz and the observed (aliased) frequency is 10 Hz (dashed line).

Indeed, in order to have minimum certainty that the observed phenomenon is real, this phenomenon must be observed at least twice, which provides only one degree of freedom. For example, if an annual cycle was observed over a period of one year only, there would be no indication that it would occur again during a second year (i.e. no degree of freedom). A similar reasoning applies to the shortest period (or highest, Nyquist frequency) detectable in the observational window. For example, if the observed phenomenon exhibits monthly variation (e.g. oscillations between maximum and minimum values over one month), two observations a month would be the absolute minimum required for identifying the presence of this cycle.

Most methods described in the present Chapter are limited to the observational window. However, some methods are mathematically capable of going beyond the upper limit (in terms of periods) of the window, because they can fit incomplete cycles of sine and cosine functions to the data series. This is the case of Schuster's periodogram (Section 12.4) and derived forms, including spectral analysis (Section 12.5). A significant period found in this region (e.g. a 3-month period in a data series 4 months long) should be interpreted with care. It only indicates that a longer time series should be observed and analysed (e.g. > 1 year of data) before drawing ecological conclusions.

Aliasing

There exists another constraint, which is also related to the observational window. This constraint follows from a phenomenon known as *aliasing*. It may happen that the observed variable exhibits fluctuations whose frequency is *higher than the Nyquist frequency*. This occurs when a period T or wavelength λ of the observed variable is smaller than 2Δ. Undersampling of high-frequency fluctuations may generate an artificial signal in the series, whose frequency is *lower than the Nyquist frequency* (Fig. 12.1). Researchers unaware of the phenomenon could attempt to interpret this artificial low frequency in the series; this would obviously be improper. In order to avoid aliasing, the sampling design must provide at least four data points per cycle of the *shortest* important period or wavelength of the variable under study. The latter period or wavelength may be determined either from theory or from a pilot study.

The sections that follow explore various aspects of series analysis. The methods discussed are those best adapted to ecological data. Additional details may be found in the biologically-oriented textbook of Diggle (1990) and the review paper of Fry *et al.* (1981), or in other textbooks on time series analysis, e.g. Jenkins & Watts (1968), Bloomfield (1976), Box & Jenkins (1976), Brillinger (1981), Priestley (1981a, b), Kendall *et al.* (1983), Chatfield (1989), and Kendall & Ord (1990). Methods for analysing time series of ecological and physiological chronobiological data have been reviewed by Legendre & Dutilleul (1992).

12.1 Characteristics of data series and research objectives

Signal
Trend
Noise

Observed data series may be decomposed into various components, which can be studied separately since they have different statistical and ecological meanings. Figure 12.2 shows an artificial data series constructed by adding three components: a periodic signal, a trend, and a noise component. Series may be analysed in terms of *deterministic* change (trend), *systematic* (periodic) variability, and *random* fluctuations (noise). Data series may be recorded with different objectives in mind (Table 12.2), which correspond to different methods of time series analysis. The following presentation of objectives is largely drawn from Legendre & Dutilleul (1992).

Trend

Objective 1 — Ecological data series often exhibit a deterministic component, known as the *trend*. The trend may be linear, polynomial, cyclic, etc. This

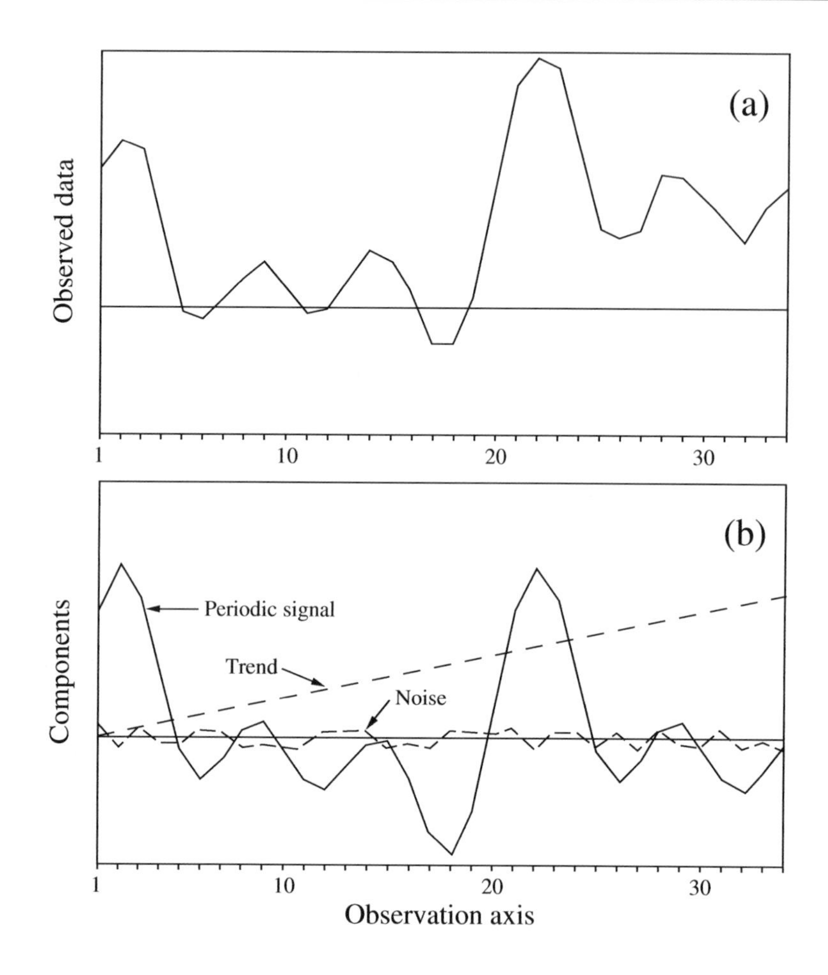

Figure 12.2 Artificial data series (a) constructed by adding the three components shown in (b), i.e. a periodic signal and a noise component, whose combination gives a stationary series (not illustrated), and a linear trend. The periodic signal is the same as in Fig. 12.13. There are n = 34 data points sampled at regular intervals. The overall mean of the noise signal is zero, by definition.

deterministic component underlies the evolution of the series (Fig. 12.2a). It must be extracted as the first step of the analysis.

In some cases, determining the *trend* is the chief objective of the study. For example, progressive changes in the characteristics of an ecosystem, over several years, may be used to assess whether this system is responding or not to anthropogenic effects. In such a case, the problem would be to characterize the long-term trend, so that the annual cycle as well as the high-frequency noise component would be of no interest. Long-term trends in data series may be modelled by regression (Section 10.3). Linear regression is used when the trend is (or seems to be) linear; this method allows one to formulate the most parsimonious hypothesis and estimate smallest number of parameters (see Ocham's razor in Subsection 10.3.3). In other cases, the ecological hypothesis or a preliminary examination of the series may indicate that the trend is of some other mathematical form (e.g. logistic), in which case the methods of polynomial or nonlinear regression should be used (e.g. Ross, 1990).

In contrast, ecologists primarily interested in the *periodic* component of data series (Objective 2) consider the long-term trend as a nuisance. Even though the trend may not be of ecological interest, it must be extracted from the series because most methods of analysis require that the series be *stationary*, i.e. that the mean, variance, and other statistical properties of the distribution be constant over the series. In the numerical example of Fig. 12.2, the observed data series (a) is obviously not stationary. It becomes so if the linear trend shown in (b) is removed by subtraction; this operation is called *detrending*. The trend may be estimated, in this case, by linear regression over a reasonably long segment of data; detrending consists in calculating the regression residuals. In practice, the analysis of series only requires *weak*, or *second-order*, or *covariance stationarity*, i.e. the mean and variance are constant and finite along the series and the autocovariance (or autocorrelation) function depends only on the distance between observations along the series; two observations separated by a given interval have the same autocovariance no matter where they occur in the series. Extracting trends may be done in various ways, which are detailed in Section 12.2.

Stationarity

Detrending

Some *low-frequency periodic* components may also be considered as trends, especially when these are both trivial and known *a priori* (e.g. annual cycle). A long-term trend as well as these periodic components may be extracted, in order to focus the analysis on finer components of the data series. Again, regression or other statistical methods (Section 12.2) may be used to model the low-frequency components and to compute residuals on which the analysis could be carried out.

Objective 2 — Identifying *characteristic periods* is a major objective of series analysis in ecology. It is generally done for one variable at a time, but it is also possible to study multidimensional series (i.e. several variables, often analysed two at a time). Ecological series always exhibit irregular and unpredictable fluctuations, called *noise* (Fig. 12.2b), which are due to non-permanent perturbation factors. The larger the noise, the more difficult it is to identify characteristic periods when analysing (stationary) series. Table 12.3 summarizes the methods available to do so; several of these are described in Sections 12.3 to 12.5.

Objective 3 — One method for identifying characteristic periods is spectral analysis. In this analysis, the variance of the data series is partitioned among frequencies (or wavenumbers) in order to estimate a *variance spectrum*. Section 12.5 shows that the spectrum is a global characteristic of the series, and presents examples where the spectra are interpreted as reflecting ecological processes.

Objective 4 — There are data series that do not behave in a periodic manner. This may be because only one or even part of a cycle has been sampled or, alternatively,

Table 12.2 Analysis of data series: research objectives and related numerical methods. Adapted from Legendre & Legendre (1984b) and Legendre & Dutilleul (1992).

Research objective	Numerical methods
1) Characterize the trend	• Regression (linear or polynomial)* • Moving averages • Variate difference method
2) Identify characteristic periods	→ Details in Table 12.3
3) Characterize series by spectrum	• Spectral analysis
4) Detect discontinuities in multivariate series	• Clustering the data series (with or without constraint) • Hawkins & Merriam or Webster segmentation methods
5) Correlate variations in a series with changes in other series	
5.1) Univariate target series	• Regression*: simple / multiple linear, nonlinear, splines • Cross-correlation
5.2) Multivariate target series	• Canonical analysis** • Mantel test*
6) Formulate a forecasting model	• Box-Jenkins modelling

Methods described in * Chapter 10 or ** Chapter 11.

because the variables under study are not under the control of periodic processes. Such series may exhibit structures other than periodic, along time or a spatial direction. In particular, one may wish to identify *discontinuities* along multidimensional data series. Such discontinuities may, for example, characterize *ecological succession.* A commonly-used method for finding discontinuities is cluster analysis. To make sure that the multidimensional series gets divided into blocks, each one containing a set of temporally contiguous observations, authors have advocated to constrain clustering algorithms so that they are forced to group only the observations that are contiguous. Various methods to do so are discussed in Section 12.6.

Objective 5 — Another objective is to *correlate variations* in the data series of interest (i.e. the *target* or *response variable*) with variations in series of some potentially *explanatory variable(s)*, with a more or less clearly specified model in mind. There are several variants. (1) When the sampling interval between observations is large, the effect of the explanatory variables on the target variable may be considered as instantaneous. In such a case, various forms of regression analysis may be used. When no explicit model is known by hypothesis, spline regression may be used to describe temporal changes in the target variable as a function of another variable (e.g. Press *et al.*, 1986). These methods are explained in Section 10.3. (2) When the interval between consecutive data is short compared to periods in the

Table 12.3 Analysis of data series: methods for identifying characteristic periods. The approaches best suited to *short* data series are: the contingency periodogram, Dutilleul's modified periodogram, and maximum entropy spectral analysis. Adapted from Legendre & Legendre (1984b) and Legendre & Dutilleul (1992).

Type of series	Methods	
	Quantitative variables only	All precision levels
1) A single variable	• Autocorrelogram	• Spatial correlogram* (quantitative, qualitative), Mantel correlogram*
	• Periodograms (Whittaker & Robinson, Schuster, Dutilleul) • Spectral analysis	• Contingency periodogram for qualitative data • Kedem's spectral analysis for binary data
2) Two variables	• Parametric cross-correlation • Coherence and phase spectra	• Nonparametric cross-correlation • Lagged contingency analysis
3) Multivariate series	• Multivariate spectral analysis	• Mantel correlogram* (data expressed as a distance matrix)

* Method described in Chapter 13

target variable, it is sometimes assumed that the target variable responded to events that occurred at some previous time, although the exact delay (*lag*) may not be known. In such a case, the method of cross-correlation may be used to identify the time lag that maximises the correlation between the explanatory and target variables (Section 12.3). When the optimal lag has been found for each of the explanatory variables in a model, multiple regression can then be used, each explanatory variable being lagged by the appropriate number of sampling intervals. (3) The previous cases apply to situations where there is a single target variable in the series under study. When there are several target variables, the target series is multivariate; the appropriate methods of data analysis are globally called canonical analysis (Chapter 11). Two forms are of special interest here: redundancy analysis and canonical correspondence analysis. (4) Finally, the relationship between two distance matrices based on two multivariate data sets can be analysed using the Mantel test or its derived forms (Sections 10.5 and 13.1).

Objective 6 — A last objective is to formulate a model to *forecast* the future behaviour of the target series. Following the tradition in economics, one way of doing this is to model the data series according to its own past behaviour (Section 12.7).

The first problem encountered when analysing series is to decide whether a *trend* is present or not. Visual examination of the series, which may be combined with previous knowledge about the process at work, is often sufficient to detect one or several trends. These may be monotonic (e.g. gradient in latitude, altitude, or water depth) or not (e.g. daily, lunar, or annual cycles). Three methods have been proposed to test for the presence of trends. (1) The numbers of positive and negative *differences between successive values* in the series are counted. These are then subjected to a *sign test* (Section 5.2), where the null hypothesis (H_0) is that the plus and minus signs are sampled from a population in which the two signs are present in equal proportions. Rejecting H_0 is indication of a trend. (2) All values in the series are *ranked* in increasing (or decreasing) order. *Kendall's rank correlation coefficient* (τ) (Section 5.2) is used to assess the degree of resemblance between the rank-ordered series and the original one; this is equivalent to computing the Kendall correlation between the original data series and the observation rank labels: 1, 2, 3, …, *n*. When τ is significantly different from zero, it is concluded that the series exhibits a *monotonic* trend. These two methods are described in Kendall & Ord (1990, pp. 21-22). The approach based on Kendall's τ is preferable to the sign test because it considers the whole series and not only differences between neighbouring values. (3) Another nonparametric test, called the *up and down runs* test, is well suited to detect the presence of various types of trends. Consider again *n* values and, for each one, the sign of the difference from the previous value. The $(n - 1)$ signs would all be the same if the observations were monotonically increasing or decreasing. Cyclical data, on the other hand, would display more runs of "+" or "–" signs than expected for random data. A *run* is a set of like signs, preceded and followed (except at the end of the series) by opposite signs. Count the number of runs in the data series, including those of length 1 (e.g. a single "+" sign, preceded and followed by a "–"). The up and down runs test, described for instance in Sokal & Rohlf (1995), allows one to compare this number to the number of runs expected from a same-length sequence of random numbers.

Up and down runs test

When there is a *trend* in the series, it must be extracted using one of the methods discussed in Section 12.2. If, after *detrending*, the mean of the series is still not stationary, a second trend must be searched for and removed. When the series does not exhibit any trend, or after detrending, one must decide, before looking for periodic variability (Sections 12.3 to 12.5), whether the stationary series presents some kind of systematic variability or if, on the contrary, it simply displays the kind of variation expected from a random process. In other words, one must test whether the series is simply *random*, or if it exhibits *periodic variability* that could be analysed.

Test of series randomness

In some instances, as in Fig. 12.3, it is useless to conduct sophisticated tests, because the random or systematic character of the series is obvious. Randomness of a series may be tested as follows: identify the *turning points* (i.e. the peaks and troughs) in the series and record the distribution of the *number of intervals* (*phase length*) between successive turning points. It is then possible to test whether these values correspond or not to those of a random series (Kendall & Ord, 1990, p. 20). This procedure actually tests the same null hypothesis as the up and down runs test described above. In practice, any ecological series with an average phase longer than two intervals may be considered non-random.

The overall procedure for analysing data series is summarized in Fig. 12.4. The following sections describe the most usual methods for extracting trends, as well as various approaches for analysing stationary series. It must be realized that, in some

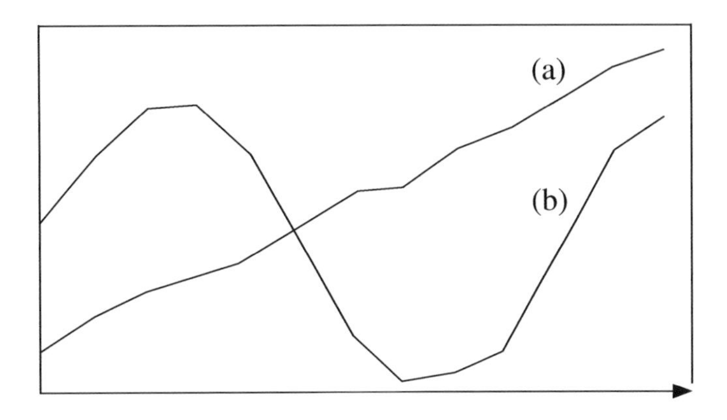

Figure 12.3 Two artificial series; (a) would be random if the linear trend was extracted, whereas (b) displays a cyclic trend.

instances, variations in stationary series may be so small that they cannot be analysed, because they are of the same order of magnitude as the background noise.

If parametric statistical tests are to be conducted during the course of the analysis, *normality* must be checked (Section 4.7) and, if the data are not normally distributed, they must be *transformed* as explained in Subsection 1.5.6. In addition, several of the methods discussed in the following sections require that observations in the series be *equally spaced*. If they are not, data may be eliminated to make them *equispaced*, or else, missing data may be calculated by regression or other interpolation methods (e.g. Section 1.6); most methods of series analysis cannot handle missing values. Obviously, it is preferable to consider the requirement of equispaced data when designing the sampling program than to have to modify the data at the stage of analysis.

In addition to the numerical methods discussed in the following sections, ecologists may find it useful to have a preliminary look at the data series, using the techniques of exploratory data analysis described by Tukey (1977, Chapters 7 and 8). These are based on simple arithmetic and easy-to-draw graphs, which may help decide which numerical treatments would be best suited for analysing the series.

12.2 Trend extraction and numerical filters

When there is a *trend* in a series (which is not always the case), it must be extracted from the data prior to further numerical analyses. As explained in the previous section, this is because most methods of analysis require that the series be *stationary*.

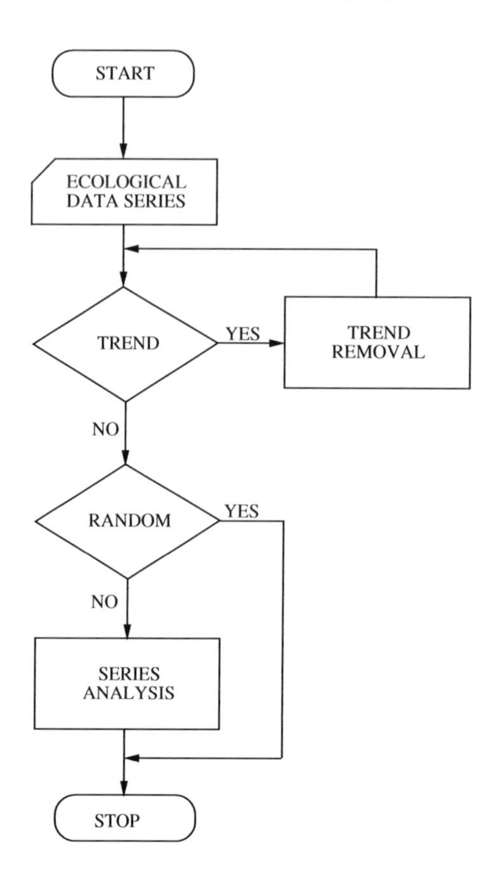

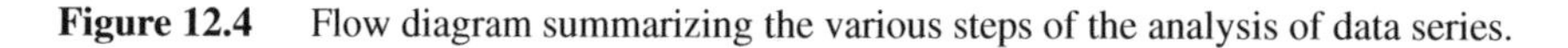
Figure 12.4 Flow diagram summarizing the various steps of the analysis of data series.

When the trend itself is of interest, it can be analysed in ecological terms (Objective 1, above). For example, Fortier *et al.* (1978) interpreted a cyclical trend in temporal changes of estuarine phytoplankton in terms of physical oceanography. When the study goes beyond the identification of a trend (Objectives 2 *et seq.*), the analysis is normally conducted on the *residual* (or *detrended*) data series. The residual (i.e. stationary) series is obtained by subtracting the value estimated by the trend function from that of each data point, observed at positions (i.e. objects) x_i along the series:

Residuals

$$\text{residual data } (\hat{y}_i) = \text{observed data } (y_i) - \text{value of the trend at } x_i \qquad \textbf{(12.1)}$$

There are cases where several trends of different natures must be extracted successively before reaching stationarity. However, because each trend extraction distorts the residuals, one must proceed with caution with detrending. The success of trend extraction may be assessed by looking at the resulting trend (Objective 1), or at the detrended series (e.g. Objective 2).

The simplest, but perhaps not so accurate way of estimating a trend in a data series is the *freehand method*. It consists in the freehand fitting of a curve corresponding to the general evolution of the series. Although the simplest, it is also the method most prone to error. The *average freehand method* is also graphical, but the trend is then represented by a broken line located halfway between two lines joining, respectively, the peaks and the troughs of the series. A continuous curve can be fitted, by freehand, to the broken line.

Moving averages

Window

The method of *moving averages* is often used to estimate trends. One calculates successive arithmetic averages over $2m + 1$ contiguous data as one moves along the data series. The interval $(2m + 1)$ over which a moving average is computed is called *window*. For example, with $m = 2$, the first moving average $\bar{y}_3$ is computed over the first 5 values y_1 to y_5, the second moving average $\bar{y}_4$ is calculated over values y_2 to y_6, the third one ($\bar{y}_5$) is the average of values y_3 to y_7, and so forth. Each average value is positioned at the centre of the window. For a series of n observations, there are $(n - 2m)$ moving averages:

x_1	x_2	x_3	x_4	...	x_{n-2}	x_{n-1}	x_n
y_1	y_2	y_3	y_4	...	y_{n-2}	y_{n-1}	y_n
moving averages		$\bar{y}_3 = \frac{1}{5}\sum_{h=1}^{5} y_h$	$\bar{y}_4 = \frac{1}{5}\sum_{h=2}^{6} y_h$	...	$\bar{y}_{n-2} = \frac{1}{5}\sum_{h=n-4}^{n} y_h$		

The general formula for moving averages is thus:

$$\bar{y}_i = \frac{1}{2m+1} \sum_{h=-m}^{m} y_{(i+h)} \tag{12.2}$$

The h values corresponding to the above example, where $m = 2$, would be: –2, –1, 0, +1, and +2, respectively.

Moving averages may also be *weighted*. In such a case, each of the $2m + 1$ values within the window is multiplied by a weight w_h. Generally, values closer to the centre of the window receive larger weights. The general formula for the weighted moving average corresponding to any position (or object) x_i is:

$$\bar{y}_i = \sum_{h=-m}^{m} y_{(i+h)} w_h / \sum_{h=-m}^{m} w_h \tag{12.3}$$

Choosing values for the weights depends on the underlying hypothesis. Kendall & Ord (1990, p. 3) give coefficients to be used under hypotheses of polynomial trend of the second, third, fourth, and fifth degrees. Another, simple method for assigning weights is that of *repeated moving averages*. After calculating a first series of non-weighted

Table 12.4 Calculation of repeated moving averages. Development of the numerator for the first and second series of averages.

x_1	x_2	x_3	x_4	...	x_i	...
y_1	y_2	y_3	y_4	...	y_i	...
	$\bar{y}'_2 = y_1 + y_2 + y_3$	$\bar{y}'_3 = y_2 + y_3 + y_4$	$\bar{y}'_4 = y_3 + y_4 + y_5$	...	$\sum_{h=-1}^{1} y_{(i+h)} w_h$ $w_0 = 1,\ w_{\pm 1} = 1$	...
		$\bar{y}''_3 = \bar{y}'_2 + \bar{y}'_3 + \bar{y}'_4$	$\bar{y}''_4 = \bar{y}'_3 + \bar{y}'_4 + \bar{y}'_5$	...	$\sum_{h=-2}^{2} y_{(i+h)} w_h$	...
		$\bar{y}''_3 = y_1 + 2y_2 + 3y_3 + 2y_4 + y_5$	...	...	$w_0 = 3,\ w_{\mp 1} = 2,\ w_{\mp 2} = 1$	

moving averages (eq. 12.2), a second series of moving averages is calculated using values from the first series. Thus calculation of three successive series of moving averages produces the following results ($\bar{y}_i$), using weights w_h (Table 12.4):

first series ($m = 1$) $\quad \bar{y}_i = \frac{1}{3}\sum_{h=-1}^{1} y_{(i+h)} w_h \qquad w_0 = 1,\ w_{\pm 1} = 1$

second series ($m = 2$) $\quad \bar{y}_i = \frac{1}{9}\sum_{h=-2}^{2} y_{(i+h)} w_h \qquad w_0 = 3,\ w_{\pm 1} = 2,\ w_{\pm 2} = 1$

third series ($m = 3$) $\quad \bar{y}_i = \frac{1}{27}\sum_{h=-3}^{3} y_{(i+h)} w_h \qquad w_0 = 7,\ w_{\pm 1} = 6,\ w_{\pm 2} = 3,\ w_{\pm 3} = 1$

It is easy to check the above values by simple calculations, as shown in Table 12.4.

When using moving averages for estimating the trend of a series, one must choose the *width of the window* (i.e. choose m) as well as the *shape* of the moving average (i.e. the degree of the polynomial or the number of iterations). These choices are not simple. They depend in part on the goal of the study, namely the ecological interpretation of the *trend* itself or the subsequent analysis of *residuals* (i.e. detrended series). To estimate a cyclic trend, for instance, it is recommended to set the window width ($2m + 1$) equal to the period of the cyclic fluctuation.

Cyclic trend

Slutzky-Yule effect

Trend extraction by moving averages may add to the detrended series an artificial periodic component, which must be identified before analysing the series. This phenomenon is called the *Slutzky-Yule effect*, because these two statisticians independently drew attention to it in 1927. According to Kendall (1976, pp. 40-45) and Kendall *et al.* (1983, pp. 465-466), the average period of this artificial component (T) is calculated using the $(2m + 1)$ weights w_i of the moving average formula (eq. 12.3)*:

$$T = 2\pi/\theta \quad \text{for angle } \theta \text{ in radians, or} \quad T = 360°/\theta \quad \text{for angle } \theta \text{ in degrees,}$$

$$\text{where} \qquad \cos\theta = \left| \sum_{h=1}^{2m+1} (w_{h+1} - w_h)(w_h - w_{h-1}) \right| / \sum_{h=1}^{2m+2} (w_h - w_{h-1})^2 \qquad \textbf{(12.4)}$$

The values of the weights located outside the window are zero: $w_0 = 0$ and $w_{2m+2} = 0$. For example, using the weights of the second series of repeated moving averages above ($m = 2$):

$$[w_h] = [1 \;\; 2 \;\; 3 \;\; 2 \;\; 1]$$

gives

$$\cos\theta = \frac{|(2-1)(1-0) + (3-2)(2-1) + (2-3)(3-2) + (1-2)(2-3) + (0-1)(1-2)|}{(1-0)^2 + (2-1)^2 + (3-2)^2 + (2-3)^2 + (1-2)^2 + (0-1)^2} = \frac{3}{6}$$

from which it follows that $\qquad \theta = 1.047 \text{ rad} = 60°$

and thus: $\qquad T = 2\pi/1.047 = 360°/60° = 6$

After detrending by this method of *repeated moving averages*, if the analysis of the series resulted in a period $T \approx 6$, this period would probably be a by-product of the moving average procedure. It would not correspond to a component of the original data series, so that one should not attempt to interpret it in ecological terms. If a period $T \approx 6$ was hypothesized to be of ecological interest, one should use different weights for trend extraction by moving average analysis.

Analytical method

The most usual approach for estimating trends is the *analytical method*. It consists in fitting a model (*regression*) to the whole series, using the least squares approach or any other algorithm. The matter was fully reviewed in Section 10.3. Smoothing methods such as splines and LOWESS may also be used (Subsection 10.3.8). The model for the trend may be linear, polynomial, exponential, logistic, etc. The main advantages of trend extraction based on regression are: the explicit choice of a model

* In Kendall (1976) and Kendall *et al.* (1983) and previous editions of *The Advanced Theory of Statistics, Vol. 3*, there is a printing error in the formula for the Slutzky-Yule effect. In the first parenthesis of the last term of their numerator, the printed sign for the second weight (w_{2m+1}) is positive; this sign should be negative, as in eq. 12.4, giving (0 – 1) in our numerical example. However, their numerical example is correct, i.e. it is computed with $-w_{2m+1}$, not $+w_{2m+1}$.

by the investigator and the ease of calculation using a statistical package. The main problem is that a new regression must be calculated upon addition of one or several observations to the data series, which may generate different values for the regression coefficients. However, as the series gets longer, estimates of the regression coefficients become progressively more independent of the length of the series.

Variate difference

Contrary to the above methods, where the trend was first estimated and then subtracted from the observed data (eq. 12.1), the *variate difference method* directly detrends the series. It consists in replacing each value y_i by the difference $(y_{i+1} - y_i)$. As in the case of repeated moving averages, differences may be calculated not only on the original data, but also on data resulting from previous detrending. If this is repeated on progressively more and more detrended series, the variance of the series usually stabilizes rapidly. The variate difference method, when applied once or a few times to a series, can successfully remove any polynomial trend. Only exponential or cyclic trends may sometimes resist the treatment. The method may be used to remove any cyclic trend whose period T is known, by using differences $(y_{i+T} - y_i)$; however, this is fully successful only in cases where T is an integer multiple of the sampling interval Δ.

Cyclic trend

In some instances, ecologists may also wish to eliminate the random *noise* component from the data series, in order to better evidence the ecological phenomenon under study. This operation, whose aim is to remove high-frequency variability from the series, is called *filtration*. In a sense, filtration is the complement of trend extraction, since trends are low-frequency components of the series. Several specialists of series analysis apply the term *filter* to any preliminary treatment of the series, whether extraction of low frequencies (trend) or removal of high frequencies (noise). Within the context of spectral analysis (Section 12.5), filtration of the series is often called “prewhitening”. This refers to the fact that filtration flattens the spectrum of a series and makes it similar to the spectrum of white light. The reciprocal operation (called “recolouring”) fits the spectrum (calculated on the filtered series) in such a way as to make it representative of the nonfiltered series. The sequence of operations — prewhitening of the series, followed by computation of the spectrum on the filtered series, and finally recolouring of the resulting spectrum — finds its justification in the fact that spectra that are more flat are also more precisely estimated.

Filtration
Filter

In addition to filters, which aim at extracting low frequencies (trends), computer programs for series analysis offer a variety of numerical filters that allow the removal, or at least the reduction, of any component located outside a given frequency band (passband). It is thus possible, depending on the objective of the study, to select the high or the low frequencies, or else a band of intermediate frequencies. It is also possible to eliminate a band of intermediate frequencies, which is the converse of the latter filter. Generally, these numerical filters are found in programs for spectral analysis (Section 12.5), but they may also be used to filter series prior to analyses using the methods described in Sections 12.3 and 12.4. In most cases, filtering data series (including trend extraction) requires solid knowledge of the techniques, because filtration always distorts the original series and thus influences further calculations. It is therefore better to do it under the supervision of an experienced colleague.

12.3 Periodic variability: correlogram

The systematic component of stationary series is called *periodic variability*. There are several methods available for analysing this type of variability. Those discussed in the present section, namely the autocovariance and autocorrelation (serial correlation) and the cross-covariance and cross-correlation, are all extensions, to the analysis of data series, of statistical techniques described in earlier chapters. These methods have been extensively used in ecology.

At this stage of series analysis, it is assumed that the data series is *stationary*, either because it originally exhibited no trend or as the result of trend extraction (Section 12.2). It is also assumed that variability is large enough to emerge from random *noise*.

A general approach for analysing periodic variability is derived from the concepts of covariance and correlation, which were defined in Chapter 4. The methods are called *autocovariance* and *autocorrelation*. The approach is to quantify the relationships between successive terms of the data series. These relationships reflect the pattern of periodic variability.

1 — Autocovariance and autocorrelation

Autocovariance measures the covariance of the series with itself, computed as the series is progressively shifted with respect to itself (Fig. 12.5). When there is no shift (i.e. lag of zero unit; $k = 0$), the covariance of the series with itself is equal to its variance (Section 4.1):

$$s_{yy} = s_y^2$$

When the series is shifted relative to itself by one unit (lag $k = 1$), the left-hand copy of the series in Fig. 12.5 loses observation y_1 and the right-hand copy loses observation y_n. The two truncated series, each of length $(n - 1)$, are compared. For a lag of k units, the means $\bar{y}'$ and $\bar{y}''$ of the $(n - k)$ terms remaining in the two truncated series are:

$$\bar{y}' = \frac{1}{n-k} \sum_{i=k+1}^{n} y_i \qquad \bar{y}'' = \frac{1}{n-k} \sum_{i=1}^{n-k} y_i$$

These means are used to compute the *autocovariance* $s_{yy}(k)$ of the series, for lag k, using an equation similar to that of the covariance (eq. 4.4):

Auto-covariance

$$s_{yy}(k) = \frac{1}{n-k-1} \sum_{i=1}^{n-k} (y_{i+k} - \bar{y}')(y_i - \bar{y}'') \qquad \textbf{(12.5)}$$

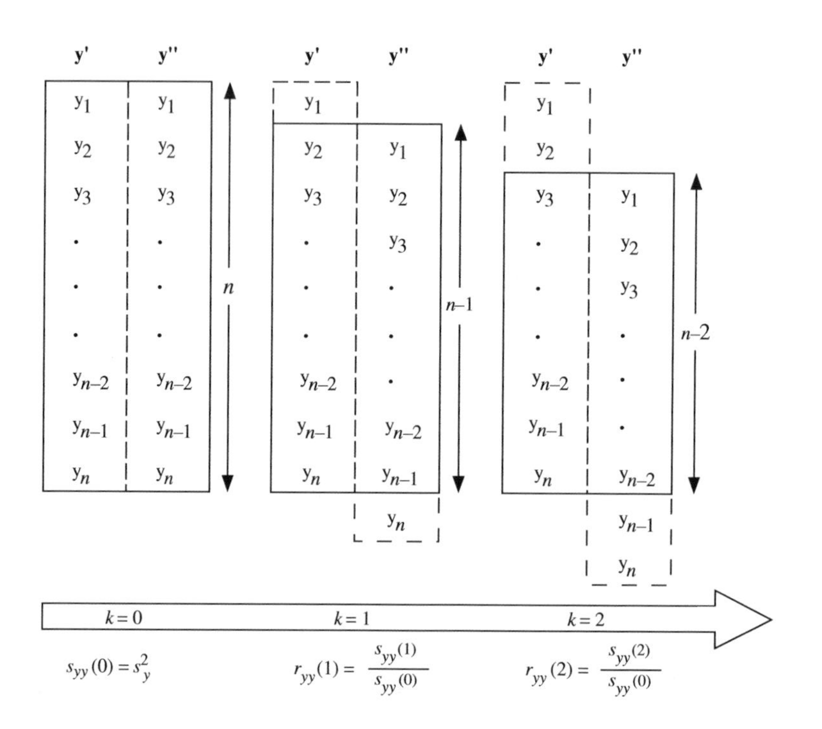

Figure 12.5 Calculation of autocovariance (s_{yy}) and autocorrelation (r_{yy}). Stepwise shift of a data series relative to itself, with successive lags of k units. The number of terms involved in the calculation $(n - k)$ decreases as k increases.

When $k = 0$ (no shift), eq. 12.5 gives, as it should, the variance (eq. 4.3) of the series:

$$s_{yy}(0) = \frac{1}{n-1}\sum_{i=1}^{n}(y_i - \bar{y})^2 = s_{yy} = s_y^2$$

In correlograms (below), the autocovariance is estimated for several successive lags k. In specific applications, researchers may decide on biological grounds how long the lag should be when computing the autocovariance of a variable of interest.

One way to calculate the *linear correlation* between two variables, y_j and y_k, is to divide their covariance by the product of their two standard deviations (eq. 4.7):

$$r_{jk} = \frac{s_{jk}}{s_j s_k}$$

In a similar way, the *autocorrelation* of a series $r_{yy}(k)$ may be defined as the ratio of its autocovariance $s_{yy}(k)$ to its variance $s_y^2 = s_{yy}(0)$:

Auto-correlation

$$r_{yy}(k) = \frac{s_{yy}(k)}{s_y s_y} = \frac{s_{yy}(k)}{s_y^2} = \frac{s_{yy}(k)}{s_{yy}(0)} \qquad \textbf{(12.6)}$$

Equation 12.6 is valid only when $(n - k)$ is reasonably large. The autocorrelation is also called *serial correlation*. It measures the average dependence of the values in the series on values found at a distance of k lags. Jenkins and Watts (1968) discuss the relative advantages of alternative formulas for estimating autocorrelation coefficients.

The autocorrelation $r_{yy}(k)$ may also be directly computed by calculating the *linear correlation* between terms y_i and y_{i+k} of the series, for the $n - k$ pairs of corresponding values in the observed and shifted series (Fig. 12.5). The corresponding formula is:

$$r_{yy}(k) = \frac{s_{yy}(k)}{\sqrt{s_{y'}^2 s_{y''}^2}} \qquad \textbf{(12.7)}$$

Equation 12.7 is not recommended, however, because the means used for computing $s_{y'}^2$ and $s_{y''}^2$, respectively, change with lag k, so that $r_{yy}(k)$ is not a satisfactory estimate when considering a set of autocorrelation coefficients (Jenkins & Watts, 1968).

Since the number of terms $(n - k)$ involved in the calculation of the autocovariance or autocorrelation decreases as k increases, it follows that, as k increases, the precision of the estimate and the number of degrees of freedom available decrease, so that the maximum lag is generally taken to be $k_{max} \le n/4$. Table 12.5 gives the values of autocovariance and autocorrelation for the artificial stationary series of Fig. 12.2b.

Autocorrel-ogram

The coefficients of autocorrelation (or autocovariance) are plotted as a function of lag k (abscissa), in a graph called *autocorrelogram* (or *correlogram*, for simplicity). Autocorrelation coefficients range between +1 and –1. The scale factor between the autocorrelation and autocovariance coefficients is the variance of the series (eq. 12.6). In Fig. 12.6, this factor is $s_{yy}(0) = 3.16$; it is given in Table 12.5 at lag $k = 0$.

The interpretation of correlograms is based on the following reasoning. At lag $k = 0$, the two copies of the series (**y**' and **y**") have the exact same values facing each other (Fig. 12.5), so that $r_{yy}(0) = +1$. With increasing lag k, corresponding values in the series **y**' and **y**" move farther apart and $r_{yy}(k)$ decreases. This is what is happening, in the numerical example, for lags up to $k = 4$ (Table 12.5 and Fig. 12.6). In series where periodic variability is present (with period T_p), increasing k eventually brings similar values to face each other again (at lag $k = T_p$), with peaks facing peaks and troughs facing troughs, hence a high positive $r_{yy}(k)$. $r_{yy}(k = T_p)$ is always $< +1$, however, because there is noise in the data and also because natural periodic phenomena seldom repeat themselves perfectly. Often, negative autocorrelation reaches its maximum at $k = T_p/2$, because the signals in **y**' and **y**" are then maximally out of phase.

Table 12.5 Autocovariance and autocorrelation coefficients for the artificial series of Fig. 12.2b, after detrending (i.e. periodic signal + noise components only). For each successive lag, the series is shifted by one sampling interval. Values corresponding to odd lags are not shown. The autocovariance and autocorrelation coefficients are plotted against lag in Fig. 12.6.

Lag	Autocovariance $s_{yy}(k)$	Autocorrelation $r_{yy}(k)$
0	3.16	1.00
2	1.17	0.37
4	–1.17	–0.37
6	–0.26	–0.08
8	–0.31	–0.10
10	–1.13	–0.36
12	–0.48	–0.15
14	–0.47	–0.15
16	–1.20	–0.38
18	0.74	0.23
20	3.12	0.99
22	1.86	0.59

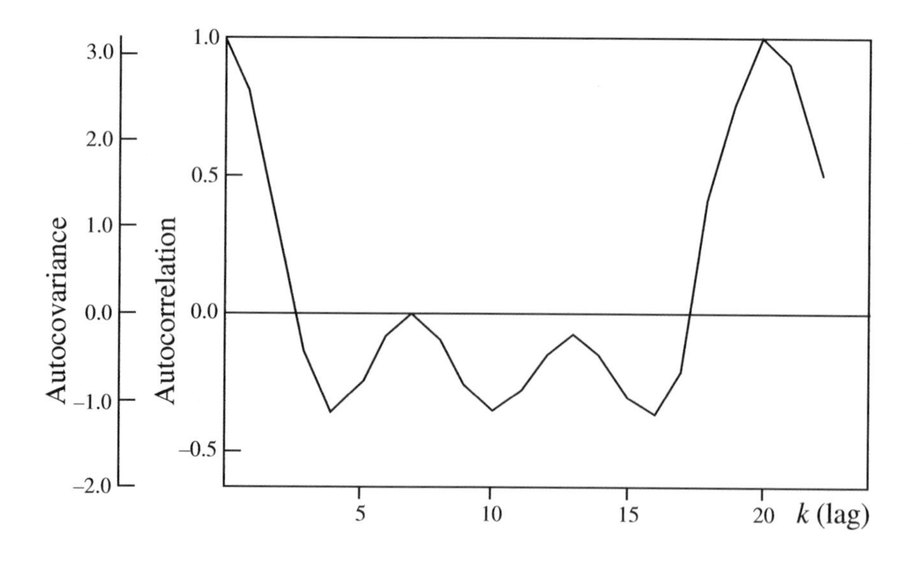

Figure 12.6 Correlogram (autocovariance and autocorrelation; values from Table 12.5) for the artificial series of Fig. 12.2b, after detrending (i.e. periodic signal + noise components only).

A practical problem occurs when there are several periodic signals in a series; this may increase the complexity of the correlogram. Nevertheless, high positive values in a correlogram may generally be interpreted as indicative of the presence of periodic variability in the series. For the numerical example, Fig. 12.6 indicates that there is a major periodicity at $k = 20$, corresponding to period $T = 20$; this interpretation is supported by the low value of $r_{yy}(10)$. Period $T = 20$ is indeed the distance between corresponding peaks or troughs in the series of Fig. 12.2b. Other features of the correlogram would be indicative of additional periods (which is indeed the case here; see Fig. 12.13) or may simply be the result of random noise.

Under the hypothesis of normality, a confidence interval can be computed and drawn on a correlogram in order to identify the values that are significantly different from zero. The confidence interval is usually represented on the correlogram as a two-standard-error band. The test formula for the Pearson correlation coefficient (eq. 4.13) cannot be used here because the data are not independent of one another, being autocorrelated (Section 1.1). According to Bartlett (1946), the variance of each term in the correlogram is a function of all the autocorrelation values in the series:

$$s^2[r_{yy}(k)] \approx \frac{1}{n} \sum_{h=-\infty}^{+\infty} [\rho^2(h) + \rho^2(h-k)\rho^2(h+k) + 2\rho^2(h)\rho^2(k) - 4\rho^2(h)\rho^2(k)\rho^2(h-k)]$$

In practice, (a) the series has a finite length, so that the maximum lag is $k_{max} \leq n/4$; (b) what is known are not the parameters ρ but their estimates r; and (c) the last three terms in the above equation are small. For these reasons, the variance of the correlogram is generally estimated using the following simplified formula, where $s(r_{yy})$ is the standard error of each autocorrelation coefficient r_{yy}:

$$s^2(r_{yy}) = \frac{1}{n}\left[1 + 2\sum_{h=1}^{k_{max}} r_{yy}^2(i)\right] \tag{12.8}$$

For $\alpha = 0.05$, when the number of observations in the series, n, is larger than 50, the confidence interval of r_{yy} is $\pm 1.96\, s(r_{yy})$. Some computer programs calculate $s(r_{yy})$ using a different formula.

To illustrate how to use eq. 12.8, let us modify the numerical example and assume that the $r_{yy}(k)$ values in Table 12.5 correspond to lags $k = 1, 2, \ldots, 11$, and that $n = 4 \times k_{max} = 4 \times 11 = 44$ observations. Equation 12.8 would then give:

$$s^2(r_{yy}) \approx \frac{1}{44}\{1 + 2[(0.37)^2 + (-0.37)^2 + \ldots + (0.99)^2 + (0.59)^2]\} = 0.11$$

so that $1.96 s(r_{yy}) = 1.96 \times \sqrt{0.11} = 1.96 \times 0.34 = 0.66$. Thus, in the correlogram of Fig. 12.6, only the values of r_{yy} larger than +0.66 or smaller than –0.66 would be significantly different from zero, i.e. in this case $r_{yy}(1)$, $r_{yy}(19)$, $r_{yy}(20)$ and $r_{yy}(21)$.

Harmonic

When the series is long, its correlogram may exhibit significant values for *harmonics* (integer multiples) of the period present in the signal (T_{series}). This is a normal phenomenon, which is generally not indicative of additional periodicity in the data series. However, when a value of the correlogram statistic is noticeably larger for a harmonic period than for the basic period, one can conclude that the harmonic is also a true period of the series. Except for very strong periodic components, *short series* should not be analysed using autocorrelograms. This is because the test of significance is not very powerful in this case, i.e. the probability of rejecting the null hypothesis of no autocorrelation is small when a real periodic component is present in short series. In the same way, when there is *more than one periodic component*, correlograms should generally not be used even with long series, because components of different periods may interfere with one another and prevent the correlogram from showing significance (see also the next paragraph). Finally, when the data are *not equispaced* and one does not wish to interpolate, methods developed for *spatial autocorrelation* analysis, which do not require equal spacing of the data, may be used (Section 13.1). Special forms of spatial autocorrelation coefficients also allow the analysis of series of *qualitative* data.

It may happen that periods present in the series do not appear in a correlogram, because they are concealed by other periods accounting for larger fractions of the variance of the series. When one or several periods have been identified using a first correlogram, one may remove these periods from the series using one of the methods recommended in Section 12.2 for cyclic trends and compute a new correlogram for the detrended series. It could bring out previously concealed periods. This is not without risk, however, because successively extracting trends rapidly distorts the residuals. Approaches better adapted to series containing several periods are discussed in Sections 12.4 and 12.5.

The following numerical example and ecological applications illustrate the computation and use of correlograms.

Numerical example. Consider the following series of 16 data points (quantitative variable):

2, 2, 4, 7, 10, 5, 2, 5, 8, 4, 1, 2, 5, 9, 6, 3

Table 12.6 illustrates the computation of the autocorrelation coefficients. These could be plotted as a function of lag (k) to form a correlogram, as in Figs. 12.6 and 12.7b. The coefficients clearly point to a dominant period at $k = 5$, for which autocorrelation is positive and maximum. This approximately corresponds to the average distance separating successive maximum values, as well as successive minima, along the data series.

Ecological application 12.3a

In order to study the spatial variability of coastal marine phytoplankton, Platt et al. (1970) measured chlorophyll *a* along a transect 8 nautical miles long, at 10 m depth and intervals of 0.1 naut. mi. (1 naut. mi. = 1852 m). The resulting 80 values are shown in Fig. 12.7a.

The series exhibited a clear linear *trend*, which was extracted at the beginning of the analysis. Autocorrelation coefficients were computed from the residual series, up to lag $k = 10$,

Table 12.6 Computation of the autocorrelation coefficients for the data of the numerical example. Boxes delimit the values included in each calculation. Note how the highest values are facing each other at lag 5, where the autocorrelation coefficient is maximum.

Lag	Data series																		Autocorrelation $r_{yy}(k)$
k=0	2	2	4	7	10	5	2	5	8	4	1	2	5	9	6	3			1.000
	2	2	4	7	10	5	2	5	8	4	1	2	5	9	6	3			
k=1	2	2	4	7	10	5	2	5	8	4	1	2	5	9	6	3			0.326
		2	2	4	7	10	5	2	5	8	4	1	2	5	9	6	3		
k=2	2	2	4	7	10	5	2	5	8	4	1	2	5	9	6	3			–0.603
			2	2	4	7	10	5	2	5	8	4	1	2	5	9	6	3	
k=3	2	2	4	7	10	5	2	5	8	4	1	2	5	9	6	3			–0.562
				2	2	4	7	10	5	2	5	8	4	1	2	5	9	…	
k=4	2	2	4	7	10	5	2	5	8	4	1	2	5	9	6	3			0.147
					2	2	4	7	10	5	2	5	8	4	1	2	5	…	
k=5	2	2	4	7	10	5	2	5	8	4	1	2	5	9	6	3			0.502
						2	2	4	7	10	5	2	5	8	4	1	2	…	
k=6	2	2	4	7	10	5	2	5	8	4	1	2	5	9	6	3			–0.178
							2	2	4	7	10	5	2	5	8	4	1	…	
etc.								etc.											etc.

because the series was quite short (Fig. 12.7b). The position of the first *zero* in the *correlogram* was taken as indicative of the average apparent *radius* of phytoplankton patches along the transect. The model underlying this interpretation is that of circular patches, separated by average distances equal to their average diameter. In such a case, it is expected that the second zero would occur at a lag three times that of the first zero, as is indeed observed on the correlogram. In the present case, the average *diameter* of phytoplankton patches and the distance separating them appear to be ca. 0.5 naut. mi.

Ecological application 12.3b

Steven & Glombitza (1972) sampled tropical phytoplankton and chlorophyll at a site off Barbados, during nearly three years. Sampling was approximately fortnightly. The physical environment there is considered to be very stable. The most abundant phytoplankton species, in surface waters, is the filamentous cyanobacterium *Trichodesmium thiebaudii*. Data were concentrations of chlorophyll *a* and of *Trichodesmium* filaments.

The raw data were subjected to two transformations: (1) computation of *equispaced* data at intervals of 15 days, by interpolation, and (2) *filtration* intended to reduce the importance of non-dominant variations. The filtered data are shown in Fig. 12.8a, where synchronism between

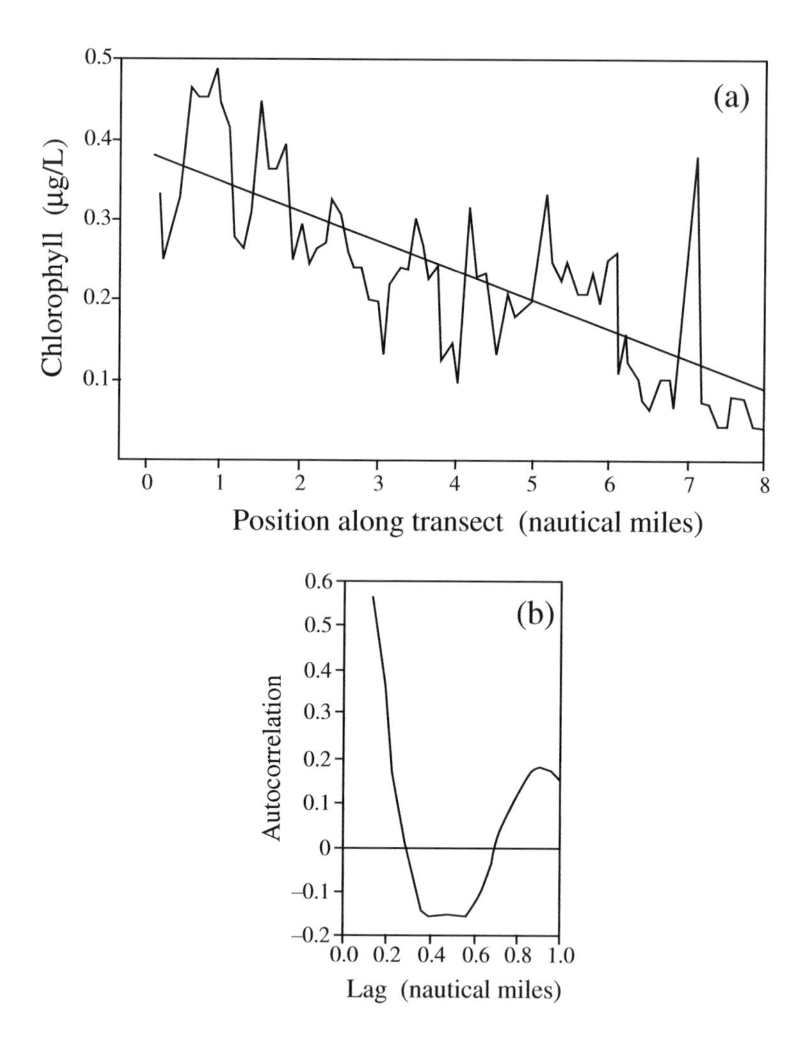

Figure 12.7 Chlorophyll *a* concentrations in a coastal marine environment, along a transect 8 naut. miles long in St. Margaret's Bay (Nova Scotia, Canada). (a) Data series exhibiting a linear trend, and (b) correlogram of the detrended series where lags (abscissa) are given as distances along the transect. After Platt *et al.* (1970).

the two sampled variables is obvious. Correlograms for the nonfiltered (Fig. 12.8b) and filtered (Fig. 12.8c) series clearly show the same periodic signal, of ca. 8 lags × (15 days lag^{-1}) = 120 days. Nonfiltered data provide the same information as the filtered series, but not quite as clearly. According to the authors, these periodic variations could be an example of free oscillations, since they seem independent of any control by the environment which is stable all the year round. The same ecological application will be used again to illustrate the calculation of cross-correlation (next Subsection) and Schuster's periodogram (Section 12.4).

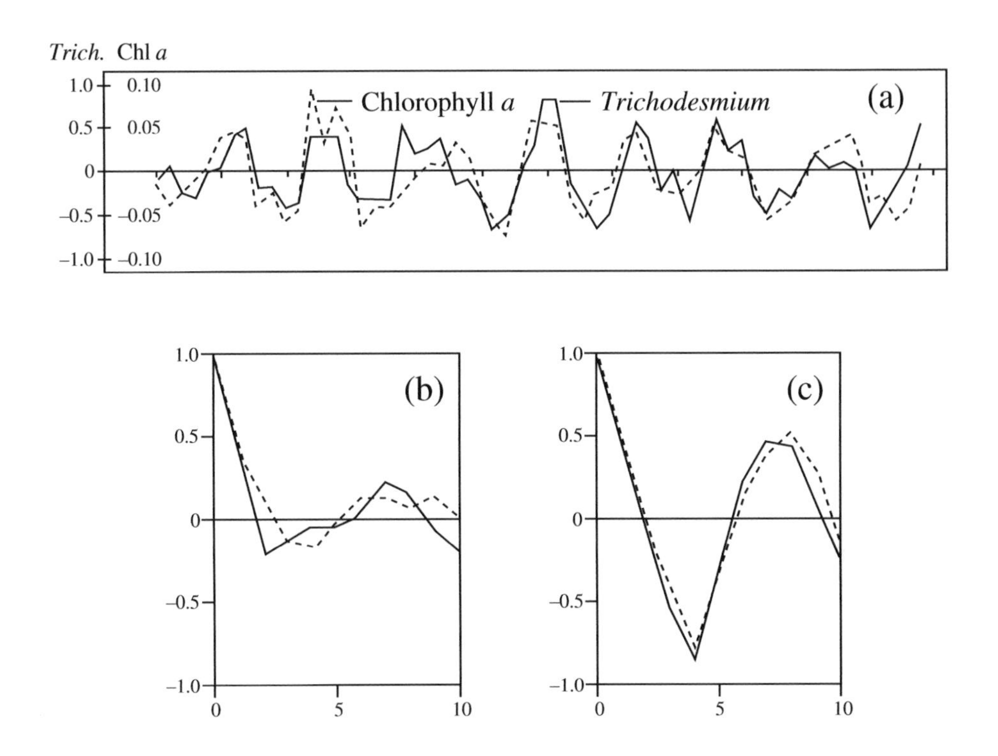

Figure 12.8 (a) Filtered time series of chlorophyll *a* and *Trichodesmium* in tropical surface waters off Barbados. Marks along the abscissa are spaced by 75 days. On the ordinate, units are 10^3 filaments *Trichodesmium* L^{-1} and μg chlorophyll *a* L^{-1}. Correlograms of (b) the nonfiltered series and (c) the filtered series. After Steven & Glombitza (1972).

2 — *Cross-covariance and cross-correlation*

In order to determine the extent to which two data series exhibit concordant periodic variations, one can use a method closely related to autocovariance and autocorrelation. This method has two variants called *cross-covariance* and *cross-correlation* (or *lag correlation*).

Consider two series, $\mathbf{y}_j$ and $\mathbf{y}_l$, of identical lengths. One is progressively shifted with respect to the other, with lags $k = 1, 2, \ldots$ As the lag increases, the zone of overlap of the two series shortens. *Cross-covariance* of order k is computed in a way which is analogous to autocovariance, using the means of the two truncated series. For a lag of k units, the means $\bar{y}_j$ and $\bar{y}_l$ of the $(n - k)$ terms of the two truncated series are:

$$\bar{y}_j = \frac{1}{n-k} \sum_{i=k+1}^{n} y_{ij} \qquad \bar{y}_l = \frac{1}{n-k} \sum_{i=1}^{n-k} y_{il}$$

As in eq. 12.5 for autocovariance, these mean values are used to compute the cross-covariance $s_{jl}(k)$ between the two series, for lag k:

Cross-covariance

$$s_{jl}(k) = \frac{1}{n-k-1} \sum_{i=1}^{n-k} [y_{(i+k)j} - \bar{y}_j]\,[y_{il} - \bar{y}_l] \qquad \textbf{(12.9)}$$

When $k = 0$ (no shift), eq. 12.5 gives the covariance (eq. 4.4) of the two variables:

$$s_{jl}(0) = \frac{1}{n-1} \sum_{i=1}^{n} (y_{ij} - \bar{y}_j)\,(y_{il} - \bar{y}_l) = s_{jl}$$

Equation 12.9 stresses an important difference between *cross-covariance* and *autocovariance*, namely that the relative direction in which a series is shifted with respect to the other must be taken into account. Indeed, shifting series $\mathbf{y}_j$ "to the right" with respect to series $\mathbf{y}_l$ is not equivalent to shifting it "to the left":

y_{1j} — — | — — — — — y_{nj} |
← k → | y_{1l} — — — — — | — y_{nl}

← k → | y_{1j} — — — — — | — y_{nj}
y_{1l} — — | — — — — — y_{nl} |

The value of cross-covariance for lag k would be different if $\mathbf{y}_j$ and $\mathbf{y}_l$ were interchanged in eq. 12.9; in other words, generally $s_{jl}(k) \neq s_{lj}(k)$. In order to distinguish between the two sets of cross-covariances, one set of shifts is labelled as positive and the other as negative. The choice of the positive and negative directions is arbitrary and without consequence. One must specify which variable leads the other, however. In eq. 12.9, if the cross-covariance of $\mathbf{y}_j$ relative to $\mathbf{y}_l$ is identified as $s_{jl}(k)$, the converse would be labelled $s_{jl}(-k)$. No distinction was made between the two relative shift directions in autocovariance (eq. 12.4) because $s_{yy}(+k) = s_{yy}(-k)$.

Cross-covariance is generally plotted as a function of the positive and negative lags k, to the right and left of $k = 0$. The alternative is to plot the two sets on the positive side of the abscissa using different symbols. Maximum cross-covariance does not necessarily occur at $k = 0$. Sometimes, the interaction between the two series is maximum at a lag $k \neq 0$. In predator-prey interactions for example, cross-covariance is maximum for a lag corresponding to the response time of the predator population (*target variable*) to changes in the number of prey (*predictor variable*). One then says that the target variable *lags* the causal variable.

Cross-covariance may be transformed into *cross-correlation*. To do so, the cross-covariance $s_{jl}(k)$ is divided by the product of the corresponding standard deviations:

Cross-correlation

$$r_{jl}(k) = \frac{s_{jl}(k)}{s_j(k)\,s_l(k)} = \frac{\sum_{i=1}^{n-k} [y_{(i+k)j} - \bar{y}_j]\,[y_{il} - \bar{y}_l]}{\sqrt{\sum_{i=1}^{n-k} [y_{(i+k)j} - \bar{y}_j]^2 \sum_{i=1}^{n-k} [y_{il} - \bar{y}_l]^2}} \qquad (12.10)$$

Cross-correlogram

As for cross-covariance, cross-correlation is defined for $+k$ and $-k$. Values are plotted as a function of k in a *cross-correlogram*. In the same vein, Fortier & Legendre (1979) used Kendall's τ (Section 5.3) instead of Pearson's r for computing cross-correlations between series of quantitative variables which were *not linearly related*. They called this measure *Kendall's cross-correlation*. It may also be applied to series of *semiquantitative* data; Spearman's r (Section 5.3) could be used instead of Kendall's τ. Extending this approach to *qualitative* data was proposed by Legendre & Legendre (1982) under the name *cross-contingency*. In this case, contingency statistics (X^2 or uncertainty coefficients; Section 6.2) are computed for the two series, as these are progressively shifted with respect to each other.

When several ecological variables are observed simultaneously, the resulting *multidimensional series* may be analysed using cross-covariance or cross-correlation. Such methods are obviously of interest in ecology, where variations in one variable are often interpreted in terms of variations in others. However, eqs. 12.9 and 12.10 consider only two series at a time; for multidimensional data series, it is sometimes useful to extend the concept of *partial correlation* (Sections 4.5 and 5.3) to the approach of cross-correlation. In Ecological application 12.3d, Fréchette & Legendre (1982) used Kendall's partial (partial τ; Section 5.3) cross-correlation to analyse an ecological situation involving three variables.

Ecological application 12.3c

In their study of the temporal variability of tropical phytoplankton (Ecological application 12.3b), Steven & Glombitza (1972) compared the variations in concentrations of chlorophyll *a* and *Trichodesmium*, using cross-correlations (Fig. 12.9). The cross-correlogram shows that changes in the two variables were in phase, with a period of 8 lags × 15 days lag^{-1} = 120 days. Filtration of the data series brought but a small improvement to the cross-correlation. These results confirm the conclusions drawn from the correlograms (Fig. 12.8) and show that variations of chlorophyll *a* concentration, in surface waters, were due to changes in the

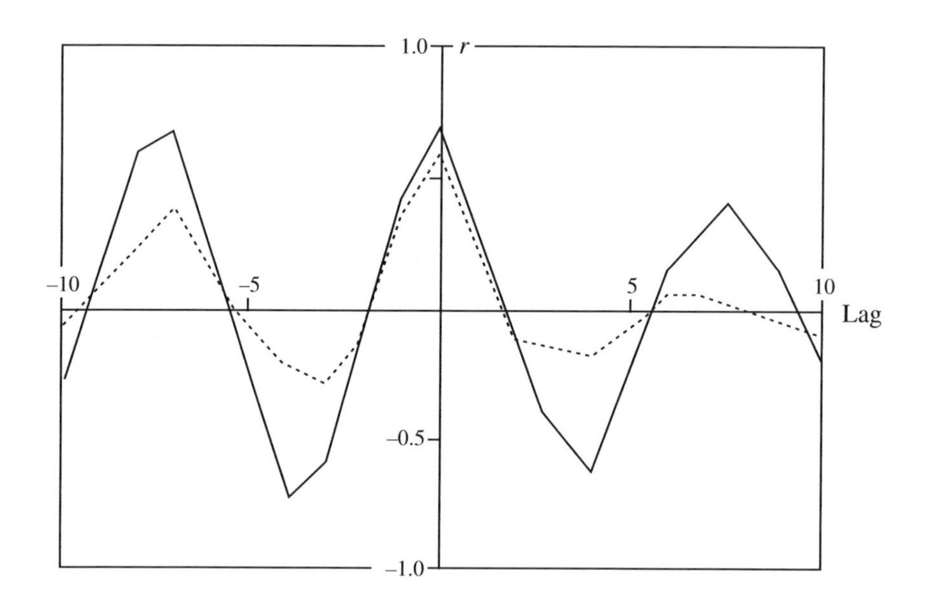

Figure 12.9 Cross-correlations between temporal changes in concentrations of chlorophyll *a* and *Trichodesmium*, in tropical surface waters, computed on nonfiltered (solid line) and filtered (dotted line) data series. After Steven & Glombitza (1972).

concentration of *Trichodesmium* filaments. This same application will be further discussed in Section 12.4 (Ecological application 12.4d).

Ecological application 12.3d

At an anchor station in the St. Lawrence Estuary (Québec), Fréchette & Legendre (1982) determined the photosynthetic capacity of phytoplankton (P^B_{max}) hourly, during six consecutive days. The area is subjected to internal tides, which drove changes in two important physical variables: (1) vertical oscillations of the water mass (characterized in this study by the depth of isopycnal $\sigma_t = 22$, i.e. the depth where the density of water is 1022 kg m^{-3}), and (2) variations in the vertical stability of the upper water column, estimated as the density gradient between 1 and 25 m. Two hypotheses could explain the observed effects of the internal tides on P^B_{max}: (1) upwelling, to the depths where sampling took place, of deeper water containing phytoplankton with lower P^B_{max}, under the effect of incoming internal tides, or (2) adaptation of P^B_{max} to changes in the vertical stability of the upper water column. Since the two physical variables were controlled by the same mechanism (i.e. internal tides), it was not easy to identify their specific contributions to phytoplankton photosynthesis. This was achieved by computing two Kendall's partial cross-correlations (partial τ): (1) between P^B_{max} and the depth of $\sigma_t = 22$, controlling for the effect of vertical stability, and (2) between P^B_{max} and stratification, controlling for vertical displacement. When calculating the partial cross-correlations, the response variable (P^B_{max}) was shifted relative to the two potentially causal (physical) variables until a maximum value was reached. The authors concluded that the photosynthetic activity of phytoplankton responded to changes in the vertical stability of the water column, driven by internal tides. This was interpreted as an adaptation of the cells to periodic variations in their light environment.

Mantel correlograms may be used to detect periodic phenomena in quantitative, semiquantitative or qualitative data series, or else in multivariate data series involving either quantitative variables alone, or a mixture of variables of different precision levels. This type of correlogram, in Subsection 13.1.5, is computed from a similarity or distance matrix among the observations in the series.

12.4 Periodic variability: periodogram

In addition to the relatively simple methods discussed in the previous section, there is another general approach to the study of periodic variability, called *harmonic analysis*. This approach is mathematically more complex than correlogram analysis, but it is often better adapted to the study of ecological data series. Results of harmonic analysis are generally plotted in a graph called *periodogram*.

1 — Periodogram of Whittaker and Robinson

The simplest way to approach harmonic analysis is to examine a *Buys-Ballot table*. Assume that a series of n quantitative observations is characterized by a period T_{series}. If $T = T_{series}$ is known, the series can be split into n/T sequences, each containing T observations. A Buys-Ballot table (Table 12.7) is a double-entry table whose rows contain the $r = n/T$ sequences of T observations. The number of columns corresponds to the known or assumed period of the data series. If $T = T_{series}$, the r successive rows in the table are repetitions of the same oscillation, although the actual values in any column (j) are generally not identical because of noise. Calculating the mean value for each column ($\bar{y}_{T,j}$) and comparing these means is a way of characterizing the variation within period T_{series}.

When there exists a hypothesis concerning the value of T_{series} (e.g. a diurnal cycle), Buys-Ballot tables may be constructed for this value and also for neighbouring lower and higher values T_k. As the period of the table (T_k) approaches that of the series (T_{series}), values within each column become more similar, so that all maximum values tend to be located in one column and all minimum values in another. As a result, the difference between the highest and lowest mean values is maximum when period T_k of the table is the same as period T_{series} of the series. The *amplitude* of a Buys-Ballot table is some measure of the variation found among the columns of the table. It may be measured by the *range* of the column means (Whittaker & Robinson, 1924):

Amplitude

Range

$$[\bar{y}_{max} - \bar{y}_{min}] \qquad \textbf{(12.11)}$$

Table 12.7 Buys-Ballot table. Allocation of data from a series containing n observations to the rows of the table.

	1	2	2	...	T
1	y_1	y_2	y_3	...	y_T
2	y_{T+1}	y_{T+2}	y_{T+3}	...	y_{2T}
.	.	.	.	...	.
.	.	.	.	...	.
.	.	.	.	...	.
r	$y_{(r-1)T+1}$	$y_{(r-1)T+2}$	$y_{(r-1)T+3}$	...	$y_{rT} = y_n$
$\bar{y}_T$	$\bar{y}_{T,1}$	$\bar{y}_{T,2}$	$\bar{y}_{T,3}$	...	$\bar{y}_{T,T}$

or by the *standard deviation* of the column means (Enright, 1965):

Standard deviation

$$\sqrt{\frac{1}{T_k}\sum_{j=1}^{T_k}(\bar{y}_{T_k,j}-\bar{y}_{T_k})^2}, \quad \text{where} \quad \bar{y}_{T_k} = \frac{1}{T_k}\sum_{j=1}^{T_k}\bar{y}_{T_k,j} \qquad \textbf{(12.12)}$$

When the period T of interest is not an integer multiple of the interval between two observations, a problem occurs in the construction of the Buys-Ballot table. The solution proposed by Enright (1965) is to construct the table with a number of columns equal to the largest integer which is less than or equal to the period of interest, T. Observations are attributed to the columns in sequence, as usual, leaving out an observation here and there in such a way that the average rate of advance in the series, from row to row of the Buys-Ballot table, is T. This is done, formally, by using the following formula for the mean of each column j:

$$\bar{y}_{T,j} = \frac{1}{r}\sum_{i=1}^{r} y_{[(i-1)T+j]} \qquad \textbf{(12.13)}$$

where r is the number with data, in column j of the table. The subscript of y is systematically rounded to the next integer. Thus, for example, if $T = 24.5$, $\bar{y}_{T,1}$ would be estimated from values y_1, y_{26}, y_{50}, y_{75}, y_{99}, y_{124}, etc.; in other words, intervals of 24 and 25 units would be successively used, to give an average period $T = 24.5$. This modified formula is required to understand Ecological application 12.4a, where fractional periods are being used.

When studying an empirical data series, the period T_{series} is not known *a priori*. Even when some hypothesis is available concerning the value of T_{series}, one may want to check whether the hypothesized value is the one that best emerges from the data. In both situations, estimating T_{series} becomes the purpose of the analysis. The values of amplitude, computed for different periods T, may be plotted together as a periodogram in order to determine which period best characterizes the data series.

Periodogram

The *periodogram of Whittaker & Robinson* is a graph in which the measures of amplitude (eq. 12.11 or 12.12) are plotted as a function of periods T_k. According to Enright (1965), periodograms based on the statistic of eq. 12.12 are more internally consistent than those based on eq. 12.11. Various ways have been proposed for testing the significance of statistic 12.12 (reviewed by Sokolove and Bushell, 1978); these tests are only asymptotically valid, so that they are not adequate for short time series.

Numerical example. Consider again the series (2, 2, 4, 7, 10, 5, 2, 5, 8, 4, 1, 2, 5, 9, 6, 3) used in Subsection 12.3.1 to compute Table 12.6. In order to examine period $T_k = 4$, for instance, the series is cut into segments of length 4 as follows:

2, 2, 4, 7;   10, 5, 2, 5;   8, 4, 1, 2;   5, 9, 6, 3

and distributed in the successive rows of the table. The Buys-Ballot tables for periods $T_k = 4$ and 5 are constructed as follows:

$T = 4$	1	2	3	4
Row 1	2	2	4	7
Row 2	10	5	2	5
Row 3	8	4	1	2
Row 4	5	9	6	3
Means	6.25	5	3.25	4.25

Range = 3, standard deviation = 1.0951

$T = 5$	1	2	3	4	5
Row 1	2	2	4	7	10
Row 2	5	2	5	8	4
Row 3	1	2	5	9	6
Row 4	3				
Means	2.75	2	4.67	8	6.67

Range = 6, standard deviation = 2.2708

The range is calculated using eq. 12.11 and the standard deviation with eq. 12.12. Repeating the calculations for $k = 2$ to 8 produces the periodogram in Fig. 12.10.

Interpretation of the periodogram may be quite simple. If one and only one oscillation is present in the series, the period with maximum amplitude is taken as the best estimate for the true period of this oscillation. Calculation of the periodogram is made under the assumption that there is *a single stable period* in the series. If several periods are present, the periodogram may be so distorted that its interpretation could lead to erroneous conclusions. Enright (1965) provides examples of such distortions, using artificial series. Other methods, discussed below, are better adapted to series with several periods.

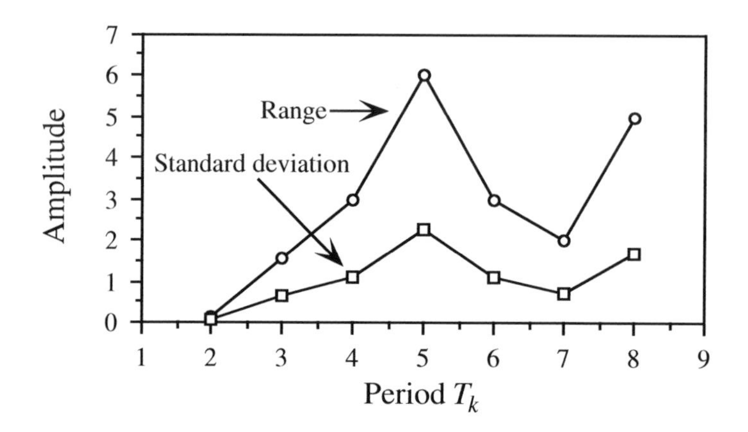

Figure 12.10 Periodogram of Whittaker and Robinson for the artificial data series. The periodogram may be based on either the range or standard deviation amplitude statistics.

Ecological application 12.4a

Enright (1965) re-examined 17 time series taken from the literature, which described the activity of animals as diverse as the chaffinch, laboratory rat, crayfish, oyster, quahog (mollusc), and fiddler crab. The purpose of Enrigh's study was to determine, using periodograms, whether the cycles of activity described by the authors of the original papers (solar, i.e. 24 h, or lunar, i.e. 24.8 h) could withstand rigorous numerical analysis.

The approach is exemplified here by a series of 28 days of observations on the perch-hopping activity of a chaffinch, a European songbird, kept under constant light conditions. The periodogram shown in Fig. 12.11a is clearly dominated by a period of 21.8 h. Figures 12.11b-d display the mean values $\bar{y}_{T,j}$ of the columns of the Buys-Ballot tables constructed for some of the time periods investigated: $T_k = 21.8$, 24.0 and 24.8 h. (The values $\bar{y}_{T,j}$ of Fig. 12.11b-d were used to calculate the amplitudes of the periodogram, Fig. 12.11a.) Similar figures could be drawn for each point of the periodogram, since a Buys-Ballot table was constructed for each period considered. Without the array of values in the periodogram, examination of, say, the sole Buys-Ballot table for $T_k = 24$ h (Fig. 12.11c) could have led to the conclusion of a diurnal rhythm. Similarly, examination of the table for $T_k = 24.8$ h (Fig. 12.11d) could have suggested a lunar rhythm. In the present case, the periodogram allowed Enright to (1) reject periods that are intuitively interesting (e.g. $T_k = 24$ h) but whose amplitude is not significantly high, and (2) identify a somewhat unexpected 21-h rhythm, which seems to be of endogenous nature.

The 17 data series re-examined by Enright (1965) had been published with the objective of demonstrating the occurrence of diurnal or tidal cycles. Enright's periodogram analyses confirmed the existence of diurnal cycles for *only two* of the series: one for the rat locomotor activity and one for the quahog shell-opening activity. *None* of the published series actually exhibited a tidal (lunar) cycle. This stresses the usefulness of periodogram analysis in ecology and the importance of using appropriate numerical methods when dealing with data series.

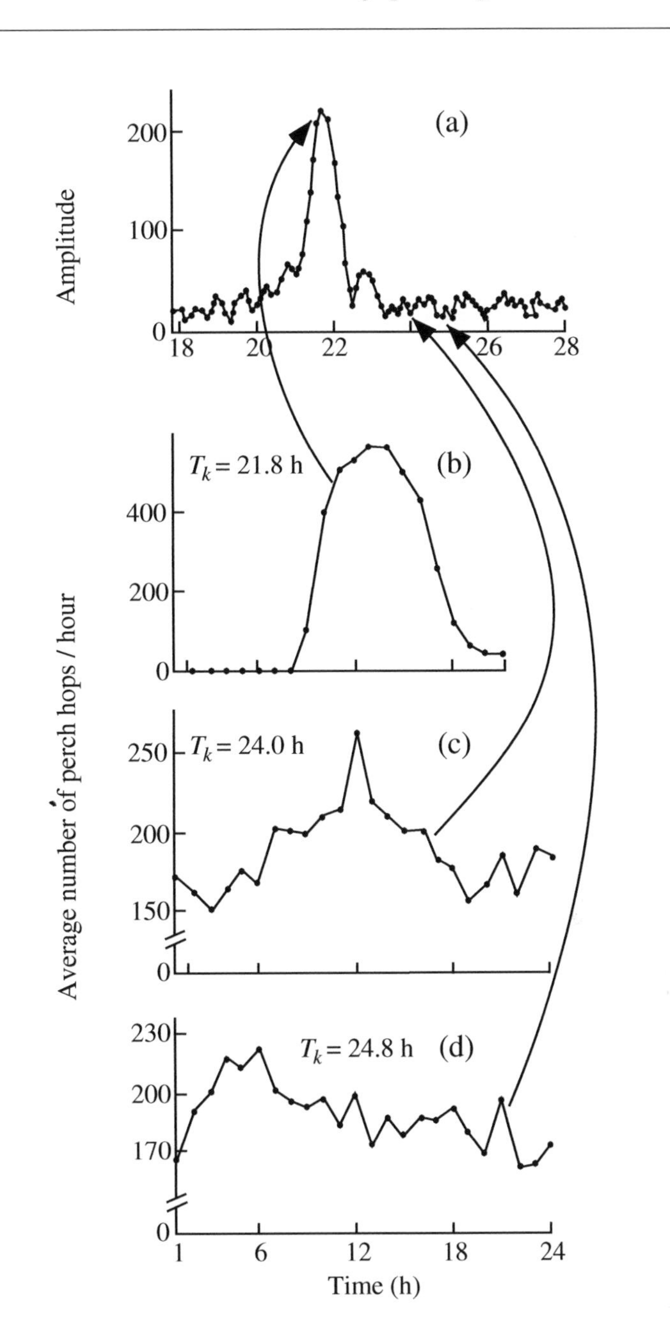

Figure 12.11 (a) Periodogram for the chaffinch perch-hopping activity series (n = 672 data points). The amplitude was calculated using Enright's formula (eq. 12.13). The three lower panels illustrate examples of values from which the amplitudes in (a) were calculated. These graphs show the means $\bar{y}_{T,j}$ of the columns in the Buys-Ballot tables, as functions of time, for periods T_k of (b) 21.8 h, (c) 24.0 h, and (d) 24.8 h. After Enright (1965).

2 — Contingency periodogram of Legendre et al.

Another type of periodogram has been proposed by Legendre *et al.* (1981) to identify rhythms in series of *qualitative* ecological data. In this *contingency periodogram*, the Buys-Ballot table is replaced by a *contingency table* (Section 6.2). The columns of the table (Colwell, 1974) are the same as in a Buys-Ballot table, but the rows are the r states of the qualitative descriptor under study. Values in the table are frequencies f_{ij} of the states of the descriptor (rows i), observed at the various times (columns j) of period T_k. As in the periodogram of Whittaker & Robinson (above), a different table is constructed for each period T_k considered in the periodogram.

Information statistic

Information (H) as to the states of the qualitative variable of interest (S), which is accounted for by a given period T_k, is the information in common between S and the sampling axis X (most often, time). This amount of information is computed as the intersection between S and X, for period T_k:

$$H(\mathrm{S} \cap \mathrm{X}) = H(\mathrm{S}) + H(\mathrm{X}) - H(\mathrm{S}, \mathrm{X}) \tag{12.14}$$

Equation 12.14 is the same as eq. 6.10, used for calculating the information shared by two descriptors (statistic B), so that $H(\mathrm{S} \cap \mathrm{X}) = \mathrm{B}$.

The *contingency periodogram* is a graph of the values $H(\mathrm{S} \cap \mathrm{X}) = \mathrm{B}$ on the ordinate, as a function of periods T_k. Periodograms, as well as spatial correlograms (Section 13.1), are often read from left (shortest periods or lags) to right (larger periods or lags). This is the case when the process that may have generated the periodic or autocorrelated structure of the data, if any, is assumed to be stronger at small lags and to generate short periods before these are combined into long periods.

Section 6.2 has shown that statistic B is related to Wilks' X^2_{W} statistic:

$$X^2_{\mathrm{W}} = 2n\mathrm{B} \quad \text{(when B in nats; eq. 6.13)}$$

$$\text{or } X^2_{\mathrm{W}} = 2n\mathrm{B}\,\log_e 2 = n\mathrm{B}\,\log_e 4 \quad \text{(when B in bits; eq. 6.14).}$$

Because X^2_{W} can be tested for significance, critical values may be drawn on the periodogram. The critical value of B is found by replacing X^2_{W} in eq. 6.13 by the critical value $\chi^2_{\alpha,\nu}$:

$$\mathrm{B}_{\text{critical}} = \chi^2_{\alpha,\nu}/2n \quad \text{(for B in nats)}$$

$\chi^2_{\alpha,\nu}$ is read from a χ^2 table, for significance level α and $\nu = (r-1)(T_k-1)$. For the periodogram, an alternative to B is to plot the χ^2_{W} statistic as a function of periods T_k; the critical value to be use is then $\chi^2_{\alpha,\nu}$ directly. As one proceeds from left (smaller periods) to right (larger periods) in the periodogram, T_k and ν increase; as a consequence, the critical value, $\chi^2_{\alpha,\nu}$ or $\mathrm{B}_{\text{critical}}$, monotonically increases from left to right in this type of periodogram, as will be shown in the numerical example below.

Since multiple tests are performed in a contingency periodogram, a correction must be made on the critical values of B (Box 1.3). The simplest type is the Bonferroni correction, where significance level α is replaced by $\alpha' = \alpha/$(number of simultaneous tests). In a periodogram, the number of simultaneous tests is the number of periods T_k for which the statistic (B or X^2_W) has been computed. Since the maximum number of periods that can be investigated is limited by the observational window (Section 12.0), the maximum number of simultaneous tests is $[(n/2) - 1]$ and the strongest Bonferroni correction that can be made is $\alpha' = \alpha/[(n/2) - 1]$. This is the correction recommended by Oden (1984) to assess the global significance of spatial correlograms (Section 13.1). In practice, when analysing long data series, one usually does not test the significance past some arbitrarily chosen point; if there are h statistics that have been tested for significance, the Bonferroni method would call for a corrected significance level $\alpha' = \alpha/h$.

Progressive Bonferroni correction

There are two problems with the Bonferroni approach applied to periodograms and spatial correlograms. The first one is that the correction varies in intensity, depending on the number of periods (in periodograms) or lags (in spatial correlograms) for which statistics have been computed and tested. The second problem is that the interest in the results of tests of significance decreases as the periods (or lags) get longer, especially in long data series; when a basic period has been identified, its harmonics are of lesser interest. These problems can be resolved by resorting to a *progressive Bonferroni* correction, proposed by P. Legendre in the Hewitt *et al.* (1997) paper. In this method, the first periodogram or spatial correlogram statistic is tested against the α significance level; the second statistic is tested against the Bonferroni-corrected level $\alpha' = \alpha/2$ because, at this point, two tests have been performed; and so forth until the k-th statistic, which is tested against the Bonferroni-corrected level $\alpha' = \alpha/k$. This approach also solves the problem of "where to stop computing a periodogram or spatial correlogram"; one goes on as long as significant values are likely to emerge, considering the fact that the significance level becomes progressively more stringent.

Numerical example. Consider the following series of qualitative data ($n = 16$), for a qualitative variable with 3 states (from Legendre *et al.*, 1981):

1, 1, 2, 3, 3, 2, 1, 2, 3, 2, 1, 1, 2, 3, 3, 1

To analyse period $T_k = 4$, for instance, the series is cut into segments of length 4 as follows:

1, 1, 2, 3;   3, 2, 1, 2;   3, 2, 1, 1;   2, 3, 3, 1

and distributed in the successive rows of the table. The first four data go into columns 1 to 4 of the contingency table, each one in the row corresponding to its code; similarly, observations 5 to 8 are placed into the columns of the table, each in the row corresponding to its code; and so forth. When the operation is completed, the number of occurrences of observations are counted in each cell of the table, so that the resulting table is a contingency table containing frequencies

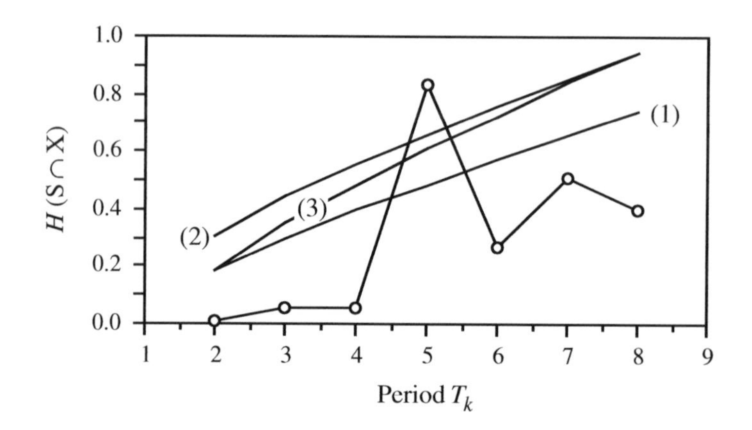

Figure 12.12 Contingency periodogram for the artificial data series (circles). The contingency statistic used here is B = H(S ∩ X). (1) Uncorrected critical values. (2) Bonferroni-corrected critical values, correcting for 7 simultaneous tests in the observational window. (3) Progressive Bonferroni correction; the first value ($T_k = 2$) is without correction, while the last ($T_k = 8$) receives the full Bonferroni correction.

f_{ij}. As an exercise, readers should try to reproduce the contingency tables shown below for $T_i = 4$ and $T_i = 5$. The values of X^2_W and B (in *nats*) are given for these two periods:

$T = 4$	1	2	3	4
State 1	1	1	2	2
State 2	1	2	1	1
State 3	2	1	1	1

B (in *nats*) = 0.055, X^2_W = 1.76

$T = 5$	1	2	3	4	5
State 1	3	3	0	0	0
State 2	1	0	3	0	1
State 3	0	0	0	3	2

B (in *nats*) = 0.835, X^2_W = 26.72

Repeating the calculations for $k = 2$ to 8 produces the periodogram shown in Fig. 12.12. Only X^2_W = 26.72 ($T = 5$) is larger than the corresponding critical value, which may be computed in various ways (as explained above), depending on the need:

- Uncorrected critical value: $\alpha = 0.05$, $\nu = (3-1)(5-1) = 8$, critical $\chi^2_{\alpha,\nu} = 15.5$. $B_{critical} = 15.5/(2 \times 16) = 0.484$.

- Bonferroni correction for 7 simultaneous tests: $\alpha' = \alpha/(n/2 - 1) = 0.05/7$, $\nu = 8$, critical $\chi^2_{\alpha',\nu} = 21.0$. $B_{critical} = 21.0/32 = 0.656$.

- Progressive Bonferroni correction. Example for the 4th test: $\alpha' = \alpha/4 = 0.05/4$, $\nu = 8$, critical $\chi^2_{\alpha',\nu} = 19.5$. $B_{critical} = 19.5/32 = 0.609$.

Thus, the only significant period in the data series is $T_k = 5$.

The contingency periodogram can be directly applied to qualitative descriptors. Quantitative or semiquantitative descriptors must be divided into states before analysis with the contingency periodogram. A method to do so is described in Legendre *et al.* (1981).

In their paper, Legendre *et al.* (1981) established the robustness of the contingency periodogram in the presence of strong random variations, which often occur in ecological data series, and its ability to identify hidden periods in series of non-quantitative ecological data. Another advantage of the contingency periodogram is its ability to analyse very short data series.

One of the applications of the contingency periodogram is the analysis of multivariate series (e.g. multi-species; Ecological application 12.4b). Such series may be transformed into a single qualitative variable describing a partition of the observations, found through a clustering method. With the contingency periodogram, it is possible to analyse the data series, now transformed into a single nonordered variable corresponding to the partition of the observations. The only alternative approach would be to carry out the analysis on the multivariate distance matrix among observations, using the Mantel correlogram described in Subsection 13.1.5.

Ecological application 12.4b

Phytoplankton was enumerated in a series of 175 water samples collected hourly at an anchor station in the St. Lawrence Estuary (Québec). Using the contingency periodogram, Legendre *et al.* (1981) analysed the first 80 h of that series, which corresponded to neap tides. The original data consisted of six functional taxonomic groups. The *six-dimensional quantitative descriptor* was transformed into a *one-dimensional qualitative descriptor* by clustering the 80 observations (using flexible clustering; Subsection 8.5.10). Five clusters of "hours" were obtained; each hour of the series was attributed to one of them. Each cluster thus defined a state of the new qualitative variable resulting from the classification of the hourly data.

When applied to the qualitative series, the contingency periodogram identified a significant period $T = 3$ h, which suggested rapid changes in surface waters at the sampling site. The integer multiples (harmonics) of the basic period (3 h) in the series also appeared in the contingency periodogram. Periods $T = 6$ h, 9 h, and so on, had about the same significance as the basic period, so that they did not indicate the presence of additional periods in the series.

3 — Periodograms of Schuster and Dutilleul

Harmonic analysis

For *quantitative* serial variables, there exists another method for calculating a periodogram, which is mathematically more complex than the periodogram of Whittaker and Robinson (Subsection 12.4.1) but is also more powerful. It is sometimes called *harmonic analysis* or *periodic regression*. This method is based on the fact that the periodic variability present in series of quantitative data can often be represented by a sum of periodic terms, involving combinations of sines and cosines (Fig. 12.13):

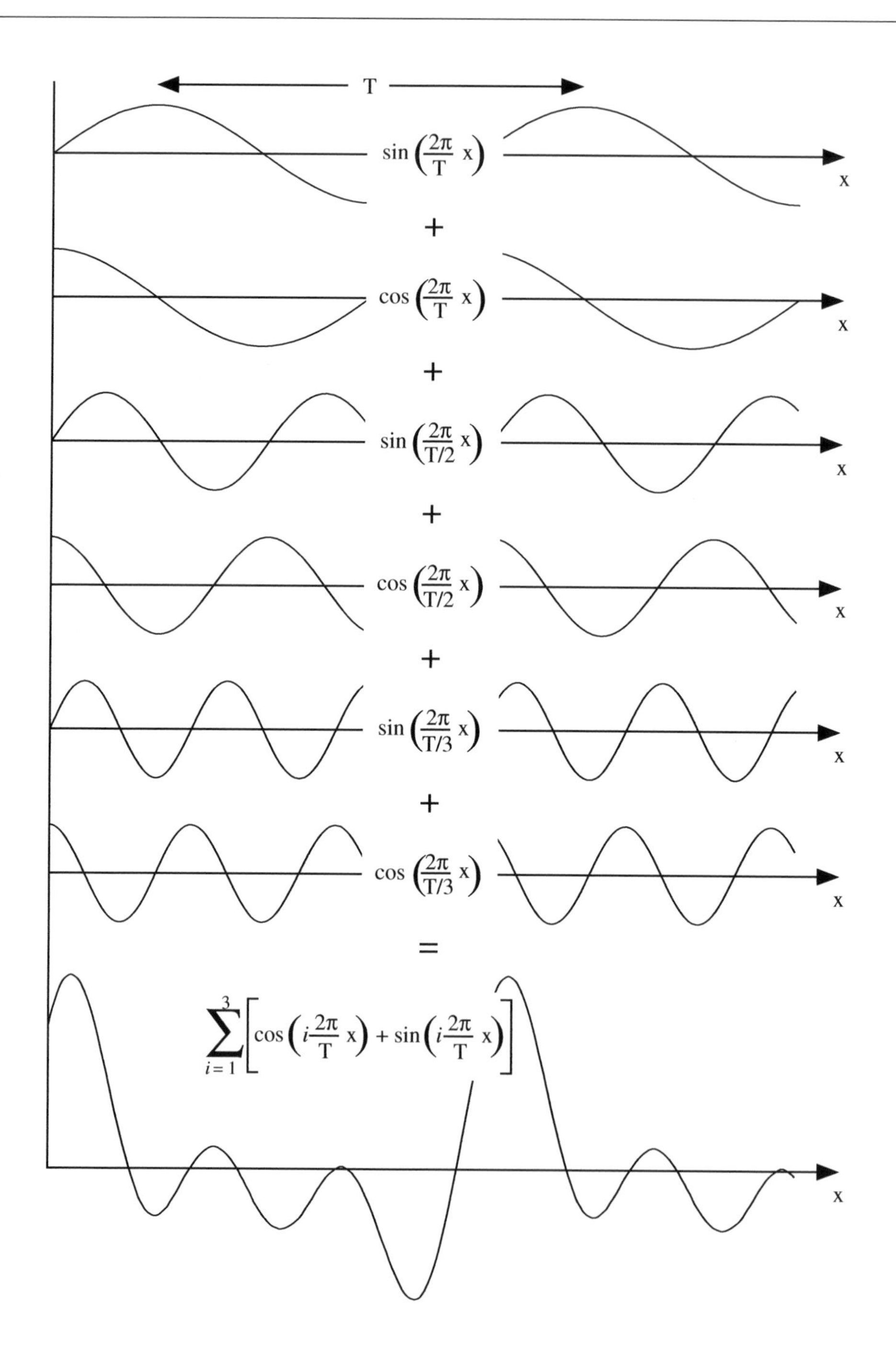

Figure 12.13 Fourier series. The periodic variation (bottom series) results from the sum of three sines and three cosines, which make up a harmonic sequence ($T_k = T$, $T/2$ and $T/3$). The mean of the series is 0 ($a_0 = 0$) and the amplitude of each sine and cosine is equal to 1 ($a_k = b_k = 1$).

Fourier series

$$y(x) = a_0 + \sum_k \left[a_k \cos\left(\frac{2\pi}{T_k}x\right) + b_k \sin\left(\frac{2\pi}{T_k}x\right) \right] \quad \textbf{(12.15)}$$

Equation 12.15 is called a *Fourier series*. Constant a_0 is the mean of the series, whereas parameters a_k and b_k determine the importance of a given period T_k in the resulting signal. Using eq. 12.15, any periodic signal can be partitioned into a sequence of superimposed oscillations (Fig. 12.13). The term $[(2\pi/T_k)/x]$ transforms the explanatory variable x into a cyclic variable. Periods T_k are generally chosen in such a way that the sines and cosines, which model the data series, are *harmonics* (Section 12.0) of a fundamental period T_0:$T_k = T_0/k$ (where $k = 1, 2, \ldots, n/2$). Periods T_k become shorter as k increases. Equation 12.15 may be rewritten as:

$$y(x) = a_0 + \sum_{k=1}^{n/2} \left[a_k \cos\left(k\frac{2\pi}{T_0}x\right) + b_k \sin\left(k\frac{2\pi}{T_0}x\right) \right]$$

Generally, T_0 is taken equal to the length of the series ($T_0 = n\Delta$, where Δ is the interval between data points), so that:

$$y(x) = a_0 + \sum_{k=1}^{n/2} \left[a_k \cos\left(k\frac{2\pi}{n\Delta}x\right) + b_k \sin\left(k\frac{2\pi}{n\Delta}x\right) \right] \quad \textbf{(12.16)}$$

The purpose of Fourier analysis is not to determine the values of coefficients a_k and b_k, but to find out which periods, among all periods T_k, best explain the variance observed in the response variable $y(x)$. After estimating the values of a_k and b_k, the *amplitude* of the periodogram for each period T_k is computed as the *fraction of the variance* of the series that is explained by the given period. This quantity, which is a coefficient of determination, combines the estimates of coefficients a_k and b_k as follows:

$$s^2(T_k) = (a_k^2 + b_k^2)/2 \quad \textbf{(12.17)}$$

Values in the periodogram are thus calculated by fitting to the data series (by least squares) a finite number of sine and cosine functions with different periods. There are $n/2$ such functions in the harmonic case. The shortest period considered is 2Δ ($T_{k\,max} = T_0/(n/2) = n\Delta/(n/2) = 2\Delta$). It corresponds to the limit of the observational window (Section 12.0). The amplitude is computed for each period T_k independently.

Schuster periodogram

Plotting the amplitudes from eq. 12.17 as a function of periods T_k produces the *periodogram of Schuster* (1898), which is used to identify significant periods in data series. Frequencies T_k are in general harmonics of T_0, but it is also possible to choose them to correspond to values of particular interest in the study. Contrary to the periodogram of Whittaker & Robinson, which does not refer to an underlying mathematical model, Schuster's periodogram is based on Fourier series (eqs. 12.15 and 12.16). Indeed, Kendall & Ord (1990, p. 158) have shown that any time series may be decomposed into a set of cycles based on the harmonic frequencies, even if the series does not display periodicity.

One advantage of Schuster's periodogram is that it can handle series with several periods, contrary to the periodogram of Whittaker and Robinson which is limited to series with only one stable period (see above). Values in Schuster's periodogram may be tested for significance by reference to a critical value, which is calculated using a formula derived from Anderson (1971, p. 110 *et seq.*):

$$-(2/n)\log_e\left(1-\sqrt[m]{1-\alpha}\right) \qquad \textbf{(12.18)}$$

where n is the number of observations in the series, m is the largest computed harmonic period (usually, $m = n/2$), and α is the probability level.

Ecological application 12.4c

Demers & Legendre (1981) used Schuster's periodogram to analyse a 76-h series of oceanographic data. For a significance level $\alpha = 0.05$, the critical value for the periodogram was:

$$-(2/76)\log_e\left(1-\sqrt[38]{1-0.05\alpha}\right) = 0.174 = 17.4\%$$

Thus, any period explaining more than 17.4% of the variance of the series was significantly different from zero at level $\alpha = 0.05$.

Ecological application 12.4d

The time series of chlorophyll *a* and *Trichodesmium* filaments in tropical waters (Steven & Glombitza, 1972), already discussed in Ecological applications 12.3b and 12.3c, were subjected to harmonic analysis. Results are reported in Table 12.8. Each column of results could also be plotted as a periodogram. The period $T = 120$ days, already evidenced by autocorrelation (Fig. 12.8) and cross-correlation (Fig. 12.9), was also clearly identified by harmonic analysis.

Periods that do not correspond to an integer number of cycles in the series are usually not computed in Schuster's periodogram, although there is nothing that prevents it mathematically except the fact that the test of statistical significance of individual values (eq. 12.18) is only asymptotically valid with fractional frequencies. As a consequence, Schuster's periodogram is poorly adapted to the analysis of short time series, in which the periods of interest are likely to be fractional. A rule of thumb is to only analyse series that are at least 10 times as long as the longest hypothesized period.

Dutilleul (1990) proposed to modify Schuster's periodogram, in order to compute the portion of total variance associated with periods that do not correspond to integer fractions of the fundamental period T_0 (i.e. *fractional periods*). The method allows a more precise detection of the periods of interest and is especially useful with *short data series*.

Table 12.8 Harmonic analysis of time series of chlorophyll *a* and *Trichodesmium* filaments, in tropical marine waters. The Table reports the amplitudes corresponding to harmonic periods. The dominant period (T_k = 120) is in italics. After Steven & Glombitza (1972).

Harmonic	Period	Nonfiltered series		Filtered series	
k	T_k = 840 days/k	Chl *a*	*Trichodesmium*	Chl *a*	*Trichodesmium*
4	210	0.007	67	0.010	75
5	168	0.007	178	0.006	168
6	140	0.022	113	0.019	129
7	*120*	*0.039*	*318*	*0.038*	*311*
8	105	0.017	147	0.016	162
9	93	0.018	295	0.019	291
10	84	0.020	123	0.020	144

Dutilleul periodogram

The statistic in *Dutilleul's modified periodogram* is the exact fraction of the total variance of the time series explained by regressing the series on the sines and cosines corresponding to one or several periodic components. In contrast, Schuster's periodogram is estimated *for a single period at a time*, i.e. each period T_k in eq. 12.15. It follows that, when applied to short series, Schuster's periodogram generally only provides an approximation of the explained fraction of the variance. In general, the number of periodic components actually present in a series is unknown *a priori*, but it may be estimated using a stepwise procedure proposed by Dutilleul (1990; see also Dutilleul, 1998). The modified periodogram thus offers two major extensions over Schuster's: (1) it may be computed for *several periods at a time* (i.e. it is *multifrequential*) and (2) its maximization over the continuous domain of possible periods provides the maximization of the sum of squares of the corresponding trigonometric model fitted by least squares to the series. Both periodograms lead to the same estimates when computed for a single period over a long data series, or when the period corresponds to an integer fraction of T_0. In all other cases, the modified periodogram has better statistical properties (Dutilleul, 1990; see also Legendre & Dutilleul, 1992; Dutilleul & Till, 1992; Dutilleul, 1998):

• The explained fraction of the variance tends to be maximum for the true periods present in the time series, even when these are fractional, because the periodogram statistic exactly represents the sum of squares of the trigonometric model fitted by least squares to the series at the frequencies considered, whether these are integers or not (when expressed in number of cycles over the series).

• Assuming normality for the data series, the periodogram statistic is distributed like χ^2 for all periods in small or large samples, which leads to exact tests of significance.

With Schuster's periodogram, this is only the case for periods corresponding to integer fractions of T_0 or, outside these periods, only for large samples.

• When the number of frequencies involved in the computation corresponds to the true number of periodic components in the series, the frequencies maximizing the periodogram statistic are unbiased estimates of the true frequencies. The stepwise procedure mentioned above allows the estimation of the number of periodic components present in the series.

In order to compare Dutilleul's periodogram to Schuster's, Legendre & Dutilleul (1992) created a test data series of 30 simulated observations containing two periodic components, which jointly accounted for 70.7% of the total variance in the series, with added noise. The true periods were $T = 12$ and 15 units. Schuster's periodogram brought out only one peak, because the two components were close to each other and Schuster's periodogram statistic was estimated for only one period at a time. When estimated for a single period, Dutilleul's modified periodogram shared this drawback. However, when estimated for the correct number of periods (i.e. two, as found by the stepwise procedure mentioned above), the modified periodogram showed maxima near the two constructed periods, i.e. at $T = 11.3$ and 14.4 units. The authors also compared the results of Dutilleul's method to those obtained with the stepwise procedure of Damsleth & Spjøtvoll (1982), which is based on Schuster's periodogram. Results from the latter (estimated periods $T = 10.3$ and 13.5) were not as good as with Dutilleul's modified periodogram. Dutilleul (1998) also showed the better performance of the modified periodogram over autocorrelograms in the context of scale analysis.

Dutilleul & Till (1992) published an application of the modified periodogram to the analysis of long dendrochronological series. Dutilleul's periodogram clearly detected the annual solar signal in cedar tree-ring series in the Atlas, a sub-tropical region where, typically, the annual dendrochronological signal is weak. An application to a series of moderate length (river discharge) was published by Tardif *et al.* (1998).

4 — Harmonic regression

Legand (1958) proposed to use the first term of the Fourier series (eq. 12.15) to analyse ecological periodic phenomena with known *sinusoidal* periodic variability (e.g. circadian). This method is called *harmonic regression.* As in the case of Fourier series (see above), the explanatory variable x (e.g. time of day) is transformed into a cyclic variable:

$$x' = \cos\left[\frac{2\pi}{T}(x + c)\right] \quad \text{(angles in radians)} \qquad \textbf{(12.19)}$$

In the above expression, which is the first term of a Fourier series, T is the period suggested by hypothesis (e.g. 24 hours); x is the explanatory variable (e.g. local time); and 2π is replaced by 360° when angles are in degrees. Constant c fits the position of the cosine along the abscissa, so that it corresponds to the time of minimum and maximum values in the data set. The regression coefficients are estimated, as usual, by the least-squares method:

$$\hat{y} = b_0 + b_1 x'$$

Ecological application 12.4e

Angot (1961) studied the diurnal cycle of marine phytoplankton production near New Caledonia, in the South Pacific. Values of primary production exhibited regular diurnal cyclic variations, which might reflect physiological rhythms. After logarithmic transformation of the primary production values, the author found *significant* harmonic regressions, with $T = 24$ h and $c = 3$ h, the explanatory variable x being the local time. Coefficients of regression b_0 and b_1 were used to compare different sampling sites.

Ecological application 12.4f

Taguchi (1976) used harmonic regression to study the short-term variability of marine phytoplankton production for different irradiance conditions and seasons. Data, which represented a variety of coastal conditions, were first transformed into ratios of production to chlorophyll *a*. The explanatory variable x was local time, $c = 4$ h, and T was generally 24 h. The intercept b_0 represented the mean production and b_1 was the slope of the regression line. The two coefficients decreased with irradiance and varied with seasons. The author interpreted the observed changes of regression coefficients in terms of photosynthetic dynamics.

Periodogram analysis is of interest in ecology because calculations are relatively simple and interpretation is direct. The correlogram and periodogram approaches, however, often give way to *spectral analysis* (next Section). Spectral analysis is more powerful than correlogram or periodogram analyses, but it is also a more complex method for studying series. For simple problems where spectral analysis would be an unnecessary luxury, ecologists should rely on correlograms or, better, periodograms.

12.5 Periodic variability: spectral analysis

Spectral analysis is the most advanced approach to analyse data series. The general concepts upon which spectral analysis is founded are described below and illustrated by ecological applications. However, the analysis cannot be conducted without taking into account a number of theoretical and practical considerations, whose discussion exceeds the scope of the present book. Interested readers should refer, for instance, to the review papers by Platt & Denman (1975) and Fry *et al.* (1981). They may also consult the book of Bendat & Piersol (1971) and the references provided at the end of Section 12.0. Ecologists wishing to use spectral analysis are advised to consult a colleague with *practical experience* of the method. Up to now, spectral analysis has been used mostly by engineers, physicists, and Earth scientists, but applications to ecological data series are rapidly increasing in number.

1 — Series of a single variable

In the previous section, calculation of the periodogram involved least-squares fitting of a Fourier series to the data (eq. 12.15):

$$y(x) = a_0 + \sum_{k=1}^{n/2} \left[a_k \cos\left(\frac{2\pi}{T_k}x\right) + b_k \sin\left(\frac{2\pi}{T_k}x\right) \right]$$

When calculating Schuster's periodogram, the Fourier series was constructed using periods T_k. In spectral analysis, frequencies $f_k = 1/T_k$ are used instead of periods T_k. Thus, eq. 12.15 is rewritten as:

$$y(x) = a_0 + \sum_{k=1}^{n/2} [a_k \cos(2\pi f_k x) + b_k \sin(2\pi f_k x)] \qquad \textbf{(12.20)}$$

Using a formula similar to eq. 12.17, the *intensity* of the periodogram, at frequency f_k, is computed using the least-squares estimates of coefficients a_k and b_k:

$$I(f_k) = n(a_k^2 + b_k^2)/2 \qquad \textbf{(12.21)}$$

The intensity of the periodogram is defined only for *harmonic* frequencies $k/n\Delta$. It is possible, however, to turn the intensity of the periodogram into a *continuous* function over all frequencies from zero to the Nyquist frequency (see Section 12.0). This defines the *spectrum* of the series:

Spectrum

$$S_{yy}(f) = n(a_f^2 + b_f^2)/2 \qquad 0 \le f \le f_{n/2} \qquad \textbf{(12.22)}$$

Power, or variance spectrum

The spectrum is thus a *continuous* function of frequencies, whereas the periodogram is discontinuous. Calculation and interpretation of spectra is the object of *spectral analysis*. Because of its origin in the field of electricity and telecommunications, the spectrum is sometimes called "power spectrum" or "energy spectrum". As shown below, it is also a "variance spectrum", which is the terminology used in ecology.

Fast Fourier Transform

In algebra, there exist mathematically equivalent pairs of equations, which are used to go from one independent variable to another. Two mathematically equivalent equations, one being a function of x and the other a function of frequency $f = 1/x$, are called a pair of *Fourier transforms*. It can be shown that the *autocovariance* or *autocorrelation* function (eqs. 12.5-12.7) and the *spectral density* function (eq. 12.22) are a pair of Fourier transforms. Therefore, both the correlogram (Section 12.3) and periodogram analyses (Section 12.4), when they are generalized, lead to spectral analysis (Fig. 12.14). Classically, the spectrum is computed by Fourier transformation of the autocorrelation, followed by smoothing. There is another method, called *Fast Fourier Transform* (FFT), which is faster than the previous approach (shorter

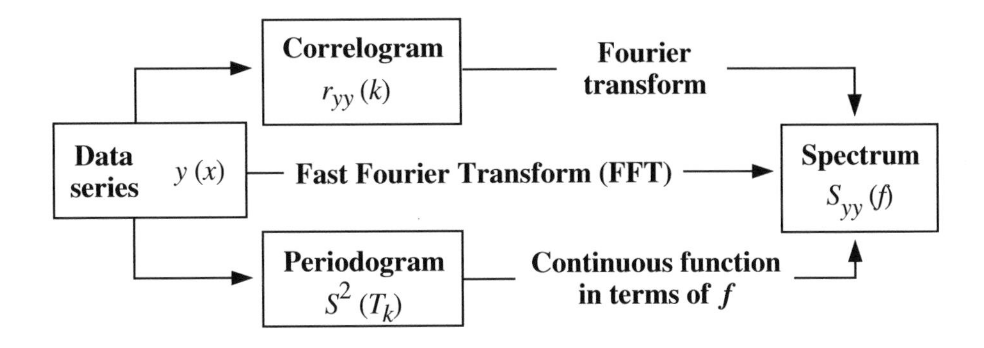

Figure 12.14 Relationships between a data series, its correlogram and periodogram, and its variance spectrum.

computing time) and efficiently computes the pair of Fourier transforms written in discrete form. This last method has the advantage of computational efficiency, but it involves a number of constraints which can only be fully mastered after acquiring some *practical* experience of spectral analysis.

Smoothing window

The spectrum computed from a correlogram or autocovariance function is an unbiased estimate of the true spectrum. However, the standard error of this spectral estimate is 100%, whatever the length of the series. It follows that the computed spectrum must be *smoothed* in order to reduce its variance. Smoothing is done using a *window*, which is a function by which one multiplies the spectrum itself (spectral window), or the autocovariance estimates (lag window) prior to Fourier transformation. The two types of windows lead to the same result. The main problem of *smoothing* is that reduction of the standard error of the spectral estimates, on the ordinate, always leads to spreading of the variance on the abscissa. As a result, the spectral estimate, at any given frequency, may become contaminated by variance that is "leaking" from neighbouring frequencies. This *leakage* may result in biased smoothed spectral estimates. The various windows found in the literature (e.g. Bartlett, Daniell, de la Valle-Poussin or Parzen, Hamming, von Han, Tukey) provide different compromises between reduction of the standard error of spectral estimates and loss of resolution between adjacent frequencies. In the literature for ecologists, Platt & Denman (1975) mainly discuss Bartlett's window, whereas Laurec (1982, pp. 242-245) compares the windows of Bartlett, Daniell, Parzen (or de la Valle-Poussin), and Tukey. As already stressed above, the practical aspects of spectral analysis, including the choice of windows, filters (Section 12.2), and so on, most often necessitate the help of an experienced colleague.

The ecological interpretation of spectra is not necessarily the same as that of correlograms or periodograms. First, the spectrum is a true *partition of the variance* of the series among frequencies. Therefore, spectral analysis is a third type of variance decomposition, in addition to the usual partitioning among experimental factors or sampling axes (ANOVA) and the partition among principal axes (Sections 4.4 and 9.1). The *units* of spectral density are [variance × frequency^{-1}], i.e. [(units of the response variable y)^2 × (units of the explanatory variable x)]. Therefore, the *variance* that corresponds to a frequency band is the *area under the curve* between the upper and lower frequencies, i.e. the integration of [variance × frequency^{-1}] over the frequency band. Spectra may be computed to identify harmonics in the data series or they may be regarded as characteristics of whole series, whether they are true sums of harmonics or not (Kendall & Ord, 1990, p. 158). Following this idea, Platt & Denman (1975) suggested the concept of *spectroscopy of ecosystems*, i.e. the characterization of ecosystems according to their spatio-temporal scales, which could provide a global approach to the study of ecological systems. These concepts should become clearer with the following Ecological applications.

Ecological application 12.5a

At an anchor station in the Gulf of St. Lawrence, Platt (1972) continuously recorded *in vivo* fluorescence in surface waters as an estimate of phytoplankton chlorophyll *a*. Spectral analysis of the detrended data series (Fourier transform of autocorrelation) resulted in a spectrum characterized by a slope of –5/3, over frequencies ranging between ca. 0.01 and 1 cycle min^{-1}. The average current velocity being ca. 20 cm s^{-1} (10 m min^{-1}), the time series covered spatial scales ranging between ca. 1000 and 10 m (wavelength = speed × frequency^{-1}). This is illustrated in Fig. 12.15.

Interpretation of the spectrum was based on the fact that spectral analysis is a type of variance decomposition in which the total variance of the series is partitioned among the frequencies considered in the analysis (here: 0.03 cycle min^{-1} $< f <$ 1.5 cycle min^{-1}). The slope –5/3 corresponds to that of turbulent processes. This led the author to hypothesize that the local concentration of phytoplankton could be mainly controlled by turbulence. In a subsequent review paper, Platt & Denman (1975) cite various studies, based on spectral analysis, whose results confirm the hypothesis that the mesoscale spatial organization of phytoplankton is controlled by physical processes, in both marine and freshwater environments. This is in fact a modern version of the model proposed in 1953 by Kierstead & Slobodkin, which is discussed in Ecological applications 3.2d and 3.3a. Other references on spectral analysis of *in vivo* fluorescence series include Demers *et al.* (1979), Denman (1976, 1977), Denman & Platt (1975, 1976), Denman *et al.* (1977), Fashman & Pugh (1976), Legendre & Demers (1984), Lekan & Wilson (1978), Platt (1978), Platt & Denman (1975), and Powell *et al.* (1975), among others.

Ecological application 12.5b

Campbell & Shipp (1974) tried to explain the migrations of an Australian cricket from observations on rhythms of locomotor activity of the males and females. One summer migration was followed during 100 days, starting in mid-February. In addition, locomotor activity rhythms of the males and females were observed in the laboratory during ca. 100 days. Fig. 12.16 shows smoothed spectra for numbers of migrating crickets and locomotor activity, for both sexes.

Peaks corresponding to periods of ca. 2.5, 5, 10, and 20 days were observed in one or several spectra, which suggested a long-term biological rhythm with several harmonics. It followed

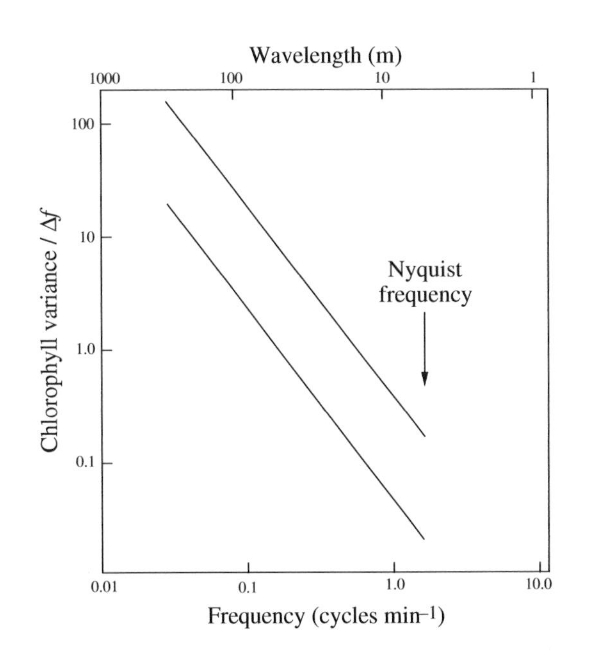

Figure 12.15 Horizontal distribution of chlorophyll *a* (*in vivo* fluorescence; arbitrary units) in surface waters of the Gulf of St. Lawrence. The two parallel lines on the variance spectrum show the envelope of calculated spectral densities. The Nyquist frequency is 1.5 cycle min^{-1}. After Platt (1972).

from spectral analysis that the migratory waves might be explained by synchronization of the locomotor activity cycles of individuals in the population. Migrations of the males appeared to follow a 20-day cycle, whereas those of females seemed to follow a cycle of ca. 10 days. The authors suggested that, during these periods, males attract females to their burrows and form relatively stable couples.

2 — Multidimensional series

Spectral analysis may be used not only with univariate but also with *multidimensional* series, when several ecological variables have been recorded simultaneously. This analysis is an extension of *cross-covariance* or *cross-correlation*, in the same way as the variance spectrum is a generalization of autocovariance or autocorrelation (Fig. 12.14).

Two-dimensional series

From two data series, $\mathbf{y}_j$ and $\mathbf{y}_l$, one can compute a pair of smoothed spectra S_{jj} and S_{ll} and a cross-correlation function $r_{jl}(k)$. These are used to define the *co-spectrum* (K_{jl}) and the *quadrature spectrum* (Q_{jl}):

Co-spectrum

$$K_{jl}(f) = \text{Fourier transform of } [r_{jl}(k) + r_{jl}(-k)]/2 \qquad \textbf{(12.23)}$$

Quadrature s.

$$Q_{jl}(f) = \text{Fourier transform of } [r_{jl}(k) - r_{jl}(-k)]/2 \qquad \textbf{(12.24)}$$

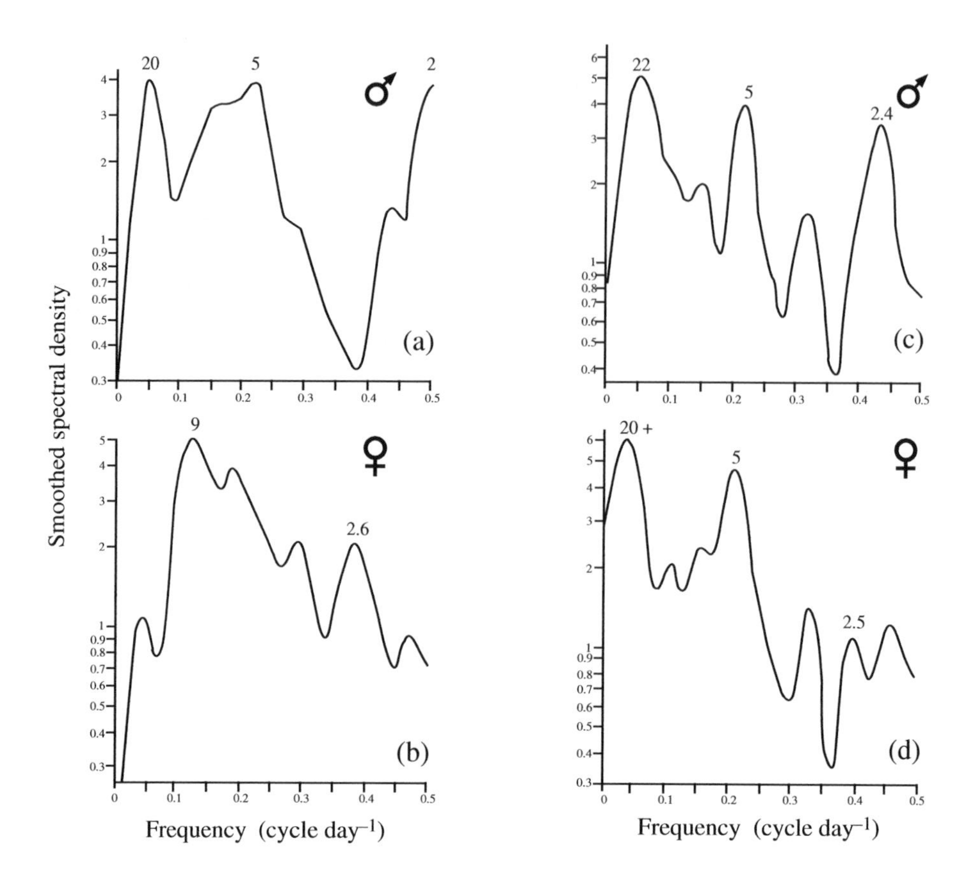

Figure 12.16 Estimates of smoothed spectra for numbers of migrating (a) male and (b) female crickets and for the locomotor activity of (c) male and (d) female crickets in the laboratory. The Nyquist frequency is 0.5 cycle day^{-1}. Periods corresponding to the main peaks are indicated above the curve, in order to facilitate interpretation; periods are the inverse of frequencies (abscissa). After Campbell & Shipp (1974).

The *co-spectrum* (eq. 12.23) measures the distribution, as a function of frequencies, of the covariance between those components of the two series that are in phase, whereas the *quadrature spectrum* (eq. 12.24) provides corresponding information for a phase shift of 90° between the same components. For example, a sine and cosine function are in perfect quadrature. These spectra are used, below, to compute the *coherence*, *phase*, and *gain*.

The *cross-amplitude spectrum* is defined as:

Cross-amplitude spectrum

$$\sqrt{K_{jl}^2(f) + Q_{jl}^2(f)} \tag{12.25}$$

The spectra for $\mathbf{y}_j$ and $\mathbf{y}_l$ are used to compute the (squared) *coherence spectrum* (C_{jl}) and the *phase spectrum* (Φ_{jl}):

$$C^2_{jl}(f) = \frac{K^2_{jl}(f) + Q^2_{jl}(f)}{S_{jj}(f)\,S_{ll}(f)} \tag{12.26}$$

Phase spectrum

$$\Phi_{jl}(f) = \arctan\left(\frac{-Q_{jl}(f)}{K_{jl}(f)}\right) \tag{12.27}$$

The *squared coherence* (eq. 12.26) is a dimensionless measure of the correlation of the two series in the frequency domain; $C^2_{jl}(f) = 1$ indicates perfect correlation between two series whereas $C^2_{jl}(f) = 0$ implies the opposite, for frequency *f*. The *phase spectrum* (eq. 12.27) shows the phase shift between the two series. When the phase is a regular function of the frequency, the squared coherence is usually significantly different from zero; when the phase is very irregular, the squared coherence is generally low and not significant.

Gain spectrum

In order to assess the causal relationships between two variables, one can use the *gain spectrum* (R^2_{jl}), which is analogous to a coefficient of simple linear regression. One can determine the response of $\mathbf{y}_j$ to $\mathbf{y}_l$:

$$R^2_{jl}(f) = \frac{S_{jj}(f)\,C^2_{jl}(f)}{S_{ll}(f)} \tag{12.28}$$

or, alternatively, the response of $\mathbf{y}_l$ to $\mathbf{y}_j$:

$$R^2_{lj}(f) = \frac{S_{ll}(f)\,C^2_{jl}(f)}{S_{jj}(f)} \tag{12.29}$$

Ecological application 12.5c

In a study of the spatial variability of coastal marine phytoplankton, Platt *et al.* (1970) repeated, in 1969, the sampling programme of 1968 described in Ecological application 10.3a. This time, data were collected not only on chlorophyll *a* but also on temperature and salinity at 80 sites along a transect. Figure 12.17 shows the coherence spectra for the three pairs of series, recorded on 24 June. Strong coherence between temperature and salinity indicates that these variables well-characterized the water masses encountered along the transect. Significant coherence between the series of chlorophyll *a* and those of temperature and salinity, at ca. 3 cycles (naut. mi.)$^{-1}$, were consistent with the hypothesis that the spatial distribution of phytoplankton was controlled to some extent by the physical structure of the environment.

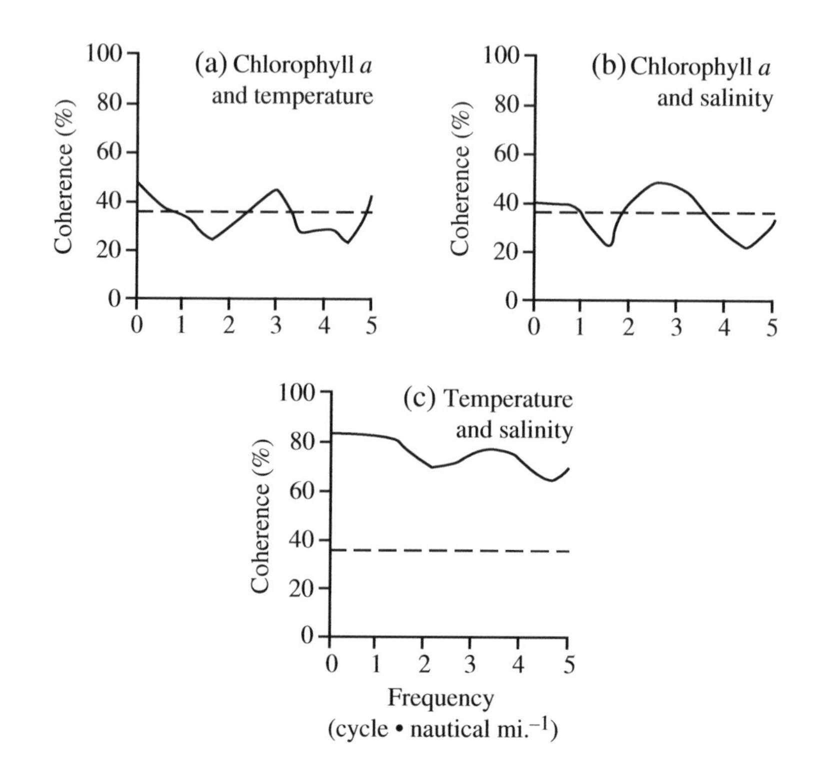

Figure 12.17 Coherence spectra between pairs of variables sampled along a transect 8 nautical miles long in St. Margaret's Bay (Nova Scotia, Canada). Dashed lines: approximate 95% confidence limits. After Platt *et al.* (1970).

Ecological application 12.5d

In order to identify the factors controlling the spatial heterogeneity of marine phytoplankton (patchiness), Denman & Platt (1975) analysed values of chlorophyll *a* and temperature, recorded continuously along a transect in the St. Lawrence Estuary. Two pumping systems were towed, at depths of 5 and 9 m, over a distance of 16.6 km (10 nautical miles). The sampling interval was 1 s, which corresponds to 3.2 m given the speed of the ship. After detrending, computations were carried out using the Fast Fourier Transform. Four coherence and phase spectra were calculated, as shown in Fig. 12.18.

For a given depth (Fig. 12.18a: 5 m; b: 9 m), the coherence between temperature and chlorophyll *a* was high at low frequencies and the phase was relatively constant. At higher frequencies, the coherence decreased rapidly and the phase varied randomly. The lower panels of Fig. 12.18 indicate the absence of covariation between series from different depths. The authors concluded that physical processes played a major role in the creation and control of phytoplankton heterogeneity at intermediate scales (i.e. from 50 m to several kilometres). Weak coherence between series from the two depths, which were separated by a vertical distance of only 4 m, suggested the presence of a strong vertical gradient in the physical structure. Such

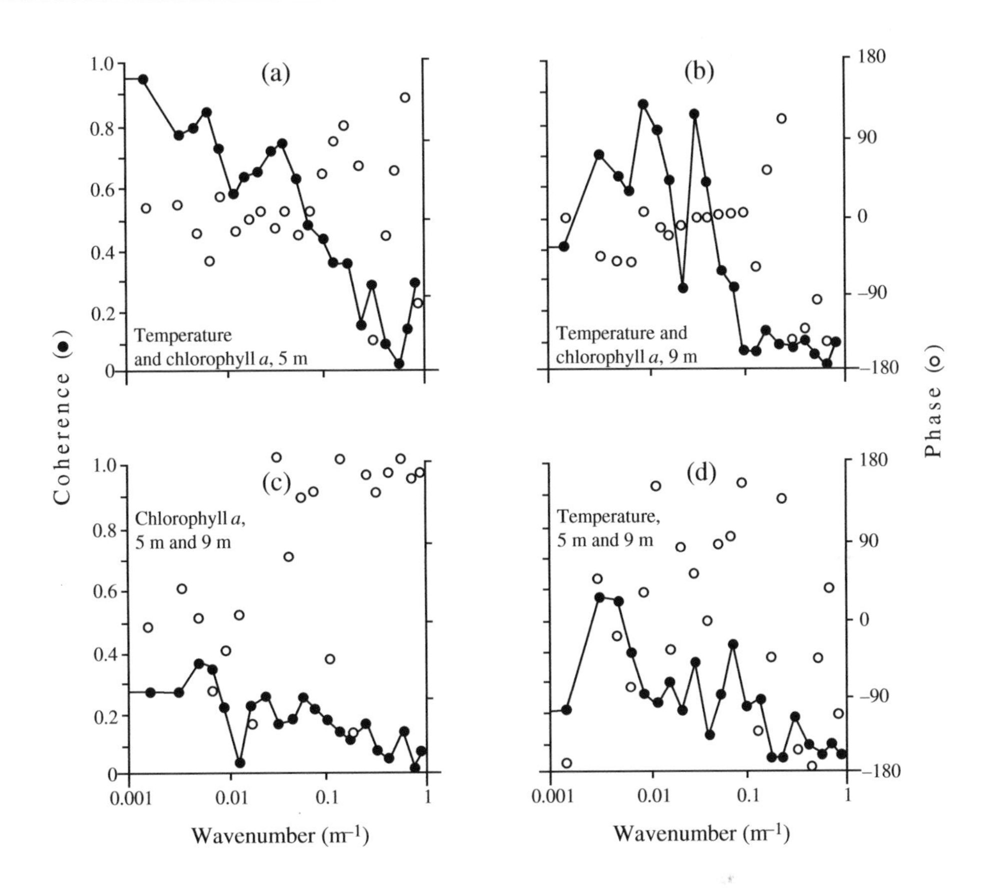

Figure 12.18 Values of coherence (solid lines) and phase (open circles), for pairs of spatial series continuously recorded in the St. Lawrence Estuary. Abscissa: *wavenumber* (= 2π/wavelength = 2π frequency/speed). Adapted from Denman & Platt (1975).

gradients are known to favour the propagation of internal waves (analogous to the propagation of waves at the air-water discontinuity). The authors proposed that the strong coherence between temperature and chlorophyll a, at each of the sampled depths, could reflect the presence of internal waves.

Multidimensional series

In the last paragraphs, the approach to multidimensional situations was to consider two series at a time. Brillinger (1981) provides the mathematical bases for processing multidimensional series using methods that are fully multivariate. When a stochastic series is a time-invariant function of several other series, the method recommended is *frequency regression*. It is analogous to multiple linear regression (Subsection 10.3.3), computed in the frequency domain. More generally, the method to study relationships among several series is that of *principal components in the frequency domain* (see Ecological application 12.5e). In this case, a spectrum is computed for each of the principal components, which are linear combinations of the serial variables (Section 9.1). The method has been adapted by Laurec (1979), who explained how to use it in ecology.

Another approach to the analysis of multivariate data series is *Mantel's correlogram* (Subsection 13.1.5). This type of correlogram is based upon a similarity or distance matrix among observations (Chapter 7), so that it is suitable for multivariate data. It may also be used to analyse series of *semiquantitative*, *qualitative*, or even *binary* data, like the species presence-absence data often collected by ecologists. The geostatistical literature proposes other approaches to the analysis of multivariate data series. Some of these are mentioned in Subsections 1.6.3 and 13.1.4.

Ecological application 12.5e

Arfi *et al.* (1982, pp. 359-363) report results from a study on the impact of the main sewage effluent of the city of Marseilles on coastal waters in the Western Mediterranean. During the study, 31 physical, chemical, and biological variables were observed simultaneously, at an anchor station 1 km offshore, every 25 min during 24 h ($n = 58$). Spectra for individual series (detrended) all show a strong peak at $T =$ ca. 6 h. Comparing the 31 data series two at a time did not make sense because this would have required $(31 \times 30)/2 = 465$ comparisons. Thus, the 31-dimensional data series was subjected to principal component analysis in the frequency domain. Figure 12.19 shows the 31 variables, plotted in the plane of the first two principal components (as in Fig. 9.6), for $T = 6$ h. The long arrows pointing towards the upper left-hand part of the graph correspond to variables that were indicative of the effluent (e.g. dissolved nutrients, bacterial concentrations) whereas the long arrows pointing towards the lower right-hand part of the ordination plane correspond to variables that indicated unperturbed marine waters (e.g. salinity, dissolved O_2, phytoplankton concentrations). The positions of the two groups of variables in the plane show that their variations were out of phase by ca. 180°, for period $T = 6$ h. This was interpreted as a periodic increase in the effluent every 6 h. This periodicity corresponds to the general activity rhythm of the adjacent human population (wake-up, lunch, end of work day, and bedtime).

3 — Maximum entropy spectral analysis

As explained in Subsection 1 above, estimating spectra requires the use of spectral or lag *windows*. Each type of window provides a compromise between reduction of the standard error of the spectral estimates and loss of resolution between adjacent frequencies. As an alternative to windows, Burg (1967) proposed to improve the spectral resolution by *extrapolating* the autocorrelation function beyond the maximum lag (k_{max}), whose value is limited by the length of the series (Subsection 12.3.1). For each extrapolated lag ($k_{max} + k$), he suggested to calculate an autocorrelation value $r_{yy}(k_{max} + k)$ that *maximizes the entropy* (Chapter 6) of the probability distribution of the autocorrelation function. Burg's (1967) method will not be further discussed here, because a different algorithm (Bos, 1971; see below) is now used for computing this *maximum entropy spectral analysis* (MESA). Estimation of the spectrum, in MESA, does not require spectral or lag windows. An additional advantage, especially for ecologists, is that it allows one to compute spectra for very short series

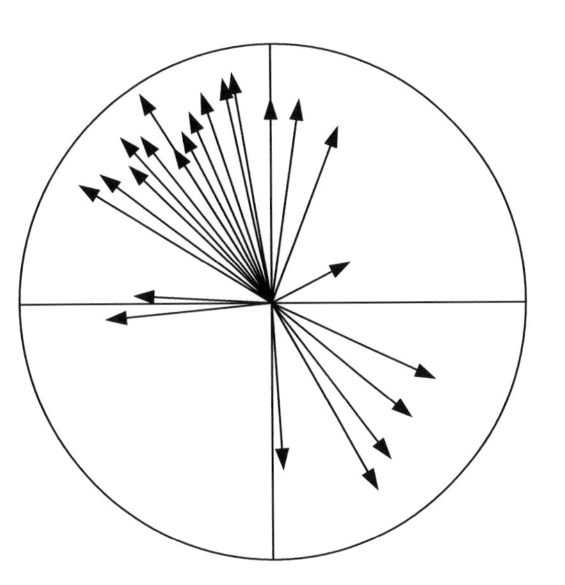

Figure 12.19 Principal component analysis in the frequency domain of 31 simultaneous series of physical, chemical, and biological variables, obtained at an anchor station in the Western Mediterranean. Plot of the 31 variables (arrows), in the plane of the first two principal components, for period $T = 6$ h. Adapted from Arfi *et al.* (1982).

AR model

Data series may be mathematically described as stochastic linear processes. A corresponding mathematical model is the *autoregressive model* (also called *AR model* or *all-pole model*), where each observation in the series $\tilde{y}_t$ (centred on the mean $\bar{y}$ of the series: $\tilde{y}_t = y_t - \bar{y}$) is represented as a function of the q preceding observations:

$$\tilde{y}_t = \phi_1 \tilde{y}_{t-1} + \phi_2 \tilde{y}_{t-2} + \ldots + \phi_q \tilde{y}_{t-q} + a_t \qquad \textbf{(12.30)}$$

White noise

q specifies how many steps back one takes into account to forecast value $\tilde{y}_t$. This is called the *order* of the process. The *autoregression coefficients* ϕ are estimated using the observations of the data series itself. Residual values a_t must be independent of one another; the series of residual values is called *white noise*. Their overall variance is noted s_a^2. This type of model will be further discussed in Section 12.7.

Concerning maximum entropy spectral analysis, Bos (1971) has shown that the maximum entropy method (MEM) proposed by Burg (1967) is equivalent to a least-squares fitting of an AR model to the data series. Using the autoregression coefficients ϕ, it is possible to compute the same spectral densities as those resulting from the entropy calculation of Burg (1967). Thus, the spectrum is estimated directly from the autoregression coefficients ϕ of the AR model, which are themselves estimated from the values $\tilde{y}_t$ of the data series. The spectral density for each frequency f is:

$$S(f) = \frac{2s_a^2\Delta}{\left|1 - \sum_{j=1}^{q} \phi_j \exp(-2i\pi fj)\right|^2} \qquad \textbf{(12.31)}$$

where $i = \sqrt{-1}$. Generally, the sampling interval $\Delta = 1$ time or space unit.

Maximum entropy spectral analysis is not entirely free of problems. Some of these are still the subject of specialized papers. A first practical problem is choosing the *order* q of the AR model for an empirical data series. Various criteria for determining q have been reviewed by Berryman (1978) and Arfi & Dumas (1990). Another problem concerns the estimation of the coefficients of the AR model (see, for instance, Ulrych & Clayton, 1976). A third problem, also discussed by Ulrych & Clayton (1976), is that other processes may fit the data series better than the AR model; for example, an autoregressive-moving average model (ARMA; Section 12.7). Fitting such models, however, may raise other practical problems. The criteria for deciding to use models other than AR are partly intuitive (Section 12.7).

Ulrych & Bishop (1975) briefly reviewed the theoretical bases underlying the algorithms of Burg (1967) and Bos (1971). They provide FORTRAN subroutines for estimating the autoregression coefficients ϕ (also called *autoregressive coefficients* or *prediction error filter coefficients*; G in the subroutines) and the variance of residuals s_a^2 (PM in the subroutines), using algorithms developed either by Yule-Walker or Burg. These subroutines also offer the method of Akaike for determining the order q of the best-fitting AR model. It is then easy to compute the values of the spectrum (eq. 12.31). Barrodale & Erikson (1980a) propose another algorithm for estimating the coefficients ϕ of the AR model, based on least squares, which provides a more precise estimation of the spectrum frequencies. The same authors criticize, on an empirical basis, the method of Akaike, and they propose a different approach. Barrodale & Erikson (1980b) provide a FORTRAN program to compute the maximum entropy spectrum using their algorithm.

Maximum entropy spectral analysis (summarized by Ables, 1974) has been successfully used in astronomy, geophysics, meteorology, and so on. Its popularity for the analysis of ecological data series is increasing. MESA can handle short series as well as series with data exhibiting measurement errors (Ables, 1974). It may also be used to analyse series with missing data (Ulrych & Clayton, 1976). Arfi & Dumas (1990) compared MESA to the classical Fourier approach, using simulated and real oceanographic data series. For long series ($n = 450$), the two approaches have the same efficiency when noise is low, but MESA is more efficient when noise is high. For short ($n = 49$ to 56) and very short ($n = 30$) series, MESA is systematically more efficient. For long data series with low noise, it may often be simpler to compute the spectrum in the traditional way (Berryman, 1978). However, for many ecological data series,

MESA would be the method of choice. The maximum entropy approach can be generalized to handle multivariate series, since coherence and phase spectra can be computed (Ulrych & Jensen, 1974).

Spectral analysis and, thus, *Objective 3* of the analysis of data series (Table 12.2), are presently restricted to *quantitative data*. The only exception is the computation of spectra for long (i.e. $n > 500$ to 1000) series of *binary variables*, using the method of Kedem (1980). Since MESA is not very demanding as to the precision of the data, it could probably be used as well for analysing series of *semiquantitative data* coded using several states.

Ecological application 12.5f

Colebrook & Taylor (1984) analysed the temporal variations of phytoplankton and zooplankton series recorded monthly in the North Atlantic Ocean and in the North Sea during 33 consecutive years (1948 to 1980). Similar series were also available for some environmental variables (e.g. surface water temperature). The series were analysed using MESA. In addition, coherence spectra were computed between series of some physical variables and the series representing the first principal component calculated for the plankton data. For the plankton series, one spectrum was computed for each species in each of 12 regions, after which the spectra were averaged over the species in each region. The resulting 12 species-averaged spectra exhibited a number of characteristic periods, of which some could be related to periods in the physical environment using coherence spectra. For example, a 3 to 4-year periodicity in plankton abundances was associated to heat exchange phenomena at the sea surface. Other periods in the spectra of the physical and biological variables could not easily be explained. Actually, 33-year series are relatively short compared with the long-term meteorological or oceanographic variations, so that some of the identified periods may turn out not to be true cycles.

12.6 Detection of discontinuities in multivariate series

Succession

Detection of discontinuities in *multivariate data series* is a problem familiar to ecologists (*Objective 4* in Section 12.2 and Table 12.2). For example, studies on changes in species assemblages over time often refer to the concept of *succession*. According to Margalef (1968), the theory of species succession within ecosystems plays the same role in ecology as does evolutionary theory in general biology.

The simplest way to approach the identification of discontinuities in multivariate series is by *visual inspection* of the curves depicting changes with time (or along a spatial direction) in the abundance of the various taxa or/and in the values of the environmental variables. For example, in Ecological application 8.9c, inspection of Fig. 8.21 was sufficient to determine that a succession, from diatoms to dinoflagellates, took place after the spring bloom. In most instances, however, simple visual examination of a set of curves does not allow one to unambiguously identify discontinuities in multivariate series. Numerical techniques must be used.

Methods of series analysis described in Sections 12.3 to 12.5 are not appropriate for detecting discontinuities in multivariate series, because the presence of discontinuities is not the same as periodicity in the data. Four types of methods are summarized here.

Instead of dividing multivariate series into subsets, Orlóci (1981) proposed a multivariate method for identifying successional trends and separating them into monotonic and cyclic components. This method may be viewed as complementary to those described below.

1 — Ordinations in reduced space

Several authors have used *ordinations in reduced space* (Chapter 9) for representing multispecies time series in low-dimensional space. To help identify the discontinuities, successive observations of the time series are connected with lines, as in Figs. 9.5 and 12.23. When several observations corresponding to a bloc of time are found in a small part of the reduced space, they may be thought of as a "step" in the succession. Large jumps in the two-dimensional ordination space are interpreted as discontinuities. This approach has been used, for example, by Williams *et al.* (1969; vegetation, principal coordinates), Levings (1975; benthos, principal coordinates), Legendre *et al.* (1984a; benthos, principal components), Dessier & Laurec (1978; zooplankton, principal components and correspondence analysis), and Sprules (1980; nonmetric multidimensional scaling; zooplankton; Ecological application 9.3). In studies of annual succession in temperate or polar regions, using ordination in reduced space, one expects the observations to form some kind of a circle in the plane of the first two axes, since successive observations are likely to be close to each other in the multidimensional space, due to autocorrelation (Section 1.1), and the community structure is expected to come back to its original structure after one year; the rationale for this null model of succession is developed in Legendre *et al.* (1985, Appendix D). Departures from a regular circular pattern are thus interpreted as evidence for the existence of subsets in the data series. In simple situations, such subsets are indeed observed in the plane of the first two ordination axes (e.g. Fig. 9.5). When used *alone*, however, this approach has two major drawbacks.

• Plotting a multivariate data series in two or three dimensions only is not the best way of using the multivariate information. In most studies, the first two principal axes used to represent the data series account together for only 10 to 50% of the multivariate information. In such cases, distances from the main clusters of observations to isolated objects (which are in some particular way different from the major groups) are likely to be expressed by some minor principal axes which are orthogonal (i.e. perpendicular in the multidimensional space) to the main projection plane. As a consequence, these objects may well be projected, in the reduced-spaced ordination, within a group from which they are actually quite different. Moreover, it has been observed that the "circle" of observations (see previous paragraph) may be deformed in a spoon shape so that groups that are distinct in a third or higher dimension may well be packed together in some part of the two-dimensional ordination plane. These problems are common to all ordinations when used alone for the purpose of group recognition. They are not as severe for ordinations obtained by nonmetric multidimensional scaling, however, because that method is, by definition, more efficient than others at flattening multidimensional phenomena into a user-determined small number of dimensions (Section 9.3). The best way to eliminate this first drawback is to associate ordination to clustering results, as explained in Section 10.1. This was the approach of Allen *et al.* (1977) in a study of the phytoplankton succession in Lake Wingra. See also Fig. 12.23.

• The second drawback is the lack of a criterion for assigning observations to groups in an ordination diagram. As a consequence, groups delineated on published ordination diagrams often look rather arbitrary.

2 — Segmenting data series

Hawkins & Merriam (1973, 1974) proposed a method for segmenting a multivariate data series into homogeneous units, by *minimizing the variability* within segments in the same way as in *K*-means partitioning (Section 8.8). Their work followed from the introduction of a contiguity constraint in the grouping of data by Fisher (1958), who called it *restriction* in space or time. The method of Hawkins & Merriam has been advocated by Ibanez (1984) for studying successional steps.

Contiguity constraint

The method has three interesting properties. (a) The multidimensional series is partitioned into homogeneous groups using an *objective clustering criterion.* (b) The partitioning is done with a *constraint of contiguity* along the data series. Within the context of series analysis, contiguity means that only observations that are neighbours along the series may be grouped together. The notion of contiguity has been used by several authors to resolve specific clustering problems: temporal contiguity (Subsection 4, below) or spatial contiguity (Subsection 13.3.2). (c) The observations do not have to be equispaced.

A first problem with Hawkins & Merriam's method is that users must determine the number of segments that the method is requested to identify. To do so, the increase in explained variation relative to the increase in the number of segments is used as a guide. Any one of the stopping rules used with *K*-means partitioning could also be used here (end of Section 8.8). A second and more serious problem, with ecological data, is that strings of zeros in multispecies series are likely to result in segments that are determined by the simultaneous absence of species.

3 — Webster's method

Window

Webster (1973) proposed a rather simple method to detect discontinuities in data series. He was actually working with spatial transects, but his method is equally applicable to time series. Draw the sampling axis as a line and imagine a window that travels along this line, stopping at the mid-points between adjacent observations (if these are equispaced). Divide the window in two equal parts (Fig. 12.20a). There are observations in the left-hand and right-hand halves of the window. Calculate the

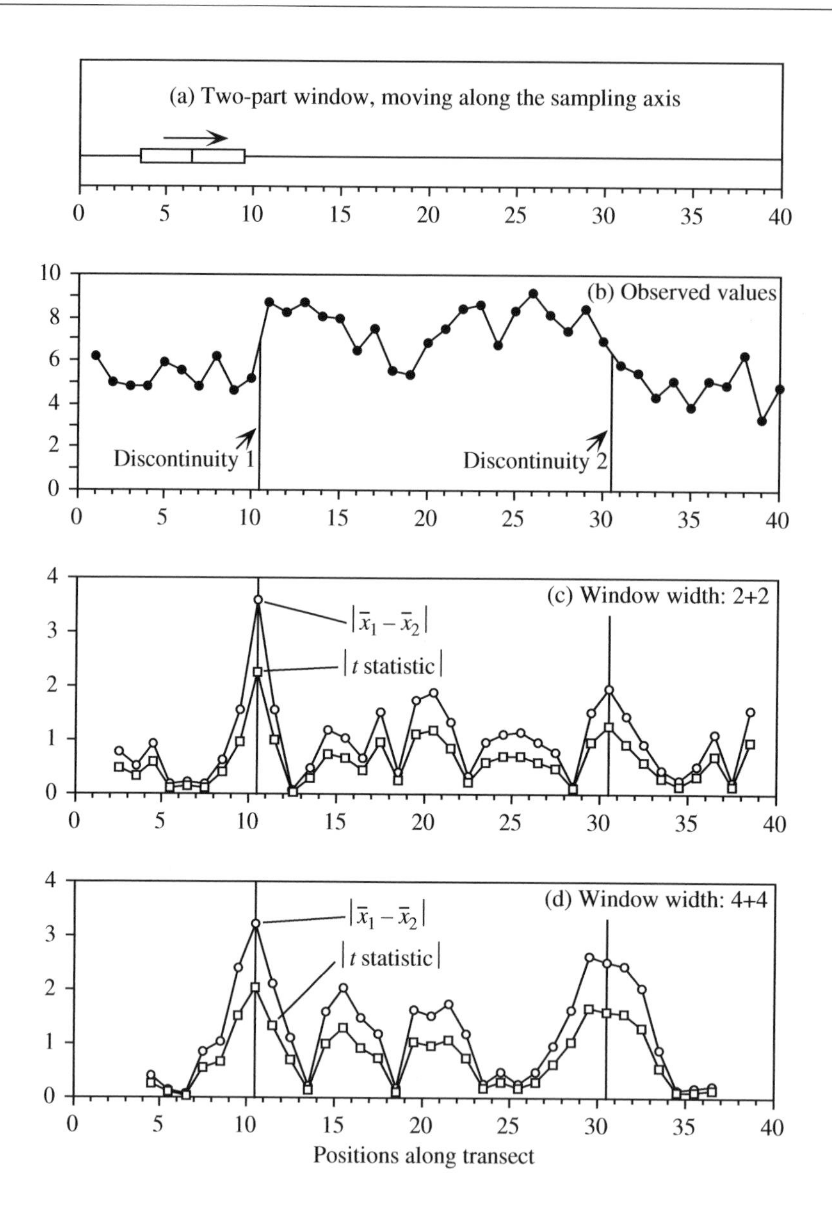

Figure 12.20 Webster's method for detecting discontinuities in data series. (a) Principle of the method. (b) Numerical example (see text). Results using a window (c) 4 observations wide, or (d) 8 observations wide.

difference (see below) between the points located in the left-hand and right-hand halves and plot these differences in a graph, as the window is moved from one end of the series to the other (Fig. 12.20c, d). The principle of the method is that the difference should be large at points where the left-hand and right-hand halves of the window contain values that are appreciably different, i.e. where discontinuities occur in the series. Various statistics may be used in the computations:

• With univariate data, calculate the absolute value of the difference between the means of the values in the left-hand and right-hand halves of the window: Statistic $= |\bar{x}_1 - \bar{x}_2|$.

• With univariate data again, one may choose to compute the absolute value of a t statistic comparing the two halves of the window: Statistic $= |\bar{x}_1 - \bar{x}_2| / s_x$. If one uses the standard deviation of the whole series as the best estimate of the standard deviations in the two halves (assuming homoscedasticity), this statistic is linearly related to the previous one. Alternatively, one could use the regular t-statistic formula for t-tests, estimating the variance in each window from the few values that it contains; this may lead to very unstable estimates when windows are narrow, which is often the case with this method.

• For multivariate series, compare the two halves of the window using either the Mahalanobis generalized distance (D_5 or D_5^2, eq. 7.40), which is the multivariate equivalent of a t statistic, or the coefficient of racial likeness (D_{12}, eq. 7.52).

The width of the window is an empirical decision made by the investigator. It is recommended to try different window widths and compare the results. The window width is limited, of course, by the spacing of observations, considering the approximate interval between the expected discontinuities. Webster's method works best with equispaced observations, but some departure from equal spacing, or missing data points, are allowed, because of the empirical nature of the method.

Numerical example. A series of 40 observations was generated using a normal pseudo-random number generator $N(5,1)$. The values of observations 11 to 30 were increased by adding 3 to the generated values in order to artificially create discontinuities between observations 10 and 11, on the one hand, and observations 30 and 31, on the other. It so happened that the first of these discontinuities was sharp whereas the second was rather smooth (Fig. 12.20b).

Webster's method for univariate data series was used with two window widths. The first window had a width of 4 observations, i.e. 2 observations in each half; the second window had a width of 8 observations, i.e. 4 in each half. Both the absolute value of the difference between means and the absolute value of the t statistic were computed. The overall standard deviation of the series was used as the denominator of t, so that this statistic was a linear transformation of the difference-between-means statistic. Results (Fig. 12.20c, d) are reported at the positions occupied by the centre of the window.

The sharp discontinuity between observations 10 and 11 was clearly identified by the two statistics and window widths. This was not the case for the second discontinuity, between observations 30 and 31. The narrow window (Fig. 12.20c) estimated its position correctly, but did not allow one to distinguish it from other fluctuations in the series, found between observations 20 and 21 for instance (remember, observations are randomly-generated numbers; so there is no structure in this part of the series). The wider window (Fig. 12.20d) brought out the second discontinuity more clearly (higher values of the statistics), but its exact position was not estimated precisely.

D_5^2 to the centroid

Window

Ibanez (1981) proposed a related method to detect discontinuities in multivariate records (e.g. simultaneous records of temperature, salinity, *in vivo* fluorescence, etc. in aquatic environments). He called the method D_5^2 *to the centroid*. For every sampling site, the method computes a generalized distance D_5^2 (eq. 7.40) between the new multivariate observation and the centroid (i.e. multidimensional mean) of the m previously recorded observations, m defining the width of a window. Using simulated and real multivariate data series, Ibanez showed that changes in D_5^2 to the centroid, drawn on a graph like Figs. 12.20c or d, allowed one to detect discontinuities. For multi-species data, however, the method of Ibanez suffers from the same drawback as the segmentation method of Hawkins & Merriam: since the simultaneous absence of species is taken as an indication of similarity, it could prevent changes occurring in the frequencies of other species from producing high, detectable distances.

McCoy *et al.* (1986) proposed a segmentation method somewhat similar to that of Webster, for species occurrence data along a transect. A matrix of Raup & Crick similarities is first computed among sites (S_{27}, eq. 7.33) from the species presence-absence data. A "+" sign is attached to a similarity found to be significant in the upper tail (i.e. when a_{hi} is significantly larger than expected under the random sprinkling hypothesis) and a "–" sign to a similarity which is significant in the lower tail (i.e. when a_{hi} is significantly smaller than expected under that null hypothesis). The number of significant pluses and minuses is analysed graphically, using a rather complex empirical method, to identify the most informative boundaries in the series.

4 — Chronological clustering

Temporal contiguity

Combining some of the best aspects of the methods described above, Gordon & Birks (1972, 1974) and Gordon (1973) introduced a constraint of temporal contiguity in a variety of clustering algorithms to study pollen stratigraphy. Analysing bird surveys repeated at different times during the breeding season, North (1977) also used a constraint of temporal contiguity to cluster bird presence locations on a geographic map and delineate territories. Recent applications of time-constrained clustering to palaeoecological data (where a spatial arrangement of the observations corresponds to a time sequence) are Bell & Legendre, 1987, Hann *et al.* (1994) and Song *et al.* (1996). Algorithmic aspects of constrained clustering are discussed in Subsection 13.3.2.

Succession model

Using the same concept, Legendre *et al.* (1985) developed the method of *chronological clustering*, based on hierarchical clustering (Chapter 8). The algorithm was designed to identify discontinuities in multi-species time series. It has also been successfully used to analyse spatial transects (e.g. Galzin & Legendre, 1987; Ardisson *et al.*, 1990; Tuomisto & Ruokolainen, 1994: Ecological application 12.6b). When applied to *ecological succession*, chronological clustering corresponds to a well-defined *model*, in which succession proceeds by steps and the transitions between steps are rapid (see also Allen *et al.*, 1977, on this topic). Broad-scale successional steps

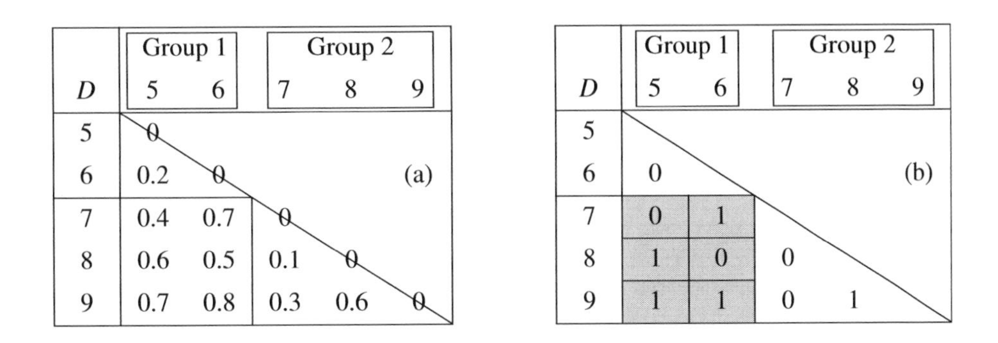

Figure 12.21 Numerical example. (a) Distance matrix for two contiguous groups from a multidimensional time series (used also in Fig. 10.22). The lower half of the symmetric matrix is shown. (b) 50% of the distances, i.e. those with the highest values, are coded 1; the others are coded 0.

contain finer-scale steps, which may be identified using a finer analysis if finer-scale data are available. Chronological clustering takes into account the sampling sequence, imposing a constraint of temporal contiguity to the clustering activity.

The method also permits the elimination of *singletons* (in the game of *bridge*, a card which is the only one of a suit in the hand of a player). Such singular observations often occur in ecological series. In nature, singletons are the result of random fluctuations, migrations, or local changes in external forcing. In an aquatic system studied at a fixed location (Eulerian approach, Section 12.0), such changes may be due to temporary movements of water masses. Singletons may also result from improper sampling or inadequate preservation of specimens.

Agglomerative clustering (Section 8.5) proceeds from an association matrix ($n \times n$) among the observations of the data series (length n), computed using an appropriately chosen similarity or distance coefficient (Chapter 7). Any method of agglomerative clustering may be used, although Legendre *et al.* (1985) used intermediate linkage clustering (Subsection 8.5.3). The clustering is modified to include the contiguity constraint; Fig. 13.23 shows how a constraint of spatial or temporal contiguity can be introduced into any agglomerative clustering algorithm. Each clustering step is subjected to a permutation test (Section 1.2) before the fusion of two objects or groups is authorized.

Permutation test

Consider two adjacent groups of objects pertaining to some data series (Fig. 12.21). The first group ($n_1 = 2$) includes objects 5 and 6 and the second ($n_2 = 3$) contains objects 7, 8 and 9. Assume that an agglomerative clustering algorithm now proposes that these two groups are the next pair to join. Distances among the five objects are given in Fig. 12.21a. Before applying the permutation test of cluster fusion, the distances are divided in two groups: the 50% of the distances (5 in this example) that have the highest values are called "high distances" and are coded 1 (Fig. 12.21b) whereas the other 50% are called "low distances" and are coded 0. The test statistic is the number of high distances (h) in the between-group matrix (shaded area); $h = 4$ in this example. Under the null hypothesis, the objects in the two groups are drawn from the same statistical population and, consequently, it is only an artefact of the agglomerative clustering algorithm that they temporarily form two groups. If the null hypothesis is true, the number of high distances ($h = 4$) presently found in the between-group matrix should be comparable to that found among all possible permutations of the five objects in two groups with $n_1 = 2$ and $n_2 = 3$ objects. If the null hypothesis is false and the two groups come from different statistical populations (i.e. different steps of the succession), the number of high distances presently found in the between-group matrix should be *higher* than most of the values found after permutation of the objects into two groups with $n_1 = 2$ and $n_2 = 3$ objects. This calls for a one-tailed test. After setting a significance level α, the permutations are performed and results that are higher than or equal to h are counted. The number of distinguishable combinations of the objects in two groups of sizes n_1 and n_2 is $(n_1 + n_2)!/(n_1!\, n_2!)$. If this number is not too large, all possible permutations can be examined; otherwise, permutations may be selected at random to form the reference distribution for significance testing. The number of permutations producing a result as large as or larger than h, divided by the number of permutations performed, gives an estimate of the probability p of observing the data under the null hypothesis.

- If $p > \alpha$, the null hypothesis is accepted and fusion of the two groups is carried out.
- If $p \le \alpha$, the null hypothesis is rejected and fusion of the groups is prevented.

This test may actually be reformulated as a Mantel test (Section 10.5.1) between the matrix of recoded distances (Fig. 12.21b) and another matrix of the same size containing 1's in the among-group rectangle and 0's elsewhere.

Internal validation criterion

This "test of significance" is actually an internal validation clustering criterion (Section 8.12), because the alternative hypothesis (H_1: the two groups actually found by the clustering method differ) is not independent of the data that are used to perform the test; it comes from the data through the agglomerative clustering algorithm. Legendre *et al.* (1985) have shown, however, that this test criterion has a correct probability of type I error; when testing on randomly generated data (Monte Carlo simulations) at significance level α, the null hypothesis is rejected in a proportion of the cases which is approximately equal to α.

Resolution

Significance level α used in the test of cluster fusion determines how easy it is to reject the null hypothesis. When α is small (close to 0), the null hypothesis is almost always accepted and only the sharpest discontinuities in the time or space series are identified. Increasing the value of α actually makes it easier to reject the null hypothesis, so that more groups are formed; the resulting groups are thus smaller and bring out more discontinuities in the data series. So, changing the value of α actually changes the resolution of the clustering results.

Singleton

A singleton is defined as a single observation whose fusion has been rejected with the groups located to its right and left in the series. When the test leads to the discovery of a singleton, it is temporarily removed from the series and the clustering procedure is started again from the beginning. This is done because the presence of a singleton can disturb the whole clustering geometry, as a result of the contiguity constraint.

The end result of chronological clustering is a *nonhierarchical partition* of the series into nonoverlapping homogeneous groups. Within the context of ecological succession, these groups correspond to the steps of a succession. *A posteriori* tests are used to assess the relationships between distant groups along the series as well as the origin of singletons. Plotting the clusters of observations onto an ordination diagram in reduced space may help in the overall interpretation of the results.

Legendre (1987b) showed that time-constrained clustering possesses some interesting properties. On the one hand, applying a constraint of spatial or temporal contiguity to an agglomerative clustering procedure forces different clustering methods to produce approximately the same results; without the constraint, the methods may lead to very different clustering results (Chapter 8), except when the spatial or temporal structure of the data (patchiness, gradient: Chapter 13) is very strong. Using autocorrelated simulated data series, he also showed that, if patches do exist in the data, constrained clustering always recovers a larger fraction of the structure than the unconstrained equivalent.

Ecological application 12.6a

In May 1977, the Société d'Énergie de la Baie James impounded a small reservoir (ca. 7 km^2), called Desaulniers, in Northern Québec (77°32' W, 53°36' N). Ecological changes occurring during the operation were carefully monitored in order to use them to forecast the changes that would take place upon impoundment of much larger hydroelectric reservoirs in the same region. Several sampling sites were visited before and after the flooding. Effects of flooding on the zooplankton community of the deepest site (max. depth: 13 m), located ca. 800 m from the dam, were studied by Legendre *et al.* (1985) using chronological clustering. Before flooding, the site was located in a riverbed and only zooplankton drifting from lakes located upstream were found there (i.e. there was no zooplankton community indigenous to the river). Changes observed are thus an example of primary succession.

After logarithmic normalization of the data (eq. 1.14), the Canberra metric (D_{10}, Chapter 7) was used to compute distances among all pairs of the 47 observations. Homogeneous groups of observations were identified along the data series, using a time-constrained algorithm for intermediate linkage clustering (Subsection 8.5.3) and the permutation test of cluster fusion described above. Results of chronological clustering are shown in Fig. 12.22 for different levels of resolution α. Plotting the groups of observations from Fig. 12.22, for α = 0.25, on an ordination diagram obtained by nonmetric multidimensional scaling (Fig. 12.23), led to the following conclusions concerning changes in the zooplankton community. In 1976, as mentioned above, zooplankton were drifting randomly from small lakes located upstream. This was evidenced by low species numbers and highly fluctuating evenness (eq. 6.44), which indicated that no stable community was present. After impoundment of the reservoir, the community departed rapidly from the river status (Fig. 12.23) and formed a fairly well-

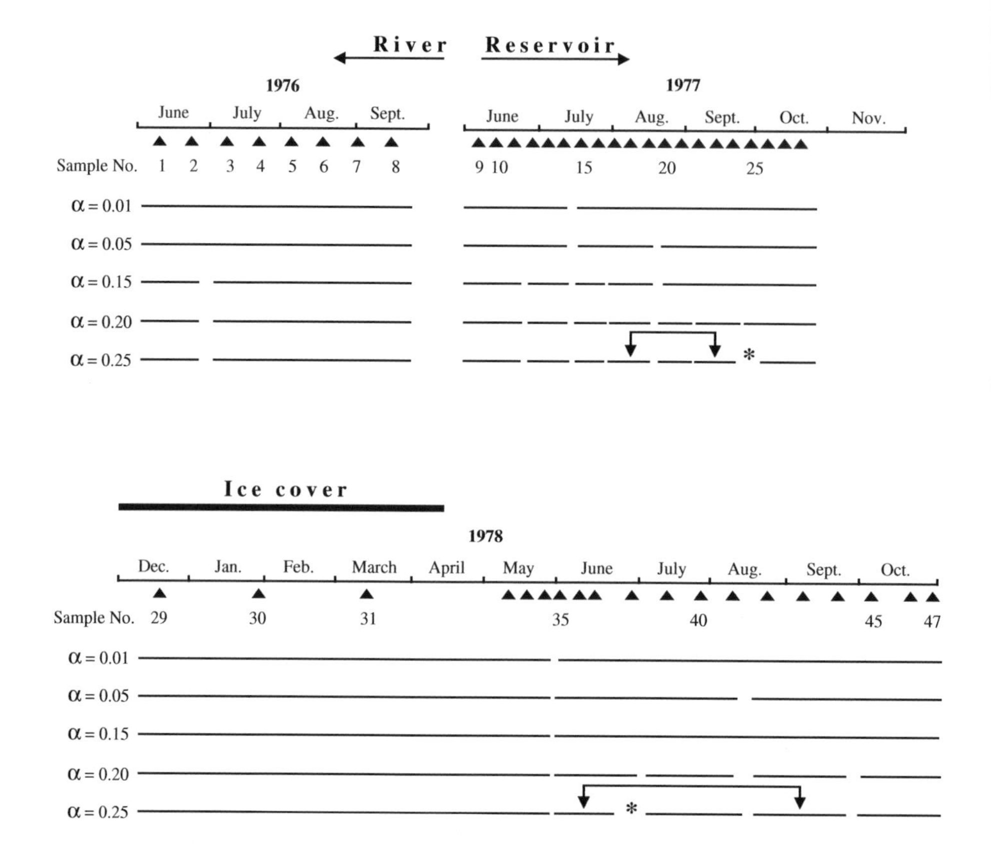

Figure 12.22 Chronological clustering: zooplankton time series. Results for different levels of resolution (α). For α = 0.25, the double arrows identify *a posteriori* tests with probabilities of fusion larger than α. Asterisks (*) identify singletons. Modified from Legendre *et al.* (1985).

developed assemblage, with 13 to 20 species in the summer of 1977, despite large chemical and water-level fluctuations. After the autumn overturn and during the 1977-1978 winter period, the community moved away from the previous summer's status. When spring came (observation 35), the community had reached a zone of the multidimensional scaling plane quite distinct from that occupied in summer 1977. Zooplankton was then completely dominated by rotifers, which increased from 70 to 87% in numbers and from 18 to 23% in biomass between 1977 and 1978, with a corresponding decrease in crustaceans, while the physical and chemical conditions had stabilized (Pinel-Alloul *et al.*, 1982). When the succession was interrupted by the 1978 autumn overturn, the last group in the series (observations 45-47) was found (Fig. 12.23) near the position of the previous winter's observations (29-34), indicating that the following year's observations might resemble the 1978 succession.

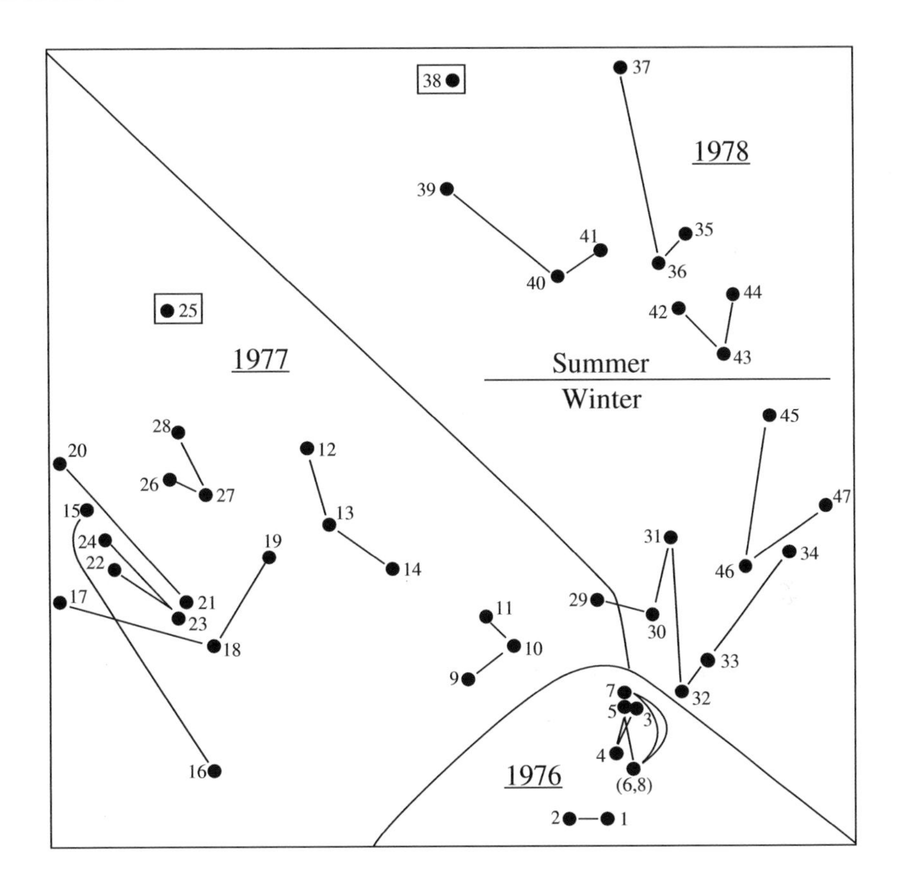

Figure 12.23 Chronological clustering: zooplankton time series. Nonmetric multidimensional scaling plot showing groups of observations from Fig. 12.22, for α = 0.25. The groups are the sets of observations that are connected by lines materializing the sampling sequence. Objects in boxes are singletons. From Legendre *et al.* (1985).

Ecological application 12.6b

Tuomisto & Ruokolainen (1994) studied species assemblages of *Pteridophyta* (ferns; 40 species in the study) and *Melastomataceae* (a family of shrubs, vines, and small trees restricted to the American tropics; 22 species in the study) along two spatial transects (replicates) in a non-flooded area of the Amazonian rain forest in Peru, covering an edaphic and topographic gradient from clay soil on level ground, to quarzitic sand on a hill top. The two 700-m-long and 5-m-wide, parallel transects were 50 m apart. Chronological clustering was applied to the edaphic and floristic variables separately, using different similarity coefficients and three levels of resolution (parameter α). In all cases, the transects could be divided into distinct sections; the results of constrained clustering were more readily interpretable than the unconstrained equivalent. The groups of plants selected proved adequate for the rapid assessment of changes in the floristic composition of the rain forest.

12.7 Box-Jenkins models

Forecasting

Objective 6 of time series analysis in ecology (Section 12.1 and Table 12.2) is to *forecast* future values. The Foreword explained that ecological modelling is not, as such, within the scope of numerical ecology. In ecological studies, however, *Box-Jenkins modelling* is often conducted together with other forms of series analysis; this is why it is briefly presented here. This type of technique has already been mentioned within the context of maximum entropy spectral analysis (MESA or MEM, Section 12.5). The present section summarizes the principles that underlie the approach. Interested readers may refer to Box & Jenkins (1976), Cryer (1986), and Bowerman & O'Connell (1987) for the theory and to user's manuals of computer packages for actual implementation.

MA model

Stochastic linear models (processes) described here are based on the idea that, in a series where data within a small window are strongly interrelated, the observed values are generated by a number of "shocks" a_t. These shocks are independent of each other and their distribution is purely random (mean zero and variance s_a^2). Such a series (a_t, a_{t-1}, a_{t-2}, ...) is called *white noise*. In the *moving average* (*MA*) *model*, each observations in the series ($\tilde{y}_t = y_t - \bar{y}$, i.e. data centred on the mean $\bar{y}$ of the series) can be represented as a weighted sum of the values of process a:

$$\tilde{y}_t = a_t - (\theta_1 a_{t-1} + \theta_2 a_{t-2} + \ldots + \theta_q a_{t-q}) \qquad \textbf{(12.32)}$$

where θ are the weights and q is the *order* of the model. The name *moving average* for this process comes from the fact that eq. 12.32 is somewhat similar to that of the moving average (see the last column of Table 12.4). The weights θ are estimated by numerical iteration, using techniques that are described in the above references and available in some computer packages.

When the above model does not fit the series adequately (see below), another possibility is to represent an observation by a weighted sum of the q previous observations plus a random shock:

$$\tilde{y}_t = \phi_1 \tilde{y}_{t-1} + \phi_2 \tilde{y}_{t-2} + \ldots + \phi_q \tilde{y}_{t-q} + a_t$$

AR model

This is the *autoregressive model* (*AR*, or *all-pole* model) already defined in eq. 12.30. In this model (of *order q*), q successive terms of the series are used to forecast term $(q + 1)$, with error a_t. When estimating the autocorrelation coefficients φ (by least squares), it is easy to compute residual errors $a_t = y_t - \tilde{y}_t$. Residual errors, as specified above for all Box-Jenkins models, must be independent of one another; this means that a correlogram of the series of residuals a_t should display no significant value. The residuals must also be normally distributed.

ARMA model

Combining the above two models gives the *autoregressive-moving average model* (*ARMA model*), whose general form is:

$$\tilde{y}_t = \phi_1\tilde{y}_{t-1} + \phi_2\tilde{y}_{t-2} + \dots + \phi_q\tilde{y}_{t-q} + a_t - (\theta_1 a_{t-1} + \theta_2 a_{t-2} + \dots + \theta_q a_{t-q}) \quad \mathbf{(12.33)}$$

An important advantage of ARMA models is that they can be fitted to data series using a small number of parameters (i.e. the coefficients ϕ and θ). However, such models may only be estimated for strictly *stationary* series (Section 12.1 and 12.2).

One approach described in Section 12.2 for extracting the trend from a series is the *variate difference method.* In the computation, each value y_t is replaced by $(y_t - y_{t-T})$ where T is the period of the trend:

$$\tilde{y}_t = y_t - y_{t-T} \quad \mathbf{(12.34)}$$

Since $\tilde{y}_t$ results from a *difference*, y_t is called the *integrated form* of $\tilde{y}_t$. When an ARMA model is applied to a series of values computed with eq. 12.34, it is called an *autoregressive-integrated-moving average model* (*ARIMA model*).

ARIMA model

Box-Jenkins analysis normally proceeds in four steps. (1) Identification of the *type of model* to be fitted to the data series (i.e. MA, AR, ARMA, or ARIMA). Even though Box & Jenkins (1976) described some statistical properties of the series (e.g. shape of the autocorrelation) that can guide this choice, identification of the proper model remains a somewhat intuitive step (e.g. Ibanez, 1982). (2) Estimation of the *parameters* of the model. For each case, various methods are generally available, so that one is confronted with a choice. (3) The *residuals* must be independent and normally distributed. If not, either the model is not adequate for the data series, or the parameters were not properly estimated. In such a case, step (2) can be repeated with a different method or, if this does not improve the residuals, a different model must be chosen at step (1). Steps (1) through (3) may be repeated as many times as necessary to obtain a good fit. The procedure of identification of the appropriate model is therefore *iterative*. (4) Using the model, values can be *forecasted* beyond the last observation.

It may happen that the data series is under external influences, so that the models described above cannot be used as such. For example, in the usual ARIMA model, the state of the series at time t is a function of the previous q observations ($\tilde{y}$) and of the random errors (a). In order to account for the additional effect of external variables, some computer programs allow the inclusion of a *transfer function* into the model (if the external forcing variable is also a random variable) and/or an *intervention component* (if the external variable is binary and not random).

It is possible to extend the forecasting to *multidimensional* data series. References to conduct the analysis are Whittle (1963) and Jones (1964).

It is important to remember that the models discussed here are *forecasting* and not *predictive* models. Indeed, the purpose of Box-Jenkins modelling is to *forecast* values of the series beyond the last observation, using the preceding data. Such forecasting is only valid as long as the environmental conditions that characterize the population under study (demographic rates, migrations, etc.) as well as the anthropogenic effects (exploitation methods, pollution, etc.) remain essentially the same. In order to *predict* with some certainty the fate of the series, causal relationships should be determined and modelled; for example, between the observed numbers of organisms, on the one hand, and the main environmental conditions, population characteristics, or/and anthropogenic factors, on the other. This requires extensive knowledge of the system under study. Forecasting models often prove quite useful in ecology, but one must be careful not to use them beyond their limits.

Ecological application 12.7

Boudreault *et al.* (1977) tried to forecast lobster landings in Îles-de-la-Madeleine (Gulf of St. Lawrence, Québec), using various methods of series analysis. In a first step, they found that an *autoregressive model* (of order 1) accounted for ca. 40% of the variance in the series of landings. This relatively low percentage could be explained by the fact that observations in the series were not very homogeneous. In a second step, external physical variables were added to the model and the data were analysed using *regression on principal components* (Section 10.3). The two external variables were: water temperature in December, 8.5 years before the fishing season, and average winter temperature 3.5 years before. This increased to 90% the variance explained by the model. Lobster landings in a given year would thus depend on: the available stock (autocorrelated to landings during the previous year), the influence of water temperature on larval survival (lobster *Homarus americanus* around Îles-de-la-Madeleine reach commercial size when ca. 8 years old), and the influence of water temperature at the time the animals reached sexual maturity (at the age of ca. 5 years).

12.8 Computer programs

Procedures available in statistical packages presently on the market are not systematically reviewed here, because the task would be beyond the scope of the present book and the information would rapidly become obsolete. Sources of information about computer programs are given in Section 1.3. However, because ecologists are often not very familiar with series analysis, some information about the main commercial computer packages that offer programs for time series analysis is assembled in Table 12.9. In addition, three university-based packages are mentioned because they are the only ones, for the time being, that offer programs for some of the methods discussed in the previous sections. Information in Table 12.9 is not exhaustive. It was up-to-date at the time this book was completed (1998), but new software rapidly appears on the market.

Table 12.9 Computer programs available for methods of time series analysis discussed in Chapter 12.

Program	Methods
Mainframe computers	
BMDP	Correlogram, Schuster periodogram, spectral analysis, Box-Jenkins models
SPSSX	Correlogram, Schuster periodogram, spectral analysis, Box-Jenkins models
SAS/ETS	Correlogram, Schuster periodogram, spectral analysis, Box-Jenkins models
R Package*	Contingency periodogram, Mantel correlogram, chronological clustering
MS-DOS/Windows machines	
BMDP	Correlogram, Schuster periodogram, spectral analysis, Box-Jenkins models
SAS-PC/ETS	Correlogram, Schuster periodogram, spectral analysis, Box-Jenkins models
STATISTICA**	Filters, correlogram, Schuster periodogram, spectral analysis, ARIMA
PASSTEC***	Trend analysis, filters, correlogram, harmonic analysis, spectral analysis (univariate, bivariate), contingency periodogram, chronological clustering, etc.
ITSM ****	Filters, correlogram, Schuster periodogram, spectral analysis (univariate, bivariate), autoregressive models (univariate, multivariate)
Macintosh and MS-DOS/Windows machines	
SYSTAT	Correlogram
SPSS	Correlogram, Schuster periodogram, spectral analysis, Box-Jenkins models
STATISTICA**	Filters, correlogram, Schuster periodogram, spectral analysis, ARIMA
R Package*	Contingency and Dutilleul modified periodograms, Mantel correlogram, chronological clustering

* See Table 13.4.

** In STATISTICA, numerical filters (Section 12.2) are called "smoothing methods".

*** The PASSTEC package is distributed by F. Ibanez and M. Étienne, Observatoire Océanologique, Station Zoologique, B.P. 28, F-06230 Villefranche-sur-Mer, France. PASSTEC means: *Programme d'Analyse des Séries Spatio-Temporelles en Écologie Côtière.*

**** The *Interactive Time Series Modelling* package is provided on diskettes with the book of Brockwell & Davis (1991b). This book briefly describes the theory and gives examples of how to use the programs. It was designed as companion to the book of Brockwell & Davis (1991a).

Chapter

13 Spatial analysis

13.0 Spatial patterns

The analysis of spatial patterns is of prime interest to ecologists because most ecological phenomena investigated by sampling geographic space are structured by forces that have spatial components. Spatial patterns are studied through surveys (called *mensurative experiments* by Hurlbert, 1984), whereas underlying processes can be studied by *manipulative experiments* (Subsection 10.2.3). Ecological processes may give rise to spatially recognizable structures which may display spatial patterns and be the subject of spatial analysis. Most ecological patterns may be described as either patches (such as tree groves, phytoplankton patches, and animal herds) or gradients. The latter may be linear or not.

Experiment

Gradient
Patch

Ecologists examine the spatial patterns of species or assemblages in order to understand the mechanisms that control species distributions. Patchiness is found at all spatial scales — from micrometres to continent and ocean-wide scales. Displaying the spatial variation of an ecological variable in the form of a map shows whether the structure is smoothly continuous or marked by sharp discontinuities. Most field studies cover only a part of any variable's spatial structure. So, gradients or patches displayed by maps may only be interpreted with respect to the scale of the sampling programme, which should be compared to the scale of the phenomenon under study.

It is now understood that species distributions result from the combined action of several forces, some of which are external whereas others are intrinsic to the community. According to the environmental control model (Whittaker, 1956; Bray & Curtis, 1957; Hutchinson, 1957), environmental characteristics are the external forces which control species distributions. The internal forces relate to population dynamics or to top-down or bottom-up biotic interactions within the community (Lindeman, 1942; Southwood, 1987). Both types of forces generate spatial patterns within species or communities. Historical events (Sousa, 1979; Pickett & White, 1985; Reynolds, 1987) are other possible sources of spatial patterns; examples are given in Subsection 13.5.2. The mechanisms that create spatial structures and, hence, autocorrelation in the data, have been briefly discussed in Section 1.1.

Historical
events

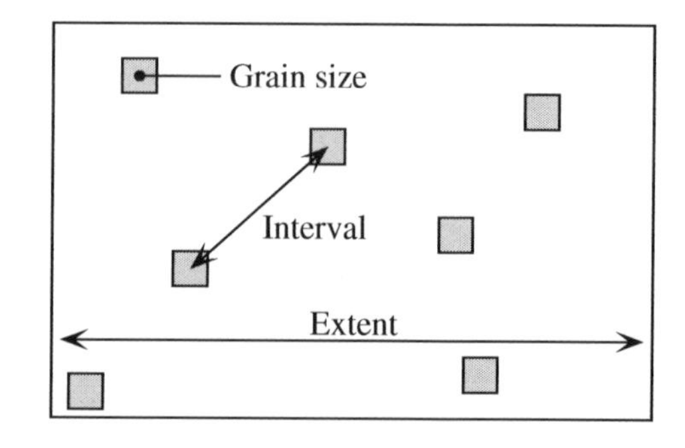

Figure 13.1 Components of a sampling design are the grain size, sampling interval, and extent. In the Figure, the sampling units are represented by squares.

The analysis of spatial structures is a fairly recent subject. It is still plagued with conceptual and methodological problems. The present Chapter is not a tutorial discussing all possible questions of interest. Its scope is more modest; it will describe some methods that allow the investigation of some of the questions of interest in spatial pattern analysis. Other avenues will be mentioned where developments are expected from adjacent fields such as geostatistics. A fundamental question will be left unanswered in this Chapter: that of designing efficient sampling programmes for studying and analysing spatial patterns. The theory of spatial ecological sampling has to be re-written to provide meaningful answers to this question.

Scale

Scale is a key concept in both sampling design and the analysis of spatial (or temporal) patterns. It includes several characteristic spatial (or temporal) properties of random variables. Definition of these properties, which follows, depends on context.

In sampling theory (Fig. 13.1), spatial scale encompasses three elements of the sampling design (Wiens, 1989; Allen & Hoekstra, 1991; He *et al.*, 1994):

• *Grain size* is the size of the elementary sampling units. It may be expressed as the diameter, surface or volume of matter supporting the measurements. In time series analysis, it is the duration over which measurements are integrated. The *resolution* of a study (Schneider, 1994) is equal to the grain size of its sampling design.

• *Sampling interval* is the average distance between neighbouring sampling units. It is called *lag* in time series analysis. For fixed extent, the sampling interval is a function of n, the number of sampling units. In turn, n is determined by the total effort that can be allocated to sampling.

• *Extent* is the total length, area or volume included in the study, or the total duration of the time series. It is called *range* by Schneider (1994) who also defines the *scope* as the ratio of the extent to the grain size. Since extent and grain size are expressed in the same units, scope is a dimensionless variable (Section 3.1).

It may happen that the data consist of contiguous sampling areas that completely cover the extent, instead of small sampling units distant from one another. This may

Pixel

occur in a variety of circumstances where a map is divided into contiguous "picture cells" or *pixels*, which include satellite data, video analysis of a transect, and modelling. The linear measurement of grain size is equal to the sampling interval in such a case. The same thing may happen with time series.

The spatial scale of patterns or processes is described as follows:

• How big is a unit object, or how much space is disturbed by a unit process? This amount of space, which is equivalent to grain size, is called the ecological neighbourhood (Addicott *et al.*, 1987) or the area of resolution of individuals (Wiens, 1989). *Unit objects* may be individual plants or animals, bacterial colonies, etc. Examples of measurable structures resulting from *unit processes* are: the neighbourhood occupied by a territorial animal, the width of the wetland zone along a stream or of a tidal sand flat, the size of the patch of soil modified by the root system of a plant, and the size of phytoplankton patches which result from the combined action of primary production and diffusion (see Ecological applications 3.2d and 3.3a).

• What is the average distance between unit objects or processes? This distance is equivalent to the sampling interval.

• Over how much space does this type of object, or this process, occur? This amount of space is equivalent to the extent. For some processes, the extent may be an ocean or the whole planet.

The same notions may be applied to temporally occurring patterns or processes. While they are readily applicable to patterns that concern the distribution of objects, they may sometimes be applied as well to processes.

Sampling design

The scale of the sampling design should follow from what is known (e.g. from a pilot study) about the scale of the pattern or process, and from the ecological question being addressed. A well-focused question generally reduces the difficulty of choosing the type (simple random, systematic, stratified, etc.) as well as the scale components (grain, interval, extent) of the sampling design.

• The sampling grain should be larger than a unit object (e.g. an individual organism) and the same as, or preferably smaller than, the structures resulting from a unit process (e.g. a patch) which is to be detected by the sampling design.

• The sampling interval should be smaller than the average distance between the structures resulting from a unit process to be detected by the sampling design.

• The sampling extent may, in some cases, be the same as the total area covered by the type of objects or by the process under study. In other cases, it is limited to a smaller area, determined by the total allowable effort (n) and the maximum interval that one wishes to maintain between adjacent sampling units. For constant n, the sampling extent can be maximized by turning the sampling area into a transect (see Ecological application 13.1b).

The extent and grain define the observation window in spatial pattern analysis. No structure can be detected which is smaller than the grain or larger than the extent of a study. Wiens (1989) compares them to the overall size and mesh size of a sieve, respectively.

In quantitative ecology, the term "scale" is generally used in a sense opposite to that of cartography. For cartographers, the scale is the ratio between the linear size of an object on a map and its size in nature, so that a small-scale map (e.g. 1:100000) is less detailed than a large-scale map (e.g. 1:25000). For ecologists, scale generally refers to the unit of measurement, e.g. the kilometre sampling scale is bigger than the centimetre scale and weekly observations are broader-scaled than hourly observations. Confusion is avoided by using "broad scale" for phenomena with large extents and "fine scale" for those with small extents (Wiens, 1989)*. In any case, these terms only have comparative values.

Broad scale
Fine scale

In many instances, not one but several scales may be pertinent for the study of a pattern or process. Different processes are often at work, depending on the scale, to determine spatial patterns. As a consequence, conclusions derived for a spatial scale often cannot be extrapolated to other scales. The scale chosen for any particular study may be considered as a variable-sized window through which one can study nature. For example, He *et al.* (1994) have shown how species diversity changes as a function of different components of scale (grain size, sampling interval, and extent). The techniques described in Section 13.1, in particular, allow researchers to describe how spatial correlation changes as a function of the sampling interval.

Scale is an important reference to help understand the difference between environmental management problems and the answers that may be found in ecological studies. Most studies are conducted at scales (extents) finer than those of natural or anthropogenic disturbances (Fig. 13.2). As a consequence, environmental problems usually involve scales broader than the information available from field studies — surveys or field experiments. Scaling up from studies to environmental problems is a challenge that ecologists are often facing. New concepts and statistical tools must be developed to do so (Thrush *et al.*, 1997). Spatial analysis of the results of surveys conducted across several spatial scales is one means towards this end.

Heterogeneity

An important concept is that of *heterogeneity* (Kolasa & Rollo, 1991; Dutilleul & Legendre, 1993). With reference to spatial patterns, heterogeneity is the opposite of *homogeneity* which means the absence of variation. In everyday's language, heterogeneous means "composed of unlike elements or parts". Pitard (1992) distinguishes *constitution heterogeneity*, which is a property of the objects under study, from *distribution heterogeneity* which can be altered by mixing. In spatial pattern analysis, heterogeneous refers to variation in the measurements, in some general sense that applies to quantitative, semiquantitative, or qualitative variables

* Unfortunately these two terms are not antonymic. *Broad* scale refers to the extent; its antonym is *narrow*. *Fine* scale refers to the grain; its antonym is *coarse*.

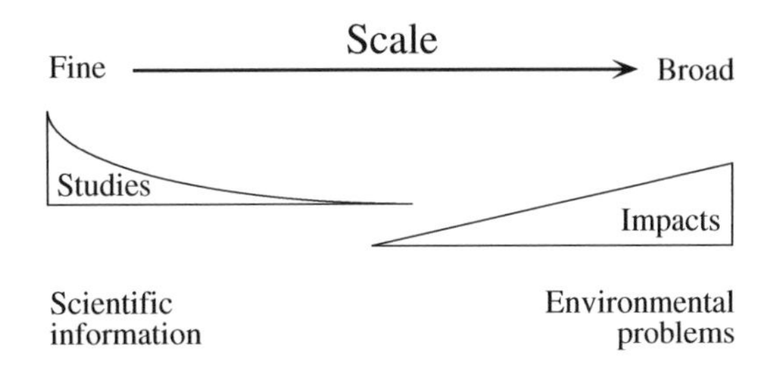

Figure 13.2 Scale differences between environmental management problems and the answers that may be found in ecological studies. From S. F. Thrush (pers. comm.).

(Subsection 1.4.1). The concept of heterogeneity may also be applied to the time dimension, considering repeated observations made at a single point in space. Heterogeneity can be measured in a univariate (e.g. the variance of a singe variable) or a multivariate way (e.g. the trace of a dispersion matrix). It can be decomposed into orthogonal components (as in PCA, Section 9.1) or with respect to spatial or temporal distance classes (e.g. correlograms for spatial survey data, Section 13.1, or for time series, Section 12.3). Kolasa & Rollo (1991) recognize that "measured heterogeneity", which reflects the observer's perspective, may be inadequate in that it may differ from the "functional heterogeneity" which is the heterogeneity that influences the organisms. Functional heterogeneity may not be the same for different groups of organisms because the processes that are important for different groups may act at different temporal or spatial scales. In the sea, for instance, the doubling time of organisms is of the order of 1 day for phytoplankton, 10 to 40 days for zooplankton, 100 to 900 days for fish, and 120 to 500 days for mussels. Spatially, the horizontal scales that characterize patches are of the order of 0.1 to 1 km for phytoplankton and zooplankton, and 1 to 100 km for fish (Legendre *et al.*, 1986). Measured heterogeneity converges towards functional heterogeneity as our knowledge of a system increases and, with it, our ability to use our measures to characterize important properties of the system (Kolasa & Rollo, 1991; Dutilleul & Legendre, 1993).

Point pattern

The analysis of spatial ecological patterns comprises two families of methods. *Point pattern analysis* is concerned with the distribution through space of individual objects — for instance individual plants or animals. Its chief purpose is to determine whether the geographic distribution of data points is random or not and to describe the type of pattern, in order to infer what kind of process may have generated it. In this family of methods, the quadrat-density and nearest-neighbour approaches have been widely used in vegetation science (Galiano 1982; Carpenter & Chaney 1983). Point pattern analysis will not be discussed further in this Chapter. It has been authoritatively reviewed by a number of authors, including Pielou (1977), Cicéri *et al.* (1977), Getis & Boots (1978), Ripley (1981, 1987), and Upton & Fingleton (1985).

Regionalized variable
Surface

Surface pattern

Values of a variable observed over a delimited geographic area form a *regionalized variable* (Matheron, 1965) or simply a *surface* (Oden *et al.*, 1993; Legendre & McArdle, 1997) if the sites where the variable has been observed may be viewed as a sample from an underlying continuous surface. The second family of methods, called *surface pattern analysis*, deals with the study of spatially continuous phenomena. The spatial distributions of the variables are known, as usual, through sampling at discrete sampling sites. One or several variables are observed or measured at the observation sites, each site representing its surrounding portion of the geographic space. The analysis of continuous surfaces, where pixels cover the whole map (including data obtained by echolocation or remote sensing), is not specifically discussed here.

Surface pattern analysis includes a large number of methods developed to answer a variety of questions (Table 13.1). Several of these methods are discussed in the present Chapter. General references are: Cliff & Ord (1981), Ripley (1981), Upton & Fingleton (1985, 1989), Griffith (1987), Legendre & Fortin (1989), and Rossi *et al.* (1992). The geostatistical literature is briefly reviewed in Subsection 13.2.2. The comparison of surfaces, i.e. univariate measures over the same area repeated at two or more sampling times, has been discussed by Legendre & McArdle (1997). Section 13.7 provides a list of computer programs available from researchers; most methods for surface pattern analysis are not available in the major statistical packages.

Line pattern

Geographers have also developed *line pattern analysis* which is a topological approach to the study of networks of connections among points. Examples are: roads, telephone lines, and river networks.

For a point pattern, heterogeneity refers to the distribution of individuals across space; one often compares the observed density variation of organisms to that expected for randomly distributed objects. For a surface pattern, heterogeneity refers to the variability of quantitative or qualitative descriptors across space. Dutilleul & Legendre (1993) provide a summary of the main statistical tools available to ecologists to quantify spatial heterogeneity in both the point pattern and the surface pattern cases. Dutilleul (1993) describes in more detail how experimental designs can be accommodated to the spatial heterogeneity found in nature; spatial heterogeneity may be a nuisance for the experimenter, or a characteristic of interest. The analysis of spatial patterns is the study of organized arrangements of [ecological] heterogeneity across space.

13.1 Structure functions

Ecologists are interested in describing spatial structures in quantitative ways and testing for the presence of spatial autocorrelation in data. The primary objective is to:

Table 13.1 Surface pattern analysis: research objectives and related numerical methods. Modified from Legendre & Fortin (1989).

Research objective	Numerical methods
1) Description of spatial structures and testing for the presence of spatial autocorrelation (Descriptions using structure functions should always be complemented by maps.)	Univariate structure functions: correlogram, variogram, etc. (Section 13.1)
	Multivariate structure functions: Mantel correlogram (Section 13.1)
	Testing for a gradient in multivariate data: (1) constrained (canonical) ordination between the multivariate data and the geographic coordinates of the sites (Section 13.4). (2) Mantel test between ecological distances (computed from the multivariate data) and geographic distances (Subsection 10.5.1)
2) Mapping; estimation of values at given locations	Univariate data: local interpolation map; trend-surface map (global statistical model) (Sect. 13.2)
	Multivariate data: clustering with spatial contiguity constraint, search for boundaries (Section 13.3); interpolated map of the 1st (2nd, etc.) ordination axis (Section 13.4); multivariate trend-surface map obtained by constrained ordination (canonical analysis) (Section 13.4)
3) Modelling species-environment relationships while taking spatial structures into account	Raw data tables: partial canonical analysis (Section 13.5)
	Distance matrices: partial Mantel analysis (Section 13.6)
4) Performing valid statistical tests on autocorrelated data	Subsection 1.1.1

• either support the null hypothesis that no significant spatial autocorrelation is present in a data set, or that none remains after detrending (Subsection 13.2.1), thus insuring valid use of the standard univariate or multivariate statistical tests of hypotheses.

• or reject the null hypothesis and show that significant spatial autocorrelation is present in the data, in order to use it in conceptual or statistical models.

Tests of spatial autocorrelation coefficients may only support or reject the null hypothesis of the absence of significant spatial structure. When significant spatial structure is found, it may correspond, or not, to spatial autocorrelation (Section 1.1, model b) — depending on the hypothesis of the investigator.

Spatial structures may be described through *structure functions*, which allow one to quantify the spatial dependency and partition it amongst distance classes. Interpretation of this description is usually supported by maps of the univariate or multivariate data (Sections 13.2 to 13.4). The most commonly used structure functions are correlograms, variograms, and periodograms.

Map

Spatial correlogram

A *correlogram* is a graph in which autocorrelation values are plotted, on the ordinate, against *distance classes* among sites on the abscissa. Correlograms (Cliff & Ord 1981) can be computed for single variables (Moran's I or Geary's c autocorrelation coefficients, Subsection 1) or for multivariate data (Mantel correlogram, Subsection 5); both types are described below. In all cases, a test of significance is available for each individual autocorrelation coefficient plotted in a correlogram.

Variogram

Similarly, a *variogram* is a graph in which semi-variance is plotted, on the ordinate, against *distance classes* among sites on the abscissa (Subsection 3). In the geostatistical tradition, semi-variance statistics are not tested for significance, although they could be through the test developed for Geary's c, when the condition of second-order stationarity is satisfied (Subsection 13.1.1). Statistical models may be fitted to variograms (linear, exponential, spherical, Gaussian, etc.); they allow the investigator to relate the observed structure to hypothesized generating processes or to produce interpolated maps by kriging (Subsection 13.2.2).

Because they measure the relationship between pairs of observation points located a certain distance apart, correlograms and variograms may be computed either for preferred geographic directions or, when the phenomenon is assumed to be isotropic in space, in an all-directional way.

2-D periodogram

A *two-dimensional Schuster* (1898) *periodogram* may be computed when the structure under study is assumed to consist of a combination of sine waves propagated through space. The basic idea is to fit sines and cosines of various periods, one period at a time, and to determine the proportion of the series' variance (r^2) explained by each period. In periodograms, the abscissa is either a period or its inverse, a frequency; the ordinate is the proportion of variance explained. Two-dimensional periodograms may be plotted for all combinations of directions and spatial frequencies. The technique is described Priestley (1964), Ripley (1981), Renshaw and Ford (1984) and Legendre & Fortin (1989). It is not discussed further in the present book.

1 — Spatial correlograms

In the case of quantitative variables, spatial autocorrelation may be measured by either Moran's I (1950) or Geary's c (1954) spatial autocorrelation statistics (Cliff & Ord, 1981):

Moran's I:

$$I(d) = \frac{\frac{1}{W}\sum_{h=1}^{n}\sum_{i=1}^{n} w_{hi}(y_h - \bar{y})(y_i - \bar{y})}{\frac{1}{n}\sum_{i=1}^{n}(y_i - \bar{y})^2} \quad \text{for } h \neq i \qquad \mathbf{(13.1)}$$

Geary's c:

$$c(d) = \frac{\frac{1}{2W}\sum_{h=1}^{n}\sum_{i=1}^{n} w_{hi}(y_h - y_i)^2}{\frac{1}{(n-1)}\sum_{i=1}^{n}(y_i - \bar{y})^2} \quad \text{for } h \neq i \qquad \mathbf{(13.2)}$$

The y_h's and y_i's are the values of the observed variable at sites h and i. Before computing spatial autocorrelation coefficients, a matrix of geographic distances $\mathbf{D} = [D_{hi}]$ among observation sites must be calculated. In the construction of a correlogram, spatial autocorrelation coefficients are computed, in turn, for the various distance classes d. The weights w_{hi} are Kronecker deltas (as in eq. 7.20); the weights take the value $w_{hi} = 1$ when sites h and i are at distance d and $w_{hi} = 0$ otherwise. In this way, only the pairs of sites (h, i) within the stated distance class (d) are taken into account in the calculation of any given coefficient. This approach is illustrated in Fig. 13.3. W is the sum of the weights w_{hi} for the given distance class, i.e. the number of pairs used to calculate the coefficient. For a given distance class, the weights w_{ij} are written in a $(n \times n)$ matrix $\mathbf{W}$. Jumars *et al.* (1977) present ecological examples where the distance^{-1} or distance^{-2} among adjacent sites is used for weight instead of 1's.

The numerators of eqs. 13.1 and 13.2 are written with summations involving each pair of objects twice; in eq. 13.2 for example, the terms $(y_h - y_i)^2$ and $(y_i - y_h)^2$ are both used in the summation. This allows for cases where the distance matrix $\mathbf{D}$ or the weight matrix $\mathbf{W}$ is asymmetric. In studies of the dispersion of pollutants in soil, for instance, drainage may make it more difficult to go from A to B than from B to A; this may be recorded as a larger distance from A to B than from B to A. In spatio-temporal analyses, an observed value may influence a later value at the same or a different site, but not the reverse. An impossible connection may be coded by a very large value of distance. In most applications, however, the geographic distance matrix among sites is symmetric and the coefficients may be computed from the half-matrix of distances; the formulae remain the same, in that case, because W, as well as the sum in the numerator, are half the values computed over the whole distance matrix $\mathbf{D}$ (except $h = i$).

One may use distances along a network of connections (Subsection 13.3.1) instead of straight-line geographic distances; this includes the "chess moves" for regularly-spaced points as obtained from systematic sampling designs: rook's, bishop's, or king's connections (see Fig. 13.19). For very broad-scale studies, involving a whole ocean for instance, "great-circle distances", i.e. distances along earth's curved surface, should be used instead of straight-line distances through the earth crust.

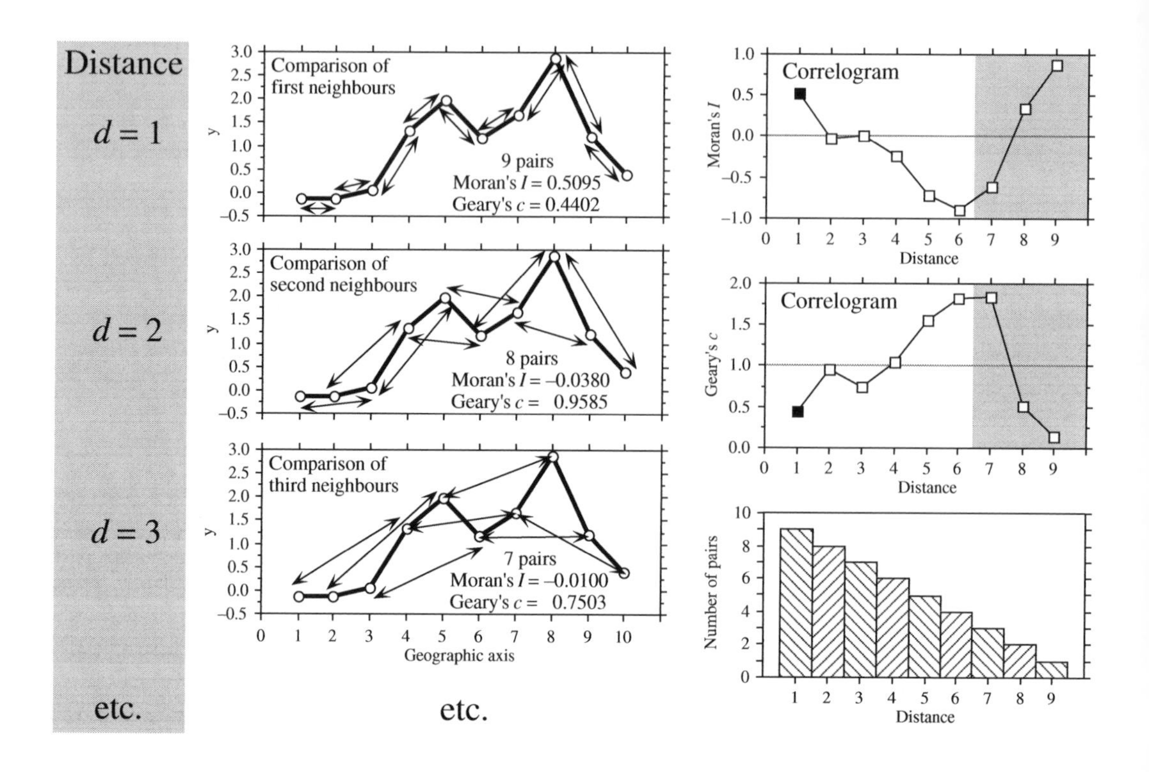

Figure 13.3 Construction of correlograms. Left: data series observed along a single geographic axis (10 equispaced observations). Moran's I and Geary's c statistics are computed from pairs of observations found at preselected distances ($d = 1$, $d = 2$, $d = 3$, etc.). Right: correlograms are graphs of the autocorrelation statistics plotted against distance. Dark squares: significant autocorrelation statistics ($p \leq 0.05$). Lower right: histogram showing the number of pairs in each distance class. Coefficients for the larger distance values (grey zones in correlograms) should not be considered in correlograms, nor interpreted, because they are based on a small number of pairs (test with low power) and only include the pairs of points bordering the series or surface.

Moran's I formula is related to Pearson's correlation coefficient; its numerator is a covariance, comparing the values found at all pairs of points in turn, while its denominator is the maximum-likelihood estimator of the variance (i.e. division by n instead of $n - 1$); in Pearson's r, the denominator is the product of the standard deviations of the two variables (eq. 4.7), whereas in Moran's I there is only one variable involved. Moran's I mainly differs from Pearson's r in that the sums in the numerator and denominator of eq. 13.1 do not involve the same number of terms; only the terms corresponding to distances within the given class are considered in the numerator whereas all pairs are taken into account in the denominator. Moran's I usually takes values in the interval [–1, +1] although values lower than –1 or higher than +1 may occasionally be obtained. Positive autocorrelation in the data translates into positive values of I; negative autocorrelation produces negative values.

Readers who are familiar with correlograms in time series analysis will be reassured to know that, when a problem involves equispaced observations along a single physical dimension, as in Fig. 13.3, calculating Moran's I for the different distance classes is nearly the same as computing the autocorrelation coefficient of time series analysis (Fig. 12.5, eq. 12.6); a small numeric difference results from the divisions by $(n-k-1)$ and $(n-1)$, respectively, in the numerator and denominator of eq. 12.6, whereas division is by $(n-k)$ and (n), respectively, in the numerator and denominator of Moran's I formula (eq. 13.1).

Geary's c coefficient is a distance-type function; it varies from 0 to some unspecified value larger than 1. Its numerator sums the squared differences between values found at the various pairs of sites being compared. A Geary's c correlogram varies as the reverse of a Moran's I correlogram; strong autocorrelation produces high values of I and low values of c (Fig. 13.3). Positive autocorrelation translates in values of c between 0 and 1 whereas negative autocorrelation produces values larger than 1. Hence, the reference 'no correlation' value is $c = 1$ in Geary's correlograms.

For sites lying on a surface or in a volume, geographic distances do not naturally fall into a small number of values; this is true for regular grids as well as random or other forms of irregular sampling designs. Distance values must be grouped into distance classes; in this way, each spatial autocorrelation coefficient can be computed using several comparisons of sampling sites.

Numerical example. In Fig. 13.4 (artificial data), 10 sites have been located at random into a 1-km^2 sampling area. Euclidean (geographic) distances were computed among sites. The number of classes is arbitrary and left to the user's decision. A compromise has to be made between resolution of the correlogram (more resolution when there are more, narrower classes) and power of the test (more power when there are more pairs in a distance class). Sturge's rule is often used to decide about the number of classes in histograms; it was used here and gave:

$$\text{Number of classes} = 1 + 3.3\log_{10}(m) = 1 + 3.3\log_{10}(45) = 6.46 \qquad \textbf{(13.3)}$$

where m is, in the present case, the number of distances in the upper (or lower) triangular matrix; the number was rounded to the nearest integer (i.e. 6). The distance matrix was thus recoded into 6 classes, ascribing the class number (1 to 6) to all distances within a class of the histogram.

An alternative to distance classes with equal widths would be to create distance classes containing the same number of pairs (notwithstanding tied values); distance classes formed in this way are of unequal widths. The advantage is that the tests of significance have the same power across all distance classes because they are based upon the same number of pairs of observations. The disadvantages are that limits of the distance classes are more difficult to find and correlograms are harder to draw.

Spatial autocorrelation coefficients can be tested for significance and confidence intervals can be computed. With proper correction for multiple testing, one can determine whether a significant spatial structure is present in the data and what are the distance classes showing significant positive or negative autocorrelation. Tests of significance require, however, that certain conditions specified below be fulfilled.

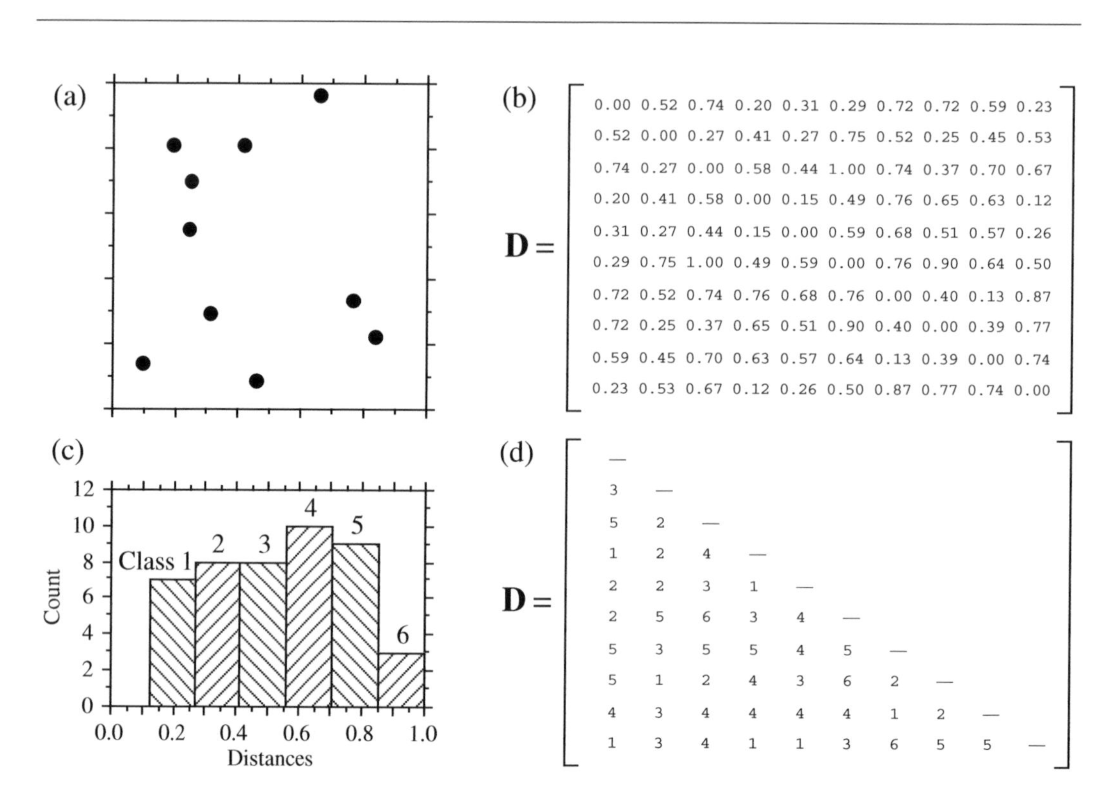

Figure 13.4 Calculation of distance classes, artificial data. (a) Map of 10 sites in a 1-km^2 sampling area. (b) Geographic distance matrix (**D**, in km). (c) Frequency histogram of distances (classes 1 to 6) for the upper (or lower) triangular portion of **D**. (d) Distances recoded into 6 classes.

Second-order stationarity

The tests require that the condition of *second-order stationarity* be satisfied. This rather strong condition states that the expected value (mean) and spatial covariance (numerator of eq. 13.1) of the variable is the same all over the study area, and the variance (denominator of eq. 13.1) is finite. The value of the autocorrelation function depends only on the length and orientation of the vector between any two points, not on its position in the study area (David, 1977).

Intrinsic assumption

A relaxed form of stationarity hypothesis, the *intrinsic assumption*, states that the differences $(y_h - y_i)$ for any distance d (in the numerator of eq. 13.2) must have zero mean and constant and finite variance over the study area, independently of the location where the differences are calculated. Here, one considers the *increments* of the values of the regionalized variable instead of the values themselves (David, 1977). As shown below, the variance of the increments is the variogram function. In layman's terms, this means that a single autocorrelation function is adequate to describe the entire surface under study. An example where the intrinsic assumption does not hold is a region which is half plain and half mountains; such a region should be divided in two subregions in which the variable "altitude" could be modelled by separate autocorrelation functions. This condition must always be met when variograms or correlograms (including multivariate Mantel correlograms) are computed, even for descriptive purpose.

Cliff & Ord (1981) describe how to compute confidence intervals and test the significance of spatial autocorrelation coefficients. For any normally distributed statistic *Stat*, a confidence interval at significance level α is obtained as follows:

$$Pr\left(Stat - z_{\alpha/2}\sqrt{\mathrm{Var}\,(Stat)} < Stat < Stat + z_{\alpha/2}\sqrt{\mathrm{Var}\,(Stat)}\right) = 1 - \alpha \qquad \textbf{(13.4)}$$

For significance testing with large samples, a one-tailed critical value $Stat_{\alpha}$ at significance level α is obtained as follows:

$$Stat_{\alpha} = z_{\alpha}\sqrt{\mathrm{Var}\,(Stat)} + \text{Expected value of } Stat \text{ under } \mathrm{H}_0 \qquad \textbf{(13.5)}$$

It is possible to use this approach because both I and c are asymptotically normally distributed for data sets of moderate to large sizes (Cliff & Ord, 1981). Values $z_{\alpha/2}$ or z_{α} are found in a table of standard normal deviates. Under the hypothesis (H_0) of random spatial distribution of the observed values y_i, the expected values (E) of Moran's I and Geary's c are:

$$\mathrm{E}(I) = -(n-1)^{-1} \quad \text{and} \quad \mathrm{E}(c) = 1 \qquad \textbf{(13.6)}$$

Under the null hypothesis, the expected value of Moran's I approaches 0 as n increases. The variances are computed as follows under a randomization assumption, which simply states that, under H_0, the observations y_i are independent of their positions in space and, thus, are exchangeable:

$$\mathrm{Var}\,(I) = \mathrm{E}\,(I^2) - [\mathrm{E}\,(I)]^2 \qquad \textbf{(13.7)}$$

$$\mathrm{Var}\,(I) = \frac{n\,[\,(n^2-3n+3)\,S_1 - nS_2 + 3W^2] - b_2\,[\,(n^2-n)\,S_1 - 2nS_2 + 6W^2]}{(n-1)\,(n-2)\,(n-3)\,W^2} - \frac{1}{(n-1)^2}$$

$$\mathrm{Var}\,(c) = \frac{(n-1)\,S_1\,[n^2-3n+3-(n-1)\,b_2]}{n\,(n-2)\,(n-3)\,W^2} \qquad \textbf{(13.8)}$$

$$+ \frac{-0.25\,(n-1)\,S_2\,[n^2+3n-6-(n^2-n+2)\,b_2] + W^2\,[n^2-3+(-(n-1)^2)\,b_2]}{n\,(n-2)\,(n-3)\,W^2}$$

In these equations,

- $S_1 = \frac{1}{2}\sum_{h=1}^{n}\sum_{i=1}^{n}(w_{hi} + w_{ih})^2$ (there is a term of this sum for *each cell* of matrix **W**);

- $S_2 = \sum_{i=1}^{n}(w_{i+} + w_{+i})^2$ where w_{i+} and w_{+i} are respectively the sums of row i and column i of matrix **W**;

- $b_2 = n\sum_{i=1}^{n}(y_i - \bar{y})^4 \Big/ \left[\sum_{i=1}^{n}(y_i - \bar{y})^2\right]^2$ measures the kurtosis of the distribution;

- W is as defined in eqs. 13.1 and 13.2.

In most cases in ecology, tests of spatial autocorrelation are one-tailed because the sign of autocorrelation is stated in the ecological hypothesis; for instance, contagious biological processes such as growth, reproduction, and dispersal, all suggest that ecological variables are positively autocorrelated at short distances. To carry out an approximate test of significance, select a value of α (e.g. $\alpha = 0.05$) and find z_α in a table of the standard normal distribution (e.g. $z_{0.05} = +1.6452$). Critical values are found as in eq. 13.5, with a correction factor that becomes important when n is small:

- $I_\alpha = z_\alpha\sqrt{\mathrm{Var}(I)} - k_\alpha(n-1)^{-1}$ in all cases, using the value in the upper tail of the z distribution when testing for positive autocorrelation (e.g. $z_{0.05} = +1.6452$) and the value in the lower tail in the opposite case (e.g. $z_{0.05} = -1.6452$).

- $c_\alpha = z_\alpha\sqrt{\mathrm{Var}(c)} + 1$ when $c < 1$ (positive autocorrelation), using the value in the lower tail of the z distribution (e.g. $z_{0.05} = -1.6452$).

- $c_\alpha = z_\alpha\sqrt{\mathrm{Var}(c)} + 1 - k_\alpha(n-1)^{-1}$ when $c > 1$ (negative autocorrelation), using the value in the upper tail of the z distribution (e.g. $z_{0.05} = +1.6452$).

The value taken by the correction factor k_α depends on the values of n and W. If $4(n - \sqrt{n}) < W \le 4(2n - 3\sqrt{n} + 1)$, then $k_\alpha = \sqrt{10\alpha}$; otherwise, $k_\alpha = 1$. If the test is two-tailed, use $\alpha^* = \alpha/2$ to find z_{α^*} and k_{α^*} before computing critical values. These corrections are based upon simulations reported by Cliff & Ord (1981, section 2.5).

Other formulas are found in Cliff & Ord (1981) for conducting a test under the assumption of normality, where one assumes that the y_i's result from n independent draws from a normal population. When n is very small, tests of I and c should be conducted by randomization (Section 1.2).

Moran's I and Geary's c are sensitive to extreme values and, in general, to asymmetry in the data distributions, as are the related Pearson's r and Euclidean distance coefficients. Asymmetry increases the variance of the data. It also increases the kurtosis and hence the variance of the I and c coefficients (eqs. 13.7 and 13.8); this makes it more difficult to reach significance in statistical tests. So, practitioners usually attempt to normalize the data before computing correlograms and variograms.

Statistical testing in correlograms implies multiple testing since a test of significance is carried out for each autocorrelation coefficient. Oden (1984) has developed a Q statistic to test the global significance of spatial correlograms; his test is an extension of the Portmanteau Q-test used in time series analysis (Box & Jenkins, 1976). An alternative global test is to check whether the correlogram contains at least one autocorrelation statistic which is significant at the Bonferroni-corrected significance level (Box 1.3). Simulations in Oden (1984) show that the power of the Q-test is not appreciably greater than the power of the Bonferroni procedure, which is computationally a lot simpler. A practical question remains, though: how many distance classes should be created? This determines the number of simultaneous tests that are carried out. More classes mean more resolution but fewer pairs per class and, thus, less power for each test; more classes also mean a smaller Bonferroni-corrected α' level, which makes it more difficult for a correlogram to reach global significance.

When the overall test has shown global significance, one may wish to identify the individual autocorrelation statistics that are significant, in order to reach an interpretation (Subsection 2). One could rely on Bonferroni-corrected tests for all individual autocorrelation statistics, but this approach would be too conservative; a better solution is to use Holm's correction procedure (Box 1.3). Another approach is the *progressive Bonferroni correction* described in Subsection 12.4.2; it is only applicable when the ecological hypothesis indicates that significant autocorrelation is to be expected in the smallest distance classes and the purpose of the analysis is to determine the extent of the autocorrelation (i.e. which distance class it reaches). With the progressive Bonferroni approach, the likelihood of emergence of significant values decreases as one proceeds from left to right, i.e. from the small to the large distance classes of the correlogram. One does not have to limit the correlogram to a small number of classes to reduce the effect of the correction, as it is the case with Oden's overall test and with the Bonferroni and Holm correction methods. This approach will be used in the examples that follow.

Autocorrelation coefficients and tests of significance also exist for qualitative (nominal) variables (Cliff & Ord 1981); they have been used to analyse for instance spatial patterns of sexes in plants (Sakai & Oden 1983; Sokal & Thomson 1987). Special types of spatial autocorrelation coefficients have been developed to answer specific problems (e.g. Galiano 1983; Estabrook & Gates 1984). The paired-quadrat variance method, developed by Goodall (1974) to analyse spatial patterns of ecological data by random pairing of quadrats, is related to correlograms.

2 — *Interpretation of all-directional correlograms*

Isotropy

Anisotropy

When the autocorrelation function is the same for all geographic directions considered, the phenomenon is said to be *isotropic*. Its opposite is *anisotropy*. When a variable is isotropic, a single correlogram may be computed over all directions of the

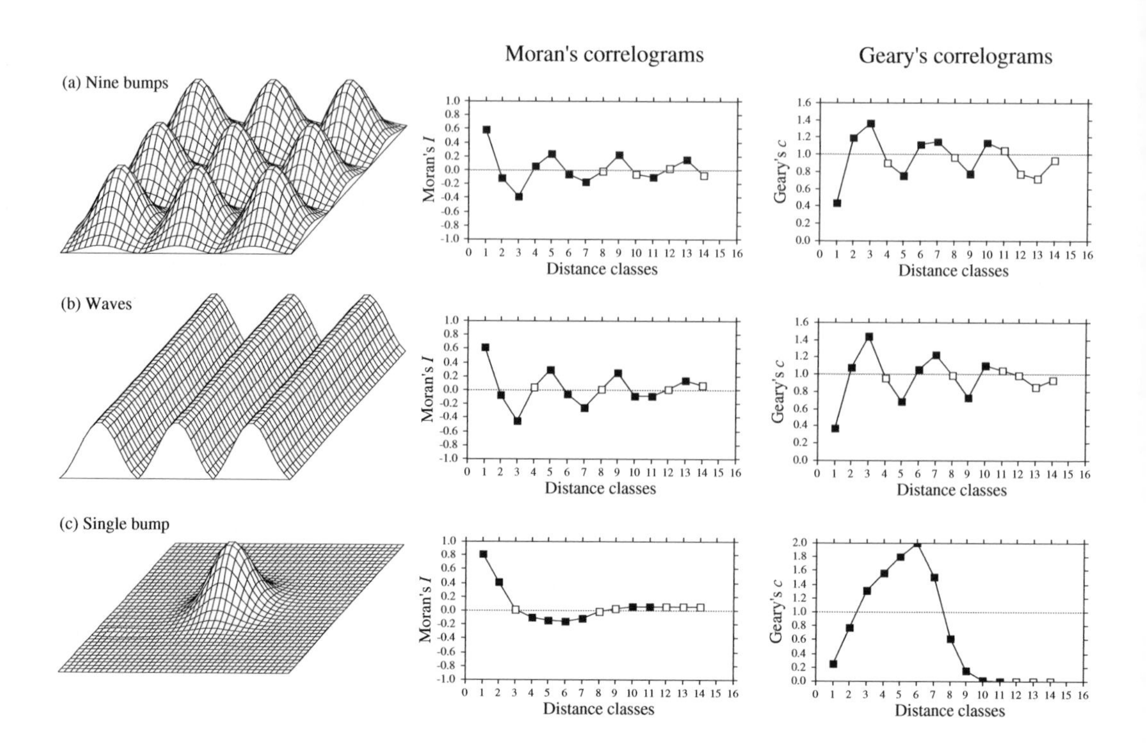

Figure 13.5 Spatial autocorrelation analysis of artificial spatial structures shown on the left: (a) nine bumps; (b) waves; (c) a single bump. Centre and right: all-directional correlograms. Dark squares: autocorrelation statistics that remain significant after progressive Bonferroni correction ($\alpha = 0.05$); white squares: non-significant values.

study area. The correlogram is said to be *all-directional* or *omnidirectional*. Directional correlograms, which are computed for a single direction of space, are discussed together with anisotropy and directional variograms in Subsection 3.

Correlograms are analysed mostly by looking at their shapes. Examples will help clarify the relationship between spatial structures and all-directional correlograms. The important message is that, although correlograms may give clues as to the underlying spatial structure, the information they provide is not specific; a blind interpretation may often be misleading and should always be supported by maps (Section 13.2).

Numerical example. Artificial data were generated that correspond to a number of spatial patterns. The data and resulting correlograms are presented in Fig. 13.5.

• Nine bumps — The surface in Fig. 13.5a is made of nine bi-normal curves. 225 points were sampled across the surface using a regular 15 × 15 grid (Fig. 13.5f). The "height" was noted at each sampling point. The 25200 distances among points found in the upper-triangular portion of the distance matrix were divided into 16 distance classes, using Sturge's rule (eq. 13.3), and

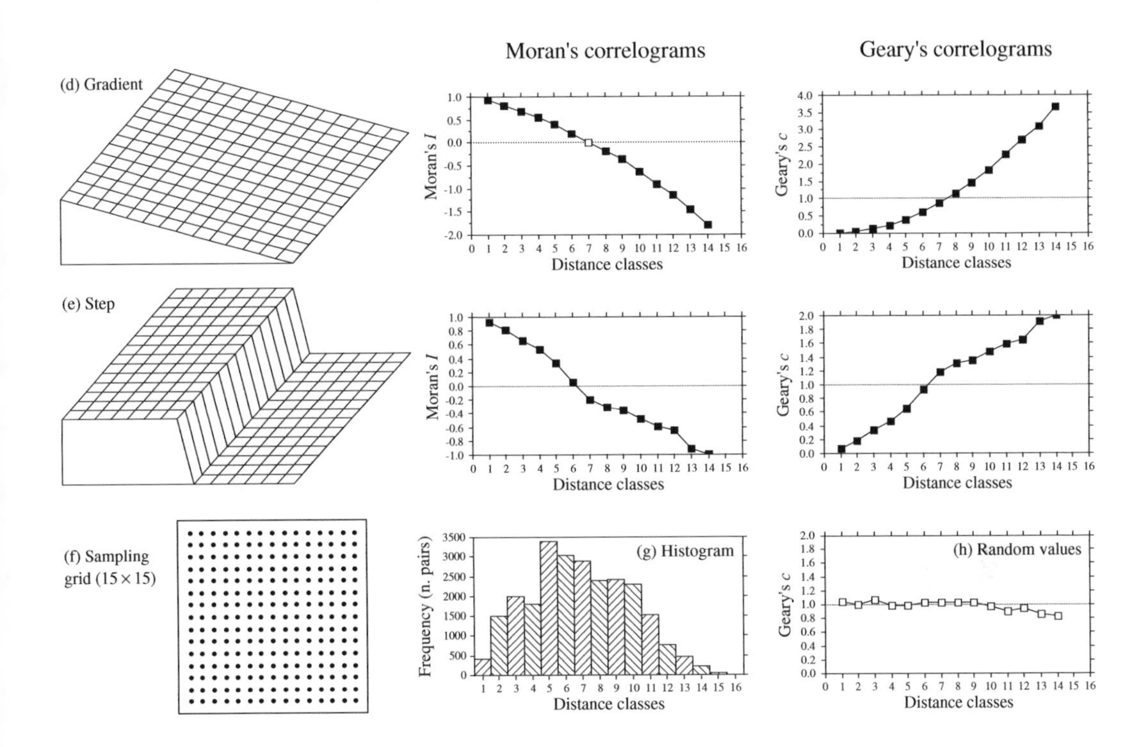

Figure 13.5 **(continued)** Spatial autocorrelation analysis of artificial spatial structures shown on the left: (d) gradient; (e) step. (h) All-directional correlogram of random values. (f) Sampling grid used on each of the artificial spatial structures to obtain 225 "observed values" for spatial autocorrelation analysis. (g) Histogram showing the number of pairs in each distance class. Distances, from 1 to 19.8 in units of the sampling grid, were grouped into 16 distance classes. Spatial autocorrelation statistics (I or c) are not shown for distance classes 15 and 16; see text.

correlograms were computed. According to Oden's test, the correlograms are globally significant at the $\alpha = 5\%$ level since several individual values are significant at the Bonferroni-corrected level $\alpha' = 0.05/16 = 0.00312$. In each correlogram, the progressive Bonferroni correction method was applied to identify significant spatial autocorrelation coefficients: the coefficient for distance class 1 was tested at the $\alpha = 0.05$ level; the coefficient for distance class 2 was tested at the $\alpha' = 0.05/2$ level; and, more generally, the coefficient for distance class k was tested at the $\alpha' = 0.05/k$ level. Spatial autocorrelation coefficients are not reported for distance classes 15 and 16 (60 and 10 pairs, respectively) because they only include the pairs of points bordering the surface, to the exclusion of all other pairs.

There is a correspondence between individual significant spatial autocorrelation coefficients and the main elements of the spatial structure. The correspondence can clearly be seen in this example, where the data generating process is known. This is not the case when analysing field data, in which case the existence and nature of the spatial structures must be confirmed by mapping the data. The presence of several equispaced patches produces an alternation of significant positive and negative values along the correlograms. The first spatial autocorrelation coefficient, which is above 0 in Moran's correlogram and below 1 in Geary's, indicates positive spatial autocorrelation in the first distance class; the first class contains the 420 pairs of points that are at distance 1 of each other on the grid (i.e. the first neighbours in the N-S or E-W directions of the map). Positive and significant spatial autocorrelation in the first distance class confirms that the distance between first neighbours is smaller than the patch size; if the distance between first neighbours in this example was larger than the patch size, first neighbours would be dissimilar in values and autocorrelation would be negative for the first distance class. The next peaking positive autocorrelation value (which is smaller than 1 in Geary's correlogram) occurs at distance class 5, which includes distances from 4.95 to 6.19 in grid units; this corresponds to positive autocorrelation between points located at similar positions on neighbouring bumps, or neighbouring troughs; distances between successive peaks are 5 grid units in the E-W or N-S directions. The next peaking positive autocorrelation value occurs at distance class 9 (distances from 9.90 to 11.14 in grid units); it includes value 10, which is the distance between second-neighbour bumps in the N-S and E-W directions. Peaking negative autocorrelation values (which are larger than 1 in Geary's correlogram) are interpreted in a similar way. The first such value occurs at distance class 3 (distances from 2.48 to 3.71 in grid units); it includes value 2.5, which is the distance between peaks and troughs in the N-S and E-W directions on the map. If the bumps were unevenly spaced, the correlograms would be similar for the small distance classes, but there would be no other significant values afterwards.

The main problem with all-directional correlograms is that the diagonal comparisons are included in the same calculations as the N-S and E-W comparisons. As distances become larger, diagonal comparisons between, say, points located near the top of the nine bumps tend to fall in different distance classes than comparable N-S or E-W comparisons. This blurs the signal and makes the spatial autocorrelation coefficients for larger distance classes less significant and interpretable.

• Wave (Fig. 13.5b) — Each crest was generated as a normal curve. Crests were separated by five grid units; the surface was constructed in this way to make it comparable to Fig. 13.5a. The correlograms are nearly indistinguishable from those of the nine bumps. All-directional correlograms alone cannot tell apart regular bumps from regular waves; directional correlograms or maps are required.

• Single bump (Fig. 13.5c) — One of the normal curves of Fig. 13.5a was plotted alone at the centre of the study area. Significant negative autocorrelation, which reaches distance classes 6 or 7, delimits the extent of the "range of influence" of this single bump, which covers half the study area. It is not limited here by the rise of adjacent bumps, as this was the case in (a).

• Linear gradient (Fig. 13.5d) — The correlogram is monotonic decreasing. Nearly all autocorrelation values in the correlograms are significant.

True, false gradient

There are actually two kinds of gradients (Legendre, 1993). "True gradients", on the one hand, are deterministic structures. They correspond to generating model 2 of Subsection 1.1.1 (eq. 1.2) and can be modelled using trend-surface analysis (Subsection 13.2.1). The observed values have independent error terms, i.e. error terms which are not autocorrelated. "False gradients", on the other hand, are structures that may look like gradients, but actually correspond to autocorrelation generated by some spatial process (model 1 of Subsection 1.1.1; eq. 1.1). When the sampling area is small relative to the range of influence of the generating process, the data generated by such a process may look like a gradient.

In the case of "true gradients", spatial autocorrelation coefficients should not be tested for significance because the condition of second-order stationarity is not satisfied (definition in previous Subsection); the expected value of the mean is not the same over the whole study area. In the case of "false gradients", however, tests of significance are warranted. For descriptive purposes, correlograms may still be computed for "true gradients" (without tests of significance) because the intrinsic assumption is satisfied. One may also choose to extract a "true gradient" using trend-surface analysis, compute residuals, and look for spatial autocorrelation among the residuals. This is equivalent to trend extraction prior to time series analysis (Section 12.2).

How does one know whether a gradient is "true" or "false"? This is a moot point. When the process generating the observed structure is known, one may decide whether it is likely to have generated spatial autocorrelation in the observed data, or not. Otherwise, one may empirically look at the *target population* of the study. In the case of a spatial study, this is the population of potential sites in the larger area into which the study area is embedded, the study area representing the *statistical population* about which inference can be made. Even from sparse or indirect data, a researcher may form an opinion as to whether the observed gradient is deterministic ("true gradient") or is part of a landscape displaying autocorrelation at broader spatial scale ("false gradient").

• Step (Fig. 13.5e) — A step between two flat surfaces is enough to produce a correlogram which is indistinguishable, for all practical purposes, from that of a gradient. Correlograms alone cannot tell apart regular gradients from steps; maps are required. As in the case of gradients, there are "true steps" (deterministic) and "false steps" (resulting from an autocorrelated process), although the latter is rare. The presence of a sharp discontinuity in a surface generally indicates that the two parts should be subjected to separate analyses. The methods of boundary detection and constrained clustering (Section 13.3) may help detect such discontinuities and delimit homogeneous areas prior to spatial autocorrelation analysis.

• Random values (Fig. 13.5h) — Random numbers, drawn from a standard normal distribution, were generated for each point of the grid and used as the variable to be analysed. Random data are said to represent a "pure nugget effect" in geostatistics. The autocorrelation coefficients were small and non-significant at the 5% level. Only the Geary correlogram is presented.

Sokal (1979) and Cliff & Ord (1981) describe, in general terms, where to expect significant values in correlograms, for some spatial structures such as gradients and large or small patches. Their summary tables are in agreement with the test examples above. The absence of significant coefficients in a correlogram must be interpreted with caution, however:

• It may indicate that the surface under study is free of spatial autocorrelation at the study scale. Beware: this conclusion is subject to type II (or β) error. Type II error depends on the power of the test which is a function of (1) the α significance level, (2) the size of effect (i.e. the minimum amount of autocorrelation) one wants to detect, (3) the number of observations (n), and (4) the variance of the sample (Cohen, 1988):

$$\text{Power} = (1 - \beta) = f(\alpha, \text{size of effect}, n, s_x)$$

Is the test powerful enough to warrant such a conclusion? Are there enough observations to reach significance? The easiest way to increase the power of a test, for a given variable and fixed α, is to increase n.

• It may indicate that the sampling design is inadequate to detect the spatial autocorrelation that may exist in the system. Are the grain size, extent and sampling interval (Section 13.0) adequate to detect the type of autocorrelation one can hypothesize from knowledge about the biological or ecological process under study?

Ecologists can often formulate hypotheses about the mechanism or process that may have generated a spatial phenomenon and deduct the shape that the resulting surface should have. When the model specifies a value for each geographic position (e.g. a spatial gradient), data and model can be compared by correlation analysis. In other instances, the biological or ecological model only specifies process generating the spatial autocorrelation, not the exact geographic position of each resulting value. Correlograms may be used to support or reject the biological or ecological hypothesis. As in the examples of Fig. 13.5, one can construct an artificial model-surface corresponding to the hypothesis, compute a correlogram of that surface, and compare the correlograms of the real and model data. For instance, Sokal *et al.* (1997a) generated data corresponding to several gene dispersion mechanisms in populations and showed the kind of spatial correlogram that may be expected from each model. Another application concerning phylogenetic patterns of human evolution in Eurasia and Africa (space-time model) is found in Sokal *et al.* (1997b).

Bjørnstad & Falck (1997) and Bjørnstad *et al.* (1998) proposed a spline correlogram which provides a continuous and model-free function for the spatial covariance. The spline correlogram may be seen as a modification of the nonparametric covariance function of Hall and co-workers (Hall & Patil, 1994; Hall *et al.*, 1994). A bootstrap algorithm estimates the confidence envelope of the entire correlogram or derived statistics. This method allows the statistical testing of the similarity between correlograms of real and simulated (i.e. model) data.

Ecological application 13.1a

During a study of the factors potentially responsible for the choice of settling sites of *Balanus crenatus* larvae (Cirripedia) in the St. Lawrence Estuary (Hudon *et al.*, 1983), plates of artificial substrate (plastic laminate) were subjected to colonization in the infralittoral zone. Plates were positioned vertically, parallel to one another. A picture was taken of one of the plates after a 3-month immersion at a depth of 5 m below low tide, during the summer 1978. The picture was divided into a (10 × 15) grid, for a total of 150 pixels of 1.7 × 1.7 cm. Barnacles were counted by C. Hudon and P. Legendre for the present Ecological application (Fig. 13.6a; unpublished *in op. cit.*). The hypothesis to be tested is that barnacles have a patchy distribution. Barnacles are gregarious animals; their larvae are chemically attracted to settling sites by arthropodine secreted by settled adults (Gabbott & Larman, 1971).

A gradient in larval concentration was expected in the top-to-bottom direction of the plate because of the known negative phototropism of barnacle larvae at the time of settlement (Visscher, 1928). Some kind of border effect was also expected because access to the centre of the plates located in the middle of the pack was more limited than to the fringe. These large-scale effects create violations to the condition of second-order stationarity. A trend-surface equation (Subsection 13.2.1) was computed to account for it, using only the Y coordinate (top-

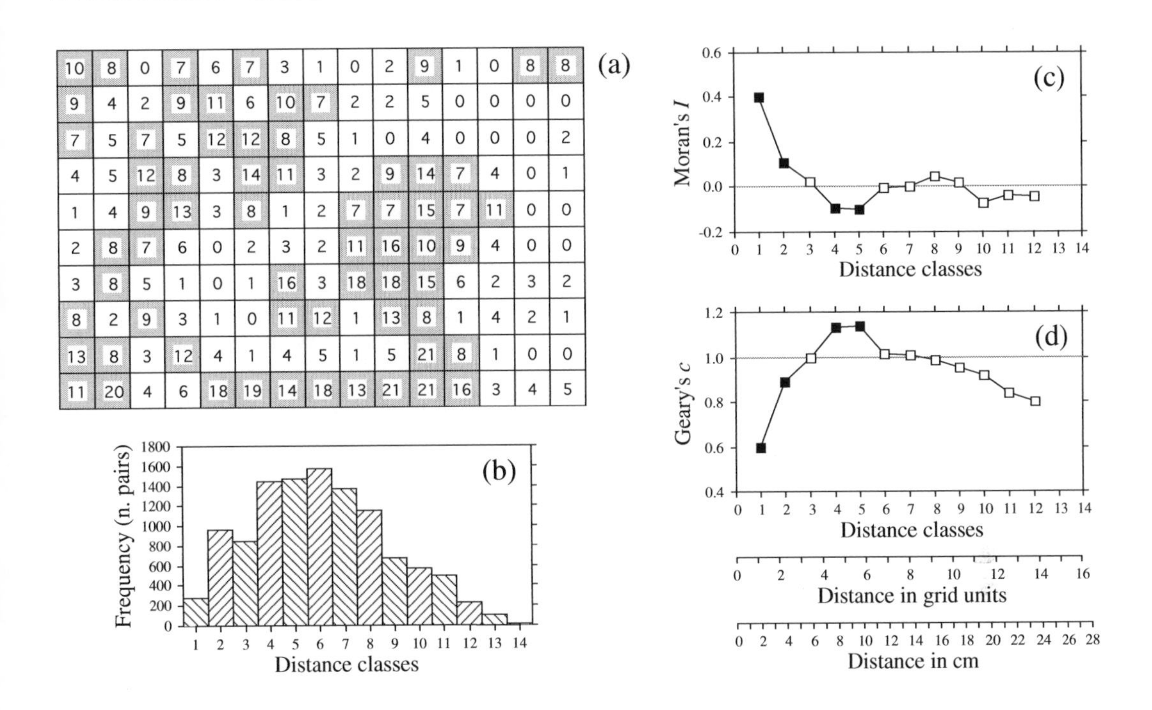

Figure 13.6 (a) Counts of adult barnacles in 150 (1.7×1.7 cm) pixels on a plate of artificial substrate (17×25.5 cm). The mean concentration is 6.17 animals per pixel; pixels with counts ≥ 7 are shaded to display the aggregates. (b) Histogram of the number of pairs in each distance class. (c) Moran's correlogram. (d) Geary's correlogram. Dark squares: autocorrelation statistics that remain significant after progressive Bonferroni correction ($\alpha = 0.05$); white squares: non-significant values. Coefficients for distance classes 13 and 14 are not given because they only include the pairs of points bordering the surface. Distances are also given in grid units and cm.

to-bottom axis). Indeed, a significant trend surface was found, involving Y and Y^2, that accounted for 10% of the variation. It forecasted high barnacle concentration in the bottom part of the plate and near the upper and lower margins. Residuals from this equation were calculated and used in spatial autocorrelation analysis.

Euclidean distances were computed among pixels; following Sturge's rule (eq. 13.3), the distances were divided into 14 classes (Fig. 13.6b). Significant positive autocorrelation was found in the first distance classes of the correlograms (Fig. 13.6c, d), supporting the hypothesis of patchiness. The size of the patches, or "range of influence" (i.e. the distance between zones of high and low concentrations), is indicated by the distance at which the first maximum negative autocorrelation value is found. This occurs in classes 4 and 5, which corresponds to a distance of about 5 in grid units, or 8 to 10 cm. The patches of high concentration are shaded on the map of the plate of artificial substrate (Fig. 13.6a).

In anisotropic situations, directional correlograms should be computed in two or several directions. Description of how the pairs of points are chosen is deferred to Subsection 3 on variograms. One may choose to represent either a single, or several of these correlograms, one for each of the aiming geographic directions, as seems fit for the problem at hand. A procedure for representing in a single figure the directional correlograms computed for several directions of a plane has been proposed by Oden & Sokal (1986); Legendre & Fortin (1989) give an example for vegetation data. Another method is illustrated in Rossi *et al.* (1992).

Another way to approach anisotropic problems is to compute two-dimensional spectral analysis. This method, described by Priestley (1964), Rayner (1971), Ford (1976), Ripley (1981) and Renshaw & Ford (1984), differs from spatial autocorrelation analysis in the structure function it uses. As in time-series spectral analysis (Section 12.5), the method assumes the data to be stationary (second-order stationarity; i.e. no "true gradient" in the data) and made of a combination of sine patterns. An autocorrelation function $r_{dX,dY}$ for all combinations of lags (dX, dY) in the two geographic axes of a plane, as well as a periodogram with intensity I for all combinations of frequencies in the two directions of the plane, are computed. Details of the calculations are also given in Legendre & Fortin (1989), with an example.

3 — Variogram

Like correlograms, semi-variograms (called *variograms* for simplicity) decompose the spatial (or temporal) variability of observed variables among distance classes. The structure function plotted as the ordinate, called *semi-variance*, is the numerator of eq. 13.2:

$$\gamma(d) = \frac{1}{2W} \sum_{h=1}^{n} \sum_{i=1}^{n} w_{hi} (y_h - y_i)^2 \quad \text{for } h \neq i \tag{13.9}$$

or, for symmetric distance and weight matrices,

$$\gamma(d) = \frac{1}{2W} \sum_{h=1}^{n-1} \sum_{i=h+1}^{n} w_{hi} (y_h - y_i)^2 \tag{13.10}$$

$\gamma(d)$ is thus a non-standardized form of Geary's c coefficient. γ may be seen as a measure of the error mean square of the estimate of y_i using a value y_h distant from it by d. The two forms lead to the same numerical value in the case of symmetric distance and weight matrices. The calculation is repeated for different values of d. This provides the *sample variogram*, which is a plot of the empirical values of variance $\gamma(d)$ as a function of distance d.

The equations usually found in the geostatistical literature look a bit different, but they correspond to the same calculations:

$$\gamma(d) = \frac{1}{2W(d)} \sum_{i=1}^{W(d)} (y_i - y_{i+d})^2 \quad \text{or} \quad \gamma(d) = \frac{1}{2W(d)} \sum_{(h,i)\,|\,d_{hi} \approx d}^{W(d)} (y_h - y_i)^2$$

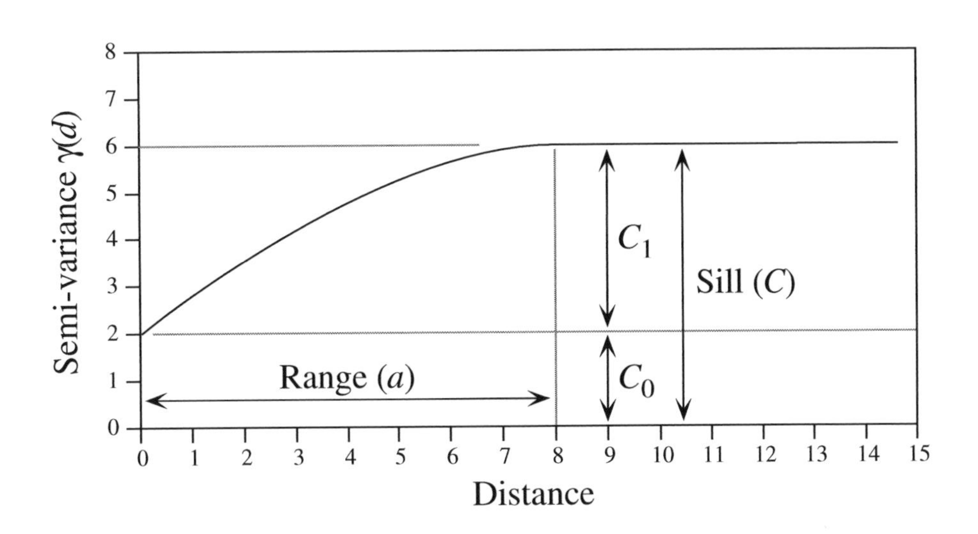

Figure 13.7 Spherical variogram model showing characteristic features: nugget effect ($C_0 = 2$ in this example), spatially structured component ($C_1 = 4$), sill ($C = C_0 + C_1 = 6$), and range ($a = 8$).

Both of these expressions mean that pairs of values are selected to be at distance d of each other; there are $W(d)$ such pairs for any given distance class d. The condition $d_{hi} \approx d$ means that distances may be grouped into distance classes, placing in class d the individual distances d_{hi} that are approximately equal to d. In directional variograms (below), d is a directional measure of distance, i.e. taken in a specified direction only. The semi-variance function is often called the variogram in the geostatistical literature. When computing a variogram, one assumes that the autocorrelation function applies to the entire surface under study (intrinsic hypothesis, Subsection 13.1.1).

Generally, variograms tend to level off at a *sill* which is equal to the variance of the variable (Fig. 13.7); the presence of a sill implies that the data are second-order stationary. The distance at which the variance levels off is referred to as the *range* (parameter a); beyond that distance, the sampling units are not spatially correlated. The discontinuity at the origin (non–zero intercept) is called the *nugget effect*; the geostatistical origin of the method transpires in that name. It corresponds to the local variation occurring at scales finer than the sampling interval, such as sampling error, fine-scale spatial variability, and measurement error. The nugget effect is represented by the error term ε_{ij} in spatial structure model 1b of Subsection 1.1.1. It describes a portion of variation which is not autocorrelated, or is autocorrelated at a scale finer than can be detected by the sampling design. The parameter for the nugget effect is C_0 and the spatially structured component is represented by C_1; the sill, C, is equal to $C_0 + C_1$. The *relative nugget effect* is $C_0/(C_0 + C_1)$.

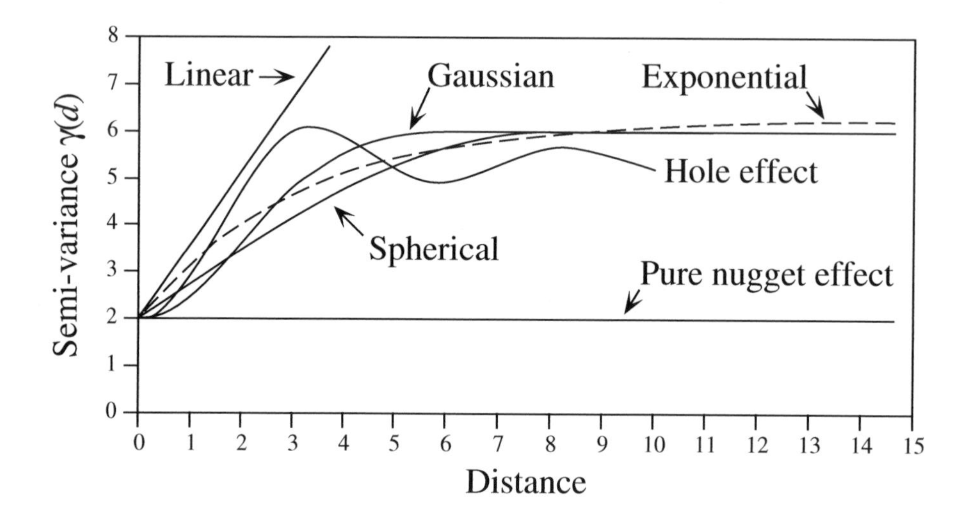

Figure 13.8 Commonly used variogram models.

Although a sample variogram is a good descriptive summary of the spatial contiguity of a variable, it does not provide all the semi-variance values needed for kriging (Subsection 13.2.2). A model must be fitted to the sample variogram; the model will provide values of semi-variance for all the intermediate distances. The most commonly used models are the following (Fig. 13.8):

- Spherical model: $\gamma(d) = C_0 + C_1\left[1.5\frac{d}{a} - 0.5\left(\frac{d}{a}\right)^3\right]$ if $d \le a$; $\gamma(d) = C$ if $d > a$.

- Exponential model: $\gamma(d) = C_0 + C_1\left[1 - \exp\left(-3\,\frac{d}{a}\right)\right]$.

- Gaussian model: $\gamma(d) = C_0 + C_1\left[1 - \exp\left(-3\,\frac{d^2}{a^2}\right)\right]$.

- Hole effect model: $\gamma(d) = C_0 + C_1\left[1 - \frac{\sin(ad)}{ad}\right]$. An equivalent form is $\gamma(d) = C_0 + C_1\left[1 - \frac{a'\sin(d/a')}{d}\right]$ where $a' = 1/a$. $(C_0 + C_1)$ represents the value of γ towards which the dampening sine function tends to stabilize. This equation would adequately model a variogram of the periodic structures in Fig. 13.5a-b (variograms only differ from Geary's correlograms by the scale of the ordinate).

• Linear model: $\gamma(d) = C_0 + bd$ where b is the slope of the variogram model. A linear model with sill is obtained by adding the specification: $\gamma(d) = C$ if $d \geq a$.

• Pure nugget effect model: $\gamma(d) = C_0$ if $d > 0$; $\gamma(d) = 0$ if $d = 0$. The second part applies to a point estimate. In practice, observations have the size of the sampling grain (Section 13.0); the error at that scale is always larger than 0.

Other less-frequently encountered models are described in geostatistics textbooks. A model is usually chosen on the basis of the known or assumed process having generated the spatial structure. Several models may be added up to fit any particular sample variogram. Parameters are fitted by weighted least squares; the weights are function of the distance and the number of pairs in each distance class (Cressie, 1991).

Anisotropy

As mentioned at the beginning of Subsection 2, anisotropy is present in data when the autocorrelation function is not the same for all geographic directions considered (David, 1977; Isaaks & Srivastava, 1989). In *geometric anisotropy*, the variation to be expected between two sites distant by d in one direction is equivalent to the variation expected between two sites distant by $b \times d$ in another direction. The range of the variogram changes with direction while the sill remains constant. In a river for instance, the kind of variation expected in phytoplankton concentration between two sites 5 m apart across the current may be the same as the variation expected between two sites 50 m apart along the current even though the variation can be modelled by spherical variograms with the same sill in the two directions. Constant b is called the *anisotropy ratio* ($b = 50/5 = 10$ in the river example). This is equivalent to a change in distance units along one of the axes. The anisotropy ratio may be represented by an ellipse or a more complex figure on a map, its axes being proportional to the variation expected in each direction. In *zonal anisotropy*, the sill of the variogram changes with direction while the range remains constant. An extreme case is offered by a strip of land. If the long axis of the strip is oriented in the direction of a major environmental gradient, the variogram may correspond to a linear model (always increasing) or to a spherical model with a sill larger than the nugget effect, whereas the variogram in the direction perpendicular to it may show only random variation without spatial structure with a sill equal to the nugget effect.

Directional variogram and correlogram

Directional variograms and correlograms may be used to determine whether anisotropy (defined in Subsection 2) is present in the data; they may also be used to describe anisotropic surfaces or to account for anisotropy in kriging (Subsection 13.2.2). A direction of space is chosen (i.e. an angle θ, usually by reference to the geographic north) and a search is launched for the pairs of points that are within a given distance class d in that direction. There may be few such pairs perfectly aligned in the aiming direction, or none at all, especially when the observed sites are not regularly spaced on the map. More pairs can usually be found by looking within a small neighbourhood around the aiming line (Fig. 13.9). The neighbourhood is determined by an angular tolerance parameter φ and a parameter κ that sets the tolerance for distance classes along the aiming line. For each observed point $Ø_h$ in turn, one looks for other points $Ø_i$ that are at distance $d \pm \kappa$ from it. All points found

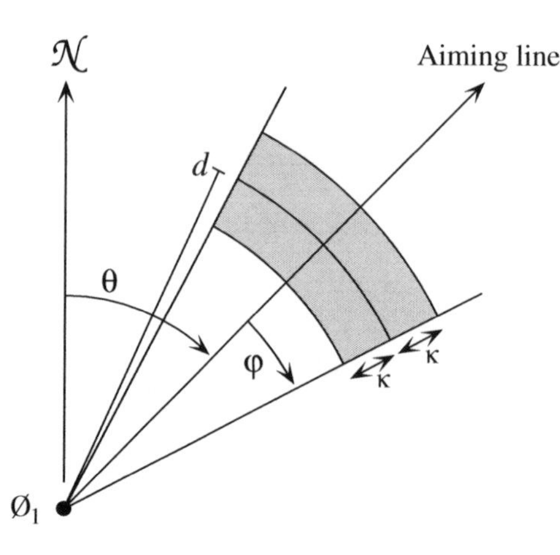

Figure 13.9 Search parameters for pairs of points in directional variograms and correlograms. From an observed study site Ø_1, an aiming line is drawn in the direction determined by angle θ (usually by reference to the geographic north, indicated by $\mathcal{N}$). The angular tolerance parameter φ determines the search zone (grey) laterally whereas parameter κ sets the tolerance along the aiming line for each distance class d. Points within the search window (in gray) are included in the calculation of $I(d)$, $c(d)$ or $\gamma(d)$.

within the search window are paired with the reference point Ø_h and included in the calculation of semi-variance or spatial autocorrelation coefficients for distance class d. In most applications, the search is bi-directional, meaning that one also looks for points within a search window located in the direction opposite (180°) the aiming direction. Isaaks & Srivastava (1989, Chapter 7) propose a way to assemble directional measures of semi-variance into a single table and to produce a contour map that describes the anisotropy in the data, if any; Rossi *et al.* (1992) have used the same approach for directional spatial correlograms.

Numerical example. An artificial data set was produced containing random autocorrelated data. The data were generated using the turning bands method (David, 1977; Journel & Huijbregts, 1978); random normal deviates were autocorrelated following a spherical model with a range of 5. Pure spatial autocorrelation, as described in the spatial structure model 1b of Subsection 1.1.1, generates continuity in the data (Fig. 13.10a). The variogram (without test of significance) and spatial correlograms (with tests) are presented in Figs. 13.10b-d. In this example, the data were standardized during data generation, prior to spatial autocorrelation analysis, so that the denominator of eq. 13.2 is 1; therefore, the variogram and Geary's correlogram are identical. The variogram suggests a spherical model with a range of 6 units and a small nugget effect (Fig. 13.10b).

Besides the description of spatial structures, variograms are used for several other purposes in spatial analysis. In Subsection 13.2.2, they will be the basis for interpolation by kriging. In addition, structure functions (variograms, spatial

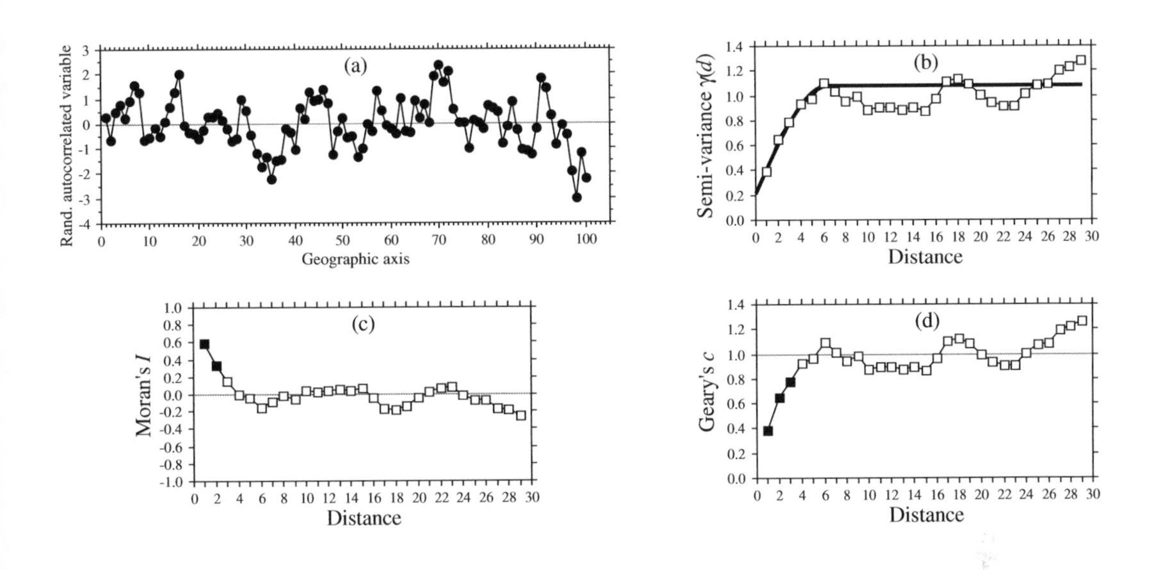

Figure 13.10 (a) Series of 100 equispaced random, spatially autocorrelated data. (b) Variogram, with spherical model superimposed (heavy line). Abscissa: distances between points along the geographic axis in (a). (c) and (d) Spatial correlograms. Dark squares: autocorrelation statistics that remain significant after progressive Bonferroni correction ($\alpha = 0.05$); white squares: non-significant values.

correlograms) may prove extremely useful to help determine the grain size of the sampling units and the sampling interval to be used in a survey, based upon the analysis of a pilot study. They may also be used to perform change-of-scale operations and predict the type of autocorrelation and variance that would be observed if the grain size of the sampling design was different from that actually used in a field study (Bellehumeur *et al.*, 1997).

4 — Spatial covariance, semi-variance, correlation, cross-correlation

This Subsection examines the relationships between spatial covariance, semi-variance and correlation (including cross-correlation), under the assumption of second-order stationarity, leading to the concept of cross-correlation. This assumption (Subsection 13.2.1) may be restated as follows:

- The first moment (mean of points i) of the variable exists:

$$\mathrm{E}[y_i] = \frac{1}{n}\sum_{i=1}^{n} y_i = m_i \tag{13.11}$$

Its value does not depend on position in the study area.

- The second moment (covariance, numerator of eq. 13.1) of the variable exists:

$$C(d) = \left[\frac{1}{W(d)} \sum_{(h,i)\,|\,d_{hi} \approx d}^{W(d)} y_h y_i \right] - m_h m_i \qquad (13.12)$$

$$C(d) = \mathrm{E}[y_h y_i] - m^2 \quad \text{for } h, i \,|\, d_{hi} \approx d \qquad (13.13)$$

The value of $C(d)$ depends only on d and on the orientation of the distance vector, but not on position in the study area. To understand eq. 13.12 as a measure of covariance, imagine the elements of the various pairs y_h and y_i written in two columns as if they were two variables. The equation for the covariance (eq. 4.4) may be written as follows, using a final division by n instead of $(n - 1)$ (maximum-likelihood estimate of the covariance, which is standard in geostatistics):

$$s_{y_h y_i} = \frac{\sum y_h y_i}{n} - \frac{\sum y_h \sum y_i}{n \quad n} = \frac{\sum y_h y_i}{n} - m_h m_i$$

The overall variance (Var, with division by n instead of $n - 1$) also exists since it is the covariance calculated for $d = 0$:

$$\mathrm{Var}[y_i] = \mathrm{E}[y_i - m_i]^2 = C(0) \qquad (13.14)$$

When computing the semi-variance, one only considers pairs of observations distant by d. Eqs. 13.9 and 13.10 are re-written as follows:

$$\gamma(d) = \frac{1}{2}\mathrm{E}[y_h - y_i] \quad \text{for } h, i \,|\, d_{hi} \approx d \qquad (13.15)$$

A few lines of algebra obtain the following formula:

$$\gamma(d) = \frac{\sum y_i^2 - \sum y_h y_i}{W(d)} = C(0) - C(d) \quad \text{for } h, i \,|\, d_{hi} \approx d \qquad (13.16)$$

Two properties are used in the derivation: (1) $\Sigma y_h = \Sigma y_i$, and (2) the variance (Var, eq. 13.14) can be estimated using any subset of the observed values if the hypothesis of second-order stationarity is verified.

The correlation is the covariance divided by the product of the standard deviations (eq. 4.7). For a spatial process, the (auto)correlation is written as follows (leading to a formula which differs from eq. 13.1):

$$r(d) = \frac{C(d)}{s_h s_i} = \frac{C(d)}{\mathrm{Var}[y_i]} = \frac{C(d)}{C(0)} \qquad (13.17)$$

Consider the formula for Geary's c (eq. 13.2), which is the semi-variance divided by the overall variance. The following derivation:

$$c(d) = \frac{\gamma(d)}{\mathrm{Var}[y_i]} = \frac{C(0) - C(d)}{C(0)} = 1 - \frac{C(d)}{C(0)} = 1 - r(d)$$

leads to the conclusion that Geary's c is one minus the coefficient of spatial (auto)correlation. In a graph, the semi-variance and Geary's c coefficient have exactly the same shape (e.g. Figs. 13.10b and d); only the ordinate scales may differ. An autocorrelogram plotted using $r(d)$ has the exact reverse shape as a Geary correlogram. An important conclusion is that the plots of semi-variance, covariance, Geary's c coefficient, and $r(d)$, are equivalent to characterize spatial structures under the hypothesis of second-order stationarity (Bellehumeur & Legendre, 1998).

Cross-covariances may also be computed from eq. 13.12, using values of *two different variables* observed at locations distant by d (Isaaks & Srivastava, 1989). Eq. 13.17 leads to a formula for cross-correlation which may be used to plot cross-correlograms; the construction of the correlation statistic is the same as for time series (eq. 12.10). With transect data, the result is similar to that of eq. 12.10. However, the programs designed to compute spatial cross-correlograms do not require the data to be equispaced, contrary to programs for time-series analysis. The theory is presented by Rossi *et al.* (1992), as well as applications to ecology.

Ecological application 13.1b

A survey was conducted on a homogeneous sandflat in the Manukau Harbour, New Zealand, to identify the scales at which spatial heterogeneity could be detected in the distribution of adult and juvenile bivalves (*Macomona liliana* and *Austrovenus stutchburyi*), as well as indications of adult-juvenile interactions within and between species. The results were reported by Hewitt *et al.* (1997); see also Ecological application 13.2. Sampling was conducted along transects established at three sites located within a 1-km^2 area; there were two transects at each site, forming a cross. Sediment cores (10 cm diam., 13 cm deep) were collected using a nested sampling design; the basic design was a series of cores 5 m apart, but additional cores were taken 1 m from each of the 5-m-distant cores. This design provided several comparison in the short distance classes (1, 4, 5, and 6 m). Using transects instead of rectangular areas allowed relatively large distances (150 m) to be studied, given the allowable sampling effort. Nested sampling designs have also been advocated by Fortin *et al.* (1989) and by Bellehumeur & Legendre (1998).

Spatial correlograms were used to identify scales of variation in bivalve concentrations. The Moran correlogram for juvenile *Austrovenus*, computed for the three transects perpendicular to the direction of tidal flow, displayed significant spatial autocorrelation at distances of 1 and 5 m (Fig. 13.11a). The same pattern was found in the transects parallel to tidal flow. Figure 13.11a also indicates that the range of influence of autocorrelation was about 15 m. This was confirmed by plotting bivalve concentrations along the transects: LOWESS smoothing of the graphs (Subsection 10.3.8) showed patches of about 25-30 m in diameter (Hewitt *et al.*, 1997, Figs. 3 and 4).

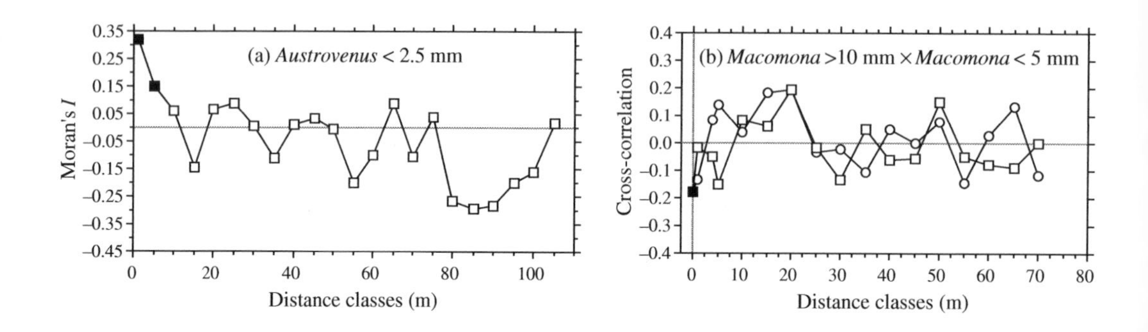

Figure 13.11 (a) Spatial autocorrelogram for juvenile *Austrovenus* densities. (b) Cross-correlogram for adult-juvenile *Macomona* interactions, folded about the ordinate: circles = positive lags, squares = negative lags. Dark symbols: correlation statistics that are significant after progressive Bonferroni correction ($\alpha = 0.05$). Redrawn from Hewitt *et al.* (1997).

Cross-correlograms were computed to detect signs of adult-juvenile interactions. In the comparison of adult (> 10 mm) to juvenile *Macomona* (< 5 mm), a significant negative cross-correlation was identified at 0 m in the direction parallel to tidal flow (Fig. 13.11b); correlation was not significant for the other distance classes. As in time series analysis, the cross-correlation function is not symmetrical; the correlation obtained by comparing values of $\mathbf{y}_1$ to values of $\mathbf{y}_2$ located at distance d on their right is not the same as when values of $\mathbf{y}_2$ are compared to values of $\mathbf{y}_1$ located at distance d on their right, except for $d = 0$. In Fig. 13.11b, the cross-correlogram is folded about the ordinate (compare to Fig. 12.9). Contrary to time series analysis, it is not useful in spatial analysis to discuss the direction of lag of a variable with respect to the other unless one has a specific hypothesis to test.

5 — Multivariate Mantel correlogram

Sokal (1986) and Oden & Sokal (1986) found an ingenious way to compute a correlogram for multivariate data, using the normalized Mantel statistic r_M and test of significance (Subsection 10.5.1). This method is useful, in particular, to describe the spatial structure of species assemblages.

The principle is to quantify the ecological relationships among sampling sites by means of a matrix $\mathbf{Y}$ of multivariate similarities or distances (using, for instance, coefficients S_{17} or D_{14} in the case of species abundance data), and compare $\mathbf{Y}$ to a *model matrix* $\mathbf{X}$ (Subsection 10.5.1) which is different for each geographic distance class (Fig. 13.12).

• For distance class 1 for instance, pairs of neighbouring stations (that belong to the first class of geographic distances) are coded 1, whereas the remainder of matrix $\mathbf{X}_1$ contains zeros. A first Mantel statistic (r_{M1}) is calculated between $\mathbf{Y}$ and $\mathbf{X}_1$.

• The process is repeated for the other distance classes d, building each time a model-matrix $\mathbf{X}_d$ and recomputing the normalized Mantel statistic. Matrix $\mathbf{X}_d$ may contain 1's

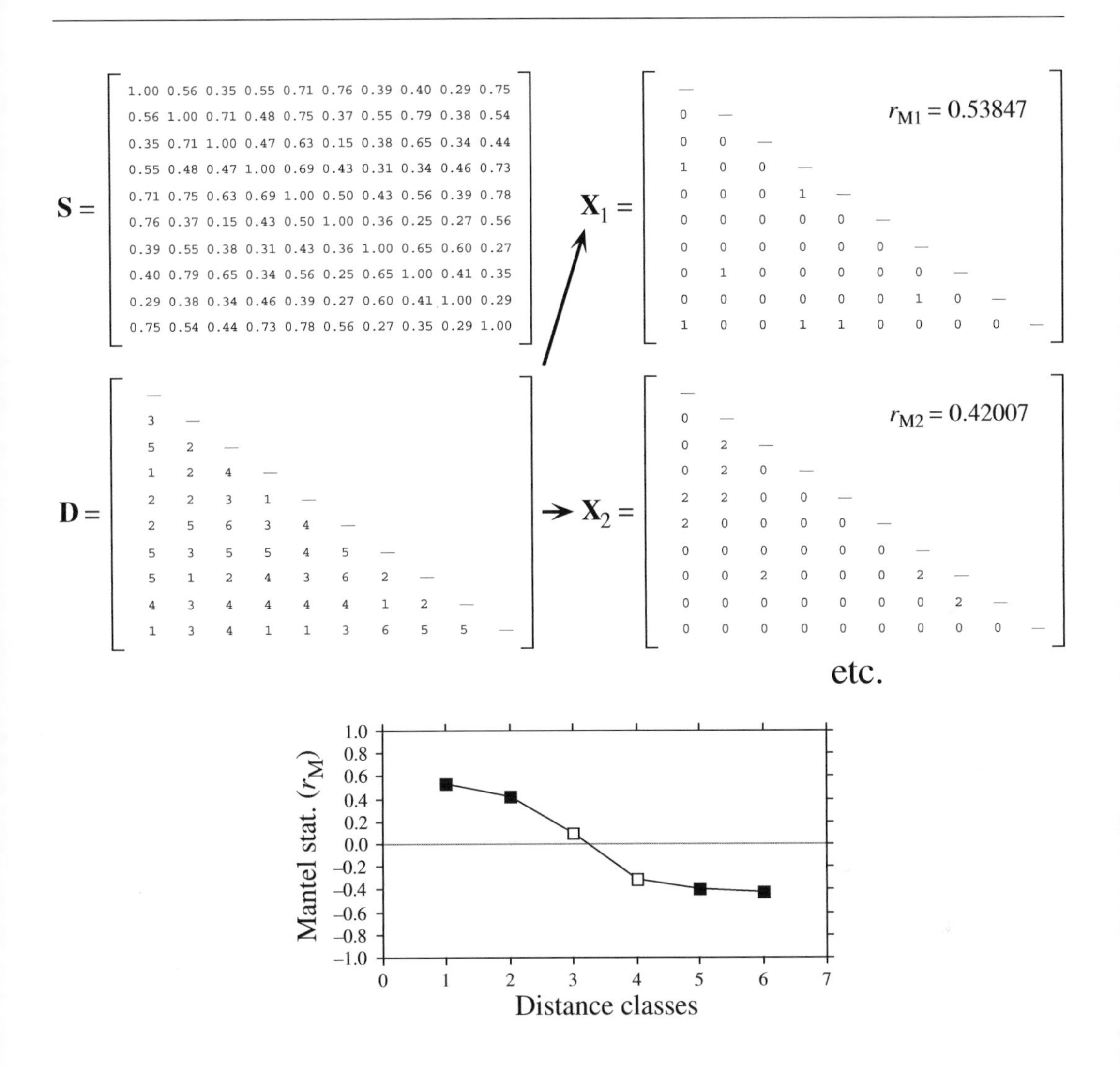

Figure 13.12 Construction of a Mantel correlogram for a similarity matrix **S** ($n = 10$ sites). The matrix of geographic distance classes **D**, from Fig. 13.4, gives rise to model matrices $\mathbf{X}_1$, $\mathbf{X}_2$, etc. for the various distance classes d. These are compared, in turn, to matrix $\mathbf{Y} = \mathbf{S}$ using standardized Mantel statistics (r_{Md}). Dark symbols in the correlogram: Mantel statistics that are significant after progressive Bonferroni correction ($\alpha = 0.05$).

for pairs that are in the given distance class, or the code value for that distance class (d), or any other value different from zero; all coding methods lead to the same value of the normalized Mantel statistic r_M.

The Mantel statistics, plotted against distance classes, produce a multivariate correlogram. Each value is tested for significance in the usual way, using either permutations or Mantel's normal approximation (Box 10.2). Computation of standardized Mantel statistics assumes second-order stationarity. As in the case of univariate correlograms (above), one is advised to use some form of correction for multiple testing before interpreting Mantel correlograms.

Numerical example. Consider again the 10 sampling sites of Fig. 13.4. Assume that species assemblage data were available and produced similarity matrix **S** of Fig. 13.12. Matrix **S** played here the role of **Y** in the computation of Mantel statistics. Were the species data autocorrelated? Distance matrix **D**, already divided into 6 classes in Fig. 13.4, was recoded into a series of model matrices $\mathbf{X}_d$ (d = 1, 2, etc.). In each of these, the pairs of sites that were in the given distance class received the value d, whereas all other pairs received the value 0. Mantel statistics were computed between **S** and each of the $\mathbf{X}_d$ matrices in turn; positive and significant Mantel statistics indicate positive autocorrelation in the present case. The statistics were tested for significance using 999 permutations and plotted against distance classes d to form the Mantel correlogram. The progressive Bonferroni method was used to account for multiple testing because interest was primarily in detecting autocorrelation in the first distance classes.

Before computing the Mantel correlogram, one must assume that the condition of second-order stationarity is satisfied. This condition is more difficult to explain in the case of multivariate data; it means essentially that the surface is uniform in (multivariate) mean and variance at broad scale. The correlogram illustrated in Fig. 13.12 suggests the presence of a gradient. If the condition of second-order-stationarity is satisfied, this means that the gradient detected by this analysis is a part of a larger, autocorrelated spatial structure. This was called a "false gradient" in the numerical example of Subsection 2, above.

When **Y** is a similarity matrix and distance classes are coded as described above, positive Mantel statistics correspond to positive autocorrelation; this is the case in the numerical example. When the values in **Y** are distances instead of similarities, or if the 1's and 0's are interchanged in matrix **X**, the signs of all Mantel statistics are changed. One should always specify whether positive autocorrelation is expressed by positive or negative values of the Mantel statistics when presenting Mantel correlograms. The method was applied to vegetation data by Legendre & Fortin (1989).

13.2 Maps

The most basic step in spatial pattern analysis is the production of maps displaying the spatial distributions of values of the variable(s) of interest. Furthermore, maps are essential to help interpret spatial structure functions (Section 13.1).

Several methods are available in mapping programs. The final product of modern computer programs may be a contour map, a mesh map (such as Figs. 13.13b and 13.16b), a raised contour map, a shaded relief map, and so on. The present Section is not concerned with the graphic representation of maps but instead with the way the mapped values are obtained. Spatial interpolation methods have been reviewed by Lam (1983).

Geographic information systems (GIS) are widely used nowadays, especially by geographers, to manage complex data corresponding to points, lines and surfaces in space. The present Section is not an introduction to these complex systems. It only aims at presenting the most widespread methods for mapping univariate data (i.e. a single variable y). The spatial analysis of multivariate data (multivariate matrix **Y**) is deferred to Sections 13.3 to 13.5.

Beware of non-additive variables such as pH, logarithms of counts of organisms, diversity measures, and the like (Subsection 1.4.2). Maps of such variables, produced by trend-surface analysis or interpolation methods, should be interpreted with caution; the interpolated values only make sense by reference to sampling units of the same size as those used in the original sampling design. Block kriging (Subsection 2) for blocks representing surfaces or volumes that differ from the grain of the observed data simply does not make sense for non-additive variables.

1 — Trend-surface analysis

Trend-surface analysis is the oldest method for producing smoothed maps. In this method, estimates of the variable at given locations are not obtained by interpolation, as in the methods presented in Subsection 2, but through a regression model calibrated over the entire study area.

In 1914, Student proposed to express observed values as a polynomial function of time and mentioned that it could be done for spatial data as well. This is also one of the most powerful tools of spatial pattern analysis, and certainly the easiest to use. The objective is to express a response variable y as a nonlinear function of the geographic coordinates X and Y of the sampling sites where the variable was observed:

$$y = f(\mathrm{X}, \mathrm{Y})$$

In most cases, a polynomial of X and Y with cross-product terms is used; trend-surface analysis is then an application of polynomial regression (Subsection 10.3.4) to spatially-distributed data. For example a relatively complex, but smooth surface might be fitted to a variable using a third-order polynomial with 10 parameters (b_0 to b_9):

$$\hat{y} = f(\mathrm{X}, \mathrm{Y}) = b_0 + b_1\mathrm{X} + b_2\mathrm{Y} + b_3\mathrm{X}^2 + b_4\mathrm{XY} + b_5\mathrm{Y}^2 + b_6\mathrm{X}^3 + b_7\mathrm{X}^2\mathrm{Y} + b_8\mathrm{XY}^2 + b_9\mathrm{Y}^3 \quad \textbf{(13.18)}$$

Note the distinction between the response variable y, which may represent a physical or biological variable, and the Cartesian geographic coordinate Y. Using polynomial regression, trend-surface analysis produces an equation which is linear in its parameters, although the response of y to the explanatory variables in matrix $\mathbf{X}$ = [X, Y] may be nonlinear. If variables y, X and Y have been centred on their respective means prior to model fitting, the model has an intercept of 0 by construct; therefore parameter b_0 does not have to be fitted and it can be removed from the model.

Numerical example. The data from Table 10.5 are used here to illustrate the method of trend-surface analysis. The dependent variable of the analysis, y, is Ma, which was the log-

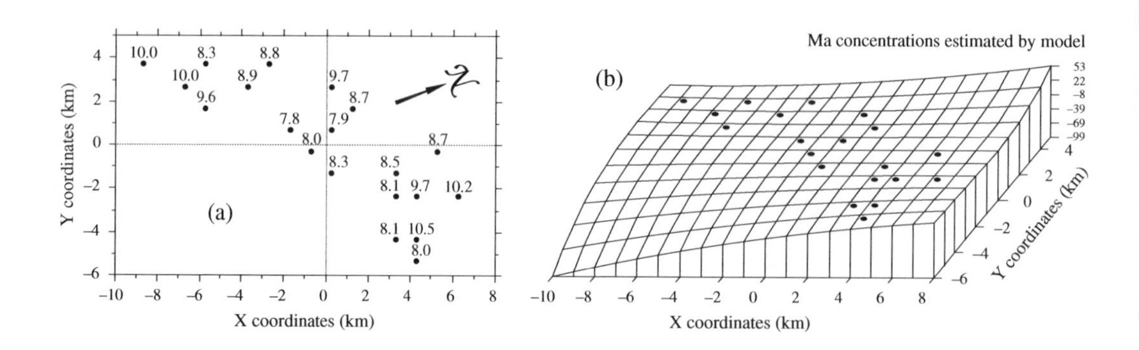

Figure 13.13 Variable Ma (log-transformed concentrations of aerobic heterotrophic bacteria growing on marine agar at salinity of 34 psu) at 20 sites in the Thau lagoon on 25 October 1988. (a) Map of the sampling sites with respect to arbitrary geographic coordinates X and Y. The observed values of Ma, from Table 10.5, are also shown. The $\mathcal{N}$ arrow points to the north. (b) Trend-surface map; the vertical axis gives the values of Ma estimated by the polynomial regression equation. Dots represent the sampling sites.

transformed ($\ln(x + 1)$) concentrations of aerobic heterotrophic bacteria growing on marine agar at salinity of 34 psu. The explanatory variables are the X and Y geographic coordinates of the sampling sites (Fig. 13.13a). The steps of the calculations are the following:

• Centre the geographic coordinates on their respective means. The reason for centring X and Y is given in Subsection 10.3.4; the amount of variation explained by a trend-surface equation is not changed by a translation (centring) of the spatial coordinates across the map.

• Determine the order of the polynomial equation to be used. A first-degree regression equation of Ma as a function of the geographic coordinates X and Y alone would only represent the linear variation of Ma with respect to X and Y; in other words, a flat surface, possibly sloping with respect to X, Y, or both. With the present data, the first-degree regression equation was not significant ($R^2 = 0.02$), meaning that there was no significant linear geographic trend to be described in the data. A regression equation incorporating the second-degree monomials (X^2, XY and Y^2) together with X and Y would be appropriate to model a surface presenting a single large bump or trough. Again, this did not seem to be the case with the present data since the second-degree equation was not significant ($R^2 = 0.39$). An equation incorporating the third-degree, fourth-degree, etc. terms would be able to model structures of increasing complexity and refinement. The cost, however, is a loss of degrees of freedom for every new monomial in the equation; trend-surface analysis using high-order equations thus requires a large number of observed sampling sites. In the present example, the polynomial was limited to the third degree, for a total of 9 terms; this is a large number of terms, considering that the data only contained 20 sampling sites.

• Calculate the various terms of the third-degree polynomial, by combining the variables X and Y as follows: X^2, $X{\times}Y$, Y^2, X^3, $X^2{\times}Y$, $X{\times}Y^2$, Y^3.

• Compute the multiple regression equation. The model obtained using all 9 regressors had $R^2 = 0.87$, but several of the partial regression coefficients were not significant.

• Remove nonsignificant terms. The linear terms may be important to express a linear gradient; the quadratic and cubic terms may be important to model more complex surfaces. Nonsignificant terms should not be left in the model, however, except when they are required for comparison purpose. Nonsignificant terms were removed one by one (backward elimination, Subsection 10.3.3) until all terms (monomials) in the polynomial equation were significant. The resulting trend-surface equation was highly significant ($R^2 = 0.81$, $p < 0.0001$):

$$\hat{y} = 8.13 - 0.16\ XY - 0.09\ Y^2 + 0.04\ X^2Y + 0.14\ XY^2 + 0.10\ Y^3$$

Remember, however, that tests of significance are too liberal with autocorrelated data, due to the non-independence of residuals (Subsection 1.1.1).

• Lay out a regular grid of points (X', Y') and, using the regression equation, compute forecasted values ($\hat{y}'$) for these points. Plot a map (Fig. 13.13b) using the file with (X', Y', and $\hat{y}'$). Values estimated by a trend-surface equation at the observed study sites do not coincide with the values observed at these sites; regression is not an exact interpolator, contrary to kriging (Subsection 2).

Different features could be displayed by rotating the Figure. The orientation chosen in Fig. 13.13b does not clearly show that the values along the long axis of the Thau lagoon are smaller near the centre than at the ends. It displays, however, the wavy structure of the data from the lower left-hand to the upper right-hand corner, which is roughly the south-to-north direction. The Figure also clearly indicates that one should refrain from interpreting extrapolated data values, i.e. values located outside the area that has actually been sampled. In the present example, the values forecasted by the model in the lower left-hand and the upper right-hand corners (–99 and +53, respectively) are meaningless for log bacterial concentrations. Within the area where real data are available, however, the trend-surface model provides a good visual representation of the broad-scale spatial variation of the response variable.

Examination of the residuals is essential to make sure that the model is not missing some salient feature of the data. If the trend-surface model has extracted all the spatially-structured variation of the data, given the scale of the study, residuals should look random when plotted on a map and a correlogram of residuals should be non-significant. With the present data, residuals were small and did not display any recognizable spatial pattern.

A cubic trend-surface model is often appropriate with ecological data. Consider an ecological phenomenon which starts at the mean value of the response variable y at the left-hand border of the sampled area, increases to a maximum, then goes down to a minimum, and comes back to the mean value at the right-hand border. The amount of space required for the phenomenon to complete a full cycle — whatever the shape it may take — is its extent (Section 13.0). Using trend-surface analysis, such a phenomenon would be correctly modelled by a third-degree trend surface equation. A polynomial equation is a more flexible mathematical model than sines or cosines, in that it does not require symmetry or strict periodicity.

The degree of the polynomial which is appropriate to model a phenomenon is predictable to a certain extent. If the extent is of the same order as the size of the study area (say, in the X direction), the phenomenon will be correctly modelled by a polynomial of degree 3 which has two extreme values, a minimum and a maximum. If the extent is larger than the study area, a polynomial of degree less than 3 is sufficient; degree 2 if there is only one maximum, or one minimum, in the sampling window; and degree 1 if the study area is limited to the increasing, or decreasing, portion of the phenomenon. Conversely, if the scale of the phenomenon controlling the variable is smaller than the study area, more than two extreme values (minima and maxima) will be found, and a polynomial of order larger than 3 is required to model it correctly. The same reasoning applies to the X and Y directions when using a polynomial combining the X and Y geographic coordinates. So, using a polynomial of degree 3 acts as a filter: it is a way of looking for phenomena that are of the same extent, or larger, than the study area.

An assumption must be made when using the method of trend-surface analysis: that all observations form a single statistical population, subjected to one and the same generating process, and can consequently be modelled using a single polynomial equation of the geographic coordinates. Evidence to that effect may be available prior to the analysis. When this is not the case, the hypothesis of homogeneity may be supported by examining the regression residuals (Subsection 10.3.1). When there are indications that values in different regions of the geographic space obey different processes (e.g. different geology, action of currents or wind, or influence of other physical variables), the study area should be divided into regions, to be modelled by separate trend-surface equations.

Polynomial regression, used in the numerical example above, is a good first approach to fitting a model to a surface when the shape to be modelled is unknown, or known to be simple. In some instances, however, it may not provide a good fit to the data; trend-surface analysis must then be conducted using nonlinear regression (Subsection 10.3.6), which requires that an appropriate numerical model be provided to the estimation program. Consider the example of the effect of some human-generated environmental disturbance at a site, the indicator variable being the number of species. The response, in this case, is expected to be stronger near the impacted site, tapering off as one gets farther away from it. Assume that data were collected along a transect (a single geographic coordinate X) and that the impacted site is near the centre of the transect. A polynomial equation would not be appropriate to model an inverse-squared-distance diffusion process (Fig. 13.14a), whereas an equation of the form:

$$\hat{y} = b_0 + \frac{b_1 X^2}{b_2 X^2 + 1}$$

would provide a much better fit (Fig. 13.14b). The minimum of this equation is b_0; it is obtained when $X = 0$. The maximum, b_1/b_2, is reached asymptotically as X becomes large in either the positive or negative direction. For data collected in different directions around the impacted site, a nonlinear trend-surface equation with similar properties would be of the form:

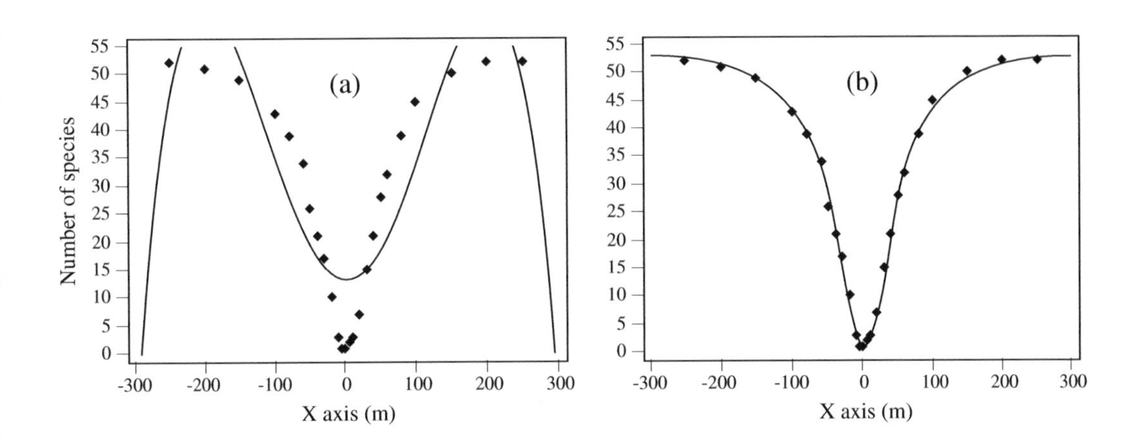

Figure 13.14 (a) Artificial data representing the number of species around the site of an environmental disturbance (located at X = 0) are not well-fitted by a 4th-order polynomial equation of the X coordinates ($R^2 = 0.7801$). (b) They are well-fitted by the following inverse-squared-distance diffusion equation: $\hat{y} = 1 + [0.0213\text{X}^2 / (0.0004\text{X}^2 + 1)]$ ($R^2 = 0.9975$).

$$\hat{y} = b_0 + \frac{b_1\text{X}^2 + b_2\text{Y}^2}{b_3\text{X}^2 + b_4\text{Y}^2 + 1}$$

where X and Y are the coordinates of the sites in geographic space.

Trend-surface analysis is appropriate for describing broad-scale spatial trends in data. It does not produce accurate fine-grained maps of the spatial variation of a variable, however. Other methods described in this Chapter may prove useful to model variation at finer scales. In some studies, the broad-scale trend itself is of interest; this is the case in the numerical example above and in Ecological application 13.2. In other instances, and especially in studies that cover large geographic expanses, the broad-scale trend may be already known and understood; students of geographic variation patterns may want to conduct analyses on detrended data, i.e. data from which the broad-scale trend has been removed. Detrending a variable may be achieved by computing the residuals from a trend-surface equation of sufficient order, as in time-series analysis (Section 12.2).

If there is replication at each point, it is possible to perform a test of goodness-of-fit of a trend-surface model (Draper and Smith, 1981; Legendre & McArdle, 1997). By comparing the observed error mean square after fitting the trend surface to the error mean square estimated by the among-replicate within-location variation, one can test if the model fits the data properly. The among-replicate within-location variation is computed from the deviations from the means at the various locations; it is actually the

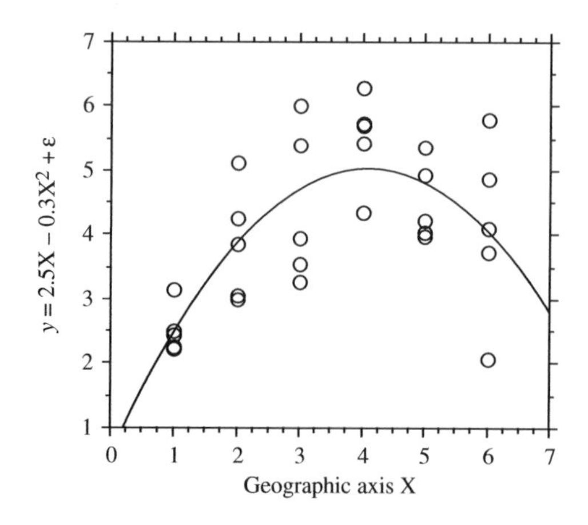

Figure 13.15 Artificial data representing sampling along a geographic axis X with 5 replicates at each site; $n = 30$. The F test of goodness-of-fit indicates that the trend-surface equation $\hat{y} = 0.562 + 2.184X - 0.267X^2$ ($R^2 = 0.4899$) fits the data properly.

residual mean square of an ANOVA among locations. These error mean squares are not much different if the trend surface goes through the expected values at the various locations, so that the F ratio of the two mean squares is not significant. If, on the contrary, the fitted surface does not follow the major features of the variation among locations, the deviations of the data from the fitted trend-surface values are likely to be larger than expected from our knowledge of the sampling error; the F statistic is then significantly larger than 1, indicating that the trend surface is misrepresenting the variation among locations.

Numerical example. Consider the artificial data in Fig. 13.15. Variable X represents a geographic axis along which sampling has taken place at 6 sites. Variable y was constructed using equation $y = 2.5X - 0.3X^2 + \varepsilon$, where ε is a random standard normal deviate [N(0, 1)]. A quadratic trend-surface model of X was fitted to the data. The residual mean square, or "error mean square after fitting the trend surface", was $MS_1 = 0.84909$ ($\nu = 27$). An analysis of variance was conducted on y using the grouping into 6 sites as the classification criterion. The residual mean square obtained from the ANOVA was $MS_2 = 0.87199$ ($\nu = 24$). The ratio of these two mean squares gave an F statistic:

$$F = \frac{MS_1}{MS_2} = \frac{0.84909}{0.87199} = 0.97374$$

which was tested against $F_{\alpha=0.05(27,\ 24)} = 1.959$. The F statistic was not significantly different from 1 ($p = 0.5308$), which indicated that the model fitted the data properly.

The trend-surface analysis was recomputed using a linear model of X. The model obtained was $\hat{y} = 3.052 + 0.316X$ ($R^2 = 0.1941$). MS_1 in this case was 1.29358 ($\nu = 28$). The F ratio

$MS_1/MS_2 = 1.29358/0.87199 = 1.48348$. The reference value was $F_{0.05(28, 24)} = 1.952$. The probability associated with the F ratio, $p = 0.1651$, indicated that this model still fitted the data, which were constructed to contain a linear term (2.5X in the construction equation) as well as a quadratic trend (term $-0.3X^2$), but the fit was poorer than with the quadratic polynomial model which was capable of accounting for both the linear and quadratic trends.

This numerical example shows that trend-surface analysis may be applied to data collected along a transect; the "trend surface" is one-dimensional in that case. The numerical example at the end of Subsection 10.3.4 is another example of a trend-surface analysis of a dependent variable, salinity, with respect to a single geographic axis (Fig. 10.9). Trend-surface analysis may also be used to model data in three-dimensional geographic space (geographic coordinates X, Y and Z, where Z is either altitude or depth) or with one of the dimensions representing time. Section 13.5 will show how the analysis may be extended to a multivariate dependent data matrix **Y**.

Haining (1987) described alternative methods for estimating the parameters of a trend-surface model when the residuals are spatially autocorrelated; in that case, least-squares estimation of the parameters is inefficient and standard errors as well as tests of significance are biased. Haining's methods allow one to recognize three components of spatial variation corresponding to the site, local, and regional scales.

Ecological application 13.2

A survey was conducted at 200 locations within a fairly homogeneous 12.5 ha rectangular sandflat area in Manukau Harbour, New Zealand, to identify factors that control the spatial distributions of the two dominant bivalves, *Macomona liliana* Iredale and *Austrovenus stutchburyi* (Gray), and to look for evidence of adult-juvenile interactions within and between species. Results are reported in Legendre *et al.* (1997). Most of the broad-scale spatial structure detected in the bivalve counts (two species, several size classes) was explained by the physical and biological variables. Results of principal component analysis and spatial regression modelling suggested that different factors controlled the spatial distributions of adults and juveniles. Larger size classes of both species displayed significant spatial structures, with physical variables explaining some but not all of this variation; the spatial patterns of the two species differed, though. Smaller organisms were less strongly spatially structured; virtually all of their spatial structure was explained by physical variables.

Highly significant trend-surface equations were found for all bivalve species and size classes (log-transformed data), indicating that the spatial distributions of the organisms were not random, but highly organised at the scale of the study site. The trend-surface models for smaller animals had much smaller coefficients of determination (10-20%) than for larger animals (30-55%). The best models, i.e. the models with the highest coefficients of determination (R^2), were for the *Macomona* > 15 mm and *Austrovenus* > 10 mm. The coefficients of determination were consistently higher for *Austrovenus* than for *Macomona*, despite the fact that *Macomona* were usually far more numerous than *Austrovenus*. A map illustrating the trend-surface equation is presented for the largest *Macomona* size class (Fig. 13.16); the field counts are also given for comparison.

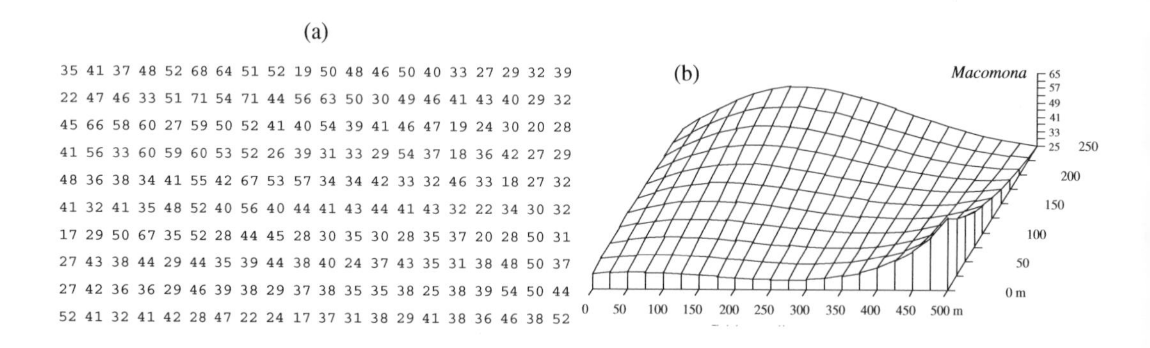

35	41	37	48	52	68	64	51	52	19	50	48	46	50	40	33	27	29	32	39
22	47	46	33	51	71	54	71	44	56	63	50	30	49	46	41	43	40	29	32
45	66	58	60	27	59	50	52	41	40	54	39	41	46	47	19	24	30	20	28
41	56	33	60	59	60	53	52	26	39	31	33	29	54	37	18	36	42	27	29
48	36	38	34	41	55	42	67	53	57	34	34	42	33	32	46	33	18	27	32
41	32	41	35	48	52	40	56	40	44	41	43	44	41	43	32	22	34	30	32
17	29	50	67	35	52	28	44	45	28	30	35	30	28	35	37	20	28	50	31
27	43	38	44	29	44	35	39	44	38	40	24	37	43	35	31	38	48	50	37
27	42	36	36	29	46	39	38	29	37	38	35	35	38	25	38	39	54	50	44
52	41	32	41	42	28	47	22	24	17	37	31	38	29	41	38	36	46	38	52

Figure 13.16 *Macomona* > 15 mm at 200 sites in Manukau Harbour, New Zealand, on 22 January 1994. (a) Actual counts at sampling sites in 200 regular grid cells; in the field, sites were not perfectly equispaced. (b) Map of the trend-surface equation explaining 32% of the spatial variation in the data. The values estimated from the trend-surface equation (log-transformed data) were back-transformed to raw counts before plotting. Modified from Legendre *et al.* (1997).

2 — *Interpolated maps*

In this family of methods, the value of the variable at a point location on a map is estimated by local interpolation, using only the observations available in the vicinity of the point of interest. In this respect, interpolation mapping differs from trend surface analysis (Subsection 1), where estimates of the variable at given locations were not obtained by interpolation, as in the present Subsection, but through a statistical model calibrated over the entire study area. Fig. 13.17 illustrates the principle of interpolation mapping. A regular grid of nodes (Fig. 13.17c) is defined over the area that contains the study sites Ø_i (Fig. 13.17a, b). Interpolation assigns a value to each point of that grid. This is the single most important step in mapping. Following that, results may be represented in the form of contours (Fig. 13.17d) with or without colours or shades, or three-dimensional constructs such as Fig. 13.16b.

Assigning a value to each grid node may be done in different ways. Different interpolation methods may produce maps that look different; this is also the case when using different parameters with a same method (e.g. different exponents in inverse-distance weighting).

The most simple rule would be to give, to each node of the grid, the value of the observation which is the closest to it. The end result is a division of the map into Voronoï polygons (Subsection 13.3.1) displaying a "zone of influence" drawn around each observation. Another simple solution consists in dividing the map into Delaunay triangles (Subsection 13.3.1). There is an observed value y_i at each site Ø_i. A triangular portion of plane, adjusted to the points Ø_i that form the apices (corners) of a Delaunay triangle, provides interpolated values for all points lying within the triangle. Maps obtained using these solutions are shown in Chapter 11 of Isaaks & Srivastava (1989).

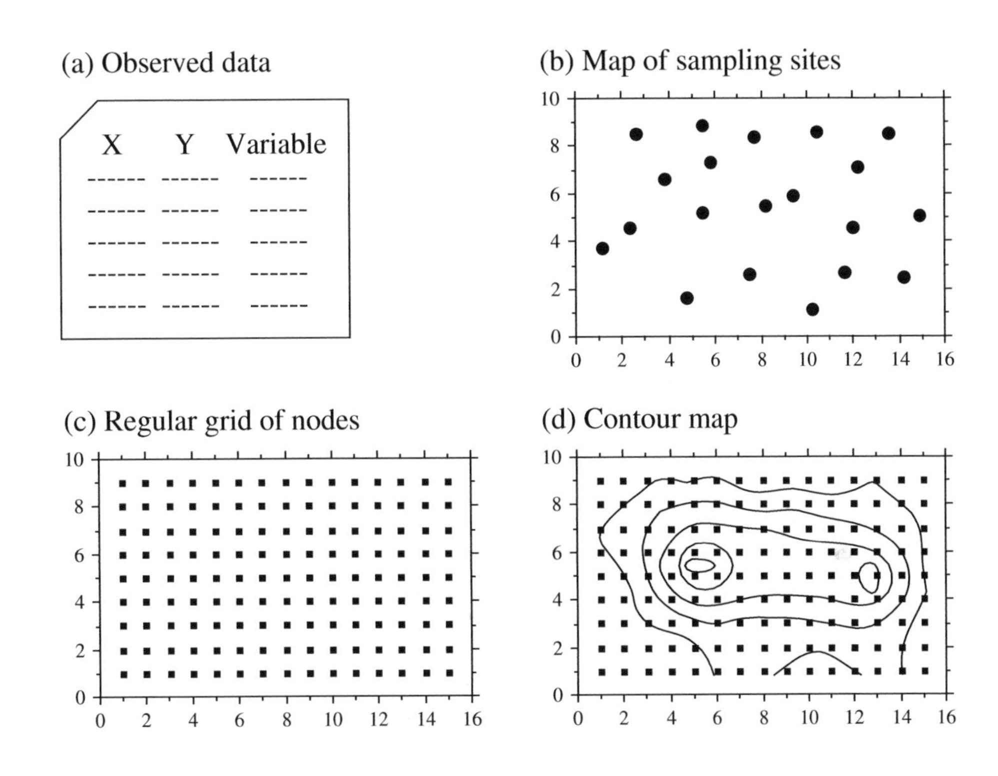

Figure 13.17 Summary of the interpolation procedure.

Alternatively, one may draw a "search circle" (or an ellipsoid for anisotropic data) around each grid node (Fig. 13.18). The radius of the circle may be determined in either of two ways. (1) One may fix a minimum number of observed points that must be included in the interpolation for each grid node; or (2) one may use the "distance of influence of the process" found by correlogram or variogram analysis (Section 13.1). The estimation procedure is repeated for each node of the grid. Several methods of interpolation may be used.

• Mean — Consider all the observed study sites found within the circle; assign the mean of these values to the grid node. This method does not produce smooth maps; discontinuities in neighbouring grid node values occur as observed points move in or out of the search circle.

• Inverse-distance weighting — Consider the observation sites found within the circle and calculate a weighted mean value, using the formula:

$$\hat{y}_{\text{Node}} = \sum_i w_i y_i \tag{13.19}$$

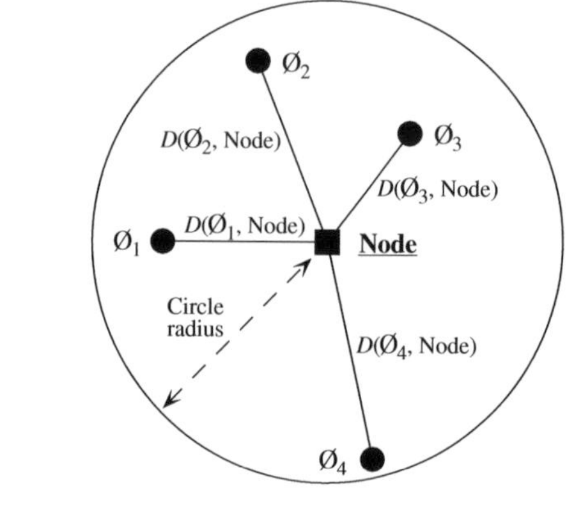

Figure 13.18 To estimate the value at a grid node (square), draw a search circle around it and consider the observed points (Ø_i) found within the circle. Observed points are separated from the node by distances $D(\text{Ø}_i, \text{Node})$.

where y_i is the value observed at point Ø_i and weight w_i is the inverse of the distance (D) from point Ø_i to the grid node to be estimated. The inverse distances, to some power k, are scaled by the sum of the weights for all points Ø_i in the estimation, so as to produce values that are consistent with the values observed at points Ø_i (unbiasedness condition):

$$w_i = \left(\frac{1}{D(\text{Ø}_i, \text{Node})^k}\right) \Big/ \sum_i \frac{1}{D(\text{Ø}_i, \text{Node})^k} \tag{13.20}$$

A commonly-used exponent is $k = 2$. This corresponds, for instance, to the decrease in energy of waves dispersing across a two-dimensional surface. The greater the value of k, the less influence distant data points have on the value assigned to the grid node. This method produces smooth values over the grid of nodes. The range of estimated values is smaller than the range of observed data so that, contrary to trend-surface analysis (Fig. 13.13b), inverse-distance weighting does not produce meaningless values in the parts of the map beyond the area that was actually sampled. When the observation sites Ø_i do not form a regular or nearly regular grid, however, this interpolation method may generate features in maps that have little to do with reality. As a consequence, inverse-distance weighting is not recommended in that situation.

• Weighted polynomial fitting — In this method, a trend-surface equation (Subsection 13.2.1) is adjusted to the observed data points within the search circle, weighting each observation Ø_i by the inverse of its distance (using some appropriate power k) to the grid node to be estimated. A first or second-order polynomial equation is usually used. z_{Node} is taken to be the value estimated by the polynomial equation for the coordinates of the grid node. This method suffers from the same problem as inverse distance weighting with respect to observation sites Ø_i that do not form a regular or nearly regular grid of points.

Kriging

• Kriging — This is the mapping tool in the toolbox of geostatisticians. The method was named by Matheron after the South African geostatistician D. G. Krige, who was the first to develop formal solutions to the problem of estimating ore reserves from sampling (core) data (Krige, 1952, 1966). Geostatistics was developed by Matheron (1962, 1965, 1970, 1971, 1973) and co-workers at the *Centre de morphologie mathématique* of the *École des Mines de Paris*. Geostatistics comprises the estimation of variograms (Subsection 13.1.6), kriging, validation methods for kriging estimates, and simulations methods for geographically distributed ("regionalized") data. Major textbooks have been written by former students of Matheron: David (1977) and Journel & Huijbregts (1978). Other useful references are Clark (1979), Rendu (1981), Verly *et al.* (1984), Armstrong (1989), Isaaks & Srivastava (1989), and Cressie (1991). Applications to environmental sciences and ecology have been discussed by Gilbert & Simpson (1985), Robertson (1987), Armstrong *et al.* (1989), Legendre & Fortin (1989), Soares *et al.* (1992), and Rossi *et al.* (1992). Geostatistical methods can be implemented using the software library of Deutsch & Journel (1992).

As in inverse-distance weighting (eq. 13.19), the estimated value for any grid node is computed as:

$$\hat{y}_{\text{Node}} = \sum_i w_i y_i$$

The chief difference with inverse-distance weighting is that, in kriging, the weights w_i applied to the points $\varnothing_i$ used in the estimation are not standardized inverses of the distances to some power k. Instead, the weights are based upon the covariances (semi-variances, eq. 13.9 and 13.10) read on a variogram model (Subsection 13.1.6). They are found by linear estimation, using the equation:

$$\mathbf{C} \cdot \mathbf{w} = \mathbf{d}$$

$$\begin{bmatrix} c_{11} & \dots & c_{1n} & 1 \\ \cdot & \dots & \cdot & 1 \\ \cdot & \dots & \cdot & 1 \\ c_{n1} & \dots & c_{nn} & 1 \\ 1 & \dots & 1 & 0 \end{bmatrix} \begin{bmatrix} w_1 \\ \cdot \\ \cdot \\ w_n \\ \lambda \end{bmatrix} = \begin{bmatrix} d_1 \\ \cdot \\ \cdot \\ d_n \\ 1 \end{bmatrix} \tag{13.21}$$

where **C** is the covariance matrix among the n points $\varnothing_i$ used in the estimation, i.e. the semi-variances corresponding to the distances separating the various pair of points, as read on the variogram model; **w** is the vector of weights to be estimated (with the constraint that the sum of weights must be 1); and **d** is a vector containing the covariances between the various points $\varnothing_i$ and the grid node to be estimated. This is where a variogram model becomes essential; it provides the weighting function for the entire map and is used to construct matrix **C** and vector **d** for each grid node to be estimated. λ is a Lagrange parameter (as in Section 4.4) introduced to minimize the variance of the estimates under the constraint $\Sigma w_i = 1$ (unbiasedness condition). The solution to this linear system is obtained by matrix inversion (Section 2.8):

$$\mathbf{w} = \mathbf{C}^{-1}\,\mathbf{d} \qquad \textbf{(13.22)}$$

Vector **d** plays a role similar to the weights in inverse-distance weighting since the covariances in vector **d** decrease with distance. Using covariances, the weights are statistical in nature instead of geometrical.

Kriging takes into account the grouping of observed points $Ø_i$ on the map. When two points $Ø_i$ are close to each other, the value of the corresponding coefficient c_{ij} in matrix **C** is high; this contributes to lowering their respective weights w_i. In this way, the redundancy of information introduced by dense groups of sampling sites is taken into account.

When anisotropy is present, kriging can use two, four, or more variogram models computed for different geographic directions and combine their estimates when calculating the covariances in matrix **C** and vector **d**. In the same way, when estimation is performed for sampling sites in a volume, a separate variogram can be used to describe the vertical spatial variation. Kriging is the best interpolation method for data that are not on a regular grid or display anisotropy. The price to pay is increased mathematical complexity during interpolation.

Among the interpolation methods, kriging is the only one that provides a measure of the error variance for each value estimated at a grid node. For each grid node, the error variance, called *ordinary kriging variance* (s^2_{OK}), is calculated as follows (Isaaks & Srivastava, 1989), using vectors **w** and **d** from eq. 13.21:

$$s^2_{\text{OK}} = \text{Var}\,[y_i] - \mathbf{w}^{-1}\mathbf{d} \qquad \textbf{(13.23)}$$

where Var$[y_i]$ is the maximum-likelihood estimate of the variance of the observed values y_i (eq. 13.14). Equation 13.23 shows that s^2_{OK} only depends on the variogram model and the local density of points, and not on the values observed at points $Ø_i$. The ordinary kriging variance may be used to construct confidence intervals around the grid node estimates at some significance level α, using eq. 13.4. It may also be mapped directly. Regions of the map with large values s^2_{OK} indicate that more observations should be made because sampling intensity was too low.

Kriging, as described above, provides point estimates at grid nodes. Each estimate actually applies to a "point" whose size is the same as the grain of the observed data. The geostatistical literature also describes how *block kriging* may be used to obtain estimates for blocks (i.e. surfaces or volumes) of various sizes. Blocks may be small, or cover the whole map if one wishes to estimate a resource over a whole area. As mentioned in the introductory remarks of the present Section, additive variables only may be used in block kriging. Block kriging programs always assume that the variable is *intensive*, e.g. the concentration of organisms (Subsection 1.4.2). For *extensive* variables, such as the number of individual trees, one must multiply the block estimate by the ratio (block size / grain size of the original data).

3 — Measures of fit

Different measures of fit may be used to determine how well an interpolated map represents the observed data. With most methods, some measure may be constructed of the closeness of the estimated (i.e. interpolated) values $\hat{y}_i$ to the values y_i observed at sites $Ø_i$. Four easy-to-use measures are:

- The mean absolute error: $MAE = \frac{1}{n}\sum_i |y_i - \hat{y}_i|$

- The mean squared error: $MSE = \frac{1}{n}\sum_i (y_i - \hat{y}_i)^2$

- The Euclidean distance: $D_1 = \sqrt{\sum_i (y_i - \hat{y}_i)^2}$

- The correlation coefficient (r) between values y_i and $\hat{y}_i$ (eq. 4.7). In the case of a trend-surface model, the square of this correlation coefficient is the coefficient of determination of the model.

In the case of kriging, the above measures of fit cannot be used because the estimated and observed values are equal, at all observed sites $Ø_i$. The technique of cross-validation may be used instead (Isaaks & Srivastava, 1989, Chapter 15). One observation, say $Ø_1$, is removed from the data set and its value is estimated using the remaining points $Ø_2$ to $Ø_n$. The procedure is repeated for $Ø_2$, $Ø_3$, …, $Ø_n$. One of the measures of fit described above may be used to measure the closeness of the estimated to the observed values. If replicated observations are available at each sampling site (a situation which does not often occur), the test of goodness-of-fit described in Subsection 1 can be used with all interpolation methods.

13.3 Patches and boundaries

Multivariate data may be condensed into spatially-constrained clusters. These may be displayed on maps, using different colours or shades. The present Section explains how clustering algorithms can be constrained to produce groups of spatially contiguous sites; study of the boundaries between homogeneous zones is also discussed. Prior to clustering, one must state unambiguously which sites are neighbours in space; the most common solutions to this are presented in Subsection 1.

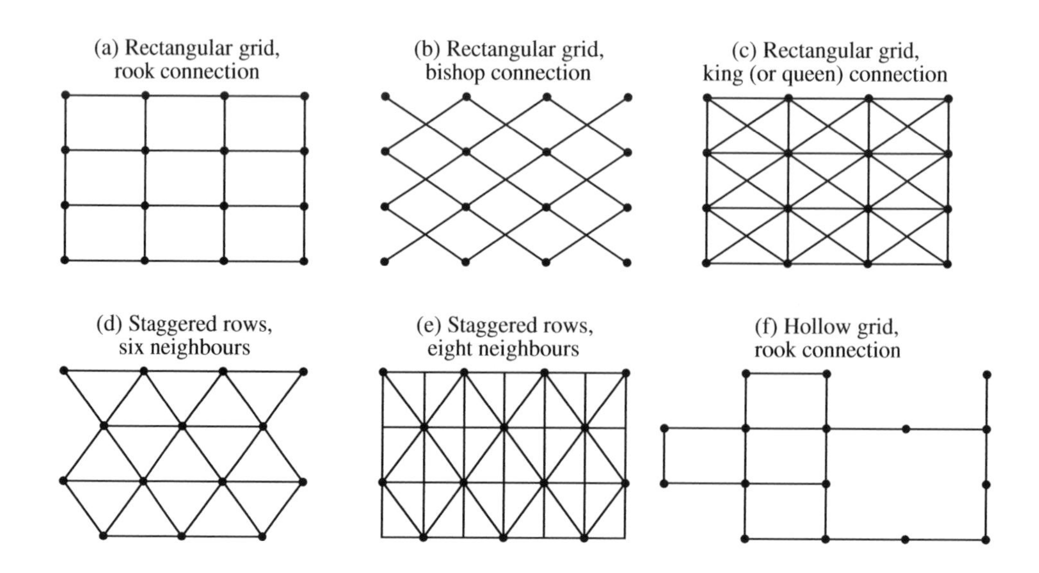

Figure 13.19 Connecting schemes for regular grids of points. See text.

1 — Connection networks

When sampling has been conducted on a regular rectangular grid, neighbouring points may be linked using simple connecting schemes whose names are derived from the game of chess (Cliff & Ord, 1981): rook's (rectangular: Fig. 13.19a), bishop's (diagonal: Fig. 13.19b), or king's connections (also called queen's: both rectangular and diagonal, Fig. 13.19c). Sampling in staggered rows leads to connecting each point (except borders) to six (Fig. 13.19d) or eight neighbours (Fig. 13.19e). Algorithms may allow the construction of regular grids with missing points (Fig. 13.19f). When the objects represent irregularly-shaped land units covering a geographic area (e.g. electoral units), parcels sharing a common boundary are regarded as contiguous.

When the localities are positioned in an irregular manner, geometric connecting schemes may be used, such as Delaunay triangulation, Gabriel graph, relative neighbourhood graph or minimum spanning tree. There exists an inclusion relationship among the four connecting schemes: all edges that are members of a minimum spanning tree also obey the relative neighbourhood graph criterion; these are all members of a Gabriel graph, which in turn are all included in a Delaunay triangulation (Toussaint, 1980; Matula & Sokal, 1980; Gordon, 1996c):

$$\text{Minimum spanning tree} \subseteq \text{Relative neighbourhood gr.} \subseteq \text{Gabriel gr.} \subseteq \text{Delaunay triangulation}$$

Delaunay triangulation

• Delaunay triangulation — The Delaunay triangulation criterion (Dirichlet, 1850; Upton & Fingleton, 1985) is illustrated in Fig. 13.20. For any triplet of points A, B and C, the three edges (i.e. lines) connecting these points are included in the triangulation

A B C

Point identifiers	Coordinates X	Y
1	0	3
2	1	5
3	2	2
4	2	1
5	4	4
6	5	2
7	8	0
8	7.5	3
9	8	5

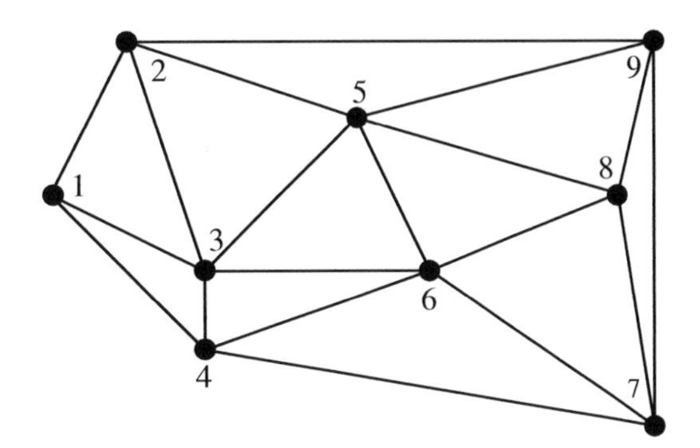

19 edges form the Delaunay triangulation:

1–2	1–3	1–4	2–3	2–5	2–9	3–4
3–5	3–6	4–6	4–7	5–6	5–8	5–9
6–7	6–8	7–8	7–9	8–9		

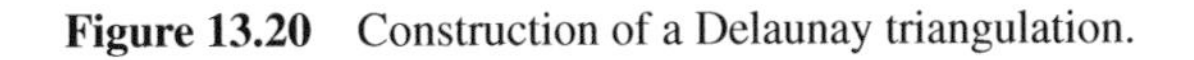

Figure 13.20 Construction of a Delaunay triangulation.

if and only if the circumscribed circle (i.e. the circle passing through the three points; on the left in the Figure) includes no other point. For example, the file of coordinates shown in the central part of the Figure gives rise to the triangulation on the right. The triangulation is fully described by a list of pairs of points corresponding to its edges; this is how the information can be passed on to a computer program for constrained clustering (Subsection 2).

Long edges may be created at the outskirts of a set of points, simply because there is no other point located farther away in the sampling design; this is called a *border effect*. For example, edges 2–9 and 7–9 might have been removed from the triangulation in Fig. 13.20 by the presence of other points in the circumscribed circles of triangles (2, 5, 9) and (7, 8, 9) had the sampling extent been broader. Long peripheral edges may be removed by hand or by the computer algorithm.

Gabriel graph

• Gabriel graph — The Gabriel graph criterion (Gabriel & Sokal, 1969) differs from that of the Delaunay triangulation (Fig. 13.21a). Draw a line between two points A and B. This line is part of the Gabriel graph if and only if no other point C lies inside the circle *whose diameter is that line*. In other words, the edge between A and B is part of the Gabriel graph if $D^2(\text{A, B}) < D^2(\text{A, C}) + D^2(\text{B, C})$ for all other points C in the study, where $D^2(\text{A, B})$ is the square of the geographic distance between points A and B. Another way of expressing this criterion is the following: if CENTRE represents the middle point between A and B, the edge connecting A to B is part of the Gabriel graph if $D(\text{A, B})/2 < D(\text{CENTRE, C})$ for any other point C in the study.

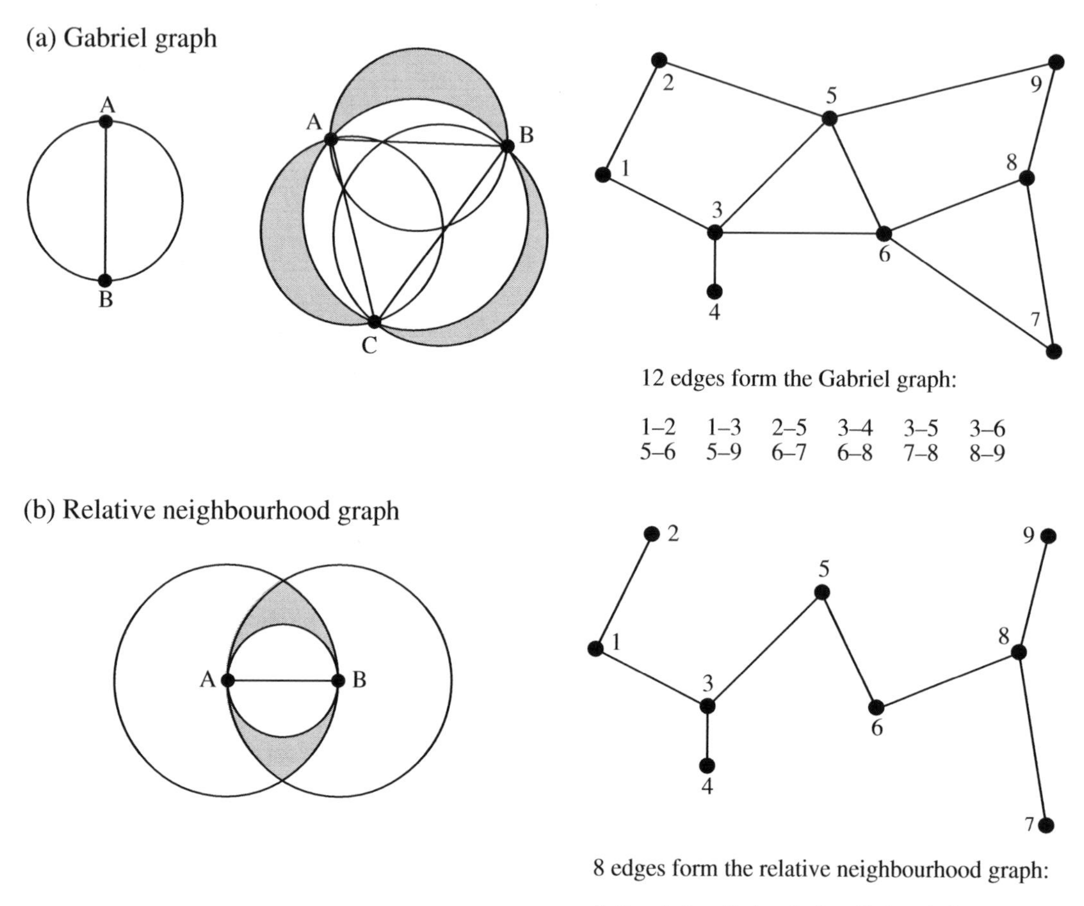

Figure 13.21 (a) Left: geometric criterion for constructing a Gabriel graph. Centre: the zone of exclusion of the Gabriel criterion, here for three points (grey zones + white inner circle), is larger than that of the Delaunay criterion (white inner circle). Right: Gabriel graph for the nine points of Fig. 13.20. (b) Left: geometric criterion for constructing a relative neighbourhood graph. The zone of exclusion of the relative neighbourhood criterion, here for two points (grey zones + white inner circle), is larger than that of the Gabriel criterion (white inner circle). Right: relative neighbourhood graph for the nine points of Fig. 13.20.

The Gabriel graph in Fig. 13.21a is constructed for the same points as the Delaunay triangulation in Fig. 13.20. The 12 edges forming the Gabriel graph are a subset of the 19 edges of the Delaunay triangulation. Indeed, as shown by the sketch in the centre of the Figure, the exclusion zone formed by the three circles corresponding to the Gabriel criterion (which have for diameters the edges A–B, B–C and A–C) may contain, in the shadowed areas outside the Delaunay circle (white inner circle), some points that the Delaunay criterion circle does not exclude. This is why some edges that are authorized by the Delaunay criterion are excluded from the Gabriel graph.

Relative neighbourhood graph

• Relative neighbourhood graph — The relative neighbourhood criterion is as follows (Toussaint, 1980; Fig. 13.21b). Draw a line between two points A and B. Draw a first circle centred over A and a second one centred over B, each one having the line from A to B as its radius. This line is part of the graph if no other point C in the study lies inside *the intersection of the two circles.* In other words, the edge from A to B is part of the relative neighbourhood graph if and only if $D(\text{A, B}) \leq \max\,[D(\text{A, C}), D(\text{B, C})]$ for all other points C in the study. For points forming an equilateral triangle, for instance, the three edges are included in the relative neighbourhood graph.

The relative neighbourhood graph in Fig. 13.21b is constructed for the same set of points as in Figs. 13.20 and 13.21a. The number of edges in a relative neighbourhood graph is $(n - 1)$. The 8 edges forming the relative neighbourhood graph are a subset of the 12 edges of the Gabriel graph. Indeed, as shown by the sketch on the left of the Figure, the exclusion zone at the intersection of the two circles corresponding to the relative neighbourhood criterion (which have for radius the edge A–B) may contain, in the shadowed zone outside the Gabriel circle (white inner circle), some points that the Gabriel criterion circle does not exclude. This is why some edges authorized by the Gabriel criterion are excluded from the relative neighbourhood graph.

Minimum spanning tree

• Minimum spanning tree — In this tree, which connects all points of a study, the sum of the edge lengths is minimum. Its construction is described at the end of Section 8.2; one way of obtaining it is to list the edges forming the primary connections of a single-linkage dendrogram. For points forming an equilateral triangle, for example, only two of the edges are included in the minimum spanning tree, whereas the three edges are included in a relative neighbourhood graph; the choice of the edge to leave out is arbitrary. The edges of a minimum spanning tree are either the same as, or a subset of, the edges of a relative neighbourhood graph of the same points. For the example data set, the edges that form the minimum spanning tree are the same as those of the relative neighbourhood graph of Fig. 13.21b.

Another approach is to select a distance threshold and connect all points that are within that distance of each other. One possible criterion to choose the distance threshold is to make it equal to the range of a variogram model (Fig. 13.7).

The list of connecting edges (Figs. 13.20 and 13.21) may be written out to a file. The file may be modified to take into account other information that researchers may have about the study area. For example, one may wish to eliminate edges that do not make sense in terms of gene flow because they cross unsuitable areas (e.g. a sea or a mountain range, in the case of terrestrial mammals). Or, one may wish to add connections that are potentially of interest although they do not imply first neighbours; for example, plants or animals may cross water bodies (lake, sea) and settle in non-

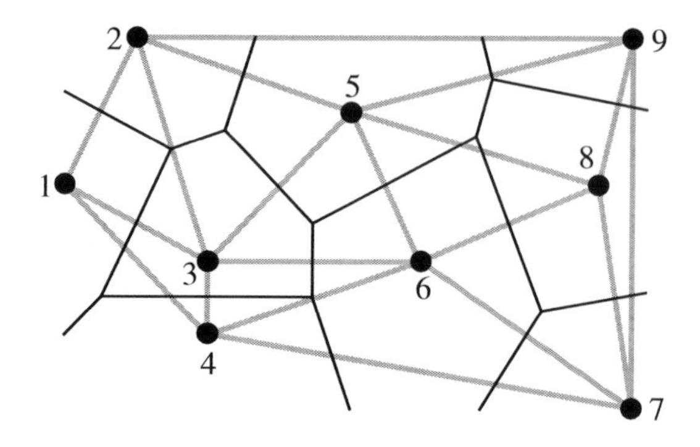

Figure 13.22 Delaunay triangulation (grey lines) and influence polygons (black lines) for the nine points of Fig. 13.20.

contiguous sites, which should nevertheless be considered contiguous because there is a direct path between them. Users of constrained clustering methods should not hesitate to modify lists of connections obtained from geometric criteria such as described above, to make the list of edges a better description of potential flow among sites, given the problem under study.

Influence polygon

It may be interesting to determine the geometric zone of influence of each point on a map. The zone of influence of a point A includes all the other points of the surface that are closer to A than to any other point in the study. The zones of influence so defined have the shape of polygons, also called tiles, tessellae, or tesserae (singular: tessella or tessera). The resulting picture is called a mosaic or tessellation (adjective: tessellated); it may be referred to as a Dirichlet tessellation (1850), Voronoï polygons (1909), or Thiessen polygons (1911), from the names of the authors who first described these mathematical structures.

Polygons are easily constructed from a Delaunay triangulation (Fig. 13.22). Draw the perpendicular bisector of each segment in the triangulation; the crossing points of the bisectors delimit the polygons (tiles). Computer algorithms may be used to calculate the surface area of each polygon, at least those that are closed; peripheral tiles may be open. Upton & Fingleton (1985) and Isaaks & Srivastava (1989) propose various applications of tessellations to spatial analysis.

2 — Constrained clustering

The delineation of clusters of contiguous objects has been discussed in Section 12.6 for time series and spatial transects. The method of chronological clustering, in particular, was described in Subsection 12.6.4; it proceeds by imposing to a clustering algorithm a constraint of contiguity along the time series. Constraints of contiguity have been applied to spatial clustering by several authors, including Lefkovitch (1978,

1980), Monestiez (1978), Lebart (1978), Roche (1978), Perruchet (1981) and Legendre & Legendre (1984c). In the present Subsection, it is generalized to two- or three-dimensional spatial data and to spatio-temporal data.

Constrained clustering differs from its unconstrained counterpart in the following way.

• Unconstrained clustering methods (Chapter 8) only use the information in the similarity or distance matrix computed among the objects. In hierarchical methods, a local criterion is optimized at each step; in all methods included in the Lance and Williams general model, for instance, the objects or groups clustered at each step are those with the largest fusion similarity or the smallest fusion distance. In partitioning methods, a global criterion is optimized; in K-means, for instance, the algorithm looks for K groups that feature the smallest sum of within-group sums-of-squares E_K.

• Constrained clustering methods take into account more information than the unconstrained approaches. In the case of spatial or temporal contiguity, the only admissible clusters are those that obey the contiguity relationship. Spatial contiguity may be described by one of the connecting schemes of Subsection 1. The criterion to be optimized during clustering is relaxed to give priority to the constraint of spatial contiguity. It is no surprise, then, that a constrained solution may be less optimal than its unconstrained counterpart in terms of the clustering criterion, e.g. E_K. This is balanced by the fact that the solution is likely to more readily interpretable.

It is fairly easy to modify clustering algorithms to incorporate a constraint of spatial contiguity (Fig. 13.23). As an example, consider the clustering methods included in the Lance and Williams general clustering model (Subsection 8.5.9). At the beginning of the clustering process, the vector of group membership has each object as a different group. Proceed as follows:

1. Compute a similarity matrix among objects, using the non-geographic information.

2. Choose a connecting scheme (Subsection 1) and produce a list of connection edges as in Figs. 13.20 and 13.21. Read in the file of edges and transform it into a *contiguity matrix* containing 1's for connected sites and 0's elsewhere.

3. Compute the Hadamard product of these two matrices. The Hadamard product of two matrices is the product element by element. The resulting matrix contains similarity values in the cells where the contiguity matrix contained 1's, and 0's elsewhere.

4. The largest similarity value in the matrix resulting from step 3 determines the next pair of objects or groups (h and i) to be clustered. Modify the vector of group membership (right of the Figure), giving the same group label to all members of former groups h and i.

5. Update the similarity matrix using eq. 8.11.

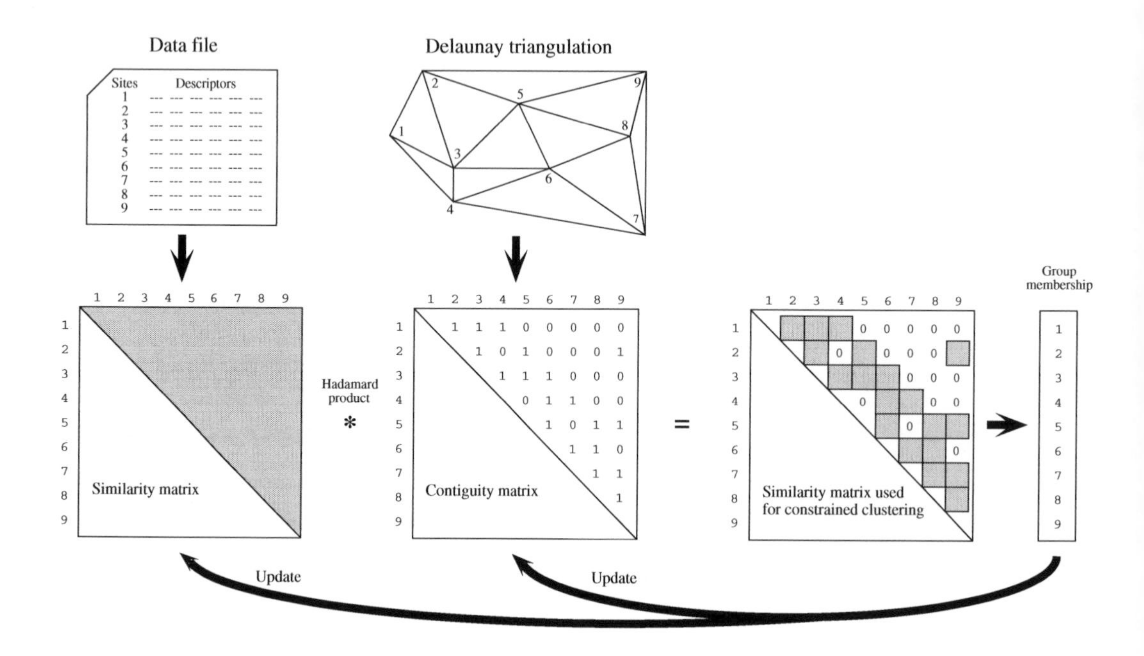

Figure 13.23 Summary of the spatially-constrained clustering procedure for methods included in the Lance and Williams general clustering model. The vector of group membership is represented at the start of the clustering iterations; see text. Locations of the points are the same as in Fig. 13.20.

6. Update also the contiguity matrix. All objects that were neighbours to h are now also neighbours to i and vice versa.

7. Go back to step 3. Iterate until all objects are members of a single group.

Ferligoj & Batagelj (1982) have shown, however, that the introduction of relational constraints (e.g. spatial contiguity) may occasionally produce reversals with any of the hierarchical clustering methods included in the Lance & Williams algorithm (Subsection 8.5.9], except complete linkage. Additional constraints may be added to the algorithm, for example to limit the size or composition of any group (Gordon, 1996c). K-means partitioning algorithms (Section 8.8) may also be constrained by the contiguity matrix shown in Fig. 13.23.

Spatially constrained clustering is useful in a variety of situations. Here are some examples.

• In many studies, there are compelling reasons to force the clusters to be composed of contiguous sites; for instance, when delineating ecological regions, political voting units, or resource distribution networks.

• One may wish to relate the results of clustering to geographically-located potential causal factors that are known to be spatially autocorrelated, e.g. geological data.

• One may wish to cluster sites based upon physical variables, using a constraint of spatial contiguity, in order to design a stratified biological sampling program to study community composition.

• To test the hypothesis that neighbouring sites are ecologically similar, one may compare unconstrained and constrained clustering solutions using the modified Rand index (Subsection 8.11.2). De Soete *et al.* (1987) give other examples where such comparisons may help test hypotheses in the fields of molecular evolution, psycholinguistics, cognitive psychology, and evolution of languages.

• Constrained solutions are less variable than unconstrained clustering results, which may differ in major ways among clustering methods. Indeed, the constraint of spatial contiguity reduces the number of possible solutions and forces different clustering algorithms to converge onto largely similar clusters (Legendre *et al.*, 1985).

Constrained clustering may also be used for three-dimensional or spatio-temporal sampling designs (e.g. Planes *et al.*, 1993). As long as the three-dimensional or spatio-temporal contiguity of the observations can be accurately described as a file of edges as in Figs. 13.20 and 13.21, constrained clustering programs have no difficulty in computing the solution; the only difficulty is the representation of the results as three-dimensional or spatio-temporal maps. Higher-dimensional extensions of the geometric connecting schemes presented in Subsection 1 are available if required.

Legendre (1987b) suggested a way of introducing spatial proximity into clustering algorithms which is less stringent than the methods described above. The method consists in weighting the values in the ecological similarity or distance matrix by some function of the geographic distances among points, before clustering. The idea was taken up by Bourgault *et al.* (1992) who proposed to use a multivariate variogram or covariogram as spatial weighting function prior to clustering. Large ecological distances between sites that are close in space are downweighted to some extent by this procedure. It is then easier for clustering algorithms to incorporate somewhat diverging sites into neighbourhood clusters. Oliver & Webster (1989) suggested to use a univariate variogram for the same purpose.

Constrained classification methods have recently been reviewed by Gordon (1996c). Formal aspects have been discussed by Ferligoj & Batagelj (1982, 1983). Algorithms have been surveyed by Murtagh (1985). Generalized forms of constrained clustering have been described by De Soete *et al.* (1987).

Numerical example. An artificial set of 16 sites was constructed to represent staggered-row sampling of a distribution with two peaks. From the geographic positions of the sites, a Delaunay triangulation (35 edges) was computed (Fig. 13.24a). A single variable was attributed to the sites. For three groups, the unconstrained *K*-means solution has a sum of within-group sums-of-squares $E_K = 53$ (Fig. 13.24b). The constrained *K*-means solution, for three groups,

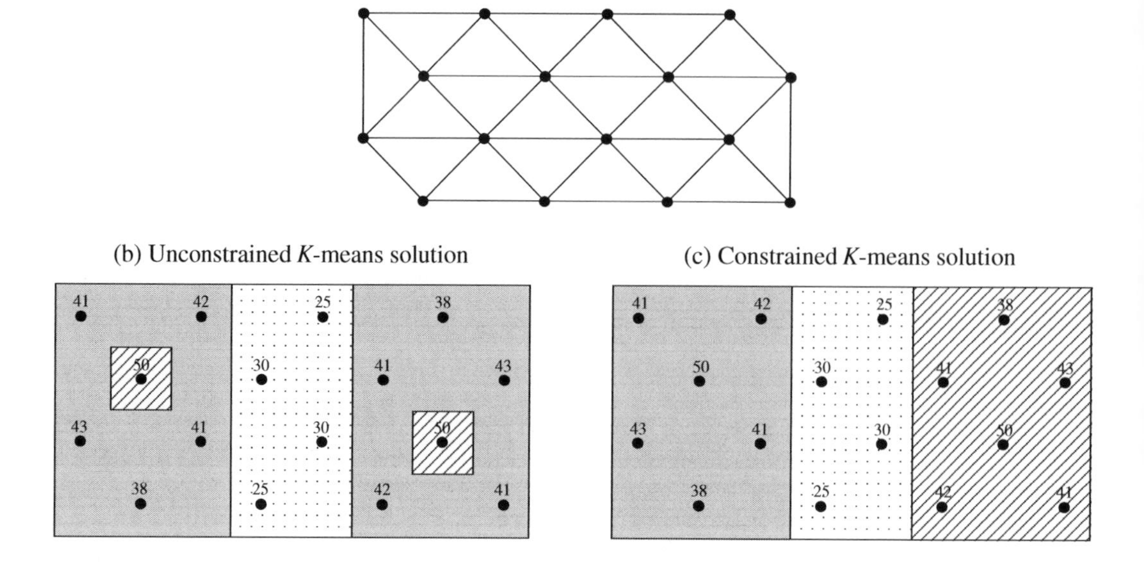

Figure 13.24 Numerical example showing the difference between the unconstrained (b) and constrained (c) clustering solutions. (a) Delaunay triangulation with 35 edges; they were used as constraint in (c). The values of the artificial variable are given in panels (b) and (c); the three groups obtained by unconstrained and constrained *K*-means are identified by shadings.

has a value $E_K = 188$ (Fig. 13.24c) which is higher than that of the unconstrained solution, for reasons explained above. The two partitions are interesting in different ways. The unconstrained solution identifies sites with similar values, whereas the constrained solution brings out the two peaks plus a region of lower values forming a valley between the peaks.

Spatially constrained clustering has been applied to a variety of ecological situations. Some applications to two-dimensional map data are: Legendre & Legendre (1984c), Legendre & Fortin (1989), Legendre *et al.* (1989), and Lapointe & Legendre (1994). Applications to transect data and paleoecology (stratigraphic data) have been listed in Subsection 12.6.4.

3 — Ecological boundaries

Detection of boundaries is a complementary problem to the detection of homogeneous regions of space. Boundaries appear on maps as a by-product of constrained clustering, for instance. Most methods of clustering delineate groups even in gradient situations; a boundary between groups does not have to correspond to a sharp discontinuity in the data. In any case, boundaries detected by clustering may be partly interpolated. Other methods have been developed that focus on boundary elements; they do not pretend to completely isolate regions of space, however.

For univariate or multivariate transect data, boundaries may be detected using one of the methods described in Section 12.6 (especially Subsection 12.6.3). Detection of boundaries of various sorts on maps is more complex. This is a well-studied topic in the field of image analysis; it has been reviewed by Davis (1975), Peli & Malah (1982) and Huang & Tseng (1988); see also Hobbs & Mooney (1990). The present Section briefly summarizes the efforts made to detect boundaries in ecological data sets, using a technique called *wombling*, and to statistically assess their significance.

Wombling is a technique for detecting zones of rapid spatial change in a set of regionalized variables. It was developed by Womble (1951) and Barbujani *et al.* (1989) for gene frequencies and morphological measurements, and refined by Fortin and co-authors (Oden *et al.*, 1993; Fortin, 1994, 1997; Fortin & Drapeau, 1995; Fortin *et al.*, 1996) with emphasis on ecological data. The original form of wombling (*lattice wombling*) could only be applied to quantitative variables observed at sites forming a regular, rectangular grid of points. Recent developments include *categorical wombling* for qualitative variables (Oden *et al.*, 1993) and *triangulation wombling* for sites linked by a Delaunay triangulation which do not necessarily correspond to a regular sampling grid (Fortin, 1994). The latter is a common situation in ecology.

A boundary is defined as a set of spatially adjacent locations where the variable under study shows high rates of change (Fortin, 1994). Triangulation wombling proceeds as follows:

• Link the observed sites by a Delaunay triangulation (Subsection 13.3.1).

• Consider a quantitative variable measured at three sites $Ø_i$ forming a Delaunay triangle. Each site has geographic coordinates (X_i, Y_i) and an observed value y_i. The plane to be fitted to these points is a linear function $y = f(X, Y) = b_0 + b_1X + b_2Y$ whose parameters are calculated by matrix inversion (Section 2.8):

$$\begin{bmatrix} b_0 \\ b_1 \\ b_2 \end{bmatrix} = \begin{bmatrix} 1 & X_1 & Y_1 \\ 1 & X_2 & Y_2 \\ 1 & X_3 & Y_3 \end{bmatrix}^{-1} \begin{bmatrix} y_1 \\ y_2 \\ y_3 \end{bmatrix}$$

• Find the direction of maximum slope of the triangle. The slope varies with the direction considered (arrows in Fig. 13.25a). Using the *b* coefficients calculated above, the *maximum slope* of the triangle is:

$$m = \sqrt{\left[\frac{\partial f(X, Y)}{\partial X}\right]^2 + \left[\frac{\partial f(X, Y)}{\partial Y}\right]^2} = \sqrt{b_1^2 + b_2^2} \qquad \textbf{(13.24)}$$

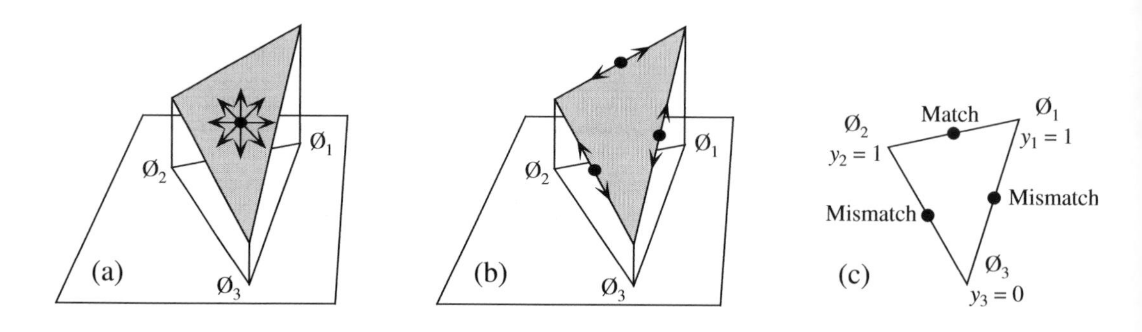

Figure 13.25 Boundary detection methods. Consider three sites $Ø_i$, with coordinates (X_i, Y_i), forming a Delaunay triangle. (a, b) Variable y measured at sites $Ø_i$ is quantitative (values shown as heights). (a) Find the direction of maximum slope of the triangle; allocate this slope value to the triangle centroid (dot). (b) Compute slopes along the edges connecting adjacent sites; allocate the values to the edge centres (dots). (c) For a qualitative variable y (with 2 states in this example), comparison between adjacent sites is made in terms of matches (0-0 or 1-1) or mismatches (0-1); allocate the matches and mismatches to the edge centres (dots).

Allocate this value of slope (m) to the centroid of the triangle, which is the point with coordinates:

$$[X, Y]_{centroid} = \left[\frac{X_1 + X_2 + X_3}{3}, \frac{Y_1 + Y_2 + Y_3}{3}\right] \qquad \textbf{(13.25)}$$

• If several variables are considered (i.e. several species), calculate the mean slope ($\bar{m}$) of the variables at the centroid of each Delaunay triangle.

• Create an ordered list of the slope values. Starting at the top of the list (highest slopes), mark the corresponding triangle centroids on the map; they become *boundary elements*. Going down the list, mark a pre-determined proportion of the slopes (e.g. 10%), or go down to a preselected value of slope. Other strategies are possible, e.g. going down the list to the value of the mean plus one or two standard deviations.

Boundary

• A *boundary* is defined as any set of one or more contiguous boundary elements.

An alternative would be to compute the slopes of the edges between adjacent sites (Fig. 13.25b). For univariate data, the rate of change would simply be the absolute value of the difference between values at sites $Ø_h$ and $Ø_i$: $|y_h - y_i|$. For multivariate data, any of the distance functions of Chapter 7 could be used. The disadvantage of this method is that slopes calculated along the edges of the Delaunay triangle do not have the same value as the *maximum* slope of the triangle, which is computed by eq. 13.24. To alleviate this problem, Dufrêne & Legendre (1991) calculated multivariate distances in four directions between pixels of a map; for each pixel, they used the largest of the distances to delineate boundaries.

Computation of statistics along the edges between adjacent sites is the option used in categorical wombling which is appropriate for species presence-absence data. The basic statistic is to record a match or a mismatch between adjacent observed sites (Fig. 13.25c). For multivariate qualitative data, one may count both the positive and negative matches and embed this number into one of the symmetrical binary coefficients of Subsection 7.3.1 (e.g. the simple matching coefficient); for species presence-absence data, one may count the positive matches only and embed this number into one of the asymmetrical binary coefficients of Subsection 7.3.2 (e.g. the Jaccard coefficient).

Tests of significance, based on permutations (Section 1.2), have been proposed by Fortin and co-authors (Oden *et al.*, 1993; Fortin, 1994, 1997; Fortin & Drapeau, 1995; Fortin *et al.*, 1996) to answer the following questions:

- Are the boundaries found by this analysis similar to random boundaries in terms of the number of separate boundaries, their maximum or mean lengths, or other boundary or graph-theoretic statistics?

- Are the boundaries found by wombling the same as borders stated by hypothesis, or found by clustering methods, or obtained using different data for the same locations?

These papers also present applications of the method to real and simulated data. A computer program for wombling is commercially available (GBAS, Table 13.4).

4 — Dispersal

Individuals, populations, and communities may cross ecological boundaries; these crossings occur on different time scales. The routes taken by species when they invade a territory after a perturbation event (long-term: e.g. glaciation; short term: e.g. pollution) is a question of interest in biogeographic analysis. Dispersal routes may be easier to identify if, as a first step in the analysis, one delineates regions that are largely homogeneous in species composition. Regions may be delimited using prior hypotheses, by unconstrained or constrained cluster analysis, or using boundary detection methods.

Legendre & Legendre (1984c) have developed coefficients to measure the likelihood of species dispersal between geographically contiguous regions, for species presence-absence as well as abundance data. The assumptions of these coefficients are that the species arrived by migration and that past dispersal has left traces in present-day distributions. For presence-absence data, adjacent regions $\mathbf{x}_1$ and $\mathbf{x}_2$ can be compared using the same quantities a, b, and c as in the similarity coefficients of Subsection 7.3.1 and 7.3.2: a is the number of species that two regions have in common; b is the number of species found in $\mathbf{x}_1$ but not in $\mathbf{x}_2$; c is the number of species found in $\mathbf{x}_2$ but not in $\mathbf{x}_1$. The *combination* of the following indications is evidence for species dispersal from region $\mathbf{x}_1$ to $\mathbf{x}_2$:

- The number of species common to the two zones is high, i.e. a is large.

- b is substantially larger than c; c larger than b would support the hypothesis of dispersal from $\mathbf{x}_2$ to $\mathbf{x}_1$.

The basic form of the *coefficient of species dispersal direction* (*DD*) is thus $a(b-c)$. To make the values of the coefficient comparable for faunas of different richness, each term is standardized by dividing it by the richness of the fauna or flora of the two regions combined:

$$DD_1(\mathbf{x}_1 \rightarrow \mathbf{x}_2) = \frac{a}{(a+b+c)} \frac{(b-c)}{(a+b+c)} \qquad \textbf{(13.26)}$$

This coefficient is signed. It measures the likelihood that species have dispersed from $\mathbf{x}_1$ to $\mathbf{x}_2$. A negative sign indicates that, if dispersal occurred, species have migrated from $\mathbf{x}_2$ to $\mathbf{x}_1$ instead.

The asymmetric portion of this coefficient may be tested for significance using a McNemar test. Under the null hypothesis of no asymmetry (H_0: $b = c$), the test statistic

$$X_P^2 = \frac{(b-c)^2}{(b+c)}$$

is distributed as χ^2 with one degree of freedom. The test may be one-tailed if one has specific hypotheses about the direction of dispersal; otherwise, use a two-tailed test. The log-linear form of this statistic is:

$$X_W^2 = 2\{b\ \ln b + c\ \ln c - (b+c)\ln[(b+c)/2]\}$$

If any of the values b or c is 0, the corresponding term $(x \ln x)$ is 0 since $\lim_{x \to 0} x \ln x = 0$.

The first portion of DD_1 is easily recognized as the Jaccard coefficient of similarity (eq. 7.10). One may prefer to give double weight to the common species, as in the coefficient of Sørensen (eq. 7.11):

$$DD_2(\mathbf{x}_1 \rightarrow \mathbf{x}_2) = \frac{2a}{(2a+b+c)} \frac{(b-c)}{(a+b+c)} \qquad \textbf{(13.27)}$$

Two other forms of the coefficient use species abundance data instead of presence-absence:

$$DD_3(\mathbf{x}_1 \rightarrow \mathbf{x}_2) = \frac{W(A-B)}{(A+B-W)^2} \quad \textbf{(13.28)}$$

and

$$DD_4(\mathbf{x}_1 \rightarrow \mathbf{x}_2) = \frac{2W}{(A+B)} \frac{(A-B)}{(A+B-W)} \quad \textbf{(13.29)}$$

where *W*, *A*, and *B* are as in the Steinhaus similarity coefficient (eq. 7.24). Coefficient DD_4 gives double weight to the abundances of the species in common and is thus the counterpart of DD_2, whereas DD_3 gives these species single weight, as in DD_1. These two coefficients take the following indications as evidence for dispersal from $\mathbf{x}_1$ to $\mathbf{x}_2$:

• The number of species common to the two zones and their abundances are high, i.e. *W* is large.

• *A* is substantially larger than *B*; *B* larger than *A* would produce a negative coefficient, supporting the hypothesis of dispersal from $\mathbf{x}_2$ to $\mathbf{x}_1$.

Legendre & Legendre (1984c) used the above coefficients and tests of significance to reconstruct plausible routes taken by freshwater fishes to reinvade the Québec peninsula after the last glaciation. Borcard *et al.* (1995) used the same method in a finer-scale study, showing possible patterns of migration of Oribatid mites between zones of an exploited peat bog in the Swiss Jura.

13.4 Unconstrained and constrained ordination maps

Section 13.3 has shown how clustering methods may help produce maps for multivariate data; these maps consist of discontinuous areas. For continuous variables, however, maps can only be produced for single variables, using the techniques of Section 13.2. The present Section shows how continuously-varying maps can be produced for multivariate data sets, through various types of ordination methods. The relationship between univariate or multivariate structure functions (Section 13.1) and maps has been stressed in the introductory paragraph of Section 13.2.

The simplest method consists in analysing a data table, using one of the ordination methods of Chapter 9, and map the first few ordination axes. For example:

• Decompose the variation of a (sites × species) presence-absence or abundance table into successive ordination axes, using principal coordinate or correspondence analysis.

• Consider the ordination of sites along the first axis. This axis is a new, synthetic quantitative variable for the variation among sites. Associate it to a table of the (X, Y) geographic coordinates of the sites. Produce a map using one of the methods in Section 13.2. An example of such a map is given in Fig. 9.19 for correspondence analysis axis I of a vegetation data table.

• Repeat the operation, producing maps for ordination axes II, III, etc. as long as interesting or significant spatial variation can be detected. The R^2 of trend surface analysis (Subsection 13.2.1) may be used as criterion for deciding which of the ordination axes should be mapped.

Simple ordination analysis leaves it to chance to find spatially-structured components of variation. One may decide instead to look directly for such components, by forcing the analysis to bring out axes of variation that are related to the X and Y coordinates, or combinations of X and Y into a spatial polynomial equation. The spatial polynomial is constructed as in Subsection 13.2.1. Ordination analysis of a species data table, constrained to be related to a spatial polynomial, can be done by canonical analysis (Chapter 11), as suggested by Legendre (1990). Canonical analysis then becomes an extension to multivariate data tables of the method of trend surface analysis. The method will be described with the help of a numerical example. Another example (vegetation data) is found in Legendre (1990).

Numerical example. Data from the Thau lagoon are used again here (Tables 10.5 and 13.2). To facilitate mapping, the X and Y geographic coordinates of the sampling sites were rotated by principal component analysis (PCA using the covariance matrix; the eigenvectors were normalized to lengths 1); Table 13.2 shows the rotated coordinates. A third-degree spatial polynomial of these new X and Y coordinates was created (Subsection 13.2.1) and subjected to the "forward selection of environmental variables" procedure of program CANOCO; in this procedure, variables from the matrix containing the spatial monomials were selected one by one, in a stepwise manner, for their capacity to significantly contribute to the explanation of the "species data" (here, the Bna and Ma bacterial variables). The following five terms of the spatial polynomial were retained by the selection procedure: X^2, X^3, X^2Y, XY^2, and Y^3.

Redundancy analysis (Section 11.1) produced two canonical axes ($\lambda_1 = 0.622$, $\lambda_2 = 0.111$), as expected from Table 11.1 for two dependent variables (Bna, Ma). The canonical relationship accounted for 73.4% of the variation in the bacterial data; it was globally significant (p = 0.001 after 999 permutations of residuals under the full model, using CANOCO); so was the first canonical eigenvalue (62% of the variance in the bacterial variables; p = 0.001). 81% of the variance of Ma was expressed along axis I, but only 7% of the variance of Bna. The second canonical eigenvalue (11% of the variance of the two bacterial variables) did not reach significance at $\alpha = 0.05$ (p = 0.076) although 42% of the variance of Bna was expressed on this axis. There are two non-canonical axes representing the non-spatially-structured variation of the response variables; they represent 16% and 10% of the variation of matrix **Y**, respectively. Canonical axis I differs from the first principal component of matrix **Y**, which would express the variation in the response variables (Bna, Ma) without the constraint of being a linear combination of the spatial variables.

For axis I, the "site scores" and the "fitted site scores" ("site scores that are linear combinations of the environmental variables" in program CANOCO) were mapped (Fig. 13.26). Maps were obtained by kriging (Subsection 13.2.2) using program OKB2D of the GSLIB

Table 13.2 Data from Table 10.5. There are two dependent variables (Bna and Ma, forming matrix **Y**) and three environmental variables (NH_4, phaeopigments, and bacterial production, forming matrix **X**). Five spatial variables (X^2, X^3, X^2Y, XY^2, and Y^3, included in matrix **W**) were derived from the X and Y coordinates, reported in the Table, that were obtained by PCA rotation of the geographic coordinates of Table 10.5. The variables are further described in Numerical example 1 of Subsection 10.3.5.

Station No.	Bna y_1	Ma y_2	NH_4 x_1	Phaeo. *a* x_2	Prod. x_3	X after PCA rotation	Y
1	4.615	10.003	0.307	0.184	0.274	–9.4173	–1.2516
2	5.226	9.999	0.207	0.212	0.213	–7.1865	–1.0985
3	5.081	9.636	0.140	0.229	0.134	–5.8174	–1.4528
4	5.278	8.331	1.371	0.287	0.177	–6.8322	0.2706
5	5.756	8.929	1.447	0.242	0.091	–4.6014	0.4238
6	5.328	8.839	0.668	0.531	0.272	–4.2471	1.7929
7	4.263	7.784	0.300	0.948	0.460	–1.8632	–0.2848
8	5.442	8.023	0.329	1.389	0.253	–0.4940	–0.6391
9	5.328	8.294	0.207	0.765	0.235	0.8751	–0.9934
10	4.663	7.883	0.223	0.737	0.362	–0.1398	0.7300
11	6.775	9.741	0.788	0.454	0.824	–1.1546	2.4534
12	5.442	8.657	1.112	0.395	0.419	0.2145	2.0992
13	5.421	8.117	1.273	0.247	0.398	4.9824	–2.0562
14	5.602	8.117	0.956	0.449	0.172	3.9676	–0.3328
15	5.442	8.487	0.708	0.457	0.141	3.4602	0.5289
16	5.303	7.955	0.637	0.386	0.360	6.3515	–2.4105
17	5.602	10.545	0.519	0.481	0.261	5.8441	–1.5488
18	5.505	9.687	0.247	0.468	0.450	4.8293	0.1746
19	6.019	8.700	1.664	0.321	0.287	4.6762	2.4054
20	5.464	10.240	0.182	0.380	0.510	6.5527	1.1894

library (Deutsch & Journel, 1992); all-directional spherical variogram models were fitted to the empirical variograms prior to kriging (Subsection 13.1.3). Figure 13.26a shows the "site scores" expressing the point-by-point variation of the original data (matrix **Y** of bacterial variables) projected onto axis I (eq. 11.12). Figure 13.26b maps the "fitted site scores" which are the fitted values of the multiple regressions (matrix $\hat{\mathbf{Y}}$) projected onto canonical axis I (eq. 11.13). The data projected onto axis I in Fig. 13.26a are with residuals ($\mathbf{Y} = \hat{\mathbf{Y}} + \mathbf{Y}_{res}$; Fig. 11.2) whereas the residuals are excluded in Fig. 13.26b. The trend surface equation that produced the "fitted site scores" for the 20 sampling stations (Fig. 13.26b) is written out by program CANOCO under the heading "Regression/canonical coefficients for standardized variables":

$$\widehat{\text{Axis I}} = 1.0526X^2 + 1.1881X^3 - 0.5225X^2Y - 0.7674XY^2 + 0.7167Y^3$$

The spatial variables were standardized before computing this equation.

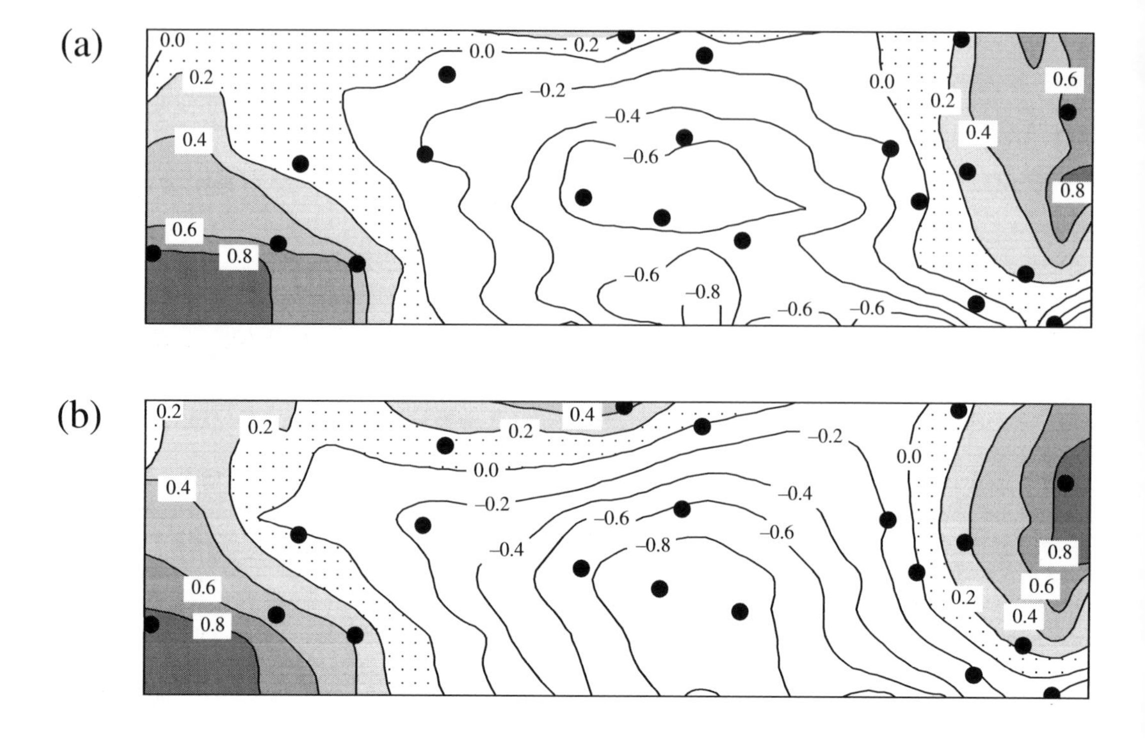

Figure 13.26 Map of the first canonical axis of the Thau lagoon bacterial variables constrained by the spatial polynomial. (a) Site scores. (b) Fitted site scores. Dots represent the 20 sampling sites. The north-south direction is nearly parallel to the vertical axis of the maps; compare the point positions with Fig. 13.13.

Interpretation of the maps is rather simple in this example: examination of Table 13.2 shows that the sites with the highest scores along canonical axis I (i.e. sites 1-3, 11, 17-20; grey areas in Fig. 13.26) possessed the highest concentrations of aerobic heterotrophic bacteria growing on marine agar (variable Ma); in the canonical analysis output, Ma was also identified as the variable dominating canonical axis I.

Thioulouse *et al.* (1995) proposed a different approach to mapping, which combines connection networks, decomposition of the variation into local and global components, eigenvalue decomposition, and mapping. The neighbouring relationships among sites are represented by some appropriate connection network (e.g. Delaunay triangulation for a homogeneous two-dimensional sampling area, or neighbouring relationships for sites along a river network) which is translated into a contiguity matrix **M** (Fig. 13.23). **M** is standardized to **P** by division by the total number of pairs of neighbours. A diagonal matrix **D** describes the degree of connectedness of the sites. Using matrices **P** and **D**, the authors define principal component and correspondence analysis for the total, local, or global components of variation; each fraction is decomposed into orthogonal axes, which may be mapped to facilitate interpretation. The paper presents applications to simulated and real ecological data (bird survey).

13.5 Causal modelling: partial canonical analysis

The significance of spatial heterogeneity for the functioning of ecosystems was discussed in Section 1.1. Models of ecosystem processes may fall short of being optimal unless they include the spatial organization of the players — populations and communities — among the predictor variables. Although this type of modelling is still in its infancy, two main approaches have been proposed. The first one, described in the present Section, consists in modelling the spatial variation of the variables of interest as a linear combination of the environmental variables and the geographic coordinates of the sites. This approach makes use of methods discussed in previous Chapters and Sections: partial regression analysis (univariate: Subsection 10.3.5) and partial canonical analysis (multivariate: Section 11.3) on the one hand; trend surface analysis (univariate: Subsection 13.2.1) and constrained ordination mapping (multivariate: Section 13.4) on the other. In the second approach (Section 13.6), the spatial structure is conveniently represented by a matrix of geographic distances among sites.

In both Sections, the analysis considers three data sets, as in partial regression modelling: **Y** contains the response variables; **X** is the set of explanatory environmental variables; **W** is the set of explanatory spatial variables (based on the spatial coordinates of the sampling sites, transformed in various ways as described below). There are two motivations for analysing data in this way:

• Spatial structures are a major source of false correlations, which are not indicative of causal relationships. Spatial autocorrelation may cause correlations to appear between autocorrelated variables, due to spatial structures that are generated by the autocorrelation in the data; examples of such structures are the "false gradients" described in Subsection 13.1.2. The question is the following: is there a significant amount of correlation between the response and explanatory variables, besides some common spatial structure which may be non-causal? The interest here is in fraction [a] of Figs. 10.10 and 13.27, which corresponds to the "explanation" (in the sense of Subsection 10.2.1) of the response data (**Y**) by the explanatory environmental variables (**X**) after the common spatial structure (fraction [b]) has been controlled for. If the analysis of the spatial structure of **Y** is also of interest, it should be carried out independently of that of **X**. If fraction [b] is large, the hypothesis that it indicates false correlations between autocorrelated variables in matrices **Y** and **X** is supported if new data, obtained by sampling at a different spatial scale, fail to produce a large fraction [b]. Nested sampling designs, as used for instance in Ecological application 13.1b, are a way of obtaining data sets for modelling at more than one spatial scale.

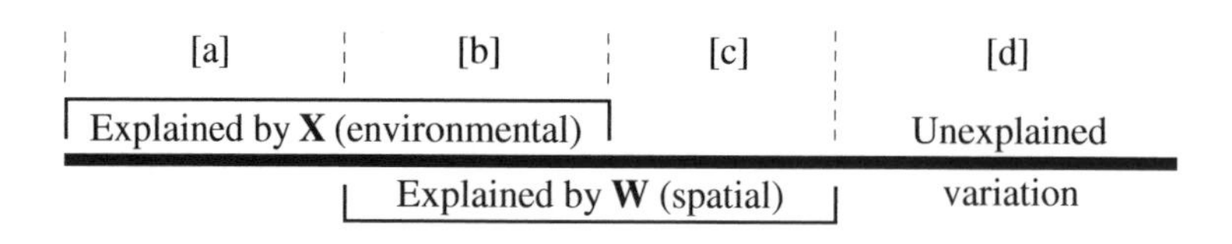

Figure 13.27 Partition of the variation of a response matrix **Y** between environmental (matrix **X**) and spatial (matrix **W**) explanatory variables. The length of the horizontal line corresponds to 100% of the variation in **Y**. Compare to Fig. 10.10. (Adapted from Borcard *et al.*, 1992, and Legendre, 1993).

• If both the spatial and non-spatial structures of the explanatory variables are assumed to be causal to the spatial variation of **Y**, fraction [a + b] (instead of fraction [a] alone, as above) empirically measures the support to the hypothesis of causal relationship between the response and explanatory variables. One may or may not be interested in breaking it down into a purely environmental fraction ([a] in Fig. 10.10 and 13.27) and a spatially-structured environmental effect (fraction [b]). One may be interested, however, in determining whether the response data (**Y**) have some spatial structure of their own (fraction [c]) which is not explained by the explanatory variables in the model. If such a structure is found, reasons for its existence are worth investigating. It may be caused by environmental variables that have not been included in the analysis, by historical events, or by population or community dynamic processes that may be masked by some dominant environmental effect (Borcard & Legendre, 1994). In this approach, any spatial structure identified in the response data set is considered an indication of the presence of some process generating it. Mapping fraction [c] of the variation may help generate hypotheses about the processes responsible for the observed residual spatial pattern.

It is important to note that the approaches described above are essentially correlative, like regression analysis. They differ from the analysis of variance, which estimates the variation associated with well-defined effects in structured sampling or manipulative experiments. In the initial stages of ecological research, correlative methods are routinely used to sort out hypotheses centring on broad correlative patterns among groups of variables, before specific hypotheses can be experimentally tested. In particular, the analyses presented in this Section allow researchers to consider different groups of explanatory variables (environmental, spatial, or temporal) and examine their capacity to explain patterns in the multivariate response variables (species or others) that are of interest in a study; they further allow one to measure the degree of overlap that exists among these groups of explanatory variables with regard to that capacity (Anderson & Gribble, 1998). The correlations brought out by the analyses are only interpretable insofar as hypotheses can be formulated about the processes that may have generated the observed patterns. This approach is related to regression (Section 10.3) and path analysis (Section 10.4), in which a large number of plausible relationships may be hypothesized and sorted out by statistical analysis.

1 — Partitioning method

A method for implementing the approach described above was proposed by Borcard *et al.* (1992). It consists in partitioning the variation of a response matrix **Y** between the environmental (**X**) and spatial (**W**) matrices of explanatory variables, in the way described in Subsection 10.3.5. Partial RDA or CCA are used if **Y** contains several response variables, or partial linear regression if there is a single variable in **Y** and a linear relationship is assumed between **Y** and the variables in **X** and **W**. Calculation of the fractions of variation ([a] to [c] in Fig. 10.10 and 13.27) is described in Subsection 10.3.5. Estimation of the vectors corresponding to the fractions of variation is also described in that Subsection; these vectors may be used to produce maps.

If matrix **Y** is multivariate, the variation in **Y** is decomposed using partial canonical analysis (Section 11.3). Borcard *et al.* (1992) applied this decomposition to three ecological data sets: soil mites, forest vegetation, and aquatic heterotrophic bacteria. The decomposition may be carried out in many different ways, which are equivalent, using three of the five following analyses:

(1) A simple RDA (redundancy analysis: Section 11.1) or CCA (canonical correspondence analysis: Section 11.2) of the matrix of response variables **Y**, constrained by matrix **X**, extracts fraction [a + b].

(2) A simple RDA or CCA of matrix **Y**, constrained by matrix **W**, extracts fraction [b + c].

(3) A simple RDA or CCA of matrix **Y**, constrained by a matrix combining the variables of matrices **X** and **W**, extracts fraction [a + b + c]. The residual variation corresponds to fraction [d].

(4) A partial RDA or CCA (Section 11.3) of matrix **Y**, constrained by matrix **X** and using **W** as the matrix of covariates, extracts fraction [a].

(5) A partial RDA or CCA of matrix **Y**, constrained by matrix **W** and using **X** as the matrix of covariates, extracts fraction [c].

For each of these analyses, the sum of all canonical eigenvalues, divided by the sum of all (unconstrained) eigenvalues, gives the corresponding fraction of the variation explained by the analysis. The sum of the (unconstrained) eigenvalues in **Y** is the sum of all the eigenvalues of a regular PCA (principal component analysis, Section 9.1) of matrix **Y** if RDA is used for the decomposition, or the sum of all the eigenvalues of a regular CA (correspondence analysis, Section 9.4) of matrix **Y** if the decomposition is obtained by CCA. The sum of all canonical eigenvalues of an analysis can be tested for significance (Subsection 11.3.2). Each analysis decomposes the canonical variation into canonical axes, and each axis may be tested for significance. For each significant axis, either the "site scores" or the "fitted site scores" may be used to produce maps as in Fig. 13.26.

Analyses (4) and (5) must be carried out only if one is interested in estimating the vectors of fitted values corresponding to fractions [a] and [c], so as to map them for instance. Otherwise, fraction [a] can be obtained from the results of analyses (2) and (3) which are simpler to carry out. Indeed, because the fractions of variation are additive (Subsection 10.3.5), it follows that:

$$[a] = [a + b + c] - [b + c]$$

where [a] represents the fraction of variation of **Y**, explained by the environmental variables, which is not spatially structured. In the same way, fraction [c] may be obtained from analyses (1) and (3):

$$[c] = [a + b + c] - [a + b]$$

where [c] is the spatially-structured fraction of variation of **Y** which is not explained by the environmental variables.

[b] is the fraction of variation of **Y**, explained by the environmental variables, which is spatially structured. It may be equally attributed to the environmental variables **X** or the spatial variables **W**. It can only be obtained by subtraction:

$$[b] = [a + b] + [b + c] - [a + b + c]$$

or

$$[b] = [a + b] - [a]$$

or

$$[b] = [b + c] - [c]$$

or

$$[b] = [a + b + c] - [a] - [c]$$

No combination of simple or partial canonical analyses can produce fraction [b]; it is not a fitted variance component, but the difference between the variances explained by two models which have no structural relationship to each other. As a result, the measured quantity cannot be described as variance explained by a specific linear model of some predictors. Interpretation of fraction [b] remains interesting, however, as explained in Subsection 10.3.5. The estimation of vectors of site scores corresponding approximately to fraction [b] is a difficult problem; some solutions have been proposed by Méot *et al.* (1998).

Fraction [d] is the residual variation of analysis (3). It can be calculated as:

$$[d] = 1 - [a + b + c]$$

The amount of variation associated with fraction [d] is the residual sum of squares (RSS) in the denominator of the F statistics (eq. 11.19) used to test the overall significance of the simple and partial canonical analyses described above, or the significance of particular canonical axes (eq. 11.20).

The broad-scale spatial variation in matrix **Y** may be modelled by a polynomial of the geographic coordinates of the sampling sites, as in trend surface analysis

(Subsection 13.2.1) and constrained ordination mapping (Section 13.4); in that case, component [a] of the variation corresponds to the more regional or local variation, as well as the non-spatially-structured environmental variation. Alternatively, more local spatial variation can be captured using a nearest-neighbour autocorrelation model (Legendre and Borcard, 1994); an example is given by He *et al.* (1994), who applied this method to explain the spatial variation of tree density, species richness and Shannon diversity in a tropical rain forest.

Numerical example. The data from Table 13.2 (Thau lagoon) are reanalysed here. In the numerical example of Section 13.4, the variable selection procedure retained the following terms of the spatial polynomial: X^2, X^3, X^2Y, XY^2, and Y^3; the same terms are used in the present example. The five analyses described above were carried out to obtain the fractions of variation:

Fractions of variation	Proportion of variation of Y	Probability (999 perm.)	Canonical λ_1	Probability (999 perm.)
[a + b]	0.450	0.005*	0.359	0.025*
[b + c]	0.734	0.001*	0.622	0.001*
[a + b + c]	0.784	0.001*	0.632	0.001*
[a]	0.051	0.549	0.042	0.561
[b]	0.399	-----	-----	-----
[c]	0.334	0.011*	0.304	0.004*
[d]	0.216	-----	-----	-----
[a + b + c + d]	1.0000			

Significant fractions are identified by asterisks ($\alpha = 0.05$). Eigenvalues of the first canonical axis (canonical λ_1) are reported as fractions of the total variance in **Y**. The second canonical axis was never significant; since each analysis only produced two canonical eigenvalues in this example, the portions of variation corresponding to λ_1 are the differences between columns 2 and 4 from the left. Fraction [b] is not an independently-calculated component of the variation; hence, it cannot be tested for significance nor decomposed into canonical axes (see Méot *et al.*, 1998, for alternative solutions).

The analysis decomposed the total explained variation [a + b + c] into a significant environmental component [a + b] and a significant component [c] which estimates the spatially-structured variation of **Y** not explained by the environmental variables. The table shows that [a + b], which is the variation of **Y** explained by the environmental variables, is mostly spatially structured since [b] represents 89% of [a + b] and [a] is not significant.

Figure 13.28 shows maps of the "fitted site scores" of the first canonical axis of fraction [a + b + c] and of its two components, [a + b] and [c]. These maps were obtained by kriging (Subsection 13.2.2) using program OKB2D of the GSLIB library (Deutsch & Journel, 1992); all-directional spherical variogram models were fitted to the empirical variograms prior to kriging (Subsection 13.1.3). While the proportions of variation of [a + b] and [c] add up to that of [a + b + c] (0.450 + 0.334 = 0.784), this is not the case for the proportions of variation represented by the first canonical axes: $\lambda_{1[a+b]} + \lambda_{1[c]} \neq \lambda_{1[a+b+c]}$. This is because the partition

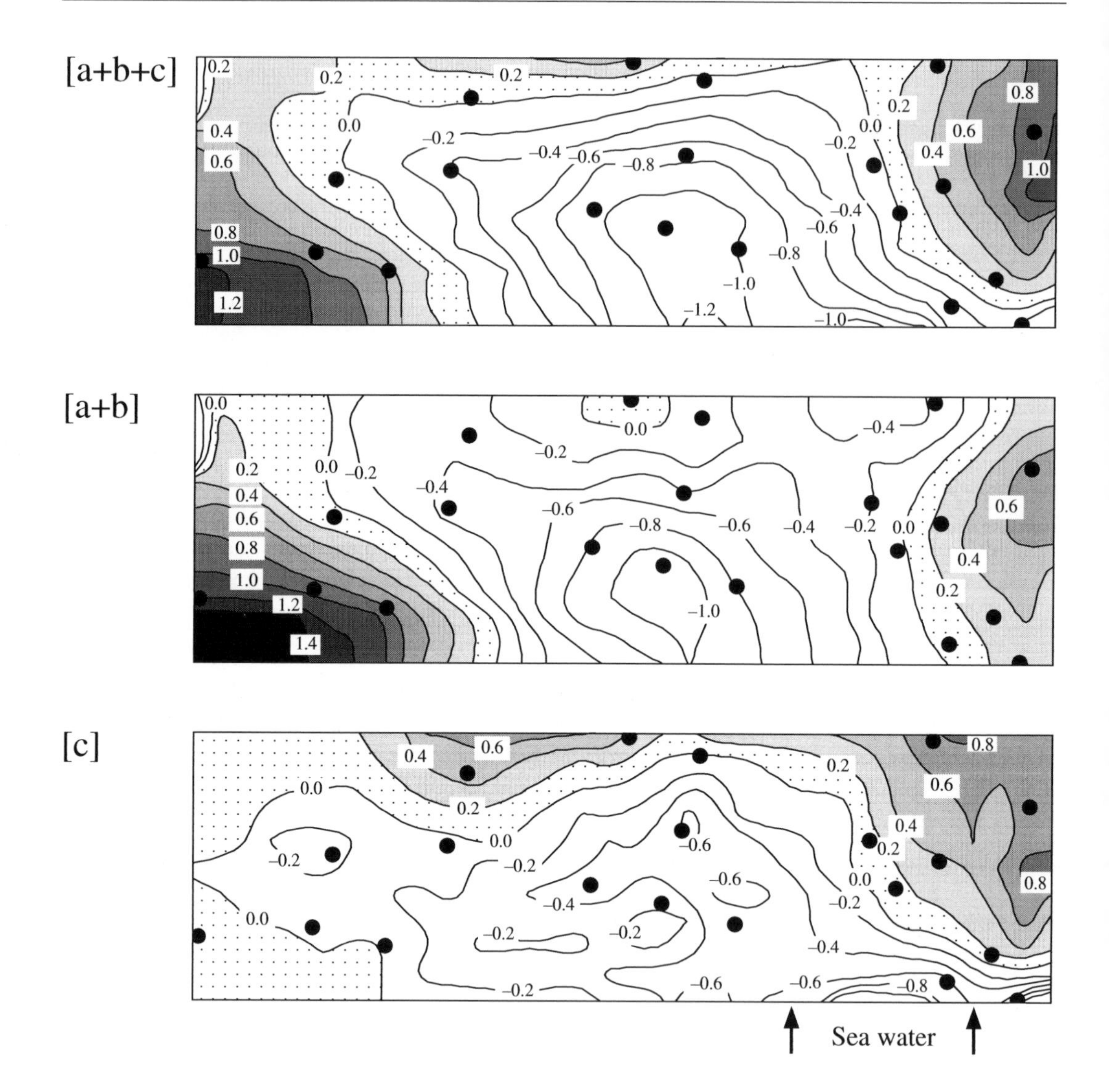

Figure 13.28 Bacterial variables: map of the "fitted site scores" of the first canonical axes of three fractions of the variation: top [a + b + c], middle [a + b], bottom [c]. Dots represent the 20 sampling sites. North is nearly parallel to the vertical axis of the maps. Compare with Fig. 13.26b which represents fraction [b + c]. Arrows at the base of map [c], labelled "sea water", indicate the positions of connections of the Thau lagoon with the Mediterranean Sea.

of fraction [a + b + c] into canonical axes is done independently of the partitions of [a + b] or [c]. As a consequence, maps of a given axis of variation (e.g. axis I of the various fractions, mapped in Fig. 13.28) do not exactly add up with this method; they only add up approximately.

Fraction [b + c] (73.4% of the variation in the bacterial data) is the one extracted by the canonical analysis, for the same data, in the numerical example of Section 13.4; Fig. 13.26 presents two maps of this fraction of the variation. In this example, the maps of axis I of fraction [b + c] (Fig. 13.26) are very similar to the map of axis I of [a + b + c] (Fig. 13.28) because [a] is very small.

Before mapping, all signs of axis I of fractions [a + b] and [c] were reversed to make them agree with the signs of the site scores of axis I of [a + b + c]; signs obtained in unconstrained or constrained ordination analyses are arbitrary. With signs reversed, axis I of [a + b], which is the fraction extracted by the environmental variables, is negatively correlated to variables NH_4 and phaeopigments.

The map of axis I of fraction [a + b + c] (63% of the variation in the response bacterial variables) and [a + b] (36%) are quite similar, whereas the map of axis I of fraction [c] (33% of the variation) is quite different. The trend surface equation that produced the "fitted site scores" for the 20 sampling stations is written out by program CANOCO under the heading "Regression/canonical coefficients for standardized variables":

$$\widehat{\text{Axis I of [c]}} = 1.8017X^2 + 2.2817X^3 - 1.0809X^2Y - 1.3064XY^2 + 1.5563Y^3$$

In this equation, the spatial variables are residuals of the standardized terms of the spatial polynomial after controlling for the effect of the three environmental variables. Examination of the map of fraction [c] suggests a hypothesis for the origin of this fraction of variation: that of a marine influence, which had not been included among the explanatory variables in the analysis. Indeed, the negative values on the map form a plume originating at the connections of the Thau lagoon with the sea and extending westwards. To "explain away" fraction [c], i.e. to make it become non-significant, another analysis could be conducted that would include variables quantifying the marine influence on the stations of the lagoon among the environmental variables. Such variables could be derived from a hydrodynamic model of the lagoon.

Applications of this method cover a wide range of ecological problems. Here is a selected list of fields and papers: palaeoecology (Zeeb *et al.*, 1994; see also Ecological application 10.3), vegetation (Heikkinen & Birks, 1996), periphyton (Cattaneo *et al.*, 1993), protozoa (Buttler *et al.*, 1996), zooplankton (Pinel-Alloul, 1995), aquatic macroinvertebrates (Pinel-Alloul *et al.*, 1996), fish (Rodríguez & Magnan, 1995), and birds (Bersier & Meyer, 1994).

Variation partitioning has been applied to more than two explanatory data sets. (1) Pinel-Alloul *et al.* (1995) tested the hypothesis that biotic and abiotic factors, as well as spatial structuring, explain together the broad-scale spatial heterogeneity of zooplankton assemblages among lakes. The explanatory variables comprised abiotic (physics and chemistry, morphometry) and biotic factors (phytoplankton and fish assemblages); these factors were analysed separately and together, using four approaches described in the paper. (2) Quinghong & Bråkenhielm (1995) explained the spatial patterns of epiphytic green algae and lichens using climatic, pollution, and geographic variables. They showed how to isolate the seven components of variation resulting from crossing three sets of explanatory variables. (3) Anderson & Gribble (1998) extended the variation partitioning method to three matrices of explanatory variables representing the environmental, spatial and temporal components, respectively. They also showed how to isolate the seven components of variation. Using this approach, they were able to resolve the confounding of space and time which is often encountered when sampling is conducted over an extensive period, because of the large size of the area to be surveyed.

2 — *Interpretation of the fractions*

In simple regression or canonical analysis modelling, one is interested in the variation of the response data (vector **y** or matrix **Y**) which is accounted for by the explanatory variables (matrix **X**) according to a model of causal relationships, vague or precise, stated prior to the analysis. The fraction of variation explained by the model is measured by the coefficient of determination (R^2, eq. 10.19) in multiple regression and by the ratio [sum of all canonical eigenvalues/trace of matrix **Y**] in canonical analysis (Subsection 11.3.2). The residual variance is assumed to be a random error component.

In the introduction to Section 13.5, two motivations were discussed for the decomposition of the variation of a multivariate response matrix **Y** into the additive components estimated in Subsection 1.

• If the spatial structure is considered to be a source of false correlations which are not indicative of causal relationships, fraction [b] measures the interference of the spatial variables with the analysis of the relationship between **Y** and **X**. Fractions [b] and [c] should not be interpreted separately, although one may still be interested in modelling the spatial structure of **Y** (fraction [b + c]) separately from **X**.

• If both the spatial and non-spatial structures of the explanatory variables are considered causal to the spatial variation of **Y**, fraction [a + b] estimates the amount of variation of **Y** explained by **X**. In such a case, the residuals of the analysis of **Y** by **X** are assumed to contain two identifiable fractions: [c] which is spatially structured and [d] which is the random error component. A test of significance allows one to determine, at some confidence level α, whether fraction [c] may be attributed to random variation. If this is not the case, one should try to interpret fraction [c]. The next step is to "explain away" fraction [c], if possible. In other words, one should try to make fraction [c] disappear by adding variables to matrix **X** and recomputing the model. The numerical example of the previous Subsection has shown how maps of the site scores for the significant canonical axes of fraction [c] may help identify the processes responsible for this fraction of variation. Borcard & Legendre (1994) developed an application showing how fraction [c] may be explained away by adding environmental variables in a stepwise manner to the set of explanatory variables.

In statistical analysis, causality, if invoked, resides in the hypotheses of the researcher. This point has been stated repeatedly throughout the book. The objective of causal statistical modelling is to verify how much of the observed variation can be explained by a consistent body of hypotheses (i.e. a set of compatible hypotheses). Problems of interpretation may occur, however, when important causal factors are left out of the model. The amount of variation of **Y** explained by the model may be reduced and, if these factors are causally anterior to both the variables in **Y** and some of the variables in set **X**, false correlations may appear in the model; this is also the case in path analysis (Section 10.4).

In community analysis, researchers are faced with a multiplicity of potential causal agents acting at a variety of spatial and temporal scales, thus creating a network of interactions that may be difficult to untangle. Section 13.0 mentioned three general models often invoked to explain community variation: the environmental control model (ECM), the biotic control model (BCM), and historical dynamics (HD). The latter refers to past natural events, such as isolation by geographic barriers and disturbances of various kinds (storms, forest fires, volcanic eruptions, landfalls, etc.), and to anthropogenic causes such as agriculture, logging, constructions of various sizes, etc. They are usually not explicitly represented by variables in matrix **X**. Some of these events may be traced by researchers (e.g. tornadoes, forest fires, logging, past agricultural plots) and explicitly included in a second round of modelling, while others cannot and may only be invoked in general terms to account for community variation. Table 13.3 summarizes the interpretation of the various fractions of variation of Fig. 13.27. Examples follow of factors that may intervene to explain community variation in a temperate forest; they illustrate the statements in Table 13.3.

[a] The environmental and biotic factors that are explicitly represented by variables in matrix **X** usually have fine-scale variation. They may explain part of the local variation of the forest community which is not taken into account by the broader-scale spatial polynomial model. Besides these factors, local variation in unobserved soil chemistry variables or other factors may be responsible for part of the variation in the community structure (matrix **Y**) and in the explanatory variables included in **X**, a case which would lead to covariation between **X** and **Y** (false correlation). For example, localized infestation by pest insects may have occurred in the past, leaving variation at some sites in the forest that persisted throughout the years; such a historical event may also have left traces in the variables of matrix **X**, leading to causal or non-causal correlations.

[b] The environmental and biotic factors that are explicitly represented by variables in matrix **X** often have broad-scale variation, detectable by the spatial polynomial model, which may explain part of the variation of the forest community. Besides these factors, broad-scale variation in unobserved environmental factors may be responsible for part of the variation in the community structure (matrix **Y**) and in the explanatory variables included in **X**, a case which would lead to covariation between **X** and **Y** (false correlation). For example, past occupation of the territory under study by agriculture may have left spatially-structured variation in the forest community; it may also have left traces in the measured soil variables of matrix **X**, leading to causal or non-causal correlations. Autocorrelation in both the response and explanatory variables may cause covariation between matrices **Y** and **X**, hence inflating fraction [b].

[c] Part of the spatial structure of the community may be caused by environmental or biotic factors that were not included in the analysis; for instance, a humidity gradient

Table 13.3 Causal factors invoked to explain the various fractions of variation, and in particular the correlations between environmental variables (matrix **X**) and community composition (matrix **Y**). The following hypotheses are invoked: the environmental control model (ECM), the biotic control model (BCM), historical dynamics (HD), and spatial autocorrelation. Bullets: factors explicitly stated in the model; asterisks: factors not explicitly spelled out. Arrows: causal relationships. Modified from Borcard & Legendre (1994).

Fraction	Causal factors	Process	Causal model[1]
[a]	• Non-spatially-structured component of environmental or biotic factors	ECM BCM	E → C
	* Non-spatially-structured environmental or biotic factors not included in the analysis	ECM BCM	F → E, F → C
	* Historical events without spatial structure at the scale of the study	HD	(same as above)
[b]	• Spatially-structured component of biotic or environmental factors included in the analysis	ECM BCM	E → C
	* Spatially-structured environmental or biotic factors not included in the analysis	ECM BCM	F → E, F → C
	* Spatially-structured historical events	HD	(same as above)
	* Spatial autocorrelation in **X** and **Y**	Autocorrelation	E↻ C↻
[c]	* Spatially-structured environmental or biotic factors not included in the analysis	ECM BCM	F → C
	* Spatially-structured historical events	HD	(same as above)
	* Spatial autocorrelation in matrix **Y**	Autocorrelation	C↻
[d]	* Environmental or biotic factors not included in analysis and not spatially structured at scale of study	ECM BCM	
	* Historical events not included in analysis and not spatially structured at scale of study	HD	
	• Random variation, sampling error, etc.	Noise	

[1] C: community structure (matrix **Y**)
E: factor explicitly represented by explanatory variable(s) in the analysis (in matrix **X**)
F: factor not represented by explanatory variable(s) in the analysis

or the effect of grazers may not have been measured. In other types of communities, competition for resources may play an important role but may have been left unmeasured. A windfall may have occurred in the past, creating a clearing in the forest which was then recolonized and has left a detectable broad-scale spatial structure in the forest community. Community processes such as growth and reproduction are a major source of spatial autocorrelation, which is responsible for part of the observed spatially-structured variation observed in communities although it cannot be explained by external factors.

[d] This fraction represents the unexplained variation of matrix **Y** which does not have broad-scale spatial structure. Perhaps some of it could be explained by the fine-scale component of additional factors that have not been included in the analysis; for instance, local patches of grazers. The rest is random variation and sampling error.

These examples illustrate the fact that, in some cases, it is of no use to try to increase the fraction of explained variation by incorporating more environmental variables into the model. Fraction [c], which may represent an important proportion of the unexplained variation, may often only be explained by population or community-based spatial processes (autocorrelation, biotic interactions) or by past events that may sometimes be documented, but often cannot.

Partitioning the spatial variation of communities into components, and mapping them, allows researchers to find interesting correlations supporting models of causal relationships. It also allows one to quantify and map fraction [c] which measures by how much preconceived models may fall short of accounting for the observed data. The same type of analysis may be conducted on time series. Ecologists may use insights obtained by analysing fraction [c] to formulate better ecological models, before going back to the field.

13.6 Causal modelling: partial Mantel analysis

In parallel with causal modelling by partial canonical analysis (Section 13.5), another approach has been proposed to include spatial relationships as predictors in statistical models on an equal footing with the set of explanatory environmental variables. The spatial structure may be represented by a matrix of geographic distances among sampling sites — or some modification of such a matrix, for instance some kind of contiguity matrix (e.g. Fig. 13.23). In the typical case, geographic (Euclidean) distances are computed for all pairs of sampling stations, from their geographic coordinates, and assembled into a "spatial distance" matrix. Similarly, resemblance matrices may be computed for the response (matrix **Y**) and the explanatory variables (matrix **X**) using appropriate similarity or distance measures (Chapter 7).

1 — Partial Mantel correlations

Legendre & Troussellier (1988) proposed to apply to resemblance matrix modelling the predictions made about the values of simple and partial correlations for causal models involving three variables (Subsection 4.5.5; Fig. 4.11). This led to *causal modelling on resemblance matrices*, mentioned in Subsection 10.5.2. Calculations involve a resemblance matrix describing community structure or some other set of response variables (**Y**), a matrix of ecological distances computed for the environmental variables (**X**), and a matrix of geographic distances (**W**). Using the approach of Subsection 4.5.5 is straightforward because the computations of simple and partial Mantel statistics (Subsection 10.5.1 and 10.5.2) are identical to those of simple and partial Pearson correlation coefficients. The algebra is the same; the only difference is that the values involved in the calculations come from resemblance matrices instead of vectors. Remember, however, that a correlation between two distance matrices is not equivalent to the correlation between two data tables; a Mantel correlation $r_{M(\mathbf{AB})}$ between matrices **A** and **B** measures the extent to which the variations in the *similarities* or *distances* of **A** correspond to the variations in **B**.

Ecological applications of causal modelling on resemblance matrices to models that include geographic distances are found in Burgman (1987), Legendre & Troussellier (1988), Villeneuve *et al.* (1991), Legendre & Fortin (1989), Nantel & Neumann (1992), Fromentin *et al.* (1993), Leduc *et al.* (1992), Mandrak (1995), and Bjørnstad *et al.* (1995). Other interesting applications of the partial Mantel test to anthropology and population genetics are found in Sokal (1986) and Sokal *et al.* (1987). Let us examine two ecological applications.

Ecological application 13.6a

In the paper of Legendre & Troussellier (1988), the question was whether the well-established relationship (model 1 of Fig. 4.11) between aquatic heterotrophic bacteria and phytoplankton biomass (estimated by chlorophyll *a*, variable Chl *a*) held for two identifiable components of the bacterial heterotrophic community of a marine lagoon: the aerobic heterotrophs growing on low-salinity medium (variable Bna), which were presumably of continental origin, and the aerobic heterotrophs growing on high-salinity medium (variable Ma), expected to be mostly of marine origin; see the numerical example of Subsection 10.3.5 for a more detailed description of the bacterial variables. The two bacterial variables were well correlated to Chl *a*, but this pattern could be the result of a common spatial structure created by currents (model 2 of Fig. 4.11). The question thus belonged to the first type mentioned in the introduction of Section 13.5: was there a significant correlation between the response variables (Bna, Ma) and the explanatory variable (Chl *a*), besides some common spatial structure?

The spatial structure was represented by a matrix of geographic (i.e. Euclidean) distances among the 63 sampling sites, called **SPACE**. Each biological variable was turned into a distance matrix by computing Euclidean distances among the sites, thus producing matrices **BNA**, **MA** and **CHL A**; abbreviations are as in the original publication. The Mantel and partial Mantel statistics, computed to decide between the two models, are reported in Fig. 13.29. The nonsignificant partial Mantel relationship between **BNA** and **CHL A**, when the effect of **SPACE** is controlled for, points to model 2 in Fig. 4.11, which contradicts the hypothesis of control of the variation of continental heterotrophs by phytoplankton variation. In contrast, the nonsignificant partial Mantel relationship between **MA** and **SPACE**, when **CHL A** is held constant, points to model 1, which supports the hypothesis of control of the variation of marine heterotrophs by phytoplankton.

(a) Analysis of the **BNA-CHL A-SPACE** relationships

	BNA	CHL A	SPACE
BNA	----	0.258*	0.521*
CHL A	–0.006	-----	0.505*
SPACE	0.468*	0.449*	-----

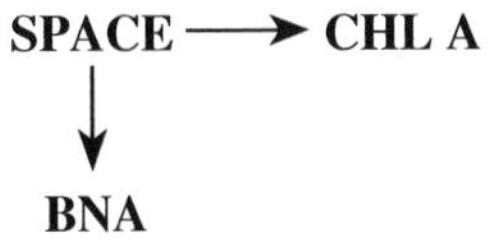

SPACE ⟶ CHL A
SPACE ↓ BNA

(b) Analysis of the **MA-CHL A-SPACE** relationships

	MA	CHL A	SPACE
MA	----	0.325*	0.223*
CHL A	0.252*	-----	0.505*
SPACE	0.073	0.469*	-----

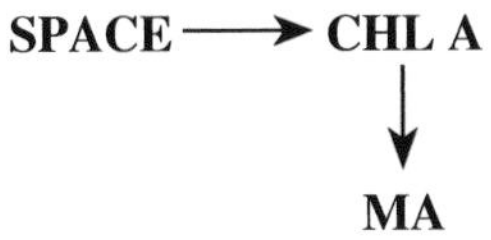

SPACE ⟶ CHL A
CHL A ↓ MA

Figure 13.29 Mantel analysis of the relationships between matrices representing bacterial variables (a: **BNA**, b: **MA**, in matrix form), **CHL A**, and **SPACE**. Left tables, above the diagonals: simple Mantel statistics; below: partial Mantel statistics controlling for the effect of the third matrix. Asterisks (*) indicate significance ($\alpha = 0.05$); tests of significance are one-tailed. Right: causal models supported by the results. Adapted from Legendre & Troussellier (1988).

Ecological application 13.6b

Leduc *et al.* (1992) analysed the relationship between environmental conditions and the spatial distributions of adult trees and saplings, for 12 tree species in 198 vegetation quadrats from a 0.5 km^2 forested area. Distance matrices were computed for the adults and saplings of each species separately. A matrix of geographic distances ("Space") was also computed among sites. "Environment" is a matrix computed among sites for the 6 drainage, soil and geomorphology variables, using the Estabrook-Rogers coefficient (S_{16}, Chapter 7). Partial Mantel tests were used in conjunction with path analysis computed from the Mantel statistics, as in Ecological application 10.4b. The question here belonged to the second type mentioned in the introduction of Section 13.5: does the spatial structure of the environmental variables fully explain the spatial structure of the vegetation, or is there a part of the spatial structure of the vegetation which is not explained by the environmental variables in the model?

After analysing each species separately, the authors found that the spatial patterns of species associated with hydric conditions were largely explained by the spatial distributions of the environmental variables (Fig.13.30c, d), whereas mesic-site species still displayed significant spatial patterns after controlling for the environmental variables (Fig. 13.30a, b). Past events (disturbances) and forest cover dynamics were suggested as explanations for the latter.

Residual distance matrices, as computed in partial Mantel analysis, provide the basis for interesting illustrations of the relationships among sites, using ordination by principal coordinate analysis (PCoA, Section 9.2) or nonmetric multidimensional scaling (MDS, Sections 9.3).

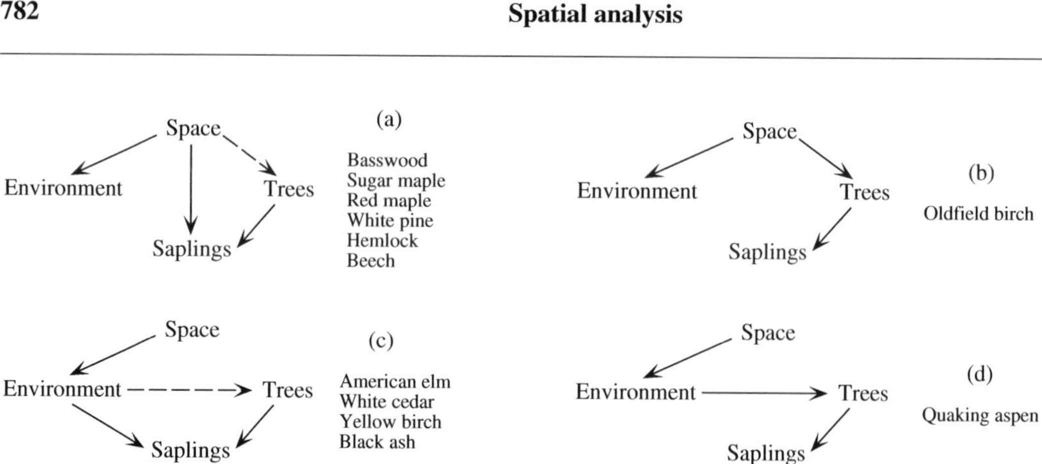

Figure 13.30 Models of relationships derived from partial Mantel tests for each tree species. (a, b) Species associated with mesic sites (i.e. moderate moisture conditions); (c, d) species associated with hydric conditions. Dashed lines indicate that, for some species in the group, the relationship, although high, did not reach the Bonferroni-corrected significance level. Species pertaining to group (a) are: basswood (*Tilia americana*), sugar maple (*Acer saccharum*), red maple (*Acer rubrum*), white pine (*Pinus strobus*), hemlock (*Tsuga canadensis*), American beech (*Fagus grandifolia*); group (b): oldfield birch (*Betula populifolia*); group (c): American elm (*Ulmus americana*), white cedar (*Thuja occidentalis*), yellow birch (*Betula alleghaniensis*), black ash (*Fraxinus nigra*); group (d): aspen (*Populus tremuloides*). Modified from Leduc *et al.* (1992).

• On the one hand, the matrix of residuals $\mathbf{R}_{\mathbf{Y}|\mathbf{W}}$ obtained by partialling out the effect of the geographic distances (**W**) from the matrix of community similarities or distances (**Y**) may be used as the basis for an ordination. It could be interesting to compare it to an ordination of the original matrix **Y**. Alternatively, the site scores along the various ordination axes could be mapped as in Fig. 13.28. Matrix $\mathbf{R}_{\mathbf{Y}|\mathbf{W}}$ is analogous to fraction [a + d] of the partial canonical analysis approach.

• On the other hand, the matrix of residuals $\mathbf{R}_{\mathbf{Y}|\mathbf{X}}$ obtained by partialling out the effect of the environmental variables (resemblance matrix **X**) from the matrix of ecological similarities or distances (**Y**) may be used for ordination. This approach has been used by Fromentin *et al.* (1993, Fig. 13). Alternatively, the site scores for the various ordination axes could be mapped as in Fig. 13.28. Matrix $\mathbf{R}_{\mathbf{Y}|\mathbf{X}}$ is analogous to fraction [c + d] of the partial canonical analysis approach.

There is still progress to be made in this elementary form of modelling. Spatial relationships can be construed in other ways than a Euclidean distance matrix. Transformations such as the inverse of the Euclidean distance, or the inverse of its square, have been used; they give more importance to the small distances (Jumars *et al.*, 1977). In order to limit the effect of large distances on Mantel correlations, Leduc *et al.* (1992) truncated their geographic distance matrix; all distances larger than the range of the variogram were made equal to that distance. Other authors prefer to use distances along connection schemes such as rook's or king's connections for regular grids of points, instead of regular or transformed Euclidean distances, and Gabriel graphs or Delaunay triangulations for irregularly spaced points (Subsection 13.3.1).

2 — Multiple regression approach

In several instances, ecological distances are not linearly, nor even monotonically related to the geographic distances. *Multiple regression on resemblance matrices* (Subsection 10.5.2; Fig. 10.21) would allow one to take such nonlinear relationships into account in modelling. This method is as an extension of partial Mantel analysis to cases involving several explanatory matrices. The method is briefly outlined as follows:

• From the matrix of geographic distances among sites, create a set of model matrices corresponding to distance classes, $\mathbf{X}_1$, $\mathbf{X}_2$, …, as in Fig. 13.12.

• Regress the matrix of ecological distances $\mathbf{Y}$ on the set of model matrices, using multiple regression on resemblance matrices. The regression model is:

$$\hat{\mathbf{Y}} = b_0 + b_1\mathbf{X}_1 + b_2\mathbf{X}_2 + \ldots \quad \mathbf{(13.30)}$$

• Other matrices of environmental variables $\mathbf{W}_1$, $\mathbf{W}_2$, … may also be incorporated into the analytical model:

$$\hat{\mathbf{Y}} = b_0 + b_1\mathbf{X}_1 + b_2\mathbf{X}_2 + \ldots + b_{k+1}\mathbf{W}_1 + b_{k+2}\mathbf{W}_2 + \ldots \quad \mathbf{(13.31)}$$

This approach looks promising for studying complex models involving spatial or temporal structures in connection with different groups of environmental variables, especially when the response variables naturally come in the form of similarity or distance matrices. No ecological application of this method has been published yet.

3 — Comparison of methods

Both the partial canonical analysis (Section 13.5) and matrix approaches (partial Mantel analysis and multiple regression; present Section) have been successful in enriching our understanding of spatial processes in ecosystems. Which one should be preferred? At least three aspects may be considered.

• Resemblance matrix — The only resemblance measures available in the partial canonical analysis approach are the Euclidean distance (in partial RDA) and the chi-square distance (in partial CCA). The matrix approach is certainly to be preferred when some other resemblance measure (Chapter 7) is justified, or when the dependent data naturally present themselves in the form of a resemblance or proximity matrix (e.g. genetic distances obtained by DNA or RNA pairing; preference or dominance data in behaviour studies, serology). Resemblance coefficients often provide greater

Table 13.4 Computer programs available for the methods of surface pattern analysis discussed in Chapter 13. The list is not exhaustive.

Program	Methods	Source
CANOCO	(Partial) canonical ordination methods	See ter Braak (1988b) who wrote the program. Available for several platforms from Scientia Publishing, P.O. Box 658. H-1365 Budapest, Hungary, and Micro-computer Power, 113 Clover Lane, Ithaca, NY 14850, USA; http://www.microcomputerpower.com
C2D	Directional spatial correlograms	BioMedware Inc., 516 North State Street, Ann Arbor, Michigan 48104-1236, USA; http://ic.net/~biomware/
GBAS	Geographic boundary analysis	BioMedware (see C2D).
GEODAT	Mapping and digitizing, including base maps of the world	BioMedware (see C2D).
GEO-EAS	Variogram, kriging; contour mapping	Developed by US Environmental Protection Agency. Available from ACOGS, P.O. Box 44247, Tucson, Arizona 85733-4247, USA; ftp://math.arizona.edu/incoming/unix.geoeas/
GEOSTAT	Variogram, kriging; contour mapping	Geostat Systems Internat., 4385 Saint-Hubert, Suite 1, Montréal (Québec) H2J 2X1, Canada.
GS+	Spatial autocorrelogram (Moran's *I*), variogram, kriging; contour mapping	BioMedware (see C2D).
GSLIB	Geostatistical software library	See Deutsch & Journel (1992). WWWeb site: http://ekofisk.stanford.edu/scrf/software.html
ISATIS	Variogram, kriging; contour mapping.	Geovariances, 38 avenue Franklin-Roosevelt, F-77210 Avon, France. WWWeb site: http://cg.ensmp.fr/HomePageEnglish.html
Kellogg's	Variogram, kriging; contour mapping	Computer Laboratory, W.K. Kellogg Biological Station, Michigan State University, Hickory Corners, Michigan 49060, USA.
MACGRIDZO	Inverse distance and weighted least squares interpolation; contour maps	RockWare Inc., 4251 Kipling St., Suite 595, Wheat Ridge, Colorado 80033, USA.
NTSYS-PC	Simple Mantel test	Developed by F. J. Rohlf. Available from Exter Software Inc., 100 North Country Road, Bldg. B, Setauket, New York 11733, USA; http://www.exeterSoftware.com
R Package	Spatial autocorrelograms (Moran's *I* and Geary's *c*), simple and partial Mantel tests, Mantel correlogram, clustering with space and time contiguity constraint. ANOVA for spatially autocorrelated regional data; connection networks	Developed by P. Legendre. Available for several platforms from the WWWeb site: http://www.fas.umontreal.ca/BIOL/legendre/
SAAP	Spatial autocorrelograms (Moran's *I* and Geary's *c*)	D. Wartenberg, Department of Environmental and Community Medicine, Robert Wood Johnson Medical School, 675 Hoes Lane, Piscataway, New Jersey 08854, USA.
SASP	Two-dimensional spectral analysis	E. Renshaw, Department of Statistics, Univ. of Edinburgh, King's Buildings, Mayfield Road, Edinburgh EH9 3JZ, Scotland.
STAT!	Point and surface pattern statistics	BioMedware (see C2D).
SURFER	Kriging; other interpolation methods; contour mapping	Golden Software Inc., P.O. Box 281, Golden, Colorado 80402, USA.
UNIMAP	Variogram, kriging; other interpolation methods; contour mapping	European Software Contractors A/S, Nørregade, DK-2800 Lyngby, Denmark.
VARIOWIN	Variogram	Information available on the WWWeb site: http://www-sst.unil.ch/geostatistics.html

flexibility for handling mixed-type data. This was the case in Ecological application 13.6b, where the environmental data matrix was computed from a mixture of ordered and nonordered variables. Ecological application 13.6a could just as well have used the partial canonical analysis approach.

The partial canonical analysis approach may actually be used even in situations where a resemblance function other than Euclidean or chi-square is preferred for either the response or the explanatory data. In order to do so, (1) the resemblance matrix is transformed back into a table of quantitative axes using principal coordinate analysis (PCoA, Section 9.2) or nonmetric multidimensional scaling (MDS, Sections 9.3), and (2) partial canonical analysis is computed using the matrix of ordination scores.

• Statistical testing — The partial canonical analysis approach (Section 13.5) allows for statistical testing and provides estimates of the contributions of the explanatory and response variables to the canonical axes. The matrix approach, at its present stage of development, focuses on statistical testing and largely disregards parameter estimation. Multiple regression on resemblance matrices (Subsection 2), however, offers a way of estimating the importance of the contributions of various subsets of the explanatory data to the explanation of the response matrix **Y**.

• Scale — Partial Mantel analysis focuses on the broad-scale spatial structure (i.e. trend). Following a decomposition of the geographic distances into distance classes, multiple regression on distance matrices allows one to model both the fine-scale (i.e. autocorrelation) and broad-scale (i.e. trend) spatial structures. The partial canonical analysis approach, as developed so far, is mostly used to model trends. Fine-scale spatial autocorrelation can also be modelled with this method (He *et al.*, 1994; Legendre & Borcard, 1994), but the analysis becomes quite cumbersome when there are many response variables.

13.7 Computer programs

Most of the methods described in this Chapter cannot be implemented using the major statistical packages available in 1998. Table 13.4 lists computer programs available either commercially or from researchers. The list is not exhaustive.

Bibliography

Numbers in brackets: pages of the book where the reference is cited

Aart, P. J. M. (van der) 1973. Distribution analysis of wolfspiders (Araneae, Lycosidae) in a dune area by means of principal component analysis. *Neth. J. Zool.* **23**: 266-329. [415, 416, 479]

Aart, P. J. M. (van der) & N. Smeenk-Enserink. 1975. Correlations between distributions of hunting spiders (Lycosidae, Ctenidae) and environmental characteristics in a dune area. *Neth. J. Zool.* **25**: 1-45. [615, 616]

Ables, J. G. 1974. Maximum entropy spectral analysis. *Astron. Astrophys. Suppl.* **15**: 383-393. [690]

Addicott, J. F., J. M. Aho, M. F. Antolin, M. F. Padilla, J. S. Richardson & D. A. Soluk. 1987. Ecological neighborhoods: scaling environmental patterns. *Oikos* **49**: 340-346. [709]

Ali, M. A. [ed.]. 1992. *Rhythms in fishes.* NATO ASI Series. Vol. A-236. Plenum Press, New York. viii + 348 pp. [638]

Allen, T. F. H., S. M. Bartell & J. F. Koonce. 1977. Multiple stable configurations in ordination of phytoplankton community change rates. *Ecology* **58**: 1076-1084. [693, 696]

Allen, T. F. H. & T. W. Hoekstra. 1991. Role of heterogeneity in scaling of ecological systems under analysis. 47-68 *in*: J. Kolasa & S. T. A. Pickett [eds.] *Ecological heterogeneity.* Springer-Verlag, New York. [708]

Allen, T. F. H. & T. B. Starr. 1982. *Hierarchy – Perspectives for ecological complexity.* Univ. of Chicago Press, Chicago. xvi + 310 pp. [8]

Amanieu, M., P. Legendre, M. Troussellier & G.-F. Frisoni. 1989. Le programme Écothau: théorie écologique et base de la modélisation. *Oceanol. Acta* **12**: 189-199. [527, 529]

Anderberg, M. R. 1973. *Cluster analysis for applications.* Academic Press, New York. xiii + 359 pp. [352]

Anderson, M. J. 1996. *Tests of ecological hypotheses in intertidal estuarine assemblages.* Ph.D. thesis, University of Sydney, New South Wales, Australia. x + 335 pp. [433]

Anderson, M. J. & N. A. Gribble. 1998. Partitioning the variation among spatial, temporal and environmental components in a multivariate data set. *Aust. J. Ecol.* **23**: 158-167. [770, 775]

Anderson, M. J. & P. Legendre. 1999. An empirical comparison of permutation methods for tests of partial regression coefficients in a linear model. *J. Statist. Comput. Simulation* (in press). [559, 607, 611]

Anderson, T. W. 1958. *An introduction to multivariate statistical analysis.* J. Wiley & Sons, New York. xii + 374 pp. [612]

Anderson, T. W. 1971. *The statistical analysis of time series.* John Wiley & Sons, New York. 704 pp. [676]

Anderson, T. W. 1984. *An introduction to multivariate statistical analysis. 2nd edition.* Wiley, New York. xvii + 675 pp. [131]

Angot, M. 1961. Analyse quantitative du cycle diurne de la production primaire dans le Pacifique subtropical près de la Nouvelle-Calédonie. *Bull. Inst. Océanogr. (Monaco)* (1200): 1-34. [679]

Ardisson, P.-L., E. Bourget & P. Legendre. 1990. Multivariate approach to study species assemblages at large spatiotemporal scales: the community structure of the epibenthic fauna of the Estuary and Gulf of St. Lawrence. *Can. J. Fish. Aquat. Sci.* **47**: 1364-1377. [441, 442, 696]

Arfi, R., F. Blanc & D. Calmet. 1982. Étude d'impact en milieu marin: échantillonnage et traitement des données. 341-364 *in*: S. Frontier [ed.] *Stratégies d'échantillonnage en écologie.* Masson, Paris et Les Presses de l'Université Laval, Québec. [688, 689]

Arfi, R. & P. Dumas. 1990. Séries chronologiques: analyse spectrale de Fourier et par maximisation d'entropie – Présentation, simulations, applications. 105-126 *in*: S. Frontier [ed.] *Biométrie et océanographie.* Actes de Colloques, 10, IFREMER, Brest. [690]

Armstrong, M. [ed.]. 1989. *Geostatistics. Vol. 1 and 2.* Kluwer Academic Publishers, Dordrecht. xxix + 491 pp., xvii + 546 pp. [749]

Armstrong, M., D. Renard & P. Berthou. 1989. *Applying geostatistics to the estimation of a population of bivalves.* ICES C. M. 1989/K37. 22 pp. [749]

Bach, P., P. Legendre, M. Amanieu & G. Lasserre. 1992. Strategy of eel (*Anguilla anguilla* L.) exploitation in the Thau lagoon. *Estuarine Coastal Shelf Sci.* **35**: 55-73. [171]

Barbalat, S. & D. Borcard. 1997. Distribution of four beetle families (Coleoptera: Buprestidae, Cerambycidae, phytophagous Scarabaeidae and Lucanidae) in different forest ecotones in the Areuse Gorges (Neuchâtel, Switzerland). *Écologie* **28**: 199-208. [371]

Barbujani, G., N. L. Oden & R. R. Sokal. 1989. Detecting regions of abrupt change in maps of biological variables. *Syst. Zool.* **38**: 377-389. [761]

Bare, B. B. & D. W. Hann. 1981. Applications of ridge regression in forestry. *For. Sci.* **27**: 339-348. [524]

Barnes, H. 1952. The use of transformations in marine biological statistics. *J. Cons. Cons. Int. Explor. Mer* **18**: 61-71. [42]

Barrodale, I. & R. E. Erickson. 1980a. Algorithms for least-square linear prediction and maximum entropy spectral analysis. Part I: Theory. *Geophysics* **45**: 420-432. [690]

Barrodale, I. & R. E. Erickson. 1980b. Algorithms for least-square linear prediction and maximum entropy spectral analysis. Part II: Fortran program. *Geophysics* **45**: 433-446. [690]

Bartlett, M. S. 1938. Further aspects of the theory of multiple regression. *Proc. Camb. Phil. Soc.* **34**: 33-40. [623]

Bartlett, M. S. 1946. On the theoretical specification and sampling properties of autocorrelated time-series. *Suppl. J. Roy. Statist. Soc.* **8**: 27-41. [657]

Bartlett, M. S. 1948. Internal and external factor analysis. *Brit. J. Psychol. Stat. Sect.* **1**: 73-81. [623]

Bartlett, M. S. 1950. Tests of significance in factor analysis. *Brit. J. Psychol. Stat. Sect.* **3**: 77-85. [409]

Bartlett, M. S. 1954. A note on the multiplying factors for various chi-squared approximations. *J. Roy. Statist. Soc. Ser. B* **16**: 296-298. [144]

Bartlett, M. S. 1978. Nearest neighbour models in the analysis of field experiments. *J. Roy. Statist. Soc. Ser. B* **40**: 147-174. [15]

Barton, D. E. & F. N. David. 1956. Some notes on ordered random intervals. *J. Roy. Statist. Soc. Ser. B* **18**: 79-94. [244]

Baum, B. R. 1992. Combining trees as a way of combining data for phylogenetic inference and the desirability of combining gene trees. *Taxon* **41**: 3-10. [379]

Beach, C. M. & J. G. MacKinnon. 1978. A maximum likelihood procedure for regression with autocorrelated errors. *Econometrica* **46**: 51-58. [15]

Belbin, L. 1991. Semi-strong hybrid scaling, a new ordination algorithm. *J. Veg. Sci.* **2**: 491-496. [444]

Belbin, L. & C. McDonald. 1993. Comparing three classification strategies for use in ecology. *J. Veg. Sci.* **4**: 341-348. [348]

Bell, M. A. & P. Legendre. 1987. Multicharacter chronological clustering in a sequence of fossil sticklebacks. *Syst. Zool.* **36**: 52-61. [696]

Bellehumeur, C. & P. Legendre. 1998. Multiscale sources of variation in ecological variables: modelling spatial dispersion, elaborating sampling designs. *Landscape Ecology* **13**: 15-25. [735]

Bellehumeur, C., P. Legendre & D. Marcotte. 1997. Variance and spatial scales in a tropical rain forest: changing the size of sampling units. *Plant Ecol.* **130**: 89-98. [733]

Bendat, J. S. & A. G. Piersol. 1971. *Random data – Analysis and measurement procedures.* Wiley-Interscience, New York. xv + 407 pp. [679]

Benzécri, J. P. 1969. Statistical analysis as a tool to make patterns emerge from data. 35-60 *in*: S. Watanabe [ed.] *Methodologies of pattern recognition.* Academic Press, New York. [451]

Benzécri, J. P. et coll. 1973. *L'analyse des données. Tome I: La taxinomie. Tome II: L'analyse des correspondances.* Dunod, Paris. viii + 615, vii + 619 pp. [390, 451, 462, 466]

Bergmann, C. 1847. Ueber die Verhältnisse der Wärmeökonomie der Thiere zu ihrer Grösse. *Gottinger Studien* **3**: 595-708. [498]

Bernstein, B. B. & J. Zalinski. 1983. An optimum sampling design and power tests for environmental biologists. *J. Environ. Manag.* **16**: 35-43. [249]

Berryman, J. G. 1978. Choice of operator length for maximum entropy spectral analysis. *Geophysics* **43**: 1384-1391. [690, 690]

Bersier, L.-F. & D. Meyer. 1994. Bird assemblages in mosaic forests: the relative importance of vegetation structure and floristic composition along the sucessional gradient. Acta Œcologica **15**: 561-576. [775]

Bertalanffy, L. 1968. *General system theory: foundations, development, applications.* Braziller, New York. xv + 289 pp. [209]

Besag, J. & P. Clifford. 1989. Generalized Monte Carlo significance tests. *Biometrika* **76**: 633-642. [15]

Best, D. J. 1974. *Tables for Kendall's tau and an examination of the normal approximation.* Division of Mathematical Statistics, Technical paper no. 39. CSIRO, Australia. 15 pp. [835]

Beum, C. O. J. & E. G. Brundage. 1950. A method for analyzing the sociomatrix. *Sociometry* **13**: 141-145. [372]

Bezdek, J. C. 1987. Some non-standard clustering algorithms. 225-287 *in*: P. Legendre & L. Legendre [eds.] *Developments in numerical ecology.* NATO ASI Series, Vol. G-14. Springer-Verlag, Berlin. [304, 305]

Binet, D., M. Gaborit, A. Dessier & M. Roux. 1972. Premières données sur les copépodes pélagiques de la région congolaise. II. Analyse des correspondances. *Cah. O. R. S. T. O. M. Sér. Océanogr.* **10**: 125-137. [465]

Birks, H. J. B. 1993. Is the hypothesis of survival on glacial nunataks necessary to explain the present-day distributions of Norwegian mountain plants? *Phytocoenologia* **23**: 399-426. [536]

Birks, H. J. B. 1995. Quantitative palaeoenvironmental reconstructions. 161-254 *in*: D. Maddy & J. S. Brew [eds.] *Statistical modelling of Quaternary science data.* Quaternary Research Association, Technical Guide 5. Cambridge University Press, Cambridge. [603, 604]

Birks, H. J. B. 1996. Statistical approaches to interpreting diversity patterns in the Norwegian mountain flora. *Ecography* **19**: 332-340. [535, 536]

Birks, H. J. B., S. Juggins & J. M. Line. 1990a. Lake surface-water chemistry reconstructions from palaeolimnological data. 301-313 *in*: B. J. Mason [ed.] *The surface waters acidification programme.* Cambridge University Press, Cambridge. [603]

Birks, H. J. B., J. M. Line, S. Juggins, A. C. Stevenson & C. J. F. ter Braak. 1990b. Diatoms and pH reconstruction. *Philos. Trans. R. Soc. Lond. B* **327**: 263-278. [604]

Birks, H. J. B., S. M. Peglar & H. A. Austin. 1994. *An annotated bibliography of canonical correspondence analysis and related constrained ordination methods 1986-1993.* Available from H. J. B. Birks, Botanical Institute, University of Bergen, Allégaten 41, N-5007 Bergen, Norway. Also available from the WWWeb site <http://alize.ere.umontreal.ca/~casgrain/cca_bib_93/> [579, 603]

Bishop, Y. M. M., S. E. Fienberg & P. W. Holland. 1975. *Discrete multivariate analysis – Theory and practice.* MIT Press, Cambridge, Mass. x + 557 pp. [217, 223, 228, 232, 233]

Bivand, R. 1980. A Monte Carlo study of correlation coefficient estimation with spatially autocorrelated observations. *Quaest. Geogr.* **6**: 5-10. [13]

Bjørnstad, O. N. & W. Falck. 1997. An extension of the spatial correlogram and the x-intercept for genetic data. P9 *in*: O. N. Bjørnstad. *Statistical models for fluctuating populations – Patterns and processes in time and space.* Dr. Philos. Dissertation, University of Oslo. ISBN 82-90934-57-2. [726]

Bjørnstad, O. N., A. Iversen & M. Hansen. 1995. The spatial structure of the gene pool of a viviparous population of *Poa alpina* – Environmental controls and spatial constraints. *Nord. J. Bot.* **15**: 347-354. [780]

Bjørnstad, O. N., N. C. Stenseth & T. Saitoh. 1998. Scaling of population dynamics of voles and mice in northern Japan. *Ecology* (in press). [726]

Blackith, R. E. & F. O. Albrecht. 1959. Morphometric differences between the eye-stripe polymorphs of the red locust. *Scient. J. Roy. Coll. Sci.* **27**: 13-27. [615]

Blackith, R. E. & R. A. Reyment. 1971. *Multivariate morphometrics.* Academic Press, London. ix + 412 pp. [615]

Blanc, F., P. Chardy, A. Laurec & J.-P. Reys. 1976. Choix des métriques qualitatives en analyse d'inertie. Implications en écologie marine benthique. *Mar. Biol. (Berl.)* **35**: 49-67. [252]

Blashfield, R. K. & M. S. Aldenderfer. 1978. The literature on cluster analysis. *Multivar. Behav. Res.* **13**: 271-295. [306]

Bloom, S. A. 1981. Similarity indices in community studies: potential pitfalls. *Mar. Ecol. Prog. Ser.* **5**: 125-128. [287, 298]

Bloomfield, P. 1976. *Fourier analysis of time series – An introduction.* Wiley, New York. xiii + 258 pp. [641]

Bock, H. H. 1989. Probabilistic aspects in cluster analysis. 12-44 *in*: O. Opitz [ed.] Conceptual and numerical analysis of data. Springer-Verlag, Berlin. [378]

Bock, H. H. 1996. Probability models and hypotheses testing in partitioning cluster analysis. 377-453 *in*: P. Arabie, L. J. Hubert & G. De Soete [eds.] *Clustering and Classification.* World Scientific Publ. Co., River Edge, New Jersey. [378]

Boggs, P. T. & J. E. Rogers. 1990. Orthogonal distance regression. *Contemp. Math.* **112**: 183-194. [517]

Bollen, K. A. 1989. *Structural equations with latent variables.* Wiley, New York. 514 pp. [480]

Boltzmann, L. 1898. *Vorlesungen über Gastheorie, Vol. II.* J. A. Barth, Leipzig. [209]

Borcard, D. 1996. Typologie des assemblages d'espèces d'Oribates (Acari, Oribatei) de la tourbière du Cachot (Jura suisse): espèces indicatrices ou groupements caractéristiques? *Bull. Soc. Neuchâtel. Sci. Nat.* **119**: 63-73. [371]

Borcard, D. & A. Buttler. 1997. *Écologie numérique.* Instituts de Biologie, Université de Neuchâtel, Switzerland. 235 pp. [594]

Borcard, D., W. Geiger & W. Matthey. 1995. Oribatid mite assemblages in a contact zone between a peat-bog and a meadow in the Swiss Jura (Acari, Oribatei): influence of landscape structures and historical processes. *Pedobiologia* **39**: 318-330. [765]

Borcard, D. & P. Legendre. 1994. Environmental control and spatial structure in ecological communities: an example using oribatid mites (Acari, Oribatei). *Environ. Ecol. Stat.* **1**: 37-53. [770, 775, 778]

Borcard, D., P. Legendre & P. Drapeau. 1992. Partialling out the spatial component of ecological variation. *Ecology* **73**: 1045-1055. [770, 771]

Borcard, D. & C. Vaucher-von Ballmoos. 1997. Oribatid mites (Acari, Oribatida) of a primary peat bog-pasture transition in the Swiss Jura mountains. *Écoscience* **4**: 470-479. [371]

Borgman, L. E. & W. F. Quimby. 1988. Sampling for tests of hypothesis when data are correlated in space and time. 25-43 *in*: L. H. Keith [ed.] *Principles of environmental sampling.* ACS Professional Reference Book. American Chemical Society. [16]

Bos, A. (van den) 1971. Alternative interpretation of maximum entropy spectral analysis. *IEEE Trans. Inf. Theory* **17**: 493-494. [688, 689, 690]

Boudoux, M. & C.-H. Ung. 1979. Applications de la régression pseudo-orthogonale en recherche forestière. *Biom-Praxim.* **19**: 59-89. [524]

Boudreault, F. R., J. D. Dupont & C. Sylvain. 1977. Modèles linéaires de prédiction des débarquements de homard aux îles de la Madeleine (Golfe du Saint-Laurent). *J. Fish. Res. Board Can.* **34**: 379-383. [704]

Bourbaki, N. 1960. *Eléments d'histoire des mathématiques.* Hermann, Paris. 277 pp. [51]

Bourbeau, L., F. Ouellette & F. Pinard. 1984. *Le système BLOPS 2.0: un dictionnaire morphologique informatisé du français et sa logithèque.* Université de Montréal, Montréal. [216]

Bourgault, G., D. Marcotte & P. Legendre. 1992. The multivariate (co)variogram as a spatial weighting function in classification methods. *Math. Geol.* **24**: 463-478. [759]

Bowerman, B. L. & R. T. O'Connell. 1987. *Time series forecasting.* Duxbury Press, Boston. xi + 540 pp. [702]

Box, G. E. P. & D. R. Cox. 1964. An analysis of transformations. *J. Roy. Statist. Soc. Ser. B* **26**: 211-243. [43]

Box, G. E. P. & G. M. Jenkins. 1976. *Time series analysis – Forecasting and control. Revised edition.* Holden-Day, San Francisco. xxi + 575 pp. [15, 641, 702, 703, 721]

Bray, R. J. & J. T. Curtis. 1957. An ordination of the upland forest communities of southern Wisconsin. *Ecol. Monogr.* **27**: 325-349. [287, 535, 569, 600, 633, 707]

Brill, R., H. Christanson, G. F. Estabrook, H. S. Fleming, B. Handley, P. Legendre, F. Ouellette, D. J. Rogers & M. Wirth. 1972. *Program CHARANAL. User's manual.* H. S. Fleming & S. G. Appan [eds.] Gulf Universities Research Corp., NASA/Mississippi Test Facility, Bay St-Louis. [344]

Brillinger, D. R. 1981. *Time series – Data analysis and theory. Expanded edition.* Holden-Day, San Francisco. xii + 540 pp. [641, 687]

Brillouin, L. 1956. *Science and information theory.* Academic Press, New York. 320 pp. [216, 241]

Brockwell, P. J. & R. A. Davis. 1991a. *Time series – Theory and methods. 2nd edition.* Springer-Verlag, New York. xvi + 577 pp. [705]

Brockwell, P. J. & R. A. Davis. 1991b. *ITSM – An interactive time series modelling package for the PC.* Springer-Verlag, New York. ix + 104 pp. + 3 diskettes. [705]

Bronson, R. 1989. *Schaum's outline of matrix operations.* McGraw-Hill, New York. 230 p. [51]

Brown, M. B. 1976. Screening effects in multidimensional contingency tables. *Appl. Statist.* **25**: 37-46. [227]

Brown, B. M. & J. S. Maritz. 1982. Distribution-free methods in regression. *Aust. J. Stat.* **24**: 318-331. [25]

Buckingham, E. 1914. On physically similar systems; illustrations of the use of dimensional equations. *Phys. Rev. (2nd series)* **4**: 345-376. [105]

Burg, J. P. 1967. *Maximum entropy spectral analysis.* Paper presented at the 37th Annual Meeting of Exploration Geophysics, October 31, Oklahoma City, Okla. [688, 689, 690]

Burgman, M. A. 1987. An analysis of the distribution of plants on granite outcrops in southern Western Australia using Mantel tests. *Vegetatio* **71**: 79-86. [556, 780]

Burgman, M. A. 1988. Spatial analysis of vegetation patterns in southern Western Australia: implications for reserve design. *Aust. J. Ecol.* **13**: 415-429. [556]

Burt, C. 1952. Tests of significance in factor analysis. *Brit. J. Psychol. Stat. Sect.* **5**: 109-133. [410, 478]

Buttler, A., B. G. Warner, P. Grosvernier & Y. Matthey. 1996. Vertical patterns of testate amoebae (Protozoa: Rhizopoda) and peat-forming vegetation on cutover bogs in the Jura, Switzerland. *New Phytol.* **134**: 371-382. [775]

Cade, B. S. & J. D. Richards. 1996. Permutation tests for least absolute deviation regression. *Biometrics* **52**: 886-902. [609]

Cadoret, L., P. Legendre, M. Adjeroud & R. Galzin. 1995. Répartition spatiale des Chaetodontidae dans différents secteurs récifaux de l'île de Moorea, Polynésie française. *Écoscience* **2**: 129-140. [472, 473, 602]

Cailliez, F. 1983. The analytical solution of the additive constant problem. *Psychometrika* **48**: 305-308. [434]

Cailliez, F. & J.-P. Pagès. 1976. *Introduction à l'analyse des données.* Société de Mathématiques appliquées et de Sciences humaines, Paris. xxii + 616 pp. [34, 251, 437, 438]

Cain, A. J. & G. A. Harrison. 1958. An analysis of the taxonomist's judgement of affinity. *Proc. Zool. Soc. Lond.* **131**: 85-98. [38]

Campbell, D. J. & E. Shipp. 1974. Spectral analysis of cyclic behaviour with examples from the field cricket *Teleogryllus commodus* (Walk.). *Anim. Behav.* **22**: 862-875. [682, 684]

Carlier, A. & P. M. Kroonenberg. 1996. Decompositions and biplots in three-way correspondence analysis. *Psychometrika* **61**: 355-373. [251]

Carpenter, S. R. & J. E. Chaney. 1983. Scale of spatial pattern: four methods compared. *Vegetatio* **53**: 153-160. [711]

Carroll, J. B. 1957. Biquartimin criterion for rotation to oblique simple structure in factor analysis. *Science (Wash. D. C.)* **126**: 1114-1115. [478]

Carroll, J. D. 1987. Some multidimensional scaling and related procedures devised at Bell Laboratories, with ecological applications. 65-138 *in*: P. Legendre & L. Legendre [eds.] *Developments in numerical ecology.* NATO ASI Series, Vol. G-14. Springer-Verlag, Berlin. [444, 448, 563]

Casgrain, P., P. Legendre, J.-L. Sixou & C. Mouton. 1996. A graph-theory method to establish serological relationships within a bacterial taxon, with example from *Porphyromonas gingivalis. J. Microbiol. Methods* **26**: 225-236. [60]

Cassie, R. M. & A. D. Michael. 1968. Fauna and sediments of an intertidal mud flat: a multivariate analysis. *J. Exp. Mar. Biol. Ecol.* **2**: 1-23. [411]

Cattaneo, A., P. Legendre & T. Niyonsenga. 1993. Exploring periphyton unpredictability. *J. North Am. Benthol. Soc.* **12**: 418-430. [775]

Cattell, R. B. 1952. *Factor analysis — An introduction and manual for the psychologist and social scientist.* Harper, New York. 462 pp. [248]

Cattell, R. B. 1966. The data box: its ordering of total resources in terms of possible relational systems. 67-128 *in*: R. B. Cattell [ed.] *Handbook of multivariate experimental psychology.* Rand McNally & Co., Chicago. [248, 249]

Cattell, R. B. & J. L. Muerle. 1960. The "maxplane" program for factor rotation to oblique simple structure. *Educ. Psychol. Measur.* **20**: 569-590. [479]

Cavalli-Sforza, L. L. & A. W. F. Edwards. 1967. Phylogenetic analysis: models and estimation procedures. *Evolution* **21**: 550-570. [312]

Chalmond, B. 1986. Régression avec résidus spatialement autocorrélés et recherche de la tendance spatiale. *Statist. Anal. Données* **11**: 1-25. [13]

Chambers, J. M. 1977. *Computational methods for data analysis.* Wiley, New York. xi + 268 pp. [543]

Chambers, J. M., W. S. Cleveland, B. Kleiner & P. A. Tukey. 1983. *Graphical methods for data analysis.* Wadsworth International Group, Belmont, California. xiv + 395 pp. [544]

Chambers, J. M. & B. Kleiner. 1982. Graphical techniques for multivariate data and for clustering. 209-244 *in*: P. R. Krishnaiah & L. N. Kanal [eds.] *Handbook of statistics. Vol. 2.* North-Holland Publ. Co., Amsterdam. [381]

Chatfield, C. 1989. *The analysis of time series. 4th edition.* Chapman & Hall, London. xiii + 241 pp. [641]

Cheetham, A. H. & J. E. Hazel. 1969. Binary (presence-absence) similarity coefficients. *J. Paleontol.* **43**: 1130-1136. [252]

Chipman, J. S. 1979. Efficiency of least squares estimation of linear trend when residuals are autocorrelated. *Econometrica* **47**: 115-128. [15]

Chodorowski, A. 1959. Ecological differentiation of turbellarians in Harsz-Lake. *Pol. Arch. Hydrobiol.* **6**: 33-73. [238]

Cicéri, M.-F., B. Marchand & S. Rimbert. 1977. *Introduction à l'analyse de l'espace.* Collection de Géographie applicable. Masson, Paris. ix + 173 pp. [711]

Clark, I. 1979. *Practical geostatistics.* Elsevier Applied Sciences, London. xii + 129 pp. [749]

Clark, P. J. 1952. An extension of the coefficient of divergence for use with multiple characters. *Copeia* **1952**: 61-64. [283]

Clarke, K. R. 1988. Detecting change in benthic community structure. 131-142 *in*: R. Oger [ed.] *Proceedings of invited papers, fourteenth international biometric conference, Namur, Belgium.* Société Adolphe Quételet, Gembloux, Belgium. [560, 561, 562, 563]

Clarke, K. R. 1993. Non-parametric multivariate analyses of changes in community structure. *Aust. J. Ecol.* **18**: 117-143. [560, 561, 562, 563]

Clarke, K. R. & M. Ainsworth. 1993. A method of linking multivariate community structure to environmental variables. *Mar. Ecol. Prog. Ser.* **92**: 205-219. [555]

Clarke, K. R. & R. M. Warwick. 1994. *Change in marine communities – An approach to statistical analysis and interpretation.* Plymouth Marine Laboratory, UK. 144 pp. [555, 560, 563]

Cleveland, W. S. 1979. Robust locally weighted regression and smoothing scatterplots. *J. Amer. Statist. Assoc.* **74**: 829-836. [544]

Cleveland, W. S. 1985. *The elements of graphing data.* Wadsworth, Monterey, California. xii + 323 pp. [544]

Cliff, A. D. & J. K. Ord. 1973. *Spatial autocorrelation.* Pion, London. 178 pp. [15]

Cliff, A. D. & J. K. Ord. 1975. The comparison of means when samples consist of spatially autocorrelated observations. *Environment and Planning A* **7**: 725-734. [14]

Cliff, A. D. & J. K. Ord. 1981. *Spatial processes – Models and applications.* Pion, London. 266 pp. [12, 13, 15, 712, 714, 719, 720, 721, 725, 752]

Clifford, H. T. & D. W. Goodall. 1967. A numerical contribution to the classification of the Poaceae. *Aust. J. Bot.* **15**: 499-519. [315, 361, 363, 364]

Clifford, H. T. & W. Stephenson. 1975. *An introduction to numerical classification.* Academic Press, New York. xii + 229 pp. [252]

Clifford, P., S. Richardson & D. Hémon. 1989. Assessing the significance of the correlation between two spatial processes. *Biometrics* **45**: 123-134. [13, 15]

Clint, M. & A. Jennings. 1970. The evaluation of eigenvalues and eigenvectors of real symmetric matrices by simultaneous iteration. *Computer J.* **13**: 76-80. [418, 464]

Cochran, W. G. 1954. Some methods for strengthening the common χ^2 tests. *Biometrics* **10**: 417-451. [218]

Cochran, W. G. 1977. *Sampling techniques. 3rd edition.* Wiley, New York. [16, 228]

Cochrane, D. & G. H. Orcutt. 1949. Application of least squares regression to relationships containing autocorrelated error terms. *J. Amer. Statist. Assoc.* **44**: 32-61. [15]

Cohen, J. 1988. *Statistical power analysis for the behavioral sciences. 2nd edition.* Lawrence Erlbaum Assoc., Publ., Hillsdale, New Jersey. xxi + 567 pp. [725]

Cole, L. C. 1949. The measurement of interspecific association. *Ecology* **30**: 411-424. [252]

Cole, L. C. 1957. The measurement of partial interspecific association. *Ecology* **38**: 226-233. [252]

Colebrook, J. M. & A. H. Taylor. 1984. Significant time scale of long-term variability in the plankton and the environment. *Rapp. P.-V. Réun. Cons. Int. Explor. Mer.* **183**: 20-26. [691]

Coleman, J. S. 1964. *Introduction to mathematical sociology.* The Free Press of Glencoe, Collier-Macmillan Ltd., New York. xiv + 554 pp. [60]

Colwell, R. K. 1974. Predictability, constancy and contingency of periodic phenomena. *Ecology* **55**: 1148-1153. [670]

Conover, W. J. 1980. *Practical nonparametric statistics. 2nd edition.* Wiley, New York. xiv + 493 pp. [537]

Cook, D. G. & S. J. Pocock. 1983. Multiple regression in geographical mortality studies, with allowance for spatially correlated errors. *Biometrics* **39**: 361-371. [15]

Cooley, W. W. & P. R. Lohnes. 1971. *Multivariate data analysis.* John Wiley & Sons, New York. 364 pp. [132]

Cordier, B. 1965. *Sur l'analyse factorielle des correspondances.* Thèse de doctorat, Université de Rennes, France. [451]

Cox, D. R. 1957. Note on grouping. *J. Amer. Statist. Assoc.* **52**: 543-547. [229]

Cramér, H. 1946. *Mathematical methods of statistics.* Princeton Univ. Press. xvi + 575 pp. [1]

Cressie, N. A. C. 1991. *Statistics for spatial data.* Wiley, New York. xx + 900 pp. [731, 749]

Crowder, M. J. & D. J. Hand. 1990. *Analysis of repeated measures.* Chapman and Hall, London. [13]

Cryer, J. D. 1986. *Time series analysis.* PWS-Kent Pub. Co., Boston. xi + 286 pp. [702]

Cullis, B. R. & A. C. Gleeson. 1991. Spatial analysis of field experiments. An extension to two dimensions. *Biometrics* **47**: 1449-1460. [15]

Czekanowski, J. 1909. Zur Differentialdiagnose der Neandertalgruppe. *Korrespondenz-Blatt deutsch. Ges. Anthropol. Ethnol. Urgesch.* **40**: 44-47. [xiii, 265, 282, 371]

Czekanowski, J. 1913. *Zarys metod statystycznych w zastosowaniu do antropologii.* Travaux de la Société des Sciences de Varsovie. III. Classe des sciences mathématiques et naturelles, no. 5. iv + 228 pp. [265]

Daget, J. 1976. *Les modèles mathématiques en écologie.* Collection d'Écologie, No. 8. Masson, Paris. viii + 172 pp. [252]

Daget, P. 1980. Le nombre de diversité de Hill, un concept unificateur dans la théorie de la diversité écologique. *Acta Oecol. Oecol. Gen.* **1**: 51-70. [241, 242, 243]

Dagnelie, P. 1960. Contribution à l'étude des communautés végétales par l'analyse factorielle. *Bull. Serv. Carte phytogéogr. B* **5**: 7-71, 93-195. [252]

Dagnelie, P. 1965. L'étude des communautés végétales par l'analyse statistique des liaisons entre les espèces et les variables écologiques: principes fondamentaux. *Biometrics* **21**: 345-361. [406]

Dagnelie, P. 1975. *L'analyse statistique à plusieurs variables.* Les Presses agronomiques de Gembloux, Gembloux (Belgique). 362 pp. [184]

D'Agostino, R. B. 1971. An omnibus test of normality for moderate and large sample sizes. *Biometrika* **58**: 341-348. [183]

D'Agostino, R. B. 1972. Small sample probability points for the *D* test of normality. *Biometrika* **59**: 219-221. [183]

D'Agostino, R. B. 1982. Departures from normality, tests for. 315-324 *in*: S. Kotz & N. L. Johnson [eds.] *Encyclopedia of statistical sciences. Vol. 2.* Wiley, New York. [178, 170, 181, 183]

Damsleth, E. & E. Spjøtvoll. 1982. Estimation of trigonometric components in time series. *J. Amer. Statist. Assoc.* **77**: 381-387. [678]

David, M. 1977. *Geostatistical ore reserve estimation.* Developments in Geomathematics, 2. Elsevier Scient. Publ. Co., Amsterdam. xix + 364 pp. [16, 718, 731, 732, 749]

Davies, P. T. & M. K.-S. Tso. 1982. Procedures for reduced-rank regression. *Appl. Statist.* **31**: 244-255. [605]

Davis, L. S. 1975. A survey of edge detection techniques. *Computer Graphics Image Process* **4**: 248-270. [761]

Day, W. H. E. 1977. Validity of clusters formed by graph-theoretic cluster methods. *Math. Biosci.* **36**: 299-317. [359, 374]

Day, W. H. E. 1983. Distribution of distances between pairs of classifications. 127-131 *in*: J. Felsenstein [ed.] *Numerical taxonomy.* NATO ASI Series, Vol. G-1. Springer-Verlag, Berlin. [489]

Day, W. H. E. 1986. Analysis of quartet dissimilarity measures between undirected phylogenetic trees. *Syst. Zool.* **35**: 325-333. [489]

de Boor, C. 1978. *A practical guide to splines.* Springer-Verlag, Berlin. xxiv + 392 pp. [543]

de Gruijter, J. J. & C. J. F. ter Braak. 1990. Model-free estimation from spatial samples: a reappraisal of classical sampling theory. *Math. Geol.* **22**: 407-415. [6]

De Neufville, R. & J. H. Stafford. 1971. *Systems analysis for engineers and managers.* McGraw-Hill, New York. xiii + 353 pp. [xiii, 167]

De Soete, G., J. D. Carroll & W. S. DeSarbo. 1987. Least squares algorithms for constructing constrained ultrametric and additive tree representations of symmetric proximity data. *J. Classif.* **4**: 155-173. [759, 759]

Deevey, E. S. 1969. Coaxing history to conduct experiments. *BioScience* **19**: 40-43. [238]

Demers, S., P. E. Lafleur, L. Legendre & C. L. Trump. 1979. Short-term covariability of chlorophyll and temperature in the St. Lawrence Estuary. *J. Fish. Res. Board Can.* **36**: 568-573. [682]

Demers, S. & L. Legendre. 1981. Mélange vertical et capacité photosynthétique du phytoplancton estuarien (estuaire du Saint-Laurent). *Mar. Biol (Berl.)* **64**: 243-250. [676]

Denman, K. L. 1976. Covariability of chlorophyll and temperature in the sea. *Deep-Sea Res.* **23**: 539-550. [682]

Denman, K. L. 1977. Short-term variability in vertical chlorophyll structure. *Limnol. Oceanogr.* **22**: 434-441. [682]

Denman, K. L., A. Okubo & T. Platt. 1977. The chlorophyll fluctuation spectrum in the sea. *Limnol. Oceanogr.* **22**: 1033-1038. [682]

Denman, K. L. & T. Platt. 1975. Coherences in the horizontal distributions of phytoplankton and temperature in the upper ocean. *Mém. Soc. R. Sci. Liège, Ser. 6.* **7**: 19-30. [682, 686, 687]

Denman, K. L. & T. Platt. 1976. The variance spectrum of phytoplankton in a turbulent ocean. *J. Mar. Res.* **34**: 593-601. [682]

Dessier, A. & A. Laurec. 1978. Le cycle annuel du zooplancton à Pointe-Noire (RP Congo). Description mathématique. *Oceanol. Acta* **1**: 285-304. [692]

Deutsch, C. V. & A. G. Journel. 1992. *GSLIB – Geostatistical software library and user's guide.* Oxford University Press, New York. [767, 773]

Dévaux, J. & G. Millerioux. 1976a. Possibilité de l'utilisation de la cotation d'abondance de Frontier (1969) pour l'analyse multivariable des populations phytoplanctoniques. *C. R. Hebd. Séances Acad. Sci., Sér. D Sci. Nat.* **283**: 41-44. [30, 412]

Dévaux, J. & G. Millerioux. 1976b. Méthode d'estimation de la biomasse totale du phytoplancton à partir des nombres de cellules, issus d'une cotation d'abondance. *C. R. Hebd. Séances Acad. Sci., Sér. D Sci. Nat.* **283**: 927-930. [30]

Dévaux, J. & G. Millerioux. 1977. Sur la possibilité d'un calcul de la diversité spécifique de populations phytoplanctoniques à partir de dénombrements issus d'une cotation d'abondance. *C. R. Hebd. Séances Acad. Sci., Sér. D Sci. Nat.* **284**: 1569-1571. [30, 189]

Dice, L. R. 1945. Measures of the amount of ecologic association between species. *Ecology* **26**: 297-302. [257, 294]

Dickman, M. 1968. Some indices of diversity. *Ecology* **49**: 1191-1193. [239]

Dietz, E. J. 1983. Permutation tests for association between two distance matrices. *Syst. Zool.* **32**: 21-26. [552, 554]

Digby, P. G. N. & R. A. Kempton. 1987. *Multivariate analysis of ecological communities.* Chapman & Hall, London. viii + 206. pp. [60]

Diggle, P. J. 1990. *Time series – A biostatistical introduction.* Oxford University Press, New York. xi + 257 pp. [641]

Dirichlet, G. L. 1850. Über die Reduktion der positiven quadratischen Formen mit drei unbestimmten ganzen Zahlen. *Journal für die reine und angewandte Mathematik* **40**: 209-234. [752, 756]

Dixon, W. J. [ed.] 1981. *BMDP statistical software 1981.* Univ. California Press, Berkeley. x + 725 pp. [217, 218, 227, 228]

Dolédec, S. & D. Chessel. 1994. Co-inertia analysis: an alternative method for studying species-environment relationships. *Freshwater Biol.* **31**: 277-294. [616]

Downing, J. A. 1979. Aggregation, transformation, and the design of benthos sampling programs. *J. Fish. Res. Board Can.* **36**: 1454-1463. [44]

Draper, N. & H. Smith. 1981. *Applied regression analysis. 2nd edition.* Wiley, New York. xiv + 709 pp. [498, 525, 743]

Dufrêne, M. & P. Legendre. 1991. Geographic structure and potential ecological factors in Belgium. *J. Biogeogr.* **18**: 257-266. [762]

Dufrêne, M. & P. Legendre. 1997. Species assemblages and indicator species: the need for a flexible asymmetrical approach. *Ecol. Monogr.* **67**: 345-366. [369, 370]

Dutilleul, P. 1990. *Apport en analyse spectrale d'un périodogramme modifié et modélisation des séries chronologiques avec répétitions en vue de leur comparaison en fréquence.* Doctoral Dissertation, Univ. Cath. Louvain, Louvain-la-Neuve, Belgium. vi + 304 pp. [676, 677]

Dutilleul, P. 1993a. Modifying the *t* test for assessing the correlation between two spatial processes. *Biometrics* **49**: 305-314. [13, 15]

Dutilleul, P. 1993b. Spatial heterogeneity and the design of ecological field experiments. *Ecology* **74**: 1646-1658. [16, 712]

Dutilleul, P. 1998. Incorporating scale in ecological experiments: data analysis. 387-425 in: D. L. Peterson & V. T. Parker [eds.] *Ecological scale – Theory and applications.* Columbia University Press, New York. [677, 678]

Dutilleul, P. & P. Legendre. 1992. Lack of robustness in two tests of normality against autocorrelation in sample data. *J. Statist. Comput. Simulation* **42**: 79-91. [13, 178, 183]

Dutilleul, P. & P. Legendre. 1993. Spatial heterogeneity against heteroscedasticity: an ecological paradigm versus a statistical concept. *Oikos* **66**: 152-171. [16, 710, 711, 712]

Dutilleul, P. & C. Till. 1992. Evidence of periodicities related to climate and planetary behaviors in ring-width chronologies of Atlas cedar (*Cedrus atlantica*) in Morocco. *Can. J. For. Res.* **22**: 1469-1482. [677, 678]

Eber, H. W. 1966. Toward oblique simple structure maxplane. *Multivar. Behav. Res.* **1**: 112-125. [479]

Eckart, C. & G. Young. 1936. The approximation of one matrix by another of lower rank. *Psychometrika* **1**: 211-218. [94]

Edgington, E. S. 1987. *Randomization tests. 2nd edition.* Marcel Dekker Inc., New York. xvii + 341 pp. [516]

Edgington, E. S. 1995. *Randomization tests. 3rd edition.* Marcel Dekker, Inc., New York. xxii + 409 pp. [20, 21, 22, 24, 25, 26, 607, 609]

Edwards, A. W. F. & L. L. Cavalli-Sforza. 1965. A method for cluster analysis. *Biometrics* **21**: 362-375. [345]

Efron, B. 1979. Bootstrap methods: another look at the jackknife. *Ann. Stat.* **7**: 1-26. [26]

Efron, B. & R. J. Tibshirani. 1993. *An introduction to the bootstrap.* Chapman & Hall, New York. xvi + 436 pp. [26]

Eisner, F. 1931. Das Wiederstands Problem. *in*: *3ième Congrès international de Mécanique appliquée*, Stockholm. [106]

Enright, J. T. 1965. The search for rhythmicity in biological time-series. *J. Theor. Biol.* **8**: 426-468. [666, 667, 668, 669]

Escofier-Cordier, B. 1969. L'analyse factorielle des correspondances. *Cah. Bur. Univ. Rech. Opér. Univ. Paris* **13**: 25-59. [451]

Estabrook, G. F. 1966. A mathematical model in graph theory for biological classification. *J. Theor. Biol.* **12**: 297-310. [311, 374]

Estabrook, G. F. & B. Gates. 1984. Character analysis of the *Banisteriopsis campestris* complex (Malpighiaceae), using spatial autocorrelation. *Taxon* **33**: 13-25. [721]

Estabrook, G. F. & D. J. Rogers. 1966. A general method of taxonomic description for a computed similarity measure. *Bioscience* **16**: 789-793. [260, 261, 263, 375]

Eubank, R. L. 1988. *Spline smoothing and nonparametric regression.* Marcel Dekker, New York. xvii + 438 pp. [543]

Everitt, B. S. 1980. *Cluster Analysis. 2nd edition.* Halsted Press, New York. 136 pp. [332, 352]

Fager, E. W. 1957. Determination and analysis of recurrent groups. *Ecology* **38**: 586-595. [292, 294, 358, 359]

Fager, E. W. 1963. Communities of organisms. 415-437 *in*: M. N. Hill [ed.] *The sea. Vol. 2.* Interscience Publ., New York. [356]

Fager, E. W. & J. A. McGowan. 1963. Zooplankton species groups in the North Pacific. *Science (Wash. D. C.)* **140**: 453-460. [294, 356, 358, 359]

Faith, D. P. 1983. Asymmetric binary similarity measures. *Oecologia (Berl.)* **57**: 287-290. [258]

Faith, D. P., P. R. Minchin & L. Belbin. 1987. Compositional dissimilarity as a robust measure of ecological distance. *Vegetatio* **69**: 57-68. [444, 449]

Falconer, D. S. 1960. *Introduction to quantitative genetics.* Ronald Press Inc., New York. ix + 365 pp. [29]

Fasham, M. J. R. 1977. A comparison of nonmetric multidimensional scaling, principal component and reciprocal averaging for the ordination of simulated coenoclines and coenoplanes. *Ecology* **58**: 551-561. [450]

Fasham, M. J. R. & P. R. Pugh. 1976. Observations on the horizontal coherence of chlorophyll *a* and temperature. *Deep-Sea Res.* **23**: 527-538. [682]

Fausett, L. 1994. *Fundamentals of neural networks – Architectures, algorithms and applications.* Prentice Hall, Englewood Cliffs, New Jersey. xvi + 461 pp. [304]

Ferligoj, A. & V. Batagelj. 1982. Clustering with relational constraint. *Psychometrika* **47:** 413-426. [758, 759]

Ferligoj, A. & V. Batagelj. 1983. Some types of clustering with relational constraints. *Psychometrika* **48**: 541-552. [759]

Field, J. G. 1969. The use of the information statistic in the numerical classification of heterogeneous systems. *J. Ecol.* **57**: 565-569. [341]

Field, J. G., K. R. Clarke & R. M. Warwick. 1982. A practical strategy for analysing multispecies distribution patterns. *Mar. Ecol. Prog. Ser.* **8**: 37-52. [483]

Field, J. G. & F. T. Robb. 1970. Numerical methods in marine ecology. 2. Gradient analysis of rocky shore samples from False Bay. *Zool. Afr.* **5**: 191-210. [440, 441]

Fienberg, S. E. 1970. The analysis of multidimensional contingency tables. *Ecology* **51**: 419-433. [230]

Fienberg, S. E. 1980. *The analysis of cross-classified categorical data. 2nd edition.* MIT Press, Cambridge, Mass. xiv + 198 pp. [218, 227, 228, 229, 497]

Finn, J. D. 1974. *A general model for multivariate analysis.* Holt, Rinehart and Winston, New York. xiii + 423 pp. [518, 582]

Fisher, R. A. 1935. *The design of experiments.* Oliver and Boyd, Edinburgh. 252 pp. [20]

Fisher, R. A. 1936. The use of multiple measurements in taxonomic problems. *Annals of Eugenics* **7:** 179-188. [617]

Fisher, R. A. 1940. The precision of discriminant functions. *Annals of Eugenics* **10**: 422-429. [451]

Fisher, R. A. 1954. *Statistical methods for research workers. 12th edition.* Oliver & Boyd, Edinburgh. 356 pp. [18]

Fisher, W. D. 1958. On grouping for maximum homogeneity. *J. Amer. Statist. Assoc.* **53**: 789-798. [693]

Flos, J. 1976. Seston superficial de la zona de afloramiento del NW de Africa. *Oecologia Aquatica.* **2**: 27-39. [324]

Ford, E. D. 1976. The canopy of a Scots pine forest: description of a surface of complex roughness. *Agric. Meteorol.* **17**: 9-32. [728]

Fortier, L. & L. Legendre. 1979. Le contrôle de la variabilité à court terme du phytoplancton estuarien: stabilité verticale et profondeur critique. *J. Fish. Res. Board Can.* **36**: 1325-1335. [663]

Fortier, L., L. Legendre, A. Cardinal & C. L. Trump. 1978. Variabilité à court terme du phytoplancton de l'estuaire du Saint-Laurent. *Mar. Biol. (Berl.)* **46**: 349-354. [648]

Fortin, M.-J. 1994. Edge detection algorithms for two-dimensional ecological data. *Ecology* **75**: 956-965. [761, 763]

Fortin, M.-J. 1997. Effects of data types on vegetation boundary delineation. *Can. J. For. Res.* **27**: 1851-1858. [761, 763]

Fortin, M.-J. & P. Drapeau. 1995. Delineation of ecological boundaries: comparison of approaches and significance tests. *Oikos* **72**: 323-332. [761, 763]

Fortin, M.-J., P. Drapeau & G. M. Jacquez. 1996. Quantification of the spatial co-occurrence of ecological boundaries. *Oikos* **77**: 51-60. [761, 763]

Fortin, M.-J., P. Drapeau & P. Legendre. 1989. Spatial autocorrelation and sampling design in plant ecology. *Vegetatio* **83**: 209-222. [735]

Fourier, J. 1822. Théorie analytique de la chaleur. *in*: G. Darboux [ed.] *1888, Oeuvres de Fourier. Tome 1.* Gauthier-Villars, Paris. [97]

François-Bongarçon, D. 1991. Geostatistical determination of sample variances in the sampling of broken gold ores. *CIM Bulletin* **84** (950): 46-57. [16]

Fréchet, A. 1990. Catchability variations of cod in the marginal ice zone. *Can. J. Fish. Aquat. Sci.* **47**: 1678-1683. [230]

Fréchette, M., & L. Legendre. 1982. Phytoplankton photosynthetic response to light in an internal tide dominated environment. *Estuaries* **5**: 287-293. [663, 664]

Freedman, D. & D. Lane. 1983. A nonstochastic interpretation of reported significance levels. *J. Bus. Econ. Statist.* **1**: 292-298. [609, 611]

Freund, R. J. & P. D. Minton. 1979. *Regression methods – A tool for data analysis.* Statistics: Textbooks and Monographs, Vol. 30. Marcel Dekker Inc., New York. xi + 261 pp. [522]

Friedman, M. 1937. The use of ranks to avoid the assumption of normality implicit in the analysis of variance. *J. Amer. Statist. Assoc.* **32**: 675-701. [203]

Fromentin, J.-M., F. Ibanez & P. Legendre. 1993. A phytosociological method for interpreting plankton data. *Mar. Ecol. Prog. Ser.* **93**: 285-306. [780]

Frontier, S. 1969. Sur une méthode d'analyse faunistique rapide du zooplancton. *J. Exp. Mar. Biol. Ecol.* **3**: 18-26. [30]

Frontier, S. 1973. Evaluation de la quantité totale d'une catégorie d'organismes planctoniques dans un secteur néritique. *J. Exp. Mar. Biol. Ecol.* **12**: 299-304. [30]

Frontier, S. 1976. Étude de la décroissance des valeurs propres dans une analyse en composantes principales: comparaison avec le modèle du bâton brisé. *J. Exp. Mar. Biol. Ecol.* **25**: 67-75. [410, 837]

Frontier, S. & F. Ibanez. 1974. Utilisation d'une cotation d'abondance fondée sur la progression géométrique, pour l'analyse en composantes principales en écologie planctonique. *J. Exp. Mar. Biol. Ecol.* **14**: 217-224. [30, 412]

Frontier, S. & D. Viale. 1977. Utilisation d'une cotation d'abondance mise au point en planctologie pour l'évaluation des troupeaux de cétacés en mer. *J. Rech. Océanogr.* **2**: 15-22. [30]

Fry, J. C., N. C. B. Humphrey & T. C. Iles. 1981. Time-series analysis for identifying cyclic components in microbiological data. *J. Appl. Bacteriol.* **50**: 189-224. [641, 679]

Furnas, G. W. 1984. The generation of random, binary unordered trees. *J. Classif.* **1**: 187-233. [25]

Gabbott, P. A. & V. N. Larman. 1971. Electrophoretic examination of partially purified extracts of *Balanus balanoides* containing a settlement inducing factor. 143-153 *in*: D. J. Crisp [ed.] *Fourth european marine biology symposium.* Cambridge Univ. Press, Cambridge. [726]

Gabriel, K. R. 1971. The biplot graphic display of matrices with application to principal component analysis. *Biometrika* **58**: 453-467. [403]

Gabriel, K. R. 1982. Biplot. 263-271 *in:* S. Kotz & N. L. Johnson [eds.] *Encyclopedia of statistical sciences.* Vol. 1. Wiley, New York. [403]

Gabriel, K. R. & R. R. Sokal. 1969. A new statistical approach to geographic variation analysis. *Syst. Zool.* **18**: 259-278. [753]

Galiano, E. F. 1982. Pattern detection in plant populations through the analysis of plant-to-all-plants distances. *Vegetatio* **49**: 39-43. [711, 721]

Galton, F. 1889. *Natural inheritance.* Macmillan & Co., London. ix + 259 pp. [499]

Galzin, R. & P. Legendre. 1987. The fish communities of a coral reef transect. *Pac. Sci.* **41**: 158-165. [696]

Garland, T. Jr. 1983. The relation between maximal running speed and body mass in terrestrial mammals. *J. Zool. (Lond.)* **199**: 157-170. [110]

Gauch, H. G. Jr. 1982. *Multivariate analysis in community ecology.* Cambridge University Press, Cambridge. x + 298 pp. [466, 470]

Gauch, H. G. Jr., R. H. Whittaker & T. R. Wentworth. 1977. A comparative study of reciprocal averaging and other ordination techniques. *J. Ecol.* **65**: 157-174. [464]

Gause, G. F. 1935. *Vérification expérimentale de la théorie mathématique de la lutte pour la vie.* Actual. Sci. Ind. no 277. Hermann Éditeur, Paris. [463]

Gauss, K. F. 1809. *Theoria motus corporum coelestium in sectionibus conicis solem ambientium.* Frid. Perthes et I. H. Besser, Hamburg. [501]

Geary, R. C. 1954. The contiguity ratio and statistical mapping. *Incorp. Statist.* **5**: 115-145. [714]

Getis, A. & B. Boots. 1978. *Models of spatial processes – An approach to the study of point, line and area patterns.* Cambridge Univ. Press, Cambridge. [711]

Gifi, A. 1990. *Nonlinear multivariate analysis.* John Wiley & Sons, Chichester. xx + 579 pp. [34]

Gilbert, R. O. & J. C. Simpson. 1985. Kriging for estimating spatial pattern of contaminants: potential and problems. *Environ. Monit. Assess.* **5**: 113-135. [749]

Gittins, R. 1985. *Canonical analysis – A review with applications in ecology.* Springer-Verlag, Berlin. 351 pp. [579, 612, 616, 626, 631]

Glansdorff, P. & I. Prigogine. 1971. *Structure, stabilité et fluctuations.* Masson, Paris. 288 pp. [31, 638]

Gleick, J. 1987. *Chaos: making a new science.* Viking, New York, xi + 352 p. [2]

Gokhale, D. V. & S. Kullback. 1978. *The information in contingency tables.* Marcel Dekker Inc., New York. x + 365 pp. [223]

Gold, H. J. 1977. *Mathematical modelling of biological systems – An introductory guidebook.* John Wiley & Sons, New York. xii + 357 pp. [xiii]

Goldstein, M. & W. R. Dillon. 1978. *Discrete discriminant analysis*. John Wiley & Sons, New York. x + 186 pp. [492]

Goodall, D. W. 1954. Objective methods for the classification of vegetation. III. An essay in the use of factor analysis. *Aust. J. Bot.* **2**: 304-324. [387, 466]

Goodall, D. W. 1964. A probabilistic similarity index. *Nature (Lond.)* **203**: 1098. [269]

Goodall, D. W. 1966a. A new similarity index based on probability. *Biometrics* **22**: 882-907. [269, 269]

Goodall, D. W. 1966b. Deviant index: a new tool for numerical taxonomy. *Nature (Lond.)* **210**: 216. [346]

Goodall, D. W. 1974. A new method for the analysis of spatial pattern by random pairing of quadrats. *Vegetatio* **29**: 135-146. [721]

Goodman, L. A. & W. H. Kruskal. 1954. Measures of association for cross classifications. *J. Amer. Statist. Assoc.* **49**: 732-764. [252]

Goodman, L. A. & W. H. Kruskal. 1959. Measures of association for cross classifications. II. Further discussion and references. *J. Amer. Statist. Assoc.* **54**: 123-163. [252]

Goodman, L. A. & W. H. Kruskal. 1963. Measures of association for cross classifications. III. Approximate sampling theory. *J. Amer. Statist. Assoc.* **58**: 310-364. [252]

Gordon, A. D. 1973. Classification in the presence of constraints. *Biometrics* **29**: 821-827. [696]

Gordon, A. D. 1994. Identifying genuine clusters in a classification. *Comput. Statist. Data Anal.* **18**: 561-581. [378, 379, 563]

Gordon, A. D. 1996a. Hierarchical classification. 65-121 *in*: P. Arabie, L. J. Hubert & G. De Soete [eds.] *Clustering and Classification*. World Scientific Publ. Co., River Edge, New Jersey. [334, 378]

Gordon, A. D. 1996b. Null models in cluster validation. 32-44 *in*: W. Gaul & D. Pfeifer [eds.] *From data to knowledge*. Springer-Verlag, Berlin. [378]

Gordon, A. D. 1996c. A survey of constrained classification. *Comput. Statist. Data Anal.* **21**: 17-29. [752, 758, 759]

Gordon, A. D. & H. J. B. Birks. 1972. Numerical methods in Quaternary palaeoecology. I. Zonation of pollen diagrams. *New Phytol.* **71**: 961-979. [696]

Gordon, A. D. & H. J. B. Birks. 1974. Numerical methods in Quaternary palaeoecology. II. Comparison of pollen diagrams. *New Phytol.* **73**: 221-249. [696]

Gosselin, M., L. Legendre, S. Demers, J.-C. Therriault & M. Rochet. 1986. Physical control of the horizontal patchiness of sea-ice microalgae. *Mar. Ecol. Prog. Ser.* **29**: 289-295. [550]

Gower, J. C. 1966. Some distance properties of latent root and vector methods used in multivariate analysis. *Biometrika.* **53**: 325-338. [347, 407, 412, 424, 425, 429, 431, 449]

Gower, J. C. 1967. A comparison of some methods of cluster analysis. *Biometrics* **23**: 623-637. [314, 322, 324, 326, 344, 346]

Gower, J. C. 1971a. A general coefficient of similarity and some of its properties. *Biometrics* **27**: 857-871. [258, 266]

Gower, J. C. 1971b. Statistical methods of comparing different multivariate analyses of the same data. 138-149 *in*: F. R. Hodson, D. G. Kendall & P. Tautu [eds.] *Mathematics in the archaeological and historical sciences*. Edinburgh University Press, Edinburgh. [563]

Gower, J. C. 1975. Generalized Procrustes analysis. *Psychometrika* **40**: 33-51. [563]

Gower, J. C. 1982. Euclidean distance geometry. *Math. Scientist* **7**: 1-14. [425, 431, 432]

Gower, J. C. 1983. Comparing classifications. 137-155 *in*: J. Felsenstein [ed.] *Numerical taxonomy.* NATO ASI Series, Vol. G-1. Springer-Verlag, Berlin. 644 pp. [376]

Gower, J. C. 1984. Ordination, multidimensional scaling and allied topics. 727-781 *in*: W. Lederman [ed.] *Handbook of Applicable Mathematics*. Vol. VI: E. Lloyd [ed.] *Statistics*. Wiley, Chichester. [387, 390]

Gower, J. C. 1985. Measures of similarity, dissimilarity, and distance. 397-405 *in*: S. Kotz & N. L. Johnson [eds.] *Encyclopedia of statistical sciences. Vol. 5.* Wiley, New York. [252, 432]

Gower, J. C. 1987. Introduction to ordination techniques. 3-64 *in*: P. Legendre & L. Legendre [eds.] *Developments in numerical ecology.* NATO ASI Series, Vol. G14. Springer-Verlag, Berlin. [444, 563]

Gower, J. C. & P. Legendre. 1986. Metric and Euclidean properties of dissimilarity coefficients. *J. Classif.* **3**: 5-48. [252, 275, 298, 433]

Gower, J. C. & G. J. S. Ross. 1969. Minimum spanning trees and single linkage cluster analysis. *Appl. Statist.* **18**: 54-64. [312, 483]

Graybill, F. A. 1983. *Matrices with applications in statistics. 2d edition.* Wadsworth, Belmont. xi + 461. [51, 132]

Green, P. G. & J. D. Carroll. 1976. *Mathematical tools for applied multivariate analysis.* Academic Press, New York. 376 pp. [52]

Green, R. H. 1979. *Sampling design and statistical methods for environmental biologists.* John Wiley & Sons, New York. xi + 257 pp. [16, 228, 249]

Greenacre, M. J. 1983. *Theory and applications of correspondence analysis.* Academic Press, London. xi + 364 pp. [390, 451]

Greenberg, J. H. 1956. The measurement of linguistic diversity. *Language* **32**: 109-115. [242]

Griffith, D. A. 1978. A spatially adjusted ANOVA model. *Geogr. Anal.* **10**: 296-301. [15]

Griffith, D. A. 1987. *Spatial autocorrelation – A primer.* Association of American Geographers, Washington, D. C. 10 + 82 pp. [15, 712]

Griffith, D. A. 1988. *Advanced Spatial Statistics.* Kluwer, Dordrecht. xiv + 273 pp. [12, 13]

Grondona, M. O. & N. Cressie. 1991. Using spatial considerations in the analysis of experiments. *Technometrics* **33**: 381-392. [15]

Guille, A. 1970. Bionomie benthique du plateau continental de la côte catalane française. II. Les communautés de la macrofaune. *Vie Milieu* **21**: 149-280. [373]

Günther, B. 1975. Dimensional analysis and theory of biological similarity. *Physiol. Rev.* **55**: 659-699. [110, 129]

Gutierrez, L. T. & W. R. Fey. 1980. *Ecosystem succession – A general hypothesis and a test model of a grassland.* MIT Press, Cambridge. xiv + 231 pp. [238]

Haining, R. 1987. Trend-surface models with regional and local scales of variation with an application to aerial survey data. *Technometrics* **29**: 461-469. [745]

Haining, R. 1990. *Spatial data analysis in the social and environmental sciences.* Cambridge University Press, Cambridge. xxi + 409 pp. [13]

Hajdu, L. J. 1981. Geographical comparison of resemblance measures in phytosociology. *Vegetatio* **48**: 47-59. [297, 298]

Hájek, J. 1969. *A course in nonparametric statistics.* Holden-Day, San Francisco. viii + 184 pp. [191]

Hall, A. V. 1965. The peculiarity index, a new function for use in numerical taxonomy. *Nature (Lond.)* **206**: 952. [346]

Hall, P., N. I. Fisher & B. Hoffmann. 1994. On the nonparametric estimation of covariance functions. *Ann. Stat.* **22**: 2115-2134. [726]

Hall, P. & P. Patil. 1994. Properties of nonparametric estimators of autocovariance for stationary random fields. *Probability Theory and Related Fields* **99**: 399-424. [726]

Hall, P. & D. M. Titterington. 1989. The effect of simulation order on level accuracy and power of Monte Carlo tests. *J. Roy. Statist. Soc. Ser. B* **51**: 459-467. [610]

Hann, B. J., P. R. Leavitt & P. S. S. Chang. 1994. Cladocera community response to experimental eutrophication in Lake 227 as recorded in laminated sediments. *Can. J. Fish. Aquat. Sci.* **51**: 2312-2321. [696]

Harris, C. W. & H. F. Kaiser. 1964. Oblique factor analytic solutions by orthogonal transformations. *Psychometrika* **29**: 347-362. [479]

Harris, R. E. & W. A. G. Charleston. 1977. An examination of the marsh microhabitats of *Lymnaea tomentosa* and *L. columella* (Mollusca: Gastropoda) by path analysis. *N. Z. J. Zool.* **4**: 395-399. [550]

Hartigan, J. A. 1975. *Cluster algorithms.* John Wiley & Sons, New York. xiii + 349 pp. [352]

Harvey, A. C. 1981. *The econometric analysis of time series.* Wiley, New York. xi + 384 pp. [15]

Harvey, A. C. & G. D. A. Phillips. 1979. Maximum likelihood estimation of regression models with autoregressive-moving average disturbances. *Biometrika* **66**: 49-58. [15]

Hatcher, A. & C. A. Frith. 1985. The control of nitrate and ammonium concentrations in a coral reef lagoon. *Coral Reefs* **4**: 101-110. [112]

Hatcher, B. G., J. Imberger & S. V. Smith. 1987. Scaling analysis of coral reef systems: an approach to problems of scale. *Coral Reefs* **5**: 171-181. [112]

Hatheway, W. H. 1971. Contingency-table analysis of rain forest vegetation. 271-313 *in*: G. P. Patil, E. C. Pielou & W. E. Waters [eds.] *Statistical ecology.* Vol. 3: *Many species populations, ecosystems, and systems analysis.* Pennsylvania State University Press, University Park and London. [451]

Hawkins, D. M. & D. F. Merriam. 1973. Optimal zonation of digitized sequential data. *Math. Geol.* **5**: 389-395. [693]

Hawkins, D. M. & D. F. Merriam. 1974. Zonation of multivariate sequences of digitized geologic data. *Math. Geol.* **6**: 263-269. [693]

Hawksworth, F. G., G. F. Estabrook & D. J. Rogers. 1968. Application of an information theory model for character analysis in the genus *Arceuthobium* (Viscaceae). *Taxon* **17**: 605-619. [221]

He, F., P. Legendre, C. Bellehumeur & J. V. LaFrankie. 1994. Diversity pattern and spatial scale: a study of a tropical rain forest of Malaysia. *Environ. Ecol. Stat.* **1**: 265-286. [32, 539, 708, 710, 773, 785]

He, F., P. Legendre & J. V. LaFrankie. 1996. Spatial pattern of diversity in a tropical rain forest of Malaysia. *J. Biogeogr.* **23**: 57-74. [240, 539]

He, F., P. Legendre & J. V. LaFrankie. 1997. Distribution patterns of tree species in a Malaysian tropical rain forest. *J. Veg. Sci.* **8**: 105-114. [539, 540]

Hecky, R. E. & P. Kilham. 1988. Nutrient limitation of phytoplankton in freshwater and marine environments. A review of recent evidence on the effects of enrichment. *Limnol. Oceanogr.* **33**: 796-822. [7]

Heikkinen, R. K. & H. J. B. Birks. 1996. Spatial and environmental components of variation in the distribution patterns of subarctic plant species at Kevo, N. Finland. A case study at the meso-scale level. *Ecography* **19**: 341-351. [775]

Hendrickson, A. E. & P. O. White. 1964. Promax: a quick method for rotation to oblique simple structure. *Brit. J. Stat. Psychol.* **17**: 65-70. [479]

Hermy, M. 1987. Path analysis of standing crop and environmental variables in the field layer of two Belgian riverine forests. *Vegetatio* **70**: 127-134. [550]

Hewitt, J. E., P. Legendre, B. H. McArdle, S. F. Thrush, C. Bellehumeur & S. M. Lawrie. 1997. Identifying relationships between adult and juvenile bivalves at different spatial scales. *J. Exp. Mar. Biol. Ecol.* **216**: 77-98. [671, 735, 736]

Hill, A. V. 1950. The dimensions of animals and their muscular dynamics. *Sci. Prog.* **38**: 209-230. [107, 109]

Hill, M. O. 1973a. Diversity and evenness: a unifying notation and its consequences. *Ecology* **54**: 427-432. [239, 242, 243]

Hill, M. O. 1973b. Reciprocal averaging: an eigenvector method of ordination. *J. Ecol.* **61**: 237-249. [418, 451, 463, 464, 475]

Hill, M. O. 1974. Correspondence analysis: a neglected multivariate method. *Appl. Statist.* **23**: 340-354. [390, 451, 455, 463, 464]

Hill, M. O. 1979. *DECORANA – A FORTRAN program for detrended correspondence analysis and reciprocal averaging.* Section of Ecology and Systematics, Cornell University, Ithaca, New York. 52 pp. [468, 470]

Hill, M. O. 1979. *TWINSPAN – A FORTRAN program for arranging multivariate data in an ordered two-way table by classification of the individuals and attributes.* Section of Ecology and Systematics, Cornell University, Ithaca, New York. iv + 49 pp. [347, 368]

Hill, M. O. & H. G. Gauch Jr. 1980. Detrended correspondence analysis, an improved ordination technique. *Vegetatio* **42**: 47-58. [466, 467, 470]

Hirschfeld, H. O. 1935. A connection between correlation and contingency. *Proc. Camb. Phil. Soc.* **31**: 520-524. [451]

Hobbs, R. J. & H. A. Mooney [eds.] 1990. *Remote sensing of biosphere functioning.* Springer-Verlag, New York. 350 pp. [761]

Hochberg, Y. 1988. A sharper Bonferroni procedure for multiple tests of significance. *Biometrika* **75**: 800-802. [18]

Hocking, R. R. 1976. The analysis and selection of variables in linear regression. *Biometrics* **32**: 1-49. [521]

Hoerl, A. E. 1962. Application of ridge analysis to regression problems. *Chem. Eng. Prog.* **58** (3): 54-59. [523, 524]

Hoerl, A. E. & R. W. Kennard. 1970a. Ridge regression: biased estimation for nonorthogonal problems. *Technometrics.* **12**: 55-67. [523]

Hoerl, A. E. & R. W. Kennard. 1970b. Ridge regression: applications to nonorthogonal problems. *Technometrics* **12**: 69-82. [523]

Hollander, M. & D. A. Wolfe. 1973. *Nonparametric statistical methods.* Wiley, New York. [537]

Holm, S. 1979. A simple sequentially rejective multiple test procedure. *Scand. J. Stat.* **6**: 65-70. [18]

Hope, A. C. A. 1968. A simplified Monte Carlo significance test procedure. *J. Roy. Statist. Soc. Ser. B* **30**: 582-598. [22]

Hotelling, H. 1931. The generalization of Student's ratio. *Ann. Math. Statist.* **2**: 360-378. [281]

Hotelling, H. 1933. Analysis of a complex of statistical variables into principal components. *J. Educ. Psychol.* **24**: 417-441, 498-520. [xiii, 391]

Hotelling, H. 1936. Relations between two sets of variates. *Biometrika* **28**: 321-377. [612]

Hotelling, H. & M. R. Pabst. 1936. Rank correlation and tests of significance involving no assumption of normality. *Ann. Math. Statist.* **7**: 29-43. [202]

Huang, J. S. & D. H. Tseng. 1988. Statistical theory of edge detection. *Computer Vision Graphics Image Process* **43**: 337-346. [761]

Hubálek, Z. 1982. Coefficients of association and similarity, based on binary (presence-absence) data: an evaluation. *Biol. Rev.* **57**: 669-689. [298]

Hubert, L. J. & P. Arabie. 1985. Comparing partitions. *J. Classif.* **2**: 193-218. [376]

Hubert, L. J. & F. B. Baker. 1977. The comparison and fitting of given classification schemes. *J. Math. Psychol.* **16**: 233-253. [379]

Hubert, L. J. & R. G. Golledge. 1981. A heuristic method for the comparison of related structures. *J. Math. Psychol.* **23**: 214-226. [559]

Hudon, C., E. Bourget & P. Legendre. 1983. An integrated study of the factors influencing the choices of the settling site of *Balanus crenatus* cyprid larvae. *Can J. Fish. Aquat. Sci.* **40**: 1186-1194. [726]

Hudon, C. & Lamarche, G. 1989. Niche segregation between American lobster *Homarus americanus* and rock crab *Cancer irroratus*. *Mar. Ecol. Prog. Ser.* **52**: 155-168. [556]

Huet, S., E. Jolivet & A. Messéan. 1992. *La régression non-linéaire – Méthodes et applications en biologie.* Institut National de la Recherche Agronomique, Paris. 247 pp. [537]

Huntley, H. E. 1967. *Dimensional analysis.* Dover, New York. 153 pp. [97]

Hurlbert, S. H. 1971. The nonconcept of species diversity: a critique and alternative parameters. *Ecology* **52**: 577-586. [238, 238, 240, 242, 243]

Hurlbert, S. H. 1984. Pseudoreplication and the design of ecological field experiments. *Ecol. Monogr.* **54**:187-211. [7, 131, 495, 707]

Hutchinson, G. E. 1957. Concluding remarks. *Cold Spring Harbor Symp. Quant. Biol.* **22**: 415-427. [3, 245, 253, 356, 463, 569, 707]

Hutchinson, G. E. 1965. *The ecological theater and the evolutionary play.* Yale Univ. Press, New Haven. xiii + 139 pp. [3, 245]

Ibanez, F. 1971. Effet des transformations des données dans l'analyse factorielle en écologie planctonique. *Cah. Océanogr.* **23**: 545-561. [411]

Ibanez, F. 1972. Interprétation de données écologiques par l'analyse des composantes principales: écologie planctonique de la mer du Nord. *J. Cons. Cons. Int. Explor. Mer* **34**: 323-340. [413]

Ibanez, F. 1973. Méthode d'analyse spatio-temporelle du processus d'échantillonnage en planctologie, son influence dans l'interprétation des données par l'analyse en composantes principales. *Ann. Inst. Océanogr. (Paris)* **49**: 83-111. [409]

Ibanez, F. 1976. Contribution à l'analyse mathématique des événements en écologie planctonique. Optimisations méthodologiques; étude expérimentale en continu à petite échelle de l'hétérogénéité du plancton côtier. Thèse de doctorat d'état ès sciences naturelles, Université Paris VI. *Bull. Inst. Océanogr. (Monaco)* **72** (1431): 1-96. [479]

Ibanez, F. 1981. Immediate detection of heterogeneities in continuous multivariate, oceanographic recordings. Application to time series analysis of changes in the bay of Villefranche sur Mer. *Limnol. Oceanogr.* **26**: 336-349. [696]

Ibanez, F. 1982. L'échantillonnage en continu en océanographie. 365-384 *in*: S. Frontier [ed.] *Stratégies d'échantillonnage en écologie.* Collection d'Écologie, No. 17. Masson, Paris et les Presses de l'Université Laval, Québec. [703]

Ibanez, F. 1984. Sur la segmentation des séries chronologiques planctoniques multivariables. *Oceanol. Acta* **7**: 481-491. [693]

Ibanez, F. & G. Seguin. 1972. Etude du cycle annuel du zooplancton d'Abidjan. Comparaison de plusieurs méthodes d'analyse multivariable: composantes principales, correspondances, coordonnées principales. *Invest. Pesq.* **36**: 81-108. [451, 465, 478]

Ifrah, G. 1981. *Histoire universelle des chiffres.* Éditions Seghers, Paris. 568 pp. [59]

Iman, R. L. & W. J. Conover. 1979. The use of the rank transformation in regression. *Technometrics* **21**: 499-509. [30]

Iman, R. L. & W. J. Conover. 1983. *A modern approach to statistics.* Wiley, New York. xxiii + 497 pp. [537]

Ipsen, D. C. 1960. *Units, dimensions and dimensionless numbers.* McGraw-Hill, New York. xii + 236 pp. [97]

Isaaks, E. H. & R. M. Srivastava. 1989. *Applied geostatistics.* Oxford Univ. Press, New York. xix + 561 pp. [731, 732, 735, 746, 749, 750, 751, 756]

Jaccard, P. 1900. Contribution au problème de l'immigration post-glaciaire de la flore alpine. *Bull. Soc. Vaudoise Sci. Nat.* **36**: 87-130. [xiii, 256]

Jaccard, P. 1901. Etude comparative de la distribution florale dans une portion des Alpes et du Jura. *Bull. Soc. Vaudoise Sci. nat.* **37**: 547-579. [256]

Jaccard, P. 1908. Nouvelles recherches sur la distribution florale. *Bull. Soc. Vaudoise Sci. nat.* **44**: 223-270. [256]

Jackson, D. A. 1993. Stopping rules in principal components analysis: a comparison of heuristical and statistical approaches. *Ecology* **74**: 2204-2214. [409, 410]

Jackson, D. A. 1995. PROTEST: a PROcrustean randomization TEST of community environment concordance. *Écoscience* **2**: 297-303. [564]

Jackson, D. A. & K. M. Somers. 1989. Are probability estimates from the permutation model of Mantel's test stable? *Can. J. Zool.* **67**: 766-769. [25]

Jackson, D. A. & K. M. Somers. 1991a. The spectre of 'spurious' correlations. *Oecologia* **86**: 147-151. [37]

Jackson, D. A. & K. M. Somers. 1991b. Putting things in order: the ups and downs of detrended correspondence analysis. *Am. Nat.* **137**: 704-712. [349, 471]

Jackson, D. A., K. M. Somers & H. H. Harvey. 1992. Null models and fish communities: evidence of nonrandom patterns. *Am. Nat.* **139**: 930-951. [357]

Jackson, R. C. & T. J. Crovello. 1971. A comparison of numerical and biosystematic studies in *Haplopappus*. *Brittonia* **23**: 54-70. [483, 484]

Jain, A. K. & R. C. Dubes. 1988. *Algorithms for clustering data.* Prentice Hall, Englewood Cliffs, New Jersey. xiv + 320 pp. [312, 313, 331, 335, 343, 352, 378]

Jambu, M. & M.-O. Lebeaux. 1983. *Cluster analysis and data analysis.* Elsevier-North-Holland, Amsterdam. xxiv + 898 pp. [315]

Jardine, N. & R. Sibson. 1968. The construction of hierarchic and non-hierarchic classifications. *Comput. J.* **11**: 177-184. [359]

Jardine, N. & R. Sibson. 1971. *Mathematical taxonomy.* Wiley, London. xviii + 286 pp. [359]

Jenkins, G. M. & D. G. Watts. 1968. *Spectral analysis and its applications.* Holden-Day, San Francisco. xviii + 525 pp. [641, 655]

Jenkins, S. H. 1975. Food selection by beavers. A multidimensional contingency table analysis. *Oecologia* (Berl.) **21**: 157-173. [230]

Jolicoeur, P. 1959. Multivariate geographical variation in the wolf *Canis lupus* L. *Evolution* **13**: 283-299. [619]

Jolicoeur, P. 1973. Imaginary confidence limits of the slope of the major axis of a bivariate normal distribution: a sampling experiment. *J. Amer. Statist. Assoc.* **68**: 866-871. [507]

Jolicoeur, P. 1975. Linear regression in fishery research: some comments. *J. Fish. Res. Board Can.* **32**: 1491-1494. [511]

Jolicoeur, P. 1990. Bivariate allometry: interval estimation of the slopes of the ordinary and standardized normal major axes and structural relationship. *J. Theor. Biol.* **144**: 275-285. [508, 511, 513, 515, 515]

Jolicoeur, P. & J. E. Mosimann. 1960. Size and shape variation in the painted turtle. A principal component analysis. *Growth* **24**: 339-354. [403]

Jolicoeur, P. & J. E. Mosimann. 1968. Intervalles de confiance pour la pente de l'axe majeur d'une distribution normale bidimensionnelle. *Biom-Praxim.* **9**: 121-140. [508, 511]

Jolivet, E. 1982. *Introduction aux modèles mathématiques en biologie.* INRA Actualités Scientifiques et Agronomiques, No. 11, Masson, Paris. 151 pp. [xiii]

Jones, R. H. 1964. Prediction of multivariate time series. *J. Appl. Meteorol.* **3**: 285-289. [703]

Jørgensen, S. E. [ed.] 1983. *Application of ecological modelling in environmental management. Part A.* Developments in Environmental Modelling, 4A. Elsevier Scientific Publ. Co., Amsterdam, The Netherlands. viii + 735 pp. [xiii]

Journel, A. G. & C. J. Huijbregts. 1978. *Mining geostatistics.* Academic Press, London. x + 600 pp. [32, 732, 749]

Jumars, P. A., D. Thistle & M. L. Jones. 1977. Detecting two-dimensional spatial structure in biological data. *Oecologia (Berl.)* **28**: 109-123. [715, 782]

Kaiser, H. F. 1958. The varimax criterion for analytic rotation in factor analysis. *Psychometrika* **23**: 187-200. [478]

Kaiser, H. F. & K. W. Dickman. 1959. *Analytic determination of common factors.* Univ. Illinois. [478]

Kedem, B. 1980. *Binary time series.* Lecture notes in pure and applied mathematics, Vol. 39. Marcel Dekker, New York. ix + 140 pp. [691]

Kempthorne, O. 1952. *The design and analysis of experiments.* Robert E. Krieger Publ. Co., Huntington, N. Y. xix + 631 pp. [6, 608]

Kendall, D. G. 1988. Seriation. 417-424 *in*: S. Kotz & N. L. Johnson [eds.] *Encyclopedia of statistical sciences. Vol. 8.* John Wiley & Sons, New York. [371]

Kendall, D. G. 1971. Seriation from abundance matrices. 215-252 *in*: F. R. Hodson, D. G. Kendal & P. Tautu [eds.] *Mathematics in the archaeological and historical sciences*. Edinburgh Univ. Press, Edinburgh. [466]

Kendall, M. G. 1938. A new measure of rank correlation. *Biometrika* **30**: 81-93. [376]

Kendall, M. G. 1948. *Rank correlation methods.* Charles Griffin & Co., London. vii + 160 pp. [195, 202, 203, 204]

Kendall, M. G. 1976. *Time series.* Charles Griffin & Co., London. ix + 197 pp. [651]

Kendall, M. G. & W. R. Buckland. 1960. *A dictionary of statistical terms. 2nd edition.* Oliver and Boyd, Edinburgh. xi + 575 pp. [221]

Kendall, M. G. & J. K. Ord. 1990. *Time series. 3rd edition.* Edward Arnold, Sevenoaks, Kent. x + 296 pp. [641, 646, 649, 675, 682]

Kendall, M. G. & A. Stuart. 1963. *The advanced theory of statistics. Vol. 1. 2nd edition.* Charles Griffith & Co., London. xii + 433 pp. [329]

Kendall, M. G. & A. Stuart. 1966. *The advanced theory of statistics. Vol. 3.* Hafner Publ. Co., New York. ix + 552 pp. [506, 612, 615]

Kendall, M. G., A. Stuart & J. K. Ord. 1983. *The advanced theory of statistics. Vol. 3. 4th edition.* Charles Griffin & Co., London. x + 780 pp. [641, 651]

Kenkel, N. C. & L. Orlóci. 1986. Applying metric and nonmetric multidimensional scaling to ecological studies: some new results. *Ecology* **67**: 919-928. [471]

Kennedy, P. E. 1995. Randomization tests in econometrics. *J. Bus. Econ. Statist.* **13**: 85-94. [559, 611]

Kent, M. & P. Coker. 1992. *Vegetation description and analysis – A practical approach.* John Wiley & Sons, New York. x + 363 p. [348]

Kermack, K. A. & J. B. S. Haldane. 1950. Organic correlation and allometry. *Biometrika* **37**: 30-41. [507]

Kierstead, H. & L. B. Slobodkin. 1953. The size of water masses containing plankton blooms. *J. Mar. Res.* **12**: 141-147. [113, 114, 124, 682]

Kim, J. O. & C. W. Mueller. 1978a. *Introduction to factor analysis – What is it and how to do it.* Sage University Paper Series on Quantitative Applications in the Social Sciences, 07-013. Sage Pubns., Beverly Hills, Calif. 80 pp. [476, 477]

Kim, J. O. & C. W. Mueller. 1978b. *Factor analysis – Statistical methods and practical issues.* Sage University Paper Series on Quantitative Applications in the Social Sciences, 07-014. Sage Pubns., Beverly Hills, Calif. 88 pp. [476, 479]

Kluge, A. G. & J. S. Farris. 1969. Quantitative phyletics and the evolution of anurans. *Syst. Zool.* **18**: 1-32. [46]

Koch, G. G. & D. B. Gillings. 1983. Inference, design based vs. model based. 84-88 *in*: S. Kotz & N. L. Johnson [eds.] *Encyclopedia of statistical sciences. Vol. 4.* Wiley, New York. [6]

Kolasa, J. & S. T. A. Pickett [eds.] 1991. *Ecological heterogeneity.* Springer-Verlag, New York. xi + 332 pp. [16]

Kolasa, J. & C. D. Rollo. 1991. Introduction: The heterogeneity of heterogeneity – A glossary. 1-23 *in*: J. Kolasa & S. T. A. Pickett [eds.] *Ecological heterogeneity.* Springer-Verlag, New York. [710, 711]

Kotz, S. & N. L. Johnson. 1982. Degrees of freedom. 293-294 *in*: S. Kotz & N. L. Johnson [eds.] *Encyclopedia of statistical science. Vol. 2.* Wiley, New York. [14]

Krackhardt, D. 1988. Predicting with networks: Nonparametric multiple regression analysis of dyadic data. *Soc. Networks* **10**: 359-381. [559]

Krige, D. G. 1952. A statistical analysis of some of the borehole values in the Orange Free State goldfield. *J. Chem. Metall. Min. Soc. S. Afr.* **53**: 47-70. [749]

Krige, D. G. 1966. Two-dimensional weighted moving average trend surfaces for ore evaluation. 13-38 *in*: *Proceedings of the symposium on mathematical statistics and computer applications in ore valuation*, Johannesburg. [749]

Kroonenberg, P. M. 1983. *Three-mode principal component analysis – Theory and applications.* DSWO Press, Leiden. 398 pp. [251]

Kroonenberg, P. M. 1996. *3WayPack user's manual. A package of three-way programs.* Department of Education, Leiden University, Leiden, The Netherlands. [251]

Kruskal, J. B. 1964a. Multidimensional scaling by optimizing goodness of fit to a nonmetric hypothesis. *Psychometrika* **29**: 1-27. [444, 446]

Kruskal, J. B. 1964b. Nonmetric multidimensional scaling: a numerical method. *Psychometrika* **29**: 115-129. [444, 446, 447]

Kruskal, J. B. & M. Wish. 1978. *Multidimensional scaling.* Sage University Paper series on Quantitative Applications in the Social Sciences, 07-011. Sage Publications, Beverly Hills. 93 pp. [444, 448]

Kruskal, W. H. & F. Mosteller. 1988. Representative sampling. 77-81 *in*: S. Kotz & N. L. Johnson [eds.] *Encyclopedia of statistical sciences. Vol. 8.* Wiley, New York. [6]

Krylov, V. V. 1968. Species association in plankton. *Oceanology* (translated from *Okeanologiya*, in Russian) **8**: 243-251. [295, 358, 359, 359]

Kulczynski, S. 1928. Die Pflanzenassoziationen der Pieninen. *Bull. Int. Acad. Pol. Sci. Lett. Cl. Sci. Math. Nat. Ser. B*, Suppl. II (1927): 57-203. [257, 266, 306, 371, 372, 373]

Kullback, S. 1959. *Information theory and statistics.* John Wiley & Sons, New York. xvii + 395 pp. [223, 623]

Kuusipalo, J. 1987. Relative importance of factors controlling the success of *Oxalis acetosella*: an example of linear modelling in ecological research. *Vegetatio* **70**: 171-179. [550]

Lam, N. S.-N. 1983. Spatial interpolation methods: a review. *Am. Cartographer* **10**: 129-149. [738]

Lambshead, P. J. D. & G. L. J. Paterson. 1986. Ecological cladistics. An investigation of numerical cladistics as a method for analysing ecological data. *J. Nat. Hist.* **20**: 895-909. [357]

Lance, G. N. & W. T. Williams. 1965. Computer programs for monothetic classification («association analysis»). *Comput. J.* **8**: 246-249. [344]

Lance, G. N. & W. T. Williams. 1966a. A generalized sorting strategy for computer classifications. *Nature (Lond.)* **212**: 218. [319, 333, 335, 336]

Lance, G. N. & W. T. Williams. 1966b. Computer programs for hierarchical polythetic classification («similarity analyses»). *Comput. J.* **9**: 60-64. [269, 336, 338, 341]

Lance, G. N. & W. T. Williams. 1966c. Computer programs for classification. *Proc. ANCCAC Conference*, Canberra, May 1966, Paper 12/3. [282]

Lance, G. N. & W. T. Williams. 1967a. Mixed-data classificatory programs. I. Agglomerative systems. *Aust. Comput. J.* **1**: 15-20. [282]

Lance, G. N. & W. T. Williams. 1967b. Mixed-data classificatory programs. II. Divisive systems. *Aust. Comput. J.* **1**: 82-85. [343]

Lance, G. N. & W. T. Williams. 1967c. A general theory of classificatory sorting strategies. I. Hierarchical systems. *Computer J.* **9**: 373-380. [312, 316, 319, 322, 333, 335, 336, 337]

Lance, G. N. & W. T. Williams. 1967d. A general theory of classificatory sorting strategies. II. Clustering systems. *Comput. J.* **10**: 271-277. [315, 351, 352, 361]

Lance, G. N. & W. T. Williams. 1968. Note on a new information-statistic classificatory program. *Comput. J.* **11**: 195. [344]

Langhaar, H. L. 1951. *Dimensional analysis and theory of models.* Wiley, New York. xi + 166 p. [97, 114]

Lapointe, F.-J. 1998. How to validate phylogenetic trees? A stepwise procedure. 71-88 in: C. Hayashi, N. Oshumi, K. Yajima, Y. Tanaka, H. H. Bock & Y. Baba [eds.] *Data science, classification, and related methods.* Springer-Verlag, Tokyo. [378]

Lapointe, F.-J. & P. Legendre. 1990. A statistical framework to test the consensus of two nested classifications. *Syst. Zool.* **39**: 1-13. [489, 559]

Lapointe, F.-J. & P. Legendre. 1991. The generation of random ultrametric matrices representing dendrograms. *J. Classif.* **8**: 177-200. [489, 559]

Lapointe, F.-J. & P. Legendre. 1992a. A statistical framework to test the consensus among additive trees (cladograms). *Syst. Biol.* **41**: 158-171. [489, 559]

Lapointe, F.-J. & P. Legendre. 1992b. Statistical significance of the matrix correlation coefficient for comparing independent phylogenetic trees. *Syst. Biol.* **41**: 378-384. [489]

Lapointe, F.-J. & P. Legendre. 1994. A classification of pure malt Scotch whiskies. *Appl. Statist.* **43**: 237-257. [489, 760]

Lapointe, F.-J. & P. Legendre. 1995. Comparison tests for dendrograms: a comparative evaluation. *J. Class.* **12**: 265-282. [489]

Larntz, K. 1978. Small sample comparisons of exact levels for chi-square goodness-of-fit statistics. *J. Amer. Statist. Assoc.* **73**: 253-263. [218]

Laurec, A. 1979. *Analyse des données et modèles prévisionnels en écologie marine.* Ph.D. Thesis. Univ. Aix-Marseille. 405 pp. + annexes. [688]

Laurec, A. 1982. Traitement des signaux quantitatifs et implications dans l'échantillonnage. 217-270 *in*: S. Frontier [ed.] *Stratégies d'échantillonnage en écologie.* Collection d'Écologie No 17. Masson, Paris et les Presses de l'Université Laval, Québec. [681]

Laws, E. A. & J. W. Archie. 1981. Appropriate use of regression analysis in marine biology. *Mar. Biol. (Berl.)* **65**: 13-16. [516]

Lebart, L. 1978. Programme d'agrégation avec containtes (C. A. H. contiguïté). *C. Anal. Données* **3**: 275-287. [757]

Lebart, L. & J. P. Fénelon. 1971. *Statistique et informatique appliquées.* Dunod, Paris. 426 pp. [285, 455]

Lebart, L., A. Morineau & J.-P. Fénelon. 1979. *Traitement des données statistiques – Méthodes et programmes.* Dunod, Paris. xiii + 510 pp. [30, 412]

Leclerc, B. & G. Cucumel. 1987. Consensus en classification: une revue bibliographique. *Math. Sci. Humaines* **100**: 109-128. [380]

Leduc, A., P. Drapeau, Y. Bergeron & P. Legendre. 1992. Study of spatial components of forest cover using partial Mantel tests and path analysis. *J. Veg. Sci.* **3**: 69-78. [551, 780, 781, 782]

Lefebvre, J. 1980. *Introduction aux analyses statistiques multidimensionnelles. 2nd ed.* Masson, Paris. xviii + 259 pp. [136. 289]

Lefkovitch, L. P. 1976. Hierarchical clustering from principal coordinates: an efficient method for small to very large numbers of objects. *Math. Biosci.* **31**: 157-174. [346, 485]

Lefkovitch, L. P. 1978. Cluster generation and grouping using mathematical programming. *Math. Biosci.* **41**: 91-110. [756]

Lefkovitch, L. P. 1980. Conditional clustering. *Biometrics* **36**: 43-58. [757]

Legand, M. 1958. Variations diurnes du zooplancton autour de la Nouvelle-Calédonie. *O. R. S. T. O. M., Inst. Fr. Océanie Sect. Océanogr. Rapp. Sci.* (6): 1-42. [678]

Le Gendre, A. M. 1805. *Nouvelles méthodes pour la détermination des orbites des comètes.* Courcier, Paris. [501]

Legendre, L. 1971a. *Phytoplankton structure in Baie des Chaleurs.* Ph.D. dissertation, Institute of Oceanography, Dalhousie University, Halifax, Canada. 137 pp. [361, 363]

Legendre, L. 1971b. Production primaire dans la Baie-des-Chaleurs (Golfe Saint-Laurent) *Nat. Can. (Qué.)* **98**: 743-773. [102, 368]

Legendre, L. 1973. Phytoplankton organization in Baie des Chaleurs (Gulf of St-Lawrence). *J. Ecol.* **61**: 135-149. [245, 293, 356, 362, 366, 367]

Legendre, L. 1987a. Multidimensional contingency table analysis as a tool for biological oceanography. *Biol. Oceanogr.* **5**: 13-28. [225, 226, 229]

Legendre, L., M. Aota, K. Shirasawa, M. J. Martineau & M. Ishikawa. 1991. Crystallographic structure of sea ice along a salinity gradient and environmental control of microalgae in the brine cells. *J. Mar. Syst.* **2**: 347-357. [550, 631, 633]

Legendre, L. & S. Demers. 1984. Towards dynamic biological oceanography and limnology. *Can. J. Fish. Aquat. Sci.* **41**: 2-19. [682]

Legendre, L., S. Demers & D. Lefaivre. 1986. Biological production at marine ergoclines. 1-29 *in*: J.-C. Nihoul [ed.] *Marine interfaces ecohydrodynamics.* Elsevier, Amsterdam. [711]

Legendre, L., M. Fréchette & P. Legendre. 1981. The contingency periodogram: a method of identifying rhythms in series of nonmetric ecological data. *J. Ecol.* **69**: 965-979. [670, 671, 673]

Legendre, L., R. G. Ingram & Y. Simard. 1982. A periodic changes of water column stability and phytoplankton in an Arctic coastal embayment, Manitounuk Sound, Hudson Bay. *Nat. Can. (Qué.)* **109**: 775-786. [234]

Legendre, L. & P. Legendre. 1978. Associations. 261-272 *in*: A. Sournia [ed]. *Phytoplankton manual. Monographs on oceanographic Methodology. Vol. 6.* Unesco [356]

Legendre, L. & P. Legendre. 1983a. *Numerical ecology.* Developments in environmental modelling, 3. Elsevier Scientific Publ. Co., Amsterdam, The Netherlands. xvi + 419 pp. [xii, xiv, 239]

Legendre, L. & P. Legendre. 1983b. Partitioning ordered variables into discrete states for discriminant analysis of ecological classifications. *Can. J. Zool.* **61**: 1002-1010. [229, 230]

Legendre, L. & P. Legendre. 1984a. *Écologie numérique. 2nd ed. 1. Le traitement multiple des données écologiques.* Masson, Paris et les Presses de l'Université du Québec. xv + 260 pp. [xiv]

Legendre, L. & P. Legendre. 1984b. *Écologie numérique. 2nd ed. 2. La structure des données écologiques.* Masson, Paris et les Presses de l'Université du Québec. viii + 335 pp. [xiv, 644, 645]

Legendre, P. 1976. An appropriate space for clustering selected groups of western North American *Salmo. Syst. Zool.* **25**: 193-195. [312, 483]

Legendre, P. 1987b. Constrained clustering. 289-307 *in*: P. Legendre & L. Legendre [eds.] *Developments in numerical ecology.* NATO ASI series, Vol. G-14. Springer-Verlag, Berlin. [699, 759]

Legendre, P. 1990. Quantitative methods and biogeographic analysis. 9-34 *in*: D. J. Garbary & R. G. South [eds.] *Evolutionary biogeography of the marine algae of the North Atlantic.* NATO ASI Series, Vol. G 22. Springer-Verlag, Berlin. [766]

Legendre, P. 1993. Spatial autocorrelation: Trouble or new paradigm? *Ecology* **74**: 1659-1673. [9, 170, 532, 559, 724, 770]

Legendre, P. & M. J. Anderson. 1999. Distance-based redundancy analysis: testing multi-species responses in multi-factorial ecological experiments. *Ecol. Monogr.* (in press). [189, 436, 557, 605, 606]

Legendre, P. & A. Beauvais. 1978. Niches et associations de poissons des lacs de la Radissonie québécoise. *Nat. Can. (Qué.)* **105**: 137-158. [359, 360]

Legendre, P. & D. Borcard. 1994. Rejoinder. *Environ. Ecol. Stat.* **1**: 57-61. [773, 785]

Legendre, P. & A. Chodorowski. 1977. A generalization of Jaccard's association coefficient for *Q* analysis of multi-state ecological data matrices. *Ekol. Pol.* **25**: 297-308. [262, 263, 267, 271, 308, 1977, 484, 485]

Legendre, P., S. Dallot & L. Legendre. 1985. Succession of species within a community: chronological clustering, with applications to marine and freshwater zooplankton. *Am. Nat.* **125**: 257-288. [297, 692, 696, 697, 698, 699, 700, 701, 759]

Legendre, P. & P. Dutilleul. 1991. Comments on Boyle's "Acidity and organic carbon in lake water: variability and estimation of means". *J. Paleolimnol.* **6**: 103-110. [14]

Legendre, P. & P. Dutilleul. 1992. Introduction to the analysis of periodic phenomena. 11-25 *in*: M. A. Ali [ed.] *Rhythms in fishes.* NATO ASI Series, Vol. A-236. Plenum, New York. [641, 644, 645, 677, 678]

Legendre, P. & M.-J. Fortin. 1989. Spatial pattern and ecological analysis. *Vegetatio* **80**: 107-138. [9, 553, 556, 712, 713, 714, 728, 728, 738, 749, 760, 780]

Legendre, P., R. Galzin & M. Harmelin-Vivien. 1997. Relating behavior to habitat: solutions to the fourth-corner problem. *Ecology* **78**: 547-562. [565, 566, 572, 745, 746]

Legendre, P., F.-J. Lapointe & P. Casgrain. 1994. Modeling brain evolution from behavior: a permutational regression approach. *Evolution* **48**: 1487-1499. [559]

Legendre, P. & L. Legendre. 1982. Échantillonnage et traitement des données. 163-216 in: S. Frontier [ed.] *Stratégies d'échantillonnage en écologie.* Collection d'Écologie, No. 17. Masson, Paris, and Les Presses de l'Université Laval, Québec. [189, 190, 192, 663]

Legendre, P. & V. Legendre. 1984c. Postglacial dispersal of freshwater fishes in the Québec peninsula. *Can. J. Fish. Aquat. Sci.* **41**: 1781-1802. [757, 763, 765]

Legendre, P., F. Long, R. Bergeron & J. M. Levasseur. 1978. Inventaire aérien de la faune dans le Moyen Nord québecois. *Can. J. Zool.* **56**: 451-462. [222, 263]

Legendre, P. & B. H. McArdle. 1997. Comparison of surfaces. *Oceanol. Acta* **20**: 27-41. [712, 743]

Legendre, P., N. L. Oden, R. R. Sokal, A. Vaudor & J. Kim. 1990. Approximate analysis of variance of spatially autocorrelated regional data. *J. Classif.* **7**: 53-75. [13, 15, 569]

Legendre, P., D. Planas & M.-J. Auclair. 1984a. Succession des communautés de gastéropodes dans deux milieux différant par leur degré d'eutrophisation. *Can. J. Zool.* **62**: 2317-2327. [405, 692]

Legendre, P. & D. J. Rogers. 1972. Characters and clustering in taxonomy: a synthesis of two taximetric procedures. *Taxon* **21**: 567-606. [262, 305, 312, 344]

Legendre, P. & M. Troussellier. 1988. Aquatic heterotrophic bacteria: modeling in the presence of spatial autocorrelation. *Limnol. Oceanogr.* **33**: 1055-1067. [559, 779, 780, 781]

Legendre, P., M. Troussellier & B. Baleux. 1984b. Indices descriptifs pour l'étude de l'évolution des communautés bactériennes. 71-84 *in*: A. Bianchi [ed.] *Bactériologie marine – Colloque international no 331* Editions du CNRS, Paris. [245]

Legendre, P., M. Troussellier, V. Jarry & M.-J. Fortin. 1989. Design for simultaneous sampling of ecological variables: from concepts to numerical solutions. *Oikos* **55**: 30-42. [760]

Legendre, P., S. F. Thrush, V. J. Cummings, P. K. Dayton, J. Grant, J. E. Hewitt, A. H. Hines, B. H. McArdle, R. D. Pridmore, D. C. Schneider, S. J. Turner, R. B. Whitlatch & M. R. Wilkinson. 1997. Spatial structure of bivalves in a sandflat: scale and generating processes. *J. Exp. Mar. Biol. Ecol.* **216**: 99-128. [745, 746]

Legendre, P. & A. Vaudor. 1991. *The R Package – Multidimensional analysis, spatial analysis.* Département de sciences biologiques, Université de Montréal. iv + 142 pp. [302]

Lehn, W. H. 1979. Atmospheric refraction and lake monsters. *Science (Wash. D. C.)* **205**: 183-185. [212]

Lehn, W. H. & I. Schroeder. 1981. The Norse merman as an optical phenomenon. *Nature (Lond.)* **289**: 362-366. [212]

Lekan, J. F. & R. E. Wilson. 1978. Spatial variability of phytoplankton biomass in the surface waters of Long Island Sound. *Estuarine Coastal Mar. Sci.* **6**: 239-251. [682]

Levings, C. D. 1975. Analyses of temporal variation in the structure of a shallow-water benthic community in Nova Scotia. *Int. Revue ges. Hydrobiol.* **60**: 449-470. [692]

Lilliefors, H. W. 1967. The Kolmogorov-Smirnov test for normality with mean and variance unknown. *J. Amer. Statist. Assoc.* **62**: 399-402. [181]

Lindeman, R. L. 1942. The trophic-dynamic aspect of ecology. *Ecology* **23**: 399-418. [707]

Line, J. M., C. J. F. ter Braak & H. J. B. Birks. 1994. WACALIB version 3.3 – A computer program to reconstruct environmental variables from fossil assemblages by weighted averaging and to derive sample-specific errors of prediction. *J. Paleolimnol.* **10**: 147-152. [604]

Lingoes, J. C. 1971. Some boundary conditions for a monotone analysis of symmetric matrices. *Psychometrika* **36**: 195-203. [434]

Lipschutz, S. 1968. *Theory and problems of linear algebra.* Schaum's Outline Series, McGraw-Hill, New York. 334 pp. [527]

Little, R. J. A. & D. B. Rubin. 1987. *Statistical analysis with missing data.* Wiley, New York. xiv + 278 pp. [48, 49]

Lloyd, M. & R. J. Ghelardi. 1964. A table for calculating the «equitability» component of species diversity. *J. Anim. Ecol.* **33**: 217-225. [243, 244, 244, 836]

Lloyd, M., R. F. Inger & F. W. King. 1968. On the diversity of reptile and amphibian species in a Bornean rain forest. *Am. Nat.* **102**: 497-515. [241]

Loehle, C. 1983. Evaluation of theories and calculation tools in ecology. *Ecol. Model.* **19**: 239-247. [xiii]

Lukaszewicz, J. 1951. Sur la liaison et la division des points d'un ensemble fini. *Colloq. Math.* **2**: 282-285. [312, 316]

MacArthur, R. H. 1957. On the relative abundance of bird species. *Proc. Natl. Acad. Sci. USA* **43**: 293-295. [244]

Macnaughton-Smith, P., W. T. Williams, M. B. Dale & L. G. Mockett. 1964. Dissimilarity analysis: a new technique of hierarchical sub-division. *Nature (Lond.)* **202**: 1034-1035. [346]

MacQueen, J. 1967. Some methods for classification and analysis of multivariate observations. 281-297 *in*: L. M. Le Cam & J. Neyman [eds.] *Proceedings of the fifth Berkeley symposium on mathematical statistics and probability. Vol. 1.* University of California Press, Berkeley. [352]

Madansky, A. 1959. The fitting of straight lines when both variables are subject to error. *J. Amer. Statist. Assoc.* **54**: 173-205. [506]

Magnan, P., M. A. Rodriguez, P. Legendre & S. Lacasse. 1994. Dietary variation in a freshwater fish species: Relative contributions of biotic interactions, abiotic factors, and spatial structure. *Can. J. Fish. Aquat. Sci.* **51:** 2856-2865. [46]

Mahalanobis, P. C. 1936. On the generalized distance in statistics. *Proc. Natl. Inst. Sci. India* **2**: 49-55. [280]

Mandrak, N. E. 1995. Biogeographic patterns of fish species richness in Ontario lakes in relation to historical and environmental factors. *Can. J. Fish. Aquat. Sci.* **52**: 1462-1474. [780]

Manly, B. F. J. 1986. Randomization and regression methods for testing for associations with geographical, environmental, and biological distances between populations. *Res. Popul. Ecol. (Kyoto)* **28**: 201-218. [559]

Manly, B. J. F. 1997. *Randomization, bootstrap and Monte Carlo methods in biology. 2nd edition.* Chapman and Hall, London. xix + 399 pp. [20, 22, 24, 25, 26, 608, 609, 611]

Mantel, N. 1967. The detection of disease clustering and a generalized regression approach. *Cancer Res.* **27**: 209-220. [375, 552, 554]

Marcotorchino, J. F. & P. Michaud. 1979. *Optimisation en analyse ordinale des données.* Masson, Paris. xii + 211 pp. [251]

Margalef, R. 1958. Information theory in ecology. *General Systems* **3**: 36-71. [240, 241, 243, 245]

Margalef, R. 1968. *Perspectives in ecological theory.* Univ. Chicago Press, Chicago. viii + 111 pp. [2, 238, 691]

Margalef, R. 1974. *Ecologia.* Ediciones Omega, Barcelona. xv + 951 pp. [31, 237, 239, 243]

Margalef, R. & F. González Bernáldez. 1969. Grupos de especies asociadas en el fitoplancton del mar Caribe (NE de Venezuela). *Invest. Pesq.* **33**: 287-312. [413]

Marquardt, D. W. & R. D. Snee. 1975. Ridge regression in practice. *Am. Statist.* **29**: 3-20. [524]

Matheron, G. 1962. *Traité de géostatistique appliquée. Tomes 1 et 2.* Éditions Technip, Paris. 334 pp., 172 pp. [712, 749]

Matheron, G. 1965. *Les variables régionalisées et leur estimation – Une application de la théorie des fonctions aléatoires aux sciences de la nature.* Masson, Paris. 305 pp. [749]

Matheron, G. 1970. *La théorie des variables régionalisées, et ses applications.* Les Cahiers du Centre de Morphologie Mathématique. Fontainebleau, fascicule 5. 212 pp. [749]

Matheron, G. 1971. *The theory of regionalised variables and its applications.* Les Cahiers du Centre de Morphologie Mathématique, Fasc. 5, ENSMP, Paris. 211 pp. [749]

Matheron, G. 1973. The intrinsic random functions and their applications. *Adv. Appl. Prob.* **5**: 439-468. [749]

Matula, D. W. & R. R. Sokal. 1980. Properties of Gabriel graphs relevant to geographic variation research and the clustering of points in the plane. *Geogr. Anal.* **12**: 205-222. [752]

Maxwell, J. C. 1871. Remarks on the mathematical classification of physical quantities. *Proc. Lond. Math. Soc.* **3**: 224-232. [97, 99]

McArdle, B. 1988. The structural relationship: regression in biology. *Can. J. Zool.* **66**: 2329-2339. [507, 510, 511, 513, 515, 517]

McBratney, A. B. & R. Webster. 1981. The design of optimal sampling schemes for local estimation and mapping of regionalized variables – II. *Comput. Geosci.* **7**: 335-365. [16]

McBratney, A. B., R. Webster & T. M. Burgess. 1981. The design of optimal sampling schemes for local estimation and mapping of regionalized variables – I. *Comput. Geosci.* **7**: 331-334. [16]

McCammon, R. B. 1970. *Minimum entropy criterion for analytic rotation.* Univ. Kansas Computer Contribution 43. [478]

McCoy, E. D., S. S. Bell & K. Walters. 1986. Identifying biotic boundaries along environmental gradients. *Ecology* **67**: 749-759. [273, 696]

McCullagh, P. & J. A. Nelder. 1983. *Generalized linear models.* Chapman and Hall, London. 261 pp. [539]

McCune, B. 1997. Influence of noisy environmental data on canonical correspondence analysis. *Ecology* **78**: 2617-2623. [597]

McCune, B. & T. F. H. Allen. 1985. Will similar forest develop on similar sites? *Can. J. Bot.* **63**: 367-376. [556]

McGeoch, M. A. & S. L. Chown. 1998. Scaling up the value of bioindicators. *Trends Ecol. Evol.* **13**: 46-47. [370, 371]

McIntyre, R. M. & R. K. Blashfield. 1980. A nearest-centroid technique for evaluating the minimum-variance clustering procedure. *Multivar. Behav. Res.* **15**: 225-238. [380]

Mead, R. 1988. *The design of experiments – Statistical principles for practical applications.* Cambridge University Press, Cambridge. xiv + 620 pp. [495]

Mendelssohn, R. & P. Cury. 1987. Fluctuations of a fortnightly abundance index of the Ivoirian coastal pelagic species and associated environmental conditions. *Can. J. Fish. Aquat. Sci.* **44**: 408-421. [50]

Méot, A., P. Legendre & D. Borcard. 1998. Partialling out the spatial component of ecological variation: questions and propositions in the linear modeling framework. *Environ. Ecol. Stat.* **5**: 1-26. [772]

Mesplé, F., M. Troussellier, C. Casellas & P. Legendre. 1996. Evaluation of simple statistical criteria to qualify a simulation. *Ecol. Model.* **88**: 9-18. [505, 516]

Meulman, J. 1982. *Homogeneity analysis of incomplete data.* DSWO Press, Leiden. 168 pp. [451]

Milligan, G. W. 1979. Ultrametric hierarchical clustering algorithms. *Psychometrika* **44:** 343-346. [343]

Milligan, G. W. 1980. An examination of the effect of six types of error perturbation on fifteen clustering algorithms. *Psychometrika* **45**: 325-342. [378]

Milligan, G. W. 1981. A Monte Carlo study of thirty internal criterion measures for cluster analysis. *Psychometrika* **46**: 187-199. [379]

Milligan, G. W. 1996. Clustering validation – Results and implications for applied analyses. 341-375 *in*: P. Arabie, L. J. Hubert & G. De Soete [eds.] *Clustering and Classification.* World Scientific Publ. Co., River Edge, New Jersey. [316, 355, 378]

Milligan, G. W. & M. C. Cooper. 1985. An examination of procedures for determining the number of clusters in a data set. *Psychometrika* **50**: 159-179. [355]

Milligan, G. W. & M. C. Cooper. 1987. Methodological review: clustering methods. *Appl. Psychol. Meas.* **11**: 329-354. [352]

Milligan, G. W. & M. C. Cooper. 1988. A study of standardization of variables in cluster analysis. *J. Classif.* **5**: 181-204. [39]

Minchin, P. R. 1987. An evaluation of the relative robustness of techniques for ecological ordination. *Vegetatio* **69**: 89-107. [471]

Monestiez, P. 1978. Méthodes de classification automatique sous contraintes spatiales. 367-379 *in*: J. M. Legay & R. Tomassone [eds.] *Biométrie et écologie.* Inst. nat. Rech. agronomique, Jouy-en-Josas. [757]

Montgomery, D. C. & E. A. Peck. 1982. *Introduction to linear regression analysis.* Wiley, New York. 504 pp. [543, 544]

Moran, P. A. P. 1950. Notes on continuous stochastic phenomena. *Biometrika* **37**: 17-23. [714]

Moreau, G. & L. Legendre. 1979. Relation entre habitat et peuplements de poissons: essai de définition d'une méthode numérique pour des rivières nordiques. *Hydrobiologia* **67**: 81-87. [283]

Morice, E. 1968. *Dictionnaire de statistique.* Dunod, Paris. ix + 196 pp. [221]

Morrall, R. A. A. 1974. Soil microfungi associated with aspen in Saskatchewan: synecology and quantitative analysis. *Can. J. Bot.* **52**: 1803-1817. [417]

Morrison, D. F. 1990. *Multivariate statistical methods. 3rd edition.* McGraw-Hill, New York. xvii + 495. [1, 10, 132, 134, 390, 391]

Motyka, J. 1947. O zadaniach i metodach badan geobotanicznych. Sur les buts et les méthodes des recherches géobotaniques. *Annales Universitatis Mariae Curie-Sklodowska (Lublin, Polonia), Sectio C, Supplementum I.* viii + 168 pp. [265]

Muirhead, R. J. 1982. *Aspects of multivariate statistical theory.* Wiley, New York. xix + 673 pp. [131]

Mulaik, S. A. 1972. *The foundations of factor analysis.* McGraw-Hill, New York. xvi + 453 pp. [476]

Murtagh, F. 1985. A survey of algorithms for contiguity-constrained clustering and related problems. *Comput. J.* **28**: 82-88. [759]

Myers, D. E. 1982. Matrix formulation of co-kriging. *Math. Geol.* **14**: 249-257. [50]

Myers, D. E. 1983. Estimation of linear combinations and co-kriging. *Math. Geol.* **15**: 633-637. [50]

Myers, D. E. 1984. Co-kriging — New developments. 295-305 *in*: G. Verly, M. David, A. G. Journel & A. Marechal [eds.] *Geostatistics for natural resources characterization, Part 1. Vol. C 122.* NATO ASI Series. D. Reidel Publ. Co., Dordrecht. [50]

Nantel, P. & P. Neumann. 1992. Ecology of ectomycorrhizal-basidiomycete communities on a local vegetation gradient. *Ecology* **73**: 99-117. [550, 780]

Nemec, A. F. L. & R. O. Brinkhurst. 1988. Using the bootstrap to assess statistical significance in the cluster analysis of species abundance data. *Can. J. Fish. Aquat. Sci.* **45**: 965-970. [379]

Neu, C. W., C. R. Byers & J. M. Peek. 1974. A technique for analysis of utilization-availability data. *J. Wildl. Manag.* **38**: 541-545. [233]

Nie, N. H., C. H. Hull, J. G. Jenkins, K. Steinbrenner & D. H. Bent. 1975. *SPSS - Statistical package for the social sciences. 2 edition.* McGraw-Hill, New York. xxiv + 675 p. pp. [546, 547, 549]

Nishisato, S. 1980. *Analysis of categorical data – Dual scaling and its applications.* Mathematical expositions No. 24. University of Toronto Press, Toronto. xiii + 276 pp. [451]

Norusis, M. J. 1990. *SPSS advanced statistics user's guide.* SPSS Inc., Chicago. 285 pp. [541]

O'Neill, R. V., R. H. Gardner, B. T. Milne, M. G. Turner & B. Jackson. 1991. Heterogeneity and spatial hierarchies. 85-96 *in*: J. Kolasa & S. T. A. Pickett [eds.] *Ecological heterogeneity.* Springer-Verlag, New York. [8]

Obenchain, R. L. 1977. Classical F-tests and confidence regions for ridge regression. *Technometrics* **19**: 429-439. [524]

Ochiai, A. 1957. Zoogeographic studies on the soleoid fishes found in Japan and its neighbouring regions. *Bull. Jpn. Soc. Sci. Fish.* **22**: 526-530. [257]

Oden, N. L. 1984. Assessing the significance of a spatial correlogram. *Geogr. Anal.* **16**: 1-16. [671, 721]

Oden, N. L. & R. R. Sokal. 1986. Directional autocorrelation: an extension of spatial correlograms to two dimensions. *Syst. Zool.* **35**: 608-617. [728, 736]

Oden, N. L. & R. R. Sokal. 1992. An investigation of three-matrix permutation tests. *J. Classif.* **9**: 275-290. [559]

Oden, N. L., R. R. Sokal, M.-J. Fortin & H. Goebl. 1993. Categorical wombling: detecting regions of significant change in spatially located categorical variables. *Geogr. Anal.* **25**: 315-336. [712, 761, 763]

Odum, E. P. 1950. Bird populations of the Highlands (North Carolina) plateau in relation to plant succession and avian invasion. *Ecology* **31**: 587-605. [287]

Okubo, A. 1987. Fantastic voyage into the deep – Marine biofluids mechanics. 32-47 *in*: E. Teramoto & M. Yamaguti [eds.] *Mathematical topics in population biology, morphogenesis and neurosciences.* Lecture Notes in Biomathematics. Springer-Verlag, Berlin. [117]

Olea, R. A. 1991. *Geostatistical glossary and multilingual dictionary.* Oxford University Press. New York. [32]

Oliver, M. A. & R. Webster. 1989. A geostatistical basis for spatial weighting in multivariate classification. *Math. Geol.* **21**: 15-35. [759]

Orlóci, L. 1966. Geometric models in ecology. I. The theory and application of some ordination methods. *J. Ecol.* **54**: 193-215. [441]

Orlóci, L. 1967a. Data centering: a review and evaluation with reference to component analysis. *Syst. Zool.* **16**: 208-212. [412]

Orlóci, L. 1967b. An agglomerative method for classification of plant communities. *J. Ecol.* **55**: 193-205. [279]

Orlóci, L. 1975. *Multivariate analysis in vegetation research.* Dr. W. Junk B. V., The Hague. ix + 276 pp. [252, 451]

Orlóci, L. 1978. *Multivariate analysis in vegetation research. 2nd edition.* Dr. W. Junk B. V., The Hague. ix + 451 pp. [252, 269, 270, 278, 280, 288, 371, 461]

Orlóci, L. 1981. Probing time series vegetation data for evidence of succession. *Vegetatio* **46**: 31-35. [692]

Paloheimo, J. E. & L. M. Dickie. 1965. Food and growth of fishes. I. A growth curve derived from experimental data. *J. Fish. Res. Board. Can.* **22**: 521-542. [115, 125]

Parker, R. H. 1975. *The study of benthic communities – A model and a review.* Elsevier Sci. Publ. Co., Amsterdam. x + 279 pp. [478]

Patrick, R. 1949. A proposed biological measure of stream conditions, based on a survey of the Conestoga basin, Lancaster County, Pennsylvania. *Proc. Acad. Nat. Sci. Phila.* **101**: 277-341. [189, 240]

Patten, B. C. 1962. Species diversity in net phytoplankton of Raritan Bay. *J. Mar. Res.* **20**: 57-75. [244]

Pearson, K. 1900. On the criterion that a given system of deviations from the probable in the case of a correlated system of variables is such that it can be reasonably supposed to have arisen from random sampling. *Philos. Mag., Ser. 5* **50**: 157-172. [217]

Pearson, K. 1901. On lines and planes of closest fit to systems of points in space. *Philos. Mag., Ser. 6* **2**: 559-572. [507]

Pearson, K. 1926. On the coefficient of racial likeness. *Biometrika* **18**: 105-117. [283]

Peet, R. K. 1974. The measurement of species diversity. *Annu. Rev. Ecol. Syst.* **5**: 285-307. [239, 239, 243]

Peet, R. K., R. G. Knox, J. S. Case & R. B. Allen. 1988. Putting things in order: the advantages of detrended correspondence analysis. *Am. Nat.* **131**: 924-934. [471]

Peli, T & D. Malah. 1982. A study of edge detection algorithms. *Computer Graphics Image Process.* **20**: 1-21. [761]

Perruchet, C. 1981. Classification sous contrainte de contiguïté continue. 71-92 *in*: *Classification automatique et perception par ordinateur.* Séminaires de l'Institut national de Recherche en Informatique et en Automatique (C 118), Rocquencourt. [757]

Perruchet, C. 1983a. Significance tests for clusters: overview and comments. 199-208 *in*: J. Felsenstein [ed.] *Numerical taxonomy.* NATO Advanced Study Institute Series G (Ecological Sciences), No. 1. Springer-Verlag, Berlin. [378]

Perruchet, C. 1983b. Une analyse bibliographique des épreuves de classifiabilité en analyse des données. *Statist. Anal. Données* **8**: 18-41. [378]

Peters, R. H. 1983. *The ecological implications of body size.* Cambridge Univ. Press, Cambridge. xii + 329 pp. [498]

Petrie, W. M. F. 1899. Sequences in prehistoric remains. *J. Anthropol. Inst.* **29**: 295-301. [371]

Piazza, A. & L. L. Cavalli-Sforza. 1975. Spectral analysis of patterned covariance matrices and evolutionary relationships. 76-105 *in*: G. F. Estabrook [ed.] *Proceedings of the eight international conference on numerical taxonomy.* W. H. Freeman, San Francisco. [346]

Pickett, S. T. A. & P. S. White [eds.] 1985. *The ecology of natural disturbance and patch dynamics.* Academic Press, New York. xiv + 472 pp. [707]

Pielou, E. C. 1966. The measurement of diversity in different types of biological collections. *J. Theor. Biol.* **13**: 131-144. [241, 243]

Pielou, E. C. 1969. *An introduction to mathematical ecology.* John Wiley & Sons, New York. viii + 286 pp. [239, 242]

Pielou, E. C. 1975. *Ecological diversity.* John Wiley & Sons, New York. viii + 165 pp. [209, 215, 238, 239, 239, 244, 245]

Pielou, E. C. 1977. *Mathematical ecology. 2nd edition.* Wiley, New York. x + 385 pp. [711]

Pinel-Alloul, B. 1995. Spatial heterogeneity as a multiscale characteristic of zooplankton community. *Hydrobiologia* **300/301**: 17-42. [775]

Pinel-Alloul, B., E. Magnin & G. Codin-Blumer. 1982. Effet de la mise en eau du réservoir Desaulniers (Territoire de la Baie de James) sur le zooplancton d'une rivière et d'une tourbière réticulée. *Hydrobiologia* **86**: 271-296. [700]

Pinel-Alloul, B., C. Méthot, G. Verreault & Y. Vigneault. 1990. Phytoplankton in Quebec lakes: variation with lake morphometry, and with natural and anthropogenic acidification. *Can. J. Fish. Aquat. Sci.* **47**: 1047-1057. [336]

Pinel-Alloul, B., G. Méthot, L. Lapierre & A. Willsie. 1996. Macroinvertebrate community as a biological indicator of ecological and toxicological factors in Lake Saint-François (Québec). *Environ. Pollut.* **91**: 65-87. [775]

Pinel-Alloul, B., T. Niyonsenga & P. Legendre. 1995. Spatial and environmental components of freshwater zooplankton structure. *Écoscience* **2**: 1-19. [775]

Pinty, J. J. & C. Gaultier. 1971. *Dictionnaire pratique de mathématiques et statistiques en sciences humaines.* Editions universitaires, Paris. 298 pp. [215]

Pitard, F. F. 1992. *Pierre Gy's sampling theory and sampling practice. Volume I: Heterogeneity and sampling.* CRC Press Inc., Boca Raton, Florida. 214 pp. [710]

Plackett, R. L. 1974. *The analysis of categorical data.* Griffin's statistical monographs and courses, no. 35. Griffin, London. viii + 159 pp. [223]

Planes, S., A. Lefèvre, P. Legendre & R. Galzin. 1993. Spatio-temporal variability in fish recruitment to a coral reef (Moorea, French Polynesia). *Coral Reefs* **12**: 105-113. [759]

Platt, T. 1969. The concept of energy efficiency in primary production. *Limnol. Oceanogr.* **14**: 653-659. [102]

Platt, T. 1972. Local phytoplankton abundance and turbulence. *Deep-Sea Res.* **19**: 183-187. [682, 683]

Platt, T. 1978. Spectral analysis of spatial structure in phytoplankton populations. 73-84 *in*: J. H. Steele [ed.] *Spatial pattern in plankton communities.* Plenum Press, New York. [682, 682]

Platt, T. 1981. Thinking in term of scale: introduction to dimensional analysis. 112-121 *in*: T. Platt, K. H. Mann & R. E. Ulanowicz [eds.] *Mathematical models in biological oceanography.* The Unesco Press, Paris. [113, 115, 118]

Platt, T. & K. L. Denman. 1975. Spectral analysis in ecology. *Annu. Rev. Ecol. Syst.* **6**: 189-210. [638, 679, 681, 682]

Platt, T., L. M. Dickie & R. W. Trites. 1970. Spatial heterogeneity of phytoplankton in a near-shore environment. *J. Fish. Res. Board Can.* **27**: 1453-1473. [658, 660, 685, 686]

Platt, T. & D. V. Subba Rao. 1970. Energy flow and species diversity in a marine phytoplankton bloom. *Nature (Lond.).* **227**: 1059-1060. [102]

Platt, T. & D. V. Subba Rao. 1973. Some current problems in marine phytoplankton productivity. *Fish. Res. Board Can. Tech. Rep.* **307**: 1-90. [111, 112]

Powell, T. M., P. J. Richerson, T. M. Dillon, B. A. Agee, B. J. Dozier, D. A. Godden & L. O. Myrup. 1975. Spatial scales of current speed and phytoplankton biomass fluctuations in Lake Tahoe. *Science (Wash. D. C.)* **189**: 1088-1090. [682]

Press, S. J. 1972. *Applied multivariate analysis.* Holt, Rinehart and Winston, New York. xix + 521 pp. [132]

Press, W. H., B. P. Flannery, S. A. Teukolsky & W. T. Vetterling. 1986. *Numerical recipes – The art of scientific computing.* Cambridge Univ. Press, Cambridge. xx + 818 pp. [77, 94, 95, 418, 435, 448, 476, 644]

Priestley, M. B. 1964. The analysis of two dimensional stationary processes with discontinuous spectra. *Biometrika* **51**: 195-217. [714, 728]

Priestley, M. B. 1981a. *Spectral analysis and time series. 1. Univariate series.* Academic Press, London. xviii + 653 pp. [641]

Priestley, M. B. 1981b. *Spectral analysis and time series. 2. Multivariate series, prediction and control.* Academic Press, London. xviii + 237 pp. [641]

Prim, R. C. 1957. Shortest connection networks and some generalizations. *Bell System Technical Journal* **36**: 1389-1401. [312]

Quenouille, M. H. 1950. *Introductory statistics.* London. [42]

Quinghong, L. & S. Bråkenhielm. 1995. A statistical approach to decompose ecological variation. *Water Air Soil Pollut.* **85**: 1587-1592. [775]

Ragan, M. A. 1992. Phylogenetic inference based on matrix representation of trees. *Mol. Phylogenet. Evol.* **1**: 53-58. [379]

Rajski, C. 1961. Entropy and metric spaces. 44-45 *in:* C. Cherry [ed.] *Information theory.* Butterworths, London. [220]

Rand, W. M. 1971. Objective criteria for the evaluation of clustering methods. *J. Amer. Statist. Assoc.* **66**: 846-850. [376]

Rao, C. R. 1948. The utilization of multiple measurements in problems of biological classification. *J. Roy. Statist. Soc. B* **10**: 159-203. [617]

Rao, C. R. 1951. An asymptotic expansion of the distribution of Wilks' criterion. *Bull. Internat. Stat. Inst.* **33**: 177-181. [623]

Rao, C. R. 1952. *Advanced statistical methods in biometric research.* John Wiley & Sons, New York. xvii + 390 pp. [617]

Rao, C. R. 1964. The use and interpretation of principal component analysis in applied research. *Sankhyaá, Ser. A* **26**: 329-358. [347, 392, 412, 425, 579]

Rao, C. R. 1973. *Linear statistical inference and its applications. 2nd edition.* Wiley, New York. 625 pp. [579, 583]

Rao, C. R. 1995. A review of canonical coordinates and an alternative to correspondence analysis using Hellinger distance. *Qüestiió* (*Quaderns d'Estadística i Investigació Operativa*) **19**: 23-63. [286]

Ratkowsky, D. A. 1983. *Nonlinear regression modeling – A unified practical approach.* Marcel Dekker Inc., New York. viii + 276 pp. [498, 537]

Raup, D. M. & R. E. Crick. 1979. Measurement of faunal similarity in paleontology. *J. Paleontol.* **53**: 1213-1227. [273, 274]

Rendu, J.-M. 1981. *An introduction to geostatistical methods of mineral evaluation.* South African Institute of Mining and Metallurgy, Johannesburg. 84 pp. [749]

Renshaw, E. & E. D. Ford. 1984. The description of spatial pattern using two-dimensional spectral analysis. *Vegetatio* **56**: 75-85. [714, 728, 760]

Rényi, A. 1961. On measures of entropy and information. 547-561 *in:* J. Neyman [ed.] *Proceedings of the fourth Berkeley symposium on mathematical statistics and probability.* University of California Press, Berkeley. [239]

Reynolds, C. S. 1987. Community organization in the freshwater plankton. Ch. 14. *in*: J. H. R. Gee & P. S. Giller [eds.] *Organization of communities: past and present.* Blackwell Scient. Publ., Oxford. [707]

Reyssac, J. & M. Roux. 1972. Communautés planctoniques dans les eaux de Côte d'Ivoire. Groupes d'espèces associées. *Mar. Biol. (Berl.)* **13**: 14-33. [294, 413, 465]

Ricker, W. E. 1973. Linear regression in fishery research. *J. Fish. Res. Board Can.* **30**: 409-434. [507]

Rigler, F. H. 1982. The relation between fisheries management and limnology. *Trans. Am. Fish. Soc.* **111**: 121-132. [212]

Ripley, B. D. 1981. *Spatial statistics.* Wiley, New York. x + 252 pp. [711, 712, 714, 728]

Ripley, B. D. 1987. Spatial point pattern analysis in ecology. 407-429 *in:* P. Legendre, & L. Legendre [eds.] *Developments in numerical ecology.* NATO ASI Series, Vol. G 14. Springer-Verlag, Berlin. [711]

Robertson, G. P. 1987. Geostatistics in ecology: interpolating with known variance. *Ecology* **68**: 744-748. [749]

Robinson, G. K. 1982. Behrens-Fisher problem. 205-209 *in:* S. Kotz & N. L. Johnson [eds.] *Encyclopedia of statistical sciences. Vol. 1.* Wiley, New York. [20]

Roche, C. 1978. Exemple de classification hiérarchique avec contrainte de contiguïté. Le partage d'Aix-en-Provence en quartiers homogènes. *C. Anal. Données* **3**: 289-305. [757]

Rodríguez, M. A. & P. Magnan. 1995. Application of multivariate analyses in studies of the organization and structure of fish and invertebrate communities. *Aquat. Sci.* **57**: 199-216. [775]

Rogers, D. J. & T. T. Tanimoto. 1960. A computer program for classifying plants. *Science (Wash. D. C.)* **132**: 1115-1118. [255]

Rohlf, F. J. 1963. Classification of *Aedes* by numerical taxonomic methods (Diptera: Culicidae). *Ann. Entomol. Soc. Am.* **56**: 798-804. [319]

Rohlf, F. J. 1970. Adaptive hierarchical clustering schemes. *Syst. Zool.* **19**: 58-82. [372, 483]

Rohlf, F. J. 1972. An empirical comparison of three ordination techniques in numerical taxonomy. *Syst. Zool.* **21**: 271-280. [390]

Rohlf, F. J. 1974. Methods for comparing classifications. *Annu. Rev. Ecol. Syst.* **5**: 101-113. [376]

Rohlf, F. J. 1978. A probabilistic minimum spanning tree algorithm. *Information Processing Letters* **7**: 44-48. [315]

Rohlf, F. J. 1982a. Single linkage clustering algorithms. 267-284 *in:* P. R. Krishnaih [ed.] *Handbook of Statistics.* North-Holland, Amsterdam. [315]

Rohlf, F. J. 1982b. Consensus indices for comparing classifications. *Math. Biosci.* **59:** 131-144. [376, 380]

Ross, G. J. S. 1990. *Nonlinear estimation.* Springer-Verlag, New York. viii + 189 pp. [498, 537, 643]

Rossi, R. E., D. J. Mulla, A. G. Journel & E. H. Franz. 1992. Geostatistical tools for modeling and interpreting ecological spatial dependence. *Ecol. Monogr.* **62**: 277-314. [712, 728, 732, 735, 749]

Roux, G. & M. Roux. 1967. À propos de quelques méthodes de classification en phytosociologie. *Rev. Stat. Appl.* **15**: 59-72. [451]

Roux, M. & J. Reyssac. 1975. Essai d'application au phytoplancton marin de méthodes statistiques utilisées en phytosociologie terrestre. *Ann. Inst. Océanogr. (Paris)* **51**: 89-97. [283]

Royston, J. P. 1982a. An extension of Shapiro and Wilk's *W* test for normality to large samples. *Appl. Statist.* **31**: 115-124. [183]

Royston, J. P. 1982b. Algorithm AS 177. Expected normal order statistics (exact and approximate). *Appl. Statist.* **31**: 161-165. [183]

Royston, J. P. 1982c. Algorithm AS 181. The *W* test for normality. *Appl. Statist.* **31**: 176-177. [183]

Rubenstein, D. I. & M. A. R. Koehl. 1977. The mechanisms of filter feeding: some theoretical considerations. *Am. Nat.* **111**: 981-994. [117]

Russell, P. F. & T. R. Rao. 1940. On habitat and association of species of anopheline larvae in south-eastern Madras. *J. Malar. Inst. India* **3**: 153-178. [257]

Sabatier, R., J.-D. Lebreton & R. Chessel. 1989. Principal component analysis with instrumental variables as a tool for modelling composition data. 341-352 in: R. Coppi & S. Bolasso [eds.] *Multiway data analysis.* Elsevier Science Publishers B. V., North Holland, the Netherlands. [606]

Sakai, A. K. & N. L. Oden. 1983. Spatial pattern of sex expression in silver maple (*Acer saccharinum* L.): Morisita's index and spatial autocorrelation. *Am. Nat.* 122: 489-508. [721]

Sale, P. F. 1978. Coexistence of coral reef fishes – A lottery for living space. *Environ. Biol. Fishes* **3**: 85-102. [570]

Sanders, H. L. 1960. Benthic studies in Buzzards Bay. III. The structure of the soft-bottom community. *Limnol. Oceanogr.* **5**: 138-153. [373]

Sanders, H. L. 1968. Marine benthic diversity: a comparative study. *Am. Nat.* **102**: 243-282. [240]

Särndal, C.-E. 1978. Design-based and model-based inference in survey sampling. *Scand. J. Stat.* **5**: 27-52. [6]

SAS Institute Inc. 1985. *SAS user's guide: statistics. Version 5 edition.* SAS Institute Inc., Cary, North Carolina. xvi + 956 pp. [332, 352]

Saunders, D. R. 1961. The rationale for an «oblimax» method of transformation in factor analysis. *Psychometrika* **26**: 317-324. [478]

Scheider, W. & P. Wallis. 1973. An alternate method of calculating the population density of monsters in Loch Ness. *Limnol. Oceanogr.* **18**: 343. [212]

Scherrer, B. 1982. Techniques de sondage en écologie. 63-162 *in*: S. Frontier [Ed.] *Stratégies d'échantillonnage en écologie.* Collection d'Écologie, No. 17. Masson, Paris et les Presses de l'Université Laval, Québec. [16]

Schneider, D. C. 1994. *Quantitative ecology – Spatial and temporal scaling.* Academic Press, San Diego. xv + 395 pp. [97, 112, 708]

Schnell, G. D. 1970. A phenetic study of the suborder Lari (Aves). I. Methods and results of principal components analyses. *Syst. Zool.* **19**: 35-57. [483]

Schoener, T. W. 1970. Nonsynchronous spatial overlap of lizards in patchy habitats. *Ecology* **51**: 408-418. [230]

Schumacker, R. E. & R. G. Lomax. 1996. *A beginner's guide to structural equation modeling.* Lawrence Erlbaum Associates, Mahwah, New Jersey. 288 pp. [480]

Schuster, A. 1898. On the investigation of hidden periodocities with application to a supposed 26 day period of meteorological phenomena. *Terrestrial Magnetism* **3**: 13-41. [714]

Searle, S. R. 1966. *Matrix algebra for the biological sciences (including applications in statistics).* John Wiley & Sons, New York. xii + 296 pp. [51]

Searle, S. R. 1987. *Linear models for unbalanced data.* Wiley. xxiv + 536 pp. [525]

Shannon, C. E. 1948. A mathematical theory of communications. *Bell System Technical Journal* **27**: 379-423. [209, 212]

Shao, K. & F. J. Rohlf. 1983. Sampling distributions of consensus indices when all bifurcating trees are equally likely. 132-137 *in*: J. Felsenstein [ed.] *Numerical taxonomy.* NATO ASI Series, Vol. G-1. Springer-Verlag, Berlin. [489]

Shao, K. & R. R. Sokal. 1986. Significance tests of consensus indices. *Syst. Zool.* **35**: 582-590. [489]

Shapiro, S. S. & M. B. Wilk. 1965. An analysis of variance test for normality (complete samples). *Biometrika* **52**: 591-611. [181, 183]

Sharpe, D. M., G. R. Guntenspergen, C. P. Dunn, L. A. Leitner & F. Stearns. 1987. Vegetation dynamics in a southern Wisconsin agricultural landscape. 137-155 *in*: M. G. Turner [ed.] *Landscape heterogeneity and disturbance.* Ecological Studies 64. Springer-Verlag, New York. [472, 474]

Sheldon, R. W. & S. R. Kerr. 1972. The population density of monsters in Loch Ness. *Limnol. Oceanogr.* **17**: 796-798. [212]

Shepard, R. N. 1962. The analysis of proximities: multidimensional scaling with an unknown distance function. *Psychometrika* **27**: 125-139. [376, 389, 444]

Shepard, R. N. 1966. Metric structures in ordinal data. *J. Math. Psychol.* **3**: 287-315. [444]

Shumway, R. H. & D. S. Stoffer. 1982. An approach to time series smoothing and forecasting using the EM algorithm. *J. Time Ser. Anal.* **3**: 253-264. [50]

Siegel, S. 1956. *Nonparametric statistics for the behavioral sciences.* McGraw-Hill Series in Psychology, McGraw-Hill, New York. xvii + 312 pp. [19, 26, 191, 192, 202, 203, 218]

Siegel, S. and N. J. Castellan, Jr. 1988. *Nonparametric statistics for the behavioral sciences. 2nd edition.* McGraw-Hill, New York. xxiii + 399 pp. [19, 191]

Simon, H. A. 1962. The architecture of complexity. *Proc. Am. Philos. Soc.* **106**: 467-482. [8]

Simpson, E. H. 1949. Measurement of diversity. *Nature (Lond.)* **163**: 688. [242]

Slutzky, E. 1927. The summation of random causes as the source of cyclic processes. In Russian. Translation revised by the author in 1937. *Econometrica* **5**: 105-146. [651]

Smouse, P. E., J. C. Long & R. R. Sokal. 1986. Multiple regression and correlation extensions of the Mantel test of matrix correspondence. *Syst. Zool.* **35**: 627-632. [558, 559, 611]

Sneath, P. H. A. 1957. The application of computers to taxonomy. *J. Gen. Microbiol.* **17**: 201-226. [308]

Sneath, P. H. A. 1966. A comparison of different clustering methods as applied to randomly-spaced points. *Classification Soc. Bull.* **1**: 2-18. [318, 319]

Sneath, P. H. A. & R. R. Sokal. 1973. *Numerical taxonomy – The principles and practice of numerical classification.* W. H. Freeman, San Francisco. xv + 573 pp. [30, 31, 38, 252, 269, 312, 314, 315, 316, 319, 322, 326, 335, 339, 375, 381, 483]

Snedecor, G. W. & W. G. Cochran. 1967. *Statistical methods*. 6th Iowa State Univ. Press, Ames. xiv + 593 pp. [20]

Soares, A., J. Távora, L. Pinheiro, C. Freitas & J. Almeida. 1992. *Predicting probability maps of air pollution concentration – A case study on Barreiro/Seixal industrial area.* Fourth international geostatistics congress, 13-18 september 1992, Troya, Portugal. [749]

Sokal, R. R. 1979. Ecological parameters inferred from spatial correlograms. 167-196 *in*: G. P. Patil & M. L. Rosenzweig [eds.] *Contemporary quantitative ecology and related ecometrics. Vol. 12.* Statistical Ecology Series. International Co-operative Publ. House, Fairland, Maryland. [725]

Sokal, R. R. 1986. Spatial data analysis and historical processes. 29-43 *in:* E. Diday *et al.* [eds.] *Data analysis and informatics, IV.* North-Holland, Amsterdam. [736, 780]

Sokal, R. R., I. A. Lengyel, P. A. Derish, M. C. Wooten & N. L. Oden. 1987. Spatial autocorrelation of ABO serotypes in mediaeval cemeteries as an indicator of ethnic and familial structure. *J. Archaeol. Sci.* **14**: 615-633. [25, 780]

Sokal, R. R. & C. D. Michener. 1958. A statistical method for evaluating systematic relationships. *Univ. Kans. Sci. Bull.* **38**: 1409-1438. [255, 306, 321]

Sokal, R. R., N. L. Oden & B. A. Thomson. 1997a. A simulation study of microevolutionary inferences by spatial autocorrelation analysis. *Biol. J. Linn. Soc.* **60**: 73-93. [726]

Sokal, R. R., N. L. Oden, J. Walker & D. M. Waddle. 1997b. Using distance matrices to choose between competing theories and an application to the origin of modern humans. *J. Hum. Evol.* **32**: 501-522. [726]

Sokal, R. R. & F. J. Rohlf. 1962. The comparison of dendrograms by objective methods. *Taxon* **11**: 33-40. [313, 375]

Sokal, R. R. & F. J. Rohlf. 1981. Taxonomic congruence in Leptopodomorpha re-examined. *Syst. Zool.* **30**: 309-325. [42]

Sokal, R. R. & F. J. Rohlf. 1995. *Biometry – The principles and practice of statistics in biological research. 3rd edition.* W. H. Freeman, New York. xix + 887 pp. [18, 20, 26, 42, 43, 44, 179, 181, 191, 217, 218, 223, 233, 498, 507, 508, 511, 512, 522, 546, 547, 549, 553, 555, 562, 646]

Sokal, R. R. & P. H. A. Sneath. 1963. *Principles of numerical taxonomy.* W. H. Freeman, San Francisco. xvi + 359 pp. [xiii, 252, 255, 256, 257]

Sokal, R. R. & J. D. Thomson. 1987. Applications of spatial autocorrelation in ecology. 431-466 *in*: P. Legendre & L. Legendre [eds.] *Developments in numerical ecology.* NATO ASI Series, Vol. G-14. Springer-Verlag, Berlin. [721]

Sokoleve, P. G. & W. N. Bushell. 1978. The chi square periodogram: its utility for analysis of circadian rhythms. *J. Theor. Biol.* **72**: 131-160. [667]

Somers, K. M. & R. H. Green. 1993. Seasonal patterns in trap catches of the crayfish Cambarus bartoni and *Orconectes virilis* in six south-central Ontario lakes. *Can. J. Zool.* **71**: 1136-1145. [556]

Song, C. Q., B. Y. Wang & X. J. Sun. 1996. Implication of paleovegetational changes in Diaojiao Lake, Inner Mongolia. (In Chinese). *Acta Bot. Sin.* **38**: 568-575. [696]

Sørensen, T. 1948. A method of establishing groups of equal amplitude in plant sociology based on similarity of species content and its application to analysis of the vegetation on Danish commons. *Biol. Skr.* **5**: 1-34. [256, 316, 317]

Sousa, W. P. 1979. Experimental investigations of disturbance and ecological succession in a rocky intertidal algal community. *Ecol. Monogr.* **49**: 227-254. [707]

Southwood, T. R. E. 1966. *Ecological methods with particular reference to the study of insect populations.* Chapman and Hall, London. xviii + 391 pp. [44, 292]

Southwood, T. R. E. 1987. The concept and nature of the community. 3-27 *in*: J. H. R. Gee & P. S. Giller [eds.] *Organization of communities: past and present.* Blackwell Scientific Publ., Oxford. [707]

Soyer, J. 1970. Bionomie benthique du plateau continental de la côte catalane française. III. Les peuplements de copépodes harpacticoïdes (Crustacea). *Vie Milieu 21: 337-511.* [373]

Späth, H. 1975. *Cluster-Analyse-Algorithmen zur Objektklassifizierung und Datenreduktion.* R. Oldenbourg Verlag, München. 217 pp. [352]

Späth, H. 1980. *Cluster analysis algorithms.* Ellis Horwood, Chichester. 226 pp. [352]

Spearman, C. 1904. "General intelligence", objectively determined and measured. *Am. J. Psychol.* **15**: 201-293. [xiii, 306, 476]

Sprules, W. G. 1980. Nonmetric multidimensional scaling analyses of temporal variation in the structure of limnetic zooplankton communities. *Hydrobiologia* **69**: 139-146. [450, 692]

Steel, M. A. & D. Penny. 1993. Distributions of tree comparison metrics. Some new results. *Syst. Biol.* **42**: 126-141. [489]

Steiner, D., K. Baumberger & H. Maurer. 1969. Computer-processing and classification of multi-variate information from remote sensing imagery – A review of the methodology as applied to a sample of agricultural crops. 895-907 *in*: *Proc. sixth int. Symp. on Remote Sensing of Environment. Vol. II.* Willow Run Laboratories, Inst. of Science and Technology, Univ. of Michigan. [631]

Stephens, M. A. 1974. EDF statistics for goodness of fit and some comparisons. *J. Amer. Statist. Assoc.* **69**: 730-737. [179, 181, 183, 834]

Stephenson, W., W. T. Williams & S. D. Cook. 1972. Computer analyses of Petersen's original data on bottom communities. *Ecol. Monogr.* **42**: 387-415. [283]

Stephenson, W., W. T. Williams & S. D. Cook. 1974. The benthic fauna of soft bottoms, southern Moreton Bay. *Mem. Queensl. Mus.* **17**: 73-123. [278]

Steven, D. M. & R. Glombitza. 1972. Oscillatory variation of a phytoplankton population in a tropical ocean. *Nature (Lond.)* **237**: 105-107. [659, 661, 663, 664, 676]

Stewart-Oaten, A., W. M. Murdoch & K. R. Parker. 1986. Environmental impact assessment: 'pseudoreplication' in time? *Ecology* **67**: 929-940. [249]

Stone, L. 1993. Period-doubling reversals and chaos in simple ecological models. *Nature* **365**: 617-620. [2]

Strömgren, T., R. Lande & S. Engen. 1973. Intertidal distribution of the fauna on muddy beaches in the Borgenfjord area. *Sarsia* **53**: 49-70. [238]

Student [W. S. Gosset]. 1914. The elimination of spurious correlation due to position in time or space. *Biometrika* **10**: 179-180. [739]

Swan, J. M. A. 1970. An examination of some ordination problems by use of simulated vegetational data. *Ecology* **51**: 89-102. [466, 471]

Taguchi, S. 1976. Short-term variability of photosynthesis in natural marine phytoplankton populations. *Mar. Biol. (Berl.)* **37**: 197-207. [679]

Tardif, J., P. Dutilleul & Y. Bergeron. 1998. Variations in periodicities of the ring width of black ash (*Fraxinus nigra* Marsh.) in relation to flooding and ecological site factors at Lake Duparquet in northwestern Quebec. *Biol. Rhythm Res.* **29**: 1-29. [678]

Tatsuoka, M. M. 1971. *Multivariate analysis.* John Wiley & Sons, New York. 310 pp. [132]

Taylor, L. R. 1961. Aggregation, variance, and the mean. *Nature* **189**: 732-735. [44]

Teissier, G. 1948. La relation d'allométrie: sa signification statistique et biologique. *Biometrics* **4**: 14-53. [507, 510]

ter Braak, C. J. F. 1985. Correspondence analysis of incidence and abundance data: properties in terms of a unimodal response model. *Biometrics* **41**: 859-873. [349, 463]

ter Braak, C. J. F. 1986. Canonical correspondence analysis: a new eigenvector technique for multivariate direct gradient analysis. *Ecology* **67**: 1167-1179. [594, 600, 601, 615]

ter Braak, C. J. F. 1987a. The analysis of vegetation-environment relationships by canonical correspondence analysis. *Vegetatio* **69**: 69-77. [594, 600]

ter Braak, C. J. F. 1987b. Calibration. 78-90 *in*: R. H. G. Jongman, C. J. F. ter Braak & O. F. R. van Tongeren [eds.] *Data analysis in community and landscape ecology.* Pudoc, Wageningen, The Netherlands. Reissued in 1995 by Cambridge Univ. Press, Cambridge, England. [604]

ter Braak, C. J. F. 1987c. Ordination. 91-173 *in*: R. H. G. Jongman, C. J. F. ter Braak & O. F. R. van Tongeren [eds.] *Data analysis in community and landscape ecology.* Pudoc, Wageningen, The Netherlands. Reissued in 1995 by Cambridge Univ. Press, Cambridge, England. [390, 418, 419, 456, 457, 462, 466, 470, 474, 475, 486, 583, 592, 594, 596, 601, 617, 634]

ter Braak, C. J. F. 1988a. Partial canonical correspondence analysis. 551-558 *in*: H. H. Bock [ed.] *Classification and related methods of data analysis.* North-Holland, Amsterdam. [605]

ter Braak, C. J. F. 1988b. CANOCO – an extension of DECORANA to analyze species-environment relationships. *Vegetatio* **75**: 159-160. [474, 579, 594]

ter Braak, C. J. F. 1988c. *CANOCO – a FORTRAN program for canonical community ordination by [partial] [detrended] [canonical] correspondence analysis, principal component analysis and redundancy analysis (version 2.1).* Agricultural Mathematics Group, Ministry of Agriculture and Fisheries, Wageningen. ii + 95 pp. [474, 594, 607]

ter Braak, C. J. F. 1990. *Update notes: CANOCO version 3.10.* Agricultural Mathematics Group, Wageningen. [16, 456, 457, 474, 569, 594, 607, 608, 609, 610]

ter Braak, C. J. F. 1992. Permutation versus bootstrap significance tests in multiple regression and ANOVA. 79-86 *in:* K.-H. Jöckel, G. Rothe & W. Sendler [eds.] *Bootstrapping and related techniques.* Springer-Verlag, Berlin. [610]

ter Braak, C. J. F. 1994. Canonical community ordination. Part I: Basic theory and linear methods. *Écoscience* **1**: 127-140. [403, 585]

ter Braak, C. J. F. 1995. Non-linear methods for multivariate statistical calibration and their use in palaeoecology: a comparison of inverse (*k*-nearest neighbours, partial least squares and weighted averaging partial least squares) and classical approaches. *Chemometrics Intelligent Lab. Syst.* **28**: 165-180. [604, 635]

ter Braak, C. J. F. & S. Juggins. 1993. Weighted averaging partial least squares regression (WA-PLS): an improved method for reconstructing environmental variables from species assemblages. *Hydrobiologia* **269**: 485-502. [604]

ter Braak, C. J. F. & C. W. N. Looman. 1987. Regression. 29-77 *in*: R. H. G. Jongman, C. J. F. ter Braak & O. F. R. van Tongeren [eds.] *Data analysis in community and landscape ecology.* Pudoc, Wageningen, The Netherlands. Reissued in 1995 by Cambridge Univ. Press, Cambridge, England. [525, 541]

ter Braak, C. J. F. & I. C. Prentice. 1988. A theory of gradient analysis. *Adv. Ecol. Res.* **18**: 271-317. [451, 604] ter Braak, C. J. F. & P. Smilauer. 1998. *CANOCO release 4 reference manual and user's guide to Canoco for Windows – Software for canonical community ordination.* Microcomputer Power, Ithaca, New York. [474, 594]

ter Braak, C. J. F. & H. van Dam. 1989. Inferring pH from diatoms: a comparison of old and new calibration methods. *Hydrobiologia* **178**: 209-223. [604]

ter Braak, C. J. F. & P. F. M. Verdonschot. 1995. Canonical correspondence analysis and related multivariate methods in aquatic ecology. *Aquat. Sci.* **57**: 255-289. [597]

ter Braak, C. J. F. & J. Wiertz. 1994. On the statistical analysis of vegetation change: a wetland affected by water extraction and soil acidification. *J. Veg. Sci.* **5**: 361-372. [606]

Thiessen, A. W. 1911. Precipitation averages for large areas. *Monthly Weather Review* **39**: 1082-1084. [756]

Thioulouse, J., D. Chessel & S. Champely. 1995. Multivariate analysis of spatial patterns: a unified approach to local and global structures. *Environ. Ecol. Stat.* **2**: 1-14. [768]

Thioulouse, J., D. Chessel, S. Dolédec & J.-M. Olivier. 1996. ADE-4: a multivariate analysis and graphical display software. *Stat. Comp.* **7**: 75-83. [579]

Thompson, S. K. 1992. *Sampling*. Wiley, New York. 368 pp. [228]

Thorrington-Smith, M. 1971. West Indian Ocean phytoplankton: a numerical investigation of phytohydrographic regions and their characteristic phytoplankton associations. *Mar. Biol. (Berl.)* **9**: 115-137. [294, 357]

Thrush, S. F., D. C. Schneider, P. Legendre, R. B. Whitlatch, P. K. Dayton, J. E. Hewitt, A. H. Hines, V. J. Cummings, S. M. Lawrie, J. Grant, R. D. Pridmore, S. J. Turner & B. H. McArdle. 1997. Scaling-up from experiments to complex ecological systems: where to next? *J. Exp. Mar. Biol. Ecol.* 216: 243-254. [710]

Thurstone, L. L. 1947. *Multiple-factor analysis - A development and expansion of the vectors of mind.* Chicago Univ. Press, Illinois. xix + 535 pp. [479]

Torgerson, W. S. 1958. *Theory and methods of scaling.* Wiley, New York. 460 pp. [425]

Toussaint, G. 1980. The relative neighbourhood graph of a finite planar set. *Pattern Recogn.* **12**: 261-268. [752, 755]

Tranter, D. J. & P. E. Smith. 1968. Filtration performance. 27-56 *in:* Zooplankton sampling. Monographs on Oceanographic Methodology, 2. Unesco. [117]

Trexler, J. C. & J. Travis. 1993. Nontraditional regression analyses. *Ecology* **74**: 1629-1637. [542, 544]

Troussellier, M. & P. Legendre. 1981. A functional evenness index for microbial ecology. *Microb. Ecol.* **7**: 283-296. [244]

Troussellier, M. & P. Legendre. 1989. Dynamics of fecal coliform and culturable heterotroph densities in an eutrophic ecosystem: stability of models and evolution of these bacterial groups. *Microb. Ecol.* **17**: 227-235. [524]

Troussellier, M., P. Legendre & B. Baleux. 1986. Modeling of the evolution of bacterial densities in an eutrophic ecosystem (sewage lagoons). *Microb. Ecol.* **12**: 355-379. [524, 550]

Tukey, J. W. 1958. Bias and confidence in not-quite large samples. (Abstract). *Ann. Math. Statist.* **29**: 614. [26]

Tukey, J. W. 1977. *Exploratory data analysis.* Addison-Wesley, Reading, Mass. xvi + 688 pp. [647]

Tuomisto, H. & K. Ruokolainen. 1994. Distribution of *Pteridophyta* and Melastomataceae along an edaphic gradient in an Amazonian rain forest. *J. Veg. Sci.* **5**: 25-34. [696, 701]

Ulrych, T. J. & T. N. Bishop. 1975. Maximum entropy spectral analysis and autoregressive decomposition. *Rev. Geophys. Space Phys.* **13**: 183-200. [690]

Ulrych, T. J. & R. W. Clayton. 1976. Time series modelling and maximum entropy. *Phys. Earth Planet. Inter.* **12**: 188-200. [690]

Ulrych, T. J. & O. Jensen. 1974. Cross-spectral analysis using maximum entropy. *Geophysics* **39**: 353-354. [691]

Underwood, A. J. 1991. Beyond BACI: experimental designs for detecting human environmental impacts on temporal variations in natural populations. *Aust. J. Mar. Freshwater Res.* **42**: 569-587. [249]

Underwood, A. J. 1992. Beyond BACI: the detection of environmental impacts on populations in the real, but variable, world. *J. Exp. Mar. Biol. Ecol.* **161**: 145-178. [249]

Underwood, A. J. 1994. On beyond BACI: sampling designs that might reliably detect environmental disturbances. *Ecol. Appl.* **4**: 3-15. [249]

Underwood, A. J. 1997. *Experiments in ecology – Their logical design and interpretation using analysis of variance.* Cambridge University Press, Cambridge, England. xviii + 504 pp. [495]

Upton, G. J. G. 1978. *The analysis of cross-tabulated data.* John Wiley & Sons, New York. xii + 148 pp. [223]

Upton, G. J. G. & B. Fingleton. 1985. *Spatial data analysis by example. Vol. 1: Point pattern and quantitative data.* Wiley, Chichester. xi + 410 pp. [553, 711, 712, 752, 756]

Upton, G. J. G. & B. Fingleton. 1989. *Spatial data analysis by example. Vol. 2: categorical and directional data.* Wiley, Chichester. xi + 416 pp. [712]

van der Meer, J. 1991. Exploring macrobenthos-environment relationship by canonical correlation analysis. *J. Exp. Mar. Biol. Ecol.* **148**: 105-120. [616]

van Rijckevorsel, J. L. A. & J. de Leeuw. 1988. *Component and correspondence analysis – Dimension, reduction by functional approximation.* John Wiley & Sons, Chichester. xiii + 146 pp. [451]

Varis, O. 1991. Associations between lake phytoplankton community and growth factors. A canonical correlation analysis. *Hydrobiologia* **210**: 209-216. [616]

Venrick, E. L. 1971. Recurrent groups of diatoms in the North Pacific. *Ecology* **52**: 614-625. [359]

Verdonschot, P. F. M & C. J. F. ter Braak. 1994. An experimental manipulation of oligochaete communities in mesocosms treated with chlorpyrifos or nutrient additions: multivariate analyses with Monte Carlo permutation tests. *Hydrobiologia* **278**: 251-266. [606]

Verly, G., M. David, A. G. Journel, & A. Marechal. 1984. *Geostatistics for natural resources characterization. Parts 1 and 2.* Reidel, Dordrecht. xxi + 585 pp., xii + 506 pp. [749]

Villeneuve, N., M. M. Grandtner & J. A. Fortin. 1991. The coenological organization of ectomycorrhizal macrofungi in the Laurentide Mountains of Quebec. *Can. J. Bot.* **69**: 2215-2224. [780]

Visscher, J. P. 1928. Reactions of the cyprid larvae of barnacles at the time of attachment. *Biol. Bull. (Woods Hole)* **54**: 327-335. [726]

Voronoï, G. F. 1909. Recherches sur les parallélloèdres primitifs. *Journal für die reine und angewandte Mathematik* **136**: 67-179. [756]

Walliser, B. 1977. *Systèmes et modèles.* Seuil, Paris. [32]

Ward, J. H. 1963. Hierarchical grouping to optimize an objective function. *J. Amer. Statist. Assoc.* **58**: 236-244. [329, 381]

Wartenberg, D., S. Ferson & F. J. Rohlf. 1987. Putting things in order: a critique of detrended correspondence analysis. *Am. Nat.* **129**: 434-448. [466, 471]

Watson, L., W. T. Williams & G. N. Lance. 1966. Angiosperm taxonomy: a comparative study of some novel numerical techniques. *J. Linn. Soc. Lond. Bot.* **59**: 491-501. [286]

Webster, R. 1973. Automatic soil-boundary location from transect data. *Math. Geol.* **5**: 27-37. [693]

Webster, R. & T. M. Burgess. 1984. Sampling and bulking strategies for estimating soil properties in small regions. *J. Soil Sci.* **35**: 127-140. [16]

Wegman, E. J. & I. W. Wright. 1983. Splines in statistics. *J. Amer. Statist. Assoc.* **78**: 351-365. [543]

Whittaker, E. T. & G. Robinson. 1924. *The calculus of observations – A treatise on numerical mathematics.* Blackie & Son, London. xvi + 396 pp. [665]

Whittaker, J. 1984. Model interpretation from the additive elements of the likelihood function. *Appl. Statist.* **33**: 52-64. [534, 535]

Whittaker, R. H. 1952. A study of summer foliage insect communities in the Great Smoky Mountains. *Ecol. Monogr.* **22**: 1-44. [282]

Whittaker, R. H. 1956. Vegetation of the Great Smoky Mountains. *Ecol. Monogr.* **26**: 1-80. [535, 569, 600, 633, 707]

Whittaker, R. H. 1960. Vegetation of the Siskiyou Mountains, Oregon and California. *Ecol. Monogr.* **30**: 279-338. [463]

Whittaker, R. H. 1962. Classification of natural communities. *Bot. Rev.* **28**: 1-239. [304, 355]

Whittaker, R. H. 1967. Gradient analysis of vegetation. *Biol. Rev. (Camb.)* **42**: 207-264. [253, 439, 463]

Whittaker, R. H. & H. G. Gauch. 1973. Evaluation of ordination techniques. 287-321 *in*: R. H. Whittaker [ed.] *Handbook of vegetation science. Part V.* Dr. W. Junk, The Hague. [413]

Whittington, H. B. & C. P. Hughes. 1972. Ordovician geography and faunal provinces deduced from trilobite distribution. *Philos. Trans. R. Soc. London, Ser. B* **263**: 235-278. [450]

Whittle, P. 1963. On the fitting of multivariate autoregressions and the approximate canonical factorization of a spectral density matrix. *Biometrika* **50**: 129-154. [703]

Wiens, J. A. 1989. Spatial scaling in ecology. *Funct. Ecol.* **3**: 385-397. [708, 709, 710]

Wieser, W. 1960. Benthic studies in Buzzards Bay. II. The meiofauna. *Limnol. Oceanogr.* **5**: 121-137. [373, 374]

Wilhm, J. L. 1968. Use of biomass units in Shannon's formula. *Ecology* **49**: 153-156. [239]

Wilhm, J. L. & T. C. Dorris. 1968. Biological parameters for water quality criteria. *BioScience* **18**: 477-481. [238, 239]

Wilkinson, G. N., S. R. Eckert, T. W. Hancock & O. Mayo. 1983. Nearest neighbour (NN) analysis of field experiments. *J. R. Stat. Soc. Ser. B* **45**: 151-211. [15]

Wilks, S. S. 1932. Certain generalizations in the analysis of variance. *Biometrika* **24**: 471-494. [623]

Wilks, S. S. 1935. The likelihood test of independence in contingency tables. *Ann. Math. Statist.* **6**: 190-196. [217]

Williams, B. K. 1983. Some observations on the use of discriminant analysis in ecology. *Ecology* **64**: 1283-1291. [542]

Williams, B. K. & K. Titus. 1988. Assessment of sampling stability in ecological applications of discriminant analysis. *Ecology* **69**: 1275-1285. [634]

Williams, D. A. 1976. Improved likelihood ratio tests for complete contingency tables. *Biometrika* **63**: 33-37. [218, 225]

Williams, E. J. 1952. Use of scores for the analysis of association in contingency tables. *Biometrika* **39**: 274-289. [461]

Williams, W. T. 1976b. Hierarchical divisive strategies. *In*: W. T. Williams [ed.] *Pattern analysis in agricultural science.* CSIRO, Melbourne, Australia. [346]

Williams, W. T., H. J. Clay & J. S. Bunt. 1982. The analysis, in marine ecology, of three-dimensional data matrices with one dimension of variable length. *J. Exp. Mar. Biol. Ecol.* **60**: 189-196. [251]

Williams, W. T. & M. B. Dale. 1965. Fundamental problems in numerical taxonomy. *Adv. Bot. Res.* **2**: 35-68. [250, 252, 346]

Williams, W. T. & J. M. Lambert. 1959. Multivariate methods in plant ecology. I. Association-analysis in plant communities. *J. Ecol.* **47**: 83-101. [343]

Williams, W. T. & J. M. Lambert. 1961. Multivariate methods in plant ecology. III. Inverse association-analysis. *J. Ecol.* **49**: 717-729. [345]

Williams, W. T., J. M. Lambert & G. N. Lance. 1966. Multivariate methods in plant ecology. V. Similarity analyses and information-analysis. *J. Ecol.* **54**: 427-445. [336, 341]

Williams, W. T., G. N. Lance, L. J. Webb, J. G. Tracey & M. B. Dale. 1969. Studies in the numerical analysis of complex rain-forest communities. III. The analysis of successional data. *J. Ecol.* **57**: 515-535. [692]

Williams, W. T., G. N. Lance, M. B. Dale & H. T. Clifford. 1971. Controversy concerning the criteria for taxonometric strategies. *Computer J.* **14**: 162-165. [307]

Williams, W. T. & W. Stephenson. 1973. The analysis of three-dimensional data (sites x species x times) in marine ecology. *J. Exp. Mar. Biol. Ecol.* **11**: 207-227. [251, 278]

Wirth, M., G. F. Estabrook & D. J. Rogers. 1966. A graph theory model for systematic biology, with an example for the Oncidiinae (Orchidaceae). *Syst. Zool.* **I5**: 59-69. [375, 381]

Wishart, D. 1978. Treatment of missing values in cluster analysis. 281-287 *in*: *Proc. COMPSTAT 1978.* Physica-Verlag Wien. [48]

Wishart, D. 1985. Estimation of missing values and diagnosis using hierarchical classification. *Comput. Stat. Quarterly* **2**: 125-134. [48]

Wold, S. 1974. Spline functions in data analysis. *Technometrics* **16**: 1-11. [543]

Wolda, H. 1981. Similarity indices, sample size and diversity. *Oecologia (Berl.)* **50**: 296-302. [298]

Wollenberg, A. L. van den. 1977. Redundancy analysis. An alternative for canonical correlation analysis. *Psychometrika* **42**: 207-219. [579]

Womble, W. H. 1951. Differential systematics. *Science* **114**: 315-322. [761]

Wright, S. 1921. Correlation and causation. *J. Agric. Res.* **20**: 557-585. [546]

Wright, S. 1960. Path coefficients and path regressions: alternative or complementary concepts? *Biometrics* **16**: 189-202. [546]

Wright, S. P. 1992. Adjusted P-values for simultaneous inference. *Biometrics* **48**: 1005-1013. [18]

Yates, F. E., D. J. Marsh & A. S. Iberall. 1972. Integration of the whole organism – A foundation for a theoretical biology. 110-132 *in*: *Challenging biological problems: directions towards their solutions.* Oxford Univ. Press, New York. [638]

Young, F. W. 1985. Multidimensional scaling. 649-659 *in*: S. Kotz & N. L. Johnson [eds.] *Encyclopedia of statistical sciences. Vol. 5.* Wiley, New York. [444]

Yule, G. U. 1927. On a method of investigating periodicities in disturbed series, with special reference to Wolfer's sunspot numbers. *Phil. Trans. A* **226**: 267-298. [651]

Zeeb, B. A., C. E. Christie, J. P. Smol, D. L. Findlay, H. J. Kling & H. J. B. Birks. 1994. Responses of diatom and chrysophyte assemblages in Lake 227 sediments to experimental eutrophication. *Can. J. Fish. Aquat. Sci.* **51**: 2300-2311. [775]

Tables

Table A Critical values of D in the Kolmogorov-Smirnov goodness-of-fit test of normality.

Table B Critical values of Kendall's rank-order correlation coefficient.

Table C Species diversity as a function of the number of species, according to the broken stick model.

Table D Percentage of the total variance of a principal component analysis associated with the successive eigenvalues, according to the broken stick model.

Table A Critical values of D in the Kolmogorov-Smirnov goodness-of-fit test of normality of distributions, recomputed following Stephens (1974), for mean and variance estimated from the sample data. When the maximum deviation D between cumulative relative frequencies and cumulative normal distribution exceeds the critical value in the Table, one rejects H_0: the sample data were drawn from a normal population. Significance levels given at the top of the columns.

n	$\alpha = 0.15$	0.10	0.05	0.025	0.01
4	0.321	0.339	0.371	0.412	0.429
5	0.297	0.314	0.343	0.382	0.397
6	0.278	0.294	0.321	0.357	0.371
7	0.262	0.277	0.303	0.336	0.350
8	0.248	0.263	0.287	0.319	0.332
9	0.237	0.250	0.273	0.304	0.316
10	0.227	0.239	0.262	0.291	0.303
11	0.218	0.230	0.251	0.279	0.290
12	0.209	0.221	0.242	0.269	0.280
13	0.202	0.214	0.234	0.260	0.270
14	0.196	0.207	0.226	0.251	0.261
15	0.190	0.201	0.219	0.244	0.254
16	0.184	0.195	0.213	0.237	0.246
17	0.179	0.190	0.207	0.230	0.240
18	0.175	0.185	0.202	0.224	0.233
19	0.171	0.180	0.197	0.219	0.228
20	0.167	0.176	0.192	0.214	0.222
21	0.163	0.172	0.188	0.209	0.218
22	0.159	0.168	0.184	0.205	0.213
23	0.156	0.165	0.180	0.200	0.209
24	0.153	0.162	0.177	0.197	0.204
25	0.150	0.159	0.173	0.193	0.201
26	0.147	0.156	0.170	0.189	0.197
27	0.145	0.153	0.167	0.186	0.193
28	0.142	0.150	0.164	0.183	0.190
29	0.140	0.148	0.162	0.180	0.187
30	0.138	0.146	0.159	0.177	0.184
31	0.136	0.143	0.157	0.174	0.181
32	0.134	0.141	0.154	0.172	0.179
33	0.132	0.139	0.152	0.169	0.176
34	0.130	0.137	0.150	0.167	0.173
35	0.128	0.135	0.148	0.164	0.171
36	0.126	0.134	0.146	0.162	0.169
37	0.125	0.132	0.144	0.160	0.167
38	0.123	0.130	0.142	0.158	0.164
39	0.122	0.129	0.140	0.156	0.162
40	0.120	0.127	0.139	0.154	0.160
41	0.119	0.126	0.137	0.152	0.159
42	0.117	0.124	0.136	0.151	0.157
43	0.116	0.123	0.134	0.149	0.155
44	0.115	0.121	0.133	0.147	0.153
45	0.114	0.120	0.131	0.146	0.152
46	0.112	0.119	0.130	0.144	0.150
47	0.111	0.118	0.128	0.143	0.149
48	0.110	0.116	0.127	0.141	0.147
49	0.109	0.115	0.126	0.140	0.146
50	0.108	0.114	0.125	0.139	0.144
>50	0.775/S	0.819/S	0.895/S	0.955/S	1.035/S

where $S = \sqrt{n} - 0.01 + (0.85/\sqrt{n})$

Table B Critical values of Kendall's rank-order correlation coefficient τ_a for given numbers of objects n. A value of $|\tau_a|$ larger than or equal to the tabulated value is significant at level α shown in the header of the Table (first row: one-tailed test; second row: two-tailed test). Derived from Table 1 of Best (1974), with permission of the author.

α (one-tailed) =	0.10	0.05	0.025	0.01	0.005
α (two-tailed) =	0.20	0.10	0.05	0.02	0.01
n					
4	1.00000	1.00000	—	—	—
5	0.80000	0.80000	1.00000	1.00000	—
6	0.60000	0.73333	0.86667	0.86667	1.00000
7	0.52381	0.61905	0.71429	0.80952	0.90476
8	0.42857	0.57143	0.64286	0.71429	0.78571
9	0.38889	0.50000	0.55556	0.66667	0.72222
10	0.37778	0.46667	0.51111	0.60000	0.64444
11	0.34545	0.41818	0.49091	0.56364	0.60000
12	0.30303	0.39394	0.45455	0.54545	0.57576
13	0.30769	0.35897	0.43590	0.51282	0.56410
14	0.27473	0.36264	0.40659	0.47253	0.51648
15	0.27619	0.33333	0.39048	0.46667	0.50476
16	0.25000	0.31667	0.38333	0.43333	0.48333
17	0.25000	0.30882	0.36765	0.42647	0.47059
18	0.24183	0.29412	0.34641	0.41176	0.45098
19	0.22807	0.28655	0.33333	0.39181	0.43860
20	0.22105	0.27368	0.32632	0.37895	0.42105
21	0.20952	0.26667	0.31429	0.37143	0.40952
22	0.20346	0.26407	0.30736	0.35931	0.39394
23	0.20158	0.25692	0.29644	0.35178	0.39130
24	0.19565	0.24638	0.28986	0.34058	0.37681
25	0.19333	0.24000	0.28667	0.33333	0.36667
26	0.18769	0.23692	0.28000	0.32923	0.36000
27	0.17949	0.23077	0.27066	0.32194	0.35613
28	0.17989	0.22751	0.26455	0.31217	0.34392
29	0.17241	0.22167	0.26108	0.31034	0.33990
30	0.17241	0.21839	0.25517	0.30115	0.33333
31	0.16559	0.21290	0.25161	0.29462	0.32473
32	0.16532	0.20968	0.24597	0.29032	0.32258
33	0.16288	0.20455	0.24242	0.28788	0.31439
34	0.15865	0.20143	0.23708	0.27986	0.31194
35	0.15630	0.19664	0.23361	0.27731	0.30420
36	0.15238	0.19365	0.23175	0.27302	0.30159
37	0.15015	0.19219	0.22823	0.26727	0.29730
38	0.14936	0.18919	0.22333	0.26316	0.29161
39	0.14710	0.18758	0.21997	0.26046	0.28745
40	0.14359	0.18462	0.21795	0.25641	0.28462
41	0.14146	0.18049	0.21463	0.25366	0.28049
42	0.14053	0.17770	0.21254	0.24971	0.27526
43	0.13843	0.17608	0.20930	0.24695	0.27353
44	0.13742	0.17336	0.20719	0.24313	0.26850
45	0.13535	0.17172	0.20404	0.24040	0.26667
46	0.13237	0.16908	0.20193	0.23865	0.26377
47	0.13228	0.16744	0.19889	0.23589	0.25994
48	0.12943	0.16667	0.19681	0.23227	0.25709
49	0.12925	0.16327	0.19558	0.22959	0.25340
50	0.12653	0.16245	0.19184	0.22776	0.25061

Table C Species diversity $M(n')$ as a function of the number of species n', according to the broken stick model. This Table may be used (1) to estimate the broken stick diversity, M, corresponding to the observed number of species $n = n'$, or (2) to find the number of species n' predicted by the model, for a computed diversity $H(n) = M(n')$. From Lloyd & Ghelardi (1964) by permission of Blackwell Scientific Publications, Oxford. See Subsection 6.5.2 for explanations.

n'	$M(n')$	n'	$M(n')$	n'	$M(n')$	n'	$M(n')$
1	0.0000	51	5.0941	102	6.0792	205	7.0783
2	0.8113	52	5.1215	104	6.1069	210	7.1128
3	1.2997	53	5.1485	106	6.1341	215	7.1466
4	1.6556	54	5.1749	108	6.1608	220	7.1796
5	1.9374	55	5.2009	110	6.1870	225	7.2118
6	2.1712	56	5.2264	112	6.2128	230	7.2434
7	2.3714	57	5.2515	114	6.2380	235	7.2743
8	2.5465	58	5.2761	116	6.2629	240	7.3045
9	2.7022	59	5.3004	118	6.2873	245	7.3341
10	2.8425	60	5.3242	120	6.3113	250	7.3631
11	2.9701	61	5.3476	122	6.3350	255	7.3915
12	3.0872	62	5.3707	124	6.3582	260	7.4194
13	3.1954	63	5.3934	126	6.3811	265	7.4468
14	3.2960	64	5.4157	128	6.4036	270	7.4736
15	3.3899	65	5.4378	130	6.4258	275	7.5000
16	3.4780	66	5.4594	132	6.4476	280	7.5259
17	3.5611	67	5.4808	134	6.4691	285	7.5513
18	3.6395	68	5.5018	136	6.4903	290	7.5763
19	3.7139	69	5.5226	138	6.5112	295	7.6008
20	3.7846	70	5.5430	140	6.5318	300	7.6250
21	3.8520	71	5.5632	142	6.5521	310	7.6721
22	3.9163	72	5.5830	144	6.5721	320	7.7177
23	3.9779	73	5.6027	146	6.5919	330	7.7620
24	4.0369	74	5.6220	148	6.6114	340	7.8049
25	4.0937	75	5.6411	150	6.6306	350	7.8465
26	4.1482	76	5.6599	152	6.6495	360	7.8870
27	4.2008	77	5.6785	154	6.6683	370	7.9264
28	4.2515	78	5.6969	156	6.6867	380	7.9648
29	4.3004	79	5.7150	158	6.7050	390	8.0022
30	4.3478	80	5.7329	160	6.7230	400	8.0386
31	4.3936	81	5.7506	162	6.7408	410	8.0741
32	4.4381	82	5.7681	164	6.7584	420	8.1087
33	4.4812	83	5.7853	166	6.7757	430	8.1426
34	4.5230	84	5.8024	168	6.7929	440	8.1757
35	4.5637	85	5.8192	170	6.8099	450	8.2080
36	4.6032	86	5.8359	172	6.8266	460	8.2396
37	4.6417	87	5.8524	174	6.8432	470	8.2706
38	4.6792	88	5.8687	176	6.8596	480	8.3009
39	4.7157	89	5.8848	178	6.8758	490	8.3305
40	4.7513	90	5.9007	180	6.8918	500	8.3596
41	4.7861	91	5.9164	182	6.9076	550	8.4968
42	4.8200	92	5.9320	184	6.9233	600	8.6220
43	4.8532	93	5.9474	186	6.9388	650	8.7373
44	4.8856	94	5.9627	188	6.9541	700	8.8440
45	4.9173	95	5.9778	190	6.9693	750	8.9434
46	4.9483	96	5.9927	192	6.9843	800	9.0363
47	4.9787	97	6.0075	194	6.9992	850	9.1236
48	5.0084	98	6.0221	196	7.0139	900	9.2060
49	5.0375	99	6.0366	198	7.0284	950	9.2839
50	5.0661	100	6.0510	200	7.0429	1 000	9.3578

Table D Percentage of the total variance of a principal component analysis associated with the successive eigenvalues λ_i, according to the broken stick model, for $p = 2$ to 20 principal axes. See Subsection 9.1.6 and Table 9.4. Further values may be computed using eq. 6.49. From Frontier (1976), with permission of the author and Elsevier Biomedical Press, Amsterdam.

$p =$	2	3	4	5	6	7	8	9	10
λ_1	75.00	61.11	52.08	45.67	40.83	37.04	33.97	31.43	29.29
λ_2	25.00	27.78	27.08	25.67	24.17	22.76	21.47	20.32	19.29
λ_3		11.11	14.58	15.67	15.83	15.61	15.22	14.77	14.29
λ_4			6.25	9.00	10.68	10.85	11.06	11.06	10.96
λ_5				4.00	6.11	7.28	7.93	8.28	8.46
λ_6					2.78	4.42	5.43	6.06	6.46
λ_7						2.04	3.35	4.21	4.79
λ_8							1.56	2.62	3.36
λ_9								1.23	2.11
λ_{10}									1.00

$p =$	11	12	13	14	15	16	17	18	19	20
λ_1	27.45	25.86	24.46	23.23	22.12	21.13	20.23	19.42	18.67	17.99
λ_2	18.36	17.53	16.77	16.08	15.45	14.88	14.35	13.86	13.41	12.99
λ_3	13.82	13.36	12.92	12.51	12.12	11.75	11.41	11.08	10.78	10.49
λ_4	10.79	10.58	10.36	10.13	9.90	9.67	9.45	9.23	9.02	8.82
λ_5	8.51	8.50	8.44	8.34	8.23	8.11	7.98	7.84	7.71	7.57
λ_6	6.70	6.83	6.90	6.92	6.90	6.86	6.80	6.73	6.65	6.57
λ_7	5.18	5.44	5.62	5.73	5.79	5.82	5.82	5.81	5.78	5.74
λ_8	3.88	4.25	4.52	4.71	4.84	4.92	4.98	5.01	5.03	5.02
λ_9	2.75	3.21	3.56	3.81	4.00	4.14	4.25	4.32	4.37	4.40
λ_{10}	1.74	2.29	2.70	3.02	3.26	3.45	3.59	3.70	3.78	3.84
λ_{11}	0.83	1.45	1.93	2.30	2.60	2.82	3.00	3.15	3.26	3.34
λ_{12}		0.69	1.23	1.65	1.99	2.26	2.47	2.64	2.78	2.89
λ_{13}			0.59	1.06	1.43	1.73	1.98	2.18	2.34	2.47
λ_{14}				0.51	0.92	1.25	1.53	1.75	1.93	2.09
λ_{15}					0.44	0.81	1.11	1.35	1.56	1.73
λ_{16}						0.39	0.71	0.98	1.21	1.40
λ_{17}							0.35	0.64	0.88	1.09
λ_{18}								0.31	0.57	0.79
λ_{19}									0.28	0.51
λ_{20}										0.25

Subject index

N

O